Plant Tolerance to Environmental Stress

Role of Phytoprotectants

Plant Tolerance to Environmental Stress

Role of Phytoprotectants

Edited by
Mirza Hasanuzzaman
Masayuki Fujita
Hirosuke Oku
M. Tofazzal Islam

CRC Press
Taylor & Francis Group
Boca Raton London New York

CRC Press is an imprint of the
Taylor & Francis Group, an **informa** business

CRC Press
Taylor & Francis Group
6000 Broken Sound Parkway NW, Suite 300
Boca Raton, FL 33487-2742

Printed on acid-free paper

International Standard Book Number-13: 978-1-138-55917-2 (Hardback)

Visit the Taylor & Francis Web site at
http://www.taylorandfrancis.com

and the CRC Press Web site at
http://www.crcpress.com

Contents

Preface

The population of the world is estimated to increase from 7.3 to 9.7 billion by the year 2050. To keep up with the pace of population growth, it has recently been estimated that food production will need to be increased by 50% by 2030 and by 70–100% by 2050 for a well-fed world population. This situation will be exacerbated by a decrease in available arable land coupled with reduced crop yields, both of which are predicted to arise from climate change. Global climate change affects crop production not only through altered weather patterns, but also via increased environmental stresses such as soil salinity, drought, flooding, metal/metalloid toxicity, environmental pollution, low and high temperature, and the emergence of new diseases and insect-pests. It is estimated that these stresses will reduce the crop yield of staple food up to 70%. As sessile organisms, plants cannot avoid all these predicted abiotic stresses, but can evolve sophisticated mechanisms to adapt to the changing environment. Therefore, a better understanding of plant responses to abiotic stresses is fundamental in designing biorational strategies to improve crop plants and obtain a sustainable crop production. Over decades, agronomic, physiological, genetics, genomics and other molecular biological studies have generated a large body of knowledge on reactions of plants to various abiotic stresses. Considerable success has also been achieved in the improvement of plant tolerance to abiotic stresses through the utilization of new knowledge generated by research. The application of various omics approaches is shedding light on molecular crosstalks between plants and abiotic stresses; this should promote the development of stress-tolerant crop plants by the application of exogenous and endogenous regulators.

The application of various phytoprotectants has become one of the most effective and plausible approaches in enhancing plant tolerance to abiotic stresses. These phytoprotectants include but are not limited to osmoprotectants (compatible solutes), antioxidants, phytohormones, nitric oxide, polyamines, amino acids, and nutrient elements of plants. A large body of literature is available on how these protective agents exert beneficial effects on plants, helping them tolerate abiotic stresses. However, the actual dose, timing, and methods of practical application of phytoprotectants need to be finely tuned. Several lines of evidence suggest that exogenous osmolytes, phytohormones, signaling molecules and nutrient elements enhance the antioxidant defense system of plants. This, too, augments their tolerance to abiotic stresses. Additionally, the application of phytoprotectants activates reactive oxygen species (ROS) signaling in plants.

This book outlines the recent updates of our understanding of the effects that various abiotic and biotic agents (commonly known as phytoprotectants) have on plant tolerance to major abiotic stresses. It includes 26 chapters contributed by 127 leading experts, including diverse areas of life sciences such as agronomy, plant physiology, cell biology, environmental sciences, and biotechnology. Chapter 1 describes the morphological and physiological changes undergone by plants under various abiotic stresses. The roles of phytoprotectants on abiotic stress signaling in plants are discussed in Chapter 2. Chapter 3 focuses on the improvement of abiotic stress tolerance in plants brought by seed priming. The enhancement of abiotic stress tolerance in plants through the application of various osmolytes is discussed in Chapter 4. Proline is a small molecule biosynthesized in plants which plays a significant role in their tolerance to salinity and drought. Chapter 5 comprehensively reviews the functions of proline in conferring salinity and drought tolerance to plants. Phytohormones are involved in various physiological processes including plant adaptation to harsh environments. Chapter 6 discusses the role of phytohormones in improving abiotic stress tolerance in plants. The enhancement of drought tolerance in plants through the ROS-scavenging system of phytohormones is illustrated in Chapter 7.

Stigalactones are chemical compounds biosynthesized in roots and are known as branching-inhibition hormones in plants. Their molecules have profound roles in plant–fungi (e.g., mycorrhiza) interactions and signaling systems in plants. Chapter 8 describes the role of strigalactones as mediators of abiotic stress responses and parasitic attraction in plants. The effects of non-enzymatic antioxidants in improving the tolerance of plants to abiotic stresses are presented in Chapter 9. Nitric oxide is a signaling molecule involved in various physiological processes in plants and other organisms. It also plays an important role in plant tolerance to abiotic stresses through antioxidants and ROS responses. Chapter 10 updates the regulatory role of nitric oxides in plant tolerance to abiotic stresses. The improvement of abiotic stress tolerance in plants through exogenous hydrogen peroxide and nitric oxide is covered in Chapter 11. The application of amino acids, calcium, balanced sulfur nutrition, silicon, and selenium in the enhancement of plants tolerance to abiotic stresses is described in Chapters 12, 13, 14, 15, and 16, respectively. The role of bioorganic fertilizers and biochar in the improvement of soil health and mitigation of stress-induced damages in plants is discussed in Chapters 17 and 18, respectively.

Plant-associated beneficial microorganisms (such as plant probiotic bacteria) play important roles in promoting abiotic stress tolerance in plants. Chapters 19–22 focus on the effects of various beneficial microorganisms in protecting plants from abiotic stresses and increasing the yield of crops. The long road to the development of novel priming products to increase crop yield under stressful environments is covered by Chapter 24. Finally, the alleviation of the adverse effects of abiotic stresses to plants through plant-derived smoke and magnetopriming is illustrated in Chapters 25 and 26, respectively.

This book represents a cooperative effort from editors and contributors representing many different countries. The editors gratefully acknowledge the authors who contributed to this book project. We are thankful for the enthusiasm and collegial spirit they demonstrated. Our profound thanks are also due to Lecturer Dr. Mahbub Alam of the Department of Agriculture of Noakhali Science and Technology University for his valuable support in formatting and incorporating all editorial changes in the manuscripts. We would like to thank Randy Brehm (Senior Editor) and Laura Piedrahita (Editorial Assistant) of CRC Press, who made suggestions on improving this book in view of our audience. Our thanks are also due to other editorial staffs for their precious help in formatting and incorporating editorial changes in the manuscripts. We believe researchers who work on plants tolerance to abiotic stress will find this book an essential reference.

Mirza Hasanuzzaman
Dhaka, Bangladesh

Masayuki Fujita
Kagawa, Japan

Hirosuke Oku
Okinawa, Japan

Tofazzal Islam
Gazipur, Bangladesh

About the Editors

Mirza Hasanuzzaman is a Professor of Agronomy at Sher-e-Bangla Agricultural University, Dhaka, Bangladesh. In 2012, he received his PhD on 'Plant Stress Physiology and Antioxidant Metabolism' from the United Graduate School of Agricultural Sciences, Ehime University, Japan with a Japanese Government (MEXT) Scholarship. Later, he completed his postdoctoral research in the Center of Molecular Biosciences (COMB), University of the Ryukyus, Okinawa, Japan with a 'Japan Society for the Promotion of Science (JSPS)' postdoctoral fellowship. Subsequently, he joined as Adjunct Senior Researcher at the University of Tasmania with an Australian Government's Endeavour Research Fellowship. He joined as a Lecturer in the Department of Agronomy, Sher-e-Bangla Agricultural University in June 2006. He was promoted to Assistant Professor, Associate Professor and Professor in June 2008, June 2013, and June 2017, respectively. Prof. Hasanuzzaman has been devoting himself to research in the field of Crop Science, especially focused on Environmental Stress Physiology, since 2004. He has been performing as team leader/principal investigator of different projects funded by World Bank, FAO, University Grants Commission of Bangladesh, Ministry of Science and Technology (Bangladesh), and so on.

Prof. Hasanuzzaman published over 100 articles in peer-reviewed journals and books. He has edited 2 books and written 35 book chapters on important aspects of plant physiology, plant stress responses, and environmental problems in relation to plant species. These books were published by internationally renowned publishers (Springer, Elsevier, CRC Press, Wiley, etc.). His publications received over 2,000 citations with h-index: 23 (according to Scopus). Prof. Mirza Hasanuzzaman is a research supervisor of undergraduate and graduate students and has supervised 20 M.S. students so far. He is an editor and reviewer of more than 50 peer-reviewed international journals and a recipient of the Publons Peer Review Award 2017. Dr. Hasanuzzaman is an active member of about 40 professional societies and acting as Publication Secretary of the Bangladesh Society of Agronomy. He has been honored by different authorities due to his outstanding performance in different fields, such as research and education. He received the World Academy of Science (TWAS) Young Scientist Award 2014. He attended and presented 25 papers and posters at national and international conferences in several countries (United States, United Kingdom, Germany, Australia, Japan, Austria, Sweden, Russia, etc.).

Masayuki Fujita is a Professor in the Laboratory of Plant Stress Responses, Faculty of Agriculture, Kagawa University, Kagawa, Japan. He received his B.Sc. in Chemistry from Shizuoka University, Shizuoka, and his M.Agr. and PhD in plant biochemistry from Nagoya University, Nagoya, Japan. His research interests include physiological, biochemical, and molecular biological responses based on secondary metabolism in plants under various abiotic and biotic stresses; phytoalexin, cytochrome P450, glutathione *S*-transferase, and phytochelatin; beside redox reaction and antioxidants. In the last decade his works have focused on oxidative stress and antioxidant defense in plants under environmental stress. His group investigates the role of different exogenous protectants in enhancing antioxidant defense and methylglyoxal detoxification systems in plants. He has acted as main supervisor for 4 M.S. students and PhD students. He has about 150 publications in journals and books and has edited four books.

Hirosuke Oku is a Professor in the Center of Molecular Biosciences at the Tropical Biosphere Research Center in the University of the Ryukyus, Okinawa, Japan. He obtained his Bachelor of Science in Agriculture from the University of the Ryukyus in 1980. He received his PhD in Biochemistry from Kyushu University, Japan in 1985. In the same year he started his career as Assistant Professor in the Faculty of Agriculture, University of the Ryukyus. He became a professor in 2009. He received several prestigious awards and medals including the Encouragement Award of Okinawa Research (1993) and the Encouragement Award of the Japanese Society of Nutrition and Food Science (1996). Prof. Oku is the group leader of the Molecular

Biotechnology Group of the Center of Molecular Biosciences at the University of the Ryukyus. His research focused on lipid biochemistry, molecular aspects of phyto-medicine, secondary metabolites biosynthesis and abiotic stress tolerance of tropical forest trees. He has about 10 PhD students and over 20 M.S. students. Prof. Oku has over 50 peer-reviewed publications.

M. Tofazzal Islam is a Professor of the Department of Biotechnology of Bangabandhu Sheikh Mujibur Rahman Agricultural University in Bangladesh. He received his MS and PhD in Applied Biosciences from Hokkaido University in Japan. Dr. Islam attended postdoctoral research experiences at Hokkaido University, University of Goettingen, University of Nottingham and West Virginia University under the JSPS, Alexander von Humboldt, Commonwealth, and Fulbright Fellowships, respectively. He has published articles in many international journals and book series (over 200 peer-reviewed articles, total citation 1,664, h-index 22, i10-index 48; RG score 39.06). Dr. Islam was awarded many prizes and medals including the Bangladesh Academy of Science Gold Medal in 2011, University Grants Commission Bangladesh Awards in 2004 and 2008, and Best Young Scientist Award 2003 from the JSBBA. Prof. Islam is the chief editor of a book series, Bacillus and Agrobiotechnology, published by Springer. His research interests include genomics, genome editing, plant probiotics and novel biologicals, and bioactive natural products.

List of Contributors

Najam Abbas
Department of Environmental Sciences & Engineering
Government College University
Faisalabad, Pakistan

Nabil Abumhadi
AgroBioInstitute
Agricultural Academy
Sofia, Bulgaria

Manuel Acosta
Department of Plant Physiology
University of Murcia, Campus de Espinardor
Murcia, Spain

José Ramón Acosta-Motos
Universidad Católica San Antonio de Murcia
Campus de los Jeronimos
Guadalupe, Spain

Muhammad Adnan
Agriculture Department
The University of Swabi
Kyber Pakhtunkhwa, Pakistan

Shakeel Ahmad
Department of Agronomy
Bahauddin Zakariya University
Multan, Pakistan

Niaz Ahmed
Department of Soil Science
Bahauddin Zakariya University
Multan, Pakistan

Saleh M.S. Al-Garni
Department of Biological Sciences
King Abdulaziz University
Jeddah, Kingdom of Saudi Arabia

Hesham Alharby
Department of Biological Sciences
King Abdulaziz University
Jeddah, Kingdom of Saudi Arabia

Aqsa Ali
Department of Botany
Government College University
Faisalabad, Pakistan

Ehsan Ali
Hubei Insect Resources Utilization and Sustainable Pest Management Key Laboratory
College of Plant Science and Technology
Huazhong Agricultural University
Wuhan, P.R. China

Muhammad Arif Ali
Department of Soil Science
Bahauddin Zakariya University
Multan, Pakistan

Qasim Ali
Department of Botany
Government College University
Faisalabad, Pakistan

Usman Ali
National Key Laboratory of Crop Genetic Improvement
Huazhong Agricultural University
Wuhan, P.R. China

Amanullah
Department of Agronomy
The University of Agriculture
Peshawar, Pakistan

Misbah Amir
Institute of Pure and Applied Biology
Bahauddin Zakariya University
Multan, Pakistan

Rabia Amir
Atta-ur-Rahman School of Applied Biosciences
National University of Sciences and Technology (NUST)
Islamabad, Pakistan

Muhammad Akbar Anjum
Department of Horticulture
Bahauddin Zakariya University
Multan, Pakistan

Yasir Anwar
Department of Biological Sciences
King Abdulaziz University
Jeddah, Kingdom of Saudi Arabia

Muhammad Arif
Department of Agronomy
The University of Agriculture
Peshawar, Pakistan

Muhammad Saleem Arif
Department of Environmental Sciences & Engineering
Government College University Faisalabad
Faisalabad, Pakistan

Saroj Arora
Department of Botanical and Environmental Sciences
Guru Nanak Dev University
Amritsar, India

Muhammad Arslan Ashraf
Department of Botany
Government College University Faisalabad
Faisalabad, Pakistan

Riffat Ashraf
Department of Botany
Government College University
Faisalabad, Pakistan

Nosheen Aslam
Department of Biochemistry
Government College University
Faisalabad, Pakistan

Habib-ur-Rehman Athar
Institute of Pure and Applied Biology
Bahauddin Zakariya University
Multan, Pakistan

Ahmed Bahieldin
Department of Biological Sciences
King Abdulaziz University
Jeddah, Kingdom of Saudi Arabia
and
Department of Genetics
Ain Shams University
Cairo, Egypt

Tamara I. Balakhnina
Institute of Basic Biological Problems
Russian Academy of Sciences
Pushchino, Russia

Shagun Bali
Plant Stress Physiology Lab
Guru Nanak Dev University
Amritsar, India

Palak Bakshi
Plant Stress Physiology Lab
Department of Botanical and Environmental Sciences
Guru Nanak Dev University
Amritsar, India

Celaleddin Barutcular
Department of Field Crops
Cukurova University
Adana, Turkey

Renu Bhardwaj
Department of Botanical and Environmental Sciences
Guru Nanak Dev University
Amritsar, India

Fatima Bibi
Department of Botany
PMAS Arid Agriculture University
Rawalpindi, Pakistan

Andrés A. Borges
Grupo de Activadores Químicos de las Defensas de la Planta
Instituto de Productos Naturales y Agrobiología – CSIC
Canary Islands, Spain

Estefanía Carrillo-Perdomo
Agroécologie, AgroSup Dijon, Institut National de la Recherche Agronomique (INRA)
Université Bourgogne Franche-Comté
Dijon, France

Bishwanath Chakraborty
Department of Botany
University of North Bengal
Siliguri, India

Usha Chakraborty
Department of Botany
University of North Bengal
Siliguri, India

Subhash Chander
ICAR-Directorate of Weed Research
Jabalpur, India

C.R. Chethan
ICAR-Directorate of Weed Research
Jabalpur, India

Sikander Pal Choudhary
Department of Botany
University of Jammu
Jammu and Kashmir, India

V.K. Choudhary
ICAR-Directorate of Weed Research
Jabalpur, India

Pedro Diaz-Vivancos
Group of Fruit Tree Biotechnology
Campus de Espinardo
Murcia, Spain

Shaghef Ejaz
Department of Horticulture
Bahauddin Zakariya University
Multan, Pakistan

Shah Fahad
National Key Laboratory of Crop Genetic Improvement
Huazhong Agricultural University
Wuhan, Hubei, P.R. China
and
Agriculture Department
The University of Swabi
Kyber Pakhtunkhwa, Pakistan
and
College of Life Science
Linyi University
Linyi, P.R. China

Samar Fatima
Department of Environmental Sciences & Engineering
Government College University
Faisalabad, Pakistan

Gábor Feigl
Department of Plant Biology
University of Szeged
Szeged, Hungary

Patrick Michael Finnegan
School of Biological Sciences
University of Western Australia
Perth, Australia

Francisco J. García-Machado
Grupo de Activadores Químicos de las Defensas de la Planta
Instituto de Productos Naturales y Agrobiología – CSIC
La Laguna, Tenerife, Spain
and
Departamento de Botánica, Ecología y Fisiología Vegetal
Universidad de La Laguna
Santa Cruz de Tenerife, Spain

Vandana Guatum
Department of Botanical and Environmental Sciences
Guru Nanak Dev University
Amritsar, India

Ghader Habibi
Department of Biology
Payame Noor University (PNU)
Tehran, Iran

Muhammad Sajjad Haider
Department of Forestry
University College of Agriculture, University of Sargodha
Sargodha, Pakistan

Muhammad Zulqurnain Haider
Department of Botany
Government College University
Faisalabad, Pakistan

Khalid Rehman Hakeem
Department of Biological Sciences
King Abdulaziz University
Jeddah, Kingdom of Saudi Arabia

Neha Handa
Department of Botanical and Environmental Sciences
Guru Nanak Dev University
Amritsar, India

Hanan A. Hashem
Department of Botany
Ain Shams University
Cairo, Egypt

José Antonio Hernandez
Group of Fruit Tree Biotechnology
Campus de Espinardo
Murcia, Spain

Iqbal Hussain
Department of Botany
Government College University Faisalabad
Faisalabad, Pakistan

Qaiser Hussain
Department of Soil Science and SWC
PMAS Arid Agricultural University
Rawalpindi, Pakistan

Sajjad Hussain
Department of Horticulture
Bahauddin Zakariya University
Multan, Pakistan

Syed Murtaza Hussain
Department of Botany
Government College University
Faisalabad, Pakistan

Syed Sarfraz Hussain
Department of Biological Sciences
Forman Christian College (A Chartered University)
Lahore, Pakistan
and
School of Agriculture, Food & Wine, Waite Campus
University of Adelaide
Adelaide, Australia

Noshin Ilyas
Department of Botany
PMAS Arid Agriculture University
Rawalpindi, Pakistan

Muhammad Iqbal
Department of Environmental Sciences & Engineering
Government College University
Faisalabad, Pakistan

Tooba Iqbal
Atta-ur-Rahman School of Applied Biosciences
National University of Sciences and Technology (NUST)
Islamabad, Pakistan

Meeta Jain
School of Biochemistry
Devi Ahilya University
Indore, India

David Jiménez-Arias
Grupo de Activadores Químicos de las Defensas de la Planta
Instituto de Productos Naturales y Agrobiología – CSIC
Canary Islands, Spain

Muhammad Kamran
College of Agronomy
Key Laboratory of Crop PhysioEcology and Tillage in Northwestern Loess Plateau, Minister of Agriculture
Northwest A&F University
Yangling, P.R. China
and
Institute of Water Saving Agriculture in Arid Areas of China
Northwest A&F University
Yangling, P.R. China

Dhriti Kapoor
School of Biotechnology and Biosciences
Lovely Professional University
Jalandhar, India

Sunita Kataria
School of Biochemistry
Devi Ahilya University
Indore, India

Parminder Kaur
Department of Botanical and Environmental Sciences
Guru Nanak Dev University
Amritsar, India

Rajinder Kaur
Department of Botanical and Environmental Sciences
Guru Nanak Dev University
Amritsar, India

Ravdeep Kaur
Plant Stress Physiology Lab
Department of Botanical and Environmental Sciences
Guru Nanak Dev University
Amritsar, India

Rumana Keyani
Bio Sciences Department
Comsats Institute of Information Technology (CIIT)
Islamabad, Pakistan

Muhammad Fasih Khalid
Department of Horticulture
Bahauddin Zakariya University
Multan, Pakistan

Maryam Khan
Atta-ur-Rahman School of Applied Biosciences
National University of Sciences and Technology (NUST)
Islamabad, Pakistan

Md. Mohibul Alam Khan
Department of Biological Sciences
King Abdulaziz University
Jeddah, Kingdom of Saudi Arabia

Shahbaz Ali Khan
Department of Environmental Sciences & Engineering
Government College University
Faisalabad, Pakistan

Kanika Khanna
Plant Stress Physiology Lab
Department of Botanical and Environmental Sciences
Guru Nanak Dev University
Amritsar, India

Renu Khanna-Chopra
Stress Physiology and Biochemistry Laboratory
Water Technology Centre, Indian Agricultural Research Institute (IARI)
New Delhi, India

Sukhmeen Kaur Kohli
Department of Botanical and Environmental Sciences
Guru Nanak Dev University
Amritsar, India

Zsuzsanna Kolbert
Department of Plant Biology
University of Szeged
Szeged, Hungary

Bhumesh Kumar
ICAR-Directorate of Weed Research
Jabalpur, Madhya Pradesh, India

Vinod Kumar
Department of Botany
DAV University
Sarmastpur, India

Nita Lakra
School of Life Sciences
Stress Physiology and Molecular Biology Laboratory
Jawaharlal Nehru University
New Delhi, India

Juan C. Luis
Departamento de Botánica, Ecología y Fisiología Vegetal
Universidad de La Laguna
Santa Cruz de Tenerife, Spain

Sajid Mahmood
Department of Arid Land Agriculture
King Abdulaziz University
Jeddah, Kingdom of Saudi Arabia

Hamid Manzoor
Institute of Molecular Biology and Biotechnology
Bahauddin Zakariya University
Multan, Pakistan

Bilal Ahmad Mir
Department of Botany
School of Life Sciences, Satellite Campus Kargil
University of Kashmir
Kargil, India

Muhammad Salman Mubarik
Centre of Agricultural Biochemistry and Biotechnology (CABB)
University of Agriculture
Faisalabad, Pakistan

Faiza Munir
Atta-ur-Rahman School of Applied Biosciences
National University of Sciences and Technology (NUST)
Islamabad, Pakistan

Muhammad Naeem
Department of Agronomy
University of Agriculture
Faisalabad, Pakistan

Jazia Naseem
Department of Botany
Government College University
Faisalabad, Pakistan

Mariela Odjakova
Department of Biochemistry
Sofia University "St Kliment Ohridski"
Sofia, Bulgaria

Puja Ohri
Department of Zoological Sciences
Guru Nanak Dev University
Amritsar, India

Ashwani Pareek
School of Life Sciences
Stress Physiology and Molecular Biology Laboratory
Jawaharlal Nehru University
New Delhi, India

Rizwan Rasheed
Department of Botany
Government College University Faisalabad
Faisalabad, Pakistan

Sumaira Rasul
Institute of Molecular Biology and Biotechnology
Bahauddin Zakariya University
Multan, Pakistan

Muhammad Riaz
Department of Environmental Sciences & Engineering
Government College University Faisalabad
Faisalabad, Pakistan

Ayman EL Sabagh
Department of Agronomy
Kafrelsheikh University
Kafr El Sheikh, Egypt

Poonam Saini
Plant Stress Physiology Lab
Department of Botanical and Environmental Sciences
Guru Nanak Dev University
Amritsar, India

Jayanwita Sarkar
Department of Botany
University of North Bengal
Siliguri, West Bengal, India

Shah Saud
College of Horticulture
Northeast Agricultural University
Harbin, P.R. China

Vimal Kumar Semwal
Stress Physiology and Biochemistry Laboratory
Water Technology Centre, Indian Agricultural Research Institute (IARI)
New Delhi, India

Sumera Shabir
Department of Botany
PMAS Arid Agriculture University
Rawalpindi, Pakistan

Sumreena Shahid
Department of Botany
Government College University
Faisalabad, Pakistan

Anket Sharma
Department of Botany
DAV University
Jalandhar, India

Faisal Shehzad
Department of Botany
Government College University
Faisalabad, Pakistan

Sher Muhammad Shehzad
Department of Soil and Environmental Science
University College of Agriculture, University of Sargodha
Sargodha, Pakistan

Muhammad Siddique
Department of Environmental Sciences & Engineering
Government College University
Faisalabad, Pakistan

Réka Szőllősi
Department of Plant Biology
University of Szeged
Szeged, Hungary

Hafiz Muhammad Tauqeer
Department of Environmental Sciences & Engineering
Government College University
Faisalabad, Pakistan
and
Department of Environmental Sciences
University of Gujrat
Gujrat City, Pakistan

Denitsa Teofanova
Department of Biochemistry
Sofia University "St Kliment Ohridski"
Sofia, Bulgaria

A.K. Thukral
Plant Stress Physiology Lab
Department of Botanical and Environmental Sciences
Guru Nanak Dev University
Amritsar, India

Veysel Turan
Department of Soil Science and Plant Nutrition
Bingöl University
Bingöl, Turkey

Abid Ullah
National Key Laboratory of Crop Genetic Improvement
Huazhong Agricultural University
Wuhan, Hubei, P.R. China

Francisco Valdés-González
Departamento de Botánica, Ecología y Fisiología Vegetal
Universidad de La Laguna
Santa Cruz de Tenerife, Spain

Adarsh Pal Vig
Department of Botanical and Environmental Sciences
Guru Nanak Dev University
Amritsar, India

Poonam Yadav
Department of Botanical and Environmental Sciences
Guru Nanak Dev University
Amritsar, India

Tahira Yasmeen
Department of Environmental Sciences & Engineering
Government College University
Faisalabad, Pakistan

Lyuben Zagorchev
Department of Biochemistry
Sofia University "St Kliment Ohridski"
Sofia, Bulgaria

Iqra Zakir
Department of Agronomy
Bahauddin Zakariya University
Multan, Pakistan

Afia Zia
Department of Agricultural Chemistry
The University of Agriculture
Peshawar, Pakistan

1 Impacts of Abiotic Stresses on Growth and Development of Plants

Muhammad Fasih Khalid, Sajjad Hussain, Shakeel Ahmad, Shaghef Ejaz, Iqra Zakir, Muhammad Arif Ali, Niaz Ahmed, and Muhammad Akbar Anjum

CONTENTS

1.1 INTRODUCTION

Plants are frequently subjected to unfavorable conditions such as abiotic stresses, which play a major part in determining their yields (Boyer, 1982) as well as in the distribution of different plants species in distinctive environments (Chaves et al., 2003). Plants can face several kinds of abiotic stresses, such as low amounts of available water, extreme temperatures, insufficient availability of soil supplements and/or increase in toxic ions, abundance of light and soil hardness, which restrict plant growth and development (Versulues et al., 2006). A plant's ability to acclimate to diverse atmospheres is related to its adaptability and strength of photosynthetic process, in combination with other types of metabolism involved in its growth and development (Chaves et al., 2011). When plants are adapted to abiotic stresses, they activate different enzymes, complex gene interactions and crosstalk with molecular pathways (Basu, 2012; Umezawa et al., 2006).

The major abiotic stresses (cold, heat, drought and salinity) negatively affect survival, yield and biomass production of crops by as much as 70% and threaten food safety around the world. Desiccation is the major factor in plant growth, development and productivity, mainly occurring due to salt, drought and heat stress (Thakur et al., 2010). When a plant is exposed to abiotic stresses, it can face a number of problems (Figure 1.1).

Since resistance and tolerance to this problem in plants is of great importance in nature (Collins et al., 2008), breeders face an enormous challenge in attempting to manipulate genetic modification in plants to overcome the issue. Conventional plant breeding approaches have had limited effectiveness in improving resistance and tolerance to these stresses (Flowers et al., 2000).

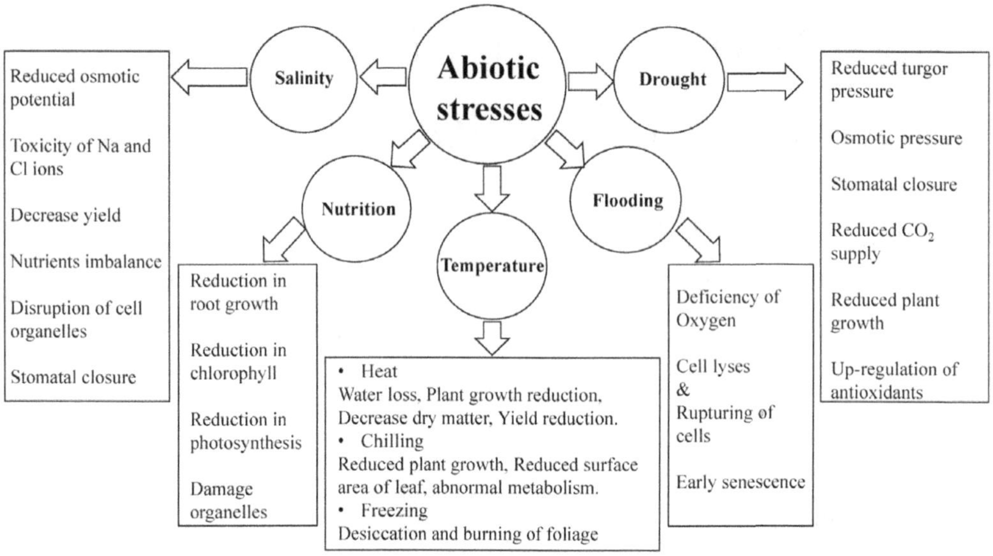

FIGURE 1.1 Effect of different abiotic stresses (salinity, nutrition, temperature, flooding and drought) on plant.

1.2 ABIOTIC STRESS AND PLANT DEVELOPMENT

1.2.1 Temperature

1.2.1.1 Chilling

Chilling injury in plants depends on the sensitivity of their different components to low temperatures. Cellular membrane integrity is the basic component that directly relates to the plants that are sensitive to chilling stress (Levitt, 1980). Lipids present in cell membranes change their state from liquid to solid when plants are brought into light at low temperatures, which mainly depends upon the amount of unsaturated fatty acids they contain (Quinn, 1988). In order for plants to tolerate frost or chilling stress, there must be some alterations in the classes of these unsaturated fatty acids, since plants that have more fatty acids in their cell membranes can tolerate more chilling or frost stress. Different enzyme contents and activities also increase or decrease under extremely low temperatures (Figure 1.2). In plant cell membranes, several changes in physicochemical states occur that compensate for the effect of chilling or frost by increasing cell membrane permeability and also cause ionic and pH imbalance, ultimately decreasing ATP (Levitt, 1980).

1.2.1.2 Freezing

Plants significantly vary in their capacities to manage freezing temperatures. Plants grown under tropical and subtropical conditions (i.e., maize, cotton, soybean, rice, mango, tomato, etc.) are more sensitive to freezing. Plants grown in a temperate climate can tolerate low temperatures, although the degree of tolerance varies from species to species. Moreover, the extreme freezing resistance of these plants is not inherent, as at low temperatures plants activate different physiological and biochemical processes to cope with freezing in a process known as 'cold acclimation'. For example, in one study, rye plants exposed to -5°C without prior acclimation to cold were not able to survive, but when cold acclimatized at 2°C for 7–14 days, they were able to survive temperatures as low as -30°C (Fowler et al., 1977).

Previous research on cold acclimation of plants was designed to examine what happens at low temperatures, helping improve resistance under these conditions. In earlier studies, it was shown that plants activate their enzymatic and non-enzymatic antioxidant defense mechanism to cope with freezing temperatures (Levitt, 1980; Sakai & Larcher, 1987). To prevent freezing injury (cellular membrane damage) under low-temperature conditions, plants produce different osmolytes and osmoprotectants, such as lipids, proline, glycine betaine, and sugars, which help in decreasing membrane damage. Along with solutes and polypeptides, studies have shown that many genes are involved in the activation of cold acclimation processes, ultimately improving freezing resistance. For example, all the chromosomes are involved in coping with freezing injury in hexaploid wheat.

Cold acclimatization is a combination of physiological, biochemical and hereditary processes. Changes do occur in plants exposed to freezing temperatures, but it is not yet clear which changes are involved in the cold acclimation process and which are involved in other resistance processes. However, freezing-tolerance mechanisms are not related to any individual gene that copes with the freezing injury. Many efforts have been

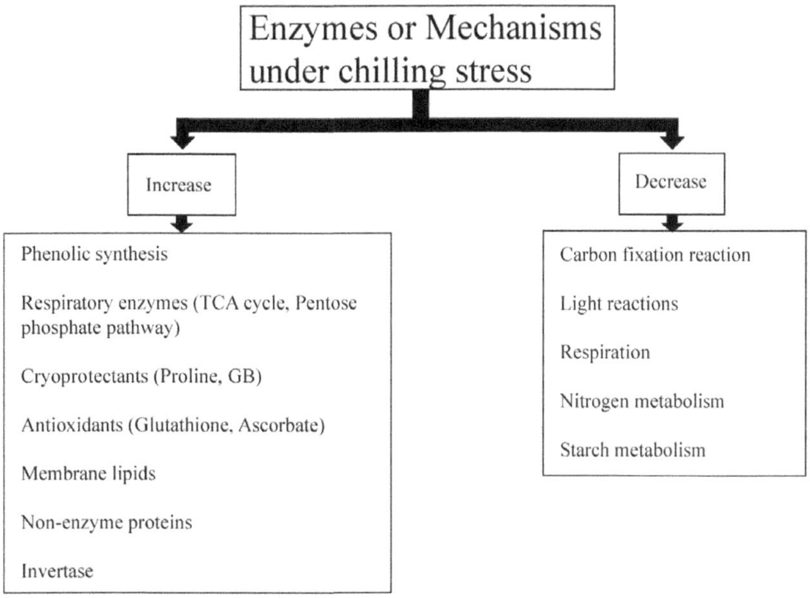

FIGURE 1.2 Enzymes and mechanisms which increase or decrease under chilling stress.

made to isolate and evaluate the genes that are induced under extremely low temperature. The latest studies confirm that *ICE2* is involved in the acclimation process in *Arabidopsis*, ultimately increasing the tolerance of plants to low-temperature stress. The recent discovery of the CBF cold-response pathway describes the involvement of clod acclimation in resistance to low temperatures.

1.2.1.3 Heat

Plants subjected to high temperatures undergo numerous distinctive changes and adjustments. Adaptation to high temperatures occurs on diverse time scales and levels of plant organization. Exposure of plants to extremely high temperatures for long periods of time may cause severe injury, leading to death. Under such temperatures, different plant parts are affected in different ways. The nature of the injury depends on the growth stage of the plant, its susceptibility and cellular processes taking place at that time. However, in extremely high temperatures, heat not only damages the plant at a cellular level but also affects many complex processes and structures that ultimately cause death of the plant. When plants are exposed to high temperatures, several changes occur at the cellular level, including modification of lipids (which become more fluid and develop osmotic pressure) and production of reactive oxygen species (ROS) that disturb the cell membrane metabolism. Proteins work efficiently at a specific temperature, but when temperatures exceed their optimal zone, they are deactivated, causing changes in the activity of enzymes and increasing the production of active oxygen species (AOS) and ROS. Several enzymatic (e.g., SOD, POD, CAT) and non-enzymatic antioxidants (e.g., proline, GB) are produced to cope with ROS, but antioxidants have very little effect on the activity of AOS, which are prevalent under extremely high temperatures. The effects of AOS, including reduced photosynthesis, increased oxidative stress, and changes in movement of assimilates, significantly damage the plant and ultimately cause its death (Hall, 2001). However, many processes and genes are involved in plant tolerance under heat stress. Commonly, under field conditions, plants are exposed to heat with a combination of other stresses, such as heat and drought stress or heat and radiation stress. Similar damage may occur via other stresses when combined with heat stress. Plants respond to different stresses in similar ways. Under heat stress, plants produce proteins to cope with the effects of high temperatures at a cellular level. These are called heat-shock proteins (HSPs; Boston et al., 1996). Different pathways are involved in mitigating heat stress in which HSPs are the only component of heat tolerance. When a plant is exposed to extremely high temperatures, different signaling genes become involved in resistance mechanisms by signaling and activating several metabolic processes. More effort is required to identify and characterize the genes that play a vital role in plant tolerance to extremely high temperatures. When the pathways involved in the resistance and tolerance of plants to heat stress are clearer, it will be easier to recognize the interaction between high-temperature stress and other stresses, which in turn will aid in devising strategies to make plants more tolerant (Nobel, 1991). The main challenge here is to recognize the cellular and metabolic pathways affected by heat stress in order to devise strategies to reduce adverse effects on crop yield.

1.2.2 Salinity

Salt stress is the major problem in arid and semiarid regions, limiting agricultural production and crop yields worldwide. About 20% of cultivated and 50% of irrigated land is directly affected by salinity, which reduces crop yield and productivity (Flowers, 2004; Munns, 2002). At present, improvement in plant growth and yield in salt-affected soils is the basic research priority. Salt tolerance in crop plants can be enhanced by conventional breeding methods as well as molecular breeding techniques (Flowers, 2004). Crop-yield stability under salt-stress conditions is also an important factor. There is only limited understanding about how the different cellular metabolisms of cell division and differentiation are directly affected under salt-stress conditions. This ultimately leads to reduced plant growth and yield (Zhu, 2001). In the past, it was proposed that plant adjustment to salinity most likely includes phenological reactions that are imperative for plant-health management in a saline environment but may negatively affect yield (Bressan et al., 1990). Yet, in some cases, high crop yield can be attained under saline conditions by following proper crop management practices (Flowers, 2004).

Some species of tomatoes, mainly wild species and primitive cultivars, possess genes for salt tolerance (Cuartero et al., 1992; Jones, 1986). Cases where adequate hereditary diversity for salt resistance exists or is available in cultivars ordinarily used in programs of breeding are risky (Flowers, 2004; Yeo & Flowers, 1986). The varieties chosen against the capacity for resistance to salt have a limited gene pool in which negative linkages exist between loci mindful for tall abdicate and those which are important for salt resistance. Recent studies on distinguishing proof of salt-resistance determinants and separation of the integration between salt resistance, plant growth and development offer some considerations for plant breeding and biotechnological strategies to enhance yield. Molecular marker-based breeding techniques will distinguish loci dependable for salt resistance and encourage the division of physically connected loci that adversely impact the yield (Flowers et al., 2000; Foolad et al., 2001; Foolad & Lin, 1997). Monogenic introgression of salt-resistance determinants can be done directly into high-yielding recent crop genotypes to improve yield. Seed germination is the basic phase in seedling establishment, determining effective crop generation (Almansouri et al., 2001). The establishment of crops relies on the interaction between seed quality and seedbed environment (Khajeh-Hosseini et al., 2003). Factors unfavorably affecting seed germination include susceptibility to water-deficit conditions and salt resilience. Earlier development stages are more susceptible to salinity than advanced ones, and the growth development and yield of plants are adversely affected when humidity is reduced. Due to constrained precipitation at sowing time, poor and unsynchronized seedlings are produced if soil moisture is low in seedbeds at the time of sowing (Mwale et al., 2003), influencing the consistency of the plant population with a negative impact on crop productivity.

Additionally, salt stress delays seed germination and reduces the rate and speed of germination, resulting in decreased plant growth, improper development and reduced final-crop productivity. Seeds are powerless to cope with stress, especially stress experienced between sowing and seedling foundation, while plant salt resistance usually increases with plant ontogeny. Soil salinity may influence the germination of seeds either by increasing osmotic potential outside the seeds avoiding water uptake, or through the toxic effects of Na^+ and Cl^- ions on germinating seeds (Khajeh-Hosseini et al., 2003).

1.2.3 Water Stress

1.2.3.1 Drought

Living creatures have two distinctive features: a cellular organization and a need for water. In spite of the fact that the cellular origin of life can be discussed, especially due to evolutionary science developments, the necessity for water reigns supreme. Life forms are able to exploit any biological specialty, no matter how extraordinary, if free water is accessible. Water is very important for plants, as it performs many vital functions. According to Kramer and Boyer (1995), herbaceous plants contain 90% of water in their fresh weight. Water maintains plant cell turgor, thus facilitating the respiration process. Water also has several biophysical characteristics (i.e., high-temperature vaporization, increased surface tension) that make it a good solvent. These characteristics allow water to remain in liquid form even in extreme temperature ranges and to act as a solvent for many molecules, minerals, ions and elements. In addition, water plays a critical role in many biochemical processes, such as serving as primary electron donor in photosynthesis. According to Boyer (1982), scarcity of water is one of the major issues in plant production, decreasing plant health, productivity and distribution of plant species. On Earth, about 35% of land is arid or semi-arid. Here, the only source of water is rainfall. Areas having sufficient amounts of rainfall but uneven distribution throughout the year still face water-deficit conditions that decrease crop yields. Drought causes approximately 50% of global yield loss. Almost all cultivated regions in the world are facing water-deficit conditions. Drought stress or water-deficit conditions are mostly unpredictable, but in some areas 'dry seasons' are expected. In the 21st century, developing plants that can resist drought stress or withstand water-deficit conditions for a long time and maintain their health and yield constitute one of the main research areas in agriculture. Thus, more studies are required to understand the cell physiology of plants under drought stress. Such studies will help us to increase plant growth and yield under water-deficit conditions.

1.2.3.2 Flooding

Flooding of land mainly occurs due to over-irrigation, poor drainage and heavy rainfall (Kozlowski & Pallardy, 1997). Waterlogging is currently a significant concern, not only in rainfall areas but also in areas where irrigation water is used. In some countries, about 0.7 million acres are affected by flooding, and 60 thousand acres are always waterlogged due to poor drainage and leakage of water through water channels. Sodium also causes waterlogging conditions in some soils, and sodicity can cause infiltration in various types of soil. When the amount of Na ion is increased in the soil, it restricts the soil pores, limiting the movement of air and water, leading to waterlogging. Fruit trees also vary in flood-resistance capacity, as found by multiple studies of citrus rootstocks under flood conditions. For example, 80% *Citrus jambhiri* plants can withstand waterlogging for two months, whereas only 10% of *Citrus aurantium* plants can tolerate one month. Under flood stress, plants are affected in many ways (i.e., decreased photosynthesis, decrement in stomatal conductance, reduced chlorophyll and rubisco or RuBisCO; Vu & Yelenosky, 1991).

Under waterlogging conditions, plants act differently depending on their stages. When a plant is in the development stage, waterlogging conditions severely decrease its yield and productivity. However, when a plant is dormant, the effect is very small and seen only for a short time (Kozlowski & Pallardy, 1997). When plants are exposed to waterlogging stress, they are affected in many ways, such as reticence of seed germination, decrease in vegetative and reproductive growth, alteration in plant structure and accelerated senescence. Plant responses under waterlogging conditions also vary with plant age, genotype, duration of stress and properties of water (Kozlowski, 1984). Unfavorable impacts of flooding frequently lead to changes in forest dissemination and composition (Oliveira-Filho et al., 1994). Trees and herbs that can be grown in short- and long-term flooded soil are listed in Figure 1.3.

1.2.4 Nutrition

The transport of metals into roots is increased in corrosive soils. Heavy metals (iron, manganese, copper and zinc) can damage the growth of plants at higher concentrations, affecting root development, reducing photosynthesis and inhibiting several enzymes which could also lead to cell damage. However, numerous plant species have created hereditary and physiological resistance to

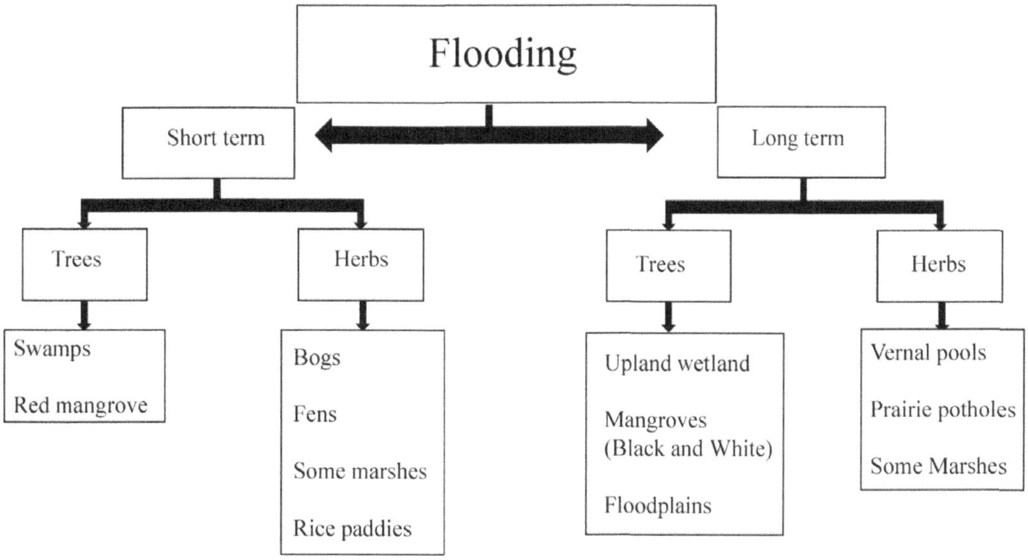

FIGURE 1.3 Trees and herbs grown in short- and long-term flooded areas.

survive in common metal-rich soils or in soils subjected to more heavy-metal contamination (Herrera-Estrella et al., 1999). The plants or crops possessing the ability to tolerate heavy metals can be divided into two distinct pathways: excluders or gatherers. In plants with an exclusion mechanism, the transport of heavy metals is controlled, leading to a relatively consistently low level of metallic particles in sprouts over an extensive period of outside concentrations. In contrast, gatherers take up the metals and store them in their non-food parts or apply a chemical transformation to metal salts, thus reducing their toxicity. In the excluder methodology, metals can be kept outside the root via root exudation of chelating substances or by the actuation of membrane transporters that drive metal particles back into the soil.

1.3 PLANT DEVELOPMENT AND MORPHOLOGY UNDER ABIOTIC FACTORS

1.3.1 Germination and Seedling Emergence

The development of plants is a complex mechanism influenced by both biotic and abiotic factors. Among these, soil moisture is an important feature that influences the germination and development of seedlings. This feature is of more critical importance in species grown in riparian zones. In fact, the provision of oxygen could be an essential feature in the activation of physiological processes for seed germination. Soil moisture may confine oxygen accessibility to the embryo, thus accelerating or delaying seed germination (Martin et al., 2011). The germination rate and respiration rates are higher in some species that grow at a normal partial pressure of oxygen comparable with that of air. A decreased partial pressure of oxygen could result in failure of the seeds to germinate and marked decline in growth (Bradford et al., 2007; Finch-Savage et al., 2005).

1.3.2 Vegetative Growth

1.3.2.1 Leaves

Plant responses to various stresses are multilateral in nature; however, leaves are thought to respond early. Although roots are most susceptible to abiotic stresses, they are also responsible for poor growth of leaves, shoots and other parts. Abiotic stress unfavorably influences shoot development in numerous woody plants by suppressing extension of leaves, internodes and leaf initiation, causing untimely leaf senescence that could further lead to dieback (Vinocur & Altman, 2005).

1.3.2.2 Shoots

As previously mentioned, abiotic stress unfavorably influences shoot development. Scientists around the globe have presented various hypotheses for this, many of which have been given serious consideration.

Salinity diminishes shoot development by halting leaf formation and development of internodes and by quickening leaf abscission (de Lacerda et al., 2003; Kozlowski & Pallardy, 2002). Reduced development could be linked to accumulation of chlorides that might result in shedding of leaves in all types of plants, including gymnosperms and angiosperms (de Lacerda et al., 2003; Hatfield & Prueger, 2015).

1.3.2.3 Roots

Soil moisture decreases root development in several timber crops through various mechanisms that might include hindering root establishment and branching along with stopping the growth of already mature roots. Moreover, such phenomena could result in enhanced susceptibility of the plant to various pathological diseases (Achuo et al., 2006). Shallow, spreading root frameworks are characteristic of areas with high water tables (Roering et al., 2003). Since root development regularly decreases more than stem development, the root/shoot proportion is reduced. It is interesting to observe that when floods subside, the overflowed plants tend to be more susceptible to abiotic stresses owing to their reduced root network.

1.3.2.4 Reproductive Growth

The effects of constant high temperatures are multifarious in nature, but grain development, yield and fertilization are more likely to be affected by heat stress. Unfortunately, diminished fertility and reduced grain filling have not been thoroughly researched, and many areas need to be explored owing to constant variations over time. Such problems include malfunctioning in male meiosis, pollen germination, pollen tube development, or mega gametophyte surrenders. Moreover, effects on bloom generation, grain set, endosperm division, source photosynthesis, and absorption transport and dividing can all contribute to extreme seed yield and weight. Indeed, attempts to characterize the heat-sensitive forms of distinctive plant species continue (Hatfield & Prueger, 2015; Liu et al., 2016). Drastic changes could be seen in the starch content of wheat. In spite of the fact that transcript levels of three isoforms of starch synthase were significantly reduced by the heat treatment, this effect was not reflected in the rate of starch accumulation. The general time to grain-fill was seriously shortened, and the starch granule type changed.

In this regard, Zahedi et al. (2003) research to study the influence of heat stress on the enzymes responsible for starch synthesis. They concluded that temperature could drastically reduce the activity of enzymes, thus leading to poor filling of grains and poor yield. Soil moisture frequently represses blossom bud start, anthesis, natural product set, and natural product extension in flood-intolerant species. It also actuates early abscission of blossoms and natural products. The degree of the changes in regenerative development varies as a function of plant genotype along with intensity of floods. Saltiness antagonistically affects several aspects of regenerative development, including blossoming, fertilization, natural product advancement, yield and quality, and seed generation. The regenerative development of *Citrus* is particularly sensitive to saline flooding (Arbona et al., 2017; Syvertsen & Garcia-Sanchez, 2014). As already discussed, when plants are exposed to environmental stress, they face numerous problems. When *Citrus scinensis* is exposed to salinity, the stress reduces flowering, which decreases the fruit set and number of fruits. Fruit trees exposed to salt-stress conditions display delays in fruiting and yield (Howie & Lloyd, 1989).

1.4 CONCLUSION

Plant growth and development are affected by various factors, including abiotic factors. Abiotic stresses include temperature extremities, nutritional imbalance, salinity, and water deficit and excess. These stresses alter physiological (reduction in photosynthesis, stomatal conductance, transpiration) and biochemical processes (denaturing of proteins, production of ROS, activation of antioxidants) in plants. The effects of these abiotic stresses overlap. When plants are exposed to high-temperature stress, eventually drought stress follows. Similarly, under drought or salinity stress, nutritional imbalance also occurs. Tolerant genotypes have been produced to cope with these abiotic stresses, but more efforts are required for genetic modification of plant genomes that can help us to maintain plant growth and ultimately increase yield to meet the food requirements of the increasing world population.

REFERENCES

Achuo, E.A., Prinsen, E., Höfte, M. (2006) Influence of drought, salt stress and abscisic acid on the resistance of tomato to *Botrytis cinerea* and *Oidium neolycopersici*. *Plant Pathol*ogy 55(2): 178–186.

Almansouri, M., Kinet, J.-M., Lutts, S. (2001) Effect of salt and osmotic stresses on germination in durum wheat (*Triticum drum Desf.*). *Plant and S*oil 231(2): 243–254.

Arbona, V., Zandalinas, S.I., Manzi, M., González-Guzmán, M., Rodriguez, P.L., Gómez-Cadenas, A. (2017) Depletion of abscisic acid levels in roots of flooded Carrizo citrange (*Poncirus trifoliata* L. Raf.× *Citrus sinen*sis L. Osb.) plants is a stress-specific response associated to the differential expression of PYR/PYL/RCAR receptors. *Plant Molecular Biol*ogy 93(6): 623–640.

Basu, U. (2012) Identification of molecular processes underlying abiotic stress plants adaptation using "omics" technologies. In *Sustainable Agriculture and New Technolog*ies, ed. Benkeblia, N., 149–171. Boca Raton: CRC Press.

Boston, R.S., Viitanen, P.V., Vierling, E. (1996) Molecular chaperones and protein folding in plants. *Plant Molecular Biol*ogy 32(1–2): 191–222.

Boyer, J.S. (1982) Plant productivity and the environment. *Sci*ence 218(4571): 443–448.

Bradford, K.J., Côme, D., Corbineau, F. (2007) Quantifying the oxygen sensitivity of seed germination using a population-based threshold model. *Seed Science Resea*rch 17(1): 33–43.

Bressan, R.A., Nelson, D.E., Iraki, N.M., LaRosa, P.C., Singh, N.K., Hasegawa, P.M., Carpita, N.C. (1990) Reduced cell expansion and changes in cell walls of plant cells adapted to NaCl. In *Environmental Injury to Plants*, ed. Kattermann, F., 137–171. New York: Academic Press.

Chaves, M.M., Costa, J.M., Saibo, N.J.M. (2011) Recent advances in photosynthesis under drought and salinity. In *Advances in Botanical Research*, ed. Turkan, I., 50–83. San Diego: Elsevier, Ltd.

Chaves, M.M., Maroco, J.P., Pereira, J.S. (2003) Understanding plant responses to drought from genes to the whole plant. *Functional Plant Biol*ogy 30(3): 239–264.

Collins, N.C., Tardieu, F., Tuberosa, R. (2008) Quantitative trait loci and crop performance under abiotic stress: Where do we stand? *Plant Physiol*ogy 147(2): 469–486.

Cuartero, J., Yeo, A.R., Flowers, T.J. (1992) Selection of donors for salt-tolerance in tomato using physiological traits. *New Phytologist* 121(1): 63–69.

de Lacerda, C.F., Cambraia, J., Oliva, M.A., Ruiz, H.A., Prisco, J.T. (2003) Solute accumulation and distribution during shoot and leaf development in two sorghum genotypes under salt stress. *Environmental and Experimental Botany* 49(2): 107–120.

Finch-Savage, W.E., Coˆme, D., Lynn, J.R., Corbineau, F. (2005) Sensitivity of *Brassica oleracea* seed germination to hypoxia: A QTL analysis. *Plant Sci*ence 169(4): 753–759.

Flowers, T.J. (2004) Improving crop salt tolerance. *Journal of Experimental Botany* 55(396): 307–319.

Flowers, T.J., Koyama, M.L., Flowers, S.A., Sudhakar, C., Singh, K.P., Yeo, A.R. (2000) QTL: Their place in engineering tolerance of rice to salinity. *Journal of Experimental Botany* 51(342): 99–106.

Foolad, M.R., Lin, G.Y. (1997) Genetic potential for salt tolerance during germination in Lycopersicon species. *Horticultural Science* 32: 296–300.

Foolad, M.R., Zhang, L.P., Lin, G.Y. (2001) Identification and validation of QTLs for salt tolerance during vegetative growth in tomato by selective genotyping. *Genome* 44(3): 444–454.

Fowler, D.B., Dvorak, J., Gusta, L.V. (1977) Comparative Cold Hardiness of Several Triticum species and *Secale cereale* L. *Crop Scie*nce 17: 941–943.

Hall, A.E. (2001) Crop Responses to the Environment. Boca Raton, Florida: CRC Press.

Hatfield, J.L., Prueger, J.H. (2015) Temperature extremes: Effect on plant growth and development. *Weather and Climate Extremes* 10: 4–10.

Herrera-Estrella, L., Guevara-Garcı´a, A., Lo´pez-Bucio, J. (1999) Heavy metal adaptation. In *Encyclopedia of Life Scien*ces, 1–5. London: Macmillan Publishers.

Howie, H., Lloyd, J. (1989) Response of orchard 'Washington navel' orange, Citrus sinensis (L.) Osbeck to saline irrigation water. 2. Flowering, fruit set and fruit growth. *Australian Journal of Agricultural Research* 40(2): 371–380.

Jones, R.A. (1986) High salt tolerance potential in Lycopersicon species during germination. *Euphytica* 35(2): 575–582.

Khajeh-Hosseini, M., Powell, A.A., Bingham, I.J. (2003) The interaction between salinity stress and seed vigour during germination of soybean seeds. *Seed Science and Technology* 31(3): 715–725.

Kozlowski, T.T. (1984) Responses of woody plants to flooding. In *Flooding and Plant Growth*, ed. Kozlowski, T.T., 129–163. Orlando: Academic Press.

Kozlowski, T.T., Pallardy, S.G. (1997) Growth control in woody plants. San Diego: Academic Press.

Kozlowski, T.T., Pallardy, S.G. (2002) Acclimation and adaptive responses of woody plants to environmental stresses. *The Botanical Review* 68(2): 270–334.

Kramer, P.J., Boyer, J.S. (1995) Water Relations of Plants and Soils. San Diego: Academic Press.

Levitt, J. (1980) Responses of plants to environmental stresses. II. Water, radiation, salt, and other stresses. New York: Academic Press.

Liu, B., Asseng, S., Liu, L., Tang, L., Cao, W., Zhu, Y. (2016) Testing the responses of four wheat crop models to heat stress at anthesis and grain filling. *Global Change Biology* 22(5): 1890–1903.

Martin, C.G., Mannion, C., Schaffer, B. (2011) Leaf gas exchange and growth responses of green buttonwood and swingle citrumelo to *Diaprepes abbreviatus* (Coleoptera: Curculionidae) larval feeding and flooding. *Florida Entomologist* 94(2): 279–289.

Munns, R. (2002) Comparative physiology of salt and water stress. *Plant, Cell and Environment* 25(2): 239–250.

Mwale, S.S., Hamusimbi, C., Mwansa, K. (2003) Germination, emergence and growth of sunflower (*Helianthus annus L.*) in response to osmotic seed priming. *Seed Science and Technology* 31(1): 199–206.

Nobel, P.S. (1991) Physicochemical and Environmental Plant Physiology. San Diego: Academic Press.

Oliveira-Filho, A.T., Vilela, E.A., Gavilanes, M.L., Carvalho, D.A. (1994) Effect of flooding regime and understorey bamboos on the physiognomy and tree species composition of a tropical semideciduous forest in southeastern Brazil. *Vegetatio* 113: 99–124.

Quinn, P.J. (1988) Effects of temperature on cell membranes. *Symposia of Society for Experimental Bio*logy 42: 237–258.

Roering, J.J., Schmidt, K.M., Stock, J.D., Dietrich, W.E., Montgomery, D.R. (2003) Shallow landsliding, root reinforcement, and the spatial distribution of trees in the Oregon Coast Range. *Canadian Geotechnical Journal* 40(2): 237–253.

Sakai, A., Larcher, W. (1987) Frost Survival of Plants: Responses and Adaptation to Freezing Stress. Berlin: Springer-Verlag.

Syvertsen, J.P., Garcia-Sanchez, F. (2014) Multiple abiotic stresses occurring with salinity stress in citrus. *Environmental and Experimental Botany* 103: 128–137.

Thakur, P., Kumar, S., Malik, J.A., Berger, J.D., Nayyar, H. (2010) Cold stress effects on reproductive development in grain crops: An overview. *Environmental and Experimental Botany* 67(3): 429–443.

Umezawa, T., Fujita, M., Fujita, Y., Yamaguchi-Shinozaki, K., Shinozaki, K. (2006) Engineering drought tolerance in plants: Discovering and tailoring genes to unlock the future. *Current Opinion in Biotechnology* 17(2): 113–122.

Versulues, P.E., Agarwal, M., Katiyar-Agarwal, S., Zhu, J., Zhu, J.K. (2006) Methods and concepts in quantifying resistance to drought, salt and freezing, abiotic stresses that affect plant water status. *The Plant Journal: for Cell and Molecular Bio*logy 45(4): 523–539.

Vinocur, B., Altman, A. (2005) Recent advances in engineering plant tolerance to abiotic stress: Achievements and limitations. *Current Opinion in Biotechnology* 16(2): 123–132.

Vu, J.C.V., Yelenosky, G. (1991) Photosynthetic responses of citrus trees to soil flooding. *Physiologia Plantarum* 81: 7–14.

Yeo, A.R., Flowers, T.J. (1986) The physiology of salinity resistance in rice (*Oryza sativa* L.) and a pyramiding approach to breeding varieties for saline soils. *Australian Journal of Plant Physiology* 13(1): 161–173.

Zahedi, M. Sharma, R., Jenner, C.F. (2003) Effects of high temperature on grain growth and on the metabolites and enzymes in the starch-synthesis pathway in the grains of two wheat cultivars differing in their responses to temperature. *Functional Plant Biology* 30(3): 291–300.

Zhu, J.K. (2001) Plant salt tolerance. *Trends in Plant Science* 6(2): 66–71.

2 Influence of Phytoprotectants on Abiotic Stress Signaling in Plants

Rabia Amir, Tooba Iqbal, Maryam Khan, Faiza Munir, and Rumana Keyani

CONTENTS

2.1 INTRODUCTION: ABIOTIC STRESS SIGNALING IN PLANTS

Plants as multicellular organisms communicate to bring coordination in their physiological and molecular responses. They have a complex cellular signaling network, as it involves various responses to stimuli from the outside and inside world after their reception and transduction (Gill et al., 2016). Being sessile organisms, plants have to face constant environmental harshness that interferes with their optimum growth. In case of defense signaling, plants react to stresses by regulating gene expression, which acts as a molecular control mechanism (Akpinar et al., 2012). Two major classes of genes are induced under stress conditions: 1) structural genes which confer tolerance to abiotic stress and 2) regulatory genes which control downstream processing and expression of stress-responsive genes (Hirayama and Shinozaki, 2010; Nakashima et al., 2012).

In plants, the perception of abiotic stress by receptors, for example receptor-like kinases (RLKs), hormones, G-protein-coupled receptors (GPCRs), phytochromes, etc., leads to signal transduction which generates secondary signaling molecules. Consequently, these secondary molecules, such as ROS, inositol phosphate and abscisic acid (ABA), facilitate Ca^{2+} flux for the initiation of protein phosphorylation in order to generate a stress response via transcription factors (Boguszewska and Zagdańska, 2012). For example, the SOS protein kinase complex, through Ca^{2+} mediated cell signaling, maintains the homeostasis of Na^{+} in the cytoplasm of plant cells during salt stress (Chinnusamy et al., 2004). In another example, ABA triggered by osmotic or cold stress reduces water loss by ROS and calcium-mediated signaling (Roychoudhury et al., 2013). Therefore, transcription factors (TFs), reactive oxygen species (ROS), phytohormones, calcium (Ca^{2+}) and protein kinases play vital roles in abiotic stress signaling in plants (Boursiac et al., 2008). Plants produce numerous chemicals such as polyamines, antioxidants, osmoprotectants, and trace elements that trigger adaptive immune responses. These molecules have also been reported to act as immunoboosters after their exogenous application on plants under stress conditions. Increased photosynthesis rate, yield and antioxidative capacities are some of the underlying mechanisms of phytoprotectants (Ahmad and Wani, 2013). Hormonal signaling, ROS signaling, Ca^{2+} mediated cell signaling and transcriptional networking, which form an integrative signaling network in abiotic stress, will be discussed in reference to phytochemicals in the following sub-sections. A concise overview of abiotic stress signaling is illustrated in Figure 2.1.

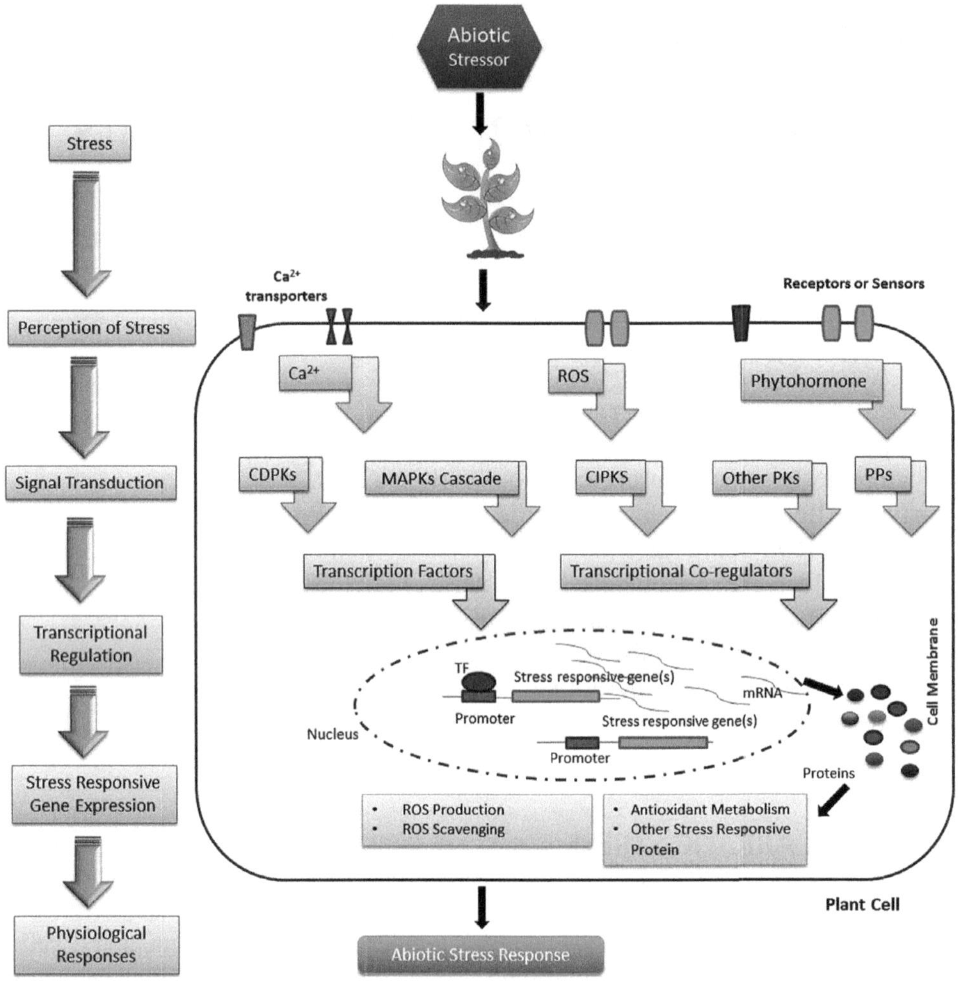

FIGURE 2.1 Overview of abiotic stress signaling components from stress perception to stress response. Where ROS–Reactive oxygen species; PK– Protein kinase; PP–Protein phosphatase; CBL–Calcineurin B-like proteins; CDPKs–Ca^{2+} dependent protein kinases; MAPK–mitogen-activated protein kinase and TF–Transcription factors (Wang et al., 2016; You and Chan, 2015).

2.2 PHYTOPROTECTANTS AFFECTING PHYTO-HORMONAL SIGNALING

Phytohormones are specialized chemicals produced through different pathways which help plants fine-tune their responses with respect to environmental conditions. They integrate numerous stress signals and control the transcription of stress-responsive genes. The interconnected mode of action of one hormonal signaling pathway with the other makes them an exceptionally important and useful resource for helping plants to cope with different abiotic stresses (Chapman and Estelle, 2009). There are several plant hormones, such as Gibberellins (GA), auxins, abscisic acid (ABA), jasmonates (JA), strigolactones (SLs), cytokinins (CK), ethylene (ET), salicylic acid (SA), brassinosteroids (BR). The list is getting longer with time. Plant hormones play an important role in growth, development, nutrient allocation and sink/source transitions of plants. But our main focus here is on how they regulate various signal transduction pathways, and how a network of hormones correlate with each other to exchange information and produce a robust response to cope with environmental stresses.

Synthesis of ABA is one of the fastest responses of plants to various abiotic stresses. ABA significantly contributes to the regulation and stimulation of adaptive responses, e.g. in drought stress ABA triggers the expression of ABA-inducible genes, which cause stomatal closure leading to a reduced rate of transpiration and consequently a reduced growth rate of the plant (Schroeder et al., 2001). ABA signaling causes changes in gene expression at transcriptional and post-transcriptional level (Cutler et al., 2010). It has been observed that ABA functions as a connecting center for primary metabolism and environmental adaptation in plants. ABA elicits the transcriptional programming of different cellular mechanisms underlying abiotic stresses and also changes the expression of genes which control lipid and carbohydrate metabolism. This indicates that ABA functions as an interface for abiotic stress regulation and also monitors primary metabolism in plants (Hey et al., 2010; Li et al., 2006; Seki et al., 2002).

Analysis of ABA-inducible genes revealed that the expression of these genes contains numerous cis-regulatory elements called ABA-responsive elements or ABREs (PyACGTGG/TC) (Giraudat et al., 1994; Umezawa et al., 2010). Proteins which bind to ABA-responsive elements are called ABRE binding factors. The expression of AREB1/ABF2 is up-regulated during different abiotic stresses such as salinity and dehydration. An overexpression of these binding factors resulted in increased tolerance to drought (Arasimowicz and Floryszak-Wieczorek, 2007; Fujita et al., 2005).

In the presence of ABA, hormonal signals are received by specific cellular receptors of the PYR/PYL/RCAR family (PYrabactin Resistance-Like/Regulatory component of ABA receptors). ABA binds these receptors and inactivates PP2Cs (type 2C protein phosphatases) such as ABI2 and ABI1. Upon inactivation of PP2C, another class of proteins called SnRK2 is activated (SNF1-related protein kinases) (Ma et al., 2009; Park et al., 2009). SnRK2 proteins regulate transcription factors such as binding factors (ABFs) and ABA-responsive promoter elements (ABREs), which are involved in the activation of ABA-responsive genes and ABA-dependent physiological processes (Umezawa et al., 2009; Vlad et al., 2009).

Another class of receptors involved in ABA signaling is GTGs (G-protein-coupled receptor-type G protein) localized in the plasma membrane (Pandey et al., 2009). The function of GTG proteins as ABA receptors was confirmed in *Arabidopsis* when the GTG1/GTG2 absent mutants were found hyposensitive to ABA (Pandey et al., 2009).

There is increasing evidence for the role of CHLH/ABAR (H subunit of Mg-chelatase) in ABA perception. Incorporation of CHLH/ABAR in the ABA signaling cascade at cellular level as a chloroplastic receptor and by plastid-to-nucleus regressive signaling via the ABA-responsive nucleo-cytoplasmic transcription repressor WRKY40 has been reported. This evidence suggests that chloroplast-mediated pathway also controls cellular ABA signaling (Shang et al., 2010; Shen et al., 2006).

Furthermore, studies indicate that ABA-mediated stress signaling is also modulated through interaction with other key hormone regulators (CK, SA, ET, and JA) associated with plant growth and development. The complex cascades of exogenous and endogenous signals which plants experience during environmental fluctuation and development are linked to each other through some convergence points between their signal transduction pathways, called crosstalk. This is predominant in modulating ABA signaling during stress and developmental transitions (Golldack et al., 2013). In *Arabidopsis*, GA signaling is modulated by the binding of GA to specific receptors called GID1a/b/c, which are orthologs of GA receptors in rice, *OsGID1 (GA-Insensitive Dwarf)* (Ueguchi-Tanaka et al., 2005). GA signals mediate the binding of DELLA proteins to GID1, which is followed by a conformational conversion of DELLA proteins. The modified DELLAs are recognized by the F-box protein SLEEPY1 (SLY1) in *Arabidopsis*. Subsequently, DELLAs are polyubiquitinated by the SCFSLY1/GID2 ubiquitin E3 ligase complex and degraded via the 26S proteasome pathway, thus activating GA-mediated responses (Dill et al., 2004; Silverstone et al., 2001). DELLA proteins, which consist of RGA-LIKE1 (RGL1), GA-insensitive repressor of GA1–3, RGL2, and RGL3, act as an interface linking GA-controlled developmental responses and ABA-mediated abiotic stress signaling (Achard et al., 2006). In addition, RING-H2 gene *XERICO* regulates tolerance to drought and ABA biosynthesis in *Arabidopsis*, as it is a transcriptional target of DELLAs downstream. This clearly indicates XERICO functions as an assembler of plant development and abiotic stress responses by linking ABA and GA signaling pathways.

Keeping the role of XERICO in mind, there is an increasing number of reports which suggest a crosstalk occurs between GA, ABA and jasmonate, another regulator in the response to drought stress. Jasmonates have signaling functions in biotic stress responses (Golldack et al., 2014); however, it was recently reported that JA receptor proteins such as *OsCOI1a* (Coronatineinsensi-Tive 1) and *JAZ* (jasmonic acid ZIM-domain proteins) are transcriptionally regulated in response to drought stress, which shows that JA signaling has a role in abiotic stress responses as well, as illustrated in Figure 2.2. Besides, the expression of the DELLA protein RGL3 responds to JA, and RGL3 interacts with JAZ proteins (Wild et al., 2012). These research advances indicate that DELLA functions as an interface between GA, JA and ABA signaling. There is also a pivotal functional involvement of lipid-related signaling in abiotic stress responses (Golldack et al., 2014).

2.3 ROLE OF PHYTOPROTECTANTS IN CALCIUM-MEDIATED CELL SIGNALING

Calcium is known as a strict spatiotemporal regulator for its critical role as a key player in the signaling network as well as in plant growth and development.

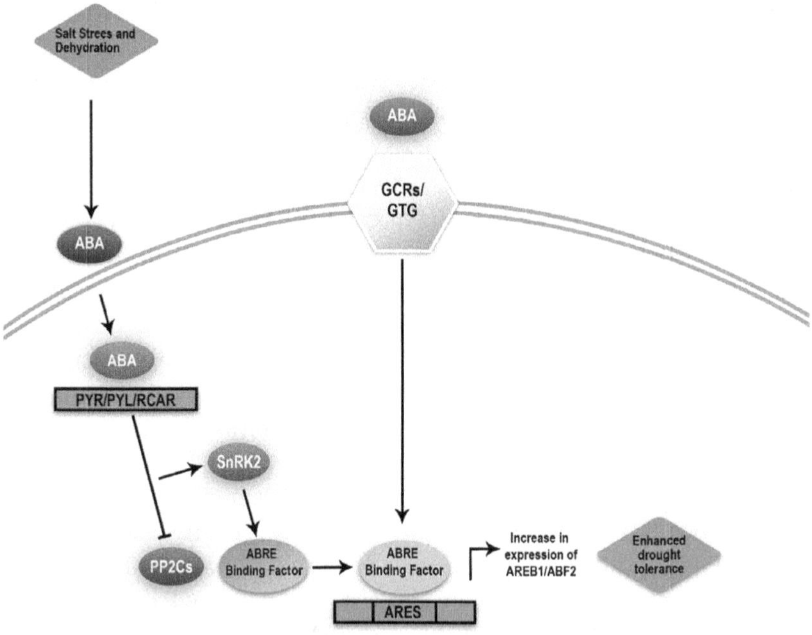

FIGURE 2.2 Schematic representation of ABA-mediated abiotic stress signaling. ABA accumulation induced by abiotic stresses cues cascade of signaling events which induce expression of stress-responsive genes.

Abiotic stresses are perceived by plants through calcium-mediated signaling which consecutively triggers stress-induced transcription factors, ROS signaling and lipid signaling. The specificity of the signaling response to abiotic stresses depends on the calcium level in a cell and its ability to metabolize hydroxyl radical signaling (Wilkins et al., 2016). Plants maintain their Ca^{2+} homeostasis through carriers, pumps and channels in their cellular membranes under the influences of various stimuli (Kudla et al., 2010). Furthermore, plants have an abundance of Ca^{2+} binding proteins termed as Ca^{2+} sensors, which have Ca^{2+} buffering capacity (Dodd et al., 2010). Such sensors include calcineurin B-like proteins (CBL) and calmodulins (CaMs) that transmit a signal to Ca^{2+} dependent protein kinases (CDPKs) or Ca^{2+} and calmodulin-dependent protein kinases (CCaMKs) through direct calcium binding. They decipher Ca^{2+} signals and encourage Ca^{2+} mediated modification of specific proteins (Huang et al., 2013; Tuteja and Mahajan, 2007). Various transcription factors, protein kinases, phosphatases, channels, antiporters, pumps, metabolic enzymes and other functional proteins are the target proteins for CBL and CaM, which one way or the other respond to environmental stresses (Zeng et al., 2015).

Calcium-mediated signaling has been elucidated in plant defense responses against chilling, heat shock and salinity stresses (Reddy and Reddy, 2004). Ca^{2+} mediated stress signaling in plants under salt stress has been described in Figure 2.3. Stress-induced expressions of Ca^{2+} sensors by multiple genes have been reported in soybeans (DeFalco et al., 2010) and *A. thaliana* (Reddy and Reddy, 2004). Recent studies have suggested that calcium-binding TFs have a vital position in stress signaling (Reddy et al., 2011). Such DNA-binding TFs maintain the homeostasis of ROS and regulate other intracellular signaling networks (Nookaraju et al., 2012; Zeng et al., 2015). Ca^{2+} mediated signaling is highly robust and evolved due to its ability to process multiple stimuli at the same time, and shares tightly regulated crosstalk with other signaling networks due to its architectural structures (Dodd et al., 2010). Plants regulate their Ca^{2+} signaling to manipulate the biochemical and molecular processes that influence their physiological, developmental and stress-related responses (Nookaraju et al., 2012). However, it has not yet been clarified how Ca^{2+} signaling brings specificity to the response and how it identifies its downstream target proteins (Zeng et al., 2015). The need of the hour is to enhance our knowledge regarding Ca^{2+} mediated signaling and understand how it helps cope with environmental stresses so that stress-tolerant crop can be developed.

When plants are exposed to stress, a transient fluctuation in calcium levels affects the signaling network. Such fluctuations create unique stress-associated calcium signatures that are deciphered by signal transduction pathways. For the reestablishment of normal Ca^{2+} concentration, there are several cytosolic Ca^{2+} buffering mechanisms, including Ca^{2+}/ATPase pumps and Ca^{2+}/

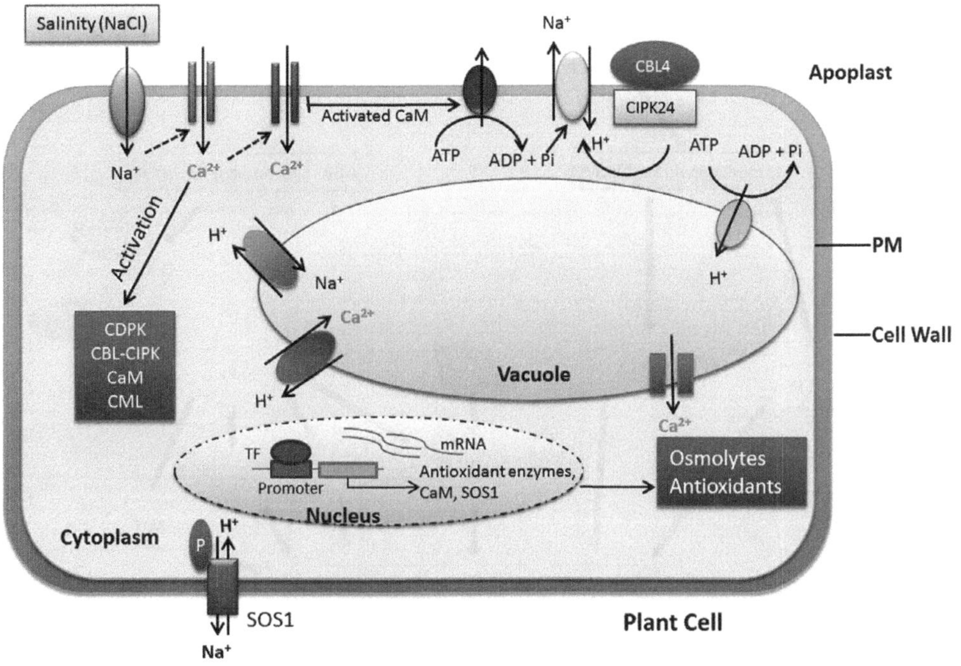

FIGURE 2.3 Ca^{2+} mediated signaling in plants under salt stress. Plant perceives salt stress by increased level of Na^+ and causes Ca^{2+} accumulation which cause hydroxyl radical formation and activation of CBL and CDPKs which leads to Na+efflux. Where, PM=Plasma membrane; DACC=Depolarization activated cation channel; HACCs=Hyperpolarization activated cation channel; cGMP=Cyclic guanosine monophosphate; cAMP=Cyclic adenosine monophosphate; InsP3R=inositol 1,4,5-trisphosphate receptor-like channel; cADPR=cyclic ADP-ribose-activator ryanodine receptor-like channel; CDPKs=Ca2+dependent protein kinases; CaM=Calmodulin; CIPKs=CBL-interacting protein kinases; and CBL=Calcineurin B-like proteins (Kurusu et al., 2015; Wilkins et al., 2016).

H^+ antiporter, which terminate Ca^{2+} mediated signaling (Bose et al., 2011). Several phytochemicals help plants restore their health, either by normalizing Ca^{2+} concentration, after the stress, in a feedback mechanism, or by elevating Ca^{2+} concentration to create a prompt response against stress. Take the example of polyamines (PA) and H_2O_2, which are the representative of phytoprotectants. PA, synergistically, affects Ca^{2+} efflux under stress conditions, whereas H_2O_2 activates various Ca^{2+} channels that affect cytosolic ionic homeostatic (Demidchik et al., 2002; Pei et al., 2000). Moreover, there is a cross-talk between PA and ROS, which results in the oxidation of PA into H_2O_2 and OH^-, which, ultimately, causes Ca^{2+} influx across the plasma membrane (Moschou et al., 2008).

2.4 EFFECT OF PHYTOPROTECTANTS ON TRANSCRIPTIONAL NETWORKING

Plants tolerate abiotic stresses through one of the principal stress controllers known as transcription factors (TFs). TFs are DNA-binding regulatory proteins mainly encoded by early stress-responsive genes, and comprise approximately 7% of the coding genome of plants (Udvardi et al., 2007). TFs regulate the transcription of other proteins by either blocking or recruiting RNA polymerase to DNA in a sequence-specific manner (Riechmann and Ratcliffe, 2000). They take part in almost all physiological, developmental and defense mechanism of plants, thus playing vital roles in plant survival and adaptation (Lindemose et al., 2013). Due to extensive diversity, TFs have been classified into gene families: myelocytomatosis oncogene (MYC)/myeloblastosis oncogene regulon (MYB), APETALA 2/ethylene responsive element binding factor (AP2/ERF), basic leucine zipper (bZIP), heat shock factor (HSF), NAC,WRKY, Cys2Hizinc fingers (C2H2 ZF), MADS-box, nuclear factor Y (NFY) and ten others (Hirayama and Shinozaki, 2010; Reguera et al., 2012; Yamaguchi-Shinozaki and Shinozaki, 2006). A noteworthy fraction of TFs, including bZIP, C2H2 ZF, AP2/ERF, MYB, WRKY, NAC and bHLH, has been characterized as coordinator of abiotic stress signaling, which confers tolerance to plants in order to promote growth and development, as depicted in Figure 2.4 (Lindemose et al., 2013).

bZIP TFs are a member of the vast family of dimerizing TFs that is present in all the eukaryotes and plays

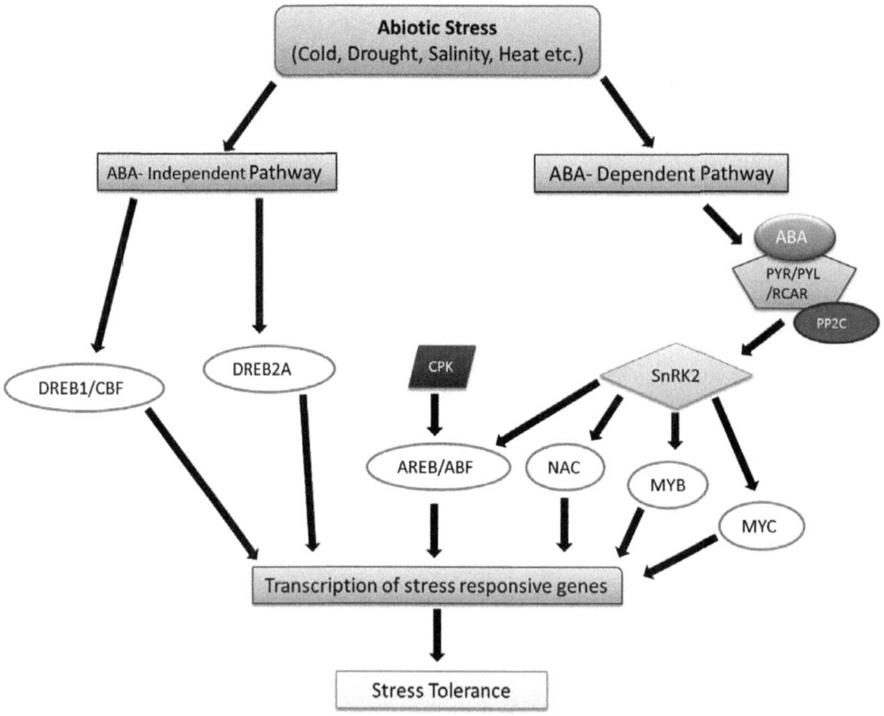

FIGURE 2.4 Transcriptional networking under abiotic stress (Joshi et al., 2016; Kumar, 2014).

an essential role in ABA signaling. ABA-mediated phosphorylation leads to activation of bZIP that control the ABA-responsive genes expression by interacting with their promoter region, which contains ABA-responsive elements (ABREs) (Soma et al., 2017). Elevated bZIP expression in response to ABA, salinity, low temperature and drought has been observed in *Glycine max, A. thaliana*, wheat and rice (Gao et al., 2011; Lindemose et al., 2013).

C2H2 ZF proteins are transcriptional repressors, having an ERF-associated amphiphilic repression (EAR) domain which regulates stress responses (Ciftci-Yilmaz et al., 2007). Improved tolerance against salinity, osmotic pressure and heat stress has been observed in plants with an enhanced level of C2H2 ZF proteins (Mittler et al., 2006). These proteins transcriptionally repress plant growth under stress condition by inhibiting auxin-responsive growth-related genes (Kodaira et al., 2011). While under an unstressed condition, they downregulate ABA-responsive genes (Jiang et al., 2008). It has been observed that C2H2 ZF overexpression can induce tolerance in plants against drought and salt stress by manipulating H_2O_2 homeostasis (Huang et al., 2009).

Dehydration-responsive element binding-protein (DREB), AP2, ERF and ABI3/VP1 (RAV) are the four main subfamilies of plant AP2/ERF TFs. Among these TFs, DREBs are well characterized transcriptional regulators of ABA-dependent abiotic stress responses (Mizoi et al., 2012). DREBs consist of six members that interact with dehydration-responsive element/C-repeat element (DRE/CRT) to induce defense responses against cold, heat, salinity and drought (Haake et al., 2002; Liu et al., 1998). DREB TFs have the capability to enhance drought tolerance in plants, but several reports have observed deformation in the growth of plants due to overexpression of these TFs (Yang et al., 2010).

MYBs are characterized by the presence of R1, R2 and R3 sequence repeats of the MYB domain (Dubos et al., 2010). Out of them, the R2R3 subfamily has occupied a vital position in ABA-dependent abiotic stress response. MYBs have an antagonistic role between ABA and Jasmonic acid (JA)-mediated signaling (Jung et al., 2010). Induced MYB expression in plants has been observed in response to ABA, osmotic stress, cold stress and drought stress by joining auxin and ABA signals (Seo et al., 2009).

WRKY TFs have been divided into three groups based on the number of interacting domains and their association with a zinc finger-like motif (Li et al., 2011). Earlier, these were only associated with biotic stress, but now evidence is available for their role in the abiotic stress response. For instance, ABA sensitivity, heat stress tolerance, drought tolerance and salt sensitivity have been observed in plants with overexpression of WRKY (Jiang and Deyholos, 2009). WRKY68 involves ABRE bZIP, a factor that plays a central role in ABA-dependent plant responses. Transcriptional regulation by WRKYs is complex because of their antagonistic and agonistic roles in various situations. For example, WRKY40 enhances ABA sensitivity while WRKY60 reduces ABA sensitivity (Liu et al., 2012).

NAC is a widely spread plant TF family which reprograms the transcription of plant stress-responsive genes. The word NAC has been derived from three genes with a conserved domain: no apical meristem (NAM), ATAF, and cup-shaped cotyledon (CUC2) (Aida et al., 1997; Souer et al., 1996). NAC TFs have an N-terminus conserved domain consisting of approximately 150–160 amino acids whose functions are linked with DNA binding, homo/hetro-dimerization and nuclear localization (Olsen et al., 2005; Ooka et al., 2003). During stress conditions, NAC genes contribute to the formation of a complex signaling network that makes them potential nominees for conferring stress tolerance (Nuruzzaman et al., 2013).

bHLH proteins are positive regulators of ABA-dependent or independent stress-responsive genes (Bailey et al., 2003). Among bHLH TFs, MYC2 has a principal role in the crosstalk among various cellular signaling pathways, including salicylic acid (SA), JA, ABA, auxin and Gibberellin signaling pathways (Kazan and Manners, 2013). Drought, salinity, mannitol and cold tolerance have been reported in bHLH92 overexpressing plants (Jiang et al., 2009).

Molecular studies have revealed that plants regulates their physiological, developmental and defense response through fine-tuning transcriptional networking. Phytoprotectants directly or indirectly involve various TFs that regulate the expression of stress-responsive genes. For example, both glutathione (GSH) and nitric oxide (NO) are effective entrants of phytoprotectants that work through various TFs under stress. GSH offsets stress-induced oxidation by changing gene expression directly or with the help of transcription factors. It has an important role in signal transduction and ROS signaling at multiple levels. It also acts as redox sensor and helps plants tolerate oxidative stress (Srivalli and Khanna-Chopra, 2008). Thus, GSH is one of the powerful phytoprotectants that can confer tolerance to plants against abiotic stresses. NO is a gaseous biologically active molecule that emerged as a significant antioxidant and signaling molecule. NO triggers many kinds of redox-associated gene expressions to establish tolerance against plant stress (Sung and Hong, 2010). Furthermore, plant hormones as phytoprotectants mediate plant adaptive responses to biotic or abiotic stresses. Hormones recognize stress signals that stimulate transcriptional network to produce plant adaptive responses under stress (Pandey et al., 2017). In a nutshell, phytoprotectants greatly affect the transcriptional networking that helps plants overcome abiotic stress.

2.5 PHYTOPROTECTANTS AFFECTING OXIDATIVE STRESS MECHANISM (ROS PRODUCTION)

Due to abundant molecular oxygen in plants environment, all plant cells confront conditions such as environmental stresses or UV radiations when some toxic chemical entities called reactive oxygen species accumulate in them. ROS include hydroxyl radical (HO•), superoxide anion ($O_2^{\cdot -}$) and hydrogen peroxide (H_2O_2) which, if uncontrollably produced, can lead to far-reaching damage to the cell by degrading proteins, inactivating enzymes or altering gene actions (Choudhury et al., 2013; Mittler et al., 2004).

However, with time, plants have evolved mechanisms to use ROS as chemical signals to mitigate the effects of abiotic stress by regulating ROS network genes, which are comprised of almost 152 genes. Such ROS-regulatory networks involve redox-sensitive transcription factors (TFs), receptor proteins and inhibition of phosphatases by ROS. Thus at any given time, an active balance between ROS-producing and ROS-scavenging pathways promotes cellular well-being (Choudhury et al., 2013).

Upon exposure to numerous abiotic stresses, plants display unique expression patterns of ROS-scavenging and producing enzymes. These changes have been observed in many forms, such as: altered levels of byproducts of lipid peroxidation, increase in enzymes such as peroxidases, glutathione-*S*-transferase, and CAT, and accumulation of phytoprotectants which act as antioxidants, such as ascorbate, phenolic compounds, carotenoids, alkaloids, sucrose and trehalose (Choudhury et al., 2017).

Among the various chemicals which act as phytoprotectants in abiotic stress conditions, soluble sugars which, by definition, are mono and di saccharides, display dual roles, acting both as mediators of ROS production like in mitochondrial respiration and as antioxidants in oxidative pentose phosphate pathways (PPP). The current topic comprehends how ROS production is mediated by phytoprotectants such as soluble sugars and affects oxidative mechanisms that cause damage to plants (Couée et al., 2006). In plants, ROS are accumulated in several different cellular sources, such as NADPH oxidase located in cell membrane, electron transport chain in chloroplast and mitochondria, β-oxidation of fatty acids and the glycolate oxidase stage of photorespiration in peroxisomes and respiration in mitochondria, respectively (Doudican et al., 2005; Møller, 2001).

Soluble sugars have the effect of increasing ROS production in plants during an increase in photosynthetic activity. In contrast, a decrease in soluble sugars negatively regulates the expression of photosynthetic genes even in normal daylight conditions, especially the genes involved in the expression of the Calvin cycle. Such type of regulation of gene expression by soluble sugar and light simultaneously is better understood thanks to the elucidation of the relationship between light and sugar accumulation. This situation is compounded in abiotic stress conditions such as chilling stress, where sugar accumulation is considered to act as cold protectant (Ciereszko et al., 2001; Havaux and Kloppstech, 2001).

Similarly, a condition of fluctuation in carbohydrate levels or carbohydrate starvation at specific developmental stages may also increase ROS production. This is due to the fact that ADP regeneration is significantly decreased and the electron transport flow, through cytochrome c oxidase, results in increased ROS in mitochondria (Dutilleul et al., 2003). Sugar starvation is also thought to activate lipid mobilization and β-oxidation in peroxisomes. This involves the stimulation of acyl co-A oxidase, the protein, and mRNA activity levels, as illustrated in Figure 2.3. Such ROS activation facts are also confirmed by transcriptomic analysis where sugar stress activates oxidative enzymes such as catalases (Contento et al., 2004).

Besides, soluble sugars and the interactive roles of some phytohormones (auxin, brassinosteroid and ABA) have also been observed for ROS production. Auxins can induce the production of ROS and regulate ROS homeostasis, suggesting a relationship between auxin signaling and oxidative stress. For example, auxins activate a Rho-GTPase (RAC/ROP) that interacts with NADPH oxidases, resulting in apoplastic ROS production (Duan et al., 2010). On the contrary, ROS activate a MAPK signaling cascade which inhibits auxin-dependent signaling and triggers oxidative signaling cascades (Kovtun et al., 2000). Auxin-induced changes in cellular redox status, brought about by auxin-induced ROS production, regulate the plant cell cycle (Vivancos et al., 2011). Although plant cells are equipped with numerous other protectants which help in scavenging the ROS produced and mitigate the consequences of abiotic stresses, the part played by such protectants as dual-role entities is also helpful in understanding valuable ROS-production-mediated effects imparted to plant cells during normal activities. An overview of this mechanism is depicted in Figure 2.5.

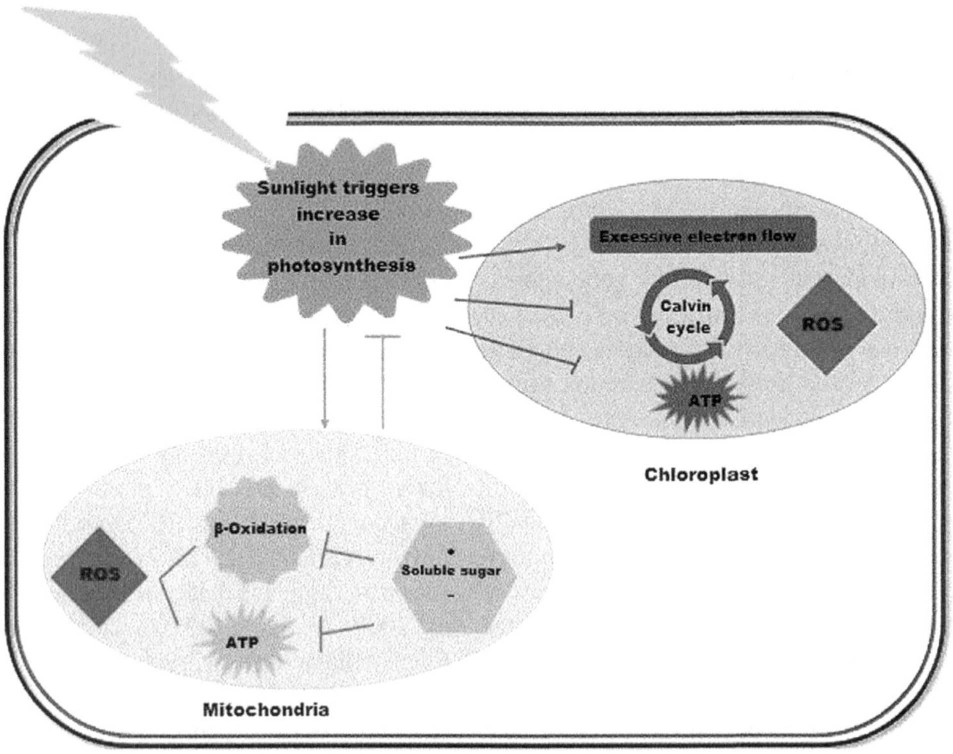

FIGURE 2.5 High photosynthetic activities stimulate accumulation of soluble sugars which in turn downregulate the expression of photosynthetic genes and increase electron flow leading to ROS generation.

2.6 PHYTOPROTECTANTS AS ROS SCAVENGERS

With the span of evolution, as discussed above, green plants have learned to mitigate the effects of ROS through different phytoprotectants that can be non-enzymatic, like carotenoids, ascorbate (AsA), tocopherol and glutathione, or enzymatic, like catalase (CAT), superoxide dismutase (SOD) and glutathione-*S*-transferase. They act as redox buffers and influence the expression of the genes involved in abiotic stresses (Foyer and Noctor, 2005). Here, we will elaborate on the roles of such protectants in mitigating oxidative stress through ROS scavenging and helping plants cope with abiotic stresses.

Ascorbic acid (AsA) is a powerful antioxidant which is present in all plant tissues, particularly in actively photosynthesizing tissues such as meristems. It works as a reductant for numerous free radicals such as $O_2^{\bullet}$, $HO^{\bullet}$ and H_2O_2. This forms the basis of its action as an antioxidant. Plants contain sufficient amounts of AsA. It not only scavenges ROS but also maintains other antioxidant components such as α-tocopherols through the AsA-GSH cycle (Smirnoff, 2000). AsA metabolism and recycling involve enzymes such as mono-dehydroascorbate reductase (MDHAR) and dehydroascorbate reductase (DHAR) (Meena et al., 2017), which are increased during certain abiotic stresses such as high temperature and confer tolerance to plants against stress. The AsA recycling system ensures the maintenance of sufficient amounts of AsA, which confer heat stress tolerance to plants. Experiments carried out on *Arabidopsis* under high-temperature stress support the fact that overexpression of DHAR in cellular compartments substantially increases AsA levels 2–4.25 times, which lowers membrane damage and improves chlorophyll content as compared to normal plants (Wang et al., 2010).

Another important protectant is glutathione (GSH), which occurs abundantly in almost all subcellular compartments such as mitochondria, endoplasmic reticulum and cytosol, whereby it manifests substantial ROS-scavenging capacity. It can act as ROS scavenger either directly by reacting with free radicals such as $HO^{\bullet}$ and $O_2^{\bullet}$, or else by indirectly maintaining the levels of other antioxidants such as tocopherol and zeaxanthin in a reduced state. Additionally, it is a substrate for some other enzymatic antioxidants such as glutathione-*S*-transferases (GST) and glutathione peroxidase (GPX) (Hasanuzzaman et al., 2013). The mechanism of action of GSH is such that the onset of abiotic stress stimulates its accumulation. Increased GSH concentrations counterbalance the stress-initiated oxidation of GSH and cause changes in gene expression directly or through interaction with regulatory proteins and/or transcription factors. This increase is equally important in signal transduction and defense against ROS and works through a multilevel control mechanism, which includes the coordinated activation of genes encoding GSH biosynthetic enzymes and GR (Srivalli and Khanna-Chopra, 2008). Thus, GSH acts as a redox sensor of environmental cues, and the increase in GSH helps plants tolerate oxidative stress.

As amphiphilic antioxidants and protectants, Tocopherols contribute to ROS scavenging in photosynthetic membranes. They limit the extent of lipid peroxidation by reducing free radicals such as lipid peroxyl ($LOO^{\bullet}$) to their respective hydroperoxides (Maeda et al., 2005). They are also part of numerous ROS-controlled signaling networks, phytohormones and some other antioxidants and therefore are appropriate candidates for influencing abiotic stress signaling. Studies indicate that α tocopherols are not important for plant survival under optimal conditions, but an adequate amount of redox-state tocopherol in chloroplasts helps in imparting tolerance to plants against abiotic stresses. Exogenous application of α-tocopherol on *H. annuus* seeds grown under salt stress markedly enhanced activities of antioxidative enzymes and mineral nutrient content, and minimized salt-induced leaf senescence (Hasanuzzaman et al., 2013).

Taken together, all cellular compartments are equipped with antioxidants to scavenge ROS immediately at the site of production by local antioxidants. However, if the stress is severe or the antioxidant capacity is not sufficient to cope with it, then free radicals (H_2O_2) can leak into the cytosol and move to other cellular compartments. Cells are also equipped with mechanisms which allow them to combat the production of excessive H_2O_2 by transporting it in vacuoles for detoxification. Vacuoles are rich in protectants such as flavonoids and ascorbate, which scavenge them there and help plants stabilize themselves on the onset of abiotic stresses (Gechev et al., 2006).

2.7 PHYSIOLOGICAL AND MOLECULAR ADAPTATIONS IN RESPONSE TO PHYTOPROTECTANTS

The utmost requirement of every plant on the inception of an environmental stress is to allocate its energy in such a way that it adapts itself better to the environment while maintaining its growth and productivity. These functions demand a change in physiology within plants, such as the activation of many metabolic reactions, ion homeostasis and plant hormonal signaling leading to the expression of stress-responsive genes (Ahmad and Wani, 2013). Phytoprotectants act as chemical barriers to resist environmental constraints and help plants remain physiologically stable. However, depending on the severity of the stress encountered, the physiological adaptation may not be robust enough to resist the stress and the plant may suffer. We will explore different types of physiological and molecular adaptation which occur in plants due to phytoprotectants and improve the effects of a protectant in producing stress-tolerant crops.

Abiotic stresses such as drought, cold and salinity induce several alterations in plants physiology, in the form of low water and nutrient availability, accumulation of toxic concentrations of salts, lower seed germination rate, early senescence and, in severe cases, plant mortality (Ahmad and Wani, 2013). Several potential low-molecular-weight organic compounds, including compatible solutes produced in plants under osmotic stress, have a potential role in the adaptation of plant physiology with respect to external conditions.

Osmolytes are important phytoprotectants comprising amines (polyamines), amino acids (Pro), sugars (sucrose trehalose), and sugar alcohols (sorbitol, mannitol), which have a role in maintaining cell turgor, lessening ionic toxicity and protecting cell structures. Enhanced levels of sugar in drought- and salt-tolerant rice varieties suggest that these protective compounds can contribute to stress tolerance in rice (Ahmad and Wani, 2013).

Reports suggest that, alternative protective functions of osmolytes have a role in adapting plant physiology to environmental conditions. Rice plants grown under salt stress accumulate soluble nitrogenous compounds such as polyamines (PA), betaines, imides and amino acids along with certain proteins. Salt stress has been shown to increase the content of a polyamine called putrescene (put), which is involved in developing tolerance to salt (Do et al., 2014). Exogenously applied PA has also shown an ability to overcome the damaging effects of salinity in several other plant species (Mansour, 2000). The physiological role of PA-induced adaptation in salt stress has been extensively studied. Due to their polycationic nature, PA can directly interact with the surface of membranes or indirectly affect some membrane-binding enzymes maintaining their structure. They also act as ROS scavengers and ammonia detoxifiers, but due to their low number, their contribution is less felt in osmotic adjustments as compared to other osmoprotectants (Mansour, 2000) (Kushad and Dumbroff, 1991).

A common alleviator of many stresses is the amino acid proline (pro), which is elevated during abiotic stress conditions such as salinity, drought, intense radiation and oxidative stress. Due to its antioxidative potential, it maintains redox balance, stabilizes enzymes and protects cellular structures (Ahmad and Wani, 2013). Studies on transgenic *Arabidopsis* with 90% lower contents of proline than the wild type ascertained that they produce significantly more ROS and lipid-peroxidation products (Székely et al., 2008). Proline is also considered as an osmotic adjustment agent under abiotic stresses and functions by lowering cellular osmotic potential, permitting the reabsorption of water to take place. Proline also has a role in stabilizing membrane structures and protecting cell membrane against salt-induced injuries. On a molecular level, this membrane stabilization involves a decreased accumulation of Na^+ and Cl^- in shoots and thus enhances growth in response to proline treatment on salt-affected plants (Mansour, 2000)

Plants accumulate non-reducing sugars such as trehalose in high concentrations under distinct abiotic stresses such as cold, salinity and drought. Stress moderately increases its levels, which helps stabilizing proteins and membranes (Ahmad and Wani, 2013). Trehalose is also a precursor of glucose and is catabolized by trehalose to give glucose (Brodmann et al., 2002; Goddijn et al., 1997). Trehalose treatments cause an increase in the transcription of antioxidant enzyme genes such as superoxide dismutase, ascorbate peroxidase, and catalase in salt-stressed rice plants. Trehalose-treated plants recover immediately compared to non-treated plants (Nounjan et al., 2012).

2.8 PHYTOPROTECTANTS AFFECTING STRESS SIGNALING IN DIFFERENT CROPS

Abiotic stresses are considered as one of the utmost constraints to crop production across the globe. It has been estimated that more than 50% of yield reduction is the direct result of abiotic stresses. This is becoming a threatening scenario for food security all over the world (Rodríguez et al., 2005). Numerous signaling compounds such as phytohormones, Nitric oxide (NO), sugars and Hydrogen Peroxide display expedient effects in crop plants to combat stresses. In this section, we will describe a few such signaling protectants, which are induced upon stress perception and regulate the expression of genes through signal transduction pathways which enable plants to better adapt to environmental conditions.

Nitric oxide (NO) is a significant signaling molecule and antioxidant which is induced under several environmental conditions and helps plants acclimatize to environmental adversities. NO has been reported to play a significant role in alleviating adverse effects in wheat seedlings. The exogenous application of NO continuously improved the antioxidant system of plants by increasing the level of antioxidants and antioxidative enzymes such as AsA, GSH, APX, GST and MDHAR, which further leads to the generation of a cascade of events that cause physiological adaptations in the plant in numerous forms, such as reduced lipid peroxidation and H_2O_2 content (Nahar et al., 2015). In grapevines, NO accumulation was observed under drought stress, which suggested its potential signaling role in drought. This role was also strengthened when similar results were obtained in other crops, such as maize seedlings grown under stress. In this case, an exogenous treatment of NO scavenger (cPTIO) increased NO scavenging, and an application of NO-donor (SNP) reduced NO scavenging, which strengthens the role of NO as a cellular messenger to mediate adaptive responses in plants against drought stress (Hao et al., 2008).

It was generally thought that hydrogen peroxide (H_2O_2) is a byproduct of aerobic metabolism and a harmful free radical entity. But recent studies suggested its role as redox-signaling molecule and mediator of plant adaptive responses under stress. An exogenous treatment of plants with H_2O_2 imparted tolerance to salinity stress by enhancing the activities of antioxidants and minimizing membrane lipid peroxidation in roots and leaves of maize as an acclimation process (de Azevedo Neto et al., 2005).

Furthermore, accumulation of H_2O_2 at low levels can induce tolerance to high-temperature stress in a few plants. Rice seedling, when pre-treated with a low amount of H_2O_2 (less than 10 mM), were found to have greater quanta yield from photosystem II and thus a better survival rate of green tissues as compared to untreated seedlings. Additionally, this enhanced the activities of antioxidative enzymes and the expression of stress-related genes such as those encoding small heat-shock protein 2, Δ-pyrroline-5-carboxylate synthase and sucrose-phosphate synthase, which have vital signaling roles in enhancing tolerance to high-temperature stress in rice seedlings (Uchida et al., 2002). The role of phytoprotectants in developing tolerance in green plants is unquestionable but a robust response to stress imparts multiple signaling cascades.

2.9 ROLE OF PHYTOPROTECTANTS IN CROSSTALK MECHANISMS

Plants induce the production of unique metabolites comprising numerous signaling molecules when subjected to stresses. Stress perception by specific plant receptors initiate stress responses by triggering particular signal transduction pathways that ensure the survival and well-being of the stressed plant. Oxidative burst, ROS, ion efflux and influx, especially that of Ca^{2+} through Ca^{2+} signaling, acidification or alkalinization of cytoplasm, nitric oxide, abscisic acid signaling, jasmonates signaling, lipid signaling and cyclic nucleotides including cyclic guanosine monophosphate (cGMP) and cyclic adenosine monophosphate (cAMP), are among the signaling components that directly or indirectly participate in plant abiotic stress responses. Integration of different signaling pathways by transcription factors or other signaling components gives rise to the essential crosstalk of signaling pathways that is crucial for plant welfare under stress conditions (Zhao et al., 2005).

Crosstalk of signaling pathways helps plants regulate the expression of arrays of genes in a spatiotemporal manner to create a wide range of defense responses against abiotic stresses. Crosstalk of signaling pathways can occur either at transcription, translation, RNA splicing/editing or post-translational modification level. Transcription factors (TFs) are converging points for almost all the signaling pathways and they can be synthesized or activated either directly through stress signals/signal transduction pathways or indirectly by the feedback mechanism regulated by other TFs (Liu et al., 1999). Association of various signaling pathways through TFs has been observed under the influence of drought, cold, wounding, salinity, pathogens and hormonal treatment (Denekamp and Smeekens, 2003; Eulgem et al., 2000; Liu et al., 1998).

ROS production is central to oxidative stress that changes redox status within a cell. ROS regulate the wide range of genes involved in defense and antioxidant responses (Kandlbinder et al., 2004). Application of phytoprotectants mainly affects ROS signaling by interfering in the antioxidant system, thus accelerating ROS scavenging by enhancing ascorbate peroxidase and catalase activities. They manipulate lipid signaling by upregulating lipid peroxidation under osmotic stress (Cruz et al., 2013). Furthermore, they change membrane permeability by improvising potential gradient of ions across cellular membranes in stress conditions. Moreover, they crosstalk with nitrogen metabolic pathways, photosynthetic pathways and hormonal pathways to bring coordination in plant growth and development under stress (Zhang et al., 2013). Several phytoprotectants have been reported in the following examples that mediate crosstalk with nitric oxide (NO) signaling, lipid signaling, hormonal signaling and $Ca2^{+}$ signaling through ROS signaling.

Nitric oxide (NO) is one of the potential phytoprotectants that crosstalk with ROS signaling. NO can work either agonistically or antagonistically with ROS in plant defensive response (Neill et al., 2002). NO signaling also causes the accumulation of cGMP and cAMP though the help of Ca2+signaling, thus creating specific defense responses in stressed plants (Lamotte et al., 2004).

Salicylic acid (SA) is a phytohormone that affects plant growth and development. It has a major part in biotic stresses but its role in abiotic stresses has been discovered recently. SA has also been included in the list of phytoprotectants because of its anti-stress program. It causes the accumulation of IAA and ABA but does not influence cytokinin levels (Sakhabutdinova et al., 2003). SA has been reported to induce crosstalk of signaling pathways, which takes place at the level of NO and ROS production under salt stress. It has also been suggested as a functional link to control various stressors (Gémes et al., 2011).

Tocopherols are amphiphilic antioxidants having four isomers that lessen the ROS level in photosynthetic membranes and limit lipid peroxidation (Garg and Manchanda, 2009). Furthermore, they take part in an intricate signaling network regulated by antioxidants, phytohormones and ROS (Munne-Bosch, 2005). Polyamines, on the other hand, are low-molecular amines that regulate a wide range of plant adaptive immune responses. They show high biological activities as they are involved in plant growth and development, membrane stabilization, gene expression and adaptation to abiotic stresses. Polyamines also have a role in signaling molecules as they are an essential signal for crosstalk between hormonal signaling and ABA signaling, and induce NO in plants (Alcázar et al., 2010; Tun et al., 2006). Thus, tocopherols and polyamines are the kind of molecules that can serve as candidates to influence plant cellular signaling. The interaction of different transcription factors under abiotic stress has been shown in Figure 2.6 for further understanding.

2.10 PHYTOPROTECTANTS AND SIGNALING PATHWAY ENGINEERING

Humans have struggled for food security since the beginning of their existence. Climate change and population growth has led to resource depletion, ultimately threatening food security worldwide. But due to advancements in genetic engineering in the current era, it has become possible to overcome these threats (Koning, 2017). Continued efforts are being made to develop new biotechnology approaches for improving crop varieties in an efficient and effective manner. The production of *Bacillus thuringiensis* (*Bt*) bacterial proteins in plants is a landmark in genetically engineered crops (Haq et al.,

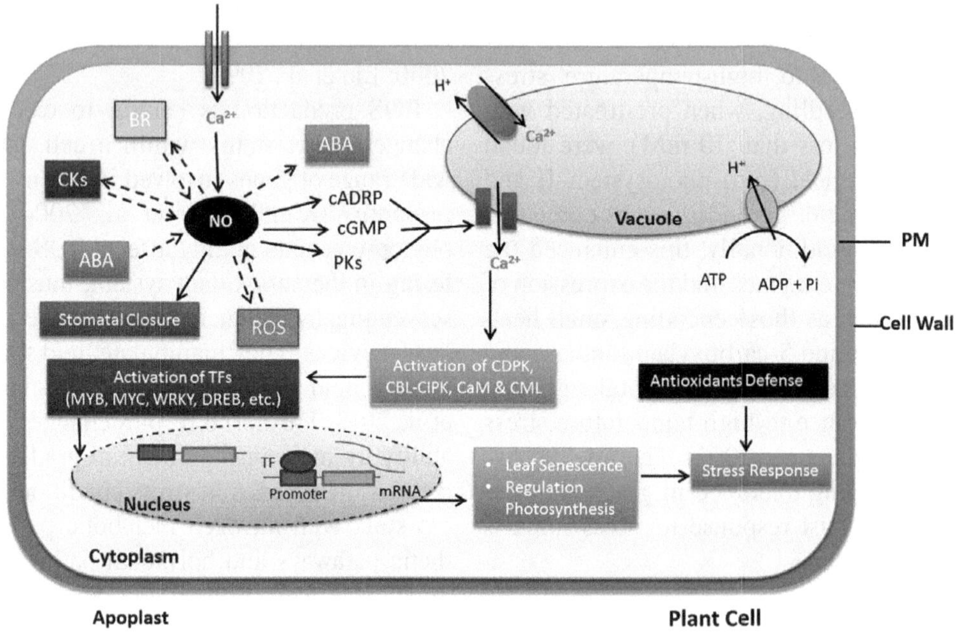

FIGURE 2.6 Phytoprotectants including nitric oxide and polyamine interact with abiotic stress signaling to generate stress responses in stressed plants. CKs cause the production of NO which induces ABA signaling, thus controlling the physiochemical reaction in a stressed plant. Furthermore, application of PAs and BR with NO promotes plant defense response against abiotic stresses. Where, NO=Nitric oxide; PAs=Polyamines; ABA=abscisic acid; ROS=Reactive Oxygen Species; CDPKs=Ca2+dependent protein kinases; CBL=Calcineurin B-like proteins. BR=Brassinosteroid; and CKs=Cytokinins CaM=Calmodulin; and CIPKs=CBL-interacting protein kinases (Asgher et al., 2017; Besson-Bard et al., 2008).

2004). Many studies have been conducted on various phytoprotectants regarding their application in plants, their role in the stress signaling network and metabolic response. A genetic transformation approach has been exploited to support plant phytoprotection under stress conditions and improved performance of crops has been observed under suboptimal conditions. Activation or inhibition of the biosynthetic and catabolic pathways can lead to an accumulation of these protectants.

Glycine betaine (Du et al.), an amine, protects plants by activating the ROS detoxification system. In genetically modified tomatoes, a 10–30% increase in production is evidence that GB is a potential candidate to protect plants against drought and cold (Park et al., 2007). Transgenic maize tolerant to drought and cold has confirmed the suitability of GB as a target for genetic engineering (He et al., 2013). γ-aminobutyric acid (GABA), an amino acid, has been associated with carbon-nitrogen pool and ROS scavenging (Liu et al., 2011). Trehalose is a non-reducing sugar playing a significant role in stress responses. Genetic analysis has demonstrated that GABA is a potent candidate for the role of phytoprotectant for biosynthetic pathway engineering to produce transgenic plants against salt and drought stress (Nounjan et al., 2012; Renault et al., 2010).

Polyols and simple sugar have also been targeted for genetic engineering because of their role as osmoprotectants. Polyols act as a molecular chaperone. and thus play a supportive role in ROS scavenging. Accumulation of straight chain polyols such as mannitol increases the tolerance of plants such as *A. thaliana*, poplar, wheat and tobacco to salinity and drought stresses (Ahmad and Wani, 2013), while a higher expression of sorbitol has been reported as being toxic due to inference in carbon metabolism (Llanes et al., 2013). Thus, mannitol biosynthetic pathway genes are a strong candidate for pathway engineering. With every passing day, research is being conducted to exploit biotechnology approaches to create plant crops with stress tolerance traits. But still, a better understanding is needed to fill the gaps in pathway engineering.

2.11 CONCLUSION AND FUTURE PROSPECTIVE

Global climate deterioration and various abiotic stresses such as heat, drought and cold adversely affect plant growth and development. Plants exert different immune responses by coordinating their physiological and molecular responses, which are important defense strategies. However, adjustment to environmental conditions involves different signaling cascades simultaneously, which makes this adaptation process more complex. A rapidly changing environment demands novel approaches to overcome threats to plant development. Using phytoprotectants as immuno-boosters under stress conditions is an interesting approach as it enhances photosynthetic rate, yield and antioxidative capacity.

Phytoprotectants such as proline and sugar can function as signaling molecules and have an obvious influence on various physiological and metabolic processes. There exists high integration between these protectants and the transcriptional activation of stress-responsive genes, ROS scavenging and calcium-mediated protective responses against environmental stresses. Extensive research has been conducted in order to understand plant abiotic stress signaling pathways mediated by phytoprotectants through the aid of powerful molecular tools including transcriptome and proteome analyses. To acquire more insights into the underlying molecular mechanism(s), a genetic transformation approach can be used to enhance the endogenous production of protectants, which will result in better performing crops under suboptimal conditions. Furthermore, novel pathways can be established in plants by introducing genes from other species. More sophisticated and high throughput techniques can be employed in genetic engineering to improve the networks of stress signaling mediated by phytoprotectants in plants, thus helping them combat abiotic stresses in a better way.

REFERENCES

Achard, P., Cheng, H., De Grauwe, L., Decat, J., Schoutteten, H., Moritz, T., Van Der Straeten, D., Peng, J., Harberd, N. P. (2006) Integration of plant responses to environmentally activated phytohormonal signals. *Science* 311(5757): 91–94.

Ahmad, P., Wani, M. R. (2013) *Physiological mechanisms and adaptation strategies in plants under changing environment, (Vol. 2)*: Springer, New York.

Aida, M., Ishida, T., Fukaki, H., Fujisawa, H., Tasaka, M. (1997) Genes involved in organ separation in Arabidopsis: An analysis of the cup-shaped cotyledon mutant. *The Plant Cell* 9(6): 841–857.

Akpinar, B. A., Avsar, B., Lucas, S. J., Budak, H. (2012) Plant abiotic stress signaling. *Plant Signaling & Behavior* 7(11): 1450–1455.

Alcázar, R., Altabella, T., Marco, F., Bortolotti, C., Reymond, M., Koncz, C., Carrasco, P.,Tiburcio, A. F. (2010) Polyamines: Molecules with regulatory functions in plant abiotic stress tolerance. *Planta* 231(6): 1237–1249.

Arasimowicz, M., Floryszak-Wieczorek, J. (2007) Nitric oxide as a bioactive signalling molecule in plant stress responses. *Plant Science* 172(5): 876–887.

Asgher, M., Per, T. S., Masood, A., Fatma, M., Freschi, L., Corpas, F. J., Khan, N. A. (2017) Nitric oxide signaling and its crosstalk with other plant growth regulators in plant responses to abiotic stress. *Environmental Science and Pollution Research International* 24(3): 2273–2285.

Bailey, P. C., Martin, C., Toledo-Ortiz, G., Quail, P. H., Huq, E., Heim, M. A., Jakoby, M., Werber, M., Weisshaar, B. (2003) Update on the basic helix-loop-helix transcription factor gene family in Arabidopsis thaliana. *The Plant Cell* 15(11): 2497–2502.

Besson-Bard, A., Pugin, A., Wendehenne, D. (2008) New insights into nitric oxide signaling in plants. *Annual Review of Plant Biology* 59: 21–39.

Boguszewska, D., Zagdańska, B. (2012) ROS as signaling molecules and enzymes of plant response to unfavorable environmental conditions. In *Oxidative Stress-Molecular Mechanisms and Biological Effects,* eds. Luschak, V., Semchyshyn, H. InTech: 342–362.

Bose, J., Pottosin, I. I., Shabala, S. S., Palmgren, M. G., Shabala, S. (2011) Calcium efflux systems in stress signaling and adaptation in plants. Frontiers in Plant Science 2: 85

Boursiac, Y., Boudet, J., Postaire, O., Luu, D. T., Tournaire-Roux, C. and Maurel, C. (2008) Stimulus-induced downregulation of root water transport involves reactive oxygen species-activated cell signalling and plasma membrane intrinsic protein internalization. *The Plant Journal* 56(2): 207–218.

Brodmann, D., Schuller, A., Ludwig-Müller, J., Aeschbacher, R. A., Wiemken, A., Boller, T., Wingler, A. (2002) Induction of trehalase in Arabidopsis plants infected with the trehalose-producing pathogen *Plasmodiophora brassicae. Molecular Plant–Microbe Interactions* 15: 693–700.

Chapman, E. J., Estelle, M. (2009) Mechanism of auxin-regulated gene expression in plants. *Annual Review of Genetics* 43: 265–285.

Chinnusamy, V. Schumaker, K., Zhu, J. K. (2004) Molecular genetic perspectives on cross-talk and specificity in abiotic stress signalling in plants. *Journal of Experimental Botany* 55(395): 225–236.

Choudhury, F. K., Rivero, R. M., Blumwald, E., Mittler, R. (2017) Reactive oxygen species, abiotic stress and stress combination. *The Plant Journal: for Cell and Molecular Biology* 90(5): 856–867.

Choudhury, S., Panda, P., Sahoo, L., Panda, S. K. (2013) Reactive oxygen species signaling in plants under abiotic stress. *Plant Signaling & Behavior* 8(4): e23681.

Ciereszko, I., Johansson, H., Kleczkowski, L. A. (2001) Sucrose and light regulation of a cold-inducible UDP-glucose pyrophosphorylase gene via a hexokinase-independent and abscisic acid-insensitive pathway in Arabidopsis. *The Biochemical Journal* 354(1): 67–72.

Ciftci-Yilmaz, S., Morsy, M. R., Song, L., Coutu, A., Krizek, B. A., Lewis, M. W., Warren, D., *et al.* (2007) The EAR-motif of the Cys2/His2-type zinc finger protein Zat7 plays a key role in the defense response of Arabidopsis to salinity stress. *The Journal of Biological Chemistry* 282(12): 9260–9268.

Contento, A. L., Kim, S. J., Bassham, D. C. (2004) Transcriptome profiling of the response of Arabidopsis suspension culture cells to Suc starvation. *Plant Physiology* 135(4): 2330–2347.

Couée, I., Sulmon, C., Gouesbet, G., El Amrani, A. (2006) Involvement of soluble sugars in reactive oxygen species balance and responses to oxidative stress in plants. *Journal of Experimental Botany* 57(3): 449–459.

Cruz, F. J. R., Castro, G. L. S., Silva Júnior, D. D., Festucci-Buselli, R. A., Pinheiro, H. A. (2013) Exogenous glycine betaine modulates ascorbate peroxidase and catalase activities and prevent lipid peroxidation in mild water-stressed Carapa guianensis plants. *Photosynthetica* 51(1): 102–108.

Cutler, S. R., Rodriguez, P. L., Finkelstein, R. R., Abrams, S. R. (2010) Abscisic acid: Emergence of a core signaling network. *Annual Review of Plant Biology* 61: 651–679.

de Azevedo Neto, A. D., Prisco, J. T., Enéas-Filho, J., Medeiros, J. V. R., Gomes-Filho, E. (2005) Hydrogen peroxide pre-treatment induces salt-stress acclimation in maize plants. *Journal of Plant Physiology* 162(10): 1114–1122.

DeFalco, T. A., Chiasson, D., Munro, K., Kaiser, B. N., Snedden, W. A. (2010) Characterization of GmCaMK1, a member of a soybean calmodulin-binding receptor-like kinase family. *FEBS Letters* 584(23): 4717–4724.

Demidchik, V. Bowen, H. C., Maathuis, F. J., Shabala, S. N., Tester, M. A., White, P. J., Davies, J. M. (2002) Arabidopsis thaliana root non-selective cation channels mediate calcium uptake and are involved in growth. *The Plant Journal: for Cell and Molecular Biology* 32(5): 799–808.

Denekamp, M., Smeekens, S. C. (2003) Integration of wounding and osmotic stress signals determines the expression of the AtMYB102 transcription factor gene. *Plant Physiology* 132(3): 1415–1423.

Dill, A., Thomas, S. G., Hu, J., Steber, C. M., Sun, T. P. (2004) The Arabidopsis F-box protein SLEEPY1 targets gibberellin signaling repressors for gibberellin-induced degradation. *The Plant Cell* 16(6): 1392–1405.

Do, P. T., Drechsel, O., Heyer, A. G., Hincha, D. K., Zuther, E. (2014) Changes in free polyamine levels, expression of polyamine biosynthesis genes, and performance of rice cultivars under salt stress: A comparison with responses to drought. *Frontiers in Plant Science* 5: 182.

Dodd, A. N., Kudla, J., Sanders, D. (2010) The language of calcium signaling. *Annual Review of Plant Biology* 61: 593–620.

Doudican, N. A., Song, B., Shadel, G. S., Doetsch, P. W. (2005) Oxidative DNA damage causes mitochondrial genomic instability in Saccharomyces cerevisiae. *Molecular and Cellular Biology* 25(12): 5196–5204.

Duan, Q., Kita, D., Li, C., Cheung, A. Y., Wu, H. M. (2010) FERONIA receptor-like kinase regulates Rho GTPase signaling of root hair development. *Proceedings of the National Academy of Sciences of the United States of America* 107(41): 17821–17826.

Dubos, C., Stracke, R., Grotewold, E., Weisshaar, B., Martin, C., Lepiniec, L. (2010) MYB transcription factors in Arabidopsis. *Trends in Plant Science* 15(10): 573–581.

Dutilleul, C., Garmier, M., Noctor, G., Mathieu, C., Chétrit, P., Foyer, C. H., de Paepe, R. (2003) Leaf mitochondria modulate whole cell redox homeostasis, set antioxidant capacity, and determine stress resistance through altered signaling and diurnal regulation. *The Plant Cell* 15(5): 1212–1226.

Eulgem, T., Rushton, P. J., Robatzek, S., Somssich, I. E. (2000) The WRKY superfamily of plant transcription factors. *Trends in Plant Science* 5(5): 199–206.

Foyer, C. H. Noctor, G. (2005) Redox homeostasis and antioxidant signaling: A metabolic interface between stress perception and physiological responses. *The Plant Cell* 17(7): 1866–1875.

Fujita, Y., Fujita, M., Satoh, R., Maruyama, K., Parvez, M. M., Seki, M., Hiratsu, K., *et al.* (2005) AREB1 is a transcription activator of novel ABRE-dependent ABA signaling that enhances drought stress tolerance in Arabidopsis. *The Plant Cell* 17(12): 3470–3488.

Gao, S. Q., Chen, M., Xu, Z. S., Zhao, C. P., Li, L., Xu, H. J., Tang, Y. M., Zhao, X., Ma, Y. Z. (2011) The soybean GmbZIP1 transcription factor enhances multiple abiotic stress tolerances in transgenic plants. *Plant Molecular Biology* 75(6): 537–553.

Garg, N., Manchanda, G. (2009) ROS generation in plants: Boon or bane? *Plant Biosystems - an International Journal Dealing with All Aspects of Plant Biology* 143(1): 81–96.

Gechev, T. S., Van Breusegem, F., Stone, J. M., Denev, I., Laloi, C. (2006) Reactive oxygen species as signals that modulate plant stress responses and programmed cell death. *BioEssays: News and Reviews in Molecular, Cellular and Developmental Biology* 28(11): 1091–1101.

Gémes, K., Poór, P., Horváth, E., Kolbert, Z., Szopkó, D., Szepesi, A., Tari, I. (2011) Cross-talk between salicylic acid and NaCl-generated reactive oxygen species and nitric oxide in tomato during acclimation to high salinity. *Physiologia Plantarum* 142(2): 179–192.

Gill, S. S., Anjum, N. A., Gill, R., Tuteja, N. (2016) Abiotic stress signaling in plants – an overview. *Abiotic Stress Response in Plants* 3: 1–12.

Giraudat, J., Parcy, F., Bertauche, N., Gosti, F., Leung, J., Morris, P. C., Bouvier-Durand, M., Vartanian, N. (1994) Current advances in abscisic acid action and signalling. *Plant Molecular Biology* 26(5): 1557–1577.

Goddijn, O. J., Verwoerd, T. C., Voogd, E., Krutwagen, R. W., de Graaf, P. T., van Dun, K., Poels, J., *et al.* (1997) Inhibition of trehalase activity enhances trehalose accumulation in transgenic plants. *Plant Physiology* 113(1): 181–190.

Golldack, D., Li, C., Mohan, H., Probst, N. (2013) Gibberellins and abscisic acid signal crosstalk: Living and developing under unfavorable conditions. *Plant Cell Reports* 32(7): 1007–1016.

Golldack, D., Li, C., Mohan, H., Probst, N. (2014) Tolerance to drought and salt stress in plants: Unraveling the signaling networks. *Frontiers in Plant Science* 5: 151.

Haake, V., Cook, D., Riechmann, J. L., Pineda, O., Thomashow, M. F., Zhang, J. Z. (2002) Transcription factor CBF4 is a regulator of drought adaptation in Arabidopsis. *Plant Physiology* 130(2): 639–648.

Hao, G. P. Xing, Y., Zhang, J. H. (2008) Role of nitric oxide dependence on nitric oxide synthase-like activity in the water stress signaling of maize seedling. *Journal of Integrative Plant Biology* 50(4): 435–442.

Haq, S. K., Atif, S. M., Khan, R. H. (2004) Protein proteinase inhibitor genes in combat against insects, pests, and pathogens: Natural and engineered phytoprotection. *Archives of Biochemistry and Biophysics* 431(1): 145–159.

Hasanuzzaman, M., Nahar, K., Fujita, M. (2013) Plant response to salt stress and role of exogenous protectants to mitigate salt-induced damages. *Ecophysiology and Responses of Plants under Salt Stress* 25–87: Springer.

Havaux, M., Kloppstech, K. (2001) The protective functions of carotenoid and flavonoid pigments against excess visible radiation at chilling temperature investigated in *Arabidopsis* npq and tt mutants. *Planta* 213(6): 953–966.

He, C., He, Y., Liu, Q., Liu, T., Liu, C., Wang, L., Zhang, J. (2013) Co-expression of genes ApGSMT2 and ApDMT2 for glycinebetaine synthesis in maize enhances the drought tolerance of plants. *Molecular Breeding* 31(3): 559–573.

Hey, S. J., Byrne, E., Halford, N. G. (2010) The interface between metabolic and stress signalling. *Annals of Botany* 105(2): 197–203.

Hirayama, T., Shinozaki, K. (2010) Research on plant abiotic stress responses in the post-genome era: Past, present and future. *The Plant Journal: for Cell and Molecular Biology* 61(6): 1041–1052.

Huang, S. S., Kirchoff, B. K., Liao, J. P. (2013) Effect of heat shock on ultrastructure and calcium distribution in *Lavandula pinnata* L. glandular trichomes. *Protoplasma* 250(1): 185–196.

Huang, X. -Y., Chao, D. -Y., Gao, J. -P., Zhu, M. -Z., Shi, M., Lin, H. -X. (2009) A previously unknown zinc finger protein, DST, regulates drought and salt tolerance in rice via stomatal aperture control. *Genes & Development* 23(15): 1805–1817.

Jiang, C. J., Aono, M., Tamaoki, M., Maeda, S., Sugano, S., Mori, M., Takatsuji, H. (2008) SAZ, a new SUPERMAN-like protein, negatively regulates a subset of ABA-responsive genes in *Arabidopsis. Molecular Genetics and Genomics: MGG* 279(2): 183–192.

Jiang, Y., Deyholos, M. K. (2009) Functional characterization of Arabidopsis NaCl-inducible WRKY25 and WRKY33 transcription factors in abiotic stresses. *Plant Molecular Biology* 69(1–2): 91–105.

Jiang, Y., Yang, B., Deyholos, M. K. (2009) Functional characterization of the Arabidopsis bHLH92 transcription factor in abiotic stress. *Molecular Genetics and Genomics: MGG* 282(5): 503–516.

Joshi, R., Wani, S. H., Singh, B., Bohra, A., Dar, Z. A., Lone, A. A., Pareek, A., Singla-Pareek, S. L. (2016) Transcription factors and plants response to drought stress: Current understanding and future directions. *Frontiers in Plant Science* 7: 1029.

Jung, C., Shim, J. S., Seo, J. S., Lee, H. Y., Kim, C. H., Do Choi, Y. D., Cheong, J. J. (2010) Non-specific phytohormonal induction of AtMYB44 and suppression of jasmonate-responsive gene activation in *Arabidopsis thaliana*. *Molecules and Cells* 29(1): 71–76.

Kandlbinder, A., Finkemeier, I., Wormuth, D., Hanitzsch, M., Dietz, K. J. (2004) The antioxidant status of photosynthesizing leaves under nutrient deficiency: Redox regulation, gene expression and antioxidant activity in Arabidopsis thaliana. *Physiologia Plantarum* 120(1): 63–73.

Kazan, K., Manners, J. M. (2013) MYC2: The master in action. *Molecular Plant* 6(3): 686–703.

Kodaira, K. S., Qin, F., Tran, L. S. P., Maruyama, K., Kidokoro, S., Fujita, Y., Shinozaki, K., Yamaguchi-Shinozaki, K. (2011) *Arabidopsis* Cys2/His2 zinc-finger proteins AZF1 and AZF2 negatively regulate abscisic acid-repressive and auxin-inducible genes under abiotic stress conditions. *Plant Physiology* 157(2): 742–756.

Koning, N. (2017) *Food Security, Agricultural Policies and Economic Growth: Long-Term Dynamics in the Past, Present and Future*. Oxford, Routledge.

Kovtun, Y., Chiu, W. L., Tena, G., Sheen, J. (2000) Functional analysis of oxidative stress-activated mitogen-activated protein kinase cascade in plants. *Proceedings of the National Academy of Sciences of the United States of America* 97(6): 2940–2945.

Kudla, J. Batistič, O., Hashimoto, K. (2010) Calcium signals: The lead currency of plant information processing. *The Plant Cell* 22(3): 541–563.

Kumar, R. (2014) Role of microRNAs in biotic and abiotic stress responses in crop plants. *Applied Biochemistry and Biotechnology* 174(1): 93–115.

Kurusu, T., Kuchitsu, K., Tada, Y. (2015) Plant signaling networks involving Ca^{2+} and Rboh/Nox-mediated ROS production under salinity stress. *Frontiers in Plant Science* 6: 427.

Kushad, M., Dumbroff, E. (1991) Metabolic and physiological relationships between the polyamine and ethylene biosynthetic pathways. In *Biochemistry and Physiology of Polyamines in Plants*, eds. Slocum, R D., Flores, H. E. Boca Raton, CRC Press: 77–92.

Lamotte, O., Gould, K., Lecourieux, D., Sequeira-Legrand, A., Lebrun-Garcia, A., Durner, J., Pugin, A., Wendehenne, D. (2004) Analysis of nitric oxide signaling functions in tobacco cells challenged by the elicitor cryptogein. *Plant Physiology* 135(1): 516–529.

Li, S., Fu, Q., Chen, L., Huang, W., Yu, D. (2011) Arabidopsis thaliana WRKY25, WRKY26, and WRKY33 coordinate induction of plant thermotolerance. *Planta* 233(6): 1237–1252.

Li, Y., Lee, K. K., Walsh, S., Smith, C., Hadingham, S., Sorefan, K., Cawley, G., Bevan, M. W. (2006) Establishing glucose-and ABA-regulated transcription networks in Arabidopsis by microarray analysis and promoter classification using a Relevance Vector Machine. *Genome Research* 16(3): 414–427.

Lindemose, S., O'Shea, C., Jensen, M. K., Skriver, K. (2013) Structure, function and networks of transcription factors involved in abiotic stress responses. *International Journal of Molecular Sciences* 14(3): 5842–5878.

Liu, C., Zhao, L., Yu, G. (2011) The dominant glutamic acid metabolic flux to produce γ-amino butyric acid over proline in Nicotiana tabacum leaves under water stress relates to its significant role in antioxidant activity. *Journal of Integrative Plant Biology* 53(8): 608–618.

Liu, L. White, M. J., MacRae, T. H. (1999) Transcription factors and their genes in higher plants. *The FEBS Journal* 262: 247–257.

Liu, Q., Kasuga, M., Sakuma, Y., Abe, H., Miura, S., Yamaguchi-Shinozaki, K., Shinozaki, K. (1998) Two transcription factors, DREB1 and DREB2, with an EREBP/AP2 DNA binding domain separate two cellular signal transduction pathways in drought-and low-temperature-responsive gene expression, respectively, in Arabidopsis. *The Plant Cell ONLINE* 10(8): 1391–1406.

Liu, Z. Q., Yan, L., Wu, Z., Mei, C., Lu, K., Yu, Y. T., Liang, S., *et al.* (2012) Cooperation of three WRKY-domain transcription factors WRKY18, WRKY40, and WRKY60 in repressing two ABA-responsive genes ABI4 and ABI5 in Arabidopsis. *Journal of Experimental Botany* 63(18): 6371–6392.

Llanes, A., Bertazza, G., Palacio, G., Luna, V. (2013) Different sodium salts cause different solute accumulation in the halophyte Prosopis strombulifera. *Plant Biology* 15: 118–125.

Ma, Y., Szostkiewicz, I., Korte, A., Moes, D., Yang, Y., Christmann, A., Grill, E. (2009) Regulators of PP2C phosphatase activity function as abscisic acid sensors. *Science* 324(5930): 1064–1068.

Maeda, H., Sakuragi, Y., Bryant, D. A., DellaPenna, D. (2005) Tocopherols protect Synechocystis sp. strain PCC 6803 from lipid peroxidation. *Plant Physiology* 138(3): 1422–1435.

Mansour, M. M. F. (2000) Nitrogen containing compounds and adaptation of plants to salinity stress. *Biologia Plantarum* 43(4): 491–500.

Meena, K. K., Sorty, A. M., Bitla, U. M., Choudhary, K., Gupta, P., Pareek, A., Singh, D. P., *et al.* (2017) Abiotic stress responses and microbe-mediated mitigation in plants: The omics strategies. *Frontiers in Plant Science* 8: 172.

Mittler, R., Kim, Y., Song, L., Coutu, J., Coutu, A., Ciftci-Yilmaz, S., Lee, H., Stevenson, B., Zhu, J. -K. (2006) Gain and loss of function mutations in Zat10 enhance the tolerance of plants to abiotic stress. *FEBS Letters* 580(28–29): 6537–6542.

Mittler, R., Vanderauwera, S., Gollery, M., Van Breusegem, F. (2004) Reactive oxygen gene network of plants. *Trends in Plant Science* 9(10): 490–498.

Mizoi, J., Shinozaki, K., Yamaguchi-Shinozaki, K. (2012) AP2/ERF family transcription factors in plant abiotic stress responses. *Biochimica et Biophysica Acta* 1819(2): 86–96.

Møller, I. M. (2001) Plant mitochondria and oxidative stress: Electron transport, NADPH turnover, and metabolism of reactive oxygen species. *Annual Review of Plant Physiology and Plant Molecular Biology* 52: 561–591.

Moschou, P. N., Paschalidis, K. A., Roubelakis-Angelakis, K. A. (2008) Plant polyamine catabolism: The state of the art. *Plant Signaling & Behavior* 3(12): 1061–1066.

Munne-Bosch, S. (2005) The role of α-tocopherol in plant stress tolerance. *Journal of Plant Physiology* 162(7): 743–748.

Nahar, K., Hasanuzzaman, M., Ahamed, K. U., Hakeem, K. R., Ozturk, M., Fujita, M. (2015) *Plant Responses and Tolerance to High Temperature Stress: Role of Exogenous Phytoprotectants Crop Production and Global Environmeental Issues*, 385–435: Springer, Cham.

Nakashima, K., Takasaki, H., Mizoi, J., Shinozaki, K., Yamaguchi-Shinozaki, K. (2012) NAC transcription factors in plant abiotic stress responses. *Biochimica et Biophysica Acta* 1819(2): 97–103.

Neill, S. J., Desikan, R., Clarke, A., Hurst, R. D., Hancock, J. T. (2002) Hydrogen peroxide and nitric oxide as signalling molecules in plants. *Journal of Experimental Botany* 53(372): 1237–1247.

Nookaraju, A., Pandey, S. K., Upadhyaya, C. P., Heung, J. J., Kim, H. S., Chun, S. C., Kim, D. H., Park, S. W. (2012) Role of Ca^{2+}- mediated signaling in potato tuberization: An overview. *Botanical Studies* 53(2): 177–189.

Nounjan, N., Nghia, P. T., Theerakulpisut, P. (2012) Exogenous proline and trehalose promote recovery of rice seedlings from salt-stress and differentially modulate antioxidant enzymes and expression of related genes. *Journal of Plant Physiology* 169(6): 596–604.

Nuruzzaman, M., Sharoni, A. M., Kikuchi, S. (2013) Roles of NAC transcription factors in the regulation of biotic and abiotic stress responses in plants. *Frontiers in Microbiology* 4: 248.

Olsen, A. N., Ernst, H. A., Leggio, L. L., Skriver, K. (2005) NAC transcription factors: Structurally distinct, functionally diverse. *Trends in Plant Science* 10(2): 79–87.

Ooka, H., Satoh, K., Doi, K., Nagata, T., Otomo, Y., Murakami, K., Matsubara, K., *et al.* (2003) Comprehensive analysis of NAC family genes in Oryza sativa and Arabidopsis thaliana. *DNA Research: an International Journal for Rapid Publication of Reports on Genes and Genomes* 10(6): 239–247.

Pandey, N., Iqbal, Z., Pandey, B. K., Sawant, S. V. (2017) Phytohormones and drought stress: Plant responses to transcriptional regulation. *Mechanism of Plant Hormone Signaling under Stress*, Vol. 2, ed. Pandey, G. K. Hoboken, John Wiley & Sons: 477–504.

Pandey, S., Nelson, D. C., Assmann, S. M. (2009) Two novel GPCR-type G proteins are abscisic acid receptors in Arabidopsis. *Cell* 136(1): 136–148.

Park, E. J., Jeknić, Z., Chen, T. H., Murata, N. (2007) The codA transgene for glycinebetaine synthesis increases the size of flowers and fruits in tomato. *Plant Biotechnology Journal* 5(3): 422–430.

Park, S. Y., Fung, P., Nishimura, N., Jensen, D. R., Fujii, H., Zhao, Y., Lumba, S., *et al.* (2009) Abscisic acid inhibits type 2C protein phosphatases via the PYR/PYL family of START proteins. *Science* 324(5930): 1068–1071.

Pei, Z. M., Murata, Y., Benning, G., Thomine, S., Klüsener, B., Allen, G. J., Grill, E., Schroeder, J. I. (2000) Calcium channels activated by hydrogen peroxide mediate abscisic acid signalling in guard cells. *Nature* 406(6797): 731–734.

Reddy, A. S., Ali, G. S., Celesnik, H., Day, I. S. (2011) Coping with stresses: Roles of calcium-and calcium/calmodulin-regulated gene expression. *The Plant Cell* 23(6): 2010–2032.

Reddy, V. S., Reddy, A. S. (2004) Proteomics of calcium-signaling components in plants. *Phytochemistry* 65(12): 1745–1776.

Reguera, M., Peleg, Z., Blumwald, E. (2012) Targeting metabolic pathways for genetic engineering abiotic stress-tolerance in crops. *Biochimica et Biophysica Acta* 1819(2): 186–194.

Renault, H., Roussel, V., El Amrani, A., Arzel, M., Renault, D., Bouchereau, A., Deleu, C. (2010) The Arabidopsis pop2-1 mutant reveals the involvement of GABA transaminase in salt stress tolerance. BMC Plant Biology 10: 20.

Riechmann, J. L., Ratcliffe, O. J. (2000) A genomic perspective on plant transcription factors. *Current Opinion in Plant Biology* 3(5): 423–434.

Rodríguez, M., Canales, E., Borrás-Hidalgo, O. (2005) Molecular aspects of abiotic stress in plants. *Biotecnología Aplicada* 22: 1–10.

Roychoudhury, A., Paul, S., Basu, S. (2013) Cross-talk between abscisic acid-dependent and abscisic acid-independent pathways during abiotic stress. *Plant Cell Reports* 32(7): 985–1006.

Sakhabutdinova, A., Fatkhutdinova, D., Bezrukova, M., Shakirova, F. (2003) Salicylic acid prevents the damaging action of stress factors on wheat plants. *Bulgarian Journal of Plant Physiology* 21: 314–319.

Schroeder, J. I., Kwak, J. M., Allen, G. J. (2001) Guard cell abscisic acid signalling and engineering drought hardiness in plants. *Nature* 410(6826): 327–330.

Seki, M., Narusaka, M., Ishida, J., Nanjo, T., Fujita, M., Oono, Y., Kamiya, A., *et al.* (2002) Monitoring the expression profiles of 7000 Arabidopsis genes under drought, cold and high-salinity stresses using a full-length cDNA microarray. *The Plant Journal: for Cell and Molecular Biology* 31(3): 279–292.

Seo, P. J., Xiang, F., Qiao, M., Park, J. Y., Lee, Y. N., Kim, S. G., Lee, Y. H., Park, W. J., Park, C. M. (2009) The MYB96 transcription factor mediates abscisic acid signaling during drought stress response in Arabidopsis. *Plant Physiology* 151(1): 275–289.

Shang, Y., Yan, L., Liu, Z. Q., Cao, Z., Mei, C., Xin, Q., Wu, F. Q., *et al.* (2010) The Mg-chelatase H subunit of Arabidopsis antagonizes a group of WRKY transcription repressors to relieve ABA-responsive genes of inhibition. *The Plant Cell* 22(6): 1909–1935.

Shen, Y. Y., Wang, X. F., Wu, F. Q., Du, S. Y., Cao, Z., Shang, Y., Wang, X. L., *et al.* (2006) The Mg-chelatase H subunit is an abscisic acid receptor. *Nature* 443(7113): 823–826.

Silverstone, A. L., Jung, H. S., Dill, A., Kawaide, H., Kamiya, Y., Sun, T. P. (2001) Repressing a repressor gibberellin-induced rapid reduction of the RGA protein in Arabidopsis. *The Plant Cell* 13(7): 1555–1566.

Smirnoff, N. (2000) Ascorbic acid: Metabolism and functions of a multi-facetted molecule. *Current Opinion in Plant Biology* 3(3): 229–235.

Soma, F., Mogami, J., Yoshida, T., Abekura, M., Takahashi, F., Kidokoro, S., Mizoi, J., Shinozaki, K., Yamaguchi-Shinozaki, K. (2017) ABA-unresponsive SnRK2 protein kinases regulate mRNA decay under osmotic stress in plants. Nature Plants 3: 16204.

Souer, E., van Houwelingen, A., Kloos, D., Mol, J., Koes, R. (1996) The no apical meristem gene of Petunia is required for pattern formation in embryos and flowers and is expressed at meristem and primordia boundaries. *Cell* 85(2): 159–170.

Srivalli, S., Khanna-Chopra, R. (2008) *Role of Glutathione in Abiotic Stress Tolerance Sulfur Assimilation and Abiotic Stress in Plants*, 207–225: Springer.

Sung, C. H., Hong, J. K. (2010) Sodium nitroprusside mediates seedling development and attenuation of oxidative stresses in Chinese cabbage. *Plant Biotechnology Reports* 4(4): 243–251.

Székely, G., Ábrahám, E., Cséplő, A., Rigó, G., Zsigmond, L., Csiszár, J., Ayaydin, F., *et al.* (2008) Duplicated P5CS genes of Arabidopsis play distinct roles in stress regulation and developmental control of proline biosynthesis. *The Plant Journal: for Cell and Molecular Biology* 53(1): 11–28.

Tun, N. N., Santa-Catarina, C., Begum, T., Silveira, V., Handro, W., Floh, E. I. S., Scherer, G. F. (2006) Polyamines induce rapid biosynthesis of nitric oxide (NO) in *Arabidopsis thaliana* seedlings. *Plant and Cell Physiology* 47(3): 346–354.

Tuteja, N., Mahajan, S. (2007) Further characterization of calcineurin B-like protein and its interacting partner CBL-interacting protein kinase from Pisum sativum. *Plant Signaling & Behavior* 2(5): 358–361.

Uchida, A., Jagendorf, A. T., Hibino, T., Takabe, T., Takabe, T. (2002) Effects of hydrogen peroxide and nitric oxide on both salt and heat stress tolerance in rice. *Plant Science* 163(3): 515–523.

Udvardi, M. K., Kakar, K., Wandrey, M., Montanari, O., Murray, J., Andriankaja, A., Zhang, J. Y., *et al.* (2007) Legume transcription factors: Global regulators of plant development and response to the environment. *Plant Physiology* 144(2): 538–549.

Ueguchi-Tanaka, M., Ashikari, M., Nakajima, M., Itoh, H., Katoh, E., Kobayashi, M., Chow, T. Y., *et al.* (2005) Gibberellin INSENSITIVE DWARF1 encodes a soluble receptor for gibberellin. *Nature* 437(7059): 693–698.

Umezawa, T., Nakashima, K., Miyakawa, T., Kuromori, T., Tanokura, M., Shinozaki, K., Yamaguchi-Shinozaki, K. (2010) Molecular basis of the core regulatory network in ABA responses: Sensing, signaling and transport. *Plant and Cell Physiology* 51(11): 1821–1839.

Umezawa, T., Sugiyama, N., Mizoguchi, M., Hayashi, S., Myouga, F., Yamaguchi-Shinozaki, K., Ishihama, Y., Hirayama, T., Shinozaki, K. (2009) Type 2C protein phosphatases directly regulate abscisic acid-activated protein kinases in Arabidopsis. *Proceedings of the National Academy of Sciences of the United States of America* 106(41): 17588–17593.

Vivancos, P. D., Driscoll, S. P., Bulman, C. A., Ying, L., Emami, K., Treumann, A., Mauve, C., Noctor, G., Foyer, C. H. (2011) Perturbations of amino acid metabolism associated with glyphosate-dependent inhibition of shikimic acid metabolism affect cellular redox homeostasis and alter the abundance of proteins involved in photosynthesis and photorespiration. *Plant Physiology* 157(1): 256–268.

Vlad, F., Rubio, S., Rodrigues, A., Sirichandra, C., Belin, C., Robert, N., Leung, J., *et al.* (2009) Protein phosphatases 2C regulate the activation of the Snf1-related kinase OST1 by abscisic acid in Arabidopsis. *The Plant Cell* 21(10): 3170–3184.

Wang, G. P., Zhang, X. Y., Li, F., Luo, Y., Wang, W. (2010) Overaccumulation of glycine betaine enhances tolerance to drought and heat stress in wheat leaves in the protection of photosynthesis. *Photosynthetica* 48(1): 117–126.

Wang, H., Wang, H., Shao, H., Tang, X. (2016) Recent advances in utilizing transcription factors to improve plant abiotic stress tolerance by transgenic technology. *Frontiers in Plant Science* 7: 67.

Wild, M., Davière, J. M., Cheminant, S., Regnault, T., Baumberger, N., Heintz, D., Baltz, R., Genschik, P., Achard, P. (2012) The Arabidopsis DELLA RGA-LIKE3 is a direct target of MYC2 and modulates jasmonate signaling responses. *The Plant Cell* 24(8): 3307–3319.

Wilkins, K. A., Matthus, E., Swarbreck, S. M., Davies, J. M. (2016) Calcium-mediated abiotic stress signaling in roots. *Frontiers in Plant Science* 7: 1296.

Yamaguchi-Shinozaki, K., Shinozaki, K. (2006) Transcriptional regulatory networks in cellular responses and tolerance to dehydration and cold stresses. *Annual Review of Plant Biology* 57: 781–803.

Yang, S., Vanderbeld, B., Wan, J., Huang, Y. (2010) Narrowing down the targets:The production of Bacillus thuringiensis (Bt) bacterial proteins into plants is a landmark Towards successful genetic engineering of drought-tolerant crops. *Molecular Plant* 3(3): 469–490.

You, J., Chan, Z. (2015) ROS Regulation during abiotic stress responses in crop plants. *Frontiers in Plant Science* 6: 1092.

Zeng, H., Xu, L., Singh, A., Wang, H., Du, L., Poovaiah, B. W. (2015) Involvement of calmodulin and calmodulin-like proteins in plant responses to abiotic stresses. *Frontiers in Plant Science* 6: 600.

Zhang, Y., Hu, X. H., Shi, Y., Zou, Z. R., Yan, F., Zhao, Y. Y., Zhang, H., Zhao, J. Z. (2013) Beneficial role of exogenous spermidine on nitrogen metabolism in tomato seedlings exposed to saline-alkaline stress. *Journal of the American Society for Horticultural Science* 138: 38–49.

Zhao, J., Davis, L. C., Verpoorte, R. (2005) Elicitor signal transduction leading to production of plant secondary metabolites. *Biotechnology Advances* 23(4): 283–333.

3 Effect of Seed Priming on Abiotic Stress Tolerance in Plants

V.K. Choudhary, Subhash Chander, C.R. Chethan, and Bhumesh Kumar

CONTENTS

3.1 INTRODUCTION

Environmental stress such as drought and submergence, nutrient excess and deficiency, salinity, temperature extremes, etc. negatively affects crop growth and productivity throughout the world. Salinity and drought, the most pervasive types of abiotic stress, are very common, and less than 10% arable land throughout the world is free from these. The soil of 34 million hectares of irrigated land is salt-affected worldwide. Waterlogging and related salinity affect 60–80 million hectares (FAO, 2011). High salinity levels take 1.5 million hectares of land out of production each year (Munns and Tester, 2008; Pitman and Lauchli, 2002). The potential yield losses by individual abiotic stress are estimated to be 17% by drought, 20% by salinity, 40% by high temperature, 15% by low temperature and 8% by other factors (Ashraf and Harris, 2004). Water deficit is a major constraint, limiting crop production worldwide especially during the germination stage, resulting in a decline or complete inhibition of seedling emergence and crop stand establishment (Kaya et al., 2006). The reductions in germination or in stand establishment were due to the drop of water potential, under moisture stress, which results in a decline in water uptake (Farooq et al., 2009a). The exposure of plants to drought stress, a type of oxidative damage caused by the overproduction of reactive oxygen species (ROS), is another major problem (Gill and Tuteja, 2010). Thus, it is necessary to alleviate the adverse effects of drought stress in order to achieve good crop yields (Ashraf and Rauf, 2001).

Irrigated lands are more affected by salinity, mainly due to inappropriate water management (irrigation and drainage), low rainfall, very high evaporation and irrigation with poor-quality saline waters. Worldwide, about 34 M ha of irrigated lands are salt-affected. High salinity levels take 1.5 M ha of land out of cultivation every year (Munns and Tester, 2008). Hence, the majority of cultivable land will be lost by the end of the 21 st century. Germination and establishment of seedlings are very sensitive to abiotic stress and this may delay the germination of high-quality seeds (Fazalali et al., 2013).

Plant health and soil fertility are being decided by the interaction between beneficial soil microbes and plants. In sustainable agriculture, rhizospheric bacteria are used to help plants achieve an easy nutrient uptake and to further the solubilization of fixed nutrients such as phosphorus (Hayat et al., 2010), as well as to minimize dependency on the use of external inputs. Symbiotic bacteria such as *Rhizobium* spp. and *Frankia* spp. and asymbiotic bacteria such as *Azotobacter, Azospirillum, Bacillus* and *Klebsiella* spp. are widely used throughout the world to increase crop growth and yield (Staley and Drahos, 1994).

3.2 SEED PRIMING

Optimum seed germination and stand establishment are prerequisite factors affecting crop production under adverse conditions. Enhancement of seed germination and stand establishment significantly contribute to seed saving and reduction in costs incurred for the same. The method used to produce a change in the germination process is known as seed priming (Paparella et al., 2015). The process of seed priming involves, first, exposure to an eliciting factor, which makes the plant more tolerant to future stress (Tanou et al., 2012). Priming is a procedure that hydrates the seed in a specific environment. Subsequently, the seed is dried to bring out its initial moisture content, so that the germination processes can begin, but restrict radicle emergence (Giri and Schillinger, 2003). Seed priming improves germination mechanisms, especially a defense mechanism that empowers seeds to compete against adverse conditions while germinating (Farooq et al., 2009b). Seed priming mechanisms include the occurrence of epigenetic changes as well as the accumulation of transcription factors and inactive forms of signaling proteins. These mechanisms are modulated upon exposure to stress and developed into a more efficient defense mechanism (Bruce et al., 2007; Tanou et al., 2012). At the sub-cellular level, seed priming was found to improve germination and seedling vigor by conferring protection to cellular proteins (Varier et al., 2010), repairing DNA damage during seed storage (Thornton et al., 1993), improving the functioning of protein synthesis machinery (Soeda et al., 2005) and increasing energy status by improving mitochondrial integrity. This led to uniform germination and seedling emergence, and better crop establishment under abiotic stress conditions. Seed priming is also used to lighten certain dormancy conditions (Bewley et al., 2013).

Seed priming is generally used to enhance germination under different conditions (Jisha et al., 2013). It is an easy, low-cost and low-risk technique generally used to conquer adverse conditions (Maiti and Pramanik, 2013). The effect of seed priming is more visible under adverse conditions than in normal conditions (Chen, 2011). Seed priming has an important role in increasing the yield of different crops. It has produced increases in the region of 37, 40, 70, 22, 31, 56, 50 and 20.6% in wheat, barley, upland rice, maize, sorghum, pearl millet, and chickpea respectively (Harris et al., 2005). The positive effects of seed priming in crops are shown in Table 3.1, while different types of seed priming and their effect on crops are shown in Tables 3.2 and 3.3.

TABLE 3.1
Positive Effect of Seed Priming on Different Crops

Crops	References
Tomato (*Solanum lycopersicum* L.)	Pradhan et al. (2014)
Hot pepper (*Capsicum annuum* var. *acuminatum* L.)	Khan et al. (2009a,b)
Pepper (*Capsicum annuum* L.)	Aloui et al. (2014)
Lettuce (*Lactuca sativa* L.)	Nasri et al. (2011)
Maize (*Zea mays* L.)	Abraha and Yohannes (2013); Tabatabaei (2014)
Okra (*Abelmoschus esculentus* L.)	Dkhil et al. (2014)
Pea (*Pisum sativum* L.)	Naz et al. (2014)
Milk thistle (*Silybum marianum* L.)	Zavariyan et al. (2015)
Soybean (*Glycine max* L.)	Miladinov et al. (2015)

Among abiotic stresses, drought is a major limiting factor for crop productivity all over the world. Drought affects almost every aspect of the physiology and biochemistry of plants, significantly reducing yield (Munns and Tester, 2008; Parida and Das, 2005).

3.2.1 Seed Priming for Drought Alleviation

Agricultural drought is defined as a period or periods during the life cycle of the crop when the supply of water is too small to meet the evaporative demand for so long that it causes the reduction in yield to be economically unacceptable. Drought occurs mostly due to the variation in quantity and distribution of rainfall. Initial, terminal and intermittent droughts during crop growing periods can be due to a delay in the onset of monsoons, as well as to their early withdrawal or intermittent breaks in their pattern.

Drought stress is a major abiotic agent that seriously decreases crop productivity in the arid and semiarid regions of the world (Lipiec et al., 2013; Yang et al., 2010). Drought stress severely inhibits seed germination, early establishment and seedling growth in the majority of crops by creating low osmotic potential, preventing water uptake (Kaya et al., 2006). Seed priming with H_2O, KNO_3 and urea all stimulated the germination

TABLE 3.2
Types of Seed Priming

Priming	What it is	Advantages
Hydro-priming	Hydro-priming means soaking the seeds in tap water before sowing and may or may not be followed by air drying of the seeds.	Hydro-priming generally enhances seed germination and seedling emergence under saline and non-saline conditions and also has a beneficial effect on enzyme activity required for rapid germination.
Halo-priming	Halo-priming refers to soaking seeds in a solution of inorganic salts, i.e. NaCl, KNO_3, $CaCl_2$ and $CaSO_4$, etc.	Halo-priming improves seed germination, seedling emergence and establishment and final crop yield in salt affected soil.
Osmo-priming	Soaking of seeds for a certain period of time in a solution of sugar, PEG, etc. followed by air drying before sowing.	Osmo-priming not only improves seed germination but also enhances crop performance under non-saline or saline conditions.
Hormonal priming	Hormonal priming is a pre-seed treatment involving several hormones such as GA_3, kinetin, abscisic acid (ABA), salicylic acid (SA) and ascorbic acid.	Hormonal priming reduces the severity of the effect of salinity. GA_3 treatment enhances vegetative growth and the deposition of Na^+ and Cl^-in both root and shoots. It also causes a significant increase in photosynthesis at the vegetative stage of the crops.
Bio-priming	Bio-priming is a new technique of seed treatment based on the inoculation of seeds with beneficial organism to protect them. Seed priming with living bacterial inoculums is termed bio-priming. This involves the application of plant-growth-promoting rhizobacteria.	Bio-priming increases speed and uniformity of germination; it also ensures a rapid, uniform and high establishment of crops; hence, it improves harvest quality and yield. Seed bio-priming allows bacteria to enter/adhere to seeds and promotes acclimatization of bacteria in the prevalent conditions.

TABLE 3.3
Type of Seed Priming in Various Crops

Crops	Type of Seed Priming	References
Wheat	Ascorbic acid and potassium salt	Farooq et al. (2013)
Lentil	Hydro-priming	Saglam et al. (2010)
Sunflower	KNO_3 and hydro-priming	Kaya et al. (2006)
Barley & Rice	Polyethylene glycol (PEG) and hydro-priming	Rouhollah (2013); Tabatabaei (2013); Sun et al. (2010); Choudhary (2015); Choudhary et al. (2017)

traits by modulating CAT, POD and SOD activities and the levels of proline and soluble sugar.

Among various strategies adopted to improve plant drought tolerance, seed priming is thought to be an easily applied, low-cost and effective approach (Ashraf and Foolad, 2005). The positive effects of priming are associated with a wide range of metabolic and physiological improvements (Shehab et al., 2010). Among them, activating protective enzymes, such as SOD, POD and CAT and accumulating osmoprotectants such as proline, soluble sugar and soluble protein are the typical stress-avoidance responses (Farhad et al., 2011). Compatible solutes also prevent destabilization during drought by replacing the water molecules around the component. Activation of enzymatic antioxidants can reduce ROS-induced oxidative damages (Posmyk et al., 2009). Osmoprotectants help enhance water uptake by improving the water status (Farooq et al., 2009a). Sugar accumulation in drought stress conditions helps maintain the stability of the membrane and protein (Lipiec et al., 2013).

In drought conditions, plants accelerate the production of ROS, and an excess of ROS causes oxidative damages to plant macromolecules. To control this excessive production and protect macromolecules, plants have evolved stronger ROS-scavenging systems including enzymatic antioxidants (SOD, CAT and POD) and non-enzymatic antioxidants (flavonoids and proline) (Gill and Tuteja, 2010; Hossain et al., 2011; Sharma et al., 2012).

Drought stress delayed or inhibited seed germination and seedling growth by causing a low osmotic potential, preventing water uptake (Kaya et al., 2006). Moreover, one of the mechanisms plants use to withstand drought stress is regulating the osmotic potential of cells, especially if drought stress increases gradually from mild to severe (Lipiec et al., 2013; Naghavi et al., 2013). Under these conditions, plants usually accumulate organic materials, such as proline and soluble sugar, to counter osmotic pressure (Farhad et al., 2011; Liu et al., 2011; Moaveni, 2011). Compatible solutes also prevent destabilization during drought by replacing the water molecules around the component. As a compatible osmolyte, proline exerts a protective function by regulating osmotic potential and scavenging free radicals (Hasegawa et al., 2000). Sugar accumulation in drought stress conditions helps maintain the stability of the membrane and protein (Lipiec et al., 2013). It has been noticed that proline and total soluble sugar contents significantly increase with increasing drought stress. However, priming enhances sugar and proline contents, which helps increase germination traits.

3.2.2 Seed Priming for Nutrient Alleviation

Plants require certain essential nutrients for optimum growth and development. These elements meet plants from the soil; thus, if any elements are lacking in the soil, growth suppression or even complete inhibition may take place. Micronutrients often act as co-factors in enzyme systems and participate in redox reactions, in addition to having several other vital functions in crops. Most micronutrients are involved in the key physiological processes of photosynthesis and respiration.

A deficiency of micronutrients can impede these vital physiological processes, thus limiting yield gain. Boron (B) deficiency, for instance, can substantially reduce yield in wheat (Rerkasem and Jamjod, 2004), chickpea (Johnson et al., 2005) and lentil (Srivastava et al., 2000); while for rice, zinc (Zn) deficiency is a major yield-limiting factor in several Asian countries (Farooq et al., 2012).

In crops, micronutrients may be applied to the soil, administered through foliar sprays or added as seed treatments. The required amounts of micronutrients can be supplied by any of these methods, but foliar sprays have been more effective in yield improvement and grain enrichment (Johnson et al., 2005). Seed treatment is a better option from an economic perspective as less micronutrient is needed, it is easy to apply and seedling growth is improved (Singh et al., 2003). Seeds may be primed with micronutrients either by soaking them in nutrient solution of a specific concentration for a specific duration or by coating them with micronutrients. Invigoration is a relatively new term and has been interchangeably used for both methods of seed treatment (Farooq et al., 2009b).

Micro-organisms can alter the root uptake of various toxic ions and nutrients by changing host physiology or by directly reducing foliar accumulation of toxic ions (Na^+ and Cl^-) while improving the nutritional status of both macro (N, P and K) and micronutrients (Zn, Fe, Cu and Mn), mostly via unknown mechanisms. Potassium plays a key role in plant water stress tolerance and has been found to be the cationic solute responsible for stomatal movements in response to changes in bulk leaf water status (Caravaca et al., 2004). Various researchers suggested the advantages and limitations of micronutrients as priming in Table 3.4.

The application of micronutrients with priming can improve stand establishment, growth, and yield; furthermore, the enrichment of grain with micronutrients is also reported in most cases (Farooq et al., 2012). Researchers highlighted the potential of nutrient priming in improving wheat, rice and forage legumes. Among micronutrients, Zn, B, Mo, Mn, Cu and Co are highly used as seed treatments for most field crops (Peeran and Natanasabapathy, 1980; Sherrell, 1984; Wilhelm et al., 1988). Seed treatment with micronutrients is a potentially low-cost way to improve the nutrition of crops. Farmers in South Asia have responded positively to seed treatment, which is a simple technique involving soaking seeds in water overnight before planting them (Harris et al., 2001). Seed priming with zinc salts is used to increase growth and disease resistance in seedlings. Silicon (Si) is an important element which has been known to augment plant defense against biotic and abiotic pressures. Maize is a Si accumulator and relatively susceptible to alkaline stress.

Plants under alkaline stress conditions show a reduction in morpho-physiological growth, relative water content (RWC), photosynthetic pigment contents, soluble sugars, total phenols and potassium ion (K^+), as well as potassium/sodium ion (K^+/Na^+) ratio. On the other hand, alkaline stress improved the contents of soluble proteins, total free amino acids, proline, Na^+ and malondialdehyde (MDA), as well as the activities of superoxide dismutase (SOD), catalase (CAT) and peroxidase (POD) in stressed plants (Abdel et al., 2016). Seed priming with Si improves the growth of plants, which is accompanied by an enhancement in RWC and levels of photosynthetic pigments, soluble sugars, soluble proteins, total free amino acids and K^+, as well as the activities of SOD, CAT, and POD enzymes (Abdel et al., 2016). Taken

TABLE 3.4
Advantages and Limitations of Micro-Nutrient Priming

Micro-nutrient Priming	Deficiency	Advantages	References
Zinc (Zn)	Poor growth and small brown spots on leaves in rice and maize plants. Zn-deficient fruit trees may have profuse growth at shoot tips with a rosette-like appearance. Interveinal leaf chlorosis and leaf mottling have also been observed in Zn-deficient citrus trees.	In barley, 44% water use efficiency. In chickpea, 10–122% more yield. 27% yield increment over non-primed seeds. Improved grain Zn content by 12% in wheat and 29% in chickpea.	Ajouri et al.(2004), Harris et al. (2005), Harris et al. (2007), Harris et al. (2008)
Boron (B)	Severe reductions in crop yield by deficiency of B, as it causes severe disturbances in B-involving metabolic processes, such as metabolism of nucleic acid, carbohydrate, protein and indole acetic acid, cell wall synthesis, membrane integrity and function, and phenol metabolism.	Grain yield increment by 8.42%. Improvement in rates of leaf emergence, leaf elongation and tiller appearance in seedlings raised from seeds primed in 0.001% B solution.	Rehman et al. (2012)
Molybdenum (Mo)	Mo is involved in nitrogen nutrition and assimilation. Mo serves an additional function, helping root nodule bacteria to fix atmospheric N.	Common bean seeds primed in sodium molybdate improved nodulation, dry matter accumulation, nitrogen fixation and yield. The yield increase in chickpea was 20% more than in non-primed seeds.	Khanal et al. (2005)
Manganese (Mn)	Mn plays a vital role in nitrogen metabolism and photosynthesis, and forms several other compounds required for plant metabolism Its deficiency is usually indicated by interveinal chlorosis; however, in some species, maturity is delayed.	In wheat, improved growth, grain yield and grain Mn contents.	
Copper (Cu)	Carbon assimilation and nitrogen metabolism are governed by Cu; hence its deficiency results in severe growth retardation. It is also involved in lignin biosynthesis, which provides strength to cell walls and prevents wilting.	Increased grain yield, but suppressed seedling emergence. Increases in seed yield and stand establishment by 43% in wheat.	Malhi (2009), Foti et al. (2008)
Cobalt (Co)	Co in legumes is required for nitrogen fixation and affects metabolism and plant growth. It is an essential component of several enzymes and co-enzymes.		

together, Si plays a pivotal role in alleviating the negative effects of alkaline stress on growth by improving water status, enhancing photosynthetic pigments, accumulating osmoprotectants rather than proline, activating the antioxidant machinery and maintaining the balance of K^+/Na^+ ratio. Thus, our findings demonstrate that seed priming with Si is an effective strategy that can be used to boost the tolerance of maize plants to alkaline stress.

3.2.3 Seed Priming for Salt Stress Alleviation

Salinity is another type of abiotic stress which significantly affects crop yield, especially in arid and semi-arid areas. Salinization spreads more in irrigated lands. This is due to inappropriate management of irrigation and drainage, low precipitation, high evaporation and irrigation with saline waters (Munns and Tester, 2008).

TABLE 3.5
Salt Affected Soils Occur in All Continents and Under Almost All Climatic Conditions (Paul and Lade, 2014)

	Total Area	Saline Soils		Sodic Soils	
Regions	**M ha**	**M ha**	**%**	**M ha**	**%**
Africa	1,899	39	2.0	34	1.8
Asia, the Pacific and Australia	3,107	195	6.3	249	8.0
Europe	2,011	7	0.3	73	3.6
Latin America	2,039	61	3.0	51	2.5
Near East	1,802	92	5.1	14	0.8
North America	1,924	5	0.2	15	0.8
Total	**12,781**	**397**	**3.1**	**434**	**3.4**

TABLE 3.6
Excess Soil Salinity Causes Poor and Spotty Stands of Crops, Uneven and Stunted Growth and Poor Yields, the Extent Depending on the Degree of Salinity

Soil Salinity Status	Conductivity of the Saturation Extracts (dS/m)	Effect on Crop Plants
Non saline	0–2	Salinity effects negligible
Slightly saline	2–4	Yields of sensitive crops may be restricted
Moderately saline	4–8	Yields of many crops are restricted
Strongly saline	8–16	Only tolerant crop yield satisfactorily
Very strongly saline	>16	Only a few very tolerant crop yield satisfactorily

Salinity affects 6% of the world's total land area, which is approximately 800 million hectares (FAO, 2012) and occurs in all continents and under almost all climatic conditions (Table 3.5). The soil of 34 million hectares of irrigated land is salt-affected worldwide (FAO, 2011). High salinity levels take 1.5 million hectares of land out of production each year (Munns and Tester, 2008; Pitman and Lauchli, 2002). Thus, 50% of cultivable lands will be lost by the middle of the 21 st century (Wang et al., 2003). High soil salinity causes poor and spotty stands of crops, uneven and stunted growth and poor yields (Table 3.6). Salt stress causes adverse physiological and biochemical changes in germinating seeds. It can affect seed germination and stand establishment through osmotic stress, ion-specific effects and oxidative stress. Salinity delays or prevents seed germination through various factors, such as a reduction in water availability and changes in the mobilization of stored reserves and affects the structural organization of proteins. Various techniques can improve emergence and stand establishment under salt conditions. Priming stimulates many of the metabolic processes involved with the early phases of germination. Seedlings from primed seeds emerge faster, grow more vigorously and perform better in adverse conditions (Cramer, 2002). However, solution composition and osmotic potential affect seed priming responses (Chinnusamy et al., 2005). It has been shown that seed priming by NaCl could be used as an adaptive method to improve salt tolerance in seeds.

Crops under salt stress alter metabolic processes and morphological characteristics to adapt. This process induces the expression of cell membranes and cell wall-related proteins, changes cell shapes, induces various kinds of osmoprotectants, such as late-embryogenesis abundant (LEA) proteins, chaperones and detoxification enzymes; protects cells, cell organelles and proteins and stimulates related stress signal transduction pathways (Seki et al., 2003; Shinozaki et al., 2003).

Similarly, Cheng et al. (2016) stated that the adaptive responses of crops to salt stress might include the following: (a) increasing ATP metabolism; (b) promoting cell softening and cell expansion; (c) repairing chloroplast structure and enhancing chlorophyll synthesis; (d) enhancing disease/defense tolerance; (e) improving cytomembrane stability; (f) strengthening transcription and signal transduction; (g) promoting the selective absorption and transportation of organic and inorganic molecules; and (h) controlling the transport of cellular materials in and out of cells, and promoting metabolism.

Seed germination and seedling growth are the two critical stages for the establishment of crops, and they are sensitive to abiotic stress. Germination of high-quality seeds may be delayed or prevented by various abiotic stresses. Among abiotic stresses, increasing salt concentration decreases germination percentage and prolongs germination time in planting zones. Lauchli and Grattan (2007) proposed the idea that delays in germination might be caused by an increase in the concentration of salt content in irrigation water. According to this, salt concentration not only decreases seed germination but also extends germination time (Kaveh et al., 2011;

TABLE 3.7
Negative Effect of Salinity on Different Crops

Crops	References
Cabbage (*Brassica oleracea* L.var. *capitata)*	Sarker et al. (2014)
Okra (*Abelmoschus esculentus* L.)	Dkhil et al. (2014)
Cowpea (*Vigna unguiculata* L. Walp.)	Thiam et al. (2013); Farhad and El-Shaieny (2015)
Celery (*Apium graveolens* L.), Fennel (*Foeniculum vulgare* Mill.) and Parsley (*Petroselinum crispum* Mill.)	–
Radish (*Raphanus sativus* var. *radicula* L.)	Sarker et al. (2014)
Water spinach (*Impomoea aquatic* Forsk.)	Sarker et al. (2014)
Milk thistle (*Silybum marianum* L.)	Zavariyan et al. (2015)

Thiam et al., 2013). Lower salt concentration induces the seed for dormancy, whereas higher concentrations, can inhibit germination and germination percentage (Khan and Weber, 2008). Seeds are generally sown in the soil at a maximum depth of 10 cm-. This layer is more saline than the lower one. Therefore, seeds show non-uniform germination and deficient seedling development.

Salt stress leads to damage to the plant cell membrane and hence increases its permeability, resulting in electrolyte leakage and accumulation of it in the surrounding tissues. The majority of crops are sensitive to salinity and are severely affected by high salt content, as shown in Table 3.7. Inoculation with *Rhizobium* and *Pseudomonas* in *Zea mays* has been reported to lower electrolyte leakage (Bano and Fatima, 2009; Sandhya et al., 2010). Similar observations were made by Shukla et al. (2012) in *Arachis hypogaea*, suggesting that PGPR protect the integrity of the plant cell membrane from the detrimental effect of salt.

Seed priming treatments in different crops are given in Table 3.8. Seed priming up-regulates the functioning of many germination-related genes, which are necessary for biophysical and biochemical processes and accelerate seedling emergence. Seed priming reprograms the gene expression for antioxidant synthesis. This defends the cell against oxidative damage and lipid peroxidation (Kubala et al., 2015a; Wahid et al., 2008). Seed priming with salt helps with the accumulation of inorganic or synthesis of organic solutes such as free proline, glycine, betaine, and free amino acids, through a process known as osmotic adjustment in response to the decreased external water potential. It also accumulates soluble sugar and promotes the activity of protective enzymes, such as superoxide dismutase, peroxidase and catalase, which are important indicators that enhance crop resistance against salinity (Kubala et al., 2015a,b; Zavariyan et al., 2015). Therefore, the accumulation of organic solutes such as sugar, compatible with plant metabolism, is an important mechanism of salt tolerance in plants.

Iqbal et al. (2006) reported that priming with NaCl and KCl was helpful in removing the deleterious effects of salts. Similarly, Kadiri and Hussaini (1999), priming with $CaCl_2$ or KNO_3 solution, increased the activity of total amylase and proteases in germinating seeds in sorghum, under salt stress. Jyotsna and Srivastava (1998) stated that $CaCl_2$ or KNO_3 generally exhibited improvement in proteins, free amino acid and soluble sugars during germinating under salt stress in pigeon pea.

3.2.4 Seed Priming for Flood Alleviation

Direct seeding of rice under the rainfed low-land condition is economically beneficial and results in very good seedling establishment. However, seedling establishment is extremely poor if seeding is followed soon by substantial rain. There are a few varieties in rice, viz. *Swarna SUB1, IR64 SUB1,* that can alleviate the consequences of flooding (Neeraja et al., 2007; Singh et al., 2009). An experiment was conducted by Sarkar (2012), showing that seed priming with water and 2% jamun (*Syzygium cumini*) leaf extract improves seedling establishment under flooding. The acceleration of growth occurred due to seed pre-treatment, which resulted in longer seedling and greater accumulation of biomass. Seed priming greatly hastened the activities of amylase and alcohol dehydrogenase in *Swarna SUB1* comprared to *Swarna*. *Swarna SUB1* outperformed *Swarna* when plants were cultivated under flooding condition. Priming had a positive effect on yield and yield-attributing parameters both under non-flooding and early flooding conditions.

3.2.5 Hormonal Priming for Stress Alleviation

Hormonal priming is a pre-seed treatment with hormones such as GA3, kinetin, ascorbate, etc., which promote growth and development in seedlings. Hormonal priming plays a crucial role in the alleviation of abiotic stress: treatment with abscisic acid (ABA), salicylic acid (SA) and ascorbic acid in wheat grown under normal and saline conditions (15 dS m^{-1}) found that germination time was reduced by 50% and final germination was increased. It was also reported that there was an improvement in fresh and dry weight of wheat (Afzal et al., 2006). This treatment has reduced the severity of the

TABLE 3.8
Seed Priming Treatments for Some Crops Under Salinity Conditions

Crop	Seed Priming Treatments	References
Field crops		
Wheat	Choline chloride (0, 5 and 10 mM) for 24 h	Salama et al. (2011), Salama and Mansour (2015)
Soybean	KNO_3 (1%), H_2O_2 (0.1%), and H_2O with the duration of 6 h, 12 h, 18 h, and 24 h.	Miladinov et al. (2015)
Green gram	β-amino butyric acid solution (0, 0.5, 1.0, 1.5, 2.0, and 2.5 mM) for 6 h	Jisha and Puthur, (2016)
Rape	Polyethylene glycol (PEG) 6000 solution (osmotic potential −1.2 MPa) during 7 days at 25°C in the darkness	Kubala et al. (2015b)
Maize	Salicylic acid (25 ppm) and NaCl (−10 bar)	Tabatabaei (2014)
	NaCl solution (5 g/L) for 12 h at room temperature	Abraha and Yohannes (2013)
Pea	KCl (250 ppm) and KOH (500 ppm)	Naz et al. (2014)
Sunflower	Polyethylene glycol 6000 (−2 bar) and hydro-priming	Moghanibashi et al. (2013)
Vegetables		
Faba bean	Melatonin (100 and 500 mM) for 12 h at 25°C	Dawood and EL-Awadi (2015)
Alfalfa	GA_3 (0, 3, 5 and 8 mM) for 24 h at room temperature	Younesi and Moradi (2015)
Capsicum	Seeds primed with three solutions of NaCl, KCl and $CaCl_2$ (0, 10, 20 and 50 mM) for (12, 24 and 36 h)	Aloui et al. (2014)
	Glycinebetaine (0, 1, 5, 10 or 25 mM) for 24 h	Korkmaz and Şirikçi (2011)
	NaCl solution (1 mM)	Khan et al. (2009a)
	Salicylic acid (0.8 mM) and acetylsalicylic acid (0.2 mM) for 48 h at 25°C	Khan et al. (2009b)
Okra	KCl 4%, mannitol 0.75 M or $CaCl_2$ 10 mM for 24 h at 20°C	Dkhil et al. (2014)
Tomato	Polyethylene glycol 6000 (−0.5, −1.0, −1.5 and −2.0 MPa) and hydro-priming for 48 h at 25°C	Pradhan et al. (2014)
	NaCl (300 mM) or distilled water for 24 h at 25°C in the dark	Nakaune et al. (2012)
	Seeds were mixed with sand particles containing 4% (v/w) water, sealed in a plastic box, for 72 h at 25°C	Nawaz et al. (2012)
Cucumber	2 mM $CaCl_2$ solution for 24	Joshi et al. (2013)
Pumpkin	Ascorbic acid (0, 500 and 1000 M) for 24 h at 20°C.	Fazalaliet al. (2013)
Lettuce	KNO_3 (0.05%) for 2 h at 25°C in the dark	Nasri et al. (2011)
	GA_3 (0, 3, 4.5 and 6 mM) for 12 h at room temp.	Hela et al. (2012)
Muskmelon	100 mmol NaCl solution for 36 h at 20°C	Farhoudi et al. (2011)
Sweet fennel	Gibberellic acid 20 mg/l, 25% Manitol, 3% NaCl and distilled water for 24 h	Sedghi et al. (2010)

effects of salinity, but an even better effect was noticed with 50 ppm SA and 50 ppm ascorbic acid. Treatment with GA3 enhanced the vegetative growth of two wheat cultivars. It enhanced the deposition of Na^+ and Cl^- in both shoots and roots and there was also an increase in photosynthetic activities at the vegetative stage. However, this largely depends on plant species, priming media and its concentration, priming duration, temperature and storage conditions, etc. Assefa et al. (2010) reported that seed priming with GA3 enhances the emergence and germination rate of soybean. Cytokinins can also be used as a priming agent as they are mainly involved in the breakdown of dormancy in some seeds (Arteca, 1996).

Bassi (2005) found that priming with GA enhanced germination and emergence (94 and 82%, respectively) as compared to non-primed seed treatment (85 and 77%, respectively). Bassi et al. (2011) reported that priming with GA_3 at 50 ppm for 2 hours enhanced emergence, germination and speed of germination in soybean as compared to non-primed seed lots. Sarika et al. (2013) reported that in French bean, chemo priming with GA and Ethrel improved seed quality and showed improved seedling length and seedling dry weight, which in turn improved seedling vigor index, germination speed and mean germination time. Significant improvements in root and shoot length, seedling vigor index and dry seedling weight were also observed. Kata et al. (2014) revealed that seed priming for 12 hours in seven priming media (salicylic acid 50 ppm, ascorbic acid 200 ppm, citric acid 200 ppm, proline 0.2%, calcium chloride 2%, Na_2HPO_4 100 ppm and distilled water) resulted in improved germination properties of paddies in ascorbic acid and salicylic acid pre-treatment at 200 ppm and 50 ppm respectively under heat stress conditions due to their higher antioxidant capacity over others.

3.2.6 Bio-Priming on Abiotic Stress Alleviation

Bio-priming has significant effects against drought; it was found that two strains, *A. brasilense* and *B. amyloliquefaciens*, up-regulated the genes related to stress in wheat and increased tolerance against drought (Kasim et al., 2013). PGPR supplies nutrients to crops, stimulates plant growth by producing phytohormones, provides bio-control of phytopathogens, and improves soil structure, bio-accumulation of organic compounds and bio-remediation of metal-contaminated soils. It has been found that bio-priming has a significant effect on tolerance against salinity stress. *Bacillus* is the prominent genus used in abiotic stress tolerance, especially in potato (Gururani et al., 2013), radish (Kaymak et al., 2009) rice, mungbean and chickpea (Chakraborty et al., 2011).

Gururani et al. (2013) found that *Bacillus pumilus, B. furmus* works with ACC-deaminase activity, IAA production, phosphate solubilization, phytate mineralization and siderophor reproduction, which are responsible for salinity, drought and heavy metal stress tolerance and lead to increase in plant height, number of leaves/plant, number of tubers/plant and tuber yield/plant in potato. Chakraborty et al. (2011) stated that *Bacillus cereus* regulates phosphate solubilization, IAA, catalase, protease, chitinase and siderophor reproduction, nitrate reduction and starch hydrolysis, which are responsible for salinity tolerance and an increase in seedling height, number and length of leaves, and root and shoot biomass in rice, greengram and chickpea. Kaymak et al. (2009) found that *Agrobacterium rubi, Burkholderia gladioli, P. putida, B. subtilis* and *B. megaterium* improved seed germination under saline conditions in radish.

PGPR-inoculated plants have shown an increase in K^+ concentration, which in turn resulted in a high K^+/Na^+ ratio leading to their effectiveness in salinity tolerance (Kohler et al., 2009; Nadeem et al., 2013; Rojas-Tapias et al., 2012). Ashraf et al. (2004) found that *Azospirillum* could restrict Na^+ influx into roots. In addition, high K^+/Na^+ ratios were found in salt-stressed maize in which selectivity for Na^+, K^+ and Ca^{2+} was altered in favor of the plant, upon inoculation with *Azospirillum* (Hamdia et al., 2004).

Salinity not only reduces Ca^{2+} and K^+ availability in plants but also reduces Ca^{2+} and K^+ mobility and transport to the growing parts of plants. However, Fu et al. (2010) reported significantly increased Ca^{2+} in shoots of eggplants inoculated with *Pseudomonas* when compared to non-inoculated eggplant under saline conditions. Yao et al. (2010) demonstrated that PGPRs are involved in significantly increasing cotton's absorbability of Mg^{2+} and Ca^{2+} and decreasing the absorption of Na^+. It has also been shown that Ca^{2+} plays a major role as an early signaling molecule at the onset of salinity.

Strains primed in different crops and PGP activities have significant effect on crops. Bio-fertilizers containing *Bacillus lentus, B. subtilis, Pseudomonas fluorescens, P. putida* and *Azospirillum* spp. have several benefits on agro-morphological traits in wheat. These also reduce nitrogen and phosphorus requirements in crops. In a laboratory study, priming with PGPR increased the fresh weight of stems and seedlings of maize. Similarly, bio-priming of different strains of *Azotobacter* and *Azospirillum* in maize significantly increased crop growth rate, dry matter accumulation and grain yield (Sharifi, 2011). Bio-priming with a consortium of *Azotobacter chroococcum* and *Azospirillum lipoferum* along with 80 kg N and 60 kg P_2O_5/ha increased dry matter accumulation, yield attributes and biological yield in barley (Mirshekari et al., 2012). In safflower, *Pseudomonas* strain significantly increased growth, yield attributes and yield (Sharifi, 2012). Bio-priming and its potential effects have been expanding at a considerable rate. Bio-stimulants have a significant role in inducing growth promotion and nutrient availability incrops (Table 3.9).

3.2.6.1 Role of Bio-Priming in Different PGP Activities

Mirshekari et al. (2012) reported that *Azotobacter chroococcum* and *A. lipoferum* improve test weight, dry biomass, grain and biological yield of barley. Sharifi (2012) found that *Pseudomonas* spp. increase branches, head diameter, grain number, test weight, grain and oil yield of safflower. Whereas *Azotobacter chroococcum, Azospirillum lipoferum, A. chroococcum* and *A. lipoferum* improve dry biomass, crop growth and grain yield of maize. Sharifi and Khavazi (2011) noticed an increase in growth, number of grains and rows/cob and grain yield of maize treated with *Azotobacter* and *Azospirillum* spp. Moeinzadeh et al. (2010) found that shoot height, root length and seedling weight of sunflower increased with *Pseudomonas fluorescens*. Bennett et al. (2009) reported that *Clonostachys rosea, P. chlororaphis, P. fluorescens, T. harzianum* and *T. viride* help increase the emergence and yield of carrot and onion.

TABLE 3.9
Biostimulants Ameliorating the Effects of Abiotic Stress

Biostimulants	References
Protein hydrolysates	Colla et al. (2015)
Seaweed extracts	Battacharyya et al. (2015)
Silicon	Savvas and Ntatsi (2015)
Chitosan	Pichyangkura and Chadchawan (2015)
Humic and fulvic acids	Canellas et al. (2015)
The role of phosphite	Gómez-Merino and Trejo-Tellez (2015)
Arbuscular mycorrhizal fungi	Rouphael et al. (2015)
Trichoderma	López-Bucio et al. (2015)
Plant growth-promoting rhizobacteria	Ruzzi and Aroca (2015)

Mahmood et al. (2016) stated that there are certain advantages and limitations to different application methods, which are as follows:

3.2.6.1.1 Carrier-Based Inoculation

This is a low-cost technique, easy to prepare and readily available, but has limitation that it may get contaminated with unwanted inoculums from carriers, such as peat. It also has limited storage ability with non-uniformity of carriers.

3.2.6.1.2 Seed Coating and Covering

This can be easily applied. It does not require any specific machinery and it is being practised by farmers in cases of pesticide application. However, the sticking agent may be harmful to bacteria with less flexibility in seeding.

3.2.6.1.3 Pelleting

It is easy to apply, flexible and liked by farmers, but the survival of bacteria is hindered mainly due to low moisture levels. Machinery is required to prepare it, therefore costs are higher.

3.2.6.1.4 Direct Soil Application

Injection in the root zone is possible, easy and simple. However, when it is exposed to the sun, desiccation problems may occur. Moreover, it is required in higher volumes.

3.2.6.1.5 Root Dipping

This is simple and easy but requires a nursery. Large amounts of liquid media and bacterial cells are needed, and they might carry contaminants from the environment.

3.3 SEED COATING AND PELLETING

Seed coating is the process of covering seeds with external materials to improve handling, protection and, to a lesser extent, to enhance germination and plant establishment (Pedrini et al., 2017). The materials used in the seed coating process can be broadly categorized according to their function as binders, fillers and active ingredients. Binders and fillers must be compatible with active compounds, and not adversely impact the ability of a seed to germinate and grow.

According to the characteristics of the natural seed coat (testa), applied compounds can be dissolved and transmitted into the seed via imbibing water, or if the testa is impermeable to those substances, through uptake from the emerging radicle and root system (Salanenka and Taylor, 2008, 2011).

3.3.1 Protectants

The most regularly used active ingredients in seed coatings include fungicides, pesticides, insecticides and nematicides. These protectants slightly promote germination and emergence, and sometimes adversely affect the rate of germination (Yang et al., 2014). However, protectants enhance plant growth and yield through decreasing predation and infection by insect pest. In spite of many benefits, sometimes protectants do have an undesirable off-target ecological effect. For example, neonicotinoids, a commonly used insecticidal complex (Jeschke et al., 2011) in crop seed coating, have been shown to have harmful effects on wild-bee diversity and distribution (Rundlof et al., 2015), with indirect impacts on honeybee health (Alburaki et al., 2015). Furthermore, fungicidal and insecticidal coating have indirect effects on the soil seed bank, potentially interfering with agroecosystem processes (Smith et al., 2016).

3.3.2 Nutrients

Analysis of nutrient alterations in seed coating and their effects on germination, growth, and yield is usually positive. However, although the application of macronutrients such as phosphorus (Peltonen-Sainio et al., 2006) and potassium (Tavares et al., 2013) can improve growth and yield, there is a possibility of harmful impacts on germination and emergence (Masauskas et al., 2008; Peltonen-Sainio et al., 2006) caused by nutrient-induced osmotic stress (Scott, 1989). The majority of nutrient alterations have instead focused on the delivery of micro-nutrients such as zinc (Adhikari et al., 2016; Oliveira et al., 2014), boron (Rehman et al., 2012;

Rehman and Farooq, 2013), molybdenum (Hara, 2013), copper (John et al., 2005; Wiatrak, 2013b) and manganese (Wiatrak, 2013a). These alterations have been used to indemnify for soil deficiencies of micro-nutrients (Farooq et al., 2012). The amalgamation of seed biology, plant physiology, and soil science could be used by the seed industry to optimize the use of seed coating as a way of delivering nutrients, ultimately allowing for the cultivation of varieties with predefined micro-nutrient requirements tailored to soil types with different trace element deficiencies.

3.3.3 Symbionts

Rhizobia are symbiotic organisms that are most commonly used for seed coating and the inoculation of legumes (Deaker et al., 2004). Their use leads to an improvement in seedling growth and, to a lesser extent, germination. On the other hand, the incorporation of inocula in an artificial seed coat results in loss of microbial viability and coated seeds unable to be stored for extended periods (Scott, 1989). The artificial seed coat usually has an antagonistic effect on rhizobia, generally due to desiccation (McIntyre et al., 2007) and osmotic stress (John et al., 2010). Protectant compounds could pose a threat to the survival and biological activity of symbiotic bacteria (Scott, 1989). Thus, the identification of rhizobia-friendly coating formulations, along with the selection of desiccation and osmotic stress-resistant bacteria, could enhance symbiotic bacteria survival and increase storage life.

3.3.4 Soil Adjuvants

In seed coating, hydrophilic materials (hydrogels or hydro absorbers) are commonly used compounds due to their innate capability to draw and retain water in the vicinity of the seed (Gorim and Asch, 2012; Mangold and Sheley, 2007; Serena et al., 2012). A new strategy to enhance water availability to seeds and seedlings in water-repellent soil is to apply a soil surfactant within the seed-coating material (Honglu and Guomei, 2008; Madsen et al., 2013). Polymer coatings have been used to postpone germination through influencing water absorption, in a sense creating an artificial dormancy. Such type of coating hampers germination when climatic conditions (water availability) are not favorable (Archer and Gesch, 2003). It generally provides protection from pathogens, fungi, and predators (Johnson et al., 1999). This method of seed coating is suitable for early sowing and it relies on the coat to trigger the germination process when optimal moisture is available (Archer and Gesch, 2003; Stokes, 2001). Thus, it can improve seedling emergence in no-tillage soil (Gesch et al., 2012). The delay is usually achieved through temperature-activated polymers that regulate water uptake at predefined temperature thresholds.

3.3.5 Phytoactive Promoters

Phytoactive promoters encompass an array of compounds if coated on to seeds. They potentially enhance germination, encourage growth and improve stress resistance. A very small number of promoters has been considered or disclosed in publications. Looking at commercial products, we can see how phytoactive promoters have gone mostly unnoticed. However, the benefits of tested phytoactive compounds suggest a potential for significant improvement in crop performance. Further investigations of these compounds are needed to better understand the efficiency of delivery through the seed coat. The use of promoters has the potential to improve seedling and plant vigor, resistance to biotic/abiotic stresses and performance under water, salinity and temperature stress conditions. Innovation in the use of phytoactive compounds via seed coats and pellets could be a key part in making farming possible in degraded areas or areas adversely affected by climatic change (Pedrini et al., 2017).

3.4 CONCLUSION

It can be concluded that seed priming enhances germination and seed performance under different abiotic conditions. It develops defense mechanism in seeds, such as the antioxidant defense system and osmotic adjustment. These mechanisms form a 'primary memory' in seeds, which can be recruited upon a later exposure to abiotic stress and help achieve greater tolerance in germination and establishment. Further systematic research is required on the interaction of various abiotic stresses, which can occur one at a time or simultaneously.

Similarly, Cheng et al. (2016) stated that the adaptive responses of crops to abiotic stresses might include (a) increasing ATP metabolism; (b) promoting cell softening and cell expansion; (c) repairing chloroplast structure and enhancing chlorophyll synthesis; (d) improving cytomembrane stability; (e) strengthening transcription and signal transduction; (f) promoting the selective absorption and transportation of organic and inorganic molecules; (g) controlling the transport of cellular materials in and out of cells; and (h) promoting metabolism and enhancing disease/defense tolerance.

REFERENCES

Abdel Latef, A.A., Tran, L.S. (2016) Impacts of priming with silicon on the growth and tolerance of maize plants to alkaline stress. *Frontiers in Plant Science* 7: 243.

Abraha, B., Yohannes, G. (2013) The role of seed priming in improving seedling growth of maize (*Zea mays* L.) under salt stress at field conditions. *Agricultural Sciences* 4: 666–672.

Adhikari, T., Kundu, S., Rao, A.S. (2016) Zinc delivery to plants through seed coating with nano-zinc oxide particles. *Journal of Plant Nutrition* 39: 136–146.

Afzal, S., Nadeem, A., Zahoor, A., Qaiser, M. (2006) Role of seed priming with zinc in improving the hybrid maize (*Zea mays*) yield. *American-Eurasian Journal of Agricultural & Environmental Sciences* 13: 301–306.

Ajouri, A., Asgedom, H., Becker, M. (2004) Seed priming enhances germination and seedling growth of barley under conditions of P and Zn deficiency. *Journal of Plant Nutrition and Soil Science* 167: 630–636.

Alburaki, M., Boutin, S., Mercier, P.L., Loublier, Y., Chagnon, M., Derome, N. (2015) Neonicotinoid-coated *Zea mays* seeds indirectly affect honeybee performance and pathogen susceptibility in field trials. *PLOS ONE* 10: e0125790.

Aloui, H., Souguir, M., Latique, S., Hannachi, C. (2014) Germination and growth in control and primed seeds of pepper as affected by salt stress. Cercetari Agronomice in Moldova 47: 83–95.

Archer, D.W., Gesch, R.W. (2003) Value of temperature-activated polymer-coated seed in the northern corn belt. *Journal of Agricultural and Applied Economics* 35: 625–637.

Ashraf, M., Foolad, M.R. (2005) Pre-sowing seed treatment – a shotgun approach to improve germination, plant growth, and crop yield under saline and non-saline conditions. Advances in Agronomy 88: 223–271.

Ashraf, M., Harris, P.J.C. (2004) Potential biochemical indicators of salinity tolerance in plants. Plant Science 166: 3–16.

Arteca, R.N. (1996) Brassinosteroids. In Plant Hormones, Physiology, Biochemistry and Molecular Biology. ed. Davies, P.J., 206–213.New York: Huwer Academic Publishers.

Ashraf, M., Hasnain, S., Berge, O., Mahmood, T. (2004) Inoculating wheat seedlings with exopolysaccharide-producing bacteria restricts sodium uptake and stimulates plant growth under salt stress. *Biology and Fertility of Soils* 40: 157–162.

Ashraf, M., Rauf, H. (2001) Inducing salt tolerance in maize (*Zea mays* L.) through seed priming with chloride salts, growth and ion transport at early growth stages. Acta Physiologiae Plantarum 23: 407–414.

Assefa, M.K., Hunje, R., Koti, R.V. (2010) Enhancement of seed quality in soybean following priming treatment. *Karnataka Journal of Agricultural Sciences* 23: 787–789.

Bano, A., Fatima, M. (2009) Salt tolerance in *Zea mays* (L) following inoculation with Rhizobium and Pseudomonas. *Biology and Fertility of Soils* 45: 405–413.

Bassi, G. (2005) Seed priming for invigourating late sown wheat (*Triticum aestivum*). *Crop Improvement* 32: 121–123.

Bassi, G., Sharma, S., Gill, B.S. (2011) Pre-sowing seed treatment and quality in-vigouration in soybean [*Glycine max* (L) Merrill]. *Seed Research* 31: 81–84.

Battacharyya, D., Babgohari, M.Z., Rathor, P., Prithiviraj, B. (2015) Seaweed extracts as biostimulants in horticulture. *Scientia Horticulturae* 196: 39–48.

Bennett, A.J., Mead, A., Whipps, J.M. (2009) Performance of carrot and onion seed primed with beneficial microorganisms in glasshouse and field trials. *Biological Control* 51: 417–426.

Bewley, J.D., Bradford, K.J., Hilhorst, H.W.M., Nonogaki, H. (2013) Seeds physiology of development. In *Germination and Dormancy*, 3rd edn. Springer: New York. 392 pp.

Bruce, T.J.A., Matthes, M.C., Napier, J.A., Pickett, J.A. (2007) Stressful memories of plants: Evidence and possible mechanisms. *Plant Science* 173: 603–608.

Canellas, L.P., Olivares, F.L., Aguiar, N.O., Jones, D.L., Nebbioso, A., Mazzei, P., Piccolo, A. (2015) Humic and fulvic acids as biostimulants in horticulture. *Scientia Horticulturae* 196: 15–27.

Caravaca, F., Figueroa, D., Barea, J.M., Azcon-Aguilar, C., Roldan, A. (2004) Effect of mycorrhizal inoculation on nutrient acquisition, gas exchange, and nitrate reductase activity of two Mediterraneanautochthonous shrub species under drought stress. Journal of Plant Nutrition 27: 57–74.

Chakraborty, U., Roy, S., Chakraborty, A.P., Dey, P., Chakraborty, B. (2011) Plant growth promotion and amelioration of salinity stress in crop plants by a salt-tolerant bacterium. *Recent Research in Science and Technology* 3: 61–70.

Chen, K. (2011) Antioxidants and Dehydrin Metabolism Associated with Osmopriming-Enhanced Stress Tolerance of Germinating Spinach (*Spinacia oleracea* L. cv. Bloomsdale) Seeds. PhD, Iowa State University, Ames, Iowa, USA.

Cheng, X., Deng, G., Sub, Y., Liu, J.J., Yang, Y., Du, G.H., Chen, Z.Y., Liu, F.H. (2016) Protein mechanisms in response to NaCl-stress of salt-tolerant and salt-sensitive industrial hemp based on iTRAQ technology. Industrial Crops and Products 83: 444–452.

Chinnusamy, V., Jagendorf, A., Zhu, J.K. (2005) Understanding and improving salt tolerance in plants. *Crop Science* 45: 437–448.

Choudhary, V.K. (2015) Surface drainage in transplanted rice: Productivity, relative water and leaf rolling, root behaviour and weed dynamics. *Proceeding of the National Academy of Sciences, India Section B. (Biological Sciences)* 87: 869–876.

Choudhary, V.K., Choudhury, B.U., Bhagawati, R. (2017) Seed priming and in situ moisture conservation measures in increasing adaptive capacity of rain-fed upland rice to moisture stress at Eastern Himalayan Region of India. *Paddy and Water Environment* 15: 343–357.

Colla, G. Nardi, S., Cardarelli, M., Ertani, A., Lucini, L., Canaguier, R., Rouphael, Y. (2015) Protein hydrolysates as biostimulants in horticulture. *Scientia Horticulturae* 196: 28–38.

Cramer, G.R. (2002) Sodium-calcium interactions under salinity stress. In Salinity: Environment-Plants-Molecules, 205–227. Dordrecht: Springer.

Dawood, M.G., EL-Awadi, M.E. (2015) Alleviation of salinity stress on Vicia faba L. plants via seed priming with melatonin. *Acta Biológica Colombiana* 20: 223–235.

Deaker, R., Roughley, R.J., Kennedy, I.R. (2004) Legume seed inoculation technology—A review. *Soil Biology and Biochemistry* 36: 1275–1288.

Dkhil, B.B., Issa, A., Denden, M. (2014) Germination and seedling emergence of primed okra (*Abelmoschus esculentus* L.) seeds under salt stress and low temperature. *American Journal of Plant Physiology* 9: 38–45.

FAO (2011) *The State of the World's Land and Water Resources for Food and Agriculture (SOLAW)—Managing Systems at Risk*. Rome: Food and Agriculture Organization of the United Nations and London: Earthscan Publications.

FAO (2012) *Land and Plant Nutrition Management Service*. Rome: Food and Agriculture Organization of the United Nations and London: Earthscan Publications.

Farhad, W., Cheema, M.A., Saleem, M.F., Saqib, M. (2011) Evaluation of drought tolerance in maize hybrids. *International Journal of Agricultural and Biological Engineering* 13(4): 523–528.

Farhad, W. El-Shaieny, A.A.H. (2015) Seed germination percentage and early seedling establishment of five (*Vigna unguiculata* L.) genotypes under salt stress. *European Journal of Experimental Biology* 5: 22–32.

Farhoudi, R., Saeedipour, S., Mohammadreza, D. (2011) The effect of NaCl seed priming on salt tolerance, antioxidant enzyme activity, proline and carbohydrate accumulation of muskmelon (*Cucumismelo* L.) under saline condition. African Journal of Agricultural Research 6: 1363–1370.

Farooq, M., Basra, S.M.A., Wahid, A., Ahmad, N., Saleem, B.A. (2009a) Improving the drought tolerance in rice (*Oryza sativa* L.) by exogenous application of salicylic acid. *Journal of Agronomy and Crop Science* 195: 237–246.

Farooq, M., Irfan, M., Aziz, T., Ahmad, I., Cheema, S.A. (2013) Seed priming with ascorbic acid improves drought resistance of wheat. *Journal of Agronomy and Crop Science* 199: 12–22.

Farooq, M., Nawaz, A., Iqbal, S., Rehman, A. (2012) Optimizing the boron seed coating treatments for improving the germination and early seedling growth of fine grain rice. *International Journal of Agriculture & Biology* 14: 453–456.

Farooq, M., Wahid, A., Kobayashi, N., Fujita, D., Basra, S.M.A. (2009b) Plant drought stress: Effects, mechanisms and management. *Agronomy for Sustainable Development* 29: 185–212.

Farooq, M., Wahid, A., Siddique, K.H.M. (2012) Micronutrient application through seed treatments—A review. *Journal of Soil Science and Plant Nutrition* 12: 125–142.

Fazalali, R., Asli, D.E., Moradi, P. (2013) The effect of seed priming by ascorbic acid on bioactive compounds of naked seed pumpkin (*Cucurbita pepo* var. *styriaca*) under salinity stress. *International Journal of Farming and Allied Science* 2: 587–590.

Foti, R., Abureni, K., Tigere, A., Gotosa, J., Gere, J. (2008) The efficacy of different seed priming osmotica on the establishment of maize (*Zea mays* L.) caryopses. *Journal of Arid Environments* 72: 1127–1130.

Fu, Q.L., Liu, C., Ding, N.F., Lin, Y.C., Guo, B. (2010) Ameliorative effects of inoculation with the plant growth-promoting rhizobacterium Pseudomonas sp. DW1 on growth of eggplant (*Solanum melongena* L.) seedlings under salt stress. *Agricultural Water Management* 97: 1994–2000.

Gesch, R.W., Archer, D.W., Spokas, K. (2012) Can using polymer-coated seed reduce the risk of poor soybean emergence in no-tillage soil? *Field Crops Research* 125: 109–116.

Gill, S.S., Tuteja, N. (2010) Reactive oxygen species and antioxidant machinery in abiotic stress tolerance in crop plants. *Plant Physiology and Biochemistry: PPB* 48: 909–930.

Giri, G.S., Schillinger, W.F. (2003) Seed priming winter wheat for germination, emergence and yield. Crop Science 43: 2135–2141.

Gómez-Merino, F.C., Trejo-Tellez, L.I. (2015) Biostimulant activity of phosphite in horticulture. *Scientia Horticulturae* 196: 82–90.

Gorim, L., Asch, F. (2012) Effects of composition and share of seed coatings on the mobilization efficiency of cereal seeds during germination. *Journal of Agronomy and Crop Science* 198: 81–91.

Gururani, M.A., Upadhyaya, C.P., Baskar, V., Venkatesh, J., Nookaraju, A., Park, S.W. (2013) Plant growth promoting rhizobacteria enhance abiotic stress tolerance in *Solanum tuberosum* through inducing changes in the expression of ROS-scavenging enzymes and improved photosynthetic performance. *Journal of Plant Growth Regulation* 32: 245–258.

Hamdia, M.A.E., Shaddad, M.A.K., Doaa, M.M. (2004) Mechanisms of salt tolerance and interactive effects of *Azospirillum brasilense* inoculation on maize cultivars grown under salt stress conditions. *Plant Growth Regulation* 44: 165–174.

Hara, Y. (2013) Improvement of rice seedling establishment in sulfate-applied submerged soil by application of molybdate. Plant Production Science 16: 61–68.

Harris, D., Raghuwanshi, B.S., Gangwar, J.S., Singh, S.C., Joshi, K.D., Rashid, A., Hollington, P.A. (2001) Participatory evaluation by farmers of 'on-farm' seed priming in wheat in India, Nepal and Pakistan. *Experimental Agriculture* 37: 403–415.

Harris, D., Rashid, A., Arif, M., Yunas, M. (2005) Alleviating micronutrient deficiencies in alkaline soils of the North-West Frontier Province of Pakistan: On-farm seed priming with zinc in wheat and chickpea. In *Micronutrients in South and South East Asia*, ed. Andersen, P., Tuladhar, J.K., Karki, K.B., Maskey, S.L., 143–151. Kathmandu: ICIMOD.

Harris, D., Rashid, A., Miraj, G., Arif, M., Shah, H. (2007) A On-farm seed priming with zinc sulphate solution - A cost-effective way to increase the maize yields of resource-poor farmers. *Field Crops Research* 102: 119–127.

Harris, D., Rashid, A., Miraj, G., Arif, M., Yunas, M. (2008) A 'On-farm' seed priming with zinc in chickpea and wheat in Pakistan. *Plant and Soil* 306: 3–10.

Hasegawa, P.M., Bressan, R.A., Zhu, J.K., Bohnert, H.J. (2000) Plant cellular and molecular responses to high salinity. *Annual Review of Plant Physiology and Plant Molecular Biology* 51: 463–499.

Hayat, R., Ali, S., Amara, U., Khalid, R., Ahmed, I. (2010) Soil beneficial bacteria and their role in plant growth promotion: A review. *Annals of Microbiology* 60: 579–598.

Hela, M., Hanen, Z., Imen, T., Olfa, B., Nawel, N., Raouia, B.M., Maha, Z., *et al.* (2012) Combined effect of hormonal priming and salt treatments on germination percentage and antioxidant activities in lettuce seedlings. *African Journal of Biotechnology* 11: 10373–10380.

Honglu, X., Guomei, X. (2008) Suspension Property of Gemini Surfactant in SeedCoating Agent. Journal of Dispersion Science and Technology 29: 496–501.

Hossain, M.A., Hasanuzzaman, M., Fujita, M. (2011) Coordinate induction of antioxidant defense and glyoxalase system by exogenous proline and glycinebetaine is correlated with salt tolerance in mung bean. Frontiers of Agriculture in China 5: 1–14.

Iqbal, M., Ashraf, M., Jamil, A., Rehmaan, S. (2006) Does seed priming induce changes in the level of some endogenous plant hormones in hexaploid wheat plant under salt stress? *Journal of Integrative Plant Biology* 48: 181–189.

Jeschke, P., Nauen, R., Schindler, M., Elbert, A. (2011) Overview of the status and global strategy for neonicotinoids. *Journal of Agricultural and Food Chemistry* 59: 2897–2908.

Jisha, K.C., Puthur, J.T. (2016) Seed priming with BABA (β-amino butyric acid): Acost-effective method of abiotic stress tolerance in *Vigna radiata* (L.) Wilczek. *Protoplasma* 253: 277–289.

Jisha, K.C., Vijayakumari, K., Puthur, J.T. (2013) Seed priming for abiotic stress tolerance: an overview. Acta Physiologiae Plantarum 35: 1381–1396.

John, S.S., Bharathi, A., Natesan, P., Raja, K. (2005) Seed film coating technology for maximizing the growth and productivity of maize. *Karnataka Journal of Agricultural Sciences* 18: 349.

John, R.P., Tyagi, R.D., Brar, S.K., Prevost, D. (2010) Development of emulsion from rhizobial fermented starch industry wastewater for application as *Medicago sativa* seed coat. *Engineering in Life Sciences* 10: 248–256.

Johnson, G.A., Hicks, D.H., Stewart, R.F., Duan, X. (1999) Use of temperature-responsive polymer seed coating to control seed germination. *Acta Horticulturae*. 504: 229–236.

Johnson, S.E., Lauren, J.G., Welch, R.M., Duxbury, J.M. (2005) A comparison of the effects of micronutrient seed priming and soil fertilization on the mineral nutrition of chickpea (*Cicer arietinum*), lentil (*Lens culinaris*), rice (*Oryza sativa*) and wheat (*Triticum aestivum*) in Nepal. Experimental Agriculture 41: 427–448.

Joshi, N., Jain, A., Arya, K. (2013) Alleviation of salt stress in *Cucumissativus* L. through seed priming with calcium chloride. Indian Journal of Applied Research 3: 22–25.

Jyotsna, V., Srivastava, A.K. (1998) Physiological basis of salt stress resistance in pigeon pea (*Cajanus cajan* L.)- pre-sowing seed soaking treatment in regulating early seeding metabolism during seed germination. *Plant Physiology and Biochemistry* 25: 89–94.

Kadiri, M., Hussaini, M.A. (1999) Effect of hardening pretreatment on vegetative growth, enzyme activities and yield of Pennisetum americanum and *Sorghum bicolour* L. *Global Journal of Pure and Applied Science* 5: 179–183.

Kasim, W.A., Osman, M.E., Omar, M.N., El-Daim, I.A., Bejai, S., Meijer, J. (2013) Control of drought stress in wheat using plant growth-promoting bacteria. *Journal of Plant Growth Regulation* 32: 122–130.

Kata, L.P., Bhaskaran, M., Umarani, R. (2014) Influence of priming treatments on stress tolerance during seed germination of rice. *International Journal of Agriculture, Environment & Biotechnology* 7: 225–232.

Kaveh, H., Nemati, H., Farsi, M., Jartoodeh, S.V. (2011) How salinity affects germination and emergence of tomato lines. Journal of Biological and Environmental Sciences 5: 159–163.

Kaya, M.D., Okcu, G., Atak, M., Cıkıhı, Y., Kolsarıcı, Ö (2006) Seed treatments to overcome salt and drought stress during germination in sunflower (*Helianthus annuus* L.). *European Journal of Agronomy* 24: 291–295.

Kaymak, H.C., Guvenç, I., Yarali, F., Donmez, M.F. (2009) The effects of bio-priming with PGPR on germination of radish (*Raphanus sativus* L.) seeds under saline conditions. *Turkish Journal of Agriculture and Forestry* 33: 173–179.

Khan, H.A., Ayub, C.M., Pervez, M.A., Bilal, R.M., Shahid, M.A., Ziaf, K. (2009a) Effect of seed priming with NaCl on salinity tolerance of hot pepper (*Capsicum annuum* L.) at seedling stage. *Soil & Environment* 28: 81–87.

Khan, H.A., Pervez, M.A., Ayub, C.M., Ziaf, K., Bilal, R.M., Shahid, M.A., Akhtar, N. (2009b) Hormonal priming alleviates salt stress in hot pepper (*Capsicum annuum* L.). *Soil & Environment* 28: 130–135.

Khan, M.A., Weber, D.J. (2008) *Ecophysiology of high salinity tolerant plants (Tasks for Vegetation Science)*, 1st edn. Springer: Amsterdam.

Khanal, N., Joshi, D., Harris, D., Chand, S.D. (2005) Effect of micronutrient loading, soil application, and foliar sprays of organic extracts on grain legumes and vegetable crops under marginal farmers' conditions in Nepal. In *Micronutrients in South and South East Asia*, ed. Andersen, P. Tuladhar, J.K., Karki, K.B., Maskey, S.L. Proceedings of an International Workshop Held in Kathmandu, Nepal, 8–11 September, 2004. The International Centre for Integrated Mountain Development, Kathmandu, Sri Lanka, 121–132.

Kohler, J., Hernandez, J.A., Caravaca, F., Roldan, A. (2009) Induction of antioxidant enzymes is involved in the greater effectiveness of a PGPR versus AM fungi with respect to increasing the tolerance of lettuce to severe salt stress. *Environmental and Experimental Botany* 65: 245–252.

Korkmaz, A., Şirikçi, R. (2011) Improving salinity tolerance of germinating seeds by exogenous application of glycinebetaine in pepper. *Seed Science and Technology* 39: 377–388.

Kubala, S., Garnczarska, M., Wojtyla, Ł., Clippe, A., Kosmala, A., Zmienko, A., Lutts, S., Quinet, M. (2015a) Deciphering priming-induced improvement of rapeseed (*Brassica napus* L.) germination through an integrated transcriptomic and proteomic approach. *Plant Science: an International Journal of Experimental Plant Biology* 231: 94–113.

Kubala, S., Wojtyla, L., Quinet, M., Lechowska, K., Lutts, S., Garnczarska, M. (2015b) Enhanced expression of the proline synthesis gene P5CSA in relation to seed osmopriming improvement of Brassica napus germination under salinity stress. Plant Science 183: 1–12.

Lauchli, A., Grattan, S.R. (2007) Plant growth and development under salinity stress. In *Advances in molecular breeding towards drought and salt tolerant crops*, ed. Jenks, M.A., Hasegawa, P.M., Mohan, J.S., 1–32. Springer: Berlin.

Lipiec, J., Doussan, C., Nosalewicz, A., Kondracka, K. (2013) Effect of drought and heat stresses on plant growth and yield: a review. International Agrophysics 27: 463–477.

Liu, C., Liu, Y., Guo, K., Fan, G., Li, G., Zheng, Y. (2011) Effect of drought on pigments, osmotic adjustment, and antioxidant enzymes in six woody plant species in karst of southwestern china. *Environmental and Experimental Botany* 71: 174–183.

López-Bucio, J., Pelagio-Flores, R., Herrera-Estrella, A. (2015) Trichoderma as biostimulant: Exploiting the multilevel properties of a plant beneficial fungus. *Scientia Horticulturae* 196: 109–123.

Madsen, M.D., Kostka, S.J., Hulet, A., Mackey, B.E., Harrison, M.A., McMillan, M.F. (2013) Surfactant seed coating–A strategy to improve turfgrass establishment on water repellent soils. *International Symposium on Adjuvants for Agrochemicals*: 205–210.

Mahmood, A., Turgay, O.C., Farooq, M., Hayat, R. (2016) Seed biopriming with plant growth promoting rhizobacteria: A review. *FEMS Microbiology Ecology* 92: 1–14.

Maiti, R., Pramanik, K. (2013) Vegetable seed priming: A low cost, simple and powerful technique for farmers' livelihood. *International Journal of Bio-Resource and Stress Management* 4: 475–481.

Malhi, S.S. (2009) Effectiveness of seed-soaked Cu, autumn-versus spring applied Cu, and Cu-treated P fertilizer on seed yield of wheat and residual nitrate-N for a Cu-deficient soil. *Canadian Journal of Plant Science* 89: 1017–1030.

Mangold, J.M., Sheley, R.L. (2007) Effects of soil texture, watering frequency, and a hydrogel on the emergence and survival of coated and uncoated crested wheat grass seeds. Ecological. Restoration 25: 6–11.

Masauskas, V., Masauskiene, A., Repsiene, R., Skuodiene, R., Braziene, Z., Peltonen, J. (2008) Phosphorus seed coating as starter fertilization for spring malting barley. *Acta Agriculturae Scandinavica, Section B - Plant Soil Science* 58: 124–131.

McIntyre, H.J., Davies, H., Hore, T.A., Miller, S.H., Dufour, J.P., Ronson, C.W. (2007) Trehalose biosynthesis in *Rhizobium leguminosarum* bv. *trifolii* and its role in desiccation tolerance. *Applied and Environmental Microbiology* 73: 3984–3992.

Miladinov, Z.J., Balesevic-Tubic, S.N., Dor–devic, V.B., Dukic, V.H., Ilic, A.D., Cobanovic, L.M. (2015) Optimal time of soybean seed priming and primer effect under salt stress conditions. *Journal of Agricultural Sciences, Belgrade* 60: 109–117.

Mirshekari, B., Hokmalipour, S., Sharifi, R.S., Farahvash, F., Gadim, A.E. (2012) Effect of seed biopriming with plant growth promoting rhizobacteria (PGPR) on yield and dry matter accumulation of spring barley (*Hordeum vulgare* L.) at various levels of nitrogen and phosphorus fertilizers. *Journal of Food, Agriculture & Environment* 10: 314–320.

Moaveni, P. (2011) Effect of water deficit stress on some physiological traits of wheat (*Triticum aestivum*). Agricultural Science Research Journal 1: 64–68.

Moeinzadeh, A., Sharif-Zadeh, F., Ahmadzadeh, M., Tajabadi, F. (2010) Bio-priming of Sunflower (*Helianthus annuus* L.) Seed with *Pseudomonas fluorescens* for improvement of seed invigoration and seedling growth. Australian Journal of Crop Science 4: 564.

Moghanibashi, M., Karimmojeni, H., Nikneshan, P. (2013) Seed treatment to overcome drought and salt stress during germination of sunflower (*Helianthus annuus* L.). Journal of Agrobiology 30: 89–96.

Munns, R., Tester, M. (2008) Mechanism of salinity tolerance. *Annual Review of Plant Biology* 59: 651–681.

Nadeem, S.M., Zahir, Z.A., Naveed, M., Nawaz, S. (2013) Mitigation of salinity induced negative impact on the growth and yield of wheat by plant growth-promoting rhizobacteria in naturally saline conditions. Annals of Microbiology 63: 225–232.

Naghavi, M.R., Aboughadareh, A.P., Khalili, M. (2013) Evaluation of drought tolerance indices for screening some of corn (*Zea mays* L.) cultivars under environmental conditions. Notulae Scientia Biologicae 5: 388–393.

Nakaune, M., Hanada, A., Yin, Y.G., Matsukura, C., Yamaguchi, S., Ezura, H. (2012) Molecular and physiological dissection of enhanced seed germination using short-term low-concentration salt seed priming in tomato. *Plant Physiology and Biochemistry: PPB* 52: 28–37.

Nawaz, A., Amjad, M., Jahangir, M.M., Khan, S.M., Cui, H., Hu, J. (2012) Induction of salt tolerance in tomato (*Lycopersicon esculentum* Mill.) seeds through sand priming. Australian Journal of Crop Science 6: 1199–1203.

Nawel, N., Rym, K., Hela, M., Olfa, B., Najoua, B., Mokhtar, Ll (2011) The effect of osmo-priming on germination, seedling growth and phosphatase activities of lettuce under saline condition. *African Journal of Biotechnology* 10: 14366–14372.

Naz, F., Gul, H., Hamayun, M., Sayyed, A., Khan, H., Sherwani, S. (2014) Effect of NaCl stress on *P. sativum* germination and seedling growth with the influence of seed priming with potassium (KCL and KOH). *American-Eurasian Journal of Agricultural & Environmental* 14: 1304–1311.

Neeraja, C., Maghirang-Rodriguez, R., Pamplona, A.M., Heuer, S., Collard, B., Septiningsih, E., Vergara, G., *et al.* (2007) A marker-assisted backcross approach for developing submergence-tolerance rice cultivars. *Theoretical and Applied Genetics* 115: 767–776.

Oliveira, S.D., Tavares, L.C., Lemes, E.S., Brunes, A.P., Dias, I.L., Meneghello, G.E. (2014) Tratamento de sementes de *Avena sativa* L. com zinco: Qualidade fisiológica e desempenho inicial de plantas. *Semina: Ciencias Agrarias* 35: 1131–1142.

Paparella, S., Araujo, S.S., Rossi, G., Wijayasinghe, M., Carbonera, D., Balestrazzi, A. (2015) Seed priming: State of the art and new perspectives. *Plant Cell Reports* 34: 1281–1293.

Parida, A.K., Das, A.B. (2005) Salt tolerance and salinity effects on plants: A review. *Ecotoxicology and Environmental Safety* 60: 324–349.

Paul, D., Lade, H. (2014) Plant-growth-promoting rhizobacteria to improve crop growth in saline soils: a review. Agronomy for Sustainable Development 34: 737–752.

Pedrini, S., Merritt, D.J., Stevens, J., Dixon, K. (2017) Seed coating: Science or marketing spin? *Trends in Plant Science* 22: 106–116.

Peeran, S.N., Natanasabapathy, S. (1980) Potassium chloride pretreatment on rice seeds. International Rice Research Newsletter 5: 19.

Peltonen-Sainio, P., Kontturi, M., Peltonen, J. (2006) Phosphorus seed coating enhancement on early growth and yield components in oat. *Agronomy Journal* 98: 206–211.

Pichyangkura, R., Chadchawan, S. (2015) Biostimulant activity of chitosan in horticulture. Scientia Horticulturae 196: 49–65.

Pitman, M.G., Lauchli, A. (2002) Global impact of salinity and agricultural ecosystem. In: *Salinity: Environment-Plants-Molecules*, ed. Lauchli, A., Luttge, U., Lauchli, A., Luttge, U., 3–20. Netherlands: Kluwer Academic Publishers.

Posmyk, M.M., Kontek, R., Janas, K.M. (2009) Antioxidant enzymes activity and phenolic compounds content in red cabbage seedlings exposed to copper stress. Ecotoxicology and Environmental Safety 72: 596–602.

Pradhan, N., Prakash, P., Tiwari, S.K., Manimurugan, C., Sharma, R.P., Singh, P.M. (2014) Osmo-priming of tomato genotypes with polyethylene glycol 6000 induces tolerance to salinity stress. Trends in Biosciences 7: 4412–4417.

Rehman, A.U., Farooq, M. (2013) Boron application through seed coating improves the water relations, panicle fertility, kernel yield, and biofortification of fine grain aromatic rice. Acta Physiologiae Plantarum 35: 411–418.

Rehman, A., Farooq, M., Cheema, Z.A., Wahid, A. (2012) Role of boron in leaf elongation and tillering dynamics in fine grain aromatic rice. *Journal of Plant Nutrition* 36: 42–54.

Rerkasem, B., Jamjod, S. (2004) Boron deficiency in wheat: a review. *Field Crops Research* 89: 173–186.

Rojas-Tapias, D., Moreno-Galvan, A., Pardo-Diaz, S., Obando, M., Rivera, D., Bonilla, R. (2012) Effect of inoculation with plant growth-promoting bacteria (PGPB) on amelioration of saline stress in maize (*Zea mays*). *Applied Soil Ecology* 61: 264–272.

Rouhollah, A. (2013) Drought stress tolerance of barley (*Hordeum vulgare* L.) affected by priming with PEG. *International Journal of Farming and Allied Sciences* 2: 803–808.

Rouphael, Y., Franken, P., Schneider, C., Schwarz, D., Giovannetti, M., Agnolucci, M., DePascale, S.D., Bonini, P., Colla, G. (2015) Arbuscular mycorrhizal fungi act as biostimulants in horticultural crops. *Scientia Horticulturae* 196: 91–108.

Rundlof, M., Andersson, G.K., Bommarco, R., Fries, I., Hederstrom, V., Herbertsson, L., Jonsson, O., *et al.* (2015) Seed coating with a neonicotinoid insecticide negatively affects wild bees. *Nature* 521: 77–80.

Ruzzi, M., Aroca, R. (2015) Plant growth-promoting rhizobacteria act as biostimulants in horticulture. *Scientia Horticulturae* 196: 124–134.

Saglam, S., Day, S., Kaya, G., Gurbuz, A. (2010) Hydropriming increases germination of lentil (*Lens culinaris* Medik.) under water stress. Notulae Scientia Biologicae 2: 103–106.

Salama, K.H.A., Mansour, M.M.F. (2015) Choline priming-induced plasma membrane lipid alterations contributed to improved wheat salt tolerance. *Acta Physiologiae Plantarum* 37: 1–7.

Salama, K.H.A., Mansour, M.M.F., Hassan, N.S. (2011) Choline priming improves salt tolerance in wheat (*Triticum aestivum* L.). *Australian Journal of Basic and Applied Sciences* 5: 126–132.

Salanenka, Y.A., Taylor, A.G. (2008) Seed Coat Permeability and Uptake of Applied Systemic Compounds. Acta Horticulturae 782: 151–154.

Salanenka, Y.A., Taylor, A.G. (2011) Seed coat permeability: Uptake and post-germination transport of applied model tracer compounds. Horticultural Science 46: 622–626.

Sandhya, V., Ali, S.Z., Grover, M., Reddy, G., Venkateswarlu, B. (2010) Effect of plant growth promoting *Pseudomonas* spp. on compatible solutes, antioxidant status and plant growth of maize under drought stress. Plant Growth Regulation 62: 21–30.

Sarika, G., Basavaraju, G.V., Bhanuprakash, K., Chaanakeshava, V., Paramesh, R., Radha, B.N. (2013) Investigation on seed viability and vigour of aged seed by priming in French bean. *Vegetable Science* 40: 169–173.

Sarkar, R.K. (2012) Seed priming improves agronomic trait performance under flooding and non-flooding conditions in rice with QTL SUB1. *Rice Science* 19: 286–294.

Sarker, A., Hossain, m.D.I., Abul Kashem, M.D. (2014) Salinity (NaCl) tolerance of four vegetable crops during germination and early seedling growth. *International Journal of Latest Research in Science and Technology* 3: 91–95.

Savvas, D., Ntatsi, G. (2015) Biostimulant activity of silicon in horticulture. *Scientia Horticulturae* 196: 66–81.

Scott, J.M. (1989) Seed coatings and treatments and their effects on plant establishment. *Advances in Agronomy* 42: 43–83.

Sedghi, M., Nemati, A., Esmaielpour, B. (2010) Effect of seed priming on germination and seedling growth of two medicinal plants under salinity. Emirates Journal of Food and Agriculture 22: 130–139.

Seki, M., Kamei, A., Yamaguchi-Shinozaki, K., Shinozaki, K. (2003) Molecular responses to drought, salinity and frost: common and different paths for plant protection. Current Opinion in Biotechnology 14: 194–199.

Serena, M., Leinauer, B., Sallenave, R., Schiavon, M., Maier, B. (2012) Turfgrass establishment from polymer-coated seed under saline irrigation. *Horticulture Science* 47: 1789–1794.

Sharifi, R.S. (2011) Study of grain yield and some of physiological growth indices in maize (*Zea mays* L.) hybrids under seed biopriming with plant growth promoting rhizobacteria (PGPR). Journal of Food, Agriculture and Environment 189: 3–4.

Sharifi, R.S. (2012) Study of nitrogen rates effects and seed biopriming with PGPR on quantitative and qualitative yield of Safflower (*Carthamus tinctorius* L.). Technical Journal of Engineering and Applied Sciences 2: 162–166.

Sharifi, R.S., Khavazi, K. (2011) Effects of seed priming with plant growth promotion rhizobacteria (PGRP) on yield and yield attribute of maize (*Zea mays* L.) hybrids. Journal of Food, Agriculture and Environment 9: 496–500.

Sharma, P., Jha, A.B., Dubey, R.S., Pessarakli, M. (2012) Reactive oxygen species, oxidative damage, and antioxidative defense mechanisms in plants under stressful conditions. *Journal of Botany* 2012: 1–26 doi:10.1155/2012/217037.

Shehab, G.G., Ahmed, O.K., El-Beltagi, H.S. (2010) Effects of various chemical agents for alleviation of drought stress in rice plants (*Oryza sativa* L.). Notulae Botanicae Horti Agrobotanici Cluj-Napoca 38: 139–148.

Sherrell, C.G. (1984) Effect of molybdenum concentration in the seed on the response of pasture legumes to molybdenum. New Zealand Journal of Agricultural Research 27: 417–423.

Shinozaki, K., Yamaguchi-Shinozaki, K., Seki, M. (2003) Regulatory network of gene expression in the drought and cold stress responses. Current Opinion in Biotechnology 6: 410–417.

Shukla, P.S., Agarwal, P.K., Jha, B. (2012) Improved salinity tolerance of *Arachis hypogaea* (L.) by the interaction of halotolerant plant growth-promoting rhizobacteria. Journal of Plant Growth Regulation 31: 195– 206.

Singh, S., Mackill, D.J., Ismail, A.M. (2009) Responses of SUB1 rice introgression lines to submergence in the field: Yield and grain quality. Field Crops Research 113: 12–23.

Singh, B., Natesan, S.K.A., Singh, B.K., Usha, K. (2003) Improving zinc efficiency of cereals under zinc deficiency. Current Science 88: 36–44.

Smith, R.G., Atwood, L.W., Morris, M.B., Mortensen, D.A., Koide, R.T. (2016) Evidence for indirect effects of pesticide seed treatments on weed seed banks in maize and soybean. *Agriculture, Ecosystems & Environment* 216: 269–273.

Soeda, Y., Konings, M.C.J.M., Vorst, O., van Houwelingen, A.M., Stoopen, G.M., Maliepaard, C.A., Kodde, J., *et al.* (2005) Gene expression programs during *Brassica olearacea* seed maturation, osmo-priming and germination are indicators of progression of the germination process and the stress tolerance level. *Plant Physiology* 137: 354–368.

Srivastava, S.P., Bhandari, T.M.S., Yadav, C.R., Joshi, M., Erskine, W. (2000) Boron deficiency in lentil: Yield loss and geographic distribution in a germplasm collection. Plant & Soil 219: 147–151.

Staley, T.E., Drahos, D.J. (1994) Marking soil bacteria with lacZY. In *Methods of soil analysis. part 2. Microbiological and biochemical properties*, ed. Weaver, R.W., Angel, J.S., Bottomley, P.J., 689–706. Madison: Soil Science Society of America.

Stokes, T. (2001) Coating enables early planting. Trends in Plant Science 6: 243.

Sun, Y.Y., Sun, Y.J., Wang, M.T., Li, X.Y., Guo, X., Hu, R., Ma, J. (2010) Effects of seed priming on germination and seedling growth under water stress in rice. *Acta Agronomica Sinica* 36: 1931–1940.

Tabatabaei, S.A. (2013) Effect of osmo-priming on germination and enzyme activity in barley (*Hordeum vulgare* L.) seeds under drought stress conditions. Journal of Stress Physiology and Biochemistry 9: 25–31.

Tabatabaei, S.A. (2014) The effect of priming on germination indexes and seed reserve utilization of maize seeds under salinity stress. Journal of Seed Science and Technology 3: 44–51.

Tanou, G., Fotopoulos, V., Molassiotis, A. (2012) Priming against environmental challenges and proteomics in plants: update and agricultural perspectives. Frontiers in Plant Science 3: 216.

Tavares, L.C., Madruga de Tunes, L., Pich Brunes, A., Robe Fonseca, D.A., de Araujo Rufino, C., Souza Albuquerque Barros, A.C. (2013) Potássio via recobrimento de sementes de soja: efeitos na qualidade fisiológica e no rendimento. *Ciencia Rural* 43: 1196–1202.

Thiam, M., Champion, A., Diouf, D., MameOureye, S.Y. (2013) NaCl effects on in-vitro germination and growth of some Senegalese cowpea (*Vigna unguiculata* (L.) Walp.) cultivars. *ISRN Biotechnology* 2013: 1–11.

Thornton, J.M., Collins, A.R.S., Powell, A.A. (1993) The effect of aerated hydration on DNA synthesis in embryos of *Brassica oleracea* L. *Seed Science Research* 3: 195–199.

Varier, R.A., Vari, A.K., Dadlani, M. (2010) The subcellular basis of seed priming. *Current Science* 99: 450–456.

Wahid, A., Noreen, A., Basra, S., Gelani, S., Farooq, M. (2008) Priming induced metabolic changes in sunflower (*Helianthus annuus*) achenes improve germination and seedling growth. Botanical Studies 49: 343–350.

Wang, W., Vinocur, B., Altman, A. (2003) Plant responses to drought, salinity and extreme temperatures: toward genetic engineering for stress tolerance. Planta 218: 1–14.

Wiatrak, P. (2013a) Effect of polymer seed coating with micronutrients on soybeans in southeastern coastal plains. *American Journal of Agricultural and Biological Sciences* 8: 302–308.

Wiatrak, P. (2013b) Influence of seed coating with micronutrients on growth and yield of winter wheat in southeastern coastal plains. *American Journal of Agricultural and Biological Sciences* 8: 230–238.

Wilhelm, N.S., Graham, R.D., Rovira, A.D. (1988) Application of different sources of manganese sulphate decreases take-all (*Gaeumannomyces graminis* var. tritici) of wheat grown in a manganese deficient soil. Australian Journal of Agricultural Research 39: 1–10.

Yang, S., Vanderbeld, B., Wan, J., Huang, Y. (2010) Narrowing down the targets: Towards successful genetic engineering of drought-tolerant crops. *Molecular Plant* 3: 469–490.

Yang, D., Wang, N., Yan, X., Shi, J., Zhang, M., Wang, Z., Yuan, H. (2014) Micro-encapsulation of seed-coating tebuconazole and its effects on physiology and biochemistry of maize seedlings. *Colloids and Surfaces B: Biointerfaces* 114: 241–246.

Yao, L.X., Wu, Z.S., Zheng, Y.Y., Kaleem, I., Li, C. (2010) Growth promotion and protection against salt stress by Pseudomonas putida Rs-198 on cotton. European Journal of Soil Biology 46: 49–54.

Younesi, O., Moradi, A. (2015) Effect of priming of seeds of *Medicago sativa* 'bami' with gibberellic acid on germination, seedlings growth and antioxidant enzymes activity under salinity stress. Journal of Horticultural Research 22: 167–174

Zavariyan, A., Rad, M., Asghari, M. (2015) Effect of seed priming by potassium nitrate on germination and biochemical indices in *Silybum marianum* L. under salinity stress.International Journal of Life Science 9: 23–29.

4 Application of Osmolytes in Improving Abiotic Stress Tolerance in Plant

Shaghef Ejaz, Sajjad Hussain, Muhammad Akbar Anjum, and Shakeel Ahmad

CONTENTS

4.1 INTRODUCTION

Environmental stresses on plants determine their diversity, productivity and distribution. Unfavorable environmental conditions retard growth, development and reproductive performance in plants. Salinity, water deficiency, water logging, nutrient toxicities and deficiencies and cold or heat stresses are major environmental constraints to plant productivity worldwide. Almost 45% of agricultural land is facing drought and 6% of the world total land is affected by salinity, thus affecting more than 38% of human population (Bot et al., 2000). Low-quality irrigation water, poor cultural practices, frequent drought conditions, unpredictable weather and overall climate change are the main contributors to major environmental stresses leading toward low agricultural productivity of lands and food insecurity, especially in developing countries. Plant survival against these stresses depends mainly on how efficiently plants utilize their various defense mechanisms. The plant uses stress sensors, signaling pathways and metabolism to respond to a stress. One such metabolic response upon exposure of the plant to stress is synthesis and accumulation of low-molecular-weight compounds called osmolytes.

4.2 OSMOLYTE

Generally, osmolytes are highly soluble in water, do not interact with normal metabolic activities, carry no net charge at physiological pH and are not toxic even at high concentrations. Therefore, these are also known as 'compatible solutes' (Slama et al., 2015).

Broadly speaking, stress physiologists identify osmolytes on the basis of three types of information: (i) an increase in accumulation of organic molecules under decreased water potential in cells, (ii) physicochemical properties of the putative osmolyte when studied *in vitro* and (iii) comparative physiology as a pointer to gain insight into whether osmolytes confer protection in desiccation-tolerant species or in dehydration tolerant stages of development such as during seed maturation (Ingram and Bartels, 1996).

Osmolytes are a class of phytoprotectants that can alter the solubility of water, adjust osmotic potential, stabilize folded proteins and protect membrane structures when climatic conditions are unfavorable (Yancey, 2005). Owing to their ability to protect cellular components from osmotic and oxidative damage and dehydration, these are called osmoprotectants.

Osmolytes can be divided into the following major groups:

Amino acids: In higher plants, amino acids (alanine, arginine, glycine and proline), amides (asparagine and glutamine) and non-protein amino acids (γ-aminobutyric acid, citrulline, ornithine and pipecolic acid) accumulate under abiotic stress.

Quaternary ammonium compounds: Plants accumulate quaternary ammonium compounds such as β-alanine betaine, choline-O-sulphate, glycinebetaine, hydroxyproline betaine, pipecolate betaine and proline betaine in unfavorable environmental conditions.

Sugars: Sugars such as fructan, sucrose and trehalose.

Sugar alcohol (Polyols): D-Ononitol, inositol, mannitol, pinitol and sorbitol.

Researchers found that higher accumulations of osmolytes regulate osmotic adjustments during salinity and drought stresses (Singh et al., 2015). However, it is not always so. For example, in Arabidopsis, proline accumulation is so low that it is unable to facilitate osmotic adjustment in cells (Ghars et al., 2008).

4.3 DIVERSITY OF OSMOLYTES IN PLANTS

Where most of the osmolytes are ubiquitous in the plant kingdom, studies have confirmed that there are species-specific and environment-specific osmolytes (Slama et al., 2015). Similarly, concentration or accumulation of osmolytes vary with genetics and environments. For example, quaternary ammonium compound β-alanine betaine accumulates among a few species of Plumbaginaceae, while amino acid proline occurs in taxonomically diverse plants (Gupta and Huang, 2014). Generally, plant species having a history of continuous exposure to environmental stress have evolved to accumulate osmolytes. Below are a few examples of plant families that accumulate various osmolytes under abiotic stress (Hasanuzzaman et al., 2013; Slama et al., 2015).

Amino acids are accumulated by Aizoaceae, Asteraceae, Amaranthaceae, Anacardiaceae, Brassicaceae, Casuarinaceae, Cucurbitaceae, Cupressaceae, Cymodoceaceae, Fabaceae, Iridaceae, Juncaginaceae, Malvaceae, Myrtaceae (methylated proline), Plantaginaceae, Plumbaginaceae, Poaceae, Portulaceae, Posidoniaceae, Rhizophoraceae, Solanaceae and Zosteraceae.

Quaternary ammonium compounds are accumulated by Asteraceae, Acanthaceae, Amaranthaceae, Brassicaceae, Plumbaginaceae and Solanaceae.

Sugars are accumulated by Asteraceae, Amaranthaceae, Brassicaceae, Cyperaceae, Juncaceae, Juncaginaceae, Plumbaginaceae, Poaceae, Posidoniaceae, Rhizophoraceae, Solanaceae and Zosteraceae.

Sugar alcohol is accumulated by Apiaceae, Asteraceae, Combretaceae, Cymodoceaeceae, Fabaceae and Plantaginaceae.

4.4 ROLE OF OSMOLYTES IN THE NORMAL FUNCTIONING OF PLANT CELLS

Osmolyte gradient triggers water influx necessary for maintaining plant cell turgor where plasma membrane behaves as an osmotic barrier. The osmolyte system in the cytosol of plant cells is an aqueous gel system that maintains cell volume and fluid balance. An amino acid such as L-glutamine and sugars such as D-glucose present in cytosol contribute to turgor dynamics. Such osmolyte complexes lead to higher turgor pressure required for plant movements (Argiolas et al., 2016).

Sugars sustain the growth of plant tissues and modulate the signaling systems that up- or down-regulate the transcription of genes involved in the processes of photosynthesis, respiration and make or break of starch and sucrose (Hare et al., 1998). Free amino acids act as a precursor to proteins or nitrogen containing molecules (nucleic acid) and are constituents of proteins. Proline also alleviates cytoplasmic acidosis and sustains $NADP^+$ to NADPH ratios necessary for metabolism (Hare et al., 1998).

Glycinebetaine (GB) is synthesized by almost all living organisms including algae, animals, bacteria, cyanobacteria, fungi and plants. GB is synthesized from choline via betaine aldehyde pathway to GB catalyzed by choline monooxygenase and betaine aldehyde dehydrogenase. GB stabilizes thylakoid membrane structure in the chloroplast, primary place of its accumulation, thus maintaining photosynthesis efficiency.

4.5 MODE OF ACTION OF OSMOLYTES UNDER ABIOTIC STRESS

Abiotic stresses in plants are due to the elements of the environment that reach perturbation level to initiate a response in plants. Abiotic stresses that are ubiquitous around the globe and are a threat to plant productivity and distribution include drought, salinity, heavy metal toxicity, chilling, freezing and heat stress.

4.5.1 Osmotic Adjustment

Water deficiency is the major abiotic stress hampering plant growth and productivity by decreasing photosynthetic performance due to changes in the relevant biochemical apparatuses and processes. Photosystem II is the most sensitive component. Maintaining turgor within plant cells under abiotic stress is critical for stomata regulation and photosynthetic activity, and in such conditions osmolyte accumulation controls turgor dynamics and reduces water potential in cells to maintain normal functioning of photosynthetic machinery. Osmolytes accumulation conserves water content in plant tissues by generating low water potential in cytosol though osmotic adjustment.

4.5.2 Protein and Membrane Stability

Stabilization of the membrane integrity in plant cells during stress is pivotal for the metabolic processes of photosynthesis and respiration. An increase in the concentration of osmolytes is protective for membrane structures owing to the exclusion of osmolytes from the surface of proteins. This hydrophobic environment forces proteins to fold, as polypeptides adopt folded conformation which have the least exposed surface area (Khan et al., 2010). Osmolytes shift equilibrium toward folding of membrane structures, protecting protein backbones during stress.

4.5.3 Reactive Oxygen Species Scavenging

Osmotic and oxidative stress and ion toxicity during abiotic stress enhance the production of reactive oxygen species (ROS) such as peroxide, hydrogen peroxide, hydroxyl ion and singlet oxygen. ROS damages macromolecules and membrane structures such as mitochondria, chloroplast and other lipid membrane cell structures that cause electron leakage from electron transport chains. Osmolytes mitigate stress-induced damage directly by protecting membrane structures and indirectly by reducing ROS through modulating antioxidant mechanisms within plant cells. Accumulation of osmolytes such as GB increases the activities of antioxidative enzymes such as superoxide dismutase (SOD) during salt stress. SOD is a relatively more active antioxidant enzyme that scavenges ROS and thus protects cell membrane structure, especially that of mitochondria and chloroplast. Methylglyoxal (MG) is a cytotoxic compound that is accumulated under stress. Detoxification of MG is carried out through enzymes glyoxalase I and II. Proline and betaine have been found to detoxify MG by enhancing glyoxalase activities (Hoque et al., 2008).

4.6 ABIOTIC STRESS MITIGATION THROUGH EXOGENOUS OSMOLYTES APPLICATION

4.6.1 Salt Stress

In higher plants, proline is known to occur and accumulate in remarkable concentrations in response to salt and drought stresses. Osmolyte accumulation in salinity-affected plants is a well-known adaptive mechanism that adjusts cell osmotic potential, stabilizes membrane structure, induces the expression of salt-stress-responsive proteins and increases antioxidant enzymes activities (Ashraf and Foolad, 2007; Khedr et al., 2003). During osmotic stresses, proline plays an important role in protecting electron transport complex II, proteins and enzymes such as RuBPCO (Hamilton and Heckathorn, 2001). Proline is also considered as a biochemical marker of stress tolerance in plants and has shown great potential as a compatible osmoprotectant since its first ever reported accumulation in ryegrass during water stress (Kemble and MacPherson, 1954).

Foliar application of proline mitigates salt-induced oxidative stress in sunflower by increasing the production of sugar and maintaining water and ion homeostasis, thus protecting the photosynthetic system (Khan et al., 2014). Hossain and Fujita (2010) suggested that proline and glycinebetaine protect mung bean seedlings against salt-induced oxidative stress by reducing hydrogen peroxide and lipid peroxidation levels and enhancing antioxidant activities and methylglyoxal detoxification systems.

Proline increases salt tolerance in cultured tobacco cells by protecting cells against oxidative stress through antioxidant mechanisms, oxidation of protein side chains and methylglyoxal detoxification systems (Hoque et al., 2008). Similarly, in tobacco, proline and glycinebetaine both effectively decreased ROS accumulation and lipid peroxidation, and suppressed nuclear deformation and chromatin condensation induced by severe salt stress (Banu et al., 2009). Ben Ahmed et al. (2011) suggested that proline increased salt tolerance in an olive tree by harnessing antioxidant enzymatic activities and protecting the photosynthesis apparatus, consequently improving plant growth and water status.

Similarly, application of proline and betaine to tobacco cell cultures suspended under salt stress increased cell growth and improved the reduced activities of catalase (CAT) and peroxidase (POD) but not of SOD (Hoque et al., 2007). Salt-stressed citrus (*Citrus sinensis* 'Valencia late') cell cultures also showed increased growth rate in response to exogenous application of proline (Lima-Costa et al., 2010).

Salt (especially NaCl) stress reduces growth, increases Na^+/K^+ ratio, and up-regulates proline synthesis genes thus enhancing proline production, which increases H_2O_2 production and antioxidant enzymatic activities. Also, exogenous proline application reduces Na^+/K^+ ratio, increases endogenous proline and up-regulates transcription of genes encoding CAT, SOD, POD and ascorbate peroxidase (APX) (Nounjan et al., 2012). Explants of the seed of *Phaseolus vulgaris* consisting of cotyledon and embryonic axis were cultured *in vitro* in a medium pre-treated with L-asparagine and L-glutamine. Low concentrations (1 and 2 mM) of asparagine or glutamine led to an increase in total carbohydrates, chlorophyll a and b, carotenoids, amide nitrogen, total nitrogen, ions content (K^+, Ca^{2+} and Mg^{2+}) and protein, whereas peptide nitrogen, ammonia nitrogen and total soluble nitrogen decreased during seedling and vegetative stages. Low concentrations of asparagine and glutamine also increased auxin, gibberellin and cytokinin levels (Haroun et al., 2010).

Pre-sowing treatment of *Phaseolus vulgaris* seeds with arginine stimulated germination of unstressed and NaCl-stressed seeds. Additionally, arginine led to an increase in the content of putrescine (Pts), spermidine (Smd) and spermine (Spm) in germinating seeds. However, with the increase in the growth of seedling, Pts content decreased and total soluble sugars increased. The ratio of Na^+ to K^+ decreased in the leaves, indicating the role of arginine in mitigation of adverse effects of salt stress (Zeid, 2009). Exogenous foliar application of ornithine to induce proline accumulation via δ-ornithine amino transferase pathway was studied by Da Rocha et al. (2012) in salt-stressed cashew. The researchers suggested that ornithine considerably induced proline accumulation in intact plants and dissected leaf discs.

Glycinebetaine can potentially protect plant cells against salt stress mainly through osmotic adjustment, enzyme (RuBPCO) stabilization, photosynthetic machinery protection and reducing reactive oxygen species (Cha-Um and Kirdmanee, 2010; Chen and Murata, 2008). During salt-induced oxidative stress, proline and glycinebetaine reduce hydrogen peroxide and lipid peroxidation levels and enhance antioxidant activities and methylglyoxal detoxification systems in mung bean seedlings (Hossain and Fujita, 2010). Exogenous application on salt (NaCl) stressed cowpea plants led to increases in total soluble sugars (Manaf, 2016), while on salt-stressed maize plants it led to an increase in photosynthetic activity (Nawaz and Ashraf, 2010). Application of glycinebetaine reduced electrolyte leakage, proline content and Na^+/K^+ ratio in salt-stressed perennial ryegrass (Hu and Hu, 2012). This effect was attributed mainly to higher CAT, SOD and APX activities and the repair of the cell membrane achieved by reducing lipid oxidation.

Glycinebetaine application to snap bean mitigated salt-induced stress by stabilizing membrane structure, increasing total soluble sugars and total soluble proteins concentration, number of leaves and pods per plant, pod moisture and pod fresh weight, thus also increasing green pod yield (Osman and Saleem, 2016). Salt (30, 50, 100 mM NaCl) stressed rice plants grown in phytotron, when supplemented with 1 mM glycinebetaine, showed higher survival percentages and growing abilities than un-supplemented plants owing to reduced Na^+ accumulation and better absorption and translocation of K^+ (Lutts, 2000). Similarly, foliar spray of glycinebetaine to 150 mM NaCl-stressed rice plants increased plant height, stabilized chlorophyll and maintained water use efficiency (Cha-Um and Kirdmanee, 2010).

Research has been conducted to genetically engineer low GB producing plants such as wheat to enhance salt tolerance by overcoming low GB accumulation. Transgenic lines of wheat have been developed through genetic modification to increase GB synthesis by introducing betaine aldehyde dehydrogenase genes. Wheat transgenic lines, under salt stress, have shown slightly higher carotenoid and chlorophyll contents, high Hill reaction activities and increased Ca^{2+}-ATPase activity compared to wild-type. Changes in lipids and fatty acid compositions in the thylakoid membrane may increase salt stress tolerance of wheat transgenic lines (Tian et al., 2017).

It is suggested that trehalose functions through scavenging reactive oxygen species, protecting the protein synthesis machinery (Chang et al., 2014). Its exogenous application reduces Na^+/K^+ ratio and potentially up-regulates proline synthesis genes and transcription of antioxidant enzymes genes in rice seedling. However, exogenously applied trehalose did not improve rice seedlings growth inhibited during salt stress (Nounjan et al., 2012). Exogenously supplementing trehalose on *Catharanthus roseus* in saline conditions largely alleviated the inhibitory effects of salinity and increased biomass accumulation, total leaf area/plant, relative water content, photosynthetic rate and potassium in leaves. Trehalose also led to high levels of sugars (glucose, fructose, sucrose), free amino acids (arginine, proline, glutamate and threonine) and alkaloids. Researchers suggested that trehalose acts as a signaling molecule during salt stress and increases the accumulation of internal compatible solutes (soluble sugars and free amino acids) to control water loss, ionic flow and leaf gas exchange (Chang et al., 2014).

Seed priming of maize seeds with 10 mM trehalose mitigated salt stress at seedling stage by increasing chlorophyll and nucleic acid contents, stabilizing plasma membranes and decreasing ion leakage and the Na^+/K^+ ratio (Zeid, 2009). Transgenic plants with enhanced trehalose biosynthesis and accumulation have shown better performance during salinity stress (López-Gómez and Lluch, 2012). Abdallah et al. (2016) reported an increase in chlorophyll a and b, total soluble solids, total carbohydrates and CAT activity, and a decrease in proline concentration, SOD and POD activities in salt-stressed rice plants.

Polyols act as a radical scavenger, protein and membrane structure stabilizer and photosynthesis machinery protector during various abiotic stresses. Studies conducted by Conde et al. (2011) show that olive trees cope with salt stress and drought conditions by coordinating mannitol transport with intracellular metabolism. Wheat cannot synthesize mannitol. However, wheat seedling grown in mannitol and NaCl-added hydroponic medium showed increase in root growth, enzymatic antioxidative activities SOD, POD, CAT, APX, and glutathione reductase (GR). Mannitol also reduced lipid peroxidation which would otherwise increase under salt stress (Seckin et al., 2009). Palma et al. (2013) suggested that higher concentrations of myo-inositol and pinitol in the lucerne (*Medicago sativa*) nodules could be an adaptive response to salt stress. Although exogenously applied mannitol reduced the activities of CAT, SOD, POD, and polyphenol oxidase (PPO) in salt-treated maize plants, it increased K^+, Ca^{2+}, and P concentration and decreased Na^+ concentration (Kaya et al., 2013).

4.6.2 Water Stress

4.6.2.1 Drought

Under water-limited conditions, proper crop establishment is hindered at germination and seedling emergence stages. Water deficiency during seed germination causes poor seed germination leading to uneconomical crop establishment.

During recent times, seed priming has been employed on large scale to enhance the rate and uniformity of germination and seedling emergence under water-deficient conditions in many crop plants (Atreya et al., 2009; Yagmur and Kaydan, 2008). One of the most important techniques of seed priming is osmopriming, whereby seeds are soaked in solutions of salts, sugars and sugar alcohols. Osmopriming of chickpea seeds with mannitol has shown rapid and uniform germination (Elkoca et al., 2008). Similarly, rice seeds primed with GB showed better growth in terms of fresh and dry weights (Farooq et al., 2008), mainly owing to an improvement in photosynthetic capacity and up-regulation of the antioxidative defense mechanism. However, pre-treating sunflower seeds with GB did not affect osmotic and turgor potentials and leaf water contents under water stress conditions (Iqbal et al., 2008).

Pre-soaking of radish seeds with trehalose was effective in increasing vascular bundle areas, leaf epidermis thickness, midrib thickness and a number of vascular bundles under water stress and non-stress conditions. Also, pre-sowing trehalose-treated plants exhibited a marked increase in chlorophyll a, photosynthesis rate, water use efficiency, total soluble sugars, free proline contents and SOD activity (Akram et al., 2016b). Apart from seed priming, foliar application of osmolytes has been widely reported to mitigate water stress-induced damages. In water-stressed sunflowers, foliar application of GB led to improvements in capitulum diameter and number, weight, oil content and yield of achenes (Hussain et al., 2008).

GB-treated drought-stressed maize plants stabilized the higher gas exchange rate and improved chlorophyll synthesis, which resulted in improved growth and yield of maize plants (Anjum et al., 2011). Ali and Ashraf (2011a) applied GB through foliage to maize plants grown in water-deficit conditions and found a considerable increase in seed moisture, sugar, oil, protein, ash, fiber and GB contents. Moreover, an increase in seed oil oleic and linolenic acid contents, lipophilic compounds, macro- and micro-nutrients and seed oil DPPH scavenging activity was observed.

Growth reduction of rice plants due to water stress showed marked reversal when GB was exogenously applied (Farooq et al., 2008). These protective and recovery tendencies shown by various plant species in response to GB application may be attributed to an improvement in osmotic adjustment, antioxidant mechanisms, and photosynthetic machinery. GB increases CO_2 assimilation rates in drought-stressed olives (Denaxa et al., 2012). Mahouachi et al. (2012) suggested that pre-drought application of GB modulates abscisic acid, jasmonic acid, and proline accumulation through controlling stomata movement and enhancing the availability of other osmolytes, which resulted in better growth, leaf water status, and photosynthetic capacity. Foliar application of GB at either vegetative or reproductive growth stage increased leaf water and turgor potentials and considerably reduced water stress-induced deficiency in yields of achene per sunflower plant (Iqbal et al., 2008).

Many crop plants have shown a correlation between endogenous proline accumulation and their ability to resist drought. Accumulation of osmolytes such as proline conserves tissue water and protects protein and cellular membranes from oxidative and osmotic stresses (Ashraf and Foolad, 2007). Foliar application of proline markedly increases endogenous accumulation of proline and soluble sugars in Arabidopsis, which increase scavenging in reactive oxygen species as indicated by the decreased lipid peroxidation measured as malondialdehyde. Drought-stressed Arabidopsis plants, when sprayed with proline, were tolerant to photoinhibition as indicated by Fv/Fm values close to those shown by healthy leaves by maintaining more than 98% of photosystem II function. Proline also protects chloroplast structures and shows higher quantum efficiency of PSII photochemistry and decreased excitation pressure in stressed leaves (Moustakas et al., 2011).

Foliar-applied proline in maize plants grown under water-limited conditions showed a remarkable increase in seed moisture, sugar, oil, protein, ash, fibre. Moreover, increases in seed oil oleic, and linolenic acid, tocopherols, flavonoids, phenolics, carotenoids and seed oil DPPH scavenging activities were observed (Ali et al., 2013). Foliar-applied trehalose considerably increased maize plant biomass production and photosynthetic activity, adjusting plant water status through the endogenous accumulation of osmolytes. Furthermore, activation of plant defense mechanisms through increased activities of CAT and POD along with the enhanced production of phenolics and tocopherols reduced water stress-induced damages, probably by stabilizing membrane structures (Ali and Ashraf, 2011b). Exogenously applied trehalose changed seed composition in maize grown under water stress conditions by increasing oil oleic, oil antioxidant activity (DPPH assay) and linolenic acid contents, but there was a subsequent decrease in linoleic acid content (Ali et al., 2012).

Drought-stressed perennial ryegrass plants were exogenously treated with γ-aminobutyric acid. These plants had higher relative water content, turf quality and POD, but also lower wilting rates, electrolyte leakage, canopy temperature depression and lipid peroxidation. However, γ-aminobutyric acid application had no significant effect on the activity of CAT and SOD under drought conditions (Krishnan et al., 2013). Trehalose, applied through foliar spray, was effective in increasing leaf vascular bundle area, epidermis thickness, midrib thickness, leaf parenchyma cell area and vascular bundles in both water stress and non-stress conditions (Akram et al., 2016a). Similarly, trehalose application in water-deficit radish caused a remarkable increase in fresh and dry plant biomass, chlorophyll a, total soluble sugars, photosynthesis rate, water use efficiency, free proline contents and SOD activity (Akram et al., 2016b).

4.6.2.2 Waterlogging

Many economically important plants such as cotton are sensitive to waterlogging conditions. The major adverse effect of waterlogging is oxygen scarcity in submerged plant tissues. Waterlogging affects osmolytes such as sucrose metabolism in cotton by decreasing the activities of ribulose-1,5-bisphosphate carboxylase-oxygenase and cytosolic fructose-1,6-bisphosphatase activity. However, the activities of sucrose synthase (SuS) and sucrose phosphate synthase (SPS) first increased during the initial days of waterlogging and then decreased after 6 days. It was also noticed that genes encoding SuS and SPS were up-regulated with the increase in the duration of waterlogging (Kuai et al., 2014).

Organic acids are an important class of osmolytes (Roychoudhury and Banerjee, 2016) and their levels in plants are affected by abiotic stresses. Chen et al. (2017) reported that malate contents decreased in cotton fiber with an increase in the duration of waterlogging; similarly, sucrose levels considerably decreased in cotton fiber under waterlogging conditions.

4.6.3 Elemental Stress

4.6.3.1 Nutrient Deficiency

Fluctuation in the accumulation of osmolytes by plants under nutrient deficiency is part of the adaptive mechanism designed to cope with stress. Nitrogen deficiency in *Phaseolus vulgaris* led to a decline in proline levels owing to the stimulation of the proline dehydrogenase enzyme. Nonetheless, when nitrogen was at adequate levels, proline contents increased, probably due to the activation of the enzyme ornithine δ-aminotransferase (Sánchez et al., 2002).

Foliar application of glycinebetaine on drought-stressed wheat caused an increase in the uptake of nitrogen, phosphorus, potassium and calcium but a decreased uptake of sodium ions, hence increasing K^+/Na^+ ratio (Raza et al., 2015).

4.6.3.2 Heavy Metal Toxicity

Proline and GB confer protective effects upon plants under environmental stresses such as heavy metal stress that, otherwise, would cause oxidative damage to plant cells. Cadmium (Cd) toxicity is one of the leading contributors to heavy metal stress. Cd toxicity causes growth inhibition. However, exogenous application of proline and GB led to endogenous accumulation of both osmolytes, decrease in lipid peroxidation and Cd accumulation and increase in CAT activity in Cd-stressed cultured tobacco cells (Islam et al., 2009a).

Proline also protects enzymes against heavy metal toxicity. For example, proline mitigated the Cd and Zn stress-induced inhibition of glucose-6-phosphate dehydrogenase and nitrate reductase enzymes *in vitro* cell cultures (Sharma et al., 1998). Hydrogen peroxide, superoxide radical and malondialdehyde act as oxidative stress markers in plant cells. Arsenate hampers the growth of plants and causes accumulation of arsenic in toxic levels. However, provision of proline to arsenate-stressed eggplants diminished arsenic levels. Exogenous proline also triggered the endogenous proline biosynthesis enzyme Δ1-pyrroline-5-carboxylate synthetase (Singh et al., 2015).

Selenium (Se) application at high concentrations (4 and 6 ppm) to hydroponically grown common bean restricted its growth and elevated lipid peroxidation along with hydrogen peroxide levels. However, exogenous application of proline caused an increase in its endogenous concentrations and enzymatic and non-enzymatic antioxidants and improved growth of seedlings under Se stress. In particular, proline application greatly stimulated the ascorbate-glutathione cycle (Aggarwal et al., 2011). Heavy metal stress due to Cd, manganese and nickel (Ni) inhibited the growth of *Scenedesmus armatus* cells and increased CAT and POD activities. However, proline containing cultures showed a somewhat higher growth rate and sustained it better than non-proline-treated cultures (El-Enany and Issa, 2001).

Some plants accumulate significant amounts of specific heavy metals and are more prone to hyperaccumulation of such heavy metals, as for example, *Solanum nigrum* is a Cd hyper-accumulator. However, pretreatment with proline reduces ROS levels and stabilizes plasma membrane structures of cultured *Solanum nigrum* cells under Cd stress. This Cd tolerance was found to be correlated with an increase in CAT and SOD activities and intracellular glutathione levels (Xu et al., 2009).

Proline application to Cd-stressed chickpea plants alleviated the inhibition of chickpea plant growth by negating the adverse effects of metal exposure. Proline increased the activity of carbonic anhydrase, antioxidative enzymes, photosynthesis and yield attributes (Hayat et al., 2013). Exogenous application of proline and GB restored the membrane structure and enhanced the activities of ascorbate-glutathione cycle enzymes under Cd stress in tobacco cultured cells. However, proline was more helpful in efficiently protecting plant cell integrity against Cd stress than GB (Islam et al., 2009b). Potted wheat plants also showed signs of Cd stress in terms of reduced fresh and dry weights of shoot, leaf chlorophyll, pigments such as carotenoids and anthocyanins and other polyphenolic compounds, against increased levels of malondialdehyde and hydrogen peroxide. However, upon application of proline and GB, plants showed significantly higher shoot and root fresh weights and polyphenolic compounds and protected chlorophyll from degradation by accumulating fewer quantities of malondialdehyde and hydrogen peroxide (Rasheed et al., 2014).

Similarly, in a pot experiment carried out on Pb-stressed wheat, GB foliar application increased biomass and length of wheat shoot and root, chlorophyll content and K^+ ions. Therefore, application of proline and GB enhances protection against stress caused by metal toxicity (Bhatti et al., 2013). Along with quaternary ammonium compounds, sugars also provide a cushion to plants against heavy metal stress, sometimes by themselves and sometimes in combination with other osmolytes. Duckweed plants were exposed to Cd stress and later treated with GB and trehalose. Exogenous application of both osmolytes confers protection against Cd stress by protecting the photosynthesis apparatus, increasing endogenous proline production and enhancing antioxidant performance in duckweed plant cells (Duman et al., 2011).

Simultaneous cytotoxicity can be caused by more than one stress, thus profoundly suppressing normal plant metabolism and even collapsing whole survival systems in plants. In such scenario, a plant utilizes all its survival options by activating various defense mechanisms. Pea plants exposed to salinity (NaCl) and heavy metal (Ni) stress substantially inhibited growth by hindering chlorophyll content, photosynthesis, stomata conductance, intercellular carbon dioxide, relative water content and membrane stability index. Exogenously applied natural (*Lolium perenne* leaf extract) and synthetic proline markedly mitigated the damages caused by dual stresses by suppressing the lipid peroxidation and electrolyte leakage, accelerating the activities of synthetic enzymes for polyamine and improving the membrane stability index, leaf polyamines and osmolytes (proline, GB, free amino acids, soluble sugars), total phenols and tocopherol contents (Shahid et al., 2014).

Anthropogenic activities are catalyzing the accumulation of Chromium (Cr), one of the most toxic metals for plants, in agricultural soils especially in peri-urban areas. Cr-stressed wheat plants when supplied with mannitol showed an increase in biomass, chlorophyll and antioxidant enzymes, against a decrease in Cr acquisition, translocation and accumulation (Adrees et al., 2015).

4.6.4 Temperature Stress

4.6.4.1 Cold Stress

Accumulation of osmolytes under extreme temperatures is recognized as an effective plant defense strategy to protect membrane structures and mitigate stress effects. However, the mode of action of osmolytes in heat or cold stress is poorly understood. Cold stress tolerance is linked to the increase in unsaturated fatty acids ratio in plant membranes. GB might induce the enzymes to manufacture unsaturated fatty acids such as fatty acid desaturase and lipoxygenase, which protect tomato plants against cold stress. This could be due to the desaturation of lipids, which leads to stability of membranes and induction of the genes linked with stress defense mechanisms (Karabudak et al., 2014).

Exogenous application of GB in cold-stored peach fruits increased endogenous accumulation of GB, γ-aminobutyric acid and proline contents, resulting in an improvement in the activities of metabolic enzymes such as betaine aldehyde hydrogenase, glutamate decarboxylase, D1-pyrroline-5-carboxylate synthetase and ornithine d-aminotransferase. Hence, GB promoted chilling tolerance by reducing chilling injury and maintaining low pulp firmness in peaches during cold storage (Shan et al., 2016).

Some plants are chilling sensitive and do not accumulate the required osmoprotectants necessary for their survival. However, exogenous provision of osmoprotectants in such plants has shown promising outcomes in terms of better growth and yield. For example, tomato is a chilling-sensitive plant and does not accumulate GB. However, exogenous foliar application of GB improved chilling tolerance as tomato leaves actively uptake GB and translocate it to various organs such as meristemic tissues (shoot apices and floral buds), accumulating the maximum content of GB. These plants show increased CAT activity and decreased H_2O_2 levels after a couple of days (Park et al., 2006).

Chilling stress reduces crop stand in sugarcane. Pretreating sprouting nodal buds with proline and GB markedly minimizes H_2O_2 production and enhances K^+ and Ca^{2+} ratios, free proline and GB levels and soluble sugars content, resulting in induction of chilling tolerance in sugarcane bud chips (Rasheed et al., 2010). Application of antifreeze protein and GB on strawberry grown under chilling temperature (–2°C to –3°C) led to maintenance of membrane integrity and permeability and an increase in stomatal conductance. These positive effects on low-temperature-stressed strawberry plants indicate induction of chilling tolerance by antifreeze protein and GB (Aras and Eşitken, 2013).

Seeds of mung bean were hydroprimed with proline. Hydropriming with proline increased bean seed germination and subsequently also seedling growth at 5°C. Later, seedlings were exposed to chilling to induce chilling injury. Seedling hydroprimed with proline showed better growth during the chilling period and after rewarming at 25°C also grew better (Posmyk and Janas, 2007). During cold stress, tea buds can maintain higher thiol to disulfide ratio if exposed to proline and GB. These osmolytes enhance glutathione transferase and glutathione reductase activities. Similarly, detoxification of methylglyoxal, a toxic compound for plant cells, was induced by proline and GB by protecting and activating glyoxalase I and glyoxalase II, respectively. Thus, proline and GB may induce cold tolerance in tea by regulating methylglyoxal and peroxidation of lipids and activating or maintaining antioxidant biomolecules and enzymes (Kumar and Yadav, 2009).

4.6.4.2 Heat Stress

Pre-treatment of trehalose subsequently protected thylakoid membrane structures and photosynthetic systems in wheat cells under heat-stressed growing conditions. Furthermore, pre-treated trehalose seedlings showed a lower content of malondialdehyde, superoxide and H_2O_2, and reduced electrolyte leakage and lipoxygenase activity (Luo et al., 2010). Exogenous application of GB or proline modified the shape of the polyphasic fluorescence transient curve of heated barley leaves that led to thermostability of photosystem II. This seems to play an important role in heat tolerance of oxygen-evolving complexes. Therefore, GB and proline promote heat stress tolerance in plants (Oukarroum et al., 2012).

Foliar application of GB on heat-stressed marigold reduced the levels of hydrogen peroxide, lipid peroxidation and superoxide and cell death by reducing photoinhibition and increasing assimilation rate, stomata conductance and transpiration rate and, thus, resulting in heat stress tolerance in marigold (Sorwong and Sakhonwasee, 2015).

4.6.5 Radiation Stress

Accumulation of osmolytes such as proline, as in other stresses, also protects plants against UV-radiation-induced peroxidative processes (Jain et al., 2001). For example, if the shoots of rice, mustard and mung bean at the seedling stage are exposed to UV radiations, the proline content in the seedlings is substantially increased depending upon exposure time. UV radiations enhanced lipid peroxidation as indicated by the proliferated production of malondialdehyde in these plants. UV radiations increased peroxidation in linolenic acid micelles as well. The presence of proline along with linolenic acid micelles during UV exposure caused a considerable reduction in the production of malondialdehyde (Saradhi et al., 1995). Ashraf and Foolad (2007) also pointed out that GB and proline are two major osmolytes that accumulate in a variety of plant species in response to environmental stresses such as UV radiation.

4.7 CONCLUSION

Plants naturally produce osmolytes, a class of phytoprotectants that are also called 'compatible solutes' or 'osmoprotectants'. However, certain valuable crops such as wheat, maize, and barley do not accumulate osmolytes in higher concentrations. Similarly, within a species, only stress tolerant genotypes accumulate levels of osmolytes high enough to cope with stress compared to susceptible genotypes. Therefore, many studies have shown that exogenous application of osmolytes can overcome the deficiency in osmolyte accumulation and lead to better growth in plants. Even within the same accumulating species, plants combat environmental stress better if fed with exogenously applied osmolytes than non-fed plants. Moreover, exogenous application of a specific osmolyte or combinations of osmolytes alleviates the adverse effects of a specific stress condition in pls. Therefore, through their studies, researchers have recommended the exogenous application of specific osmolyte to a certain crop to cope with specific stress.

REFERENCES

Abdallah, M.M.-S., Abdelgawad, Z.A., El-Bassiony, H.M.S. (2016) Alleviation of the adverse effects of salinity stress using trehalose in two rice varieties. *South African Journal of Botany* 103: 275–282.

Adrees, M., Ali, S., Iqbal, M., Bharwana, S.A., Siddiqi, Z., Farid, M., Ali, Q., Saeed, R., Rizwan, M. (2015) Mannitol alleviates chromium toxicity in wheat plants in relation to growth, yield, stimulation of anti-oxidative enzymes, oxidative stress and Cr uptake in sand and soil media. *Ecotoxicology and Environmental Safety* 122: 1–8.

Aggarwal, M., Sharma, S., Kaur, N., Pathania, D., Bhandhari, K., Kaushal, N., Kaur, R., *et al.* (2011) Exogenous proline application reduces phytotoxic effects of selenium by minimising oxidative stress and improves growth in bean (*Phaseolus vulgaris* L.) seedlings. *Biological Trace Element Research* 140(3): 354–367.

Akram, N.A., Shafiq, S., Ashraf, M., Aisha, R., Sajid, M.A. (2016a) Drought-induced anatomical changes in radish (*Raphanus sativus* L.) leaves supplied with trehalose through different modes. *Arid Land Research and Management* 30(4): 412–420.

Akram, N.A., Waseem, M., Ameen, R., Ashraf, M. (2016b) Trehalose pretreatment induces drought tolerance in radish (*Raphanus sativus* L.) plants: Some key physio-biochemical traits. *Acta Physiologiae Plantarum* 38(1): 1–10.

Ali, Q., Anwar, F., Ashraf, M., Saari, N., Perveen, R. (2013) Ameliorating effects of exogenously applied proline on seed composition, seed oil quality and oil antioxidant activity of maize (*Zea mays* L.) under drought stress. *International Journal of Molecular Sciences* 14(1): 818–835.

Ali, Q., Ashraf, M. (2011a) Exogenously applied glycinebetaine enhances seed and seed oil quality of maize (*Zea mays* L.) under water deficit conditions. *Environmental and Experimental Botany* 71(2): 249–259.

Ali, Q., Ashraf, M. (2011b) Induction of drought tolerance in maize (*Zea mays* L.) due to exogenous application of trehalose: Growth, photosynthesis, water relations and oxidative defence mechanism. *Journal of Agronomy and Crop Science* 197(4): 258–271.

Ali, Q., Ashraf, M., Anwar, F., Al-Qurainy, F. (2012) Trehalose-induced changes in seed oil composition and antioxidant potential of maize grown under drought stress. *Journal of the American Oil Chemists' Society* 89: 1485–1493.

Anjum, S.A., Farooq, M., Wang, L.C., Xue, L.L., Wang, S.G., Wang, L., Zhang, S., Chen, M. (2011) Gas exchange and chlorophyll synthesis of maize cultivars are enhanced by exogenously-applied glycinebetaine under drought conditions. *Plant, Soil and Environment* 57(7): 326–331.

Aras, S., Eşitken, A. (2013) Effects of antifreeze proteins and glycine betaine on strawberry plants for resistance to cold temperature. In *International Conference on Agriculture and Biotechnology*. Singapore: IACSIT Press, *Vol. 60*, 107–111.

Argiolas, A., Puleo, G.L., Sinibaldi, E., Mazzolai, B. (2016) Osmolyte cooperation affects turgor dynamics in plants. Scientific Reports 6: 30139.

Ashraf, M., Foolad, M.R. (2007) Roles of glycine betaine and proline in improving plant abiotic stress resistance. *Environmental and Experimental Botany* 59(2): 206–216.

Atreya, A., Vartak, V., Bhargava, S. (2009) Salt priming improves tolerance to desiccation stress and extreme salt stress in Bruguiera cylindrical. International Journal of Integrative Biology 6(2): 68–73.

Banu, N.A., Hoque, A., Watanabe-Sugimoto, M., Matsuoka, K., Nakamura, Y., Shimoishi, Y., Murata, Y. (2009) Proline and glycinebetaine induce antioxidant defense gene expression and suppress cell death in cultured tobacco cells under salt stress. *Journal of Plant Physiology* 166(2): 146–156.

Ben Ahmed, Ch, Magdich, S., Ben Rouina, B., Sensoy, S., Boukhris, M., Ben Abdullah, F. (2011) Exogenous proline effects on water relations and ions contents in leaves and roots of young olive. *Amino Acids* 40(2): 565–573.

Bhatti, K.H., Anwar, S., Nawaz, K., Hussain, K., Siddiqi, E., Usmansharif, R., Talat, A., Khalid, A. (2013) Effect of exogenous application of glycinebetaine on wheat (*Triticum aestivum* L.) under heavy metal stress. *Middle East Journal of Scientific Research* 14: 130–137.

Bot, A.J., Nachtergaele, F.O., Young, A. (2000) Land resource potential and constraints at regional and country levels. *World Soil Resources Reports* 90. Rome: Land and Water Development Division, FAO.

Chang, B., Yang, L., Cong, W., Zu, Y., Tang, Z. (2014) The improved resistance to high salinity induced by trehalose is associated with ionic regulation and osmotic adjustment in *Catharanthus roseus. Plant Physiology and Biochemistry: PPB* 77: 140–148.

Cha-Um, S., Kirdmanee, C. (2010) Effect of glycinebetaine on proline, water use, and photosynthetic efficiencies, and growth of rice seedlings under salt stress. *Turkish Journal of Agriculture and Forestry* 34: 517–527.

Chen, T.H., Murata, N. (2008) Glycinebetaine: An effective protectant against abiotic stress in plants. *Trends in Plant Science* 13(9): 499–505.

Chen, Y., Wang, H., Hu, W., Wang, S., Wang, Y., Snider, J.L., Zhou, Z. (2017) Combined elevated temperature and soil waterlogging stresses inhibit cell elongation by altering osmolyte composition of the developing cotton (*Gossypium hirsutum* L.) fiber. *Plant Science: an International Journal of Experimental Plant Biology* 256: 196–207.

Conde, A., Silva, P., Agasse, A., Conde, C., Gerós, H. (2011) Mannitol transport and mannitol dehydrogenase activities are coordinated in *Olea europaea* under salt and osmotic stresses. *Plant and Cell Physiology* 52(10): 1766–1775.

Da Rocha, I.M., Vitorello, V.A., Silva, J.S., Silva, S.L.F., Viegas, R.A., Silva, E.N., Silveira, J.A.G. (2012) Exogenous ornithine is an effective precursor and the delta-ornithine amino transferase pathway contributes to proline accumulation under high N recycling in salt-stressed cashew leaves. *Journal of Plant Physiology* 169(1): 41–49.

Denaxa, N.-K., Roussos, P.A., Damvakaris, T., Stournaras, V. (2012) Comparative effects of exogenous glycine betaine, kaolin clay particles and Ambiol on photosynthesis, leaf sclerophylly indexes and heat load of olive cv. Chondrolia Chalkidikis under drought. *Scientia Horticulturae* 137: 87–94.

Duman, F., Aksoy, A., Aydin, Z., Temizgul, R. (2011) Effects of exogenous glycinebetaine and trehalose on cadmium accumulation and biological responses of an aquatic plant (*Lemna gibba* L.). *Water, Air, and Soil Pollution* 217(1–4): 545–556.

El-Enany, A.E., Issa, A.A. (2001) Proline alleviates heavy metal stress in Scenedesmus armatus. *Folia Microbiologica* 46(3): 227–230.

Elkoca, E., Kantar, F., Fiahin, F. (2008) Influence of nitrogen fixing and phosphorus solubilizing bacteria on the nodulation, plant growth, and yield of chickpea. *Journal of Plant Nutrition* 31(1): 157–171.

Farooq, M., Basra, S.M.A., Wahid, A., Cheema, Z.A., Cheema, M.A., Khaliq, A. (2008) Physiological role of exogenously applied glycinebetaine to improve drought tolerance in fine grain aromatic rice (*Oryza sativa* L.). *Journal of Agronomy and Crop Science* 194(5): 325–333.

Ghars, M.A., Parre, E., Debez, A., Bordenave, M., Richard, L., Leport, L., Bouchereau, A., Savouré, A., Abdelly, C. (2008) Comparative salt tolerance analysis between *Arabidopsis thaliana* and *Thellungiella halophila*, with special emphasis on K+/Na+ selectivity and proline accumulation. *Journal of Plant Physiology* 165(6): 588–599.

Gupta, B., Huang, B. (2014) Mechanism of salinity tolerance in plants: physiological, biochemical, and molecular characterization. International Journal of Genomics 2014: 701596.

Hamilton, E.W., Heckathorn, S.A. (2001) Mitochondrial adaptation to NaCl. Complex I is protected by antioxidants and small heat shock proteins, whereas complex II is protected by proline and betaine. *Plant Physiology* 126(3): 1266–1274.

Hare, P.D., Cress, W.A., van Staden, J. (1998) Dissecting the roles of osmolyte accumulation during stress. *Plant, Cell and Environment* 21(6): 535–553.

Haroun, S.A., Shukry, W.M., Sawy, O.E. (2010) Effect of asparagine or glutamine on growth and metabolic changes in *Phaseolus vulgaris* under in vitro conditions. *Bio-Science Research* 7(1): 01–21.

Hasanuzzaman, M., Nahar, K., Fujita, M. (2013) Plant response to salt stress and role of exogenous protectants to mitigate salt-induced damages. In *Ecophysiology and Responses of Plants under Salt Stress*, ed. Ahmad, P., Azooz, M.M. and Prasad, M.N.V., 25–87. New York: Springer.

Hayat, S., Hayat, Q., Alyemeni, M.N., Ahmad, A. (2013) Proline enhances antioxidative enzyme activity, photosynthesis and yield of *Cicer arietinum* L. exposed to cadmium stress. *Acta Botanica Croatica* 72(2): 323–335.

Hoque, M.A., Banu, M.N.A., Nakamura, Y., Shimoishi, Y., Murata, Y. (2008) Proline and glycinebetaine enhance antioxidant defense and methylglyoxal detoxification systems and reduce NaCl-induced damage in cultured tobacco cells. *Journal of Plant Physiology* 165(8): 813–824.

Hoque, M.A., Banu, M.N., Okuma, E., Amako, K., Nakamura, Y., Shimoishi, Y., Murata, Y. (2007) Exogenous proline and glycinebetaine increase NaCl-induced ascorbate–glutathione cycle enzyme activities, and proline improves salt tolerance more than glycinebetaine in tobacco Bright Yellow-2 suspension-cultured cells. *Journal of Plant Physiology* 164(11): 1457–1468.

Hossain, M.A., Fujita, M. (2010) Evidence for a role of exogenous glycinebetaine and proline in antioxidant defense and methylglyoxal detoxification systems in mung bean seedlings under salt stress. *Physiology and Molecular Biology of Plants: an International Journal of Functional Plant Biology* 16(1): 19–29.

Hu, L., Hu, T. (2012) Exogenous glycine betaine ameliorates the adverse effect of salt stress on perennial ryegrass. *Journal of the American Society for Horticultural Science* 137: 38–46.

Hussain, M., Malik, M.A., Farooq, M., Ashraf, M.Y., Cheema, M.A. (2008) Improving drought tolerance by exogenous application of glycinebetaine and salicylic acid in sunflower. *Journal of Agronomy and Crop Science* 194(3): 193–199.

Ingram, J., Bartels, D. (1996) The molecular basis of dehydration tolerance in plants. Annual Review of Plant Physiology and Plant Molecular Biology 47: 377–403.

Iqbal, N., Ashraf, M., Ashraf, M.Y. (2008) Glycinebetaine, an osmolyte of interest to improve water stress tolerance in sunflower (*Helianthus annuus* L.): Water relations and yield. *South African Journal of Botany* 74(2): 274–281.

Islam, M.M., Hoque, M.A., Okuma, E., Banu, M.N., Shimoishi, Y., Nakamura, Y., Murata, Y. (2009a) Exogenous proline and glycinebetaine increase antioxidant enzyme activities and confer tolerance to cadmium stress in cultured tobacco cells. *Journal of Plant Physiology* 166(15): 1587–1597.

Islam, M.M., Hoque, M.A., Okuma, E., Jannat, R., Banu, M.N., Jahan, M.S., Nakamura, Y., Murata, Y. (2009b) Proline and glycinebetaine confer cadmium tolerance on tobacco bright yellow-2 cells by increasing ascorbate-glutathione cycle enzyme activities. *Bioscience, Biotechnology, and Biochemistry* 73(10): 2320–2323.

Jain, M., Mathur, G., Koul, S., Sarin, N.B. (2001) Ameliorative effects of proline on salt stress-induced lipid peroxidation in cell lines of groundnut (*Arachis hypogaea* L.). *Plant Cell Reports* 20(5): 463–468.

Karabudak, T., Bor, M., Özdemir, F., Türkan, İ. (2014) Glycine betaine protects tomato (*Solanum lycopersicum*) plants at low temperature by inducing fatty acid desaturase7 and lipoxygenase gene expression. *Molecular Biology Reports* 41(3): 1401–1410.

Kaya, C., Sonmez, O., Aydemir, S., Ashraf, M., Dikilitas, M. (2013) Exogenous application of mannitol and thiourea regulates plant growth and oxidative stress responses in salt-stressed maize (*Zea mays* L.). *Journal of Plant Interactions* 8(3): 234–241.

Kemble, A.R., MacPherson, H.T. (1954) Liberation of amino acids in perennial rye grass during wilting. *The Biochemical Journal* 58(1): 46–49.

Khan, S.H., Ahmad, N., Ahmad, F., Kumar, R. (2010) Naturally occurring organic osmolytes: from cell physiology to disease prevention. *IUBMB Life* 62(12): 891–895.

Khan, A., Iqbal, I., Ahmad, I., Nawaz, H., Nawaz, M. (2014) Role of proline to induce salinity tolerance in sunflower (*Helianthus annuus* L.). *Science, and Technology and for Development* 33(2): 88–93.

Khedr, A.H.A., Abbas, M.A., Wahid, A.A.A., Quick, W.P., Abogadallah, G.M. (2003) Proline induces the expression of salt stress responsive proteins and may improve the adaptation of *Pancratium maritimum* L. to salt stress. *Journal of Experimental Botany* 54(392): 2553–2562.

Krishnan, S., Laskowski, K., Shukla, V., Merewitz, E.B. (2013) Mitigation of drought stress damage by exogenous application of a non-protein amino acid ɣ–aminobutyric acid on perennial ryegrass. *Journal of the American Society for Horticultural Science* 138: 358–366.

Kuai, J., Liu, Z., Wang, Y., Meng, Y., Chen, B., Zhao, W., Zhou, Z., Oosterhuis, D.M. (2014) Waterlogging during flowering and boll forming stages affects sucrose metabolism in the leaves subtending the cotton boll and its relationship with boll weight. *Plant Science: an International Journal of Experimental Plant Biology* 223: 79–98.

Kumar, V., Yadav, S.K. (2009) Proline and betaine provide protection to antioxidant and methylglyoxal detoxification systems during cold stress in *Camellia sinensis* (L.) O. Kuntze. *Acta Physiologiae Plantarum* 31(2): 261–269.

Lima-Costa, M.E., Ferreira, S., Duarte, A., Ferreira, A.L. (2010) Alleviation of salt stress using exogenous proline on a citrus cell line. *Acta Horticulturae* 868: 109–112.

López-Gómez, M., Lluch, C. (2012) Trehalose and abiotic stress tolerance. In *Abiotic stress responses in plants: Metabolism, productivity and sustainability*, ed. Ahmad, P. and Prasad, M.N.V., 253–265. New York: Springer.

Luo, Y., Li, F., Wang, G.P., Yang, X.H., Wang, W. (2010) Exogenously-supplied trehalose protects thylakoid membranes of winter wheat from heat-induced damage. *Biologia Plantarum* 54(3): 495–501.

Lutts, S. (2000) Exogenous glycinebetaine reduces sodium accumulation in salt-stressed rice plants. *International Rice Research Notes* 25(2): 39–40.

Mahouachi, J., Argamasila, R., Gomes-Cadenas, A. (2012) Influence of exogenous glycine betaine and abscisic acid on papaya in responses to water-deficit stress. *Journal of Plant Growth Regulation* 31(1): 1–10.

Manaf, H.H. (2016) Beneficial effects of exogenous selenium, glycine betaine and seaweed extract on salt stressed cowpea plant. *Annals of Agricultural Sciences* 61(1): 41–48.

Moustakas, M., Sperdouli, I., Kouna, T., Antonopoulou, C., Therios, I. (2011) Exogenous proline induces soluble sugar accumulation and alleviates drought stress effects on photosystem II functioning of Arabidopsis thaliana leaves. *Plant Growth Regulation* 65(2): 315–325.

Nawaz, K., Ashraf, M. (2010) Exogenous application of glycinebetaine modulates activities of antioxidants in maize plants subjected to salt stress. *Journal of Agronomy and Crop Science* 196(1): 28–37.

Nounjan, N., Nghia, P.T., Theerakulpisut, P. (2012) Exogenous proline and trehalose promote recovery of rice seedlings from salt-stress and differentially modulate antioxidant enzymes and expression of related genes. *Journal of Plant Physiology* 169(6): 596–604.

Osman, H.S., Saleem, B.B.M. (2016) Influence of exogenous application of some phytoprotectants on growth, yield and pod quality of snap bean under NaCl salinity. *Annals of Agricultural Sciences* 61(1): 1–13.

Oukarroum, A., El Madidi, S., Strasser, R.J. (2012) Exogenous glycine betaine and proline play a protective role in heat-stressed barley leaves (*Hordeum vulgare* L.): A chlorophyll a fluorescence study. *Plant Biosystems - an International Journal Dealing with All Aspects of Plant Biology* 146(4): 1037–1043.

Palma, F., Tejera, N.A., Lluch, C. (2013) Nodule carbohydrate metabolism and polyols involvement in the response of Medicago sativa to salt stress. *Environmental and Experimental Botany* 85: 43–49.

Park, E.J., Jeknic, Z., Chen, T.H. (2006) Exogenous application of glycinebetaine increases chilling tolerance in tomato plants. *Plant and Cell Physiology* 47(6): 706–714.

Posmyk, M.M., Janas, K.M. (2007) Effects of seed hydropriming in presence of exogenous proline on chilling injury limitation in Vigna radiata L. seedlings. *Acta Physiologiae Plantarum* 29(6): 509–517.

Rasheed, R., Ashraf, M.A., Hussain, I., Haider, M.Z., Kanwal, U., Iqbal, M. (2014) Exogenous proline and glycinebetaine mitigate cadmium stress in two genetically different spring wheat (*Triticum aestivum* L.) cultivars. *Brazilian Journal of Botany* 37(4): 399–406.

Rasheed, R., Wahid, A., Ashraf, M., Basra, S.M. (2010) Role of proline and glycinebetaine in improving chilling stress tolerance in sugarcane buds at sprouting. *International Journal of Agriculture and Biology* 12: 1–8.

Raza, M.A.F., Saleem, M.F.S., Khan, I.H. (2015) Combined application of glycinebetaine and potassium on the nutrient uptake performance of wheat under drought stress. *Pakistan Journal of Agricultural Sciences* 52(1): 19–26.

Roychoudhury, A., Banerjee, A. (2016) Endogenous glycine betaine accumulation mediates abiotic stress tolerance in plants. *Tropical Plant Research* 3: 105–111.

Sánchez, E., Garcia, P.C., Lefebre, L.R.L., Rivero, R.M., Ruiz, J.M., Romero, L. (2002) Proline metabolism in response to nitrogen deficiency in French Bean plants (Phaseolus vulgaris L. cv Strike). *Plant Growth Regulation* 36(3): 261–265.

Saradhi, P.P., Alia, A.S., Arora, S., Prasad, K.V. (1995) Proline accumulates in plants exposed to UV radiation and protects them against UV induced peroxidation. *Biochemical and Biophysical Research Communications* 209(1): 1–5.

Seckin, B., Sekmen, A.H., Türkan, İ (2009) An enhancing effect of exogenous mannitol on the antioxidant enzyme activities in roots of wheat under salt stress. *Journal of Plant Growth Regulation* 28(1): 12–20.

Shahid, M.A., Balal, R.M., Pervez, M.A., Abbas, T., Aqeel, M.A., Javaid, M.M., Sanchez, F.G. (2014) Exogenous proline and proline-enriched *Lolium perenne* leaf extract protects against phytotoxic effects of nickel and salinity in Pisum sativum by altering polyamine metabolism in leaves. *Turkish Journal of Botany* 38(5): 914–926.

Shan, T., Jin, P., Zhang, Y., Huang, Y., Wang, X., Zheng, Y. (2016) Exogenous glycine betaine treatment enhances chilling tolerance of peach fruit during cold storage. *Postharvest Biology and Technology* 114: 104–110.

Sharma, S.S., Schat, H., Vooijs, R. (1998) In vitro alleviation of heavy metal-induced enzyme inhibition by proline. *Phytochemistry* 49(6): 1531–1535.

Singh, M., Pratap Singh, V., Dubey, G., Mohan Prasad, S. (2015) Exogenous proline application ameliorates toxic effects of arsenate in *Solanum melongena* L. seedlings. *Ecotoxicology and Environmental Safety* 117: 164–173.

Slama, I., Abdelly, C., Bouchereau, A., Flowers, T., Savouré, A. (2015) Diversity, distribution and roles of osmoprotective compounds accumulated in halophytes under abiotic stress. *Annals of Botany* 115(3): 433–447.

Sorwong, A., Sakhonwasee, S. (2015) Foliar application of glycine betaine mitigates the effect of heat stress in three marigold (*Tagetes erecta*) cultivars. *The Horticulture Journal* 84(2): 161–171.

Tian, F., Wang, W., Liang, C., Wang, X., Wang, G., Wang, W. (2017) Overaccumulation of glycine betaine makes the function of the thylakoid membrane better in wheat under salt stress. *The Crop Journal* 5(1): 73–82.

Xu, J., Yin, H., Li, X. (2009) Protective effects of proline against cadmium toxicity in micropropagated hyperaccumulator, *Solanum nigrum* L. *Plant Cell Reports* 28(2): 325–333.

Yagmur, M., Kaydan, D. (2008) Alleviation of osmotic stress of water and salt in germination and seedling growth of triticale with seed priming treatments. African Journal of Biotechnology 7: 2156–2162.

Yancey, P.H. (2005) Organic osmolytes as compatible, metabolic and counteracting cytoprotectants in high osmolarity and other stresses. *The Journal of Experimental Biology* 208(15): 2819–2830.

Zeid, I.M. (2009) Effect of arginine and urea on polyamines content and growth of bean under salinity stress. *Acta Physiologiae Plantarum* 31(1): 65–70.

5 Proline – A Key Regulator Conferring Plant Tolerance to Salinity and Drought

Renu Khanna-Chopra, Vimal Kumar Semwal, Nita Lakra, and Ashwani Pareek

CONTENTS

5.1 INTRODUCTION

Plant growth and productivity are greatly affected by osmotic or water stress caused by drought or high salinity (Golldack et al., 2014; Khanna-Chopra and Singh, 2015). Plants respond and adapt to osmotic stress using intricate physiological mechanisms enabling survival. Many plants accumulate compatible osmolytes, such as proline (Pro), glycine betaine, or sugar alcohols, when they are subjected to osmotic stress (Delauney and Verma, 1993; Kishor et al., 2005; Szabados and Savoure, 2010). Among these, Pro is the most diversely used osmolyte accumulated under osmotic stress conditions in plants. Stress-induced accumulation of Pro was first observed in wilting ryegrass (*Lolium* sp.) (Kemble and Macpherson, 1954) and was later observed in other species in response to different environmental stresses including drought, high salinity, high light, UV irradiation, heavy metals, oxidative stress, and in response to biotic stresses (Choudhary et al., 2005; Fabro et al., 2004; Saradhi et al., 1995; Schat et al., 1997; Yang et al., 2009; Yoshiba et al., 1995). Pro is accumulated not only in plants but also in eubacteria, protozoa, marine invertebrates, and algae (Delauney and Verma, 1993). Proline is a low-molecular-weight cyclic amino acid and plays significant roles as a sink of energy or reducing power, as a source of carbon and nitrogen compounds, as a hydroxyl radical scavenger, and in protecting plasma membrane integrity in plants under osmotic stress (Kishor et al., 2015; Szabados and Savoure, 2010). Proline plays crucial roles in cellular metabolism as a component of proteins with significant contributions to protein synthesis, protein folding and structure, and is able to stabilize proteins, DNA and membranes due to its ability to scavenge reactive oxygen species (Alia et al., 1997; Wang et al., 2014). In addition, hydroxyproline and Pro-rich proteins are important components of cell wall proteins and have been shown to play key roles in cell wall signal transduction cascades, and are thus linked to stress tolerance (Kishor et al., 2015). Pro also induces expression of some drought and salt-stress-induced responsive genes (Khedr et al., 2003).

In stressed plants, Pro concentration can reach mM level, which is several folds higher than its background level depending upon the species and the severity and duration of stress (Delauney and Verma, 1993). In many plants, a strong correlation has been established between Pro accumulation and abiotic stress tolerance (Hare and Cress, 1997; Kishor et al., 1995, 2005; Saradhi et al., 1995; Siripornadulsil et al., 2002). However, in barley, Pro accumulation appears to play no role in salinity tolerance, instead of representing a symptom of susceptibility (Chen et al., 2007). Thus, the relationship of Pro accumulation and stress tolerance remains somewhat controversial (Hare and Cress, 1997). However, detailed studies using transgenic plants and mutants clearly show the multifunctional roles of Pro metabolism in growth, development, and stress responses (Kishor and Sreenivasulu, 2014).

Proline may also play a role in flowering and development both as a metabolite and a signaling molecule. Developmental accumulation of Pro in reproductive organs seems to be a widespread phenomenon among plant species (Mattioli et al., 2009a). An obvious function of Pro in development may be the protection of developing cells from osmotic damages, especially in developmental processes such as pollen development and embryogenesis where tissues undergo spontaneous dehydration. Also, Pro has been proposed to provide energy to sustain metabolically demanding programs of plant reproduction as the oxidation of one molecule of Pro yields 30 ATP equivalents. The upregulation of Pro catabolic genes typically observed in flowers, siliques, and seeds is consistent with the need to provide the plant with energy throughout the whole reproductive phase. Lehmann et al. (2010) observed that many flowers secrete Pro-rich nectar. A high level of Pro in plant nectars is thought to work as an attractant since several insect species prefer Pro-rich nectars (Bertazzini et al., 2010). Proline has also shown its involvement in aroma synthesis in aromatic rice (Yoshihashi et al., 2002) and plays a crucial role in regulating general protein synthesis and cell cycle in maize (Wang et al., 2014).

In this chapter, we will focus on Pro, its biosynthesis and transport and its various functions which contribute toward enhancing drought and salt tolerance in plants. The molecular basis of Pro metabolism and the importance of Pro homeostasis under stress will also be discussed.

5.2 PROLINE METABOLISM AND TRANSPORT

In plants, Pro biosynthesis takes place via two pathways, namely, glutamate and ornithine pathways (Figure 5.1). In glutamate pathways, Pro biosynthesis begins with the phosphorylation of glutamate to form γ-glutamyl phosphate, which is reduced to an intermediate glutamic-5-semialdehyde (GSA) by the action of bifunctional enzyme pyrroline-5-carboxylate synthetase (P5CS) which is spontaneously cyclized into pyrroline-5-carboxylate (P5C). P5CS requires both ATP and NADPH. P5C reductase (P5CR) further reduces the P5C intermediate to Pro. In ornithine pathways, ornithine can be transaminated

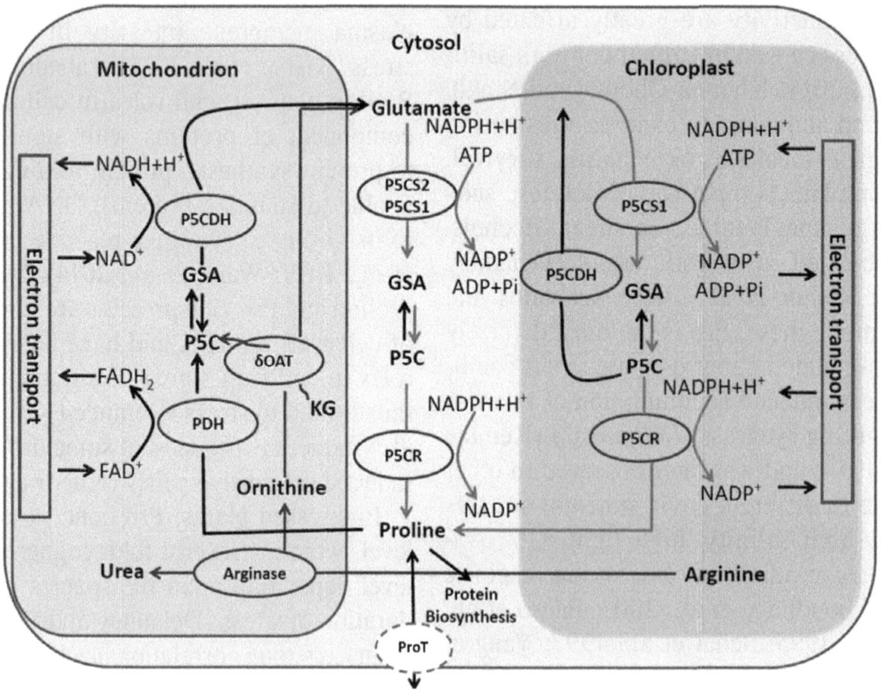

FIGURE 5.1 Proline metabolism in plants: Proline synthesis occurs in the cytosol and chloroplast. Proline degradation occurs in mitochondria. P5C - δ-pyrroline-5-carboxylate, P5CR - pyrroline-5-carboxylate reductase, P5CS - pyrroline-5-carboxylate synthase, GSA - glutamic semialdehyde, PDH - Proline dehydrogenase, OAT - Ornithine aminotransferase, KG - Ketoglutarate, ProT-Proline transporter.

to GSA by the activity of ornithine ∂-aminotransferase (∂-OAT) located in mitochondria, and subsequently gets converted to Pro via P5C (Szabados and Savoure, 2010). It has, however, been shown in *Arabidopsis thaliana* that δ-OAT generates P5C exclusively for the catabolic branch, which ultimately gives glutamate as an end product (Funck et al., 2008). Therefore, in Arabidopsis, biosynthesis of Pro occurs exclusively through the glutamate pathway. Under normal conditions, the biosynthesis of Pro takes place in the cytosol; whereas, under stress conditions, Pro synthesis occurs also in chloroplasts and mainly from glutamate (Rejeb et al., 2014). In most plant species, P5CS is encoded by two genes and P5CR is encoded by one (Strizhov et al., 1997; Verbruggen et al., 1993). Proline catabolism occurs in mitochondria via the sequential action of Pro dehydrogenase (PDH) or Pro oxidase (POX) producing P5C from Pro using flavin adenine dinucleotide (FAD) as a cofactor. Pyrroline-5-carboxylate dehydrogenase (P5CDH) converts P5C to glutamate. *PDH* is encoded by two genes, whereas a single *P5CDH* gene has been identified in Arabidopsis and tobacco (Deuschle et al., 2001; Ribarits et al., 2007). The biosynthetic enzymes P5CS1, P5CS2 and P5CR are predicted to be localized in the cytosol whereas a mitochondrial localization is predicted for the enzymes involved in Pro catabolism such as PDH1/ERD5, PDH2, P5CDH, and OAT. Using GFP (green fluorescent protein) technology P5CS1 and P5CS2 were shown to be localized in the cytosol of leaf mesophyll cells and when exposed to salt or osmotic stress only P5CS1 accumulated in the chloroplasts (Rayapati et al., 1989). P5CR protein and activity have been detected in the cytosol and plastid fraction of soybean leaf (Szoke et al., 1992) while under the osmotic condition it was localized in chloroplasts in pea (Rayapati et al., 1989). Intracellular Pro content is determined by biosynthesis, catabolism, and transport between cells and different cellular compartments. The production of Pro is promoted and its degradation suppressed during salt or drought stress. Proline biosynthesis normally occurs in the cytosol and it is controlled by the *P5CS2* gene in Arabidopsis. During osmotic stress, Pro biosynthesis is augmented in chloroplast, which is controlled by the stress-induced *P5CS1* gene in Arabidopsis. Hence Pro can be synthesized in different subcellular compartments depending on environmental conditions. Of two genes, *PDH1* and *PDH2*, identified in *A. thaliana* (Funck et al., 2010), PDH1 is considered the main isoform involved in the degradation of Pro. The expression of *PDH1* has been shown to be downregulated under water stress (Kiyosue et al., 1996). It is generally believed that the level of Pro is mainly regulated by P5CS and PDH enzymes. Transport of amino acids is regulated by both endogenous and environmental signals. The amino acid transporter family includes five sub-classes of transporters: amino acid permeases, lysine, histidine, Pro (ProT) and the putative auxin transporters. Eight different amino acid transporter clones have been isolated and characterized in Arabidopsis using yeast mutants. Among these transporters, two encoding specific Pro transporters (*ProT1* and *ProT2*) were distantly related to the amino acid/auxin permease (AAP) family. So, at present, ProT, which belongs to the ATF or AAAP family, has been shown to be localized in the plasma membrane and is likely involved in intercellular and long-distance Pro transport (Rentsch et al., 2007). Genes encoding transporters of Pro have been isolated from various plant species such as Arabidopsis, rice, tomato and barley. The expression of genes encoding two specific Pro transporters (*ProT1* and *ProT2*) was found to be enhanced during stressful conditions in *A. thaliana*. *ProT1* expressed ubiquitously, but in *Arabidopsis thaliana* plants exposed to salinity stress, higher levels of *ProT1* were recorded in roots, stems and flowers. Young flowers showed highest expression, particularly in floral stalk phloem. Under water or salinity stress, a strong expression of *ProT2* was recorded, indicating that it plays an important role in transport of nitrogen, especially Pro under water stress (Rentsch et al., 1996). In the halophyte species *Limonium latifolium*, Pro was sequestered to vacuoles in non-stressed plants, whereas high Pro content was detected in the cytosol of salt-stressed plants, suggesting the importance of *de novo* Pro biosynthesis as well as transport for Pro accumulation (Gagneul et al., 2007). Proline transport has also been reported to contribute to Pro accumulation in some tissues under stress conditions. For example, high Pro concentration in phloem sap indicated Pro transport to growing tissues in drought-stressed alfalfa (Girousse et al., 1996). Similarly, Raymond and Smirnoff (2002) presented evidence for Pro transport to the root tips of maize at low water potentials. These studies suggest that Pro transport is increased under drought-stressed conditions (Kishor et al., 2005). However, little research has been devoted to Pro accumulation mechanisms and Pro transport in plants that are relieved from stressed conditions.

5.3 PROLINE IS A KEY COMPONENT OF DROUGHT AND SALT STRESS TOLERANCE IN PLANTS

5.3.1 Role of Proline During Osmotic Stress

The accumulation of compatible solutes is a prerequisite for the adaptation of plants to drought stress. Proline is one of the major osmolytes in plants. It accumulates in the cells to balance external osmotic pressure and maintains cellular turgor under drought stress conditions. The term osmotic adjustment (OA) is used for this balance of cellular turgidity (Blum, 2017). Osmotic adjustment helps in maximizing water uptake under drought conditions and is a major protective mechanism in plants. Blum (2017) argued about the importance of compatible solutes and OA in sustaining crop yields and discussed 26 cases so far studied in 12 cultivated crops of prime importance where OA was directly compared with the stability of yield and related traits under drought stress. Out of the 26 cases he reviewed 24 cases that showed significant positive correlation of OA with crop yield indicating the importance of OA and osmolyte accumulation in plant adaptation and agricultural productivity under drought stress.

Drought stress is inherently associated with and characterized by reduced leaf water potential in plants. Many reports in plants showed high Pro accumulation under drought stress leading to maintenance of cellular turgor (Verslues and Sharma, 2010). PEG-induced drought stress resulted in a four-fold increase in cellular Pro content compared to non-stressed rice plants, the accumulation being higher in sensitive genotypes compared to the tolerant genotype (Basu et al., 2010). Karamanos (1995) showed that Pro was the highest accumulating free amino acid compared to others under drought stress in wheat and fava bean genotypes, accounting for 70% of total amino compounds under severe stress conditions. By comparing wheat genotypes and fava bean genotypes differing in drought resistance potential, it has been proved that there is a positive association between drought tolerance and Pro accumulation, as the tolerant genotypes accumulated higher Pro content compared to the susceptible ones (Karamanos, 1995). In sorghum, the younger and actively growing plant's parts accumulated higher amounts of Pro and showed faster recovery after stress. The speed of drought-stressed plants recovery was also associated with higher Pro concentration in sorghum leaves (Blum and Ebercon, 1976). High Pro accumulation in pollen alsoimproved pollen desiccation tolerance (Hare et al., 1998).

Several plants have been identified in which the Pro content increases in response to salt tolerance, and it is considered as a major marker for detecting salt stress tolerance in plants (Roychoudhury and Chakraborty, 2013). In sunflower accumulation of compatible solutes, i.e. Pro, glycinebetaine, and sucrose, contributed to osmoregulation during salt stress (Jabeen and Ahmad, 2013). In *C. paludosum* and sea aster, an increase in ion concentrations and Pro accumulation largely contributed to a decrease in leaf osmotic potential. This adaptation mechanism allowed plants to keep a positive cell turgor by continuing water uptake so that seedlings could grow under salt stress (Matsumura et al., 1998).

Drought and salt stress are inherently associated with impaired cellular functions, leading to accelerated accumulation of toxic metabolic byproducts such as reactive oxygen species (ROS) in different cellular compartments and cytosol (Khanna-Chopra, 2012). Besides its role as a compatible solute, Pro can act in the cell as a direct scavenger of a few of the most damaging ROS, i.e. singlet oxygen and hydroxyl radicals (Kaul et al., 2008; Smirnoff and Cumbes, 1989). Alia et al. (2001) reported that under high illumination the formation of singlet oxygen in isolated thylakoids of *Brassica juncea* was significantly reduced by Pro. It has been reported that free as well as polypeptide-bound Pro can produce free radical adducts of Pro by reacting with H_2O_2, singlet oxygen and hydroxyl radical (Floyd and Nagy, 1984; Kaul et al., 2008). A direct reaction of Pro with H_2O_2 does not contribute significantly to H_2O_2 scavenging and Pro has not been shown to directly react with a superoxide radical (Liang et al., 2013). Recent evidence that Pro cannot quench $^1O_2$ in the aqueous buffer may lead to reconsider the assumption about a likely scavenging role of Pro against $^1O_2$ in plants under stress. These authors proposed a Pro-Pro cycle that may act in scavenging OH^-. In this cycle, Pro captures two OH yielding P5C. P5C is then recycled back to Pro through the action of the P5CR/NADPH enzymatic system (Signorelli et al., 2014). However, the above observation needs to be confirmed under physiological conditions.

Exogenously applied Pro to *Arabidopsis* roots reduced ROS levels (Cuin and Shabala, 2007). Exogenous application of Pro resulted in enhanced activities of major ROS scavenging enzymes including SOD, CAT, POX and AsA/GSH cycle enzymes APX, DHAR and MDHAR in tobacco cell cultures under oxidative stress (Hoque et al., 2008). Proline application resulted in scavenging of ROS leading to lower membrane damage and better photosynthetic rate under drought stress in leaves of *Arabidopsis thaliana* (Moustakas et al., 2011). Foliar spray of Pro under long-term drought stress in pea resulted in a marked increase in non-enzymatic antioxidant contents and enhanced activities of SOD, APX, and CAT, leading to yield stability compared to non-sprayed drought-stressed plants (Osman, 2015). External application of Pro also resulted in enhanced GSH content and increased activity of glutathione-s-transferase and lower H_2O_2 accumulation in lentil seedlings under drought stress (Molla et al., 2014). In grapevine, Pro ameliorates inactivation of antioxidant defense enzymes, namely SOD, APX, and CAT, under drought stress (Ozden et al., 2009).

Interestingly, in a comparative study on mustard, plants that were treated with foliar spray of 1 mM Pro before being exposed to drought stress had lower endogenous H_2O_2 levels and less membrane damage coupled with higher activities of ROS scavenging enzymes and better redox state of glutathione compared to drought-stressed plants not pre-treated with Pro. This points toward an involvement of the drought acclamatory response in mustard (Hossain et al., 2014).

External application of Pro also resulted in lower ROS generation and/or higher activities of antioxidant defense enzymes such as SOD, CAT, APX, and POX in many plants including grapes, ice plant, mung bean, olive, and rice, etc., under salt stress (Ben Ahmed et al., 2010; Hossain and Fujita, 2010; Nounjan et al., 2012; Ozden et al., 2009; Shevyakova et al., 2009). Several studies have shown parallel changes in GSH pools with stress-induced biosynthesis of Pro in plants (Miller et al., 2009; Szabados and Savoure, 2010). GSH together with AsA forms major cellular antioxidant metabolite pools. Proline and GSH biosynthesis pathways are interconnected through the intermediate metabolite γ-glutamyl phosphate, hence there is cross talk between the pathways. This cross talk has been extensively studied in animals (reviewed in Liang et al., 2013). Some studies have shown evidence that an initial increase in Pro leads to increased levels of ROS. Exogenous Pro application leads to an increase in ROS in *Arabidopsis* and its *P5CDH* mutant plants (Miller et al., 2009). Hare et al. (1999) have shown that downstream signaling by ROS accompanied by enhanced Pro accumulation is more important than Pro accumulation and its antioxidant role.

Higher Pro accumulating transgenic plants overexpressing the *P5CS* gene exhibited lower oxidative stress and/or better antioxidant defense under drought, salt and osmotic stresses in wheat, chickpea, citrumelo and other plants (Tables 5.1 and 5.2; Amini et al., 2015; Bhatnagar-Mathur et al., 2009; de Campos et al., 2011; Vendruscolo et al., 2007). In contrast, *P5CS*-deficient Arabidopsis mutants accumulated lower Pro levels resulting in high levels of ROS-mediated damage and diminished antioxidant enzyme activities (Szekely et al., 2008; Szekely, 2004).

Proline metabolism is involved in regulation of cellular redox potential and storage and transfer of energy and reducing power (Giberti et al., 2014). Evidence is available to connect Pro metabolism with NADPH/$NADP^+$ redox balance, where Pro acts as redox shuttle (Giberti et al., 2014; Rejeb et al., 2014). Under drought stress, stomata are closed, which limits CO_2 uptake and hence decreases NADPH consumption. Proline synthesis in chloroplasts requires two NADPH to be oxidized to $NADP^{+,}$ hence it is a way of recycling $NADP^+$ under drought stress. By generating $NADP^+$ electron acceptor, Pro reduces electron transfer to oxygen and lowers generation of ROS (Hare et al., 1998; Szekely et al., 2008). Proline synthesis possibly increases NADP/NADPH ratio under stress. NADPH is regenerated and one CO_2 is produced in pentose-phosphate pathways when glucose-6-phosphate is converted to ribulose-5-phosphate. In stress conditions, regeneration of CO_2 allows carbon reduction to continue and NAD(P)H can be used in the biosynthesis of Pro and to prevent photoinhibition and excess ROS production in chloroplasts (Verslues and Sharma, 2010). Hamilton and Heckathorn (2001) found that Pro protects complex II in mitochondria under salt stress. Proline also works as a substrate for mitochondrial respiration under salinity stress (Hasegawa et al., 2000). Proline acts as a source of carbon, nitrogen, and energy during recovery from stress (Kishor et al., 2015).

Proline accumulation has been shown as an indicator of drought and salt stress tolerance in crop plants and thus it is a useful tool as a criterion for drought/salt resistance. Besides acting as a compatible solute, nitrogen source and essential amino acid, Pro acts as direct ROS scavenging metabolite as well as a signaling molecule by stimulating antioxidant synthesis and improving the activities of ROS scavenging enzymes. Hence, Pro plays a significant and essential role in antioxidant defense under drought stress in plants.

5.3.2 Enhancing Proline Accumulation in Plants Using a Transgenic Approach

5.3.2.1 Salinity Tolerance

Several attempts have been made to increase the accumulation of Pro in plants by transferring genes associated with Pro metabolism and checking salt tolerance in transgenic plants for many years (Hmida-Sayari et al., 2005; Karthikeyan et al., 2011; Kishor et al., 1995; Kumar et al., 2010; Surekha et al., 2014). Initial attempts were made by Kishor et al. (1995) for overexpression of moth bean *P5CS* in tobacco and the resultant transgenics synthesized 10–18-fold more Pro, high root biomass than control plants and were more tolerant to salinity. Earlier it was suggested that Pro accumulation in plants under stress might involve the loss of feedback regulation due to a conformational change in the P5CS protein, therefore Hong et al. (2000) removed the feedback inhibition using site-directed mutagenesis to replace the Phe residue at position 129 in *V. aconitifolia* P5CS with an Ala residue, and the resulting mutated enzyme (P5CS129A) was therefore no longer subjected to feedback inhibition. Consequently, the removal of this feedback inhibition resulted in two times more Pro accumulation in *P5CSF129A* transgenics as compared to wild-type, and better stress protection was observed (Hong et al., 2000). Further, the same mutagenized form of this gene was used and the resultant transgenic plants exhibited better growth performance under salt stress due to elevated Pro accumulation (Ahmed et al., 2015; Kumar et al., 2010; Reddy et al., 2015; Surekha et al., 2014). The antisense lines of *P5CS* showed morphological abnormalities in epidermal and parenchymatous cells, which indicates the role of Pro as a major constituent of cell wall proteins (Nanjo et al., 1999). Over-expression of *P5CS* in many other plants, such as *N. tabacum*, *Triticum aestivum*, *Dacus carota*, sugarcane and rice plants showed an increase in Pro levels and provided salt tolerance (Anoop and Gupta, 2003; Guerzoni et al., 2014; Sawahel and Hassan, 2002; Siripornadulsil et al., 2002). Transgenic cell lines of carrot with mothbean *P5CS* gene showed a sixfold increase in the degree of tolerance to salinity. Over-expression of *P5CS* of *S. torvum* (*StP5CS*) increased salt tolerance in transgenic soybeans (Zhang et al., 2015) and *P5CS* genes from Arabidopsis showed higher Pro accumulation, enhanced biomass production and flower development under salt stress in transgenic *S. tuberosum* (Hmida-Sayari et al., 2005).

In addition to P5CS, δ-OAT enzyme has also been reported to play a role in increasing Pro biosynthesis and take part in tolerance to salt stress. It has been found that overexpression of δ-OAT gene can increase Pro biosynthesis and subsequently increase biomass and germination rates under osmotic stress conditions (Roosens et al., 2002). Overexpressing Arabidopsis *δ-OAT* gene in *O. sativa* plants resulted in an increase in Pro content (5–15 fold) and higher tolerance to drought and salt stress. It also showed an improved yield in transgenic rice plants under salinity and drought conditions (Wu et al., 2003). Apart from the synthesis and accumulation of Pro, its degradation has also been found to participate in the resistance of plants to salt stress. To this end, suppression of Pro degradation by generating antisense transgenic Arabidopsis plants with an *AtProDH* cDNA was reported to improve freezing and high salinity tolerance in transgenic plants as compared to wild-type plants (Nanjo et al., 1999).

P5CS1 mutants studies showed a strong reduction in Pro accumulation in response to stress, concomitantly with a reduced root growth, enhanced production of ROS and a lower NADP1/NADPH ratio (Sharma et al., 2011; Szekely et al., 2008). Table 5.1 summarizes the examples of genetic modifications by transferring the genes associated with Pro biosynthesis (*P5CS1*, *P5CS2*, and *P5CR*) and its degradation (*ProDH*). Though Pro transgenics showed better tolerance to a variety of stresses, only the transgenics generated for enhanced salinity stress tolerance have been discussed.

5.3.2.2 Drought Stress Tolerance

Many reports showed that constitutive or stress-induced overexpression of *P5CS* gene in transgenic plants resulted in enhanced drought stress tolerance. Some of these studies are summarized in Table 5.2. In the majority of the transgenics developed for overexpression of *P5CS*, Pro accumulation was many folds higher than in wild-type plants. The majority of these experiments used *P5CS* or *P5CSF129A* cDNA from drought adaptive legume mothbean (*Vigna aconitifolia*). Overexpression of the *P5CS* gene resulted in higher survival rate, delayed wilting, higher shoot/root biomass, better osmotic adjustment, improved antioxidant defense, higher photosynthesis rate, and improved grain yield under drought stress as compared to non-transformed wild-type plants in many crops and other plants viz. rice, wheat, chickpea, tobacco, petunia and orange etc. (Table 5.2; Bhatnagar-Mathur et al., 2009; Borgo et al., 2015; Gubis et al., 2007; Kishor et al., 1995; Molinari et al., 2004; Pospisilova et al., 2011; Su and Wu, 2004; Vendruscolo et al., 2007; Wu et al., 2003; Yamada et al., 2005; Zhu et al., 1998). Diminishing *P5CS* expression by using antisense technology resulted in impaired drought response in plants. Severe stress effects have been reported in *P5CS* antisense soybean plants under drought stress compared to wild-type plants (Table 5.2; De Ronde et al., 2000, 2004). Su and Wu (2004) used stress-inducible promoter *AIPC* which responds to ABA for the overexpression of the *P5CS* gene. ABA accumulation is a key response in plants under drought stress and other stresses leading to osmotic stress. All other studies used constitute promoter *CaMV35S*. It is known that overexpression of genes under stress-inducible promoters results in minimizing undesirable side effects in transgenic plants under normal conditions.

Overall overexpression of the *P5CS* gene has resulted in improved drought tolerance in many crops and other plants, while *P5CS* antisense plants exhibited drought susceptibility. Hence it was proved, by using these technologies, that high Pro accumulation results in improved drought tolerance in plants.

5.3.2.3 Effect of Exogenous Proline in Alleviating Salt Stress Damage

The exposure of plants to salt stress impairs various physiological and biochemical functions such as

TABLE 5.1
List of Transgenic Plants Developed from Proline Biosynthesis Genes That Provide Salinity Stress Tolerance

Gene/ Source	Transgenic plant/growth stage and stress treatment	Tolerance mechanism	References
P5CS/ Vigna aconitifolia	*Nicotiana tabacum/6-week-old seedlings/250–400 mM*	Increased biomass production, enhanced flower and seed development under salinity stress	Kishor et al. (1995)
P5CS/ Vigna aconitifolia	*Triticum aestivum/2-week-old plants/5–200 mM*	High proline level	Sawahel and Hassan (2002)
P5CS/ Vigna aconitifolia	*Oryza sativa/6-week-old seedlings/200 mM NaCl*	Better root growth, biomass and high proline content	Anoop and Gupta (2003)
P5CS/ Arabidopsis	*Solanum tuberosum/2–3-week-old plants/100 mM*	High proline level, less altered tuber yield and weight under salinity	Hmida-Sayari et al. (2005)
P5CS/ Vigna aconitifolia	*Oryza sativa/15-day-old plants/200 mM NaCl*	Increase proline level in transgenic plants	Karthikeyan et al. (2011)
P5CS/ Arabidopsis	*Nicotiana tabacum/50–250 mM*	Transgenics showed better salt stress tolerance by enhanced proline accumulation	Rastgar et al., (2011)
P5CS/ Vigna aconitifolia	*Cicer arietinum/2-week-old plants/250 mM NaCl*	Maintains chlorophyll stability, less electrolyte leakage and no effect on yield under salinity stress	Ghanti et al., (2011)
P5CS/ Vigna aconitifolia	*Sugarcane/4-month-old plants/100–200 mM*	Enhanced proline content, lesser lipid peroxidation and increase salt tolerance	Guerzoni et al. (2014)
P5CS1/ Phaseolus vulgaris	*Arabidopsis/7-day-old plants/100–300 mM*	Increase proline content and more biomass production	Chen et al. (2010, 2013)
P5CSF129A/ Vigna aconitifolia	*Oryza sativa/21-day-old plants /150 mM*	Increase proline, lower lipid peroxidation, better growth under salt stress	Kumar et al. (2010)
P5CSF129A/ Vigna aconitifolia	*Pigeon pea/2–3-week-old plants/200 mM*	Enhanced proline accumulation, enhanced growth performance, more chlorophyll and relative water content under high salinity and lower levels of lipid peroxidation	Surekha et al. (2014)
P5CSF129A/ Vigna aconitifolia	*Sorghum bicolor/40-day-old plants/100 mM*	Transgenics showed better salt stress tolerance by protecting photosynthetic and antioxidant enzymes (SOD,CAT and GR)	Reddy et al., (2015)
OAT/Arabidopsis	*Nicotiana Plumbaginifolia/4-week-old plants/200mM*	Increased proline level	Roosens et al., (2002)
OAT/Arabidopsis	*Oryza sativa/3 or 8 weeks old plants/0.1–0.4 mol/L*	Increased proline content and improved yield under salt and drought stress	Wu et al. (2003)
P5CR/Triticum aestivum	*Arabidopsis/10-day-old plants/150mM*	Enhanced root growth under salt stress and decreased lipid peroxidation under salt, drought, and ABA stress	Ma et al. (2008)
P5CR/Ipomoea batatas	*Ipomoea batatas/4-week-old plants/100 mM*	Proline content and superoxide dismutase and photosynthetic activities were significantly increased, whereas malonaldehyde content was significantly decreased in the transgenic plants	Liu et al., (2014)
ProDH/Arabidopsis	*Arabidopsis/4-week-old plants 600 mM*	Antisense suppression accumulate more proline and provide tolerance to salinity	Nanjo et al., (1999)
ProDH/Arabidopsis	*Nicotiana tabacum/6–10-week-old plants*	Antisense suppression accumulates more proline	Kochetov et al., (2004)
At-PDH/Arabidopsis	*Arabidopsis/2-week-old plants*	Antisense suppression accumulates more proline	Mani et al., (2002)

TABLE 5.2
List of Transgenics Plants Developed from Proline Biosynthetic Pathway Genes and Their Response to Drought Stress

Gene/Source	Transgenic plant/ growth stage and stress treatment	Effect on drought tolerance	References
P5CS/ mothbean	Tobacco/ 6-week-old seedling subjected to drought stress for 10 days	Delayed wilting of transgenic plants under drought stress (DS) compared to non-transformed plants (WT). Higher root biomass and longer roots in transgenic plants compared to WT under DS	Kishor et al. (1995)
P5CS/ mothbean	Rice/ 8-week-old seedlings were subjected to 4 cycles of 6 days DS and 1 day recovery and observation taken at the end of 4th stress cycle	Delayed leaf wilting, increased growth and higher shoot biomass in transgenic plants compares to WT plants under DS	Zhu et al. (1998)
Delta-OAT/ *Arabidopsis*	Rice/ mature plants subjected to drought stress	Transgenic plants maintained higher yield than WT plants under stress treatment	Wu et al. (2003)
P5CS/ mothbean	Rice/ seedling subjected to DS	Significantly higher shoot and root weight in transgenic plants compared to WT plants under DS	Su and Wu (2004)
P5CSF129A/ mothbean	Orange/ mature plants were subjected to DS for 15 days	Transgenic plants showed superior osmotic adjustment and higher photosynthesis under DS compared to WT plants	Molinari et al. (2004)
P5CS/ *Arabidopsis*	Petunia/ 7-day-old pots-grown seedling were subjected DS for 14 days	Transgenic plants exhibited higher survival rate and better vegetative growth compared to WT plants under DS	Yamada et al. (2005)
P5CS/ mothbean	Wheat/ 65-day-old pots-grown seedlings at booting stage were subjected to DS for 4 days	Transgenic plants maintained better antioxidant defense, less membrane damage and high photosynthesis compared to WT under DS	Vendruscolo et al. (2007)
P5CSF129A/ mothbean	Tobacco/ seedlings at 5-leaf stage were subjected to DS for 12 or 18 days	Transgenic plants exhibited higher biomass and high pigment levels under DS compared to WT plants	Gubis et al. (2007)
P5CSF129A/ mothbean	Chickpea/ seedling stage subjected to DS	Transgenic plants had higher tolerance to DS due to lower ROS levels compared to WT plants	Bhatnagar-Mathur et al. (2009)
P5CSF129A/ mothbean	Tobacco/ seedling were subjected to DS for 7 days	No significant difference in photosynthesis rate, or contents of photosynthetic pigments was observed between transgenic plants and WT plants under DS	Pospisilova et al. (2011)
P5CS/ mothbean	Tobacco/ 6-week-old plants were subjected to DS for 12 days	Transgenic plants maintained higher photosynthesis up to 6 days of DS only compared to WT plants. High level of proline accumulated under DS in transgenic does not lead to any OA	Borgo et al. (2015)
P5CS/ soybean	Soyabean/seedlings subjected to drought stress at 37°C for 6 days	Antisense plants could not survive 6 days drought stress at 37°C; however, WT plants survived. Antisense plants produced fewer seeds compares to WT plants subjected to DS	De Ronde et al. (2000)
P5CS/ soyabean	Soyabean/seedlings subjected to simultaneous drought stress and heat stress	Antisense plants showed severe stress effects while WT plants exhibited only mild stress symptoms under stress combination. $NADP^+$ levels declined in antisense plants and increased in WT plants under WS.	De Ronde et al. (2004)

growth rate, photosynthetic efficiency, and cell turgor, and damages cellular components through an excessive production of ROS (Khan et al., 2012). It also disrupts and alters molecular homeostasis in plants. Proline significantly alleviated the effect of salt stress by reducing the effects of ROS, working as a molecular chaperone, protecting the integrity of cell membranes, stabilizing proteins, DNA and enzymes (Matysik et al., 2002; Szabados and Savoure, 2010). A number of studies showed that exogenous Pro reduces the effect of salt stress and thus improves tolerance, as shown in Table 5.3. Exogenous application of Pro significantly increased plant salt tolerance in both salt-stressed halophyte *Allenrolfea occidentalis* plants (Chrominski et al., 1989) and glycophyte rice cultivars (Teh et al., 2016). Hasanuzzaman et al. (2014) showed that Pro improved salinity tolerance in both salt-sensitive and salt-tolerant rice cultivars. Ben Ahmed et al. (2010) reported that photosynthetic activity, leaf water relations and antioxidant systems were enhanced in the olive tree by Pro treatment during salt stress. Moreover, exogenous Pro application to *Vicia faba* significantly increased leaf water potential and alleviated salinity-induced injury to membranes (Gadallah, 1999). In *Cordia myxa*, seedlings showed an increase in chlorophyll, Pro stem height, diameter, secondary branching and leaf area during salinity stress (Mayahi and Fayadh, 2015). Addition of Pro to the nutrient medium drastically decreased the oxidative damage to membranes caused by salinity in *Mesembryanthemum crystallinum* L., *Arachis hypogaea* and *Cucumis melo* (Jain et al., 2001; Shevyakova et al., 2009; Yan et al, 2013) and increased dry weight and free Pro content in the callus cells of *Medicago sativa* (Ehsanpour and Fatahian, 2003). It is suggested that exogenous Pro application upregulates genes/proteins responsible for stress tolerance and mitigates the damages caused by salt in *Saccharum officinarum*, and in desert plants *Pancratium maritimum* and *Oryza sativa*, under salt stress conditions (Khedr et al., 2003; Patade et al., 2014; Teh et al., 2016). The aforesaid studies clearly demonstrated that, under stress conditions, Pro modulates multiple stress response pathways and thus makes plants more tolerant to stress.

However, some studies also show that exogenously applied Pro benefits plants under salt stress conditions only at low concentrations, whereas it becomes toxic at high concentrations. Roy et al. (1993, 2014) validated that Pro at a low concentration (20, 30 mM) mitigates the adverse effects of salinity through increasing germination and seedling growth, while higher concentrations (40, 50 mM) of Pro resulted in toxic effects and reduced growth of rice seedlings and also lowered the K^+/Na^+ ratio. Similarly, a lower concentration of Pro (50 mM), in contrast to 100 mM, effectively alleviated salinity-induced deleterious impacts on the growth of groundnut (Jain et al., 2001). Further, Deivanai et al. (2011) demonstrated that soaking rice seeds in low concentrations of Pro (1 mM) overcame the adverse effects of NaCl stress, while a higher concentration (10 mM) was injurious. When Pro was added to the culture medium at low concentrations, it effectively alleviated salinity-induced decline in fresh weight and reduced peroxidative damage to lipid membranes in groundnut (*Arachis hypogea*); in contrast, higher Pro concentrations did not prove beneficial (Jain et al., 2001). Seed priming with exogenous Pro under normal growth conditions improved plant growth, photosynthesis, and antioxidative systems in two different cultivars of mustard (Wani et al., 2012) under varying concentrations of Pro (10–50 mM). It was observed that 20 mM Pro for 8 h was most effective, and also the response of both cultivars to Pro was greatly varied. The findings confirmed that the effect of Pro is concentration-dependent, and genotypes of the same species vary in their response to exogenous Pro (Garg et al., 2002; Wani et al., 2012). In agreement with this, higher concentrations of Pro had harmful effects and were genotype-dependent in *Medicago sativa* callus (Ehsanpour and Fatahian, 2003).

Despite the beneficial effects of endogenous and exogenous Pro with respect to oxidative stress tolerance, over-accumulation or excessive application of Pro was found to have negative consequences for plants. This had no direct correlation with the salt tolerance of the species (Chen et al., 2007; Lv et al., 2011; Roy et al., 1993; Widodo et al., 2009). A recent study on 46 switch grass lines showed that, under salinity, Pro concentration increased 5000-fold in salt sensitive lines while salt-tolerant lines exhibited a lower increase in Pro in response to salt treatment (Kim et al., 2016). Salt tolerant *Paulownia imperialis* accumulated the greatest Pro content at low salt stress (40 mM NaCl), whereas at higher NaCl concentrations (60, 80, 160 mM), a significant decline in Pro levels was observed (Ayala-Astorga and Alcaraz-Melendez, 2010). Widodo et al. (2009) observed that salt sensitive barley roots showed an increase in Pro, whereas salt-tolerant types did not. In rice, the Pro content was found to be higher in salt sensitive cultivars in comparison to salt-tolerant ones (Momayezi et al., 2009). Another study in soybean showed that salt sensitive cultivars accumulated more Pro than tolerant ones under high salinity stress (Phang et al., 2008). Furthermore, greater accumulation of Pro in cashew leaves was related to higher salt sensitivity, as Pro accumulation was found to result from salt-induced proteolysis (Silveira et al.,

TABLE 5.3
Studies Reporting the Role of Exogenous Application of Proline in Plants Under Salinity Stress

Plant species	Metabolism affected	Phenotype/tolerance mechanism	References
Pancratium maritimum	Osmoregulation	Improved tolerance by increasing dehydrin proteins (stress protective proteins)	Khedr et al. (2003)
Nicotiana tabacum	Antioxidant metabolism	Provided protection by increasing APX, CAT, POX, DHAR and GR activity	Hoque et al. (2007)
Cucumis sativus	Antioxidant metabolism	Alleviated plant growth inhibition through higher activity of POD, low MDA level and maintaining RWC	Huang et al. (2009)
Nicotiana tabacum	Antioxidant metabolism	Decrease in ROS accumulation and lipid peroxidation as well as improvement in membrane integrity, CAT and POX activity	Banu et al. (2009)
Vigna radiata	Antioxidant metabolism	Reduced oxidative stress by increasing GSH content, GST, GPX, GR, GlyI and Gly II activity	Hossain and Fujita (2010)
Olea europaea	Ion homeostasis	Improved plant growth by maintaining low Na^+ and high K^+ and Ca^{2+} content	Ben Ahmed et al. (2011)
Oryza sativa	Antioxidant metabolism	Alleviated plant growth inhibition by increasing activity of SOD, CAT, APX, POX	Nounjan and Theerakulpisut (2012)
Oryza sativa	Ion homeostasis	Increase K^+/Na^+ ratio	Sobahan et al. (2012)
Cucumis melo	Antioxidant metabolism	Increased plant growth, chlorophyll content, net photosynthetic rate and actual efficiency of photosystem II, enhanced the activity of SOD, POD, CAT, APX, DHAR and GR	Yan et al. (2013)
Solanum melongena	Water use efficiency	Counteracted the adverse effects on plant growth and fresh weight	Shahbaz et al. (2013)
Oryza sativa	Antioxidant metabolism	Increased tolerance to oxidative damage by upregulating the antioxidant defense system (AsA, GSH and GSH/GSSG, APX, MDHAR, DHAR, GR, GPX, CAT, and Gly I activities)	Hasanuzzaman et al. (2014)
Oryza sativa	Carbohydrate metabolism	Alleviated inhibitory effects on seed germination (improved alpha amylase activity, beta-amylase isoenzyme)	Hua-long et al. (2014)
Solanum lycopersicum	Photosynthesis	Increased plant growth, chlorophyll a fluorescence	Kahlaoui et al. (2014)
Saccharum officinarum	Antioxidant metabolism	Increased plant growth rate (SOD, GR, APX, GPX, CAT)	Patade et al. (2014)
Oryza sativa	Plant height, root length	Increased plant growth	Teh et al. (2016)
Helianthus annuus	Photosynthetic pigments, plant height, number of leaves, and fresh and dry weight of shoot	Improved plant growth by chlorophyll a, chlorophyll b and carotenoids	Sadak and Mostafa (2015)
Zea mays	Ion homeostasis	Increased K^+/Na^+ ratio and P uptake	Alam et al. (2016)
Onobrychisviciaefolia Scop	Ion homeostasis	Maintained Na^+/K^+ and increase proline accumulation	Wu et al. (2017)

2003). All together, the above-mentioned studies imply a negative correlation between Pro contents and salt tolerance in various plants, suggesting Pro accumulation as a symptom of damage. It also appears that the mitigating effect of Pro is species or even cultivar specific, which cautions against over-simplification of Pro role in salt tolerance. Therefore, it is necessary to determine the optimal concentration of Pro that has beneficial effects in different plant species before recommending it for agricultural practices.

5.3.2.4 Effect of Exogenous Proline in Alleviating Drought Stress Damage

Plant tolerance against environmental stresses can be increased by the exogenous application of certain growth enhancers such as Pro, amino acids, ABA, glycine betaine (GB), BAP, silicon, soluble sugars, humic acid, and potassium (Farooq et al., 2009). Appropriate concentrations of these synthetic enhancers could promote growth in plants and ameliorate water deficit stress by interfering with metabolic and photosynthesis processes through osmotic adjustment, ROS scavenging, and increasing enzymatic and non-enzymatic antioxidants and proteins (Bohnert and Jensen, 1996). Exogenous foliar spray of growth enhancers at the critical growth stages of wheat (tillering, booting, heading, milking) is one of the most significant treatments that increase antioxidant status against reactive oxygen species under water-deficit conditions (Yasmeen et al., 2013). Proline treatment elevated osmotic adjustment to improve turgor pressure and promoted accumulation of antioxidants to detoxify ROS, thus maintaining the integrity of membrane structures, enzymes, and other macromolecules during drought conditions (Anjum et al., 2011; Nawaz et al., 2016). The exogenous application of Pro alone and/or in combination with GB as a foliar spray has also been found to be effective in reducing the adverse effects of drought stress on cotton (Noreen et al., 2013; Ullah et al., 2017). Ali et al. (2007) reported that exogenous Pro applied as spray treatment at the seedling and/or at vegetative stage of *Zea mays* resulted in enhanced growth in a water deficient environment. Proline applied as presowing seed-soaking treatment alleviated the adverse effects generated by drought stress in *Triticum aestivum*, resulting in enhanced growth and yield characteristics (Kamran et al., 2009). Exogenous application of Pro enhanced growth and also maintained nutrient status by promoting the uptake of K^+, Ca^+, P, and N in *Zea mays* plants exposed to drought stress (Ali et al., 2008).

5.3.3 Regulation of Proline Metabolism

The core of Pro metabolism involves two enzymes catalyzing Pro synthesis from glutamate in the cytoplasm or chloroplast, two enzymes catalyzing Pro catabolism back to glutamate in the mitochondria, as well as an alternative pathway of Pro synthesis via ornithine (Figure 5.1). The interconversion of Pro and glutamate is sometimes referred to as the "proline cycle." Although Pro metabolism has been studied for more than 40 years in plants, little is known about the signaling pathways involved in its regulation. The transcriptional upregulation of Pro synthesis from glutamate and downregulation of Pro catabolism during stress are thought to control Pro levels, although exceptions to this pattern have been observed. High Pro levels in *V. vinifera* fruit were not correlated with P5CS levels, which remained largely invariant during fruit development (Stines et al., 1999).

Proline biosynthesis is controlled by the activity of two *P5CS* genes in plants, encoding one housekeeping and one stress-specific P5CS isoform. Although the duplicated *P5CS* genes share a high level of sequence homology in coding regions, their transcriptional regulation is different (Xue et al., 2009). Arabidopsis *P5CS1* is induced by osmotic and salt stresses and is activated by an abscisic acid (ABA)-dependent and ABA-insensitive 1 (ABI1)-controlled regulatory pathway and H_2O_2 derived signals (Savouré et al., 1997; Verslues et al., 2007). Induction of *P5CS* transcription by ABA was described in *A. thaliana*, rice and *Opuntia streptacantha* (Fichman et al., 2015). Rice treated with NaCl or ABA showed a higher level of *OsP5CS1* and *OsP5CR* transcript expression than that of *OsP5CS2* (Sripinyowanich et al., 2013). It is suggested that the salt-stress-induced *OsP5CS1* transcript expression in rice is controlled, at least in part, by an ABA-dependent pathway (Bray et al., 2000; Zhu, 2002). Salt-induced activation of *AtP5CS1* is controlled through ABI1-dependent signaling (Strizhov et al., 1997). The ABI1 2C protein phosphatase (PP2C), which is a component of the ABA receptor complex, regulates gene expression through the control of phosphorylation of ABA-responsive element (ABRE)-binding factor-(ABF)-type transcription factors by the sucrose non-fermenting-1-related kinases (SnRK2) (Fujii et al., 2009). Recent studies support the possibility that (i) the signaling pathway mediated by TFs of the bZIP family and *bHLH* family is involved in the regulation of Pro biosynthesis in ABA-dependent pathways and (ii) cis-regulatory element DREB2 and NAC transcription factors are involved in the regulation of Pro biosynthesis by ABA-independent signaling pathway (Zarattini and Forlani, 2017).

AtP5CS1 is induced by cold and osmotic stress in ABA-dependent and independent mechanisms and the induction of this gene by low water potential is less sensitive to ABA than that of other control stress markers (Sharma and Verslues, 2010), suggesting that the transcription of *P5CS1* and the accumulation of Pro are controlled by several signaling pathways. Moreover, P5CS1 activation and Pro accumulation are promoted by light and repressed by brassinosteroids (Abrahám et al., 2003). Calcium signaling and Phospholipase C (PLC) trigger *P5CS* transcription and Pro accumulation during salt stress (Parre et al., 2010; Sripinyowanich et al., 2013). Ca^{2+}, which is involved in lipid signaling, was shown to control downstream events that activate *AtP5CS1* expression. Annexins are evolutionarily conserved Ca^{2+}- and phospholipid-binding proteins that influence stress responses in a light-dependent manner. While an *A. thaliana* double *ANNAt1 ANNAt4* annexin mutant had enhanced drought and salt tolerance with elevated *AtP5CS1* expression, overexpression of *ANNAt4* led to hypersensitivity to these stresses with reduced *AtP5CS1* expression and Pro accumulation (Huh et al., 2010). Annexin 1 was recently shown to function as a plasma membrane Ca^{2+} transporter, controlling cytosolic free Ca^{2+} levels in response to ROS, which are known signals of stress responses (Laohavisit et al., 2010). In the halophyte *Thellungiella halophila*, Phospholipase D functions as a positive regulator, whereas Phospholipase C exerts a negative control on Pro accumulation (Ghars et al., 2008). Light-dependent fluctuations of *AtP5CS1* transcript levels were correlated with daily changes in Pro levels (Hayashi et al., 2000), and illumination was important for salt-induced *AtP5CS1* transcription and Pro accumulation (Abrahám et al., 2003). In *A. thaliana*, a Ca^{2+}-dependent calmodulin interaction with the myeloblastosis (MYB2) transcription factor could induce *AtP5CS1* transcription and Pro accumulation (Yoo et al., 2005). Exposure to H_2O_2 activated P5CS and δ-OAT and downregulated ProDH activity in *Z. mays* seedlings, leading to Pro accumulation (Yang et al., 2009). *P5CS* transcription and Pro accumulation are therefore influenced by several environmental factors such as salt, drought and light, and mediated by a complex network of ROS, lipid and Ca^{2+} signals.

The conversion of glutamate to pyrroline-5 carboxylate P5C is a rate-limiting step in Pro biosynthesis. P5CS1 is subjected to feedback inhibition by Pro (Hu et al., 1992; Zhang et al., 1995). In bacteria, change of a conserved aspartate (at position 107) to asparagine rendered γ-glutamyl kinase much less sensitive to Pro inhibition (Csonka et al., 1988). However, in plants, this conserved aspartate (at position 128) was not involved in feedback inhibition. Instead, a *F129A* (phenylalanine129alanine) mutant *V. aconitifolia* P5CS enzyme had decreased feedback inhibition (Zhang et al., 1995). Tobacco plants ectopically expressing this feedback-insensitive *P5CS* mutant gene had higher levels of Pro than either untransformed plants or plants expressing the unmutated *P5CS* (Hong et al., 2000), suggesting that reduced feedback inhibition can lead to higher Pro *in vivo*.

In *Arabidopsis*, *P5CS2* is the housekeeping gene, which is active in dividing, meristematic tissues, such as the shoot and root tips, inflorescences and cell cultures (Gagneul et al., 2007; Strizhov et al., 1997; Szekely et al., 2008). Both *P5CS* genes are active in floral shoot apical meristems, and supply Pro for flower development (Mattioli et al., 2009b). In Arabidopsis, the *P5CS2* gene was identified as one of the targets of CONSTANS (CO), a transcriptional activator that promotes flowering in response to day length, thereby suggesting its involvement in flower development (Samach et al., 2000).

Whereas Pro biosynthesis is upregulated by light and osmotic stresses, Pro catabolism is activated in the dark and during stress relief and is controlled by PDH and P5CDH (Figure 5.1). PDH transcription is activated by rehydration and Pro but repressed by dehydration, thus preventing Pro degradation during abiotic stress (Kiyosue et al., 1996; Verbruggen et al., 1996). *PDH1* transcription is repressed during daylight and induced in darkness; therefore illumination has opposite effects on *P5CS1* and *PDH1* transcription (Gagneul et al., 2007; Hayashi et al., 2000). Promoter analysis of *PDH1* identified the Pro and hypo-osmolarity-responsive element (PRE) motif ACTCAT, which is necessary for the activation of the *PDH* gene (Satoh et al., 2002). Basic leucine zipper protein (bZIP) transcription factors (AtbZIP-2, -11, -44, -53) have been identified as candidates for binding to this motif and controlling *PDH* expression (Weltmeier et al., 2006). The *P5CDH* gene is expressed at a low basal level in all Arabidopsis tissues and can be upregulated by Pro (Deuschle et al., 2001). A short sequence similar to the PRE motif has been identified on the promoters of *P5CDH* genes in Arabidopsis and cereals (Ayliffe et al., 2005). Natural antisense overlapping of the 30 UTR regions of *P5CDH* and the salt-induced similar to RCD ONE 5 (*SRO5*) gene generate endogenous small interfering RNA (24-nt and 21-nt siRNA) in Arabidopsis, which cleaves the *P5CDH* RNA and reduces transcript levels during stress. Transcriptional gene silencing is therefore important in the control of *P5CDH* gene activity (Borsani et al., 2005).

Several lines of evidence suggest that Pro has certain regulatory roles under dehydration stress and acts as a signaling molecule. Oono et al. (2003) revealed that about one-third of the rehydration responsive genes could be induced by Pro. Most of these rehydration and Pro responsive genes have putative PRE *cis*-acting elements in their promoter region, which is a target of bZIP-type transcription factors (Oono et al., 2003). Many transgenic lines overexpressing bZIP proteins are responsive to Pro and more resistant to drought stress (Zhang et al., 2016). It is reported that transgenic *Arabidopsis* carrying soybean *GmbZIP110* or wheat *TabZIP60* accumulated significantly higher amounts of Pro (Xu et al., 2016; Zhang et al., 2015). Some studies suggest that WRKY members have regulatory roles in Pro metabolism. Wheat *TaWRKY10* overexpression resulted in enhanced drought tolerance in tobacco due to enhanced levels of Pro (Wang et al., 2013). Rice plants overexpressing *AtWRKY57* showed higher transcript levels of *P5CS1* under dehydration stress (Jiang et al., 2016). Some of the MYB TF recognition sequences have been identified in the promoter region of Pro biosynthesis genes in both rice and *Arabidopsis* (Fichman et al., 2015; Zarattini and Forlani, 2017). High expression of MYB family genes results in high Pro accumulation (Shukla et al., 2015). Overexpression of *MYB2* resulted in enhanced accumulation of Pro in *Arabidopsis*, wheat and rice (Mao et al., 2011; Yang et al., 2012; Yoo et al., 2005).

5.3.4 Proline Homeostasis Is the Key to Survival Under Stress

Proline is a multifunctional amino acid playing an important role in mitigating the deleterious effects of abiotic stresses on plant growth, metabolism, and productivity. The level of Pro accumulation varies between plant species and may even go to 100 times more than the non-stressed conditions reaching cytosol concentrations of 120–230 mM (Ashraf and Foolad, 2007). Proline accumulation under various abiotic stresses and studies from transgenic plants or mutants have demonstrated that Pro metabolism is positively correlated with stress tolerance, including salinity (Choudhary et al., 2005; Yazici et al., 2007; Szekely et al., 2008; Man et al., 2011; Saeedipour, 2013; Sripinyowanich et al., 2013; Theocharis et al., 2012; Witt et al., 2012; Xu et al., 2012). Correlation between Pro accumulation and its proposed roles in salt adaptation, however, have not been clearly confirmed in several plant species. In addition, the studies relating to Pro functions and plant salt and drought tolerance are always carried out in growth chambers and are not successfully verified in field conditions (Mansour and Ali, 2017). Further, stress tolerance is a complex trait, and studies based solely on Pro accumulation do not adequately explain its functions in stress tolerance.

Proline concentration also increases significantly under normal physiological conditions in meristematic cells and tissues undergoing senescence besides playing an important role in reproductive tissues during seed maturation (Mattioli et al., 2009a,b). Whether this is part of the normal developmental process or is induced by external stress needs to be examined. Also, Pro accumulation may be toxic to certain tissues if it gets partially catabolized because it then induces an increase in P5C concentration, which in turn leads to apoptosis (Fabro et al., 2004). Hence the debate whether Pro accumulation is on its own correlated with stress tolerance or Pro homeostasis is a more important issue under stress. Many authors have credited enhanced Pro content for the upregulation of P5CS coupled with a downregulation of ProDH under stress conditions (Sharma and Verslues, 2010; Verslues et al., 2007), and again upregulation of the latter after the stress is relieved (Satoh et al., 2002). This revealed that the enhanced Pro content during abiotic stresses is not only due to the Pro synthesis but also to the perturbation of Pro homeostasis caused by the deactivation of Pro catabolism that ultimately leads to elevated levels of Pro under stress (Kishor and Sreenivasulu, 2014). Studies have also suggested that Pro homeostasis is related to the energy demand of young dividing cells and the resumed growth post-stress relief. Proline synthesis in chloroplast under stress helps in regenerating NADP for electron transport; hence ROS generation is avoided (Szabados and Savoure, 2010). Beside active Pro synthesis in stressed shoots and roots, Pro is also transported from shoots via phloem to roots, leading to Pro accumulation, which is subsequently catabolized in mitochondria generating NADPH required for root growth under moisture stress (Lee et al., 2009). Hence Pro homeostasis achieved by both intercellular compartment coordination and long-distance transport seems to be essential for normal plant growth and development under abiotic stresses.

5.4 CONCLUSION AND FUTURE PERSPECTIVES

Proline is an important compatible solute that exhibits numerous roles during plant growth and development and under abiotic stresses including drought and salinity. Proline protects plants against these stresses mainly by maintaining osmotic adjustment, ROS scavenging, and modulating major enzymatic components of the antioxidant defense system. Proline also stabilizes proteins and protein complexes in the chloroplast and cytosol and protects the photosynthetic apparatus and the enzymes involved in detoxification of ROS during stress. The enhanced rate of Pro biosynthesis in chloroplasts can contribute to the stabilization of redox balance and maintenance of cellular homeostasis by dissipating the excess of reducing potential when electron transport is saturated during adverse conditions. Proline catabolism in the mitochondria is connected to oxidative respiration and provides energy for resumed growth after stress. Moreover, Pro oxidation can regulate mitochondrial ROS levels and influence programmed cell death. Proline appears to function as a metabolic signal that regulates metabolite pools and redox balance, controls the expression of numerous genes and influences plant growth and development.

Pragmatic evidence for the protective functions of Pro is provided by studies using mutants and transgenic plants. Enhanced Pro accumulation achieved by upregulation of rate-limiting Pro biosynthesis genes in transgenics resulted in enhanced tolerance of plants to drought and salinity stress (Per et al., 2017). Inhibition of Pro degradation can also lead to Pro accumulation. However, using this approach, contrasting results for stress tolerance were obtained in different studies. Whereas antisense *PDH* was reported to improve salt tolerance in one study (Nanjo et al., 1999), other studies reported unchanged stress tolerance, accompanied by abnormal seed and plant development and increased Pro hypersensitivity of transgenic lines and mutants (Mani et al., 2002). These results suggest that it might not be the actual Pro content, but the enhanced rate of Pro biosynthesis that is an important factor for stress adaptation. Engineering strategies could, therefore, target the Pro biosynthetic pathway, and aim to accelerate Pro biosynthesis in chloroplasts.

Improvement in drought or salt tolerance of crop plants via engineering Pro metabolism is an existing possibility and should be explored more extensively. However, the way Pro accumulation influences particular regulatory pathways in response to stress is still not clear and requires further research. Therefore, identifying the genes reported to regulate Pro biosynthesis and their modulation can benefit sustainable agriculture by improving drought and salt stress tolerance in plants. Furthermore, the identification of signal transduction pathways associated with Pro synthesis, degradation and the coordination of gene expression events, promoter elements and other transcription factors under stress and stress recovery is of paramount importance. More detailed analyses might reveal novel and interesting links between metabolism and transport, which may contribute to a better understanding of the role and regulation of Pro homeostasis. Modern and more powerful metabolic profiling tools might be helpful in understanding the regulation of Pro signaling in plants. In the context of the present climate changes, further study on Pro metabolism under abiotic stress conditions will certainly supplement the knowledge regarding abiotic stress tolerance in crop plants.

ACKNOWLEDGMENTS

RK Chopra acknowledges the financial support received from the Indian National Science Academy. VKS acknowledges the support of CSIR for the RA fellowship. A.P. would like to acknowledge the financial support received from UPOE-II, International Atomic Energy Agency (Vienna), India-NWO, DBT and Indo-US Science and Technology Forum (IUSSTF), New Delhi. N. Lakra is thankful to the University Grant Commission for providing the Dr. D. S. Kothari Fellowship and Jawaharlal Nehru University (JNU), New Delhi for hosting the same.

REFERENCES

Abrahám, E., Rigó, G., Székely, G., Nagy, R., Koncz, C., Szabados, L. (2003) Light-dependent induction of proline biosynthesis by abscisic acid and salt stress is inhibited by brassinosteroid in Arabidopsis. *Plant Molecular Biology* 51(3):363–372.

Ahmed, A.A.M., Roosens, N., Dewaele, E., Jacobs, M., Angenon, G. (2015) Overexpression of a novel feedback-desensitized Δ1-pyrroline-5-carboxylate synthetase increases proline accumulation and confers salt tolerance in transgenic Nicotiana plumbaginifolia. *Plant Cell, Tissue and Organ Culture* 122(2):383–393.

Alam, R., Das, D.K., Islam, M.R., Murata, Y., Hoque, M.A. (2016) Exogenous proline enhances nutrient uptake and confers tolerance to salt stress in maize (*Zea mays* L.). *Progressive Agriculture* 27(4):409–417.

Ali, Q., Ashraf, M., Athar, H.U.R. (2007) Exogenously applied proline at different growth stages enhances growth of two maize cultivars grown under water deficit conditions. *Pakistan Journal of Botany* 39:1133–1144.

Ali, Q., Ashraf, M., Shahbaz, M., Humera, H. (2008) Ameliorating effect of foliar applied proline on nutrient uptake in water stressed maize (*Zea mays* L.) plants. *Pakistan Journal of Botany* 40(1):211–219.

Alia, Mohanty, P., Matysik, J. (2001) Effect of proline on the production of singlet oxygen. *Amino Acids* 21(2):195–200.

Alia, Saradhi, P.P., Mohanty, P. (1997) Involvement of proline in protecting thylakoid membranes against free radical-induced photodamage. *Journal of Photochemistry and Photobiology-Part B* 38:253–257.

Amini, S., Ghobadi, C., Yamchi, A. (2015) Proline accumulation and osmotic stress: An overview of P5CS gene in plants. *Journal of Plant Molecular Breeding* 3(2), 44–55.

Anjum, S.A., Wang, L.C., Farooq, M., Hussain, M., Xue, L.L., Zou, C.M. (2011) Brassinolide application improves the drought tolerance in maize through modulation of enzymatic antioxidants and leaf gas exchange. *Journal of Agronomy and Crop Science* 197(3):177–185.

Anoop, N., Gupta, A.K. (2003) Transgenic indica rice cv IR-50 overexpressing Vigna aconitifolia deltapyrroline-5-carboxylate synthetase cDNA shows tolerance to high salt. *Journal of Plant Biochemistry and Biotechnology* 12(2):109–116.

Ashraf, M., Foolad, M.A. (2007) Improving plant abiotic-stress resistance by exogenous application of osmoprotectants glycine betaine and proline. *Environmental and Experimental Botany* 59:206–216.

Ayala-Astorga, G.I., Alcaraz-Melendez, L. (2010) Salinity effects on protein content, lipid peroxidation, pigments, and proline in Paulownia imperialis (Siebold & Zuccarini) and Paulownia fortunei (Seemann & Hemsley) grown in vitro. *Electronic Journal of Biotechnology* 13(5):1–15.

Ayliffe, M.A., Mitchell, H.J., Deuschle, K., Pryor, A.J. (2005) Comparative analysis in cereals of a key proline catabolism gene. *Molecular Genetics and Genomics: MGG* 74(5):494–505.

Banu, N.A., Hoque, A., Watanabe-Sugimoto, M., Matsuoka, K., Nakamura, Y., Shimoishi, Y., Murata, Y. (2009) Proline and glycinebetaine induce antioxidant defense gene expression and suppress cell death in cultured tobacco cells under salt stress. *Journal of Plant Physiology* 166(2):146–156.

Basu, S., Roychoudhury, A., Saha, P.P., Sengupta, D.N. (2010) Differential antioxidative responses of indica rice cultivars to drought stress. *Plant Growth Regulation* 60(1):51–59.

Ben Ahmed, C., Ben Rouina, B., Sensoy, S., Boukhriss, M., Ben Abdullah, F. (2010) Exogenous proline effects on photosynthetic performance and antioxidant defense system of young olive tree. *Journal of Agricultural and Food Chemistry* 58(7):4216–4222.

Ben Ahmed, Ch, Magdich, S., Ben Rouina, B., Sensoy, S., Boukhris, M., Ben Abdullah, F. (2011) Exogenous proline effects on water relations and ions contents in leaves and roots of young olive. *Amino Acids* 40(2):565–573.

Bertazzini, M., Medrzycki, P., Bortolotti, L., Maistrello, L., Forlani, G. (2010) Amino acid content and nectar choice by forager honeybees (*Apis mellifera* L.). *Amino Acids* 39(1):315–318.

Bhatnagar-Mathur, P., Vadez, V., Devi, M.J., Lavanya, M., Vani, G., Sharma, K.K. (2009) Genetic engineering of chickpea (*Cicer arietinum* L.) with the P5CSF129A gene for osmoregulation with implications on drought tolerance. *Molecular Breeding* 23(4):591–606.

Blum, A. (2017) Osmotic adjustment is a prime drought stress adaptive engine in support of plant production. *Plant, Cell & Environment* 40(1):4–10.

Blum, A., Ebercon, A. (1976) Genotypic responses in sorghum to drought stress. III. Free proline accumulation and drought resistance. *Crop Science* 16(3):428–431.

Bohnert, H.J., Jensen, R.G. (1996) Strategies for engineering water-stress tolerance in plants. *Trends in Biotechnology* 14(3):89–97.

Borgo, L., Marur, C.J., Vieira, L.G.E. (2015) Effects of high proline accumulation on chloroplast and mitochondrial ultrastructure and on osmotic adjustment in tobacco plants. *Acta Scientiarum Agronomy* 37(2):191–199.

Borsani, O., Zhu, J., Verslues, P.E., Sunkar, R., Zhu, J.K. (2005) Endogenous siRNAs derived from a pair of natural cis-antisense transcripts regulate salt tolerance in *Arabidopsis. Cell* 123(7):1279–1291.

Bray, E.A., Bailey-Serres, J., Weretilnyk, E. (2000) Responses to abiotic stresses. In *Biochemistry and Molecular Biology of Plants*, eds. Buchanan, B., Gruissem, W., Jones, R., pp 1158–1203. American Society of Plant Physiologists, Rockville.

Chen, Z., Cuin, T.A., Zhou, M., Twomey, A., Naidu, B.P., Shabala, S. (2007) Compatible solute accumulation and stress-mitigating effects in barley genotypes contrasting in their salt tolerance. *Journal of Experimental Botany* 58(15–16):4245–4255.

Chen, J.B., Yang, J.W., Zhang, Z.Y., Feng, X.F., Wang, S.M. (2013) Two P5CS genes from common bean exhibiting different tolerance to salt stress in transgenic Arabidopsis. *Journal of Genetics* 92(3):461–469.

Choudhary, N.L., Sairam, R.K., Tyagi, A. (2005) Expression of delta1-pyrroline-5-carboxylate synthetase gene during drought in rice (*Oryza sativa* L.). *Indian Journal of Biochemistry and Biophysics* 42(6):366–370.

Chrominski, A., Halls, S., Weber, D.J., Smith, B.N. (1989) Proline effects ACC to ethylene conversion under salt and water stress in the halophytic *Allenrolfea occidentalis. Environmental and Experimental Botany* 29(3):359–363.

Csonka, L.N., Gelvin, S.B., Goodner, B.W., Orser, C.S., Siemieniak, D., Slightom, J.L. (1988) Nucleotide sequence of a mutation in the proB gene of Escherichia coli that confers proline overproduction and enhanced tolerance to osmotic stress. *Gene* 64(2):199–205.

Cuin, T.A., Shabala, S. (2007) Compatible solutes reduce ROS-induced potassium efflux in Arabidopsis roots. *Plant, Cell and Environment* 30(7):875–885.

de Campos, M.K.F., de Carvalho, K., de Souza, F.S., Marur, C.J., Pereira, L.F.P., Bespalhok Filho, J.C.B., Vieira, L.G.E. (2011) Drought tolerance and antioxidant enzymatic activity in transgenic 'Swingle' citrumelo plants over-accumulating proline. *Environmental and Experimental Botany* 72(2):242–250.

De Ronde, J.A., Cress, W.A., Krüger, G.H.J., Strasser, R.J., Van Staden, J. (2004) Photosynthetic response of transgenic soybean plants, containing an Arabidopsis P5CR gene, during heat and drought stress. *Journal of Plant Physiology* 161(11):1211–1224.

De Ronde, J.A., Spreeth, M.H., Cress, W.A. (2000) Effect of antisense L-Δ1-pyrroline-5-carboxylate reductase transgenic soybean plants subjected to osmotic and drought stress. *Plant Growth Regulation* 32(1): 13–26.

Deivanai, S., Xavier, R., Vinod, V., Timalata, K., Lim, O.F. (2011) Role of exogenous proline in ameliorating salt stress at early stage in two rice cultivars. *Journal of Stress Physiology and Biochemistry* 7:157–174.

Delauney, A.J., Verma, D.P.S. (1993) Proline biosynthesis and osmoregulation in plants. *The Plant Journal* 4(2):215–223.

Deuschle, K., Funck, D., Hellmann, H., Daschner, K., Binder, S., Frommer, W.B. (2001) A nuclear gene encoding mitochondrial Δ1-pyrroline-5-carboxylate dehydrogenase and its potential role in protection from proline toxicity. *The Plant Journal: for Cell and Molecular Biology* 27(4):345–356.

Ehsanpour, A.A., Fatahian, N. (2003) Effects of salt and proline on Medicago sativa callus. *Plant Cell, Tissue and Organ Culture* 73(1):53–56.

Fabro, G., Kovacs, I., Pavet, V., Szabados, L., Alvarez, M.E. (2004) Proline accumulation and AtP5CS2 gene activation are induced by plant-pathogen incompatible interactions in Arabidopsis. *Molecular Plant-Microbe Interactions: MPMI* 17(4):343–350.

Farooq, M., Wahid, A., Kobayashi, N., Fujita, D., Basra, S.M.A. (2009) Plant drought stress: Effects, mechanisms and management. *Agronomy for Sustainable Development* 29(1):185–212.

Fichman, Y., Gerdes, S.Y., Kovács, H., Szabados, L., Zilberstein, A., Csonka, L.N. (2015) Evolution of proline biosynthesis: Enzymology, bioinformatics, genetics, and transcriptional regulation. *Biological Reviews of the Cambridge Philosophical Society* 90(4):1065–1099.

Floyd, R.A., Nagy, I.Z. (1984) Formation of long-lived hydroxyl free radical adducts of proline and hydroxyproline in a Fenton reaction. *Biochimica et Biophysica Acta (BBA)-Protein Structure and Molecular Enzymology* 790(1):94–97.

Fujii, H., Chinnusamy, V., Rodrigues, A., Rubio, S., Antoni, R., Park, S.Y., Cutler, S.R., *et al.* (2009) In vitro reconstitution of an abscisic acid signalling pathway. *Nature* 462(7273):660–664.

Funck, D., Eckard, S., Müller, G. (2010) Non-redundant functions of two proline dehydrogenase isoforms in Arabidopsis. *BMC Plant Biology* 10:70.

Funck, D., Stadelhofer, B., Koch, W. (2008) Ornithine-d-aminotransferase is essential for arginine catabolism but not for proline biosynthesis. *BMC Plant Biology* 8(1):40.

Gadallah, M.A.A. (1999) Effect of proline and glycinebetaine on Vicia faba responses to salt stress. *Biologia Plantarum* 42:247–249.

Gagneul, D., Aïnouche, A., Duhazé, C., Lugan, R., Larher, F.R., Bouchereau, A. (2007) A reassessment of the function of the so-called compatible solutes in the halophytic Plumbaginaceae *Limonium latifolium*. *Plant Physiology* 144(3):1598–1611.

Garg, A.K., Kim, J.K., Owens, T.G., Ranwala, A.P., Choi, Y.D., Kochian, L.V., Wu, R.J. (2002) Trehalose accumulation in rice plants confers high tolerance levels to different abiotic stresses. *Proceedings of the National Academy of Sciences of the United States of America* 99(25):15898–15903.

Ghanti, S.K.K., Sujata, K.G., Vijay Kumar, B.M., Nataraja Karba, N., Janardhan Reddy, K., Srinath Rao, M., Kishor, P.B.K. (2011) Heterologous expression of P5CS gene in chickpea enhances salt tolerance without affecting yield. *Biologia Plantarum* 55(4):634–640.

Ghars, M.A., Parre, E., Leprince, A.S., Bordenave, M., Lefebvre-De Vos, D., Richard, L., Abdelly, C., Savouré, A. (2008) Opposite lipid signaling pathways tightly control proline accumulation in *Arabidopsis thaliana* and *Thellungiella halophila*. In *Biosaline Agriculture and High Salinity Tolerance*, (Abdelly, C., Öztürk, M., Ashraf, M., Grignon, C., Eds.) pp 317–324. Birkhäuser Verlag, Basel.

Giberti, S., Funck, D., Forlani, G. (2014) Δ1-pyrroline-5-carboxylate reductase from *Arabidopsis thaliana*: Stimulation or inhibition by chloride ions and feedback regulation by proline depend on whether NADPH or NADH acts as co-substrate. *The New Phytologist* 202(3):911–919.

Girousse, C., Bournoville, R., Bonnemain, J.L. (1996) Water deficit-induced changes in concentrations in proline and some other amino acids in the phloem sap of alfalfa. *Plant Physiology* 111(1):109–113.

Golldack, D., Li, C., Mohan, H., Probst, N. (2014) Tolerance to drought and salt stress in plants: Unraveling the signaling networks. *Frontiers in Plant Science* 5:151.

Gubis, J., Vaňková, R., Červená, V., Dragúňová, M., Hudcovicová, M., Lichtnerová, H., Dokupil, T., Jureková, Z. (2007) Transformed tobacco plants with increased tolerance to drought. *South African Journal of Botany* 73(4):505–511.

Guerzoni, J.T.S., Belintani, N.G., Moreira, R.M.P., Hoshimo, A.A., Domingues, D.S., Filho, J.C.B., Vieira, L.G.E. (2014) Stress-induced Δ1-pyrroline-5-carboxylate synthetase (P5CS) gene confers tolerance to salt stress in transgenic sugarcane. *Acta Physiologiae Plantarum* 36(9):2309–2319.

Hamilton, E.W., Heckathorn, S.A. (2001) Mitochondrial adaptations to NaCl. Complex I is protected by anti-oxidants and small heat shock proteins, whereas complex II is protected by proline and betaine. *Plant Physiology* 126(3):1266–1274.

Hare, P.D., Cress, W.A. (1997) Metabolic implications of stress-induced proline accumulation in plants. *Plant Growth Regulation* 21(2):79–102.

Hare, P.D., Cress, W.A., Van Staden, J. (1998) Dissecting the roles of osmolyte accumulation during stress. *Plant, Cell & Environment* 21(6):535–553.

Hare, P.D., Cress, W.A., Van Staden, J. (1999) Proline synthesis and degradation: A model system for elucidating stress-related signal transduction. *Journal of Experimental Botany* 50(333):413–434.

Hasanuzzaman, M., Alam, M.M., Rahman, A., Hasanuzzaman, M., Nahar, K., Fujita, M. (2014) Exogenous proline and glycinebetaine mediated upregulation of antioxidant defense and glyoxalase systems provides better protection against salt-induced oxidative stress in two rice (*Oryza sativa* L.) varieties. *Biomedical Research International* 2014:1–17.

Hasegawa, P.M., Bressan, R.A., Zhu, J.K., Bohnert, H.J. (2000) Plant cellular and molecular responses to high salinity. *Annual Review of Plant Physiology and Plant Molecular Biology* 51:463–499.

Hayashi, F., Ichino, T., Osanai, M., Wada, K. (2000) Oscillation and regulation of proline content by P5CS and ProDH gene expressions in the light/dark cycles in *Arabidopsis thaliana* L. *Plant and Cell Physiology* 41(10):1096–1101.

Hmida-Sayari, A., Gargouri-Bouzid, R., Bidani, A., Jaoua, L., Savoure, A., Jaoua, S. (2005) Overexpression of Δ1-pyrroline-5-carboxylate synthetase increases proline production and confers salt tolerance in transgenic potato plants. *Plant Science* 169(4):746–752.

Hong, Z., Lakkineni, K., Zhang, Z., Verma, D.P.S. (2000) Removal of feedback inhibition of D1-pyrroline-5-carboxylate synthetase results in increased proline accumulation and protection of plants from osmotic stress. *Plant Physiology* 122(4):1129–1136.

Hoque, M.A., Banu, M.N.A., Nakamura, Y., Shimoishi, Y., Murata, Y. (2008) Proline and glycinebetaine enhance antioxidant defense and methylglyoxal detoxification systems and reduce NaCl-induced damage in cultured tobacco cells. *Journal of Plant Physiology* 165(8):813–824.

Hoque, M.A., Banu, M.N.A., Okuma, E., Amako, K., Nakamura, Y., Shimoishi, Y., Murata, Y. (2007) Exogenous proline and glycinebetaine increase NaCl-induced ascorbate glutathione cycle enzyme activities, and proline improves salt tolerance more than glycinebetaine in tobacco Bright Yellow-2 suspension cultured cells. *Journal of Plant Physiology* 164(11):1457–1468.

Hossain, M.A., Fujita, M. (2010) Evidence for a role of exogenous glycinebetaine and proline in antioxidant defense and methylglyoxal detoxification systems in mung bean seedlings under salt stress. *Physiology and Molecular Biology of Plants: an International Journal of Functional Plant Biology* 16(1):19–29.

Hossain, M.A., Mostofa, M.G., Burritt, D.J., Fujita, M. (2014) Modulation of reactive oxygen species and methylglyoxal detoxification systems by exogenous Glycinebetaine and proline improves drought tolerance in mustard (*Brassica juncea* L.). *International Journal of Plant Biology and Research* 2(2):1014.

Hu, C.A., Delauney, A.J., Verma, D.P. (1992) A bifunctional enzyme (delta 1-pyrroline-5-carboxylate synthetase) catalyzes the first two steps in proline biosynthesis in plants. *Proceedings of the National Academy of Sciences* 89(19):9354–9358.

Hua-long, L., Han-jing, S., Jing-guo, W., Yang, L., De-tang, Z., Hong-wei, Z. (2014) Effect of seed soaking with exogenous proline on seed germination of rice under salt stress. *Journal of Northeast Agricultural University* 21(3):1–6.

Huang, Y., Bie, Z., Liu, Z., Zhen, A., Wang, W. (2009) Protective role of proline against salt stress is partially related to the improvement of water status and peroxidise enzyme activity in cucumber. *Soil Science and Plant Nutrition* 55(5):698–704.

Huh, S.M., Noh, E.K., Kim, H.G., Jeon, B.W., Bae, K., Hu, H.C., Kwak, J.M., Park, O.K. (2010) Arabidopsis annexins AnnAt1 and AnnAt4 interact with each other and regulate drought and salt stress responses. *Plant and Cell Physiology* 51(9):1499–1514.

Jabeen, N., Ahmad, R. (2013) The activity of antioxidant enzymes in response to salt stress in safflower (*Carthamus tinctorius* L.) and sunflower (*Helianthus annuus* L.) seedlings raised from seed treated with chitosan. *Journal of the Science of Food and Agriculture* 93(7):1699–1705.

Jain, M., Mathur, G., Koul, S., Sarin, N. (2001) Ameliorative effects of proline on salt stress-induced lipid peroxidation in cell lines of groundnut (*Arachis hypogaea* L.). *Plant Cell Reports* 20(5):463–468.

Jiang, Y., Qiu, Y., Hu, Y., Yu, D. (2016) Heterologous expression of AtWRKY57 confers drought tolerance in *Oryza sativa*. *Frontiers in Plant Science* 7:145.

Kahlaoui, B., Hachicha, M., Rejeb, S., Rejeb, M.N., Hanchi, B., Misle, E. (2014) Response of two tomato cultivars to field-applied proline under irrigation with saline water: Growth, chlorophyll fluorescence and nutritional aspects. *Photosynthetica* 52(3):421–429.

Kamran, M., Shahbaz, M., Ashraf, M., Akram, N.A. (2009) Alleviation of drought-induced adverse effects in spring wheat (*Triticum aestivum* L.) using proline as a pre-sowing seed treatment. *Pakistan Journal of Botany* 41(2):621–632.

Karamanos, A.J. (1995) The involvement of proline and some metabolites in water stress and their importance as drought resistance indicators. *Bulgarian Journal of Plant Physiology* 21(2–3):98–110.

Karthikeyan, A., Pandian, S.K., Ramesh, M. (2011) Transgenic indica rice cv. ADT 43 expressing a Δ1-pyrroline-5-carboxylate synthetase (P5CS) gene from Vigna aconitifolia demonstrates salt tolerance. *Plant Cell, Tissue and Organ Culture* 107(3):383–395.

Kaul, S., Sharma, S.S., Mehta, I.K. (2008) Free radical scavenging potential of L-proline: Evidence from in vitro assays. *Amino Acids* 34(2):315–320.

Kemble, A.R., Macpherson, H.T. (1954) Liberation of amino acids in perennial ray grass during wilting. *The Biochemical Journal* 58(1):46–49.

Khan, N.A., Nazar, R., Iqbal, N., Anjum, N.A. (2012) *Phytohormones and Abiotic Stress Tolerance in Plants.* 10.1007/978-3-642-25829-9. Springer, Berlin.

Khanna-Chopra, R. (2012) Leaf senescence and abiotic stresses share reactive oxygen species-mediated chloroplast degradation. *Protoplasma* 249(3):469–481.

Khanna-Chopra, R., Singh, K. (2015) Drought resistance in crops: Physiological and genetic basis of traits for crop productivity. In *Stress Responses in Plants*, eds. Tripathi B.N., Muller M., pp 1–35. Springer, Netherlands.

Khedr, A.H.A., Abbas, M.A., Wahid, A.A.A., Quick, W.P., Abogadallah, G.M. (2003) Proline induces the expression of salt-stress-responsive proteins and may improve the adaptation of *Pancratium maritimum* L. to salt-stress. *Journal of Experimental Botany* 54(392):2553–2562.

Kim, J., Liu, Y., Zhang, X., Zhao, B., Childs, K.L. (2016) Analysis of salt-induced physiological and proline changes in 46 switchgrass (*Panicum virgatum*) lines indicates multiple response modes. *Plant Physiology and Biochemistry: PPB* 105:203–212.

Kishor, P.B.K., Hima Kumari, P., Sunita, M.S., Sreenivasulu, N. (2015) Role of proline in cell wall synthesis and plant development and its implications in plant ontogeny. *Frontiers in Plant Science* 6:544.

Kishor, P.B.K., Hong, Z., Miao, G.H., Hu, C.-A., Verma, D.P.S. (1995) Over expression of D1-pyrroline-5-carboxylate synthetase increases proline overproduction and confers osmtolerance in transgenic plants. *Plant Physiology* 108(4):1387–1394.

Kishor, P.B.K., Sangam, S., Amrutha, R.N., Sri Laxmi, P., Naidu, K.R., Rao, K.R.S.S. (2005) Regulation of proline biosynthesis, degradation, uptake and transport in higher plants: Its implications in plant growth and abiotic stress tolerance. *Current Science* 88:424–438.

Kishor, P.B.K., Sreenivasulu, N. (2014) Is proline accumulation per se correlated with stress tolerance or is proline homeostasis a more critical issue? *Plant, Cell and Environment* 37(2):300–311.

Kiyosue, T., Yoshiba, Y., Yamaguchi-Shinozaki, K., Shinozaki, K. (1996) A nuclear gene encoding mitochondrial proline dehydrogenase, an enzyme involved in proline metabolism, is upregulated by proline but downregulated by dehydration in Arabidopsis. *The Plant Cell* 8(8):1323–1335.

Kochetov, A.V., Titov, S.E., Kolodyazhnaya, Y.S., Komarova, M.L., Koval, V.S., Makarova, N.N., Il'yinskyi, Y.Y., Trifonova, E.A., Shumny, V.K. (2004) Tobacco transformants bearing antisense suppressor of proline dehydrogenase gene are characterized by higher proline content and cytoplasm osmotic pressure. *Russian Journal of Genetics* 40(2):216–218.

Kumar, V., Shriram, V., Kishor, P.B.K., Jawali, N., Shitole, M.G. (2010) Enhanced proline accumulation and salt stress tolerance of transgenic indica rice by over expressing P5CSF129A gene. *Plant Biotechnology Reports* 4(1):37–48.

Laohavisit, A., Brown, A.T., Cicuta, P., Davies, J.M. (2010) Annexins: Components of the calcium and reactive oxygen signaling network. *Plant Physiology* 152(4):1824–1829.

Lee, B.R., Jin, Y.L., Avice, J.C., Cliquet, J.B., Ourry, A., Kim, T.H. (2009) Increased proline loading to phloem and its effects on nitrogen uptake and assimilation in water-stressed white clover (*Trifolium repens*). *The New Phytologist* 182(3):654–663.

Lehmann, S., Funck, D., Szabados, L., Rentsch, D. (2010) Proline metabolism and transport in plant development. *Amino Acids* 39(4):949–962.

Liang, X., Zhang, L., Natarajan, S.K., Becker, D.F. (2013) Proline mechanisms of stress survival. *Antioxidants & Redox Signaling* 19(9):998–1011.

Liu, D., He, S., Zhai, H., Wang, L., Zhao, Y., Wang, B., Li, R., Liu, Q. (2014) Overexpression of IbP5CR enhances salt tolerance in transgenic sweet potato. *Plant Cell, Tissue and Organ Culture* 117(1):1–16.

Lv, W.T., Lin, B., Zhang, M., Hua, X.J. (2011) Proline accumulation is inhibitory to Arabidopsis seedlings during heat stress. *Plant Physiology* 156(4):1921–1933.

Ma, L., Zhou, E., Gao, L., Mao, X., Zhou, R., Jia, J. (2008) Isolation, expression analysis and chromosomal location of P5CR gene in common wheat (*Triticum aestivum* L.). *South African Journal of Botany* 74(4):705–712.

Man, D., Bao, Y.X., Han, L.B., Zhang, X. (2011) Drought tolerance associated with proline and hormone metabolism in two tall fescue cultivars. *Horticultural Science* 46:1027–1032.

Mani, S., Van de Cotte, B., Van Montagu, M., Verbruggen, N. (2002) Altered levels of proline dehydrogenase cause hypersensitivity to proline and its analogs in Arabidopsis. *Plant Physiology* 128(1):73–83.

Mansour, M.M.F., Ali, E.F. (2017) Evaluation of proline functions in saline conditions. *Phytochemistry* 140:52–68.

Mao, X., Jia, D., Li, A., Zhang, H., Tian, S., Zhang, X., Jia, J., Jing, R. (2011) Transgenic expression of TaMYB2A confers enhanced tolerance to multiple abiotic stresses in Arabidopsis. *Functional & Integrative Genomics* 11(3):445–465.

Matsumura, T., Kanechi, M., Inagaki, N., Maekawa, S. (1998) The effects of salt stress on ion uptake, accumulation of compatible solutes, and leaf osmotic potential in safflower, Chrysanthemum paludosum and sea aster. *Engei Gakkai Zasshi* 67(3):426–431

Mattioli, R., Costantino, P., Trovato, M. (2009a) Proline accumulation in plants-not. *Plant Signaling & Behavior* 4(11):1016–1018.

Mattioli, R., Falasca, G., Sabatini, S., Altamura, M.M., Costantino, P., Trovato, M. (2009b) The proline biosynthetic genes P5CS1 and P5CS2 play overlapping roles in Arabidopsis flower transition but not in embryo development. *Physiologia Plantarum* 137(1):72–85.

Matysik, J., Bhalu, B., Mohanty, P. (2002) Molecular mechanisms of quenching of reactive oxygen species by proline under stress in plants. *Current Science* 82:525–532.

Mayahi, A.M.Z., Fayadh, M.H. (2015) Effect of exogenous proline application on salinity tolerance of *Cordia myxa* L. Seedlings. Effect on vegetative and physiological characteristics. *Journal of Natural Sciences Research* 5(24):118–125.

Miller, G., Honig, A., Stein, H., Suzuki, N., Mittler, R., Zilberstein, A. (2009) Unraveling Δ1-pyrroline-5-carboxylate-proline cycle in plants by uncoupled expression of proline oxidation enzymes. *The Journal of Biological Chemistry* 284(39):26482–26492.

Molinari, H.B.C., Marur, C.J., Bespalhok Filho, J.C.B., Kobayashi, A.K., Pileggi, M., Júnior, R.P.L., Pereira, L.F.P., Vieira, L.G.E. (2004) Osmotic adjustment in transgenic citrus rootstock Carrizo citrange (*Citrus sinensis* Osb. × *Poncirus trifoliata* L. Raf.) overproducing proline. *Plant Science* 167(6):1375–1381.

Molla, M.R., Ali, M.R., Hasanuzzaman, M., Al-Mamun, M.H., Ahmed, A., Nazim-ud-Dowla, M.A.N., Rohman, M.M. (2014) Exogenous proline and betaine-induced upregulation of glutathione transferase and glyoxalase I in lentil. *Notulae Botanicae Horti Agrobotanici Cluj-Napoca* 42:73–80.

Momayezi, M.R., Zaharah, A.R., Hanafi, M.M., Mohd Razi, I. (2009) Agronomic characteristics and proline accumulation of Iranian rice genotypes at early seedling stage under sodium salts stress. *Malaysian Journal of Soil Science* 13:59–75.

Moustakas, M., Sperdouli, I., Kouna, T., Antonopoulou, C.I., Therios, I. (2011) Exogenous proline induces soluble sugar accumulation and alleviates drought stress effects on photosystem II functioning of Arabidopsis thaliana leaves. *Plant Growth Regulation* 65(2):315–325.

Nanjo, T., Kobayashi, M., Yoshiba, Y., Kakubari, Y., Yamaguchi-Shinozaki, K., Shinozaki, K. (1999) Antisense suppression of proline degradation improves tolerance to freezing and salinity in *Arabidopsis thaliana. FEBS Letters* 461(3):205–210.

Nawaz, H., Yasmeen, A., Anjum, M.A., Hussain, N. (2016) Exogenous application of growth enhancers mitigate water stress in wheat by antioxidant elevation. *Frontiers in Plant Science* 7:597.

Noreen, S., Athar, H., Ashraf, M. (2013) Interactive effects of watering regimes and exogenously applied osmoprotectants on earliness indices and leaf area index in cotton (*Gossypium hirsutum* L.) crop. *Pakistan Journal of Botany* 45(6):1873–1881.

Nounjan, N., Nghia, P.T., Theerakulpisut, P. (2012) Exogenous proline and trehalose promote recovery of rice seedlings from salt-stress and differentially modulate antioxidant enzymes and expression of related genes. *Journal of Plant Physiology* 169(6):596–604.

Oono, Y., Seki, M., Nanjo, T., Narusaka, M., Fujita, M., Satoh, R., Satou, M., et al. (2003) Monitoring expression profiles of Arabidopsis gene expression during rehydration process after dehydration using ca. 7000 full-length cDNA microarray. *The Plant Journal: for Cell and Molecular Biology* 34(6):868–887.

Osman, H.S. (2015) Enhancing antioxidant-yield relationship of pea plant under drought at different growth stages by exogenously applied glycine betaine and proline. *Annals of Agricultural Sciences* 60(2):389–402.

Ozden, M., Demirel, U., Kahraman, A. (2009) Effects of proline on antioxidant system in leaves of grapevine (*Vitis vinifera* L.) exposed to oxidative stress by H_2O_2. *Scientia Horticulturae* 119(2):163–168.

Parre, E., de Virville, J., Cochet, F., Leprince, A.S., Richard, L., Lefebvre-De Vos, D., Ghars, M.A., *et al.* (2010) A new method for accurately measuring Δ 1-pyrroline-5-carboxylate synthetase activity. *Methods in Molecular Biology* 639:333–340.

Patade, V.Y., Lokhande, V.H., Suprasanna, P. (2014) Exogenous application of proline alleviates salt induced oxidative stress more efficiently than glycinebetaine in sugarcane cultured cells. *Sugar Tech* 16(1):22–29.

Per, T.S., Khan, N.A., Reddy, P.S., Masood, A., Hasanuzzaman, M., Khan, M.I.R., Anjum, N.A. (2017) Approaches in modulating proline metabolism in plants for salt and drought stress tolerance: Phytohormones, mineral nutrients and transgenics. *Plant Physiology and Biochemistry: PPB* 115:126–140.

Phang, J.M., Donald, S.P., Pandhare, J., Liu, Y. (2008) The metabolism of proline, a stress substrate, modulates carcinogenic pathways. *Amino Acids* 35(4):681–690.

Pospisilova, J., Haisel, D., Vankova, R. (2011) Responses of transgenic tobacco plants with increased proline content to drought and/or heat stress. *American Journal of Plant Sciences* 02(3):318–324.

Rastgar, J.F., Yamchi, A., Hajirezaei, M., Abbasi, A.R., Karkhane, A.A. (2011) Growth assessments of *Nicotiana tabaccum* cv. Xanthitrans formed with *Arabidopsis thaliana* P5CS under salt stress. *African Journal of Biotechnology* 10:8539–8552.

Rayapati, P.J., Stewart, C.R., Hack, E. (1989) Pyrroline-5-carboxylate reductase is in pea (*Pisum sativum* L.) leaf chloroplasts. *Plant Physiology* 91(2):581–586.

Raymond, M.J., Smirnoff, N. (2002) Proline metabolism and transport in maize seedlings at low water potential. *Annals of Botany* 89:813–823.

Reddy, P.S., Jogeswar, G., Rasineni, G.K., Maheswari, M., Reddy, A.R., Varshney, R.K., Kishor, P.B.K. (2015) Proline over-accumulation alleviates salt stress and protects photosynthetic and antioxidant enzyme activities in transgenic sorghum [*Sorghum bicolor* (L.) Moench]. *Plant Physiology and Biochemistry: PPB* 94:104–113.

Rejeb, K.B., Abdelly, C., Savoure, A. (2014) How reactive oxygen species and proline face stress together. *Plant Physiology and Biochemistry: PPB* 80:278–284.

Rentsch, D., Hirner, B., Schmelzer, E., Frommer, W.B. (1996) Salt stress induced proline transporters and salt stress-repressed broad specificityamino acid permeases identified by suppression of a yeast amino acid permease-targeting mutant. *The Plant Cell* 8(8):1437–1446.

Rentsch, D., Schmidt, S., Tegeder, M. (2007) Transporters for uptake and allocation of organic nitrogen compounds in plants. *FEBS Letters* 581(12):2281–2289.

Ribarits, A., Abdullaev, A., Tashpulatov, A., Richter, A., Heberle-Bors, E., Touraev, A. (2007) Two tobacco proline dehydrogenases are differentially regulated and play a role in early plant development. *Planta* 225(5):1313–1324.

Roosens, N.H., Al Bitar, F., Loenders, K., Angenon, G., Jacobs, M. (2002) Overexpression of ornithine-delta-aminotransferase increases proline biosynthesis and confers osmotolerance in transgenic plants. *Molecular Breeding* 9(2):73–80.

Roy, D., Basu, N., Bhunia, A., Banerjee, S.K. (1993) Counteraction of exogenous l-proline with NaCl in salt-sensitive cultivar of rice. *Biologia Plantarum* 35(1):69–72.

Roy, S.J., Negrao, S., Tester, M. (2014) Salt resistant crop plants. *Current Opinion in Biotechnology* 26:115–124.

Roychoudhury, A., Chakraborty, M. (2013) Biochemical and molecular basis of varietal difference in plant salt tolerance. *Annual Review and Research in Biology* 3:422–454.

Sadak, M.S., Mostafa, H.A.M. (2015) Physiological role of pre-sowing seed with proline on some growth, biochemical aspects, yield quantity and quality of two sunflower cultivars grown under seawater salinity stress. *Scientia Agriculturae* 9(1):60–69.

Saeedipour, S. (2013) Relationship of grain yield, ABA and proline accumulation in tolerant and sensitive wheat cultivars as affected by water stress. *Proceedings of the National Academy of Sciences, India Section B: Biological Sciences* 83(3):311–315.

Samach, A., Onouchi, H., Gold, S.E., Ditta, G.S., Schwarz-Sommer, Z., Yanofsky, M.F., Coupland, G. (2000) Distinct roles of CONSTANS target genes in reproductive development of Arabidopsis. *Science* 288(5471):1613–1616.

Saradhi, P.P., Alia, Arora, S., Prasad, K.V. (1995) Proline accumulates in plants exposed to UV radiation and protects them against UV induced peroxidation. *Biochemical and Biophysical Research Communications* 209(1):1–5.

Satoh, R., Nakashima, K., Seki, M., Shinozaki, K., Yamaguchi-Shinozaki, K. (2002) ACTCAT, a novel cis-acting element for proline-and hypoosmolarity-responsive expression of the ProDH gene encoding proline dehydrogenase in Arabidopsis. *Plant Physiology* 130(2):709–719.

Savouré, A., Hua, X.J., Bertauche, N., Van Montagu, M., Verbruggen, N. (1997) Abscisic acid-independent and abscisic acid-dependent regulation of proline biosynthesis following cold and osmotic stresses in Arabidopsis thaliana. *Molecular & General Genetics: MGG* 254(1):104–109.

Sawahel, W.A., Hassan, A.H. (2002) Generation of transgenic wheat plants producing high levels of the osmoprotectant proline. *Biotechnology Letters* 24(9):721–725.

Schat, H.S., Sharma, S., Vooijs, R. (1997) Heavy metal-induced accumulation of free proline in a metal-tolerant and a non-tolerant ecotype of Silene vulgaris. *Physiologia Plantarum* 101:477–482.

Shahbaz, M., Mushtaq, Z., Andaz, F., Masood, A. (2013) Does proline application ameliorate adverse effects of salt stress on growth, ions and photosynthetic ability of eggplant (*Solanum melongena* L.)? *Scientia Horticulturae* 164:507–511.

Sharma, S., Verslues, P.E. (2010) Mechanisms independent of abscisic acid (ABA) or proline feedback have a predominant role in transcriptional regulation of proline metabolism during low water potential and stress recovery. *Plant, Cell & Environment* 33(11):1838–1851.

Sharma, S., Villamor, J.G., Verslues, P.E. (2011) Essential role of tissue-specific proline synthesis and catabolism in growth and redox balance at low water potential. *Plant Physiology* 157(1):292–304.

Shevyakova, N.I., Bakulina, E.A., Kuznetsov, V.V. (2009) Proline antioxidant role in the common ice plant subjected to salinity and paraquat treatment inducing oxidative stress. *Russian Journal of Plant Physiology* 56(5):663–669.

Shukla, P.S., Gupta, K., Agarwal, P., Jha, B., Agarwal, P.K. (2015) Overexpression of a novel SbMYB15 from Salicornia brachiata confers salinity and dehydration tolerance by reduced oxidative damage and improved photosynthesis in transgenic tobacco. *Planta* 242(6):1291–1308.

Signorelli, S., Coitiño, E.L., Borsani, O., Monza, J. (2014) Molecular mechanisms for the reaction between ·OH radicals and proline: Insights on the role as reactive oxygen species scavenger in plant stress. *The Journal of Physical Chemistry B* 118(1):37–47.

Silveira, J.A., Viégas, Rde A., da Rocha, I.M., Moreira, A.C., Moreira, Rde A., Oliveira, J.T. (2003) Proline accumulation and glutamine synthetase activity are increased by salt-induced proteolysis in cashew leaves. *Journal of Plant Physiology* 160(2):115–123.

Siripornadulsil, S., Traina, S., Verma, D.P., Sayre, R.T. (2002) Molecular mechanisms of proline-mediated tolerance to toxic heavy metals in transgenic microalgae. *The Plant Cell* 14(11):2837–2847.

Smirnoff, N., Cumbes, Q.J. (1989) Hydroxyl radical scavenging activity of compatible solutes. *Phytochemistry* 28(4):1057–1060.

Sobahan, M.A., Akter, N., Ohno, M., Okuma, E., Hirai, Y., Mori, I.C., Nakamura, Y., Murata, Y. (2012) Effects of exogenous proline and glycinebetaine on the salt tolerance of rice cultivars. *Bioscience, Biotechnology, and Biochemistry* 76(8):1568–1570.

Sripinyowanich, S., Klomsakul, P., Boonburapong, B., Bangyeekhun, T., Asami, T., Gu, H., Buaboocha, T., Chadchawan, S. (2013) Exogenous ABA induces salt tolerance in indica rice (*Oryza sativa* L.): The role of OsP5CS1 and OsP5CR gene expression during salt stress. *Environmental and Experimental Botany* 86:94–105.

Stines, A.P., Naylor, D.J., Høj, P.B., Van Heeswijck, R. (1999) Proline accumulation in developing grapevine fruit occurs independently of changes in the levels of Δ1-pyrroline-5-carboxylate synthetase mRNA or protein. *Plant Physiology* 120(3):923.

Strizhov, N., Ábrahám, E., Ökrész, L., Blickling, S., Zilberstein, A., Schell, J., Koncz, C., Szabados, L. (1997) Differential expression of two P5CS genes controlling proline accumulation during salt-stress requires ABA and is regulated by ABA1, ABI1 and AXR2 in Arabidopsis. *The Plant Journal: for Cell and Molecular Biology* 12(3):557–569.

Su, J., Wu, R. (2004) Stress-inducible synthesis of proline in transgenic rice confers faster growth under stress conditions than that with constitutive synthesis. *Plant Science* 166(4):941–948.

Surekha, C., Kumari, K.N., Aruna, L.V., Suneetha, G., Arundhati, A.K., Kishor, P.B.K. (2014) Expression of the Vigna aconitifolia P5CSF129A gene in transgenic pigeonpea enhances proline accumulation and salt tolerance. *Plant Cell, Tissue and Organ Culture* 116(1):27–36.

Szabados, L., Savoure, A. (2010) Proline: A multifunctional amino acid. *Trends in Plant Science* 15(2):89–97.

Szekely, G. (2004) The role of proline in *Arabidopsis thaliana* osmotic stress. *Acta Biologica Szegediensis* 48:81.

Szekely, G., Abraham, E., Cseplo, A., Rigo, G., Zsigmond, L., Csiszar, J., Ayaydin, F., et al. (2008) Duplicated P5CS genes of Arabidopsis play distinct roles in stress regulation and developmental control of proline biosynthesis. *The Plant Journal: for Cell and Molecular Biology* 53(1):11–28.

Szoke, A., Miao, G.H., Hong, Z., Verma, D.P.S. (1992) Subcellular location of δ1-pyrroline-5-carboxylate reductase in root/nodule and leaf of soybean. *Plant Physiology* 99(4):1642–1649.

Teh, C., Shaharuddin, N.A., Ho, C., Mahmood, M. (2016) Exogenous proline significantly affects the plant growth and nitrogen assimilation enzymes activities in rice (*Oryza sativa*) under salt stress. *Acta Physiologiae Plantarum* 38(6):151.

Theocharis, A., Clement, C., Barka, E.A. (2012) Physiological and molecular changes in plants grown at low temperatures. *Planta* 235(6):1091–1105.

Ullah, A., Sun, H., Yang, X., Zhang, X. (2017) Drought coping strategies in cotton: Increased crop per drop. *Plant Biotechnology Journal* 15(3):271–284.

Vendruscolo, E.C.G., Schuster, I., Pileggi, M., Scapim, C.A., Molinari, H.B.C., Marur, C.J., Vieira, L.G.E. (2007) Stress-induced synthesis of proline confers tolerance to water deficit in transgenic wheat. *Journal of Plant Physiology* 164(10):1367–1376.

Verbruggen, N., Hua, X.J., May, M., VanMontagu, M. (1996) Environmental and developmental signals modulate proline homeostasis: Evidence for a negative transcriptional regulator. *Proceedings of the National Academy of Sciences of the United States of America* 93(16):8787–8791.

Verbruggen, N., Villarroel, R., Montagu, M.V. (1993) Osmoregulation of a pyrroline-5-carboxylate reductase gene in Arabidopsis thaliana. *Plant Physiology* 103(3):771–781.

Verslues, P.E., Kim, Y.S., Zhu, J.K. (2007) Altered ABA, proline and hydrogen peroxide in an Arabidopsis glutamate: Glyoxylate aminotransferase mutant. *Plant Molecular Biology* 64(1–2):205–217.

Verslues, P.E., Sharma, S. (2010) Proline metabolism and its implications for plant-environment interaction. *The Arabidopsis Book* 8:e0140.

Wang, C., Deng, P., Chen, L., Wang, X., Ma, H., Hu, W., Yao, N., et al. (2013) A wheat WRKY transcription factor TaWRKY10 confers tolerance to multiple abiotic stresses in transgenic tobacco. *PLOS ONE* 8(6):e65120.

Wang, G., Zhang, J., Wang, G., Fan, X., Sun, X., Qin, H., Xu, N., et al. (2014) Proline responding1 plays a critical role in regulating general protein synthesis and the cell cycle in maize. *The Plant Cell* 26(6):2582–2600 doi:10.1105/tpc.114.125559.

Wani, A.S., Irfan, M., Hayat, S., Ahmad, A. (2012) Response of two mustard (*Brassica juncea* L.) cultivars differing in photosynthetic capacity subjected to proline. *Protoplasma* 249(1):75–87.

Weltmeier, F., Ehlert, A., Mayer, C.S., Dietrich, K., Wang, X., Schütze, K., Alonso, R., *et al.* (2006)Combinatorial control of Arabidopsis proline dehydrogenase transcription by specific heterodimerisation of bZIP transcription factors. *The EMBO Journal* 25(13):3133–3143.

Widodo, J.H.P., Patterson, J.H., Newbigin, E., Tester, M., Bacic, A., Roessner, U. (2009) Metabolic responses to salt stress of barley (*Hordeum vulgare* L.) cultivars, Sahara and Clipper, which differ in salinity tolerance. *Journal of Experimental Botany* 60(14):4089–4103.

Witt, S., Galicia, L., Lisec, J., Cairns, J., Tiessen, A., Araus, J.L., Palacios-Rojas, N., Fernie, A.R. (2012) Metabolic and phenotypic responses of greenhouse-grown maize hybrids to experimentally controlled drought stress. *Molecular Plant* 5(2):401–417

Wu, L., Fan, Z., Guo, L., Li, Y., Zhang, W., Qu, L.J., Chen, Z. (2003) Over-expression of an Arabidopsis δ-OAT gene enhances salt and drought tolerance in transgenic rice. *Chinese Science Bulletin* 48(23):2594–2600.

Wu, G.Q., Feng, R.J., Li, S.J., Du, Y.Y. (2017) Exogenous application of proline alleviates salt-induced toxicity in sainfoin seedlings. *The Journal of Animal and Plant Sciences* 27(1):246–251.

Xu, Z., Ali, Z., Xu, L., He, X., Huang, Y., Yi, J., Shao, H., Ma, H., Zhang, D. (2016) The nuclear protein GmbZIP110 has transcription activation activity and plays important roles in the response to salinity stress in soybean. *Scientific Reports* 6:20366.

Xu, F.Y., Wang, X.L., Wu, Q.X., Zhang, X.R., Wang, L.H. (2012) Physiological responses differences of different genotype sesames to flooding stress. *Advance Journal of Food Science and Technology* 4:352–356.

Xue, X., Liu, A., Hua, X. (2009) Proline accumulation and transcriptional regulation of proline biothesynthesis and degradation in Brassica napus. *BMB Reports* 42(1):28–34.

Yamada, M., Morishita, H., Urano, K., Shiozaki, N., Yamaguchi-Shinozaki, K., Shinozaki, K., Yoshiba, Y. (2005) Effects of free proline accumulation in petunias under drought stress. *Journal of Experimental Botany* 56(417):1975–1981.

Yan, Z., Guo, S., Shu, S., Sun, J., Tezuka, T. (2013) Effects of proline on photosynthesis, root reactive oxygen species (ROS) metabolism in two melon cultivars (*Cucumis melo* L.) under NaCl stress. *African Journal of Biotechnology* 10:18381–18390.

Yang, A., Dai, X., Zhang, W.H. (2012) A R2R3-type MYB gene, OsMYB2, is involved in salt, cold, and dehydration tolerance in rice. *Journal of Experimental Botany* 63(7):2541–2556.

Yang, S.L., Llan, S.S., Gong, M. (2009) Hydrogen peroxide-induced proline and metabolic pathway of its accumulation in maize seedlings. *Journal of Plant Physiology* 166(15):1694–1699.

Yasmeen, A., Basra, S.M.A., Farooq, M., Rehman, Hu, Hussain, N., Athar, HuR. (2013) Exogenous application of Moringa leaf extract modulates the antioxidant enzyme system to improve wheat performance under saline conditions. *Plant Growth Regulation* 69(3):225–233.

Yazici, I., Türkan, I., Sekmen, A.H., Demiral, T. (2007) Salinity tolerance of purslane (*Portulaca oleraceae* L.) is achieved by enhanced antioxidative system, lower level of lipid peroxidation and proline accumulation. *Environmental and Experimental Botany* 61(1):49–57.

Yoo, J.H., Park, C.Y., Kim, J.C., Do Heo, W.D., Cheong, M.S., Park, H.C., Kim, M.C., et al. (2005) Direct interaction of a divergent CaM isoform and the transcription factor, MYB2, enhances salt tolerance in Arabidopsis. *The Journal of Biological Chemistry* 280(5):3697–3706.

Yoshiba, Y., Kiyosue, T., Katagiri, T., Ueda, H., Mizoguchi, T., Yamaguchi-Shinozaki, K., Wada, K., Harada, Y., Shinozaki, K. (1995) Correlation between the induction of a genefor delta 1-pyrroline-5-carboxylate synthetase and the accumulation of proline in *Arabidopsis thaliana* under osmotic stress. *The Plant Journal: for Cell and Molecular Biology* 7(5):751–760.

Yoshihashi, T., Huong, N.T.T., Inatomi, H. (2002) Precursors of 2-acetyl-1-pyrroline, a potent flavor compound of an aromatic rice variety. *Journal of Agricultural and Food Chemistry* 50(7):2001–2004.

Zarattini, M., Forlani, G. (2017) Toward unveiling the mechanisms for transcriptional regulation of proline biosynthesis in the plant cell response to biotic and abiotic stress conditions. *Frontiers in Plant Science* 8:927.

Zhang, C.S., Lu, Q., Verma, D.P.S. (1995) Removal of feedback inhibition of Δ1-pyrroline-5-carboxylate synthetase, a bifunctional enzyme catalyzing the first two steps of proline biosynthesis in plants. *The Journal of Biological Chemistry* 270(35):20491–20496.

Zhang, W., Yang, G., Mu, D., Li, H., Zang, D., Xu, H., Zou, X., Wang, Y. (2016) An ethylene-responsive factor BpERF11 negatively modulates salt and osmotic tolerance in Betula platyphylla. *Scientific Reports* 6:23085.

Zhang, G.C., Zhu, W.L., Gai, J.Y., Zhu, Y.L., Yang, L.F. (2015) Enhanced salt tolerance of transgenic vegetable soybeans resulting from overexpression of a novel D1-pyrroline-5-carboxylate synthetase gene from *Solanum torvum* Swartz. *Horticulture, Environment, and Biotechnology* 56(1):94–104.

Zhu, J.K. (2002) Salt and drought stress signal transduction in plants. *Annual Review of Plant Biology* 53(1):247–273.

Zhu, B., Su, J., Chang, M., Verma, D.P.S., Fan, Y.L., Wu, R. (1998) Overexpression of a Δ1-pyrroline-5-carboxylate synthetase gene and analysis of tolerance to water and salt-stress in transgenic rice. *Plant Science* 139(1):41–48.

6 Phytohormones in Improving Abiotic Stress Tolerance in Plants

Parminder Kaur, Poonam Yadav, Shagun Bali, Vandana Guatum, Sukhmeen Kaur Kohli, Dhriti Kapoor, Saroj Arora, Adarsh Pal Vig, Rajinder Kaur, and Renu Bhardwaj

CONTENTS

6.1 INTRODUCTION

The rapid increase in world population and environmental pollution has resulted in a decline in food security. The need of a substantial increase in food production through agriculture by restoring the sustainability of resources has emerged as a challenging task (Hussain et al., 2015). To cope with the world's demand for food with an ever-growing human population, crop production needs to be increased by 70% (FAO, 2009). Changing environmental conditions (including abiotic and biotic stress) act as limiting factors in the productivity of crops. Plants face several abiotic stresses such as chilling and freezing, high temperature, heat, drought, salinity, heavy metals, etc. (Saud et al., 2013). Plant mechanisms to endure various stresses include a series of cellular and molecular responses. Plants use many strategies to adapt to abiotic stress, which ultimately enhances plant growth and productivity. The response of plants varies with the nature and intensity of stress. Inhibition in germination rate, stunted growth and reduced yield of plants are seen as sudden effects in plants exposed to abiotic stress (Parida and Das, 2005). Adaptation to all these stresses is accompanied by metabolic adjustments that lead to the accumulation of several organic solutes such as sugars, polyols, betaines and proline, protection of cellular machinery, maintenance of ionic homeostasis, scavenging of free radicals, expression of certain proteins and upregulation of their genes beside induction of phytohormones (Munns and Tester, 2008; Parida and Das, 2005; Tuteja, 2007).

Phytohormones are plant growth regulators synthesized in plants, which either act at the same site of synthesis or are transported to some other part of the plant to regulate growth and development processes under normal as well as stressed conditions (Peleg and Blumwald, 2011). Plant hormones such as abscisic acid, gibberellins, ethylene, auxins, cytokinins etc. interact with each other to regulate growth, development, physiological and biochemical processes in plants (Iqbal et al., 2014). Phytohormones thus have an important function in regulating plant response to abiotic stress, by which the plant may endeavor to escape or endure stressful conditions and may result in declined growth so that the plant can focus its resources on withstanding the stress (Skirycz and Inzé, 2010). To meet the demand of food worldwide, crop production needs novel and potent approaches. Engineering of phytohormones proved to be a great method for biotechnologies to enhance crop productivity and stress tolerance in plants. These hormones work as chemical messengers to coordinate various signal transduction pathways during abiotic stress conditions. The challenges caused by abiotic stress and the role of various phytohormones in abiotic stress tolerance, as well as the use of engineering techniques in conferring abiotic stress tolerance in plants have been discussed in this chapter.

6.2 ABIOTIC STRESS: A GLOBAL THREAT TO PLANTS IN A CHANGING ENVIRONMENT

Climate changes are intensified by various anthropogenic activities which have an adverse effect on the ecosystem. These changes are comprised of intense alterations in environmental drivers consisting of elevation in atmospheric CO_2 level and temperatures, changes in precipitation patterns and increase in the level of pollutants which lead to the degradation of the ecosystem (Field et al., 2014). Due to the shifting regime in the environment, plants encounter multiple abiotic stresses, which restrain their growth and yield. It was reported by Goel and Madan (2014) that abiotic stresses caused more than 50% decline in crop production.Drought, salinity, waterlogging, heavy metals, extreme temperatures (chilling, heat and freezing), light (high and low) and radiations (UV-A/B) have become a serious menace to crop yield and quality (Hasanuzzaman et al., 2013; Vardhini and Anjum, 2015). These conditions adversely influence plant growth, photosynthetic activity, pigment content, membrane integrity, osmotic balance and water relations. Under stress, plants confront cellular dehydration and this leads to the formation of reactive oxygen species (ROS), which eventually affect membrane potentials adversely, and macromolecules such as carbohydrates, DNA and proteins (Doltchinkova et al., 2013). Thus, to produce the required amounts of food, crop production must be enhanced numerically. For a better crop yield, plant biotechnologists are investigating various approaches in order to generate abiotic stress-tolerant species. Genetic engineering introduced a new tool for the generation of abiotic stress-tolerant species. Singh et al. (2012) introduced transgenic tobacco, which is tolerant to heavy metal stress and salinity. Singh and his coworkers transferred a gene from rice to tobacco. It's a necessity to investigate how abiotic stresses influence plant growth at every developmental stage and generate new strategies for the improvement of crops.

6.3 PHYTOHORMONES: THEIR ROLE AS MAIN MEDIATORS IN ABIOTIC STRESS IMPROVEMENT

Plants must mediate their growth and development by countering several external and internal stimuli (Wolters and Jurgens, 2009). Plant hormones, a divergent group of signaling molecules present in minute quantities in cells, regulate these responses. Their crucial roles in triggering plant reconciliation to ever-changing environments by regulating growth, source/sink transitions, development and nutrient distribution have been well studied (Fahad et al., 2015). Though plants feedback to abiotic stresses depends on several factors, plant hormones are recognized as the most pivotal endogenous substances for mediating physiological and molecular responses, a vital requirement for plant endurance as sessile organisms (Fahad et al., 2015). They act either at their site of biosynthesis or somewhere else in plants through their subsequent transportation (Peleg and Blumwald, 2011). They consist of auxin (IAA), abscisic acid (ABA), cytokinins (CKs), gibberellins (GAs), ethylene (ET), salicylic acid (SA), brassinosteroids (BRs) and jasmonates (JAs). Various studies showed that phytohormones have ameliorative properties against abiotic stresses. IAA treatment is identified as a significant component of defense responses through regulation of various genes and negotiation of the interaction between biotic and abiotic stress responses (Fahad et al., 2015). IAA is an important, multifaceted plant hormone, and apart from its crucial role in plant growth and development it also mediates responses against stress (Kazan, 2013). IAA enhanced root and shoot growth of wheat and *Brassica napus* plants under salt or cadmium stress (Egamberdieva, 2009; Sheng and Xia, 2006).

ABA is recognized as an important messenger in the adaptive mechanisms against abiotic stress in plants. Under stress, endogenous ABA concentration was enhanced rapidly, stimulating particular signal transduction pathways and modifying expression levels of stress-responsive genes (O'Brien and Benkova, 2013). ABA is concerned in vigorous root growth and architectural modifications in maize under drought stress (Giuliani et al., 2005). It modulates the expression of various stress-responsive genes and the biosynthesis of LEA proteins, dehydrins and various stress-protective proteins (Sreenivasulu et al., 2012; Verslues et al., 2006). ABA is involved in the maintenance of cell turgor pressure and production of osmoprotectants and antioxidant enzymes under drought stress (Chaves et al., 2003). Zhang et al. (2006) reported that ABA concentration was enhanced under salt and drought stress.

CKs are recognized as master regulators and play a significant role in growth and developmental processes in plants (Kang et al., 2012; Nishiyama et al., 2011). Changes in the endogenous concentration of CKs in response to stress highlight their role under abiotic stress such as drought and salinity in *Arabidopsis thaliana* (Kang et al., 2012; Nishiyama et al., 2011; O'Brien and

Benkova, 2013). There was an antagonistic interaction between CKs and ABA (Pospíšilová, 2003). In water-stressed plants, reduction in CK content and accumulation of ABA lead to an enhanced ABA/CK ratio in water-stressed plants. A decreased CK concentration leads to an enhancement in apical dominance and mediates ABA regulation of stomatal aperture, which helps plants acclimatize under drought stress (O'Brien and Benkova, 2013).

GAs have an essential role in abiotic stress response and acclimatization (Colebrook et al., 2014). This study was conducted to explore the role of GAs in *Arabidopsis thaliana* seedlings under osmotic stress (Claeys et al., 2012). GAs crosstalk with other plant hormones during various developmental and stimulus-responsive processes (Munteanu et al., 2014). The interaction between GAs and ET has both an antagonistic and a synergistic nature which depends upon the tissue and signaling cascade (Munteanu et al., 2014).

Apart from the physiological role in plants, JAs mediate plant defense responses against abiotic stress and pathogen attack (Pauwels et al., 2009; Seo et al., 2011). These are important signaling molecules triggered by diverse environmental stresses such as salinity, drought and UV irradiation (Demkura et al., 2010; Du et al., 2013a; Pauwels et al., 2009; Seo et al., 2011). Treatment with Methyl Jasmonate (MeJA) significantly decreased salinity stress in soybean seedlings (Yoon et al., 2009). JAs ameliorated Cd stress in *Capsicum frutescens* var. fasciculatum seedlings by stimulating the antioxidant machinery (Yan et al., 2013).

SA, a naturally occurring phenolic compound found in plants, is involved in the regulation of pathogenesis-related protein transcripts (Miura and Tada, 2014). SA reduced the adverse effects caused by drought stress in Arabidopsis (Miura et al., 2013), salt stress in maize (Khodary, 2004), chilling stress in peach fruit (Yang et al., 2012), heat stress in barley (Fayez and Bazaid, 2014) and heavy metal stress in *Brassica juncea* (Kohli et al., 2018). In *Phillyrea angustifolia*, there was an enhancement in the endogenous concentration of SA under drought conditions (Munne-Bosch and Penuelas, 2003). SA content was increased about two-fold under water-deficit conditions in *Phillyrea angustifolia* (Bandurska and Stroinski, 2005).

BRs are the most influential category among phytohormones. They improve plant growth and regulate various responses against abiotic and biotic stresses (Fariduddin et al., 2013; Houimli et al., 2010). Treatment with 24-epibrassinolide (EBL) improved germination percentage, shoot length and root length in *Oryza sativa* under chromium (Cr) stress (Sharma et al., 2016). The study conducted by Sharma et al. (2017) reported that seed-soaking treatment of EBL significantly lowered oxidative stress and imidacloprid (IMI) residues by modulating the expression of antioxidative defense-related genes in *Brassica juncea* under pesticide (IMI) stress. Phytohormones could be used to generate abiotic stress-tolerant species which will be a promising tool for improving crop yield and may help to achieve the target of food security. The use and role of phytohormones under various abiotic stresses is discussed below.

6.3.1 Extreme Temperature

Low temperature restricts the growth and geographic distribution of plants. Low-temperature stress can be classified into chilling (0–15°C) and freezing, which cause disorders in subtropical and tropical plants by necrosis, chlorosis, changes in enzyme activity, cytoplasm viscosity and membrane damage (Ruelland and Zachowski, 2010). These physiological and biochemical changes induce a series of events leading to changes in gene expression and production of proteins to help with cold stress tolerance. Various phytohormones also play vital roles in cold acclimation by inducing gene expression or activating the defense system. It has been reported that SA biosynthesized growth-promoting and protective molecules and thus decreased the negative impact of chilling stress in common beans, leading to enhanced growth. Six common bean varieties were observed in optimum temperatures (25°C) and cold conditions (15°C). Seed germination and growth of seedlings were found to decrease under cold conditions. Treating seeds with exogenous SA enhanced germination rate in comparison to seeds of untreated control as well as of chilling conditions (Gharib and Hegazi, 2010). JA has been reported to enhance chilling stress tolerance by causing the production of cryoprotective agents, polyamines, ABA, antioxidants and proteinase inhibitors and lowering the activity of lipoxygenase (Cao et al., 2009; Zhao et al., 2013). It has been found that, under cold stress, the level of endogenous JA increases. Studies showed that in cold stressed rice seedlings, genes related to JA biosynthesis were upregulated. It has been found in *Arabidopsis* that ET signaling negatively controls cold stress tolerance (Shi et al., 2012). EIN3, which is a regulator of ethylene signaling pathways, represses the CBF pathway, and thus acts antagonistically to JA. The study of Zhu et al., (2011a) reported that there is crosstalk between ethylene and JA signaling pathways through the interaction of JAZ and EIN3/EIL1. Ethylene is needed for the stabilization of EIN3/EIL1, while JA helps releasing these from JAZ degradation. Thus, the regulation of these two leads to an amelioration of stress-induced changes on plant growth and an increase in stress tolerance. It has been reported that intracellular IAA response controlled by local IAA gradient plays a role in the regulation of plant growth under low-temperature conditions. Intracellular trafficking pathways, which control multiple signaling pathways, regulate protein homeostasis and the interactions of different hormones were found to participate in the regulation of IAA under low-temperature stress (Rahman, 2013). ABA, a naturally occurring growth hormone, reduces oxidative damage due to the enhanced activities of enzymatic and nonenzymatic antioxidants under cold stress. Enhancement in leaf water content and reduction in oxidative stress leading to improvement in cold tolerance has been observed by Kumar et al. (2012). BRs have been reported to be effective in reducing oxidative damage produced under different types of abiotic stress, including high and low-temperature stresses, by activating the antioxidative defense system.

Temperature above the threshold value causes a negative impact on the growth of plants by causing injury or permanent damage. It has been suggested in reports of IPCC (2014) that by the end of the twenty-first century, the global temperature will rise by 4–5 °C. High-temperature stress impairs cell homeostasis by affecting plant processes and cellular machinery (Bokszczanin et al., 2013). Plants produce heat-shock proteins (HSPs) in response to high temperature. ABA is the primary hormone involved in the regulation of plant growth and development (Pospisilova et al., 2009). It has been reported that under heat stress, exogenous application of ABA reduces oxidative stress by activating the defense system and thus helps in redox homeostasis (Ding et al., 2010). Several studies that have reported the effects of various hormones are listed in Table 6.1.

6.3.2 Water Stress (Drought/Flooding)

Drought is regarded as the most dangerous stress, as it limits agricultural production globally. It affects plant growth, membrane integrity, pigment levels, water relations, photosynthetic apparatus and osmotic adjustment (Pathak et al., 2014; Sanghera et al., 2011) Phytohormones, molecules produced in plants, regulate various cellular processes. Various phytohormones have shown positive protective functions against different abiotic stresses (Voß et al., 2014). ABA plays a crucial role in water-deficit conditions and indicates shoots regarding stressful conditions around roots, thus resulting in stomatal closure, reduced antitranspirant activity and decreased leaf expansions (Wilkinson et al., 2012). Root growth and other architectural changes under drought stress have been controlled by ABA (Giuliani et al., 2005). CKs are considered antagonist to ABA. Under drought conditions, CKs content decreases while ABA content was found to increase leading to enhanced ABA/CKs ratio. The decrease in CKs increases apical dominance, which along with ABA regulates stomatal aperture and helps adapt to water-deficit conditions (O'Brien and Benkova, 2013). It has been reported that SA and ABA together take part in the regulation of drought responses (Miura and Tada, 2014). Under water-deficit conditions, in *Phillyrea angustifolia* the endogenous content of SA increased five times (Munne-Bosch and Penuelas, 2003). Similarly, in barley roots SA levels showed a two-fold increase under drought conditions (Bandurska and Stroinski, 2005). SA inducible Pathogenesis-related genes (*PR1* and *PR2*) are induced in water-deficit conditions (Miura et al., 2013). An auxin efflux carrier gene, *OsPIN3t,* has been identified by Zhang et al. (2012) in rice plants. This gene under drought stress helps in auxin transport. The downregulation of these genes causes root abnormalities at seedlings stage and their upregulation enhances the tolerance of plants to drought conditions. Table 6.2 shows the recent studies on phytohormones reported to affect different plants under water stress (drought/flooding) conditions.

Flooding can happen in various ecosystems including salt marshes, dune slacks, bogs, river floodplains, etc. Episodic anaerobic condition is one of the characteristic features of flooding. Shortage of oxygen or hypoxia is a major growth limiting factor in floods in vulnerable regions. Flooding may also cause accumulation of toxic compounds in soil, reduced light availability and induction of free radicals on re-aeration. All these factors affect the growth of plants. Flooding is one of the prominent abiotic factors that affect crop yield worldwide. Few crops can withstand waterlogging for a few hours, while other tolerant crops can endure partial or complete submergence for days or months (Bailey-Serres and Voesenek, 2008). The enhanced production of ethylene is a metabolic response toward flooding in tolerant species (Trinh et al., 2014). It has been noted that plant response to flooding is linked with compartmentalization of enzymes associated with synthesis of ethylene. The enhancement of ACC synthesis occurs in roots followed by the initiation of ACC synthase genes under hypoxia conditions (Grichko and Glick, 2001). Reports showing the effect of various phytohormones on different plant species have been listed in Table 6.2.

6.3.3 Salt

Salinity has emerged as one of the most concerning problems for plants in this changing environment. Soil salinity is increasing at an alarming rate globally, affecting

TABLE 6.1
Effect of Various Plant Hormones on Different Plant Species Under Temperature Stress (Freezing and High-Temperature Stress)

Stress	Plant species	Hormone(s)	Effect on plant system	References
Low temperature stress	*Hordeum vulgare*	SA	Induced activities of antioxidant enzymes, ice nucleation activity and the patterns of apoplastic proteins	Mutlu et al. (2013)
	Triticum aestivum	SA	Decrease in ABA and increase in CKs, IAA and GAs	Kosová et al. (2012)
	Solanum lycopersicum	GAs	Decreased electrolyte leakage, MDA level, increased proline level and improvement in antoxidant enzyme activities	Ding et al. (2015)
	Cucumis melo	ABA	Increased root, shoot length, fresh weight and stem diameter. Also improved endogenous GAs and SA levels	Kim et al. (2016)
	Citrus limon	SA and MeJA	Decreased lipid peroxidation and ROS accumulation, increased total phenolics and phenylalanine ammonia lyase activity (PAL) and inhibited the activity of polyphenol oxidase (PPO) and peroxidase (POD)	Siboza et al. (2014)
	Capsicum annuum	ABA	Increase in the activities o f monodehydroascorbate reductase (DHAR), dehydroascorbate reductase, glutathione reductase, guaiacol peroxidase, ascorbate peroxidase, ascorbate, glutathione and related gene expression	Guo et al. (2012)
	Cucumis sativus	GAs	Reduced lipid peroxidation, increase in SOD, CAT, APX, GPX activity	Li et al. (2011)
	Elymus nutans	ABA	Increase in GSH, AsA, total glutathione, and total ascorbate concentrations, as well as SOD, CAT, APX and GR activities	Fu et al. (2017)
	Cucumis sativus	BRs	Increased photosynthesis, reduced ROS accumulation, increased the activities of SOD and APX.	Hu et al. (2010)
	Triticum aestivum	GAs	Increased seed germination rate, germination index, weight and length of radical and coleoptiles, seed respiration rate while decreasing mean germination time	Li et al. (2013)
	Piper nigrum	BRs	Improvement in photoinhibition and PSII efficiency, increase in the level of proline, soluble sugars and protein and enhanced activity of defense enzymes	Li et al. (2015)
	Oryza sativa	BRs, SA	Reduced MDA content, increased proline content, leaf area, seedling dry weight, seedling g height, increased activity of antioxidative enzymes	Mo et al. (2016)
	Arabidopsis	JA	Plant freezing tolerance, CBF/DREB1 signaling pathway upregulated	Hu et al. (2013a)
Heat	*Solanum lycopersicum*	BRs	Improved recovery of *PN*, stomatal conductance and maximum carboxylation rate of Rubisco, electron transport rate, relative quantum efficiency of PSII photochemistry, photochemical quenching and increased NPQ	Hu et al. (2010)
	Cicer arietinum	ABA	Decrease in lipid peroxidation and hydrogen peroxide content	Kumar et al. (2012)
	Solanum melongena	BRs	Promoted plant growth and photosynthesis, increased antioxidative enzyme activities, proline, soluble sugars and protein levels	Wu et al. (2014)
	Phragmites communis	ABA	Decrease in lipid peroxidation and hydrogen peroxide content, increase in activities of superoxide dismutase, catalase, ascorbate peroxidase and peroxidise	Ding et al. (2010)

(*Continued*)

TABLE 6.1 (CONTINUED)
Effect of Various Plant Hormones on Different Plant Species Under Temperature Stress (Freezing and High-Temperature Stress)

Stress	Plant species	Hormone(s)	Effect on plant system	References
	Cucumis melo	BRs	Enhanced chlorophyll content, net photosynthetic rate, stomatal conductance, water use efficiency, reduced leaf transpiration rate, non-photochemical quenching, thermal dissipation	Zhang et al. (2013a)
	Cucumis melo	BRs	Enhanced chlorophyll content, net photosynthetic rate, stomatal conductance, water use efficiency, reduced leaf transpiration rate, non-photochemical quenching, decreased MDA content and enhanced soluble proteins, proline levels and activities of SOD, POD, CAT, APX	Zhang et al. (2014)
	Leymus chinensis	BRs	Reduced MDA content, increased accumulation of proline, sugars and protein, upregulated activities of SOD, POD, APX, CAT, GR.	Niu et al. (2016)
	Brassica juncea	BRs	Enhanced activity of SOD, CAT, GPOX, APOX, reduced accumulation of MDA, H_2O_2 and NO	Sirhindi et al. (2017)
	Vitis	SA	Increase in photosynthetic rate, Rubisco activation state and improvement in PSII	Wang et al. (2010)
	Oryza sativa	BRs	Maintained net photosynthetic rate, stomatal conductance, water use efficiency	Thussagunpanit et al. (2015)
	Snap Beans	BRs	Increase in growth, yield, quality of pods, free amino acids and phenolic acids level	El-Bassiony et al. (2012)
	Medicago sativa	ABA	Decrease in electrolyte leakage and stomatal conductance, and increase in growth and leaf water potential	An et al. (2014)
	Triticum aestivum	SA	Decrease in 1-aminocyclopropane carboxylic acid (ACC) synthase (ACS) acivity, improvement in proline metabolism, N assimilation and photosynthesis	Khan et al. (2013)
	Arabidopsis	JA	Helps in maintenance of cell viability	Clarke et al. (2009)

more than 10% of productive land and reducing the average yield of major crops by 50% (Wang et al., 2009). Salinity affects plant growth and development by affecting various biochemical reactions and physiological processes. Plants acclimatize to the adverse effects of these salinity conditions by changing with various metabolic processes which result in accumulation of compatible solutes such as sugars, polyphenols, proline, etc. Some phytohormones have also been shown to influence salinity tolerance by modulating several physiological processes (Fatma et al., 2013; Khan et al., 2013; Khan and Khan, 2013; Table 6.3). The endogenous level of various phytohormones shows enhanced tolerance capability in plants as stress-induced proteins are synthesized by various phytohormones (Hamayun et al., 2010). Moreover, exogenous application of phytohormones in managing salt stress by reducing adverse effects has been reported by Iqbal et al. (2012), Sharma et al. (2013), Iqbal and Ashraf (2013a, b) and Amjad et al. (2014).

ABA has been reported as endogenous stress messenger and is often reported to regulate water status in plants by upregulating genes controlling guard cells movement (Narusaka et al., 2003; Zhu, 2000). It was reported by Zhang et al. (2006) that plants under salt stress showed an increased synthesis of ABA generally related to leaf and soil water potential. ABA was also reported to decrease Na^+ and Cl^- content and Na^+/K^+ ratio, increased K^+ and Ca^{2+} content, proline accumulation, soluble sugar content and grain yield in *Oryza sativa* (Gurmani et al., 2013).

IAA is most commonly known for regulating salinity in plants. A genetic link between IAA signaling and salt stress management has been reported by Jung and Park (2011). A membrane-bound transcription factor (*NTM2*) was reported to be responsible for seed germination under salt stress. It was also reported by Park et al. (2011) that the signaling pathway to germination was controlled by upregulation of the *IAA30* gene of *NTM2*. The expression of IAA responsive genes (*AtMEKK1*, *AtRSH3*, *Cat1*, *Fer1*) was reported to be downregulated and overexpression of *NIT1* and *NIT2* was observed in *Arabidopsis thaliana* to maintain the content of IAA in

TABLE 6.2
Role of Various Phytohormones Under Water Stress (Drought and Flooding) in Different Plant Species

Stress	Plant	Hormone(s)	Effect on plant system	References
Drought	*Arabidopsis*	ABA	Upregulation of genes encoding for LHCB (light harvesting chlorophyll binding proteins)	Xu et al. (2012)
	Oryza sativa	ABA	Increased xanthophylls content, Improved PSII efficiency	Du et al. (2010)
	Lycium chinese	ABA	Improvement in NPQ (Non-photo chemical quenching) NAND reduced decrease in PSII efficiency	Guan et al. (2015)
	Arabidopsis	IAA	Increased shoot branching, modified rosette shape, enhanced PSII efficiency, increase in electron transfer rate and photochemical quenching	Tognetti et al. (2010)
	Salvia officinalis	SA	Reduction in chlorophyll levels and induced leaf senescence	Abreu and Munné-Bosch (2008)
	Arabidopsis	SA	Enhanced survival rate, upregulation of photosynthesis related genes	Van Ha et al. (2014)
	Cucumis sativus	BRs	Increased Rubisco activity, enhanced soluble sugars, starch levels, higher NPQ, improved PSII efficiency	Yu et al. (2004)
	Nicotiana tabacum	CKs	Genes related to Chl synthesis, light reactions, the Calvin-Benson cycle and photorespiration, PSII upregulated	Rivero et al. (2010)
	Arabidopsis	SA	Increased photosynthetic rate, maximal efficiency and quantum yield of PSII	He et al. (2014)
	Capsicum annuum	BRs	Improved photosynthesis, increased CO_2 assimilation capacity	Hu et al. (2013b)
	Lycopersicon esculentum	SA	Increase in photosynthetic parameters, chlorophyll content, activities of nitrate reductase, carbonic anhydrase, realative water content and leaf water potential	Hayat et al. (2008)
	Oryza sativa	SA	Increase in photosynthetic pigment content, net photosynthetic rate, water use efficiency, enhanced activity of SOD and perxoidases	Li and Zhang (2012)
Flooding	*Rumex palustris*	ET, IAA and GAs	Shoot elongation	Voesenek et al. (2003)
	Glycine max	ABA	Control of energy conservation via glycolytic system, regulate proteins, cell division cycle 5 protein and transduction	Komatsu et al. (2013)
	Glycine max	GAs	Improvement of secondary metabolism- and cell-related proteins, and proteins involved in protein degradation/synthesis	Won Oh et al. (2014)
	Gylcine max	JA and SA	Increased MDHAR activity	Kamal and Komatsu (2016)
	Morus	ET	Upregulation of genes from AP2/ERF family, *MaERF*-B2-1 and *MaERF*-B2-2	Shang et al. (2014)
	Glycine max	ET	Protein phosphorylation	Yin et al. (2014)

response to salt stress tolerance (Fang and Yang, 2002). Similarly, in *Arabidopsis thaliana,* the gene responsible for IAA signaling, *ARX1,* was observed under salt stress by Tirkayi (2007). CKs act opposite to ABA in regulating major metabolic processes and increasing salt tolerance in plants. A decline in the content of CK, with an alleviation of the adverse effect of salinity on plant growth, has been reported by Barciszewski et al. (2000). However, exogenous application of CKs was reported to increase salt tolerance by reversing the aging effects caused by a decline of ABA content in plants (Iqbal et al., 2006). Similarly, the role of ET in salt tolerance was reported by Khan et al. (2012) and Iqbal et al. (2013). The ET specific receptor gene *NTHK1* was observed to be overexpressed in *Nicotiana tabaccum* on salt exposure suggesting its role in salt tolerance (Cao et al., 2007; Zhang et al., 2001). It was reported by Zhao and Schaller (2004) that change in expression of ET receptors corresponds to salt tolerance in *Arabidopsis.* A reduction in the expression of *ETR1* (an ethylene receptor) levels was observed, which increases the sensitivity of plants to ethylene. *ERF* activation of transcriptional factors of *ERF* improves tolerance

TABLE 6.3
Role of Various Phytohormones Under Salt Stress in Different Plant Species

Plant	Hormone(s)	Effect on plant system	References
Nicotiana tabacum	ET	Improved photosynthetic rate	Wi et al. (2010)
Oryza sativa, Brassica oleracea	ABA	Enhanced PSII efficiency and photochemistry	Zhu et al. (2011b)
Oryza sativa,	IAA	Prevention of photosynthetic pigment loss	Anuradha and Rao (2003)
Solanum melongena	CKs	Enhaced growth, pigment content, photosynthethetic activity and reduced oxidative stress	Wu et al. (2014)
Lycopersicon	ET	Overexpression *JERF1,* (ERF protein), enhanced tolerance of transgenic tobacco to high salt	Zhang et al. (2010)
Brassica juncea	SA	Enhanced photosynthesis, improved activity of nitrate reductase (NR) and ATP sulfurylase (ATPS)	Nazar et al. (2011)
Arabidopsis	SA	Decline in salt-induced K^+ loss via Guard cell Outward Rectifying Channel (GORK)	Jayakannan et al. (2013)
Arabidopsis	BRs	Enhanced antioxidative system	Divi et al. (2010)
Oryza sativa	BRs	Increased proline content	Sharma et al. (2013)
Indica rice	ABA	Upregulation of gene *OsP5CS1* regulating proline accumulation	Sripinyowanich et al. (2013)
Triticum aestivum	Gas	Increase in grain yield and decrease in putrescine (Put) and spermidine (Spd) leaf content	Iqbal and Ashraf (2013b)
Arabidopsis thaliana	ABA	Decline in stress-susceptible transcription factor.	Suzuki et al. (2016)
Solanum lycopersicum L.	Gas	Modifies hormone balance in plants and improves growth	Khalloufi et al. (2017)

to salinity by inhibiting ROS production in response to salt stress (Wu et al., 2008).

GAs is also reported to promote normal growth of *Glycine max* under salt by regulating the levels of various other plant hormones (Hamayun et al., 2010). Co-application of GAs along with N helps overcome salt stress in *Brassica juncea* (Siddiqui et al., 2008). The role of BRs in the induction of salinity tolerance has also been reported by Kagale et al. (2007). It was reported that EBL (epibrassinolide) treatment enhances *Brassica napus* tolerance to salinity by activating the antioxidative defense system in plants (Özdemir et al., 2004). The role and effect of various hormones in salt stress tolerance have been presented in Table 6.3.

6.3.4 Heavy Metal

Heavy metals are natural non-biodegradable constituents of the Earth's crust which accumulate and persist forever in the ecosystem as a result of anthropogenic activities. Since the Industrial Revolution, the concentration of lead, mercury, cadmium, arsenic and zinc has progressively contaminated soil and water resources, triggering considerable losses in plant yields. These issues have become a significant concern of scientific interest (Bücker-Neto et al., 2017). Presently, the contamination of natural ecosystems by heavy metals is a global environmental concern, endangering agricultural systems (Micó et al., 2006). This problem has arisen from the long-term use of untreated wastewater for irrigation in developing countries, leading to amplified concentrations of heavy metals in soils (Lu et al., 2015). This causes a reduction in the maximum genetic potential of plants for growth, development and reproduction.

ABA is associated with tolerance to adverse environmental conditions and its signaling pathway is a central regulator of abiotic stress response in plants (Danquah et al., 2014). The concentration of ABA in plant tissues is known to enhance after exposure to heavy metals, which suggests an involvement of this phytohormone in the induction of protective mechanisms against heavy metal toxicity (Hollenbach et al., 1997). It was found that Cd treatment leads to enhanced endogenous ABA levels in the roots of *Typha latifolia* and *Phragmites australis* (Fediuc et al., 2005), in potato tubers (Stroiński et al., 2010) and also in rice plants (Kim et al., 2014). When mercury (Hg), cadmium (Cd) and copper (Cu) solutions were applied to wheat seeds during germination, ABA levels were found to enhance (Munzuro et al., 2008). In cucumbers, seed germination was reduced and ABA content was increased under Cu^{2+} and zinc (Zn^{2+}) stress (Wang et al., 2014).

Among these extreme environmental situations faced by plants, the interaction between IAA homeostasis and heavy metal toxicity is of great interest. This phytohormone is often described as an important mediator in several aspects of plant growth and development. Despite the detrimental effects of heavy metals in IAA metabolism, it has been reported that exogenous application of these phytohormones can save the endogenous levels of IAA. Exogenous supply of IAA has also been found to improve the growth of *Brassica juncea* exposed to As (Srivastava et al., 2013). In the same way, the application of different levels of L-TRP (a precursor of auxin) to the roots of rice seedlings growing in contaminated soil increased plant growth and yield under Cd stress, when compared to untreated seedlings in Cd-contaminated pots without this auxin precursor (Farooq et al., 2015). Tandon et al. (2015) analyzed six concentrations of two representative natural auxins (IAA and IBA), and a synthetic auxin (1-Naphathaleneacetic acid), in wetland and non-wetland plant species in a water environment. The authors observed that exogenous supply of IAA enhanced phytoremediation efficiency in wastewater treatment. Similarly, Pandey and Gupta (2015) analyzed the effect of the co-application of selenium (Se) and auxin on morphological and biochemical characteristics in rice seedlings exposed to arsenic (As) stress.

Plant steroids (BRs) regulate cell expansion and elongation, photomorphogenesis, flowering, male fertility and seed germination (Mandava, 1988). The heavy metal nickel (Ni) is a very important environmental contaminant. High doses of Ni^{2+} ions can bind to proteins and lipids inducing oxidative damage (Bal and Kasprzak, 2002). Exogenous application of 24- epiBL has been observed to overcome Ni-stress in *Brassica juncea* by increasing the activity of antioxidant enzymes (Kanwar et al., 2013). In the same way, enhanced POD, CAT and SOD activity by exogenous application of 28-homoBL protects wheat against Ni toxicity (Yusuf et al., 2011). Increased antioxidant activity in response to Ni was also found in *Raphanus sativus* and *Vigna radiate* pretreated with 24-epiBL (Yusuf et al., 2012).

The plant growth hormone ethylene plays a significant role in many developmental processes, such as the "triple response" in seedlings. Under heavy metal exposure, the enzymes of ethylene biosynthesis, ACS2 and ACS6, are phosphorylated by MAPKs, which in turn increase their half-life. Both phosphorylated and native ACS forms are functional. However the former is more stable and active compared to the latter (Skottke et al., 2011).

6.3.5 Xenobiotics

Xenobiotics are chemical compounds or other substances which are not found naturally in the ecosystem and are therefore foreign in nature to organisms. Herbicides and pesticides are general xenobiotics which plants generally get exposed to (Croom, 2012). These pesticide and herbicide remains from the agricultural system put human health at risk globally in the absence of proper metabolism. Xenobiotics may get deposited in higher concentrations inside cells, which can be a toxic condition to life forms. A variety of xenobiotics are recognized as factors responsible for ROS production in cells. ROS stop several cellular proteins and genes from working properly, thereby modifying responses and causing oxidative stress (Ramel et al., 2012). Even though the detoxification systems for xenobiotics have been comprehensively explored in mammalian cells, we still need to fully understand the directive set of connections in plants. Different enzymes, their specific cofactors and energy are necessary for the metabolism of those compounds in the cell (Zhou et al., 2015). Xenobiotics metabolism in plants usually presents three stages: transformation, conjugation and compartmentalization (Sandermann, 1992). The enzymes responsible for the metabolism of xenobiotics are conceivably demarcated into step-I, step-II and carrier enzymes. Step-I enzymes make available sites for conjugation reactions, thus metabolizing xenobiotics by transforming them into more polar, more water-soluble, more chemically reactive and occasionally more biologically active derivatives by introducing new, different functional groups in the xenobiotic molecule or the exposure of pre-existing functional groups (Komives and Gullner, 2005). Step-II enzymes carry forward the conjugation process by interacting either directly with xenobiotics or with metabolites produced by step-I enzymes. Finally these metabolites are removed by means of transportation through carrier enzymes. The mode of transport can be either active or passive, or both. Generally, a large number of different complicated pathways and enzymes are involved in the elimination process of xenobiotics. The efficiency of the process depends upon the amount of xenobiotics present, magnitude and kinship of enzymes and accessibility of cofactors in the organism (Croom, 2012). The individual detoxification systems of plants for diminishing xenobiotics residues involve various phytohormones (Varshney et al., 2012). Investigations involving mutants have been widely used to enlighten the fact that xenobiotic responses in plants are associated with hormone pathways. Mutants of IAA signaling pathways have been extensively used for the investigation of auxin-like analogues (Gleason et al., 2011). Herbicides have an effect on genes related to ethylene biosynthesis (Ramel et al., 2007). The phenotypic mutants that match up to the main phases of the signaling pathway of ethylene exposed the fact that diverse xenobiotic reactions were in any case connected to the ethylene pathway to a degree (Weisman et al., 2010). Those phases include the ethylene precursor biosynthesis enzyme, signal transduction proteins, negative regulators and ethylene receptors. Phenanthrene hinders the biosynthesis and signal transduction of ethylene (Weisman et al., 2010). Ethylene pathways also play a vital role in the stimulation of atrazine tolerance in *Arabidopsis* seedlings (Sulmon et al., 2006). The significance of mitogen-activated protein kinases in the stress signaling mechanisms is highlighted by the participation of the Raf-like Ser/Thr kinase *CTR1*(Nakagami et al., 2005). The salicylic acid pathway also participates in the xenobiotics responses in *Arabidopsis* as being independent of the systemic acquired resistance-related gene; in cooperation with TGA factor-dependent and -independent pathways, salicylic acid intervenes in the responses to safeners to a certain extent in the "sid2-2" (salicylic acid deficient) and "sai" (NON-EXPRESSOR of PR-1 (NPR1)-deficient) mutants (Behringer et al., 2011). Xenobiotics stress alleviation through detoxification is carried out by jasmonates and jasmonate-related oxylipins in *Arabidopsis thaliana* (Ramel et al., 2007). Signs of a very robustly decreased sensitivity to xenobiotics were shown by a triple mutant *Arabidopsis thaliana* (fad3-2/fad7-2/fad8) that is unable to generate a precursor of oxylipin (linolenic acid) (Skipsey et al., 2011). GAs single-handedly or in a combination with cytokinin overturned the phytotoxic outcome of the herbicide glyphosate in *Vicia faba* plants (Shaban et al., 1987). In rice plants, brassinosteroids are reported to trim down the injurious impact of herbicides and enhance resistance against pretilachlor, simazine and butachlor (Takematsu et al., 1986). BRs also reduce the residue levels of various pesticides and promote their metabolism in *Cucumis sativus* L. (Xia et al., 2009). Zhou et al. (2015) published the very first genetic proof showing that pesticide dilapidation by the enzymatic system in higher plants is regulated by the phytohormone signaling. The external treatment of brassinosteroids to tomato, rice, tea, broccoli, cucumber, strawberry and other plants decreased pesticide residues by 30%–70%. The pesticides tested were common organophosphorus, organochlorine and carbamate pesticides. A genome-wide microarray analysis was conducted which revealed that 301 genes were co-upregulated by brassinosteroids (naturally occurring plant hormone) and chlorothalonil (fungicide). Those genes contained a set of detoxifying genes encoding cytochrome P-450, oxidoreductase, hydrolase and transferase in tomato plants. The concentration of brassinosteroids was intimately associated to glutathione biosynthesis and redox homeostasis, respiratory burst oxidase 1 (RBOH1)-encoded NADPH oxides-reliant generation of hydrogen peroxide and the action of glutathione-S-transferase. Further, it was demonstrated by gene silencing experiments that pesticide residues were decreased as the brassinosteroids enhanced the metabolism of pesticides via a signaling pathway connecting brassinosteroids-activated hydrogen peroxide generation and cellular redox modification (Zhou et al., 2015). To reduce the risk of accumulation of xenobiotics in plants, the latest scientific approaches, such as genetic engineering and signaling, could be adopted rather than foliar application.

6.3.6 UV-Radiation

Ultraviolet (UV) radiations are present in sunlight in the spectral band between 10 and 400 nm, contributing about 10% of the total sunlight output. Different types of ultraviolet radiations and their wavelength ranges are given in Table 6.4. These are electromagnetic in nature, longer than X-rays but shorter than visible light (Andrady et al., 2006). The concentrations of ultraviolet radiations have increased in the atmosphere because of the depletion of the stratospheric ozone layer caused by anthropogenic activities (Rowland, 2006). Plants cannot avoid exposure to ultraviolet light as they are sessile and require sunlight for photosynthesis. The increased concentrations of ultraviolet radiations cause increased incidences of disturbance in photosynthesis and transpiration; injury to membranes, proteins and DNA;

TABLE 6.4
Different Types of Ultraviolet Radiations and their Wavelength Ranges (Mainster, 2006)

Name	Wavelength(nm)	Remarks
Ultraviolet A (UV-A)	315–400	Long-wave, black light, not absorbed by the ozone layer
Ultraviolet B (UV-B)	280–315	Medium-wave, mostly absorbed by the ozone layer
Ultraviolet C (UV-C)	100–280	Short-wave, germicidal, completely absorbed by the ozone layer and atmosphere
Near ultraviolet (N-UV)	300–400	Visible to birds, insects and fish
Middle ultraviolet (M-UV)	200–300	–
Far ultraviolet (F-UV)	122–200	–

and alteration in development, growth and morphology of plants (Hectors et al., 2007). Plants have developed various defense reactions as there are many secondary metabolites that provide a shield against UV radiations. Plant tolerance against tUV radiations is dependent upon the production and management of UV radiation-shielding metabolites. A vital role of plant hormones in this process has been reported. Abscisic acid (ABA) induces radiation stress tolerance as it considerably increases the concentrations of kaempferol, quercetin and flavonols, which absorb UV-B radiations (Mazid et al., 2011; Tossi et al., 2012). In addition, ABA also activates the plant's defense system against UV radiations stress by regulating processes such as commencement and continuation of seed and bud dormancy, enhancement of Ca^{2+} amount in the cytosol, modification of membrane properties. It also causes an increase in caeffic acid, hydroxycinnamic acid and ferulic acid concentrations, as well as alkalinization of the cytosol. Therefore, ABA plays an important role in activating the plants defence system against UV radiation stress (Berli et al., 2010).

Hectors et al. (2012) reported that the IAA element of the regulatory structure is in command of UV-mediated amassing of flavonoids and UV-induced morphogenesis. UV-B radiation causes down-curling of leaves in *Arabidopsis thaliana.* Fierro et al. (2015) reported that an integral IAA signaling apparatus is essential for the emergence of leaf curling phenotypes in UV-B light as epinasty depends upon changed auxin allocation, keeping auxin at the leaf-blade boundaries in the presence of UV-B. Foliar spray of SA improved growth and oil production in *Thymus vulgaris* L. and *Thymus daenensis* Celak plants under UV stress (Yadegari, 2017).

6.4 THE INTERACTION BETWEEN DIFFERENT PHYTOHORMONES DURING ABIOTIC STRESS

Different phytohormones interact with each other for signaling responses in plants defense against environmental stress. The point of convergence in signal transduction of various hormones is considered as crosstalk, and the interaction forms a signaling network. It has been shown in various reports that the interaction between various plant hormones and their respective regulating functions helps improve stress tolerance in plants (Kohli et al., 2013). Out of the various plant hormones, ABA, JA and SA are considered as the most important in the regulation of the signaling response. ABA is considered as the prime controller of various responses. It is involved in the generation of a plant's sudden response to abiotic stress (Liang et al., 2014). The closing of stomata as an abrupt change is controlled primarily by ABA. The latter also regulates various long-term development processes such as growth and stomatal movement by controlling the expression of genes. Under water-deficient processes, ABA interacts with various other plant hormones such as JA, SA, NO and BRs to promote the closing of stomata and maintenance of osmotic balance (Hossain et al., 2011). However, antagonistic reports were given by Daszkowska-Golec and Szarejko (2013), who affirmed that stomatal opening is enhanced by CKs and IAA while leaf senescence is inhibited by GAs, CKs and IAA.

ABA also regulates the gene expression that includes the initiation of the response of ET, CKs and IAA. JA synthesized under stress is also reported to interact with ABA for the closing of stomata either by inhibiting the entrance of external Ca^{+2} or by activating the signaling of H_2O_2/NO. Leaf senescence is also promoted by ET and it acts opposite to ABA in regulating stomatal opening and growth under water-deficient conditions (Tanaka et al., 2005; Yin et al., 2015). Under flooding conditions, ABA and ethylene act opposite to each other to control plant response (Voesenek and Bailey-Serres, 2015). The ET regulated response to flooding is simultaneously controlled by GAs, BRs and IAA (Ayano et al., 2014; Cox et al., 2006; van Veen et al., 2013). These reports indicated that a plant's response to abiotic stress is controlled by interactions among various hormones as shown in Figure 6.1.

6.5 ENGINEERING PHYTOHORMONES FOR PRODUCING ABIOTIC STRESS-TOLERANT PLANTS

The role of various hormones as main mediators of plant growth and development process in response to abiotic processes has been well established (Sreenivasulu et al., 2012). The function of endogenous as well as exogenous phytohormones in enhancing plant tolerance has become evident. The use of mutants and transgenic phytohormones in genetic engineering for the successful manipulation of plant transformation methods helps improve the response of plants to various stress conditions. The genetic engineering techniques available for transferring genes include the improvement of specific regulatory components either by direct expression or through the use of stress-inducible promoters (Hardy, 2010). The other techniques involved in the manipulation factors would only affect the subset of stress-responsive genes (Umezawa et al., 2006). The role of ABA as the main regulator of stress tolerance has been transgenically investigated. It was reported by Park et al. (2008) that

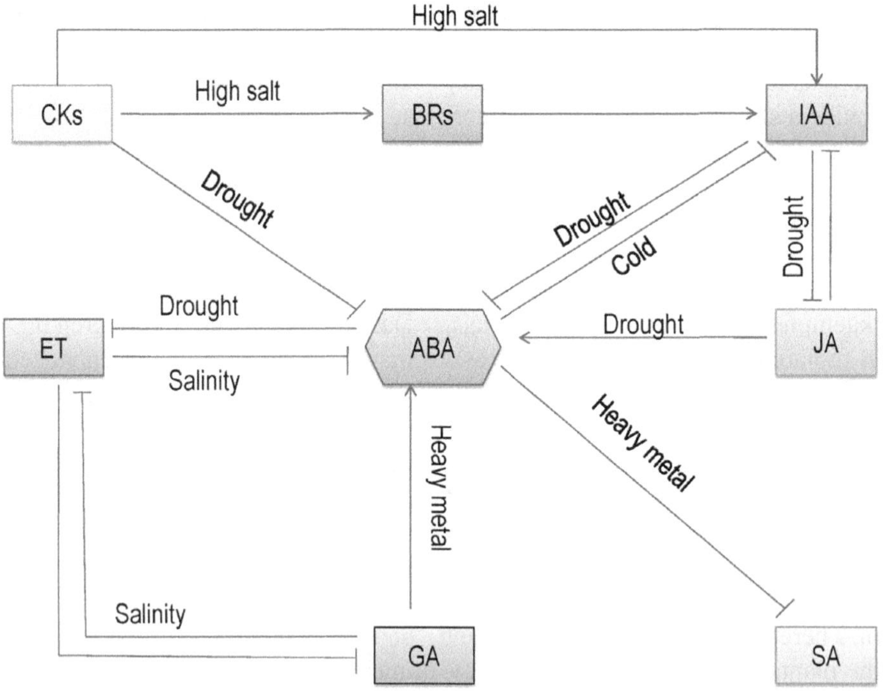

FIGURE 6.1 Interactions of various hormones in plant during abiotic stress tolerance.

Arabidopsis showed increased tolerance to abiotic stress by upregulating ABA-responsive stress-regulated genes. The expression of dehydration-responsive element binding (DREB) and C-repeat binding factor (CBF) has been changed to maintain high tolerance to drought stress. Increased tolerance in *Lycopersicon esculentum* for stress has been observed due to upregulation of *DREB/CRF1* genes from *Arabidopsis*. In soybean plants, the overexpression of the *MoCo* sulfurase gene results in enhanced production of biomass and yield with better drought tolerance. This caused an accumulation of ABA, with declined water loss and activation of the antioxidative defense system.

The limitations of using a transgenic approach for enhanced tolerance have been reported in some studies. The overexpression of genes involved in ABA biosynthesis and enhanced tolerance to drought results in abnormal growth due to induction of promoters (Hwang et al., 2010). Overexpression of *CRK45*, a stress-inducible kinase, involved in ABA signaling was overexpressed in Arabidopsis transgenic plants which showed better tolerance to drought. The upregulation of *CRK45* indicates its role in controlling the levels of ABA in plants (Zhang et al., 2013b). Similar results were observed in *Agrostis stolonifera* where *IPT* (isopenetenyltransferase) controlling CK synthesis was expressed in the presence of a promoter to ignore the pleiotropic effects. This leads to enhancement in the content of CK along with activation of antioxidants to scavenge free radicals, better growth and enhanced yield (Xu et al., 2016). The increased drought tolerance was also observed in rice mutants, *CONSTITUTIVELY WILTED1*-deficent in *YUCCA* homolog (*YUCCA is* member of flavinmonooxygenase-like proteins involved in the tryptophan-dependent IAA synthesis pathway). The activation of *YUCCA7* in *Arabidopsis* showed enhanced drought tolerance (Lee et al., 2012). A tolerance to drought in potato plants due to overexpression of *AtYUC6* was also reported (Im Kim et al., 2013). The expression of various genes in making plants with engineered phytohormones synthesis in various studies is shown below in Table 6.5.

6.6 CONCLUSION

Overall, it has been well documented that various phytohormones and their interactions help maintain sustainable yield and growth of plants in changing environmental conditions. Phytohormones interplay with each other to form a signaling network which provides necessary adaptations of plants toward abiotic stress tolerance. The generation of engineered plants by altering the gene expression of phytohormones provides a promising insight into the future. This approach shows the potential of the use of phytohormones in developing crops with tolerance to various abiotic stresses.

TABLE 6.5
Gene Expression of Various Plants in Synthesis of Engineering Phytohormones

Plant	Hormone	Gene	Expression	Function	Effect on Plant	References
Hordeum vulgare	ABA	*AtLOS5*	Overexpression	ABA Biosynthesis	Salt stress tolerane	Zhang et al. (2016a)
Petunia	ABA	*NCED*	Overexpression	ABA Biosynthesis	Increased level of ABA and drought tolerance	Estrada-Melo et al. (2015)
Medicago Sativa	ABA	*MsZEP*	Overexpression	ABA Biosynthesis	Salt and drought tolerance	Zhang et al. (2016b)
Triticum aestivum	ABA	*SnRK2.4*	Overexpression	Important in ABA signaling	Induce salt, drought and chilling stress tolerance	Mao et al. (2010)
Arabidopsis	IAA	*YUCCA6*	Overexpression	IAA Biosysnthesis	Higher IAA content and drought tolerance	Im Kim et al. (2013)
Oryza sativa	IAA	*OsIAA6*	Overexpression	Part of IAA gene family	Auxin regulation and biosynthesis	Jung et al. (2015)
Solanum Lycopersicum	CKs	*SlIPT3*	Overexpression		Increased salinity tolerance	Žižková et al. (2015)
Arabidopsis	CKs	*CKX*	Overexpression	CK dehydrogenase	Increased drought tolerance	Werner et al. (2010)
Hordeum vulgare	CKs	*AtCKX1*	Overexpression	CK dehydrogenase	Enhanced drought tolerance	Pospíšilová, et al. (2016)
Oryza sativa	CKs	*ERF-1*	Overexpression	Response factor for jasmonates and ethylene	Increased drought tolerance	Du et al. (2013b)
Zea mays	ET	*ACC synthase*	Gene silencing	Ethylene-biosynthesis	Better drought tolerance	Habben et al. (2014)
Arabidopsis thaliana	ET	*ZmARGOS*	Overexpression	Negative regulator ethylene signal transduction	Better drought tolerance	Shi et al. (2015)
Arabidopsis thaliana	BRs	*AtHSD1*	Overexpression	BR biosynthesis	Increases growth and salinity tolerance	Li et al. (2007)
Brachypodium distachyon	BRs	*BdBR1*	Downregulation	BR-receptor	Enhanced growth tolerance	Feng et al. (2015)

REFERENCES

Abreu, M. E., Munné-Bosch, S. (2008) Salicylic acid may be involved in the regulation of drought-induced leaf senescence in perennials: A case study in field-grown *Salvia officinalis* L. plants. *Environmental and Experimental Botany* 64(2): 105–112.

Amjad, M., Akhtar, J., Anwar-ul-Haq, M., Yang, A., Akhtar, S. S., Jacobsen, S. (2014) Integrating role of ethylene and ABA in tomato plants adaptation to salt stress. *Scientia Horticulturae* 172: 109–116.

An, Y., Zhou, P., Liang, J. (2014) Effects of exogenous application of abscisic acid on membrane stability, osmotic adjustment, photosynthesis and hormonal status of two Lucerne (*Medicago sativa* L.) genotypes under high temperature stress and drought stress. *Crop and Pasture Science* 65(3): 274–286.

Andrady, A., Aucamp, P. J., Bais, A. F. (2006) Environmental effects of ozone depletion and its interactions with climate change: progress report. *Photochemical and Photobiological Sciences* 5: 13–24.

Anuradha, S., Rao, S. S. R. (2003) Application of brassinosteroids to rice seeds (*Oryza sativa* L.) reduced the impact of salt stress on growth, prevented photosynthetic pigment loss and increased nitrate application of brassinosteroids to rice seeds (*Oryza sativa* L.) reduced the impact of salt stress on growth, prevented photosynthetic pigment loss and increased nitrate reductase activity. *Plant Growth Regulation* 40(1): 29–32.

Ayano, M., Kani, T., Kojima, M., Sakakibara, H., Kitaoka, T., Kuroha, T., Angeles-Shim, R. B., *et al.* (2014) Gibberellin biosynthesis and signal transduction is essential for internode elongation in deepwater rice. *Plant, Cell & Environment* 37(10): 2313–2324.

Bailey-Serres, J., Voesenek, L. A. C. J. (2008) Flooding stress: Acclimations and genetic diversity. *Annual Review of Plant Biology* 59: 313–339.

Bal, W., Kasprzak, K. S. (2002) Induction of oxidative DNA damage by carcinogenic metals. *Toxicology Letters* 127(1–3): 55–62.

Bandurska, H., Stroinski, A. (2005) The effect of salicylic acid on barley response to water deficit. *Acta Physiologiae Plantarum* 27(3): 379–386.

Barciszewski, J., Siboska, G., Rattan, S. I. S., Clark, B. F. C. (2000) Occurrence, biosynthesis and properties of kinetin (N6-furfuryladenine). *Plant Growth Regulation* 32(2/3): 257–265.

Behringer, C., Bartsch, K., Schaller, A. (2011) Safeners recruit multiple signalling pathways for the orchestrated induction of the cellular xenobiotic detoxification machinery in Arabidopsis. *Plant, Cell and Environment* 34(11): 1970–1985.

Berli, F. J., Moreno, D., Piccoli, P., Hespanhol-Viana, L., Silva, M. F., Bressan-Smith, R., Cavagnaro, J. B., Bottini, R. (2010) Abscisic acid is involved in the response of grape (*Vitis vinifera* L.) cv. Malbec leaf tissues to ultraviolet-B radiation by enhancing ultraviolet absorbing compounds, antioxidant enzymes and membrane sterols. *Plant, Cell & Environment* 33(1): 1–10.

Bokszczanin, K.L., Fragkostefanakis, S., Bostan, H., Bovy, A., Chaturvedi, P., Chiusano, M.L., Firon, N., Iannacone, R., Jegadeesan, S., Klaczynskid, K., Li, H. (2013) Perspectives on deciphering mechanisms underlying plant heat stress response and thermotolerance. *Frontiers in Plant Science* 4: 315.

Bücker-Neto, L., Paiva, A.L.S., Machado, R.D., Arenhart, R.A., Margis-Pinheiro, M. (2017) Interactions between plant hormones and heavy metals responses. *Genetics and Molecular Biology* 40 (1): 373–386.

Cao, W. H., Liu, J., He, X. J., Mu, R. L., Zhou, H. L., Chen, S. Y., Zhang, J. S. (2007) Modulation of ethylene responses affects plant salt stress responses. *Plant Physiology* 143(2): 707–719.

Cao, S., Zheng, Y., Wang, K., Jin, P., Rui, H. (2009) Methyl jasmonate reduces chilling and enhances antioxidant enzyme activity in postharvest loquat fruit. *Food Chemistry* 115(4): 1458–1463.

Chaves, M. M., Maroco, J. P., Pereira, J. S. (2003) Understanding plant responses to drought – from genes to the whole plant. *Functional Plant Biology* 30(3): 239–264.

Claeys, H., Skirycz, A., Maleux, K., Inzé, D. (2012) DELLA signaling mediates stress-induced cell differentiation in Arabidopsis leaves through modulation of anaphase-promoting complex/cyclosome activity. *Plant Physiology* 159(2): 739–747.

Clarke, S. M., Cristescu, S. M., Miersch, O., Harren, F. J., Wasternack, C., Mur, L. A. (2009) Jasmonates act with salicylic acid to confer basal thermotolerance in *Arabidopsis thaliana. The New Phytologist* 182(1): 175–187.

Colebrook, E. H., Thomas, S. G., Phillips, A. L., Hedden, P. (2014) The role of gibberellin signaling in plant responses to abiotic stress. *The Journal of Experimental Biology* 217(1): 67–75.

Cox, M. C., Peeters, A. J., Voesenek, L. A. (2006) The stimulating effects of ethylene and auxin on petiole elongation and on hyponastic curvature are independent processes in submerged *Rumex palustris. Plant, Cell & Environment* 29(2): 282–290.

Croom, E. (2012) Metabolism of xenobiotics of human environments. *Progress in Molecular Biology and Translational Science* 112: 31–88.

Danquah, A., de Zelicourt, A., Colcombet, J., Hirt, H. (2014) The role of ABA and MAPK signaling pathways in plant abiotic stress responses. *Biotechnology Advances* 32(1): 40–52.

Daszkowska-Golec, A., Szarejko, I. (2013) Open or close the gate—Stomata action under the control of phytohormones in drought stress conditions. *Frontiers in Plant Science* 4: 138.

Demkura, P. V., Abdala, G., Baldwin, I. T., Ballaré, C. L. (2010) Jasmonate-dependent and-independent pathways mediate specific effects of solar ultraviolet B radiation on leaf phenolics and antiherbivore defense. *Plant Physiology* 152(2): 1084–1095.

Ding, Y., Sheng, J., Li, S., Nie, Y., Zhao, J., Zhu, Z., Tang, Z., Tang, X. (2015) The role of gibberellins in the mitigation of chilling injury in cherry tomato (*Solanum lycopersicum* L.) fruit. *Postharvest Biology and Technology* 101: 88–95.

Ding, W., Song, L., Wang, X., Bi, Y. (2010) Effect of abscisic acid on heat stress tolerance in the calli from two ecotypes of *Phragmites communis. Biologia Plantarum* 54(4): 607–613.

Divi, U. K., Rahman, T., Krishna, P. (2010) Brassinosteroid-mediated stress tolerance in Arabidopsis shows interaction with absicisic acid, ethylene and salicylic acid pathways. BMC Plant Biology 10: 151.

Doltchinkova, V., Angelova, P., Ivanova, E., Djilianov, D., Moyankova, D., Konstantinova, T., Atanassov, A. (2013) Surface electric charge of thylakoid membrane from genetically modified tobacco plants under freezing stress. *Journal of Photochemistry and Photobiology B, Biology* 119: 22–30.

Du, H., Liu, H., Xiong, L. (2013a) Endogenous auxin and jasmonic acid levels are differentially modulated by abiotic stresses in rice. *Frontiers in Plant Science* 4: 397.

Du, H., Wang, N., Cui, F., Li, X., Xiao, J., Xiong, L. (2010) Characterization of the β-carotene hydroxylase gene DSM2 conferring drought and oxidative stress resistance by increasing xanthophylls and abscisic acid synthesis in rice. *Plant Physiology* 154(3): 1304–1318.

Du, Z., Li, H., Wei, Q., Zhao, X., Wang, C., Zhu, Q., Yi, X., Xu, W., Liu, X.S., Jin, W., Su, Z. (2013b) Genome-wide analysis of histone modifications: H3K4me2, H3K4me3, H3K9ac, and H3K27ac in Oryza sativa L. Japonica. *Molecular Plant* 6(5): 1463–1472.

Egamberdieva, D. (2009) Alleviation of salt stress by plant growth regulators and IAA producing bacteria in wheat. *Acta Physiologiae Plantarum* 31(4): 861–864.

El-Bassiony, A. M., Ghoname, A. A., El-Awadi, M. E., Fawzy, Z. F., Gruda, N. (2012) Ameliorative effects of brassinosteroids on growth and productivity of Snap Beans grown under high temperature. *Gesunde Pflanzen* 64(4): 175–182. 10.1007/s10343-012-0286-x.

Estrada-Melo, A. C., Reid, M. S., Jiang, C. Z. (2015) Overexpression of an ABA biosynthesis gene using a stress-inducible promoter enhances drought resistance in petunia. *Horticulture Research* 2: 15013.

Fahad, S., Hussain, S., Bano, A., Saud, S., Hassan, S., Shan, D., Khan, F. A., *et al.* (2015) Potential role of phytohormones and plant growth-promoting rhizobacteria in abiotic stresses: Consequences for changing environment. *Environmental Science and Pollution Research International* 22(7): 4907–4921.

Fang, B., Yang, L. J. (2002) Evidence that the auxin signaling pathway interacts with plant stress response. Acta Botanica Sinica 44: 532–536.

Fariduddin, Q., Khalil, R. R., Mir, B. A., Yusuf, M., Ahmad, A. (2013) 24-Epibrassinolide regulates photosynthesis, antioxidant enzyme activities and proline content of *Cucumis sativus* under salt and/or copper stress. *Environmental Monitoring and Assessment* 185(9): 7845–7856.

Farooq, H., Asghar, H. N., Khan, M. Y., Saleem, M., Zahir, Z. A. (2015) Auxin-mediated growth of rice in cadmium-contaminated soil. *Turkish Journal of Agricultural Engineering* 39: 272–276.

Fatma, M., Khan, M. I. R., Masood, A., Khan, N. A. (2013) Coordinate changes in assimilatory sulphate reduction are correlated to salt tolerance: Involvement of phytohormones. *Annual Review and Research in Biology* 3: 267–295.

Fayez, K. A., Bazaid, S. A. (2014) Improving drought and salinity tolerance in barley by application of salicylic acid and potassium nitrate. *Journal of the Saudi Society of Agricultural Sciences* 13(1): 45–55.

Fediuc, E., Lips, S. H., Erdei, L. (2005) O-acetylserine (thiol) lyase activity in Phragmites and Typha plants under cadmium and NaCl stress conditions and the involvement of ABA in the stress response. *Journal of Plant Physiology* 162(8): 865–872.

Feng, Y., Yin, Y., Fei, S. (2015) Down-regulation of BdBRI1, a putative brassinosteroid receptor gene produces a dwarf phenotype with enhanced drought tolerance in Brachypodium distachyon. *Plant Science: an International Journal of Experimental Plant Biology* 234: 163–173.

Field, C. B. (2014) *Climate Change 2014: Impacts, Adaptation, and Vulnerability. Part A: Global and Sectoral Aspects. Contribution of Working Group II to the Fifth Assessment Report of the Intergovernmental Panel on Climate Change.* Cambridge: Cambridge University Press.

Fierro, A. C., Leroux, O., De Coninck, B., Cammue, B. P., Marchal, K., Prinsen, E., Van Der Straeten, D., Vandenbussche, F. (2015) Ultraviolet-B radiation stimulates downward leaf curling in *Arabidopsis thaliana*. *Plant Physiology and Biochemistry: PPB* 93: 9–17.

Food and Agricultural Organization (FAO) (2009) Land and Plant Nutrition Management Service, www.fao.org/wsfs/forum2050/

Fu, J., Wu, Y., Miao, Y., Xu, Y., Zhao, E., Wang, J., Sun, H., *et al.* (2017) Improved cold tolerance in Elymus nutans by exogenous application of melatonin may involve ABA-dependent and ABA-independent pathways. *Scientific Reports* 7: 39865.

Gharib, F. A., Hegazi, A. Z. (2010) Salicylic acid ameliorates germination, seedling growth, phytohormone and enzymes activity in bean (*Phaseolus vulgaris* L.) under cold stress. Journal of American Science 6: 675–683.

Giuliani, S., Sanguineti, M. C., Tuberosa, R., Bellotti, M., Salvi, S., Landi, P. (2005) Root-ABA1 a major constitutive QTL affects maize root architecture and leaf ABA concentration at different water regimes. *Journal of Experimental Botany* 56(422): 3061–3070.

Gleason, C., Foley, R. C., Singh, K. B. (2011) Mutant analysis in Arabidopsis provides insight into the molecular mode of action of auxinic herbicide dicamba. *PLOS ONE* 6: e17245.

Goel, S., Madan, B. (2014) Genetic engineering of crop plants for abiotic stress tolerance. In *Emerging Technologies and Managements of Crop Stress Tolerance*, eds. Ahmed, P., Rasool S., *Vol. 1.* Amsterdam: Elsevier.

Grichko, V. P., Glick, B. R. (2001) Ethylene and flooding stress in plants. *Plant Physiology and Biochemistry* 39(1): 1–9.

Guan, C., Ji, J., Zhang, X., Li, X., Jin, C., Guan, W., Wang, G. (2015) Positive feedback regulation of a Lycium chinense-derived VDE gene by drought-induced endogenous ABA, and over-expression of this VDE gene improve drought-induced photo-damage in Arabidopsis. Journal of Plant Physiology 175: 26–36.

Guo, W. L., Chen, R. G., Gong, Z. H., Yin, Y. X., Ahmed, S. S., He, Y. M. (2012) Exogenous abscisic acid increases antioxidant enzymes and related gene expression in pepper (*Capsicum annuum*) leaves subjected to chilling stress. *Genetics and Molecular Research: GMR* 11(4): 4063–4080.

Gurmani, A. R., Bano, A., Ullah, N., Khan, H., Jahangir, M., Flowers, T. J. (2013) Exogenous abscisic acid (ABA) and silicon (Si) promote salinity tolerance by reducing sodium (Na^+) transport and bypass flow in rice ('*Oryza sativa*' indica). *Australian Journal of Crop Science* 7(9): 1219.

Habben, J. E., Bao, X., Bate, N. J., DeBruin, J. L., Dolan, D., Hasegawa, D., Helentjaris, T. G., *et al.* (2014) Transgenic alteration of ethylene biosynthesis increases grain yield in maize under field drought-stress conditions. *Plant Biotechnology Journal* 12(6): 685–693.

Hamayun, M., Khan, S. A., Khan, A. L., Shin, J. H., Ahmad, B., Shin, D. H., Lee, I. J. (2010) Exogenous gibberellic acid reprograms soybean to higher growth and salt stress tolerance. *Journal of Agricultural and Food Chemistry* 58(12): 7226–7232.

Hardy, A. (2010) Candidate stress response genes for developing commercial drought tolerant crops. *Basic Biotechnology* 6: 54–58.

Hasanuzzaman, M., Gill, S. S., Fujita, M. (2013) Physiological role of nitric oxide in plants grown under adverse environmental conditions. In *Plant acclimation to environmental stress*, eds. Gill, S. S., Tuteja, N., 269–322. New York: Springer.

Hayat, S., Hasan, S. A., Fariduddin, Q., Ahmad, A. (2008) Growth of tomato (*Lycopersicon esculentum*) in response to salicylic acid under water stress. *Journal of Plant Interactions* 3(4): 297–304.

He, Q., Zhao, S., Ma, Q., Zhang, Y., Huang, L., Li, G., Hao, L. (2014) Endogenous salicylic acid levels and signaling positively regulate Arabidopsis response to polyethylene glycol-simulated drought Stress. *Journal of Plant Growth Regulation* 33(4): 871–880.

Hectors, K., Prinsen, E., De Coen, W., Jansen, M. A., Guisez, Y. (2007) *Arabidopsis thaliana* acclimated to low dose rates of ultraviolet B radiation show specific changes in morphology and gene expression in the absence of stress symptoms. *The New Phytologist* 175(2): 255–270.

Hectors, K., van Oevelen, S., Guisez, Y., Prinsen, E., Jansen, M. A. (2012) The phytohormone auxin is a component of the regulatory system that controls UV-mediated accumulation of flavonoids and UV-induced morphogenesis. *Physiologia Plantarum* 145(4): 594–603.

Hollenbach, B., Schreiber, L., Hartung, W., Dietz, K. J. (1997) Cadmium leads to stimulated expression of the lipid transfer protein genes in barley: Implications for the involvement of lipid transfer proteins in wax assembly. *Planta* 203(1): 9–19.

Hossain, M. A., Munemasa, S., Uraji, M., Nakamura, Y., Mori, I. C., Murata, Y. (2011) Involvement of endogenous abscisic acid in methyl jasmonate-induced stomatal closure in Arabidopsis. *Plant Physiology* 156(1): 430–438.

Houimli, S. I. M., Denden, M., Mouhandes, B. D. (2010) Effects of 24-epibrassinolide on growth, chlorophyll, electrolyte leakage and proline by pepper plants under NaCl-stress. *EurAsian Journal of Biosciences* 4: 96–104.

Hu, Y., Jiang, L., Wang, F., Yu, D. (2013a) Jasmonate regulates the inducer of CBF expression-C-repeat binding factor/DRE binding factorl cascade and freezingtolerance in *Arabidopsis. The Plant Cell* 25(8): 2907–2924.

Hu, W. H., Wu, Y., Zeng, J. Z., He, L., Zeng, Q. M. (2010) Chill-induced inhibition of photosynthesis was alleviated by 24-epibrassinolide pretreatment in cucumber during chilling and subsequent recovery. *Photosynthetica* 48(4): 537–544.

Hu, W. H., Yan, X. H., Xiao, Y. A., Zeng, J. J., Qi, H. J., Ogweno, J. O. (2013b) 24-Epibrassinosteroid alleviate drought-induced inhibition of photosynthesis in *Capsicum annuum. Scientia Horticulturae* 150: 232–237.

Hussain, S., Peng, S., Fahad, S., Khaliq, A., Huang, J., Cui, K., Nie, L. (2015) Rice management interventions to mitigate greenhouse gas emissions: A review. *Environmental Science and Pollution Research International* 22(5): 3342–3360.

Hwang, S. G., Chen, H. C., Huang, W. Y., Chu, Y. C., Shii, C. T., Cheng, W. H. (2010) Ectopic expression of rice OsNCED3 in Arabidopsis increases ABA level and alters leaf morphology. *Plant Science* 178(1): 12–22.

Im Kim, J. I., Baek, D., Park, H. C., Chun, H. J., Oh, D. H., Lee, M. K., Cha, J. Y., *et al.* (2013) Overexpression of Arabidopsis YUCCA6 in potato results in high-auxin developmental phenotypes and enhanced resistance to water deficit. *Molecular Plant* 6(2): 337–349.

IPCC (2014) Summary for policymakers. In *Climate Change 2014: Impacts, Adaptation, and Vulnerability. Part A: Global and Sectoral Aspects. Contribution of Working Group II to the Fifth Assessment Report of the Intergovernmental Panel on Climate Change*, eds. Field, C. B., Barros, V. R., Dokken, D. J., Mach, K. J., Mastrandrea, M. D., Bilir, T. E., Chatterjee, M., Ebi, K. L., *et al.* 1–32. Cambridge, UK; New York, NY: Cambridge University Press.

Iqbal, M., Ashraf, M. (2006) Wheat seed priming in relation to salt tolerance: growth, yield and levels of free salicylic acid and polyamines. In *Annales Botanici Fennici*. 250–259. Finnish Zoological and Botanical Publishing Board.

Iqbal, M., Ashraf, M. (2013a) Gibberellic acid mediated induction of salt tolerance in wheat plants: Growth, ionic partitioning, photosynthesis, yield and hormonal homeostasis. *Environmental and Experimental Botany* 86: 76–85.

Iqbal, M., Ashraf, M. (2013b) Salt tolerance and regulation of gas exchange and hormonal homeostasis by auxin-priming in wheat. *Pesquisa Agropecuária Brasileira* 48(9): 1210–1219.

Iqbal, N., Masood, A., Khan, N. A. (2012) Phytohormones in salinity tolerance: Ethylene and gibberellins cross talk. In *Phytohormones and Abiotic Stress Tolerance in Plants*, eds. Khan, N. A., Nazar, R., Iqbal N., Anjum, N. A., 77–98. Berlin: Springer.

Iqbal, N., Umar, S., Khan, N. A., Khan, M. I. R. (2014) A new perspective of phytohormones in salinity tolerance: Regulation of proline metabolism. *Environmental and Experimental Botany* 100: 34–42.

Jayakannan, M., Bose, J., Babourina, O., Rengel, Z., Shabala, S. (2013) Salicylic acid improves salinity tolerance in Arabidopsis by restoring membrane potential and preventing salt-induced K^+ loss via a GORK channel. *Journal of Experimental Botany* 64(8): 2255–2268.

Jung, H., Lee, D. K., Do Choi, Y. D., Kim, J. K. (2015) OsIAA6, a member of the rice Aux/IAA gene family, is involved in drought tolerance and tiller outgrowth. *Plant Science: an International Journal of Experimental Plant Biology* 236: 304–312.

Jung, J. H., Park, C. M. (2011) Auxin modulation of salt stress signaling in Arabidopsis seed germination. *Plant Signaling & Behavior* 6(8): 1198–1200.

Kagale, S., Divi, U. K., Krochko, J. E., Keller, W. A., Krishna, P. (2007) Brassinosteroid confers tolerance in *Arabidopsis thaliana* and *Brassica napus* to a range of abiotic stresses. *Planta* 225(2): 353–364.

Kamal, A. H. M., Komatsu, S. (2016) Jasmonic acid induced protein response to biophoton emissions and flooding stress in soybean. *Journal of Proteomics* 133: 33–47.

Kang, N. Y., Cho, C., Kim, N. Y., Kim, J. (2012) Cytokinin receptor-dependent and receptor-independent pathways in the dehydration response of *Arabidopsis thaliana*. *Journal of Plant Physiology* 169(14): 1382–1391.

Kanwar, M. K., Bhardwaj, R., Chowdhary, S. P., Arora, P., Sharma, P., Kumar, S. (2013) Isolation and characterization of 24-Epibrassinolide from *Brassica juncea* L. and its effects on growth, Ni ion uptake, antioxidant defence of *Brassica* plants and in vitro cytotoxicity. *Acta Physiologiae Plantarum* 35(4): 1351–1362.

Kazan, K. (2013) Auxin and the integration of environmental signals into plant root development. *Annals of Botany* 112(9): 1655–1665.

Khalloufi, M., Martínez-Andújar, C., Lachaâl, M., Karray-Bouraoui, N., Pérez-Alfocea, F., Albacete, A. (2017) The interaction between foliar GA 3 application and arbuscular mycorrhizal fungi inoculation improves growth in salinized tomato (*Solanum lycopersicum* L.) plants by modifying the hormonal balance. *Journal of Plant Physiology* 214: 134–144.

Khan, M. I. R., Khan, N. A. (2013) Salicylic acid and jasmonates: Approaches in abiotic stress tolerance. *Journal of Plant Biochemistry and Physiology* 1(4): 4.

Khan, M. I. R., Iqbal, N., Masood, A., Khan, N. A. (2012) Variation in salt tolerance of wheat cultivars: Role of glycinebetaine and ethylene. *Pedosphere* 22(6): 746–754.

Khan, M. I. R., Iqbal, N., Masood, A., Per, T. S., Khan, N. A. (2013) Salicylic acid alleviates adverse effects of heat stress on photosynthesis through change in proline production and ethylene formation. *Plant Signaling & Behavior* 8(11): e26374.

Khodary, S. E. A. (2004) Effect of salicylic acid on growth, photosynthesis and carbohydrate metabolism in salt stressed maize plants. *International Journal of Agriculture and Biology* 6: 5–8.

Kim, Y. H., Choi, K. I., Khan, A. L., Waqas, M., Lee, I. J. (2016) Exogenous application of abscisic acid regulates endogenous gibberellins homeostasis and enhances resistance of oriental melon (*Cucumis melo* L.) against low temperature. *Scientia Horticulturae* 207: 41–47.

Kim, Y. H., Khan, A. L., Kim, D. H., Lee, S. Y., Kim, K. M., Waqas, M., Jung, H. Y., *et al.* (2014) Silicon mitigates heavy metal stress by regulating P-type heavy metal ATPases, *Oryza sativa* low silicon genes, and endogenous phytohormones. *BMC Plant Biology* 14: 13.

Kohli, S. K., Handa, N., Bali, S., Arora, S., Sharma, A., Kaur, R., Bhardwaj, R. (2018) Modulation of antioxidative defense expression and osmolyte content by co-application of 24-epibrassinolide and salicylic acid in Pb exposed Indian mustard plants. *Ecotoxicology and Environmental Safety* 147: 382–393.

Kohli, A., Sreenivasulu, N., Lakshmanan, P., Kumar, P. P. (2013) The phytohormone crosstalk paradigm takes center stage in understanding how plants respond to abiotic stresses. *Plant Cell Reports* 32(7): 945–957.

Komatsu, S., Han, C., Nanjo, Y., Altaf-Un-Nahar, M., Wang, K., He, D., Yang, P. (2013) Label-free quantitative proteomic analysis of abscisic acid effect in early-stage soybean under flooding. *Journal of Proteome Research* 12(11): 4769–4784.

Komives, T., Gullner, G. (2005) Phase I xenobiotic metabolic systems in plants. *Zeitschrift Fur Naturforschung C, Journal of Biosciences* 60(3–4): 179–185.

Kosová, K., Prášil, I. T., Vítámvás, P., Dobrev, P., Motyka, V., Floková, K., Novák, O., *et al.* (2012) Complex phytohormone responses during the cold acclimation of two wheat cultivars differing in cold tolerance. *Journal of Plant Physiology* 169(6): 567–576.

Kumar, S., Kaushal, N., Nayyar, H., Gaur, P. (2012) Abscisic acid induces heat tolerance in chickpea (*Cicer arietinum* L.) seedlings by facilitated accumulation of osmoprotectants. *Acta Physiologiae Plantarum* 34(5): 1651–1658.

Lee, M., Jung, J. H., Han, D. Y., Seo, P. J., Park, W. J., Park, C. M. (2012) Activation of a flavin monooxygenase gene YUCCA7 enhances drought resistance in Arabidopsis. *Planta* 235(5): 923–938.

Li, F., Asami, T., Wu, X., Tsang, E. W., Cutler, A. J. (2007) A putative hydroxysteroid dehydrogenase involved in regulating plant growth and development. *Plant Physiology* 145(1): 87–97.

Li, X., Jiang, H., Liu, F., Cai, J., Dai, T., Cao, W., Jiang, D. (2013) Induction of chilling tolerance in wheat during germination by pre-soaking seed with nitric oxide and gibberellins. *Plant Growth Regulation* 71(1): 31–40.

Li, Q., Li, C., Yu, X., Shi, Q. (2011) Gibberellin A3 pretreatment increased antioxidative capacity of cucumber radicles and hypocotyls under suboptimal temperature. *African Journal of Agricultural Research* 6(17): 4091–4098.

Li, J., Yang, P., Gan, Y., Yu, J., Xie, J. (2015) Brassinosteroid alleviates chilling-induced oxidative stress in pepper by enhancing antioxidation systems and maintenance of photosystem II. *Acta Physiologiae Plantarum* 37(11): 222.

Li, X., Zhang, L. (2012) SA and PEG-induced priming for water stress tolerance in rice seedling. In *Information Technology and Agricultural Engineering*, 881–887. *Advances in Intelligent and Soft Computing*. Berlin, Heidelberg: Springer.

Liang, C., Wang, Y., Zhu, Y., Tang, J., Hu, B., Liu, L., Ou, S., *et al.* (2014) OsNAP connects abscisic acid and leaf senescence by fine-tuning abscisic acid biosynthesis and directly targeting senescence-associated genes in rice. *Proceedings of the National Academy of Sciences of the United States of America* 111(27): 10013–10018.

Lu, Y., Song, S., Wang, R., Liu, Z., Meng, J., Sweetman, A.J., Jenkins, A., Ferrier, R.C., Li, H., Luo, W., Wang, T. (2015) Impacts of soil and water pollution on food safety and health risks in China. *Environment International* 77: 5–15.

Mainster, M. A. (2006) Violet and blue light blocking intraocular lenses: Photoprotection versus photoreception. *The British Journal of Ophthalmology* 90(6): 784–792.

Mandava, N. B. (1988) Plant growth-promoting brassinosteroids. *Annual Review of Plant Physiology and Plant Molecular Biology* 39(1): 23–52.

Mao, X., Zhang, H., Tian, S., Chang, X., Jing, R. (2010) TaSnRK 2.4, an SNF1-type serine/threonine protein kinase of wheat (*Triticum aestivum* L.), confers enhanced multistress tolerance in Arabidopsis. *Journal of Experimental Botany* 61(3): 683–696.

Mazid, M., Khan, T. A., Mohammad, F. (2011) Role of secondary metabolites in defense mechanisms of plants. *Biology and Medicine* 3(2): 232–249.

Micó, C., Recatalá, L., Peris, M., Sánchez, J. (2006) Assessing heavy metal sources in agricultural soils of an European Mediterranean area by multivariate analysis. *Chemosphere* 65(5): 863–872.

Miura, K., Okamoto, H., Okuma, E., Shiba, H., Kamada, H., Hasegawa, P. M., Murata, Y. (2013) SIZ1 deficiency causes reduced stomatal aperture and enhanced drought tolerance via controlling salicylic acid-induced accumulation of reactive oxygen species in Arabidopsis. *The Plant Journal: for Cell and Molecular Biology* 73(1): 91–104.

Miura, K., Tada, Y. (2014) Regulation of water, salinity, and cold stress responses by salicylic acid. Frontiers in Plant Science 5: 4.

Mo, Z. W., Ashraf, U., Pan, S. G., Kanu, A. S., Li, W., Duan, M. Y., Tian, H., Tang, X. R. (2016) Exogenous application of plant growth regulators induce chilling tolerance in direct seeded super and non-super rice seedlings through modulations in morpho-physiological attributes. *Cereal Research Communications* 44(3): 524–534.

Munne-Bosch, S., Penuelas, J. (2003) Photo and antioxidative protection and a role for salicylic acid during drought and recovery in field-grown *Phillyrea angustifolia* plants. *Planta* 217(5): 758–766.

Munns, R., Tester, M. (2008) Mechanisms of salinity tolerance. *Annual Review of Plant Biology* 59: 651–681

Munteanu, V., Gordeev, V., Martea, R., Duca, M. (2014) Effect of gibberellin cross talk with other phytohormones on cellular growth and mitosis to endoreduplication transition. International Journal of Advance Research in Biological Science 1: 136–153.

Munzuro, Ö., Fikriye, K. Z., Yahyagil, Z. (2008) The abscisic acid levels of wheat (Triticum aestivum L. cv. Çakmak 79) seeds that were germinated under heavy metal (Hg++, Cd++, Cu++) stress. *Gazi University Journal of Science* 21: 1–7.

Mutlu, S., Karadağoğlu, Ö, Atici, Ö, Nalbantoğlu, B. (2013) Protective role of salicylic acid applied before cold stress on antioxidative system and protein patterns in barley apoplast. *Biologia Plantarum* 57(3): 507–513.

Nakagami, H., Pitzschke, A., Hirt, H. (2005) Emerging MAP kinase pathways in plant stress signalling. *Trends in Plant Science* 10(7): 339–346.

Narusaka, Y., Nakashima, K., Shinwari, Z. K., Sakuma, Y., Furihata, T., Abe, H., Narusaka, M., Shinozaki, K., Yamaguchi-Shinozaki, K. (2003) Interaction between two cisacting elements, ABRE and DRE, in ABA-dependent expression of Arabidopsis rd29A gene in response to dehydration and high-salinity stresses. *The Plant Journal: for Cell and Molecular Biology* 34(2): 137–148.

Nazar, R., Iqbal, N., Syeed, S., Khan, N. A. (2011) Salicylic acid alleviates decreases in photosynthesis under salt stress by enhancing nitrogen and sulfur assimilation and antioxidant metabolism differentially in two mungbean cultivars. *Journal of Plant Physiology* 168(8): 807–815.

Nishiyama, R., Watanabe, Y., Fujita, Y., Le, D. T., Kojima, M., Werner, T., Vankova, R., *et al.* (2011) Analysis of cytokinin mutants and regulation of cytokinin metabolic genes reveals important regulatory roles of cytokinins in drought, salt and abscisic acid responses, and abscisic acid biosynthesis. *The Plant Cell* 23(6): 2169–2183.

Niu, J. H., Ahmad Anjum, S., Wang, R., Li, J. H., Liu, M. R., Song, J. X., Zohaib, A., *et al.* (2016) Exogenous application of brassinolide can alter morphological and physiological traits of *Leymus chinensis* (Trin.) Tzvelev under room and high temperatures. *Chilean Journal of Agricultural Research* 76(1): 27–33.

O'Brien, J. A., Benkova, E. (2013) Cytokinin cross-talking during biotic and abiotic stress responses. Frontiers in Plant Science 4: 451.

Özdemir, F., Bor, M., Demiral, T., Türkan, İ. (2004) Effects of 24-epibrassinolide on seed germination, seedling growth, lipid peroxidation, proline content and antioxidant system of rice (*Oryza sativa* L.) under salinity stress. *Plant Growth Regulation* 42(3): 203–211.

Pandey, C., Gupta, M. (2015) Selenium and auxin mitigates arsenic stress in rice (*Oryza sativa* L.) by combining the role of stress indicators, modulators and genotoxicity assay. *Journal of Hazardous Materials* 287: 384–391.

Parida, A. K., Das, A. B. (2005) Salt tolerance and salinity effects on plants: A review. *Ecotoxicology and Environmental Safety* 60(3): 324–349.

Park, J., Kim, Y. S., Kim, S. G., Jung, J. H., Woo, J. C. W., Park, C. M. (2011) Integration of auxin and salt signals by the NAC transcription factor NTM2 during seed germination in Arabidopsis. *Plant Physiology* 156(2): 537–549.

Park, H. Y., Seok, H. Y., Park, B. K., Kim, S. H., Goh, C. H., Lee, B. H., Lee, C. H., Moon, Y. H. (2008) Overexpression of Arabidopsis ZEP enhances tolerance to osmotic stress. *Biochemical and Biophysical Research Communications* 375(1): 80–85.

Pathak, M. R., Teixeira da Silva, J. A., Wani, S. H. (2014) Polyamines in response to abiotic stress tolerance through transgenic approaches. *GM Crops and Food* 5(2): 87–96.

Pauwels, L., Inzé, D., Goossens, A. (2009) Jasmonate-inducible gene: What does it mean? *Trends in Plant Science* 14(2): 87–91.

Peleg, Z., Blumwald, E. (2011) Hormone balance and abiotic stress tolerance in crop plants. *Current Opinion in Plant Biology* 14(3): 290–295.

Pospíšilová, H., Jiskrová, E., Vojta, P., Mrízová, K., Kokáš, F., Čudejková, M. M., Bergougnoux, V., *et al.* (2016) Transgenic barley overexpressing a cytokinin dehydrogenase gene shows greater tolerance to drought stress. *New Biotechnology* 33(5 Pt B): 692–705.

Pospíšilová, J. (2003) Interaction of cytokinins and abscisic acid during regulation of stomatal opening in bean leaves. *Photosynthetica* 41(1): 49–56.

Pospíšilová, J., Synková, H., Haisel, D., Baťková, P. (2009) Effect of abscisic acid on photosynthetic parameters during ex vitro transfer of micropropagated tobacco plantlets. *Biologia Plantarum* 53 (1): 11–20.

Rahman, A. (2013) Auxin: A regulator of cold stress response. *Physiologia Plantarum* 147(1): 28–35.

Ramel, F., Sulmon, C., Cabello-Hurtado, F., Taconnat, L., Martin-Magniette, M. L., Renou, J. P., El Amrani, A., Couée, I., Gouesbet, G. (2007) Genome-wide interacting effects of sucrose and herbicide-mediated stress in *Arabidopsis thaliana*: Novel insights into atrazine toxicity and sucrose-induced tolerance. *BMC Genomics* 8: 450.

Ramel, F., Sulmon, C., Serra, A. A., Gouesbet, G., Couée, I. (2012) Xenobiotic sensing and signaling in higher plants. *Journal of Experimental Botany* 63(11): 3999–4014.

Recatala, L., Ive, J.R., Baird, I.A., Hamilton, N., Sanchez, J. (2000) Land-use planning in the Valencian Mediterranean region: using LUPIS to generate issue relevant plans. *Journal of Environmental Management* 59: 169–184.

Rivero, R. M., Gimeno, J., Van Deynze, A., Walia, H., Blumwald, E. (2010) Enhanced cytokinin synthesis in tobacco plants expressing PSARK:IPT prevents the degradation of photosynthetic protein complexes during drought. *Plant and Cell Physiology* 51(11): 1929–1941.

Rowland, F. (2006) Review: Stratospheric ozone depletion. Philosophical. *Transactions of the Royal Society B: Biology Sciences* 361: 769–790.

Ruelland, E., Zachowski, A. (2010) How plants sense temperature. *Environmental and Experimental Botany* 69(3): 225–232.

Sandermann Jr, H. (1992) Plant metabolism of xenobiotics. *Trends in Biochemical Sciences* 17(2): 82–84.

Sanghera, G.S., Wani, S.H., Hussain, W., Singh, N.B. (2011) Engineering cold stress tolerance in crop plants. *Current genomics* 12 (1): 30.

Saud, S. H. A. H., Chen, Y., Baowen, L., Fahad, S. H. A. H., Arooj, S. (2013) The different impact on the growth of cool season turf grass under the various conditions on salinity and drought stress. *International Journal of Agriculture Science and Research* 3(4): 77–84.

Seo, J. S., Joo, J., Kim, M. J., Kim, Y. K., Nahm, B. H., Song, S. I., Cheong, J. J., *et al.* (2011) OsbHLH148, a basic helix-loop-helix protein, interacts with OsJAZ proteins in a jasmonate signaling pathway leading to drought tolerance in rice. *The Plant Journal: for Cell and Molecular Biology* 65(6): 907–921.

Shaban, S. A., El-Hattab, A. H., Hassan, E. A., Abo-El Suoud, M. R. (1987) Recovery of faba bean (*Vicia faba* L.) plants as affected by glyphosate. *Journal of Agronomy and Crop Science* 158(5): 294–303.

Shang, J., Song, P., Ma, B., Qi, X., Zeng, Q., Xiang, Z., He, N. (2014) Identification of the mulberry genes involved in ethylene biosynthesis and signaling pathways and the expression of MaERF-B2-1 and MaERF-B2-2 in the response to flooding stress. *Functional and Integrative Genomics* 14(4): 767–777.

Sharma, I., Ching, E., Saini, S., Bhardwaj, R., Pati, P. K. (2013) Exogenous application of brassinosteroid offers tolerance to salinity by altering stress responses in rice variety Pusa Basmati-1. *Plant Physiology and Biochemistry: PPB* 69: 17–26.

Sharma, P., Kumar, A., Bhardwaj, R. (2016) Plant steroidal hormone epibrassinolide regulate–Heavy metal stress tolerance in *Oryza sativa* L. by modulating antioxidant defense expression. *Environmental and Experimental Botany* 122: 1–9.

Sharma, A., Thakur, S., Kumar, V., Kesavan, A. K., Thukral, A. K., Bhardwaj, R. (2017) 24-epibrassinolide stimulates Imidacloprid detoxification by modulating the gene expression of *Brassica juncea* L. *BMC Plant Biology* 17(1): 56.

Sheng, X. F., Xia, J. J. (2006) Improvement of rape (*Brassica napus*) plant growth and cadmium uptake by cadmium-resistant bacteria. *Chemosphere* 64(6): 1036–1042.

Shi, J., Habben, J. E., Archibald, R. L., Drummond, B. J., Chamberlin, M. A., Williams, R. W., Lafitte, H. R., Weers, B. P. (2015) Overexpression of ARGOS genes modifies plant sensitivity to ethylene, leading to improved drought tolerance in both Arabidopsis and maize. *Plant Physiology* 169(1): 266–282.

Shi, Y., Tian, S., Hou, L., Huang, X., Zhang, X., Guo, H., Yang, S. (2012) Ethylene signaling negatively regulates freezing tolerance by repressing expression of CBF and type-A ARR genes in Arabidopsis. *The Plant Cell* 24(6): 2578–2595.

Siboza, X. I., Bertling, I., Odindo, A. O. (2014) Salicylic acid and methyl jasmonate improve chilling tolerance in cold-stored lemon fruit (*Citrus limon*). *Journal of Plant Physiology* 171(18): 1722–1731.

Siddiqui, M. H., Khan, M. N., Mohammad, F., Khan, M. M. A. (2008) Role of nitrogen and gibberellic acid (GA3) in the regulation of enzyme activities and in osmoprotectant accumulation in *Brassica juncea* L. under salt stress. *Journal of Agronomy and Crop Science* 194(3): 214–224.

Singh, A. K., Kumar, R., Pareek, A., Sopory, S. K., Singla-Pareek, S. L. (2012) Overexpression of Rice CBS domain containing protein improves salinity, oxidative and heavy metal tolerance in transgenic tobacco. *Molecular Biotechnology* 52(3): 205–216.

Sirhindi, G., Kaur, H., Bhardwaj, R., Sharma, P., Mushtaq, R. (2017) 28-homobrassinolide potential for oxidative interface in *Brassica juncea* under temperature stress. *Acta Physiologiae Plantarum* 39(10): 228.

Skipsey, M., Knight, K. M., Brazier-Hicks, M., Dixon, D. P., Steel, P. G., Edwards, R. (2011) Xenobiotic responsiveness of Arabidopsis thaliana to a chemical series derived from a herbicide safener. *The Journal of Biological Chemistry* 286(37): 32268–32276.

Skirycz, A., Inzé, D. (2010) More from less: Plant growth under limited water. *Current Opinion in Biotechnology* 21(2): 197–203.

Skottke, K. R., Yoon, G. M., Kieber, J. J., DeLong, A. (2011) Protein phosphatase 2A controls ethylene biosynthesis by differentially regulating the turnover of ACC synthase isoforms. *PLOS Genetics* 7(4): e1001370.

Sreenivasulu, N., Harshavardhan, V. T., Govind, G., Seiler, C., Kohli, A. (2012) Contrapuntal role of ABA: Does it mediate stress tolerance or plant growth retardation under long-term drought stress? *Gene* 506(2): 265–273.

Sripinyowanich, S., Klomsakul, P., Boonburapong, B., Bangyeekhun, T., Asami, T., Gu, H., Buaboocha, T., Chadchawan, S. (2013) Exogenous ABA induces salt tolerance in indica rice (*Oryza sativa* L.): The role of *OsP5CS1* and *OsP5CR* gene expression during salt stress. *Environmental and Experimental Botany* 86: 94–105.

Srivastava, S., Verma, P.C., Chaudhry, V., Singh, N., Abhilash, P.C., Kumar, K.V., Sharma, N., Singh, N. (2013) Influence of inoculation of arsenic-resistant Staphylococcus arlettae on growth and arsenic uptake in Brassica juncea (L.) Czern. Var. R-46. *Journal of Hazardous Materials* 262: 1039–1047.

Stroiński, A., Chadzinikolau, T., Giżewska, K., Zielezińska, M. (2010) ABA or cadmium induced phytochelatin synthesis in potato tubers. *Biologia Plantarum* 54(1): 117–120.

Sulmon, C., Gouesbet, G., El Amrani, A., Couée, I. (2006) Sucrose-induced tolerance to atrazine in Arabidopsis seedlings involves activation of oxidative and xenobiotic stress responses. *Plant Cell Reports* 25: 489–498.

Suzuki, N., Bassil, E., Hamilton, J. S., Inupakutika, M. A., Zandalinas, S. I., Tripathy, D., Luo, Y., *et al.* (2016) ABA is required for plant acclimation to a combination of salt and heat stress. *PLOS ONE* 11(1): e0147625.

Takematsu, T., Takeuchi, Y., Choi, C. D. (1986) Overcoming effects of brassinosteroids on growth inhibition of rice caused by unfavourable growth conditions. *Shokucho [a journal from Japan Association for Advancement of Phytoregulators]* 20: 2–12.

Tanaka, Y., Sano, T., Tamaoki, M., Nakajima, N., Kondo, N., Hasezawa, S. (2005) Ethylene inhibits abscisic acid-induced stomatal closure in Arabidopsis. *Plant Physiology* 138(4): 2337–2343.

Tandon, S. A., Kumar, R., Parsana, S. (2015) Auxin treatment of wetland and non-wetland plant species to enhance their phytoremediation efficiency to treat municipal wastewater. *Journal of Scientific and Industrial Research* 74: 702–707.

Thussagunpanit, J., Jutamanee, K., Sonjaroon, W., Kaveeta, L., Chai-Arree, W., Pankean, P., Suksamrarn, A. (2015) Effects of brassinosteroid and brassinosteroid mimic on photosynthetic efficiency and rice yield under heat stress. *Photosynthetica* 53(2): 312–320.

Tirkayi, I. (2007) The role of auxin-signaling gene axrl in salt stress and jasmonic acid inducible gene expression in *Arabidopsis thaliana. Journal of Molecular Cell Biology* 6: 189–195.

Tognetti, V. B., Van Aken, O., Morreel, K., Vandenbroucke, K., Van De Cotte, B., De Clercq, I., Chiwocha, S., *et al.* (2010) Perturbation of indole-3-butyric acid homeostasis by the UDP-glucosyltransferase UGT74E2 modulates Arabidopsis architecture and water stress tolerance. *The Plant Cell* 22(8): 2660–2679.

Tossi, V., Cassia, R., Bruzzone, S., Zocchi, E., Lamattina, L. (2012) ABA says NO to UV-B: A universal response? *Trends in Plant Science* 17(9): 510–517.

Trinh, N. N., Huang, T. L., Chi, W. C., Fu, S. F., Chen, C. C., Huang, H. J. (2014) Chromium stress response effect on signal transduction and expression of signaling genes in rice. *Physiologia Plantarum* 150(2): 205–224.

Tuteja, N. (2007) Mechansims of high salinity tolerance in plants. *Methods in Enzymology* 428: 419–438.

Umezawa, T., Fujita, M., Fujita, Y., Yamaguchi-Shinozaki, K., Shinozaki, K. (2006) Engineering drought tolerance in plants: Discovering and tailoring genes to unlock the future. *Current Opinion in Biotechnology* 17(2): 113–122.

Van Ha, C., Leyva-González, M. A., Osakabe, Y., Tran, U. T., Nishiyama, R., Watanabe, Y., Tanaka, M., *et al.* (2014) Positive regulatory role of strigolactone in plant responses to drought and salt stress. *Proceeding of National Academy of Sciences of the United States of America* 111: 851–856.

van Veen, H., Mustroph, A., Barding, G. A., Vergeer-van Eijk, M., Welschen-Evertman, R. A., Pedersen, O., Visser, E. J., *et al.* (2013) Two *Rumex* species from contrasting hydrological niches regulate flooding tolerance through distinct mechanisms. *The Plant Cell* 25(11): 4691–4707.

Vardhini, B. V., Anjum, N. A. (2015) Brassinosteroids make plant life easier under abiotic stresses mainly by modulating major components of antioxidant defense system. *Frontiers in Environmental Science* 2: 1–16.

Varshney, S., Hayat, S., Alyemeni, M. N., Ahmad, A. (2012) Effects of herbicide applications in wheat fields: Is phytohormones application a remedy? *Plant Signaling & Behavior* 7(5): 570–575.

Verslues, P. E., Agarwal, M., Katiyar-Agarwal, S., Zhu, J., Zhu, J. K. (2006) Methods and concepts in quantifying resistance to drought, salt and freezing, abiotic stresses that affect plant water status. *The Plant Journal: for Cell and Molecular Biology* 45(4): 523–539.

Voesenek, L. A. C. J., Bailey-Serres, J. (2015) Flood adaptive traits and processes: An overview. *The New Phytologist* 206(1): 57–73.

Voesenek, L. A. C. J., Benschop, J. J., Bou, J., Cox, M. C. H., Groeneveld, H. W., Millenaar, F. F., Vreeburg, R. A. M., Peeters, A. J. M. (2003) Interactions between plant hormones regulate submergence-induced shoot elongation in the flooding-tolerant dicot *Rumex palustris*. *Annals of Botany* 91(2): 205–211.

Voß, U., Bishopp, A., Farcot, E., Bennett, M. J. (2014) Modelling hormonal response and development. *Trends in Plant Science* 19(5): 311–319.

Wang, L. J., Fan, L., Loescher, W., Duan, W., Liu, G. J., Cheng, J. S., Luo, H. B., Li, S. H. (2010) Salicylic acid alleviates decreases in photosynthesis under heat stress and accelerates recovery in grapevine leaves. *BMC Plant Biology* 10: 34.

Wang, W. B., Kim, Y. H., Lee, H. S., Kim, K. Y., Deng, X. P., Kwak, S. S. (2009) Analysis of antioxidant enzyme activity during germination of alfalfa under salt and drought stresses. *Plant Physiology and Biochemistry: PPB* 47(7): 570–577.

Wang, Y., Wang, Y., Kai, W., Zhao, B., Chen, P., Sun, L., Ji, K., *et al.* (2014) Transcriptional regulation of abscisic acid signal core components during cucumber seed germination and under Cu^{2+}, Zn^{2+}, NaCl and simulated acid rain stresses. *Plant Physiology and Biochemistry* 76: 67–76.

Weisman, D., Alkio, M., Colón-Carmona, A. (2010) Transcriptional responses to polycyclic aromatic hydrocarbon-induced stress in Arabidopsis thaliana reveal the involvement of hormone and defense signalling pathways. *BMC Plant Biology* 10: 59.

Werner, T., Nehnevajova, E., Köllmer, I., Novák, O., Strnad, M., Krämer, U., Schmülling, T. (2010) Root-specific reduction of cytokinin causes enhanced root growth, drought tolerance, and leaf mineral enrichment in *Arabidopsis* and tobacco. *The Plant Cell* 22(12): 3905–3920.

Wi, S. J., Jang, S. J., Park, K. Y. (2010) Inhibition of biphasic ethylene production enhances tolerance to abiotic stress by reducing the accumulation of reactive oxygen species in *Nicotiana tabacum*. *Molecules and Cells* 30(1): 37–49.

Wilkinson, S., Kudoyarova, G. R., Veselov, D. S., Arkhipova, T. N., Davies, W. J. (2012) Plant hormone interactions: Innovative targets for crop breeding and management. *Journal of Experimental Botany* 63(9): 3499–3509.

Wolters, H., Jurgens, G. (2009) Survival of the flexible: Hormonal growth control and adaptation in plant development. *Nature Reviews Genetics* 10(5): 305–317.

Won Oh, M. W., Nanjo, Y., Komatsu, S. (2014) Analysis of soybean root proteins affected by gibberellic acid treatment under flooding stress. *Protein and Peptide Letters* 21(9): 911–947.

Wu, X., He, J., Chen, J., Yang, S., Zha, D. (2014) Alleviation of exogenous 6-benzyladenine on two genotypes of egg plant (*Solanum melongena* Mill.) growth under salt stress. *Protoplasma* 251(1): 169–176.

Wu, X., Yao, X., Chen, J., Zhu, Z., Zhang, H., Zha, D. (2014) Brassinosteroids protect photosynthesis and antioxidant system of eggplant seedlings from high-temperature stress. *Acta Physiologiae Plantarum* 36(2): 251–261.

Wu, L., Zhang, Z., Zhang, H., Wang, X. C., Huang, R. (2008) Transcriptional modulation of ethylene response factor protein JERF3 in the oxidative stress response enhances tolerance of tobacco seedlings to salt, drought, and freezing. *Plant Physiology* 148(4): 1953–1963.

Xia, X. J., Zhang, Y., Wu, J. X., Wang, J. T., Zhou, Y. H., Shi, K., Yu, Y. L., Yu, J. Q. (2009) Brassinosteroids promote metabolism of pesticides in cucumber. *Journal of Agricultural and Food Chemistry* 57(18): 8406–8413.

Xu, Y., Burgess, P., Zhang, X., Huang, B. (2016) Enhancing cytokinin synthesis by overexpressing ipt alleviated drought inhibition of root growth through activating ROS-scavenging systems in *Agrostis stolonifera*. *Journal of Experimental Botany* 67(6): 1979–1992.

Xu, Y. H., Liu, R., Yan, L., Liu, Z. Q., Jiang, S. C., Shen, Y. Y., Wang, X. F., Zhang, D. P. (2012) Light-harvesting chlorophyll a/b-binding proteins are required for stomatal response to abscisic acid in *Arabidopsis*. *Journal of Experimental Botany* 63(3): 1095–1106.

Yadegari, M. (2017) Study of phytohormones effects on UV-B stress seeds of thyme species. *Journal of Herbal Drugs* 8(2): 109–115.

Yan, Z., Chen, J., Li, X. (2013) Methyl jasmonate as modulator of Cd toxicity in *Capsicum frutescens* var. fasciculatum seedlings. *Ecotoxicology and Environmental Safety* 98: 203–209.

Yang, Z., Cao, S., Zheng, Y., Jiang, Y. (2012) Combined salicyclic acid and ultrasound treatments for reducing the chilling injury on peach fruit. *Journal of Agricultural and Food Chemistry* 60(5): 1209–1212.

Yin, C. C., Ma, B., Collinge, D. P., Pogson, B. J., He, S. J., Xiong, Q., Duan, K. X., *et al.* (2015) Ethylene responses in rice roots and coleoptiles are differentially regulated by a carotenoid isomerase-mediated abscisic acid pathway. *The Plant Cell* 27(4): 1061–1081.

Yin, X., Sakata, K., Komatsu, S. (2014) Phosphoproteomics reveals the effect of ethylene in soybean root under flooding stress. *Journal of Proteome Research* 13(12): 5618–5634.

Yoon, J. Y., Hamayun, M., Lee, S. K., Lee, I. J. (2009) Methyl jasmonate alleviated salinity stress in soybean. *Journal of Crop Science and Biotechnology* 12(2): 63–68.

Yu, J. Q., Huang, L. F., Hu, W. H., Zhou, Y. H., Mao, W. H., Ye, S. F., Nogués, S. (2004) A role for brassinosteroids in the regulation of photosynthesis in *Cucumis sativus*. *Journal of Experimental Botany* 55(399): 1135–1143.

Yusuf, M., Fariduddin, Q., Ahmad, A. (2012) 24-Epibrassinolide modulates growth, nodulation, antioxidant system, and osmolyte in tolerant and sensitive varieties of *Vigna* radiate under different levels of nickel: A shotgun approach. *Plant Physiology and Biochemistry: PPB* 57: 143–153.

Yusuf, M., Fariduddin, Q., Hayat, S., Hasan, S. A., Ahmad, A. (2011) Protective response of 28-homobrassinolide in cultivars of *Triticum aestivum* with different levels of nickel. *Archives of Environmental Contamination and Toxicology* 60(1): 68–76.

Zhang, Y. P., He, J., Yang, S. J., Chen, Y. Y. (2014) Exogenous 24-epibrassinolide ameliorates high temperature-induced inhibition of growth and photosynthesis in *Cucumis melo*. *Biologia Plantarum* 58(2): 311–318.

Zhang, J., Jia, W., Yang, J., Ismail, A. M. (2006) Role of ABA in integrating plant responses to drought and salt stresses. *Field Crops Research* 97(1): 111–119.

Zhang, Z., Li, F., Li, D., Zhang, H., Huang, R. (2010) Expression of ethylene response factor JERF1 in rice improves tolerance to drought. *Planta* 232(3): 765–774.

Zhang, Q., Li, J., Zhang, W., Yan, S., Wang, R., Zhao, J., Li, Y., *et al.* (2012) The putative auxin efflux carrier OsPIN3t is involved in the drought stress response and drought tolerance. *The Plant Journal: for Cell and Molecular Biology* 72(5): 805–816.

Zhang, Z., Wang, Y., Chang, L., Zhang, T., An, J., Liu, Y., Cao, Y., *et al.* (2016b) MsZEP, a novel zeaxanthin epoxidase gene from alfalfa (*Medicago sativa*), confers drought and salt tolerance in transgenic tobacco. *Plant Cell Reports* 35(2): 439–453.

Zhao, M. L., Wang, J. N., Shan,W., Fan, J. G., Kuang, J. F., Wu, K. Q., Li, X., *et al.* (2013) Induction of jasmonate signalling regulators MaMYC2s and their physical interactions with MaICE1 in methyl jasmonate-induced chilling tolerance in banana fruit. *Plant, Cell and Environment* 36(1): 30–51.

Zhang, J. -S., Xie, C., Shen, Y. -G., Chen, S. -Y. (2001) A two-component gene (NTHK1) encoding a putative ethylene-receptor homolog is both developmentally and stress-regulated in tobacco. *TAG Theoretical and Applied Genetics* 102(6–7): 815–824.

Zhang, X., Yang, G., Shi, R., Han, X., Qi, L., Wang, R., Xiong, L., Li, G. (2013b) Arabidopsis cysteine-rich receptor-like kinase 45 functions in the responses to abscisic acid and abiotic stresses. *Plant Physiology and Biochemistry: PPB* 67: 189–198.

Zhang, J., Yu, H., Zhang, Y., Wang, Y., Li, M., Zhang, J., Duan, L., Zhang, M., Li, Z. (2016a) Increased abscisic acid levels in transgenic maize overexpressing AtLOS5 mediated root ion fluxes and leaf water status under salt stress. *Journal of Experimental Botany* 67(5): 1339–1355.

Zhang, Y. P., Zhu, X. H., Ding, H. D., Yang, S. J., Chen, Y. Y. (2013a) Foliar application of 24-epibrassinolide alleviates high-temperature-induced inhibition of photosynthesis in seedlings of two melon cultivars. *Photosynthetica* 51(3): 341–349.

Zhao, X. C., Schaller, G. E. (2004) Effect of salt and osmotic stress upon expression of the ethylene receptor ETR1 in Arabidopsis thaliana. *FEBS Letters* 562(1–3): 189–192.

Zhou, Y., Xia, X., Yu, G., Wang, J., Wu, J., Wang, M., Yang, Y., *et al.* (2015) Brassinosteroids play a critical role in the regulation of pesticide metabolism in crop plants. *Scientific Reports* 590: 1–8.

Zhu, J. K. (2000) Genetic analysis of plant salt tolerance using *Arabidopsis*. *Plant Physiology* 124(3): 941–948

Zhu, S. Q., Chen, M. W., Ji, B. H., Jiao, D. M., Liang, J. S. (2011b) Roles of xanthophylls and exogenous ABA in protection against NaCl-induced photodamage in rice (*Oryza sativa* L.) and cabbage (*Brassica campestris*). *Journal of Experimental Botany* 62(13): 4617–4625.

Zhu, Z., An, F., Feng, Y., Li, P., Xue, L. A. M., Jiang, Z., *et al.* (2011a) Derepression of ethylene-stabilized transcription factors (EIN3/EIL1) mediates jasmonate and ethylene signaling synergy in Arabidopsis. *Proceedings of the National Academy of Sciences of the United States of America* 108(30): 12539–12544.

Žižková, E., Dobrev, P. I., Muhovski, Y., Hošek, P., Hoyerová, K., Haisel, D., Procházková, D., *et al.* (2015) Tomato (*Solanum lycopersicum* L.) SlIPT3 and SlIPT4 isopentenyltransferases mediate salt stress response in tomato. *BMC Plant Biology* 15(1): 85.

7 Drought Tolerance in Plants

Role of Phytohormones and Scavenging System of ROS

Shah Fahad, Abid Ullah, Usman Ali, Ehsan Ali, Shah Saud, Khalid Rehman Hakeem, Hesham Alharby, Ayman EL Sabagh, Celaleddin Barutcular, Muhammad Kamran, Veysel Turan, Muhammad Adnan, Muhammad Arif, and Amanullah

CONTENTS

7.1 INTRODUCTION

To feed the overpopulating world, industries are building on agricultural lands and consequently forests are being replaced by agricultural lands, which is disturbing the water cycle. In addition, these industries are creating global warming via air, land and water pollution. It has been observed in the last few years that global warming is increasing steadily. Water is evaporating from the ground and causing drought stress (Fahad and Bano, 2012; Fahad et al., 2013, 2015a, b, c, d, 2016a, b, c, d, 2018; Ullah et al., 2015). Agriculture drought is the prolonged shortfall in moisture required for plant growth and development to complete the life cycle (Fahad et al., 2018; Farooq et al., 2012). Generally, drought stresses severely alter physiological, biochemical and molecular processes and consequently affect plant growth and development. Specifically, it affects the photosynthetic process, stomata regulation, transpiration rate, water potential and carboxylation efficiency (Fahad et al., 2015a; Saud et al., 2013, 2014, 2016; Ullah et al., 2017). Altogether, these drought-related parameters adversely affect plant yield. For example, cotton production in Pakistan declined by 34% in 2016 from 14.4 M bales to 9.68 M bales as compared to 2015 due to drought stress. In another case, drought stress caused a 67% crop loss in the United States (Comas et al., 2013). In order to feed the growing global population, it is essential for plant scientists to maintain as well as improve yield under drought conditions. In this regard, it is very important to study the mechanism of drought-tolerant plants, so that drought-tolerance adaptations can be engineered into them.

Plants have developed a complicated drought-tolerance mechanism in the form of morphological, physiological, biochemical and molecular strategies during the evolution and domestication of crops. On the basis of these strategies, the plant drought-tolerance mechanism can be divided into (1) drought avoidance, (2) drought tolerance, (3) drought escape and (4) drought recovery (Fang and Xiong, 2015).

1. Drought avoidance: During moderate drought conditions, plants maintain key physiological processes such as photosynthesis, stomata

regulation and root development (Ullah et al., 2017).
2. Drought tolerance: Plants have the capability to withstand severe drought conditions through drought-tolerance physiological activities, for instance, osmotic adjustment (Luo, 2010).
3. Drought escape: Capability of plants to shorten their life cycle under unfavorable conditions i.e., drought stress to avoid seasonal drought stress (Manavalan et al., 2009).
4. Drought recovery: Plants have the ability to recover growth and yield after exposure to severe drought stress.

Numerous organs and substances are involved in these strategies to improve drought tolerance. These organs include the closing of stomata, root development, leaf structure (shape, expansion, area, senescence, pubescence, waxiness) and cuticle (Nezhadahmadi et al., 2013). Phytohormones play a crucial role in regulating these morpho-physiological and biochemical processes to enhance drought tolerance in plants.

7.2 ROLE OF PHYTOHORMONES IN DROUGHT TOLERANCE

Upon drought stress, the resulting signal transduction triggers the production of various substances including phytohormones to respond and adapt to drought stress. Phytohormones are organic substances that are produced in low concentrations but are able to regulate plant growth, development, response to stresses (drought) and other physiological processes (Wani et al., 2016). Although some other factors are also involved in the response to biotic and abiotic stresses, phytohormones are the key endogenous substances that modulate various mopho-physiological, biochemical and molecular processes for the plant to survive under critical conditions, i.e., drought stress (Basu et al., 2016). These phytohormones include abscisic acid (ABA), jasmonic acid (JA), auxin (IAA), salicylic acid (SA), ethylene (ET), cytokinins (CKs), gibberellins (GAs) and brassinosteroids (BRs). The chemical structure of these phytohormones has been shown in Figure 7.1. Among these phytohormones, ABA is the key hormone that regulates drought resistance in plants.

7.2.1 Abscisic Acid (ABA)

ABA, a sesquiterpenoid ($C_{15}H_{2}0O_{4}$) with a 15-carbon ring (Figure 7.1), has a variety of biological functions and is found ubiquitously across several kingdoms, including cyanobacteria, sponges, algae, lichens, mosses and mammals (Fernando and Schroeder, 2016). Abscisic acid is the most important drought-tolerant phytohormone and is also known as a stress hormone. It assists plants in adapting and surviving under drought conditions. In addition, ABA has the ability to antagonize the

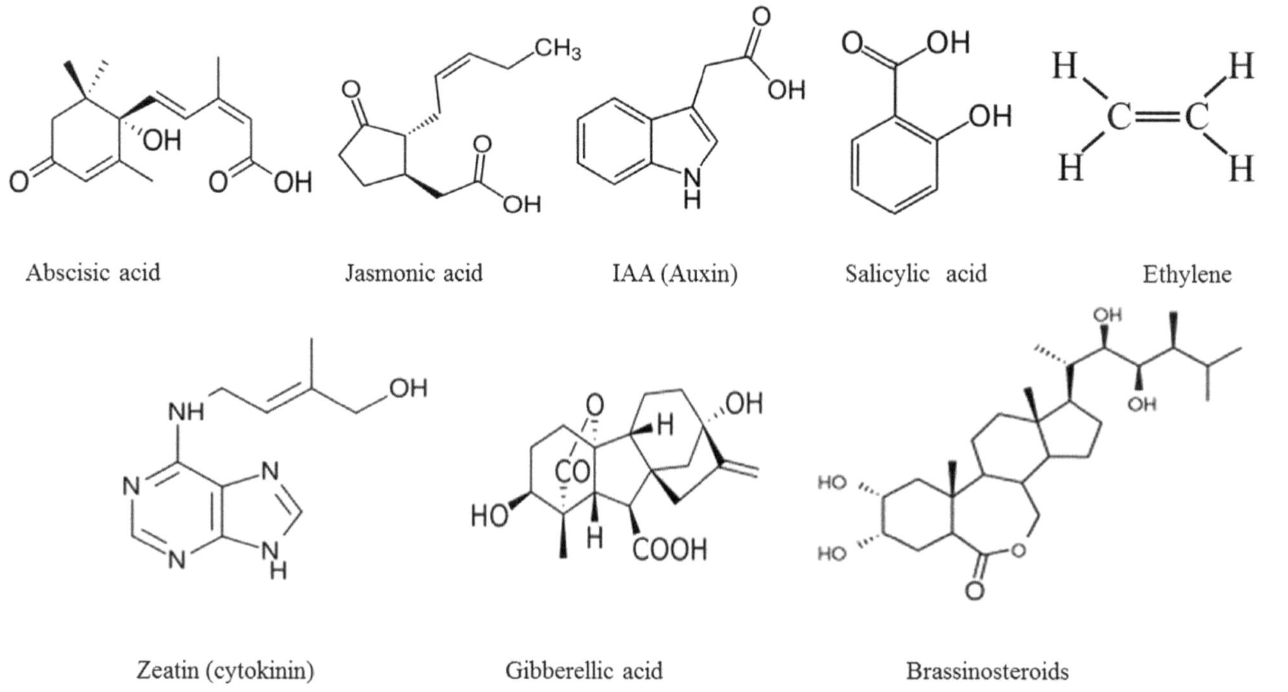

FIGURE 7.1 Chemical structures of phytohormones.

germination-promoting effects of gibberellin, regulate guard cells and regulate stress-responsive gene expression under water-deprived conditions (Fernando and Schroeder, 2016). On the other hand, ABA also regulates various other plant growth and developmental processes (Sah et al., 2016). In *Arabidopsis thaliana*, about 10% of the genome consists of ABA-regulated genes. Approximately half of these genes are ABA-induced and the rest are ABA-repressed. ABA-induced genes code for proteins that confer stress tolerance such as dehydrins, detoxifying enzymes of reactive oxygen species, regulatory proteins (transcription factors, protein kinases, phosphatases) and enzymes required for phospholipid signaling. Genes that are repressed by ABA are mostly related to growth (Cutler et al., 2010). Under drought conditions, ABA regulates stomatal conductance, root development and photosynthesis.

Endogenous ABA levels are elevated in response to osmotic stress, which in turn coordinates the plant's response to reduced water availability. Water balance is achieved through guard cell regulation and the latter role by induction of genes that encode dehydration tolerance proteins in nearly all cells (Fujita et al., 2011; Zhu, 2002). Stomata regulation plays an important role in gas exchange and production of energy. More profuse (higher root length density) and deeper root systems in the soil are often proposed as desirable characteristics for drought adaptation. For example, Liu et al. (2014) produced a transgenic cotton plant that was more tolerant to drought stress, and, in addition to other factors, root development contributed to drought resistance.

ABA also mediates root elongation to reach deep water in the soil during drought conditions (Xiong et al., 2006). Rowe et al. (2016) reported that ABA regulates root growth under osmotic stress conditions. At low water potential, ABA accumulation is essential for maize root elongation. Fluridone, an ABA synthesis inhibitor, inhibits primary root elongation at low water availability (Sharp et al., 1994). In addition, Sharp et al. (2004) found that water deficit increases ABA and ROS, while ABA acts as a suppressing agent of growth-inhibitory ROS and ethylene. In another case, lateral roots were partially mediated by ABA under drought stress. Root characteristics, such as root length, density and size play critical roles in maintaining plant productivity under drought stress, specifically when water is available deeper in the soil (Comas et al., 2013). For example, small xylem diameters can store water present in deep soil, which can be used upon crop maturity and to maintain a high yield, especially in environments with late-season water deficits. However, deep root growth with large xylem diameters is also a fruitful method for the uptake of deep water when it is available in the lower soil profile. Moreover, root growth according to the distribution of water available in the soil is another trait that increases yield under drought conditions (Wasson et al., 2012). When the surface soil dries, the deep soil may still contain water; thus, several agricultural species have developed deeper root systems to obtain access to this deep water. To use this deep water in the soil profile, root architecture, including the greater hierarchical structure, can support hydraulic lift for increased utilization of water (Comas et al., 2013; Doussan et al., 2006). Another mechanism is hydraulic redistribution, where water is available in the deep soil layers. Larger-diameter xylem vessels can increase the axial hydraulic conductivity of roots present in deeper soil profiles (Wasson et al., 2012). These processes, i.e., hydraulic lift and redistribution, may be sufficient to carry out transpiration and support of plants even in extreme drought periods (Comas et al., 2013). Thus, roots play a pivotal role in drought stress of cotton as well as other plants.

7.2.2 Jasmonic Acid

JA and its derivatives are another important phytohormone that responds to drought stress. Jasmonic acid (JA) is derived from α-linolenic acid. JA and its active derivatives, which are known as jasmonates, have a significant role in regulating stress responses of plants to various biotic as well as abiotic stresses (Ullah et al., 2017). In addition to plant growth and development, JA is also involved in root growth, fruit ripening, tendril coiling and viable pollen production (Wasternack, 2007). JA has been shown to participate in the response to drought conditions. Genome-wide functional analyses of cotton were performed to analyze the molecular mechanism of drought resistance, and they identified various genes related to JA signaling pathways (Chen et al., 2013). Similarly, several studies have shown that exogenous application of jasmonates enhances plant resistance to water-deficit conditions (Bandurska et al., 2003). Similar to ABA, various studies have shown that jasmonates also participate in the regulation of stomatal closure (Riemann et al., 2015).

JA induces drought tolerance in plants in various ways including stomata closure, the scavenging of ROS and root development. Similar to ABA, various studies have shown that jasmonates also participate in the regulation of stomata closure and stomata regulation, which have a vital role in the response to drought stress (Munemasa et al., 2007; Riemann et al., 2015). In a case, when the JA precursor, i.e., 12-OPDA was treated to Arabidopsis, stomata closure was observed. In addition, elevated levels of OPDA are associated with decreased stomatal aperture and consequently drought tolerance was enhanced in *Arabidopsis*. It was proposed that the conversion of 12-OPDA to JA was prevented by drought stress and then OPDA acted either independently or together with ABA to induce stomatal closure and enhance drought tolerance (Kazan, 2015; Savchenko et al., 2014). Enhanced scavenging of ROS was reported by Fang et al., (2016) in the transgenic plants carrying overexpressed *VaNAC26* which exhibited higher tolerance to drought stress. They revealed that JA-synthesis related genes were upregulated in the overexpressing lines during drought as well as in normal conditions. Exogenous application of JA has also been found to ameliorate the response of plants to drought stress. Numerous researchers reported that exogenous application of JA increased antioxidant activity in plants under drought conditions (Ai et al., 2008; Bandurska et al., 2003; Nafie et al., 2011; Shan et al., 2015). In a case, exogenous application of JA significantly enhanced glutathione reductase (GR), ascorbate peroxidase (APX), dehydroascorbate reductase (DHAR) and monodehydroascorbate reductase (MDHAR) under drought stress in wheat seedlings (Shan et al., 2015). Moreover, the role of JA has also been identified in modulating root hydraulic conductivity to uptake water from the soil in limited-moisture conditions (Sánchez-Romera et al., 2014). Ollas et al. (2013) affirmed that a transient accumulation of JA in roots is needed to increase ABA under drought stress. However, the functions of JA still need to be elucidated under drought stress so that plants with better drought tolerance can be attained.

Traditionally, salicylic acid (SA), JAs, and ET are associated with plant defense, whereas gibberellins (GAs), auxins (IAAs), brassinosteroids (BRs), and cytokinins are associated with plant development. Abscisic acid (ABA) is the main hormone that regulates plant responses to abiotic stresses. However, it is becoming increasingly evident that all plant hormones can have direct and/or indirect effects on multiple plant functions. For instance, SA, JA and ET are also involved in plant development and abiotic stress tolerance, whereas IAAs and GAs play roles in biotic and abiotic stress tolerance (Kazan, 2015).

7.2.3 Auxin

As a well-known plant phytohormone, auxin (Indole-3-acetic acid, IAA) is an important regulator of several processes including seed dormancy, tropistic growth, root patterning, cell differentiation and flower organ development. In recent years, evidence has increasingly suggested a potential link between auxin response and abiotic and biotic stresses (Shi et al., 2014). Recently, it has been reported that auxins are the crucial player in regulating drought stress response. The YUC (YUCCA) family of flavin monooxygenases play a crucial role in auxin biosynthesis as it converts IPA (indole-3-pyruvate) into IAA (Naser and Shani, 2016). Several YUC genes in Arabidopsis are upregulated only hours after exposure to dehydration (2 h air drying on paper). The Arabidopsis activation-tagged mutant, yuc7–1D, in which YUC7 is upregulated, had higher total auxin levels compared to WT but with similar free auxin levels. The plants exhibited phenotypes typical for an auxin overproducer, such as tall, slender stems, enhanced apical dominance, curled narrow leaves and drought resistance (high water-deficit survival rates) (Lee et al., 2012). In potato (*Solanum tuberosum*) and poplar (*Populus alba*, *P. glandulosa*), plants overexpressing Arabidopsis YUC6 displayed phenotypes such as narrow downward-curled leaves and increased height, suggesting a high auxin content. These phenotypes were accompanied by enhanced drought tolerance (water withdrawal) due to reduced water loss (Im Kim et al., 2013; Ke et al., 2015). Consistent with an increased drought resistance obtained by auxin overproduction, *Arabidopsis* iaaM-OX transgenic lines, where higher IAA levels can be obtained by constitutively expressing the *Agrobacterium tumefaciens* tryptophan monooxygenase (iaaM) gene, exhibited enhanced drought stress resistance (withdrawal of water for 21 days). At the same time, yuc1/yuc2/yuc6 triple mutants with lower endogenous IAA levels showed decreased stress resistance. Additional experiments suggest that auxin can promote drought stress resistance by modulating root architecture, ABA (abscisic acid)-responsive genes expression, and ROS (reactive oxygen species) metabolism (Shi et al., 2014). These studies revealed the positive role of auxins in drought-tolerance mechanisms.

7.2.4 Ethylene

ET (C_2H_4) is a naturally occurring gaseous hormone with multiple actions including fruit ripening, growth, senescence, seed germination, flowering, and response to various stresses (Arraes et al., 2015). It has been reported recently that ET also plays an active role in the regulation of drought tolerance in plants. In contrast to JAs, which promote stomatal closure, ET has been implicated in both stomatal opening and closure (Daszkowska-Golec and Szarejko, 2013). On the one hand, ET inhibits ABA-induced stomatal closure, and the ET overproducing mutant eto1 closes its stomata more slowly than wild-type plants during drought stress. By contrast, ET signaling components EIN3 and ETR1 do not affect stomatal closure (Tanaka et al., 2005). On the other hand, ET promotes stomatal closure by activating NADPH oxidase-mediated ROS production in stomatal guard cells (Desikan et al., 2006). ET-mediated production of antioxidant flavanols, which accumulate in guard cells under stress conditions in an EIN2-dependent manner, negatively regulates stomatal closure by inhibiting ROS production (Watkins et al., 2014). Possible effects of ET-mediated stomatal movements on drought tolerance require further investigation. Similarly to the Arabidopsis etol mutant, the rice etoll mutant with an inactive ETHYLENE OVERPRODUCER 1-LIKE (OsETOL1) protein contains elevated ET levels and shows increased tolerance to drought, while OsETOL1 overexpressing plants show reduced ET accumulation and reduced drought tolerance. Similarly to its Arabidopsis counterpart, OsETOL1 interacts with the type II ACS enzyme OsACS2 and reduces its activity (Du et al., 2014).

7.2.5 Cytokinins

CKs play an important role in the regulation of various biological processes in plants, including growth and development as well as acclimation/adaptation to environmental stresses (Ha et al., 2012). CK homeostasis in plant cells is tightly regulated by the concerted actions of both isopentenyltransferase (IPT) and CK oxidase/dehydrogenase (CKX) enzymes (Li et al., 2016). A total of nine IPT and seven CKX genes were identified in the model *Arabidopsis thaliana* (Ha et al., 2012; Kieber and Schaller, 2014). To cope with unfavorable water-deficit conditions, plants use sophisticated and complex mechanisms to adjust endogenous CK levels to appropriately control the CK-signal transduction pathway and its downstream genes. In the past several years, intensive research has been conducted to find an answer to how plants control CK homeostasis and the CK-signaling network for acclimation and adaptation on perceiving drought. A growing body of evidence has indicated the multifaceted nature of CKs; that is, CKs can have both negative and positive effects on drought tolerance (Ha et al., 2012; Zwack and Rashotte, 2015). Likewise, CK levels may decrease or increase depending on stress duration or intensity (Zwack and Rashotte, 2015). The positive effect of CKs on drought tolerance was elegantly evidenced by a carefully designed enhancement of endogenous CK levels in transgenic plants that led to improved drought tolerance through delaying senescence (Rivero et al., 2007). Regarding the negative effect of CKs on drought tolerance, studies of drought responses using Arabidopsis CK-deficient mutants, such as CKX1, CKX2, CKX3, or CKX4 overexpressors or iptl/3/5/7 mutants, or CK-signaling mutants, such as ahk2/3, ahp2/3/5, or arrl/10/12 plants, have provided evidence that CKs, and thus CK signaling, can act as negative regulators of plant drought adaptation (Nishiyama et al., 2011, 2013; Werner et al., 2010).

7.2.6 Gibberellins

Gibberellins (GAs) are tetracyclic diterpenoid carboxylic acids which function as a growth hormone and in response to abiotic stresses. They act throughout the plant life cycle to stimulate the growth of most organs through cell elongation and enhanced cell division; they also promote developmental phase transitions, juvenile and adult phase growth, and vegetative and reproductive development (Colebrook et al., 2014; Wang et al., 2008). For the first time, it came to our knowledge that GAs negatively regulate drought tolerance in plants by inhibiting the biosynthesis of GAs. GA-deficient mutants were created and checked for drought tolerance. It was found that they had enhanced stress tolerance as well as their dwarf growth habit (Gilley and Fletcher, 1998; Vettakkorumakankav et al., 1999). There has been relatively little material published on the influence of water deficit on GA metabolism, although osmotic stress, which is often used as a proxy for drought, was reported to reduce GA content in maize leaves (Wang et al., 2008). In addition to promoting tolerance to submergence, the *SublA* gene also promotes tolerance to dehydration and drought stress (Fukao et al., 2011). On de-submergence (which triggers dehydration), and after withholding water, *SublA* lines maintained a higher leaf relative water content, showed reduced ROS accumulation and oxidative stress, and recovered their growth more effectively on re-watering. This enhanced tolerance was associated with increased responsiveness to ABA, and enhanced expression of both ABA-dependent and ABA-independent drought-responsive transcripts (Jung et al., 2010).

7.2.7 Salicylic Acid

Generally, salicylic acid (SA) is involved in the regulation of pathogenesis-related protein expression, leading to plant defense against biotrophic pathogens. It also plays an important role in the regulation of plant growth, development, ripening, flowering, and responses to abiotic stresses such as drought stress (Miura and Tada, 2014). In addition to ABA and JA, SA is also involved in the regulation of drought responses. Endogenous SA levels are increased up to five-fold in the evergreen shrub *Phillyrea angustifolia* (Munne-Bosch and Penuelas, 2003). The SA content in barley roots is increased approximately two-fold by water deficit (Bandurska and Stroiński, 2005). Furthermore, the SA-inducible genes *PR1* and *PR2* are induced by drought stress (Miura et al., 2013). The induction of SA accumulation may play a role in protective mechanisms during water stress. Stomata play an important role in the uptake of CO_2 and transpiration. Because stomata are pores in the epidermis, pathogens can enter unchallenged. After an attack by a pathogen, endogenous SA levels are increased to induce SAR. An increase in endogenous SA levels promotes stomatal closure. This closure is likely caused by the generation of ROS, which is induced by SA (Melotto et al., 2006). The exogenous application of SA also induces ROS, H_2O_2 and $Ca2^+$ accumulation, leading to stomatal closure (Miura and Tada, 2014).

However, the effect of SA on drought tolerance remains to be determined because some investigators have reported enhancement of drought tolerance by SA application whereas others have reported a reduction in drought tolerance. Generally, low concentrations of applied SA increase drought tolerance and high concentrations decrease it. As described above, SA induces ROS production in photosynthetic tissues (Borsani et al., 2001). Thus, the application of a high concentration of SA may cause high levels of oxidative stress, leading to decreased abiotic stress tolerance. Both drought tolerance and plant growth are suppressed when a high concentration (2–3 mM) of SA is applied to wheat seedlings, whereas plant growth is enhanced by the application of a low concentration (0.5 mM) of SA (Kang et al., 2012).

7.2.8 Brassinosteroids

Brassinosteroids (BRs) are ubiquitous plant steroid hormones, regulating an array of physiological and developmental processes including cell elongation and division, photomorphogenesis, xylem differentiation, seed germination, seed size, photosynthesis, stem and root growth, floral initiation, male fertility, development of flowers and fruits, fruit ripening, grain filling, leaf senescence and resistance to various biotic and abiotic stresses (Ahammed et al., 2015). 24-epibrassonolide (EBR) is well recognized as an important bioactive BR having a positive effect on photosynthesis (Yu et al., 2004). EBR treatment significantly alleviates drought stress by increasing net photosynthetic rate, relative water content and activity of antioxidant enzymes in tomato (Yuan et al., 2012). The content of malondialdehyde and H_2O_2 was remarkably reduced by BR treatment under drought stress (Yuan et al., 2010). EBR alleviates drought-induced deleterious effects on photosynthesis by increasing efficiency of light utilization and dissipation of excitation energy in the PSII antennae in pepper leaves (Hu et al., 2013). When *Chorispora bungeana* was challenged with polyethylene glycol (PEG)-imposed drought stress following foliar application of 0.1 μM EBR, EBR-treated plants showed improved drought tolerance compared to non-treated control plants based on different biochemical and physiological parameters (Li et al., 2012). It was reported that *Brassica juncea* plants challenged with a weeklong drought stress at an early growth stage (8–21 days after sowing, DAS) showed pronounced inhibition in growth and photosynthesis even at 60 DAS. However, a follow-up treatment with 28-homobrassinolide (HBL, 0.01 μM) at 30 DAS resulted in significantly improved growth and photosynthesis when measured at 60 DAS (Fariduddin et al., 2009). EBR-induced drought stress tolerance also involves alterations in the gene expression that encodes both structural and regulatory proteins. For instance, EBR treatment induces expression of drought-responsive genes such as *BnCBF5* and *BnDREB* which may in part confer drought stress tolerance to *Brassica napus* seedlings (Kagale et al., 2007).

7.3 SCAVENGING SYSTEM OF ROS

Partial reduction of atmospheric O_2 leads to the production of ROS, also known as active oxygen species (AOS) or reactive oxygen intermediates (ROI). Cellular ROS basically consists of four forms, hydrogen peroxide (H_2O_2), the hydroxyl radical ($HO^\bullet$), superoxide anion radical (O_2^-), and singlet oxygen ($^1O_2$) (Ullah et al., 2017). Two of these forms are especially reactive, i.e., $HO^\bullet$ and $^1O_2$. They can harm and oxidize various components of the cell, such as lipids, proteins, DNA and RNA. Eventually, they can result in cell death if the oxidation of cellular components is not controlled (Fang and Xiong, 2015). Sub-cellular locations, such as mitochondria, plasma membrane, cell wall, chloroplast and nucleus, are responsible for the production of ROS (Gill and Tuteja, 2010). Under drought stress, the production of these ROS increases in various ways. For example, a reduction in CO_2 fixation leads to decreased $NADP^+$ regeneration during the Calvin cycle, which will reduce the activity of the photosynthetic electron transport chain. Moreover, during drought conditions, there is excessive leakage of electrons to O_2 by the Mehler reaction during photosynthesis (Carvalho, 2008). The Mehler reaction reduces O_2 to O_2^- by donation of an electron in photosystem I. O_2^- can be converted to H_2O_2 by SOD. This can be further converted to water by ascorbate peroxidase (Heber, 2002). However, it is difficult to evaluate the levels of ROS produced during the Mehler reaction compared to those generated through photorespiration. Drought conditions also enhance the photorespiratory pathway, particularly when RuBP oxygenation is high due to limited CO_2 fixation. Noctor et al. (2002) found that approximately 70% of total H_2O_2 production occurs through photorespiration under drought stress.

Plants have developed complicated scavenging mechanisms and regulatory pathways to monitor the ROS redox homeostasis to prevent excess ROS in cells. Alterations in antioxidant enzyme metabolism could influence drought tolerance in cotton. The defense mechanism against ROS has been reviewed in detail by Das and Roychoudhury (2014). An antioxidant machinery has been developed by plants to ensure survival. It has two arms: (i) enzymatic components, such as catalase (CAT), SOD, ascorbate peroxidase (APX), glutathione reductase (GR), guaiacol peroxidase (GPX), dehydroascorbate reductase (NADH) and monodehydroascorbate reductase (MDAR) and (ii) non-enzymatic antioxidants, such as reduced glutathione (GSH), ascorbic acid (AA), α-tocopherol, flavonoids, carotenoids, and the osmolyte proline. To scavenge ROS, these two arms/components work together (Das and Roychoudhury, 2014; Wu et al., 2015). APX, along with MDAR, NADH and GR, removes H_2O_2 via the Halliwell–Asada pathway (Uzilday et al., 2012). APX reduces H_2O_2 to water by oxidizing ascorbate to MDHA and thus plays a key role in the ascorbate-glutathione cycle (de Azvedo Neto et al., 2006). MDHA is then reduced to ascorbate by MDHAR. However, two molecules of MDHA can be non-enzymatically converted to MDHA and dehydroascorbate, which is further reduced to ascorbate via the NADH and GR cycle (Szalai et al., 2009). In this cycle, glutathione (GSH) is reduced by GR oxidation to oxidized glutathione at the expense of NADPH (nicotinamide adenine dinucleotide phosphate). Glutathione reductase activity increased during drought stress to keep oxidized and reduced glutathione ratios at an adequate level (Chan et al., 2013). The balance between ROS production and antioxidative enzyme activities determines whether oxidative signaling and/or damage will occur (Zhang et al., 2014). In a study, Zhang et al. (2014) experimented with cotton cultivars: drought-resistant (CCRI-60) and drought-sensitive (CCRI-27). They found that the CCRI-60 cultivar was drought-tolerant due to increased root length and vigor, antioxidant enzyme activities and significantly increased GR activity and proline content. CCRI-60 has the ability to scavenge free radicals and provides better protection compared to CCRI-27; thus, it is more resistant to drought and has increased growth.

7.4 CONCLUSIONS

Recently, drought stress has become an extremely serious issue. It is responsible for extensive crop loss and will likely become worse in the future; thus, there is international interest in increasing drought-tolerant crops. The goal of our study is to explore the functions of phytohormones under water-deficit conditions. Among phytohormones, ABA is the key hormone regulating plant responses to environmental stresses i.e., drought and salt stress. These processes include root development, stomata regulation, photosynthesis, scavenging of ROS and activation of various drought-related genes. JA also plays a crucial role in responding to abiotic stresses including drought stress. Furthermore, auxins, gibberellins, ethylene, cytokinins and brassinosteroids are also involved directly or indirectly in response to drought stress. On the other hand, drought stress overproduces ROS, which are deleterious to cell and over-all plant. However, the scavenging of these ROS is possible through enzymatic and non-enzymatic ways. Numerous genes have been reported to regulate these phytohormones, ROS scavenging and other drought-related traits. Usually, a single gene used to up or downregulate is not adequate to enhance drought tolerance in plants. If two, three or more genes are up or downregulated in plants, this can potentially enhance the drought-tolerance capability of plants according to current needs.

REFERENCES

Ahammed, G.J., Xia, X.J., Li, X., Shi, K., Yu, J.Q., Zhou, Y.H. (2015) Role of brassinosteroid in plant adaptation to abiotic stresses and its interplay with other hormones. *Current Protein and Peptide Science* 16(5): 462–473.

Ai, L., Li, Z.H., Xie, Z.X., Tian, X.L., Eneji, A.E., Duan, L.S. (2008) Coronatine alleviates polyethylene glycol-induced water stress in two rice (*Oryza sativa* L.) cultivars. *Journal of Agronomy and Crop Science* 194(5): 360–368.

Arraes, F.B., Beneventi, M.A., Lisei de Sa, M.E., Paixao, J.F., Albuquerque, E.V., Marin, S.R., Purgatto, E., Nepomuceno, A.L., Grossi-de-Sa, M.F. (2015) Implications of ethylene biosynthesis and signaling in soybean drought stress tolerance. *BMC Plant Biology* 15(1): 213.

Bandurska, H., Stroiński, A. (2005) The effect of salicylic acid on barley response to water deficit. *Acta Physiologiae Plantarum* 27(3): 379–386.

Bandurska, H., Stroiński, A., Kubis, J. (2003) The effect of jasmonic acid on the accumulation of ABA, proline and spermidine and its influence on membrane injury under water deficit in two barley genotypes. *Acta Physiologiae Plantarum* 25(3): 279–285.

Basu, S., Ramegowda, V., Kumar, A., Pereira, A. (2016) Plant adaptation to drought stress. *F1000Research* 5(F1000 Faculty Rev): 1554.

Borsani, O., Valpuesta, V., Botella, M.A. (2001) Evidence for a role of salicylic acid in the oxidative damage generated by NaCl and osmotic stress in *Arabidopsis* seedlings. *Plant Physiology* 126(3): 1024–1030.

Chan, K.X., Wirtz, M., Phua, S.Y., Estavillo, G.M., Pogson, B.J. (2013) Balancing metabolites in drought: The sulfur assimilation conundrum. *Trends in Plant Science* 18(1): 18–29.

Chen, Y., Liu, Z.H., Feng, L., Zheng, Y., Li, D.D., Li, X.B. (2013) Genome-wide functional analysis of cotton (*Gossypium hirsutum*) in response to drought. *PLOS ONE* 8(11): e80879.

Colebrook, E.H., Thomas, S.G., Phillips, A.L., Hedden, P. (2014) The role of gibberellin signalling in plant responses to abiotic stress. *The Journal of Experimental Biology* 217(1): 67–75.

Comas, L.H., Becker, S.R., Cruz, V.M.V., Byrne, P.F., Dierig, D.A. (2013) Root traits contributing to plant productivity under drought. Frontiers in Plant Science 4: 442.

Cruz de Carvalho, M.H. (2008) Drought stress and reactive oxygen species: Production, scavenging and signaling. *Plant Signaling and Behavior* 3(3): 156–165.

Cutler, S.R., Rodriguez, P.L., Finkelstein, R.R., Abrams, S.R. (2010) Abscisic acid: Emergence of a core signaling network. *Annual Review of Plant Biology* 61: 651–679.

Das, K., Roychoudhury, A. (2014) Reactive oxygen species (ROS) and response of antioxidants as ROS-scavengers during environmental stress in plants. *Frontiers in Environmental Science* 2: 53.

Daszkowska-Golec, A., Szarejko, I. (2013) Open or close the gate – Stomata action under the control of phytohormones in drought stress conditions. *Frontiers in Plant Science* 4: 138.

de Azvedo Neto, A.D., Prisco, J.T., Eneas-Filho, J., de Abreao, C.E.Bd, Gomes-Filho, E. (2006) Effect of salt stress on antioxidative enzymes and lipid peroxidation on leaves and roots of salt-tolerant and salt-sensitive maize genotypes. *Environmental and Experimental Botany* 56(1): 87–94.

Desikan, R., Last, K., Harrett-Williams, R., Tagliavia, C., Harter, K., Hooley, R., Hancock, J.T., Neill, S.J. (2006) Ethylene-induced stomatal closure in Arabidopsis occurs via AtrbohF-mediated hydrogen peroxide synthesis. *The Plant Journal: for Cell and Molecular Biology* 47(6): 907–916

Doussan, C., Pierret, A., Garrigues, E., Pagès, L. (2006) Water uptake by plant roots: II – modelling of water transfer in the soil root-system with explicit account of flow within the root system – comparison with experiments. *Plant Soil* 283: 99–117.

Du, H., Wu, N., Cui, F., You, L., Li, X., Xiong, L. (2014) A homolog of ETHYLENE OVERPRODUCER, OsETOL1, differentially modulates drought and submergence tolerance in rice. *The Plant Journal: for Cell and Molecular Biology* 78(5): 834–849.

Fahad, S., Bano, A. (2012) Effect of salicylic acid on physiological and biochemical characterization of maize grown in saline area. Pakistan journal of botany 44: 1433–1438.

Fahad, S., Chen, Y., Saud, S., Wang, K., Xiong, D., Chen, C., Wu, C., *et al.* (2013) Ultraviolet radiation effect on photosynthetic pigments, biochemical attributes, antioxidant enzyme activity and hormonal contents of wheat. *Journal of Food, Agriculture and Environment* 11: 1635–1641.

Fahad, S., Hussain, S., Bano, A., Saud, S., Hassan, S., Shan, D., Khan, F.A., *et al.* (2015a) Potential role of phytohormones and plant growth-promoting rhizobacteria in abiotic stresses: Consequences for changing environment. *Environmental Science and Pollution Research International* 22(7): 4907–4921.

Fahad, S., Hussain, S., Matloob, A., Khan, F.A., Khaliq, A., Saud, S., Hassan, S., *et al.* (2015b) Phytohormones and plant responses to salinity stress: A review. *Plant Growth Regulation* 75(2): 391–404. doi:10.1007/s10725-014-0013-y.

Fahad, S., Hussain, S., Saud, S., Hassan, S., Chauhan, B.S., Khan, F., Ihsan, M.Z., *et al.* (2016a) Responses of rapid viscoanalyzer profile and other rice grain qualities to exogenously applied plant growth regulators under high day and high night temperatures. *PLOS ONE* 11(7): e0159590. doi:10.1371/journal.pone.0159590.

Fahad, S., Hussain, S., Saud, S., Hassan, S., Ihsan, Z., Shah, A.N., Wu, C., *et al.* (2016b) Exogenously applied plant growth regulators enhance the morphophysiological growth and yield of rice under high temperature. *Frontiers in Plant Science* 7: 1250.

Fahad, S., Hussain, S., Saud, S., Hassan, S., Tanveer, M., Ihsan, M.Z., Shah, A.N., *et al.* (2016c) A combined application of biochar and phosphorus alleviates heat-induced adversities on physiological, agronomical and quality attributes of rice. *Plant Physiology and Biochemistry: PPB* 103: 191–198.

Fahad, S., Hussain, S., Saud, S., Khan, F., Hassan, S., Amanullah, Nasim, W., *et al.* (2016d) Exogenously applied plant growth regulators affect heat-stressed rice pollens. *Journal of Agronomy and Crop Science* 202(2): 139–150.

Fahad, S., Hussain, S., Saud, S., Tanveer, M., Bajwa, A.A., Hassan, S., Shah, A.N., *et al.* (2015c) A biochar application protects rice pollen from high-temperature stress. *Plant Physiology and Biochemistry: PPB* 96: 281–287.

Fahad, S., Ihsan, M.Z., Khaliq, A., Daur, I., Saud, S., Alzamanan, S., Nasim, W., *et al.* (2018)Consequences of high temperature under changing climate optima for rice pollen characteristics-concepts and perspectives. *Archives of Agronomy and Soil Science* 64(11): 1473–1488. doi:10.1080/03650340.2018.1443213.

Fahad, S., Nie, L., Chen, Y., Wu, C., Xiong, D., Saud, S., Hongyan, L., Cui, K., Huang, J. (2015d) Crop plant hormones and environmental stress. *Sustainable Agriculture Reviews* 15: 371–400.

Fang, L., Su, L., Sun, X., Li, X., Sun, M., Karungo, S.K., Fang, S., *et al.* (2016) Expression of *Vitis amurensis* NAC26 in Arabidopsis enhances drought tolerance by modulating jasmonic acid synthesis. *Journal of Experimental Botany* 67(9): 2829–2845.

Fang, Y., Xiong, L. (2015) General mechanisms of drought response and their application in drought resistance improvement in plants. *Cellular and Molecular Life Sciences: CMLS* 72(4): 673–689.

Fariduddin, Q., Khanam, S., Hasan, S.A., Ali, B., Hayat, S., Ahmad, A. (2009) Effect of 28 homobrassinolide on the drought stress induced changes in photosynthesis and antioxidant system of *Brassica juncea* L. *Acta Physiologiae Plantarum* 31(5): 889–897.

Farooq, M., Hussain, M., Wahid, A., Siddique, K.H.M. (2012) Drought stress in plants: An overview. In *Plant Responses to Drought Stress*. Springer, Berlin: 1–33.

Fernando, V.D., Schroeder, D.F. (2016) Role of ABA in Arabidopsis salt, drought, and desiccation tolerance. In *Abiotic and Biotic Stress in Plants-Recent Advances and Future Perspectives*, pp. 507–525. InTech, Rijeka.

Fujita, Y., Fujita, M., Shinozaki, K., Yamaguchi-Shinozaki, K. (2011) ABA-mediated transcriptional regulation in response to osmotic stress in plants. *Journal of Plant Research* 124(4): 509–525.

Fukao, T., Yeung, E., Bailey-Serres, J. (2011) The submergence tolerance regulator SUB1A mediates crosstalk between submergence and drought tolerance in rice. *The Plant Cell* 23(1): 412–427.

Gill, S.S., Tuteja, N. (2010) Reactive oxygen species and antioxidant machinery in abiotic stress tolerance in crop plants. *Plant Physiology and Biochemistry: PPB* 48(12): 909–930.

Gilley, A., Fletcher, R.A. (1998) Gibberellin antagonizes paclobutrazol-induced stress protection in wheat seedlings. *Journal of Plant Physiology* 153(1–2): 200–207.

Ha, S., Vankova, R., Yamaguchi-Shinozaki, K., Shinozaki, K., Tran, L.S. (2012) Cytokinins: Metabolism and function in plant adaptation to environmental stresses. *Trends in Plant Science* 17(3): 172–179.

Heber, U. (2002) Irrungen, Wirrungen? The Mehler reaction in relation to cyclic electron transport in C3 plants. *Photosynthesis Research* 73(1–3): 223–231.

Hu, W.H., Yan, X.H., Xiao, Y.A., Zeng, J.J., Qi, H.J., Ogweno, J.O. (2013) 24-Epibrassinosteroid alleviate drought-induced inhibition of photosynthesis in Capsicum annuum. Scientia Horticulturae 150: 232–237.

Im Kim, J.I., Baek, D., Park, H.C., Chun, H.J., Oh, D.H., Lee, M.K., Cha, J.Y., *et al.* (2013) Overexpression of Arabidopsis YUCCA6 in potato results in high-auxin developmental phenotypes and enhanced resistance to water deficit. *Molecular Plant* 6(2): 337–349.

Jung, K.H., Seo, Y.S., Walia, H., Cao, P., Fukao, T., Canlas, P.E., Amonpant, F., Bailey-Serres, J., Ronald, P.C. (2010) The submergence tolerance regulator Sub1A mediates stress-responsive expression of AP2/ERF transcription factors. *Plant Physiology* 152(3): 1674–1692.

Kagale, S., Divi, U.K., Krochko, J.E., Keller, W.A., Krishna, P. (2007) Brassinosteroid confers tolerance in *Arabidopsis thaliana* and *Brassica napus* to a range of abiotic stresses. *Planta* 225(2): 353–364.

Kang, G., Li, G., Xu, W., Peng, X., Han, Q., Zhu, Y., Guo, T. (2012) Proteomics reveals the effects of salicylic acid on growth and tolerance to subsequent drought stress in wheat. *Journal of Proteome Research* 11(12): 6066–6079.

Kazan, K. (2015) Diverse roles of jasmonates and ethylene in abiotic stress tolerance. *Trends in Plant Science* 20(4): 219–229.

Ke, Q., Wang, Z., Ji, C.Y., Jeong, J.C., Lee, H.S., Li, H., Xu, B., Deng, X., Kwak, S.S. (2015) Transgenic poplar expressing Arabidopsis YUCCA6 exhibits auxin-overproduction phenotypes and increased tolerance to abiotic stress. *Plant Physiology and Biochemistry: PPB* 94: 19–27.

Kieber, J.J., Schaller, G.E. (2014) Cytokinins. *The Arabidopsis Book* 12: e0168.

Lee, M., Jung, J.H., Han, D.Y., Seo, P.J., Park, W.J., Park, C.M. (2012) Activation of a Flavin monooxygenase gene YUCCA7 enhances drought resistance in Arabidopsis. *Planta* 235(5): 923–938.

Li, W., Herrera-Estrella, L., Tran, L.P. (2016) The Yin–Yang of cytokinin homeostasis and drought acclimation/adaptation. *Trends in Plant Science* 21(7): 548–550.

Li, Y.H., Liu, Y.J.,Xu, X.L., Jin, M., An, L.Z., Zhang, H. (2012) Effect of 24-epibrassinolide on drought stress-induced changes in *Chorispora bungeana*. *Biologia Plantarum* 56(1): 192–196.

Liu, G., Li, X., Jin, S., Liu, X., Zhu, L., Nie, Y., Zhang, X. (2014) Overexpression of rice NAC gene SNAC1 improves drought and salt tolerance by enhancing root development and reducing transpiration rate in transgenic cotton. *PLOS ONE* 9(1): e86895.

Luo, L.J. (2010) Breeding for water-saving and drought-resistance rice (WDR) in China. *Journal of Experimental Botany* 61(13): 3509–3517.

Manavalan, L.P., Guttikonda, S.K., Tran, L.S.P., Nguyen, H.T. (2009) Physiological and molecular approaches to improve drought resistance in soybean. *Plant and Cell Physiology* 50(7): 1260–1276.

Melotto, M., Underwood, W., Koczan, J., Nomura, K., He, S.Y. (2006) Plant stomata function in innate immunity against bacterial invasion. *Cell* 126(5): 969– 980.

Miura, K., Okamoto, H., Okuma, E., Shiba, H., Kamada, H., Hasegawa, P.M., Murata, Y. (2013) SIZ1 deficiency causes reduced stomatal aperture and enhanced drought tolerance via controlling salicylic acid-induced accumulation of reactive oxygen species in Arabidopsis. *Plant Journal* 73(1): 91–104.

Miura, K., Tada, Y. (2014) Regulation of water, salinity, and cold stress responses by salicylic acid. *Frontiers in Plant Science* 5: 4.

Munemasa, S., Oda, K., Watanabe-Sugimoto, M., Nakamura, Y., Shimoishi, Y., Murata, Y. (2007) The coronatine-insensitive 1 mutation reveals the hormonal signaling interaction between abscisic acid and methyl jasmonate in Arabidopsis guard cells. Specific impairment of ion channel activation and second messenger production. *Plant Physiology* 143(3): 1398–1407.

Munne-Bosch, S., Penuelas, J. (2003) Photo- and antioxidative protection, and a role for salicylic acid during drought and recovery in field-grown *Phillyrea angustifolia* plants. *Planta* 217(5): 758–766.

Nafie, E., Hathout, T., Mokadem, A.S.A. (2011) Jasmonic acid elicits oxidative defense and detoxification systems in *Cucumis melo* L. cells. *Brazilian Journal of Plant Physiology* 23(2): 161–174.

Naser, V., Shani, E. (2016) Auxin response under osmotic stress. *Plant Molecular Biology* 91(6): 661–672.

Nezhadahmadi, A., Prodhan, Z.H., Faruq, G. (2013) Drought tolerance in wheat. *The Scientific World Journal* 2013: 610721.

Nishiyama, R., Watanabe, Y., Fujita, Y., Le, D.T., Kojima, M., Werner, T., Vankova, R., *et al.* (2011) Analysis of cytokinin mutants and regulation of cytokinin metabolic genes reveals important regulatory roles of cytokinins in drought, salt and abscisic acid responses, and abscisic acid biosynthesis. *The Plant Cell* 23(6): 2169–2183.

Nishiyama, R., Watanabe, Y., Leyva-Gonzalez, M.A., Ha, C.V., Fujita, Y., Tanaka, M., Seki, M., *et al.* (2013) Arabidopsis AHP2, AHP3, and AHP5 histidine phosphotransfer proteins function as redundant negative regulators of drought stress response. *Proceedings of the National Academy of Sciences of the United States of America* 110(12): 4840–4845.

Noctor, G., Veljovic-Jovanovic, S., Driscoll, S., Novitskaya, L., Foyer, C.H. (2002) Drought and oxidative load in the leaves of C3 plants: A predominant role for photorespiration? *Annals of Botany* 89: 841–850.

Riemann, M., Dhakarey, R., Hazman, M., Miro, B., Kohli, A., Nick, P. (2015) Exploring jasmonates in the hormonal network of drought and salinity responses. *Frontiers in Plant Science* 6: 1077.

Rivero, R.M., Kojima, M., Gepstein, A., Sakakibara, H., Mittler, R., Gepstein, S., Blumwald, E. (2007) Delayed leaf senescence induces extreme drought tolerance in a flowering plant. *Proceedings of the National Academy of Sciences of the United States of America* 104(49): 19631–19636.

Rowe, J.H., Topping, J.F., Liu, J., Lindsey, K. (2016) Abscisic acid regulates root growth under osmotic stress conditions via an interacting hormonal network with cytokinin, ethylene and auxin. *The New Phytologist* 211(1): 225–239.

Sah, S.K., Reddy, K.R., Li, J. (2016) Abscisic acid and abiotic stress tolerance in crop plants. *Frontiers in Plant Science* 7: 571.

Sánchez-Romera, B., Ruiz-Lozano, J.M., Li, G., Luu, D.T., Martínez-Ballesta, Mdel C., Carvajal, M., Zamarreño, A.M., *et al.* (2014) Enhancement of root hydraulic conductivity by methyl jasmonate and the role of calcium and abscisic acid in this process. *Plant, Cell and Environment* 37(4): 995–1008.

Saud, S., Chen, Y., Fahad, S., Hussain, S., Na, L., Xin, L., Alhussien, S.A.A.F.E. (2016) Silicate application increases the photosynthesis and its associated metabolic activities in Kentucky bluegrass under drought stress and post-drought recovery. *Environmental Science and Pollution Research International* 23(17): 17647–17655. doi:10. 1007/s11356-016-6957-x.

Saud, S., Chen, Y., Long, B., Fahad, S., Sadiq, A. (2013) The different impact on the growth of cool season turf grass under the various conditions on salinity and draught stress. *International Journal of Agricultural Science and Research* 3: 77–84.

Saud, S., Li, X., Chen, Y., Zhang, L., Fahad, S., Hussain, S., Sadiq, A., Chen, Y. (2014) Silicon application increases drought tolerance of Kentucky bluegrass by improving plant water relations and morpho physiological functions. *The Scientific World Journal* 2014: 368694. doi:10.1155/2014/368694.

Savchenko, T., Kolla, V.A., Wang, C.Q., Nasafi, Z., Hicks, D.R., Phadungchob, B., Chehab, W.E., *et al.* (2014) Functional convergence of oxylipin and abscisic acid pathways controls stomatal closure in response to drought. *Plant Physiology* 164(3): 1151–1160.

Shan, C., Zhou, Y., Liu, M. (2015) Nitric oxide participates in the regulation of the ascorbate-glutathione cycle by exogenous jasmonic acid in the leaves of wheat seedlings under drought stress. *Protoplasma* 252(5): 1397–1405.

Sharp, R.E., Poroyko, V., Hejlek, L.G., Spollen, W.G., Springer, G.K., Bohnert, H.J., Nguyen, H.T. (2004) Root growth maintenance during water deficits: Physiology to functional genomics. *Journal of Experimental Botany* 55(407): 2343–2351.

Sharp, R.E., Wu, Y., Voetberg, G.S., Saab, I.N., LeNoble, M.E. (1994) Confirmation that abscisic acid accumulation is required for maize primary root elongation at low water potentials. *Journal of Experimental Botany* 45(Special_Issue): 1743–1751.

Shi, H., Chen, L., Ye, T., Liu, X., Ding, K., Chan, Z. (2014) Modulation of auxin content in *Arabidopsis* confers improved drought stress resistance. *Plant Physiology and Biochemistry: PPB* 82: 209–217.

Szalai, G., Kellos, T., Galiba, G., Kocsy, G. (2009) Glutathione as an antioxidant and regulatory molecule in plants under abiotic stress conditions. *Journal of Plant Growth Regulation* 28(1): 66–80.

Tanaka, Y., Sano, T., Tamaoki, M., Nakajima, N., Kondo, N., Hasezawa, S. (2005) Ethylene inhibits abscisic acid-induced stomatal closure in Arabidopsis. *Plant Physiology* 138(4): 2337–2343.

Ullah, A., Heng, S., Munis, M.F.H., Fahad, S., Yang, X. (2015) Phytoremediation of heavy metals assisted by plant growth promoting (PGP) bacteria: A review. *Environmental and Experimental Botany* 117: 28–40.

Ullah, A., Sun, H., Yang, X., Zhang, X. (2017) Drought coping strategies in cotton: Increased crop per drop. *Plant Biotechnology Journal* 15(3): 271–284. doi.org/10.1111/pbi.12688.

Uzilday, B., Turkan, I., Sekmen, A.H., Ozgur, R., Karakaya, H.C. (2012) Comparison of ROS formation and antioxidant enzymes in *Cleome gynandra* (C4) and *Cleome spinosa* (C3) under drought stress. *Plant Science* 182: 59–70.

Vettakkorumakankav, N.N., Falk, D., Saxena, P., Fletcher, R.A. (1999) A crucial role for gibberellins in stress protection of plants. *Plant and Cell Physiology* 40(5): 542–548.

Wang, C., Yang, A., Yin, H., Zhang, J. (2008) Influence of water stress on endogenous hormone contents and cell damage of maize seedlings. *Journal of Integrative Plant Biology* 50(4): 427–434.

Wani, S.H., Kumar, V., Shriram, V., Sah, S.K. (2016) Phytohormones and their metabolic engineering for abiotic stress tolerance in crop plants. *The Crop Journal* 4(3): 162–176.

Wasson, A.P., Richards, R.A., Chatrath, R., Misra, S.C., Prasad, S.V., Rebetzke, G.J., Kirkegaard, J.A., Christopher, J., Watt, M. (2012) Traits and selection strategies to improve root systems and water uptake in water-limited wheat crops. *Journal of Experimental Botany* 63(9): 3485–3498.

Wasternack, C. (2007) Jasmonates: An update on biosynthesis, signal transduction and action in plant stress response, growth and development. *Annals of Botany* 100(4): 681–697.

Watkins, J.M., Hechler, P.J., Muday, G.K. (2014) Ethylene-induced flavonol accumulation in guard cells suppresses reactive oxygen species and moderates stomatal aperture. *Plant Physiology* 164(4): 1707–1717.

Werner, T., Nehnevajova, E., Köllmer, I., Novák, O., Strnad, M., Krämer, U., Schmülling, T. (2010) Root-specific reduction of cytokinin causes enhanced root growth, drought tolerance, and leaf mineral enrichment in *Arabidopsis* and tobacco. *The Plant Cell* 22(12): 3905–3920.

Wu, S., Hu, C., Tan, Q., Li, L., Shi, K., Zheng, Y., Sun, X. (2015) Drought stress tolerance mediated by zinc-induced antioxidative defense and osmotic adjustment in cotton (*Gossypium hirsutum*). *Acta Physiologiae Plantarum* 37(8): 167.

Xiong, L., Wang, R.G., Mao, G., Koczan, J.M. (2006) Identification of drought tolerance determinants by genetic analysis of root response to drought stress and abscisic acid. *Plant Physiology* 142(3): 1065–1074.

Yu, J.Q., Huang, L.F., Hu, W.H., Zhou, Y.H., Mao, W.H., Ye, S.F., Nogues, S. (2004) A role for brassinosteroids in the regulation of photosynthesis in *Cucumis sativus*. *Journal of Experimental Botany* 55(399): 1135–1143.

Yuan, G.F., Jia, C.G., Li, Z.,Sun, B., Zhang, L.P., Liu, N., Wang, Q.M. (2010) Effect of brassinosteroids on drought resistance and abscisic acid concentration in tomato under water stress. *Scientia Horticulturae* 126(2): 103–108.

Yuan, L.Y., Yuan, Y.H., Du, J.,Sun, J., Guo, S.R. (2012) Effects of 24-epibrassinolide on nitrogen metabolism in cucumber seedlings under $Ca(NO_3)_2$ stress. *Plant Physiology and Biochemistry* 61: 29–35.

Zhang, L., Peng, J., Chen, T.T., Zhao, X.H., Zhang, S.P., Liu, S.D., Dong, H.L., *et al.* (2014) Effect of drought stress on lipid peroxidation and proline content in cotton roots. *Journal of Animal and Plant Sciences* 24: 1729–1736.

Zhu, J.K. (2002) Salt and drought stress signal transduction in plants. *Annual Review of Plant Biology* 53(1): 247–273.

Zwack, P.J., Rashotte, A.M. (2015) Interactions between cytokinin signalling and abiotic stress responses. *Journal of Experimental Botany* 66(16): 4863–4871.

8 Strigolactones
Mediators of Abiotic Stress Response and Weakness in Parasite Attraction

Denitsa Teofanova, Mariela Odjakova, Nabil Abumhadi, and Lyuben Zagorchev

CONTENTS

8.1 INTRODUCTION

8.1.1 Strigolactones – Chemical Structure and Biosynthetic Pathway

8.1.1.1 Structural Diversity and Related Compounds

Strigolactones (SL) were defined as key phytohormones only recently (Koltai, 2011), although they were first identified in the middle of the twentieth century (Cook et al., 1966). The first member, because of its role as germination stimulant for the parasitic plant *Striga lutea* Lour., was named strigol with the general chemical formula $C_{19}H_{22}O_6$. Since then, numerous other strigolactones were identified in every plant species studied (Soto et al., 2010). Beside being potent germination stimulants for parasitic plants of the Orobanchaceae family (Matusova et al., 2005), namely *Striga* spp. and *Orobanche* spp., SLs also participate in a variety of developmental processes such as inhibition of shoot branching (Dun et al., 2013, Gomez-Roldan et al., 2008), root elongation (Sun et al., 2016), response to abiotic stresses (Pandey, Sharma and Pandey, 2016) and arbuscular mycorrhizal (AM) symbiosis (Gough and Bécard, 2016). Notably, SLs can exhibit their action endogenously (Gomez-Roldan et al., 2008) or as external signal molecules, released by roots (Rasmann and Turlings, 2016). Because of their emerging role in various aspects of plant life, virtually hundreds of studies on SLs studies were reported in recent years.

Currently, SLs are classified into two major groups – canonical and non-canonical SLs (Yoneyama et al., 2017). Canonical SLs contain an ABC three-ring structure, enol-ether-linked to a D (methylbutenolide) ring (Figure 8.1). Currently, twenty-three different canonical SLs (Figure 8.2) have been characterized (Yoneyama et al., 2017). Based on the stereochemical orientation of the C ring, canonical SLs are subdivided into strigol and orobanchol type. In contrast, non-canonical SLs miss the ABC ring system (Figure 8.3), but contain the biologically essential enol-ether D ring. Currently, the

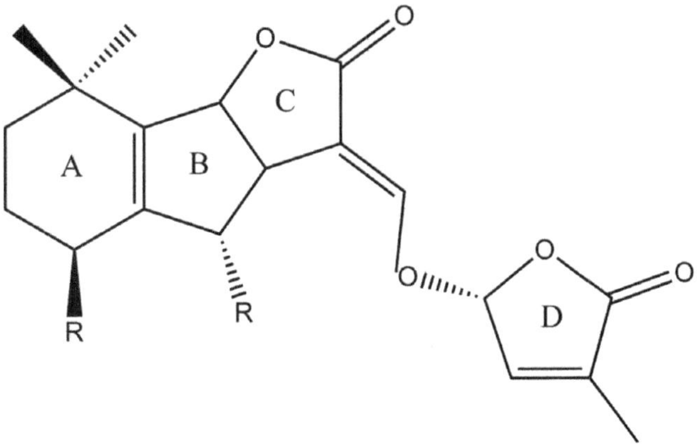

FIGURE 8.1 General structure of strigolactones, showing the A, B, C three-ring system and the enol-ether-linked D ring.

number of known non-canonical SLs is low, but it was suggested that new members of this group may be identified in the future. In addition, another type of butenolide compounds with similar signaling functions were identified in the smoke of burning plants. They were named karrikins and were shown to enhance germination and seedling growth after fires (Morffy, Faure and Nelson, 2016).SLs research was further driven by the generation of several synthetic SL analogues. Zwanenburg (Zwanenburg et al., 2009) provided a comprehensive overview of some SLs synthetic agonists and their functional activity. The best characterized synthetic SLs are GR5, GR7 and GR24 (Figure 8.4), all of them first synthesized during early research on parasitic plant germination as strigol analogues. Unlike GR5 and GR7, which lack the A and B or the A ring respectively, GR24, which share the entire ring structure with strigol, showed the highest germination potential on *Striga hermonthica* (Delile) Benth. and *Orobanche crenata* Forsk. As expected, the effective dose (ED_{50}) for *Striga* was twice lower than for *Orobanche*. When the enol-ether bridge was replaced with a CH_2 linker, as in the Carba-analogues of the GR compounds, the germination stimulatory effect was lost, while receptor binding was presumably conserved (Mangnus and Zwanenburg, 1992). Such synthetic analogues were proposed as putative means of control of parasitic weeds as they block receptors, inhibiting germination (Thuring, Nefkens and Zwanenburg, 1997). More recently, fluorescent analogues such as Yoshimulactone Green (YLG) were also developed in order to directly probe the target receptors in both fungi and parasitic plants (Prandi et al., 2011, Tsuchiya et al., 2015). The contemporary variety of synthetic SL analogues cannot be easily covered, but it is clear that even minor chemical modifications could lead to strong differences in biological effect (Kondo et al., 2007), thus giving a wide choice of methodological instruments for various research.

8.1.1.2 Strigolactone Biosynthesis

It was demonstrated that SLs are derived from the carotenoid biosynthetic pathway (López-Ráez et al., 2008, Matusova et al., 2005). The first dedicated step is the conversion of 9-cis-β-carotene into carlactone by the cleavage dioxygenases CCD7 and CCD8 (Figure 8.5). Carlactone is further converted to canonical SLs by more axillary growth 1 (MAX1) cytochrome P450 (Al-Babili and Reski, 2017). Further action of oxidase leads to the formation of the B and C rings in the structure of 4-deoxyorobanchol (Brewer et al., 2016). Alternatively, a series of reactions may lead to the synthesis of the non-canonical SLs zealactone as in *Zea mays* L. (Charnikhova et al., 2017) or methyl carlactonoate as in *Arabidopsis thaliana* L. (Brewer et al., 2016). The variety of SLs among plant species seems to be enormous and species-specific. The first obvious differences is in the prevalence of either strigol- or orobanchol-type SLs in particular species, although some plants, such as *Nicotiana tabacum* L., produce both (Xie et al., 2013). The overall picture is further complicated by the non-uniform pattern of SLs production among plant families where different species of the same family are either strigol- or orobanchol-rich (Yoneyama et al., 2017). The functions of this wide structural variety are still not well understood.

8.2 PRIMARY FUNCTIONS OF STRIGOLACTONES IN PLANTS

8.2.1 Regulation of Plant Development

The role of the internal SL signaling (Figure 8.6) was first reported in 2008, when SLs were identified as the most probable phytohormones, involved in inhibition of shoot branching (Gomez-Roldan et al., 2008, Umehara et al., 2008). The molecular mechanism was further elucidated by the finding that auxins and SLs are mutually regulated phytohormones with coordinated action in lateral bud suppression (Hayward et al., 2009). Similarly, SLs are also responsible for the inhibition of lateral roots development, driven by the same signaling mechanism and by modulation of the shoot-to-root auxin transport (Sun et al., 2014). Additionally, germination enhancement of non-parasitic plants and light-driven hypocotyl growth regulation are also proposed functions (Tsuchiya et al., 2010).

To date, one protein was identified as a putative SLs receptor in plants DWARF14-class α/β-fold hydrolase

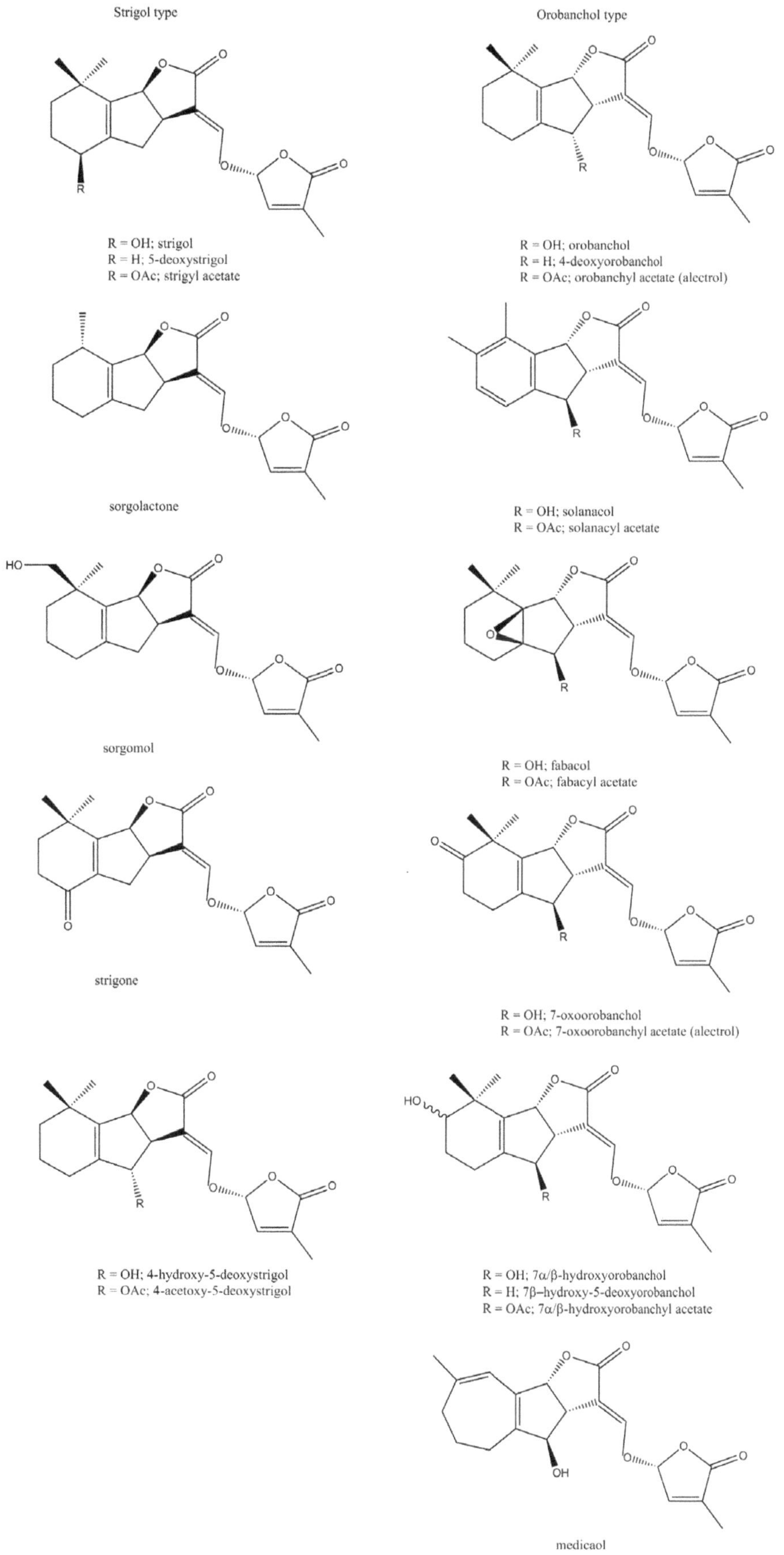

FIGURE 8.2 Structure of canonical strigolactones, divided into strigol and orobanchol types (Yoneyama et al., 2017).

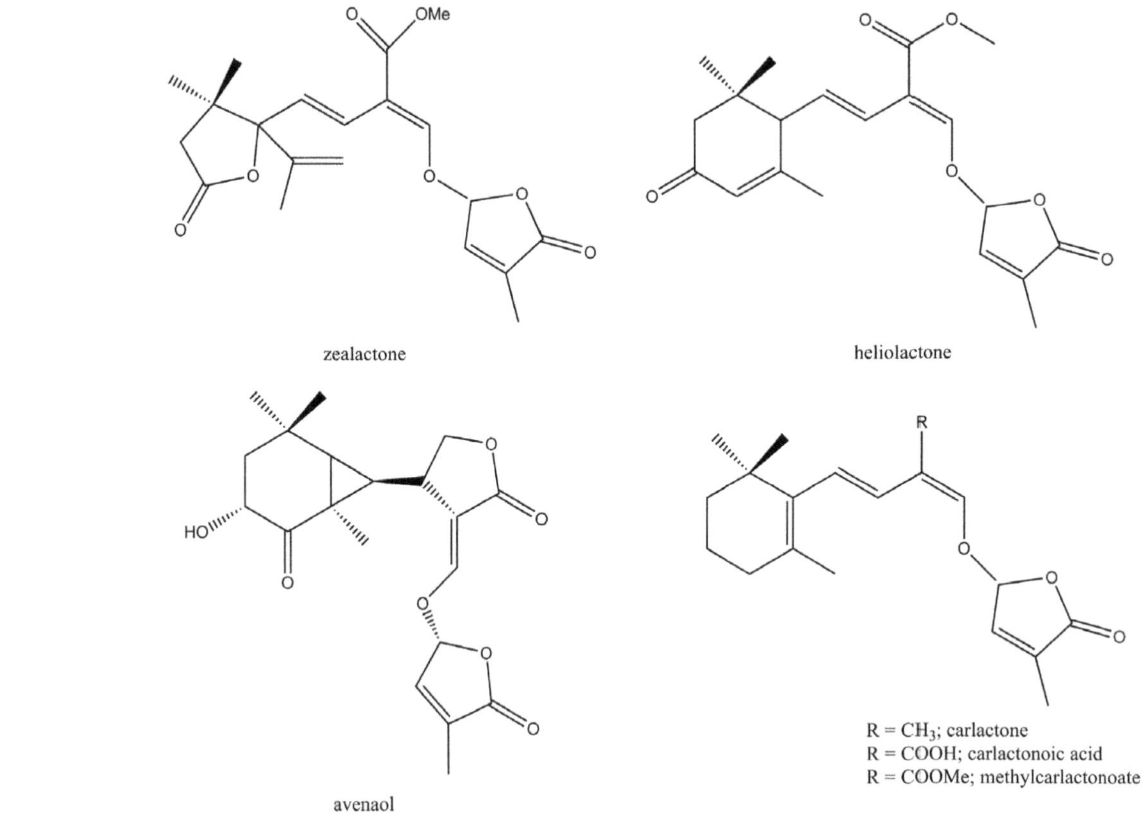

FIGURE 8.3 Structure of non-canonical strigolactones (Yoneyama et al., 2017).

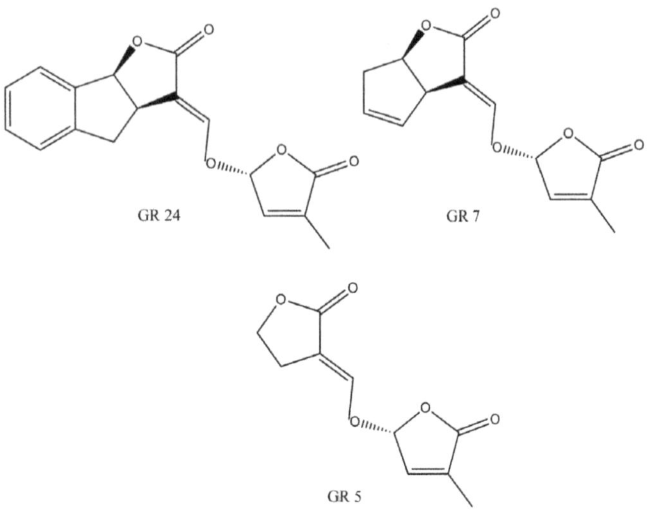

FIGURE 8.4 Structure of synthetic strigolactone analogues.

(D14), which acts with its protein ligand – MAX2 class F-box protein (Bennett et al., 2016). According to the now accepted model, D14 act as SL receptor, cleaving the enol-ether linkage and forming a covalent bond with the D ring (Lumba, Holbrook-Smith and McCourt, 2017). The resulting conformational changes in D14 lead to interaction with MAX2 and subsequent formation of ubiquitin-ligase complex (D14/MAX2/SCF complex/E2) with downstream targeted protein degradation (Bennett et al., 2016). In rice, the orthologue of MAX2 is denoted D3 (Zhou et al., 2013). The signal transduction of SLs signals is further regulated by SL-independent phloem transport of D14 and the presence or absence of both D14 and MAX2 in the particular tissue (Kameoka et al., 2016).

Several protein targets were shown to be ubiquitinilated and further degraded in response to SLs. The most probable candidate is a class I Clp ATPase named DWARF53 or D53 (Zhou et al., 2013). It was suggested that D53 is a repressor of SL function, targeted to proteolysis by SLs themselves. In *Arabidopsis*, three D53 proteins (SMLX 6–8) were identified as potential SLs targets and the mechanism was further elucidated, by proposing that D53- and D53-like proteins are repressors of transcription of SL-responsive genes (Wang et al., 2015). Conversely, *d53* mutants exhibit SLs-insensitive phenotype and extensive shoot branching (Zhou et al., 2013). Other known targets of SLs-induced degradation include brassinosteroid transcriptional effector BES1 (Wang et al., 2013) and their own receptor D14 in a feedback regulation mechanism (Chevalier et al., 2014). Besides affecting the transcriptional upregulation

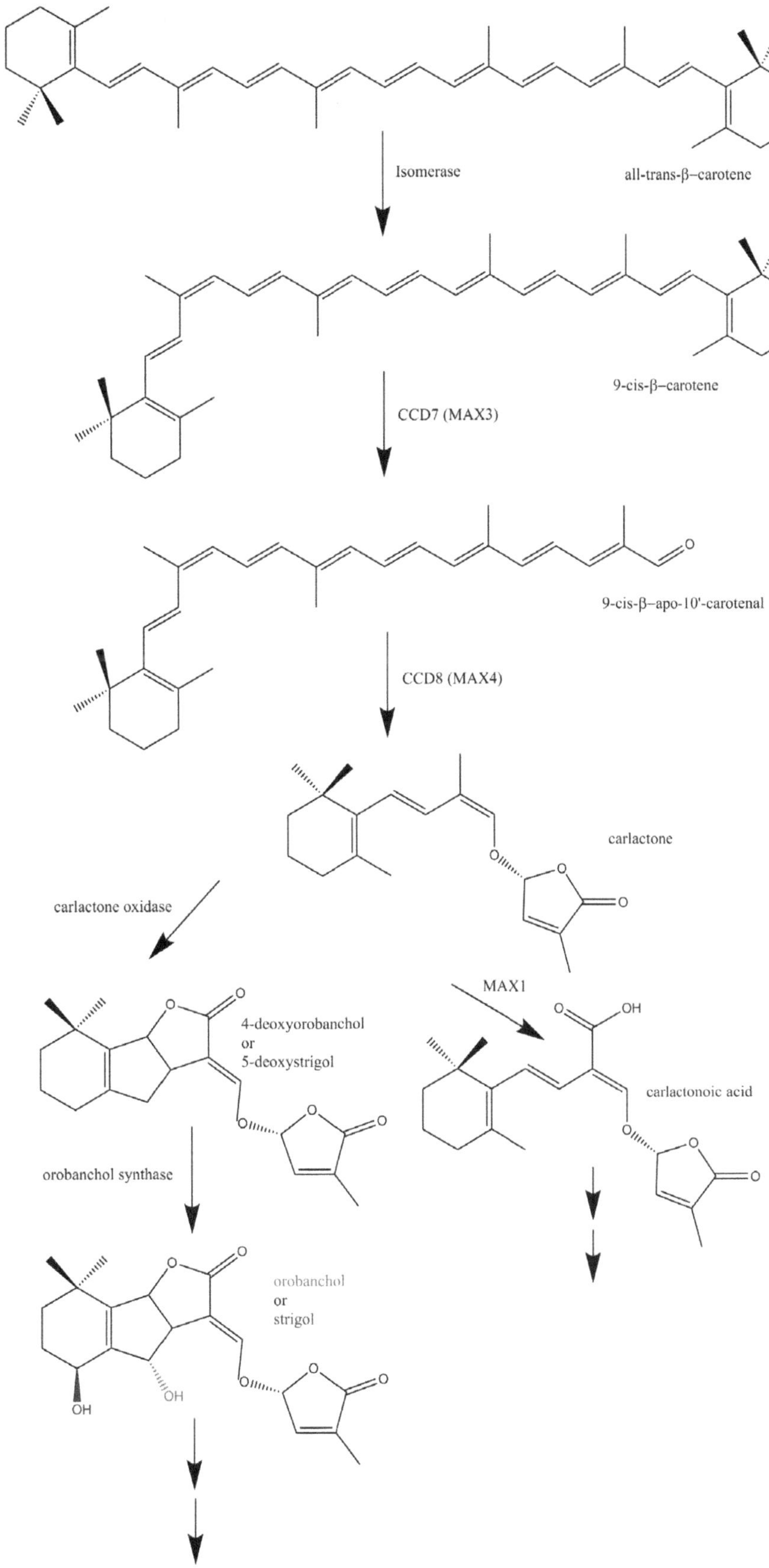

FIGURE 8.5 Biosynthetic pathway of strigolactones from β-carotene. Multiple arrows at the end represent multiple steps to other strigolactones. CCD – carotenoid cleavage dioxygenase. MAX – more axillary branching.

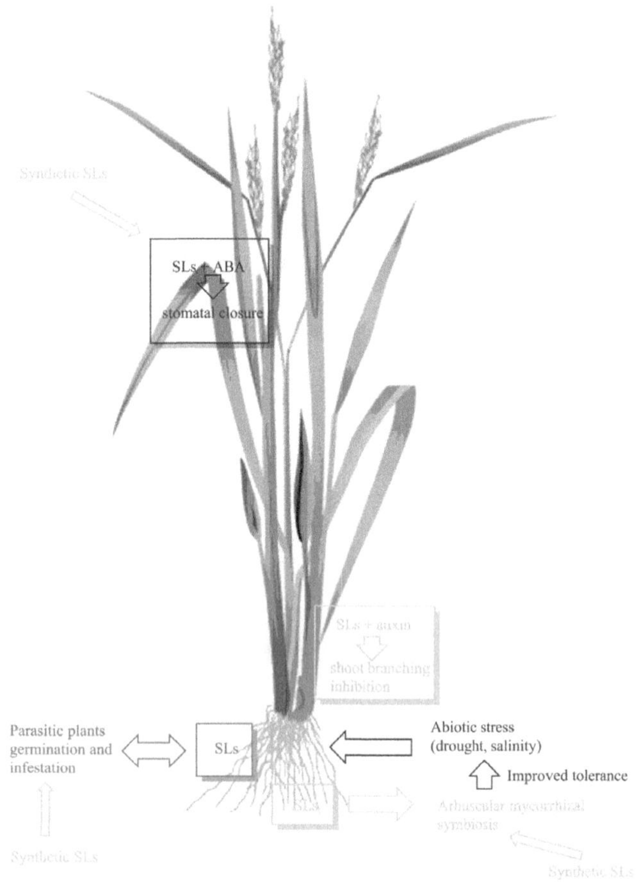

FIGURE 8.6 Overview of the multiple roles of strigolactones in plant development and arbuscular mycorrhizal symbiosis, abiotic stress response and parasitic plants germination and infestation.

of genes involved in shoot branching inhibition (Waldie, McCulloch and Leyser, 2014), SLs were also found to upregulate genes involved in photosynthesis (Mayzlish-Gati et al., 2010), which suggests that these phytohormones have a broader function.

8.2.2 Induction of Arbuscular Mycorrhiza

Arbuscular mycorrhiza is a widespread endosymbiosis, in which beneficial interrelations are established between plant roots and fungal hyphae. The primary role of the fungal partner is to uptake and supply the plant with water and nutrients, especially phosphates, while the plant partner offers photosynthates, with up to 20% of the assimilated carbon estimated to be consumed by AM symbionts (Parniske, 2008). It was proven that most if not all terrestrial plants establish AM, an evolutionary 400 million-year-old relation, providing the fungi mostly with carbohydrates e.g. hexoses and, as recently shown, with fatty acids (Roth and Paszkowski, 2017). The main benefit for plants is the enhanced phosphate uptake promoted by AM, as well as accelerated decomposition of organic material in the root area and nitrogen flow to the plant in the form of arginine and ammonia (Parniske, 2008). Thus, AM fungi are forced into relations with plants as they have no other source of organic carbon. They significantly improve plant productivity. These relations are further developed by some plant lineages that lost their ability to photosynthesize and draw organic carbon from established symbiosis through AM fungi in a case of parasitism called mycoheterotrophy (Gomes et al., 2017). Coordinated expression of specific genes, as well as extensive hormonal signaling, are required for the establishment of AM. While the positive effect of abscisic acid (ABA) was demonstrated, both ethylene and gibberellins seem to be inhibitory to AM (Fracetto, Peres and Lambais, 2017, Martín-Rodríguez et al., 2016).

Among other chemical factors released by plant roots, SLs are proven as positive regulators for recognition, hyphal growth and hyphal branching (Figure 8.6) of AM fungi (Gutjahr and Parniske, 2013). Unlike vascular plants, fungal SLs receptors have not yet been identified. It was hypothesized that in response to SLs the mitochondrial oxidative phosphorylation and fatty acid degradation of symbiotic fungi is enhanced, thus increasing their growth (Besserer et al., 2006, Besserer et al., 2008). However, the role of SLs in arbuscular symbiosis does not affect the fungal partner solely, but also induces signal transduction and symbiosis promotion in the host plant itself. The receptors and molecular mechanism of action seem to be at least partially similar to the one functioning in shoot branching regulation through D14/MAX2 (or D3) interaction. For example, rice *d3* mutants showed significant inhibition of fungal colonization (Yoshida et al., 2012). Apart from SLs being important for both fungal and plant partners, these phytohormones are also released by non-AM plants in a possible autosignaling mechanism. In *Arabidopsis*, phosphate starvation induces the release of orobanchol in root exudates (Kohlen et al., 2010). Synthetic GR24 promotes the increase of root hair length and inhibition of lateral root density (Kapulnik et al., 2011).

8.3 STRIGOLACTONES IN ABIOTIC STRESS TOLERANCE

8.3.1 Alteration of Strigolactone Synthesis Under Abiotic Stress

It will not be exaggerated to state that abiotic stress is among the factors most damaging to agriculture and abiotic stress response and tolerance are among the most studied plant traits. Traditionally, stress signaling in plants is associated with the action of four phytohormones – abscisic acid (ABA), jasmonic acid (JA), salicylic acid (SA) and ethylene (ET), which are more or less also involved in defense responses against pathogens (Fujita et al., 2006). In a broader sense, all phytohormones are involved and mutually regulated under abiotic stress (Kohli et al., 2013), but it was not until very recently when SLs were also reported to be regulated by abiotic stress factors. The first evidence of abiotic stress regulation of SLs was probably provided by the report that ABA not only shares a common precursor (β-carotene) with SLs, but also regulates their synthesis (López-Ráez et al., 2010). The crosstalk between ABA and SLs was further confirmed under osmotic stress (Liu et al., 2015) and evidence for SLs, involvement in abiotic stress responses was further accumulated (Andreo-Jimenez et al., 2015, Kapulnik and Koltai, 2014, Quain et al., 2014, Van Ha et al., 2014).

The major questions arising from the above said are whether abiotic stresses affect SLs biosynthesis in quantitative and qualitative manners and whether endogenous and root exudate levels are differentially affected. Evidence is still limited and probably further complicated by the extremely low concentrations in which SLs exhibit their action. In stimulation of fungal hyphal branching, the lower active concentration of sorgolactone was found to be 10^{-15} M (Besserer et al., 2006). The amount of SLs, released in root exudates tend to be in the range of ng g^{-1} root tissue and quantitation is highly complicated in soil (Dor et al., 2011b). In shoot tissue, SLs concentration could be 100-fold lower than the one released by roots (Liu et al., 2015). To cope with this, studies of SLs involvement in abiotic stress response either exploit SL-deficient mutants (Van Ha et al., 2014) or exogenous application of synthetic SL analogues, e.g., GR24 (Liu et al., 2015). Indirect quantitation is also provided by parasitic plant germination assay (either *Striga* or *Orobanche*) and using GR24 as a positive control (Dor et al., 2011b). We should also keep in mind that phosphate availability is the primary regulator of SLs synthesis (López-Ráez and Bouwmeester, 2008) and insignificant variability in phosphate concentrations could significantly affect the SLs released by roots.

Overall, osmotic stress was proved to inhibit SLs synthesis. This was proved in *Lotus japonicus* roots, where 5-deoxystrigol significantly decreased in root exudates under PEG treatment and independently of phosphate supply (Liu et al., 2015). Simultaneously, shoot SLs concentrations were not affected. Drought stress was also reported to differentially affect SLs synthesis in tomato (Visentin et al., 2016). Briefly, in roots, solanacol, orobanchol and didehydro orobanchol, along with CCD7 and CCD8 transcripts, decreased under drought with orobanchol and didehydro orobanchol being less affected. In contrast, CCD7 and CCD8 transcripts increased in shoots in response to drought. The pattern of SLs release in the exudate, however, could be dramatically different when an arbuscular fungal symbiotic partner is colonizing the roots. In established symbiosis, it was shown that SLs concentrations decreased and the decrease was different and depended on the fungal species (López-Ráez et al., 2011). Under both drought and salinity stress, however, colonized plants showed higher SLs concentration, which was also related to higher stress tolerance (Aroca et al., 2013, Ruiz-Lozano et al., 2016).

8.3.2 Abiotic Stress Tolerance Through Crosstalk with Stress Responsible Hormones

As previously discussed, SLs do interact with auxin signaling to provide lateral bud suppression (Hayward et al., 2009) and also with other phytohormones such as cytokinins, brassinosteroids and ethylene in various developmental processes (Cheng et al., 2013). When referring to abiotic stress response, their crosstalk with ABA signaling pathways is of major importance. The first evidence of such interplay was reported in tomato, where ABA was shown to regulate SLs production (López-Ráez et al., 2010). Further experiments suggested that decrease of root SLs and increase of shoot SLs under drought stress are needed as signaling events, priming the shoot to meet the environmental challenge. At least in tomato, SLs render stomata cells hypersensitive to ABA, thus decreasing transpiration and water loss (Visentin et al., 2016). Strigolactones-deficient and *max Arabidopsis* mutants were shown to be drought and salinity hypersensitive due to slower stomatal closure, e.g., lower ABA sensitivity (Van Ha et al., 2014). This is a comprehensive demonstration of the importance of both SLs and SL receptors in ABA-mediated stress response.

However, in mycorrhizal-colonized plants, the concentration of SLs, released by roots under stress, is increased (Aroca et al., 2013, Ruiz-Lozano et al., 2016), which correlates with higher stress tolerance. In both cases, ABA levels also increased in response to salinity and drought, and in relation to root SLs increase in AM plants. Despite the relatively higher stomatal conductance in AM plants compared to non-colonized plants, the photosynthetic efficiency was also higher in AM plants.

8.3.3 Abiotic Stress Tolerance Through Symbiotic Interactions

Considering the above, it may be rational to think that abiotic stress tolerance of AM plants is increased not because of increased SL levels in roots, but also because of additional features that the fungal symbiotic partner has to offer. The various effects, induced by the AM fungi, could indeed render drought and salinity tolerance in a wide range of plants. Numerous reports indicated differential aquaporin regulation of AM plants, thus giving more efficient water homeostasis under stress as proven in rice (Grondin et al., 2016) and maize (Quiroga et al., 2017). Especially in saline conditions, AM also regulates the expression levels of cation transporters, thus establishing a beneficially higher K^+/Na^+ ratio, improving stress tolerance (Porcel et al., 2016).

Arbuscular mycorrhiza was also shown to significantly improve both enzymatic and non-enzymatic antioxidant systems of host plants, thus reducing levels of oxidative damages (Kapoor and Singh, 2017). Improved selenium accumulation due to AM symbiosis also improves drought tolerance (Durán et al., 2016). The better performance of AM plants under abiotic stress is not restricted to salinity and drought, but also encompasses heavy metals (Miransari, 2017), cold (Pedranzani et al., 2015) and heat (Cabral et al., 2016) stress. Considering the importance of SLs in AM establishment, it is clear that strigolactones are essential players in abiotic stress tolerance beyond their endogenous crosstalk with ABA signaling.

8.3.4 Exogenous Application of Strigolactones for Abiotic Tolerance

The potential use of exogenously applied SLs in agriculture arose naturally in recent years, in concordance with the understanding of their multiple functions and signaling pathways. Because of the picomolar concentrations released naturally by plant roots, the possibility for global-scale application is largely dependent on the possibility for the efficient chemical synthesis of SL or SL analogues. Although the production process is complicated and expensive (Vurro, Prandi and Baroccio, 2016), their biological activity in very low concentrations could justify their use.

In concordance to the role of SLs as parasitic plant germination stimulants, the first efforts into application in agriculture were directed to suicide germination of *Orobanche* spp. and *Striga* spp. in the absence of a host (Zwanenburg, Mwakaboko and Kannan, 2016). The elegant idea behind this approach is to induce germination of the parasitic weed before planting the crop plant as the absence of host leads to fast depletion of storage compounds and dying of the target species. Results of extensive field trials on *Orobanche ramosa* L. infested tobacco fields with the SL analogues Nijmegen-1 and Nijmegen-1 Me varied from very promising to negative (Zwanenburg et al., 2016). Alternatively, various modified analogues such as Carba-GR24 were also synthesized and probed as potential, non-active inhibitors of germination receptors (Thuring et al., 1997), but their use may be compromised as they could equally affect non-target plants. Aside from parasitic weed control, synthetic SLs analogues are also potential crop-enhancing additives (Screpanti et al., 2016, Vurro et al., 2016). In respect to abiotic stress tolerance, the application of SLs seems to be an emerging trend in various trials.

First, exogenous application of the synthetic GR24 largely affects the root architecture of *Arabidopsis thaliana* plants (Ruyter-Spira et al., 2011). Proteomics approaches revealed nearly 150 root proteins, differentially regulated by GR24 again in *Arabidopsis*, with a correlated increase in stress-responsive flavonoids synthesis (Walton et al., 2016). Furthermore, wheat seeds pretreated with GR24 showed a variable response to saline conditions, with a notable increase in photosynthetic efficiency, but overall non-significant changes in fresh mass accumulation (Kausar and Shahbaz, 2017). Wheat, treated with a combination of GR24 and salicylic acid, showed activation of antioxidant enzymes and reduced lipid peroxidation under drought conditions (Sedaghat et al., 2017). GR24 supplementation was also proved effective in alleviation of Cd stress in *Panicum virgatum* L. by reducing Cd uptake and improving growth and photosynthesis (Tai et al., 2017).

Overall, it seems that exogenous application of synthetic SLs and especially GR24, being the most available, has a direct positive effect on stress tolerance of various model and crop plants. Another, probably more substantial role of SLs, however, should be noted. As proposed, SLs could be exogenously applied as "indirect" fertilizers, improving AM symbiosis (Vurro et al., 2016) and conversely, abiotic stress tolerance.

8.4 STRIGOLACTONES AS KEY SIGNALS FOR PATHOGENS

8.4.1 Parasitic Plants Exploitation of Strigolactones for Host Localization

Despite their important role as endogenous phytohormones and AM symbiosis stimulants, SLs were first identified as germination stimulants of parasitic root plants of the family Orobanchaceae (Yoneyama et al., 2010). In the time-course of SLs identification, they were first named after the genus name of the parasitic plant, e.g., strigol (*Striga* spp.), orobanchol (*Orobanche* spp.) and alectrol (*Alectra* spp.). It was not until recently that SLs started to be named after the producing plant taxon, e.g., zealacton (*Zea mays*), solanacol (Solanaceae), fabacol (Fabaceae), etc. In rare cases, certain parasitic plant species require particular SLs as germination stimulants (*Orobanche cumana* Wallr. requires sunflower dehydrocostus lactone [Joel et al., 2011]) and parasitic plants respond to various SLs. The germination percentage, however, could vary between different SLs and some are active at 100-fold lower concentration than others (Yoneyama et al., 2010). In other words, some SLs are superior to others in their ability to induce parasitic plants' germination, with the synthetic GR24 being the less active. Conversely, *Striga* spp. and *Orobanche* spp. express broader specialization toward predominantly strigol- and orobanchol-type SLs-producing plants respectively.

The availability of SLs as germination stimulants is crucial for root parasitic plants due to limited seed resources and absent or limited photosynthesis not allowing long-term survival in the absence of a host (Heide-Jørgensen, 2008). Recently, highly-sensitive SLs receptors were identified in *Striga hermonthica* (Toh et al., 2015) and *Orobanche aegyptiaca* Pers. (Conn et al., 2015). Studies confirmed that SL receptors in parasitic plants evolved from KAI2 (karikkin receptors) α/β-hydrolase, a germination-responsive receptor through new ligand specificity gain rather than from D14 (SL receptor) through new signaling function gain (Conn et al., 2015). Parasitic plants strongly depend on the quantitative and qualitative profile of SLs in the root exudates of the host. A comparative study of *Arabidopsis thaliana* ability to induce *Orobanche* spp. germination showed that *A. thaliana* root exudates are good stimulants for *O. aegyptiaca* and *O. ramosa*, and far worse for *O. crenata*, *O. minor* Sm. and *O. cernua* Loefl. (Westwood, 2000). Further studies indicated that the non-AM plant *A. thaliana* exuded orobanchol as the main SL and two more SLs in a lower amount (Goldwasser et al., 2008). However, host resistance to root parasitic plants may not be exclusively dependent on the induced germination percentage of the parasite, while still being dependent on the SLs profile. This was demonstrated with sorghum *Striga*-resistant and *Striga*-susceptible cultivars (Yoneyama et al., 2010). Fluorescent probing of *Striga* SL receptors showed that SLs are required not only for germination but also during the later stages of seedling elongation (Tsuchiya et al., 2015).

In any case, it is intriguing to think that abiotic stress-induced changes in SLs release by the host would cause changes in susceptibility to root parasitic plants. In non-AM plants and non-colonized AM plants, the abiotic stress-caused decrease of SLs in root exudates may render resistance to parasites. This was proved in *Arabidopsis thaliana* – *Phelipanche ramosa* interaction (Demirbas et al., 2013) and in lettuce – *Ph. ramosa* interaction (Aroca et al., 2013) under salinity. The AM symbiosis and a concomitant increase in SLs release under stress could, however, significantly alter this trend (Aroca et al., 2013). To summarize, AM colonization is beneficial for resistance to parasitic plants under optimal conditions, but has the opposite effect under abiotic stress. Accordingly, the exogenous application of SLs should be avoided in parasitic plants-infested areas.

8.4.2 Strigolactones and Other Plant Pathogens

The well-known stimulating effect of SLs on AM fungi most logically leads to the idea that they can be used as signals by non-beneficial and even detrimental fungal species in the soil. Relatively early, it was concluded that SLs are not general plant-to-fungi signal molecules and exert AM specific action (Steinkellner et al., 2007). This was later confirmed for phytopathogenic fungi such as *Fusarium* (Foo et al., 2016) and *Pythium* (Blake et al., 2016). SL-deficient plants did not prove to be more susceptible to fungal infections, and neither did the supplementation with GR 24 induce spore germination or mycelium growth of the pathogens. Furthermore, SLs and GR 24, in particular, were shown to negatively affect the growth of various pathogens such as *Fusarium, Sclerotinia, Macrophomina, Alternaria, Colletotrichum* and *Botrytis* (Dor et al., 2011a). Accordingly, SLs-deficient tomato mutants became more susceptible to various fungal infections (Torres-Vera et al., 2014). This trend, however, may not be universal in plant-fungal interaction and could differ according to the pathogen (López-Ráez, Shirasu and Foo, 2017). Currently, the molecular mechanism of the differential effects of SLs on symbiotic and pathogenic fungi has not been elucidated. The protective role of AM fungi against pathogens (Harrier and Watson, 2004), however, could also enhance the positive effects of SLs. Strigolactones are, in any case, an important player in these complex soil interactions. An interesting aspect of these interactions may be a recently reported case of misbalanced AM symbiosis between non-native plants and native AM fungi (Řezáčová et al., 2017) which resulted in intensified carbon drainage by the fungi with no compensation. Otherwise said, AM symbiosis is not beneficial per se.

Similarly, SL-deficient plant mutants showed higher susceptibility to pathogenic bacteria in various cases

(López-Ráez et al., 2017). This was explained either by the antagonistic effect of SLs on *Rhodococcus* – induced shoot branching (Stes et al., 2015) or by SL/ABA-induced stomatal closure, which blocked the major route of *Pectobacterium* and *Pseudomonas* entry (Piisilä et al., 2015). Thus, it could be concluded that SLs could be applied as general phytoprotectants not only to abiotic stresses but also to plant pathogens.

8.5 CONCLUSIONS AND FUTURE PERSPECTIVES

Initially identified as parasitic root-plants' germination stimulants, strigolactones emerged as important phytohormones with broad functions both endogenously, in plant development, and exogenously, as communication molecules in the soil. Their positive role as exogenously applied substances in the abiotic and biotic stress responses of plants could be either direct or indirect, through arbuscular mycorrhizal symbiosis. The commercial application of natural SLs, however, is largely hampered by the low concentrations in root exudates, which could be overcome by the possibility for chemical synthesis of SLs analogues. Ongoing studies on synthetic SLs derivatives could further improve their application by specifically targeting their beneficial effects and avoiding unwanted infestations by root parasitic plants of the Orobanchaceae family.

REFERENCES

Andreo-Jimenez, B., Ruyter-Spira, C., Bouwmeester, H. J., Lopez-Raez, J. A. (2015) Ecological relevance of strigolactones in nutrient uptake and other abiotic stresses, and in plant-microbe interactions below-ground. *Plant and Soil* 394: 1–19.

Aroca, R., Ruiz-Lozano, J. M., Zamarreño, A. M., Paz, J. A., García-Mina, J. M., Pozo, M. J., López-Ráez, J. A. (2013) Arbuscular mycorrhizal symbiosis influences strigolactone production under salinity and alleviates salt stress in lettuce plants. *Journal of Plant Physiology* 170: 47–55.

Bennett, T., Liang, Y., Seale, M., Ward, S., Müller, D., Leyser, O. (2016) Strigolactone regulates shoot development through a core signalling pathway. *Biology Open* 5: 1806–1820.

Besserer, A., Bécard, G., Jauneau, A., Roux, C., Séjalon-Delmas, N. (2008) GR24, a synthetic analog of strigolactones, stimulates the mitosis and growth of the arbuscular mycorrhizal fungus gigaspora Rosea by boosting its energy metabolism. *Plant Physiology* 148: 402–413.

Besserer, A., Puech-Pagès, V., Kiefer, P., Gomez-Roldan, V., Jauneau, A., Roy, S., Portais, J. C., *et al.* (2006) Strigolactones stimulate arbuscular mycorrhizal fungi by activating mitochondria. *PLOS Biology* 4: e226.

Blake, S. N., Barry, K. M., Gill, W. M., Reid, J. B., Foo, E. (2016) The role of strigolactones and ethylene in disease caused by *Pythium irregulare. Molecular Plant Pathology* 17: 680–690.

Brewer, P. B., Yoneyama, K., Filardo, F., Meyers, E., Scaffidi, A., Frickey, T., Akiyama, K., *et al.* (2016) LATERAL BRANCHING OXIDOREDUCTASE acts in the final stages of strigolactone biosynthesis in Arabidopsis. *Proceedings of the National Academy of Sciences of the United States of America* 113: 6301–6306.

Cabral, C., Ravnskov, S., Tringovska, I., Wollenweber, B. (2016) Arbuscular mycorrhizal fungi modify nutrient allocation and composition in wheat (*Triticum aestivum* L.) subjected to heat-stress. *Plant and Soil* 408: 385–399.

Charnikhova, T. V., Gaus, K., Lumbroso, A., Sanders, M., Vincken, J. P., De Mesmaeker, A., Ruyter-Spira, C. P., Screpanti, C., Bouwmeester, H. J. (2017) Zealactones. Novel natural strigolactones from maize. *Phytochemistry* 137: 123–131.

Cheng, X., Ruyter-Spira, C., Bouwmeester, H. (2013) The interaction between strigolactones and other plant hormones in the regulation of plant development. Frontiers in Plant Science 4: 199.

Chevalier, F., Nieminen, K., Sánchez-Ferrero, J. C., Rodríguez, M. L., Chagoyen, M., Hardtke, C. S., Cubas, P. (2014) Strigolactone promotes degradation of DWARF14, an α/β hydrolase essential for strigolactone signaling in *Arabidopsis. The Plant Cell* 26: 1134–1150.

Conn, C. E., Bythell-Douglas, R., Neumann, D., Yoshida, S., Whittington, B., Westwood, J. H., Shirasu, K., *et al.* (2015) PLANT EVOLUTION. Convergent evolution of strigolactone perception enabled host detection in parasitic plants. *Science* 349: 540–543.

Cook, C. E., Whichard, L. P., Turner, B., Wall, M. E., Egley, G. H. (1966) Germination of witchweed (*Striga lutea* Lour.): Isolation and properties of a potent stimulant. *Science* 154: 1189–1190.

Decker, E. L., Alder, A., Hunn, S., Ferguson, J., Lehtonen, M. T., Scheler, B., Kerres, K. L., *et al.* (2017) Strigolactone biosynthesis is evolutionarily conserved, regulated by phosphate starvation and contributes to resistance against phytopathogenic fungi in a moss, *Physcomitrella patens.The New Phytologist* 216: 455–468.

Demirbas, S., Vlachonasios, K. E., Acar, O., Kaldis, A. (2013) The effect of salt stress on *Arabidopsis thaliana* and *Phelipanche ramosa* interaction. *Weed Research* 53: 452–460.

Dor, E., Joel, D. M., Kapulnik, Y., Koltai, H., Hershenhorn, J. (2011a) The synthetic strigolactone GR24 influences the growth pattern of phytopathogenic fungi. *Planta* 234: 419–427.

Dor, E., Yoneyama, K., Wininger, S., Kapulnik, Y., Yoneyama, K., Koltai, H., Xie, X., Hershenhorn, J. (2011b) Strigolactone deficiency confers resistance in tomato line SL-ORT1 to the parasitic weeds *Phelipanche* and *Orobanche* spp. *Phytopathology* 101: 213–222.

Dun, E. A., de Saint Germain, A., Rameau, C., Beveridge, C. A. (2013) Dynamics of strigolactone function and shoot branching responses in *Pisum sativum*. *Molecular Plant* 6: 128–140.

Durán, P., Acuña, J. J., Armada, E., López-Castillo, O. M., Cornejo, P., Mora, M. L., Azcón, R. (2016) Inoculation with selenobacteria and arbuscular mycorrhizal fungi to enhance selenium content in lettuce plants and improve tolerance against drought stress. *Journal of Soil Science and Plant Nutrition* 16: 211–225.

Foo, E., Blake, S. N., Fisher, B. J., Smith, J. A., Reid, J. B. (2016) The role of strigolactones during plant interactions with the pathogenic fungus *Fusarium oxysporum*. *Planta* 243: 1387–1396.

Fracetto, G. G. M., Peres, L. E. P., Lambais, M. R. (2017) Gene expression analyses in tomato near isogenic lines provide evidence for ethylene and abscisic acid biosynthesis fine-tuning during arbuscular mycorrhiza development. *Archives of Microbiology* 199: 787–798.

Fujita, M., Fujita, Y., Noutoshi, Y., Takahashi, F., Narusaka, Y., Yamaguchi-Shinozaki, K., Shinozaki, K. (2006) Crosstalk between abiotic and biotic stress responses: A current view from the points of convergence in the stress signaling networks. *Current Opinion in Plant Biology* 9: 436–442.

Goldwasser, Y., Yoneyama, K., Xie, X., Yoneyama, K. (2008) Production of strigolactones by *Arabidopsis thaliana* responsible for *Orobanche aegyptiaca* seed germination. *Plant Growth Regulation* 55: 21–28.

Gomes, S. I., Aguirre-Gutiérrez, J., Bidartondo, M. I., Merckx, V. S. (2017) Arbuscular mycorrhizal interactions of mycoheterotrophic Thismia are more specialized than in autotrophic plants. *The New Phytologist* 213: 1418–1427.

Gomez-Roldan, V., Fermas, S., Brewer, P. B., Puech-Pagès, V., Dun, E. A., Pillot, J. P., Letisse, F., *et al.* (2008) Strigolactone inhibition of shoot branching. *Nature* 455: 189–194.

Gough, C., Bécard, G. (2016) Strigolactones and lipo-chitooligosaccharides as molecular communication signals in the arbuscular mycorrhizal symbiosis. In *Molecular Mycorrhizal Symbiosis*, ed. Martin, F., 107–124. Wiley-Blackwell, New Jersey.

Grondin, A., Mauleon, R., Vadez, V., Henry, A. (2016) Root aquaporins contribute to whole plant water fluxes under drought stress in rice (*Oryza sativa* L.). *Plant, Cell & Environment* 39: 347–365.

Gutjahr, C., Parniske, M. (2013) Cell and developmental biology of arbuscular mycorrhiza symbiosis. *Annual Review of Cell and Developmental Biology* 29: 593–617.

Harrier, L. A., Watson, C. A. (2004) The potential role of arbuscular mycorrhizal (AM) fungi in the bioprotection of plants against soil-borne pathogens in organic and/or other sustainable farming systems. *Pest Management Science* 60: 149–157.

Hayward, A., Stirnberg, P., Beveridge, C., Leyser, O. (2009) Interactions between auxin and strigolactone in shoot branching control. *Plant Physiology* 151: 400–412.

Heide-Jørgensen, H. (2008) *Parasitic Flowering Plants*. Brill, Leiden.

Joel, D. M., Chaudhuri, S. K., Plakhine, D., Ziadna, H., Steffens, J. C. (2011) Dehydrocostus lactone is exuded from sunflower roots and stimulates germination of the root parasite *Orobanche cumana*. *Phytochemistry* 72: 624–634.

Kameoka, H., Dun, E. A., Lopez-Obando, M., Brewer, P. B., de Saint Germain, A., Rameau, C., Beveridge, C. A., Kyozuka, J. (2016) Phloem transport of the receptor DWARF14 protein is required for full function of strigolactones. *Plant Physiology* 172: 1844–1852.

Kapoor, R., Singh, N. (2017) Arbuscular mycorrhiza and reactive oxygen species. In *Arbuscular Mycorrhizas and Stress Tolerance of Plants*, ed. Wu, Q.-S., 225–243. Springer, Singapore.

Kapulnik, Y., Delaux, P. M., Resnick, N., Mayzlish-Gati, E., Wininger, S., Bhattacharya, C., Séjalon-Delmas, N., *et al.* (2011) Strigolactones affect lateral root formation and root-hair elongation in *Arabidopsis*. *Planta* 233: 209–216.

Kapulnik, Y., Koltai, H. (2014) Strigolactone involvement in root development, response to abiotic stress, and interactions with the biotic soil environment. *Plant Physiology* 166: 560–569.

Kausar, F., Shahbaz, M. (2017) Influence of strigolactone (GR24) as a seed treatment on growth, gas exchange and chlorophyll fluorescence of wheat under saline conditions. *International Journal of Agriculture & Biology* 19: 321–327.

Kohlen, W., Charnikhova, T., Liu, Q., Bours, R., Domagalska, M. A., Beguerie, S., Verstappen, F., *et al.* (2010) Strigolactones are transported through the xylem and play a key role in shoot architectural response to phosphate deficiency in non-AM host *Arabidopsis thaliana*. *Plant Physiology* 110: 974–987.

Kohli, A., Sreenivasulu, N., Lakshmanan, P., Kumar, P. P. (2013) The phytohormone crosstalk paradigm takes center stage in understanding how plants respond to abiotic stresses. Plant Cell Reports 32: 945–957.

Koltai, H. (2011) Strigolactones are regulators of root development. *The New Phytologist* 190: 545–549.

Kondo, Y., Tadokoro, E., Matsuura, M., Iwasaki, K., Sugimoto, Y., Miyake, H., Takikawa, H., Sasaki, M. (2007) Synthesis and seed germination stimulating activity of some imino analogs of strigolactones. *Bioscience, Biotechnology, and Biochemistry* 71: 2781–2786.

Liu, J., He, H., Vitali, M., Visentin, I., Charnikhova, T., Haider, I., Schubert, A., *et al.* (2015) Osmotic stress represses strigolactone biosynthesis in *Lotus japonicus* roots: Exploring the interaction between strigolactones and ABA under abiotic stress. *Planta* 241: 1435–1451.

López-Ráez, J. A., Bouwmeester, H. (2008) Fine-tuning regulation of strigolactone biosynthesis under phosphate starvation. *Plant Signaling & Behavior* 3: 963–965.

López-Ráez, J. A., Charnikhova, T., Fernández, I., Bouwmeester, H., Pozo, M. J. (2011) Arbuscular mycorrhizal symbiosis decreases strigolactone production in tomato. *Journal of Plant Physiology* 168: 294–297.

López-Ráez, J. A., Charnikhova, T., Gómez-Roldán, V., Matusova, R., Kohlen, W., De Vos, R., Verstappen, F., *et al.* (2008) Tomato strigolactones are derived from carotenoids and their biosynthesis is promoted by phosphate starvation. *The New Phytologist* 178: 863–874.

López-Ráez, J. A., Kohlen, W., Charnikhova, T., Mulder, P., Undas, A. K., Sergeant, M. J., Verstappen, F., *et al.* (2010) Does abscisic acid affect strigolactone biosynthesis? *The New Phytologist* 187: 343–354.

López-Ráez, J. A., Shirasu, K., Foo, E. (2017) Strigolactones in plant interactions with beneficial and detrimental organisms: The Yin and Yang. *Trends in Plant Science* 22: 527–537.

Lumba, S., Holbrook-Smith, D., McCourt, P. (2017) The perception of strigolactones in vascular plants. *Nature Chemical Biology* 13: 599–606.

Mangnus, E. M., Zwanenburg, B. (1992) Tentative molecular mechanism for germination stimulation of Striga and Orobanche seeds by strigol and its synthetic analogs. *Journal of Agricultural and Food Chemistry* 40: 1066–1070.

Martín-Rodríguez, J. A., Huertas, R., Ho-Plágaro, T., Ocampo, J. A., Turečková, V., Tarkowská, D., Ludwig-Müller, J., García-Garrido, J. M. (2016) Gibberellin–abscisic acid balances during arbuscular mycorrhiza formation in tomato. Frontiers in plant science 7: 1273.

Matusova, R., Rani, K., Verstappen, F. W., Franssen, M. C., Beale, M. H., Bouwmeester, H. J. (2005) The strigolactone germination stimulants of the plant-parasitic Striga and Orobanche spp. are derived from the carotenoid pathway. *Plant Physiology* 139: 920–934.

Mayzlish-Gati, E., LekKala, S. P., Resnick, N., Wininger, S., Bhattacharya, C., Lemcoff, J. H., Kapulnik, Y., Koltai, H. (2010) Strigolactones are positive regulators of light-harvesting genes in tomato. *Journal of Experimental Botany* 61: 3129–3136.

Miransari, M. (2017) Arbuscular mycorrhizal fungi and heavy metal tolerance in plants. In *Arbuscular Mycorrhizas and Stress Tolerance of Plants*, ed. Wu, Q. S., 147–161. Springer, Singapore.

Morffy, N., Faure, L., Nelson, D. C. (2016) Smoke and hormone mirrors: Action and evolution of karrikin and strigolactone signaling. *Trends in Genetics: TIG* 32: 176–188.

Pandey, A., Sharma, M., Pandey, G. K. (2016) Emerging roles of strigolactones in plant responses to stress and development. *Frontiers in Plant Science* 7: 434.

Parniske, M. (2008) Arbuscular mycorrhiza: The mother of plant root endosymbioses. *Nature Reviews Microbiology* 6: 763–775.

Pedranzani, H., Tavecchio, N., Gutiérrez, M., Garbero, M., Porcel, R., Ruiz-Lozano, J. M. (2015) Differential effects of cold stress on the antioxidant response of mycorrhizal and non-mycorrhizal *Jatropha curcas* (L.) plants. *Journal of Agricultural Science* 7: 35–43.

Piisilä, M., Keceli, M. A., Brader, G., Jakobson, L., Jõesaar, I., Sipari, N., Kollist, H., Palva, E. T., Kariola, T. (2015) The F-box protein max2 contributes to resistance to bacterial phytopathogens in *Arabidopsis thaliana. BMC Plant Biology* 15: 53.

Porcel, R., Aroca, R., Azcon, R., Ruiz-Lozano, J. M. (2016) Regulation of cation transporter genes by the arbuscular mycorrhizal symbiosis in rice plants subjected to salinity suggests improved salt tolerance due to reduced Na^+ root-to-shoot distribution. *Mycorrhiza* 26: 673–684.

Prandi, C., Occhiato, E. G., Tabasso, S., Bonfante, P., Novero, M., Scarpi, D., Bova, M. E., Miletto, I. (2011) New potent fluorescent analogues of strigolactones: Synthesis and biological activity in parasitic weed germination and fungal branching. *European Journal of Organic Chemistry* 2011: 3781–3793.

Quain, M. D., Makgopa, M. E., Márquez-García, B., Comadira, G., Fernandez-Garcia, N., Olmos, E., Schnaubelt, D., Kunert, K. J., Foyer, C. H. (2014) Ectopic phytocystatin expression leads to enhanced drought stress tolerance in soybean (*Glycine max*) and *Arabidopsis thaliana* through effects on strigolactone pathways and can also result in improved seed traits. *Plant Biotechnology Journal* 12: 903–913.

Quiroga, G., Erice, G., Aroca, R., Chaumont, F., Ruiz-Lozano, J. M. (2017) Enhanced drought stress tolerance by the arbuscular mycorrhizal symbiosis in a drought-sensitive maize cultivar is related to a broader and differential regulation of host plant aquaporins than in a drought-tolerant cultivar. Frontiers in Plant Science 8: 1056.

Rasmann, S., Turlings, T. C. (2016) Root signals that mediate mutualistic interactions in the rhizosphere. *Current Opinion in Plant Biology* 32: 62–68.

Řezáčová, V., Slavíková, R., Konvalinková, T., Hujslová, M., Gryndlerová, H., Gryndler, M., Püschel, D., Jansa, J. (2017) Imbalanced carbon-for-phosphorus exchange between European arbuscular mycorrhizal fungi and non-native Panicum grasses—A case of dysfunctional symbiosis. *Pedobiologia* 62: 48–55.

Roth, R., Paszkowski, U. (2017) Plant carbon nourishment of arbuscular mycorrhizal fungi. *Current Opinion in Plant Biology* 39: 50–56.

Ruiz-Lozano, J. M., Aroca, R., Zamarreño, ÁM., Molina, S., Andreo-Jiménez, B., Porcel, R., García-Mina, J. M., Ruyter-Spira, C., López-Ráez, J. A. (2016) Arbuscular mycorrhizal symbiosis induces strigolactone biosynthesis under drought and improves drought tolerance in lettuce and tomato. *Plant, Cell & Environment* 39: 441–452.

Ruyter-Spira, C., Kohlen, W., Charnikhova, T., van Zeijl, A., van Bezouwen, L., de Ruijter, N., Cardoso, C., *et al.* (2011) Physiological effects of the synthetic strigolactone analog GR24 on root system architecture in Arabidopsis: Another belowground role for strigolactones? *Plant Physiology* 155: 721–734.

Screpanti, C., Fonné-Pfister, R., Lumbroso, A., Rendine, S., Lachia, M., De Mesmaeker, A. (2016) Strigolactone derivatives for potential crop enhancement applications. *Bioorganic & Medicinal Chemistry Letters* 26: 2392–2400.

Sedaghat, M., Tahmasebi-Sarvestani, Z., Emam, Y., Mokhtassi-Bidgoli, A. (2017) Physiological and antioxidant responses of winter wheat cultivars to strigolactone and salicylic acid in drought. *Plant Physiology and Biochemistry: PPB* 119: 59–69.

Soto, M. J., Fernández-Aparicio, M., Castellanos-Morales, V., García-Garrido, J. M., Ocampo, J. A., Delgado, M. J., Vierheilig, H. (2010) First indications for the involvement of strigolactones on nodule formation in alfalfa (*Medicago sativa*). *Soil Biology and Biochemistry* 42: 383–385.

Steinkellner, S., Lendzemo, V., Langer, I., Schweiger, P., Khaosaad, T., Toussaint, J. P., Vierheilig, H. (2007) Flavonoids and strigolactones in root exudates as signals in symbiotic and pathogenic plant-fungus interactions. *Molecules* 12: 1290–1306.

Stes, E., Depuydt, S., De Keyser, A., Matthys, C., Audenaert, K., Yoneyama, K., Werbrouck, S., Goormachtig, S., Vereecke, D. (2015) Strigolactones as an auxiliary hormonal defence mechanism against leafy gall syndrome in *Arabidopsis thaliana*. *Journal of Experimental Botany* 66: 5123–5134.

Sun, H., Bi, Y., Tao, J., Huang, S., Hou, M., Xue, R., Liang, Z., *et al.* (2016) Strigolactones are required for nitric oxide to induce root elongation in response to nitrogen and phosphate deficiencies in rice. *Plant, Cell & Environment* 39: 1473–1484.

Sun, H., Tao, J., Liu, S., Huang, S., Chen, S., Xie, X., Yoneyama, K., Zhang, Y., Xu, G. (2014) Strigolactones are involved in phosphate- and nitrate-deficiency-induced root development and auxin transport in rice. *Journal of Experimental Botany* 65: 6735–6746.

Tai, Z., Yin, X., Fang, Z., Shi, G., Lou, L., Cai, Q. (2017) Exogenous GR24 alleviates cadmium toxicity by reducing cadmium uptake in Switchgrass (*Panicum virgatum*) seedlings. International Journal of Environmental Research and Public Health 14: 852.

Thuring, J. W. J. F., Nefkens, G. H. L., Zwanenburg, B. (1997) Synthesis and biological evaluation of the strigol analogue Carba-GR24. *Journal of Agricultural and Food Chemistry* 45: 1409–1414.

Toh, S., Holbrook-Smith, D., Stogios, P. J., Onopriyenko, O., Lumba, S., Tsuchiya, Y., Savchenko, A., McCourt, P. (2015) Structure-function analysis identifies highly sensitive strigolactone receptors in Striga. *Science* 350: 203–207.

Torres-Vera, R., García, J. M., Pozo, M. J., López-Ráez, J. A. (2014) Do strigolactones contribute to plant defence? *Molecular Plant Pathology* 15: 211–216.

Tsuchiya, Y., Vidaurre, D., Toh, S., Hanada, A., Nambara, E., Kamiya, Y., Yamaguchi, S., McCourt, P. (2010) A small-molecule screen identifies new functions for the plant hormone strigolactone. *Nature Chemical Biology* 6: 741–749.

Tsuchiya, Y., Yoshimura, M., Sato, Y., Kuwata, K., Toh, S., Holbrook-Smith, D., Zhang, H., *et al.* (2015) PARASITIC PLANTS. Probing strigolactone receptors in *Striga hermonthica* with fluorescence. *Science* 349: 864–868.

Umehara, M., Hanada, A., Yoshida, S., Akiyama, K., Arite, T., Takeda-Kamiya, N., Magome, H., *et al.* (2008) Inhibition of shoot branching by new terpenoid plant hormones. *Nature* 455: 195–200.

Van Ha, C. V., Leyva-González, M. A., Osakabe, Y., Tran, U. T., Nishiyama, R., Watanabe, Y., Tanaka, M., *et al.* (2014) Positive regulatory role of strigolactone in plant responses to drought and salt stress. *Proceedings of the National Academy of Sciences* 111: 851–856.

Visentin, I., Vitali, M., Ferrero, M., Zhang, Y., Ruyter-Spira, C., Novák, O., Strnad, M., *et al.* (2016) Low levels of strigolactones in roots as a component of the systemic signal of drought stress in tomato. *The New Phytologist* 212: 954–963.

Vurro, M., Prandi, C., Baroccio, F. (2016) Strigolactones: How far is their commercial use for agricultural purposes? *Pest Management Science* 72: 2026–2034.

Waldie, T., McCulloch, H., Leyser, O. (2014) Strigolactones and the control of plant development: Lessons from shoot branching. *The Plant Journal: for Cell and Molecular Biology* 79: 607–622.

Walton, A., Stes, E., Goeminne, G., Braem, L., Vuylsteke, M., Matthys, C., De Cuyper, C., *et al.* (2016) The response of the root proteome to the synthetic strigolactone GR24 in Arabidopsis. *Molecular & Cellular Proteomics: MCP* 15: 2744–2755.

Wang, L., Wang, B., Jiang, L., Liu, X., Li, X., Lu, Z., Meng, X., *et al.* (2015) Strigolactone signaling in Arabidopsis regulates shoot development by targeting D53-like SMXL repressor proteins for ubiquitination and degradation. *The Plant Cell* 27: 3128–3142.

Wang, Y., Sun, S., Zhu, W., Jia, K., Yang, H., Wang, X. (2013) Strigolactone/MAX2-induced degradation of brassinosteroid transcriptional effector BES1 regulates shoot branching. *Developmental Cell* 27: 681–688.

Westwood, J. H. (2000) Characterization of the Orobanche–Arabidopsis system for studying parasite–host interactions. *Weed Science* 48: 742–748.

Xie, X., Yoneyama, K., Kisugi, T., Uchida, K., Ito, S., Akiyama, K., Hayashi, H., *et al.* (2013) Confirming stereochemical structures of strigolactones produced by rice and tobacco. *Molecular Plant* 6: 153–163.

Yoneyama, K., Awad, A. A., Xie, X., Yoneyama, K., Takeuchi, Y. (2010) Strigolactones as germination stimulants for root parasitic plants. *Plant and Cell Physiology* 51: 1095–1103.

Yoneyama, K., Xie, X., Yoneyama, K., Kisugi, T., Nomura, T., Nakatani, Y., Akiyama, K., McErlean, C. S. (2017) Which are major players, canonical or non-canonical strigolactones? *Journal of Experimental Botany* 69: 2231–2239.

Yoshida, S., Kameoka, H., Tempo, M., Akiyama, K., Umehara, M., Yamaguchi, S., Hayashi, H., Kyozuka, J., Shirasu, K. (2012) The D3 F-box protein is a key component in host strigolactone responses essential for arbuscular mycorrhizal symbiosis. *The New Phytologist* 196: 1208–1216.

Zhou, F., Lin, Q., Zhu, L., Ren, Y., Zhou, K., Shabek, N., Wu, F., *et al.* (2013) D14-SCFD3-dependent degradation of D53 regulates strigolactone signalling. *Nature* 504: 406–410.

Zwanenburg, B., Mwakaboko, A. S., Kannan, C. (2016) Suicidal germination for parasitic weed control. *Pest Management Science* 72: 2016–2025.

Zwanenburg, B., Mwakaboko, A. S., Reizelman, A., Anilkumar, G., Sethumadhavan, D. (2009) Structure and function of natural and synthetic signalling molecules in parasitic weed germination. *Pest Management Science* 65: 478–491.

9 The Role of Non-Enzymatic Antioxidants in Improving Abiotic Stress Tolerance in Plants

Muhammad Arslan Ashraf, Muhammad Riaz, Muhammad Saleem Arif, Rizwan Rasheed, Muhammad Iqbal, Iqbal Hussain, and Muhammad Salman Mubarik

CONTENTS

ABBREVIATIONS

SOD superoxide dismutase
CAT catalase
POD peroxidase
APX ascorbate peroxidase
ROS reactive oxygen species
$O_2^{\bullet-}$ superoxide radical
OH hydroxyl radical
H_2O_2 hydrogen peroxide
DHAR dehydroascorbate
MDHA monodehydroascorbate
GSH reduced glutathione
GSSG oxidized glutathione
GR glutathione reductase
DHAR dehydroascorbate reductase
Cd cadmium
FC field capacity
MDA malondialdehyde
RCBD randomized complete block design
PEG polyethylene glycol
Pb lead
RWC relative water content
Na^+ sodium
K^+ potassium
Ca^{2+} calcium
Cl^- chloride
Al Aluminum
N nitrogen
P phosphorous
PSI photosystem I
Ni nickel

9.1 INTRODUCTION

Reactive oxygen species (ROS) are considered as uninvited substances produced as a result of aerobic life since the introduction of molecular oxygen (O_2) in our environment through oxygen-releasing photosynthetic organisms about 2.7 billion years ago (Halliwell, 2006). Molecular oxygen produces ROS when it accepts electrons (Gill and Tuteja, 2010). ROS production also occurs as byproducts of different metabolic pathways localized in a number of compartments inside the cell, such as peroxisomes, mitochondria and chloroplast (del Río et al., 2006; Navrot et al., 2007). In algae and higher plants, photosynthesis occurs in chloroplasts with organized thylakoid membranes which contain all the components for light-harvesting photosynthetic machinery. Molecular oxygen is produced during photosynthesis and when it passes across photosystems, it accepts electrons to form ROS. Plants contain an antioxidant system that scavenges ROS and in this way plants maintain a balance between ROS and antioxidants (Foyer and Noctor, 2005). However, this equilibrium between ROS production and detoxification is disturbed in plants challenged with abiotic stresses such as air pollution, nutrient deficiency, temperature extremes, heavy metals, drought, UV radiation and salinity. These alterations in the equilibrium increase the cellular levels of ROS which in turn induce substantial damage to plant growth and development (Gill and Tuteja, 2010). Furthermore, 1–2% oxygen utilization leads to the production of ROS (Bhattachrjee, 2005). Molecular oxygen undergoes various reactions to generate different ROS such as hydrogen peroxide (H_2O_2), hydroxyl radical ($OH^{\cdot}$) and other ROS species. Different ROS are highly reactive in nature and induce great damage to nucleic acids, carbohydrates, lipids and proteins and thereby result in cell death (Gill and Tuteja, 2010). Major losses in crop productivity occur throughout the world due to overproduction of ROS in plants exposed to environment stresses (Hirt, 2009; Gill et al., 2011). A myriad of metabolic pathways are inhibited in response to higher cell concentration of ROS that causes substantial damage to nucleic acid and oxidize proteins which induce lipid peroxidation (Foyer and Noctor, 2005). It is essential to determine whether ROS levels are good for signaling, protection or inhibitory functions which depend on the equilibrium between ROS scavenging and generation at a specific time and site (Gratão et al., 2005). Apart from cellular damage, ROS can also induce gene expression. The provoked cell response is related to various factors. The subcellular localization of ROS production is important since ROS diffuse some distance before their damaging impact on biomolecule. Abiotic stress-induced accumulation of ROS is mitigated by an antioxidant system with enzymatic antioxidants such as superoxide dismutase (SOD), ascorbate peroxidase (APX), catalase (CAT) peroxidase (POD) and non-enzymatic (flavonoids, carotenoids, phenolics, anthocyanins, α-tocopherols, glutathione (GSH) and ascorbate (AsA)) low-molecular-weight antioxidant compounds (Mittler et al., 2004; Gratão et al., 2005; Gill et al., 2011). Moreover, proline is also reported to be the part of non-enzymatic antioxidants that plants need to counteract the inhibitory effects of ROS (Chen and Dickman, 2005). The above-mentioned antioxidant compounds are reported in every cellular compartment, which highlights the significance of ROS scavenging for plant survival (Gill et al., 2011). ROS also affect the expression of several genes and interfere with signaling pathways, which suggests that ROS function as signals and biological stimuli regulating important genetic stress-related processes (Dalton et al., 1999). It has been evident in the literature that plants actively generate ROS to mediate defense processes against abiotic and biotic stress factors (Gill and Tuteja, 2010).

9.2 NON-ENZYMATIC ANTIOXIDANTS

9.2.1 Ascorbic Acid (Vitamin C)

Ascorbic acid, also known as vitamin C, contains significant antioxidant activity that reduces or protects plants from ROS-induced oxidative damage. Ascorbic acid is a water-soluble non-enzymatic antioxidant (Smirnoff, 2005; Athar et al., 2008). Ascorbic acid is found in every plant tissue, with a higher presence in meristems and photosynthetic cells. Higher ascorbic acid levels have been found in mature leaves which have fully developed chloroplasts with higher chlorophyll level. Ascorbic acid is available in reduced state in chloroplasts and leaves under normal conditions (Smirnoff, 2000). AsA levels in chloroplasts may reach 50 mM, which means 30 to 40% of total cellular levels of AsA (Gill and Tuteja, 2010). Metabolism of ascorbic acid takes place in the mitochondria of plants. Mitochondria are not only able to produce AsA via L-galactono-g-lactone dehydrogenase but are also effectively involved in the regeneration of AsA from its oxidized form (Szarka et al., 2007). The regeneration of ascorbic acid is critical due to short half-life of oxidized dehydroascorbic acid which would be degraded if not reduced to AsA. Researchers have reported AsA as having significant ROS detoxification potential owing to its ability to donate electrons in different reactions in the presence or absence of enzymes. AsA protects membranes from deterioration by detoxification of $OH^{\cdot}$ and $O_2^{\cdot-}$ (superoxide radical) coupled with regeneration of α-tocopherols through tocopheroxyl radical. Ascorbic acid also dissipates excess excitation energy in chloroplast as it functions as a cofactor for violaxanthin de-epoxidase (Smirnoff, 2000). Furthermore, AsA also plays an essential part in the ascorbate-glutathione cycle by buffering the activities of enzymes with metal ions (Noctor and Foyer, 1998). L-ascorbic acid, dehydroascorbate (DHA) and monodehydroascorbate (MDHA) are important components in the ascorbate redox system. It is reported that AsA is unstable in aqueous medium when in oxidized form, while DHA undergoes reduction by GSH to produce AsA (Foyer and Halliwell, 1976). The evidence that glutathione reductase (GR), GSH and DHAR maintained foliar AsA pool came from the studies on transformed plants with enhanced expression of GR (Foyer et al., 1995). Populus and *N. tabacum, Canescens* plants had greater tolerance to oxidative stress due to greater levels of foliar ascorbate (Aono et al., 1993; Foyer et al., 1995). In another study, increased cellular levels of oxidized ascorbate were reported in *H. vulgare* plants under Cd stress (Demirevska-Kepova et al., 2006). Similarly, *P. asperata* seedlings contained elevated levels of ascorbate when subjected to high light intensity (Yang et al., 2008). In another study, *C. auriculata* had an increase in DHA and ascorbate along with greater GSH/GSSG under UV-B stress (Agarwal, 2007). In contrast, a decrease in AsA levels was evident in *Glycine max* subjected to Cd stress (Balestrasse et al., 2001). Likewise, Cd stress also resulted in substantial decline in AsA levels in leaves and chloroplast in *P. sativum* and *A. thaliana* (Schützendübel et al., 2002; Zhang et al., 2002; Skórzyńska-Polit et al., 2003; Romero-Puertas et al., 2007).

The protective role of AsA has been reported in many plant studies (Table 9.1). Aziz et al. (2018) studied the influence of orange juice (natural source of ascorbic acid) or synthetic ascorbic acid on growth and important metabolic reactions in quinoa under drought stress. Quinoa plants (14-day) were subjected to drought stress as control (100% field capacity (FC), 60% FC, 40% FC and 20% FC). Foliar applications of orange juice (25%) and 150 mg L^{-1} besides no spray and distilled water spray were done one month after the imposition of various drought conditions. After 15 days of foliar spray, plants were harvested to measure important physio-biochemical attributes. Drought stress caused significant reduction in growth, flavonoids, carotenoids, photosynthesis and relative water content. In contrast, increase in cellular levels of H_2O_2, malondialdehyde (MDA), ascorbic acid, relative membrane permeability, total free amino acids, total free proteins, phenolics, glycine betaine, activities of antioxidant enzymes (SOD and POD), and endogenous levels of sugars were registered. Changes in these parameters were more evident in plants grown under 20% and 40% FC. Exogenous application of ascorbic acid (150 mg L^{-1}) or orange juice (25%) were effective as their applications resulted in improvement in growth, flavonoids, sugars, antioxidant enzyme activities, total soluble sugars, proline, ascorbic acid, carotenoids, photosynthesis, and MDA levels. This study suggested that the use of 25% orange juice is equally effective when compared with synthetic ascorbic acid (150 mg L^{-1}) in reducing drought effects on quinoa plants. Authors recommended the use of orange juice since it served as cheaper source of vitamin C for mediating plant defense responses under drought. Likewise, Kamal et al. (2017) also studied the influence of exogenously applied ascorbic acid in mitigating heat stress in cotton. Heat stress was applied at squaring and flowering stages. The experiment was conducted in two consecutive years (2013 and 2014) in (randomized complete block design) RCBD with split plot arrangement. Each treatment was replicated three times. Treatment breakup was as H0 (control), H1 (heat at square initiation), H2 (heat imposition at flowering stage) along with exogenous application of ascorbic acid as 0, 20, 40 and 60 mg L^{-1} ascorbic acid. The results of the experiment clearly indicated substantial damage to osmotic potential and relative water content, antioxidant enzyme activities and total chlorophyll level which was related to a decrease in yield. They suggested the use of 40 and 60 mg L^{-1} ascorbic acid to mitigate heat damage in cotton. This was associated with ascorbic acid-induced improvement in antioxidant enzymes, stay green and water relations. In another study, ascorbic acid application was reported to increase mineral nutrients, carbohydrates and amino acids in maize. Previously it was known that ascorbic acid increases soluble amino acids, proteins and carbohydrates in other species. Seed priming with ascorbic acid is considered as a key process to make maize tolerant to abiotic stress. However, the influence of seed priming on grain quality was not studied. In this study, authors carried out maize yield quality analysis from plants of ascorbic acid-primed seeds. They determined mineral nutrients, carbohydrate and amino acid profile. They observed the improvement in maize quality in the form of ascorbic acid levels, allocation of boron, soluble amino acids, and carbohydrate levels in ascorbic acid-primed plants. In contrast, there was a decline in soluble methionine. They suggested that ascorbic acid could be used to improve yield quality in maize (Alcantara et al., 2017).

A field experiment was conducted to evaluate the role of foliar-applied ascorbic acid and salicylic acid (200 mg L^{-1}) in mitigating the adverse effects of water-deficit conditions in canola. The role of these chemicals was studied in canola plants irrigated at three intervals: irrigation 1 (25 day), irrigation 2 (35 day) and irrigation 3 (45 day). The results indicated a decrease in relative water content and chlorophyll with an increase in irrigation interval. Likewise, yield and yield components also decreased due to increase in irrigation interval. Exogenous application of ascorbic acid and salicylic acid alleviated the inhibitory effects of stress and improved growth and yield in canola. Authors suggested that ascorbic acid application was effective during irrigation 2. The results indicated that ascorbic acid application effectively mitigated the inhibitory effects of water-deficit conditions in canola (El-Sabagh et al., 2017). In another study, *Caralluma tuberculata* calli were exposed to PEG-induced drought in nutrient solution. They were treated with exogenous application of

TABLE 9.1
Some Reports on Ascorbic Acid Role in Mediating Plant Defense Responses Under Abiotic Stresses

Mode of Application	Concentration	Plant Species	Stress Type	Plant Response	References
Rooting medium	2 mM	Rice	Aluminum (Al)	Plants under Al stress had substantial rise in activities of antioxidant enzymes (SOD, POD, CAT and APX) along with significant increase in MDA and H_2O_2 levels. Al stress also reduced H^+-ATPase activities and endogenous levels of ascorbic acid. Exogenous application of ascorbic acid (2 mM) significantly decreased H_2O_2 and MDA levels in roots along with increase in ascorbic acid contents in rice plants under Al stress. In addition, ascorbic acid application also enhanced the activities of H^+-ATPases and antioxidant enzymes (SOD, POD, CAT and APX). It was also noticed that ascorbic acid application increased NO_3 –N uptake in rice.	Zhou et al. (2016)
Foliar spray	200 mg L^{-1}	Wheat	Water stress	Plants were grown with different amounts of water (well watered and water stress) and nitrogen levels (0, 80, 160 and 240 kg ha-1). There was a significant decrease in yield and yield components in plants under drought stress. Exogenous application of ascorbic acid mitigated water deficit stress by increasing chlorophyll and water status, and thereby yield was also increased. Furthermore, ascorbic acid-treated plants also saw a significant increase in the activities of CAT and APX under water stress. Higher nitrogen levels did not improve nitrogen use efficiency due to a decrease in grain yield. However, ascorbic acid application alleviated water stress effects by increasing antioxidant enzyme activities.	Hafez and Gharib (2016)
Foliar spray	50 and 100 mg L^{-1}	Cucumber	Water stress	Drought condition resulted in marked decline in growth, C_i/C_a, internal CO_2 levels (C_i), stomatal conductance, photosynthesis, relative water content and chlorophyll. In contrast, plants had increases in glycine betaine and proline contents and relative membrane permeability under drought stress. Exogenous application of ascorbic acid resulted in increase in plant fresh and dry biomass, internal CO_2 levels, RWC, chlorophyll and proline. It was concluded that ascorbic acid application could have enhanced protected plants in terms of greater fresh and dry masses, RWC and chlorophyll concentration under drought conditions.	Naz et al. (2016)
Foliar spray	2 mM	Olive	Salinity	Ascorbic acid application protected olive trees from salinity effects. Salinity caused a substantial decrease in growth with a concomitant increase in Na^+ and Cl^-. However, the endogenous levels of toxic ions decrease in plants treated with foliar ascorbic acid. There was also a significant decrease in K^+ and K/Na which were substantially higher in plants treated with ascorbic acid under salinity. Ascorbic acid also decreased electrolyte leakage in plants under salinity stress.	Aliniaeifard et al. (2016)
Seed priming	75 and 150 mg L^{-1}	Cauliflower	Water stress	Water stress caused decreased growth, relative water content, chlorophyll content, total soluble protein, shoot and root concentrations of K^+ and P. In contrast, drought conditions caused increase in phenolics, glycine betaine, proline, H_2O_2, ascorbic acid, relative membrane permeability and activity of SOD. However, exogenous application of ascorbic acid mitigated drought effects by lowering H_2O_2. Furthermore, ascorbic acid-treated plants had higher activities of SOD and CAT, increased growth, relative water content, proline, glycine betaine, phenolics and ascorbic acid contents. The results suggested that exogenous application of ascorbic acid protected plants from adverse effects of drought.	Latif et al. (2016)

(Continued)

TABLE 9.1 (CONTINUED)
Some Reports on Ascorbic Acid Role in Mediating Plant Defense Responses Under Abiotic Stresses

Mode of Application	Concentration	Plant Species	Stress Type	Plant Response	References
Foliar spray	200 mg L^{-1}	Barley	Cadmium (Cd)	Cd stress decreased growth and nutrients in barely cultivars. However, ascorbic acid application increased plant growth by reducing the levels of Cd uptake and subsequent accumulation in aerial parts.	Ullah et al. (2015)
Foliar spray	100 mM	Maize	Water stress	Drought stress decreased plant growth and chlorophyll with concomitant increase in activities of antioxidant enzymes (SOD and POD). The exogenous application of ascorbic acid resulted in further increase in the activities of antioxidant enzymes. Authors concluded that exogenous ascorbic acid protected plants by stimulating the oxidative defense system under drought conditions.	Noman et al. (2015)
Rooting medium	5 mM	Fescue	Water stress	Fescue plants were grown in hydroponic with or without PEG. Plants treated with ascorbic acid had greater root growth along with minimal levels of ROS (O^{-}_2, H_2O_2) under drought. Moreover, ascorbic acid application also decreased the accumulation of MDA in plants under PEG-induced drought. Ascorbic acid application significantly increased the concentration of ascorbic acid coupled with increased transcript levels of genes, namely monohydroascorbate reductase, dehydroascorbate reductase, glutathione reductase, ascorbate peroxidase, ascorbate peroxidase, catalase, and superoxide dismutase. Ascorbic acid application increased root growth in drought stressed plants by increasing endogenous levels of non-enzymatic antioxidants and lowering the endogenous levels of ROS.	Xu et al. (2015)
Foliar spray	100 mM	Maize	Salinity	Salinity stress induced significant decrease in the uptake of N, P, Ca^{2+} and K^+ with a simultaneous increase in Na^+ in different plant parts. Ascorbic acid application increased the endogenous levels of ascorbic acid in leaves and roots along with an associated increase in proline and glycine betaine. The uptake of the above-mentioned ions was also increased in plants treated with ascorbic acid. Ascorbic acid application protected plants by reducing the uptake and accumulation of toxic Na^+. Authors concluded that protection from salinity stress in ascorbic acid-treated plants might be due to alterations in the de novo synthesis of ascorbic acid or increased accumulation of ascorbic acid in leaves and roots.	Jamil et al. (2015)
Foliar spray	100 mg L^{-1}	Chickpea	Water stress	Exogenous application of ascorbic acid enhanced drought tolerance in chickpea by improving growth and protein content. In addition, ascorbic acid also water use efficiency of plants under drought.	Farjam et al. (2015)

ascorbic acid and salicylic acid. Exogenous application of ascorbic acid enhanced the activities of antioxidant enzymes (GR, APX, CAT, POD and SOD) under drought stress (Rehman et al., 2017). Wang et al. (2018) reported the damaging effects of some herbicides in wheat under low-temperature conditions. They gave pre-treatment to wheat seedlings with herbicides, namely fenoxaprop-P-ethyl, fluroxypyr and isoproturon. Afterwards, wheat plants were grown under low-temperature conditions. Use of herbicides further aggravated the adverse effects of low temperature on growth, cell membrane integrity, chlorophyll and electron transport rate. However, foliar ascorbic acid application alleviated the negative effects of herbicide use under low-temperature conditions. Ascorbic acid-treated plants had lower ROS and MDA levels and membrane damage along with enhanced activities of different antioxidant enzymes (SOD, POD and CAT). Authors further reported the significant role of foliar-applied ascorbic acid in mitigating oxidative stress due to combined abiotic stresses. Penella et al. (2017) studied the effects of foliar-applied ascorbic acid in peach trees under water-deficit conditions. Two cultivars of peach were used in this study, namely (CaroTiger and Scarletprince). Trees were given full water-to-field capacity one week after the application of ascorbic acid. It was reported that ascorbic acid application increased photosynthesis in peach trees under water-deficit conditions. There existed variation in cultivars with respect to exogenous application of ascorbic acid. Exogenous application of ascorbic acid enhanced catalase activity and membrane integrity in both peach cultivars under water-deficit conditions. Results suggested the application of ascorbic acid for mitigating water deficit effects in peach trees. Similarly, Hussain et al. (2017) reported that seed priming with ascorbic acid increased Pb stress tolerance in okra. They used two okra cultivars (Arka Anamika and Subz-Pari) in the study. Ascorbic acid priming (50 and 100 mg L^{-1}) along with hydropriming and without priming resulted in significant changes in physio-biochemical attributes in two okra cultivars challenged with Pb stress (0 and 100 mg L^{-1}). Imposition of Pb stress in okra significantly decreased growth and chlorophyll concentration, whereas increases in the activities of antioxidant enzymes (SOD, POD, CAT), ascorbic acid content, total soluble protein, proline and total free amino acids were recorded. These changes were pronounced when plants were treated with 100 mg L^{-1} Pb. Plants grown under Pb stress also suffered oxidative damage in the form of significant increase in cellular levels of MDA and H_2O_2. In contrast, hydro- and ascorbic acid-primed plants had greater growth, activities of antioxidant enzymes, ascorbic acid, total free amino acids, proteins, proline and chlorophyll. Seed priming with 50 mg L^{-1} ascorbic acid significantly increased Pb tolerance in okra plants. Cultivar Arka Anamika showed greater tolerance to Pb compared with Subz-Pari. In another study, ascorbic acid was reported as an important growth regulator of plants under salinity. Ascorbic acid application (0.5 and 1 mM) was assessed in rice callus culture under 200 mM salinity. There was a marked decrease in fresh and dry biomass, growth rate, Ca^{2+} and K^+ levels with elevated levels of Na^+ and Na^+/K^+. However, exogenous application of ascorbic acid alleviated the inhibitory effects of salinity. Antioxidant enzyme activities (SOD, POD and CAT) and proline levels increased in salinity stressed plants and exogenous application of ascorbic acid (0.5 mM) resulted in further increase in these variables. The results of the study indicated that exogenous application of ascorbic acid had the potential to substantially increase rice callus tolerance against salinity (Alhasnawi et al., 2016).

9.2.2 Flavonoids

Plants are an enriched source of flavonoids which usually occur in pollens, floral parts and leaves. Flavonoids are accumulated in vacuoles as glycosides or flavonoids and are also found as exudates on aerial parts of plants. Plants contain more than 1 mM flavonoids (Vierstra et al., 1982). There are various classes of flavonoids based on their structural variations such as anthocyanins, isoflavones, flavones and flavanols. Flavonoids are associated with a number of functions related to seed pigmentation, fruits, flowers, protection against phytopathogens and UV light, and also serve as signaling molecules in plant-microbe interactions, playing essential role in pollen germination and fertility (Olsen et al., 2010). Among plant secondary metabolites, flavonoids are considered the most bioactive compounds. The antioxidant function of flavonoids often outcompetes other well-known antioxidants such as α-tocopherols and ascorbic acid (Hernandez et al., 2009). Flavonoids are important defensive compounds inside plants because they readily detoxify ROS radicals under environmental constraints (Lovdal et al., 2010). The antioxidant function of flavonoids is attributed to the arrangement and number of OH (hydroxyl) groups attached to ring structures. Phenolic compounds and flavonoids rapidly absorb high UV irradiation and therefore significantly protect plants subjected to UV irradiation. Mutant plants with inability for flavonoid production were unable to tolerate high UV irradiation stress (Cle et al., 2008). The expression of genes related to flavonoid biosynthesis is enhanced in plants grown in stressful conditions. The endogenous levels of flavonoids substantially increase in plants subjected to abiotic and biotic stress such as wounding, nutrient deficiency, drought and metal toxicity (Winkel-Shirley, 2002). In this context, mutant plants with inability to produce flavonoids appeared more sensitive to UV-B light compared with their wild-type relative (Filkowski et al., 2004). Flavonoids also play an essential role in plants grown in nitrogen deficient environment (Solovchenko and Schmitz-Eiberger, 2003). Flavonoids protect plants against pathogen attacks (Andersen and Markham, 2005).

9.2.3 Carotenoids

Plants employ a number of pathways to dissipate the excess energy of photosynthetic membranes. Some pathways also involve the use of isoprenoid compounds. In photosynthetic organisms, tocopherols, β-carotene, carotenoids and zeaxanthin play an essential part in photoprotection either by decreasing lipid peroxidation or through active scavenging of ROS. Carotenoids are present widely in microorganism and plants. They exhibit huge diversity as there are about 600 known carotenoids in nature and they are found in various concentrations (Table 9.2). Carotenoids are a lipid-soluble antioxidant in plants with multiple functions. They also

TABLE 9.2
Some Reports from Literature Which Indicate Abiotic Stress-Induced Changes in Carotenoids in Plants

Plant Species	Stress Type	Carotenoid Contents increased/decreased	References
Tomato	Cd	Decreased	Alyemeni et al. (2018)
Radish	Water stress	Decreased	Noman et al. (2018)
Tomato	Salinity	Decreased	Rouphael et al. (2018)
Sorghum	Salinity	Increased	Forghani et al. (2018)
Olive	Salinity and drought	Increased	Ben Abdallah et al. (2018)
Wheat	Drought	Decreased	Abid et al. (2018)
Rapeseed	Cd	Decreased	Ali et al. (2018)
Sunflower	Salinity	Increased	Youssef et al. (2017)
Melon	Salinity	Decreased	Sarabi et al. (2017)
Carex leucochlora	Salinity	Increased	Ye et al. (2017)
Amsonia orientalis	Salinity	Decreased	Acemi et al. (2017)
Parsley	Salinity	Decreased	Habibi (2017)
Rice	Salinity	Decreased	Khunpona et al. (2017)
Lupinus termis	Salinity	Decreased	Abdel-Latef et al. (2017)
Sesuvium portulacastrum	Salinity	Increased	Slama et al. (2017)
Anethum graveolens	Salinity and water stress	Increased	Tsamaidi et al. (2017)
Leymus chinensis	High temperature	Decreased	Anjum et al. (2016)
Fennel	Drought	Decreased	Mirjahanmardi and Ehsanzadeh (2016)
Mung bean	Chromium	Decreased	Jabeen et al. (2016)
Pea	Salinity	Decreased	Husen et al. (2016)

protect plants from oxidative damage. Carotenoids perform three main roles in plants: first, they are the accessory pigments which absorb light of 400 and 550 nm wavelength and transfer the absorbed light to chlorophyll (Siefermann-Harms, 1987). Second, carotenoids protect the photosynthetic apparatus by efficiently scavenging ROS radicals, which are produced during photosynthesis (Collins, 2001). Third, carotenoids are essential for assembly of PSI and also for the integrity of light harvesting complex proteins. Carotenoids also play a structural role by protecting the thylakoid membrane (Niyogi et al., 2001). Moreover, they are reported to protect plants from singlet oxygen. However, photosynthetic cells deficient in carotenoids suggest the presence of other mechanisms for protection against singlet oxygen. Razinger et al. (2008) and Rai et al. (2005) reported Cd-induced decrease in the concentration of chlorophyll and carotenoids in *V. mungo* and *Phyllanthus amarus*, respectively. Cd stress was also reported to decrease the concentration of carotenoids in barley (Demirevska-Kepova et al., 2006). In another study, carotenoids were increased in plants challenged with Cd (Rai et al., 2005). Furthermore, isoprenoids, which also include tocopherols and carotenoids, are reported to play an essential part in photoprotection (Penuelas and Munne-Bosch, 2005). It has also been known that monoterpenes protect plants from oxidative damage and high-temperature stress (Loreto et al., 2004).

9.2.4 Tocopherols

Tocopherols are lipid-soluble antioxidant enzymes which have substantial potential for the degradation of ROS (Holländer-Czytko et al., 2005). Tocopherols are abundant in membranes where they perform both enzymatic and non-enzymatic functions. Tocopherols protect membranes by scavenging ROS like H_2O_2. Tocopherols are localized in thylakoid membranes inside chloroplasts. Plants have four isoforms of tocopherols (α, β, γ, δ). Among various forms of tocopherols, α-tocopherols depict the highest antioxidant activities (Kamal-Eldin and Appelqvist, 1996). Higher amounts of α-tocopherols are evident in leaves. *Artemisia ordosiea* Kraschen and *Artemisia sphaerocephala* Kraschen are widespread in arid and semiarid regions of northwest China. It was found that α-tocopherol protected from membrane damage and also decreased Na^+/K^+. In this experiment, the role of α-tocopherol in modulating the composition of membrane fatty acids was studied. The above-mentioned plant species were subjected to salinity for 7, 14 and 21 days. Higher levels of toxic Na^+ coupled with more chlorophyll and dry weight were evident in *A. sphaerocephala* compared with *A. ordosiea*. The significant higher level of unsaturated lipids in *A. sphaerocephala* was attributed to C18:2. Furthermore, there was linear increase in C18:2 and α-tocopherol levels in *A. sphaerocephala*, whereas C18:3, C18:1, Na^+ and jasmonic acid decreased. It was observed that α-tocopherol had negative correlation with C18:3 and positive with C18:2 (Chen et al., 2018). Under environmental constraints, plants usually undergo changes in endogenous levels of α-tocopherols to overcome the inhibitory effects of stress through detoxification of ROS (Ashraf and Ashraf, 2016). For example, exogenous application of α-tocopherol modulated a number of physio-biochemical attributes in *Carex leucochlora* challenged with salinity. It was observed that exogenous application of α-tocopherol (0.8 mM) exhibited significant mitigation of salinity effects on growth and loss of cell membranes integrity. Exogenous α-tocopherol significantly decreased the endogenous levels of H_2O_2 and superoxide radical in salinity stressed plants. Furthermore, plants treated with foliar α-tocopherol also showed significant increase in soluble protein, proline, carotenoids and chlorophyll b in salinity stressed plants. In contrast, α-tocopherol application did not influence soluble sugars and chlorophyll a contents in plants under salinity stress. Exogenous application of α-tocopherol mitigated the inhibitory effects of salinity in *Carex leucochlora* by improving osmotic adjustment and photosynthetic pigments (Ye et al., 2017). In another study, exogenous application of α-tocopherol mitigated salinity-induced oxidative stress in soybean and this function of α-tocopherol was attributed to its role as an important signal transmitter during salinity stress. Plants treated with exogenous α-tocopherol had maximal root length with concomitant decline in the generation of superoxide radical ($O_2^{\cdot -}$) and H_2O_2 which in turn reduced the accumulation of MDA. However, α-tocopherol treatment significantly decreased the endogenous levels of proline. Intriguingly, supplemented α-tocopherol significantly increased the cellular levels of auxins. The influence of exogenous α-tocopherol was also observed in the form of a decline in SOD and isoenzyme activity coupled with an increase in glutathione-s-transferase and peroxidase activities. In contrast, APX activity remained unaffected due to the exogenous application of this biomolecule. It was suggested that α-tocopherol functioned as an important signal molecule that enhanced auxins and antioxidant activities in roots of salinity stressed soybean plants (Sereflioglu et al., 2017).

Salinization of agricultural soil is the major concern in Egypt. In field experiments, exogenous application of α-tocopherol (100 and 200 ppm) and hand hoeing was done to find out their integrative effects on nutritional values of lupine, yield and yield-related attributes, mineral content, osmoprotectant and growth in saline soil. Imposition of salinity significantly decreased plant dry mass, leaf area and chlorophyll content. However, exogenous application of α-tocopherols significantly improved the above-mentioned attributes. Salinity stress also caused significant decline in essential amino acid levels which were substantially increased due to exogenous application of α-tocopherols in salinity stressed plants. Authors concluded that exogenous application of α-tocopherols (200 ppm) mitigated the adverse salinity effects on lupine plants in terms of growth and yield quality (Dawood et al., 2016). In another study, salinity stress resulted in a significant decrease in endogenous levels of α-tocopherols in hydroponically grown wheat cultivars (S-24 and MH-97) with contrasting salinity tolerance. Likewise, when rice plants were subjected to NaCl or Na^+ and Cl^- alone, there was significant variation in the cellular levels of α-tocopherols and ascorbic acids, which are important non-enzymatic antioxidants (Khare et al., 2015). Likewise, exogenous application of α-tocopherol was done in soybean plants under saline conditions. Three foliar applications of α-tocopherol (100 mg L^{-1}) were used in the study. Salinity stress increased Na^+, nitrogen, soluble sugars, total free amino acids, proline, chlorophyll *a*/*b*, relative growth rate and net photosynthetic rate in soybean. Conversely, salinity stress decreased K^+/Na^+, K^+, P, chlorophyll and relative water content. Exogenous application of α-tocopherol increased N levels, relative growth rate and net assimilation rate for photosynthate. Authors concluded that α-tocopherols should be added in commercial formulations to alleviate salinity effects in soybean (Mostafa et al., 2015). Semida et al. (2014) studied the effects of foliar-applied α-tocopherols in salinity stressed faba beans. Two cultivars of faba beans (Giza 429 and Giza 40) were used in the experiment. Faba bean plants subjected to salinity had substantial decline in growth, yield and physiology with significant changes in anatomy. However, plants treated with foliar α-tocopherols had better growth, yield and yield components under saline conditions. Moreover, exogenously applied α-tocopherol mitigated salinity effects on stem and leaf anatomy, nutrient relations, membrane integrity and relative water content. Greater salinity tolerance was evident in Giza 429 compared to Giza 40 in term of better growth. It was concluded that α-tocopherol application stimulated an antioxidant system that protected plants from salinity-induced oxidative damage and thereby improved physiological performance in plants. In a study, exogenous application of α-tocopherols (α-Toc) and resveratrol (Res), alone and in combination, was performed in order to alleviate salinity damages in citrus seedlings. Exogenous application of Res+α-Toc or Res, α-Toc decreased chlorophyll degradation, membrane permeability and lipid peroxidation in terms of MDA accumulation. Likewise, combined applications of both biomolecules decreased cellular H_2O_2 levels along with restoration of photosynthesis under salinity. Exogenous application of Res restricted aerial transport of Cl^-, while α-Toc+Res application significantly reduced the translocation of Cl^- and Na^+ to leaves. There existed a marked perturbation in endogenous levels of glutathione, ascorbic acid, phenol, proline and carbohydrates in citrus plants subjected to salinity. Activities of antioxidant enzymes, namely polyphenol oxidase, glutathione reductase, peroxidase, ascorbate peroxidase and superoxide dismutase were higher in leaves of salinity stressed plants whereas POD and SOD activities decreased in roots due to salinity. Studies have reported tissue-specific activation or deactivation of antioxidant enzymes in response to exogenous application of α-Toc and Res (Kostopoulou et al., 2014). Naliwajski and Skłodowska (2014) measured the activities of antioxidant enzymes, α-tocopherols and oxidative stress markers in non-acclimated and acclimated cucumber cell suspensions when exposed to 150 and 200 mM NaCl. Salinity stress caused a significant increase in glutathione peroxidase, while glutathione-S-transferase activity increased only in acclimated cell suspensions. Similarly, greater cellular levels of α-tocopherols were evident in acclimated cells due to salinity. NaCl treatment significantly increased α-tocopherols. The results of the present study suggested that the salinity tolerance of acclimated plants was higher than in non-acclimated plants. Enhanced salinity tolerance was attributed to greater levels of α-tocopherols.

Kumar et al. (2013) reported that tocopherols, apart from being powerful antioxidants, are also good signaling molecules and significantly mediate signaling. Among different forms of tocopherols, α-tocopherol is the most effective. In this study, α-tocopherol enriched transgenic *Brassica juncea* plants were grown along with their wild-type counterparts under osmotic, heavy metal and salt stress to find out the impact of overproduction of α-tocopherols on important metabolites and antioxidant enzymes. Transgenic plants with enhanced production of α-tocopherols had minimal oxidative damage reflected as higher relative water content, lower H_2O_2, and MDA levels. Furthermore, transgenic plants also had lower H_2O_2 and superoxide radical as indicated in histochemical staining under abiotic stress. Transgenic and wild-type plants did not show any differences when grown under normal conditions. However, there was a marked improvement in the activities of antioxidant enzymes, namely GR, APX, CAT and SOD in transgenic plants over wild-type plants under stressed conditions. Transgenic plants had higher reductive ratios with lower glutathione and ascorbate levels under stress conditions. The results suggested that plants with enhanced ability to produce α-tocopherols are better able to tolerate osmotic, heavy metal and salt stress with significant interaction with water-soluble antioxidant compounds. In another study, salinity stress resulted in a significant increase in α- and γ-tocopherols in sunflower plants (Noreen and Ashraf, 2010). Likewise, salinity stress enhanced the endogenous levels of gamma- and Delta-tocopherols in peas (Noreen and Ashraf, 2009).

9.2.5 Glutathione

Glutathione is an important metabolite produced in plants during stress. It protects plants from ROS-induced oxidative stress. GSH is usually found in reduced form and is present is every cell compartment such as apoplast, peroxisomes, chloroplasts, mitochondria, vacuole, endoplasmic reticulum and cytosol (Mittler and Zilinskas, 1992; Jiménez et al., 1998). GSH plays an essential part in important physiological mechanisms including detoxification of xenobiotics, conjugation of metabolites, signal transduction and regulation of sulfate transport (Xiang et al., 2001). It mediates the expression of stress-related genes (Mullineaux and Rausch, 2005) and also plays an essential part in plant growth and developmental events such as enzymatic regulation, pathogen resistance, senescence and cell death, and cell differentiation (Rausch and Wachter, 2005). GSH provides substrates to different metabolic pathways that produce GSSG. The maintenance of cellular redox state is related to the appropriate balance between GSSG and GSH (Foyer and Noctor, 2005). GSH is required to maintain a reduced cell state so as to protect cells from oxidative damage induced by ROS (Meyer, 2008). GSH is a strong scavenger of H_2O_2 and $^1O_2$ (Gill and Tuteja, 2010) and is also able to detoxify the most lethal ROS, such as $OH^·$ (Larson, 1988). In addition, GSH plays an essential part in mediating the antioxidant system through the regeneration of another water-soluble antioxidant compound, namely ascorbate, through the GSH-ascorbate cycle (Foyer and Halliwell, 1976; Gill and Tuteja, 2010). Under stressful conditions, the endogenous levels of GSH decreased, resulting in a more oxidized redox state which damages plants (Tausz et al., 2004). GSH acts as precursor in the biosynthesis of phytochelatins that mediates cell response to heavy metal levels. Researchers also use the cell concentration of GSH as an important stress marker. However, the changes in the levels of GSH play a major role in its function as an antioxidant, as they undergo significant variations in response to abiotic stress. In addition, higher endogenous levels of GSH are associated with the ability of plants to withstand heavy metal stress. Higher cell concentration of GSH protected *Phragmites australis* Trin. from oxidative damage under Cd and also protected enzymes of photosynthetic pathways (Tausz et al., 2004). Furthermore, cell concentration of GSH significantly rose in Cd-stressed *Sedum alfredii* (Sun et al., 2007) and *P. sativum* (Metwally et al., 2005).

9.2.6 Proline

In addition to its function as osmolyte, proline is also a potent antioxidant with significant ability to protect plants from oxidative damage as reflected in the form of decreased lipid peroxidation. Therefore, proline is now also referred to as a non-enzymatic antioxidant, and living organisms need this biomolecule to scavenge ROS (Chen and Dickman, 2005). Proline biosynthesis from L-glutamic acid requires enzymes, namely D1-pyrroline-5-carboxylate synthetase (P5CS), D1-pyrroline- 5-carboxylate (P5C) and D1-pyrroline-5-carboxylate reductase (P5CR) in plants (Verbruggen and Hermans, 2008). In contrast, enzymes of mitochondria P5C dehydrogenase (P5CDH) and proline dehydrogenase (oxidase) (ProDH) metabolize proline into L-glutamic acid via P5C. Literature showed the enhanced accumulation of proline in plants facing metal, drought and salinity stress. This enhanced accumulation of proline is attributed to either limited proline degradation or higher biosynthesis of proline. In addition to its role as osmoprotectant, free proline also acts as scavenger of $^1O_2$ and $OH^{\cdot}$, inhibitor of lipid peroxidation, metal chelator and protein stabilizer (Ashraf and Foolad, 2007; Trovato et al., 2008). Proline along with other osmolytes, such as myo-inositol and mannitol, functions as $OH^{\cdot}$ scavenger. However, among these osmolytes, proline has been shown to be the most effective $OH^{\cdot}$ scavenger. Therefore, it is an important ROS scavenger and molecule in redox signaling under abiotic stress (Alia and Saradhi, 1991). For instance, proline addition was reported to scavenge ROS in DARas mutant cells and hence protected from cell death (Chen and Dickman, 2005). Moreover, proline protected *C. trifolii* from H_2O_2, heat, salt and UV light stress. It also protected yeast from MV (herbicide). It was suggested that proline's function as ROS scavenger protects ROS-mediated apoptosis and thereby significantly regulates defense responses of cells under stress. Enhanced tolerance to abiotic stress such as salinity and drought is expected in plants with higher proline concentrations (Gill and Tuteja, 2010). Under stress, proline protected cells from acidosis and maintained $NADP^+$: NADPH at an appropriate rate required for metabolism (Gill and Tuteja, 2010). Proline application refills $NADP^+$ and thereby supports redox cycling, which is essential in antioxidant defense processes during abiotic stresses (Babiychuk et al., 1995). Salinity stress also caused substantial increase in proline accumulation in two rice cultivars with contrasting response to salinity (Demiral and Türkan, 2005). Evidence exists in the literature which also highlights the function of proline as potential mediator of the pentose-phosphate pathway (Hare and Cress, 1997). The pentose-phosphate pathway is an important component in the antioxidant system, requiring NADPH to maintain ascorbate and glutathione. In another study, MDA levels and reduced/oxidized GSH showed the correlation of proline with GSH redox state in algae under heavy metal stress (Siripornadulsil et al., 2002). Proline also functioned as antioxidant to protect Cd-stressed cells (Siripornadulsil et al., 2002). Similarly, the role of proline as an antioxidant was noted in pea plants exposed to Ni stress (Gajewska and Skłodowska, 2005). Proline protected plants from salinity damages by increasing the GSH content and maintaining greater glutathione redox state with enhanced activities of glyoxalase I and glyoxalase II, glutathione reductase, glutathione-S-transferase, and glutathione peroxidase enzymes associated with ROS detoxification. Proline reduced oxidative damage as reflected from lower H_2O_2 and lipid peroxidation in mung bean under salinity (Hossain and Fujita, 2010). It protected plants from oxidative damage from Cd by increasing ascorbate and GSH levels with higher GSH/GSSH, activities of antioxidant enzymes, namely CAT, GPX, GR, MDHAR, DHAR and APX. Proline decreased lipid peroxidation in the form of MDA and also endogenous H_2O_2 levels in Cd-stressed mung bean (Hossain et al., 2010).

9.3 Conclusion

Increasing abiotic stresses significantly compromise food security, and this condition is more prevalent in developing countries. Abiotic stresses cause a substantial increase in the endogenous levels of ROS that in turn damage important cellular molecules such as proteins, lipids, nucleic acid, and pigment. Plants have well-defined defense pathways in the form of enzymatic and non-enzymatic antioxidants. Ascorbic acid, glutathione, carotenoids, flavonoids, phenolics, tocopherols, and proline are important molecules that function as non-enzymatic antioxidants and protect plants from oxidative stress caused by the accumulation of ROS. The above-cited review of literature indicated the role of these non-enzymatic antioxidants as mediators of plant defense response. Plants undergo significant variations in the synthesis and degradation of these antioxidant compounds, which highlights their potential involvement in plant protection under environmental hazards such as drought, salinity, temperature extremes, heavy metal, waterlogging and UV light. Plant researchers also apply these antioxidant compounds exogenously to circumvent the negative effects of abiotic stress on plants. Therefore, there is a need to produce plants that have an ability for enhanced production of non-enzymatic antioxidants so as to better combat environmental constraints.

REFERENCES

Abdel-Latef, A.A.H., Abu-Alhmad, M.F., Abdel-Fattah, K.E. (2017) The possible roles of priming with ZnO nanoparticles in mitigation of salinity stress in lupine (*Lupinus termis*) plants. *Journal of Plant Growth Regulation* 36(1): 60–70.

Abid, M., Hakeem, A., Shao, Y., Liu, Y., Zahoor, R., Fan, Y., Suyu, J., *et al.* (2018) Seed osmopriming invokes stress memory against post-germinative drought stress in wheat (Triticum aestivum L.). *Environmental and Experimental Botany* 145: 12–20.

Acemi, A., Duman, Y., Karakuş, Y.Y., Kompe, Y.Ö, Ozen, F. (2017) Analysis of plant growth and biochemical parameters in *Amsonia orientalis* after in vitro salt stress. *Horticulture, Environment, and Biotechnology* 58(3): 231–239.

Agarwal, S. (2007) Increased antioxidant activity in Cassia seedlings under UV-B radiation. *Biologia Plantarum* 51(1): 157–160.

Alcantara, B.K., Rizzi, V., Gaziola, S.A., Azevedo, R.A. (2017) Soluble amino acid profile, mineral nutrient and carbohydrate content of maize kernels harvested from plants submitted to ascorbic acid seed priming. *Anais da Academia Brasileira de Ciências* 89(1 Suppl 0): 695–704.

Alhasnawi, A.N., Che Radziah, C.M.Z., Kadhimi, A.A., Isahak, A., Mohamad, A., Yusoff, W.M.W. (2016) Enhancement of antioxidant enzyme activities in rice callus by ascorbic acid under salinity stress. *Biologia Plantarum* 60(4): 783–787.

Ali, E., Hussain, N., Shamsi, I.H., Jabeen, Z., Siddiqui, M.H., Jiang, L.X. (2018) Role of jasmonic acid in improving tolerance of rapeseed (*Brassica napus* L.) to Cd toxicity. *Journal of Zhejiang University Science B* 19(2): 130–146.

Alia, Saradhi, P.P. (1991) Proline accumulation under heavy metal stress. *Journal of Plant Physiology* 138(5): 554–558.

Aliniaeifard, S., Hajilou, J., Tabatabaei, S.J., Sifi-Kalhor, M. (2016) Effects of ascorbic acid and reduced glutathione on the alleviation of salinity stress in olive plants. *International Journal of Fruit Science* 16(4): 395–409.

Alyemeni, M.N., Ahanger, M.A., Wijaya, L., Alam, P., Bhardwaj, R., Ahmad, P. (2018) Selenium mitigates cadmium-induced oxidative stress in tomato (*Solanum lycopersicum* L.) plants by modulating chlorophyll fluorescence, osmolyte accumulation, and antioxidant system. *Protoplasma* 255(2): 459–469.

Andersen, O.M., Markham, K.R. (2005) *Flavonoids: Chemistry, Biochemistry and Applications*. CRC Press, Florida.

Anjum, S.A., Jian-Hang, N., Ran, W., Jin-Huan, L., Mei-Ru, L., Ji-Xuan, S., Jun, L., *et al.* (2016) Regulation mechanism of exogenous 5-aminolevulinic acid on growth and physiological characters of *Leymus chinensis* (Trin.) under high temperature stress. *The Philippine Agricultural Scientist* 99(3): 253–259.

Aono, M., Kubo, A., Saji, H., Tanaka, K., Kondo, N. (1993) Enhanced tolerance to photooxidative stress of transgenic *Nicotiana tabacum* with high chloroplastic glutathione reductase activity. *Plant and Cell Physiology* 34: 129–135.

Ashraf, M.A., Ashraf, M. (2016) Growth stage-based modulation in physiological and biochemical attributes of two genetically diverse wheat (*Triticum aestivum* L.) cultivars grown in salinized hydroponic culture. *Environmental Science and Pollution Research International* 23(7): 6227–6243.

Ashraf, M., Foolad, M.R. (2007) Roles of glycine betaine and proline in improving plant abiotic stress resistance. *Environmental and Experimental Botany* 59(2): 206–216.

Athar, H.-U.-R., Khan, A., Ashraf, M. (2008) Exogenously applied ascorbic acid alleviates salt-induced oxidative stress in wheat. *Environmental and Experimental Botany* 63(1–3): 224–231.

Aziz, A., Akram, N.A., Ashraf, M. (2018) Influence of natural and synthetic vitamin C (ascorbic acid) on primary and secondary metabolites and associated metabolism in quinoa (*Chenopodium quinoa* Willd.) plants under water deficit regimes. *Plant Physiology and Biochemistry: PPB* 123: 192–203.

Babiychuk, E., Kushnir, S., Belles-Boix, E., Van Montagu, M., Inze, D. (1995) *Arabidopsis thaliana* NADPH oxidoreductase homologs confer tolerance of yeasts toward the thiol-oxidizing drug diamide. *The Journal of Biological Chemistry* 270(44): 26224–26231.

Balestrasse, K.B., Gardey, L., Gallego, S.M., Tomaro, M.L. (2001) Response of antioxidant defence system in soybean nodules and roots subjected to cadmium stress. *Functional Plant Biology* 28(6): 497–504.

Ben Abdallah, M., Trupiano, D., Polzella, A., De Zio, E., Sassi, M., Scaloni, A., Zarrouk, M., Ben Youssef, N., Scippa, G.S. (2018) Unraveling physiological, biochemical and molecular mechanisms involved in olive (*Olea europaea* L. cv. Chétoui) tolerance to drought and salt stresses. *Journal of Plant Physiology* 220: 83–95.

Bhattachrjee, S. (2005) Reactive oxygen species and oxidative burst: Roles in stress, senescence and signal transduction in plant. *Current Science* 89: 1113–1121.

Chen, C., Dickman, M.B. (2005) Proline suppresses apoptosis in the fungal pathogen *Colletotrichum trifolii*. *Proceedings of the National Academy of Sciences of the United States of America* 102(9): 3459–3464.

Chen, X., Zhang, L., Miao, X., Hu, X., Nan, S., Wang, J., Fu, H. (2018) Effect of salt stress on fatty acid and alpha-tocopherol metabolism in two desert shrub species. *Planta* 247(2): 499–511.

Cle, C., Hill, L.M., Niggeweg, R., Martin, C.R., Guisez, Y., Prinsen, E., Jansen, M.A. (2008) Modulation of chlorogenic acid biosynthesis in *Solanum lycopersicum*; consequences for phenolic accumulation and UV-tolerance. *Phytochemistry* 69(11): 2149–2156.

Collins, A.R. (2001) Carotenoids and genomic stability. *Mutation Research* 475(1–2): 21–28.

Dalton, T.P., Shertzer, H.G., Puga, A. (1999) Regulation of gene expression by reactive oxygen. *Annual Review of Pharmacology and Toxicology* 39: 67–101.

Dawood, M.G., El-Metwally, I.M., Abdelhamid, M.T. (2016) Physiological response of lupine and associated weeds grown at salt-affected soil to αtocopherol and hoeing treatments. *Gesunde Pflanzen* 68(2): 117–127.

Del Río, L.A., Sandalio, L.M., Corpas, F.J., Palma, J.M., Barroso, J.B. (2006) Reactive oxygen species and reactive nitrogen species in peroxisomes. Production, scavenging, and role in cell signaling. *Plant Physiology* 141(2): 330–335.

Demiral, T., Türkan, I. (2005) Comparative lipid peroxidation, antioxidant defense systems and proline content in roots of two rice cultivars differing in salt tolerance. *Environmental and Experimental Botany* 53(3): 247–257.

Demirevska-Kepova, K., Simova-Stoilova, L., Stoyanova, Z.P., Feller, U. (2006) Cadmium stress in barley: Growth, leaf pigment, and protein composition and detoxification of reactive oxygen species. *Journal of Plant Nutrition* 29(3): 451–468.

El-Sabagh, A.E., Abdelaal, K.A.A., Barutcular, C. (2017) Impact of antioxidants supplementation on growth, yield and quality traits of canola (*Brassica napus* L.) under irrigation intervals in north Nile delta of Egypt. *Journal of Experimental Biology and Agricultural Sciences* 5(2): 163–172.

Farjam, S., Kazemi-Arbat, H., Siosemardeh, A., Yarnia, M., Rokhzadi, A. (2015) Effects of salicylic and ascorbic acid applications on growth, yield, water use efficiency and some physiological traits of chickpea (*Cicer arietinum* L.) under reduced irrigation. *Legume Research - an International Journal* 38(1): 66–71.

Filkowski, J., Kovalchuk, O., Kovalchuk, I. (2004) Genome stability of vtc1, tt4, and tt5 Arabidopsis thaliana mutants impaired in protection against oxidative stress. *The Plant Journal : for Cell and Molecular Biology* 38(1): 60–69.

Forghani, A.H., Almodares, A., Ehsanpour, A.A. (2018) Potential objectives for gibberellic acid and paclobutrazol under salt stress in sweet sorghum (*Sorghum bicolor* [L.] Moench cv. Sofra). *Applied Biological Chemistry* 61(1): 113–124.

Foyer, C.H., Halliwell, B. (1976) The presence of glutathione and glutathione reductase in chloroplasts: A proposed role in ascorbic acid metabolism. *Planta* 133(1): 21–25.

Foyer, C.H., Noctor, G. (2005) Redox homeostasis and antioxidant signaling: A metabolic interface between stress perception and physiological responses. *The Plant Cell* 17(7): 1866–1875.

Foyer, C.H., Souriau, N., Perret, S., Lelandais, M., Kunert, K.J., Pruvost, C., Jouanin, L. (1995) Overexpression of glutathione reductase but not glutathione synthetase leads to increases in antioxidant capacity and resistance to photoinhibition in poplar trees. *Plant Physiology* 109(3): 1047–1057.

Gajewska, E., Skłodowska, M. (2005) Antioxidative responses and proline level in leaves and roots of pea plants subjected to nickel stress. *Acta Physiologiae Plantarum* 27(3): 329–340.

Gill, S.S., Khan, N.A., Anjum, N.A., Tuteja, N. (2011) Amelioration of cadmium stress in crop plants by nutrients management: Morphological, physiological and biochemical aspects. *Plant Stress* 5: 1–23.

Gill, S.S., Tuteja, N. (2010) Reactive oxygen species and antioxidant machinery in abiotic stress tolerance in crop plants. *Plant Physiology and Biochemistry: PPB* 48(12): 909–930.

Gratão, P.L., Polle, A., Lea, P.J., Azevedo, R.A. (2005) Making the life of heavy metal-stressed plants a little easier. *Functional Plant Biology* 32(6): 481–494.

Habibi, G. (2017) Selenium ameliorates salinity stress in Petroselinum crispum by modulation of photosynthesis and by reducing shoot Na accumulation. *Russian Journal of Plant Physiology* 64(3): 368–374.

Hafez, E., Gharib, H. (2016) Effect of exogenous application of ascorbic acid on physiological and biochemical characteristics of wheat under water stress. *International Journal of Plant Production* 10: 579–596.

Halliwell, B. (2006) Reactive species and antioxidants. Redox biology is a fundamental theme of aerobic life. *Plant Physiology* 141(2): 312–322.

Hare, P.D., Cress, W.A. (1997) Metabolic implications of stress-induced proline accumulation in plants. *Plant Growth Regulation* 21(2): 79–102.

Hernandez, I., Chacon, O., Rodriguez, R., Portieles, R., Lopez, Y., Pujol, M., Borras-Hidalgo, O. (2009) Black shank resistant tobacco by silencing of glutathione S-transferase. *Biochemical and Biophysical Research Communications* 387(2): 300–304.

Hirt, H. (2009) Plant Stress Biology: From Genomics to Systems Biology. Wiley, West Sussex.

Hollander-Czytko, H., Grabowski, J., Sandorf, I., Weckermann, K., Weiler, E.W. (2005) Tocopherol content and activities of tyrosine aminotransferase and cystine lyase in Arabidopsis under stress conditions. *Journal of Plant Physiology* 162(7): 767–770.

Hossain, M.A., Fujita, M. (2010) Evidence for a role of exogenous glycinebetaine and proline in antioxidant defense and methylglyoxal detoxification systems in mung bean seedlings under salt stress. *Physiology and Molecular Biology of Plants: an International Journal of Functional Plant Biology* 16(1): 19–29.

Hossain, M.A., Hasanuzzaman, M., Fujita, M. (2010) Up-regulation of antioxidant and glyoxalase systems by exogenous glycinebetaine and proline in mung bean confer tolerance to cadmium stress. *Physiology and Molecular Biology of Plants: An International Journal of Functional Plant Biology* 16(3): 259–272.

Husen, A., Iqbal, M., Aref, I.M. (2016) IAA-induced alteration in growth and photosynthesis of pea (*Pisum sativum* L.) plants grown under salt stress. Journal of Environmental Biology 37: 421.

Hussain, I., Siddique, A., Ashraf, M.A., Rasheed, R., Ibrahim, M., Iqbal, M., Akbar, S., Imran, M. (2017) Does exogenous application of ascorbic acid modulate growth, photosynthetic pigments and oxidative defense in okra (*Abelmoschus esculentus* (L.) Moench) under lead stress? *Acta Physiologiae Plantarum* 39(6): 144.

Jabeen, N., Abbas, Z., Iqbal, M., Rizwan, M., Jabbar, A., Farid, M., Ali, S., Ibrahim, M., Abbas, F. (2016) Glycinebetaine mediates chromium tolerance in mung bean through lowering of Cr uptake and improved antioxidant system. *Archives of Agronomy and Soil Science* 62(5): 648–662.

Jamil, S., Ali, Q., Iqbal, M., Javed, M.T., Iftikhar, W., Shahzad, F., Perveen, R. (2015) Modulations in plant water relations and tissue-specific osmoregulation by foliar-applied ascorbic acid and the induction of salt tolerance in maize plants. *Brazilian Journal of Botany* 38(3): 527–538.

Jimenez, A., Hernández, J.A., Pastori, G., Del Rıo, L.A., Sevilla, F. (1998) Role of the ascorbate-glutathione cycle of mitochondria and peroxisomes in the senescence of pea leaves. *Plant Physiology* 118(4): 1327–1335.

Kamal, M.A., Saleem, M.F., Shahid, M., Awais, M., Khan, H.Z., Ahmed, K. (2017) Ascorbic acid triggered physiochemical transformations at different phenological stages of heat-stressed Bt cotton. *Journal of Agronomy and Crop Science* 203(4): 323–331.

Kamal-Eldin, A., Appelqvist, L.A. (1996) The chemistry and antioxidant properties of tocopherols and tocotrienols. *Lipids* 31(7): 671–701.

Khare, T., Kumar, V., Kishor, P.B.K. (2015) Na^+ and Cl^- ions show additive effects under NaCl stress on induction of oxidative stress and the responsive antioxidative defense in rice. *Protoplasma* 252(4): 1149–1165.

Khunpona, B., Cha-Umb, S., Faiyuec, B., Uthaibutraa, J., Saengnila, K. (2017) Influence of paclobutrazol on growth performance, photosynthetic pigments, and antioxidant efficiency of Pathumthani 1 rice seedlings grown under salt stress. *ScienceAsia* 43(2): 70–81.

Kostopoulou, Z., Therios, I., Molassiotis, A. (2014) Resveratrol and its combination with alpha-tocopherol mediate salt adaptation in citrus seedlings. *Plant Physiology and Biochemistry: PPB* 78: 1–9.

Kumar, D., Yusuf, M.A., Singh, P., Sardar, M., Sarin, N.B. (2013) Modulation of antioxidant machinery in alpha-tocopherol-enriched transgenic *Brassica juncea* plants tolerant to abiotic stress conditions. *Protoplasma* 250(5): 1079–1089.

Larson, R.A. (1988) The antioxidants of higher plants. *Phytochemistry* 27(4): 969–978.

Latif, M., Akram, N.A., Ashraf, M. (2016) Regulation of some biochemical attributes in drought-stressed cauliflower (*Brassica oleracea* L.) by seed pre-treatment with ascorbic acid. *The Journal of Horticultural Science and Biotechnology* 91(2): 129–137.

Loreto, F., Pinelli, P., Manes, F., Kollist, H. (2004) Impact of ozone on monoterpene emissions and evidence for an isoprene-like antioxidant action of monoterpenes emitted by *Quercus ilex* leaves. *Tree Physiology* 24(4): 361–367.

Lovdal, T., Olsen, K.M., Slimestad, R., Verheul, M., Lillo, C. (2010) Synergetic effects of nitrogen depletion, temperature, and light on the content of phenolic compounds and gene expression in leaves of tomato. *Phytochemistry* 71(5–6): 605–613.

Metwally, A., Safronova, V.I., Belimov, A.A., Dietz, K.J. (2005) Genotypic variation of the response to cadmium toxicity in *Pisum sativum* L. *Journal of Experimental Botany* 56(409): 167–178.

Meyer, A.J. (2008) The integration of glutathione homeostasis and redox signaling. *Journal of Plant Physiology* 165(13): 1390–1403.

Mirjahanmardi, H., Ehsanzadeh, P. (2016) Iron supplement ameliorates drought-induced alterations in physiological attributes of fennel (*Foeniculum vulgare*). *Nutrient Cycling in Agroecosystems* 106(1): 61–76.

Mittler, R., Vanderauwera, S., Gollery, M., Van Breusegem, F. (2004) Reactive oxygen gene network of plants. *Trends in Plant Science* 9(10): 490–498.

Mittler, R., Zilinskas, B.A. (1992) Molecular cloning and characterization of a gene encoding pea cytosolic ascorbate peroxidase. *The Journal of Biological Chemistry* 267(30): 21802–21807.

Mostafa, M.R., Mervat, S.S., Safaa, R.E.-L., Ebtihal, M.A.E., Magdi, T.A. (2015) Exogenous α-tocopherol has a beneficial effect on *Glycine max* (L.) plants irrigated with diluted sea water. *The Journal of Horticultural Science and Biotechnology* 90(2): 195–202.

Mullineaux, P.M., Rausch, T. (2005) Glutathione, photosynthesis and the redox regulation of stress-responsive gene expression. *Photosynthesis Research* 86(3): 459–474.

Naliwajski, M.R., Skłodowska, M. (2014) The oxidative stress and antioxidant systems in cucumber cells during acclimation to salinity. *Biologia Plantarum* 58(1): 47–54.

Navrot, N., Rouhier, N., Gelhaye, E., Jacquot, J.P. (2007) Reactive oxygen species generation and antioxidant systems in plant mitochondria. *Physiologia Plantarum* 129(1): 185–195.

Naz, H., Akram, N.A., Ashraf, M. (2016) Impact of ascorbic acid on growth and some physiological attributes of cucumber (*Cucumis sativus*) plants under water-deficit conditions. *Pakistan Journal of Botany* 48: 877–883.

Niyogi, K.K., Shih, C., Soon Chow, W., Pogson, B.J., Dellapenna, D., Björkman, O. (2001) Photoprotection in a zeaxanthin- and lutein-deficient double mutant of Arabidopsis. *Photosynthesis Research* 67(1–2): 139–145.

Noctor, G., Foyer, C.H. (1998) A re-evaluation of the ATP :NADPH budget during C3 photosynthesis: A contribution from nitrate assimilation and its associated respiratory activity? *Journal of Experimental Botany* 49(329): 1895–1908.

Noman, A., Ali, Q., Maqsood, J., Iqbal, N., Javed, M.T., Rasool, N., Naseem, J. (2018) Deciphering physio-biochemical, yield, and nutritional quality attributes of water-stressed radish (*Raphanus sativus* L.) plants grown from Zn-Lys primed seeds. *Chemosphere* 195: 175–189.

Noman, A., Ali, S., Naheed, F., Ali, Q., Farid, M., Rizwan, M., Irshad, M.K. (2015) Foliar application of ascorbate enhances the physiological and biochemical attributes of maize (*Zea mays* L.) cultivars under drought stress. *Archives of Agronomy and Soil Science* 61(12): 1659–1672.

Noreen, Z., Ashraf, M. (2009) Assessment of variation in antioxidative defense system in salt-treated pea (*Pisum sativum*) cultivars and its putative use as salinity tolerance markers. *Journal of Plant Physiology* 166(16): 1764–1774.

Noreen, S., Ashraf, M. (2010) Modulation of salt (NaCl)-induced effects on oil composition and fatty acid profile of sunflower (*Helianthus annuus* L.) by exogenous application of salicylic acid. *Journal of the Science of Food and Agriculture* 90(15): 2608–2616.

Olsen, K.M., Hehn, A., Jugde, H., Slimestad, R., Larbat, R., Bourgaud, F., Lillo, C. (2010) Identification and characterisation of CYP75A31, a new flavonoid 3'5'-hydroxylase, isolated from Solanum lycopersicum. BMC Plant Biology 10: 21.

Penella, C., Calatayud, Á. Melgar, J.C. (2017) Ascorbic acid alleviates water stress in young peach trees and improves their performance after rewatering. *Frontiers in Plant Science* 8: 1627.

Penuelas, J., Munne-Bosch, S. (2005) Isoprenoids: An evolutionary pool for photoprotection. *Trends in Plant Science* 10(4): 166–169.

Rai, V., Khatoon, S., Bisht, S.S., Mehrotra, S. (2005) Effect of cadmium on growth, ultramorphology of leaf and secondary metabolites of *Phyllanthus amarus* Schum. and Thonn. *Chemosphere* 61(11): 1644–1650.

Rausch, T., Wachter, A. (2005) Sulfur metabolism: A versatile platform for launching defence operations. *Trends in Plant Science* 10(10): 503–509.

Razinger, J., Dermastia, M., Koce, J.D., Zrimec, A. (2008) Oxidative stress in duckweed (*Lemna minor* L.) caused by short-term cadmium exposure. *Environmental Pollution* 153(3): 687–694.

Rehman, R.U., Zia, M., Chaudhary, M.F. (2017) Salicylic acid and ascorbic acid retrieve activity of antioxidative enzymes and structure of *Caralluma tuberculata* calli on PEG stress. *General Physiology and Biophysics* 36: 167–174.

Romero-Puertas, M.C., Corpas, F.J., Rodríguez-Serrano, M., Gómez, M., Del Río, L.A., Sandalio, L.M. (2007) Differential expression and regulation of antioxidative enzymes by cadmium in pea plants. *Journal of Plant Physiology* 164(10): 1346–1357.

Rouphael, Y., Raimondi, G., Lucini, L., Carillo, P., Kyriacou, M.C., Colla, G., Cirillo, V., *et al.* (2018) Physiological and metabolic responses triggered by omeprazole improve tomato plant tolerance to NaCl stress. *Frontiers in Plant Science* 9: 249.

Sarabi, B., Bolandnazar, S., Ghaderi, N., Ghashghaie, J. (2017) Genotypic differences in physiological and biochemical responses to salinity stress in melon (*Cucumis melo* L.) plants: Prospects for selection of salt tolerant landraces. *Plant Physiology and Biochemistry: PPB* 119: 294–311.

Schützendubel, A., Nikolova, P., Rudolf, C., Polle, A. (2002) Cadmium and H_2O_2-induced oxidative stress in *Populus × canescens* roots. *Plant Physiology and Biochemistry* 40(6–8): 577–584.

Semida, W.M., Taha, R.S., Abdelhamid, M.T., Rady, M.M. (2014) Foliar-applied α-tocopherol enhances salt-tolerance in *Vicia faba* L. plants grown under saline conditions. *South African Journal of Botany* 95: 24–31.

Sereflioglu, S., Dinler, B.S., Tasci, E. (2017) Alpha-tocopherol-dependent salt tolerance is more related with auxin synthesis rather than enhancement antioxidant defense in soybean roots. *Acta Biologica Hungarica* 68(1): 115–125.

Siefermann-Harms, D. (1987) The light-harvesting and protective functions of carotenoids in photosynthetic membranes. *Physiologia Plantarum* 69(3): 561–568.

Siripornadulsil, S., Traina, S., Verma, D.P., Sayre, R.T. (2002) Molecular mechanisms of proline-mediated tolerance to toxic heavy metals in transgenic microalgae. *The Plant Cell* 14(11): 2837–2847.

Skórzyńska-Polit, E., Dra̧żkiewicz, M., Krupa, Z. (2003) The activity of the antioxidative system in cadmium-treated *Arabidopsis thaliana*. *Biologia Plantarum* 47(1): 71–78.

Slama, I., M'rabet, R., Ksouri, R., Talbi, O., Debez, A., Abdelly, C. (2017) Effects of salt treatment on growth, lipid membrane peroxidation, polyphenol content, and antioxidant activities in leaves of *Sesuvium Portulacastrum* L. *Arid Land Research and Management* 31(4): 404–417.

Smirnoff, N. (2000) Ascorbic acid: Metabolism and functions of a multi-facetted molecule. *Current Opinion in Plant Biology* 3(3): 229–235.

Smirnoff, N. (2005) Ascorbate, tocopherol and carotenoids: Metabolism, pathway engineering and functions. In *Antioxidants and Reactive Oxygen Species in Plants*: 53–86.

Solovchenko, A., Schmitz-Eiberger, M. (2003) Significance of skin flavonoids for UV-B-protection in apple fruits. *Journal of Experimental Botany* 54(389): 1977–1984.

Sun, Q., Ye, Z.H., Wang, X.R., Wong, M.H. (2007) Cadmium hyperaccumulation leads to an increase of glutathione rather than phytochelatins in the cadmium hyperaccumulator Sedum alfredii. *Journal of Plant Physiology* 164(11): 1489–1498.

Szarka, A., Horemans, N., Kovács, Z., Gróf, P., Mayer, M., Bánhegyi, G. (2007) Dehydroascorbate reduction in plant mitochondria is coupled to the respiratory electron transfer chain. *Physiologia Plantarum* 129(1): 225–232.

Tausz, M., Šircelj, H., Grill, D. (2004) The glutathione system as a stress marker in plant ecophysiology: Is a stress-response concept valid? *Journal of Experimental Botany* 55(404): 1955–1962.

Trovato, M., Mattioli, R., Costantino, P. (2008) Multiple roles of proline in plant stress tolerance and development. *Rendiconti Lincei* 19(4): 325–346.

Tsamaidi, D., Daferera, D., Karapanos, I., Passam, H. (2017) The effect of water deficiency and salinity on the growth and quality of fresh dill (*Anethum graveolens* L.) during autumn and spring cultivation. *International Journal of Plant Production* 11: 33–46.

Ullah, H.A., Javed, F., Wahid, A., Sadia, B. (2015) Alleviating effect of exogenous application of ascorbic acid on growth and mineral nutrients in cadmium stressed barley (*Hordeum vulgare*) seedlings. *International Journal of Agriculture and Biology* 18(1): 73–79.

Verbruggen, N., Hermans, C. (2008) Proline accumulation in plants: A review. *Amino Acids* 35(4): 753–759.

Vierstra, R.D., John, T.R., Poff, K.L. (1982) Kaempferol 3-O-galactoside, 7-O-rhamnoside is the major green fluorescing compound in the epidermis of *Vicia faba*. *Plant Physiology* 69(2): 522–525.

Wang, X., Wu, L., Xie, J., Li, T., Cai, J., Zhou, Q., Dai, T., Jiang, D. (2018) Herbicide isoproturon aggravates the damage of low temperature stress and exogenous ascorbic acid alleviates the combined stress in wheat seedlings. *Plant Growth Regulation* 84(2): 293–301.

Winkel-Shirley, B. (2002) Biosynthesis of flavonoids and effects of stress. *Current Opinion in Plant Biology* 5(3): 218–223.

Xiang, C., Werner, B.L., Christensen, E.M., Oliver, D.J. (2001) The biological functions of glutathione revisited in Arabidopsis transgenic plants with altered glutathione levels. *Plant Physiology* 126(2): 564–574.

Xu, Y., Xu, Q., Huang, B. (2015) Ascorbic acid mitigation of water stress-inhibition of root growth in association with oxidative defense in tall fescue (*Festuca arundinacea* Schreb.). *Frontiers in Plant Science* 6: 807.

Yang, Y., Han, C., Liu, Q., Lin, B., Wang, J. (2008) Effect of drought and low light on growth and enzymatic antioxidant system of *Picea asperata* seedlings. *Acta Physiologiae Plantarum* 30(4): 433–440.

Ye, Y.R., Wang, W.L., Zheng, C.S., Fu, D.J., Liu, H.W., Shen, X. (2017) Foliar-application of α-tocopherol enhanced salt tolerance of Carex leucochlora. *Biologia Plantarum* 61(3): 565–570.

Youssef, R.A., El-Azab, M.E., Mahdy, H.A., Essa, E.M., Mohammed, K.A. (2017) Effect of salicylic acid on growth, yield, nutritional status and physiological properties of sunflower plant under salinity stress. *International Journal of Pharmaceutical and Phytopharmacological Research* 7: 54–58.

Zhang, F., Shi, W., Jin, Z., Shen, Z. (2002) Response of antioxidative enzymes in cucumber chloroplasts to cadmium toxicity. *Journal of Plant Nutrition* 26(9): 1779–1788.

Zhou, X., Gu, Z., Xu, H., Chen, L., Tao, G., Yu, Y., Li, K. (2016) The effects of exogenous ascorbic acid on the mechanism of physiological and biochemical responses to nitrate uptake in two rice cultivars (*Oryza sativa* L.) under aluminum stress. *Journal of Plant Growth Regulation* 35(4): 1013–1024.

10 Nitric Oxide
A Regulator of Plant Signaling and Defense Against Abiotic Stress

Hanan A. Hashem

CONTENTS

10.1 INTRODUCTION

Nitric oxide (NO) is a ubiquitous diatomic molecule that was first discovered as a colorless and toxic gas in 1772 by Joseph Priestly, who called it "nitrous air". This classification of NO as a toxic gas and air pollutant continued until 1987 when it was shown that NO is a natural component in the body. By 1987, NO's role in regulating blood pressure and relieving heart conditions was well established (Schmidt and Walter, 1994). In 1992, due to its widespread biological significance, NO was voted "Molecule of the Year" by *Science* magazine (Culotta and Koshland, 1992). The importance of NO became front-page news in 1998 when Louis J. Ignerro, Robert F. Furchgott and Ferid Murad were awarded the Nobel Prize for Medicine and Physiology. These scientists identified NO as a signaling molecule in mammalian cells, involved in almost all areas of biology including vasorelaxation, relaxation of smooth muscles, neurotransmission and innate immune response. NO also plays important roles in a number of signaling cascades controlling diverse physiological processes in plants, ranging from seed germination to plant senescence (Mur et al., 2012; Wilson et al., 2008). In recent years, there has been much research into the diverse biological activities of NO in response to abiotic and biotic stresses, such as drought, salt, temperature (high and low), heavy metal stresses and disease resistance (Siddiqui et al., 2013; Zhang et al., 2006).

Here, we will highlight the important role of NO as a signaling molecule in regulating physiological and molecular processes in the plant under optimal and stressful environments.

10.2 NITRIC OXIDE CHEMISTRY

NO is a ubiquitous diatomic molecule with a high diffusivity ($4.8 \times 10{-}5$ cm^2 s^{-1} in H_2O), exhibiting hydrophobic properties. Thus, NO may not only easily migrate in the hydrophilic regions of the cell, such as the cytoplasm, but also freely diffuse through the lipid phase of membranes and act as an inter- and intracellular messenger in many physiological functions (Arasimowicz and Floryszak-Wieczorek, 2007). The half-life of NO in biological tissues is estimated to be < 6 s (Bethke et al., 2004). This short half-life reflects the highly reactive nature of NO, which reacts directly with metal complexes and other radicals and indirectly as a reactive nitrogen oxide species (RNS) with DNA, proteins and lipids (RNS arise from the interaction of NO with oxygen and superoxide; Wink and Mitchell, 1998).

Nitric oxide is a gaseous free radical; its chemistry implicates an interplay between the three redox-related species: NO radical ($NO^{\cdot}$), nitrosonium cation (NO^+), and nitroxyl anion (NO^-). In biological systems, $NO^{\cdot}$ reacts rapidly with atmospheric oxygen (O_2), superoxide anion ($O_2^{\cdot -}$) and transition metals. The reaction of $NO^{\cdot}$ with O_2 results in the generation of NO_x compounds (including $NO_2^{\cdot}$, N_2O_3, and N_2O_4), which can either react with cellular amines and thiols or simply hydrolyze to form the end metabolites nitrite (NO_2^-) and nitrate (NO_3^-) (Wendehenne et al., 2001). When reactive oxygen species (ROS) and NO are produced as during recognition of plant pathogens, the reaction of $NO^{\cdot}$ with $O_2^{\cdot -}$ yields peroxynitrite ($ONOO^-$), a powerful oxidant that mediates cellular injury (Bellin et al., 2013). In addition, NO mediates electrophilic attack on reactive sulfur, oxygen, nitrogen and aromatic carbon centers, with thiols being the most reactive groups. This chemical process is referred to as nitrosation. Nitrosation of sulfhydryl (S-nitrosation) centers of many enzymes or proteins has been described and the resulting chemical modification affects the activity in many cases. Such modifications are reversible and protein S-nitrosation denitrosation could represent an important mechanism for regulating signal transduction (Hayat et al., 2010).

10.3 BIOSYNTHESIS OF NITRIC OXIDE

In plants, several potential NO sources may be distinguished, with the physiological role of each depending on the species, type of tissue or cells, external conditions and potential activation of the signal pathway (Neil et al., 2003). NO synthesis in plants includes both arginine and nitrite-dependent pathways.

The enzyme responsible for NO generation in animal organisms is NO synthase (NOS, EC 1.14.13.39) catalyzing five electron oxidations of one of the atoms in L-arginine (N^{3-} to N^{2+}) with the participation of O_2 and NADPH. Although NOS-like activity has been detected widely in plants, animal-type NOS is still elusive (Arasimowicz and Floryszak-Wieczorek, 2007). Nevertheless, NO can be produced by plant extracts incubated with NOS substrate arginine. Therefore, a NOS-like activity exists in plants (Asai and Yoshioka, 2009). This activity requires the same cofactors (NADPH, calcium, calmodulin, flavin adenine dinucleotide, flavin mononucleotide, and tetrahydrobiopterin BH4) as animal NOS (Asai et al., 2010) and is often referred to as the major source of NO in plants during plant-pathogen interactions. As an alternative to oxidative routes, several reductive enzymatic systems through the action of NAD(P)H-dependent nitrate (NR, EC 1.6.6.1) or nitrite (NiR, EC 1.7.7.1) reductases using nitrite as a substrate for NO production were also reported for plants (Gupta et al., 2011).

Godber et al. (2000) suggested that xanthine oxidase, a ubiquitous molybdo-enzyme, could catalyze the reduction of nitrite to NO under hypoxia and in the presence of NADH. Xanthine oxidoreductase (XOR) is another enzyme that has been found to produce NO in plants as well as in animals. Xanthine oxidoreductase occurs in two interconvertible forms: the superoxide-producing xanthine oxidase and xanthine dehydrogenase (Palma et al., 2002). XOR can produce the free radicals $O_2^{\cdot -}$ and $NO^{\cdot}$ during its catalytic reaction, depending on whether the oxygen tensions are high or low, respectively (Harrison, 2002). This property of producing $O_2^{\cdot -}$ and $NO^{\cdot}$ radicals confers a key role to XOR as a source of the signal molecule in plant cells. Recently it has been shown that polyamines and hydroxylamine can increase oxidative NO synthesis in plant cells (Wimalasekera et al., 2011). NO can mediate effects of polyamines in plants; however, the manner in which polyamines can increase NO synthesis is uncertain (Fröhlich and Durner, 2011). Possible mechanisms include an interaction of polyamines with the NR-catalyzed NO production (Rosales et al., 2012) and the indirect effect of polyamine synthesis on L-Arginine metabolism (Zhang et al., 2011). Chandok et al. (2003) identified in tobacco another candidate for NO enzymatic production in plants – the inducible NO synthase (iNOS), a group of evolutionarily conserved cytosolic or membrane-bound isoenzymes that convert L-arginine to L-citrulline and NO. Plant mitochondria also make NO from nitrite (Planchet et al., 2005; Tischner et al., 2004). Inhibitors of mitochondria electron transport inhibit NO synthesis, suggesting that electrons from the mitochondria electron transport chain drive nitrite reduction. This raises the possibility that NO release under hypoxic or anoxic conditions depends on NR for the production of nitrite, but it is the mitochondria that produce the bulk of NO from nitrite. This would depend on the ratio of nitrite to nitrate in the cytosol (Tischner et al., 2004).

NO may also be formed non-enzymatically in a reaction between nitrogen oxides and plant metabolites or as a result of a chemical reduction of NO_2 at acid pH (Wendehenne et al., 2001). As it has been documented, a sufficiently acid medium, required for NO_2 reduction, is found in the apoplast of barley aleurone cells (Bethke et al., 2004). It also needs to be remembered that under physiological conditions plants are exposed to NO produced with the participation of soil microorganisms. The release of NO to the atmosphere occurs in the reaction of nitrification and denitrification, which may constitute an alternative source of NO for plants. Nitrification of NH_4^+ is the primary source of N_2 emitted to the atmosphere, where it oxidizes to NO and NO_2 (Wojtaszek, 2000).

10.4 THE PHYSIOLOGICAL FUNCTION OF NITRIC OXIDE IN PLANTS

Phytohormones are chemical messengers derived from plant biosynthetic pathways that act at the site of their synthesis or are transported to some other site in the plant to mediate growth and developmental responses under both optimal and stressful environments (Peleg and Blumwald, 2011). There are five groups of phytohormones: auxin, gibberellins, ethylene, cytokinin (CK) and abscisic acid (ABA). There are also other compounds that have important growth-regulating activity and function as phytohormones. These include brassinosteroids, jasmonic acid (JA) and salicylic acid (SA). Nitric oxide is considered a new member of this group as a non-traditional plant growth regulator (Fatma et al., 2016; Leterrier et al., 2012). NO is thought to modulate a variety of developmental processes such as germination (Beligni and Lamattina, 2000), root growth and development (Fernandez-Marcos et al., 2011), flower development (Kwon et al., 2012), flowering time (He et al., 2004) and apical dominance (Kwon et al., 2012).

Here, we will briefly discuss the role of NO in different processes of plants under normal conditions (unstressed plants), as the role of NO in plants under different environmental stresses is discussed in detail later in Section 10.6.

10.4.1 Effect of Nitric Oxide on Growth and Development

In the plant, NO conducts complex biological functions, either as a cytotoxin or a cytoprotectant (Beligni and Lamattina, 2001). The dual function of nitric acid as a potent oxidant or effective antioxidant mostly depends on its concentration and on the status of the environment (Beligni and Lamattina, 1999). The effect of NO on plant growth was found to be concentration dependent (Gouvea et al., 1997). Seedlings of canola, raised from seeds treated with a lower concentration of NO-donor sodium nitroprusside (SNP), had more root length and dry mass, whereas higher concentration reduced the values of these parameters (Zanardo et al., 2005). Similar dual behavior of NO-donor SNP was also noted in wheat (Tian and Lei, 2006).

NO effects in the plant are at least in part attributed to its interaction with plant hormones. Several studies demonstrated that NO might affect biosynthesis, catabolism/conjugation, transport, perception and/or transduction of different phytohormones (Freschi, 2013; Terrile et al., 2012).

NO is able to induce adventitious root (AR) development in monocot, dicot and gymnosperm plant species (Lanteri et al., 2008). A number of second messengers involved in signaling cascades regulated by NO, implicated in AR development, have been uncovered. In the last decade, a series of experiments have implicated NO as a central component in auxin-orchestrated root growth and development. Further, the accumulating data suggest that NO might also modulate the interaction of roots with microorganisms in the rhizosphere (Boscari et al., 2013; Pagnussat et al., 2004).

Cytokinin is a pivotal phytohormone in plant growth and development. Cytokinin signaling is thought to be mediated by a phosphorelay that sequentially transfers phosphoryl groups from the cytokinin receptors to histidine phosphotransfer proteins (AHPs) and response regulators (ARRs). It has been suggested that NO might negatively regulate cytokinin signaling by blunting phosphorelay activity through S-nitrosylation (Feng et al., 2013).

Some studies proposed that NO has anti-senescence properties. Exogenous application of NO in pea leaves under senescence-promoting conditions decreased ethylene levels because of the inhibition of ethylene biosynthesis (Leshem, 2000). It was demonstrated that endogenous NO and ethylene content maintain an inverse correlation during the ripening of strawberries and avocados while unripe, green fruits contain high NO and low ethylene concentrations; the maturation process is accompanied by a marked decrease in NO concomitant with an increase in ethylene (Leshem and Pinchasov, 2000). It was also observed that NO emission decreased as ethylene production increased from anthesis to senescence (Zanardo et al., 2005).

NO donors exert a protective effect against abscisic acid (ABA)-induced senescence of rice leaves by diminishing ABA-dependent effects such as leaf senescence. In the same context, the role of NO in preserving and increasing chlorophyll content in pea and potato was also proved (Leshem et al., 1997). The protective effect of NO on chlorophyll retention may reflect NO effects on iron availability. Strong evidence supporting the role of NO in iron nutrition of plants was presented by Graziano et al. (2002), as iron-deficient growth conditions normally result in chlorosis. NO treatment increased the chlorophyll content in maize leaves up to the control level.

10.4.2 Effect of NO on Photosynthesis

Photosynthesis is one of the most important physiological processes. The whole metabolism of plants directly or indirectly depends on this process; any change in photosynthetic rate will automatically affect the rest of plant metabolism. However, the role of NO in photosynthesis is poorly understood, which may be attributed to the small number of in vivo and in vitro studies carried out and the contradictory results obtained (Takahashi and Yamasaki, 2002; Yang et al., 2004). NO and its donors, such as sodium nitroprusside (SNP), S-nitroso-N acetylpenicillamine (SNAP), and S-nitrosoglutathione (GSNO) are recognized to differentially regulate photosynthetic rate. NO gas decreases net photosynthetic rate in *Avena sativa* and *Medicago sativa* leaves. Lum et al. (2005) found that SNP decreased the amount of ribulose-1,5-bisphosphate carboxylase/oxygenase (Rubisco) activase and Rubisco subunit binding G-protein β-subunit. Meanwhile, NO can reversibly bind to several sites in PSII and inhibit electron transfer (Wodala et al., 2008); within PS II complex, important binding sites of NO are the non heme iron between QA and QB (Diner and Petrouleas, 1990), and manganese (Mn) cluster of water-oxidizing complex (Schansker et al., 2002).

Chloroplasts were proved to be the crucial player in regulation of NO concentration in a plant cell. They are not only the site of NO biosynthesis either by NR or NOS-like protein, but NO, through post-translational modifications of proteins, also affects assimilatory processes of photosynthesis and thus the entire growth capacity. One other possible reason for the observed changes in the rates of photosynthesis could be due to the effect of NO on stomatal behavior. The exogenous application of NO to both monocotyledonous and dicotyledonous epidermal strips induced stomatal closure through a Ca^2-dependent process (Neill et al., 2002). In *Pisum sativum* and *Vicia faba* plants, abscisic acid increased the endogenous production of NO that was suggested to be the reason for ABA-induced stomatal closure (García-Mata and Lamattina, 2003). There is also some convergent evidence that supports the involvement of nitrate reductase through the production of NO in guard cells leading to their closure (García-Mata and Lamattina, 2003).

10.4.3 Effect of No on Respiration

NO affects the mitochondrial functionality in plant cells and reduces total cell respiration due to its inhibitory effect on the cytochrome functioning. Mitochondria are one of the earliest targets of NO which reversibly, and competitively with O_2, inhibit their cytochrome oxidase (COX), and the same was found in isolated plant mitochondria. In carrot cell suspension, NO reduced total respiration by 50%, and this effect was accompanied by a significant increase in cell death. Similarly, in soybean cotyledon mitochondria, the oxygen uptake was inhibited after NO treatment, but it was restored upon NO depletion (Zottini et al., 2002). Nitric oxide can also modulate other mitochondrial enzymes, such as tobacco aconitase, which is a constituent of the Krebs cycle. Its inactivation by NO decreases the cellular energy metabolism that may result in reduced electron flow through the mitochondrial respiratory electron transport chain and a subsequent decrease in the generation of reactive oxygen species (ROS), the natural byproduct of respiration (Navarre et al., 2000).

Although the described NO functions in plants have been increasing over the last years, the precise molecular mechanisms underlying its physiological roles are still poorly understood. Some works about the way NO works in plants demonstrated that artificially generated as well as endogenously produced NO can modulate several gene expressions, involved in stress responses and hormonal signaling.

10.5 NITRIC OXIDE SIGNALING IN PLANTS

For a signaling molecule to be effective, it needs to be produced quickly and efficiently on demand, to induce defined effects within the cell, and to be removed rapidly and effectively when no longer required. Alternatively, it is possible that signals function together. For example, it could be that continuous NO synthesis is essential for another signaling path to operate, such that removal of NO would be inhibitory, even though induction of NO synthesis was not required. Any stimulus that inhibited NO production would increase the cellular event by virtue of its inhibition of NO synthesis (Neil et al., 2003). Although there is an ever-increasing number of NO responses in plants, we still know relatively little of the signal transduction processes by which NO interaction with cells results in altered cellular activities. There are two important mechanisms intermediating NO signal transduction in plants. The first is NO-dependent alteration of transcript profile. The second is NO-dependent post-transcriptional modifications (PTMs). Initial studies to reveal the mechanism of NO signaling focused on transcriptomic assessments of NO effects. There is a lot of evidence showing that NO affects gene expression in plants. The mechanism by which NO alters the transcript profile could be through a single cascade via the synthesis of cyclic guanosine monophosphate (cGMP), salicylic acid and Ca^{2+} ions (Arasimowicz and Floryszak-Wieczorek, 2007; Neil et al., 2003). Three PTMs are proved to reveal NO signaling components and modes of action: thiol protein S-nitrosylation, tyrosine nitration and metal nitrosylation (Mur et al., 2012 and Freschi, 2013). This specific and regulated NO-dependent PTM has been implicated as potentially controlling the function of diverse cellular processes such as photosynthesis, hormone metabolism, amino acid metabolism and genetic information processing.

10.5.1 cGMP Signaling

cGMP is a well-established second messenger molecule, that is, a biologically active intracellular signaling molecule whose concentrations are transiently altered in response to external stimuli. Typically, cGMP concentrations are increased via the enhanced activity of the biosynthetic enzyme guanylyl cyclase (GC), which synthesizes cGMP from GTP (guanosine tri-phosphate). Concentrations are returned to resting values (and generally kept at low levels) by the constitutive action of phosphodiesterases (PDE). NO reacts directly with the iron in the haem moiety of GC, inducing a conformational change that results in enzyme activation (Hancock et al., 2011). Such activation is transient, persisting only for so long as NO is present. Thus, the immediate cellular effects of NO are relatively short-lived, as cGMP is rapidly degraded by PDE. The transient increases in cGMP content induced by NO, and inhibition of NO responses by GC inhibitors, mean that the enzymes required to synthesize cGMP are present, and in some cases, rapidly activated, and that the (presumably phosphodiesterase, PDE) enzymes that affect metabolic inactivation are also present, being either constitutively active or similarly activated. cGMP has been shown to be involved in plant growth and development, gas exchange, aging, maturation and plant response to biotic and abiotic environmental factors (Dubovskaya et al., 2015)

NO has been shown to activate phosphokinases (PKs) in *Arabidopsis* and tobacco (Clarke et al., 2000). It is not known if the effects of NO were direct, or via activation of cGMP that subsequently activates PKs. NO synthesis and signaling also involve regulation via protein phosphatases.

10.5.2 Cyclic Adenosine Diphosphate-Ribose (cADPR) and Calcium

Intracellular calcium can be stored in various cellular locations, release from which is predicated on the presence of specific calcium channel proteins that recognize and bind second messengers such as cADPR. Growing evidence suggests that NO regulates the signaling cascade via cADPR and Ca^{2+} mobilization (Lamotte et al., 2005). cADPR synthesis is commonly activated by NO, such activation being mediated by cGMP, potentially via activation of a cGMP-dependent kinase (Wendehenne et al., 2001). cADPR can induce intracellular Ca^{2+} – permeable channels to release Ca^{2+} from endoplasmic reticulum and vacuoles in order to elevate free cytosolic calcium levels in cells (Arasimowicz and Floryszak-Wieczorek, 2007; Leckie et al., 1998). Calcium is a core component of stomatal ABA signaling pathways, with a full response to ABA being calcium-dependent, but a partial response likely to be calcium-independent (Webb et al., 2001). It is not surprising then, that NO-induced stomatal closure requires calcium (García-Mata and Lamattina, 2003). Both calcium influx from the cell exterior and calcium release from intracellular stores are required for ABA effects (MacRobbie, 2000).

10.5.3 Post-Translational Modifications (PTMs)

Major NO-dependent protein modifications currently investigated in plants are S-nitrosylation, nitration and metal nitrosylation.

10.5.3.1 S-Nitrosylation

S-nitrosylation is the formation of S-nitrosothiols by covalent addition of an NO moiety to the sulfhydryl group of cysteine residues in target proteins, forming S-nitrosothiol (SNO) (Lindermayr and Durner, 2009). This modification does not generically target all cysteine sensor residues. This specificity seems to be driven by the presence of surrounding acidic and basic amino acids in the vicinity of the considered Cys, and the presence of this residue in a hydrophobic pocket that can favor the concentration of nitrosylating agents (Seth and Stamler, 2010). Transnitrosylation refers to the direct transfer of an NO group from another S-nitrosylated protein or from low-molecular-weight nitrosothiols such as S-nitroglutathione (GSNO), which has an important role in plant resistance. Because of their reactivity with intracellular reducing agents such as ascorbic acid or glutathione, the half-life of S-nitrosothiols is tightly regulated by the redox state of the cell and can be very brief, making protein S-nitrosylation a highly sensitive regulation mechanism. GSNO represents a stable reservoir of potential NO signal and can be reduced by GSNO reductase to ultimately produce glutathione disulfide (GSSG) and ammonia (NH_3), and GSSG can be reduced by glutathione reductase to re-enter the glutathione (GSH) pool (Mannick and Schonhoff, 2004; Mur et al., 2012). S-nitrosylation may, therefore, play important regulatory roles similar to that of protein phosphorylation: both mechanisms are highly specific, rapidly reversible, and allow a prompt modification of the protein. Modifications of key regulatory proteins affect their activities, localization, interaction with other proteins and stability (Yun et al., 2011).

10.5.3.2 Tyrosine Nitration

Tyrosine (Tyr) nitration is the process by which a nitro group (NO_2) is added to the ortho-position of Tyr residues forming 3-nitrotyrosine (Schopfer et al., 2003). The NO_2 group originates mainly from peroxynitrite (ONOO−), a powerful oxidative agent resulting from the reaction between NO and the superoxide anion. Under physiological condition, ONOO− can react with CO_2 and be further decomposed in $CO_3^{\cdot -}$ and NO_2, a powerful nitrating agent. Moreover, Tyr nitration can result from the reaction of NO with tyrosyl radicals (Gunther et al., 2002). Because ROS and NO formation occur under stress situations as well as under normal growth conditions, it can be hypothesized that ONOO- is continuously formed in healthy cells (Romero-Puertas et al., 2004).

Tyr nitration is restricted to specific target tyrosine residues (Bayden et al., 2011; Ischiropoulos, 2003) and can promote conformational changes that lead to the activation or the inhibition of the target proteins. Only a few works have been done so far to determine Tyr nitrated proteins in plants. By using a proteomic approach, eight target proteins participating in photosynthesis, ATP synthesis, Calvin cycle, glycolysis, and nitrate assimilation differentially nitrate in *Arabidopsis thaliana* plants during hypersensitive response (HR; induced immune responses, to protect plants against microbial pathogens and herbivorous insects) (Cecconi et al., 2009). In 2011, Lozano-Juste identified 127 proteins that are putatively Tyr nitrated in plants. The functional consequences of nitration have been studied for only a few proteins in plants, but Àlvarez and colleagues (2011) reported the inhibition of O-acetylserine(thiol)lyase A1 (OASA1) by Tyr nitration in *Arabidopsis thaliana*. The authors explained that inactivation of this enzyme could avoid an extra production of cysteine and/or glutathione, preventing locally the scavenging of reactive oxygen and nitrogen species, further needed in downstream signaling events for an efficient stress response. During plant immune responses, NO and ROS are considered to act concertedly. The regulation of NADPH oxidase constitutes a key example of the control of ROS by NO through nitrosylation (Zaffagnini et al., 2016).

10.5.3.3 Metal Nitrosylation

NO can directly bind transition metals generating NO-metal complexes (Thomas et al., 2015). When these metals are located in the prosthetic group (zinc, iron or copper centers) of metalloproteins, this reaction can lead to the metal-catalyzed (or metal-assisted) nitrosylation of protein thiols (Ford, 2010). The reversible formation of the metal-nitrosyl complex will induce conformational changes that impact the reactivity or the activity of the concerned target proteins (Toledo and Augusto, 2012).

The best-characterized plant protein undergoing metal nitrosylation is hemoglobin. In plants, hemoglobins (Hb) are separated in three groups based on their structural properties: symbiotic Hb, also named leghemoglobin (Lb), localized in nitrogen-fixing root nodules of leguminous plants; non-symbiotic Hb (nsHb), consisting of two classes based on oxygen affinity (i.e. class I has higher affinity for oxygen than class II); and truncated Hb(tHb) (Besson-Bard et al., 2008). In 1998, Mathieu and colleagues reported the existence of a complex between ferrous leghemoglobin (LbFeII) and NO in intact root nodules or extracts of root nodules. Over the past decade, extensive work has shown that, in plants, oxygenated class 1 nsHbs can be oxidized by NO, resulting in nitrate production (Igamberdiev and Hill, 2004; Perazzolli et al., 2004). This NO scavenging reaction is now accepted as a general mechanism modulating NO bioavailability, participating in the regulation and detoxification of NO in plants (Hill, 2012). Some studies also reported similar processes for class-2 symbiotic hemoglobins (sHb) that are able to interact with NO and to scavenge it (Sanchez et al., 2011; Sasakura et al., 2006). A few more studies report the inhibition of some target proteins through metal nitrosylation in plants such as ascorbate peroxidase, cytochrome c oxidase, lipoxygenase or catalase (Brown, 2001).

10.6 NITRIC OXIDE AND ABIOTIC STRESS IN HIGHER PLANT

NO is involved in almost every stress response investigated for NO so far. For example, NO participates in plant response to high- and low-temperature stress. It was found that high-temperature treatment of lucerne cells resulted in an increase of NO synthesis, whereas the application of exogenous NO increased cold tolerance in tomato and survival rate in wheat and maize seedlings (Neil et al., 2003). Mackerness et al. (2001) showed the participation of NO in plant response to UV-B radiation, demonstrating post-stress induction of NOS activity and an elevation in NO level. In addition, Shi et al. (2005) proved that NO-donor treatment of potato tubers before UV-B irradiation resulted in the development of almost 50% more healthy leaves in comparison to plants not subjected to NO treatment. In relation to other abiotic stresses, it was documented that exogenous NO reduces the destructive action of mechanical damage, heavy metals and herbicides on plants (Kopyra and Gwozdz, 2003).

It is well known that various abiotic stresses such as drought, salinity, low and high temperatures, heavy metal toxicity, mechanical wounding, ultraviolet radiation and ozone exposure induce the generation of ROS (Neill et al., 2002). ROS can act as oxidizing agents on proteins, lipids and nucleic acids, modifying the activity or function of these molecules. Hence the steady-state levels of ROS must be strongly regulated by scavenging systems including enzymatic and non-enzymatic antioxidants, such as catalase (CAT), which is involved in removing hydrogen peroxide radicals; superoxide dismutases (SOD), which are involved in removing superoxide radicals, and the ascorbate-glutathione cycle (ASC-GSH) made up of ascorbate peroxidase (APX), monodehydroascorbate reductase (MDHAR), dehydroascorbate reductase (DHAR), glutathione reductase (GR), reduced glutathione (GSH) and ascorbate (ASC) (Romero-Puertas et al., 2006). Several studies revealed the vital role of NO. During environmental stress, the plant relies on its properties:

10.6.1 NO Acts as an Antioxidant Agent

In fact, NO interacts rapidly with ROS, giving rise to a number of reactive oxygen species (RNS), such as (NO•) and its derivative, peroxynitrite ($ONOO^-$), nitrogen dioxide (NO_2), which degrades to nitrite (a precursor of NO), and nitrate. It was suggested that NO might serve an antioxidant function during various stresses (Beligni and Lamattina, 1999). Modulation of superoxide formation (Caro and Puntarulo, 1998) and inhibition of lipid peroxidation (Boveris et al., 2000) by NO also illustrate its potential as an antioxidant.

NO can react with lipid alcoxy (LO·) and peroxyl (LOO·) radicals to stop the propagation of radical directly mediated lipid oxidation (Laspina et al., 2005). Interestingly, a number of antioxidant enzymes have been shown to be regulated by NO. NO is able, via PTMs, to modulate ROS levels by regulating the activities of antioxidant enzymes (Chaki et al., 2015).

On the other hand, excess NO can result in nitrosative stress, so a favorable balance of ROS/NO is important.

10.6.2 NO Acts as a Signal Molecule

NO is involved in signaling pathways as a downstream element in ABA signaling pathway, jasmonic acid synthesis and upstream of H_2O_2 synthesis (Wendehenne et al., 2004). NO-dependent signals can also be modulated through protein phosphorylation upstream of intracellular Ca^{2+} release, implicating a target for protein kinase control in ABA signaling that would feed into a NO-dependent Ca^{2+} release (Sokolovski et al., 2005).

NO also regulates the expression of some abiotic stress tolerance-related genes such as ABC-transporters, cytochrome P450 genes (Zeier et al., 2004), glutathione peroxidase, glutathione-S-transferase and *LOX2* (lipoxygenase) (Huang et al., 2004) in *A. thaliana*.

10.7 CROSSTALK BETWEEN NITRIC OXIDE AND OTHER SIGNALS INVOLVED IN STRESS RESPONSE

10.7.1 NO and Abscisic Acid (ABA)

Both important "stress-related" molecules, NO and ABA intensively crosstalk during certain signaling cascades triggered by environmental challenges, such as water limitation and UV-B radiation, which ultimately leads to the induction of plant adaptive responses, such as stomatal closure and antioxidant defenses (Hancock et al., 2011).

ABA is known to play a vital role in plant response to drought stress by inducing stomatal closure. NO was found to affect ABA-induced stomatal closure in different plant species. Plasma membrane calcium-dependent anion channels and inward-rectifying K^+ channels have been identified to be NO targets during ABA-induced guard cell response. These channels are activated and deactivated, respectively, by NO as a consequence of increases in guard cell cytoplasmatic Ca^{2+} levels due to the NO-triggered release of this anion from intercellular stores (García-Mata and Lamattina, 2003). The involvement of protein phosphorylation upstream of intracellular calcium release has also been determined, implicating protein kinases as additional targets of NO action within ABA-regulated guard cell signaling (Sokolovski et al., 2005). NO modulate calcium-independent outward-rectifying K^+ channels directly by PTMs of these channels (Sokolovski and Blatt, 2004). Several studies proved the participation of MAPKs (mitogen-activated protein kinase) in NO-dependent signaling pathways (Pagnussat et al., 2004; Xu and Dong, 2005). However, these studies didn't clarify whether MAPK activation by NO occurs directly or through other messengers. Concomitantly, MAPK levels were increased in response to H_2O_2 indicating that NO does not act alone, but interacts with other signaling molecules such as H_2O_2, to effect stomatal closure.

The effect of NO on both Ca^{2+}-dependent and Ca^{2+}-independent ion channels in the plasma membrane of guard cells facilitates osmotic solute loss, thereby reducing guard cell turgor and promoting stomatal closure. Stomatal closure guarantees minimal transpiration under water-deficit conditions, enabling plants to tolerate it.

10.7.2 NO and Ethylene

Ethylene (ET) is known as a stress hormone besides its roles in the regulation of plant growth and development. Ethylene is produced by conversion of S-adenosyl methionine (SAM) to 1-aminocyclopropanr-1-carboxylic acid (ACC) via ACC synthase (ACS). ACC is then oxidized to ethylene by ACC oxidase (ACO) (Yang and Hoffman, 1984). The levels of both ethylene and its precursor ACC increased in a plant in response to abiotic stress. The antagonistic relationship was detected between ET and NO during the maturation, senescence, and abscission of a plant organ. Lindermayr et al. (2006) showed that NO directly acts by down-regulating ethylene synthesis. Exogenous NO application modulates the generation of ethylene by controlling the transcript levels of ACS and ACO. In addition, it may also regulate ACS activity via S-nitrosylation (Abat and Deswal, 2009) and influence ACO activity by direct binding to the enzyme-forming ACO-NO binary complex, which is then chelated by ACC resulting in ternary stable complex ACO-NO-ACC (Manjunatha et al., 2010).

On the other hand, a synergetic correlation between NO and ET was also detected in case of plant defense to biotic stresses and Fe deficiency. Some studies reported that NO donors, such as SNP, stimulate ethylene production in non-senescent leaf tissues of tobacco, maize and apple embryos (Ahlfors et al., 2009; Mur et al., 2008). This effect was attributed to NO-stimulated ACS expression as well as other key ethylene biosynthetic enzymes. The signal transduction elements of these regulatory mechanisms need further investigation.

10.7.3 NO and Jasmonic Acid

Jasmonic acid (JA) is a fatty acid-derived signaling molecule involved in a variety of biological processes including plant response to biotic and abiotic stresses (Reymond and Farmer, 1998). Recent evidence suggests that NO also plays a role in the JA signaling pathway, especially in response to wounding. As mentioned earlier, NO induces the expression of lipoxygenase, a key enzyme in the JA biosynthesis pathway. Mur et al. (2013) stated that, during the induction of plant defense responses against biotic stress, NO positively affected JA and SA production. In tomato, NO donors inhibited both wounding- induced H_2O_2 synthesis and JA–induced expression of defense genes (Ferrarini et al., 2008). This inhibition was independent of SA, which has been shown to antagonize JA synthesis and/or activity. Thus, NO may interact directly with the JA pathway at a point downstream of JA synthesis and upstream of H_2O_2 generation.

Additional evidence that NO crosstalks with the JA pathway comes from the demonstration that JA treatment induces NO production in sweet potato and *A. thaliana* epidermal cells (Sokolovski et al., 2005) and that exogenous NO induces all of the genes that are required for JA biosynthesis. Interestingly, NO treatment of SA-deficient plants (*NahG* mutant) resulted in the activation of JA-responsive genes and JA production, suggesting that SA negatively regulates NO-mediated JA synthesis in wildtype plants.

10.7.4 NO and Salicylic Acid (SA)

Salicylic acid (SA) is an important signal molecule in plant defense against pests and pathogens. SA acts via the induction of a large number of defense genes, of which the most-commonly described are pathogenesis-related protein (PR) genes such as *PR1* (Cao et al., 1994).

The SA signaling pathway has now been extensively characterized. The translational activator NONEXPRESSOR OF PATHOGENESIS-RELATED PROTEINS1 (NPR1), localized in an oligomeric form in the cytoplasm (Fu et al., 2012), interacts with the SA receptors NPR3 and NPR4, likely following redox changes at key cysteine residues that result in a monomeric NPR1 form which is translocated to the nucleus (Mou et al., 2003). Within the nucleus, NPR1 interacts with a range of TGA-class transcription factors which bind to TGACG motifs encoded within the promoters of SA-induced genes (Mur et al., 2013).

Several lines of evidence point to an inter-relationship between NO and salicylic acid (SA) in plant defense. Treatment of tobacco and *A. thaliana* leaves with NO induces a substantial increase in endogenous SA (Sokolovski et al., 2005). In tobacco, this increase is required for *PR1* expression and probably involves NO-dependent induction of the phenylalanine (PAL) gene (Ribeiro et al., 1999). In addition, NOS inhibitors and a NO scavengers attenuate SA-induced systemic acquired resistance (SAR) (Modolo et al., 2006). Although these results suggest that NO is involved in both SA biosynthesis and action, other studies have indicated that NO function requires SA. Durner et al. (1998) suggested a role for NO in SAR. Given that SA treatment leads to enhanced NO production (Klepper, 1991), a complex signaling relationship between H_2O_2, NO and SA during HR and SAR is likely (Song and Goodman, 2001).

NO was proved to modulate SA signaling by controlling the oligomerization status of the translational activator NONEXPRESSOR OF PATHOGENESIS-RELATED PROTEINS1 (NPR1) via S-nitrosylation at cys 156 (Tada et al., 2008). The presence of this S-nitrosylated oligomeric form of NPR1 represses JA-triggered responses (Spoel et al., 2003). This indicates that this NO-dependent PTM of NPR1 plays a key role during hormonal signaling cascades, leading to coordinated plant immunity response (Mur et al., 2013; Yu et al., 2012).

10.8 CONCLUSIONS AND FUTURE PERSPECTIVES

NO has attracted a lot of attention in the last few years due to its broad range of functionality in the plant under both normal and various stress conditions. In almost all the investigated growth, developmental processes or defense mechanisms, NO has proved to play a key role. The exact biosynthetic pathway for NO has not been fully understood, but it is well established that there are three main pathways to generate NO in a plant cell: NOS, NR and non-enzymatically reduction of chemicals.

NO is now an important player in plant defense mechanism systems due to its power as a signaling molecule that modulates the action of many other defense system elements, as well as its function as an antioxidant, scavenging ROS from the cell and hence overcoming its toxic effect on cell components and functions.

Despite a large number of studies carried out in the last years, there are still many unanswered questions. For example, NO-dependent PTMs is a promising area for a study to explore exactly when and where these protein modifications occur, and when they are reversed.

Also, most of the studies carried out depend on NO donors such as SNP and GSNO to illustrate the effect of NO in plant metabolism; much attention should be paid to determine the changes occurred in the endogenous NO levels in plant and their effect on modulating different metabolic processes and stress responses in the plant.

Although it is confirmed now NO interacts with different signaling pathways in the cell, further investigation should be done to clarify the exact mechanism underlying this interaction; particularly the interplay between ethylene, JA and SA. Transcriptomic and proteomic analyses are still needed to explore different targets for NO bulletin plant cell.

REFERENCES

Abat, J.K., Deswal, R. (2009) Differential modulation of *S*-nitrosoproteome of *Brassica juncea* by low temperature: Change in *S*-nitrosylation of RuBisCO is responsible for the inactivation of its carboxylase activity. *Proteomics* 9 (18): 4368–4380.

Ahlfors, R., Brosche, M., Kangasjarvi, J. (2009) Ozone and nitric oxide interaction in Arabidopsis thaliana: A role for ethylene? *Plant Signaling & Behavior* 4 (9): 878–879.

Àlvarez, C., Lozano-Juste, J., Romero, L.C., Garcia, I., Gotor, C., Leon, J. (2011) Inhibition of Arabidopsis O-acetylserine (thiol) lyase A1 by tyrosine nitration. *The Journal of Biological Chemistry* 286 (1): 578–586.

Arasimowicz, M., Floryszak-Wieczorek, J. (2007) Nitric oxide as a bioactive signaling molecule in plant stress responses. *Plant Science* 172 (5): 876–887.

Asai, S., Mase, K., Yoshioka, H. (2010) A key enzyme for flavin synthesis is required for nitric oxide and reactive oxygen species production in disease resistance. *The Plant Journal: For Cell and Molecular Biology* 62 (6): 911–924.

Asai, S., Yoshioka, H. (2009) Nitric oxide as a partner of reactive oxygen species participates in disease resistance to necrotrophic pathogen *Botrytis cinerea* in *Nicotiana benthamiana. Molecular Plant–Microbe Interaction* 22: 619–629.

Bayden, A.S., Yakovlev, V.A., Graves, P.R., Mikkelsen, R.B., Kellogg, G.E. (2011) Factors influencing protein tyrosine nitration—Structure-based predictive models. *Free Radical Biology & Medicine* 50 (6): 749–762.

Beligni, M.V., Lamattina, L. (1999) Nitric oxide counteracts cytotoxic processes mediated by reactive oxygen species in plant tissues. *Planta* 208 (3): 337–344.

Beligni, M.V., Lamattina, L. (2000) Nitric oxide stimulates seed germination and de-etiolation, and inhibits hypocotyl elongation, three light-inducible responses in plants. *Planta* 210 (2): 215–221.

Beligni, M.V., Lamattina, L. (2001) Nitric oxide: A non-traditional regulator of plant growth. *Trends in Plant Science* 6 (11): 508–509.

Bellin, D., Asai, S., Delledonne, M., Yoshioka, H. (2013) Nitric oxide as a mediator for defense responses. *Molecular Plant-Microbe Interactions: MPMI* 26 (3): 271–277.

Besson-Bard, A., Courtois, C., Gauthier, A., Dahan, J., Dobrowolska, G., Jeandroz, S., Pugin, A., Wendehenne, D. (2008) Nitric oxide in plants: Production and cross-talk with Ca^{2+} signaling. *Molecular Plant* 1 (2): 218–228.

Bethke, P.C., Badger, M.R., Jones, R.L. (2004) Apoplastic synthesis of nitric oxide by plant tissues. *The Plant Cell* 16 (2): 332–341.

Boscari, A., Del Giudice, J., Ferrarini, A., Venturini, L., Zaffini, A.L., Delledonne, M., Puppo, A. (2013) Expression dynamics of the *Medicago truncatula* transcriptome during the symbiotic interaction with Sinorhizobium meliloti: Which role for nitric oxide? *Plant Physiology* 161 (1): 425–439.

Boveris, A.D., Galatro, A., Puntarulo, S. (2000) *Effect of nitric oxide* and *plant antioxidants on microsomal content of lipid radicals. Biological Research* 33 (2): 159–165.

Brown, G.C. (2001) Regulation of mitochondrial respiration by nitric oxide inhibition of cytochrome c oxidase. *Biochimica et Biophysica Acta* 1504 (1): 46–57.

Cao, H., Bowling, S.A., Gordon, A.S., Dong, X.N. (1994) Characterization of an Arabidopsis mutant that is non-responsive to inducers of systemic acquired resistance. *The Plant Cell* 6 (11): 1583–1592.

Caro, A., Puntarulo, S. (1998) Nitric oxide decreases superoxide anion generation by microsomes from soybean embryonic axes. *Physiologia Plantarum* 104 (3): 357–364.

Cecconi, D., Orzetti, S., Vandelle, E., Rinalducci, S., Zolla, L., Delledonne, M. (2009) Protein nitration during defense response in *Arabidopsis thaliana. Electrophoresis* 30 (14): 2460–2468.

Chaki, M., Alvarezde Morales, P., Ruiz, C., Begara-Morales, J.C., Barroso, J.B., Corpas, F.J., Palma, J.M. (2015) Ripening of pepper (*Capsicum annuum*) fruit is characterized by an enhancement of protein tyrosinenitration. *Annals of Botany* 116 (4): 637–647.

Chandok, M.R., Ytterberg, A.J., Van Wijk, K.J., Klessig, D.F. (2003) The pathogen-inducible nitric oxide synthase (iNOS) in plants is a variant of the P protein of the glycine decarboxylase complex. *Cell* 113 (4): 469–482.

Clarke, A., Desikan, R., Hurst, R.D., Hancock, J.T., Neill, S.J. (2000) Nitric oxide and programmed cell death in *Arabidopsis thaliana* suspension cultures. *Plant Journal* 24 (5): 667–677.

Culotta, E., Koshland, D.E. (1992) NO news is good news. *Science* 258 (5090): 1862–1865.

Diner, B.A., Petrouleas, V. (1990) Formation by NO of nitrosyl adducts of redox components of the photosystem II reaction center. II. Evidence that binds to the acceptor-side non-heme iron. *Biochimica et Biophysica Acta* 1015 (1): 141–149.

Dubovskaya, L.V., Bakakina, Y.S., Volotovski, I.D. (2015) Cyclic guanosine monophosphate as a mediator in processes of stress-signal transduction in higher plants. *Biophysics* 60 (4): 559–570.

Durner, J., Wendehemme, D., Klessig, D.F. (1998) Defense gene induction in tobacco by nitric oxide, cyclic GMP and cyclic ADP-ribose. *Proceedings of National Academy of Science, United States of America* 95: 10328–10333.

Fatma, M., Masood, A., Per, T.S., Rasheed, F., Khan, N.A. (2016) Interplay between nitric oxide and sulphur assimilation in salt tolerance in plants. *The Crop Journal* 2016: 153–161.

Feng, J., Wang, C., Chen, Q., Chen, H., Ren, B., Li, X., Zuo, J. (2013) *S*-nitrosylation of phosphotransfer proteins represses cytokinin signaling. *Nature Communications* 4: 1529.

Fernandez-Marcos, M., Sanz, L., Lewis, D.R., Muday, G.K., Lorenzo, O. (2011) Nitric oxide causes root apical meristem defects and growth inhibition while reducing PIN-FORMED 1 (PIN1)-dependent acropetal auxin transport. *Proceedings of the National Academy of Sciences of the United States of America* 108 (45): 18506–18511.

Ferrarini, A., De Stefano, M., Baudouin, E., Pucciariello, C., Polverari, A., Puppo, A., Delledonne, M. (2008) Expression of Medicago truncatula genes responsive to nitric oxide in pathogenic and symbiotic conditions. *Molecular Plant-Microbe Interactions: MPMI* 21 (6): 781–790.

Ford, P.C. (2010) Reactions of NO and nitrite with heme models and proteins. *Inorganic Chemistry* 49 (14): 6226–6239.

Freschi, L. (2013) Nitric oxide and phytohormone interactions: Current status and perspectives. *Frontiers in Plant Science* 4: 398.

Fröhlich, A., Durner, J. (2011) The hunt for plant nitric oxide synthase (NOS): Is one really needed? *Plant Science: An International Journal of Experimental Plant Biology* 181 (4): 401–404.

Fu, Z.Q., Yan, S., Saleh, A., Wang, W., Ruble, J., Oka, N., Mohan, R., *et al.* (2012)NPR3 and NPR4 are receptors for the immune signal salicylic acid in plants. *Nature* 486 (7402): 228–232.

García-Mata, C., Lamattina, L. (2003) Abscisic acid, nitric oxide and stomatal closure – Is nitrate reductase one of the missing links? *Trends in Plant Science* 8 (1): 20–26.

Godber, B.L., Doel J.J., Sapkota G.P., Blake D.R., Stevens C.R., Eisenthal R., Harrison R. (2000) Reduction of nitrite to nitric oxide catalyzed by xanthine oxidoreductase. *Journal of Biological Chemistry* 275 (11): 7757–7763.

Gouvea, C.M.C.P., Souza, J.F., Magalhaes, A.C.N., Martins, I.S. (1997) NO-releasing substances that induce growth elongation in maize root segments. *Plant Growth Regulation* 21 (3): 183–187.

Graziano, M., Beligni, M.V., Lamattina, L. (2002) Nitric oxide improves internal iron availability in plants. *Plant Physiology* 130 (4): 1852–1859.

Gunther, M.R., Sturgeon, B.E., Mason, R.P. (2002) Nitric oxide trapping of the tyrosyl radical-chemistry and biochemistry. *Toxicology* 177 (1): 1–9.

Gupta, K.J., Fernie, A.R., Kaiser, W.M., Van Dongen, J.T. (2011) On the origins of nitric oxide. *Trends in Plant Science* 16 (3): 160–168.

Hancock, J.T., Neill, S.J., Wilson, I.D. (2011) Nitric oxide and ABA in the control of plant function. *Plant Science: An International Journal of Experimental Plant Biology* 181 (5): 555–559.

Harrison, R. (2002) Structure and function of xanthine oxidoreductase: Where are we now? *Free Radical Biology & Medicine* 33 (6): 774–797.

Hayat, S., Mori, M., John, P., Ahmad, A. (2010) *Nitric Oxide in Plant Physiology*, WILEY-VCH Verlag GmbH & Co. KGaA, Weinheim.

He, Y., Tang, R.H., Hao, Y., Stevens, R.D., Cook, C.W., Ahn, S.M., Jing, L., et al. (2004) Nitric oxide represses the *Arabidopsis* floral transition. *Science* 305 (5692): 1968–1971.

Hill, R.D. (2012) Non-symbiotic haemoglobins – What's happening beyond nitric oxide scavenging? *AoB Plants* 2012: pls004.

Huang, X., Stettmaier, K., Michel, C., Hutzler, P., Mueller, M.J., Durner, J. (2004) Nitric oxide is induced by wounding and influences jasmonic acid signaling in Arabidopsis thaliana. *Planta* 218 (6): 938–946.

Igamberdiev, A.U., Hill, R.D. (2004) Nitrate, NO and haemoglobin in plant adaptation to hypoxia: An alternative to classic fermentation pathways. *Journal of Experimental Botany* 55 (408): 2473–2482.

Ischiropoulos, H. (2003) Biological selectivity and functional aspects of protein tyrosine nitration. *Biochemical and Biophysical Research Communications* 305 (3): 776–783.

Klepper, L. (1991) NOx evolution by soybean leaves treated with salicylic acid and selected derivatives. *Pesticide Biochemistry and Physiology* 39 (1): 43–48.

Kopyra, M., Gwozdz, E.A. (2003) Nitric oxide stimulates seed germination and counteracts the inhibitory effect of heavy metals and salinity on root growth of *Lupinus luteus. Plant Physiology and Biochemistry* 41 (11–12): 1011–1017.

Kwon, E., Feechan, A., Yun, B.W., Hwang, B.H., Pallas, J.A., Kang, J.G., Loake, G.J. (2012) AtGSNOR1 function is required for multiple developmental programs in *Arabidopsis. Planta* 236 (3): 887–900.

Lamotte, O., Courtois, C., Barnavon, L., Pugin, A., Wendehenne, D. (2005) Nitric oxide in plants: The biosynthesis and cell signaling properties of a fascinating molecule. *Planta* 221 (1): 1–4.

Lanteri, M.L., Laxalt, A.M., Lamattina, L. (2008) Nitric oxide triggers phosphatidic acid accumulation via phospholipase D during auxin-induced adventitious root formation in cucumber. *Plant Physiology* 147 (1): 188–198.

Laspina, N.V., Groppas, M.D., Tomaro, M.L., Benavides, M.P. (2005) Nitric oxide protects sunflower leaves against Cd-induced oxidative stress. *Plant Science* 169 (2): 323–330.

Leckie, C.P., McAinsh, M.R., Allen, G.J., Sanders, D., Hetherington, A.M. (1998) Abscisic acid-induced stomatal closure mediated by cyclic ADP-ribose. *Proceedings of National Academy of Science, United States of America* 95: 155837–155842.

Leshem, Y.Y. (2000) *Nitric Oxide in Plants, in Occurrence, Function and Use.* Kluwer Academic Publishers, Dordrecht, The Netherlands.

Leshem, Y.Y., Haramaty, E., Iiuz, D., Malik, D., Sofer, Y., Roitman, L., Leshem, Y. (1997) Effect of stress nitric oxide (NO): interaction between chlorophyll fluorescence, galactolipid fluidity and lipoxygenase activity. *Plant Physiology and Biochemistry* 35: 573–579.

Leshem, Y.Y., Pinchasov, Y. (2000) Non-invasive photoacoustic spectroscopic determination of relative endogenous nitric oxide and ethylene content stoichiometry during the ripening of strawberries *Fragaria anannasa* (Duch.) and avocados *Persea americana* (Mill.). *Journal of Experimental Botany* 51 (349): 1471–1473.

Leterrier, M., Airaki, M., Palma, J.M., Chaki, M., Barroso, J.B., Corpas, F.J. (2012) Arsenic triggers the nitric oxide (NO) and *S*-nitrosoglutathione (GSNO) metabolism in *Arabidopsis. Environmental Pollution* 166: 136–143.

Lindermayr, C., Durner, J. (2009) *S*-nitrosylation in plants: Pattern and function. *Journal of Proteomics* 73 (1): 1–9.

Lindermayr, C., Saalbach, G., Bahnweg, G., Durner, J. (2006) Differential inhibition of Arabidopsis methionine adenosyltransferases by protein S-nitrosylation. *The Journal of Biological Chemistry* 281 (7): 4285–4291.

Lozano-Juste, J., Colom-Moreno, R., León, J. (2011) In vivo protein tyrosine nitration in *Arabidopsis thaliana. Journal of Experimental Botany* 62 (10): 3501–3517.

Lum, H.K., Lee, C.H., Butt, Y.K.C., Lo, S.C.L. (2005) Sodium nitroprusside affects the level of photosynthetic enzymes and glucose metabolism in *Phaseolus aureus* (mung bean). *Nitric Oxide: Biology and Chemistry* 12 (4): 220–230.

Mackerness, S.A., John, C.F., Jordan, B., Thomas, B. (2001) Early signaling components in ultraviolet-B responses: Distinct roles for different reactive oxygen species and nitric oxide. *FEBS Letters* 489 (2–3): 237–242.

MacRobbie, E.A.C. (2000) ABA activates multiple Ca2+ fluxes in stomatal guard cell triggering vacuole K^+/ Rb^+ release. *Proceeding of National Academy of Science of the United States of America* 97: 12361–12368.

Manjunatha, G., Lokesh, V., Neelwarne, B. (2010) Nitric oxide in fruit ripening: Trends and opportunities. *Biotechnology Advances* 28 (4): 489–499.

Mannick, J.B., Schonhoff, C.M. (2004) NO means no and yes: Regulation of cell signaling by protein nitrosylation. *Free Radical Research* 38 (1): 1–7.

Mathieu, C., Moreau, S., Frendo, P., Puppo, A., Davies, M.J. (1998) Direct detection of radicals in intact soybean nodules: Presence of nitric oxide- leghemoglobin complexes. *Free Radical Biology & Medicine* 24 (7–8): 1242–1249.

Modolo, L.V., Augusto, O., Almeida, I.M.G., Pinto-Maglio, C.A.F., Oliveira, H.C., Seligman, K., Salgado, I. (2006) Decreased arginine and nitrite levels in nitrate reductase- deficient Arabidopsis thaliana plants impair nitric oxide synthesis and the hypersensitive response to *Pseudomonas syringae. Plant Science* 171 (1): 34–40.

Mou, Z., Fan, W.H., Dong, X.N. (2003) Inducers of plant systemic acquired resistance regulate NPR1 function through redox changes. *Cell* 113 (7): 935–944.

Mur, L.A.J., Laarhoven, L.J.J., Harren, F.J.M., Hall, M.A., Smith, A.R. (2008) Nitric oxide interacts with salicylate to regulate biphasic ethylene production during the hypersensitive response. *Plant Physiology* 148 (3): 1537–1546.

Mur, L.A., Mandon, J., Persijn, S., Cristescu, S.M., Moshkov, I.E., Novikova, G.V., Hall, M.A., *et al.* (2012) Nitric oxide in plants: An assessment of the current state of knowledge. *AoB Plants* 5: 1–17.

Mur, L.A.J., Prats, E., Pierre, S., Hall, M.A., Hebelstrup, K.H. (2013) Integrating nitric oxide into salicylic acid and jasmonic acid/ethylene plant defense pathways. *Frontiers in Plant Science* 4: 215.

Navarre, D.A., Wendenhenne, D., Durner, J., Noad, R., Klessing, D.F. (2000) Nitric oxide modulates the activity of tobacco aconitase. *Plant Physiology* 122 (2): 573–582.

Neil, S.J., Desikan, R., Hancock, J.T. (2003) Nitric oxide signaling in plant. *New Phytologist* 159 (1): 11–35.

Neill, S.J., Desikan, R., Clarke, A., Hancock, J.T. (2002) Nitric oxide is a novel component of abscisic acid signaling in stomatal guard cells. *Plant Physiology* 128 (1): 13–16.

Pagnussat, G.C., Lanteri, M.L., Lombardo, M.C., Lamattina, L. (2004) Nitric oxide mediates the indole acetic acid induction activation of a mitogen-activated protein kinase cascade involved in adventitious root development. *Plant Physiology* 135 (1): 279–286.

Palma, J.M., Sandalio, L.M., Corpas, F.J., Romero-Puertas, M.C., McCarthy, I., Del Rio, L.A. (2002) Plant proteases, protein degradation and oxidative stress: Role of peroxisomes. *Plant Physiology and Biochemistry* 40 (6–8): 521–530.

Peleg, Z., Blumwald, E. (2011) Hormone balance and abiotic stress tolerance in crop plants. *Current Opinion in Plant Biology* 14 (3): 290–295.

Perazzolli, M., Dominici, P., Romero-Puertas, M.C., Zago, E., Zeier, J., Sonoda, M., Lamb, C., Delledonne, M. (2004) *Arabidopsis* nonsymbiotic hemoglobin AHb1 modulates nitric oxide bioactivity. *The Plant Cell* 16 (10): 2785–2794.

Planchet, E., Jagadis Gupta, K., Sonoda, M., Kaiser, W.M. (2005) Nitric oxide emission from tobacco leaves and cell suspensions: Rate limiting factors and evidence for the involvement of mitochondrial electron transport. *The Plant Journal: For Cell and Molecular Biology* 41 (5): 732–743.

Reymond, P., Farmer, E.E. (1998) Jasmonate and salicylate as global signals for defense gene expression. *Current Opinion in Plant Biology* 1 (5): 404–411.

Ribeiro, E.A., Jr., Cunha, F.Q., Tamashiro, W.M., Martins, I.S. (1999) Growth phase-dependent subcellular localization of nitric oxide synthase in maize cells. *FEBS Letters* 445 (2–3): 283–286.

Romero-Puertas, M.C., Rodríguez-Serrano, M., Corpas, F.J., Gómez, M., Del Río, L.A., Sandalio, L.M. (2004) Cd-induced subcellular accumulation of O_2- and H_2O_2 in pea leaves. *Plant, Cell and Environment* 27 (9): 1122–1134.

Rosales, E.P., Iannone, M.F., Groppa, M.D., Benavides, M.P. (2012) Polyamines modulate nitrate reductase activity in wheat leaves: Involvement of nitric oxide. *Amino Acids* 42 (2–3): 857–865.

Sanchez, C., Cabrera, J.J., Gates, A.J., Bedmar, E.J., Richardson, D.J., Delgado, M.J. (2011) Nitric oxide detoxification in the rhizobia-legume symbiosis. *Biochemical Society Transactions* 39 (1): 184–188.

Sasakura, F., Uchiumi, T., Shimoda, Y., Suzuki, A., Takenouchi, K., Higashi, S., Abe, M.A. (2006) A Class 1 hemoglobin gene from *Alnus firma* functions in symbiotic and nonsymbiotic tissues to detoxify nitric oxide. *Molecular Plant-Microbe Interactions: MPMI* 19 (4): 441–450.

Schansker, G., Goussias, C., Petrouleas, V., Rutherford, A.W. (2002) Reduction of the Mn cluster of the water-oxidizing enzyme by nitric oxide: Formation of an S; state. *Biochemistry* 41 (9): 3057–3064.

Schmidt, H.H.H.W., Walter, U. (1994) NO at work. *Cell* 78 (6): 919–925.

Schopfer, F.J., Baker, P.R., Freeman, B.A. (2003) NO-dependent protein nitration: A cell signaling event or an oxidative inflammatory response? *Trends in Biochemical Sciences* 28 (12): 646–654.

Seth, D., Stamler, J.S. (2010) The SNO-proteome: Causation and classifications. Current Opinion in Chemistry and Biology 15: 129–136.

Shi, S., Wang, G., Wang, Y., Zhang, L., Zhang, L. (2005) Protective effect of nitric oxide against oxidative stress under ultraviolet-B radiation. *Nitric Oxide: Biology and Chemistry* 13 (1): 1–9.

Siddiqui, M.H., Al-Whaibi, M.H., Ali, H.M., Sakran, A.M., Basalah, M.O., AlKhaishany, M.Y.Y. (2013) Mitigation of nickel stress by the exogenous application of salicylic acid and nitric oxide in wheat. *Australian Journal of Crop Science* 7: 1780–1788.

Sokolovski, S., Blatt, M.R. (2004) Nitric oxide block of outward-rectifying K^+ channels indicates direct control by protein nitrosylation in guard cells. *Plant Physiology* 136 (4): 4275–4284.

Sokolovski, S., Hills, A., Gay, R., Garcia-Mata, C., Lamattina, L., Blatt, M.R. (2005) Protein phosphorylation is a prerequisite for intracellular Ca^{2+} release and ion channel control by nitric oxide and abscisic acid in guard cells. *The Plant Journal: For Cell and Molecular Biology* 43 (4): 520–529.

Song, F., Goodman, R.M. (2001) Activity of nitric oxide is dependent on, but is partially required for function of, salicylic acid in the signaling pathway in tobacco synthetic acquired resistance. *Molecular Plant-Microbe Interactions* 14: 1458–1462.

Spoel, S.H., Koornneef, A., Claessens, S.M.C., Korzelius, J.P., Van Pelt, J.A., Mueller, M.J., Buchala, A.J., et al. (2003) NPR1 modulates cross-talk between salicylate- and jasmonate-dependent defense pathways through a novel function in the cytosol. *The Plant Cell* 15 (3): 760–770.

Tada, Y., Spoel, S.H., Pajerowska-Mukhtar, K., Mou, Z.L., Song, J.Q., Wang, C., Zuo, J., Dong, X. (2008) Plant immunity requires conformational charges of NPR1 via *S*-nitrosylation and thioredoxins. *Science* 321 (5891): 952–956.

Takahashi, S., Yamasaki, H. (2002) Reversible inhibition of photophosphorylation in chloroplasts by nitric oxide. *FEBS Letters* 512 (1–3): 145–148.

Terrile, M.C., Parïs, R., Calderón-Villalobos, L.I.A., Iglesias, M.J., Lamattina, L., Estelle, M., Casalongué, C.A. (2012) Nitric oxide influences auxin signaling through *S*-nitrosylation of the *Arabidopsis* TRANSPORT INHIBITOR RESPONSE 1 auxin receptor. *The Plant Journal: For Cell and Molecular Biology* 70 (3) : 492–500.

Thomas, D.D., Heinecke, J.L., Ridnour, L.A., Cheng, R.Y., Kesarwala, A.H., Switzer, C.H., McVicar, D.W., et al. (2015) Signaling and stress: The redox landscape in NOS2 biology. *Free Radical Biology and Medicine* 87: 204–225

Tian, X., Lei, Y. (2006) Nitric oxide treatment alleviates drought stress in wheat seedlings. *Biologia Plantarum* 50 (4): 775–778.

Tischner, R., Planchet, E., Kaiser, W.M. (2004) Mitochondrial electron transport as a source for nitric oxide in the unicellular green alga Chlorella sorokiniana. *FEBS Letters* 576 (1–2): 151–155

Toledo, J.C., Augusto, O. (2012) Connecting the chemical and biological properties of nitric oxide. *Chemical Research in Toxicology* 25 (5): 975–989.

Webb, A.A.R., Larman, M.G., Montgomery, L.T., Taylor, J.E., Hetherington, A.M. (2001) The role of calcium in ABA-induced gene expression and stomatal movements. *The Plant Journal: For Cell and Molecular Biology* 26 (3): 351–362.

Wendehenne, D., Gould, K., Lamotte, O., Selvi-Srinivas, M., Klinguer, A., Pugin, A. (2004) Nitric oxide is an essential component of biotic and abiotic stress-induced signaling pathways in plants. In *Nitric oxide signaling in higher plants, focus on plant molecular biology*, ed. Magalhaes, R.J., 25–64, Studium Press, Houston, USA.

Wendehenne, D., Pugin, A., Klessig, D.F., Durner, J. (2001) Nitric oxide: Comparative synthesis and signaling in animal and plant cells. *Trends in Plant Science* 6 (4): 177–183.

Wilson, I.D., Neill, S.J., Hancock, J.T. (2008) Nitric oxide synthesis and signaling in plants. *Plant, Cell and Environment* 31 (5): 622–631.

Wimalasekera, R., Villar, C., Begum, T., Scherer, G.F. (2011) Copper amine oxidase 1 (CuAO1) of Arabidopsis thaliana contributes to abscisic acid- and polyamine-induced nitric oxide biosynthesis and abscisic acid signal transduction. *Molecular Plant* 4 (4): 663–678.

Wink, D.A., Mitchell, J.B. (1998) Chemical biology of nitric oxide: Insights into regulatory, cytotoxic and cytoprotective mechanisms of nitric oxide. *Free Radical Biology and Medicine* 25 (4–5): 434–456.

Wodala, B., Deak, Z., Vass, I., Erdei, L., Altorjay, I., Horvath, F. (2008) In vivo target sites of nitric oxide in photosynthetic electron transport as studied by chlorophyll fluorescence in pea leaves. *Plant Physiology* 146 (4): 1920–1927.

Wojtaszek, P. (2000) Nitric oxide in plants: To NO or not to NO. *Phytochemistry* 54 (1): 1–4.

Xu, M., Dong, J. (2005) Nitric oxide stimulates indole alkaloid production in *Catharanthus roseus* cell suspension cultures through a protein kinasedependent signal pathway. *Enzyme and Microbial Technology* 37 (1): 49–53.

Yang, S.F., Hoffman, N.E. (1984) Ethylene biosynthesis and its regulation in higher plants. *Annual Review of Plant Physiology* 35 (1): 155–189.

Yang, J.D., Zhao, H.L., Zhang, T.H., Yun, J.F. (2004) Effects of exogenous nitric oxide on photochemical activity of photosystem II in potato leaf tissue under non-stress condition. Acta Botanica Sinica 46: 1009–1014.

Yu, M., Yun, B.W., Spoel, S.H., Loake, G.J. (2012) A sleigh ride through the SNO: Regulation of plant immune function by protein S-nitrosylation. *Current Opinion in Plant Biology* 15 (4): 424–430.

Yun, B.W., Feechan, A., Yin, M., Saidi, N.B., Le Bihan, T., Yu, M., Moore, J.W., *et al.* (2011) S-nitrosylation of NADPH oxidase regulates cell death in plant immunity. *Nature* 478 (7368): 264–268.

Zaffagnini, M., De Mia, M., Morisse, S., Di Giacinto, N., Marchand, C.H., Maes, A., Lemaire, S.D., Trost, P. (2016) Protein S-nitrosylation in photosynthetic organisms: A comprehensive overview with future perspectives. *Biochimica et Biophysica Acta* 1864 (8): 952–966.

Zanardo, D.I.L., Zanardo, F.M.L., Ferrarese, M.D.L.L., Magalhaes, J.R., Filho, O.F. (2005) Nitric oxide seed germination and peroxidase activity in canola (*Brassica napus* L.). Physiology and Molecular Biology of Plants 11: 81–86.

Zeier, J., Delledonne, M., Mishina, T., Severi, E., Sonoda, M., Lamb, C. (2004) Genetic elucidation of nitric oxide signaling in incompatible plant–pathogen interactions. *Plant Physiology* 136 (1): 2875–2886.

Zhang, Y.Y., Wang, L.L., Liu, Y.L., Zhang, Q., Wei, Q.P., Zhang, W.H. (2006) Nitric oxide enhances salt tolerance in maize seedlings through increasing activities of proton-pump and Na^+/H^+ antiport in the tonoplast. *Planta* 224 (3): 545–555.

Zhang, A., Zhang, J., Zhang, J., Ye, N., Zhang, H., Tan, M., Jiang, M. (2011) Nitric oxide mediates brassinosteroid-induced ABA biosynthesis involved in oxidative stress tolerance in maize leaves. *Plant and Cell Physiology* 52 (1): 181–192.

Zottini, M., Formentin, E., Scattolin, M., Carimi, F., Schiavo, F., Terzi, M. (2002) Nitric oxide affects plant mitochondrial functionality in vivo. *FEBS Letters* 515 (1–3): 75–78.

11 Role of Exogenous Hydrogen Peroxide and Nitric Oxide on Improvement of Abiotic Stress Tolerance in Plants

Ghader Habibi

CONTENTS

11.1 INTRODUCTION

Abiotic stresses mainly cause oxidative damage in plants due to overproduction of reactive oxygen species (ROS) including superoxide radical ($O_2^{\cdot-}$), hydroxyl radical ($OH^{\cdot}$), hydroperoxyl radical ($HO_2^{\cdot}$), peroxy radical ($ROO^{\cdot}$), singlet oxygen ($^1O_2$), and hydrogen peroxide (H_2O_2) (Ali et al., 2018; Ashraf, 2009; Demidchik, 2015). ROS can react with photosynthetic pigments, lipids, proteins and DNA (Ashraf, 2009), leading eventually to lipid peroxidation, membrane damage, inactivation of antioxidant enzymes and cell death (Gill and Tuteja, 2010). However, for the detoxification of excessively produced ROS, plants possess a developed antioxidative defense mechanism. ROS scavenging occurs through a large number of ROS detoxifying enzymes such as superoxide dismutase (SOD), ascorbate peroxidase (APX), catalase (CAT), glutathione peroxidase (GPX), and peroxiredoxin (PRX), as well as through antioxidants such as ascorbic acid (AsA) and glutathione (GSH), in order to maintain redox homeostasis (Gill and Tuteja, 2010; Mittler et al., 2004). On the other hand, plants use ROS as signal transduction molecules (Considine et al., 2015; Dietz, 2015; Foyer and Noctor, 2013; Mignolet-Spruyt et al., 2016).

Different methodologies have been developed aiming at enhancing multiple stress tolerance. Exogenous application of low concentrations of SNP (NO donor), H_2O_2, NaHS (H_2S donor), melatonin (Mel), or polyamines (PAs) can enhance plant abiotic stress tolerance without subsequent inhibition of plant growth (Li et al., 2013; Savvides et al., 2016; Shi et al., 2010). Increasing evidence suggests that initial exposure to H_2O_2 and/or NO at the appropriate level can enhance abiotic stress tolerance in plants. Nitric oxide (NO) and H_2O_2, as signaling molecules, are involved in plant developmental and physiological processes (Ge et al., 2015; Hernández-Barrera et al., 2015; Shi et al., 2015; Wang et al., 2015) as well as in response to abiotic stresses (Fan et al., 2015; Hossain et al., 2015; Liu et al., 2015a; Shan et al., 2015).

The mitigation of oxidative stress following pre-treatment of plants with exogenous H_2O_2 or NO has been found to correlate with increased transcript levels of enzymatic antioxidants (Christou et al., 2013, 2014) and/or ascorbate and glutathione biosynthesis components (Christou et al., 2014; Wei et al., 2014). Exogenous H_2O_2 and NO have been shown to increase tissue levels of free proline and/or soluble carbohydrates and/or soluble protein compared to non-primed plants when exposed to drought, salinity (Shi et al., 2014), chilling, or freezing (Savvides et al., 2016).

Therefore, it has been well established that exogenous H_2O_2 and NO are effective chemical priming agents against different abiotic stresses; however, few studies have tested the relationship between NO and H_2O_2 in plant responses to individual and combined abiotic stresses. Since the ability of exogenous H_2O_2 and NO to mitigate various environmental stresses is dependent on the plant species, treatments and/or experimental systems, we review here exogenous H_2O_2 and NO that can mitigate abiotic stresses in different plant studies. In addition, we attempt to summarize the current knowledge regarding the role of the interplay among H_2O_2 and NO in alleviating environmental stresses.

11.2 H_2O_2 AND NO HOMEOSTASIS IN PLANT CELLS

Chloroplasts, mitochondria, peroxisome, and endoplasmic reticulum are major sites of H_2O_2 production. In addition, the apoplast seems to be an important site for H_2O_2 production in response to abscisic acid (ABA) and adverse environmental conditions, such as drought and salinity (Habibi, 2014). Under environmental stress conditions, enhanced levels of H_2O_2 can cause damage to biomolecules such as lipids, proteins, and DNA. In plant cells, both H_2O_2 production and removal processes are precisely regulated and coordinated in the same or in different cellular compartments. The mechanisms of H_2O_2 scavenging are controlled by both enzymatic (e.g. catalase (CAT), ascorbate peroxidase (APX), glutathione peroxidase (GPX), glutathione reductase (GR), monodehydroascorbate reductase (MDHAR), and dehydroascorbate reductase (DHAR)) and non-enzymatic (e.g. ascorbate (vitamin C), reduced glutathione (GSH), and tocopherol (vitamin E)) antioxidants (Mittler et al., 2004).

H_2O_2 plays a dual role in plants: at low concentrations, it acts as a signal molecule (Mittler et al., 2004); however, excess H_2O_2 leads to oxidative stress and causes lipid peroxidation, membrane destruction, protein denaturation and DNA damage. To function as a signaling molecule, H_2O_2 needs to cross the inner and outer membranes of the chloroplast and peroxisomes. Later studies have revealed that H_2O_2 transport might be mediated by specific membrane aquaporin homologues of the TIP (tonoplast intrinsic protein) and PIP (plasma membrane intrinsic protein) families (Bienert et al., 2007). Moreover, it has been suggested that H_2O_2 produced by the chloroplast electron transport chain can leak out of chloroplasts in a light-intensity-dependent manner (Mubarakshina et al., 2010). In addition, the GSH-GSSG ratios are involved in the transmission of H_2O_2 signals.

It has been established that H_2O_2 is a secondary messenger in signal transduction networks. In the cell, some major factors influence H_2O_2 interaction with other signaling molecules during various stress responses. Ca^{2+} acts as a central regulator in guard cell signaling in response to versatile stimuli such as ABA, ROS and nitric oxide (NO) (Marten et al., 2008; Young et al., 2006). H_2O_2 and ABA can activate plasma membrane hyperpolarization activated Ca^{2+} channels (Hamilton et al., 2000; Lemtiri-Chlieh et al., 2003).

Recent evidence proposes that H_2O_2 and NO are involved in stimulation of stomatal closure as well as in stimulation or inhibition of root hair growth in *Arabidopsis* (Clark et al., 2010). In *Arabidopsis*, H_2O_2 induces guard cells to synthesize NO, which is responsible for the closure of the stomatal aperture (Wang et al., 2010c).

Nitric oxide participates in a wide array of biological processes such as chlorophyll level (Liu and Guo, 2013), vegetative growth (Lozano-Juste and León, 2011), symbiosis nodule formation (Hichri et al., 2015), stomatal movement (Chen et al., 2013), and Fe homeostasis (Buet and Simontacchi, 2015) through interaction with hormones, reactive oxygen species, calcium, and protein post-translational modifications. Substantiated experimental evidence demonstrated that NO is also a key regulatory molecule in the response of plants to environmental stress (León et al., 2016; Siddiqui et al., 2011). Also, NO plays an important role in plant responses to various abiotic stresses such as cold (Fan et al., 2015), heat (Yu et al., 2015), salt (Liu et al., 2015b), drought (Shan et al., 2015), UV-B (Esringu et al., 2015) and heavy metal (Alemayehu et al., 2015; Kaur et al., 2015).

NO alters the activity, function, and stability of many target proteins through regulation of proteolytic degradation or subcellular re-localization (Guerra and Callis, 2012). Interestingly, NO generated in response to abiotic stresses is prone to mediate defense responses similar to those seen following H_2O_2 generation (Wang et al., 2015). Recent findings suggest that NO is generated through enzymatic and non-enzymatic pathways: the Arg-dependent nitric oxide synthase (NOS) and nitrite-dependent nitrate reductase (NR) pathways (Niu and Liao, 2016). Once NO is formed, specific scavenging mechanisms including (i) the presence of target molecules: superoxide radical, thiols, Fe-containing molecules, and (ii) non-symbiotic hemoglobins help modulate or repress the levels of NO (Hebelstrup et al., 2013).

To determine the effects of NO in plants under environmental stresses, pharmacological experiments have been conducted using exogenous application of NO donors (e.g. sodium nitroprusside (SNP)), NO scavengers (e.g. 2-(4-carboxyphenyl)-4,4,5,5-tetramethylimidazoline-1-oxyl-3-oxide (cPTIO)) and enzyme inhibitors, where altering NO levels with donors and scavengers were achieved, by employing mutants (e.g. decreased endogenous NO levels in *nia1nia2*, *Atnoa1* and increased in *nox1/cue1* (Chun et al., 2012; Shi et al., 2012)), or developing *Arabidopsis* transformed with the Ot*NOS* gene under the control of a stress-inducible promoter (Foresi et al., 2015). These transgenic plants showed higher NO synthesis and exhibited enhanced tolerance to a range of biotic and abiotic stresses. Interestingly, Ot*NOS* transgenic lines exhibited higher germination rate as compared to wild-type *Arabidopsis* under NaCl and drought stresses (Foresi et al., 2015). Under water stress, salinity, heavy metal stress, high or low temperature extremities, and ultraviolet radiation, the oxidative status generated by reactive oxygen species in stressed plants is alleviated by NO through the improvement of antioxidant capacity and redox homeostasis (León et al., 2016).

11.3 EXOGENOUS H_2O_2 AND NO AND ABIOTIC STRESS TOLERANCE

Multiple environmental stresses influence plant metabolism, and it is important to elevate the ability of plants to survive under multiple abiotic and biotic stresses (Mittler and Blumwald, 2010). In response to abiotic and biotic stresses, plants have developed a wide range of adaptive mechanisms to maintain growth and development and ensure survival (Mittler et al., 2011). Increasing evidence suggests that an appropriate level of H_2O_2 and/or NO as signaling molecules is not only needed for better plant production but also beneficial to mitigate different kinds of biotic and abiotic stresses. A number of studies on plants have demonstrated that initial exposure to H_2O_2 and/or NO can enhance abiotic stress tolerance through the modulation of photosynthesis, and ROS detoxification (Gondim et al., 2012, 2013; Hasanuzzaman et al., 2011; Hossain and Fujita, 2013; Mostofa et al., 2014; Mostofa and Fujita, 2013; Sathiyaraj et al., 2014; Teng et al., 2014; Wang et al., 2014a). The evidence regarding common components of the mode of action of these compounds strongly demonstrates that pre-treatment of plants with these compounds will prime the plants for stronger tolerance when they are exposed to an abiotic challenge sometime later (Savvides et al., 2016). Since exogenous application of low concentrations of these molecules initially induces an increase in their endogenous concentrations, primed plants can inaugurate a mild stress cue, similar to an acclimation response. This characteristic is similar to general animal vaccines. On exposure to stress, primed plants exhibit enhanced tolerance-related responses (Figure 11.1). However, the roles of exogenous H_2O_2 and NO in abiotic stresses alleviation, and the mechanisms by which growth and development of plants are influenced by these signaling molecules, are not well determined. The role of exogenous H_2O_2 and NO in environmental stresses alleviation is described below.

11.3.1 Salt Stress

Recent studies have demonstrated that pre-treatment of plants with exogenous H_2O_2 can enhance abiotic stress tolerance including salinity. Uchida et al. (2002) reported that H_2O_2 and nitric oxide (NO) pre-treatments in rice (*Oryza sativa*) plants induce salt tolerance through the enhanced activities of antioxidants, increased photosynthetic activity and induced expression of genes encoding Δ'-pyrroline-5-carboxylate synthase, sucrose-phosphate synthase, and the small heat-shock proteins. Their findings suggest that NO and H_2O_2 act as signaling molecules that regulate salt stress tolerance by enhancing the expression of stress-related genes. Wahid et al. (2007) indicated that seedlings from H_2O_2-treated seeds of *Triticum aestivum* had more effective antioxidant systems than untreated controls.

Fedina et al. (2009) reported that *Hordeum vulgare* seedlings pre-treated with H_2O_2 (1 and 5 mM) when exposed to salt (150 mM NaCl, 4 and 7 days) showed an increase in the rates of CO_2 fixation, when compared with seedlings subjected to NaCl stress only. Gondim et al. (2010) showed that H_2O_2 could alleviate the damage of salt stress (100 mM NaCl) in maize (*Zea mays*), seedlings which was reflected by an increase in the percentage germination of seeds as well as up-regulation of APX and CAT activities. In addition, exogenously applied H_2O_2 has significant ameliorating effects against NaCl-induced oxidative damage in wheat (Li et al., 2011), maize (Gondim et al., 2012, 2013), *Suaeda fruticosa* (Hameed et al., 2012), and *Panax ginseng* (Sathiyaraj et al., 2014) seedlings through stimulation of enzymatic and non-enzymatic antioxidants.

Recently, Terzi et al. (2014) investigated the effects of H_2O_2 pre-treatment on proline, soluble sugars, polyamines and phytohormone ABA levels for improving osmotic stress tolerance of maize seedlings. The results showed that H_2O_2 pre-treatment alleviated water loss and increased osmotic stress resistance by enhancing the levels of soluble sugars, proline and polyamines, and by reducing ABA and H_2O_2 concentrations in maize seedlings under osmotic stress. Kilic and Kahraman (2016) reported that adverse effects of salt stress on barley growth were alleviated with H_2O_2 application. The result obtained by Ashfaque et al. (2014) suggested that treatment of wheat plants with H_2O_2 positively influenced

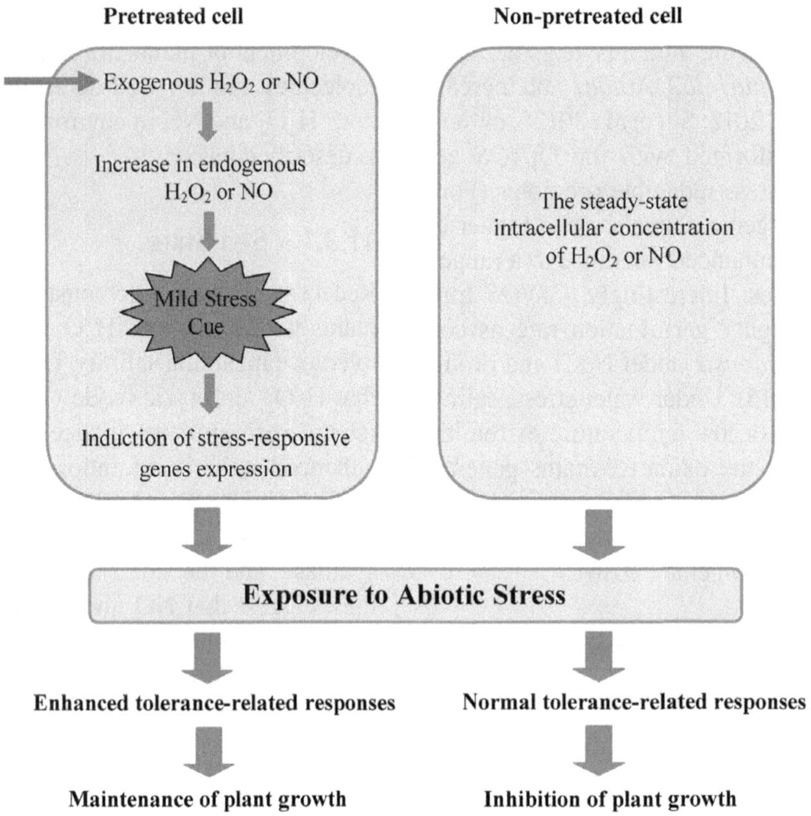

FIGURE 11.1 The mode of action of exogenous H_2O_2 or NO for enhancing stress tolerance.

plant growth under saline and non-saline conditions. They reported that the application of 50 or 100 μM H_2O_2 reduced both Na^+ and Cl^- ions levels, and increased the levels of photosynthetic pigments, proline content and N assimilation. Recently, Ellouzi et al. (2017) reported that H_2O_2 seed priming stimulated the growth and antioxidant defense of *Cakile maritima* and *Eutrema salsugineum* grown under salinity, together with significantly higher concentrations of glutathione, ascorbic acid, and proline production. Finally, these results indicated that H_2O_2 priming improved tolerance to salinity. Zhu et al. (2016) found that ethylene and H_2O_2 were involved in the BR-dependent induction of tomato salt stress tolerance.

Increased NO production by plant tissues has recently been found to occur in response to several abiotic stresses such as salinity and osmotic stress (David et al., 2010; Manai et al., 2014; Monreal et al., 2013; Valderrama et al., 2007; Zhang et al., 2006). On the other hand, salt-induced oxidative stress can be at least partially alleviated by application of exogenous NO in many plant species (Khan et al., 2012; Tanou et al., 2012; Uchida et al., 2002; Zhang et al., 2006). Exogenous application of NO enhances tolerance to salinity by improving the antioxidative defense system, osmolyte accumulation, and ionic homeostasis (Hayat et al., 2012; Khan et al., 2012; Tanou et al., 2012). Egbichi et al. (2014) proposed that the role of NO in increasing tolerance to salinity stress in soybean was attributed to an increase in antioxidant capacity by direct scavenging of H_2O_2. In addition, exogenously applied NO has significant ameliorating effects against NaCl-induced oxidative damage in the calluses of reed (*Phragmites communis*) under 200 mM NaCl treatment (Zhao et al., 2004) and sunflower seedlings (David et al., 2010) by increasing the K^+ to Na^+ ratio and ionic homeostasis. Therefore, NO improves their salt tolerance by regulation of Na^+ homeostasis and K^+ acquisition via enhancing expression of a plasma membrane Na^+/H^+ antiporter and H^+-ATPase-related genes. These findings suggest that the control of H^+-transport activity by NO contributes to ensuring an adequate inward flux of K^+ as well as to decrease K^+ loss from cells (Chen et al., 2013; Chen et al., 2014). In fact, the involvement of NO in salt tolerance depends on salt concentration as well as plant species or genotypes. Liu et al. (2007) reported that the increase in nitrate reductase-dependent NO production in red kidney bean roots during salt stress causes changes in the NADPH levels via glucose-6-phosphate dehydrogenase (G-6-PDH) activation, which results in an enhancement in the activities of antioxidant enzymes.

Manai et al. (2014) suggested that exogenous NO application is useful to alleviate salinity-induced oxidative stress in tomato plants through inducing the activity of main antioxidative enzymes including SOD, APX, glutathione reductase (GR), and peroxidase (POD), as well as stimulation of some enzymes involved in nitrogen metabolism, including nitrate reductase (NR) and nitrite reductase (NiR) activities. However, it has been reported that NO utilizes a differential modulation of the antioxidant response under conditions of salinity (Chen et al., 2014; Hasanuzzaman et al., 2011; Wang et al., 2011; Zeng et al., 2011). It is reported that the protective role of nitric oxide against salt-induced damages in crop plants is attributed to an increase in the plasma membrane H-ATPase activity, which ultimately increased the K/Na ratio in maize seedlings (Zhang et al., 2006). Exogenous NO application improved salt tolerance in rice seedlings and resulted in secondary metabolite accumulation (Bellin et al., 2013). A proteomic analysis in citrus plants revealed a crosstalk between signaling pathways of H_2O_2 and NO in acclimation to salinity (Tanou et al., 2009). Exogenous NO was recently proven to mitigate salinity stress in cucumber seedlings by regulating the content and proportions of different types of free polyamines (Fan et al., 2013). In cotton seedlings, the possible mechanisms of NO-mediated protective effects under salt stress may include enhanced plants growth and antioxidant enzyme activities, an increase in the K^+/Na^+ ratio, and a decrease in the contents of thiobarbituric acid-reactive substances (TBARS) and malondialdehyde (MDA) (Dong et al., 2014a). Recently, Ali et al. (2017) investigated the beneficial role of exogenous sodium nitroprusside (SNP) to induce salt tolerance in four wheat cultivars (Sahar-06, Punjab-11, Millat-11 and Galaxy-13). They observed that 0.1 mM SNP mitigated salinity stress in cv. Sahar-06 by increasing the activities of antioxidant enzymes such as SOD, POD and CAT, and the contents of ascorbic acid (AsA), proline, and total phenolics.

11.3.2 Drought Stress

In drought conditions, ABA is accumulated in roots and translocated to leaves to initiate adaptation of plants to drought stress. ABA causes the formation of reactive oxygen species (ROS) such as hydrogen peroxide (H_2O_2), and controls the closure of the stomatal aperture in an H_2O_2-dependent and -independent manner (Ali et al., 2018; Chan et al., 2013). Drought stress induces oxidative stress by increasing the levels of H_2O_2 and singlet oxygen (de Carvalho, 2013).

On the other hand, drought-induced oxidative stress may be at least partially alleviated by H_2O_2 priming in cucumber plants through stimulation of the antioxidant enzymes SOD, CAT, GPOX, APX, DHAR, DHAR, GR, and the levels of AsA and GSH, and maintenance of the ultrastructure of chloroplasts (Jing et al., 2009). In a similar study, exogenous application of NO enhanced soybean tolerance to salinity by improving leaf water levels, probably due to increased oligosaccharide biosynthesis (Ishibashi et al., 2011). In *Phaseolus vulgaris* L., priming seeds with H_2O_2 ameliorated the deleterious effects of drought stress on growth through maintaining the photosynthetic pigments and the total carbohydrate content (Abass and Mohamed, 2011). Liao et al. (2012a) reported that under drought conditions, exogenously applied H_2O_2 and NO significantly improved leaf chlorophyll contents, chlorophyll fluorescence parameters, and hypocotyl soluble carbohydrate and protein contents to improve drought tolerance in marigold (*Tagetes erecta* L.) plants. In mustard (*Brassica juncea* L.) seedlings, H_2O_2 pre-treated drought-stressed seedlings exhibited significantly higher APX, GR, CAT, GST, and Gly II activities, as well as a higher GSH/GSSG ratio compared with seedlings under drought only (Hossain and Fujita, 2013). Recently, Khan et al. (2017) reported that priming *Brassica napus* seeds with H_2O_2 alleviated drought-induced oxidative stress, observed as enhanced relative water content (RWC) and chlorophyll content. In addition, exogenously applied H_2O_2 was recently proven to alleviate water stress in maize seedlings (Ashraf et al., 2014) as well as osmotic stress in two cucumber (*Cucumis sativus* L.) varieties (Liu et al., 2010) by decreasing lipid membrane peroxidation through stimulation of enzymatic and non-enzymatic antioxidants.

It has been observed that exogenous NO improved drought tolerance and counteracted membrane damage and lipid peroxidation in wheat (Boyarshinov and Asafova, 2011), cucumber (Arasimowicz-Jelonek et al., 2009) and rice (Farooq et al., 2009). In guard cells, exogenous NO causes stomatal closure (Neill et al., 2008) by promoting specifically intracellular Ca^{2+} release, thus regulating Ca^{2+}-sensitive K^+ and Cl^- channels in the plasma membrane (García-Mata et al., 2003). Recently, hydrogen sulfide was described as a new component of the ABA-dependent signaling network in stomatal guard cells inducing NO production (Scuffi et al., 2014). Later studies suggest that NO signaling is integrated with CAM expression, thereby allowing plants to regulate photosynthetic plasticity, as well as adaptive responses to environmental stresses (Freschi et al., 2010). It has been reported that exogenous NO can induce CAM expression by increasing the activities of the enzymes phosphoenolpyruvate carboxylase, malate dehydrogenase, and phosphoenolpyruvate carboxykinase, as well as nocturnal accumulation of malate (Freschi et al., 2010). Ali et al. (2017) reported that the SNP-mediated mitigation of drought-induced damages was achieved by up-regulating the synthesis of proline.

11.3.3 Heavy Metal Stress

A number of studies on plants have demonstrated that H_2O_2, as a signaling molecule, interacts with phytohormones and signaling molecules ABA, SA, JA (jasmonic acid), GA (gibberellic acid), ethylene and NO, which are involved in plant tolerance to various abiotic stresses (Hasanuzzaman et al., 2017; Reczek and Chandel, 2015; Saxena et al., 2016). It has been demonstrated that exogenous application of H_2O_2 alleviates Cd toxicity in rice plants (Chao et al., 2009; Hu et al., 2009), which may be due to a stimulated antioxidant system, and increases the glutathione level and Cd sequestration. In two rice genotypes (N07-6 and N07-63), H_2O_2 pre-treatment appeared to alleviate the adverse effects of Cd stress on seedling growth by enhancing the levels of GSH and phytochelatins (PCs), as well as GST activity (Bai et al., 2011). Recently, Hasanuzzaman et al. (2017) reported that hydrogen peroxide pre-treatment alleviates cadmium-induced oxidative stress in *Brassica napus* through improvement of the antioxidant defense system (the activities of APX, MDHAR, DHAR, GR, GST, GPX, and CAT) and glyoxalase system (the activities of Gly I, and Gly II). Xu et al. (2010) reported that the beneficial role of exogenous H_2O_2 to induce tolerance in H_2O_2-primed Al-stressed wheat seedlings was attributed to higher activities of GPX, CAT, POD, MDHAR, DHAR, and GR, and higher AsA and GSH contents than seedlings subjected to Al-stress only. In another study, Yildiz et al. (2013) reported that the application of H_2O_2 counteracted the adverse effects of chromium (Cr) toxicity (50 μM) in canola plants. They observed that the ameliorating effects of H_2O_2 on chromium (Cr) toxicity were mainly attributed to improved thiol content, antioxidant enzyme activities and the growth in chlorophyll content, as well as the levels of a metallothionein protein (BnMP1). After 7 days of exposure to Cr, H_2O_2-pre-treated canola seedlings exhibited a smaller decrease in *BnMP1* expression as compared with the seedlings subjected to Cr stress only. Moreover, exogenous H_2O_2 has been reported to enhance resistance to copper (Cu) stress in maize leaves (Guzel and Terzi, 2013) by enhancing growth, water content, mineral concentration (Na^+, K^+, Ca^+, Mg^{2+}), proline content, and total soluble sugar and soluble protein contents.

Current evidence demonstrates that NO contributes to determining the efficiency of acquisition and utilization of several macro and micronutrients. NO exerts a strong effect on nitrogen acquisition efficiency through the modulation of root architecture (Yu et al., 2014; Zhang, 1998). It has been reported that nitrate supply influences primary root growth in maize and *Arabidopsis* (Trevisan et al., 2014). The addition of NO-donor SNP induces the primary and lateral root tips and the formation of cluster roots of lupine under P-deficiency conditions (Wang et al., 2010c), as well as increasing AKT1 transcripts abundance in the halophytic plant *Kandelia obovata* K-deficiency conditions (Chen et al., 2013). Additionally, NO in the confluence of signaling mechanisms may also protect cells against zinc and iron starvation conditions through its interaction with hormones, glutathione, ferritin, frataxin, and Fe-compounds (Buet and Simontacchi, 2015) the modulation of root architecture (Buet et al., 2014; Pagnussat et al., 2002). In the last decade, numerous research results have indicated that exogenous application of NO alleviated heavy metal toxicity in plants (Khairy et al., 2016; Kováčik et al., 2015; Liu et al., 2015b). Xu et al. (2011) reported that NO levels were rapidly enhanced in *Solanum nigrum*, a Cd hyperaccumulator, under Cd stress, which resulted in increased intracellular antioxidative capacity and reduced oxidative damage. Exogenous NO also induces Cd tolerance by increasing the production of Pro and total GSH and decreasing oxidative damage in the roots of *Medicago truncatula* seedlings (Xu et al., 2009). Improved growth of Cd-stressed sunflower (*Helianthus annuus* L.) plants by exogenous NO application is mainly attributed to improved chlorophyll content, and the enhancement of CAT and APX activity as well as maintained AsA and GSH levels (Laspina et al., 2005). Wang et al. (2016) reported that the addition of 300 μM SNP mitigated the effect of Cd toxicity in ryegrass plants exposed to Cd stress by increasing chlorophyll synthesis and enhancing the antioxidant defense system. They also concluded that the mitigating effect of exogenous NO was attributed to reduced Cd uptake and restricted Cd transport from roots to shoots, and enhanced the absorption of nutrient elements. Application of exogenous NO also induced Cu toxicity tolerance by regulating ROS scavenging enzymes, modulating the activity of H^+-ATPase and H^+-PPase in the plasma membrane or tonoplast, and also significantly alleviated the growth inhibition caused by Cu toxicity in tomato plants. In addition, Dong et al. (2014b) found that low concentrations of NO (100 μM) helped increase the tolerance of ryegrass to Cu stress by enhancing chlorophyll content and photosynthesis, maintaining intracellular ion equilibrium under Cu stress and restricted Cu translocation from roots to shoots, increasing antioxidant enzyme activities and protecting against Cu-induced oxidative stress. It has been reported that exogenous NO protected plants against Al toxicity by alleviating Al-induced oxidative stress (Wang et al., 2010b; Wang and Yang, 2005) via changing cell wall polysaccharides and modulating hormonal equilibrium (He et al., 2012). Another recent study reported that the exogenous application of NO reduces arsenic (As) toxicity in rice plants through modulating regulatory networks involved in As detoxification and jasmonic acid biosynthesis (Singh et al., 2017). They observed that NO regulated metal transporters, especially NIP, NRAMP, ABC, and iron transporters, stress-related genes such as CytP450, GSTs, GRXs, TFs, amino acid, hormone(s), and signaling and secondary metabolism genes involved in As detoxification.

11.3.4 Chilling and Heat Stress Tolerance

It was reported that exogenous H_2O_2 can mediate the induction of protective mechanisms against low-temperature stress tolerance in maize (Prasad et al., 1994a, b), mustard (Dat et al., 1998), and mung bean (*Vigna radiata* L.) (Hung et al., 2007; Yu et al., 2003) seedlings. The authors observed that H_2O_2-mediated chilling tolerance in mung bean plants was mainly attributed to improved GSH content that is independent of ABA. Wang et al. (2010a) reported that foliar pre-treatment with 10 mM H_2O_2 was effective in modulating chilling stress tolerance of Mascarene grass (*Zoysia tenuifolia*) and manila grass (*Zoysia matrella*). This may be explained by higher protein contents and increased the activities of APX, GPX, and CAT in *Zoysia matrella* and APX, GR, and POD activities in *Zoysia tenuifolia*, resulting in lower MDA and EL levels and, consequently, better protection against oxidative stress. In tomato plants, it has been found that exogenous H_2O_2 application could mitigate cold stress by inducing APX activity, anthocyanin levels, and proline accumulation, and the maintenance of a higher RWC under stress (İşeri et al., 2013).

Several studies have also elucidated the major role of exogenous H_2O_2 in increasing plant tolerance to heat stress (Hossain et al., 2013b). Kang et al. (2009) found that exogenously applied H_2O_2 was generally effective in countering the inhibitory effects of heat stress by enhancing the activities of APX and glucose-6-phosphated dehydrogenase (G6PDH) in cucumber and tomato seedlings. Recently, Wang et al. (2014a) showed that H_2O_2 pre-treated seedlings of ryegrass (*Lolium perenne* cv. Accent) and tall fescue (*Festuca arundinacea* cv. Barlexas) exhibited lower oxidative damage and H_2O_2 levels, and increased activities of APX, GR, GST, and GPX, as compared to untreated plants when exposed to heat stress.

Some research reports indicated that NO is involved in responses to chilling and heat stresses. Fan et al. (2014) reported that exogenous nitric oxide increased chilling tolerance in Chinese cabbage seedlings by enhancing antioxidant enzymes, chlorophyll and protein content as well as by maintaining membrane permeability. Esim and Atici (2015) showed that the pre-treatment of wheat seedlings with 0.1 mM SNP counteracted chilling-induced oxidative stress by increasing the activity of SOD, CAT and POX in wheat seedlings treated with NO. Exogenous application of NO donors has also been able to ameliorate heat-induced cellular damage in plants (Hasanuzzaman et al., 2013; Parankusam et al., 2017; Song et al., 2006). Hasanuzzaman et al. (2012) reported that pre-treatment with 0.25 mM SNP alleviated heat-induced damage in *Triticum aestivum* seedlings by up-regulating the activities of APX, MDHAR, DHAR, GR, GST, CAT and Gly. I. Li et al. (2013) found that *Zea mays* seedlings pre-treated with NO donor (0.15 mM SNP) when exposed to heat stress (48°C for 18 h) exhibited a reduced electrolyte leakage and malondialdehyde (MDA) content.

11.3.5 UV Radiation Stress

There is evidence that UV radiation can exert detrimental effects on physiological, biochemical, and molecular characteristics of plants (Frohnmeyer and Staiger, 2003; Kataria et al., 2014). Chloroplasts are very sensitive to UV-B radiation. UV-B exposure of isolated soybean chloroplasts increases the generation of reactive oxygen species (ROS) as well as enhancing lipid peroxidation and the content of carbonyl groups in proteins as compared to control chloroplasts (Galatro et al., 2001). However, the exposure of isolated chloroplasts to GSNO, as NO donor, induced a reduction in the generation rate of chloroplastic lipid radicals, as well as in the content of carbonyl groups (Jasid et al., 2006). In bean (*Phaseolus vulgaris*) leaves, maximum efficiency of PSII photochemistry (F_v/F_m) and the quantum yield of PSII (ΦPSII) were reduced under UV. Treatment with SNP exhibited Chl loss, abated F_v/F_m decrease, and alleviated the increase in carbonyl contents in thylakoid membrane proteins after UV-B irradiation in bean by enhancement of SOD, APX, and CAT activities (Shi et al., 2005). NO can increase specific isoforms of antioxidant enzymes in soybean leaves subjected to enhanced UV-B radiation (Santa-Cruz et al., 2014). As chloroplasts seem to be related to NO generation (Galatro et al., 2013; Jasid et al., 2006; Tewari et al., 2013), and NO can contract the oxidative effects of UV-B radiation, this source of NO could be operative under UV-B radiation. Tossi et al. (2011) reported that UV-B radiation increases both ROS and

NO. Then, NO decreases ROS levels and up-regulates the expression of several genes involved in flavonoid and anthocyanin synthesis (as the maize transcription factor ZmP and MYB12, its *Arabidopsis* functional homolog; as well as their target genes *Chs*, and *Chi* – chalcone isomerase). Indeed, plants accumulate flavonoids and anthocyanins for light harvesting and photoprotection, which are also involved in the removal of ROS (Tossi et al., 2012). Thus, NO decreases UV-B impact by enhancing antioxidant enzyme activities, flavonoid and anthocyanin synthesis, and by reducing oxidative stress in plants.

Recently, Wu et al. (2016) reported that exogenous H_2O_2 was involved in the UV-B-induced biosynthesis of anthocyanins in the hypocotyls of radish sprouts. They observed that H_2O_2 increased anthocyanin concentration and UV RESISTANCE LOCUS8 (UVR8) transcription. Since H_2O_2-induced anthocyanin accumulation and *UVR8* expression were significantly inhibited by co-treatment with 2-phenyl-4,4,5,5-tetramethylimidazoline-3-oxide-1-oxyl (PTIO, a NO scavenger), there was crosstalk among hydrogen peroxide, nitric oxide and the UVR8 pathway in UV-B-induced anthocyanin accumulation.

11.4 THE INTERACTION BETWEEN H_2O_2 AND NO DURING ABIOTIC STRESS

Some research reports indicated that the interaction of H_2O_2 and NO triggers a serious of physiological and biological responses in plant cells (Niu and Liao, 2016). The interaction between H_2O_2 and NO has been revealed clearly in plants, and is involved in physiological responses to various abiotic stresses (Huang et al., 2015; Shi et al., 2015). This interaction plays an important role in plant tolerance to salt stress (Niu and Liao, 2016). Tanou et al. (2009, 2010) reported an interaction between H_2O_2 and NO during salt stress response in citrus. They suggested that H_2O_2 and NO pre-treatments could ameliorate the adverse effects of salinity on protein carbonylation. Tanou et al. (2009) reported that both H_2O_2 and SNP pre-treatments strongly reduced the detrimental phenotypical and physiological effects of salinity in citrus plants. They observed that the regulation of protein carbonylation and S-nitrosylation was the main mechanism by which both H_2O_2 and SNP pre-treatments before salinity stress alleviated salinity-induced oxidative stress. These results demonstrated an overlap between H_2O_2 and NO-signaling pathways in acclimation to high salinity.

Additionally, H_2O_2 and NO pre-treatments enhanced drought tolerance in marigold explants by protecting the cell membrane against lipid peroxidation through up-regulating the antioxidant defense system, protecting the mesophyll cells' ultrastructure, and improving the photosynthetic level of leaves (Liao et al., 2012b). Similarly, the interplay between H_2O_2 and NO signaling enhanced the activity of myo-inositol phosphate synthase in mitigating drought stress (Tan et al., 2013). Importantly, the crosstalk between H_2O_2 and NO in the guard cells of *Arabidopsis* leaves and their roles in UV-B-induced stomatal closure have been determined by He et al. (2013) and Tossi et al. (2014). Thus, the crosstalk between H_2O_2 and NO may regulate stomatal movement to mitigate UV-B stress damage in plant cells.

Recently, Guo et al. (2014) investigated the relationship between H_2O_2 and NO under cold stress. They reported that a signaling crosstalk between H_2O_2 and NO may be involved in inducing S-adenosyl methionine synthetase and an increase in cold tolerance through up-regulating polyamine oxidation in *Medicago sativa* subsp. Moreover, the interaction between H_2O_2 and NO may increase cold-induced gene expression of falcata myo-inositol phosphate synthase (*MfMIPS*), which improves tolerance to cold stress (Tan et al., 2013).

Previous studies have shown that the relationship between H_2O_2 and NO may be involved in inducing heat tolerance in maize (Li et al., 2015) and *Arabidopsis* seedlings (Wang et al., 2014a). Wang et al. (2014a) examined the connection between NO and H_2O_2 in heat shock signaling. They observed the involvement of NO in H_2O_2 signaling as a downstream factor via stimulating heat shock factors (HSF), DNA-binding activity and heat-shock proteins accumulation. Thus, these data showed that H_2O_2 acts upstream of NO in thermo-tolerance. Alberto et al. (2012) reported that under copper stress the signaling interaction between H_2O_2 and NO may alter antioxidant enzyme activities and relative gene expression to improve copper tolerance of *Ulva compressa* plants. It has been demonstrated that the interplay of NO and H_2O_2 alleviates zinc toxicity in wheat plants (Chao et al., 2009; Hu et al., 2009) which may be due to a decreased lipid peroxidation as well as up-regulated resistance gene expression (Duan et al., 2015).

11.5 CONCLUSION AND FUTURE PERSPECTIVES

It has long been shown that NO and H_2O_2 are both endogenous plant molecules which are also involved in plant abiotic stress tolerance. In addition, the use of exogenous NO and H_2O_2 has been found to be very much effective in increasing plant tolerance in various crop and non-crop species against individual abiotic stresses. However, the precise physiological function and mechanism of action of these compounds under combined abiotic stresses still remain unclear. Environmental stress can increase the level of endogenous NO and H_2O_2 in both primed and non-primed plants. It appears, therefore, that pre-treatment of plants with exogenous NO and H_2O_2 can initiate a mild stress cue, similar to an acclimation response that eventually leads to increased tolerance when the plant is subjected to an abiotic challenge.

Previously, several studies have demonstrated the relationship between NO and H_2O_2 in plant responses to stress. Some studies have shown that H_2O_2 acts upstream of NO in lateral root growth, nitrogen-fixing nodule formation, drought tolerance, and stomatal movement. However, other studies have indicated that NO influences H_2O_2 accumulation. Thus, the exact mechanism, which is upstream of the other, remains unknown.

Interestingly, the relationship between H_2O_2 and NO is dependent on plant species, treatments and/or experimental systems. More importantly, crosstalk between H_2O_2 and NO in a developmental context seems to differ from that in plant responses to stress. Since the relationship between NO and H_2O_2 under combined abiotic stresses is obscure, future work will need to focus on the molecular mechanism of the interplay among H_2O_2 and NO during multiple abiotic stresses.

REFERENCES

Abass, S.M., Mohamed, H.I. (2011) Alleviation of adverse effects of drought stress on common bean (*Phaseolus vulgaris* L.) by exogenous application of hydrogen peroxide. *Bangladesh Journal of Botany* 41: 75–83.

Alemayehu, A., Zelinová, V., Boèová, B., Huttová, J., Mistrík, I., Tamás, L. (2015) Enhanced nitric oxide generation in root transition zone during the early stage of cadmium stress is required for maintaining root growth in barley. *Plant and Soil* 390(1–2): 213–222.

Ali, Q., Daud, M.K., Haider, M.Z., Ali, S., Rizwan, M., Aslam, N., Noman, A., *et al.* (2017) Seed priming by sodium nitroprusside improves salt tolerance in wheat (*Triticum aestivum* L.) by enhancing physiological and biochemical parameters. *Plant Physiology and Biochemistry: PPB* 119: 50–58.

Ali, Q., Javed, M.T., Noman, A., Haider, M.Z., Waseem, M., Iqbal, N., Waseem, M., *et al.* (2018) Assessment of drought tolerance in mung bean cultivars/lines as depicted by the activities of germination enzymes, seedling's antioxidative potential and nutrient acquisition. *Archives of Agronomy and Soil Science* 64(1): 84–102.

Arasimowicz-Jelonek, M., Floryszak-Wieczorek, J., Kubiś, J. (2009) Involvement of nitric oxide in water stress-induced responses of cucumber roots. *Plant Science* 177(6): 682–690.

Ashfaque, F., Khan, M.I.R., Khan, N.A. (2014) Exogenously applied H_2O_2 promotesproline accumulation, water relations, photosynthetic efficiency and growth of wheat (*Triticum aestivum* L.) under salt stress. *Annual Research and Review in Biology* 4(1): 105–120.

Ashraf, M. (2009) Biotechnological approach of improving plant salt tolerance using antioxidants as markers. *Biotechnology Advances* 27(1): 84–93.

Ashraf, M.A., Rasheed, R., Hussain, I., Iqbal, M., Haider, M.Z., Parveen, S. (2014) Hydrogen peroxide modulates antioxidant system and nutrient relation in maize (*Zea mays* L.) under water-deficit conditions. *Archives of Agronomy and Soil Science* 61: 507–523.

Bai, X.J., Liu, L.J., Zhang, C.H., Ge, Y., Cheng, W.D. (2011) Effect of H_2O_2 pretreatment on Cd tolerance of different rice cultivars. *Rice Science* 18(1): 29–35.

Bellin, D., Asai, S., Delledonne, M., Yoshioka, H. (2013) Nitric oxide as a mediator for defense responses. *Molecular Plant-Microbe Interactions Journal* 26: 271–277.

Bienert, G.P., Møller, A.L., Kristiansen, K.A., Schulz, A., Møller, I.M., Schjoerring, J.K., Jahn, T.P. (2007) Specific aquaporins facilitate the diffusion of hydrogen peroxide across membranes. *The Journal of Biological Chemistry* 282(2): 1183–1192.

Boyarshinov, A.V., Asafova, E.V. (2011) Stress responses of wheat leaves to dehydration: Participation of endogenous NO and effect of sodium nitroprusside. *Russian Journal of Plant Physiology* 58(6): 1034–1039.

Buet, A., Moriconi, J.I., Santa-María, G.E., Simontacchi, M. (2014) An exogenous source of nitric oxide modulates zinc nutritional status in wheat plants. *Plant Physiology and Biochemistry: PPB* 83: 337–345.

Buet, A., Simontacchi, M. (2015) Nitric oxide and plant iron homeostasis. *Annals of the New York Academy of Sciences* 1340: 39–46.

Chan, K.X., Wirtz, M., Phua, S.Y., Estavillo, G.M., Pogson, B.J. (2013) Balancing metabolites in drought: The sulfur assimilation conundrum. *Trends in Plant Science* 18(1): 18–29.

Chao, Y.-Y., Hsu, Y.T., Kao, C.H. (2009) Involvement of glutathione in heat shock- and hydrogen peroxide-induced cadmium tolerance of rice (*Oryza sativa* L.) seedlings. *Plant and Soil* 318(1–2): 37–45.

Chen, J., Xiao, Q., Wang, C., Wang, W.H., Wu, F.-H., Chen, J., He, B., *et al.* (2014) Nitric oxide alleviates oxidative stress caused by salt in leaves of a mangrove species, *Aegiceras corniculatum*. *Aquatic Botany* 117: 41–47.

Chen, J., Xiong, D.Y., Wang, W.H., Hu, W.J., Simon, M., Xiao, Q., Chen, J., *et al.* (2013) Nitric oxide mediates root K^+/Na^+ balance in a mangrove plant, Kandelia obovata, by enhancing the expression of AKT1-type K^+ channel and Na^+/H^+ antiporter under high salinity. *PLOS ONE* 8(8): e71543.

Christou, A., Manganaris, G.A., Fotopoulos, V. (2014) Systemic mitigation of salt stress by hydrogen peroxide and sodium nitroprusside in strawberry plants via transcriptional regulation of enzymatic and non-enzymatic antioxidants. *Environmental and Experimental Botany* 107: 46–54.

Christou, A., Manganaris, G.A., Papadopoulos, I., Fotopoulos, V. (2013) Hydrogen sulfide induces systemic tolerance to salinity and non-ionic osmotic stress in strawberry plants through modification of reactive species biosynthesis and transcriptional regulation of multiple defence pathways. *Journal of Experimental Botany* 64(7): 1953–1966.

Chun, H.J., Park, H.C., Koo, S.C., Lee, J.H., Park, C.Y., Choi, M.S., Kang, C.H., *et al.* (2012) Constitutive expression of mammalian nitric oxide synthase in tobacco plants triggers disease resistance to pathogens. *Molecules and Cells* 34(5): 463–471.

Clark, G., Wu, M., Wat, N., Onyirimba, J., Pham, T., Herz, N., Ogoti, J., *et al.* (2010) Both the stimulation and inhibition of root hair growth induced by extracellular nucleotides in Arabidopsis are mediated by nitric oxide and reactive oxygen species. *Plant Molecular Biology* 74(4–5): 423–435.

Considine, M.J., Sandalio, L.M., Foyer, C.H. (2015) Unravelling how plants benefit from ROS and NO reactions, while resisting oxidative stress. *Annals of Botany* 116(4): 469–473.

Dat, J.F., Foyer, C.H., Scott, I.M. (1998) Changes in salicylic acid and antioxidants during induced thermotolerance in mustard seedlings. *Plant Physiology* 118(4): 1455–1461.

David, A., Yadav, S., Bhatla, S.C. (2010) Sodium chloride stress induces nitric oxide accumulation in root tips and oil body surface accompanying slower oleosin degradation in sunflower seedlings. *Physiologia Plantarum* 140(4): 342–354.

de Carvalho, M.H.C. (2013) Drought stress and reactive oxygen species. Production, scavenging and signalling. *Plant Signaling and Behavior* 3: 156–165.

Demidchik, V. (2015) Mechanisms of oxidative stress in plants: From classical chemistry to cell biology. *Environmental and Experimental Botany* 109: 212–228.

Dietz, K.J. (2015) Efficient high light acclimation involves rapid processes at multiple mechanistic levels. *Journal of Experimental Botany* 66(9): 2401–2414.

Dong, Y.J., Jinc, S.S., Liu, S., Xu, L.L., Kong, J. (2014a) Effects of exogenous nitric oxide on growth of cotton seedlings under NaCl stress. *Journal of Soil Science and Plant Nutrition* 14: 1–13.

Dong, Y., Xu, L., Wang, Q., Fan, Z., Kong, J., Bai, X. (2014b) Effects of exogenous nitric oxide on photosynthesis, antioxidative ability, and mineral element contents of perennial ryegrass under copper stress. *Journal of Plant Interactions* 9(1): 402–411.

Duan, X., Li, X., Ding, F., Zhao, J., Guo, A., Zhang, L., Yao, J., Yang, Y. (2015) Interaction of nitric oxide and reactive oxygen species and associated regulation of root growth in wheat seedlings under zinc stress. *Ecotoxicology and Environmental Safety* 113: 95–102.

Egbichi, I., Keyster, M., Ludidi, N. (2014) Effect of exogenous application of nitric oxide on salt stress responses of soybean. *South African Journal of Botany* 90: 131–136.

Ellouzi, H., Sghayar, S., Abdelly, C. (2017) H_2O_2 seed priming improves tolerance to salinity; drought and their combined effect more than mannitol in Cakile maritima when compared to *Eutrema salsugineum*. *Journal of Plant Physiology* 210: 38–50.

Esim, N., Atici, Ö. (2015) Effects of exogenous nitric oxide and salicylic acid on chilling-induced oxidative stress in wheat (*Triticum aestivum*). *Frontiers in Life Science* 8(2): 124–130.

Esringu, A., Aksakal, O., Tabay, D., Kara, A.A. (2015) Effects of sodium nitroprusside (SNP) pretreatment on UV-B stress tolerance in lettuce (*Lactuca sativa* L.) seedlings. *Environmental Science and Pollution Research International* 23(1): 589–597.

Fan, J., Chen, K., Amombo, E., Hu, Z., Chen, L., Fu, J. (2015) Physiological and molecular mechanism of nitric oxide (NO) involved in Bermudagrass response to cold stress. *PLOS ONE* 10(7): e0132991.

Fan, H.F., Du, C.X., Guo, S.R. (2013) Nitric oxide enhances salt tolerance in cucumber seedlings by regulating free polyamine content. *Environmental and Experimental Botany* 86: 52–59.

Fan, H., Du, C., Xu, Y., Wu, X. (2014) Exogenous nitric oxide improves chilling tolerance of Chinese cabbage seedlings by affecting antioxidant enzymes in leaves. *Horticulture, Environment, and Biotechnology* 55(3): 159–165.

Farooq, M., Basra, S.M.A., Wahid, A., Rehman, H. (2009) Exogenously applied nitric oxide enhances the drought tolerance in fine grain aromatic rice (*Oryza sativa* L.). *Journal of Agronomy and Crop Science* 195(4): 254–261.

Fedina, I.S., Nedeva, D., Çiçek, N. (2009) Pre-treatment with H_2O_2 induces salt tolerance in barley seedlings. *Biologia Plantarum* 53(2): 321–324.

Foresi, N., Mayta, M.L., Lodeyro, A.F., Scuffi, D., Correa-Aragunde, N., García-Mata, C., Casalongué, C., Carrillo, N., Lamattina, L. (2015) Expression of the tetrahydrofolate-dependent nitric oxide synthase from the green alga Ostreococcus tauri increases tolerance to abiotic stresses and influences stomatal development in Arabidopsis. *The Plant Journal: For Cell and Molecular Biology* 82(5): 806–821.

Foyer, C.H., Noctor, G. (2013) Redox signaling in plants. *Antioxidants and Redox Signaling* 18(16): 2087–2090.

Freschi, L., Rodrigues, M.A., Domingues, D.S., Purgatto, E., Van Sluys, M.A., Magalhaes, J.R., Kaiser, W.M., Mercier, H. (2010) Nitric oxide mediates the hormonal control of Crassulacean acid metabolism expression in young pineapple plants. *Plant Physiology* 152(4): 1971–1985.

Frohnmeyer, H., Staiger, D. (2003) Ultraviolet-B radiation-mediated responses in plants. Balancing damage and protection. *Plant Physiology* 133(4): 1420–1428.

Galatro, A., Puntarulo, S., Guiamet, J.J., Simontacchi, M. (2013) Chloroplast functionality has a positive effect on nitric oxide level in soybean cotyledons. *Plant Physiology and Biochemistry: PPB* 66: 26–33.

Galatro, A., Simontacchi, M., Puntarulo, S. (2001) Free radical generation and antioxidant content in chloroplasts from soybean leaves exposed to ultraviolet-B. *Physiologia Plantarum* 113(4): 564–570.

García-Mata, C., Gay, R., Sokolovski, S., Hills, A., Lamattina, L., Blatt, M.R. (2003) Nitric oxide regulates K^+ and Cl^- channels in guard cells through a subset of abscisic acid-evoked signaling pathways. *Proceedings of the National Academy of Sciences of the United States of America* 100(19): 11116–11121.

Ge, X.M., Cai, H.L., Lei, X., Zou, X., Yue, M., He, J.M. (2015) Heterotrimeric G protein mediates ethylene-induced stomatal closure via hydrogen peroxide synthesis in Arabidopsis. *The Plant Journal: For Cell and Molecular Biology* 82(1): 138–150.

Gill, S.S., Tuteja, N. (2010) Reactive oxygen species and antioxidant machinery in abiotic stress tolerance in crop plants. *Plant Physiology and Biochemistry: PPB* 48(12): 909–930.

Gondim, F.A., Gomes-Filho, E., Costa, J.H., Alencar, N.L.M., Priso, J.T. (2012) Catalase plays a key role in salt stress acclimation induced by hydrogen peroxide pretreatment in maize. *Journal of Plant Physiology and Biochemistry* 56: 62–71.

Gondim, F.A., Gomes-Filho, E., Lacerda, C.F., Prisco, J.T., Neto, A.D.A., Marques, E.C. (2010) Pretreatment with H_2O_2 in maize seeds: Effects on germination and seedling acclimation to salt stress. *Brazilian Journal of Plant Physiology* 22(2): 103–112.

Gondim, F.A., Miranda, R.S., Gomes-Filho, E., Prisco, J.T. (2013) Enhanced salt tolerance in maize plants induced by H_2O_2 leaf spraying is associated with improved gas exchange rather than with non-enzymatic antioxidant system. *Theoretical and Experimental Plant Physiology* 25(4): 251–260.

González, A., Cabrera, Mde L., Henríquez, M.J., Contreras, R.A., Morales, B., Moenne, A. (2012) Cross talk among calcium, hydrogen peroxide, and nitric oxide and activation of gene expression involving calmodulins and calcium-dependent protein kinases in Ulva compressa exposed to copper excess. *Plant Physiology* 158(3): 1451–1462.

Guerra, D.D., Callis, J. (2012) Ubiquitin on the move: The ubiquitin modification system plays diverse roles in the regulation of endoplasmic reticulum- and plasma membrane-localized proteins. *Plant Physiology* 160(1): 56–64.

Guo, Z., Tan, J., Zhuo, C., Wang, C., Xiang, B., Wang, Z. (2014) Abscisic acid, H_2O_2 and nitric oxide interactions mediated cold-induced S-adenosylmethionine synthetase in Medicago sativa subsp. falcata that confers cold tolerance through up-regulating polyamine oxidation. *Plant Biotechnology Journal* 12(5): 601–612.

Guzel, S., Terzi, R. (2013) Exogenous hydrogen peroxide increases dry matter production, mineral content and level of osmotic solutes in young maize leaves and alleviates deleterious effects of copper stress. *Botanical Studies* 54(1): 26.

Habibi, G. (2014) Hydrogen peroxide (H_2O_2) generation, scavenging and signaling in plants. In *Oxidative Damage to Plants*, ed. Ahmad, P., 557–574. Amsterdam: Elsevier.

Hameed, A., Hussain, T., Gulzar, S., Aziz, I., Gul, B., Khan, M.A. (2012) Salt tolerance of a cash crop halophyte Suaeda fruticosa: Biochemical responses to salt and exogenous chemical treatments. *Acta Physiologiae Plantarum* 34(6): 2331–2340.

Hamilton, D.W., Hills, A., Köhler, B., Blatt, M.R. (2000) Ca^{2+} channels at the plasma membrane of stomatal guard cells are activated by hyperpolarization and abscisic acid. *Proceedings of the National Academy of Sciences of the United States of America* 97(9): 4967–4972.

Hasanuzzaman, M., Hossain, M.A., Fujita, M. (2011) Nitric oxide modulates antioxidant defense and the methylglyoxal detoxification system and reduces salinity-induced damage of wheat seedlings. *Plant Biotechnology Reports* 5(4): 353–365.

Hasanuzzaman, M., Nahar, K., Alam, M.M., Fujita, M. (2012) Exogenous nitric oxide alleviates high temperature induced oxidative stress in wheat (*Triticum aestivum* L.) seedlings by modulating the antioxidant defense and glyoxalase system. *Australian Journal of Agricultural Research* 6: 1314–1323.

Hasanuzzaman, M., Nahar, K., Alam, M.M., Roychowdhury, R., Fujita, M. (2013) Physiological, biochemical, and molecular mechanisms of heat stress tolerance in plants. *International Journal of Molecular Sciences* 14(5): 9643–9684.

Hasanuzzaman, M., Nahar, K., Gill, S.S., Alharby, H.F., Razafindrabe, B.H., Fujita, M. (2017) Hydrogen peroxide pretreatment mitigates cadmium-induced oxidative stress in *Brassica napus* L.: An intrinsic study on antioxidant defense and glyoxalase systems. *Frontiers in Plant Science* 8: 115.

Hayat, S., Yadav, S., Wani, A.S., Irfan, M., Alyemini, M.N., Ahmad, A. (2012) Impact of sodium nitroprusside on nitrate reductase, proline content, and antioxidant system in tomato under salinity stress. *Horticulture, Environment, and Biotechnology* 53(5): 362–367.

He, H.Y., He, L.F., Gu, M.H., Li, X.F. (2012) Nitric oxide improves aluminum tolerance by regulating hormonal equilibrium in the root apices of rye and wheat. *Plant Science: An International Journal of Experimental Plant Biology* 183: 123–130.

He, J.M., Ma, X.G., Zhang, Y., Sun, T.F., Xu, F.F., Chen, Y.P., Liu, X., Yue, M. (2013) Role and interrelationship of Gα protein, hydrogen peroxide, and nitric oxide in ultraviolet B-induced stomatal closure in *Arabidopsis* leaves. *Plant Physiology* 161(3): 1570–1583.

Hebelstrup, K.H., Shah, J.K., Igamberdiev, A.U. (2013) The role of nitric oxide and hemoglobin in plant development and morphogenesis. *Physiologia Plantarum* 148(4): 457–469.

Hernández-Barrera, A., Velarde-Buendía, A., Zepeda, I., Sanchez, F., Quinto, C., Sánchez-Lopez, R., Cheung, A.Y., Wu, H.M., Cardenas, L. (2015) Hyper, a hydrogen peroxide sensor, indicates the sensitivity of the Arabidopsis root elongation zone to aluminum treatment. *Sensors* 15(1): 855–867.

Hichri, I., Boscari, A., Castella, C., Rovere, M., Puppo, A., Brouquisse, R. (2015) Nitric oxide: A multifaceted regulator of the nitrogen-fixing symbiosis. *Journal of Experimental Botany* 66(10): 2877–2887.

Hossain, M.A., Bhattacharjee, S., Armin, S.M., Qian, P., Xin, W., Li, H.Y., Burritt, D.J., Fujita, M., Tran, L.S.P. (2015) Hydrogen peroxide priming modulates abiotic oxidative stress tolerance: Insights from ROS detoxification and scavenging. *Frontiers in Plant Science* 6: 420.

Hossain, M.A., Fujita, M. (2013) Hydrogen peroxide priming stimulates drought tolerance in mustard (*Brassica juncea* L.). *Plant Gene and Trait* 4: 109–123.

Hossain, M.A., Mostofa, M.G., Fujita, M. (2013a) Cross protection by cold-shock to salinity and drought stress-induced oxidative stress in mustard (*Brassica campestris* L.) seedlings. *Molecular Plant Breeding* 4: 50–70.

Hossain, M.A., Mostofa, M.G., Fujita, M. (2013b) Heat-shock positively modulates oxidative protection of salt and drought-stressed mustard (*Brassica campestris* L.) seedlings. *Journal of Plant Science and Molecular Breeding* 2: 1–14.

Hu, Y., Ge, Y., Zhang, C., Ju, T., Cheng, W. (2009) Cadmium toxicity and translocation in rice seedlings are reduced by hydrogen peroxide pretreatment. *Plant Growth Regulation* 59(1): 51–61.

Huang, A.X., Wang, Y.S., She, X.P., Mu, J., Zhao, J.L. (2015) Copper amine oxidase-catalysed hydrogen peroxide involves production of nitric oxide in darkness-induced stomatal closure in broad bean. *Functional Plant Biology* 42(11): 1057–1067.

Hung, S.H., Wang, C.C., Ivanov, S.V., Alexieva, V., Yu, C.W. (2007) Repetition of hydrogen peroxide treatment induces a chilling tolerance comparable to cold acclimation in mung bean. *Journal of the American Society for Horticultural Science* 132: 770–776.

İşeri, Ö.D., Körpe, D.A., Sahin, F.I., Haberal, M. (2013) Hydrogen peroxide pretreatment of roots enhanced oxidative stress response of tomato under cold stress. *Acta Physiologiae Plantarum* 35(6): 1905–1913.

Ishibashi, Y., Yamaguchi, H., Yuasa, T., Inwaya-Inoue, M., Arima, S., Zheng, S.H. (2011) Hydrogen peroxide spraying alleviates drought stress in soybean plants. *Journal of Plant Physiology* 168(13): 1562–1567.

Jasid, S., Simontacchi, M., Bartoli, C.G., Puntarulo, S. (2006) Chloroplasts as a nitric oxide cellular source, effect of reactive nitrogen species on chloroplastic lipids and proteins. *Plant Physiology* 142(3): 1246–1255.

Jing, L.Z., Kui, G.Y., Hang, L.S., Gang, B.J. (2009) Effects of exogenous hydrogen peroxide on ultrastructure of chloroplasts and activities of antioxidant enzymes in greenhouse-ecotype cucumber under drought stress. *Acta Horticulturae Sinica* 36: 1140–1146.

Kang, N.J., Kang, Y.I., Kang, K.H., Jeong, B.R. (2009) Induction of thermotolerance and activation of antioxidant enzymes in H_2O_2 pre-applied leaves of cucumber and tomato seedlings. *Journal of the Japanese Society for Horticultural Science* 78(3): 320–329.

Kataria, S., Jajoo, A., Guruprasad, K.N. (2014) Impact of increasing ultraviolet-B (UV-B) radiation on photosynthetic processes. *Journal of Photochemistry and Photobiology. B, Biology* 137: 55–66.

Kaur, G., Singh, H.P., Batish, D.R., Mahajan, P., Kohli, R.K., Rishi, V. (2015) Exogenous nitric oxide (NO) interferes with lead (Pb)-induced toxicity by detoxifying reactive oxygen species in hydroponically grown wheat (*Triticum aestivum*) roots. *PLOS ONE* 10(9): e0138713.

Khairy, A.I.H., Oh, M.J., Lee, S.M., Kim, D.S., Roh, K.S. (2016) Nitric oxide overcomes Cd and Cu toxicity in in vitro-grown tobacco plants through increasing contents and activities of RuBisCO and RuBisCO Activase. *Biochimie Open* 2: 41–51.

Khan, A., Anwar, Y., Hasan, M.M., Iqbal, A., Ali, M., Alharby, H.F., Hakeem, K.R., Hasanuzzaman, M. (2017) Attenuation of drought stress in Brassica seedlings with exogenous application of Ca^{2+} and H_2O_2. *Plants* 6(4): 20.

Khan, M.N., Siddiqui, M.H., Mohammad, F., Naeem, M. (2012) Interactive role of nitric oxide and calcium chloride in enhancing tolerance to salt stress. *Nitric Oxide: Biology and Chemistry* 27(4): 210–218.

Kilic, S., Kahraman, A. (2016) The mitigation effects of exogenous hydrogen peroxide when alleviating seed germination and seedling growth inhibition on salinity-induced stress in barley. *Polish Journal of Environmental Studies* 25(3): 1053–1059.

Kováčik, J., Klejdus, B., Babula, P., Hedbavny, J. (2015) Nitric oxide donor modulates cadmium-induced physiological and metabolic changes in the green alga *Coccomyxa subellipsoidea*. *Algal Research* 8: 45–52.

Laspina, N.V., Groppa, M.D., Tomaro, M.L., Benavides, M.P. (2005) Nitric oxide protects sunflower leaves against Cd-induced oxidative stress. *Plant Science* 169(2): 323–330.

Lemtiri-Chlieh, F., MacRobbie, E.A., Webb, A.A., Manison, N.F., Brownlee, C., Skepper, J.N., Chen, J., Prestwich, G.D., Brearley, C.A. (2003) Inositol hexakisphosphate mobilizes an endomembrane store of calcium in guard cells. *Proceedings of the National Academy of Sciences of the United States of America* 100(17): 10091–10095.

León, J., Costa, Á., Castillo, M.C. (2016) Nitric oxide triggers a transient metabolic reprogramming in *Arabidopsis*. *Scientific Reports* 6: 37945.

Li, X., Baio, G., Yun, W. (2013) Heat stress mitigation by exogenous nitric oxide application involves polyamine metabolism and PSII physiological strategies in ginger leaves. *Agricultural Sciences in China* 47: 1171–1179.

Li, Z.G., Luo, L.J., Sun, Y.F. (2015) Signal crosstalk between nitric oxide and hydrogen sulfide may be involved in hydrogen peroxide-induced thermotolerance in maize seedlings. *Russian Journal of Plant Physiology* 62(4): 507-–514.

Li, J.T., Qiu, Z.B., Zhang, X.W., Wang, L.S. (2011) Exogenous hydrogen peroxide can enhance tolerance of wheat seedlings to salt stress. *Acta Physiologiae Plantarum* 33(3): 835–842.

Li, Z.G., Yang, S.Z., Long, W.B., Yang, G.X., Shen, Z.Z. (2013b) Hydrogen sulphide may be a novel downstream signal molecule in nitric oxide-induced heat tolerance of maize (*Zea mays* L.) seedlings. *Plant, Cell & Environment* 36(8): 1564–1572.

Liao, W.B., Huang, G.B., Yu, J.H., Zhang, M.L. (2012a) Nitric oxide and hydrogen peroxide alleviate drought stress in marigold explants and promote its adventitious root development. *Plant Physiology and Biochemistry: PPB* 58: 6–15.

Liao, W.B., Zhang, M.L., Huang, G.B., Yu, J.H. (2012b) Ca^{2+} and CaM are involved in NO-and H_2O_2-induced adventitious root development in marigold. *Journal of Plant Growth Regulation* 31(2): 253–264.

Liu, F., Guo, F.Q. (2013) Nitric oxide deficiency accelerates chlorophyll breakdown and stability loss of thylakoid membranes during dark-induced leaf senescence in Arabidopsis. *PLoS One* 8(2): e56345.

Liu, Z.J., Guo, Y.K., Bai, J.G. (2010) Exogenous hydrogen peroxide changes antioxidant enzyme activity and protects ultrastructure in leaves of two cucumber ecotypes under osmotic stress. *Journal of Plant Growth Regulation* 29(2): 171–183.

Liu, W., Li, R.J., Han, T.T., Cai, W., Fu, Z.W., Lu, Y.T. (2015a) Salt stress reduces root meristem size by nitric oxide-mediated modulation of auxin accumulation and signaling in Arabidopsis. *Plant Physiology* 168(1): 343–356.

Liu, S.L., Yang, R.J., Pan, Y.Z., Ma, M.D., Pan, J., Zhao, Y., Cheng, Q.S., *et al.* (2015b) Nitric oxide contributes to minerals absorption, proton pumps and hormone equilibrium under cadmium excess in *Trifolium repens* L. plants. *Ecotoxicology and Environmental Safety* 119: 35–46.

Liu, Y., Wu, R., Wan, Q., Xie, G., Bi, Y. (2007) Glucose-6-phosphate dehydrogenase plays a pivotal role in nitric oxide-involved defense against oxidative stress under salt stress in red kidney bean roots. *Plant and Cell Physiology* 48(3): 511–522.

Lozano-Juste, J., León, J. (2011) Nitric oxide regulates DELLA content and PIF expression to promote photomorphogenesis in Arabidopsis. *Plant Physiology* 156(3): 1410–1423.

Manai, J., Gouia, H., Corpas, F.J. (2014) Redox and nitric oxide homeostasis are affected in tomato (Solanum Lycopersicum) roots under salinity-induced oxidative stress. *Journal of Plant Physiology* 171(12): 1028–1035.

Marten, H., Hyun, T., Gomi, K., Seo, S., Hedrich, R., Roelfsema, M.R.G. (2008) Silencing of NtMPK4 impairs CO_2-induced stomatal closure, activation of anion channels and cytosolic Ca^{2+} signals in Nicotiana tabacum guard cells. *The Plant Journal: For Cell and Molecular Biology* 55(4): 698–708.

Mignolet-Spruyt, L., Xu, E., Idanheimo, N., Hoeberichts, F.A., Muhlenbock, P., Brosche, M., Van Breusegem, F., Kangasjarvi, J. (2016) Spreading the news: Subcellular and organellar reactive oxygen species production and signalling. *Journal of Experimental Botany* 67(13): 3831–3844.

Mittler, R., Blumwald, E. (2010) Genetic engineering for modern agriculture: Challenges and perspectives. *Annual Review of Plant Biology* 61: 443–462.

Mittler, R., Vanderauwera, S., Gollery, M., Van Breusegem, F. (2004) Reactive oxygen gene network of plants. *Trends in Plant Science* 9(10): 490–498.

Mittler, R., Vanderauwera, S., Suzuki, N., Miller, G., Tognetti, V.B., Vandepoele, K., Gollery, M., Shulaev, V., Van Breusegem, F. (2011) ROS signaling: The new wave? *Trends in Plant Science* 16(6): 300–309.

Monreal, J.A., Arias-Baldrich, C., Tossi, V., Feria, A.B., Rubio-Casal, A., García-Mata, C., Lamattina, L., García-Mauriño, S. (2013) Nitric oxide regulation of leaf phosphoenolpyruvate carboxylase-kinase activity: Implication in sorghum responses to salinity. *Planta* 238(5): 859–869.

Mostofa, M.G., Fujita, M. (2013) Salicylic acid alleviates copper toxicity in rice (*Oryza sativa* L.) seedlings by up-regulating antioxidative and glyoxalase systems. *Ecotoxicology* 22(6): 959–973.

Mostofa, M.G., Seraj, Z.I., Fujita, M. (2014) Exogenous sodium nitroprusside and glutathione alleviate copper toxicity by reducing copper uptake and oxidative damage in rice (*Oryza sativa* L.) seedlings. *Protoplasma* 251(6): 1373–1386.

Mubarakshina, M.M., Ivanov, B.N., Naydov, I.A., Hillier, W., Badger, M.R., Krieger-Liszkay, A. (2010) Production and diffusion of chloroplastic H_2O_2 and its implication to signalling. *Journal of Experimental Botany* 61(13): 3577–3587.

Neill, S., Barros, R., Bright, J., Desikan, R., Hancock, J., Harrison, J., Morris, P., Ribeiro, D., Wilson, I. (2008) Nitric oxide, stomatal closure, and abiotic stress. *Journal of Experimental Botany* 59(2): 165–176.

Niu, L., Liao, W. (2016) Hydrogen peroxide signaling in plant development and abiotic responses: Crosstalk with nitric oxide and calcium. *Frontiers in Plant Science* 7: 230.

Pagnussat, G.C., Simontacchi, M., Puntarulo, S., Lamattina, L. (2002) Nitric oxide is required for root organogenesis. *Plant Physiology* 129(3): 954–956.

Parankusam, S., Adimulam, S.S., Bhatnagar-Mathur, P., Sharma, K.K. (2017) Nitric oxide (NO) in plant heat stress tolerance: Current knowledge and perspectives. *Frontiers in Plant Science* 8: 1582.

Prasad, T.K., Anderson, M.D., Martin, B.A., Stewart, C.R. (1994a) Evidence for chilling-induced oxidative stress in maize seedlings and a regulatory role for hydrogen-peroxide. *The Plant Cell Online* 6(1): 65–74.

Prasad, T.K., Anderson, M.D., Stewart, C.R. (1994b) Acclimation, hydrogen-peroxide, and abscisic-acid protect mitochondria against irreversible chilling injury in maize seedlings. *Plant Physiology* 105(2): 619–627.

Reczek, C.R., Chandel, N.S. (2015) ROS-dependent signal transduction. *Current Opinion in Cell Biology* 33: 8–13.

Santa-Cruz, D.M., Pacienza, N.A., Zilli, C.G., Tomaro, M.L., Balestrasse, K.B., Yannarelli, G.G. (2014) Nitric oxide induces specific isoforms of antioxidant enzymes in soybean leaves subjected to enhanced ultraviolet-B radiation. *Journal of Photochemistry and Photobiology. B, Biology* 141: 202–209

Sathiyaraj, G., Srinivasan, S., Kim, Y.J., Lee, O.R., Parvin, S., Balusamy, S.R., Khorolragchaa, A., Yang, D.C. (2014) Acclimation of hydrogen peroxide enhances salt tolerance by activating defense-related proteins in Panax ginseng CA. Meyer. *Molecular Biology Reports* 41(6): 3761–3771.

Savvides, A., Ali, S., Tester, M., Fotopoulos, V. (2016) Chemical priming of plants against multiple abiotic stresses: Mission possible? *Trends in Plant Science* 21(4): 329–340.

Saxena, I., Srikanth, S., Chen, Z. (2016) Cross talk between H_2O_2 and interacting signal molecules under plant stress response. *Frontiers in Plant Science* 7: 570.

Scuffi, D., Núñez, Á., Laspina, N., Gotor, C., Lamattina, L., García-Mata, C. (2014) Hydrogen sulfide generated by L-cysteine desulfhydrase acts upstream of nitric oxide to modulate ABA-dependent stomatal closure. *Plant Physiology* 166: 2065–2076.

Shan, C., Zhou, Y., Liu, M. (2015) Nitric oxide participates in the regulation of the ascorbate-glutathione cycle by exogenous jasmonic acid in the leaves of wheat seedlings under drought stress. *Protoplasma* 252(5): 1397–1405.

Shi, J., Fu, X.Z., Peng, T., Huang, X.S., Fan, Q.J., Liu, J.H. (2010) Spermine pretreatment confers dehydration tolerance of citrus in vitro plants via modulation of antioxidative capacity and stomatal response. *Tree Physiology* 30(7): 914–922.

Shi, H., Jiang, C., Ye, T., Tan, D.X., Reiter, R.J., Zhang, H., Chan, Z. (2014) Comparative physiological, metabolomic, and transcriptomic analyses reveal mechanisms of improved abiotic stress resistance in Bermudagrass [Cynodon dactylon (L). Pers.] by exogenous melatonin. *Journal of Experimental Botany* 66(3): 681–694.

Shi, H.T., Li, R.J., Cai, W., Liu, W., Wang, C.L., Lu, Y.T. (2012) Increasing nitric oxide content in Arabidopsis thaliana by expressing rat neuronal nitric oxide synthase resulted in enhanced stress tolerance. *Plant and Cell Physiology* 53(2): 344–357.

Shi, K., Li, X., Zhang, H., Zhang, G., Liu, Y., Zhou, Y., Xia, X., Chen, Z., Yu, J. (2015) Guard cell hydrogen peroxide and nitric oxide mediate elevated CO_2-induced stomatal movement in tomato. *The New Phytologist* 208(2): 342–353.

Shi, H., Wang, X., Ye, T., Cheng, F., Deng, J., Yang, P., Zhang, Y., Chan, Z. (2014) The cysteine2/histidine2-type transcription factor ZINC FINGER OF ARABIDOPSIS THALIANA 6 modulates biotic and abiotic stress responses by activating salicylic acid-related genes and C-REPEAT-BINDING FACTOR genes in Arabidopsis. *Plant Physiology* 165(3): 1367–1379.

Shi, S., Wang, G., Wang, Y., Zhang, L., Zhang, L. (2005) Protective effect of nitric oxide against oxidative stress under ultraviolet-B radiation. *Nitric Oxide: Biology and Chemistry* 13(1): 1–9.

Siddiqui, M.H., Al-Whaibi, M.H., Basalah, M.O. (2011) Role of nitric oxide in tolerance of plants to abiotic stress. *Protoplasma* 248(3): 447–455.

Singh, P.K., Indoliya, Y., Chauhan, A.S., Singh, S.P., Singh, A.P., Dwivedi, S., Tripathi, R.D., Chakrabarty, D. (2017) Nitric oxide mediated transcriptional modulation enhances plant adaptive responses to arsenic stress. *Scientific Reports* 7(1): 3592.

Song, L., Ding, W., Zhao, M., Sun, B., Zhang, L. (2006) Nitric oxide protects against oxidative stress under heat stress in the calluses from two ecotypes of reed. *Plant Science: An International Journal of Experimental Plant Biology* 171(4): 449–458.

Tan, J., Wang, C., Xiang, B., Han, R., Guo, Z. (2013) Hydrogen peroxide and nitric oxide mediated cold- and dehydration-induced myo-inositol phosphate synthase that confers multiple resistances to abiotic stresses. *Plant, Cell and Environment* 36(2): 288–299.

Tanou, G., Filippou, P., Belghazi, M., Job, D., Diamantidis, G., Fotopoulos, V., Molassiotis, A. (2012) Oxidative and nitrosative-based signaling and associated post-translational modifications orchestrate the acclimation of citrus plants to salinity stress. *The Plant Journal: For Cell and Molecular Biology* 72(4): 585–599.

Tanou, G., Job, C., Belghazi, M., Molassiotis, A., Diamantidis, G., Job, D. (2010) Proteomic signatures uncover hydrogen peroxide and nitric oxide cross-talk signaling network in citrus plants. *Journal of Proteome Research* 9(11): 5994–6006.

Tanou, G., Job, C., Rajjou, L., Arc, E., Belghazi, M., Diamantidis, G., Molassiotis, A., Job, D. (2009) Proteomics reveals the overlapping roles of hydrogen peroxide and nitric oxide in the acclimation of citrus plants to salinity. *The Plant Journal: For Cell and Molecular Biology* 60(5): 795–804.

Teng, K., Li, J., Liu, L., Han, Y., Du, Y., Zhang, J., Sun, H., Zhao, Q. (2014) Exogenous ABA induces drought tolerance in upland rice: The role of chloroplast and ABA biosynthesis-related gene expression on photosystem II during PEG stress. *Acta Physiologiae Plantarum* 36(8): 2219–2227.

Terzi, R., Kadioglu, A., Kalaycioglu, E., Saglam, A. (2014) Hydrogen peroxide pretreatment induces osmotic stress tolerance by influencing osmolyte and abscisic acid levels in maize leaves. *Journal of Plant Interactions* 9(1): 559–565.

Tewari, R.K., Prommer, J., Watanabe, M. (2013) Endogenous nitric oxide generation in protoplast chloroplasts. *Plant Cell Reports* 32(1): 31–44.

Tossi, V., Amenta, M., Lamattina, L., Cassia, R. (2011) Nitric oxide enhances plant ultraviolet-B protection up-regulating gene expression of the phenylpropanoid biosynthetic pathway. *Plant, Cell and Environment* 34(6): 909–921.

Tossi, V., Lamattina, L., Jenkins, G.I., Cassia, R.O. (2014) Ultraviolet-B-induced stomatal closure in Arabidopsis is regulated by the UV RESISTANCE LOCUS8 photoreceptor in a nitric oxide-dependent mechanism. *Plant Physiology* 164(4): 2220–2230.

Tossi, V., Lombardo, C., Cassia, R., Lamattina, L. (2012) Nitric oxide and flavonoids are systemically induced by UV-B in maize leaves. *Plant Science: an International Journal of Experimental Plant Biology* 193–194:103–109.

Trevisan, S., Manoli, A., Quaggiotti, S. (2014) NO signaling is a key component of the root growth response to nitrate in Zea mays L. *Plant Signaling and Behavior* 9(3): e28290.

Uchida, A., Jagendorf, A.T., Hibino, T., Takabe, T., Takabe, T. (2002) Effects of hydrogen peroxide and nitric oxide on both salt and heat stress tolerance in rice. *Plant Science* 163(3): 515–523.

Valderrama, R., Corpas, F.J., Carreras, A., Fernández-Ocaña, A., Chaki, M., Luque, F., Gómez-Rodríguez, M.V., *et al.* (2007) Nitrosative stress in plants. *FEBS Letters* 581(3): 453–461.

Wahid, A., Perveen, M., Gelani, S., Basra, S.M.A. (2007) Pretreatment of seed with H_2O_2 improves salt tolerance of wheat seedlings by alleviation of oxidative damage and expression of stress proteins. *Journal of Plant Physiology* 164(3): 283–294.

Wang, W.W., Bai, X.Y., Dong, Y.J., Chen, W.F., Song, Y.L., Tian, X.Y. (2016) Effects of application of exogenous NO on the physiological characteristics of perennial ryegrass grown in Cd-contaminated soil. *Journal of Soil Science and Plant Nutrition* 16(3): 731–744.

Wang, L., Guo, Y., Jia, L., Chu, H., Zhou, S., Chen, K., Wu, D., Zhao, L. (2014a) Hydrogen peroxide acts upstream of nitric oxide in the heat shock pathway in Arabidopsis seedlings. *Plant Physiology* 164(4): 2184–2196.

Wang, Y., Li, L., Cui, W., Xu, S., Shen, W., Wang, R. (2011) Hydrogen sulfide enhances alfalfa (*Medicago sativa*) tolerance against salinity during seed germination by nitric oxide pathway. *Plant and Soil* 351(1–2): 107–119.

Wang, Y., Li, J., Wang, J., Li, Z. (2010a) Exogenous H_2O_2 improves the chilling tolerance of manilagrass and mascarenegrass by activating the antioxidative system. *Plant Growth Regulation* 61(2): 195–204.

Wang, B.L., Tang, X.Y., Cheng, L.Y., Zhang, A.Z., Zhang, W.H., Zhang, F.S., Liu, J.Q., *et al.* (2010b) Nitric oxide is involved in phosphorus deficiency-induced cluster-root development and citrate exudation in white lupin. *The New Phytologist* 187(4): 1112–1123.

Wang, Y.S., Yang, Z.M. (2005) Nitric oxide reduces aluminum toxicity by preventing oxidative stress in the roots of Cassia tora L. *Plant and Cell Physiology* 46: 1915–1923.

Wang, L., Yang, L., Yang, F., Li, X., Song, Y., Wang, X., Hu, X. (2010c) Involvements of H_2O_2 and metallothionein in NO-mediated tomato tolerance to copper toxicity. *Journal of Plant Physiology* 167(15): 1298–1306.

Wang, Y., Zhang, J., Li, J.L., Ma, X.R. (2014b) Exogenous hydrogen peroxide enhanced the thermotolerance of *Festuca arundinacea* and *Lolium perenne* by increasing the antioxidative capacity. *Acta Physiologiae Plantarum* 36(11): 2915–2924.

Wang, P., Zhu, J.K., Lang, Z. (2015) Nitric oxide suppresses the inhibitory effect of abscisic acid on seed germination by s-nitrosylation of snrk2 proteins. *Plant Signaling & Behavior* 10(6): e1031939.

Wei, W., Li, Q.T., Chu, Y.N., Reiter, R.J., Yu, X.M., Zhu, D.H., Chen, S.Y., *et al.* (2014) Melatonin enhances plant growth and abiotic stress tolerance in soybean plants. *Journal of Experimental Botany* 66: 695–707.

Wu, Q., Su, N., Zhang, X., Liu, Y., Cui, J., Liang, Y. (2016) Hydrogen peroxide, nitric oxide and UV RESISTANCE LOCUS8 interact to mediate UV-B-induced anthocyanin biosynthesis in radish sprouts. *Scientific Reports* 6: 29164.

Xu, F.J., Jin, C.W., Liu, W.J., Zhang, Y.S., Lin, X.Y. (2010) Pretreatment with H_2O_2 alleviates aluminum-induced oxidative stress in wheat seedlings. *Journal of Integrative Plant Biology* 54: 44–53.

Xu, J., Wang, W., Sun, J., Zhang, Y., Ge, Q., Du, L., Yin, H., Liu, X. (2011) Involvement of auxin and nitric oxide in plant Cd-stress responses. *Plant and Soil* 346: 107–119.

Xu, J., Yin, H., Li, X. (2009) Protective effects of proline against cadmium toxicity in micropropagated hyperaccumulator, *Solanum nigrum* L. *Plant Cell Reports* 28(2): 325–333.

Yildiz, M., Terzi, H., Bingül, N. (2013) Protective role of hydrogen peroxide pretreatment on defense systems and BnMP1 gene expression in Cr(VI)-stressed canola seedlings. *Ecotoxicology* 22(8): 1303–1312.

Young, J.J., Mehta, S., Israelsson, M., Godoski, J., Grill, E., Schroeder, J.I. (2006) CO_2 signaling in guard cells: Calcium sensitivity response modulation, a Ca^{2+}-independent phase, and CO_2 insensitivity of the gca2 mutant. *Proceedings of the National Academy of Sciences* 103(19): 7506–7511.

Yu, M., Lamattina, L., Spoel, S.H., Loake, G.J. (2014) Nitric oxide function in plant biology: A redox cue in deconvolution. *The New Phytologist* 202(4): 1142–1156.

Yu, C.W., Murphy, T.M., Lin, C.H. (2003) Hydrogen peroxide-induced chilling tolerance in mung beans mediated through ABA-independent glutathione accumulation. *Functional Plant Biology* 30(9): 955–963.

Yu, Y., Yang, Z., Guo, K., Li, Z., Zhou, H., Wei, Y., Li, J., *et al.* (2015) Oxidative damage induced by heat stress could be relieved by nitric oxide in *Trichoderma harzianum* LTR-2. *Current Microbiology* 70(4): 618–622.

Zeng, C.-L., Liu, L., Wang, B.-R., Wu, X.-M., Zhou, Y. (2011) Physiological effects of exogenous nitric oxide on *Brassica juncea* seedlings under NaCl stress. *Biologia Plantarum* 55(2): 345–348.

Zhang, H., Forde, B.G. (1998) An Arabidopsis MADS box gene that controls nutrient-induced changes in root architecture. *Science* 279(5349): 407–409.

Zhang, Y., Wang, L., Liu, Y., Zhang, Q., Wei, Q., Zhang, W. (2006) Nitric oxide enhances salt tolerance in maize seedlings through increasing activities of proton-pump and Na^+/H^+ antiport in the tonoplast. *Planta* 224(3): 545–555.

Zhao, L., Zhang, F., Guo, J., Yang, Y., Li, B., Zhang, L. (2004) Nitric oxide functions as a signal in salt resistance in the calluses from two ecotypes of reed. *Plant Physiology* 134(2): 849–857.

Zhu, T., Deng, X., Zhou, X., Zhu, L., Zou, L., Li, P., Zhang, D., Lin, H. (2016) Ethylene and hydrogen peroxide are involved in brassinosteroid-induced salt tolerance in tomato. *Scientific Reports* 6: 35392.

12 Role of Amino Acids in Improving Abiotic Stress Tolerance to Plants

Qasim Ali, Habib-ur-Rehman Athar, Muhammad Zulqurnain Haider, Sumreena Shahid, Nosheen Aslam, Faisal Shehzad, Jazia Naseem, Riffat Ashraf, Aqsa Ali, and Syed Murtaza Hussain

CONTENTS

12.1 INTRODUCTION

Due to their sessility, plants face a wide range of environmental stresses. A huge amount of work has been cited about the effects of different biotic and abiotic stresses on plant growth (Ali et al., 2007; Ali et al., 2018; Ali and Ashraf, 2008; Ali and Ashraf, 2011a; Noman et al., 2018). Abiotic stresses include drought, salinity, extreme temperatures, nutrient deficiency, high light intensity and ozone (O_3) as well as anaerobic stresses (Ali et al., 2016; Ali et al., 2018; Ali and Ashraf, 2011b; Noman et al., 2018). It is estimated that these stresses reduce crop yield, especially in staple food, up to 70% (Thakur et al., 2010). They also disturb cellular ion homeostasis due to low availability of water for the uptake of nutrients or to the hyperaccumulation of different ions under nutrient stresses such as salinity and heavy metal stress (Munns and Tester, 2008). For example, under high intensity of salt stress, the movement patterns of Na^+ and K^+ ions changed. For normal functioning, a plant needs a high level of K^+ and a low level of Na^+ ions (Carden et al., 2003). Similar processes took place under varying levels of heavy metal stresses (Islam et al., 2009). In order to counteract the adverse effects of excessive toxic ions, plants have developed different mechanisms, such as exclusion and cellular compartmentalization of these ions (Pasapula et al., 2011; Rodríguez-Rosales et al., 2009; Ye et al., 2009).

The first and foremost effect of these stresses (drought, salinity and cold) is on the plant–water relation and disturbs the cellular turgor potential (Boudsocq and

Laurière, 2005). Depletion of plant water status directly affects photosynthesis, and ultimately reduces plant growth (Da Silva and Arrabaça, 2004). To overcome the adverse effects of these stresses, accumulation of different osmolytes (carbohydrates, betaine, proline and other amino acids) is found to be an adaptive mechanism plants use to maintain cellular turgor pressure (Da Silva and Arrabaça, 2004). Among these, the accumulation of different amino acids also plays a significant role, including stomatal opening and closing, which helps maintain cellular water balance (Kamran et al., 2009). Under abiotic stress, overproduction of reactive oxygen species (ROS) takes place due to a disturbance in the electron transport chain of PSII (Apel and Hirt, 2004). Overproduction of ROS causes damages to cells due to membrane lipid peroxidation (Imlay, 2003) that could even be fatal. To control lipid peroxidation, plants have developed various mechanisms (Vranova et al., 2002) that include the production/accumulation of different metabolites (Desikan et al., 2004). Some of the osmolytes that accumulate in cells also play a role in the scavenging of ROS and protection of the cell (Krishnan et al., 2008). Plants have developed a well-organized mechanism to counteract the adverse effects of stresses, and amino acids metabolism is one of these. Amino acids are important not only as the building blocks of proteins, but also because of the other metabolic activities they perform. It has been studied that a concentration of different amino acids increased under stressful conditions, which shows the direct or indirect role of amino acids under these conditions. For example, proline had a 4-fold increase in maize under drought stress; asparagine a 4.5-fold increase; serine and glycine a 2.2-fold increase and aspartic acid a 1.5-fold increase under stress conditions. These amino acids help plants by increasing the amination process, which enhances crop yields without the use of any other fertilizer (Shekari and Javanmardi, 2017). It has been found that the process of amination also increases the total of amino acids, fats and carbohydrates as well as cellulose yield and doubles the plant's dry weight (Osuji et al., 2011). Amino acids also play a key role in ammonium fixation and C_4 metabolism, and in the biosynthesis of different compounds including flavonoids, isoflavonoids, stilbenes, aurones, cutin, suberin, sporopolleins, catechins, proanthocyanidins, lignans, lignins, phenylpropenes, acylated polyamines and many other alkaloid derivatives (Fraser and Chapple, 2011), which are the building blocks of different cellular components. Moreover, amino acid-derived compounds help in abiotic/biotic stress tolerance by providing mechanical strength to plant cells, and in pollen viability, pest deterrence, and protection against UV, disease resistance, drought resistance and insect resistance, among many other roles (Nair and Harris, 2004). They have also has been found to be helpful in tocopherol anabolism in plants (Holländer-Czytko et al., 2005). Amino acids (AAs) play their role in abiotic stress tolerance through three mechanisms: acting as compatible osmolytes, regulating pH or acting as a nitrogen or carbon reserve.

The accumulation of AAs is the most common response of plants to abiotic stresses, and in plant tissues exposed to such stresses, concentrations of AAs may rise up to the millimolar range (Planchet et al., 2011; Singh et al., 1972). AAs have been shown to have multifarious functional roles in plant stress tolerance by acting as compatible osmolytes (Deivanai et al., 2011), being involved in pH regulation and in the detoxification of reactive oxygen species (ROS) (Rizwan et al., 2016), and acting as a nitrogen and carbon reserve, mainly for the synthesis of specific enzymes and precursors of various secondary metabolites such as flavonoids and lignins, as well as being an available stock of AAs, which are useful during recovery from stress. AAs also act as regulatory and signaling molecules (Munns, 2002). These protecting roles of AAs are also supported by the fact that stress-tolerant plants accumulate higher AAs than sensitive plants and show a correlation between stress tolerance and AA levels. The increase in the level of some AAs has been reported to lead to beneficial effects during stress acclimation. In some cases, AAs accumulate as a sink of excess nitrogen, indicating a reduction in growth rate. In the latter case, the increase of AA concentrations indicates damage to cell functioning rather than an adaptive response to environmental stresses (Munns, 2002).

Amino acids are well-known biostimulants which have positive effects on plant growth and yield and significantly mitigate the injuries caused by abiotic stresses (Kowalczyk et al., 2008). Amino acids as constituents of a variety of proteins have the potential to alleviate abiotic stresses in crop plants (Sadak et al., 2015; Sadak and Abdelhamid, 2015). In addition, metabolite analysis of the mitochondrial protein complex composition following oxidative stress suggested that enhancement of amino acid catabolism into the TCA cycle might compensate for a reduced electron supply from the TCA cycle (Obata et al., 2011). The metabolic connection of the AAs pathway with the TCA cycle has also recently emerged from a dedicated bioinformatics approach, analyzing the network expression behaviour of the entire set of Arabidopsis genes encoding metabolic enzymes and transcription factors to multiple abiotic and biotic stresses as well as to some nutritional and hormonal cues (Less et al., 2011). Plants have the ability to maintain the cellular content of specific amino acids for stress tolerance, but this is plant-species specific. Some high-yielding cultivars are an exception to that and are more tolerant through their exogenous use (Ali and Ashraf, 2011a) through varying modes. This alters the internal cellular metabolic activities in different ways. Application of amino acids as a foliar spray caused an increase in the content of total carbohydrates and polysaccharides of stressed and non-stressed plants and similar results have been reported in other plant species (Nahed et al., 2010).

In addition, application of amino acids reduced oxidative stress in plants by decreasing ROS production (Souri, 2016; Teixeira et al., 2017). Recently, the use of AAs in chelation with micro-nutrients has been increasing due to the double benefit of both partners, and its helpfulness for drought tolerance. It was reported that amino acids chelated with micro-nutrients increased the photosynthesis in plants when applied either alone or under abiotic stresses (Sadak et al., 2015; Souri, 2016). During their lifespan, plants have to deal with various adverse climatic factors that harmfully affect them in various ways. They have developed strategies regarding altered metabolism, in order to acclimatize and survive under these unfavorable conditions. These mechanisms involve some changes in gene expression and protein modification as a result altered metabolic pathways. The changed pattern in amino acids (AAs) metabolism is one of these. Mostly, plant composition is modified by changing environmental conditions and is characterized by an elevated accumulation of specific AAs that are involved in plant stress tolerance. These AAs are synthesized by various metabolic networks and accumulate differentially in plant species exposed to various stresses.

A long history is available about the role of amino acids in plants, but their effects are plant-type specific and amino acid-type specific. For example, in *Phaseolus mungo*, accumulation of aspargine and alanine (Chen and Murata, 2002); arginine in *Cryptomeria* (Agarwal et al., 2006); aspartic acid, glutamic acid and glutamine in cotton (Hanower and Brzozowska, 1975); asparagine, aspartic acid, serine and glycine in maize (Rai, 2002); aspartic acid and alanine in *Iris germinica* (Paulin, 1972) and proline, ornithine, arginine and glutamic acid in rice leaves (Yang et al., 2000). From these findings it is clear that amino acids play a critical role in plant growth and development as well as in abiotic stress tolerance. The general account, metabolism and specific roles of some important amino acids are described in detail below.

12.2 PROLINE

Proline, along with GB, is the amino acid of prime importance for the induction of stress tolerance through its metabolism. Studies regarding its role under varying abiotic stresses revealed it in detail. Its first and foremost role is in cellular osmotic adjustment for the maintenance of cellular–water relations through its accumulation (Haffani et al., 2014). It maintains not only the cellular–water relation but also ion homeostasis under ionic stress (Kamyab et al., 2016). Under NaCl stress, it decreased the uptake of Na^+ and maintained a high rate of K^+/Na^+ through its application as foliar spray. High K^+/Na^+ have also been studied through its use as seed priming or soil application (Nounjan et al., 2012). Under heavy metal stress, it also reduced the uptake of metal ions by its endogenous biosynthesis (Theriappan et al., 2011). Proline has also been found to maintain proper photosynthetic rate under varying stresses through the maintenance of cellular water content due to its role as osmoregulator/osmoprotectant by taking part in stomatal regulation. Moreover, proline protects photosynthetic units from the damaging effects of high energy free electrons under high light intensity. The protective effects of proline also act on the cellular membrane, by reducing lipid peroxidation in various ways. Proline increases the levels of various antioxidative enzymes and non-enzymatic compounds (Butt et al., 2016). For example, under abiotic stress, proline accumulation enhances the activities of antioxidative enzymes (Nayyar and Walia, 2003). Its role in non-enzymatic antioxidants has also been reported (Butt et al., 2016) in varying stresses. Not only does proline play a significant role in the improvement of antioxidative defense mechanisms, but it also functions as a potential antioxidant itself in scavenging ROS (Kaul et al., 2008). The importance of proline as an osmoprotectant/osmoregulator has also been reported in various studies (Ben-Rouina et al., 2006; Deivanai et al., 2011; Zali and Ehsanzadeh, 2018), through its exogenous use or as internal improvement. Along with maintenance in physiological mechanisms under stress, proline has been found to improve the yield of various crops (Kahlaoui et al., 2013; Siddiqui et al., 2015; Wani et al., 2010). The endogenous accumulation of proline from few to many folds has been reported depending on plant species as well as on the types of stress (Ali et al, 2007; Ben-Rouina et al., 2006; Nayyar and Walia, 2003). Exogenous use of proline through different modes under varying stresses has been reported for stress tolerance in many studies. However, the effective concentration and mode of application was plant-species and type-of-stress specific.

12.2.1 Proline Biosynthesis

Proline biosynthesis in higher plants takes place either by glutamate or ornithine pathways. Under osmotic stress or nitrogen shortage, the glutamate pathway of proline biosynthesis is thought to be the most important one for converting glutamate into proline in two-step reactions (Fichman et al., 2015). Initially, proline biosynthesis was studied in bacteria (Leisinger, 1987) and found to be similar to that of other prokaryotic and eukaryotic organisms. In this pathway, the synthesis of proline starts with ATP-dependent phosphorylation of L-glutamate, which is converted into γ -glutamyl-phosphate (γ -GP) by the γ -glutamyl-kinase enzyme (γ -GK). Subsequently, γ -GP is reduced to glutamic-γ -semi-aldehyde (GSA) by glutamic-γ-semi-aldehydedehydrogenase (GSADH), and GSA spontaneously cyclizes to pyrroline-5-Carboxylate (P5C) which is converted into L-proline by the pyrroline-5-carboxylate-reductase enzyme (P5CR). On the basis of indirect evidence, a similar proline biosynthesis pathway has been found in higher plants.

In these plants, proline is synthesized not only by the glutamate pathway but also through the ornithine pathway (Adams and Frank, 1980; Mestichelli et al., 1979), which can be transaminases either to GSA by the δ-ornithine-aminotransferasi enzyme (δ-OAT), or to α-keto-δ-aminovalerate (KAV) by the ornithine α-aminotransferase enzyme (Adams and Frank, 1980; Mestichelli et al., 1979). In the first biochemical pathway, GSA spontaneously cyclizes to P5C, which is converted into proline by P5CR. In the second pathway, KAV spontaneously cyclizes to pyrroline 2-carboxylate (P2C), which is converted into proline by P2C reductase (P2CR).

12.2.2 Role of Proline in Abiotic Stress Tolerance

Proline is the most common water-soluble amino acid, and also the most studied in relation with its role under stress. It is the only amino acid in which *a*-amino is present as a secondary amine. The metabolism of proline in plants has been extensively studied in relation to environmental stresses. The first study of proline biosynthesis under abiotic stress was conducted on ryegrass (*Lolium perenne*) by Kemble and MacPherson (1954). Later, many studies reported proline accumulation specifically in higher plants under abiotic stress including drought, salinity, high and low temperature, nutrient deficiency, anaerobiosis, heavy metals, UV irradiation and oxidative stress (Planchet et al., 2011; Verbruggen and Hermans, 2008). A large number of reviews and book chapters are available about its role in varying abiotic stresses as well as its exogenous use for induction of stress tolerance through different modes (Ashraf and Foolad, 2007). It has been observed that, under abiotic stress, proline accumulation can increase up to 100-fold as compared to non-stressed plants (Verslues and Sharma, 2010). This high accumulation of proline under abiotic stress explains its role in coping with stress conditions. However, it is still unclear why proline accumulation takes place under stress conditions. Is its accumulation a result of stress conditions? Or is it due to reduced growth of plants under abiotic stress (Verbruggen and Hermans, 2008)? There are many mechanisms for the action of proline under stressed conditions, but it depends on the amount of proline accumulated. Proline can be assimilated as a compatible osmolyte, a putative role supported by the fact that it has been reported to accumulate in the cytosol in response to hyperosmotic stresses (Theriappan et al., 2011). Accumulation of proline under stress conditions as an osmolyte helps plants maintain optimum water balance under unfavorable environmental conditions. Proline also acts as an osmoprotectant because under stress conditions it accumulates in the cytosol and chloroplast, while most of the other solutes accumulate in vacuoles (Ben-Rouina et al., 2006). Proline plays a significant role in maintaining redox homeostasis in plants under stress conditions. This alters the cellular pH and maintains $NADP^+$/NADPH ratios (Nounjan and Theerakulpisut, 2012). During dehydration conditions, proline also plays a significant role in maintaining the cellular structure as well as the cellular membrane (Szabados and Savouré, 2010; Verbruggen and Hermans, 2008). Under heavy metal stress, proline has been reported to play a role in maintaining the activities of some enzymes (for example, nitrate reductase and glucose-6-phosphate dehydrogenase) against the effects of specific toxic heavy metals by forming proline–metal complexes (Sharma and Dietz, 2006). Proline also protects plants against oxidative damage due to production of ROS under oxidative stress (Alía et al., 2001). Some authors demonstrated the capacity of proline to act as an antioxidant enzyme (Signorelli et al., 2013) (Table 12.1)

Despite its role as osmoregulator and osmoprotectant, proline also plays a significant role as a signaling molecule and acts as a part of the signaling network in plants during development (Szabados and Savouré, 2010), especially under stress conditions. Proline has been reported to induce the expression of some salt-stress-responsive genes and to improve the salt tolerance of *Pancratium maritimum* by upregulating stress-protective proteins and protecting the protein turnover machinery against stress damage (He et al., 2017). Although proline accumulation induces abiotic stress tolerance in plants, accumulation of proline due to heat stress results in sensitivity as reported in tobacco plants and *Arabidopsis* seedlings (Rizhsky et al., 2004). These results showed a

TABLE 12.1
Roles of Proline Under Various Abiotic Stresses

Type of stress	Plant species	Source of amino acid (exogenous/endogenous)	Effective concentration	Effects/ improvement	References
Salt stress	*Brassica juncea*	Foliar spray	20 mM	Increase in growth, photosynthesis, and yield parameters	Wani et al. (2010)
Salt stress	*Pancratium maritimum* L.	In nutrient solution	5 mM	Increased antioxidative activities and stress tolerance	Khedr et al. (2003)
Drought stress	*Triticum aestivum* L.	Endogenous proline accumulation	160 and 170 umol/g DW respectively in genotypes HD 2380 and C 306	Increased stress tolerance by increasing antioxidative activity	Nayyar and Walia (2003)
Salt stress	*Olea europaea* L.	Endogenous accumulation	0.84 to 2.04 μ mol mg^{-1}	Accumulation of proline was helpful to maintain an optimum photosynthetic rate	Ben-Rouina et al. (2006)
Drought stress	*Zea mays*	Exogenous (foliar)	30, 60 mM	Improved growth and photosynthetic attributes	Ali et al. (2007)
Heavy metal stress	*Nicotiana tabacum*	In rooting media	1 and 10 mM	Restored membrane integrity and increased the activities of ASC-GSH cycle enzymes under Cd stress	Islam et al. (2009)
Drought stress	*Triticum aestivum* L.	Pre-sowing	20, 40 mM	Increased biomass, WUE, photosynthesis and final yield	Kamran et al. (2009)
Cd stress	*Vigna radiata*	Hypertonic solution	5 mM	Increased antioxidative activities	Hossain et al. (2010)
Heavy metal stress	*Brassica oleracea*	Endogenous increase	Up to 2-fold increase	Decrease in heavy metal stress, seed germination and root and shoot growth, proline accumulated to induce stress tolerance	Theriappan et al. (2011)
Salt stress	*Oryza sativa*	Seed priming	1, 5, and 10 mM	Increased plant growth under stress condition, by regulating cellular osmotic potential	Deivanai et al. (2011)
Salt stress	*Oryza sativa*	Seed priming	10 mM in NaCl solution	Decrease in shoot Na^+ and Cl^- accumulation in embryo culture cells	Nounjan and Theerakulpisut (2012)
Salt stress	*Oryza sativa*	Seed priming	10 mM	Reduced Na^+/K^+ ratio and strongly decreased endogenous proline	Nounjan et al. (2012)
Salt stress	*Triticum aestivum* L.	Foliar spray	50,100 mM	Increased growth parameters and chlorophyll content	Talaat and Shawky (2013)
Salt stress	*Solanum lycopersicum*	Foliar spray	10 to 20 mg/L	Increase in leaf area, growth length and fruit yield	Kahlaoui et al. (2013)
Sea water	*Vicia faba*	Foliar spray	0, 25, 50 mM	Increase in growth and photosynthetic pigments	Dawood et al. (2014)
Salt stress	*Oryza sativa*	Foliar spray	25 ml/ plant	Increased growth, grain and straw yield	Siddiqui et al. (2015)
Salt stress	*Solanum lycopersicum*	Foliar spray	10, 20 mg L^{-1}	Increase in total soluble protein and glutamine content	Kahlaoui et al. (2015)

(Continued)

TABLE 12.1 (CONTINUED)
Roles of Proline Under Various Abiotic Stresses

Type of stress	Plant species	Source of amino acid (exogenous/endogenous)	Effective concentration	Effects/ improvement	References
Drought stress	*Pisum sativum*	Foliar spray	4 mM	Enhanced antioxidant activity	Osman (2015)
Arsenate stress	*Solanum melongena* L.	In nutrient solution	25 µM	Increased antioxidative activities	Singh et al. (2015)
Salt stress	*Capsicum annuum*	Foliar spray	0.6, 0.8, 1.0 and 1.2 mM	Increased antioxidants concentration (SOD, POD and CAT)	Butt et al. (2016)
Salt stress	*Onobrychis viciaefolia*	NaCl soultion	2.5 mM	Increase in concentration of MDA and RMP in shoots, regulation of Na^+/K^+ ratio	Wu et al. (2017)
Drought stress	*Chenopodium quinoa*	Foliar spray	12.5, 25 mM	Improved growth parameters, relative water content, photosynthetic pigments, indoleacetic acid, phenolics, TSS, proline and free amino acids in leaf tissues of quinoa plants. Also, yield, yield components and the nutritional values of the yielded seeds were improved	Elewa et al. (2017)
Salt stress	*Tetragenococcus halophilus*	Hyperhaline medium	0.5 gL^{-1}	Higher contents of intermediates involved in glycolysis, the tricarboxylic acid cycle, and the pentose phosphate pathway were observed in the cells supplemented with proline	He et al. (2017)
Natural conditions	*Foeniculum vulgare*	Foliar soray	0, 25 mM	Significant increases in carotenoids, polyphenol, chlorophyll, proline, total soluble carbohydrates and essential oil concentrations and relative water content, decrease in leaf water potential	Zali and Ehsanzadeh (2018)

contradictory relationship between proline and thermotolerance. Similarly, proline accumulated under drought stress but did not accumulate when a combination of drought and heat stress was applied. It was reported that the increased toxicity of proline to plant cells is a result of heat stress. Under these conditions, sucrose replaces proline in the cell as an osmoprotectant (Rizhsky et al., 2004). Similar observations were made in barley (Chen and Li, 2007). These findings suggest that the relationship between stress tolerance and proline accumulation is more complex than it may at first seem. Recent studies have confirmed the role of proline as a non-enzymatic antioxidant as well as an osmoprotectant (Table 12.1).

In wheat, exogenous application of proline significantly affected growth and biochemical attributes. Under drought stress, proline increases abiotic stress tolerance (drought/salinity) by decreasing the lipid peroxidation of the cellular membrane while increasing the activity of antioxidant enzymes (Nayyar and Walia, 2003). In wheat, exogenous application of proline also increased the photosynthesis and water use efficiency of the plant under water-deficit conditions (Kamran et al., 2009; Talaat and Shawky, 2013). Along with wheat, exogenous application of proline significantly increased the growth, photosynthesis and antioxidant activity of different maize cultivars as well as the final yield under drought or salinity stress (Ali et al., 2007). Furthermore, under salt stress, exogenous application of proline was helpful in maintaining the cellular osmotic adjustment (Deivanai et al., 2011) and increased plant growth. Proline was also helpful in decreasing the shoot Na^+ and Cl^- accumulation in culture cells (Nounjan and Theerakulpisut, 2012) and reduced the Na^+/K^+ ratio under salt stress (Nounjan et al., 2012), while, in tomato, exogenous application of proline was helpful in increasing the leaf area and fruit yield (Kahlaoui et al., 2013). Similar results were obtained in *Brassica juncea* under salt stress (Wani et al., 2010). In pea, sainfoin and chili, exogenous application of proline as foliar spray significantly increased the antioxidant activity of cells, which was helpful in improving plant growth under drought and salt stress respectively (Butt et al., 2016; Osman, 2015; Wu et al., 2017). Foliar application of proline under salt stress was helpful in improving the TCA cycle in *Tetragenococcus halophilus* under salt stress (He et al., 2017). Under natural environmental conditions, foliar application of proline was helpful by causing a significant increase in carotenoids, polyphenol, chlorophyll, proline, total soluble carbohydrates and essential oil content concentrations, as well as a decrease in leaf water potential (Zali and Ehsanzadeh, 2018).

Despite these significant roles of proline under varying abiotic stresses, the effect of exogenously applied proline is concentration and species-specific. Different plant species, even different cultivars of the same species, show a different response to exogenously applied concentrations. For example, under drought and salinity stress, the effective concentration of foliarly applied proline was 20–40 mM (Kamran et al., 2009), while in maize the effective concentration of exogenous application of proline was 30–60 mM (Ali et al., 2007). In rice, a concentration of 1–10 mM was found effective when proline was applied exogenously under salt stress (Deivanai et al., 2011; Nounjan et al., 2012; Nounjan and Theerakulpisut, 2012), while under drought stress a 30 mM concentration of foliarly applied proline was found to be effective for the induction of drought tolerance with enhanced yield and yield attributes (Ali et al., 2013). In tomato, exogenous application of 10–20 mg/L proline as a foliar spray was found to be effective in alleviating the adverse effects of salt stress (Kahlaoui et al., 2013; Kahlaoui et al., 2015) and increasing plant growth and fruit quality. In *Brassica junica*, *Pancratium martimum* L., *Solanum melongena* L., *Onobrychis viciifolia*, *Chenopodium quinoa* and *Tetragenococcus halophilus* the effective concentrations for the exogenous application of proline were 20 mM, 5 mM, 25 mM, 2.5 mM, 0.5 g/L and 25 mM respectively in drought or salt stress (Elewa et al., 2017; He et al., 2017; Khedr et al., 2003; Singh at al., 2015; Wani et al., 2010; Wu et al., 2017). However, under drought stress, proline concentration was increased endogenously in *Olea europaea* L., cauliflower and wheat ranging from 0.84 to 2.04 and up to many folds under salt, drought and heavy metal stress respectively (Ben-Rouina et al., 2006; Nayyar and Walia, 2003; Theriappan et al., 2011) (Table 12.1).

12.3 GLYCINE BETAINE (GB)

To tolerate adverse environmental conditions, the accumulation of different compatible solutes has great potential. These compatible solutes are water soluble, organic compounds. They include various sugars and peptides as well as amino acids, etc. Among these, glycine betaine (GB) is considered as one of the most important and effective harmonious compatible solutes. GB protects cells from the effects of various stresses by sustaining a proper osmotic balance in comparison with the surrounding stressful environment. It also stabilizes the complex protein quaternary structure. For example, it protects antioxidant enzymes activity and PSII oxygen-evolving complex under adverse environmental conditions (Sakr et al., 2012). Studies indicate that GB has an important role in plant tolerance to different environmental abiotic stresses. (LiXin et al., 2009).

12.3.1 Natural Production and Accumulation

Among many varying known and occurring quaternary ammonium compounds in plants, GB accumulates lavishly in response to dehydration stress (Mansour, 2000; Mohanty et al., 2002; Venkatesan and Chellappan, 1998; Yang et al., 2003). At cellular levels, it is rich mainly in the chloroplast, where it plays a vigorous role in the adjustment and protection of thylakoid membranes, thereby sustaining photosynthetic efficiency (Genard et al., 1991; Robinson and Jones, 1986; Yang and Lu, 2005).

Among higher plants, the chloroplast is the main site where GB is synthesized from the pathway of serine via choline, ethanolamine and betaine aldehyde (Hanson and Scott, 1980; Rhodes and Hanson, 1993). Some plants and their species such as rice, Arabidopsis, mustard and tobacco are naturally deficient in GB production both under stressed and non-stressed conditions (Rhodes and Hanson, 1993; Sakr et al., 2012). It was found that transgenics with over-production of GB exhibited a considerable improvement in tolerance level to salt, drought, cold, as well as to high temperature (Rhodes and Hanson, 1993). However, the accumulation of GB in these plant species was found to be much lower than the amount naturally found in plants under different stresses (Rhodes and Hanson, 1993). This limitation in the production of GB in high masses in transgenic plants clearly shows that there must be a low availability of choline substrate and a reduction in the transport of this choline in the chloroplast, which is the natural source for GB synthesis (Huang et al., 2000; McNeil et al., 2000; Nuccio et al., 1998).

There are certain studies that demonstrate the positive effects of GB on growth and yield attributes under different stresses in plants such as wheat, tobacco, sorghum, barley, etc. Exogenous application of GB in some other plants also mitigated the harmful effects of stresses (Ali and Ashraf, 2011a). For example, when GB was applied exogenously, a significant decrease in the adverse effects of salt stress with an increase in tolerance level in rice plants was observed (Harinasut et al., 1996; Lutts, 2000). In the case of tomato plants that were subjected to either heat or salt stress, when GB was applied exogenously, fruit yield increased about 40% (Mäkelä et al., 1998). In a comprehensive study on rice seedlings, when GB was applied exogenously, this ameliorated adverse effects on root and shoot growth as well as damages to ultra-structure (Rahman et al., 2002). Furthermore, salt-stress-induced damages to leaves, swelling, disintegration of thylakoids, inter-granular lamellae and destruction of mitochondria were overcome by use of GB. GB application causes the production of more vacuoles in root cells, which results in the accumulation of more Na^+ ions in the root zone and in a reduction of its transportation toward shoots (Hong-Bo et al., 2006) (Table 12.2).

12.3.2 Glycine Betaine Biosynthesis

GB biosynthesis is completed in two steps by the oxidation process of choline. In the first step, choline is oxidized into betaine aldehyde by the codA/betA enzymes. In the next step betain aldehyde oxidizes to glycine betaine through the help of the betaine aldehyde dehydrogenase (BADH) (Brouquisse et al., (1989). This biosynthetic process happens in the chloroplast but it is also important for the biosynthesis of GB in microorganisms, whereas the second biosynthetic pathway occurs mostly in microorganisms (Ikuta et al., 1997). It takes GB biosynthesis from glycine with the help of single-flavor enzymes (COD).

12.3.3 Accumulators and Non-Accumulators of GB

Accumulation of GB in response to various stresses has been observed in many plants (Chen and Murata, 2008; Wang et al., 2010). However, in certain plants, such as Arabidopsis, tomato and tobacco, this stress-induced component does not accumulate (Ma et al., 2007; Zhang et al., 2010). Genetically modified plants and exogenous application of GB might be the best possible way to cope with different stresses (Ali and Ashraf, 2011a; Athar et al., 2009). However, in certain plants, GB accumulation is considered too low, thus not up to the mark for stress tolerance (Su et al., 2006). To cope with this situation, exogenous application through foliar spray of GB has been suggested in order to maintain an endogenous level of GB (Harris et al., 2004, 2007). The usefulness of an exogenous application of GB depends upon its absorption in the leaves of plants and its translocation toward different organs in the plant (Ali and Ashraf, 2011a; Mäkelä et al., 1999). Various studies showed that exogenously applied GB through foliar spray was firstly absorbed into the leaf tissue and then translocated to different plant parts such as seeds, flower, stem, and roots (Ali and Ashraf, 2011a; Park et al., 2006). However, it has been reported that GB is also taken up via roots (Athar et al., 2009). Moreover, it is a significant source of carbon and nitrogen on releasing stress. GB is well known in the category of molecular chaperons protecting cellular components from the damaging effects of abiotic stresses (Khan et al., 2016).

TABLE 12.2
Roles of GB Under Various Abiotic Stresses

Type of stress	Plant species	Exogenous/endogenous	Effective concentration	Improvement/amelioration	References
Salt, cold, heat	*Oryza sativa*	Endogenous accumulation	2 to 8 mol m^{-3}	Improvement in growth	Kishitani et al. (2000)
Heat and freezing	*Arabidopsis*, *Spinacia vulgare*, *hordeum vulgare*	Endogenous accumulation	50–100 mM	Better growth	Sakamoto and Murata (2002)
Heat, cold	*Zea mays*	Endogenous accumulation	2 to 5 m mol g^{-1}	Ability to overcome stress	Quan et al. (2004)
Drought	*Brassica compestris*	Exogenous application at vegetative and flowering stage	100 mM	Increase in leaf diameter, achene weight, oil and yield	Hussain et al. (2008)
Salt stress	*Brassica napus* L.	Exogenous application at germinating stage	0, 0.1, 0.5, 1,5 mM	Growth improvement	Athar et al. (2009)
Drought	*Zea mays*	Pre-sowing, vegetative and flowering stage	20, 40, 60 ml	Osmoregulation by maintaining osmolyte accumulation	LiXin et al. (2009)
Water stress	*Oryza sativa* L.	Exogenous application during vegetative, seedling, reproductive stages	0.035 mol g^{-1}	Growth pigment and biochemical attributes	Kathuria et al. (2009)
Drought, heat	*Zea mays* L.	Over-accumulation	100 mM	Increase in photosynthesis	Wang et al. (2010)
Salt	*Solanum melongena*	Exogenous application at vegetative stage	0, 50 mM	Growth improvement and physiological process	Abbas et al. (2010)
Drought	*Zea mays*	Exogenous application as foliar spray at vegetative stage	30 mM	Nutritional quality	Ali and Ashraf (2011a)
Salt, drought stress (descrbed in review article)	Transgenic plants	Endogenous accumulation	0.035 μmol	Protection of cellular activities against different stresses	Giri (2011)
Drought stress	*Gossypium barbadense* L.	Exogenous applications as pre-sowing seed treatment	400, 600, 800	Improvement in growth attributes	Shallan et al. (2012)
Salinity	*Glycine max* L.	Exogenous application at flowering stage	0, 2.5, 5, 7.5 and 10 kg ha^{-1}	Yield attributes	Rezaei et al. (2012)
Salinity	*Brassica napus* L.	Pre-sowing vegetative and flowering stages	400 mg L^{-1}	Improvement in yield	Sakr et al. (2012)
Drought	*Zea mays* L.	Exogenous application as foliar spray	150 ppm	Improving effects on growth, photosynthetic pigments, grain yield	Miri and Armin (2013)
Salt stress	*Triticum aestivum* L.	Endogenous accumulation	24.1 to 34.3 umol g^{-1}	Improvement in yield	Rao et al. (2013)
Water stress	*Triticum aestivum* L.	Foliar application	10 mM	Improvement in growth	Aldesuquy et al. (2013)
Drought stress	*Triticum aestivum* L.	Exogenous	100 mM	Improvement in biochemical attributes	Gupta et al. (2014)

(*Continued*)

TABLE 12.2 (CONTINUED)
Roles of GB Under Various Abiotic Stresses

Type of stress	Plant species	Exogenous/endogenous	Effective concentration	Improvement/amelioration	References
Osmotic stress	*Zea mays* L.	Endogenous accumulation	38 m mol g^{-1}	Improving effects on growth, photosynthetic pigments, biochemical attributes	Moharramnejad et al. (2015)
Salt stress	*Oryza sativa* L.	In vitro	5, 10, 15, 20 mM	Improved yield components	Maziah and Teh (2016)
Salinity stress	*Triticum aestivum* L.	Exogenous application as foliar spray at vegetative stage	100 mM	Enhanced growth attributes	Khan et al. (2016)
Abiotic stress	Crop plants	Endogenous accumulation	100 m mol g^{-1}	Improved cellular homeostasis	Roychoudhury and Banerjee (2016)
Salt stress	*Triticum aestivum* L.	Endogenous accumulation	35 and 42 m mol L^{-1}	Improvement in growth, photosynthetic pigments	Tian et al. (2017)
Pathogen attack and downy mildew disease	*Pennisetum glaucum*	Exogenous application as pre-sowing seed treatment	30 mg ml^{-1}	Enhanced seed germination and seed vigour	Lavanya and Amruthesh (2017)
Drought stress	*Triticum aestivum* L.	Exogenous application at tillering and anesthesia stages	100 mM	Yield increase	Gupta and Thind (2017)

Most plant species have a well-known mechanism for the biosynthesis of GB to tolerate adverse environmental conditions. However, the biosynthesis of GB for induction of stress tolerance is species/cultivar specific. The protecting effects of GB under various stresses are multifarious and act by regulating different metabolic activities. In maize plants grown under osmotic stress, a significant increase in GB accumulation was found to be correlated with the protection of photosynthetic pigments and other biochemical attributes which were helpful to sustain better growth (Moharramnejad et al., 2015). In this study, it was found that the protective effects of GB were associated with a better cellular water content. Among the salt-stress-protective effects of GB on growth, a photosynthetic pigment has been found in wheat that showed tolerance to salt stress (Gupta et al., 2014). However, no endogenous increase of GB for stress tolerance by its accumulation was found in any plant species (Ma et al., 2007). For the induction of drought tolerance by GB metabolism, its exogenous use through different modes has been reported to be helpful especially in crop plants. Gupta et al. (2014) and Khan et al. (2016) reported that exogenous applications of GB significantly enhance growth through the improvement of biochemical attributes in wheat plants (Table 12.2).

Regarding the roles of exogenous applications of GB in growth and yield, Miri and Armin (2013) reported that foliar-applied GB in corn plants under drought was found to be effective in increasing the plant biomass production and grain yield, and was associated with enhanced photosynthesis through an increase in leaf photosynthetic pigments. In another study on water-stressed rice plants, it was found that exogenously applied GB effectively enhanced growth, which was related to improvements in photosynthetic pigments. Tian et al. (2017) reported that GB endogenous accumulation in wheat plants was found to be helpful in coping with salt stress by improving their growth, photosynthetic pigments and yield attributes. The role of GB exogenous application in the oxidative defense mechanism has also been reported (Ali and Ashraf, 2011a). Recent studies revealed that manipulation of GB biosynthetic genes in high-yielding stress-sensitive crop plants was found to be effective for the induction of stress tolerance for a better yield (Aldesuquy et al., 2013). It has been found that exogenously applied GB through foliar spray was translocated to different plant parts, where it actively took part in stress tolerance mechanisms as well as improving nutritional quality (Ali and Ashraf, 2011a). Due to persistent accumulation for a long duration, GB was not only helpful for growth improvement but also increased the seed yield and its nutritional quality (Table 12.2).

12.4 ASPARTIC ACID

Among the amino acids belonging to the aspartate family, aspartic acid has prime importance due to its role in the biosynthesis of another amino acids. Aspartic acid is involved in the biosynthesis of many amino acids, such as isoleucine, methionine and threonine in many plant species as a so-called precursor of the aspartate family metabolic pathway (Rawia et al., 2011). It is involved in various metabolic reactions taking place in plastids, such as in argininosuccinate and carbamoyl-aspartate biosynthesis, which are precursors of arginine (Katoh et al., 2006; Zrenner et al., 2006). It can act as a buffer for maintaining the pH of the cell. Saeed et al. (2005) reported the role of aspartic acid in the regulation of many processes, such as the synthesis of protein chlorophyll biosynthesis as well as the formation of photosynthetic pigments. It can also serve as the precursor of many biomolecules such as antioxidants, vitamins and co-factors that are involved in cellular homeostasis and antioxidative defense mechanisms (Azevedo et al., 2006).

12.4.1 Biosynthesis/Metabolism

Like in glutamic acid, the aspartic acid biosynthetic pathway starts from pyruvate, the end product of the glycolysis pathway. Synthesis of aspartic acid also takes place through the citric acid cycle in which pyruvate converts into citrate via the Acetyl-CoA enzyme. All steps are similar to the synthesis of glutamate but the only difference is that oxaloacetate is converted into aspartic acid with the help of an enzyme, aspartate aminotransferase, which in turn converts into α-ketoglutarate. This α-ketoglutarate is then converted into oxaloacetate. Aspartic acid biosynthesis is regulated by the aspartate aminotransferase enzyme, which catalyzes the reversible transamination between oxaloacetate and glutamate to synthesize aspartate. Under stress conditions, aspartic acid is catabolized into a number of amino acids such as asparagines, threonine, lysine, isoleucine as well as methionine.

12.4.2 Significance Under Stress

Many developmental and metabolic factors are responsible for regulating free amino acid levels in plants and to maintain their concentration within the cell. However, their accumulation in plants takes place in response to various abiotic stresses, such as temperature, water and salt as well as heavy metal stresses (Delauney and Verma, 1993; Rhodes and Hanson, 1993). The studies regarding the roles of amino acids, including their accumulation, have been reported for a variety of amino acids, especially in the aspartate family. Stewart and Larher (1980) reported the accumulation of amino acids such as aspartic acid, glutamine and serine in many plant species in response to salt stress. They reported that the accumulation of these amino acids can serve as components of salt-tolerant mechanisms and help in the biosynthesis of many other important amino acids. These amino acids play vital roles in plants. such as development and growth (Scheideler et al., 2002). Among these amino acids, aspartic acid also accumulates in response to many abiotic stresses in several plant species. Ramos et al. (2005) reported an endogenous increase in aspartic acid in two soybean cultivars under salt stress along with the accumulation of other amino acids such as proline, glutamic acid, arginine and histidine. Accumulation of aspartic acid under water stress in cotton has been reported by Hanower and Brzozowska (1975). Thakur and Rai (1982) reported a 1.5-fold increase in aspartic acid in several maize cultivars due to water stress (Table 12.3). Exogenous application (either as foliar spray or seed priming) of these amino acids can also serve as a biostimulant that has been found to be effective in increasing plant yield and growth as well as mitigating the harmful effects caused by abiotic stresses (Kowalczyk et al., 2008) (Table 12.3). Exogenous application of aspartic acid as seed priming was found to be helpful in alleviating salt stress in *Ricinus communis* L. It also decreased the harmful effects of salinity on growth and secondary metabolic compounds (Ali et al., 2008). Rizwan et al. (2017a) reported that foliar-applied aspartic acid induced Cd stress tolerance in rice plants, which resulted in an increased growth which was due to enhanced nutrient levels as well as improved photosynthetic activity. Similarly, in another study, foliar application of aspartic acid resulted in an improvement in growth parameters of salt stressed tomato plants. Aspartic acid-treated plants increased their enzymatic activities such as POD, POX and CAT, which resulted in induction of more resistance in tomato plants against salt stress as compared with controlled plants. Furthermore, accumulation of glutamate, aspartate and arginine as well as histidine was seen in tomato plants grown under saline or non-saline conditions due to foliar-applied aspartic acid (Akladious and Abbas, 2013). Among different amino acids present in plants, aspartic acid has a significant importance in plants under various stresses (Table 12.3). A range of studies was conducted on stress-induced increase in aspartic acid in many plants in relation with stress tolerance. Simon-Sarkadi and Galiba (1996) reported the effects of drought and cold stress on endogenous accumulation of aspartic acid as 14 mg/g^{-1} and 10 mg/g^{-1} respectively in wheat plants, which was found to be effective for stress tolerance. Misra et al. (2006) reported that endogenous increase in aspartic acid contents (2.5 g/g^{-1}) in green gram plants due to salt stress increased growth in relation with other amino

TABLE 12.3
Roles of Aspartate Under Various Abiotic Stresses

Type of stress	Plant species	Endogenous increase/ exogenous application	Concentration of amino acid	Effects/ response	References
Cold stress	*Triticum aestivum* L.	Endogenous increase	61.3 μmolg^{-1} dry weight	Increased amino acid contents such as aspartate, glycine betaine, proline and glutamine	Naidu et al. (1991)
Water and cold stress	*Triticum aestivum* L.	Endogenous increase	14 mg/g^{-1} (drought stress) 10 mg/g^{-1} (cold stress)	Increase was cultivar specific depending on the tolerance level	Simon-Sarkadi and Galiba (1996)
High nitrogen	*Glycine max* L.	Endogenous increase	18.4 μmol/g fw (in roots) 32.3 μmol/ml (in xylem sap)	Counteracted the nitrogen effect	Lima and Sodek (2003)
Salt stress	*Glycine max* L.	Endogenous increase	0.81 μmol g^{-1} (29 strain) and 0.98 μmol g^{-1} (CB1809)	Reduced the adverse effects of drought stress	Ramos et al. (2005)
Salt stress	*Vigna radiata*	Endogenous increase	2.5 mg/g	Increased growth	Misra et al. (2006)
Salt stress	*Ricinus communis* L.	Exogenous (seed priming)	5, 10 and 15 mM	Decreased the harmful effects of salt stress on secondary products and plant growth	Ali et al. (2008)
Salt stress	*Zea mays* l.	Endogenous increase	Increase from 53.9 to 115.3 m mol g^{-1} dry mass in salttolerant genotype and 48.2–109 m mol g^{-1} dry mass in salt sensitive genotype in leaves	Increased content of soluble protein in salt sensitive genotype; also increased the concentration of organic solutes as a result; tolerance to stress	Neto et al. (2009)
Salt stress	*Moringa oleifera*	Endogenous increase	1.82g/100g protein	Increased photosynthetic contents of plant as well as amino acid content	Hussen et al. (2013)
Salt stress	*Solanum lycopersicum*	Exogenous (foliar application)	100 mg/L	Significantly increased plant growth, the content of amino acids as well as enzymatic activities	Akladious and Abbas (2013)
Salt stress	*Vicia faba*	Exogenous (foliar spray)	(3.2 to 3.45% aspartate) in mixture of amino acids	Improved all parameters that were decreased due to salt stress	Sadak et al. (2015)
Non-stressed conditions	*Vicia faba*	Exogenous (foliar spray)	50, 100 and 150 mg L^{-1}	Higher level (150mg L^{-1}) increased photosynthetic contents, seed yield and all growth parameters	Amin et al. (2014)
Salt stress	*Glycine max* L.	Endogenous increase	47.11 mg/g	Increased the quality as well as quantity of protein	Farhangi-Abriz and Ghassemi-Golezani (2016)
Under natural conditions	*Gerbera jamesonii* L.	Exogenous (foliar spray)	5.60% in mixture of amino acids	Enhanced photosynthetic rate, number and diameter of flowers, protein contents as well as contents of proline and protein	Geshnizjani and Khosh-Khui (2016)
Cadmium (Cd)	*Oryza sativa* L.	Exogenous (foliar spray)	0, 10, 15, and 20 mg L^{-1}	Increased all growth parameters, photosynthetic contents as well as nutrients by decreasing concentration of Cd	Rizwan et al. (2017a)

acids contents, which were also increased. An endogenous increase in maize plants was reported by Neto et al. (2009) in aspartic acid contents, which resulted in an increase in the organic solutes as well as total soluble protein contents. Hussen et al. (2013) reported a salt-stress-induced increase in aspartic acid (1.82 g/100 g protein) in moringa plants, which further increased the total amino acid content and photosynthetic activity of the plants. Salinity increased endogenous aspartic acid (1.82 g/100 g protein) in soybean plants, along with other amino acids such as glutamate, methionine and lysine, which improved the quality and quantity of protein (Farhangi-Abriz and Ghassemi-Golezani, 2016). Some studies on the accumulation of amino acids in plants in response to various abiotic stresses revealed that endogenous amino acids are not enough to regulate various metabolic activities in plants. However, their exogenous supply was found to be effective for proper cellular metabolic activities and induction of drought stress as well as physiological processes (Table 12.3). Ali et al. (2008) reported a significant reduction in the harmful effects of salinity in *Riccinus communis* L. plants through exogenous application of aspartic acid as seed priming. This resulted in better growth. Exogenous application of aspartic acid as foliar spray on tomato plants was found to be helpful for the induction of salt tolerance and significantly increased plant growth, enzymatic activities and endogenous levels of other amino acids (Akladious and Abbas, 2013). In a study conducted by Sadak et al. (2015), it was reported that the induction of salt tolerance by foliar spray of aspartic acid in faba bean was related with increased growth parameters. In another study conducted on faba beans by Amin et al. (2014), it was reported that exogenous application of aspartate as foliar spray (150 mg L^{-1}) increased the photosynthetic pigment contents, seed yield and growth parameters. Furthermore, foliar application of aspartic acid in *Gerbera jamesonni* L. plants enhanced photosynthetic rate, number and diameter of flowers, and protein and proline contents (Geshnizjani and Khosh-Khui, 2016) (Table 12.3).

12.5 GLUTAMATE

Among protein amino acids, glutamate is of prime importance due to its multifarious roles in multiple metabolic activities not only under normal growth conditions but also under specific stressful conditions. The foremost role of glutamate is that it is the precursor of many other essential amino acids such as arginine, ornithine and lysine, and thus regulates many metabolic activities indirectly (Slocum, 2005). Recently, it has been found to be an important signaling molecule under various environmental conditions in many plants species (Kan et al., 2017). It is a source of carbon and nitrogen metabolism in the conversion of glutamate into α-ketoglutarate with the help of the glutamate dehyrogenase enzyme (GHD) (Forde and Lea, 2007). Glutamate is also involved in cellular osmotic adjustment, especially in stomatal guard cells, thus regulating stomatal opening and closing (Dinu et al., 2011). It is involved in controlling the antioxidative defense mechanism through the biosynthesis of glutathione, an active part of the antioxidative system (Lu, 2013). In addition to its role in carbon and nitrogen metabolism, glutamate is also involved in the synthesis of chlorophyll and vitamin B9 (functional folate) (Hanson and Gregory, 2011), thus playing its role in plant photosynthetic activity. It was found that glutamate application enhanced the leaf chlorophyll content. In the presences of biotic and abiotic stresses, glutamate is the precursor of the important biomolecule γ-aminobutyrate (GABA), a non-proteinogenic amino acid that not only plays an important role in metabolism of carbon and many signaling pathways in plants but is also involved in the biosynthesis of proline, a well-known amino acid involved in plant stress tolerance as osmolyte as well as antioxidant (Bouche and Fromm, 2004; Bown et al., 2006; Sharma and Dietz, 2006; Shelp et al., 1999). Walch-Liu et al. (2006) reported the role of glutamate in calcium signaling and stated that it served as an exogenous signal in primary root growth but did not have any effect on the growth of lateral roots. Glutamate is also involved in ammonia assimilation which describes its role in the metabolism of nitrogen (Hirel and Lea, 2001) (Table 12.4).

12.5.1 Glutamate Biosynthesis

Pyruvate is the precursor of the glutamate biosynthetic pathway, which is the end product of glycolosis. Pyruvate is converted into two molecules: the first one is oxaloacetate, which is the result of carboxylation of pyruvate; the second one is citrate (citr), which is formed by the condensation of Acetyl Coenzyme A (Ac.CoA) with oxaloacetate (OAA). Citrate is then converted into α–ketoglutarate (α-KG), which is then converted into glutamate (glu), and glu is then catalyzed by the glutamine synthase enzyme into glutamine (gln). Almost 85% of glu is converted into interconvertible molecules glutamine and GABA, while the rest is degraded into malate (mal). Malate is then converted into pyruvate to continue the TCA cycle through the Ac. CoA enzyme (Hertz et al., 2007).

TABLE 12.4
Roles of Glutamate Under Various Abiotic Stresses

Type of stress	Plant species	Source of amino acid (endogenous/exogenous)	Concentration of amino acid	Metabolic effects	References
Salt stress	*Coleus blumei* Benth.	Endogenous inclrease	2.2 nmol g^{-1}	Accumulation of nitrogen containing compounds and amino acids	Gilbert et al. (1998)
Water stress	*Arabidopsis thaliana*	Endogenous increase	1.7-fold increase (wild-type mutant)	Accumulation of free amino acid contents	Nambara et al. (1998)
Water stress	*Glycine max* L.	Endogenous increase	0.98 μmol g^{-1} (29 strain) and 1.33 μmol g^{-1} (CB1809)	Increased accumulation of amino acids under water stress	Ramos et al. (2005)
Non-stress	Arabidopsis	Exogenous (in rooting medium)	50 and 1 mM	Role as an external signal to modulate root proliferation, growth and branching	Walch-Liu et al. (2006)
Water stress	*Sporobolus stapfianus*	Endogenous increase	Increase from 20 to 30%	Improvement in desiccation tolerance but cultivar specific. Increase in concentration of glutamic acid 20–30%. The glutamate served as a source of carbon and nitrogen	Martinelli et al. (2007)
Non-stress	*Codiaeum variegatum*	Exogenous (foliar application)	100 and 200 ppm (200 ppm was more effective)	Improvement in the quality and productivity of plants, increase in all growth parameters	Mazher et al. (2011)
Temperature stress	*Triticum aestivum* L.	Endogenous increase	18–23 mg/g	Increased total free amino acid concentration, growth parameters and yield components	Hassanein et al. (2013)
Salt stress	*Moringa oleifera*	Endogenous increase	2.53/100 g protein	Increased chl. *a*, chl. *b* and carotenoids concentration, but in the case of carotenoids a lower level of salinity was more effective. Increasing salinity increased the contents of amino acids at its higher level	Hussen et al. (2013)
Salt stress	Faba bean	Exogenous (foliar application)	7.24–9.12%	Decrease in harmful effects of salinity on faba bean	Sadak et al. (2015)
Non-stress	*Triticum aestivum* L.	Exogenous (foliar application)	50 and 100 mg/l	Increased the nutritional values of the yielded grains, % age of carbohydrates, proteins, macro-nutrients, micro-nutrients, total flavonoids, antioxidant activity, total amino acids and essential amino acids in yielded grain	Abd Allah et al. (2015)
Salt stress	*Glycine max* L.	Endogenous increase	72.42 mg/g	Increase in the contents of non-essential amino acids such as aspartate, glutamic acid, tyrosine, proline, but serine and alanine remain unaffected	Farhangi-Abriz and Ghassemi-Golezani (2016)
Non-stress	*Oryza sativa*	Exogenous (in rooting medium)	10 mM	Activated the trancription of defense-related genes in roots and leaves. Induced blast resistance in rice	Kadotani et al. (2016)

(Continued)

TABLE 12.4 (CONTINUED)
Roles of Glutamate Under Various Abiotic Stresses

Type of stress	Plant species	Source of amino acid (endogenous/ exogenous)	Concentration of amino acid	Metabolic effects	References
Nitrogen deficiency	*Oryza sativa*	Exogenous glutamate (in rooting medium through hydroponic culture)	0.1–10 mM (2.5 effective)	Role in signal transduction, decrease in nitrogen deficiency	Kan et al. (2017)
Salt and osmotic stress	*Brassica campestris*	Exogenous (through injection)	50 mM	Reduced salt and osmotic stress	Stolarz and Dziubinska (2017)
Non-stress	*Glycine max* L.	Exogenous (foliar spray and seed priming)	4.2 μM (as foliar spray)/10.5 mM (as seed priming)	Increased the antioxidative activity of enzymes	Teixeira et al. (2017)
Salt and cold stress	*Brassica napus*	Exogenous	10 μM	Induced stress tolerance by activating an H_2O_2 burst and crosstalk between H_2O_2 and Ca^{2+} signaling.	Lei et al. (2017)

12.5.2 Significance Under Stress

Amino acids have direct or indirect roles in plant growth and development. The majority of these amino acids start to accumulate in response to various abiotic stresses. Like for many other amino acids, the endogenous level of glutamate also increases under various abiotic stresses in several plant species, and further regulates many physiological and metabolic functions of plants. Accumulation of glutamate has been shown in detached leaves of rice due to water stress (Thakur and Rai, 1982). Shelp et al. (1999) reported that, under abiotic stresses, glutamate is converted into GABA, a stress indicator, through the activity of an enzyme, γ-aminobutyrate (GDC). A study conducted by Martinelli et al. (2007) reported that, in the presence of water stress, endogenous glutamate concentration increased from 20% to 30% in *Sporobolus stapfinus* plants. This induced stress tolerance in *Sporobolus stapfinus* plants, but this tolerance was cultivar specific. It is reported that an increase in the concentration of glutamate can serve as a source of nitrogen and carbon. Hassanein et al. (2013) reported that increasing temperature stress increased the level of endogenous glutamate in wheat plants, which further increased the tolerance of wheat plants against temperature stress, with enhanced growth parameters and yield components (Table 12.4). Studies reveal that in all plants, endogenous glutamate levels are not high enough for the induction of stress tolerance and the regulation of other metabolic processes. In these plant species, exogenous applications of various concentrations of glutamate were found to be effective for inducing stress tolerance. For example in *Brassica napus*, an exogenously applied 10 μM level of glutamate can serve as an external signal for salt and cold tolerance through the activation of an H_2O_2 burst and crosstalk between Ca^{+2} and H_2O_2 signaling (Lei et al., 2017). Sadak et al. (2015) reported that the application of exogenous glutamate (7.24–9.12%) as foliar spray on faba beans was found to decrease the harmful effects of salinity and improve plant growth and yield. Stolarz and Dziubinska (2017) reported that exogenous application of glutamate (50 μM injection) increased osmotic and salt stress tolerance in sunflower. In rice plants, 10 mM concentration of glutamate as exogenous application (in rooting medium) increased the resistance of rice plants against various diseases (Kadotani et al., 2016) (Table 12.4). Moreover, Walch-Liu et al. (2006) reported that exogenous application (in rooting medium) of glutamate significantly affected the growth and various developmental processes of Arabidopsis plants even at a very low concentration of glutamate (50 μM). Furthermore, it has been found that different levels of glutamate (20 or 40 mM) exogenously applied increased the endogenous glutamate level in tobacco plants (Schneidereit et al., 2006). Recently,

Kan et al. (2017) reported that exogenously applied glutamate in rice plants reduced nitrogen deficiency at a concentration of 2.5 mM within 30 minutes of the glutamate application (in rooting medium through hydroponic culture) by rapidly metabolizing into nitrogen containing compounds. Like other amino acids, glutamate also plays an important role in plants under various stresses. Farhangi-Abriz and Ghassemi-Golezani (2016) reported the effects of salt stress on cellular glutamate levels in soybean plants. They reported that increasing temperature stress increased the biosynthesis of glutamate, which induced stress tolerance and also increased the contents of non-essential amino acids such as aspartate, tyrosine and proline. Ramos et al. (2005) reported an endogenous increase in soybean plants due to drought stress. In moringa plants, the biosynthesis of glutamate was increased due to salt stress which increased the biosynthesis of photosynthetic contents (chl. *a*, chl. *b* and carotenoids) and also the content of total amino acids (Hussen et al., 2013). It was reported that water stress resulted in endogenous increase in *Arabidopsis thaliana,* which further increased accumulation of free amino acids (Nambara et al., 1998). Gilbert et al. (1998) reported an increased accumulation of glutamic acid induced by salt stress in *Caleus blumei* plants, which further increased nitrogen containing metabolic compounds and amino acids. Studies reveal that the endogenous glutamate concentration is not enough to induce salt tolerance in plants. They also need an appropriate concentration of exogenous glutamate to increase stress tolerance. The putative role of glutamate in the oxidative defense mechanism has been found to be due not only to its internal metabolism but also to its exogenous use. In soybean, exogenous application of glutamate (as foliar spray and seed priming) increased the antioxidative activity of various enzymes (Teixeira et al., 2017). Abd Allah et al. (2015) reported that, in wheat, the nutritional value of grains, in terms of peroximate composition such as % age of carbohydrates, protein, micro- and macro-nutrients, was increased by exogenously applied glutamic acid at a concentration of 50 and 100 mg/l. Exogenous application (in rooting medium) of glutamate (0.5 mM) was also found to be effective in modulating root growth and branching (Walch-Liu et al., 2006). Furthermore, exogenously applied glutamate (foliar spray) improved quality and productivity and increased growth parameters in *Codiaeum variegatum* plants (Mazher et al., 2011). The information presented depicts the multifarious roles of glutamate in a number of biochemical reactions, and its significance. Although glutamate has been found to play a role in plant stress tolerance through its biosynthesis as well as due to its exogenous use, the details of how it regulates these processes are still to be understood. The missing links still need to be explored (Table 12.4).

12.6 LYSINE

Lysine is a nutritionally important essential amino acid whose level in plants is largely regulated by the rate of its synthesis (Bright et al., 1983). Wide-ranging inquiries have focused on understanding the regulatory mechanisms that control lysine accumulation in seeds. This involves complex processes including synthesis, incorporation into proteins and degradation. Because of the nutritional importance of lysine and threonine, the regulation of their metabolism has been studied extensively at biochemical, genetic, and, more recently, molecular levels. This AA was studied largely because of its importance in nutrition rather than its role in plant stress tolerance through endogenous metabolism or exogenous application. Lysine is formed by the aspartate metabolic pathway, in which three other essential amino acids – threonine, methionine and isoleucine – are also produced (Azevedo et al., 1997).

Many bacterial species and higher plants synthesize lysine from aspartate using two different branches of the aspartate family pathway (Bryan, 1980). Lysine is synthesized in plants by a specific branch of the aspartate family pathway1. This pathway is regulated by end-product feedback inhibition, with lysine inhibiting aspartate kinase (AK) and dihydrodipicolinate synthase (DHDPS). Aspartate kinase is the first enzyme in the pathway and is controlled by feedback inhibited by lysine and threonine, whereas DHDPS is specific to the lysine branch and is inhibited only by lysine. Lysine is also translocated in appreciable amounts (5% of the translocated amino acid pool) from vegetative tissues to the reproductive part. Lysine degradation is important not only for controlling free lysine levels in plant tissues. Recently, several lines of investigations have revealed that lysine degradation might be related to other physiological processes (Arruda et al., 2000). There is continuing interest in its metabolism in higher plants.

The major site of regulation appears to be at the dihydrodipicolinate synthase enzyme. There is also strong evidence for the important role of the enzymes involved in lysine catabolism for the accumulation of this amino acid, and the major advances in lysine metabolism made in more recent years have been in lysine catabolism. Lysine catabolism in higher plants was derived from the studies on amino acid metabolism using compounds labeled with 14C. In maize and barley, C14 lysine was incorporated into glutamic acid, proline and saccharopine (Arruda et al., 1982; Brandt, 1975; Sodek and Wilson, 1970), revealing that lysine catabolism in plants also occurred via the saccharopine pathway. Lysine metabolism in plants has been studied for over 50 years, since the discovery of the maize high-lysine mutant *opaque-2* (*o2*), which contains low levels of lysine-poor storage proteins (zeins) and consequently an increased lysine and tryptophan content compared with the wild-type (Mertz et al., 1964). However, enrichment of lysine levels in crops using classical genetics and breeding approaches is difficult because: (i) lysine synthesis is highly negatively regulated by a feedback inhibition loop in which lysine feedback inhibits the activity of dihydrodipicolinate synthase (DHPS), the first enzyme of the lysine biosynthesis pathway, slowing down its synthesis (Box 1); and (ii) lysine is efficiently degraded by its catabolism into the tricarboxylic (TCA) cycle, a pathway initiated by the bi-functional enzyme LKR/SDH (Box 1), which exhibits both lysine-ketoglutarate reductase (LKR) and saccharopine dehydrogenase (SDH) activities (Karchi et al., 1995). Nevertheless, it is by careful engineering of both of the above enzymes (DHPS and LKR/SDH) that Yang et al. (2016) have now developed two pyramid transgenic lines with free lysine content elevated to 25-fold compared to wild-type without changing the plant phenotype.

12.6.1 Biosynthesis

The precursor of lysine metabolism is an aspartic acid which yields β-aspartate semialdehyde which converts into 2,3-dihydrodipicolinate through the activity of an enzyme, dihydrodipicolinate synthase. The 2,3-dihydrodipicolinate converts into lysine through six enzymatic steps.

12.6.2 Significance Under Stress

The lysine and protein contents of lysine transgenic maize were detected, and stress tolerance was analyzed by measuring the free lysine and chlorophyll contents. The results showed that transgenic lines had not only increased lysine and protein contents but also showed more tolerance to stress (Wang et al., 2013). Moreover, ten stress-related gene expression changes were analyzed using quantitative RT-PCR. The saccharopine pathway is involved in the regulation of lysine levels and stress responses (Arruda and Neshich, 2012). In this pathway, LKR and SDH catalyze, in two reaction steps, the conversion of lysine to AASA and the formation of glutamate. The lysine to AASA pathways could be used to alleviate hyperosmotic stress by producing compatible solutes and preventing the accumulation of toxic aldehydes (Neshich et al., 2013). In *Silicibacterpomeroyi*, the lysine dehydrogenase (LYSDH) pathway is preferentially expressed, as compared to the saccharopine pathway, upon stress and lysine induction. The fact that the gene encoding P5CR is also upregulated in the high-osmotic medium in a lysine-independent manner corroborates the notion that lysine may fuel pipecolate and its derivatives in bacteria subjected to stress (Moulin et al., 2000, 2006). In plants, the lysine-derived from osmo-induced proteolysis increases under stress and stimulates the LKR/SDH expression, channeling lysine to AASA, which can be used to form either Acetyl-CoA or pipecolate (Moulin et al., 2000, 2006). Lysine catabolism may serve to regulate lysine homeostasis in some tissues, while efficiently converting lysine to glutamate and then to other stress-related metabolites in response to stress and certain developmental programs (Galili et al., 2001). LKR/SDH-mediated lysine catabolism provides an example in which knowledge from plant research is far more advanced than that from animals (Galili et al., 2001).

It is possible that, in senescing and stressed tissues, protein hydrolysis leads to transient increases in free lysine concentration. Alternatively, the saccharopine pathway might be responsible for the synthesis of the regulatory molecule(s) involved in developmental processes from root growth to leaf senescence in response to biotic or abiotic stress (Arruda et al., 2000). The production of aspartate kinase enzyme in processes such as abscission and response to abiotic stress is also important. It is not yet known whether these processes are related to the regulation of free lysine levels. As a consequence of the biosynthesis of lysine or its accumulation under stress as a general response to environmental stresses, the cellular energy status is profoundly affected, leading to an energy limitation. Consequently, the expression of genes encoding biosynthetic enzymes of the aspartate family pathway is partially repressed in order to preserve energy (Galili, 2011). In accordance with this suggestion, lysine catabolism has been shown to be stimulated to create additional energy.

It was found that lysine, be it exogenous or endogenous, is helpful in reducing the stress state. For example, Rizwan et al. (2017b) reported that 60 mg/kg of lysine as foliar spray on wheat was effective in reducing the adverse effects of heavy metal stress, specifically in the uptake of Cd. Recently, Noman et al. (2018) experimented with radish plants which were grown under water-deficient conditions. They applied lysine in chelation with Zn as seed priming with a concentration of 6 to 9 mg/kg and found it to be beneficial in overcoming the adverse effects of drought stress. When plants face some environmental disturbance, they activate defense mechanisms for their survival. Muttucumaru et al. (2015) reported that an endogenous increase in the level of lysine content in potatoes grown under drought stress enhanced their tolerance against water stress and stated that the effective increased concentration was about 1.462 mmol kg^{-1} (Table 12.5) It was reported by Zafari and Ebadi (2016) that an increase in lysine levels in safflower grown under water-deficient conditions was found to be effective in enhancing stress tolerance, and this was related with improved physiological traits. Wang et al. (2013) stated that when maize was grown under salinity stress, lysine production was significantly increased, which was helpful for salinity tolerance, and maize plants become capable of showing substantial growth under salt stress. Saeedipour and Moradi, (2012) reported that in different wheat cultivars grown under salt stress, lysine content increased significantly, enhancing tolerance to salt stress and viability. Along with lysine, the contents of other amino acids were also increased. There was a respective change in the concentration of lysine from 0.24 to 0.37 μmol g^{-1} FW and 0.34 to 1.5 in two cultivars respectively. Though the studies reveal a potential role of lysine in plant stress tolerance through different modes, its effective concentration as well as the tolerance mechanisms are still to be explored (Table 12.5).

12.7 METHIONINE

Of all the amino acids, methionine (Met) is the only one which has the different compositions of *Noccaea caerulescens* (NC) and *Arabidopsis halleri* (AH). Methionine biosynthesis in plants mainly takes place through the aspartate metabolic pathway through the involvement of different metabolites, using three convergent pathways. In this reaction, the S is taken from cysteine, with a backbone of aspartate, and the methyl group is taken from the β carbon of the serine amino acid. Like other amino acids, it has other functions besides its significant role in the biosynthesis of proteins. It is involved in the preparation of S-adenosylmethionine, in Transmethylation. It donates its methyl group and is involved in the biosynthesis of phytohormones such as ethylene and polyamines (Zemanová et al., 2014).

As methionine is very sensitive to oxidation processes, they immediately convert methionine into sulfoxide, which results in the altered activity of methionine as well as the biosynthesis of different proteins. Beside having oxidative properties, this amino acid is also considered as an antioxidant. According to the findings of Maxwell and Kieber (2004), AAs are highly involved in the formation of growth regulators; for instance, methionine is involved in the biosynthesis of auxins, brassinosteroids and cytokinins in plants (El-Awadi et al., 2011). Not only is the internal metabolism of AAs beneficial for the improvement of various physio-biochemical

TABLE 12.5
Roles of Lysine Under Various Abiotic Stresses

Sr. no.	Stress type	Crop type	Source of lysine	Concentration	Effects	References
1	Water stress	*Solanum tuberosum*	Endogenous increase	1.462 m mol/kg FW	Improvement in drought tolerance and yield	Muttucumaru et al. (2015)
2	Water stress	*Raphanus sativus*	Exogenous (seed priming)	6 and 9ppm	Improved drought tolerance, yield and nutritional quality	Noman et al. (2018)
3	Heavy metal stress	*Triticum aestivum* L.	Exogenous (foliar spray)	60ppm	Reduced Cd uptake	Rizwan et al. (2017b)
4	Water stress	Safflower	Endogenous increase	120 mm	Improvement in different physiological traits	Zafari and Ebadi (2016)
5	Salt stress	*Zea mays* l.	Endogenous increase	2.7g/100g	Improved salt tolerance and nutritive quality	Wang et al. (2013)
6	Nutritive stress	Grain, vegetable and weed	Endogenous increase	6g/100g	Improved nutritional quality	Andini et al. (2013)(in review article)
7	Salt stress	*Triticum aestivum* L.	Endogenous increase along with other AAs	0.24–0.37 $\mu molg^{-1}$ FW and 0.34–1.5 $\mu molg^{-1}$ FW	Increase was cultivarspecific Improved salt tolerance	Saeedipour and Moradi (2012)

attributes, but their exogenous use is also found to be helpful. In corn, the dry weight of shoots was increased by exogenous application of methionine (Chen et al., 2005), L-methionine, L-phenylalanine and L-tryptophan associated with the nutrient uptake (Chen et al., 2005). It was found that the chances of the uptake of different components such as phosphorous, potassium and nitrogen were increased by treating the corn with methionine (Chen et al., 2005).

12.7.1 Methionine Biosynthesis

Aspartate is a precursor for the biosynthesis of many other amino acids, such as methionine, lysine and isoleucine as well as threonine. Aspartate catalyzes with the help of the β-aspartyle phosphate enzyme into aspartate semialdehyde, a substrate for enzyme homoserine dehydrogenase. This enzyme converts aspartate semialdehyde into homoserine, which yields cystothionine and therionine to give methionine and isoleucine respectively.

12.7.2 Significance Under Stress

The use of amino acids through different modes has recently gained great interest in the agricultural sector. It is used through foliar spray alone or in combination with other amino acids. Recently, there has been increasing interest in chelation with different micro- and macronutrients as seed priming or foliar spray. Organic Crop Protectants Pvt. Ltd is an agricultural company providing different formulations of micro-nutrients chelated essential amino acids to fulfill the crop micro-nutrient as well as AAs requirements, under water-deficit conditions in order to reduce hidden hunger. It was found that the application of methionine (6.4 mg/g) as a foliar spray improved fruit ripening, along with having a role in stomatal regulation for a better water use through exogenous use (www.ocp.com.au). It was reported that the consumption of S-Methyl methionine through exogenous use improved the contents of stress-protective compounds such as flavonoids, phenolic and polyamines (growth regulators) under stressful conditions (Refs). An increase in the contents of these compounds showed improvements in stress tolerance in a stressful environment, which in turn provided protection (Ludmerszki et al., 2011). Under salt stress, an increase in methionine biosynthesis was observed in *Arabidopsis thaliana* (Ogawa and Mitsuya, 2012), showing its role in a stressful environment. Similarly, in pea plants, the biosynthesis of methionine was seen to have a role in seed development. Different studies also revealed the protective role of methionine on photosynthetic mechanisms against oxidative stress. It also provided chloroplast protection. However, the contents of carotenoids decreased in different varieties, showing a negative correlation between methionine and carotenoids. It was found that in drought conditions, plants show tolerance through the transformation of polyamine into methionine (Zafari and Ebadi, 2016). Furthermore, the accumulation of methionine was found to be cultivar and drought-level specific. For example, in safflower grown under different levels of stress, different cultivars showed a different response to methionine accumulation (Zafari and Ebadi, 2016). The methionine of the Golasht and Sina cultivar was decreased when evaporation was increased from 80 to 120 mm. On the other hand, methionine contents increased when evaporation was at 160 mm, as compared to normal conditions. In the Faraman safflower cultivar, the level of methionine increased when evaporation was at 160 mm and then at higher levels there was a decrease in methionine. Similarly, in canola, drought stress increased the levels of methionine (Zafari and Ebadi, 2016). It has been reported that sulfur-containing metabolic compounds including amino acids methionine and cysteine have a potential role in modifying cellular physiological and molecular pathways. The changes in these metabolic processes due to sulfur-containing metabolites have been found to be effective in the induction of stress tolerance (Mukhtar et al., 2016). Sadak et al. (2015) reported that in faba beans, exogenous use of a mixture of amino acids containing 0.23%–0.3% methionine improved stress tolerance (Sadak et al., 2015) (Table 12.6).

In wheat, enhanced biosynthesis of methionine increased salinity tolerance; this was associated with its concentration of 0.19 g/100g FW (Cornelia et al., 2011). In another study on soya bean, it was found that internal concentration of methionine increased to 10.9 mg/g, which was helpful for the induction of salt tolerance (Farhangi-Abriz and Ghassemi-Golezani, 2016). Furthermore, in cotton under drought stress, the biosynthesis of methionine increased to a level of 4.34 mmol/kg and 2.84 mmol/kg in two cultivars, IAC 13-1 and IAC 20 respectively, effectively enhancing tolerance against drought stress (Marur et al., 1994). In another study, an enhanced biosynthesis of methionine was recorded to increase tolerance under salt stress. This was associated with its role in osmotic adjustment and improvement in protein content (Neto et al., 2009). It was reported that the accumulated concentration was cultivar specific. A concentration of 14.7 mmol g^{-1} was found in cultivar BR5033 and a concentration of 17.5 mmol g^{-1} in cultivar BR5011 (Table 12.6).

TABLE 12.6
Roles of Methionine Under Various Abiotic Stresses

Type of stress	Plant species	Source of amino acid (exogenous/ endogenous)	Effective concentration of amino Acid	Effects or improvements	References
Drought	*Triticum aestivum* L.	Foliar spray in mixture	0.2 mg/ml	Improvement in plant–water relation, physio-biochemical attributes, yield and nutritional quality	Hammad and Ali (2014)
Biotic (viral infection)	*Zea mays* l.	Foliar spray (S methyl-methionine)98	0.01 g/L	Improved the photosynthetic rate and the amount of chlorophyll pigments, and significantly improved the plant's defense response to viral infection	Ludmerszki et al. (2011)
Salt stress	*Vicia faba*	Foliar spray	0.23–0.3 %	Alleviated the harmful effects of salinity	Sadak et al. (2015)
Salt stress	*Triticum aestivum* L.	Endogenous increase	Cellular level 0.19 g/100g FW	No significant effect	Cornelia et al. (2011)
Salt stress	Soybean	Endogenous increase	Internal cellular concentration 10.9 mg/g	Enhanced salinity tolerance	Farhangi-Abriz and Ghassemi-Golezani (2016)
Drought stress	*Gossypium arboretum* L.	Endogenous increase	Cellular level 4.34 m mol/kg (IAC 13–1) 2.84 m mol/kg (IAC 20)	Enhanced drought stress tolerance	Marur et al. (1994)
Drought stress	*Vigna unguiculata*	Exogenous application as foliar spray	4.0 mM	Improved growth, yield and antioxidative defence mechanism	Merwad et al. (2018)
Salt stress	*Zea mays* l.	Endogenous increase	14.7 m mol g^{-1} [salt-tolerant (BR5033)] 17.5 m mol g^{-1} [salt-sensitive (BR5011)]	Role in salt tolerance with improvement in protein content as well as role in osmotic adjustment, cultivarspecific effect	Neto et al. (2009)
Non-stress	*Chlorophyllum molybdites*	Exogenous	Foliar application of methionine at 100 mg L^{-1}. Foliar application of methionine at 200 mg L^{-1}	Increased the absorption of N, P, and K. Increased dry weight of shoot, leaf area and chlorophyll contents	Shekari and Javanmardi (2017)

Stress tolerance in plants was obtained not only through endogenous improvement but also by exogenous application through different modes with different levels. For example, in wheat, foliar spray of methionine in a mixture increased drought tolerance and was associated with an increase in plant–water relations, physio-biochemical attributes, yield attributes and nutrition quality when applied at 0.2 mg/ml (Hammad and Ali, 2014). In another study on maize, foliar spray of methionine with a concentration of 0.01 g/L enhanced physiological parameters including photosynthetic rate and chlorophyll pigment content with an improved defense mechanism (Ludmerszki et al., 2011). In another study on faba bean, foliar spray of methionine increased salinity tolerance, and it was reported that a 0.23–0.3% concentration of methionine was found to be most effective for salinity tolerance (Sadak et al., 2015). Furthermore, in cowpea, a 0.4 mM concentration of methionine was effective in enhancing stress tolerance and improving growth, yield characteristics, contents of chlorophyll, carotenoids, shoot and seed nutrients and other components (Merwad et al., 2018). In a study by Shekari and Javanmardi (2017), two levels of methionine, i.e. 100 mg L^{-1} and 200 mg L^{-1}, on green parasol plants improved the absorption of N, P and K, which was associated with increases in dry weight of shoots, leaf area and leaf chlorophyll contents (Table 12.6).

12.8 CONCLUSIONS

The recent investigations on the roles of amino acids in the metabolism, physiology and different developmental processes of higher plants showed their potential for the induction of stress tolerance. Though a lot of work has been done in order to understand this tolerance mechanism, there are still some gaps in the resistance and tolerance mechanisms. Furthermore, studies have been carried out mainly on some specific amino acids. They revealed that amino acid metabolism could be a vital component for plant abiotic stress tolerance. The knowledge of amino acid biosynthesis genes will provide help in understanding not only the precise role of amino acids in abiotic stress contexts, but also the signaling mechanism. So, more investigation is needed to devise the complex mechanism of amino acid metabolism in developing stress tolerant plants that will be better equipped to face environmental adversities.

REFERENCES

Abbas, W., Ashraf, M., Akram, N.A. (2010) Alleviation of salt-induced adverse effects in eggplant (*Solanum melongena* L.) by glycinebetaine and sugarbeet extracts. *Scientia Horticulturae* 125(3): 188–195.

Abd Allah, M.M.S., El-Bassiouny, H.M.S., Bakry, B.A., Sadak, M.S. (2015) Effect of arbuscular mycorrhiza and glutamic acid on growth, yield, some chemical composition and nutritional quality of wheat plant grown in newly reclaimed sandy soil. *Research Journal of Pharmaceutical, Biological and Chemical Sciences* 6(3): 1038–1054.

Adams, E., Frank, L. (1980) Metabolism of proline and the hydroxyprolines. *Annual Review of Biochemistry* 49(1): 1005–1061.

Agarwal, P.K., Agarwal, P., Reddy, M.K., Sopory, S.K. (2006) Role of DREB transcription factors in abiotic and biotic stress tolerance in plants. *Plant Cell Reports* 25(12): 1263–1274.

Akladious, S.A., Abbas, S.M. (2013) Alleviation of seawater stress on tomato by foliar application of aspartic acid and glutathione. *Journal of Stress Physiology & Biochemistry* 3: 282–298.

Aldesuquy, H.S., Abbas, M.A., Abo-Hamed, S.A., Elhakem, A.H. (2013) Does glycine betaine and salicylic acid ameliorate the negative effect of drought on wheat by regulating osmotic adjustment through solutes accumulation? *Journal of Stress Physiology & Biochemistry* 9(3): 5–22.

Ali, Q., Ashraf, M. (2011a) Exogenously applied glycinebetaine enhances seed and seed oil quality of maize (*Zea mays* L.) under water deficit conditions. *Environmental and Experimental Botany* 71(2): 249–259.

Ali, H.M., Saddiqui, M.H., Al-Whaibi, M.H., Basalah, M.O., Sakran, A.M., El-Zaidy, M. (2013) Effect of proline and abcisic acid on the growth and physiological performance of faba bean under water stress. *Pakistan Journal of Botany* 45(3): 933–340.

Ali, Q., Ashraf, M. (2011b) Induction of drought tolerance in maize (*Zea mays* L.) due to exogenous application of trehalose: Growth, photosynthesis, water relations and oxidative defence mechanism. *Journal of Agronomy and Crop Science* 197(4): 258–271.

Ali, Q., Ashraf, M., Athar, H.U.R. (2007) Exogenously applied proline at different growth stages enhances growth of two maize cultivars grown under water deficit conditions. *Pakistan Journal of Botany* 39(4): 1133–1144.

Ali, Q., Haider, M.Z., Iftikhar, W., Jamil, S., Tariq Javed, M., Noman, A., Iqbal, M., Perveen, R. (2016) Drought tolerance potential of *Vigna mungo* L. lines as deciphered by modulated growth, antioxidant defense, and nutrient acquisition patterns. *Brazilian Journal of Botany* 39(3): 801–812.

Ali, Q., Javed, M.T., Noman, A., Haider, M.Z., Waseem, M., Iqbal, N., Waseem, M., *et al.* (2018) Assessment of drought tolerance in mung bean cultivars/lines as depicted by the activities of germination enzymes, seedling's antioxidative potential and nutrient acquisition. *Archives of Agronomy and Soil Science* 64(1): 84–102.

Ali, R.M., Elfeky, S.S., Abbas, H. (2008) Response of salt-stressed *Ricinus communis* L. To exogenous application of glycerol and/or aspartic acid. *Journal of Biological Sciences* 8(1): 171–175.

Alía, R., Moro-Serrano, J., Notivol, E. (2001) Genetic variability of Scots pine (*Pinus sylvestris*) provenances in Spain: Growth traits and survival. *Silva Fennica* 35(1): 27–38.

Amin, A.A., Abouziena, H.F., Abdelhamid, M.T., Rashad, E.S.M., Gharib, A.F. (2014) Improving growth and productivity of faba bean plants by foliar application of thiourea and aspartic acid. *International Journal of Plant Soil Science* 3: 724–736.

Andini, R., Yoshida, S., Ohsawa, R. (2013) Variation in protein content and amino acids in the leaves of grain, vegetable and weedy types of amaranths. *Agronomy* 3(2): 391–403.

Apel, K., Hirt, H. (2004) Reactive oxygen species: Metabolism, oxidative stress, and signal transduction. *Annual Review of Plant Biology* 55: 373–399.

Arruda, P., Kemper, E.L., Papes, F., Leite, A. (2000) Regulation of lysine catabolism in higher plants. *Trends in Plant Science* 5(8): 324–330.

Arruda, P., Neshich, I.P. (2012) Nutritional-rich and stress-tolerant crops by saccharopine pathway manipulation. *Food and Energy Security* 1(2): 141–147.

Arruda, P., Sodek, L., da Silva, W.J. (1982) Lysine-ketoglutarate reductase activity in developing maize endosperm. *Plant Physiology* 69(4): 988–989.

Ashraf, M., Foolad, M.R. (2007) Roles of glycine betaine and proline in improving plant abiotic stress resistance. *Environmental and Experimental Botany* 59(2): 206–216.

Ashraf, M., Ali, Q. (2008) Relative membrane permeability and activities of some antioxidant enzymes as the key determinants of salt tolerance in canola (Brassica napus L.). *Environmental and Experimental Botany* 63(1–3): 266–273.

Athar, H.U.R., Ashraf, M., Wahid, A., Jamil, A. (2009) Inducing salt tolerance in canola (Brassica napus L.) by exogenous application of glycine betain and proline: response at initial growth stages. *Pakistan Journal of Botany* 41(3): 1311–1319.

Azevedo, R.A., Arruda, P., Turner, W.L., Lea, P.L. (1997) The biosymthesis and metabolism of aspartate derived amino acids in higher plant. *Phytochemistry* 46: 395–419.

Azevedo, R.A., Lancien, M., Lea, P.J. (2006) The aspartic acid metabolic pathway, an exciting and essential pathway in plants. *Amino Acids* 30(2): 143–162.

Ben-Rouina, B., Ben-Ahmed, C., Boukhriss, M. (2006) Water relations, proline accumulation and photosynthetic activity in olive tree (*Olea europaea* L. cv "Chemlali") in response to salt stress. *Pakistan Journal of Botany* 38(5): 1397–1406.

Bouche, N., Fromm, H. (2004) GABA in plants: Just a metabolite? *Trends in Plant Science* 9(3): 110–115.

Boudsocq, M., Laurière, C. (2005) Osmotic signaling in plants. Multiple pathways mediated by emerging kinase families. *Plant Physiology* 138(3): 1185–1194.

Bown, A.W., MacGregor, K.B., Shelp, B.J. (2006) Gamma-aminobutyrate: Defense against invertebrate pests? *Trends in Plant Science* 11(9): 424–427.

Brandt, A.B. (1975) In vivo incorporation of lysine-C14 into the endosperm of wild type and high lysine barley. *FEBS Letters* 52(2): 288–291.

Bright, S.W.J., Shewry, P.R., Kasarda, D.D. (1983) Improvement of protein quality in cereals. *Critical Reviews in Plant Sciences* 1(1): 49–93.

Brouquisse, R., Weigel, P., Rhodes, D., Yocum, C.F., Hanson, A.D. (1989) Evidence for a ferredoxin-dependent choline monooxygenase from spinach chloroplast stroma. *Plant Physiology* 90(1): 322–329.

Bryan, J.K. (1980) Synthesis of the aspartate family and branched-chain amino acids. In *Amino Acids and Derivatives*, ed. Miflin, B.J., 403–452. Academic Press: New York.

Butt, M., Ayyub, C.M., Amjad, M., Ahmad, R. (2016) Proline application enhances growth of chilli by improving physiological and biochemical attributes under salt stress. *The Pakistan Journal of Agricultural Sciences* 53(1): 43–49.

Carden, D.E., Walker, D.J., Flowers, T.J., Miller, A.J. (2003) Single-cell measurements of the contributions of cytosolic Na^+ and K^+ to salt tolerance. *Plant Physiology* 131(2): 676–683.

Chen, M., Cheng, B., Zhang, Q., Ding, Y., Yang, Z., Liu, P. (2005) Effects of applying L-methionine, L-phenylalanine and L-tryptophan on Zea mays growth and its nutrient uptake. *The Journal of Applied Ecology* 16(6): 1033–1037.

Chen, Y.L., Li, Q.Z. (2007) Prediction of apoptosis protein subcellular location using improved hybrid approach and pseudo-amino acid composition. *Journal of Theoretical Biology* 248(2): 377–381.

Chen, T.H., Murata, N. (2002) Enhancement of tolerance of abiotic stress by metabolic engineering of betaines and other compatible solutes. *Current Opinion in Plant Biology* 5(3): 250–257.

Chen, T.H., Murata, N. (2008) Glycine betaine: An effective protectant against abiotic stress in plants. *Trends in Plant Science* 13(9): 499–505.

Cornelia, P., Bogdan, A.T., Ipate, L., Chis, A., Borbely, M.V. (2011) Exogenous salicylic acid involvement in ameliorating the negative effect of salt stress in wheat (*Triticum aestivum* cv *crisana*) plants in vegetative stage. *Analele Universității din Oradea, Fascicula Protecția Mediului* 7: 137–146.

Da Silva, J.M., Arrabaça, M.C. (2004) Photosynthesis in the water-stressed C4 grass Setaria sphacelata is mainly limited by stomata with both rapidly and slowly imposed water deficits. *Physiologia Plantarum* 121(3): 409–420.

Dawood, M.G., Taie, H.A.A., Nassar, R.M.A., Abdelhamid, M.T., Schmidhalter, U. (2014) The changes induced in the physiological, biochemical and anatomical characteristics of *Vicia faba* by the exogenous application of proline under seawater stress. *South African Journal of Botany* 93: 54–63.

Deivanai, S., Xavier, R., Vinod, V., Timalata, K., Lim, O.F. (2011) Role of exogenous proline in ameliorating salt stress at early stage in two rice cultivars. *Journal of Stress Physiology & Biochemistry* 7(4): 157–174.

Delauney, A.J., Verma, D.P.S. (1993) Proline biosynthesis and osmoregulation in plants. *The Plant Journal* 4(2): 215–223.

Desikan, R., Cheung, M.K., Bright, J., Henson, D., Hancock, J.T., Neill, S.J. (2004) ABA, hydrogen peroxide and nitric oxide signalling in stomatal guard cells. *Journal of Experimental Botany* 55(395): 205–212.

Dinu, C.A., Moraru, D., Paraschiv, N.L. (2011) The physiology of glutamic acid. *Agronomy Series of Scientific Research/Lucrari Stiintifice Seria Agronomie* 54(2): 53–55.

El-Awadi, M., El-Bassiony, A., Fawzy, Z., El-Nemr, M. (2011) Response of Snap Bean (Phaseolus vulgaris L.) lants to nitrogen fertilizer and foliar application with methionine and tryptophan. *Nature and Science* 9: 87–94.

Elewa, T.A., Sadak, M.S., Saad, A.M. (2017) Proline treatment improves physiological responses in quinoa plants under drought stress. *Bio-Science Research* 14: 21–33.

Farhangi-Abriz, S., Ghassemi-Golezani, K. (2016) Improving amino acid composition of soybean under salt stress by salicylic acid and jasmonic acid. *Journal of Applied Botany and Food Quality* 89: 243–248.

Fichman, Y., Gerdes, S.Y., Kovács, H., Szabados, L., Zilberstein, A., Csonka, L.N. (2015) Evolution of proline biosynthesis: Enzymology, bioinformatics, genetics, and transcriptional regulation. *Biological Reviews of the Cambridge Philosophical Society* 90(4): 1065–1099.

Forde, B.G., Lea, P.J. (2007) Glutamate in plants: Metabolism, regulation, and signalling. *Journal of Experimental Botany* 58(9): 2339–2358.

Fraser, C.M., Chapple, C. (2011) The phenylpropanoid pathway in Arabidopsis. *The Arabidopsis Book* 9: e0152

Galili, G. (2011) The aspartate-family pathway of plants: Linking production of essential amino acids with energy and stress regulation. *Plant Signaling & Behavior* 6(2): 192–195.

Galili, G., Tang, G., Zhu, X., Gakiere, B. (2001) Lysine catabolism: A stress and development super-regulated metabolic pathway. *Current Opinion in Plant Biology* 4(3): 261–266.

Genard, H., Le Saos, J., Hillard, J., Tremolieres, A., Boucaud, J. (1991) Effect of salinity on lipid composition, glycine betaine content and photosynthetic activity in chloroplasts of Suaeda maritime. *Plant Physiology and Biochemistry* 29: 421–427.

Geshnizjani, N., Khosh-Khui, M. (2016) Promoted growth and improved quality of Gerbera jamesonni L. flowers using exogenous application of amino acids. *International Journal of Horticultural Science and Technology* 3(2): 155–166.

Gilbert, G.A., Gadush, M.V., Wilson, C., Madore, M.A. (1998) Amino acid accumulation in sink and source tissues of *Coleus blumei* Benth. during salinity stress. *Journal of Experimental Botany* 49(318): 107–114.

Giri, J. (2011) Glycinebetaine and abiotic stress tolerance in plants. *Plant Signaling & Behavior* 6(11): 1746–1751.

Gupta, N., Thind, S.K. (2017) Grain yield response of drought stressed wheat to foliar application of glycine betaine. *Indian Journal of Agricultural Research* 51(3): 287–291.

Gupta, N., Thind, S.K., Bains, N.S. (2014) Glycine betaine application modifies biochemical attributes of osmotic adjustment in drought stressed wheat. *Plant Growth Regulation* 72(3): 221–228.

Haffani, S., Mezni, M., Slama, I., Ksontini, M., Chaibi, W. (2014) Plant growth, water relations and proline content of three vetch species under water-limited conditions. *Grass and Forage Science* 69(2): 323–333.

Hammad, S.A.R., Ali, O.A.M. (2014) Physiological and Biochemical studies on drought tolerance of wheat plants by application of amino acids and yeast extract. *Annals of Agricultural Sciences* 59(1): 133–145.

Hanower, P., Brzozowska, J. (1975) Effects of osmotic stress on composition of free amino acids in cotton leaves. *Phytochemistry* 14(8): 1691–1694.

Hanson, A.D., Gregory III, J.F. (2011) Folate biosynthesis, turnover, and transport in plants. *Annual Review of Plant Biology* 62: 105–125.

Hanson, A.D., Scott, N.A. (1980) Betaine synthesis from radioactive precursors in attached, water-stressed barley leaves. *Plant Physiology* 66(2): 342–348.

Harinasut, P., Tsutsui, K., Takabe, T., Nomura, M., Takabe, T., Kishitani, S. (1996) Exogenous glycinebetaine accumulation and increased salt tolerance in rice seedlings. *Bioscience, Biotechnology, and Biochemistry* 60(2): 366–368.

Harris, D., Rashid, A., Ali, S., Hollington, P.A. (2004) On-farm seed priming with maize in Pakistan. In *Proceedings of the 8th Asian regional maize workshop: New Technologies for the New Millennium*, Bangkok: Thailand, January 3, 2001. 316–324.

Harris, K., Subudhi, P.K., Borrel, A., Jordan, D., Rosenow, D., Nguyen, H., Klein, P., Klein, R., Mullet, J. (2007) Sorghum stay-green QTL individually reduce post-flowering drought-induced leaf senescence. *Journal of Experimental Botany* 58(2): 327–338.

Hassanein, R.A., El-Khawas, S.A., Ibrahim, S.K., El-Bassiouny, H.M., Mostafa, H.A., Abdel-Monem, A.A. (2013) Improving the thermo tolerance of wheat plant by foliar application of arginine or putrescine. *Pakistan Journal of Botany* 45(1): 111–118.

He, G., Wu, C., Huang, J., Zhou, R. (2017) Effect of exogenous proline on metabolic response of Tetragenococcus halophilus under salt stress. *Journal of Microbiology and Biotechnology* 27(9): 1681–1691.

Hertz, L., Peng, L., Dienel, G.A. (2007) Energy metabolism in astrocytes: High rate of oxidative metabolism and spatiotemporal dependence on glycolysis/glycogenolysis. *Journal of Cerebral Blood Flow and Metabolism: Official Journal of the International Society of Cerebral Blood Flow and Metabolism* 27(2): 219–249.

Hirel, B., Lea, P.J. (2001) Ammonia assimilation. In *Plant Nitrogen*, ed. Lea, P.J. and Morot-Gaudry, J.F., 79–99. Springer: Berlin Heidelberg.

Holländer-Czytko, H., Grabowski, J., Sandorf, I., Weckermann, K., Weiler, E.W. (2005) Tocopherol content and activities of tyrosine aminotransferase and cystine lyase in *Arabidopsis* under stress conditions. *Journal of Plant Physiology* 162(7): 767–770.

Hong-Bo, S., Xiao-Yan, C., Li-Ye, C., Xi-Ning, Z., Gang, W., Yong-Bing, Y., Chang-Xing, Z., Zan-Min, H. (2006) Investigation on the relationship of proline with wheat anti-drought under soil water deficits. *Colloids and Surfaces. B, Biointerfaces* 53(1): 113–119.

Hossain, M.A., Hasanuzzaman, M., Fujita, M. (2010) Up-regulation of antioxidant and glyoxalase systems by exogenous glycinebetaine and proline in mung bean confer tolerance to cadmium stress. *Physiology and Molecular Biology of Plants: An International Journal of Functional Plant Biology* 16(3): 259–272.

Huang, J., Hirji, R., Adam, L., Rozwadowski, K.L., Hammerlindl, J.K., Keller, W.A., Selvaraj, G. (2000) Genetic engineering of glycinebetaine production toward enhancing stress tolerance in plants: Metabolic limitations. *Plant Physiology* 122(3): 747–756.

Hussain, M., Malik, M.A., Farooq, M., Ashraf, M.Y., Cheema, M.A. (2008) Improving drought tolerance by exogenous application of glycinebetaine and salicylic acid in sunflower. *Journal of Agronomy and Crop Science* 194(3): 193–199.

Hussen, M.M., Lobna, S., Taha, L., Rawia, A.E., Soad, M.M.I. (2013) Responses of photosynthetic pigments and amino acids content of Moringa plants to salicylic acid and salinity. *Journal of Applied Sciences Research* 9(8): 4889–4895.

Ikuta, S., Matuura, K., Imamura, S., Misaki, H., Horiuti, Y. (1977) Oxidative pathway of choline to betaine in the soluble fraction prepared from *Arthrobacter globiformis*. *Journals of Biochemst* 82: 157–163.

Imlay, J.A. (2003) Pathways of oxidative damage. *Annual Review of Microbiology* 57: 395–418.

Islam, M.M., Hoque, M.A., Okuma, E., Banu, M.N.A., Shimoishi, Y., Nakamura, Y., Murata, Y. (2009) Exogenous proline and glycinebetaine increase antioxidant enzyme activities and confer tolerance to cadmium stress in cultured tobacco cells. *Journal of Plant Physiology* 166(15): 1587–1597.

Kadotani, N., Akagi, A., Takatsuji, H., Miwa, T., Igarashi, D. (2016) Exogenous proteinogenic amino acids induce systemic resistance in rice. *BMC Plant Biology* 16(60): 60.

Kahlaoui, B., Hachicha, M., Misle, E., Fidalgo, F., Teixeira, J. (2015) Physiological and biochemical responses to the exogenous application of proline of tomato plants irrigated with saline water. *Journal of the Saudi Society of Agricultural Sciences* 17(1): 17–23.

Kahlaoui, B., Hachicha, M., Teixeira, J., Misle, E., Fidalgo, F., Hanchi, B. (2013) Response of two tamoto cultivars to field-applied proline and salt stress. *Journal of Stress Physiology and Biochemistry* 9(3): 357–365.

Kamran, M., Shahbaz, M., Ashraf, M., Akram, N.A. (2009) Alleviation of drought-induced adverse effects in spring wheat (*Triticum aestivum* L.) using proline as a pre-sowing seed treatment. *Pakistan Journal of Botany* 41(2): 621–632.

Kamyab, M., Kafi, M., Shahsavand, H., Goldani, M., Shokouhifar, F. (2016) Exploring ion homeostasis and mechanism of salinity tolerance in primary tritipyrum lines (wheat thinopyrum bessarabicum) in the presence of salinity. *Australian Journal of Crop Science* 10(7): 911–919.

Kan, C.C., Chung, T.Y., Wu, H.Y., Juo, Y.A., Hsieh, M.H. (2017) Exogenous glutamate rapidly induces the expression of genes involved in metabolism and defense responses in rice roots. *BMC Genomics* 18(1): 186.

Karchi, H., Miron, D., Ben-Yaacov, S., Galili, G. (1995) The lysine-dependent stimulation of lysine catabolism in tobacco seed requires calcium and protein phosphorylation. *The Plant Cell* 7(11): 1963–1970.

Kathuria, H., Giri, J., Nataraja, K.N., Murata, N., Udayakumar, M., Tyagi, A.K. (2009) Glycinebetaine-induced water-stress tolerance in codA-expressing transgenic indica rice is associated with up-regulation of several stress responsive genes. *Plant Biotechnology Journal* 7(6): 512–526.

Katoh, A., Uenohara, K., Akita, M., Hashimoto, T. (2006) Early steps in the biosynthesis of NAD in Arabidopsis start with aspartate and occur in the plastid. *Plant Physiology* 141(3): 851–857.

Kaul, S., Sharma, S.S., Mehta, I.K. (2008) Free radical scavenging potential of L-proline: evidence from in vitro assays. *Amino Acids* 34: 315–320.

Kemble, A.R., MacPherson, H.T. (1954) Liberation of amino acids in perennial rye grass during wilting. *The Biochemical Journal* 58(1): 46–49.

Khan, M.S., Shah, J.S., Ullah, M. (2016) Assesement of salinity stress and the protective effects of glycine betaine on local wheat varieties. *Journal of Agriculture and Biological Sciences* 11(9): 360–366.

Khedr, A.H.A., Abbas, M.A., Wahid, A.A.A., Quick, W.P., Abogadallah, G.M. (2003) Proline induces the expression of salt-stress-responsive proteins and may improve the adaptation of *Pancratium maritimum* L. to salt-stress. *Journal of Experimental Botany* 54(392): 2553–2562.

Kishitani, S., Takanami, T., Suzuki, M., Oikawa, M., Yokoi, S., Ishitani, M., Alvarez-Nakase, A.M., Takabe, T., Takabe, T. (2000) Compatibility of glycinebetaine in rice plants: Evaluation using transgenic rice plants with a gene for peroxisomal betaine aldehyde dehydrogenase from barley. *Plant, Cell & Environment* 23(1): 107–114.

Kowalczyk, K., Zielony, T., Gajewski, M. (2008) Effect of Aminoplant and Asahi on yield and quality of lettuce grown on rockwool. In *Biostimulators in Modern Agriculture Vegetable Crops*, ed. Dąbrow, Z.T., 35–43. Wieś Jutra, Warsaw: Poland.

Krishnan, V., Han, M.H., Mazei-Robison, M., Iñiguez, S.D., Ables, J.L., Vialou, V., Berton, O., et al. (2008) AKT signaling within the ventral tegmental area regulates cellular and behavioral responses to stressful stimuli. *Biological Psychiatry* 64(8): 691–700.

Lavanya, S.N., Amruthesh, K.N. (2017) Glycine betaine mediated disease resistance against Sclerospora graminicola in Pearl Millet. *Journal of Applied Biology & Biotechnology* 5(3): 45–51.

Lei, P., Pang, X., Feng, X., Li, S., Chi, B., Wang, R., Xu, H. (2017) The microbe-secreted isopeptide poly-γ-glutamic acid induces stress tolerance in *Brassica napus* L. seedlings by activating crosstalk between H_2O_2 and Ca^{2+}. *Scientific Reports* 7: 1–15.

Leisinger, T. (1987) Biosynthesis of proline. In *Escberichia coli* and *Salmonella typhimurium: Cellular and Molecular Biology*, ed. Neidhardt, F.C., Ingraham, J.L., Brooks, K., Magasanik, M., Schaechter, M., Umbarger, H.E., 345–351. Washington, DC: American Society for Microbiology.

Less, H., Angelovici, R., Tzin, V., Galili, G. (2011) Coordinated gene networks regulating *Arabidopsis* plant metabolism in response to various stresses and nutritional cues. *The Plant Cell* 23(4): 1264–1271.

Lima, J.D., Sodek, L. (2003) N-stress alters aspartate and asparagine levels of xylem sap in soybean. *Plant Science* 165(3): 649–656.

LiXin, Z., Xiu, L.S., Suo, L.Z. (2009) Differential plant growth and osmotic effects of two maize (*Zea mays*) cultivars to exogenous glycine betaine application under drought stress. *Plant, Cell and Environment* 58: 297–305.

Lu, S.C. (2013) Glutathione synthesis. *Biochimica et Biophysica Acta (BBA)-General Subjects* 1830(5): 3143–3153.

Ludmerszki, E., Rudnóy, S., Almási, A., Szigeti, Z., Rácz, I. (2011) The beneficial effects of S-methyl-methionine in maize in the case of maize dwarf mosaic virus infection. *Acta Biologica Szegediensis* 55(1): 109–112.

Lutts, S. (2000) Exogenous glycine betaine reduces sodium accumulation in salt-stressed rice plants. *International Rice Research Notes* 25: 39–40.

Ma, X.L., Wang, Y.J., Xie, S.L., Wang, C., Wang, W. (2007) Glycinebetaine application ameliorates negative effects of drought stress in tobacco. *Russian Journal of Plant Physiology* 54(4): 472–479.

Mäkelä, P., Kontturi, M., Pehu, E., Somersalo, S. (1999) Photosynthetic response of drought- and salt-stressed tomato and turnip rape plants to foliar-applied glycinebetaine. *Physiologia Plantarum* 105(1): 45–50.

Mäkelä, P., Munns, R., Colmer, T.D., Condon, A.G., Peltonen-Sainio, P. (1998) Effect of foliar applications of glycinebetaine on stomatal conductance, abscisic acid and solute concentrations in leaves of salt-or drought-stressed tomato. *Functional Plant Biology* 25(6): 655–663.

Mansour, M.M.F. (2000) Nitrogen containing compounds and adaptation of plants to salinity stress. *Biologia Plantarum* 43(4): 491–500.

Martinelli, T., Whittaker, A., Bochicchio, A., Vazzana, C., Suzuki, A., Masclaux-Daubresse, C. (2007) Amino acid pattern and glutamate metabolism during dehydration stress in the 'resurrection' plant Sporobolus stapfianus: A comparison between desiccation-sensitive and desiccation-tolerant leaves. *Journal of Experimental Botany* 58(11): 3037–3046.

Marur, C.J., Sodek, L., Magalhaes, A.C. (1994) Free amino acids in leaves of cotton plants under water deficit. *Revista Brasileira de Fisiologia Vegetal* 6(2): 103–108.

Maxwell, B., Kieber, J. (2004) Cytokinin signal transduction. In Davies, P.J., ed., *Plant Hormones, Biosynthesis, Signal Transduction, Action*, 321–349. Kluwer Academic Publishers: Dordrecht, The Netherland.

Mazher, A.A., Zaghloul, S.M., Mahmoud, S.A., Siam, H.S. (2011) Stimulatory effect of kinetin, ascorbic acid and glutamic acid on growth and chemical constituents of Codiaeum variegatum L. plants. *Am Eurasian Journal of Agriculture and Environmental Science* 10: 318–323.

Maziah, M., Teh, C.Y. (2016) Exogenous application of glycine betaine alleviates salt induced damages more efficiently than ascorbic acid in vitro rice shoots. *Australian Journal of Basic and Applied Sciences* 10(16): 58–65.

McNeil, S.D., Rhodes, D., Russell, B.L., Nuccio, M.L., Shachar-Hill, Y., Hanson, A.D. (2000) Metabolic modeling identifies key constraints on an engineered glycine betaine synthesis pathway in tobacco. *Plant Physiology* 124(1): 153–162.

Mertz, E.T., Bates, L.S., Nelson, O.E. (1964) Mutant gene that changes protein composition and increases lysine content of maize endosperm. *Science* 145(3629): 279–280.

Merwad, A.M.A., Desoky, E.M., Rady, M.M. (2018) Response of water deficit-stressed Vigna unguiculata performances to silicon, proline or methionine foliar application. *Scientia Horticulturae* 228: 132–144.

Mestichelli, L.J., Gupta, R.N., Spenser, I.D. (1979) The biosynthetic route from ornithine to proline. *The Journal of Biological Chemistry* 254(3): 640–647.

Miri, H.R., Armin, M. (2013) The interaction effect of drought and exogenous application of glycine betaine on corn (*Zea mays* L.). *European Journal of Experimental Biology* 3: 197–206.

Misra, N., Gupta, A.K., Dwivedi, U.N. (2006) Changes in free amino acid and stress protein synthesis in two genotypes of green gram under salt stress. *Journal of Plant Sciences* 1(1): 56–66.

Mohanty, A., Kathuria, H., Ferjani, A., Sakamoto, A., Mohanty, P., Murata, N., Tyagi, A.K. (2002) Transgenics of an elite indica rice variety Pusa Basmati 1 harbouring the codA gene are highly tolerant to salt stress. *Theoretical and Applied Genetics* 106(1): 51–57.

Moharramnejad, S., Sofalian, O., Valizadeh, M., Asgari, A., Shiri, M. (2015) Proline, glycine betaine, total phenolics and pigment contents in response to osmotic stress in maize seedlings. *Journal of Bioscience & Biotechnology* 4(3): 313–319.

Moulin, M., Deleu, C., Larher, F. (2000) L-lysine catabolism is osmo-regulated at the level of lysine-ketoglutarate reductase and saccharopine dehydrogenase in rapeseed leaf discs. *Plant Physiology and Biochemistry* 38(7–8): 577–585.

Moulin, M., Deleu, C., Larher, F., Bouchereau, A. (2006) The lysine-ketoglutarate reductase–saccharopine dehydrogenase is involved in the osmo-induced synthesis of pipecolic acid in rapeseed leaf tissues. *Plant Physiology and Biochemistry* 44(7–9): 474–482.

Mukhtar, I., Shahid, M.A., Khan, M.W., Balal, R.M., Iqbal, M.M., Naz, T., Ali, H.H. (2016) Improving salinity tolerance in chili by exogenous application of calcium and sulphur. *Soil & Environment* 35(1): 56–64.

Munns, R. (2002) Comparative physiology of salt and water stress. *Plant, Cell and Environment* 25(2): 239–250.

Munns, R., Tester, M. (2008) Mechanisms of salinity tolerance. *Annual Review of Plant Biology* 59: 651–681.

Muttucumaru, N., Powers, S.J., Elmore, J.S., Mottram, D.S., Halford, N.G. (2015) Effects of water availability on free amino acids, sugars, and acrylamide-forming potential in potato. *Journal of Agricultural and Food Chemistry* 63(9): 2566–2575.

Nahed, G., Abdel Aziz, A.A., Mazher, M., Farahat, M.M. (2010) Response of vegetative growth and chemical constituents of Thuja orientalis L. plant to foliar application of different amino acids at Nubaria. *Journal of American Science* 6(3): 295–301.

Naidu, B.P., Paleg, L.G., Aspinall, D., Jennings, A.C., Jones, G.P. (1991) Amino acid and glycine betaine accumulation in cold-stressed wheat seedlings. *Phytochemistry* 30(2): 407–409.

Nambara, E., Kawaide, H., Kamiya, Y., Naito, S. (1998) Characterization of an Arabidopsis thaliana mutant that has a defect in ABA accumulation: ABA-dependent and ABA-independent accumulation of free amino acids during dehydration. *Plant and Cell Physiology* 39(8): 853–858.

Nayyar, H., Walia, D.P. (2003) Water stress induced proline accumulation in contrasting wheat genotypes as affected by calcium and abscisic acid. *Biologia Plantarum* 46(2): 275–279.

Neshich, I.A., Kiyota, E., Arruda, P. (2013) Genome-wide analysis of lysine catabolism in bacteria reveals new connections with osmotic stress resistance. *The ISME Journal* 7(12): 2400–2410.

Neto, A.D.A., Prisco, J.T., Gomes-Filho, E. (2009) Changes in soluble amino-N, soluble proteins and free amino acids in leaves and roots of salt stressed maize genotypes. *Journal of Plant Interactions* 4(2): 137–144.

Noman, A., Ali, Q., Maqsood, J., Iqbal, N., Javed, M.T., Rasool, N., Naseem, J. (2018) Deciphering physio-biochemical, yield, and nutritional quality attributes of water-stressed radish (*Raphanus sativus* L.) plants grown from Zn-Lys primed seeds. *Chemosphere* 195: 175–189.

Nounjan, N., Nghia, P.T., Theerakulpisut, P. (2012) Exogenous proline and trehalose promote recovery of rice seedlings from salt-stress and differentially modulate antioxidant enzymes and expression of related genes. *Journal of Plant Physiology* 169(6): 596–604.

Nounjan, N., Theerakulpisut, P. (2012) Effects of exogenous proline and trehalose on physiological responses in rice seedlings during salt-stress and after recovery. *Plant Soil and Environment* 58(7): 309–315.

Nuccio, M.L., Russell, B.L., Nolte, K.D., Rathinasabapathi, B., Gage, D.A., Hanson, A.D. (1998) The endogenous choline supply limits glycine betaine synthesis in transgenic tobacco expressing choline monooxygenase. *The Plant Journal: For Cell and Molecular Biology* 16(4): 487–496.

Obata, T., Matthes, A., Koszior, S., Lehmann, M., Araújo, W.L., Bock, R., Sweetlove, L.J., Fernie, A.R. (2011) Alteration of mitochondrial protein complexes in relation to metabolic regulation under short-term oxidative stress in Arabidopsis seedlings. *Phytochemistry* 72(10): 1081–1091.

Ogawa, S., Mitsuya, S. (2012) S-methylmethionine is involved in the salinity tolerance of *Arabidopsis thaliana* plants at germination and early growth stages. *Physiologia Plantarum* 144(1): 13–19.

Osman, H.S. (2015) Enhancing antioxidant–yield relationship of pea plant under drought at different growth stages by exogenously applied glycine betaine and proline. *Annals of Agricultural Sciences* 60(2): 389–402.

Osuji, G.O., Brown, T.K., South, S.M., Duncan, J.C., Johnson, D. (2011) Doubling of crop yield through permutation of metabolic pathways. *Advances in Bioscience and Biotechnology* 2: 364–379.

Park, E.J., Jeknic, Z., Chen, T.H. (2006) Exogenous application of glycinebetaine increases chilling tolerance in tomato plants. *Plant and Cell Physiology* 47(6): 706–714.

Pasapula, V., Shen, G., Kuppu, S., Paez-Valencia, J., Mendoza, M., Hou, P., Chen, J., et al. (2011) Expression of an Arabidopsis vacuolar H+-pyrophosphatase gene (AVP1) in cotton improves drought- and salt tolerance and increases fibre yield in the field conditions. *Plant Biotechnology Journal* 9(1): 88–99.

Paulin, A. (1972) Influence of temporary water deficiency on nitrogen-metabolism of flowers cut from iris-germanica l. *Comptes Rendus Hebdomadaires Des Seances De L Academie Des Sciences Serie D* 275(2): 209.

Planchet, E., Rannou, O., Ricoult, C., Boutet-Mercey, S., Maia-Grondard, A., Limami, A.M. (2011) Nitrogen metabolism responses to water deficit act through both abscisic acid (ABA)-dependent and independent pathways in *Medicago truncatula* during post-germination. *Journal of Experimental Botany* 62(2): 605–615.

Quan, R., Shang, M., Zhang, H., Zhao, Y., Zhang, J. (2004) Improved chilling tolerance by transformation with betA gene for the enhancement of glycinebetaine synthesis in maize. *Plant Science* 166(1): 141–149.

Rahman, S., Miyake, H., Takeoka, Y. (2002) Effects of exogenous glycinebetaine on growth and ultrastructure of salt-stressed rice seedlings (*Oryza sativa* L.). *Plant Production Science* 5(1): 33–44.

Rai, V.K. (2002) Role of amino acids in plant responses to stresses. *Biologia Plantarum* 45(4): 481–487.

Ramos, M.L.G., Parsons, R., Sprent, J.I. (2005) Differences in ureide and amino acid content of water stressed soybean inoculated with *Bradyrhizobium japonicum* and B. elkanii. *Pesquisa Agropecuária Brasileira* 40(5): 453–458.

Rao, A., Ahmad, S.D., Sabir, S.M., Awan, S.I., Hameed, A., Abbas, S.R., Ahmad, Z. (2013) Detection of saline tolerant wheat cultivars (*Triticum aestivum* L.) using lipid peroxidation, antioxidant defense system, glycine betaine and proline contents. *Journal of Animal and Plant Sciences* 23: 1742–1748.

Rawia, A.E., Lobna, S.T., Soad, M.M.I. (2011) Alleviation of adverse effects of salinity on growth, and chemical constituents of marigold plants by using glutathione and ascorbate. *Journal of Applied Sciences Research* 7(5): 714–721.

Rezaei, M.A., Kaviani, B., Masouleh, A.K. (2012) The effect of exogenous glycine betaine on yield of soybean [*Glycine max* (L.) Merr.] in two contrasting cultivars Pershing and DPX under soil salinity stress. *Plant OMICS* 5(2): 87–93.

Rhodes, D., Hanson, A.D. (1993) Quaternary ammonium and tertiary sulfonium compounds in higher plants. *Annual Review of Plant Physiology and Plant Molecular Biology* 44(1): 357–384.

Rizhsky, L., Liang, H., Shuman, J., Shulaev, V., Davletova, S., Mittler, R. (2004) When defense pathways collide. The response of Arabidopsis to a combination of drought and heat stress. *Plant Physiology* 134(4): 1683–1696.

Rizwan, M., Ali, S., Adrees, M., Rizvi, H., Zia-ur-Rehman, M., Hannan, F., Qayyum, M.F., Hafeez, F., Ok, Y.S. (2016) Cadmium stress in rice: Toxic effects, tolerance mechanisms, and management: A critical review. *Environmental Science and Pollution Research International* 23(18): 17859–17879.

Rizwan, M., Ali, S., Akbar, M.Z., Shakoor, M.B., Mahmood, A., Ishaque, W., Hussain, A. (2017a) Foliar application of aspartic acid lowers cadmium uptake and Cd-induced oxidative stress in rice under Cd stress. *Environmental Science and Pollution Research International* 24(27): 21938–21947.

Rizwan, M., Ali, S., Hussain, A., Ali, Q., Shakoor, M.B., Zia-Ur-Rehman, M., Asma, M., Asma, M. (2017b) Effect of zinc-lysine on growth, yield and cadmium uptake in wheat (*Triticum aestivum* L.) and health risk assessment. *Chemosphere* 187: 35–42.

Robinson, S.P., Jones, G.P. (1986) Accumulation of glycinebetaine in chloroplasts provides osmotic adjustment during salt stress. *Australian Journal of Plant Physiology* 13(5): 659–668.

Rodríguez-Rosales, M.P., Gálvez, F.J., Huertas, R., Aranda, M.N., Baghour, M., Cagnac, O., Venema, K. (2009) Plant NHX cation/proton antiporters. *Plant Signaling & Behavior* 4(4): 265–276.

Roychoudhury, A., Banerjee, A. (2016) Endogenous glycine betaine accumulation mediates abiotic stress tolerance in plants. *Tropical Plant Research* 3: 105–111.

Sadak, M.S., Abdelhamid M.T. (2015) Influence of amino acids mixture application on some biochemical aspects, antioxidant enzymes and endogenous polyamines of Vicia faba plant grown under seawater salinity stress. *Gesunde Pflanze* 67: 119–129.

Sadak, M.S.H., Abdelhamid, M.T., Schmidhalter, U. (2015) Effect of foliar application of aminoacids on plant yield and some physiological parameters in bean plants irrigated with seawater. *Acta Biológica Colombiana* 20(1): 141–152.

Saeed, M.R., Kheir, A.M., Al-Sayed, A.A. (2005) Suppersive effect of some amino acids against meloidogyne incognita on soybeans. *Journal of Agricultural Sciences Mansoura Universitiy* 30(2): 1097–1103.

Saeedipour, S., Moradi, F. (2012) Stress-induced changes in the free amino acid composition of two wheat cultivars with difference in drought resistance. *African Journal of Biotechnology* 11(40): 9559.

Sakamoto, A., Murata, N. (2002) The role of glycine betaine in the protection of plants from stress: Clues from transgenic plants. *Plant, Cell & Environment* 25(2): 163–171.

Sakr, M.T., El-Sarkassy, N.M., Fuller, M.P. (2012) Osmoregulators proline and glycine betaine counteract salinity stress in canola. *Agronomy for Sustainable Development* 32(3): 747–754.

Scheideler, M., Schlaich, N.L., Fellenberg, K., Beissbarth, T., Hauser, N.C., Vingron, M., Slusarenko, A.J., Hoheisel, J.D. (2002) Monitoring the switch from housekeeping to pathogen defense metabolism in Arabidopsis thaliana using cDNA arrays. *The Journal of Biological Chemistry* 277(12): 10555–10561.

Schneidereit, J., Häusler, R.E., Fiene, G., Kaiser, W.M., Weber, A.P. (2006) Antisense repression reveals a crucial role of the plastidic 2-oxoglutarate/malate translocator DiT1 at the interface between carbon and nitrogen metabolism. *The Plant Journal: for Cell and Molecular Biology* 45(2): 206–224.

Shallan, M.A., Hassan, H.M., Namich, A.A., Ibrahim, A.A. (2012) Effect of sodium nitroprusside, putrescine and glycine betaine on alleviation of drought stress in cotton plant. *American-Eurasian Journals of Agriculture & Environmental Science* 12: 1252–1265.

Sharma, S.S., Dietz, K.J. (2006) The significance of amino acids and amino acid-derived molecules in plant responses and adaptation to heavy metal stress. *Journal of Experimental Botany* 57(4): 711–726.

Shekari, G., Javanmardi, J. (2017) Effects of foliar application pure amino acid and amino acid containing fertilizer on broccoli (*Brassica oleracea* L. var. italica) transplants. *Advances in Crop Science and Technology* 5(3): 1–4.

Shelp, B.J., Bown, A.W., McLean, M.D. (1999) Metabolism and functions of gamma-aminobutyric acid. *Trends in Plant Science* 4(11): 446–452.

Siddiqui, M.H., Al-Khaishany, M.Y., Al-Qutami, M.A., Al-Whaibi, M.H., Grover, A., Ali, H.M., Al-Wahibi, M.S., Bukhari, N.A. (2015) Response of different genotypes of faba bean plant to drought stress. *International Journal of Molecular Sciences* 16(5): 10214–10227.

Signorelli, S., Corpas, F.J., Borsani, O., Barroso, J.B., Monza, J. (2013) Water stress induces a differential and spatially distributed nitro-oxidative stress response in roots and leaves of *Lotus japonicus*. *Plant Science: An International Journal of Experimental Plant Biology* 201: 137–146.

Simon-Sarkadi, L., Galiba, G. (1996) Reflection of environmental stresses on the amino acid composition of wheat. *Periodica Polytechnica: Chemical Engineering* 40(1–2): 79.

Singh, T.N., Aspinall, D., Paleg, L.G. (1972) Proline accumulation and varietal adaptability to drought in barley; a potential metabolic measure of drought resistance. *Nature* 286: 188–190.

Singh, M., Pratap Singh, V., Dubey, G., Mohan Prasad, S. (2015) Exogenous proline application ameliorates toxic effects of arsenate in *Solanum melongena* L. seedlings. *Ecotoxicology and Environmental Safety* 117: 164–173.

Slocum, R.D. (2005) Genes, enzymes and regulation of arginine biosynthesis in plants. *Plant Physiology and Biochemistry* 43(8): 729–745.

Sodek, L., Wilson, C.M. (1970) Incorporation of leucine-14C and lysine-14C into protein in the developing endosperm of nomal and opaque-2 corn. *Archives of Biochemistry and Biophysics* 140(1): 29–38.

Souri, M.K. (2016) Aminochelate fertilizers: The new approach to the old problem; A review. *Open Agriculture* 1(1): 118–123.

Stewart, G.R., Larher, F. (1980) Accumulation of amino acids and related compounds in relation to environmental stress. In *Amino Acids and Derivatives*, ed. Miflin, B.J., 609–635. Academic Press: New York.

Stolarz, M., Dziubinska, H. (2017) Osmotic and salt stresses modulate spontaneous and glutamate-induced action potentials and distinguish between growth and circumnutation in Helianthus annuus seedlings. *Frontiers in Plant Science* 8: 1766.

Su, J., Hirji, R., Zhang, L., He, C., Selvaraj, G., Wu, R. (2006) Evaluation of the stress-inducible production of choline oxidase in transgenic rice as a strategy for producing the stress-protectant glycine betaine. *Journal of Experimental Botany* 57(5): 1129–1135.

Szabados, L., Savouré, A. (2010) Proline: A multifunctional amino acid. *Trends in Plant Science* 15(2): 89–97.

Talaat, N.B., Shawky, B.T. (2013) 24-Epibrassinolide alleviates salt-induced inhibition of productivity by increasing nutrients and compatible solutes accumulation and enhancing antioxidant system in wheat (*Triticum aestivum* L.). *Acta Physiologiae Plantarum* 35(3): 729–740.

Teixeira, W.F., Fagan, E.B., Soares, L.H., Umburanas, R.C., Reichardt, K., Neto, D.D. (2017) Foliar and seed application of amino acids affects the antioxidant metabolism of the soybean crop. *Frontiers in Plant Science* 8: 327.

Thakur, P., Kumar, S., Malik, J.A., Berger, J.D., Nayyar, H. (2010) Cold stress effects on reproductive development in grain crops: An overview. *Environmental and Experimental Botany* 67(3): 429–443.

Thakur, P.S., Rai, V.K. (1982) Dynamics of amino acid accumulation of two differentially drought resistant Zea mays cultivars in response to osmotic stress. *Environmental and Experimental Botany* 22(2): 221–226.

Theriappan, P., Gupta, A.K., Dhasarrathan, P. (2011) Accumulation of proline under salinity and heavy metal stress in cauliflower seedlings. *Journal of Applied Sciences and Environmental Management* 15(2): 251–255.

Venkatesalu, V., Chellappan, K.P. (1998) Accumulation of proline and glycinebetaine in Ipomoea pes-carpae induced by NaCl. *Biologia Plantarum* 41(2): 271–276.

Verbruggen, N., Hermans, C. (2008) Proline accumulation in plants: A review. *Amino Acids* 35(4): 753–759.

Verslues, P.E., Sharma, S. (2010) Proline metabolism and its implications for plant-environment interaction. *The Arabidopsis Book* 8: e0140.

Vranova, E., Inzé, D., Van Breusegem, F. (2002) Signal transduction during oxidative stress. *Journal of Experimental Botany* 53(372): 1227–1236.

Walch-Liu, P., Liu, L.H., Remans, T., Tester, M., Forde, B.G. (2006) Evidence that L-glutamate can act as an exogenous signal to modulate root growth and branching in *Arabidopsis thaliana*. *Plant and Cell Physiology* 47(8): 1045–1057.

Wang, M., Liu, C., Li, S., Zhu, D., Zhao, Q., Yu, J. (2013) Improved nutritive quality and salt resistance in transgenic maize by simultaneously overexpression of a natural lysine-rich protein gene, SBgLR, and an ERF transcription factor gene, TSRF1. *International Journal of Molecular Sciences* 14(5): 9459–9474.

Wang, G.P., Zhang, X.Y., Li, F., Luo, Y., Wang, W. (2010) Overaccumulation of glycine betaine enhances tolerance to drought and heat stress in wheat leaves in the protection of photosynthesis. *Photosynthetica* 48(1): 117–126.

Wani, S.H., Sofi, P.A., Gosal, S.S., Singh, N.B. (2010) In vitro screening of rice (*Oryza sativa* L.) callus for drought tolerance. *Communications in Biometry and Crop Science* 5(2): 108–115.

Wu, G.Q., Feng, R.J., Li, S.J., Du, Y.Y. (2017) Exogenous application of proline alleviates salt-induced toxicity in sainfoin seedlings. *Journal of Animal and Plant Sciences* 27: 246–251.

Yang, X., Lu, C. (2005) Photosynthesis is improved by exogenous glycinebetaine in salt-stressed maize plants. *Physiologia Plantarum* 124(3): 343–352.

Yang, J., Peng, S., Visperas, R.M., Sanico, A.L., Zhu, Q., Gu, S. (2000) Grain filling pattern and cytokinin content in the grains and roots of rice plants. *Plant Growth Regulation* 30(3): 261–270.

Yang, W.J., Rich, P.J., Axtell, J.D., Wood, K.V., Bonham, C.C., Ejeta, G., Mickelbart, M.V., Rhodes, D. (2003) Genotypic variation for glycinebetaine in sorghum. *Crop Science* 43(1): 162–169.

Yang, Q.Q., Zhang, C.Q., Chan, M.L., Zhao, D.S., Chen, J.Z., Wang, Q., Liu, Q.F., et al. (2016) Biofortification of rice with the essential amino acid lysine: Molecular characterization, nutritional evaluation, and field performance. *Journal of Experimental Botany* 67(14): 4285–4296.

Ye, H.Y. Du, H. Tang, N. Li, X.H. Xiong, L.Z. (2009) Identification and expression profiling analysis of TIFY family genes involved in stress and phytohormone responses in rice. *Plant Molecular Biology* 71(3): 291–305.

Zafari, M., Ebadi, A. (2016) Effects of water stress and brassinosteroid (24-epibrassinolide) on changes of some amino acids and pigments in safflower (*Cartamus tinctorius* L.). *Journal of Current Research in Science* 1: 711–715.

Zali, A.G., Ehsanzadeh, P. (2018) Exogenous proline improves osmoregulation, physiological functions, essential oil, and seed yield of fennel. *Industrial Crops and Products* 111: 133–140.

Zemanová, V., Pavlik, M., Pavlikova, D., Tlustos, P. (2014) The significance of methionine, histidine and tryptophan in plant responses and adaptation to cadmium stress. *Plant, Soil and Environment* 60(9): 426–432.

Zhang, H., Murzello, C., Sun, Y., Kim, M.S., Xie, X., Jeter, R.M., Zak, J.C., Dowd, S.E., Paré, P.W. (2010) Choline and osmotic-stress tolerance induced in Arabidopsis by the soil microbe *Bacillus subtilis* GB03. *Molecular Plant–Microbe Interaction* 23: 1097–1104.

Zrenner, R., Stitt, M., Sonnewold, U., Boldt, R. (2006) Pyrimidine and purine biosynthesis and degradation in plants. *Annual Review of Plant Biology* 57: 805–836.

13 Role of Calcium in Conferring Abiotic Stress Tolerance

Muhammad Naeem, Misbah Amir, Hamid Manzoor, Sumaira Rasul, and Habib-ur-Rehman Athar

CONTENTS

13.1 INTRODUCTION

Calcium is one of the major plant macronutrients, whose role in regulating various physiological roles has well been documented (Taiz et al., 2015, Yang et al., 2015). For example, it regulates various essential processes such as cytoplasmic streaming, gravitropism, thigmotropism, cell division, elongation, differentiation, cell polarization, and photomorphogenesis. It is required for structural roles in cell walls and membranes. It is also involved in regulating cellular metabolism (Marschner, 1995, Taiz and Zeiger, 2010). In addition, various environmental factors stimulate the increase of cytosolic Ca^{2+} to activate the different downstream responses that cause plant adjustment to abiotic stresses such as heat, drought, and salinity (Grattan and Grieve, 1998, Naeem et al., 2017, Naeem et al., 2018, Yang et al., 2015). Calcium also plays a significant role in biotic stress tolerance such as fungal resistance and virus resistance in cotton (Ashraf and Zafar, 2000, Marschner, 2012). In such situations, Ca^{2+} acts as an important secondary messenger in plant signaling networks (Cacho et al., 2013). In view of such diverse roles of Ca^{2+} in regulating cellular metabolism, growth, and development under normal or stressful conditions, the present study aims to assess how a single molecule regulate has such diverse functions under normal or stressful conditions. First, we look at various physiological roles of Ca^{2+} occurring in plants. Since changes in uptake and accumulation of Ca^{2+} in different abiotic stresses perturb cellular metabolism, it is necessary to update information on the mechanism of Ca^{2+} uptake and transport within the plant body, and how Ca^{2+} modulates the uptake of other nutrients directly or indirectly. It is also necessary to understand the role of Ca^{2+} in the mechanism of abiotic stress sensing and signaling. Finally, we look into the different reports demonstrating that Ca^{2+} has a role in alleviating the adverse effects of abiotic stress with the probable usage of Ca^{2+} as fertilizer or anti-stress agent.

13.2 PHYSIOLOGICAL ROLES OF CALCIUM

Calcium plays an important role in various physiological processes ranging from structural roles in cell walls to sensing and signaling (Maathuis, 2009, Riveras et al., 2015). It promptly complexes with organic compounds having negative groups, i.e., phosphates and carboxyl's phospholipids, sugars, and proteins. For example, in the cell wall it helps in cross-linking cellulose microfibrils by pectins and glycans. Carboxyl groups from contrasting pectins can be electrostatically synchronized by Ca^{2+} that consequently confers firmness to the cell wall and plant tissues. Calcium plays an equivalent role in cell membranes where Ca^{2+} organizes with phosphate groups of phospholipids. This complexion occurs largely at the outer face of the plasma membrane and thus membrane integrity under normal and abiotic stress conditions (Maathuis, 2009).

As Ca^{2+} readily forms insoluble salts with phosphates and sulfates, the free Ca^{2+} contents in the cytoplasm are largely reserved. This makes Ca^{2+} an ideal secondary messenger and a wide range of stimuli has been shown to evoke rapid changes in cytosolic free Ca^{2+} in plants (Mahouachi et al., 2006, McAinsh and Pittman, 2009). The root-shoot cell elongation needs acidification of the apoplasm and replacement of Ca from the cross-links of the pectic-chain, although this is only part of the process (Carpita et al., 2001). The increased cytosolic concentration of free Ca^{2+} trigger the synthesis of precursors for cell walls and their excretion into the apoplast. The latter process is repressed by eliminating apoplastic calcium. The elongation of root-hairs and pollen tubes also rely on the availability of apoplastic calcium. The Ca^{2+} influx is restricted from the apoplast to the top of these cells and increases confined cytosolic Ca^{2+} concentration, which acts as attention for the exocytosis of the cell wall material and creates a polarity for cell elongation (Cole and Fowler, 2006, Krichevsky et al., 2007). In root caps, apoplasmic Ca^{2+} is also required for the secretion of mucilage. Calcium also plays a role in plugging phloem cells by callose. The formation of callose is one of another illustration of a Ca^{2+}-induced secretory process. Cells produce callose (1,4 β-glucan units) when reacting to injury or the existence of toxic cations, and plants get sealed phloem cells by callose (Kartusch, 2003, Kauss, 1987, Rengel and Zhang, 2003). This switch is also initiated by a rise in cytosolic free Ca (Kauss, 1987). In germinating seeds, a high concentration of Ca^{2+} (millimolar) stimulates α-amylase enzyme activity in aleuron cells, which helps in mobilizing food reserves toward developing an embryo. Calcium is an integral part of α-amylase, which is synthesized on rough endoplasmic reticulum (ER). Transport of Ca^{2+} through the membranes of ER is improved by gibberellic acid (GA) and repressed by abscisic acid (ABA), leading to the characteristic stimulation (GA) and inhibition (ABA) of α-amylase activity in aleurone cells (Lovegrove and Hooley, 2000). Thus, Ca^{2+} is an important mineral nutrient and has a role in various physiological processes including cell wall synthesis, mitotic spindle formation during cell division, normal functioning of the plasma membrane, and environmental and hormonal signal transductions.

13.3 CALCIUM UPTAKE AND DISTRIBUTION DURING ABIOTIC STRESS

According to estimates, 60% of cultivated land suffers from nutrient deficiency. Calcium deficiency in soils causes physiological disorders in plants that result in low crop productivity. However, the effect of Ca^{2+} deficiency along with abiotic stresses causes further negative effects on the growth of plants (Marschner, 1995, Marschner, 2012). Abiotic stresses also induced Ca^{2+} deficiency by inhibiting Ca^{2+} uptake and accumulation in plants. For example, drought interrupts the nutritive status of plants by changing ion concentrations in tissues, reduces root growth, and causes little nutrient accessibility in dry soil (Samarah et al., 2004). Limited water supply causes lower transpiration rate that results in lower uptake of Ca^{2+} (Martínez-Ballesta et al., 2010). Similarly, heat stress caused a substantial reduction in root respiration, Ca^{2+} uptake, membrane integrity, and root biomass (Giri, 2013). In the same way, salt stress inhibited plant growth by inhibiting nutrient uptake, including calcium (Rewald et al., 2013). Chilling stress increases the permeability of cellular membranes that results in leakage of organic and inorganic substances. It affects the root growth through a reduction in elongation and biomass accumulation and may cause loss of leaf turgor due to a reduction in water absorption and moderate transpiration. These changes, in turn, decrease the nutrient acquisition by the plants under chilling stress (Naeem et al., 2013).

Plants absorb Ca^{2+} through roots from the soil solution (Havlin et al., 2005). It enters the root through Ca^{2+} permeable channels, some of them are selective for Ca^{2+} and others are non-selective ion channels. Concerning Ca^{2+} influx, the five particular protein families have been revealed to transport Ca in plants: cyclic nucleotide-gated channels (CNGCs) (Zelman et al., 2012); glutamate-like receptor homologs (GLRs) (Price et al., 2012); two-pore channels (TPCs) (Morgan and Galione, 2014); mechanosensitive (MC) channels (Kurusu et al., 2013); and the most recently recognized is hyperosmolality-induced Ca (OSCAs) channels (Yuan et al., 2014). OSCAs were recognized as an imperative constituent of the initial osmotic response in *Arabidopsis* and are a well-maintained family of channel proteins that exist in all eukaryotes (Edel and Kudla, 2015). It is remarkably diversified in plants. Phylogenetic studies of completely sequenced genomes show four key OSCAs clades, of which two OSCAs genes (OSCA1.1 and OCAS1.2) have been described and their potential to transport Ca^{2+} ions confirmed experimentally (Yuan et al., 2014). The middle lamella of the cell wall, ER, and vacuole contain a high concentration of Ca. Most of the water-soluble Ca^{2+} ions tend to be impounded in the big vacuole of mature cells. Calcium in the ER is linked with Ca^{2+} binding proteins. In contrast, Ca^{2+} concentration in cytosol is low (0.1–1.0 mM). Stress-induced increase in cytosolic Ca^{2+} has been reported by many scientists (Medvedev, 2005, Reddy et al., 2011). Stress-induced disturbance in cytosolic Ca^{2+} levels is distinguished by Ca-binding proteins called Ca^{2+} sensors. In plants, the Ca^{2+}dependent protein kinases (CPKs or CDPKs) characterizes a distinctive group of Ca^{2+} sensors (Harper et al., 2004). The increased level of cytosolic Ca^{2+} carried by abscisic acid was primarily triggered by Ca influx. The ABA-induced increase in cytosolic Ca^{2+} acts as a secondary messenger that plays a role in stomatal closure. The increased activity of inositol 1,4,5-triphosphate (IP3) induces an increase in cytosolic Ca^{2+} concentration, a vital step in stomatal closure (Takahashi et al., 2001). IP3 is recognized to stimulate vacuolar Ca^{2+} channels (Bokszczanin and Fragkostefanakis, 2013). In addition, chloroplast-localized Ca^{2+} sensing receptor (CAS) has been revealed to play an important part in the production of extracellular Ca^{2+}-prompted cytosolic Ca^{2+} transients and stomatal closure in *Arabidopsis* by regulating IP_3 concentrations, which causes the release of Ca^{2+} from internal stores (Han et al., 2003, Nomura et al., 2008, Tang et al., 2007).

The key transporters catalyzing Ca^{2+} effluxes from the cytosol to the apoplast and endoplasmic reticulum are Ca^{2+}-ATPases. At the membrane of the vacuole (tonoplast), both Ca-ATPases and Ca^{2+}/H^{+} anti-porters catalyze Ca-efflux from the cytosol to the vacuole. The latter is strengthened by the proton electrochemical gradient created by tonoplast H^{+}-ATPase and H^{+}-PPiase activities (McAinsh and Pittman, 2009). Chloroplasts can also enclose enormous amounts of Ca (6.5–15 mM total Ca^{2+}, generally bound to thylakoid membranes), but in the stroma of the chloroplast, the free Ca^{2+} concentrations are only in the range of 2.4–6.3 μM (Kreimer et al., 1988). A plastid Ca^{2+}ATPase catalyzes Ca^{2+} uptake by plastids (Marschner, 2012). The Ca^{2+} transporters for xylem loading still need to be identified and a fraction of xylem Ca^{2+} may reach through the apoplast (White and Krause, 2001), but Ca^{2+} mobility is also low in the plant vascular structure (Maathuis, 2009). Calcium is relatively immobile in plants and their uptake reduces in above-ground portions of plants (shoots and leaves) as well as in roots under drought conditions due to the decline in transpiration rate (Brown et al., 2006, Hu et al., 2008). The root tip area is a major site of Ca entry and these zone of Ca^{2+} uptakes are likely to be reduced under water-deficit conditions (Raza et al., 2013). Therefore, continuous supply of Ca^{2+} is required by plants for vigorous leaf and root development and overall canopy growth (Amor and Marcelis, 2003). Consequently, Ca^{2+} levels may drop below a dangerous level in fast developing tissues triggering diseases such as "blossom end rot" in tomatoes, "black heart" in celery, or "bitter-pit" in apples.

13.4 CALCIUM INTERACTION WITH OTHER MINERAL ELEMENTS DURING STRESS CONDITIONS

A recent study revealed that drought stress causes a substantial reduction in the acquisition of mineral uptake by maize. In contrast, foliar treatment of Ca^{2+} markedly increased grain N, K, Ca^{2+}, Fe, Mn, and Zn contents under water-deficit conditions (Naeem et al., 2018). The positive relations of Ca^{2+} in the uptake of micronutrients in tobacco plants have also been observed (López-Lefebre et al., 2001). High concentration of sodium (Na^{+}) under salt stress interferes with Ca^{2+} nutrition and disturbs various plant physiological processes (Tavakkoli et al., 2010). These effects can be reduced by exogenous application of Ca^{2+} as it helps the plants compete with Na^{+} ions in covering the membrane binding sites (Kaya et al., 2002). Tuna et al. (2007) revealed the effects of exogenous Ca ($CaSO_4$) on the uptake of mineral elements by tomato plants under salinity stress, and indicated an increase in leaf concentration of N, K, and Ca^{2+}, and lower concentration of Na^{+}. Patel et al. (2011) also demonstrated lower concentration of N, P, K, and Ca^{2+} in leaf tissues of *Caesalpinia crista* L. under salt stress; however, the application of Ca^{2+} restored the concentration of these nutrients.

13.5 CALCIUM SIGNALING IN PLANT ABIOTIC STRESS RESPONSES

Calcium is one of the most important secondary messengers that participates in a wide range of developmental and physiological processes. Ca^{2+} is a versatile signaling molecule that is present on the upstream of most signaling cascades. Ca^{2+} triggers the activation of different ion channels present on the plasma membrane and endomembranes. When it enters the cytoplasm, it participates in the allosteric regulation of many enzymes and proteins. Ca^{2+} can affect both directly or indirectly on signal transduction networks such as G-protein-coupled receptors. Ca^{2+} concentration in the cell is strictly controlled by Ca^{2+} pumps. The level of Ca^{2+} is different in different parts of the cells. For example, it is much higher outside the cell and in the ER as compared with the cytosol. The release of Ca^{2+} in the cytosol is used to trigger a biological response. Some signals that induce Ca^{2+} concentration change as a secondary messenger are neurotransmitters, plant hormones, and growth factors.

One of the earlier events, following the recognition of pathogen (elicitors/effectors), is the ion fluxes through the plasma membrane that have been observed after a very short time (five minutes) after elicitation. Ion fluxes include an influx of H^+ and Ca^{2+} and efflux of Cl^-, NO_3^-, and K^+, etc., and have been reported after elicitor treatments in different plant species such as parsley, tobacco or *A. thaliana* (Garcia-Brugger et al., 2006).

The Ca^{2+} and anionic fluxes trigger a depolarization of the plasma membrane. The kinetics and amplitude of depolarization depend on the nature of the elicitor (Garcia-Brugger et al., 2006). Among these ions, Ca^{2+} is considered as one of the major second messengers, acting upstream of numerous other signaling events during plant-pathogen interactions (Dodd et al., 2010). The involvement of anion fluxes in cell signaling in response to elicitors was mainly studied by using pharmacological and biochemical approaches.

Elicitins, a well-known class of elicitors, mobilize Ca^{2+} from both extracellular (apoplast) and intracellular pools (vacuole, ER). Several studies reported that upon elicitation, the intracellular concentration of free cytosolic Ca^{2+} ($[Ca^{2+}]_{cyt}$) increased from nM to mM range in a few minutes (Zhao et al., 2005). Previous studies have demonstrated that different elicitors induce different changes in $[Ca^{2+}]_{cyt}$ in the cytosol and are known as "Ca signature" (Lecourieux et al., 2005). A rapid elevation in $[Ca^{2+}]_{cyt}$ promotes the opening of other membrane channels (Blume et al., 2000, Brunner et al., 2002, Lecourieux et al., 2002, Ranf et al., 2008), or activates Ca^{2+}-dependent protein kinases (Ludwig et al., 2005). There are different channels/transporters that take part in the mobilization of Ca^{2+} into the cell, including ATPases, interchanges Ca^{2+}/H^+ channels, two pore channel1 (TPC1), glutamate receptors (GLRs), and the cyclic nucleotide-gated channel (CNGCs; (Dodd et al., 2010). TPC1, thought to participate in Ca^{2+} influx, is either located in the plasma membrane or vacuolar membrane (Hamada et al., 2012, Kadota et al., 2004, Kurusu et al., 2005, Peiter et al., 2005).

In the model plant *Arabidopsis thaliana*, the CNGCs ion channel family is composed of 20 members that are further classified into 4 groups (Mäser et al., 2001). *Cngc* channel mutants named *dnd* "defense no death" showed increase resistance to virulent and avirulent *P. syringae* strains but lacked the HR (Yu et al., 1998). Some reports showed the loss of CNGC4, along with CNGC2, resulted in increased resistance and loss of hypersensitive cell death (Balagué et al., 2003, Jurkowski et al., 2004). Ali et al. (2007) identified CNGC2 as a key Ca^{2+} permeable channel that affects plant defense response through a Ca^{2+}-dependent NO production signaling cascade. *Arabidopsis dndl* (cngc2) mutant displays no HR in response to infection by some pathogens (Ali et al., 2007). In plants, GLRs are another class of putative plasma membrane receptors and have been identified in *Arabidopsis,* rice, tomato, tobacco, poplar, and radish (Ward et al., 2009). *Arabidopsis* genome sequencing revealed the presence of 20 members of AtGLRs, which can be further grouped into 3 clades on the basis of their phylogenetic relationship (Chiu et al., 2002, Lacombe et al., 2001). Plant GLRs have been playing an important role in many different physiological processes including signal transduction, growth processes, ion transport, and response to (a)biotic stresses (Jammes et al., 2011). In two recent investigations, Kwaaitaal et al. (2011) and Vatsa et al. (2011) reported the role of GLRs in elicitor-induced plant defense signaling in tobacco and *Arabidopsis,* respectively. These studies demonstrated that treatment with different plant defense elicitors, for example, Cry, flg22, elf18, or chitin leads to an initial influx of Ca^{2+} that was partially inhibited after GLR inhibitor applications. Moreover, there were also modifications in the elicitor-mediated NO production and MAPK activation (two other important second messenger that participates actively in plant defense signaling) in response to GLR antagonists, further suggesting the role of GLRs in plant defense signaling cascade through the regulation of second messengers (Kwaaitaal et al., 2011, Vatsa et al., 2011). Interestingly, TPC1, CNGCs, and GLRs provide putative pathways for Ca^{2+} signaling leading to plant defense, although TPC1 participation is now a subject of debate (Ranf et al., 2008).

13.6 ROLE OF CALCIUM IN ABIOTIC STRESS TOLERANCE

13.6.1 Drought

Calcium regulates various physiological mechanisms, which influence plant growth and stress tolerance (Table 13.1). Calcium influences water and solutes movement, which have an influence on the functioning of stomata (McLaughlin and Wimmer, 1999). The uptake of Ca^{2+} is also decreased under drought stress, however, overall accumulation of Ca^{2+} is depressed slightly as compared with phosphorous and potassium (K^+). Jenne et al. (1958) observed that under drought stress, the accumulations of Ca^{2+}, K^+, and phosphorous is 91%, 71%, and 40% respectively in mature *Zea mays* L. plants. Therefore, continuous supply of Ca is required by plants for vigorous leaf

TABLE 13.1
Role of Exogenous Ca Application in Alleviating the Adverse Effects of Drought in Different Plant Species

Crop	Characters Improved	References
Triticum aestivum	Manipulation of activities of antioxidant enzymes.	Nayyar and Kaushal (2002)
Zea mays	Alleviates the damaging effect of drought.	Nayyar and Walia (2003)
Zoysia japonica	Improves the ability of plants to survive under drought stress.	Xu et al. (2013)
Camellia sinensis	$CaCl_2$ improves recovery potential of plants after drought stress.	Upadhyaya et al. (2011)
Lonicera japonica	Exogenous calcium regulates different physiological processes.	Qiang et al. (2012)
Phaseolus vulgaris	Seed quality and production.	Abou El-Yazied (2011)
Helianthus annuus	Alleviated the deleterious effects of drought stress by down-regulating protein expression	Hassan et al. (2011)
Arabidopsis thaliana	Induced stomatal closure and mediated protein interaction between CPK8 and CAT3	Zou et al. (2015)
Helianthus annuus	Seed weight and seed lipid contents	Hassan et al. (2011)
Vigna catjang	Increased resistance against drought	Mukherjee and Choudhuri (1985)
Glycine max		Yang and Gao (1993)
Gossypium hirsutum		Cheng et al. (1997)

TABLE 13.2
Mitigating Effects of Exogenous Ca Application on Deleterious Effects of Heat Stress in Different Plant Species

Crop	Characters Improved	References
Triticum aestivum	10 and 15 mM Ca^{2+}: Enhanced intrinsic heat tolerance of seedlings.	Bhatia and Asthir (2014)
Beta vulgaris	5–10 mM Ca^{2+}: Reduces membrane leakage.	Toprover and Glinka (1976)
Festuca arundinacea	10 mM $CaCl_2$ Improves the ability of plants to survive under heat stress.	Jiang and Huang (2001)
Poa pratensis	10 mM $CaCl_2$ improves growth quality, chlorophyll, and relative water content.	Jiang and Huang (2001)
Solanum tuberosum	Reduced thermal damage.	Coria et al. (1998)
Zea mays	Increased SOD and APX activity leading to heat tolerance.	Gong et al. (1997)
Nicotiana abacum	Heat stress elevated the cytosolic Ca^{2+} and transduced the signal transduction by mobilizing both extra-cellular and intra-cellular Ca^{2+} sources.	Gong et al. (1998)
Oryza sativa	Up-regulated *OsANN1* expression, Enhanced the level of *SOD* and *CAT* expression.	Qiao et al. (2015)

and root development and overall canopy growth (Amor and Marcelis, 2003). Calcium plays a role as a regulator of plant cell metabolism and can also contribute in the regulation mechanism of plants adjusting to adverse environmental conditions such as water deficit (Bowler and Fluhr, 2000). Application of Ca^{2+} can improve plant resistance to drought, prevent the synthesis of active oxides, protect the structure of plasma membranes, and continue normal photosynthesis along with controlling the metabolism of different plant hormones and other imperative chemicals (Table 13.1). Exogenous treatment of Ca^{2+} conferred improved tolerance to drought stress (Xu et al., 2013) by modulating antioxidant metabolism, photosynthetic effectiveness, and nitrogen assimilation (Zhu et al., 2013). Calcium alleviates the adverse effects of drought by manipulation of antioxidant activities such as glutathione reductase (GR), superoxide dismutase (SOD), catalase (CAT), peroxidase (POD), and ascorbate peroxidase (APX) reducing the membrane impairment (MDA) and aiding the plants to survive in stress conditions (Fu and Huang, 2003, Jiang and Huang, 2001, Nayyar and Kaushal, 2002). From these results, it is concluded that application of Ca at critical growth stages helps to improve the resistance of plants against adverse conditions of drought, which, in turn, may result in improved economic yield

13.6.2 Heat

Calcium also plays a regulatory role in the metabolism of plants under heat stress. High temperature lowers the cell turgor and increases membrane permeability by altering its viscosity and fluidity (Table 13.2). Heat stress changes plasma membrane fluidity which mediates signal cascades. Calcium signaling induces acclamatory responses in plants against heat stress by enhancing the expression of certain genes, such as desaturase-encoding genes (Huda et al., 2013). Similarly, Ca^{2+} regulates carbohydrate metabolism in plants under heat stress. While working wheat seedlings, Kolupaev et al. (2005) reported that exogenous application of 5 mM $CaCl_2$ prior to heat stress caused the accelerated production of reactive oxygen species, increased membrane damage, and an increase in activities of antioxidant enzymes. However, such pre-treatment with Ca^{2+} salts protected plant tissues of wheat from heat stress. They reasoned to the Ca-mediated formation of Hsps (heat-shock proteins), which are necessary for the survival of plant cells. Heat-shock proteins can be recognized as HSP9s, HSP70, and small HSPs based on their molecular mass. Likewise, while assessing the involvement of Ca^{2+} in protection against heat-induced oxidative stress in *Arabidopsis* by using Ca channel blockers and calmodulin inhibitors, Larkindale and Knight (2002) found that Ca^{2+} induced thermotolerance in *Arabidopsis* plants protected against heat stress-induced oxidative stress. Using the Ca-dependent luminescent protein, Gong et al. (1998) detected Ca transients in tobacco plants under heat stress and they concluded that Ca has a role in signaling under heat stress. Moreover, the application of inhibitors of Ca and calmodulin to maize plants under heat stress limited maize plant survival and increased membrane electrolyte leakage (Gong et al., 1997). These findings suggested that Ca^{2+} is involved in signaling pathways for the perception of heat stress and in initiating tolerance mechanisms for heat stress that may include heat-shock protein production. In contrast, in some studies, it has been reported that both Ca^{2+} deficiency and excess enhanced the heat-stress induced damage in sugar beet (Trofimova et al., 1999). Increase in Ca^{2+} helps in the recovery of heat injury and in this way recover them from adverse effects of heat. The best example is the role of Ca^{2+} in germinating seeds. During heat stress, Ca^{2+} helps in the release of alpha and beta amylase in barley (Jones and Carbonell, 1984) and maize (Laurière et al., 1992).

13.6.3 Salinity

Calcium can be used as supplement to reduce the effect of salinity (Table 13.3). Calcium can promote plant growth, reverse the effect of salinity on plasma membranes by competing with sodium ions for membrane-binding sites, reduce translocation of Na^+ to shoot from root, reduce Na^+ content, and increase Ca^{+2} content in plants (Grattan and Grieve, 1998). However, low levels of Ca causes deterioration of plasma membranes and cellular

TABLE 13.3
Role of Exogenous Ca Application in Alleviating the Adverse Effects of Salinity Stress in Different Plant Species

Calcium Salt Used and Concentration	Crop	Characters Improved	References
$CaCl_2$, $Ca(NO_3)$	*Triticum aestivum*	Improved salinity tolerance.	Genc et al. (2010)
$CaCl_2$, $CaSO_4$	*Vigna unguiculata*	Mitigate Na^+ toxicity and can improve ionic and nutritional balance.	Guimarães et al. (2012)
$CaSO_4$	*Lycopersicum esculentum*	Reduced the concentration of Na^+ in leaves.	Tuna et al. (2007)
Ca^{+2}	*Phaseolus vulgaris*	Enhanced germination and growth.	Awada et al. (1995)
Ca^{2+}	*Cajanus cajan*	Decreases uptake and accumulation of Na^+ and Cl^-.	Subbarao et al. (1990)
Ca^{+2} (1 mM)	Blueberry	Offset the damage to shoots.	Wright et al. (1993)
$CaSO_4.2H_2O$	*Caesalpinia crista*	The addition of N, P, and K restored the levels of these nutrients.	Patel et al. (2011)
2 mM Ca	*Zea mays*	Increased root viability, decreased membrane leakage, increased chlorophyll content, and increased POD activity.	Shoresh et al. (2011)
10 mM $CaCl_2$	*Nicotiana tobbaccum*	Exclusion of sodium ions enhanced and growth improved.	Maeda et al. (2005)
$CaSO_4$	Solanum tuberosum	Increased yield.	Abdullah and Ahmad (1982)
Ca^{2+}	*Bruguiera gymnorrhiza*	Mediated root ion flux.	Lu et al. (2013)
Ca^{2+}	*Kandelia candel*	Increased K^+ flux and Na^+/H^+ antiport.	Lu et al. (2013)
Ca^{2+}	*Arabidopsis thaliana*	Increased NADPH/$NADP^+$, G6PDH activity up-regulated expression of PMH^+-ATPase gene.	Li et al. (2011)

compartments, and the death of cells and tissue damage. Kaya et al. (2002) have showed that high concentrations of NaCl are found in strawberries when Ca^{+2} is low. Growth, yield, and quality of fruits have been improved when plants are supplemented with fertilizers of Ca^{2+} under saline environments (Bolat et al., 2006, Jaleel et al., 2007, Saleh et al., 2008, Sima et al., 2009, Tuna et al., 2007). Under saline conditions, the effect of blossom end rot decreased when treated with high levels of Ca. Moreover, Ca^{2+} also enhances the qualities of fruits in peppers and tomatoes leading to more marketability (Rubio et al., 2009). Calcium is able to reverse the effects of salinity on plants like decreasing germination and plant growth in seedlings of *Salvadora persica* L. Vaghela et al. (2009) also documented that Ca^{2+} enhances uptake of nitrogen, K^+, and phosphorous, and reduces accumulation of sodium in leaves. When a combination of Ca^{2+} and K^+ has been given, the negative effects of salinity diminished. Additionally, plant growth and morphological characters were improved in sorghum (Shariat Jafari et al., 2009). It is an interesting feature of supplemental Ca^{2+} that morpho-physiological activities were improved even at low concentration (Sima et al., 2009). It has been indicated that absorption of Ca^{2+} by plants is an active process. Maize has been considered as the model for the investigation of saline soils on the cells, organs, and the whole plant. Maize is used to define the competitive influx/efflux of Na and Ca^{2+} (Zidan et al., 1991). When the concentrations of Na/Ca^{2+} become too high, maize plants show growth inhibition and exhibit morphological and anatomical changes (Cramer, 1992, Evlagon et al., 1990, Maas and Grieve, 1987). Production and elongation of root cells is inhibited in maize under saline soils. But these processes can be restored partially by the addition of Ca (Cramer et al., 1988, Zidan et al., 1991). Changes in morphology and anatomy of maize plants severely affects transportation of water in primary roots (Neumann et al., 1994). Supplementation of these plants with Ca^{2+} tends to diminish the negative effects of salinity, including mitigation in reductions of hydraulic conductivity in primary roots (Azaizeh et al., 1992, Azaizeh and Steudle, 1991, Evlagon et al., 1990, Neumann et al., 1994). Maintenance of availability of Ca^{2+} under salinity serves as a crucial factor for controlling toxicity of different ions, especially in those crops that are sensitive to Na and Cl injuries (Maas, 1993). In citrus, when grown under saline environments, Ca^{2+} reduced the transportation of both chloride and sodium from roots to aerial parts, especially leaves. In this way, Ca^{2+} alleviates foliar injury. The extent to which Ca^{2+} is effective under saline conditions depends on the type of crop species, Ca^{2+} concentration, and Ca^{2+} source. Different mechanisms have been proposed that are responsible for diminishing the effect of salinity by the application of Ca^{2+} growth media. First, Ca^{2+} is able to affect permeability of membranes which leads to increased concentrations of K^+, Ca^{2+}, nitrogen, and decreased concentrations of sodium (Bolat et al., 2006). Second, Ca^{2+} increases the accumulation of proline and glycine betaine in plant leaves and calluses causing osmotic adjustments (Murugan and Sathish, 2005). Third, Ca modulates the activities of antioxidant enzymes such as CAT, SOD, and POD. Antioxidants accumulation causes enhanced protection against salinity (Shoresh et al., 2011). Calcium plays a role by causing the changes at gene level (Henriksson and Henriksson, 2005).

13.6.4 Cold

Over the past two decades, researchers have shown that Ca^{2+} has a major role in cold stress tolerance in plants. For example, exogenous application of Ca^{2+} can improve plant growth, development, morphology, and physiology by regulation of gene expression during cold stress (Gao et al., 2001). Calcium responsive protein kinases, *Arabidopsis* Ca-dependent protein kinase 6 (CDPK6) (Xu et al., 2010) and soybean calmodulin GmCAM4 (Yoo et al., 2005) positively regulate tolerance against cold stress in *Arabidopsis.* Although a number of reports are available on inducing salt and drought tolerance in plants by exogenous application of Ca^{2+} salts, a few reports are available on role exogenous Ca^{2+} application in alleviating adverse effects of cold stress on plants. Recently, Shi et al. (2014) have shown cold stress enhanced Ca^{2+} content in the leaves and roots of Bermuda grass. Moreover, they have demonstrated that such increases in Ca^{2+} was associated with a cold tolerance response. Exogenous application of $CaCl_2$ to these Bermuda grass plants enhanced the cold stress tolerance via enhancing activities of antioxidants. In addition, such exogenous application modulated the expression of several genes related with metabolites responsible for cold tolerance. Similarly, exogenous application of $CaCl_2$ improved cold tolerance in *Zoysia matrella* (Wang et al., 2009) and in *Stylosanthes guianensis* (Zhou and Guo, 2009). These studies have shown that exogenous Ca^{2+} application induces cold tolerance in plants by modulating reactive oxygen species (ROS) metabolism, nitrogen assimilation, and plant photosynthetic capacity.

13.6.5 Hypoxia

In a number of studies, it has been found that exogenous Ca^{2+} application as $CaCl_2$ or $CaNO_3$ improved the tolerance in plants against hypoxic stress or flooding stress by changes in nitrogen metabolism, photosynthetic activity, and by regulating gene expression. Exogenous Ca^{2+} application triggers the sensing and signaling pathways of flooding tolerance. The release of Ca^{2+} from mitochondria due to the decrease in cytosolic pH is an important factor when plants are exposed to hypoxia (Pandey et al., 2000, Subbaiah et al., 1998). Elevated levels of Ca^{2+} can influence substrate level oxidation and metabolic fluxes under hypoxia (Bowler and Fluhr, 2000, Gao and Guo, 2005). Calcium-calmodulin complex is involved in the transduction of an anaerobic signal. This anaerobic signal inhibits proteolysis and the release of solute. Exogenous application of Ca^{2+} is an easy and effective approach for making plants hypoxically tolerant and for improving productivity. The addition of $CaCl_2$ causes an increase in the accumulation of amino acids in the roots of rice plants in response to anaerobic conditions. Similarly, in muskmelon sensitive to hypoxia (Liu and Chen, 1987) exogenous application of Ca^{2+} enhances hypoxic tolerance (Gao and Guo, 2005). Exogenous application of Ca^{2+} plays an important role in the growth and development of muskmelon (Alarcón et al., 1999). Although various regulatory mechanisms of Ca^{2+} under hypoxic conditions are quite different, many pieces of evidence show that exogenous treatment of Ca can increase tolerance in plants by regulating the metabolism of nitrogen (Aurisano et al., 1995, Gao et al., 2011, Ma et al., 2005).

13.6.6 Heavy Metals

Heavy metals in soils such as nickel, cadmium, chromium, lead, and so on, cause poor plant growth by affecting nutrient uptake, imbalance in redox potential, and photosynthesis. For example, Nada et al. (2007) have suggested that Cd^{2+} as a heavy metal is able to reduce the levels of Ca^{+2}, K^{+}, and Mg^{+2}. Cd also decreases the K^{+} and Ca^{2+} concentration (Ghnaya et al., 2005). This decrease is observed due to the competition between Ca^{2+} and Cd^{2+}. In some investigations, it has been reported that exogenous application of Ca reversed the heavy metal toxicity in different plant species. For example, Suzuki (2005) applied 30 mM $CaCl_2$ restored the root elongation and seedling growth up to 200 uM Cd. Similarly, 5 mM $Ca(NO_3)_2$ application alleviated the toxic effects of 40 uM Cd on the growth of lentils (*Lens culinaris*) (Talukdar, 2012). In the same way, similar effect of exogenous application of Ca has been found in other plant species such as *Sedum alfredii* (Tian et al., 2011) and in leguminous crops (Rodríguez-Serrano et al., 2009; Wang and Song, 2009). Exogenous Ca application reduces the arsenic toxicity in *Pteris vittata* by lowering in uptake of toxic elements. However, organic ascorbic acid is able to increase levels of Ca^{2+} in leaves. Ca^{+2} levels provide protection from heavy metalloid toxicity. Calcium also provides a protective role against monomethylarsonic acid (MMAA). According to Shanker et al. (2005), chromium is low in plants when there is a high concentration of Ca. In addition to this, (Ali et al., 2009) studied the effects of Ni in relation to Ca in canola (*Brassica napus*). There is a negative correlation between Ni and Ca. According to Gajewska and Skłodowska (2007), Ni has similar chemical characteristics as Ca. Ni is able to compete with Ca ions at the binding site in the oxygen-evolving complex. Ni has similar chemical characteristics as Ca. Bouazizi et al. (2010) studied toxicity of Cu. They concluded that there is a high concentration of Cu when Ca is low in *Phaseolus vulgaris*. Calcium plays an important role in alleviating toxic effects of nickel on plant growth (Aziz and Khan, 2001) by increasing activities of antioxidant enzymes and by the reduction in lipid peroxidation of cell membranes (Gong et al., 1997, Jiang et al., 2001). Calcium plays a major role in the detoxification process of heavy metals (Antosiewicz and Hennig, 2004, Jáuregui-Zúñiga et al., 2005). Calcium also plays role in the stabilization of membranes, influencing the pH of cells, and in the prevention of leakage of solutes from cytoplasm (Hirschi, 2004).

13.7 CONCLUSION AND PROSPECTS

Abiotic stress tolerance in plants is a complex and multigenic trait, i.e., controlled by a number of genes. Therefore, plants generally use a wide range of adaptive mechanisms to cope with abiotic stresses. However, it has been observed that most of the abiotic stresses cause oxidative stress in chloroplast, mitochondria, cytosol, and cell walls. During this process, cytosolic Ca^{2+} increased to a high level. Increase in cytosolic activates initiates a signaling cascade for stress tolerance via Ca sensor proteins. However, such an increase in Ca in cytosol may also cause toxic effects on cellular metabolism. To counteract such a situation, Ca^{2+} cation exchangers export the Ca^{2+} outside the cell or partitioned in to the vacuole. Both events import and export Ca^{2+} from cytosol and develop specific amplitude for a specific duration and frequency, which has been decoded for downstream gene expression responsible for stress tolerance. Moreover, changes in cytosolic Ca^{2+} is a central point or hub that initiates a number of signaling pathways to initiate stress tolerance physiological and biochemical events. However, exact components/proteins/transcription factors of these signaling pathways are not known yet. The advancement of RNA sequence analysis, CRISPER-Cas genome editing tools, and proteome-based approaches can all help in identifying these components. Some recent studies focused on cytosolic changes in Ca and pH signaling under these abiotic stresses. In view of these reports, it is suggested to identify cellular Ca^{2+} and pH signature signaling components in stress tolerant and sensitive cultivars. Breeding or developing transgenics with these specific components of stress tolerance will help in developing stress tolerant crop cultivars.

REFERENCES

Abdullah, Z., Ahmad, R. (1982) Salt tolerance of *Solanum tuberosum* L. growing on saline soils amended with gypsum. *Journal of Agronomy and Crop Science* 151: 409–416.

Abou El-Yazied, A. A. (2011) Foliar application of glycine betaine and chelated calcium improves seed production and quality of common bean (*Phaseolus vulgaris* L.) under water stress conditions. *Research Journal of Agriculture and Biological Sciences* 7: 357–370.

Alarcón, A. L., Madrid, R., Egea, C., Guillén, I. (1999) Calcium deficiency provoked by the application of different forms and concentrations of Ca^{2+} to soil-less cultivated muskmelons. *Scientia Horticulturae* 81 (1): 89–102.

Ali, M. A., Ashraf, M., Athar, H. R. (2009) Influence of nickel stress on growth and some important physiological/biochemical attributes in some diverse canola (*Brassica napus* L.) cultivars. *Journal of Hazardous Materials* 172 (2–3): 964–969.

Ali, R., Ma, W., Lemtiri-Chlieh, F., Tsaltas, D., Leng, Q., von Bodman, S., Berkowitz, G. A. (2007) Death don't have no mercy and neither does calcium: *Arabidopsis* cyclic nucleotide gated channel2 and Innate Immunity. *The Plant Cell* 19 (3): 1081–1095.

Amor, F. M. D., Marcelis, L. F. M. (2003) Regulation of nutrient uptake, water uptake and growth under calcium starvation and recovery. *The Journal of Horticultural Science and Biotechnology* 78 (3): 343–349.

Antosiewicz, D. M., Hennig, J. (2004) Overexpression of LCT1 in tobacco enhances the protective action of calcium against cadmium toxicity. *Environmental Pollution* 129 (2): 237–245.

Ashraf, M., Zafar, Z. U. (2000) Effect of low and high regimes of calcium on two cultivars of cotton (*Gossypium hirsutum* L.) differing in resistance to cotton leaf curl virus (CLCuV)-growth and macronutrients. *Agrochimica* 44: 89–100.

Aurisano, N., Bertani, A., Reggiani, R. (1995) Involvement of calcium and calmodulin in protein and amino acid metabolism in rice roots under anoxia. *Plant and Cell Physiology* 36: 1525–1529.

Awada, S., Campbell, W. F., Dudley, L. M., Jurinak, J. J., Khan, M. A. (1995) Interactive effects of sodium chloride, sodium sulfate, calcium sulfate, and calcium chloride on snapbean growth, photosynthesis, and ion uptake. *Journal of Plant Nutrition* 18 (5): 889–900.

Azaizeh, H., Gunse, B., Steudle, E. (1992) Effects of NaCl and $CaCl_2$ on water transport across root cells of maize (*Zea mays* L.) seedlings. *Plant Physiology* 99 (3): 886–894.

Azaizeh, H., Steudle, E. (1991) Effects of salinity on water transport of excised maize (*Zea mays* L.) roots. *Plant Physiology* 97 (3): 1136–1145.

Aziz, I., Khan, M. A. (2001) Experimental assessment of salinity tolerance of *Ceriops tagal* seedlings and saplings from the Indus Delta, Pakistan. *Aquatic Botany* 70 (3): 259–268.

Balagué, C., Lin, B., Alcon, C., Flottes, G., Malmström, S., Köhler, C., Neuhaus, G., et al. (2003) HLM1, an essential signaling component in the hypersensitive response, is a member of the cyclic nucleotide–gated channel ion channel family. *The Plant Cell* 15 (2): 365–379.

Bhatia, S., Asthir, B. (2014) Calcium mitigates heat stress effect in wheat seeding growth by altering carbohydrate metabolism. *Indian Journal of Plant Physiology* 19 (2): 138–143.

Blume, B., Nürnberger, T., Nass, N., Scheel, D. (2000) Receptor-mediated increase in cytoplasmic free calcium required for activation of pathogen defense in parsley. *The Plant Cell* 12 (8): 1425–1440.

Bokszczanin, K. L., Solanaceae Pollen Thermotolerance Initial Training Network (SPOT-ITN) Consortium, Fragkostefanakis, S. (2013) Perspectives on deciphering mechanisms underlying plant heat stress response and thermotolerance. *Frontiers in Plant Science* 4: 315.

Bolat, I., Kaya, C., Almaca, A., Timucin, S. (2006) Calcium sulfate improves salinity tolerance in rootstocks of plum. *Journal of Plant Nutrition* 29 (3): 553–564.

Bouazizi, H., Jouili, H., Geitmann, A., El Ferjani, E. (2010) Copper toxicity in expanding leaves of *Phaseolus vulgaris* L.: Antioxidant enzyme response and nutrient element uptake. *Ecotoxicology and Environmental Safety* 73 (6): 1304–1308.

Bowler, C., Fluhr, R. (2000) The role of calcium and activated oxygens as signals for controlling cross-tolerance. *Trends in Plant Science* 5 (6): 241–246.

Brown, C. E., Pezeshki, S. R., DeLaune, R. D. (2006) The effects of salinity and soil drying on nutrient uptake and growth of *Spartina alterniflora* in a simulated tidal system. *Environmental and Experimental Botany* 58 (1–3): 140–148.

Brunner, F., Rosahl, S., Lee, J., Rudd, J. J., Geiler, C., Kauppinen, S., Rasmussen, G., Scheel, D., Nürnberger, T. (2002) Pep-13, a plant defense-inducing pathogen-associated pattern from Phytophthora transglutaminases. *The EMBO Journal* 21 (24): 6681–6688.

Cacho, M., Domínguez, A. T., Elena-Rosselló, J. A. (2013) Role of polyamines in regulating silymarin production in *Silybum marianum* (L.) Gaertn (Asteraceae) cell cultures under conditions of calcium deficiency. *Journal of Plant Physiology* 170 (15): 1344–1348.

Carpita, N., Tierney, M., Campbell, M. (2001) Molecular biology of the plant cell wall: Searching for the genes that define structure, architecture and dynamics. *Plant Molecular Biology* 47 (1–2): 1–5.

Cheng, L., Tang, L., Zhang, Y., Yan, J., Zhang, H. (1997) The effect of $CaCl_2$ on the drought resistance of young cotton seedlings. *China Cottons* 24: 17–18.

Chiu, J. C., Brenner, E. D., DeSalle, R., Nitabach, M. N., Holmes, T. C., Coruzzi, G. M. (2002) Phylogenetic and expression analysis of the glutamate-receptor–like gene family in *Arabidopsis* thaliana. *Molecular Biology and Evolution* 19 (7): 1066–1082.

Cole, R. A., Fowler, J. E. (2006) Polarized growth: Maintaining focus on the tip. *Current Opinion in Plant Biology* 9 (6): 579–588.

Coria, N. A., Sarquís, J. I., Peñalosa, I., Urzúa, M. (1998) Heat-induced damage in potato (*Solanum tuberosum*) tubers: Membrane stability, tissue viability, and accumulation of glycoalkaloids. *Journal of Agricultural and Food Chemistry* 46: 4524–4528.

Cramer, G. R. (1992) Kinetics of maize leaf elongation II. Responses of a Na-excluding cultivar and Na-including cultivar to varying Na/Ca salinities. *Journal of Experimental Botany* 43 (6): 857–864.

Cramer, G. R., Epstein, E., Läuchli, A. (1988) Kinetics of root elongation of maize in response to short-term exposure to NaCl and elevated calcium concentration. *Journal of Experimental Botany* 39 (11): 1513–1522.

Dodd, A. N., Kudla, J., Sanders, D. (2010) The language of calcium signaling. *Annual Review of Plant Biology* 61: 593–620.

Edel, K. H., Kudla, J. (2015) Increasing complexity and versatility: How the calcium signaling toolkit was shaped during plant land colonization. *Cell Calcium* 57 (3): 231–246.

Evlagon, D., Ravina, I., Neumann, P. (1990) Interactive effects of salinity and calcium on hydraulic conductivity, osmotic adjustment, and growth in primary roots of maize seedlings. *Israel Journal of Botany* 39: 239–247.

Fu, J., Huang, B. (2003) Effects of foliar application of nutrients on heat tolerance of creeping bentgrass. *Journal of Plant Nutrition* 26 (1): 81–96.

Gajewska, E., Skłodowska, M. (2007) Effect of nickel on ROS content and antioxidative enzyme activities in wheat leaves. *BioMetals: An International Journal on the Role of Metal Ions in Biology, Biochemistry, and Medicine* 20 (1): 27–36.

Gao, H. B., Guo, S. R. (2005) Influence of calcium on antioxidant system and nitrogen metabolism of muskmelon seedlings under nutrient solution hypoxia. *Acta Horticulturae* 691 (691): 321–328.

Gao, H., Jia, Y., Guo, S., Lv, G., Wang, T., Juan, L. (2011) Exogenous calcium affects nitrogen metabolism in root-zone hypoxia-stressed muskmelon roots and enhances short-term hypoxia tolerance. *Journal of Plant Physiology* 168 (11): 1217–1225.

Gao, M., Tao, R., Miura, K., Dandekar, A. M., Sugiura, A. (2001) Transformation of Japanese persimmon (*Diospyros kaki* Thunb.) with apple cDNA encoding NADP-dependent sorbitol-6-phosphate dehydrogenase. *Plant Science: An International Journal of Experimental Plant Biology* 160 (5): 837–845.

Garcia-Brugger, A., Lamotte, O., Vandelle, E., Bourque, S., Lecourieux, D., Poinssot, B., Wendehenne, D., Pugin, A. (2006) Early signaling events induced by elicitors of plant defenses. *Molecular Plant-Microbe Interactions: MPMI* 19 (7): 711–724.

Genc, Y., Tester, M., McDonald, G. K. (2010) Calcium requirement of wheat in saline and non-saline conditions. *Plant and Soil* 327 (1–2): 331–345.

Ghnaya, T., Nouairi, I., Slama, I., Messedi, D., Grignon, C., Abdelly, C., Ghorbel, M. H. (2005) Cadmium effects on growth and mineral nutrition of two halophytes: Sesuvium Portulacastrum and Mesembryanthemum crystallinum. *Journal of Plant Physiology* 162 (10): 1133–1140.

Giri, A. (2013) Effect of acute heat stress on nutrient uptake by plant roots. Master of Science, Department of Biology. Toledo, Spain: The University of Toledo.

Gong, M., Chen, S. N., Song, Y. Q., Li, Z. E. (1997) Effect of calcium and calmodulin on intrinsic heat tolerance in relation to antioxidant systems in maize seedlings. *Functional Plant Biology* 24 (3): 371–379.

Gong, M., Van der Luit, A. H., Knight, M. R., Trewavas, A. J. (1998) Heat-shock-induced changes in intracellular Ca^{2+} level in tobacco seedlings in relation to thermotolerance. *Plant Physiology* 116 (1): 429–437.

Grattan, S. R., Grieve, C. M. (1998) Salinity–mineral nutrient relations in horticultural crops. *Scientia Horticulturae* 78 (1–4): 127–157.

Guimarães, F. V. A., Lacerda, C. F., de Marques, E. C., Abreu, C. E. B., de Aquino, B. F., de Prisco, J. T., Gomes-Filho, E. (2012) Supplemental Ca^{2+} does not improve growth but it affects nutrient uptake in NaCl-stressed cowpea plants. *Brazilian Journal of Plant Physiology* 24 (1): 9–18.

Hamada, H., Kurusu, T., Okuma, E., Nokajima, H., Kiyoduka, M., Koyano, T., Sugiyama, Y., et al. (2012) Regulation of a proteinaceous elicitor-induced ca^{2+} influx and production of phytoalexins by a putative voltage-gated cation channel, ostpcl, in cultured rice cells. *The Journal of Biological Chemistry* 287 (13): 9931–9939.

Han, S., Tang, R., Anderson, L. K., Woerner, T. E., Pei, Z. M. (2003) A cell surface receptor mediates extracellular Ca^{2+} sensing in guard cells. *Nature* 425 (6954): 196–200.

Harper, J. F., Breton, G., Harmon, A. (2004) Decoding Ca^{2+} signals through plant protein kinases. *Annual Review of Plant Biology* 55 (1): 263–288.

Hassan, N. M., El-Sayed, A. K. A., Ebeid, H. T., Alla, M. M. N. (2011) Molecular aspects in elevation of sunflower tolerance to drought by boron and calcium foliar sprays. *Acta Physiologiae Plantarum* 33 (2): 593–600.

Havlin, J. L., Beaton, J. D., Tisdale, S. L., Nelson, W. L. (2005) *Soil Fertility and Fertilizers: An Introduction to Nutrient Management*, Vol. 515. Upper Saddle River, NJ: Pearson Prentice Hall.

Henriksson, E., Henriksson, K. N. (2005) Salt-stress signalling and the role of calcium in the regulation of the *Arabidopsis* ATHB7 gene. *Plant, Cell & Environment* 28 (2): 202–210.

Hirschi, K. D. (2004) The calcium conundrum. Both versatile nutrient and specific signal. *Plant Physiology* 136 (1): 2438–2442.

Hu, Y., Burucs, Z., Schmidhalter, U. (2008) Effect of foliar fertilization application on the growth and mineral nutrient content of maize seedlings under drought and salinity. *Soil Science & Plant Nutrition* 54 (1): 133–141.

Huda, K. M. K., Banu, M. S. A., Tuteja, R., Tuteja, N. (2013) Global calcium transducer P-type Ca^{2+}-ATPases open new avenues for agriculture by regulating stress signalling. *Journal of Experimental Botany* 64 (11): 3099–3109.

Jaleel, C. A., Gopi, R., Manivannan, P., Panneerselvam, R. (2007) Responses of antioxidant defense system of *Catharanthus roseus* (L.) G. Don. To paclobutrazol treatment under salinity. *Acta Physiologiae Plantarum* 29 (3): 205–209.

Jammes, F., Hu, H. C., Villiers, F., Bouten, R., Kwak, J. M. (2011) Calcium-permeable channels in plant cells. *The FEBS Journal* 278 (22): 4262–4276.

Jáuregui-Zúñiga, D., Ferrer, M. A., Calderón, A. A., Muñoz, R., Moreno, A. (2005) Heavy metal stress reduces the deposition of calcium oxalate crystals in leaves of Phaseolus vulgaris. *Journal of Plant Physiology* 162 (10): 1183–1187.

Jenne, E. A., Rhoades, H. F., Yien, C. H., Howe, O. W. (1958) Change in nutrient element accumulation by corn with depletion of soil moisture. *Agronomy Journal* 50 (2): 71–74.

Jiang, C. D., Gao, H. Y., Zou, Q. (2001) Enhanced thermal energy dissipation depending on xanthophyll cycle and D1 protein turnover in iron-deficient maize leaves under high irradiance. *Photosynthetica* 39 (2): 269–274.

Jiang, Y., Huang, B. (2001) Effects of calcium on antioxidant activities and water relations associated with heat tolerance in two cool-season grasses. *Journal of Experimental Botany* 52 (355): 341–349.

Jones, R. L., Carbonell, J. (1984) Regulation of the synthesis of barley aleurone α-amylase by gibberellic acid and calcium ions. *Plant Physiology* 76 (1): 213–218.

Jurkowski, G. I., Smith, R. K., Yu, I. C., Ham, J. H., Sharma, S. B., Klessig, D. F., Fengler, K. A., Bent, A. F. (2004) *Arabidopsis* DND2, a second cyclic nucleotide-gated ion channel gene for which mutation causes the "Defense, No Death" phenotype. *Molecular Plant-Microbe Interactions: MPMI* 17 (5): 511–520.

Kadota, Y., Furuichi, T., Ogasawara, Y., Goh, T., Higashi, K., Muto, S., Kuchitsu, K. (2004) Identification of putative voltage-dependent Ca^{2+}-permeable channels involved in cryptogein-induced Ca^{2+} transients and defense responses in tobacco BY-2 cells. *Biochemical and Biophysical Research Communications* 317 (3): 823–830.

Kartusch, R. (2003) On the mechanism of callose synthesis induction by metal ions in onion epidermal cells. *Protoplasma* 220 (3–4): 219–225.

Kauss, H. (1987) Some aspects of calcium-dependent regulation in plant metabolism. *Annual Review of Plant Physiology* 38 (1): 47–71.

Kaya, C., Ak, B. E., Higgs, D., Murillo-Amador, B. (2002) Influence of foliar-applied calcium nitrate on strawberry plants grown under salt-stressed conditions. *Australian Journal of Experimental Agriculture* 42 (5): 631–636.

Kolupaev, Y. E., Akinina, G. E., Mokrousov, A. V. (2005) Induction of heat tolerance in wheat coleoptiles by calcium ions and its relation to oxidative stress. *Russian Journal of Plant Physiology* 52 (2): 199–204.

Kreimer, G., Melkonian, M., Holtum, J. A. M., Latzko, E. (1988) Stromal free calcium concentration and light-mediated activation of chloroplast fructose-1, 6-bisphosphatase. *Plant Physiology* 86 (2): 423–428.

Krichevsky, A., Kozlovsky, S. V., Tian, G. W., Chen, M. H., Zaltsman, A., Citovsky, V. (2007) How pollen tubes grow. *Developmental Biology* 303 (2): 405–420.

Kurusu, T., Kuchitsu, K., Nakano, M., Nakayama, Y., Iida, H. (2013) Plant mechanosensing and Ca^{2+} transport. *Trends in Plant Science* 18 (4): 227–233.

Kurusu, T., Yagala, T., Miyao, A., Hirochika, H., Kuchitsu, K. (2005) Identification of a putative voltage-gated Ca^{2+} channel as a key regulator of elicitor-induced hypersensitive cell death and mitogen-activated protein kinase activation in rice. *The Plant Journal: for Cell and Molecular Biology* 42 (6): 798–809.

Kwaaitaal, M., Huisman, R., Maintz, J., Reinstädler, A., Panstruga, R. (2011) Ionotropic glutamate receptor (iGluR)-like channels mediate MAMP-induced calcium influx in *Arabidopsis thaliana*. *The Biochemical Journal* 440 (3): 355–365.

Lacombe, B., Becker, D., Hedrich, R., DeSalle, R., Hollmann, M., Kwak, J. M., Schroeder, J. I., et al. (2001) The identity of plant glutamate receptors. *Science* 292 (5521): 1486–1487.

Larkindale, J., Knight, M. R. (2002) Protection against heat stress-induced oxidative damage in *Arabidopsis* involves calcium, abscisic acid, ethylene, and salicylic acid. *Plant Physiology* 128 (2): 682–695.

Laurière, C., Doyen, C., Thévenot, C., Daussant, J. (1992) β-amylases in cereals. A study of the maize β-amylase system. *Plant Physiology* 100 (2): 887–893.

Lecourieux, D., Lamotte, O., Bourque, S., Wendehenne, D., Mazars, C., Ranjeva, R., Pugin, A. (2005) Proteinaceous and oligosaccharidic elicitors induce different calcium signatures in the nucleus of tobacco cells. *Cell Calcium* 38 (6): 527–538.

Lecourieux, D., Mazars, C., Pauly, N., Ranjeva, R., Pugin, A. (2002) Analysis and effects of cytosolic free calcium increases in response to elicitors in *Nicotiana plumbaginifolia* cells. *The Plant Cell* 14 (10): 2627–2641.

Li, J., Wang, X., Zhang, Y., Jia, H., Bi, Y. (2011) cGMP regulates hydrogen peroxide accumulation in calcium-dependent salt resistance pathway in *Arabidopsis thaliana* roots. *Planta* 234 (4): 709–722. doi:10.1007/s00425-011-1439-3.

Liu, C. S., Chen, C. Y. (1987) Callus growth, cell increase and alcohol metabolism compared between two muskmelon varieties in culture. *Plant Science* 51 (2–3): 257–261.

López-Lefebre, L. R., Rivero, R. M., García, P. C., Sanchez, E., Ruiz, J. M., Romero, L. (2001) Effect of calcium on mineral nutrient uptake and growth of tobacco. *Journal of the Science of Food and Agriculture* 81 (14): 1334–1338.

Lovegrove, A., Hooley, R. (2000) Gibberellin and abscisic acid signalling in aleurone. *Trends in Plant Science* 5 (3): 102–110.

Lu, Y., Li, N., Sun, J., Hou, P., Jing, X., Zhu, H., Deng, S., et al. (2013) Exogenous hydrogen peroxide, nitric oxide and calcium mediate root ion fluxes in two non-secretor mangrove species subjected to NaCl stress. *Tree Physiology* 33 (1): 81–95.

Ludwig, A. A., Saitoh, H., Felix, G., Freymark, G., Miersch, O., Wasternack, C., Boller, T., et al. (2005) Ethylene-mediated cross-talk between calcium-dependent protein kinase and MAPK signaling controls stress responses in plants. *Proceedings of the National Academy of Sciences of the United States of America* 102 (30): 10736–10741.

Ma, R., Zhang, M., Li, B., Du, G., Wang, J., Chen, J. (2005) The effects of exogenous Ca^{2+} on endogenous polyamine levels and drought-resistant traits of spring wheat grown under arid conditions. *Journal of Arid Environments* 63 (1): 177–190.

Maas, E. V. (1993) Salinity and citriculture. *Tree Physiology* 12 (2): 195–216.

Maas, E. V., Grieve, J. D. G. (1987) Sodium-induced calcium deficiency in salt-stressed corn. *Plant, Cell & Environment* 10: 559–564.

Maathuis, F. J. (2009) Physiological functions of mineral macronutrients. *Current Opinion in Plant Biology* 12 (3): 250–258.

Maeda, Y., Yoshiba, M., Tadano, T. (2005) Comparison of ca effect on the salt tolerance of suspension cells and intact plants of tobacco (*Nicotiana tabacum* L., cv. Bright Yellow-2). *Soil Science & Plant Nutrition* 51 (4): 485–490.

Mahouachi, J., Socorro, A. R., Talon, M. (2006) Responses of papaya seedlings (*Carica papaya* L.) to water stress and re-hydration: Growth, photosynthesis and mineral nutrient imbalance. *Plant and Soil* 281 (1–2): 137–146.

Marschner, H. (1995) *Mineral Nutrition of Higher Plants*. New York: Academic Press.

Marschner, P. (2012) *Marschner's Mineral Nutrition of Higher Plants*. 3rd edn. San Diego: Academic Press.

Martínez-Ballesta, M. C., Dominguez-Perles, R., Moreno, D. A., Muries, B., Alcaraz-López, C., Bastías, E., García-Viguera, C., Carvajal, M. (2010) Minerals in plant food: Effect of agricultural practices and role in human health. A review. *Agronomy for Sustainable Development* 30 (2): 295–309.

Mäser, P., Thomine, S., Schroeder, J. I., Ward, J. M., Hirschi, K., Sze, H., Talke, I. N., et al. (2001) Phylogenetic relationships within cation transporter families of *Arabidopsis*. *Plant Physiology* 126 (4): 1646–1667.

McAinsh, M. R., Pittman, J. K. (2009) Shaping the calcium signature. *The New Phytologist* 181 (2): 275–294.

McLaughlin, S. B., Wimmer, R. (1999) Tansley Review No. 104: calcium physiology and terrestrial ecosystem processes. *New Phytologist* 142 (3): 373–417.

Medvedev, S. S. (2005) Calcium signaling system in plants. *Russian Journal of Plant Physiology* 52 (2): 249–270.

Morgan, A. J., Galione, A. (2014) Two-pore channels (TPCs): Current controversies. *Bioessays* 36 (2): 173–183.

Mukherjee, S. P., Choudhuri, M. A. (1985) Implication of hydrogen peroxide – Ascorbate system on membrane permeability of water stressed *Vigna* seedlings. *New Phytologist* 99 (3): 355–360.

Murugan, K., Sathish, D. K. (2005) Ameliorative effect by calcium on NaCI salinity stress related to proline metabolism in the callus of *Centella asiatica* L. *Journal of Plant Biochemistry and Biotechnology* 14 (2): 205–207.

Nada, E., Ferjani, B. A., Ali, R., Bechir, B. R., Imed, M., Makki, B. (2007) Cadmium-induced growth inhibition and alteration of biochemical parameters in almond seedlings grown in solution culture. *Acta Physiologiae Plantarum* 29 (1): 57–62.

Naeem, M., Khan, M. N., Khan, M. M. A. (2013) Adverse effects of abiotic stresses on medicinal and aromatic plants and their alleviation by calcium. In N. Tuteja, S. Singh Gill (eds.) *Plant Acclimation to Environmental Stress*, 101–146. New York: Springer.

Naeem, M., Naeem, M. S., Ahmad, R., Ahmad, R. (2017) Foliar-applied calcium induces drought stress tolerance in maize by manipulating osmolyte accumulation and antioxidative responses. *Pakistan Journal of Botany* 49: 427–434.

Naeem, M., Naeem, M. S., Ahmad, R., Ihsan, M. Z., Ashraf, M. Y., Hussain, Y., Fahad, S. (2018) Foliar calcium spray confers drought stress tolerance in maize via modulation of plant growth, water relations, proline content and hydrogen peroxide activity. *Archives of Agronomy and Soil Science* 64 (1): 116–131.

Nayyar, H., Kaushal, S. K. (2002) Alleviation of negative effects of water stress in two contrasting wheat genotypes by calcium and abscisic acid. *Biologia Plantarum* 45 (1): 65–70.

Nayyar, H., Walia, D. P. (2003) Water stress induced proline accumulation in contrasting wheat genotypes as affected by calcium and abscisic acid. *Biologia Plantarum* 46 (2): 275–279.

Neumann, P. M., Azaizeh, H., Leon, D. (1994) Hardening of root cell walls: A growth inhibitory response to salinity stress. *Plant, Cell & Environment* 17 (3): 303–309.

Nomura, H., Komori, T., Kobori, M., Nakahira, Y., Shiina, T. (2008) Evidence for chloroplast control of external Ca^{2+}-induced cytosolic Ca^{2+} transients and stomatal closure. *The Plant Journal: for Cell and Molecular Biology* 53 (6): 988–998.

Pandey, S., Tiwari, S. B., Upadhyaya, K. C., Sopory, S. K. (2000) Calcium signaling: Linking environmental signals to cellular functions. *Critical Reviews in Plant Sciences* 19 (4): 291–318.

Patel, N. T., Vaghela, P. M., Patel, A. D., Pandey, A. N. (2011) Implications of calcium nutrition on the response of *Caesalpinia crista* (Fabaceae) to soil salinity. *Acta Ecologica Sinica* 31 (1): 24–30.

Peiter, E., Maathuis, F. J. M., Mills, L. N., Knight, H., Pelloux, J., Hetherington, A. M., Sanders, D. (2005) The vacuolar Ca^{2+}-activated channel TPC1 regulates germination and stomatal movement. *Nature* 434 (7031): 404–408.

Price, M. B., Jelesko, J., Okumoto, S. (2012) Glutamate receptor homologs in plants: Functions and evolutionary origins. *Frontiers in Plant Science* 3: 235.

Qiang, L., Jianhua, C., Longjiang, Y., Maoteng, L., Jinjing, L., Lu, G. (2012) Effects on physiological characteristics of Honeysuckle (*Lonicera japonica* Thunb) and the role of exogenous calcium under drought stress. *Plant OMICS: Journal of Plant Molecular Biology & Omics* 5: 1.

Qiao, B., Zhang, Q., Liu, D., Wang, H., Yin, J., Wang, R., He, M., et al. (2015) A calcium-binding protein, rice annexin OsANN1, enhances heat stress tolerance by modulating the production of H_2O_2. *Journal of Experimental Botany* 66 (19): 5853–5866.

Ranf, S., Wünnenberg, P., Lee, J., Becker, D., Dunkel, M., Hedrich, R., Scheel, D., Dietrich, P. (2008) Loss of the vacuolar cation channel, AtTPC1, does not impair Ca^{2+} signals induced by abiotic and biotic stresses. *The Plant Journal for Cell and Molecular Biology* 53 (2): 287–299.

Raza, S., Farrukh Saleem, M., Mustafa Shah, G., Jamil, M., Haider Khan, I. (2013) Potassium applied under drought improves physiological and nutrient uptake performances of wheat (*Triticum aestivum* L.). *Journal of Soil Science and Plant Nutrition* 13: 175–185.

Reddy, A. S., Ali, G. S., Celesnik, H., Day, I. S. (2011) Coping with stresses: Roles of calcium-and calcium/calmodulin-regulated gene expression. *The Plant Cell* 23 (6): 2010–2032.

Rengel, Z., Zhang, W.-H. (2003) Role of dynamics of intracellular calcium in aluminium-toxicity syndrome. *New Phytologist* 159 (2): 295–314.

Rewald, B., Shelef, O., Ephrath, J. E., Rachmilevitch, S. (2013) Adaptive plasticity of salt-stressed root systems. In P. Ahmad, M. Azooz, M. Prasad (eds.) *Ecophysiology and Responses of Plants under Salt Stress*, 169–201. New York: Springer.

Riveras, E., Alvarez, J. M., Vidal, E. A., Oses, C., Vega, A., Gutiérrez, R. A. (2015) The calcium ion is a second messenger in the nitrate signaling pathway of *Arabidopsis*. *Plant Physiology* 169 (2): 1397–1404.

Rodríguez-Serrano, M., Romero-Puertas, M. C., Pazmiño, D. M., Testillano, P. S., Risueño, M. C., Del Río, L. A., Sandalio, L. M. (2009) Cellular response of pea plants to cadmium toxicity: Cross talk between reactive oxygen species, nitric oxide, and calcium. *Plant Physiology* 150 (1): 229–243.

Rubio, J. S., García-Sánchez, F., Rubio, F., Martínez, V. (2009) Yield, blossom-end rot incidence, and fruit quality in pepper plants under moderate salinity are affected by K^+ and Ca^{2+} fertilization. *Scientia Horticulturae* 119 (2): 79–87.

Saleh, B., Allario, T., Dambier, D., Ollitrault, P., Morillon, R. (2008) Tetraploid citrus rootstocks are more tolerant to salt stress than diploid. *Comptes Rendus Biologies* 331 (9): 703–710.

Samarah, N., Mullen, R., Cianzio, S. (2004) Size distribution and mineral nutrients of soybean seeds in response to drought stress. *Journal of Plant Nutrition* 27 (5): 815–835.

Shanker, A. K., Cervantes, C., Loza-Tavera, H., Avudainayagam, S. (2005) Chromium toxicity in plants. *Environment International* 31 (5): 739–753.

Shariat Jafari, H., Kafi, M., Astaraei, A. R. (2009) Interactive effects of NaCl induced salinity, calcium and potassium on physiomorphological traits of sorghum (*Sorghum bicolor* L.). *Pakistan Journal of Botany* 41: 3053–3063.

Shi, H., Ye, T., Zhong, B., Liu, X., Chan, Z. (2014) Comparative proteomic and metabolomic analyses reveal mechanisms of improved cold stress tolerance in Bermudagrass (*Cynodon dactylon* (L.) Pers.) by exogenous calcium. *Journal of Integrative Plant Biology* 56 (11): 1064–1079.

Shoresh, M., Spivak, M., Bernstein, N. (2011) Involvement of calcium-mediated effects on ROS metabolism in the regulation of growth improvement under salinity. *Free Radical Biology and Medicine* 51 (6): 1221–1234.

Sima, N. A. K. K., Askari, H., Mirzaei, H. H., Pessarakli, M. (2009) Genotype-dependent differential responses of three forage species to calcium supplement in saline conditions. *Journal of Plant Nutrition* 32 (4): 579–597.

Subbaiah, C. C., Bush, D. S., Sachs, M. M. (1998) Mitochondrial contribution to the anoxic Ca^{2+} signal in maize suspension-cultured cells. *Plant Physiology* 118 (3): 759–771.

Subbarao, G. V., Johansen, C., Jana, M. K., Kumar Rao, J. V. D. K. (1990) Effects of the sodium/calcium ratio in modifying salinity response of pigeonpea (*Cajanus cajan*). *Journal of Plant Physiology* 136 (4): 439–443.

Suzuki, N. (2005) Alleviation by calcium of cadmium-induced root growth inhibition in *Arabidopsis* seedlings. *Plant Biotechnology* 22 (1): 19–25.

Taiz, L., Zeiger, E. (2010) *Plant Physiology*. 5th edn. Sunderland, MA: Sinauer Associates.

Taiz, L., Zeiger, E., Moller, I. S., Murphy, A. eds. (2015) *Plant Physiology and Development*. 6th edn. Sunderland, MA: Sinauer Associates.

Takahashi, S., Katagiri, T., Hirayama, T., Yamaguchi-Shinozaki, K., Shinozaki, K. (2001) Hyperosmotic stress induces a rapid and transient increase in inositol 1, 4, 5-trisphosphate independent of abscisic acid in *Arabidopsis* cell culture. *Plant and Cell Physiology* 42 (2): 214–222.

Talukdar, D. (2012) Exogenous calcium alleviates the impact of cadmium-induced oxidative stress in *Lens culinaris* medic. Seedlings through modulation of antioxidant enzyme activities. *Journal of Crop Science and Biotechnology* 15 (4): 325–334.

Tang, R. H., Han, S., Zheng, H., Cook, C. W., Choi, C. S., Woerner, T. E., Jackson, R. B., Pei, Z. M. (2007) Coupling diurnal cytosolic Ca^{2+} oscillations to the CAS-IP_3 pathway in *Arabidopsis*. *Science* 315 (5817): 1423–1426.

Tavakkoli, E., Rengasamy, P., McDonald, G. K. (2010) High concentrations of Na^+ and Cl^- ions in soil solution have simultaneous detrimental effects on growth of faba bean under salinity stress. *Journal of Experimental Botany* 61 (15): 4449–4459.

Tian, S., Lu, L., Zhang, J., Wang, K., Brown, P., He, Z., Liang, J., Yang, X. (2011) Calcium protects roots of Sedum alfredii H. against cadmium-induced oxidative stress. *Chemosphere* 84 (1): 63–69.

Toprover, Y., Glinka, Z. (1976) Calcium ions protect beet root cell membranes against thermally induced changes. *Physiologia Plantarum* 37 (2): 131–134.

Trofimova, M. S., Andreev, I. M., Kuznetsov, V. V. (1999) Calcium is involved in regulation of the synthesis of HSPs in suspension-cultured sugar beet cells under hyperthermia. *Physiologia Plantarum* 105 (1): 67–73.

Tuna, A. L., Kaya, C., Ashraf, M., Altunlu, H., Yokas, I., Yagmur, B. (2007) The effects of calcium sulphate on growth, membrane stability and nutrient uptake of tomato plants grown under salt stress. *Environmental and Experimental Botany* 59 (2): 173–178.

Upadhyaya, H., Panda, S. K., Dutta, B. K. (2011) $CaCl_2$ improves post-drought recovery potential in Camellia sinensis (L) O. Kuntze. *Plant Cell Reports* 30 (4): 495–503.

Vaghela, P. M., Patel, A. D., Pandey, I. B., Pandey, A. N. (2009) Implications of calcium nutrition on the response of *Salvadora oleoides* (Salvadoraceae) to soil salinity. *Arid Land Research and Management* 23 (4): 311–326.

Vatsa, P., Chiltz, A., Bourque, S., Wendehenne, D., Garcia-Brugger, A., Pugin, A. (2011) Involvement of putative glutamate receptors in plant defence signaling and NO production. *Biochimie* 93 (12): 2095–2101.

Wang, C. Q., Song, H. (2009) Calcium protects Trifolium repens L. seedlings against cadmium stress. *Plant Cell Reports* 28 (9): 1341–1349.

Wang, Y., Yang, Z. M., Zhang, Q. F., Li, J. L. (2009) Enhanced chilling tolerance in *Zoysia matrella* by pre-treatment with salicylic acid, calcium chloride, hydrogen peroxide or 6-benzylaminopurine. *Biologia Plantarum* 53 (1): 179–182.

Ward, J. M., Mäser, P., Schroeder, J. I. (2009) Plant ion channels: Gene families, physiology, and functional genomics analyses. *Annual Review of Physiology* 71: 59–82.

White, J. B., Krause, H. H. (2001) Short-term boron deficiency in a black spruce (*Picea mariana* [Mill.] BSP) plantation. *Forest Ecology and Management* 152 (1–3): 323–330.

Wright, G. C., Patten, K. D., Drew, M. C. (1993) Gas exchange and chlorophyll content of 'Tifblue' rabbit eye and 'Sharp blue' southern highbush blueberry exposed to salinity and supplemental calcium. *Journal of the American Society for Horticultural Science* 118: 456–463.

Xu, C., Li, X., Zhang, L. (2013) The effect of calcium chloride on growth, photosynthesis, and antioxidant responses of Zoysia japonica under drought conditions. *PLOS ONE* 8 (7): e68214.

Xu, J., Tian, Y. S., Peng, R. H., Xiong, A. S., Zhu, B., Jin, X. F., Gao, F., et al. (2010) AtCPK6, a functionally redundant and positive regulator involved in salt/drought stress tolerance in *Arabidopsis*. *Planta* 231 (6): 1251–1260.

Yang, G. P., Gao, A. L. (1993) The relation of calcium to cell permeability in water stressed soybean hypocotyls. *Plant Physiology Communications* 3: 4.

Yang, S., Wang, F., Guo, F., Meng, J. J., Li, X. G., Wan, S. B. (2015) Calcium contributes to photoprotection and repair of photosystem II in peanut leaves during heat and high irradiance. *Journal of Integrative Plant Biology* 57 (5): 486–495.

Yoo, J. H., Park, C. Y., Kim, J. C., Do Heo, W. D., Cheong, M. S., Park, H. C., Kim, M. C., et al. (2005) Direct interaction of a divergent CaM isoform and the transcription factor, MYB2, enhances salt tolerance in *Arabidopsis*. *The Journal of Biological Chemistry* 280 (5): 3697–3706.

Yu, I. C., Parker, J., Bent, A. F. (1998) Gene-for-gene disease resistance without the hypersensitive response in *Arabidopsis dnd1* mutant. *Proceedings of the National Academy of Sciences of the United States of America* 95 (13): 7819–7824.

Yuan, F., Yang, H., Xue, Y., Kong, D., Ye, R., Li, C., Zhang, J., et al. (2014) OSCA1 mediates osmotic-stress-evoked Ca^{2+} increases vital for osmosensing in *Arabidopsis*. *Nature* 514 (7522): 367–371.

Zelman, A. K., Dawe, A., Gehring, C., Berkowitz, G. A. (2012) Evolutionary and structural perspectives of plant cyclic nucleotide-gated cation channels. *Frontiers in Plant Science* 3: 95.

Zhao, J., Davis, L. C., Verpoorte, R. (2005) Elicitor signal transduction leading to production of plant secondary metabolites. *Biotechnology Advances* 23 (4): 283–333.

Zhou, B., Guo, Z. (2009) Calcium is involved in the abscisic acid-induced ascorbate peroxidase, superoxide dismutase and chilling resistance in *Stylosanthes guianensis*. *Biologia Plantarum* 53 (1): 63–68.

Zhu, X., Feng, Y., Liang, G., Liu, N., Zhu, J. K. (2013) Aequorin-based luminescence imaging reveals stimulus-and tissue-specific Ca^{2+} dynamics in *Arabidopsis* plants. *Molecular Plant* 6 (2): 444–455.

Zidan, I., Jacoby, B., Ravina, I., Neumann, P. M. (1991) Sodium does not compete with calcium in saturating plasma membrane sites regulating $^{22}Na^+$ influx in salinized maize roots. *Plant Physiology* 96 (1): 331–334.

Zou, J. J., Li, X. D., Ratnasekera, D., Wang, C., Liu, W. X., Song, L. F., Zhang, W. Z., Wu, W. H. (2015) *Arabidopsis* calcium-dependent protein kinase8 and catalase3 function in abscisic acid-mediated signaling and H_2O_2 homeostasis in stomatal guard cells under drought stress. *The Plant Cell* 27 (5): 1445–1460.

14 Sulfur Nutrition and Abiotic Stress Tolerance in Plant

Sukhmeen Kaur Kohli, Neha Handa, Parminder Kaur, Poonam Yadav, Vinod Kumar, Rajinder Kaur, Saroj Arora, Adarsh Pal Vig, and Renu Bhardwaj

CONTENTS

14.1 INTRODUCTION

Sulfur (S) is an essential element ubiquitously present in plants. It occurs in two abundant forms including cysteine (Cys) and methionine (Met). It is an important constituent of a variety of plant metabolites such as proteins, polysaccharides, vitamins, enzyme co-factors, and lipids (Nocito et al., 2007). The main source of S for plants is available sulfate ions in the rhizosphere region of the soil, which is taken up by the roots through specific transporters present in the plasma membrane of root cells (Marschner, 1995). Once the sulfate enters the plant system, it is transported via phloem and xylem routes to various sinks and is further sequestered in the leaf chloroplast and vacuoles. There are different sulfate transportation systems regulating S fluxes, efficiently fulfilling S requirements for proper growth and development of plants under stressed as well as non-stressed conditions (Hawkesford, 2000). The survival of plants under varied environmental stresses is directly dependent upon adaptation and regulation of sulfate metabolism, as several S-containing compounds are involved in plant acclimatization to abiotic stresses (Nocito et al., 2007; Rausch and Wachter, 2005). A wide array of reports are available, suggesting alterations in S-responsive gene expression, modulating sulfate assimilation, and transportation as an adaptive response to lowered S demand (Nocito et al., 2006, 2007). Several mechanisms regulating S metabolism have been suggested in the literature, including demand-driven regulation, membrane protein-assisted regulation, glutathione (GSH) biosynthesis assisted regulation, and hormonal regulation (Nazar et al., 2011; Nocito et al., 2007; Saito, 2004).

Sustainable development of crops is primarily dependent upon varied environmental stress factors (Kumar et al., 2009; Nocito et al., 2007). Fertility of soil is another imperative factor (Sogbedi et al., 2006). Alteration in the soil fertility in response to abiotic stress, specifically xenobiotics, has erupted as one of the serious concerns for agriculturists for promoting proper growth and development in plants (Turkan and Demiral, 2009).

It has been well documented that several stresses including water deficit, salinity, heat, cold, and metal stress results in activation of the oxidative signal cascade (Chinnusamy et al., 2010; Keunen et al., 2011; Miller et al., 2010). Enhancement in abiotic stress levels such as salinity results in oxidative stress due to overproduction of reactive oxygen species (ROS) (Giraud et al., 2008). High levels of ROS in plant systems result in the loss of efficiency of photosynthetic attributes, proteins, nucleic acid, and lipids in response to oxidative damage (Nazar et al., 2011). Several S-containing compounds, including Cys, Met, MTs, PCs, GSH, TRX, and GRX, are considered as significant regulatory and protective mechanisms of plants to adapt to abiotic stress (Zagorchev et al., 2013).

The aim of this chapter is to provide updated and concise information regarding S metabolism and its assimilation, regulation, and effect of abiotic stress on S metabolism and role of S-containing compounds in abiotic stress management.

14.2 SULFUR METABOLISM

14.2.1 Sulfur Uptake and Transport

S in soils is present in both inorganic and organic forms and plants primarily take up inorganic or anionic sulfates (Buchner et al., 2004). The uptake of sulfates into cells at the soil-root interface, cell to cell transport, and storage require several sulfur transporter proteins. The symport of sulfate anions (SO_4^{2-}) along with protons (H^+) in the ratio of 1:3, which is driven by a proton gradient, is thought to be the mechanism of transport across plasma membranes (Leustek and Saito, 1999). The proton gradient is maintained by ATPase. The initial uptake of S occurs in root cells from the soil and they contain two isoforms (SULTR1;1, SULTR1;2) of group 1 sulfate transporters (Buchner et al., 2004). Group 1 transporters include all high affinity sulfate transporters and are primarily present in root tips, root hair, epidermis, and even the cortex of the mature root parts (Howarth et al., 2003; Rae and Smith, 2002; Shibagaki et al., 2002; Takahashi et al., 2000; Yoshimoto et al., 2002). However, the SULTR1;3 isoform is shown to be present in sieve tubes and companion cells of phloem (Yoshimoto et al., 2003). The property of high affinity of these transporters confirms that they are able to adapt to the variations in the external concentrations of sulfate (Leustek and Saito, 1999).

Group 2 of sulfate transporters comprise low-affinity transporters and include SULTR2;1 and SULTR2;2 isoforms (Hawkesford and De Kok, 2006). It has been shown by expression studies that they are located in vascular tissues and thus, participate in long-distance sulfur distribution inside the plant body (Hawkesford and De Kok, 2006). Studies conducted on *Brassica* and *Arabidopsis* show that the expression of the two isoforms is different. The expression of SULTR2;1 in *Brassica* is predominantly in roots, stems, and leaves and SULTR2;2 is expressed in roots. However, both these isoforms are expressed in roots and leaves in the case of *Arabidopsis* (Buchner et al., 2004; Hawkesford and De Kok, 2006).

Group 3 of sulfate transporters is the largest group and SULTR3;5 was the first transporter to be characterized in this category. Other members of this group include SULTR3;1, SULTR3;2, SULTR3;3, and SULTR3;4, which are present in plastid membranes and aid in the transport of sulfate inside the plastids (Cao et al., 2013). The transporter SULTR3;5 does not aid in sulfur transport but regulates the function of SULTR2;1 by forming a heterodimer (Kataoka et al., 2004a). Also, in *Lotus japonicus*, the homolog of SULTR3;5 is present in membrane of symbiosome in the nodule, thereby helping in nitrogen fixation by transporting sulfur to the bacteria inside (Krusell et al., 2005). The other two groups, *viz.* groups 4 and 5, have been reported to be present on the tonoplast membranes of the vacuoles (Kataoka et al., 2004b).

14.2.2 Sulfur Assimilation

S in the form of sulfate ions gets activated by linking with ATP to form adenosine-5′-phosphosulfate (APS) and enter metabolic pathways. The reaction is catalyzed by ATP sulfurylase and there is a release of pyrophosphate (Reaction 1; Figure 14.1).

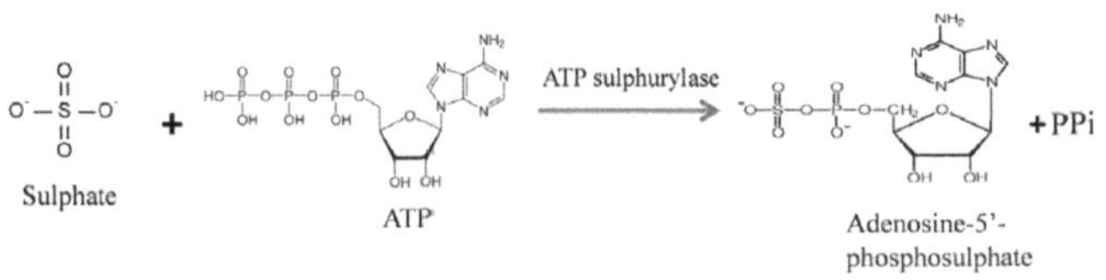

Reaction 1

APS so formed can either enter the reduction pathway or the sulfation pathway. In the reduction pathway, the sulfate gets reduced to sulfide by two enzymes in two steps. The first enzyme APS sulfotransferase, also known as ATS reductase, catalyzes the reduction of the sulfate residue of APS to sulfite (Saito, 2004). The enzyme uses thiol compounds most likely reduced glutathione (GSH) as an electron donor (Bick et al., 1998) and generates a thiol-sulfonate (*S*-sulfoglutathione in the case of GSH) (Reaction 2).

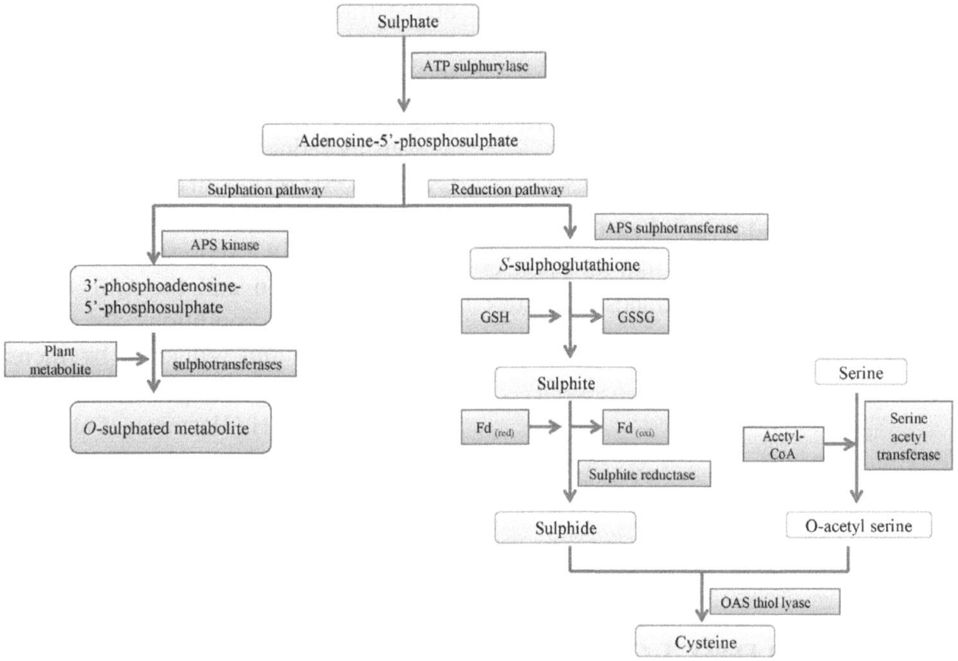

FIGURE 14.1 Schematic outline of S assimilation.

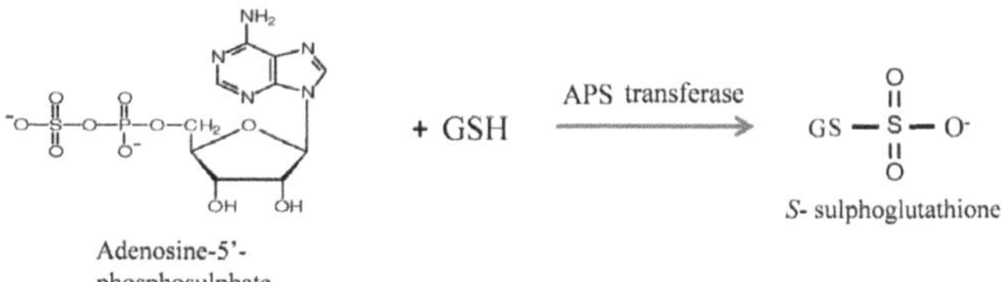

Reaction 2

The *S*-sulfoglutathione, in the presence of an excess of GSH, gets reduced and releases free sulfite in a non-enzymatic step (Reaction 3) (Schurmann and Brunold, 1980). In the next reduction step, sulfite reductase enzyme uses electrons from reduced ferredoxin and converts sulfite to sulfide (Reaction 4) (Leustek and Saito, 1999).

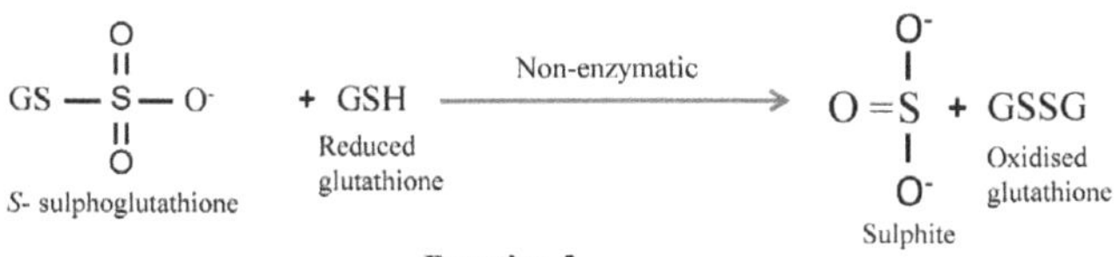

Reaction 3

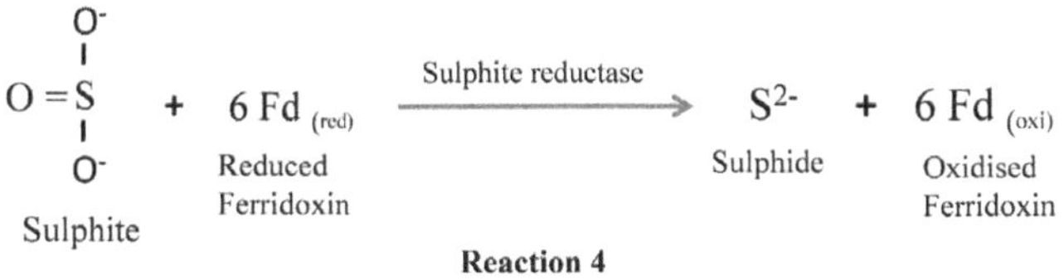

Reaction 4

The sulfide, thus formed, is ready to be incorporated in the amino acids (cysteine), which is the last step of assimilation pathway. The above-mentioned enzymes of sulfate reduction are all present in the plastids, while the enzymes for cysteine synthesis from sulfide are located in the cytosol, chloroplast, and mitochondria (Saito, 2004). In this mechanism, *O*-acetylserine is first synthesized from serine and acetyl-Co-A by the action of enzyme serine acetyltransferase (Reaction 5). The *O*-acetylserine and sulfide from the reduction step are then acted upon by *O*-acetylserine(thiol)lyase (OAS thiol lyase) to produce cysteine (Reaction 6) (Leustek and Saito, 1999; Saito, 2004).

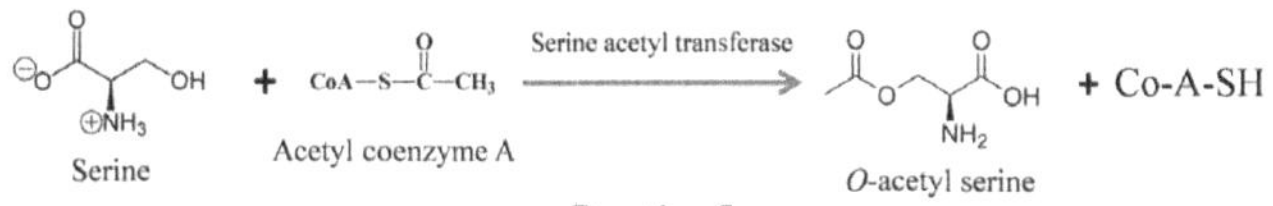

Reaction 5

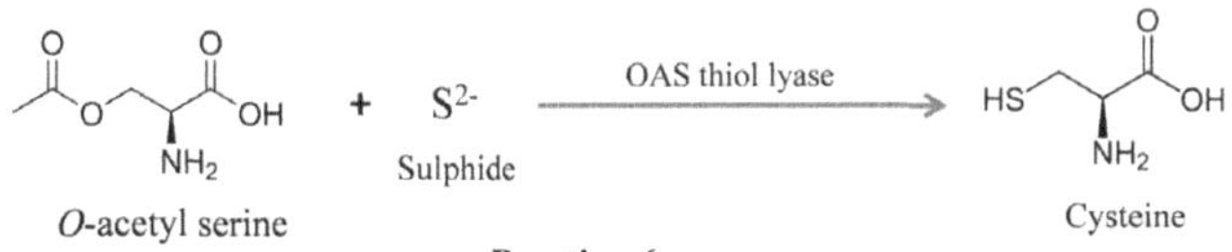

Reaction 6

The second pathway of S assimilation is the sulfation pathway where sulfate is integrated into the metabolites. In this pathway, the APS formed in the first step gets phosphorylated to form 3′-phosphoadenosine-5′-phosphosulfate (PAPS), and the reaction is catalyzed by APS kinase.

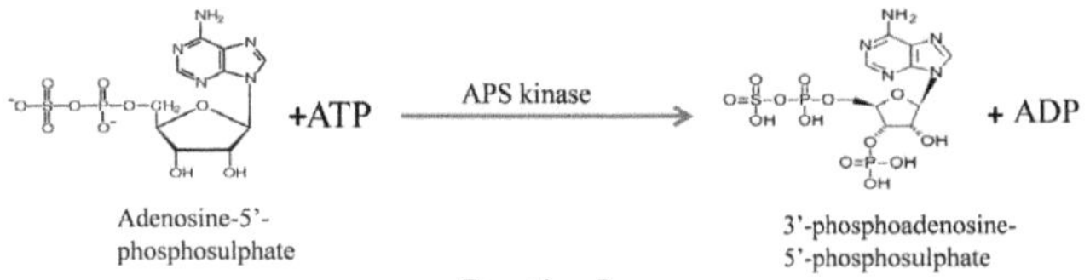

Reaction 7

PAPS is acted upon by several types of sulfotransferases and hence sulfation of metabolites like flavonoids, glucosinolates, choline, gallic acid, and glucoside occurs (Reaction 8) (Leustek and Saito, 1999; Saito, 2004).

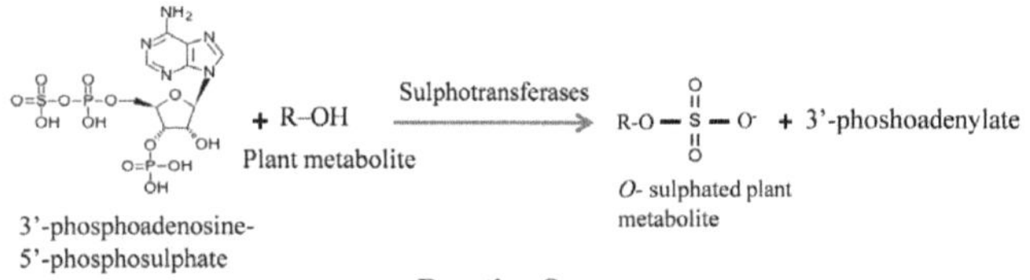

Reaction 8

14.3 REGULATION AND EFFECT OF ABIOTIC STRESSES ON S METABOLISM

14.3.1 Demand-Dependent Regulation

S assimilatory cascade is highly regulated in a demand-dependent manner (Davidian and Kopriva, 2010; Kopriva, 2006). The pathway is activated in response to increasing requirements of S for growth and development of plants (Kopriva and Rennenberg, 2004). Cys is the primary intermediate of S metabolism, its synthesis is carefully monitored to meet the demand of Cys required by plants. The demand of Cys required is continuously altered in response to varied environmental conditions to which plants are exposed. Abiotic stresses have been reported to increase the demand of several Cys derivatives, consequently resulting in enhancing the activity of the S assimilatory cascade (Nocito et al., 2007; Rausch and Wachter, 2005). A similar enhancement in the activity of the S assimilation pathway has been observed under S deficient conditions (Lappartient et al., 1999; Lappartient and Touraine, 1996). In response to lowered S from the growth medium, the levels of Cys, sulfate, and GSH in the tissue are drastically lowered, which further results in activation of key enzymes of S assimilation pathways and transportation system (Lappartient et al., 1999). Various steps of S assimilation pathway are modulated at the transcriptional level, for example, enhanced accumulation of genes encoding ATP sulfurylase, S transporters, and APS reductase (Nocito et al., 2007).

Several metabolites formed as intermediates of the S assimilation pathway and biosynthesis of GSH act as signals (positive or negative) regulating gene expression at the transcriptional level and activities of enzymes. Enhancement in the content of Cys and GSH results in repression of gene expression via negative feedback loops. This loop protects aggravation in sulfate uptake as well as reduction. However, a reduction in levels of Cys and GSH results in de-repression of gene expression and activation of the S metabolic pathway (Nocito et al., 2007). Figure 14.2 represents demand-driven regulation of S assimilation and the GSH biosynthetic pathway. In sulfate deficit conditions, the plants enhance sulfate transportation and reduction by increasing activity of APS enzyme. APS enzyme is also found in response to low levels of GSH (Hermsen et al., 2010). This indicates that GSH biosynthesis also occurs in a demand-dependent manner under abiotic stresses to scavenge high levels of ROS.

At post-transcriptional level, reversible formation of an imperative enzyme complex, that is, Ser acetyltransferase and OAS (thiol) lyase, occurs. Sulfide ions enhance the formation of OAS (thiol) lyase-Ser acetyltransferase complex resulting in a drastic decline in activity of OAS (thiol) lyase enzyme indicating huge amount of OAS (thiol) lyase responsible for the formation of Cys (Nocito et al., 2007). Synthesis of Cys and O acetyl-L-Serine (thiol) lyase (OAS-TL) is modulated by protein-protein interactions in multienzyme complexes of cysteine synthase (CSase) (Droux, 2004; Khan et al., 2014). Furthermore, the activity of serine acetyl transferase (SAT) is strongly enhanced in response to association with OAS-TL (Droux, 2003). ATP-S is considered to be the primary rate-limiting enzymes of the S assimilatory cascade (Khan et al., 2014).

14.3.2 Membrane Protein Assisted Regulation

Transportation of sulfate from the soil into the plant system involves multiple transit steps via several membranes involving plasma membrane transporters (PMTs). These transporters are present in the outermost layers of the root cells for primary uptake, translocation, and assimilation along with photosynthesis. Inside the root cells, these transporters couple with organelle transport such as plastid and vacuoles (Hawkesford, 2003; Saito, 2004). These PMTs are classified as proton/sulfate co-transporters. The sulfate co-transporters regulate active S transportation via trans-membrane protein gradient (Saito, 2004). The sulfate PMTs contain 12 membrane-spanning domains and are categorized as a large family of solutes or cation co-transporters (Hawkesford, 2003). Cao et al. (2013) reported several genes encoding sulfate transporters and were classified into five sub-families including SULTR 1, SULTR 2, SULTR 3, SULTR 4,

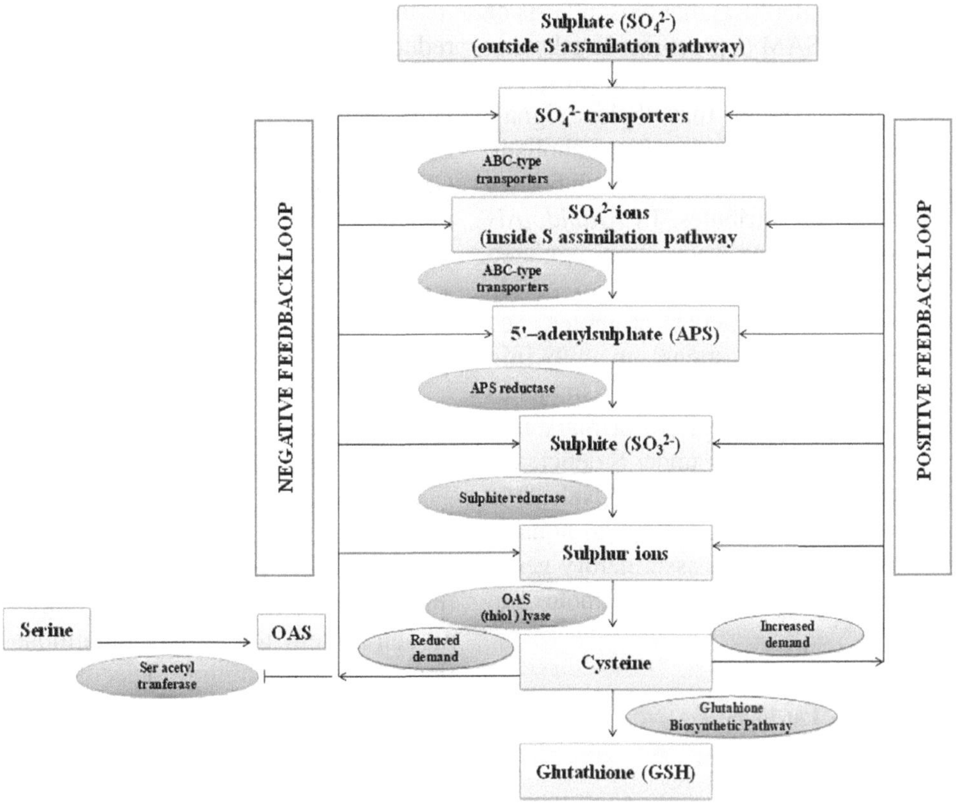

FIGURE 14.2 Schematic representation of the demand-dependent regulation of S assimilation and the GSH biosynthetic pathway.

and SULTR 5, already discussed in the previous section. Along with proton/sulfate co-transporters, anion channels, ABC type transporters may also be responsible for regulating S metabolism (Saito, 2004).

14.3.3 Glutathione Biosynthesis

GSH is considered as one of the most important thiol-containing metabolites. It is synthesized from glutamate, Cys, and glycine in the presence of two enzymes, γ-glutamyl-Cys synthetase and GSH synthetase. The activities of both the enzymes have been found in cytosol and chloroplast. The availability of Cys and inhibition of enzymes γ-glutamyl-Cys synthetase activity by GSH forms a rate-limiting factor for GSH biosynthesis (Noctor et al., 2002). Transporters of GSH and its derivatives have been recognized as proton co-transporters (Zhang et al., 2004). The uptake of several GSH derivatives associated with xenobiotics into the vacuole is regulated by ATP-energized transporters via ABC-type proteins. Another imperative transporter is S-Methyl Met formed in phloem and the cyclic conversion of Met to S-Methyl Met is responsible for short duration control of S-adenosyl-Met content (Kocsis et al., 2003). Several experimental analyses suggest the involvement of GSH, functioning as phloem-translocated signals as a regulator of S uptake and translocation (Lappartient et al., 1999; Nazar et al., 2011). Restricted GSH biosynthesis was observed in *Arabidopsis* and *Brassica* sp. in response to activation of L-buthionine (S, R) sulfoximine, an inhibitor of c-glutamyl cysteine synthetase enzyme. Consequently, it resulted in down-regulation of ATPS and mRNA associated with APR accumulation (Nazar et al., 2011; Vauclare et al., 2002). Under abiotic stress conditions, the increased requirement of GSH results in enhanced GSH biosynthesis, S assimilation, and external supply of S, improving tolerance to abiotic stress (Rennenberg et al., 2007). As a result of the higher flow of S inside cells, increase in GSH biosynthesis and APR expression is recorded in several stress conditions and channeling of sulfate into GSH synthesis pathway takes place (Scheerer et al., 2010).

14.3.4 Hormonal Regulation of S Metabolism

Nutrient use efficiency of plants is strongly regulated in response to alterations in phytohormone levels (Khan et al., 2009; Nazar et al., 2011). Gene expressions of several S metabolism encoding genes have been observed to be altered in response to plant hormones (Saito, 2004). It was suggested by Nazar et al. (2011) that biosynthesis of GSH and S-adenosyl Met, SAM (a precursor of ethylene biosynthesis) have a common precursor molecule, i.e., Cys. Therefore, it might be possible that ethylene signaling has an important role in regulating S metabolism. Cytokinins are largely reported to regulate S metabolism by modulation of molecular attributes. In *Arabidopsis* plants, sulfate transporters genes were down-regulated by cytokinin signaling (Maruyama-Nakashita et al., 2004). In *Glycine max*, S deficiency promoter, i.e., β-conglycinin, is enhanced in response to cytokinin signaling (Ohkama et al., 2002). Methyl jasmonates and auxin signaling cascades also have a primary role in regulating S assimilation pathway under S deficiency as evidenced by transcriptomic studies (Nakashita et al., 2003; Nikiforova et al., 2003). Methyl jasmonates have also induced a wide array of S assimilatory genes and jasmonate-inducible genes (Sasaki-Sekimoto and Ohta, 2004).

14.3.5 Miscellaneous Factors Regulating S Metabolism

The S assimilation pathway is extremely active in growing tissues of the plant system where high levels of Cys and MTs are required. It was reported that the gene expression of APS reductase and ATP sulfurylase is high in leaves of *Arabidopsis thaliana* plants (Rotte and Leustek, 2000). The trichome of roots has been observed to show high activities of GSH and Cys synthesizing enzymes (Gutierez-Alcala et al., 2000), probably due to an increased requirement for PCs and GSH for detoxification of several xenobiotic stresses (Choi et al., 2001). Heavy metal and other oxidative stresses largely affect the S assimilatory pathway. In response to heavy metal stress, PCs are produced from GSH via consumption of Cys (Dominguez-Solís et al., 2001; Saito, 2004). In order to alleviate oxidative stress, GSH functions as an imperative antioxidant as well as a reducing agent of several other antioxidants. As Cys production is a primary limiting factor for GSH biosynthesis, S assimilation is significantly controlled by the oxidative status of plant cells (Saito, 2004). Cys is converted to oxidized glutathione by activation of the hydrogen sulfide bond of Cys into the oxidized disulfide bond of GSH (Bick et al., 2001). Gene expression of several S transporters, Ser acetyl transferase, APS reductase, and 3-phosphoglycerate dehydrogenase are the enzymes to be activated by the onset of light as they are regulated by the circadian cycle (Harmer et al., 2000). The leaves of plants generate 3-phosphoglycerate, electrons, and ATP in response to the onset of light by enhancing photosynthesis resulting in the formation of APS, Ser formation, and sulfite reduction. The circadian cycle controls the production of Cys efficiently before the generation of various reductants and subtracts. Post-translational modification of GSH and S assimilation enzymes results in activation of GSH biosynthesis and scavenging of ROS, subsequently mitigating oxidative damage (Saito, 2004).

14.4 ROLE OF S-CONTAINING COMPOUNDS IN ABIOTIC STRESS TOLERANCE

S is recognized as one of the major essential nutrient elements, which not only plays a role in proper growth and development but also is linked to the stress tolerance (Marschner, 1995). S is present in plants as an integrated part of organic molecules and is found in thiol (-SH) groups in proteins like cysteine residues or non-protein thiols like glutathione. The concentration of thiol-containing molecules, especially glutathione, is sensitive to the oxidized surroundings, and thus behaves as a potential moderator of the stress response (Khan et al., 2013; Szalai et al., 2009). Thiols show a strong nucleophilic nature, which makes them suitable for biological redox processes. Redox processes control enzymatic action and protect plants against oxidative damage produced by abiotic stress. Plants have adopted several mechanisms involving S-containing compounds to increase tolerance to abiotic stresses. There is increasing evidence on the involvement of S-containing compounds such as Cys, MTs, GSH, PCs, TRX, and GRX in plant stress tolerance (Colville and Kranna, 2010). Cys is the primary product of sulfate metabolism and GSH is a tripeptide, low-molecular-weight thiol (Zagorchev et al., 2013). GSH, a water-soluble antioxidant, is an important long-distance storage form of S and has a significant role in the detoxification of several xenobiotics and heavy metals and modulation of the cell cycle (Notor et al., 2011). TRX and GRX are also potent protein-based protection and regulatory mechanisms (Meyer et al., 2012). A few S-containing compounds have a role in abiotic stress management, which are discussed below.

14.4.1 S-Containing Amino Acids

Cys and Met take part in abiotic stress management by chelating with heavy metals like Cd, Cu, Zn, and so on. Cys is produced in the final step of sulfate assimilation, a process through which plants take up sulfur. Thus, it acts as a S donor for the production of Met, sulfur-iron cluster, vitamins like thiamine and biotin, glutathione, and thiol-containing proteins (Hell and Wirtz, 2011).

Cys synthesis takes place via two steps. In the first step, o-acteylserine is synthesized from Ser and acetyl-CoA, and the reaction is catalyzed by serine acetyltransferase. In the second step, the reduced S (H_2S) is included by o-acteylserine lyase by eliminating the acetate residue to form Cys (Figure 14.3). Overexpression of these two enzymes under stress conditions in transgenic plants indicates that cysteine synthetase complexes are the primary enzymes of S metabolism with a role in stress response. The transgenic tobacco plants showed the involvement of cysteine synthetase complex in Cd tolerance (Kawashima et al., 2004). Increase in the free Cys concentrations in cells under abiotic stress has been reported by Ruiz and Blumwald, (2002). In many cases, the increase in free Cys content was reported to accompany enhanced glutathione levels, which leads to the conclusion that Cys is mainly required for the synthesis of S rich molecules like glutathione and stress-related proteins having anti-stress activity. It has been observed that under saline conditions, the rate of Cys synthesis increases along with the overexpression of OAS-TL, which is related to salt stress (40, 41, 43). In transgenic and wild-type canola plants, increase in the SAT was recorded under higher levels of NaCl. Under 150 mM NaCl concentration, wild-type canola plants showed 2.5 times more SAT activity than transgenic plants. This difference in SAT activity between transgenic and wild-type plants is related to Cys concentration in leaves as 1.5 times more Cys was recorded in wild-type plants (Ruiz and Blumwald, 2002). Similarly, Khan et al. (2013) have also reported increased activity of SAT and OASTL, leading to enhanced Cys biosynthesis, and thus resulting in higher accumulation of GSH under salinity stress.

Methionine is synthesized from Cys. Condensation of OPH and Cys leads to the formation of intermediate cystathione, which is transformed to homocysteine with the help of the enzyme CgS cystathionine β-lyase. The homocysteine is then converted to Met by Met synthetase. Approximately 80% of Met synthesized is further converted to *S*-adenosyl Met (SAM) by SAMS (SAM synthetase). Methionine also undergoes ROS-induced oxidation and form Met sulfoxide (MetO). MetO can be converted to Met with the help of enzymes metsulfoxide reductases, which are plastidic and cytosolic enzymes concerned with reducing oxidative damage (Cabreiro et al., 2006). Methionine is required for the synthesis of polyamine like putrescine, spermidine, and spermine, which play important roles in abiotic stress tolerance (Alcázar et al., 2010). It has been recorded that arsenic (metal stress) S interaction enhanced the levels of Met in rice plants (Dixit et al., 2015). A similar kind of increase in cysteine and Met levels was reported in macrophyte *Hydrilla verticillata* (Srivastava and D'souza, 2009). The increase in the activity of Met synthase was observed after metal treatment to the *Phytolacca Americana*, which suggests the role of Met in plant defense (Zhao et al., 2011).

14.4.2 Glutathione and Derivatives

GSH and its derivatives are widely present in all plants. GSH is a S-containing low-molecular-weight compound that is associated with the exclusion of

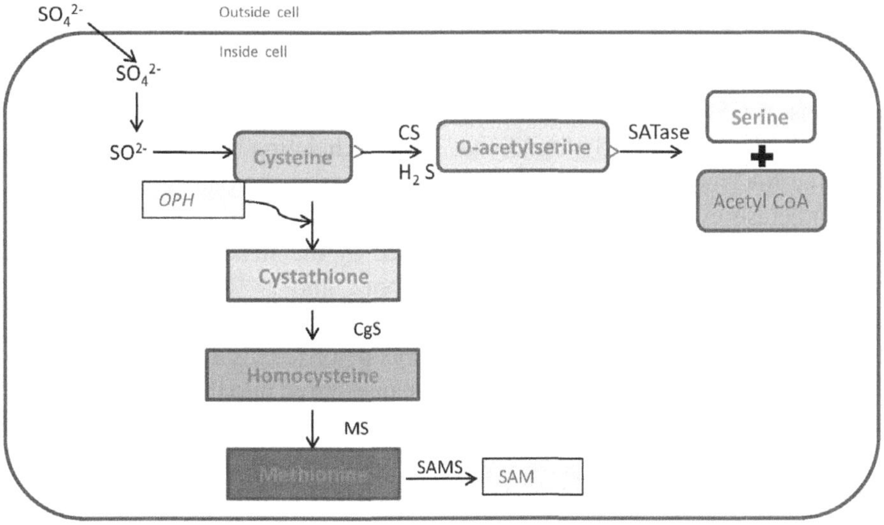

FIGURE 14.3 The biosynthetic pathway of cysteine and methionine in plants. CS, cysteine synthetase; SATase, serine acetyltransferase; OPH, *O*-phosphohomoserine; CgS, CgS. cystathionine β-lyase; MS, methionine synthase; SAMS, SAM synthetase; SAM, *S*-adenosyl methionine.

ROS (Noctor et al., 2012). GSH is biosynthesized in two steps. In the first step, an ATP dependent reaction forms γ-glutamyl Cys with the help of γ-glutamyl cysteine synthetase. The glycine is added to the γ-glutamyl Cys by the glutathione synthetase to form glutathione (Wachter et al., 2005). It has been reported that levels of GSH increase in plants under abiotic stress conditions. Exposure of plants to adverse environmental conditions induces the production of ROS leading to oxidative stress. Under oxidative stress, glutathione acts as an antioxidant as well as reducing molecules for other antioxidants like ascorbic acid (Asgher et al., 2017). GSH is a cellular redox buffer that helps the cellular environment in the reduced state. The GSH to GSSG ratio in unstressed cells is 20:1, which under stressed conditions shift to the GSSG (Szarka et al., 2012). This shift triggers various signaling pathways leading to the programmed cell death (PCD), which is an important method for stress resistance (Kranner et al., 2010). It has been observed by Semane et al. (2007) that exposure of *Arabidopsis* plants to a lower concentration of Cd (1 μM) enhanced the GSSG and reduced the GSH concentration, thus leading to increased GSSG/GSH ratio. Plants were able to cope with toxicity at a low level of metal but at higher Cd levels (10 μM), plants showed metal-induced oxidative stress.

GSH also participates in various cellular processes for the detoxification of xenobiotic and heavy metals. It has been confirmed with the help of NMR that GSH has the ability to chelate Cd (Delalande et al., 2010). GSH also activate and conjugate with xenobiotic compounds with the help of glutathione-s-transferase (Edwards and Dixon, 2005). The conjugated products are transported to vacuoles and thus protect cells from harmful effects (Yazaki, 2006). Various studies on transgenic plants and overexpressing γ-glutamyl cysteine synthetase showed that higher concentrations of GSH and PCs lead to better metal sequestration. Song et al. (2013) found that there was a two to three-fold increase in the level of γ-glutamyl cysteine synthetase in rice plants under Cu stress. It has been observed that tolerance of plants to salinity is linked with the ability to biosynthesize GSH and thus shows an important role of S nutrition (Nazar et al., 2011). The exogenous application of GSH to *Arabidopsis* plants reduced the Hg accumulation and thus confers enhanced Hg tolerance. Application of GSH significantly increased the seed germination, seedling growth, and chlorophyll content while reduced the H_2O_2 and O_2^- levels and lipid peroxidation (Kim et al., 2017). TLC and NMR studies showed that GSH binds strongly with Hg and forms more stable complexes. It has been observed that exogenous application of GSH reduced the Cu induced toxicity symptoms, decreases ROS production, and limits the Cu uptake in rice plants (Mostofa et al., 2014). Similar results were reported by Chao et al. (2011) in rice plants under Cd toxicity. The genotypic difference between rice cultivars has been linked with GSH synthesis and accumulation under stress (Cai et al., 2011). It has been seen that Cd retention and detoxification in rice roots occurs through chelation with thiol molecules and sequestration (Zhang et al., 2014). It was further reported that ABA and auxin levels also changed with an endogenous increase in GSH levels. The stress tolerance and rise in other biological impacts might be associated with increased hormone levels due to GSH treatment (Cheng et al., 2015). Increase in the GSH and cysteine levels under Cu stress in *Brassica juncea* plants has also been reported by Yadav et al. (2018, 2018).

Homo-glutathione is a homologue of glutathione and is present chiefly in legumes. In it, C– terminal of glycine is replaced by β-alanine and has the same functions as GSH (Noctor et al., 2012). It has been reported that legumes synthesize homo-phytochelatins under heavy metal stress (Sobrino-Plata et al., 2009). Cruz de Carvalho et al. (2010) reported an increase in mRNA levels in drought-tolerant cowpea cultivar encoding for the hGSH synthetase under drought and desiccation conditions. It was observed that Cd exposure causes accretion of peroxides and reduction of GSH and hGSH, which lead to redox imbalance in *Medicago sativa* plants (Ortega-Villasante et al., 2005). It has been reported by Lucini et al., (2015) that homo-glutathiones plays an important role in salinity stress tolerance along with other biomolecules like hormones, carbohydrates, and so on. Yang et al. (2012) observed a decrease in the GSH and hGSH levels under Cd toxicity in soybean.

14.4.3 Phytochelatins

Phytochelatins (PCs) are the most important metal ion chelators in the plants and few fungi as well as invertebrates. These are oligomers of reduced GSH characterized by the structure (γ-Glu-Cys)*n*-Gly, where n varies from 2 and 11. The PC synthase enzyme is responsible for their synthesis which is activated by the presence of metal ions. PCs form complexes with ions which are sequestrated into the vacuole and thereby reduce the toxic effect of metals (Pal and Rai, 2010). Overexpression of PC synthase (*alr0975*) increases abiotic stress tolerance by changing the proteomics of *Anabaena* sp. PCC 7120 (Chaurasia et al., 2017). Under drought, salt, and cold stress, the PC synthase (PCS) gene expression was upregulated in *Arabidopsis* (Seki et al., 2002). The *AtPCS1* has been expressed in transgenic *Arabidopsis* with the aim of enhancing PC synthase and metal tolerance in such plants. Transgenic PCs was found to contain 12–25 times more accumulation of *AtPCS1*and PCs formation enhanced 1.3–2.1 times more in comparison with wild-type plants under Cd stress conditions (Lee et al., 2003). The expression of *Arabidopsis AtPCS1* in *Nicotiana tabacum* plants increases the Cd tolerance in the presence of GSH (Brunetti et al., 2011). In plants, PC dependent Cd detoxification needs that PC-Cd complexes are transported into the vacuoles. *Arabidopsis thaliana* seedlings malfunctioning in the ABCC transporter *AtABCC3* have enhanced sensitivity to various concentrations of Cd and that seedlings overexpressing *AtABCC3* have enhanced Cd tolerance. The *AtABCC3* level in wild-type seedlings was less than that of *AtABCC1* and *AtABCC2* in the absence of Cd but more after Cd treatment, and even more in *atabcc1 atabcc2* mutants. The study indicated that *AtABCC3* is a transporter of PC–Cd complexes, and it was proposed that its activity is regulated by Cd and is coordinated with AtABCC1/AtABCC2 activity (Brunetti et al., 2015).

Phytochelatin transporters such as *AtABCC1* and *AtABCC2* are essential vacuolar transporters that provide tolerance to Cd and Hg, in addition to their function in the detoxification of As. These transporters present valuable techniques for genetic engineering of plants with increased metal tolerance (Park et al., 2012). PCs synthesis in *Vetiveria zizanioides* under Pb and P stress has been studied by Andra et al. (2010). From the results, it was concluded that by synthesizing PCs and Pb-PCs complexes in *V. zizanioides* is an effective mechanism for the detoxification of Pb. PCs content was studied in *Halimione portulacoides*, *Sarcocornia perennis*, and *Spartina maritime*. PCs concentration was found higher in leaves and the stems of *H. portulacoides* and *S. perennis*, whereas *S. maritime* contained higher content in roots. PC2 was synthesized by all tissues and species, and high concentration was found in large roots of *S. maritima*. The level of PC4 and PC5 were high in small roots of *S. maritima*. PC2 has a positive correlation with Zn, As, and Pb (Negrin et al., 2017). As(III) and As(V) induced the synthesis of PC in *Dunaliella salina*. PC2, PC3, and PC4 were found in the algal cells and their content was reduced slowly when exposed to As for three days. However, decreases in As uptake and PC synthesis in *D. salina* were seen as the PO_4^{-3} content in the medium enhanced. From the study, it was found that PCs play a significant role in detoxification of As in *D. salina* (Wang et al., 2017). The aquatic snail *Lymnaea stagnalis* produced PC2 and PC3 and their content was enhanced in response to sub-lethal concentrations of Cd (Gonçalves et al., 2016). *The Oryza sativa* PC synthase 1 (*OsPCS1*) plays a significant role in the detoxification of As and Cd. The *OsPCS1* mutants showed enhanced sensitivity to As and Cd in hydroponic experiments, indicating the significance of *OsPCS1*-dependent PC synthesis for rice As and Cd tolerance (Uraguchi et al., 2017). The date palm PC synthase type 1 (*PdPCS1*) catalyzes the PCs synthesis. The gene expression of *PdPCS1* was determined in seedling hypocotyls under Cu, Cr, and Cd stress (Zayneb et al., 2017).

14.4.4 Metallothionins

MTs are Cys-rich metal chelators with low molecular weight and have the potential to bind with various heavy metal ions by the generation of mercaptide bonds between Cys residues and the metals (Cobbett and Goldsbrough, 2002). MTs generally contain two metal binding Cys-rich domains that provide MTs a dumbbell conformation. They are broadly found in plants, animals, fungi, and cyanobacteria. On the basis of sequence similarity and phylogenetic relationships, these are categorized into three types (Binz and Kagi, 1999; Cherian and Chan, 1993; Zimeri et al., 2005). Type 1 MTs are commonly found in vertebrates. Type II MTs are prevalent in plants, fungi, and invertebrates. Type III MTs represents the PCs, which are enzymatically synthesized metal binding peptides. *OsMT1e-P* expression is regulated by various abiotic stresses in an organ-specific method and it is an essential gene for abiotic stress tolerance. It is projected that the *OsMT1e-P* gene helps in detoxification and cellular repair by maintaining the cellular homeostasis via reactive oxygen species scavenging directly or indirectly via other antioxidants. This gene may act together with other physiological mechanisms in the cell for selective uptake and sequestration of ions under salinity stress (Kumar et al., 2012).

MT class I protein, i.e., *CcMT1*, has been isolated from *Cajanus cajan*. Northern blot analysis of *C. cajan* under heavy metal stress indicates the enhanced contents of *CcMT1*, whereas overexpression of *CcMT1* gene in transgenic *Arabidopsis* plants showed enhanced Cu and Cd tolerance as indicated by elevated plant biomass and chlorophyll content as well as high root growth as compared with the control plants. Transgenic plants expressing the *CcMT1* gene showed increased Cu and Cd ions accumulations in the roots. The results of this study showed that *CcMT1* contributes to the detoxification of metal ions and provides increased tolerance against metal stresses. CcMT1 may be positioned as a possible candidate for increasing the metal tolerance in various crop plants (Sekhar et al., 2011). Ansarypour and Shahpiri (2017) studied the heterologous expression of the *OsMTI-1b* gene in *Saccharomyces cerevisiae* which increases Cd, H_2O_2 and ethanol tolerance. The study revealed that the *OsMTI-1b* gene increases the *S. cerevisiae* tolerance to numerous stresses. The *ScMT2-1-3* gene from sugarcane acts as a significant function in the regulation of heavy metal tolerance. Under Cu stress, expression of *ScMT2-1-3* is upregulated, whereas under Cd stress its expression is down-regulated. Real-time qPCR indicated that *ScMT2-1-3* gene expression was over 14 times more in bud and root as compared with stem and leaf. The sugarcane plantlets assay revealed that *ScMT2-1-3* is extensively concerned with Cu detoxification, whereas its functional mechanism in Cd detoxification needs further testification (Guo et al., 2013). Type I MT *ZjMT* is characterized from the *Ziziphus jujube,* and the expression of *ZjMT* gene was up-regulated by the treatments of polyethylene glycol (PEG), $CdCl_2$, and NaCl. Continuous *ZjMT* expression in wild-type *Arabidopsis* plants increased their tolerance to NaCl during germination stage. Transgenic plants accumulate more Cd^{2+} in roots but less accumulation in leaf as compared with wild-type plants, indicating that gene *ZjMT* may play a role in Cd^{2+} retention in roots and reduce the Cd^{2+} toxicity (Yang et al., 2015). MT gene expression was studied in *Solanum lycopersicum* under Cd stress conditions, and these genes were up-regulated under Cd stress. The level of MT1 and MT2 were enhanced in the roots, leaves, and the fruits, whereas MT3 and MT4 levels were varied concerning the type of the tissue (Kısa et al., 2016).

Type III MT *GhMT3a* was isolated from *Gossypium hirsutum* and northern blot analysis showed that *GhMT3a* accumulation was not only up-regulated under drought, salinity, and low-temperature conditions but also under reactive oxygen species (ROS), heavy metals, ethylene, and abscisic acid (ABA) in seedlings of cotton. *Nicotiana tabacum* expressing *GhMT3a* showed enhanced tolerance under abiotic stresses as compared with wild-type plants. The results of the *in vitro* assay indicated that *GhMT3a* has the capability to bind with the metal ions and scavenge ROS. From these results, it was found that *GhMT3a* function as efficient ROS scavenger and its expression may be regulated under abiotic stress conditions through ROS signaling (Xue et al., 2009). *SbMT-2* gene from *Salicornia brachiata* provides abiotic stress tolerance in transgenic tobacco. Transgenic tobacco lines such as L2, L4, L6, and L13 were found to significantly increase salt and osmotic and heavy metal tolerance as compared with wild-type plants. Increased expression of SOD, POD, and APX showed that the *SbMT-2* gene plays an important function in the ROS scavenging and detoxification mechanism (Chaturvedi et al., 2014). MT genes such as *IbMT1*, *IbMT2*, and *IbMT3* were cloned from *Ipomoea batatas* and their expression profiles were studied in plants under heavy metal, salt, and osmotic stresses. The expression of *IbMT1* was strongly increased under Cd and Fe, and moderately more in response to Cu, whereas the *IbMT3* gene followed the same expression level as that of *IbMT1.* Under PEG and NaCl stress conditions, the *IbMT* expression is up-regulated (Kim et al., 2014).

14.4.5 Thioredoxin and Glutaredoxin

Thioredoxin (TRX) is a widely present S-containing compound. It is 12–13 KDa heat-stable protein in plants. Due to the presence of a disulfide bond, it has a primary role in regulating the redox status of plants (Schürmann and Buchanan, 2008). It contains a pair of vicinal Cys, i.e., (-Trp-Cys-Gly-Pro-Cys-Lys-) (Holmgren, 1989). TRX is classified into two families, Family 1 and Family 2, depending upon the sequence of amino acids. Family 1 has a distinct TRX domain and Family 2 having a TRX domain (one or two) that is fused with proteins having additional domains (Gelhaye et al., 2004). Family 1 is further categorized into f, h, m, o, x, and y TRX in higher plants. TRX o is present in mitochondrial chambers, TRX f, m, x, and y are found in chloroplast. A large number of TRX have been reported in rice (about 60–70), *Arabidopsis* (about 60–70), and sorghum (11), and also the different number of genes in TRX family varies from species to species (Nuruzzaman et al., 2012). A report by Martí et al. (2011) suggests an imperative mitochondrial TRX PsTrxO1, which serves as a defense component in *Pisum sativum* plants under salt stress. An enhancement in mitochondrial PsTrxO1 activity was observed in response to an increase in salt stress. It also prevents mitochondrial oxidative damage with help of antioxidative enzymes. Zhang et al. (2011), suggested an increase in expression of Os TRXh1 gene in *Oryza sativa* plants under salinity stress which led to modulation of the redox state and development of plants. Serrato et al. (2004) reported NADP-TRX enzymes, that is, NADPH-TRX reductase (NTR) in chloroplast of *Arabidopsis thaliana* plants. Lack of NTR expression results in enhanced susceptibility of *Arabidopsis* plants to salinity stress. An *Arabidopsis* NTRC, a knock-out mutant of NTR, had retarded growth and hypersensitivity to salt stress, further suggesting the role of NTRC in promoting protection of plants against oxidative stress. NTRC was reported to undergo heat-shock oligomerization and exhibit chaperones like activity and promote tolerance to heat stress (Chae et al., 2013).

Glutaredoxin (GRX) is another imperative protein-based protection mechanism in plants under stress (Meyer et al., 2012). It is ubiquitously present in most organisms, including bacteria, Escherichia *coli*, and higher plants and animals. Similar to TRX, it is a protein thiol oxidoreductase with a pair of Cys responsible for its reducing ability. It has been reported to reduce several enzymes including S peroxidases, Met sulfoxide reductase (MSR), and metal reductase, and induces thiol-disulfide oxidoreductase activity. Abiotic stress largely elevates the levels of TRX by promoting protein content and gene expression. A report by Song et al. (2013), suggests up-regulation in the expression of both TRX and GRX genes in *Oryza sativa* plants under Cu stress. A similar enhancement in TRX and GRX gene expression was reported in salt-stressed barley plants (Fatehi et al., 2012). Another report by Sanz-Barrio et al. (2012) suggested lowered expression of TRX genes under cold stress and enhancement in expression of the same under drought stress in tobacco plants in early treatment stages. A significant family of tetratripeptide thiredoxin-like (TTL) protein has been reported to have a significant role in promoting tolerance to osmotic stress and salinity stress (Lakhssassi et al., 2012).

14.5 CONCLUSION

The ever-increasing abiotic stresses are the primary threats to soil fertility and crop productivity. Several strategies are adopted by plants to develop more tolerance to these stresses. S metabolism plays a significant role in providing structural constituents to the essential macromolecules and act as a signal under stressed conditions. The S assimilation process is largely affected by several internal and external factors. S-containing compounds are involved in initial responses to the stressed environment, alteration of redox status, and the modulation of molecular processes. The in-depth study of several S-containing compounds such as Cys, Met, PCs, MTs, TRX, and GRX, suggest the effective involvement of S in promoting tolerance to abiotic stress.

REFERENCES

Alcázar, R., Planas, J., Saxena, T., Zarza, X., Bortolotti, C., Cuevas, J., Bitrián, M., et al. (2010) Putrescine accumulation confers drought tolerance in transgenic *Arabidopsis* plants over-expressing the homologous arginine decarboxylase 2 gene. *Plant Physiology and Biochemistry: PPB* 48(7): 547–552.

Andra, S.S., Datta, R., Sarkar, D., Makris, K.C., Mullens, C.P., Sahi, S.V., Bach, S.B.H. (2010) Synthesis of phytochelatins in vetiver grass upon lead exposure in the presence of phosphorus. *Plant and Soil* 326(1–2): 171–185.

Ansarypour, Z., Shahpiri, A. (2017) Heterologous expression of a rice metallothionein isoform (OsMTI-1b) in Saccharomyces cerevisiae enhances cadmium, hydrogen peroxide and ethanol tolerance. *Brazilian Journal of Microbiology: [Publication of the Brazilian Society for Microbiology]* 48(3): 537–543.

Asgher, M., Per, T.S., Anjum, S., Khan, M.I.R., Masood, A., Verma, S., Khan, N.A. (2017) Contribution of glutathione in heavy metal stress tolerance in plants. In *Reactive oxygen species and antioxidant systems in plants: Role and regulation under abiotic stress*, ed. Khan, M., Khan, N., 297–313. Springer: Singapore.

Bick, J.A., Åslund, F., Chen, Y., Leustek, T. (1998) Glutaredoxin function for the carboxyl-terminal domain of the plant-type 5′-adenylylsulfate reductase. *Proceedings of the National Academy of Sciences of the United States of America* 95(14): 8404–8409.

Bick, J.A., Setterdahl, A.T., Knaff, D.B., Chen, Y., Pitcher, L.H., Zilinskas, B.A., Leustek, T. (2001) Regulation of the plant-type 5′-adenylylsulfate reductase by oxidative stress. *Biochemistry* 40(30): 9040–9048.

Binz, P.A., Kagi, J.H.R. (1999) *Metallothionein: Molecular evolution and classification.* In *Metallothionein*, ed. Klaassen, C.B. 4th edn, 7–13. Birkhauser Verlag: Switzerland.

Brunetti, P., Zanella, L., De Paolis, A., Di Litta, D., Cecchetti, V., Falasca, G., Barbieri, M., et al. (2015) Cadmium-inducible expression of the ABC-type transporter AtABCC3 increases phytochelatin-mediated cadmium tolerance in *Arabidopsis. Journal of Experimental Botany* 66(13): 3815–3829.

Brunetti, P., Zanella, L., Proia, A., De Paolis, A., Falasca, G., Altamura, M.M., Sanità di Toppi, L., et al. (2011) Cadmium tolerance and phytochelatin content of *Arabidopsis* seedlings over-expressing the phytochelatin synthase gene AtPCS1. *Journal of Experimental Botany* 62(15): 5509–5519.

Buchner, P., Takahashi, H., Hawkesford, M.J. (2004) Plant sulphate transporters: Co-ordination of uptake, intracellular and long-distance transport. *Journal of Experimental Botany* 55(404): 1765–1773.

Cabreiro, F., Picot, C.R., Friguet, B., Petropoulos, I. (2006) Methionine sulfoxide reductases: Relevance to Aging and Protection against Oxidative Stress. *Annals of the New York Academy of Sciences* 1067(1): 37–44.

Cai, Y., Cao, F., Cheng, W., Zhang, G., Wu, F. (2011) Modulation of exogenous glutathione in phytochelatins and photosynthetic performance against Cd stress in the two rice genotypes differing in Cd tolerance. *Biological Trace Element Research* 143(2): 1159–1173.

Cao, M.J., Wang, Z., Wirtz, M., Hell, R., Oliver, D.J., Xiang, C.B. (2013) SULTR3; 1 is a chloroplast-localized sulfate transporter in *Arabidopsis thaliana. The Plant Journal: for Cell and Molecular Biology* 73(4): 607–616.

Chae, H.B., Moon, J.C., Shin, M.R., Chi, Y.H., Jung, Y.J., Lee, S.Y., Nawkar, G.M., et al. (2013) Thioredoxin reductase type C (NTRC) orchestrates enhanced thermotolerance to *Arabidopsis* by its redox-dependent holdase chaperone function. *Molecular Plant* 6(2): 323–336.

Chao, Y.Y., Hong, C.Y., Chen, C.Y., Kao, C.H. (2011) The importance of glutathione in defence against cadmium-induced toxicity of rice seedlings. *Crop, Environment and Bioinformatics* 8: 217–228.

Chaturvedi, A.K., Patel, M.K., Mishra, A., Tiwari, V., Jha, B. (2014) The SbMT-2 gene from a halophyte confers abiotic stress tolerance and modulates ROS scavenging in transgenic tobacco. *PLOS ONE* 9(10): e111379.

Chaurasia, N., Mishra, Y., Chatterjee, A., Rai, R., Yadav, S., Rai, L.C. (2017) Overexpression of phytochelatin synthase (pcs) enhances abiotic stress tolerance by altering the proteome of transformed *Anabaena* sp. PCC 7120. *Protoplasma* 254(4): 1715–1724.

Cheng, M.C., Ko, K., Chang, W.L., Kuo, W.C., Chen, G.H., Lin, T.P. (2015) Increased glutathione contributes to stress tolerance and global translational changes in *Arabidopsis*. *The Plant Journal: for Cell and Molecular Biology* 83(5): 926–939.

Cherian, G.M., Chan, H.M. (1993) Biological functions of metallothioneins-a review. In metallothionein iii: Biological roles and medical implications, ed. Suzuki, K.T., Imura, N., Kimura, M., 87–109. Birkhauser Verlag.

Chinnusamy, V., Zhu, J.K., Sunkar, R. (2010) Gene regulation during cold stress acclimation in plants. *Methods in Molecular Biology* 639: 39–55.

Choi,Y.E., Harada, E., Wada, M., Tsuboi, H., Morita, Y., Kusano, T., Sano, H. (2001) Detoxification of cadmium in tobacco plants: Formation and active excretion of crystals containing cadmium and calcium through trichomes. *Planta* 213(1): 45–50.

Cobbett, C., Goldsbrough, P. (2002) Phytochelatins and metallothioneins: Roles in heavy metal detoxification and homeostasis. *Annual Review of Plant Biology* 53: 159–182.

Colville, L., Kranner, I. (2010) Desiccation tolerant plants as model systems to study redox regulation of protein thiols. *Plant Growth Regulation* 62(3): 241–255.

Davidian, J.C., Kopriva, S. (2010) Regulation of sulfate uptake and assimilation—The same or not the same? *Molecular Plant* 3(2): 314–325.

de Carvalho, M.H.C., Brunet, J., Bazin, J., Kranner, I., d'Arcy-Lameta, A., Zuily-Fodil, Y., Contour-Ansel, D. (2010) Homoglutathione synthetase and glutathione synthetase in drought-stressed cowpea leaves: Expression patterns and accumulation of low-molecular-weight thiols. *Journal of Plant Physiology* 167(6): 480–487

Delalande, O., Desvaux, H., Godat, E., Valleix, A., Junot, C., Labarre, J., Boulard, Y. (2010) Cadmium – Glutathione solution structures provide new insights into heavy metal detoxification. *Federation of European Biochemical Societies Journal* 277(24): 5086–5096.

Dixit, G., Singh, A.P., Kumar, A., Dwivedi, S., Deeba, F., Kumar, S., Suman, S., et al. (2015) Sulfur alleviates arsenic toxicity by reducing its accumulation and modulating proteome, amino acids and thiol metabolism in rice leaves. *Scientific Reports* 5: 16205.

Dominguez-Solís, J.R., Gutierrez-Alcalá, G., Vega, J.M., Romero, L.C., Gotor, C. (2001) The cytosolic O-acetylserine (thiol) lyase gene is regulated by heavy-metals and can function in cadmium tolerance. *The Journal of Biological Chemistry* 276(12): 9297–9302.

Droux, M. (2004) Sulfur assimilation and the role of sulfur in plant metabolism: A survey. *Photosynthesis Research* 79(3): 331–348.

Edwards, R., Dixon, D.P. (2005) Plant glutathione transferases. *Methods in Enzymology* 401: 169–186.

Fatehi, F., Hosseinzadeh, A., Alizadeh, H., Brimavandi, T. (2012) The proteome response of *Hordeum spontaneum* to salinity stress. *Cereal Research Communication*: 1–10.

Gelhaye, E., Rouhier, N., Gérard, J., Jolivet, Y., Gualberto, J., Navrot, N., Ohlsson, P.I. et al. (2004) A specific form of thioredoxin h occurs in plant mitochondria and regulates the alternative oxidase. *Proceedings of the National Academy of Sciences of the United States of America* 101(40): 14545–14550.

Giraud, E., Ho, L.H.M., Clifton, R., Carroll, A., Estavillo, G., Tan, Y.F., Howell, K.A., et al. (2008) The absence of alternative oxidase1a in *Arabidopsis* results in acute sensitivity to combined light and drought stress. *Plant Physiology* 147(2): 595–610.

Gonçalves, S.F., Davies, S.K., Bennett, M., Raab, A., Feldmann, J., Kille, P., Loureiro, S., et al. (2016) Sub-lethal cadmium exposure increases phytochelatin concentrations in the aquatic snail *Lymnaea stagnalis*. *Science of the Total Environment* 568: 1054–1058.

Guo, J., Xu, L., Su, Y., Wang, H., Gao, S., Xu, J., Que, Y. (2013) ScMT2-1-3, a metallothionein gene of sugarcane, plays an important role in the regulation of heavy metal tolerance/accumulation. *BioMed Research International*, doi: 10.1155/2013/904769.

Gutierez-Alcala, G., Gotor, C., Meyer, A.J., Fricker, M., Vega, J.M., Romero, L.C. (2000) Glutathione biosynthesis in *Arabidopsis* trichome cells. *Proceedings of the National Academy of Sciences of the United States of America* 97(20): 11108–11113.

Harmer, S.L., Hogenesch, J.B., Straume, M., Chang, H.S., Han, B., Zhu, T., Wang, X., et al. (2000) Orchestrated transcription of key pathways in *Arabidopsis* by the circadian clock. *Science* 290(5499): 2110–2113.

Hawkesford, M.J. (2000) Plant responses to sulphur deficiency and the genetic manipulation of sulphate transporters to improve S-utilization efficiency. *Journal of Experimental Botany* 51(342): 131–138.

Hawkesford, M.J. (2003) Transporter gene families in plants: The sulphatetransporter gene family: Redundancy or specialization? *Physiologia Plantarum* 117(2): 155–163.

Hawkesford, M.J., De Kok, L.J. (2006) Managing sulphur metabolism in plants. *Plant, Cell & Environment* 29(3): 382–395.

Hell, R., Wirtz, M. (2011) Molecular biology, biochemistry and cellular physiology of cysteine metabolism in *Arabidopsis thaliana*. *The Arabidopsis Book* 9: e0154.

Hermsen, C., Koprivova, A., Matthewman, C., Wesenberg, D., Krauss, G.J., Kopriva, S. (2010) Regulation of sulfate assimilation in Physcomitrella patens: Mosses are different! *Planta* 232(2): 461–470.

Holmgren, A. (1989) Thioredoxin and glutaredoxin systems. *The Journal of Biological Chemistry* 264(24): 13963–13966.

Howarth, J.R., Fourcroy, P., Davidian, J.C., Smith, F.W., Hawkesford, M.J. (2003) Cloning of two contrasting high-affinity sulfate transporters from tomato induced by low sulfate and infection by the vascular pathogen *Verticillium dahliae*. *Planta* 218(1): 58–64.

Kataoka, T., Hayashi, N., Yamaya, T., Takahashi, H. (2004a) Root-to-shoot transport of sulfate in *Arabidopsis*. Evidence for the role of SULTR3; 5 as a component of low-affinity sulfate transport system in the root vasculature. *Plant Physiology* 136(4): 4198–4204.

Kataoka, T., Watanabe-Takahashi, A., Hayashi, N., Ohnishi, M., Mimura, T., Buchner, P., Hawkesford, M.J., et al. (2004b) Vacuolar sulfate transporters are essential determinants controlling internal distribution of sulfate in *Arabidopsis*. *The Plant Cell* 16(10): 2693–2704.

Kawashima, C.G., Noji, M., Nakamura, M., Ogra, Y., Suzuki, K.T., Saito, K. (2004) Heavy metal tolerance of transgenic tobacco plants over-expressing cysteine synthase. *Biotechnology Letters* 26(2): 153–157.

Keunen, E., Remans, T., Bohler, S., Vangronsveld, J., Cuypers, A. (2011) Metal-induced oxidative stress and plant mitochondria. *International Journal of Molecular Sciences* 12(10): 6894–6918.

Khan, N.A., Anjum, N.A., Nazar, R., Iqbal, N. (2009) Increased activity of ATP sulfurylase, contents of cysteine and glutathione reduce high cadmium-induced oxidative stress in high photosynthetic potential mustard (*Brassica juncea* L.) cultivar. *Russian Journal of Plant Physiology* 56(5): 670–677.

Khan, M.I.R., Asgher, M., Iqbal, N., Khan, N.A. (2013) Potentiality of sulfur-containing compounds in salt stress tolerance. In *Ecophysiology and responses of plants under salt stress*, ed. Ahmad, P., Azooz, M.M., Prasad, M.N.V., 443–472. Springer: New York.

Khan, N.A., Khan, M.I.R., Asgher, M., Fatma, M., Masood, A., Syeed, S. (2014) Salinity tolerance in plants: Revisiting the role of sulfur metabolites. *Journal of Biochemistry and Physiology* 2: 1.

Kim, Y.O., Bae, H.J., Cho, E., Kang, H. (2017) Exogenous glutathione enhances mercury tolerance by inhibiting mercury entry into plant cells. *Frontiers in Plant Science* 8: 683.

Kim, S.H., Jeong, J.C., Ahn, Y.O., Lee, H.S., Kwak, S.S. (2014) Differential responses of three sweetpotato metallothionein genes to abiotic stress and heavy metals. *Molecular Biology Reports* 41(10): 6957–6966.

Kısa, D., Öztürk, L., Tekin, Ş. (2016) Gene expression analysis of metallothionein and mineral elements uptake in tomato (*Solanum lycopersicum*) exposed to cadmium. *Journal of Plant Research* 129(5): 989–995.

Kocsis, M.G., Ranocha, P., Gage, D.A., Simon, E.S., Rhodes, D., Peel, G.J., Mellema, S., et al. (2003) Insertional inactivation of the methionine S-methyltransferase gene eliminates the S-methylmethionine cycle and increases the methylation ratio. *Plant Physiology* 131(4): 1808–1815.

Kopriva, S. (2006) Regulation of sulfate assimilation in *Arabidopsis* and beyond. *Annals of Botany* 97(4): 479–495.

Kopriva, S., Rennenberg, H. (2004) Control of sulphate assimilation and glutathione /synthesis: Interaction with N and C metabolism. *Journal of Experimental Botany* 55(404): 1831–1842.

Kranner, I., Minibayeva, F.V., Beckett, R.P., Seal, C.E. (2010) What is stress? Concepts, definitions and applications in seed science. *The New Phytologist* 188(3): 655–673.

Krusell, L., Krause, K., Ott, T., Desbrosses, G., Krämer, U., Sato, S., Nakamura, Y., et al. (2005) The sulfate transporter SST1 is crucial for symbiotic nitrogen fixation in *Lotus japonicus* root nodules. *The Plant Cell* 17(5): 1625–1636.

Kumar, G., Purty, R.S., Sharma, M.P., Singla-Pareek, S.L., Pareek., A. (2009) Physiological responses among *Brassica* species under salinity stress show strong correlation with transcript abundance for SOS pathway-related genes. *Journal of Plant Physiology* 166(5): 507–520.

Kumar, G., Kushwaha, H.R., Panjabi-Sabharwal, V., Kumari, S., Joshi, R., Karan, R., Mittal, S., et al. (2012) Clustered metallothionein genes are co-regulated in rice and ectopic expression of OsMT1e-P confers multiple abiotic stress tolerance in tobacco via ROS scavenging. *BMC Plant Biology* 12: 107.

Lakhssassi, N., Doblas, V.G., Rosado, A., del Valle, A.E., Posé, D., Jimenez, A.J., Castillo, A.G., et al. (2012) The *Arabidopsis* tetratricopeptide thioredoxin-like gene family is required for osmotic stress tolerance and male sporogenesis. *Plant Physiology* 158(3): 1252–1266.

Lappartient, A.G., Touraine, B. (1996) Demand-driven control of root ATP sulphurylase activity and SO4 2– uptake in intact canola. *Plant Physiology* 111(1): 147–157.

Lappartient, A.G., Vidmar, J.J., Leustek, T., Glass, A.D.M., Touraine, B. (1999) Interorgan signalling in plant: Regulation of ATP sulfurylase and sulfate transporter genes expression in roots mediated by phloem-translocated compounds. *Planta* 18: 89–95.

Lee, S., Moon, J.S., Ko, T.S., Petros, D., Goldsbrough, P.B., Korban, S.S. (2003) Overexpression of *Arabidopsis* phytochelatin synthase paradoxically leads to hypersensitivity to cadmium stress. *Plant Physiology* 131(2): 656–663.

Leustek, T., Saito, K. (1999) Sulfate transport and assimilation in plants. *Plant Physiology* 120(3): 637–644.

Lucini, L., Rouphael, Y., Cardarelli, M., Canaguier, R., Kumar, P., Colla, G. (2015) The effect of a plant-derived biostimulant on metabolic profiling and crop performance of lettuce grown under saline conditions. *Scientia Horticulturae* 182: 124–133.

Marschner, H. (1995) *Mineral nutrition in higher plants*. Academic Press: London.

Martí, M.C., Florez-Sarasa, I., Camejo, D., Ribas-Carbó, M., Lázaro, J.J., Sevilla, F., Jiménez, A. (2011) Response of mitochondrial thioredoxin PsTrxo1, antioxidant enzymes, and respiration to salinity in pea (*Pisum sativum* L.) leaves. *Journal of Experimental Botany* 62(11): 3863–3874.

Maruyama-Nakashita, A., Nakamura, Y., Yamaya, T., Takahashi, H. (2004) A novel regulatory pathway of sulfate uptake in *Arabidopsis* roots: Implication of CRE1/WOL/AHK4-mediated cytokinin-dependent regulation. *The Plant Journal: for Cell and Molecular Biology* 38(5): 779–789

Meyer, Y., Belin, C., Delorme-Hinoux, V., Reichheld, J.P., Riondet, C. (2012) Thioredoxin and glutaredoxin systems in plants: Molecular mechanisms, crosstalks, and functional significance. *Antioxidants & Redox Signaling* 17(8): 1124–1160.

Meyer, A.J., Hell, R. (2005) Glutathione homeostasis and redox regulation by sulfhydryl groups. *Photosynthesis Research* 86(3): 435–457.

Miller, G., Suzuki, N., Ciftci-Yilmazi, S., Mittler, R. (2010) Reactive oxygen species homeostasis and signaling during drought and salinity stresses. *Plant, Cell and Environment* 33(4): 453–467.

Mostofa, M.G., Seraj, Z.I., Fujita, M. (2014) Exogenous sodium nitroprusside and glutathione alleviate copper toxicity by reducing copper uptake and oxidative damage in rice (*Oryza sativa* L.) seedlings. *Protoplasma* 251(6): 1373–1386.

Nazar, R., Iqbal, N., Masood, A., Syeed, S., Khan, N.A. (2011a) Understanding the significance of sulfur in improving salinity tolerance in plants. *Environmental and Experimental Botany* 70(2–3): 80–87.

Nazar, R., Iqbal, N., Syeed, S., Khan, N.A. (2011b) Salicylic acid alleviates decreases in photosynthesis under salt stress by enhancing nitrogen and sulfur assimilation and antioxidant metabolism differentially in two mungbean cultivars. *Journal of Plant Physiology* 168(8): 807–815.

Negrin, V.L., Teixeira, B., Godinho, R.M., Mendes, R., Vale, C. (2017) Phytochelatins and monothiols in salt marsh plants and their relation with metal tolerance. *Marine Pollution Bulletin* 121(1–2): 78–84. doi: 10.1016/j.marpolbul.2017.05.045.

Nikiforova, V., Freitag, J., Kempa, S., Adamik, M., Hesse, H., Hoefgen, R. (2003) Transcriptome analysis of sulfur depletion in *Arabidopsis thaliana*: Interlacing of biosynthetic pathways provides response specificity. *The Plant Journal : for Cell and Molecular Biology* 33(4): 633–650.

Nocito, F.F., Lancilli, C., Crema, B., Fourcroy, P., Davidian, J.C., Sacchi, G.A. (2006) Heavy metal stress and sulfate uptake in maize roots. *Plant Physiology* 141(3): 1138–1148.

Nocito, F.F., Lancilli, C., Giacomini, B., Sacchi, G.A. (2007) Sulfur metabolism and cadmium stress in higher plants. *Plant Stress* 1(2): 142–156.

Noctor, G., Gomez, L., Vanacker, H., Foyer, C.H. (2002) Interactions between biosynthesis, compartmentation and transport in the control of glutathione homeostasis and signaling. *Journal of Experimental Botany* 53(372): 1283–1304.

Noctor, G., Mhamdi, A., Chaouch, S., Han, Y.I., Neukermans, J., Marquez-Garcia, B.E.L.E.N., Queval, G., Foyer, C.H. (2012) Glutathione in plants: An integrated overview. *Plant, Cell and Environment* 35(2): 454–484.

Nuruzzaman, M., Sharoni, A.M., Satoh, K., Al-Shammari, T., Shimizu, T., Sasaya, T., Omura, T., Kikuchi, S. (2012) The thioredoxin gene family in rice: Genome-wide identification and expression profiling under different biotic and abiotic treatments. *Biochemical and Biophysical Research Communications* 423(2): 417–423.

Ohkama, N., Takei, K., Sakakibara, H., Hayashi, H., Yoneyama, T., Fujiwara, T. (2002) Regulation of sulfur-responsive gene expression by exogenously applied cytokinins in *Arabidopsis thaliana*. *Plant and Cell Physiology* 43(12): 1493–1501.

Ortega-Villasante, C., Rellán-Álvarez, R., Del Campo, F.F., Carpena-Ruiz, R.O., Hernández, L.E. (2005) Cellular damage induced by cadmium and mercury in *Medicago sativa*. *Journal of Experimental Botany* 56(418): 2239–2251.

Pal, R., Rai, J.P.N. (2010) Phytochelatins: Peptides involved in heavy metal detoxification. *Applied Biochemistry and Biotechnology* 160(3): 945–963.

Park, J., Song, W.Y., Ko, D., Eom, Y., Hansen, T.H., Schiller, M., Lee, T.G., et al. (2012) The phytochelatin transporters AtABCC1 and AtABCC2 mediate tolerance to cadmium and mercury. *The Plant Journal: for Cell and Molecular Biology* 69(2): 278–288.

Rae, A.L., Smith, F.W. (2002) Localisation of expression of a high-affinity sulfate transporter in barley roots. *Planta* 215(4): 565–568.

Rausch, T., Wachter, A. (2005) Sulfur metabolism: A versatile platform for launching defence operations. *Trends in Plant Science* 10(10): 503–509.

Rennenberg, H., Herschbach, C., Haberer, K., Kopriva, S. (2007) Sulphur metabolism in plants: Are tree different? *Plant Biology* 9(5): 620–637.

Rotte, C., Leustek, T. (2000) Differential subcellular localization and expression of ATP sulfurylase and APS reductase during ontogenesis of *Arabidopsis thaliana* leaves indicates that cytosolic and plastid forms of ATP sulfurylase may have specialized functions. *Plant Physiology* 124(2): 715–724.

Ruiz, J.M., Blumwald, E. (2002) Salinity-induced glutathione synthesis in *Brassica napus*. *Planta* 214(6): 965–969.

Sanz-Barrio, R., Millán, A.F.S., Carballeda, J., Corral-Martínez, P., Seguí-Simarro, J.M., Farran, I. (2012) Chaperone-like properties of tobacco plastid thioredoxins f and m. *Journal of Experimental Botany* 63(1): 365–379.

Scheerer, U., Haensch, R., Mendel, R.R., Kopriva, S., Rennenberg, H., Herschbach, C. (2010) Sulphur flux through the sulphate assimilation pathway is differently controlled by adenosine 5_-phosphosulphate reductase under stress and in transgenic poplar plants overexpressing -ECS, SO, or APR. *Journal of Experimental Botany* 61(2): 609–622.

Schurmann, P., Brunold, C. (1980) Formation of cysteine from adenosine 5′-phosphosulfate (APS) in extracts from spinach chloroplasts. *Zeitschrift für Pflanzenphysiologie* 100(3): 257–268.

Schürmann, P., Buchanan, B.B. (2008) The ferredoxin/thioredoxin system of oxygenic photosynthesis. *Antioxidants & Redox Signaling* 10(7): 1235–1274.

Sekhar, K., Priyanka, B., Reddy, V.D., Rao, K.V. (2011) Metallothionein 1 (CcMT1) of pigeonpea (*Cajanus cajan* L.) confers enhanced tolerance to copper and cadmium in *Escherichia coli* and *Arabidopsis thaliana. Environmental and Experimental Botany* 72(2): 131–139.

Seki, M., Narusaka, M., Ishida, J., Nanjo, T., Fujita, M., Oono, Y., Kamiya, A., et al. (2002) Monitoring the expression profiles of 7000 *Arabidopsis* genes under drought, cold and high-salinity stresses using a full-length cDNA microarray. *The Plant Journal: for Cell and Molecular Biology* 31(3): 279–292.

Semane, B., Cuypers, A., Smeets, K., Van Belleghem, F., Horemans, N., Schat, H., Vangronsveld, J. (2007) Cadmium responses in *Arabidopsis thaliana*: Glutathione metabolism and antioxidative defence system. *Physiologia Plantarum* 129(3): 519–528.

Serrato, A.J., Pérez-Ruiz, J.M., Spínola, M.C., Cejudo, F.J. (2004) A novel NADPH thioredoxin reductase, localized in the chloroplast, which deficiency causes hypersensitivity to abiotic stress in *Arabidopsis thaliana. The Journal of Biological Chemistry* 279(42): 43821–43827.

Shibagaki, N., Rose, A., McDermott, J.P., Fujiwara, T., Hayashi, H., Yoneyama, T., Davies, J.P. (2002) Selenate-resistant mutants of *Arabidopsis thaliana* identify Sultr1; 2, a sulfate transporter required for efficient transport of sulfate into roots. *The Plant Journal for Cell and Molecular Biology* 29(4): 475–486.

Sobrino-Plata, J., Ortega-Villasante, C., Flores-Cáceres, M.L., Escobar, C., Del Campo, F.F., Hernández, L.E. (2009) Differential alterations of antioxidant defenses as bioindicators of mercury and cadmium toxicity in alfalfa. *Chemosphere* 77(7): 946–954.

Sogbedi, J.M., van Es, H.M., Agbeko, K.L. (2006) Cover cropping and nutrient management strategies for maize production in western. *African Agronomy Journal* 98: 883–889.

Song, Y., Cui, J., Zhang, H., Wang, G., Zhao, F.J., Shen, Z. (2013) Proteomic analysis of copper stress responses in the roots of two rice (*Oryza sativa* L.) varieties differing in Cu tolerance. *Plant and Soil* 366(1–2): 647–658. doi:10.1007/s11104-012-1458-2.

Srivastava, S., D'souza, S.F. (2009) Increasing sulfur supply enhances tolerance to arsenic and its accumulation in *Hydrilla verticillata* (Lf) Royle. *Environmental Science and Technology* 43(16): 6308–6313.

Szalai, G., Kellos, T., Galiba, G., Kocsy, G. (2009) Glutathione as an antioxidant and regulatory molecule in plants under abiotic stress conditions. *Journal of Plant Growth Regulation* 28(1): 66–80.

Szarka, A., Tomasskovics, B., Bánhegyi, G. (2012) The ascorbate-glutathione-α-tocopherol triad in abiotic stress response. *International Journal of Molecular Sciences* 13(4): 4458–4483.

Takahashi, H., Watanabe-Takahashi, A., Smith, F.W., Blake-Kalff, M., Hawkesford, M.J., Saito, K. (2000) The roles of three functional sulphate transporters involved in uptake and translocation of sulphate in *Arabidopsis thaliana. The Plant Journal: for Cell and Molecular Biology* 23(2): 171–182.

Turkan, I., Demiral, T. (2009) Recent developments in understanding salinity tolerance. *Environmental and Experimental Botany* 67(1): 2–9.

Uraguchi, S., Tanaka, N., Hofmann, C., Abiko, K., Ohkama-Ohtsu, N., Weber, M., Kamiya, T., et al. (2017) Phytochelatin synthase has contrasting effects on cadmium and arsenic accumulation in rice grains. *Plant and Cell Physiology* 58(10): 1730–1742.

Vauclare, P., Kopriva, S., Fell, D., Suter, M., Sticher, L., Von Ballmoos, P., Krähenbühl, U., et al. (2002) Flux control of sulphate assimilation in *Arabidopsis thaliana*: Adenosine 5-phosphosulphate reductase is more susceptible to negative control by thiols than ATP sulphurylase. *Plant Journal* 31: 729–740.

Wachter, A., Wolf, S., Steininger, H., Bogs, J., Rausch, T. (2005) Differential targeting of GSH1 and GSH2 is achieved by multiple transcription initiation: Implications for the compartmentation of glutathione biosynthesis in the Brassicaceae. *The Plant Journal: for Cell and Molecular Biology* 41(1): 15–30.

Wang, Y., Zhang, C., Zheng, Y., Ge, Y. (2017) Phytochelatin synthesis in *Dunaliella salina* induced by arsenite and arsenate under various phosphate regimes. *Ecotoxicology and Environmental Safety* 136: 150–160.

Xue, T., Li, X., Zhu, W., Wu, C., Yang, G., Zheng, C. (2009) Cotton metallothionein GhMT3a, a reactive oxygen species scavenger, increased tolerance against abiotic stress in transgenic tobacco and yeast. *Journal of Experimental Botany* 60(1): 339–349.

Yadav, P., Kaur, R., Kanwar, M.K., Bhardwaj, R., Sirhindi, G., Wijaya, L., Alyemeni, M.N., Ahmad, P. (2018) Ameliorative role of castasterone on copper metal toxicity by improving redox homeostasis in *Brassica juncea* L. *Journal of Plant Growth Regulation* 37(2): 575–590. doi:10.1007/s00344-017-9757-8

Yadav, P., Kaur, R., Kanwar, M.K., Sharma, A., Verma, V., Sirhindi, G., Bhardwaj, R. (2018) Castasterone confers copper stress tolerance by regulating antioxidant enzyme responses, antioxidants, and amino acid balance in *B. juncea* seedlings. *Ecotoxicology and Environmental Safety* 147: 725–734.

Yang, S., Xie, J., Li, Q. (2012) Oxidative response and antioxidative mechanism in germinating soybean seeds exposed to cadmium. *International Journal of Environmental Research and Public Health* 9(8): 2827–2838.

Yang, M., Zhang, F., Wang, F., Dong, Z., Cao, Q., Chen, M. (2015) Characterization of a type 1 metallothionein gene from the stresses-tolerant plant *Ziziphus jujuba. International Journal of Molecular Sciences* 16(8): 16750–16762.

Yazaki, K. (2006) ABC transporters involved in the transport of plant secondary metabolites. *Federation of European Biochemical Societies Letters* 580(4): 1183–1191.

Yoshimoto, N., Inoue, E., Saito, K., Yamaya, T., Takahashi, H. (2003) Phloem-localizing sulfate transporter, Sultrl; 3, mediates re-distribution of sulfur from source to sink organs in *Arabidopsis*. *Plant Physiology* 131(4): 1511–1517.

Yoshimoto, N., Takahashi, H., Smith, F.W., Yamaya, T., Saito, K. (2002) Two distinct high-affinity sulfate transporters with different inducibilities mediate uptake of sulfate in *Arabidopsis* roots. *The Plant Journal: for Cell and Molecular Biology* 29(4): 465–473.

Zagorchev, L., Seal, C.E., Kranner, I., Odjakova, M. (2013) A central role for thiols in plant tolerance to abiotic stress. *International Journal of Molecular Sciences* 14(4): 7405–7432.

Zayneb, C., Imen, R.H., Walid, K., Grubb, C.D., Bassem, K., Franck, V., Hafedh, M., Amine, E. (2017) The phytochelatin synthase gene in date palm (*Phoenix dactylifera* L.): Phylogeny, evolution and expression. *Ecotoxicology and Environmental Safety* 140: 7–17.

Zhang, M., Zhao, B.C., Ge, W.N., Zhang, Y.F., Song, Y., Sun, D.Y., Guo, Y. (2011) An apoplastic h-type thioredoxin is involved in the stress response through regulation of the apoplastic reactive oxygen species in rice. *Plant Physiology* 157(4): 1884–1899.

Zhang, M.-Y., Bourbouloux, A., Cagnac, O., Srikanth, C.V., Rentsch, D., Bachhawat, A.K., Delrot, S. (2004) A novel family of transporters mediating the transport of glutathione derivatives in plants. *Plant Physiology* 134(1): 482–491.

Zhang, W., Lin, K., Zhou, J., Zhang, W., Liu, L., Zhang, Q. (2014) Cadmium accumulation, sub-cellular distribution and chemical forms in rice seedling in the presence of sulfur. *Environmental Toxicology and Pharmacology* 37(1): 348–353.

Zhao, L., Sun, Y.L., Cui, S.X., Chen, M., Yang, H.M., Liu, H.M., Chai, T.Y., Huang, F. (2011) Cd-induced changes in leaf proteome of the hyperaccumulator plant *Phytolacca americana*. *Chemosphere* 85(1): 56–66.

Zimeri, A.M., Dhankher, O.P., McCaig, B., Meagher, R.B. (2005) The plant MT1 metallothioneins are stabilized by binding cadmiums and are required for cadmium tolerance and accumulation. *Plant Molecular Biology* 58(6): 839–855.

15 Exogenous Silicon Increases Plant Tolerance to Unfavorable Environments

Tamara I. Balakhnina

CONTENTS

15.1 INTRODUCTION

15.1.1 SOIL STRESSORS AND PLANT REACTIONS

The life of plants, their growth, development, and productivity are closely related to physical and chemical processes and properties of soil (Gliñski, 2011). Transport of water, vapor, air, and chemical substances as well as capillary flow, molecular diffusion, osmosis, mass absorption/desorption, energy transport (heat conduction, convection, radiation), and some mechanical processes (impact, crushing, compression, shearing, tension) are the determining conditions of culture growth. Soil conditions based on the enumerated physical factors may create stresses for plant growth and development.

Suitability of the environment for normal plant development is also determined by the pool of oxygen stored in the soil and by the ability of its continuous supply from the atmosphere. After exhausting the soil oxygen pool, and without further oxygen supply, plants start to suffer from oxygen stress; the root system perishes, and finally, the whole plant dies (Gliñski et al., 2004). Soil oxygen is one of the most important factors, which even during a short period can severely limit plant development and nutrient uptake, thereby resulting in a significant reduction of yields (Gliñski and Stêpniewski, 1985). Soil aeration is closely connected with the ratio of air and water conditions in soils. Destruction of the dynamic equilibrium between these parameters, for instance, by flooding the soil, changes the chemical and physical soil properties, affects the biological activity of soil microorganisms, and, consequently, leads to oxygen stress. The relationship between the air and water content in soils affects the biological activity of soil organisms, mainly microorganisms that are very sensitive to oxidation or reduction processes (Gliñski and Stêpniewski, 1985). Various abiotic (soil flooding, drought, soil compaction, salinity, high temperature, or a combination of these stressors) and biotic factors result in the development of oxygen stress in the leaves and roots of plants. One of the abiotic stressors, which influences the growth and development of plants, and is recognized as an important yield-limiting factor for many crops, is soil flooding (Balakhnina et al., 2004, 2010a; Balakhnina, 2015). The most negative effects of soil overwatering are a reduction of the oxygen in the water and a decrease in the nutrient uptake, as well as disturbances in the plant respiratory metabolism. Excess water in the soil leads to a decrease of the oxygen diffusive rate in the soil and a limited supply of oxygen to the cells, which is one of the vital requirements for plants (Balakhnina, 2015; Chen et al., 2005; Pociecha et al., 2008). Oxygen deficiency in the soil, induced by flooding, affects the energy in seed germination, intensity of photosynthesis in the leaves of seedlings, and biomass accumulation, as well as a number of physiological and biochemical reactions that are connected with plant adaptation and the development of oxidative stress in the cells (Balakhnina et al., 2004; Balakhnina, 2015). Salt stress is an environmental factor that restricts plant growth and productivity worldwide. Salt causes both ionic as well as osmotic stress on plants (Hejazi Mehrizi et al., 2011; Parvaiz and Satyawati, 2008). A high concentration of Na^+ causes deficiencies in other nutrients in the soil and interacts with other environmental factors, such as drought, which exacerbate the problem (Parvaiz and Satyawati, 2008). The decline in growth observed in many plants subjected to excessive salinity is often associated with a decrease in their photosynthetic capacity (Yang et al., 2008). Conventional selection and breeding techniques have been used to improve salinity tolerance in crop plants (Parvaiz and Satyawati, 2008). Drought stress, just like flooding, results in inhibiting photosynthesis and growth (Epron and Dreyer, 1993). A consequence of extreme weather conditions and (or) insufficient field irrigation is the development of plant drought stress because of the limited amount of water in the root zone (Farkas, 2011). Drought results to crop loss worldwide, reducing average yields by more than 50% (Wang et al., 2003). The optimal supply of plants by water has positive effects on the growth processes, photosynthesis, and biomass production. Plants growing on well-moistened soil absorbs water in accordance with the requirements of transpiration. When the soil is dry, plants have difficulties in absorbing water because of its deficiency in the root zone. In this case, the rate of transpiration decreases (Shein and Pachepsky, 1995).

15.1.2 Antioxidant Enzymes and Non-Specific Plant Reactions at the Stress Development

Peroxide processes are of vital importance for living organisms and inevitably take place in the cells of all aerobes. The formation of hydrogen peroxide (H_2O_2) and other reactive oxygen species (ROS) such as superoxide radicals (O^{2-}), hydroxyl radicals (OH^-), and so on is a normal metabolic process (Alsher et al., 1997). Under unfavorable conditions of plant growth, there is the possibility of excessive accumulation of ROS, which intensify peroxide processes in the plant tissues. In this case, the antioxidant potential of cells might be insufficient and the organism suffers from oxidative damage (Ali and Alqurainy, 2006). One of the ROS formation places might be connected with the electron transport chain because of $NADP^+$ content decline. In this situation, oxygen becomes an alternative electron acceptor (Egneus et al., 1975). The result of ROS activities is peroxide oxidation of membrane lipids (LPO), as well as the oxidative destruction of proteins, pigments, and other compounds of the cells (Allen, 1995; Balakhnina et al., 2009, 2010a; Halliwell, 1987; Zakrzhevsky et al., 1995). Thanks to the evolutionarily formed system of protection against oxidative degradation, plants have the ability to withstand destructive stress.

Ascorbic acid, reduced glutathione, tocopherols, and other low molecular antioxidants, as well as enzymes of antioxidant nature, determine the ability of aerobes to quench the aggression of ROS (Larson, 1988). In the cells of living organisms there are functionally associated enzymes superoxide dismutase (SOD), catalase (CAT), guaiacol peroxidase (GPX), ascorbate peroxidase (APX), and glutathione reductase (GR), which form the basis of the enzyme antioxidant system (Asada, 1992, 2006; Gunes et al., 2007a). SOD is a primary scavenger of ROS in plants and protects cells against superoxide induced oxidative processes. However, one of the products of the superoxide anion-radical dismutation, catalyzed by SOD, is H_2O_2, which is also toxic to cells (Beyer et al., 1991). Most plants, resistant to the action of various stressors, contain enzymes CAT, GPX APX, and GR that convert H_2O_2 to water and molecular oxygen (Shim et al., 2003; Zhu et al., 2010). However, there is information that under some stress conditions when CAT activity decreased, the activities of GPXs, APX, and GR were induced by stress treatments (Shim et al., 2003). GPX is a widely prevalent enzyme in plant cells. It is found in the cell wall, cytosol, apoplast, vacuole, and extracellular medium. There is evidence that GPX is very important not only as an antioxidant, but also because of its participation in the processes of plant growth and development (Ghamsari et al., 2007). The broader substrate specifications and stronger affinity of GPX for H_2O_2 than those of CAT, explain the higher antioxidant potential of this enzyme (Brigelius-Flohe and Flohe, 2003). APX is one of the plant peroxidases, which localize in the chloroplast, microbody, and cytosol. The main function of APX is the reduction of oxidative stress in plant cells by the neutralizing of H_2O_2 in the reaction, where ascorbate is used as a specific electron donor (Ghamsari et al., 2007; Rosa et al., 2010; Shigeoka et al., 2002). It was shown that a number of specific factors such as pathogen attack, mechanical pressure, injury, ultraviolet-B (UV-B) radiation, water deficiency, salt stress, excess excitation energy, too low or too high temperature, excess oxygen after a period of anoxia, atmospheric pollution, excess metal ions, and deficiency in some mineral salts, e.g., phosphates and herbicides, can activate APX gene expression. APX activities generally increase along with activities of other antioxidant enzymes like CAT, SOD, and GR in response to various environmental stress factors (Shigeoka et al., 2002). GR is a flavoprotein that catalyzes the reduction of oxidized glutathione (GSSG) to reduced glutathione (GSH) in the NADPH-dependent reaction. A higher level of GSH concentration is needed for protection cells against oxidative processes by the neutralizing of H_2O_2 (Asada, 1992). GR, together with APX, is involved in the ascorbate-glutathione cycle, which is an important and efficient enzymatic defense system for decomposing H_2O_2 and maintaining the balance of antioxidants (Asada, 2006; Gunes et al., 2007a). APX is the first enzyme in this pathway and its major function is catalyzing the reaction of H_2O_2 to H_2O conversion. GR is the last step in the pathway, playing a crucial role in the plant antioxidant potential by maintaining a level of reduced glutathione. It was established that at the optimal metabolic conditions in the plant cells, there is a dynamic equilibrium between the activity of the antioxidant system and intensity of lipid peroxidation processes (Alsher et al., 1997). Increasing the activity level of one or more antioxidant enzymes are connected with plant resistance to stressor actions (Allen, 1995; Bennicelli et al., 2005). Because of excessive formation of ROS under abiotic stresses, the dynamic equilibrium between the activity of the antioxidant system and intensity of LPO processes is displaced to intensification of LPO processes that may lead to oxidative degradation and death of plant cells (Mittler, 2002; Molassiotis et al., 2005). Activation of the system defense against oxidative destruction might be considered a key link in the mechanism of plant tolerance to unfavorable conditions.

15.2 SILICON

15.2.1 Si in the Soil

The content of silicon, one of the most prevalent elements in soils, is determined by values from 50 to 400 g Si kg^{-1} and depends on the type of soil (Kovda, 1973). In clay soils, the content of this element varies from 200 to 350 g of Si kg^{-1} of soil, while in sandy soil, from 450 to 480 g Si kg^{-1} of soil (Kovda, 1973). The main Si-rich compounds of the soil are the inert quartz or crystalline silicates, which form the skeleton of the soil. Soluble monosilicic acids, polysilicic acids, and organosilicon compounds present the physically and chemically active Si containing substances in the soil (Matichenkov and Ammosova, 1996). Plants and microorganisms absorb soluble monosilicic acid (Yoshida, 1975). It was shown (Matichenkov et al., 2000; Sokolova, 1985) that they also control chemical and biological properties of the soil, P, Al, Fe, Mn, and heavy metal mobility, microbial activity, stability of soil organic matter, and formation of polysilicic acids and secondary minerals in the soil. Polysilicic acid has a significant influence on soil texture, the capacity of water holding, adsorption capacity, and stability of soil erosion (Matichenkov et al., 2000). Plants can absorb enough Si (Savant et al., 1997) that can determine the effect of Si on soil fertility and plants.

15.2.2 Silicon Uptake and Accumulation in Plants

Even though silicon is a very common element in soil, it practically does not occur in a form that is assimilated to plants and is always combined with other elements, usually forming oxides or silicates (Gunes et al., 2007b). It was established (Ranganathan et al., 2006) that Si is absorbed by plants in the form of uncharged silicic acid, Si $(OH)_4$, and it irreversibly turns in the plants into amorphous silicon. Therefore, although silicon is plentiful, most of the silicon-containing substances are insoluble and in a plant-unavailable form. The concentrations of silicic acid in the soil solution vary from 0.1 to 0.6 mM. The silicon concentrations in the shoot of plants on average range from 1.0 to 100.0 grams of Si kg^{-1} dry weight.

After studying more than 500 plant species, the plants were separated into groups with high and intermediate ability to accumulate silicon and plants that did not stock silicone. The groupings were based upon measurements (on a dry weight basis) of silicon and the silicon-to-calcium ratio in plant tissues (Ma et al., 2001a). The difference in Si concentration in plant species is mainly attributable to the varying ability of Si-uptake and transport. Such species as rice (Ma et al., 2001b), wheat (Rains et al., 2006), ryegrass (Jarvis, 1987), and barley (Barber and Shone, 1966) can actively uptake Si. However, some Gramineae plants such as oats (Jones and Handreck, 1967), some dicots such as cucumber, melon, strawberry, and soybean (Liang et al., 2005) take up Si passively. Despite the importance of this problem, molecular mechanisms of silicon absorption by these plants has not been studied sufficiently (Ma and Yamaji, 2006). Investigations conducted by Parry and Kelso (1975) showed that silicon interacted with polyphenols in xylem cell walls and affected lignin deposition and biosynthesis. Addition of silicon at rice growing, under conditions of water deficit induced by polyethylene glycerol, resulted in decreasing the transpiration rate and membrane permeability (Agarie et al., 1998). Application of Si increased the content of relative water and dry mass of sorghum (Sorghum bicolor Moench) plants. The increasing of drought tolerance in sorghum when adding silicon might be associated with enhancement of the plant's ability to uptake water (Hattori et al., 2005, 2007). It was shown that the addition of silicon also improved the water status and increased the dry mass of wheat plants under drought (Gong et al., 2005, 2008). Lux et al. (2002, 2003) demonstrated that Si was precipitated in the cell walls of roots, leaves, and stems. Richmond and Sussman (2003) and Ma and Yamaji (2006) have suggested that a beneficial influence of Si on plant growth during stress conditions might not be connected with Si effects on the activity of antioxidant enzymes. According to Hossain et al. (2007), silicon application resulted in the changing of the cell wall architecture, which may be responsible for improvement in the cell wall extensibility.

15.2.3 Protective Role of Silicon

In spite of the fact that silicon is not a vital element for plants, a rather large experimental material, confirming its positive influence on the growth and productivity of crops has been accumulated to date (Ahmed et al., 2011; Balakhnina et al., 2012a; Gunes et al., 2007a; Ma et al., 2004; Savant et al., 1997). Silicon is applied to improve plant growth and yield, in particular, under stress conditions (Hattori et al., 2005). Correction of nutrient imbalance, reduction of mineral toxicities, improvement of mechanical properties of plant tissues, and enhancement of resistance to other various abiotic (salt, metal toxicity, nutrient imbalance, lodging, drought, radiation, high temperature, freezing, UV) and biotic stresses are functions that silicon possesses (Epstein, 1999; Ma and Yamaji, 2006). The protective role of Si in plants may be connected with accumulation of polysilicic acids inside cells (Biel et al., 2008). This point of view was indirectly confirmed by the fact that an increase in the concentration of polysilicic acid in plant tissues led to an increase in their resistance (Matichenkov et al., 2000).

15.2.3.1 Salinity

Increasing salinity of soil leads to a malfunction of electron transport in the electron transport chains of chloroplasts and mitochondria and leads to the generation of an excess amount of ROS, which in turn intensifies phytotoxic reactions accompanied by over-oxidation of membrane lipids, protein destruction, and DNA mutation (Ali and Alqurainy, 2006). Application of Si decreased the permeability of the plasma membrane of leaf cells, and significantly improved the ultrastructure of chloroplasts, which were badly damaged by NaCl addition with the double membranes disappearing and the grannae being disintegrated in the absence of Si (Liang et al., 2003). It was found (Biel et al., 2008) that *Distichlis spicata* plants, growing under soil salinity, accumulate bigger amounts of Si in their particular parts under stressful conditions. The results of Al-Aghabary et al. (2004) showed that silicate partially removes the negative effect of NaCl stress and increases the resistance of tomato plants to salinity due to an increase in the activity of superoxide dismutase and catalase. Liang et al. (2003) demonstrated that exogenous Si significantly enhanced the activities of SOD, CAT, and GR in roots of salt-stressed plants. Increasing SOD activity under salt stress was observed by Molassiotis et al. (2005). Zhu (2001) established that application of Si led to increases of SOD, GPX, and APX activities of salt-stressed cucumbers. It was shown that at the investigation of the effect of silicon on the antioxidant capacity of two grapevines (*Vitis vinifera* L.), root stocks grown in boron toxic, saline, and boron toxic-saline soil, application of Si lowered SOD and CAT but increased APX (Soylemezoglu et al., 2009).

15.2.3.2 Drought

A large number of plants in the world often suffer from temporary or permanent water shortages in the soil due to lack of rain or poor irrigation of fields. Drought leads to a decrease in the quality and productivity of crops (Devkota and Jha, 2011; Said-Al Ahl et al., 2009). The results of numerous investigations indicate that the resistance of crops to drought is increased with enhancing antioxidant enzyme activities. Application of silicon under drought increased the activities of some antioxidant enzymes: SOD, CAT, and GR as well as the fatty acid unsaturation of lipids and the content of photosynthetic pigments, while the content of H_2O_2 was decreased and the activities of GPX and AsP remained without significant differences (Gong et al., 2005). Gong et al. (2005) suggest that the enhancing of drought resistance by silicon in wheat plants may be associated with an increase of antioxidant system activity, diminishing oxidative damage of the molecular content of cells, induced by overproduction of ROS under drought, and maintaining metabolic processes of stressed plants. Positive correlations of physiological and biochemical reactions with the amount of silicon supply were found in plants under conditions of severe stress. Gong et al. (2008) showed that the intensity of oxidative processes in the leaves of wheat was stimulated by water shortage and there was a smaller increment after application of silicon.

15.2.3.3 UV Radiation

Ultraviolet-B (UV-B) radiation negatively affects plant cells, causing generation of ROS such as superoxide anions (O^{2-}), hydrogen peroxide (H_2O_2), hydroxyl radicals (OH), and singlet oxygen (O_2) (Beckmann et al., 2012; Lizana et al., 2009; Rybus-Zajc and Kubioe, 2010; Zancan et al., 2008). Fang et al. (2011) and Li et al. (2007) have reported that Si increases plant tolerance to UV-B radiation. The experiment performed by Shen et al. (2010) showed that drought and UV-B radiation stresses caused intensification of LPO in soybean seedlings, but Si application significantly reduced the membrane damage. The CAT and SOD activities increased under the effect of UV-B radiation and significantly decreased at Si application. UV-B light had more adverse effects on growth than drought; the data also showed that Si could alleviate seedling damage under these stress conditions (Shen et al., 2010).

15.2.3.4 Flooding

We studied the influence of silicon on the growth and development of plant adaptation under optimal and soil flooding conditions—barley (*Hordeum vulgare* L.) grown on the substrate, enriched by the soluble, amorphous silica dioxide (SiO_2). The adaptive potential was evaluated by the oxidative processes intensity through the content of thiobarbituric acid reactive substances (TBARs), whereas the antioxidative capacities were evaluated by SOD, GPX, and ascorbate peroxidase (AsP) activities. Addition of amorphous SiO_2 to the soil increased the Si concentration in shoots and roots of plants, stimulated growth, and biomass accumulation under optimal soil watering. Soil flooding inhibited the growth of both $Si0_2$-treated and non-treated plants. The concentration of TBARs was lower in the tissues of the (+Si)-plants. Soil hypoxia increased SOD activity of the (–Si;+flooding) and (+Si;+flooding)-plants, but without significant differences. Under optimal soil watering, the GPX activity in the roots of (+Si)-plants was higher than in the (–Si)-ones, but there were no significant differences between (+Si)- and (–Si)-variants at the soil flooding conditions. Soil treatment with silica did not affect the activity of AsP either under optimal conditions of plant growth or soil flooding. Thus, the enrichment of soil with available silicon for plants leads to an enhancement in their growth, biomass accumulation, and a decrease in the intensity of peroxide processes under conditions of soil flooding without significant changes in the level of activity of antioxidant enzymes (Balakhnina et al., 2012a).

15.2.3.4.1 Content of Si in the Soil and Plant Samples

The content of acid-extractable Si in control (–Si)-soil was about 290 mg kg^{-1} of dry soil and remained practically unchanged at the flooding conditions (Figure 15.1a).

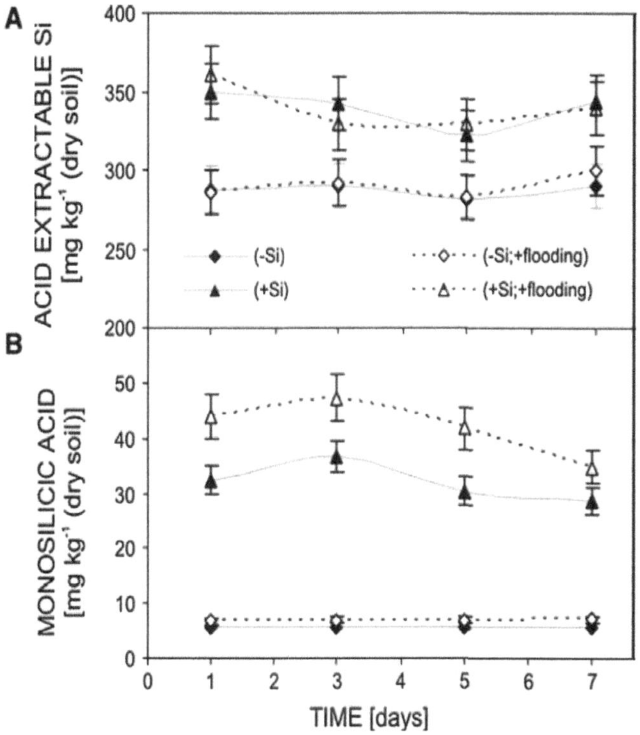

FIGURE 15.1 Concentration of acid extractable Si (a) and monosilicic acid (b) in the soil without and with application of amorphous silica under optimal soil watering [(–Si) and (+Si)] or flooding [(-Si;+flooding) and (+Si;+flooding)]. Bars indicate standard deviations (n = 3). (Source: Balakhnina et al., 2012a.)

When amorphous silica was added into the soil (+Si)-variant, the value of this parameter increased by 15–20%.

At the optimum soil watering, the concentration of mono silicon acid in (–Si), the soil was about 5.8 mg (Si) kg^{-1} of dry soil and increased to about 7.1 mg (Si) kg^{-1} dry soil, i.e., by 22% in the flooded soil (–Si; + floods) (Figure 15.1b). The concentration of monosilicic acid under optimal soil watering in the (+Si)-soil varied from 28.6 to 36.7 mg (Si) kg^{-1} dry soil (almost six-fold higher than in the (–Si)-soil). It increased to about 45.6 mg (Si) kg^{-1} of dry soil in the first days of flooding and then decreased to 34.8 mg (Si) kg^{-1} of dry soil at the end of the experiment (Figure 15.1b). The percentage Si in the dry mass of (–Si)-plants was approximately the same and varied in the roots and shoots from 2.4 to 2.6% (Figure 15.2a, b). Soil hypoxia (–Si;+flooding) resulted in a gradual increase in the Si content both in the roots and shoots. Plants grown on Si-reached soil, (+Si)-plants, contained larger amounts of Si, 3.4 and 3.6% in roots and shoots, respectively. Soil flooding in the (+Si;+flooding)-plants caused a slight tendency to increase the Si in the roots (Figure 15.2a) without significant changing of the Si content in the shoots (Figure 15.2b). It should also be

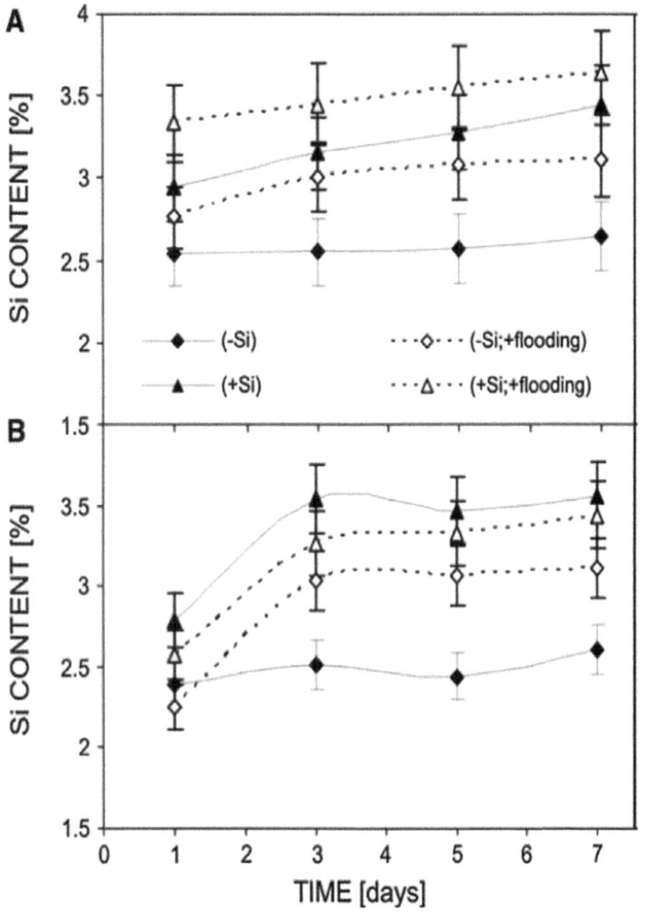

FIGURE 15.2 Content of Si in the roots (a) and shoots (b) of *Hordeum vulgare* seedlings grown in the soil without and with application of amorphous silica under optimal soil watering [(–Si) and (+Si)] or flooding [(–Si;+flooding) and (+Si;+flooding)]. Bars indicate standard deviations (n = 3). (Source: Balakhnina et al., 2012a.)

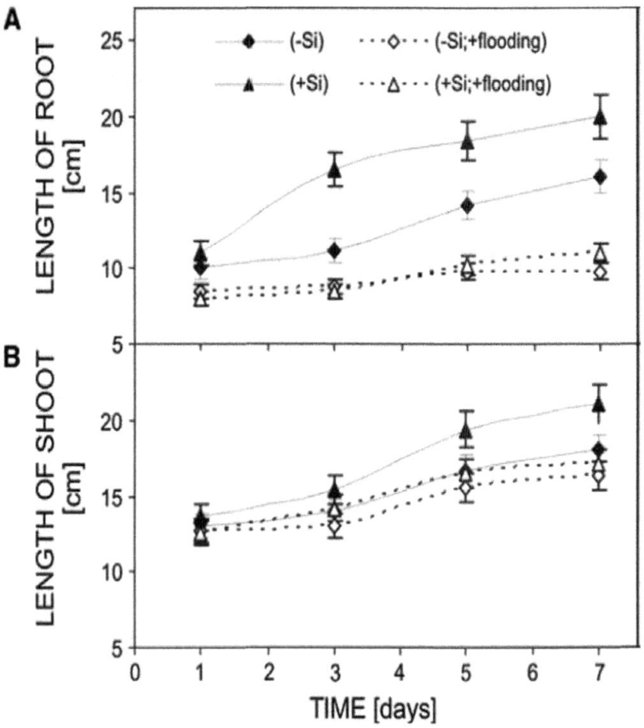

FIGURE 15.3 Length of roots (a) and shoots (b) of *Hordeum vulgare* seedlings grown in the soil without and with application of amorphous silica under optimal soil watering [(–Si) and (+Si)] or flooding [(–Si;+flooding) and (+Si;+flooding)]. Bars indicate standard deviations (n = 12). (Source: Balakhnina et al., 2012a.)

noted that the Si percentage in the (+Si;+flooding)-roots was considerably higher than in the (–Si;+flooding)-roots (Balakhnina et al., 2012a).

15.2.3.4.2 Plant Growth and Biomass Production

It was observed that after 7 days of the plant growth at the optimal soil watering, the length of roots and shoots of the (–Si)-plants increased from 10 to 16 cm and from 13 to 18 cm, respectively (Figure 15.3a, b). Addition of amorphous SiO_2 resulted in stimulation of the barley roots and shoots growth by about 33 and 15%, respectively, beginning from the third day of the experiment (Figure 15.3a, b). Soil flooding significantly inhibited the growth of barley roots of both (–Si)- and (+Si)-plants (Figure 15.3a). The five-day flooding of the soil caused a cessation of growth (–Si;+flooding)-roots in length, but they remained alive, and their biomass did not decrease. Inhibition of the shoot growth by soil flooding was not so significant (Figure 15.3b). The dry biomass of the roots and shoots of the (–Si)-plants increased during the experiment from 0.4 to 1.0 g and from 1.1 to 2.3 g per plant, respectively (Figure 15.4a, b). Application of amorphous silica stimulated biomass accumulation of roots and shoots similarly to its effect on their length, i.e., by about 34 and 18%, respectively, starting from the third day of the experiment (Figure 15.4a, b). Soil flooding significantly suppressed root and shoot biomass production of the (–Si)- and (+Si)-plants (Figure 15.4a, b).

15.2.3.4.3 Intensity of Peroxidation Processes

The content of TBARs in the leaves of the (–Si)-plants was significantly higher than in the roots (Figure 15.5a, b). Enriching of the soil by the amorphous SiO_2 did not influence the concentration of TBARs either in the roots or the leaves under optimal watering. However, under soil flooding, the intensity of the destructive processes was higher in plant tissues of the (–Si;+flooding)-plants in comparison with the (+Si;+flooding)-plants (Figure 15.5a, b).

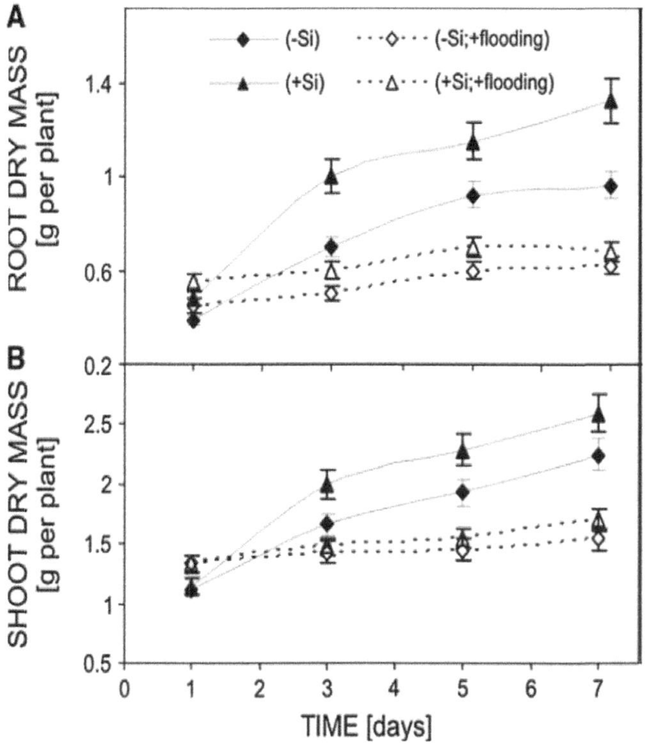

FIGURE 15.4 Biomass of roots (a) and shoots (b) of *Hordeum vulgare* seedlings grown in the soil without and with application of amorphous silica under optimal soil watering [(–Si) and (+Si)] or flooding [(–Si;+flooding) and (+Si;+flooding)]. Bars indicate standard deviations (n = 12). (Source: Balakhnina et al., 2012a.)

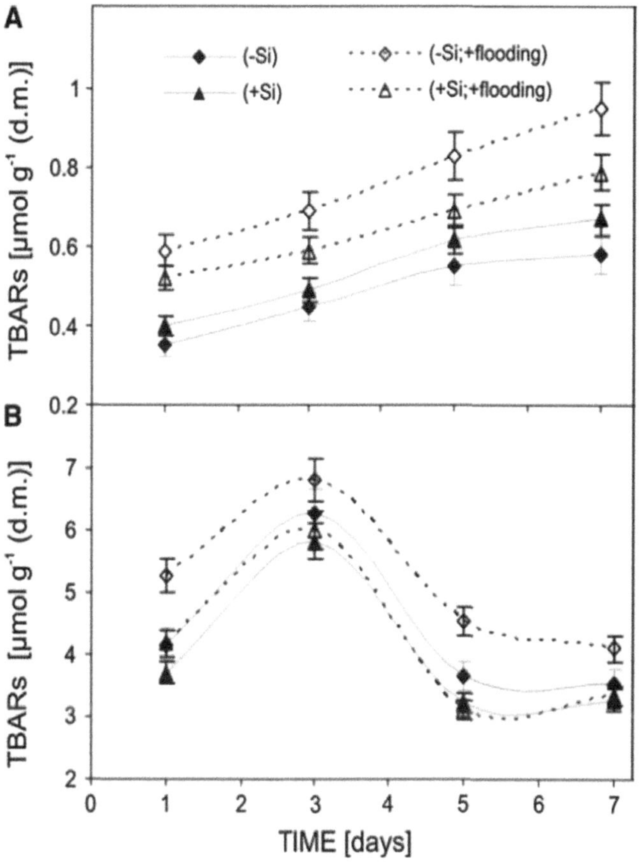

FIGURE 15.5 Concentration of TBARs in the roots (a) and leaves (b) of *Hordeum vulgare* seedlings grown in the soil without and with application of amorphous silica under optimal soil watering [(–Si) and (+Si)] or flooding [(–Si;+flooding) and (+Si;+flooding)]. Bars indicate standard deviations (n = 3). (Source: Balakhnina et al., 2012a.)

15.2.3.4.4 SOD, GPX, and AsP Activities

It was found that in the roots of the (–Si)-plants, the SOD activity was almost two times lower than in the leaves (Figure 15.6a, b) (Balakhnina et al., 2012a). After three days of the experiment, the enzyme activity in the roots of the (–Si;+flooding)-plants started to increase and became significantly (60–80%) higher than in the non-flooded ones (Figure 15.6a). In the leaves of the (–Si;+flooding)-plants, the SOD activity increased up to 160% after 1 day of soil hypoxia and then remained at the level of 115–130% higher relatively the non-stressed plants (Figure 15.6b). In the leaves of the (–Si;+flooding)-plant, the SOD activity increased up to 160% after 1 day of soil hypoxia and then remained at the level of 115–130% higher, as compared with the non-stressed plants (Figure 15.6b). Under optimal conditions of water supply, the exogenous SiO_2 application did not change SOD activity either in the roots or the leaves. However, SOD activity in the roots and the leaves of (+Si;+flooding)-plants increased and reached the values observed in the (–Si;+flooding)-ones (Figure 15.6a, b).

In contrast to SOD, the activity of GPX in the roots of the (–Si)-plants was about two times higher than the activity of this enzyme in the leaves (Figure 15.7a, b). Under soil flooding, GPX activity in the roots of (–Si)-plants increased by 26–67% and in the leaves by 23–150% (Figure 15.7a, b). Exogenous SiO_2 at the optimal soil watering increased GPX activity in the roots of the (+Si)-plants by 30–74% relative to the (–Si)-ones, but it did not effect on the GPX activity in the leaves (Figure 15.7a, b). Soil flooding resulted in increasing of GPX activity in the leaves of the (+Si;+flooding)-plants up to the values observed in the (–Si;+flooding)-ones (Figure 15.7a, b).

The activity of AsP in the leaves of (–Si)-plants was less affected by soil flooding in contrast to the GPX activity. A reliable increase in AsP activity was observed only after 5 days of flooding, and it did not exceed 50% (Figure 15.8). Exogenously available for plants, SiO_2 was

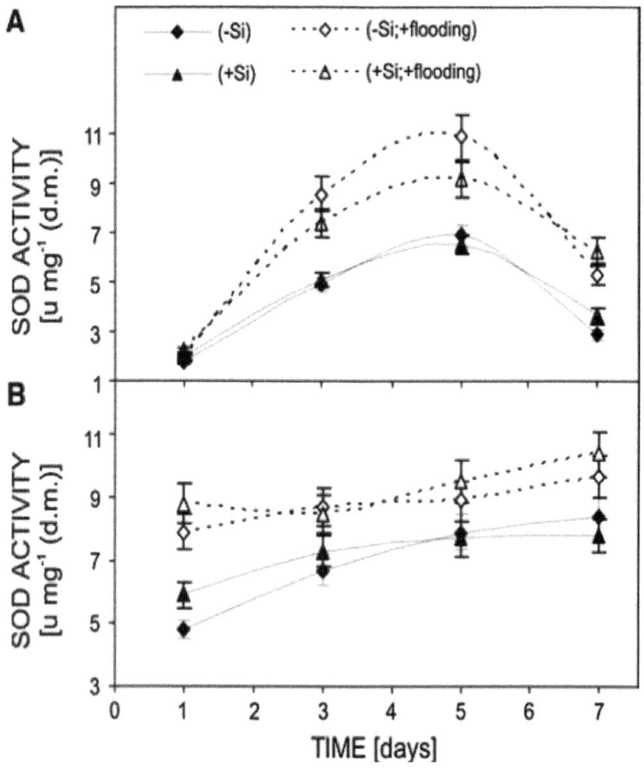

FIGURE 15.6 SOD activity in the roots (a) and leaves (b) of *Hordeum vulgare* seedlings grown in the soil without and with application of amorphous silica under optimal soil watering [(–Si) and (+Si)] or flooding [(–Si;+flooding) and (+Si;+flooding)]. Bars indicate standard deviations (n=3). (Source: Balakhnina et al., 2012a.)

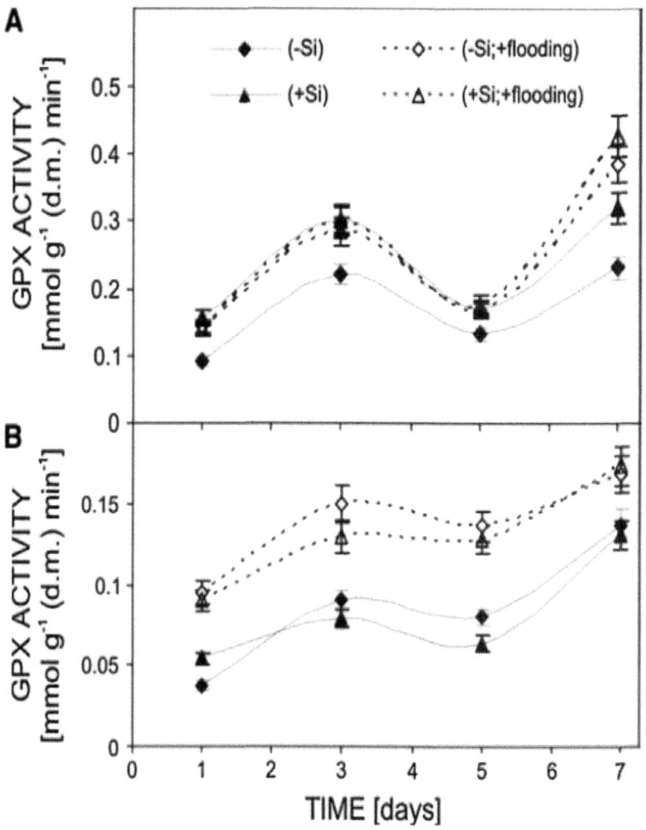

FIGURE 15.7 GPX activity in the roots (a) and leaves (b) of *Hordeum vulgare* seedlings grown in the soil without and with application of amorphous silica under optimal soil watering [(–Si) and (+Si)] or flooding [(–Si;+flooding) and (+Si;+flooding)]. Bars indicate standard deviations (n=3). (Source: Balakhnina et al., 2012a.)

not effective in relation to AsP activity either under optimal soil irrigation or flood conditions (Figure 15.8).

15.2.3.4.5 Discussion

According to data (Balakhnina et al., 2012a) concentration of the monosilicic acid under soil flooding in (–Si)-soil increased by about 22%, which can be explained by the changing equilibrium between mono- and polysilicic acids in the soil solution. Enrichment of soil by the amorphous SiO_2 resulted in the increase of the monosilicic acid content approximately six times, while the concentration of acid-extractable Si in soil only increased by 15–20%. These data show that amorphous silicon dioxide (Silicon dioxide 325 Mesh #C 9723950, Fisher Scientific) can be an effective source of monosilicic acid.

It was established that content of the Si in the roots and shoots increased at the optimal soil irrigation if the plants were grown on Si-enriched soil, compared with the (–Si)-variant (Balakhnina et al., 2012a). Under soil flooding, the Si content increased significantly in the roots of the (–Si;+flooding)-plants and did not increase in the (+Si;+flooding)-plants, relatively non-flooded (–Si)- and (+Si)-variants, respectively. This fact may be explained by because the silicon content in the (+Si)-roots was already quite high before the onset of flooding. Using other stressors showed (Biel et al., 2008; Matichenkov and Kosobryukhov, 2004) that salt toxicity, for instance, induced the accumulation of Si in plants grown on silica soil. The effect of the application of Si on the growth and biomass production of barley shoots and roots was pronounced, especially in roots, with optimal watering. However, the positive effect of Si on the intensity of plant growth processes and biomass production under flooding conditions was not observed; these processes were suppressed, especially sharply in the roots (Balakhnina et al., 2012a). The obtained data on growth and of plant biomass accumulation, as one of the main indicators of metabolic activity, are consistent with the explanation (Balakhnina et al., 2010; Chirkova, 1988; Kalashnikov et al., 1994; Yordanova and Popova, 2007) that a decrease in a metabolic activity may be one of the adaptive reactions to soil hypoxia.

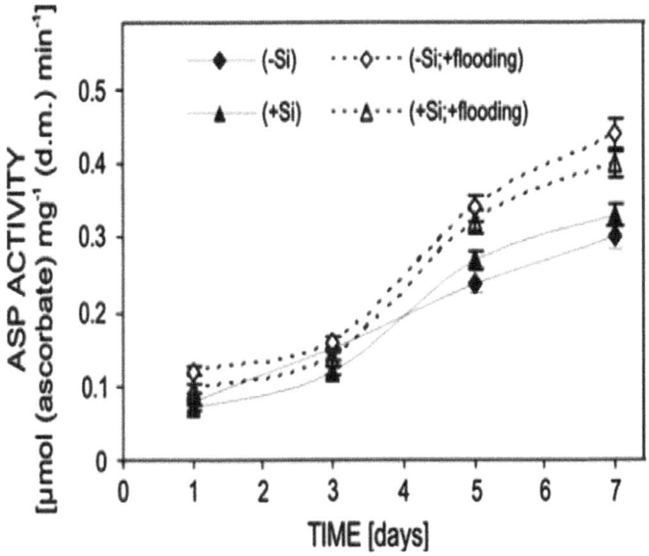

FIGURE 15.8 AsP activity in the leaves of *Hordeum vulgare* seedlings grown in the soil without and with application of amorphous silica under optimal soil watering [(–Si) and (+Si)] or flooding [(–Si;+flooding) and (+Si;+flooding)]. Bars indicate standard deviations (n = 3). (Source: Balakhnina et al., 2012a.)

However, there is evidence that in some other stressful conditions, Si stimulates plant growth and productivity (Biel et al., 2008; Gunes et al., 2007a; Matichenkov and Kosobryukhov, 2004).

The increase in TBARs concentration is considered as one of the main indicators of the intensification of peroxide processes in cells of stressed living organisms (Mittler, 2002). It was observed that the drought-induced increase of TBARs concentration in wheat leaves, and at the application of silicon, the increment of these substances was smaller (Gong et al., 2008). The application of silicon resulted in a decrease of malondialdehyde concentration induced by the soil salinity (Liang et al., 2003, 2007).

An analogical effect of Si was obtained (Balakhnina et al., 2012a): the increase in TBARs concentration during flooding was less in plants grown on Si-enriched soil. It was found that soil flooding increases stomatal resistance and thus leads to the formation of an excessive amount of ROS in the leaves, which in turn enhances LPO processes intensity (Jackson, 1991; Jackson and Drew, 1984). Blanke and Cooke (2004) found that flooding and drought caused stomatal closure in strawberry stolons and leaves. In the leaves of drought-stressed strawberry plants, stomatal conductance was strongly reduced in comparison with well-watered control plants. However, a significant but lesser decline was in flooded plants. Stomatal closure under drought was explained by increasing ABA transport from roots to leaves, while under soil flooding, stomatal closure was considered as a plant reaction to the release of stress ethylene (Blanke and Cooke, 2004).

Earlier, Zakrzhevsky et al. (1995) showed that the values of stomatal resistance in the leaves of relatively soil flooding tolerant maize increased within 12 days of soil hypoxia from 6.4 to 12.4 s cm^{-1} and in the leaves of non-resistant pea—from 2.7 to 72.4 s cm^{-1}. On the example of sorghum enriched with silicon, Hattori et al. (2003, 2005, 2007) indicated the important role of silicon in the transport of water and the growth of roots, which absorbed a larger volume of water from the dry soil and kept higher stomatal conductance. In the light of the data mentioned above, the significant difference observed by Balakhnina et al. (2012a) in LPO intensity between the (–Si;+flooding)- and (+Si;+flooding)-plants can also be connected with the action of Si on stomatal closure. This conclusion is in agreement with the concept of Si beneficial effects associated with its high deposition in plant tissues, enhancing their strength and rigidity (Ma and Yamaji, 2006). Decreased formation of secondary and tertiary lateral roots in plants lacking silicon nutrition has been demonstrated (Matichenkov and Bocharnikova, 2004). It is known that distribution of Si between plant organs may vary from 0.001% in the pulp of fruit to 10–15% in the epidermal tissues (Voronkov et al., 1978). Hodson et al. (2005) described the phylogenetic variation in the silicon composition of plants. Usually, accumulation of silicon is observed in the cell walls of root endodermis, leaf tip cells, and bundle sheath of leaves, needles, husks, and bark (Bennett, 1982a,b; Hodson, 1986; Lux et al., 2003; Ma, 2003; Ma et al., 2001a; Ranganathan et al., 2006; Snyder et al., 2006). In particular, silicon deposition in the roots, leaves, culms, floral bracts, and awns of barley was investigated in series work (Bennett, 1982a,b; Hayward and Parry, 1973, 1975, 1980). Molecules of monosilicic acid accumulating in the epidermal tissues (Hodson and Sangster, 1989) were found to form the silicon cellulose envelope (Waterkeyn et al., 1982) for the mechanical strengthening of plants (Ma, 2003) and preventing dehydration of leaves (Emadian and Newton, 1989).

Some authors showed that, compared with the non-silicon treatment, application of silicon under drought increased the activities of some antioxidant enzymes: SOD, and catalase and glutathione reductase, whereas the content of H_2O_2 was decreased, and the activities of GPX and AsP showed no significant difference (Gong et al., 2005). Later, it was found (Gong et al., 2008) that application of Si did not change the contents of H_2O_2 of drought-stressed wheat at the booting stage, but decreased it at the filling stage. There was no effect of Si on the SOD, GPX and catalase activities at the booting stage; at the filling stage, however, application of Si increased SOD activity and decreased GPX activity in drought-stressed plants (Gong et al., 2008). In salt-stressed barley, it was found that the addition of Si increased activities of SOD and GPX (Liang et al., 2003, 2007). Zhu et al. (2004) also observed that the addition of Si increased the activities of SOD, GPX, and AsP in salt-stressed cucumber. In our experiments, application of amorphous silica stimulated GPX activity in barley roots under optimal watering, but we did not find stimulatory action of silicon on the SOD, GPX, and AsP activities of the flooded plants. In summary, the results indicate that application of Si stimulates growth processes of barley shoots and roots under optimal soil watering and decreases the intensity of oxidative destruction under soil flooding without significant changes in the activities of antioxidant enzymes.

15.2.3.5 Metal Toxicity

Heavy metal stress negatively affects processes associated with biomass production and grain yield in almost all major field grown crops (Bednarek et al., 2006). Every metal interacts with a plant in a specific way, which depends on several factors such as the type of soil, growth conditions, and the presence of other ions (Rana and Masood, 2002). Hammond et al. (1995) showed that silicon treatments gave significant alleviation of the toxic effect of Al in barley plants. Aluminum uptake by roots was significantly diminished in the presence of Si. Silicon-mediated alleviation of (heavy) metal toxicity in higher plants is widely accepted. Shi et al. (2005) reported that the alleviation of Mn toxicity by Si in cucumber was attributed to a significant reduction in LPO intensity caused by excess Mn and to a significant increase in enzymatic, for example, SOD, APX, and GR, and non-enzymatic antioxidants, for example, ascorbate and glutathione. In a study conducted by Gunes et al. (2007a), unlike SOD and CAT activities, APX activity of barley was significantly higher, compared with plants growing without Si supplementation. It was concluded, from the APX results, that APX was probably more important than CAT in H_2O_2 detoxification. Such coordinated responses of APX with H_2O_2 concentrations in tissues are believed to promote tolerance to oxidative stress (Gunes et al., 2007a). Soylemezoglu et al. (2009) showed that the activities of SOD and CAT in boron stressed plants obviously increased, whereas that of APX was decreased. The results related to antioxidant enzyme responses under B toxicity were in agreement with the findings of Molassiotis et al. (2005), who reported increased SOD and CAT activity under B toxicity in apple rootstocks.

15.2.3.6 Si-Rich Mineral Zeolite Elevates Cadmium Stress

Hordeum vulgare L. cv. Bartom seedlings grown on a substrate without or with zeolite were exposed to 450 and 1,000 μmol of $Cd(NO_3)_2$. Adding the zeolite to a substrate (+Si) resulted in the accumulation of mono- and polysilicic acids in the leaves, an increase in the growth of seedlings and biomass production, a decrease of the contents of TBARs, and an increase in the activities of the antioxidant enzymes SOD, APX, and GPX. The negative effects of Cd^{2+}, which were reflected in the loss of chlorophyll contents and a decrease of the photosynthesis rate, as well as an increase of TBARs contents, were observed in the (–Si)-plants to a greater extent than in the (+Si)-plants. The activities of SOD, APX, and GPX increased in the barley leaves of (–Si)- and (+Si)-seedlings with an increase of Cd^{2+} concentration in the acting solution. At the same time, the values of the ratios between TBARs concentrations and the activities of SOD, APX, and GPX in (–Si)-plants were higher than that in (+Si)-plants both under optimal conditions and with Cd stress. This suggests that the use of Si-rich mineral zeolite for growing barley stimulates metabolism and increases the resistance of plants to cadmium stress through a shift in the dynamic equilibrium between the rate of oxidative destruction and the activity of antioxidant systems in favor of the latter.

15.2.3.6.1 Cadmium

Cadmium (Cd^{2+}) is one of the most dangerous heavy metals, which can be accumulated in cultivated plants; it is not biodegradable and cannot be substantially removed from plants (Kabata-Pendias, 2011). Cd^{2+} is regarded as a non-essential metal, which is extremely toxic to plants, animals, and humans, and it is extremely persistent in the environment (Wagner, 1993). The anthropogenic input of cadmium to soils occurs by aerial deposition of Cd-rich industrial wastes, sewage sludge, manure, and phosphate at fertilizer application (Azevedo et al., 2012). The elevated levels of Cd in the cells affect photosynthesis by inhibiting root Fe(III) reductase and the enzymes involved in CO_2 fixation, the water relations of the cells, and mineral nutrition (such as Ca, Mg, P and K) (Alcantara et al., 1994; Azevedo et al., 2012; Das et al., 1997; Nagajyoti et al., 2010). It was shown that treatment of pea seeds with $CdCl_2$ resulted in the alteration of redox and oxidative properties in seed tissues, a decrease of germination success, and inhibition of embryonic axes growth (Smiri, 2011).

Considering that Cd pollution of the cultivated area has become a global problem, the searching of the mechanisms for increasing plant resistance to Cd toxicity is one of the most important aspects of modern plant physiology and biochemistry. The important role of Si in enhancing the growth and development of plants and in the reinforcement of their resistance to biotic and abiotic stresses has been observed for many plant species (Belanger, 2005; Epstein, 1999; Gunes et al., 2007a, b; Ma et al., 2004; Ma and Takahashi, 2002). The stimulation of the antioxidant system of plants and the formation of complexes between toxic metal ions and Si-compounds are the most important mechanisms, which can explain positive influence of active silicon on the plants under heavy metal toxicity (Balakhnina and Borkowska, 2013; Belanger, 2005; Gong et al., 2005, 2008; et al., 2007; Mittler, 2002). The application of active forms of silicon to the soil resulted in a decrease of the intensity of destructive processes in the plants under soil flooding (Balakhnina et al., 2012b). Presently, the effectiveness of using silicon to improve plant stress resistance is widely recognized and has been described by many authors (Epstein, 1999; Ma et al., 2004).

Currently, the attention of researchers attracted to natural compounds is in improving the stability and productivity of plants under stress conditions. One such compound is a zeolite, a silicon-rich mineral. It has been shown (Balakhnina et al., 2015) that zeolite (brand DSP, ZeoTradeResourceLtd., Orlov region, Russia) induces growth, biomass accumulation, and resistance in barley *Hordeum vulgare* L., cv. Bartom to the subsequent effects of different doses of cadmium. This brand of zeolite consists of dense light gray granules containing SiO_2—69.0–74.0%, TiO—0.08–0.16%, Al_2O_3—11.4–14.0%, Fe_2O_3—0.60–1.8%, MnO—0.02–0.05%, CaO—1.7–3.3%, MgO—0.4–1.7%, K_2O—0.5–5%, Na_2O—0.4–0.9%, and P_2O_5—0.4%, pH 6.5.

15.2.3.6.2 Concentration of Mineral Elements in the Plants

The ICP analysis of the digested plant tissue gives a possibility to determine the element composition in the cytoplasm of the plant tissue. The obtained data (Table 15.1) showed that only concentration of Si significantly increased in the leaves of plants grown on the substrate enriched with zeolite. The content of all other tested elements (P, K, Ca, Mg, Al, and Fe) changed but insignificantly. Therefore, it is possible to conclude that in the short time experiment, the application of zeolite to the growing media has mainly influenced the silicon status of the barley.

15.2.3.6.3 Monosilicic and Polysilicic Acid Content in Plant Samples

The concentration of silicon compounds in the water-soluble fraction of the leaves was dependent on zeolite application (Table 15.2). After 2 weeks of plant growth, the concentrations of mono- and polysilicic acids in the (+Si)-plants were 30 and 41% higher than in (–Si)-plants, respectively (Table 15.2).

15.2.3.6.4 Plant Growth and Biomass Production

The rate of plant growth is one of the most important integral parameters indicating the physiological state of agriculture. The addition of zeolite to the substrate resulted in an intensification of seedling growth (Table 15.2). The biomass of shoots and roots of (+Si)-plants reached 143 and 136%, respectively, as compared with (–Si)-plants after 2 weeks of growth. The length of shoots and roots also increased up to 158 and 168%, respectively, relative to (–Si)-plants (Table 15.2). The chlorophyll content [Chl(a+b)] in the leaves of (–Si)- and (+Si)-plants did not differ significantly and consisted of about 2.62 and 2.54 mg g^{-1} FW, respectively (Table 15.3). After exposing the barley seedlings to the aqueous solutions of $Cd(NO_3)_2$ for 2 h, Chl(a+b) concentration in the leaves

TABLE 15.1

Concentration of Si, P, K, Ca, Mg, Al, and Fe in the Cytoplasm of Leaves of Barley Plants after 2 weeks of Growing without and with Zeolite in Substrate (Means ± SE, n = 5)

	Concentration, mg kg⁻¹ DW						
Substrate for Plant Growing	**Si**	**P**	**K**	**Ca**	**Mg**	**Al**	**Fe**
Coconut shavings + peat (v:v)	1198 ± 95[a]	250 ± 20[a]	3010 ± 150[a]	2050 ± 123[a]	265 ± 25[a]	184 ± 13[a]	85 ± 7[a]
Coconut shavings + zeolite (v:v)	2560 ± 153[b]	286 ± 25[a]	3240 ± 156[a]	2280 ± 122[a]	294 ± 27[a]	168 ± 11[a]	94 ± 8[a]

Using Duncan's multiple range tests, values within a column followed by the same letter are not statistically different ($P < 0.05$).
Source: Balakhnina et al., 2015.

TABLE 15.2
Concentration of Monosilicic and Polysilicic Acids in the Cytoplasm of Leaves (n = 5), Biomass and Length (n = 20) of Barley Plants after 2 weeks of Growing in Substrate Enriched by Zeolite (Means ± SE)

Substrate for Plant Growth	Monosilicic Acid, mg kg^{-1} FW (%)	Polysilicic Acids, mg kg^{-1} FW (%)	Shoot Biomass, g per plant (%)	Root Biomass, g per plant (%)	Length of Shoot, cm (%)	Length of Root, cm (%)
Coconut shavings + peat (v:v)	0.366 ± 0.02 (100)[a]	0.326 ± 0.02 (100)[a]	0.63 ± 0.03 (100)[a]	0.154 ± 0.01 (100)[a]	14.7 ± 0.7 (100)[a]	10.8 ± 0.9 (100)[a]
Coconut shavings + zeolite (v:v)	0.477 ± 0.02 (130)[b]	0.459 ± 0.02 (141)[b]	0.90 ± 0.04 (143)[b]	0.210 ± 0.02 (136)[b]	23.2 ± 1.4 (158)[b]	18.1 ± 1.2 (168)[b]

Using Duncan's multiple range tests, values within a column followed by the same letter are not statistically different ($P < 0.05$).
Source: Balakhnina et al., 2015.

TABLE 15.3
Concentration of Chl(a + b) and Photosynthetic Activity (P_N) in the Primary Leaves of Barley (Hordeum vulgare L., cv Bartom) Grown on a Substrate without (–Si) and with Zeolite (+Si) and after That Exposed to Aqueous Solutions of $Cd(NO_3)_2$ for 2 h and 24 h (n = 5)

Parameters	Time	(–Si)	(–Si)+Cd 450 µmol	(–Si)+Cd 1,000 µmol	(+Si)	(+Si)+Cd 450 µmol	(+Si)+Cd 1,000 µmol
Chl($a+b$), mg g^{-1} FW (%)	24 h	2.62 ± 0.14 (100)[a]	2.04 ± 0.1 (78)[b]	1.56 ± 0.06 (65)[c]	2.54 ± 0.1 (100)[a]	2.08 ± 0.1 (82)[b]	1.85 ± 0.08 (73)[c]
P_N, µmol CO_2 m^{-2} s^{-1} (%)	2 h	28.7 ± 1.7 (100)[a]	20.2 ± 1.4 (70)[b]	16.4 ± 1.0 (57)[c]	26.3 ± 1.6 (100)[a]	21.7 ± 1.1 (83)[b]	17.4 ± 1.3 (66)[c]
P_N, µmol CO_2 m^{-2} s^{-1} (%)	24 h	30.3 ± 1.8 (100)[a]	19.8 ± 1.3 (65)[b]	13.7 ± 0.14 (45)[c]	27.6 ± 1.4 (100)[a]	20.1 ± 1.2 (73)[b]	16.1 ± 1.5 (58)[c]

Using Duncan's multiple range tests, values within a line followed by the same letter are not statistically different ($P < 0.05$).
Source: Balakhnina et al., 2015.

remained without any significant differences (data not shown). After 24 h of exposure to $Cd(NO_3)_2$, the concentration of the pigments in the leaves of (+Si;+Cd 1,000 µmol)-plants decreased to 1.85 mg g^{-1} FW, while in the leaves of (–Si;+Cd 1,000 µmol)-plants, it decreased to 1.56 mg g^{-1} FW (Table 15.3). In the cases of 450 µmol Cd solution, a decrease in Chl(a + b) concentration in the leaves was less pronounced (Table 15.3).

After 2 weeks of growth, the photosynthetic rates of (+Si)- and (–Si)-plants not exposed to Cd stress were not different (Table 15.3). Cd suppressed photosynthesis. In particular, in leaves of (–Si;+Cd 450 µmol) and (–Si;+Cd 1,000 µmol)-plants the photosynthetic rate decreased by about 30 and 43% after 2 h of the experiment and by about 35 and 55% after 24 h, respectively (Table 15.3). A decrease of photosynthetic rate in the (+Si;+Cd 450 µmol)- and (+Si;+Cd 1,000 µmol)-plants was less pronounced, by about 17 and 34% after 2 h of the cadmium stress and 27 and 42% after 24 h, respectively in comparison with the (+Si)-variant (Table 15.3).

15.2.3.6.5 *Intensity of Oxidative Processes*

The intensities of the lipid peroxidation processes in the leaves of non-stressed plants grown on the substrate with and without zeolite were similar to each other. However, plants responses to Cd stress differed and depended on the presence of Si in the substrate. After 2 h of exposure to the aqueous solutions of Cd, an increase (6 and 23%, respectively) in the TBARs content was observed in the (–Si;+Cd 450 µmol)- and (–Si;+Cd 1,000 µmol)-plants in comparison with the (–Si)-plants (Figure 15.9). After 24 h of exposure to Cd stress, the TBARs content in the leaves of these plants increased further by 12 and 36%, respectively. Conversely, in the (+Si;+Cd 450 µmol)- and

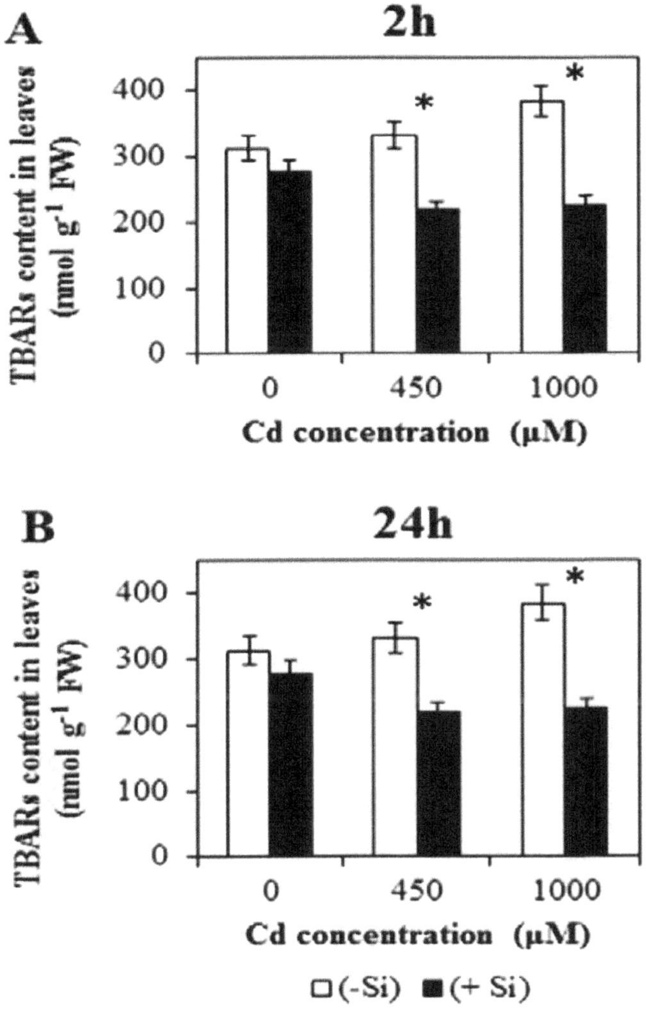

FIGURE 15.9 TBARs content in the primary leaves of barley (*Hordeum vulgare* L., cv Bartom), grown on a substrate without (–Si) and with zeolite (+Si), after 2 h and 24 h exposure in aqueous solutions $Cd(NO_3)_2$. Error bars represent the standard deviation of the mean value (n = 5). (Source: Balakhnina et al., 2015.)

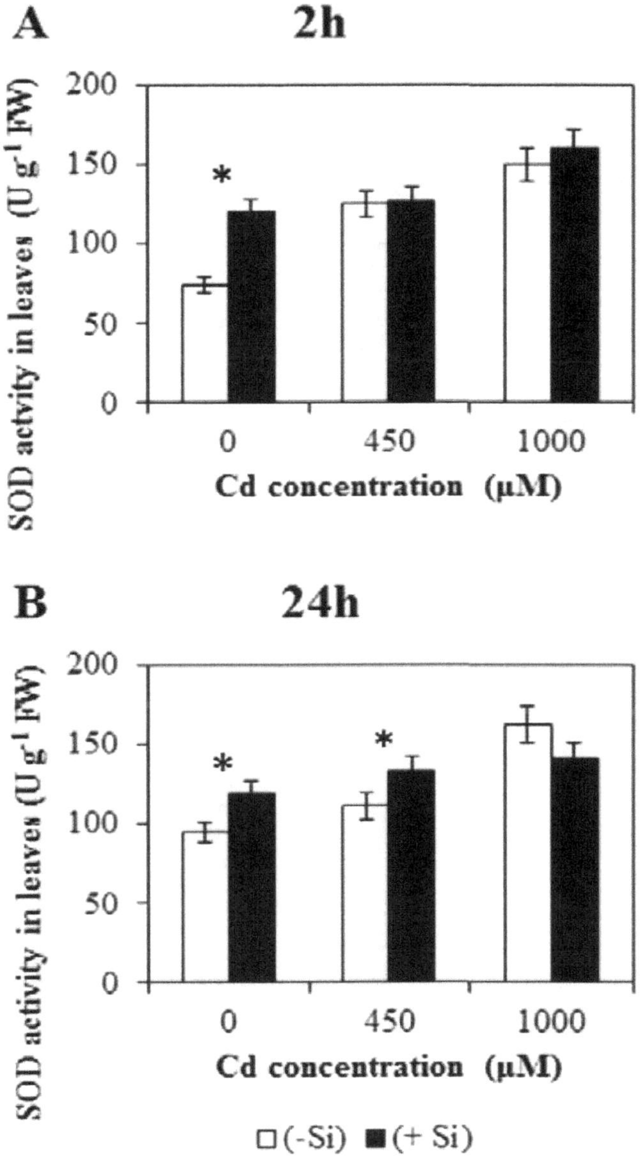

FIGURE 15.10 SOD activity in the primary leaves of barley (*Hordeum vulgare* L., cv Bartom), grown on a substrate without (–Si) and with zeolite (+Si), after 2 h and 24 h exposure in aqueous solutions $Cd(NO_3)_2$. Error bars represent the standard deviation of the mean value (n = 5).

(+Si;+Cd 1,000 µmol)-seedlings exposed to Cd solutions for 2 h, the TBARs content decreased below the value of the (+Si)-plants by 22 and 19%, respectively and remained on these levels after 24 h (Figure 15.9).

15.2.3.6.6 SOD, APX, and GPX Activities

Zeolite application stimulated SOD activity in the leaves of barley seedlings. SOD activity was 81 U g^{-1} FW in the (–Si)-plants, while in (+Si)-plants, SOD activity was significantly higher (Figure 15.10). After 2 h of exposure to the aqueous solutions of $Cd(NO_3)_2$, an increase of enzyme activity was observed up to 154 and 203%, respectively in the leaves of (–Si;+Cd 450 µmol)- and (–Si;+Cd 1,000 µmol)-plants relative to the (–Si)-plants (Figure 15.2). After 24 h of the experiment, an increase of SOD activity in these plants was not so high and amounted to 117 and 172%, respectively. Accounting the fact that SOD activity in (+Si)-plants was higher than in (–Si)-plants, it seemed that the increase of enzyme activity in response to Cd stress was lower after 2 h of stress exposure and even absent after 24 h.

APX activity in the leaves of (+Si)-plants was about 4.25 µmol g^{-1} FW min^{-1} and was 25% higher than that of the (–Si)-plants. Under the influence of Cd, the APX activity increased, and the increment of enzyme activity was dependent on the dose of the stress factor

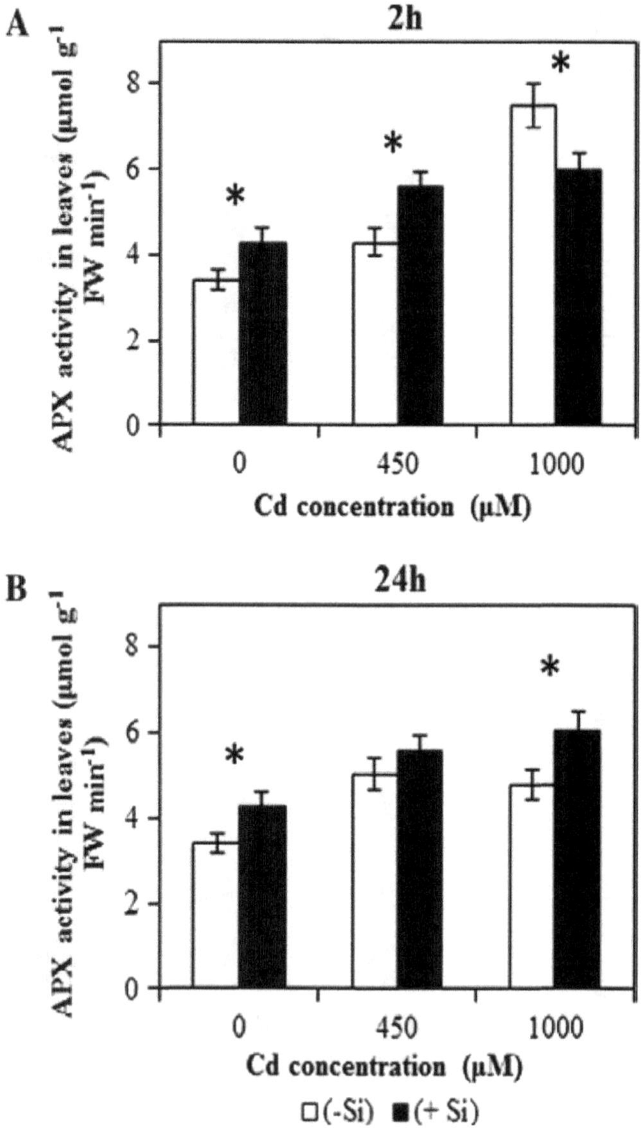

FIGURE 15.11 APX activity in primary leaves of barley (*Hordeum vulgare* L., cv Bartom), grown on a substrate without (–Si) and with zeolite (+Si), after 2 h and 24 h of exposure to aqueous solutions of $Cd(NO_3)_2$. Error bars represent the standard deviation of the mean value (n = 5). (Source: Balakhnina et al., 2015.)

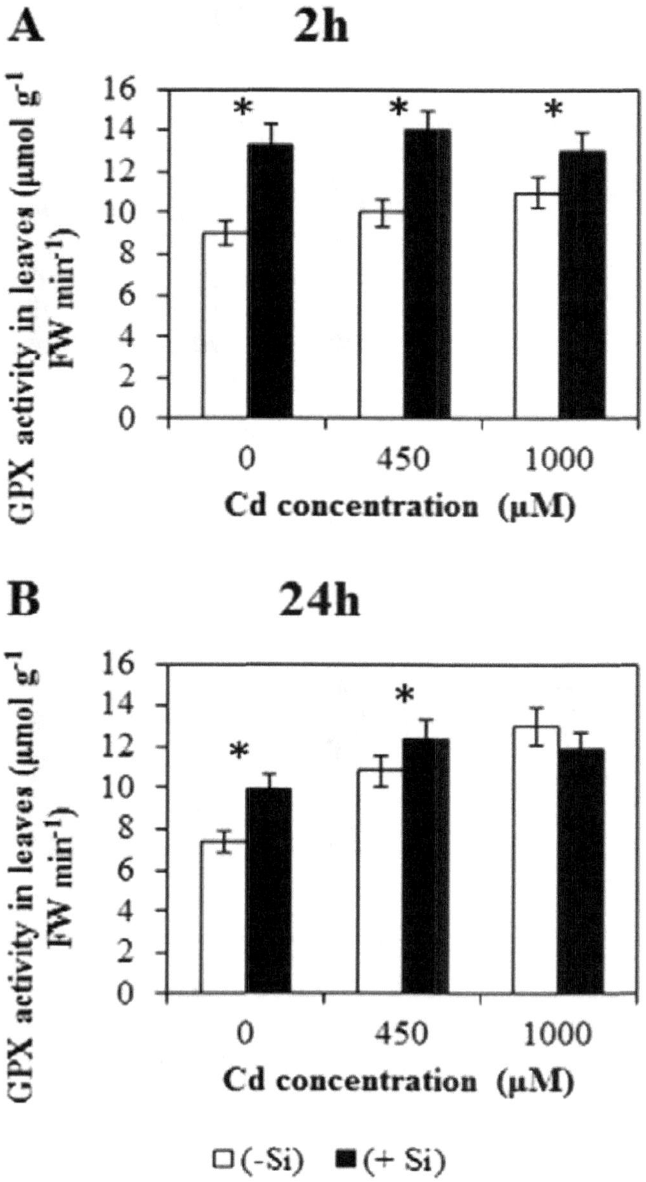

FIGURE 15.12 GPX activity in the primary leaves of barley (*Hordeum vulgare* L., cv Bartom), grown on a substrate without (–Si) and with zeolite (+Si), after 2 h and 24 h exposure in aqueous solutions $Cd(NO_3)_2$. Error bars represent the standard deviation of the mean value (n = 5). (Source: Balakhnina et al., 2015.)

(Figure 15.11). GPX activity in leaves of 2 weeks old (–Si)- and (+Si)- plants ranged between 9.0 and 7.4 μmol $g^{-1}FW\ min^{-1}$ and 13.4 and 10.0 μmol $g^{-1}FW\ min^{-1}$, respectively (Figure 11.12). In the (–Si;+Cd 450 μmol)- and (–Si;+Cd 1,000 μmol)-plants, the GPX activity increased by 11 and 22% after 2 h of stress subjection and by 46 and 75%, after 24 h, respectively (Figure 11.12). In the (+Si;+Cd 450 μmol)- and (+Si;+Cd 1,000 μmol)-plants, an increase of enzyme activity was observed only after 24 h of the experiment, making up by 25 and 19%, respectively (Figure 15.12).

15.2.3.6.7 Discussion on Underlying Mechanisms

The plant's response to various stressors, including heavy metals, depends on the dose of the stressor, the type of the plant, and the stage of the plant development. Stress responses can be divided into destructive and adaptive reactions. The latter reflects an adaptive potential of the organism that is based on its ability to use compounds against stress development. The use of Si-rich physiologically active natural compounds, organic, and mineral fertilizers promotes preserving the plant species, and maintains their homeostasis and productivity under the influence of unfavorable factors at the same levels as they would be under optimal growth conditions (Balakhnina et al., 2012b, 2015; Matichenkov et al., 2000; Matichenkov and Kosobryukhov, 2004). In particular, it was shown (Al-Busaidi et al., 2008) that zeolite treatment produced taller plants, more grains, and higher plant biomass under salt stress. Other authors (Peter et al., 2011) observed that Si could mitigate the stress reactions of plants facing drought, saline soil, or heavy metal pollution. Eshghi et al. (2010) obtained results promoting the influence of zeolite application on the shoot and root dry weight of soybean both under optimal nutrition and under Cd stress by mitigating the negative impact of Cd contamination on mineral element contents in the shoots. Zeolite application also stimulated growth processes of barley seedlings by increasing the concentration of active Si forms (monosilicic and polysilicic acids) in plant tissues (Balakhnina et al., 2015). This observation led to expections that silicon-rich plants would be less affected by the Cd stressor than the Si-non-treated plants. The rate of photosynthesis and chlorophyll content confirmed this expectation. A decrease of potential activity of photosynthetic apparatus and chlorosis were induced by Cd-stressor and manifested to the lesser degree in the silicon-rich plants than in the controls. Matoh et al. (1991) explained stress protective action of Si on photosynthesis by a Si-induced decrease in the transpiration rate. According to Ali et al. (2013), there is a correlation between the intensity of photosynthesis and the transpiration rate. Transpiration from the leaves occurs through the stomata and the cuticle. Si is deposited beneath the cuticle of the leaves forming a Si-cuticle double layer that results in a decrease of transpiration (Ali et al., 2013).

Formation of ROS appears to be an unspecific plant response to different stress factors, including heavy metal. The plants possess an evolutionary formed defensive system against damage by oxidative destruction. A significant increase in SOD, APX, and GR activities in cultured tobacco cells has been observed in response to Cd^{2+} treatment (We et al., 2012). Activation of antioxidant enzymes through silicon application enhanced plant resistance to drought, soil flooding, and salinity (Balakhnina et al., 2012b, 2015a; Hattori et al., 2005; Liang et al., 2003, 2007; Ma, 2003; Zhu et al., 2004). The TBARs contents in the leaves of control barley seedlings were expectably increased under Cd stress, while the silicon-rich Cd-stressed plants responded to Cd stress by a decrease in TBARs contents (Balakhnina et al., 2015). This may be associated with increasing activities of antioxidant enzymes (SOD, APX, and GPX) in silicon-rich plants in comparison with the controls. In other words, zeolite application enhanced the availability of the plants to stress. Zeolite reinforced the plant defense system against oxidative destructions; the activities of antioxidant enzymes in the leaves of silicon-rich plants were higher than that of the control plants, but the increase of the activity of these enzymes in Cd-stressed silicon-rich plants was lower. It was assumed that Si not only decreases the adsorption of heavy metals but also reduces the transpiration rate and binds heavy metals into complexes (Balakhnina et al., 2015). Under optimal growth conditions, intensity of the oxidative processes and the activity of the antioxidant system are known to be in the state of dynamic equilibrium (Baraboi, 1991). With the stress amplification, the equilibrium could be shifted. Earlier, we used the ratios of lipid peroxidation intensity to antioxidant activity, expressed by nmol TBARs/unit APX activity as an index of back proportional to the adaptive potential of plants (Balakhnina et al., 2009). Close inverse correlation between the lipid peroxidation intensity and the APX activity in barley tissues was observed. Stimulation of APX by the natural bioflavonoid dihydroquercetin resulted in a decrease of oxidative destruction processes (tested by TBARs contents); the index of back proportional to the adaptive potential of the plants decreased as well. In the present work, the ratios between TBARs contents and the activity of all tested antioxidant enzymes were lower in the silicon-rich plants as compared with the control plants (Table 15.4). Under Cd stress, these ratios increased with an increase of the dose of the stressor (Cd concentration and time of treatment), especially in seedlings that were not rich with silicon. In other words, silicon-rich plants displayed higher adaptive potential. It may be concluded that the use of zeolite for growing barley stimulates the growth process and enhances plant resistance to cadmium in the earlier stages of stress development via a shift in the dynamic equilibrium between the rate of oxidative destruction and the activity of the antioxidant system in favor of the latter.

15.3 CONCLUSION

Environmental stress causes huge losses in agriculture productivity worldwide. Therefore, research aimed at overcoming environmental stresses needs to be quickly and fully implemented. These reports suggest that Si has certain physiological functions in plants. Its role becomes more important under adverse environmental conditions. Increasing the content of silicon in plant tissues enhances their resistance to various stresses. The presence of silicon in the cell walls of plants increases their strength,

TABLE 15.4
The Ratio between the Content of Thiobarbituric Acid Reactive Substances (TBARs) and the Activity of Superoxide Dismutase (SOD), Ascorbate Peroxidase (APX), Gluthatione Reductase (GR), and Guaiacol Peroxidase (GPX) in the Leaves of Barley (Hordeum vulgare) Grown without (–Si) and with Zeolite (+Si) and Exposed to Different Solutions of $Cd(NO_3)_2$

Ratio of Parameters	(–Si)	(–Si)+Cd 450 µmol	(–Si)+Cd 1000 µmol	(+Si)	(+Si)+Cd 450 µmol	(+Si)+Cd 1000 µmol
2 h						
TBARs/SOD	3.9	2.7	2.3	2.3	1.7	1.4
TBARs/APX	91	77	64	65	39	35
TBARs/GR	433	346	343	357	232	235
TBARs/GPX	35	33	35	21	16	17
24 h						
TBARs/SOD	3.4	3.2	2.7	2.3	2.1	2.5
TBARs/APX	94	70	79	64	50	53
TBARs/GR	349	403	573	199	189	504
TBARs/GPX	43	33	34	27	23	30

The data presented in Figures 15.2, 15.3, 15.4, and 15.5 were used for calculations.
Source: Balakhnina et al., 2015.

as silicon increases resistance to salinity, drought tolerance, and photosynthetic activity, and promotes the active growth of roots and foliage. The results of these studies illustrate that the entry of silicon to plant tissues leads to inhibition of the oxidative destruction processes that are accompanied with increasing activity of some antioxidant enzymes that neutralize ROS induced by drought, salinity, toxic metals, and UV-B radiation. They also suggest that Si could be used as a potential growth regulator to improve plant growth and resistance under stress conditions. This may be a promising new strategy for improvement of soil properties in agriculture.

REFERENCES

Agarie, S., Uchida, H., Agata, W., Kubota, F., Kaufman, P.B. (1998) Effects of silicon on transpiration and leaf conductance in rice plants (*Oryza sativa* L.). *Plant Production Science* 1(2): 89–95.

Ahmed, M., Hassen, F., Khurshid, Y. (2011) Does silicon and irrigation have impact on drought tolerance mechanism of sorghum? *Agricultural Water Management* 98(12): 1808–1812.

Al-Aghabary, K., Zhu, Z., Shi, Q. (2004) Influence of silicon supply on chlorophyll content, chlorophyll fluorescence, and antioxidative enzyme activities in tomato plants under salt stress. *Journal of Plant Physiology* 27(12): 2101–2115.

Al-Busaidi, A., Yamamoto, T., Inoue, M., Egrinya, E.E., Mori, Y., Irshad, M. (2008) Effects of zeolite on soil nutrients and growth of barley following irrigation with saline water. In: *The 3rd international conference on water resources and arid environments and the 1st Arab Water Forum*, pp. 1–11.

Alcantara, E., Romera, F.J., Canete, M., De La Guardia, M.D. (1994) Effects of heavy metals on both induction and function of root Fe (III) reductase in Fe-deficient cucumber (*Cucumis sativus* L.) plants. *Journal of Experimental Botany* 45(12): 1893–1898.

Ali, A., Alqurainy, F. (2006) Activities of antioxidants in plants under environmental stress. In *The Lutein-Prevention and Treatment for Diseases*, ed. Motohashi, N., 187–256. Transworld Research Network, Kerala, India.

Ali, S., Farooq, M.A., Yasmeen, T., Hussain, S., Arif, M.S., Abbas, F., Bharwana, S.A., Zhang, G. (2013) The influence of silicon on barley growth, photosynthesis and ultra-structure under chromium stress. *Ecotoxicology and Environmental Safety* 89: 66–72.

Allen, R.D. (1995) Dissection of oxidative stress tolerance using transgenic plants. *Plant Physiology* 107(4): 1049–1054.

Alsher, R.G., Donahue, J.L., Cramer, C.L. (1997) Reactive oxygen species and antioxidants: Relationship in green cells. *Physiologia Plantarum* 100: 224–233.

Asada, K. (1992) Ascorbate peroxidase—Hydrogen peroxide scavenging enzyme in plants. *Physiologia Plantarum* 85: 235–224.

Asada, K. (2006) Production and scavenging of reactive oxygen species in chloroplasts and their functions. *Plant Physiology* 141(2): 391–396.

Athar, R., Ahmad, M. (2002) Heavy metal toxicity: Effect on plant growth and metal uptake by wheat, and on free living *Azotobacter. Water, Air, and Soil Pollution* 138(1/4): 165–180.

Azevedo, R.A., Gratão, P.L., Monteiro, C.C., Carvalho, R.F. (2012) What is new in the research on cadmium-induced stress in plants? *Food and Energy Security* 1(2): 133–140.

Balakhnina, T.I. (2015) Plant responses to soil flooding. In *Stress Responses in Plants, Mechanisms of Toxicity and Tolerance*, eds. Tripathi, B.N., Müller, M., 115–143. Springer International Publishing AG, Cham, Switzerland.

Balakhnina, T.I., Bennicelli, R.P., Stêpniewska, Z., Stêpniewski, W. (2004) Oxygen stress in the Root zone and plant response (some examples). In *Physics, Chemistry and Biogeochemistry in Soil and Plant Studies*, ed. Józefaciuk, G., 23–27. Institute of Agrophysics PAS Press, Lublin, Poland.

Balakhnina, T., Bennicelli, R., Stepniewska, Z., Stepniewski, W., Borkowska, A., Fomina, I. (2012b) Stress responses of spring rape plants to soil flooding. *International Agrophysics* 26(4): 347–353.

Balakhnina, T.I., Bennicelli, R.P., Stêpniewska, Z., Stêpniewski, W., Fomina, I.R. (2010) Oxidative damage and antioxidant defense system in leaves of *Vicia faba* major L. cv. Bartom during soil flooding and subsequent drainage. *Plant and Soil* 327(1–2): 293–301.

Balakhnina, T., Borkowska, A. (2013) Effects of silicon on plant resistance to environmental stresses: Review. *International Agrophysics* 27: 225–232.

Balakhnina, T.I., Bulak, P., Matichenkov, V.V., Kosobryukhov, A.A., Włodarczyk, T.M. (2015) The influence of Si-rich mineral zeolite on the growth processes and adaptive potential of barley plants under cadmium stress. *Plant Growth Regulation* 75(2): 557–565.

Balakhnina, T.I., Gavrilov, A.B., Włodarczyk, T.M., Borkowska, A., Nosalewicz, M., Fomina, I.R. (2009) Dihydroquercetin protects barley seeds against mould and increases seedling adaptive potential under soil flooding. *Plant Growth Regulation* 57(2): 127–135.

Balakhnina, T.I., Matichenkov, V.V., Wlodarczyk, T., Borkowska, A., Nosalewicz, M., Fomina, I.R. (2012a) Effects of silicon on growth processes and adaptive potential of barley plants under optimal soil watering and flooding. *Plant Growth Regulation* 67(1): 35–43.

Baraboi, V.A. (1991) Stress mechanisms and lipid peroxidation. *Uspekhi Sovremenoi Biologii* 111(6): 923–932.

Barber, D.A., Shone, M.G.T. (1966) The absorption of silica from aqueous solutions by plants. *Journal of Experimental Botany* 17(3): 569–578.

Beckmann, M., Hock, M., Bruelheide, H., Erfmeier, A. (2012) The role of UV-B radiation in the invasion of *Hieracium pilosella* – A comparison of German and New Zealand plants. *Environmental and Experimental Botany* 75: 173–180

Bednarek, W., Tkaczyk, P., Dresler, S. (2006) Heavy metals content as criterion for assessment of carrot root (in Polish). *Acta Agrophysics* 142: 779–790

Belanger, R.R. (2005) The role silicon in plant–pathogen interaction: Toward universal model. In *Proceedings of the 3rd silicon agricultural conference*, ed. Korndorfer, G.H., 34–40. Universidad Federal de Uberlandia, Uberlandia, Brazil.

Bennett, D.M. (1982a) An ultrastructural study on the development of silicified tissues in the leaf tip of barley (*Hordeum sativum* Jess.). *Annals of Botany* 50(2): 229–237.

Bennett, D.M. (1982b) Silicon deposition in the roots of *Hordeum sativum* Jess., *Arena sativa* L. and *Triticum aestivum* L. *Annals of Botany* 50(2): 239–245.

Bennicelli, R.P., Balakhnina, T.I., Szajnocha, K., Banach, A. (2005) Aerobic conditions and antioxidative system of Azolla caroliniana Willd. in the presence of Hg in water solution. *International Agrophysics* 19: 27–30.

Beyer, W., Imlay, J., Fridovich, I. (1991) Superoxide dismutase. *Program Nucleic Acid Research* 40: 221–253.

Biel, K.Y., Matichenkov, V.V., Fomina, I.R. (2008) Protective role of silicon in living systems. In *Functional Foods for Chronic Diseases*, ed. Martirosyan, D.M. D and A. Inc., Richardson Press, Dallas, pp. 208–231.

Blanke, M.M., Cooke, D.T. (2004) Effects of flooding and drought on stomatal activity, transpiration, photosynthesis, water potential and water channel activity in strawberry stolons and leaves. *Plant Growth Regulation* 42(2): 153–160.

Brigelius-Flohe, R., Flohe, L. (2003) Is there a role of glutathione peroxidases in signaling and differentiation? *BioFactors* 17(1–4): 93–102.

Chen, H., Qualls, R.G., Blank, R.R. (2005) Effect of soil flooding on photosynthesis, carbohydrate partitioning and nutrient uptake in the invasive exotic *Lepidium latifolium*. *Aquatic Botany* 82(4): 250–268.

Chirkova, T.V. (1988) *Puti adaptatsii rastenii k gipoksii i anoksii* [Pathways of plant adaptation to hypoxia and anoxia]. Leningrad State University, Saint Petersburg, Russia.

Das, P., Samantaray, S., Rout, G.R. (1997) Studies on cadmium toxicity in plants: A review. *Environmental Pollution* 98(1): 29–36.

Devkota, A., Jha, P.K. (2011) Influence of water stress on growth and yield of *Centella asiatica*. *International Agrophysics* 25: 211–214.

Egneus, H., Heber, U., Matthiesen, U., Kirk, M. (1975) Reduction of oxygen by the electron transport chain of chloroplasts during assimilation of carbon dioxide. *Biochimica et Biophysica Acta* 408(3): 252–268.

Emadian, S.F., Newton, R.J. (1989) Growth enhancement of loblolly pine (*Pinus taeda* L.) seedlings by silicon. *Journal of Plant Physiology* 134(1): 98–103.

Epron, D., Dreyer, E. (1993) Long-term effects of drought on photosynthesis of adult oak trees (*Quercus petraea* and *Q. robur*) in a natural stand. *New Phytology* 125: 381–389.

Epstein, E. (1999) Silicon – Annual review of plant physiology. *Plant Molecular Biology* 50: 641–664.

Eshghi, S., Mahmoodabadi, M.R., Abdi, G.R., Jamali, B. (2010) Zeolite ameliorates the adverse effect of cadmium contamination on growth and nodulation of soybean plant (*Glycine max* L.). *Journal of Applied Environmental and Biological Sciences* 4(10): 43–50.

Fang, C.X., Wang, Q.S., Yu, Y., Huang, L.K., Wu, X.C., Lin, W.X. (2011) Silicon and its uptaking gene Lsi1 in regulation of rice UV-B tolerance. *Acta Agronomica Sinica* 37(6): 1005–1011.

Farkas, I. (2011) Plant drought stress: Detection by image analysis. In *Encyclopedia of Agrophysics*, eds. Gliñski, J., Horabik, J., Lipiec, J. Springer Press, Dordrecht. *Polish Journal of Environmental Studies* 19(3): 565–572.

Ghamsari, L., Keyhani, E., Golkhoo, S. (2007) Kinetics properties of guaiacol peroxidase activity in *Crocus sativus* L. Corm during rooting. *Iranian Biomedical Journal* 11(3): 137–146.

Gliñski, J. (2011) Agrophysical objects (soils, plants, agricultural products, and food). In *Encyclopedia of Agrophysics*, eds. Gliñski, J., Horabik, J., Lipiec, J., 122–125. Springer Press, Dordrecht-Heidelberg-London-New York.

Gliñski, J., Stêpniewski, W. (1985) *Soil Aeration and Its Role for Plants.* CRC Press, Boca Raton, FL.

Gliñski, J., Stêpniewski, W., Ostrowski, J., Stêpniewska, Z. (2004) Spatial characteristics of soil redox conditions, ed. Albert-Ludwigs. In *Proceedings of Conference Eurosoil*, September 4–12, Freiburg, Germany.

Gong, H.J., Chen, K.M., Zhao, Z.G., Chen, G.C., Zhou, W.J. (2008) Effects of silicon on defense of wheat against oxidative stress under drought at different developmental stages. *Biologia Plantarum* 52(3): 592–596.

Gong, H., Zhu, X., Chen, K., Wang, S., Zhang, C. (2005) Silicon alleviates oxidative damage of wheat plants in pots under drought. *Plant Science* 169(2): 313–321.

Gunes, A., Inal, A., Bagci, E.G., Coban, S. (2007a) Silicon mediated changes on some physiological and enzymatic parameters symptomatic of oxidative stress in barley grown in sodic-B toxic soil. *Journal of Plant Physiology* 164(6): 807–811.

Gunes, A., Inal, A., Bagci, E.G., Coban, S., Sahin, O. (2007b) Silicon increases boron tolerance and reduces oxidative damage of wheat grown in soil with excess boron. *Biologia Plantarum* 51(3): 571–574.

Halliwell, B. (1987) Oxidative damage, lipid peroxidation and antioxidant protection in chloroplasts. *Chemistry and Physics of Lipids* 44(2–4): 327–340.

Hammond, K.E., Evans, D.E., Hodson, M.J. (1995) Aluminium/silicon interactions in barley (*Hordeum vulgare* L.) seedlings. *Plant and Soil* 173(1): 89–95.

Hattori, T., Inanaga, S., Araki, H., An, P., Morita, S., Luxova, M., Lux, A. (2005) Application of silicon enhanced drought tolerance in *Sorghum bicolor. Physiologia Plantarum* 123(4): 459–466.

Hattori, T., Inanaga, S., Tanimoto, E., Lux, A., Luxova, M., Sugimoto, Y. (2003) Silicon-induced changes in viscoelastic properties of sorghum root cell walls. *Plant and Cell Physiology* 44(7): 743–749.

Hattori, T., Sonobe, K., Inanaga, S., An, P., Tsuji, W., Araki, H., Eneji, A.E., Morita, S. (2007) Short-term stomatal responses to light intensity changes and osmotic stress in sorghum seedlings rose with and without silicon. *Environmental and Experimental Botany* 60(2): 177–182.

Hayward, D.M., Parry, D.W. (1973) Electron-probe microanalysis studies of silica distribution in barley (*Hordeum sativum* L.). *Annals of Botany* 37(3): 579–591.

Hayward, D.M., Parry, D.W. (1975) Scanning electron microscopy of silica deposition in the leaves of barley (*Hordeum sativum* L.). *Annals of Botany* 39(5): 1003–1009.

Hayward, D.M., Parry, D.W. (1980) Scanning electron microscopy of silica deposits in the culms, floral bracts and awns of barley (*Hordeum sativum* Jess.). *Annals of Botany* 46(5): 541–548.

Hejazi Mehrizi, M., Shariatmadari, H., Khoshgoftarmanesh, A.H., Zarezadeh, A. (2011) Effect of salinity and zinc on physiological and nutritional responses of rosemary. *International Agrophysics* 25: 349–353.

Hodson, M.J. (1986) Silicon deposition in the roots, culm and leaf of *Phalaris canariensis* L. *Annals of Botany* 58(2): 167–177.

Hodson, M.J., Sangster, A.G. (1989) X-ray microanalysis of the seminal root of Sorghum bicolor with particular reference to silicon. *Annals of Botany* 64(6): 659–667.

Hodson, M.J., White, P.J., Mead, A., Broadley, M.R. (2005) Phylogenetic variation in the silicon composition of plants. *Annals of Botany* 96(6): 1027–1046.

Hossain, M.T., Soga, K., Wakabayashi, K., Kamisaka, S., Fujii, S., Yamamoto, R., Hoson, T. (2007) Modification of chemical properties of cell walls by silicon and its role in regulation of the cell wall extensibility in oat leaves. *Journal of Plant Physiology* 164(4): 385–393.

Jackson, M.B. (1991) Regulation of water relationships in flooded plants by ABA from leaves, roots and xylem sap. In *Abscisic Acid: Physiology and Biochemistry*, eds. Davies, W.J., Jones, H.G., 217–226. BIOS Scientific Publishers, Oxford.

Jackson, M.B., Drew, M.C. (1984) Effects of flooding on growth and metabolism of herbaceous plants. In *Flooding and Plant Growth*, ed. Kozlowski, T.T., 47–128. Academic Press, Orlando, FL.

Jarvis, S.C. (1987) The uptake and transport of silicon by perennial ryegrass and wheat. *Plant and Soil* 97(3): 429–437.

Jones, L.H.P., Handreck, K.A. (1967) Silica in soils, plants and animals. *Advances in Agronomy* 19: 107–149.

Kabata-Pendias, A. (2011) *Trace Elements in Soils and Plants*, 4th edn. CRC Press, Boca Raton, FL, 287–288.

Kalashnikov, Y.E., Zakrzhevsky, D.A., Balakhnina, T.I. (1994) Effect of soil hypoxia on activation of oxygen and the system of protection from oxidative damage in roots and leaves of *Hordeum vulgare* L. *Russian Journal of Plant Physiology* 41: 583–588.

Kovda, V.A. (1973) The bases of learning about soils. *Nauka* 2(8) 377–428.

Larson, R.A. (1988) The antioxidants of higher plants. *Phytochemistry* 27: 969–978.

Li, B., Wei, S.C., Li, N., Zhang, J. (2007) Heterologous expression of the TsVP gene improves the drought resistance of maize. *Plant Biotechnology Journal* 6(2): 146–159.

Liang, Y., Chen, Q., Liu, Q., Zhang, W., Ding, R. (2003) Exogenous silicon (Si) increases antioxidant enzyme activity and reduces lipid peroxidation in roots of salt stressed barley (*Hordeum vulgare* L.). *Journal of Plant Physiology* 160(10): 1157–1164.

Liang, Y., Sun, W., Zhu, Y.G., Christie, P. (2007) Mechanisms of silicon mediated alleviation of abiotic stresses in higher plants. *Environmental Pollution* 147(2): 422–428.

Liang, Y.C., Si, J., Römheld, V. (2005) Silicon uptake and transport is an active process in *Cucumis sativus* L. *New Phytology* 167: 797–804.

Lizana, X.C., Hess, S., Calderini, D.F. (2009) Crop phenology modifies wheat responses to increased UV-B radiation. *Agricultural and Forest Meteorology* 149(11): 1964–1974.

Lux, A., Luxová, M., Abe, J., Morita, S., Inanaga, S. (2003) Silicification of bamboo (*Phyllostachys heterocycla* Mitf.) root and leaf. *Plant and Soil* 225: 85–91.

Lux, A., Luxová, M., Hattori, T., Inanaga, S., Sugimoto, Y. (2002) Silicification in sorghum (*Sorghum bicolor*) cultivars with different drought tolerance. *Physiologia Plantarum* 115 (1): 87–92.

Ma, J.F. (2003) Function of silicon in higher plants. In *Progress in Molecular and Subcellular Biology*, ed. Muller W.E.G., 127–147. Springer, Berlin.

Ma, C.C., Li, Q.F., Gao, Y.B., Xin, T.R. (2004) Effects of silicon application on drought resistance of cucumber plants. *Soil Science and Plant Nutrition* 50(5): 623–632.

Ma, J.F., Miyake, Y., Takahashi, E. (2001a) Silicon as a beneficial element for crop plants. In *Silicon in Agriculture*, eds. Datonoff, L., Korndorfer, G., Synder, G. Elsevier Science Press, New York, NY, pp. 17–39.

Ma, J.F., Ryan, P.R., Delhaize, E. (2001b) Aluminum tolerance in plants and the complexing role of organic acids. *Trends in Plant Science* 6(6): 273–278.

Ma, J.F., Takahashi, E. (2002) *Soil, Fertilizer, and Plant Silicon Research in Japan*. Elsevier, Amsterdam.

Ma, J.F., Yamaji, N. (2006) Silicon uptake and accumulation in lower plants. *Trends in Plant Science* 11(8): 392–397.

Matichenkov, V.V., Ammosova, J.M. (1996) Effect of amorphous silica on soil properties of a sod-podzolic soil. *Eurasian Soil Science* 28(10): 87–99.

Matichenkov, V.V., Bocharnikova, E.A. (2004) Si in horticultural industry. In *Plant Mineral Nutrition and Pesticide Management. Production Practices and Quality Assessment of Food Crops*, eds. Dris, R., Jain, S.M. 217–239. Kluwer Academic Press, Amsterdam.

Matichenkov, V.V., Calvert, D.V., Snyder, G.H. (2000) Prospective silicon fertilization for citrus in Florida. *Proceedings of the Soil and Crop Science Society of Florida* 59: 137–141.

Matichenkov, V.V., Kosobryukhov, A.A. (2004) Si effect on the plant resistance to salt toxicity. In *Proceedings of the 13th international soil conservation organization conference (ISCO)*, Brisbane, 287–295.

Matoh, T., Murata, S., Takahashi, E. (1991) Effect of silicate application on photosynthesis of rice plants. Japanese. *Journal of Soil Science and Plant Nutrition* 62: 248–252.

Mittler, R. (2002) Oxidative stress, antioxidants and stress tolerance. *Trends in Plant Science* 7(9): 405–410.

Molassiotis, A., Sotiropoulos, T., Tanou, G., Diamantidis, G., Therios, I. (2005) Boron induced oxidative damage and antioxidant and nucleolytic responses in shoot tips culture of the apple rootstock M9 (*Malus domestica* Borkh). *Environmental and Experimental Botany* 56: 54–62.

Nagajyoti, P.C., Lee, K.D., Sreekanth, T.V.M. (2010) Heavy metals, occurrence and toxicity for plants: A review. *Environmental Chemistry Letters* 8(3): 199–216.

Parry, D.W., Kelso, M. (1975) The distribution of silicon deposits in the root *Molina caerulea* (L.) Moench and *Sorghum bicolor* (L.) Moench. *Annals of Botany* 39(5): 995–1001.

Parvaiz, A., Satyawati, S. (2008) Salt stress and phytobiochemical responses of plants – A review. *Plant, Soil and Environment* 54(3): 89–99.

Peter, A., Mihaly-Cozmuta, L., Mihaly-Cozmuta, A., Nicula, C. (2011) The role of natural zeolite and of zeolite modified with ammonium ions to reduce the uptake of lead, zinc, copper and iron ions in *Hieracium aurantium* and *Rumex acetosella* grown on tailing ponds. *Analele Universitatii din Oradea: Fascicula Biologie* 18(2): 128–135.

Pociecha, E., Kooecielniak, J., Filek, W. (2008) Effects of root flooding and stage of development on the growth and photosynthesis of field bean (*Vicia faba* L. minor). *Acta Physiologiae Plantarum* 30(4): 529–535.

Rains, D.W., Epstein, E., Zasoski, R.J., Aslam, M. (2006) Active silicon uptake by wheat. *Plant and Soil* 280(1–2): 223–228.

Ranganathan, S., Suvarchala, V., Rajesh, Y.B.R.D., Prasad, M.S., Padmakumari, A.P., Voleti, S.R. (2006) Effects of silicon sources on its deposition, chlorophyll content, and disease and pest resistance in rice. *Biologia Plantarum* 50(4): 713–716.

Richmond, K.E., Sussman, M. (2003) Got silicon? The non-essential beneficial plant nutrient. *Current Opinion in Plant Biology* 6(3): 268–272.

Rosa, S.B., Caverzan, A., Teixeira, F.K., Lazzarotto, F., Silveira, J.A.G., Ferreira-Silva, S.L., Abreu-Neto, J., Margis, R., Margis-Pinheiro, M. (2010) Cytosolic APx knockdown indicates an ambiguous redox response in rice. *Phytochemistry* 71(5–6): 548–558.

Rybus-Zajc, M., Kubioe, J. (2010) Effect of UV-B radiation on antioxidative enzyme activity in cucumber cotyledons. *Acta Biologica Cracoviensia Series Botanica* 52(2): 97–102.

Said-Al Ahl, H.A.H., Omer, E.A., Naguib, N.Y. (2009) Effect of water stress and nitrogen fertilizer on herb and essential oil of oregano. *International Agrophysics* 23: 269–275.

Savant, N.K., Snyder, G.H., Datnoff, L.E. (1997) Silicon management and sustainable rice production. *Advances in Agronomy* 58: 151–199.

Shein, E.V., Pachepsky, Y.A. (1995) Influence of root density on the critical soil water potential. *Plant and Soil* 171(2): 351–357.

Shen, X., Zhou, Y., Duan, L., Li, Z., Eneji, A.E., Li, J. (2010) Silicon effects on photosynthesis and antioxidant parameters of soybean seedlings under drought and ultraviolet-B radiation. *Journal of Plant Physiology* 167(15): 1248–1252.

Shi, Q.H., Bao, Z.Y., Zhu, Z.J., He, Y., Qian, Q.Q., Yu, J.Q. (2005) Silicon mediated alleviation of Mn toxicity in *Cucumis sativus* in relation to activities of superoxide dismutase and ascorbate peroxidase. *Phytochemistry* 66(13): 1551–1559.

Shigeoka, S., Ishikawa, T., Tamoi, M., Miyagawa, Y., Takeda, T., Yabuta, Y., Yoshimura, K. (2002) Regulation and function of ascorbate peroxidase isoenzymes. *Journal of Experimental Botany* 53(372): 1305–1319.

Shim, I.S., Momose, Y., Yamamoto, A., Kim, D.W., Usui, K. (2003) Inhibition of catalase activity by oxidative stress and its relationship to salicylic acid accumulation in plants. *Plant Growth Regulation* 39(3): 285–292.

Smiri, M. (2011) Effect of cadmium on germination, growth, redox and oxidative properties in *Pisum sativum* seeds. *Journal of Environmental Chemistry and Ecotoxicology* 3: 52–59.

Snyder, G.H., Matichenkov, V.V., Datnoff, L.E. (2006) Silicon. In *Handbook of Plant Nutrition*, eds. Barker, A.V., Pibeam, D.J., 551–568. University of Massachusetts, Boston, MA.

Sokolova, T.A. (1985) *The Clay Minerals in the Humid Regions of USSR* (in Russian). Nauka Press, Novosibirsk, Russia.

Soylemezoglu, G., Demir, K., Inal, A., Gunes, A. (2009) Effect of silicon on antioxidant and stomatal response of two grapevines (*Vitis vinifera* L.) rootstocks grown in boron toxic, saline and boron toxic-saline soil. *Scientia Horticulturae* 123(2): 240–246.

Voronkov, M.G., Zelchan, G.I., Lukevits, A.Y. (1978) *Kremnii i zhizn* [*Silicon and Life*]. Zinatne, Riga.

Wagner, G.J. (1993) Accumulation of cadmium in crop plants and its consequences to human health. *Advances in Agronomy* 51(1): 173–212.

Wang, W., Vinocur, B., Altman, A. (2003) Plant responses to drought, salinity and extreme temperatures: Towards genetic engineering for stress tolerance. *Planta* 218(1): 1–14.

Waterkeyn, L., Bientait, A., Peeters, A. (1982) Callose et silice epidermiques rapports avec la transpiration culticulaire. *La Cellule* 73: 263–287.

We, J.F., Deng, M.H., Gong, M. (2012) Cd2+ stress induces two waves of H_2O_2 accumulation associated with ROS-generating system and ROS-scavenging system in cultured tobacco cells. *Australian Journal of Crop Science* 6(5): 846–853.

Yang, X., Liang, Z., Wen, X., Lu, C. (2008) Genetic engineering of the biosynthesis of glycinebetaine leads to increased tolerance of photosynthesis to salt stress in transgenic tobacco plants. *Plant Molecular Biology* 66(1–2): 73–86.

Yordanova, R.Y., Popova, L.P. (2007) Flooding induced changes in photosynthesis and oxidative status in maize plants. *Acta Physiologiae Plantarum* 29(6): 535–541.

Yoshida, S. (1975) The physiology of silicon in rice. *Technological Bulletin* 25: 24–27.

Zakrzhevsky, D.A., Balakhnina, T.I., Stepniewski, W., Stepniewska, S., Bennicelli, R.P., Lipiec, J. (1995) Oxidation and growth processes in roots and leaves of higher plants at different oxygen availability in soil. *Russian Journal of Plant Physiology* 42: 242–248.

Zancan, S., Suglia, I., La Rocca, N., Ghisi, R. (2008) Effects of UV-B radiation on antioxidant parameters of iron-deficient barley plants. *Environmental and Experimental Botany* 63(1–3): 71–79.

Zhu, J.K. (2001) Plant salt tolerance. *Trends in Plant Science* 6(2): 66–71.

Zhu, X., Song, F., Xu, H. (2010) Influence of arbuscular mycorrhiza on lipid peroxidation on lipid peroxidation and antioxidant enzyme activity of maize plants under temperature stress. *Mycorrhiza* 20(5): 325–332.

Zhu, Z., Wei, G., Li, J., Qian, Q., Yu, J. (2004) Silicon alleviates salt stress and increases antioxidant enzymes activity in leaves of salt stressed cucumber (*Cucumis sativus* L.). *Plant Science* 167(3): 527–533.

16 Selenium-Induced Abiotic Stress Tolerance in Plants

Zsuzsanna Kolbert, Réka Szőllősi, and Gábor Feigl

CONTENTS

16.1 INTRODUCTION

Selenium (Se) is a non-metal element that is special in many ways. Contrary to animals or humans, higher plants do not require it as an essential microelement, but at low concentrations, Se exerts beneficial effects on plants during normal, as well as stress conditions, which is extensively reviewed and discussed in this chapter using the latest literature and schematic figures.

16.2 BRIEF OVERVIEW OF Se UPTAKE AND ASSIMILATION BY HIGHER PLANTS

Selenium can exist in the (+6), (+4), (0), and (–2) oxidation states, which greatly affects its solubility and mobility/availability in the environment. The most oxidized forms of Se are water soluble; therefore, both selenate (SeO_4^{2-}) and selenite (SeO_3^{2-}) have a high degree of bioavailability and bioaccumulation potential in nature (Saha et al., 2017). Globally, total soil Se concentration is typically within the range 0.01–2.0 mg kg^{-1} with an overall mean of 0.4 mg kg^{-1} (Johnson et al., 2010).

Due to several mechanisms for selenite removal from water (e.g., absorption of selenite by organic and inorganic soil particles, assimilation of selenite by soil microbes), selenate is the major soluble form in water which is primarily taken up by plants (Martens, 2003). Another special feature of selenium is its chemical similarity with sulfur (S), an essential macroelement necessary for plants at high concentrations. Root cells contain sulfate transporter proteins in their plasma membrane, among which high-affinity sulfate transporters (HAST) proved to be associated with selenate uptake (Sors et al., 2005). On the other hand, selenite seems to compete with phosphate because it is (mainly) transported by high-affinity phosphate transporters (PHT, Li et al., 2008; Zhang et al., 2014) while the uptake of organic selenium compounds is not well known. The more toxic selenite is partly converted to non-toxic organic forms such as selenomethionine (SeMet) already in the root cells, while selenate mostly translocated into the aerial plant parts *via* xylem (Shrift and Ulrich, 1976; Zayed et al., 1998). Because of its similarity to S, not just the uptake but also the assimilation of selenium forms occurs through S assimilation enzymes and pathways. The assimilation takes place mainly in leaf chloroplasts. Selenate is activated by ATP sulfurylase (APS) and reduced by APS reductase, resulting in the formation of selenite, which is further reduced to selenide (Se^{2-}) either

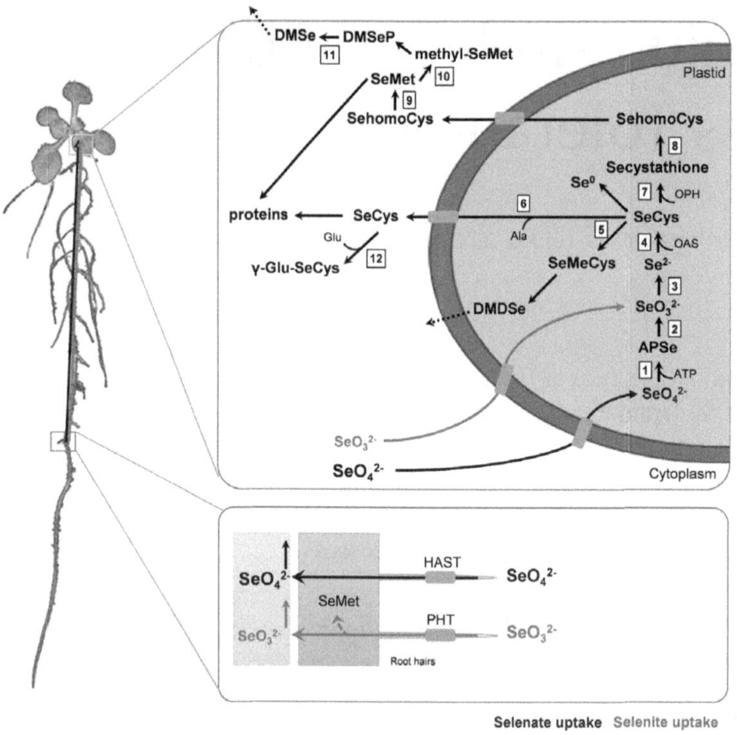

FIGURE 16.1 Uptake and assimilation of selenate and selenite by higher plants. SeO_4^{2-}, selenate; SeO_3^{2-}, selenite; HAST, high-affinity sulfate transporter; PHT, high-affinity phosphate transporter; SeMet, selenomethionine; APSe, adenosine phospho selenate; Se^{2-}, selenide; OAS, *O*-acetylserine; SeCys, selenocysteine; SeMeCys, methyl-selenocysteine; DMDSe, dimethyldiselenide; Ala, alanine; Se^0, elemental selenium; OPH, *O*-phosphohomoserine; Secystatione, selenocystatione; SehomoCys, seleno-homocysteine; methyl-SeMet, methyl-selenomethionine; DMSeP, dimethylselenoproprionate; DMSe, dimethylselenide. Numbers indicate enzymes as follows: (1) ATP sulfurylase; (2) adenosine phosphosulfate reductase; (3) sulfite reductase; (4) O-acetylserine thiol lyase; (5) selenocysteine methyltransferase; (6) selenocysteine lyase; (7) cysthathionine-γ-synthase; (8) cysthathionine-β-lyase; (9) methionine synthase; (10) methionine methyltransferase; (11) dimethylselenoproprionate lyase; (12) γ-glutamyl-cysteine synthetase (modified after Pilon-Smits and Quinn, 2010).

by sulfite reductase or by a non-enzymatic reduction with glutathione. Selenocysteine (SeCys) is formed in the reaction between selenide and O-acetylserine (OAS). SeCys can be detoxified by various alternative ways. It can be converted to the less toxic selenomethionine (SeMet) and volatile dimethyl-selenide (DMSe) or it can be converted to elemental Se (Se^0) and alanine. The formations of methyl-selenocysteine (MeSCys) and the volatile dimethyl-diselenide (DMDSe) are further possible SeCys detoxification mechanisms and are characteristic of hyperaccumulator plant species (Pilon-Smits and Quinn, 2010). The uptake and assimilation mechanisms of Se forms and the participating enzymes are depicted in Figure 16.1.

16.3 ESSENTIALITY AND BENEFICIAL EFFECTS OF Se ON PLANT GROWTH AND DEVELOPMENT

What also makes this element interesting is its essential feature in primitive plants like algae and its non-essential nature in the higher plants. According to Schiavon and Pilon-Smits (2017), the ability to use Se as an essential nutrient seems to be evolutionally lost in more advanced groups of plants perhaps because of the limited availability of Se in terrestrial habitats. Even though the lack of Se does not induce deficiency symptoms, low Se concentrations (<1 mg kg^{-1}) proved to be beneficial for plant growth and development, a feature that was first observed as early as 1925 (Levine, 1925).

Currently, Se seems to exert beneficial effects both in unstressed plants and in plants exposed to various environmental stresses mainly through its capability to regulate the production and quenching of reactive oxygen species (ROS) such as superoxide anion radical ($O_2{\cdot}^-$), singlet oxygen ($^1O_2$), hydroxyl radical (OH·), hydrogen peroxide (H_2O_2), or lipid peroxides (LOO.) (Das and Roychoudhury, 2014). In case of $O_2{\cdot}^-$ elimination, three putative ways of Se actions have been proposed including the promotion of spontaneous dismutation of $O_2{\cdot}^-$ into H_2O_2, the direct reaction between Se compounds and

$O_2^{\cdot -}$ or OH·, and the induction of antioxidant enzyme activities leading to indirect ROS scavenging. Another theoretical mechanism of Se action may be the induction of the assembly of photosystems resulting in controlled ROS production (reviewed by Feng et al., 2013).

The relatively high number of published data clearly show that low Se doses positively regulate growth processes at all three developmental stages of seed-bearing plants (seed formation and germination, vegetative, and reproductive phases) and also during senescence.

16.3.1 Se Induces Seed Germination

Plant seeds are able to take up Se from the external medium and assimilate it during germination. Exogenous application of selenium compounds (priming) breaks dormancy in species like bitter gourd (Chen and Sung, 2001) or wheat (Peng et al., 2001). In the case of *Stylosanthes humilis* seeds, Se in the form of SeMet (Barros and Freitas, 2001) or other Se species like selenium dioxide or selenourea (Pinheiro et al., 2008) breaks seed dormancy through the induction of ethylene (ET) production. Seed priming with low Se doses (15–60 μmolL^{-1}) leads to the induction of enzymes related to starch catabolism resulting in elevation of sugar content and a concomitant increase in cellular respiration producing the energy required for faster growth. These events are accompanied by greater membrane stability and intensified antioxidant capacity in Se-primed seeds which contributes to faster germination and improved seedling vigor (Khaliq et al., 2015). Se has a positive effect on seed germination, we have to consider that a significant amount of Se can be accumulated by seeds like wheat, alfalfa, or by sunflowers during germination, which might serve as an excellent Se source for human consumption (Lintschinger et al., 2000).

16.3.2 Se Induces Vegetative Growth

Similar to germination, the growth of shoots and roots of different plant species such as cucumber, wheat, or tobacco is triggered by exogenous application of Se (Haghighi and da Silva, 2016; Han et al., 2013; Hawrylak-Nowak, 2015; Peng et al., 2001). Using mungbean plants, it was found that Se up-regulates enzymes involved in carbohydrate metabolism, creating energy condition for growth (Malik et al., 2011). Low Se doses (≤4.4 mg kg^{-1}) were found to positively regulate antioxidants thus reducing lipid peroxidation and consequently inducing biomass production of tobacco (Han et al., 2013). Similarly, both selenium forms at low concentrations (1 and 5 μM sodium selenate or sodium selenite) enhanced shoot growth which was found to be associated with the enhanced activity of ascorbate peroxidase (APX) enzyme in wheat (Boldrin et al., 2016). Selenite at 20 μM concentration promoted root biomass production of both hydroponically grown 28-day-old *Arabidopsis thaliana* and 16-day-old *Brassica juncea* (Molnár et al., 2018a). In *Brassica juncea*, 20 μM selenate significantly increased the fresh weight of the shoot and root system, but in case of selenite, the same concentration proved to be inhibitory on the observed parameters (Molnár et al., 2018b). The beneficiary effect was also observable in agar-grown, non-accumulator *Arabidopsis thaliana*, where 10 μM selenite effectively increased hypocotyl length and cotyledon area (Lehotai et al., 2016a). In the case of pea grown in semi-hydroponics, 10 μM selenite remarkably increased the biomass of both organs (Lehotai et al., 2016b).

16.3.3 Se Regulates Reproductive Growth Processes

Selenite treatment at 50 and 100 μM concentrations induced the premature development of flowers in 25-day-old pea plants, while these concentrations were inhibitory with regard to vegetative growth (Lehotai et al., 2016b). Similarly, reproductive parameters such as floral bud development, the opening of flowers or podding were induced by 10 and 20 μM selenate in canola (Hajiboland and Keivanfar, 2012). It has to be noted that the exact molecular mechanisms of Se-induced flowering are still unclear and deserve more attention.

In a recent study, Zhu et al., (2017) reported that foliar pre-treatment of tomato plants with sodium selenate (1 mg L^{-1}) down-regulated ethylene biosynthetic genes (both ACC synthase and oxidase), which delayed fruit ripening and maintained fruit quality. Moreover, the Se-induced increase in antioxidant capacity may also contribute to the inhibitory effect of Se on ripening. Based on these results, selenium treatment represents a promising strategy for delaying ripening and extending the shelf life of tomato fruit (Zhu et al., 2017).

Regarding seed production, sodium selenite application was shown to increase it in *Brassica rapa* without resulting in higher total biomass. In leaves and flowers of Se-exposed *Brassica* plants, higher total respiratory activity was measured which may partly be responsible for the enhanced seed production (Lyons et al., 2009). In the case of wheat, the antioxidant capacity was increased by Se concentrations between 20 and 30 mg L^{-1} Se and this beneficial effect was the most pronounced when Se treatment was applied at heading–blooming stage, indicating that the time of Se application is a relevant parameter influencing the beneficial effect of Se.

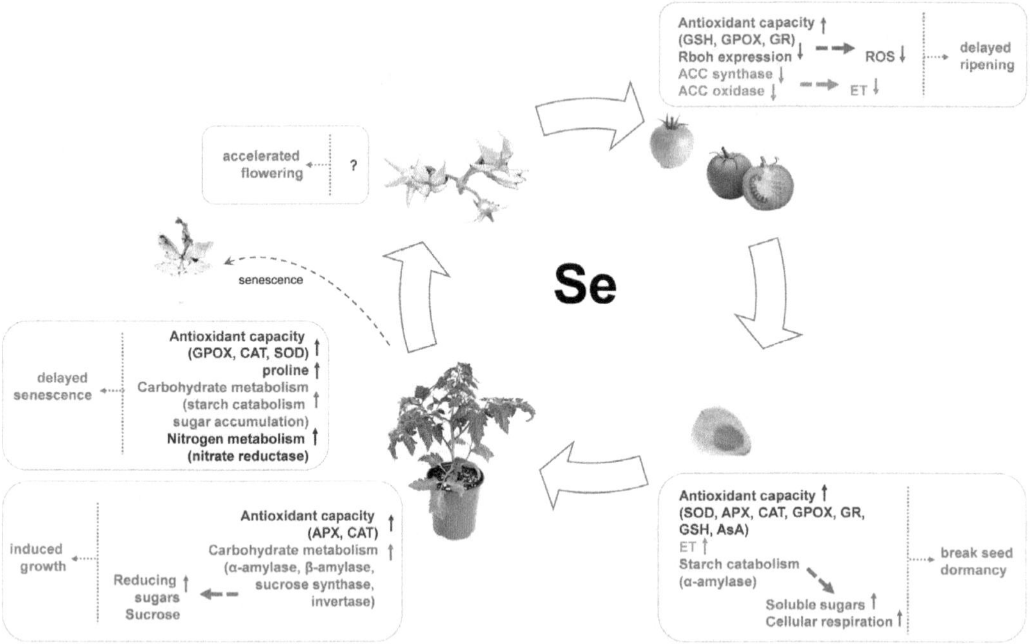

FIGURE 16.2 Growth promoting effects of selenium through the stages of plant life cycle. GSH, glutathione; GPOX, glutathione peroxidase; GR, glutathione reductase; Rboh, plant respiratory burst oxidase homolog; ET, ethylene; SOD, superoxide dismutase; APX, ascorbate peroxidase; CAT, catalase; AsA, ascorbate.

16.3.4 Se Delays Senescence

Senescence is an integral part of the plant life cycle and is thought to coincide with the overproduction of ROS and the consequent oxidative damage of membranes, proteins, and DNA (Leshem, 1988). Based on the fact that selenium is an effective antioxidant, the delay of senescence by selenium is conceivable. This hypothesis was experimentally supported in some early works with maize, tomato, and lettuce (Linchuan and Binggao, 1996; Pezzarossa et al., 1999; Xue et al., 2001). The growth of senescent lettuce was enhanced by low selenate dose, possibly through the induction of glutathione peroxidase (GPOX) and superoxide dismutase (SOD), resulting in improved antioxidant capacity. In soybean, selenium was applied through seed soaking (5 ppm) and also as foliar spray (100 ppm) and in both cases increased activities of antioxidants (SOD, GPOX, catalase), enhanced contents of proline and photosynthetic pigments, delayed leaf senescence, and improved yield were observed (Djanaguiraman et al., 2004, 2005). In a recent paper of Rahmat et al. (2017), the effects of selenium supplementation (10 μM selenate) on developmental and on induced senescence in oilseed rape were compared. Results showed that Se slows leaf senescence mainly through protection of plants from photoinhibitory effects. During developmental senescence, the delaying effect of Se was additionally associated with the regulation of carbohydrate and protein metabolism. The data obtained so far implies that in senescent plants, selenium addition is capable of reducing processes involved in oxidative damages thus delaying/slowing senescence.

Based on the above-described results, Se exerts beneficial effects on growth and development at all phases of the plant life cycle (Figure 16.2). The accumulated data shows that the main processes which are regulated by Se are the ROS scavenging through the induction of antioxidant capacity, the influencing ET biosynthesis, and the induction of carbohydrate catabolism yielding energy for growth and development. Although the growth promoting effects depend on several factors which have to be considered during both laboratory experiments and agricultural practice. Selenium tolerance of the plant species, the concentration, the chemical form of the applied Se, and the way of the treatment influences Se action.

16.4 ABIOTIC STRESS AMELIORATING EFFECTS OF Se

In the last 30 to 40 years, numerous research efforts have been dedicated to the exploration of the positive effects of Se on plants' abiotic stress tolerance. Recently, selenium has been found to ameliorate the damaging effects of various environmental stresses like the presence of heavy metals (e.g. Wu et al., 2016), drought (e.g. Nawaz et al., 2017), salinity (e.g. Jiang et al., 2017), high temperature (e.g. Djanaguiraman et al., 2010), cold (e.g. Hawrylak-Nowak et al., 2010), and UV-B irradiation (e.g. Golob et al., 2017).

16.4.1 Se and Salt Stress

Salinity is among the most severe environmental factors, exerting a negative impact on the productivity of crop plants. The presence of high concentration of salts in the soil affects yield or biomass production of most crop plants (Kronzucker and Britto, 2011; Munns and Tester, 2008; Zhang and Shi, 2013), also causing a loss in crop yield (Ahmad and Prasad, 2012; Bhatnagar-Mathur et al., 2008). In most of the cases, an increased amount of Na^+ and Cl^- ions are responsible for the negative effects of salinity (Maathuis et al., 2014); however, Cl^- is the most hazardous of the possible ions (Tavakkoli et al., 2010). High salinity may either cause hyperosmotic or hyperionic stress, possibly leading to the death of the plant (Hasanuzzaman et al., 2013). Many phenomena might stand behind the fatal outcome: imbalance of nutrients, damage of membranes, changes in the balance of growth regulators, and inhibition of enzymes of metabolic processes like photosynthesis (Hasanuzzaman et al., 2012; Mahajan and Tuteja, 2005). The exact consequence of salinity depends on many factors, like the severity of salt stress and the time of exposure, and on the plant species or genotypes, as well as on other environmental factors (Hasanuzzaman et al., 2013). According to molecular and biochemical studies of plants' salt stress responses, the levels of ROS increase significantly (Ahmad and Umar, 2011; Tanou et al., 2009).

Plants develop many different mechanisms against the negative effects of salinity to protect their ionic homeostasis, such as accumulation of different osmolytes, scavenging harmful radicals, or regulating water transport (Hasegawa et al., 2000) to maintain a reasonably even intracellular milieu (Gupta and Huang, 2014; Roy et al., 2014).

The possible interaction of Se with soil salinity has been previously reviewed by Terry et al. (2000), who concluded that sulfate salinity is able to significantly decrease uptake of selenate into plants; however, this phenomenon depends on the plant species and its relationship with selenium in general (accumulator or not). Terry et al. (2000) also stated that chloride salinity has a much less significant effect on selenate uptake; Se accumulation is only a little lower when its paired with higher salt level.

Regarding the application of low Se doses in salt-stressed plants, it was found that Se counteract the detrimental effect of salt and one general way to exert this stress-alleviating effect is that Se is able to increase the rate of photosynthesis and protect photosystem II (Diao et al., 2014), or on the other hand, Se is able to upregulate the antioxidant defense system, thus preventing lipid peroxidation, paired with the decrease of electrolyte leakage (Djanaguiraman et al., 2005; Habibi, 2014; Kong et al., 2005), to protect plant cells from the negative effect of salinity.

Similarly, in cucumber, a low amount of Se (5/10 μM) was able to promote plant growth under 5, 10, and 20 μM salt treatment by enhancing proline content and photosynthetic pigments; higher Se concentration than 20 μM, however, reduced the growth of the plants (Hawrylak-Nowak, 2009).

Also in *Cucumis sativus* 1 mg kg^{-1} Se was able to provide protection against 2000 ppm NaCl treatment by decreasing lipid peroxidation and electrolyte leakage together with the upregulation of antioxidant enzymes, like peroxidases, catalase, superoxide dismutase, or ascorbate peroxidase (Walaa et al., 2010).

Another study that supports the oxidative stress ameliorating the effect of Se was reported by Hasanuzzaman et al. (2011). In rapeseed, it was observed that 25 μM selenate was able to ameliorate the negative effect of 100 and 200 μM NaCl treatment through the decrease of H_2O_2 levels and the prevention of lipid peroxidation and the upregulation of the antioxidant defense systems, the glyoxalate pathway, and the thiol status. Diao et al. (2014) found that the application of 0.05 mM selenium was able to reverse the negative effect of 100 mM salt stress in both salinity-resistant and -sensitive tomato cultivars by reducing growth inhibition, normalizing photosynthetic parameters, decreasing hydrogen peroxide, and malondialdehyde (MDA) levels through the upregulation of both enzymatic- and non-enzymatic antioxidant system.

The differences in the effects of Se forms was evaluated by Hawrylak-Nowak (2015), who observed that 2 μM selenite was more effective in the improvement of salt-tolerance than the same concentration of selenate, while at 6 μM concentration, selenite still had a positive effect, not like selenate, which had no such property. The beneficial effects of Se, in this case, were not due to the reduction of Na^+ or Cl^- content or to the accumulation of proline, but to the upregulation of antioxidant capacity and photosynthetic activity.

Salt stress induced growth inhibition and decreased chlorophyll content of *Anethum graveolens*, which was ameliorated by 5 μM Se in hydroponic culture. Na^+ content decreased and K^+ content increased due to the Se treatment, while the salt-induced lipid peroxidation was also alleviated, because of the increased SOD and

CAT activity. The improved growth of salt-stressed dill treated with Se could be explained by the improved ion balance and osmotic adjustment, besides the increased antioxidant capacity (Shekari et al., 2017).

When tomato plants were treated with NaCl, Se supplementation was able to alleviate the adverse effect of salt stress through the improvement of cell membrane integrity and increased leaf relative water content. Physiological and biochemical data indicate that 10 μM Se is able to support the survival of the plants under higher NaCl concentrations by increasing the amount of photosynthetic pigments and free radical scavenging activity (Mozafariyan et al., 2016).

Recently, it was reported that 1 μM Se was able to alleviate the inhibitory effect of 100 mM salt stress in maize plants (Jiang et al., 2017) by restoring plant growth, increasing the photosynthetic rate, and alleviating chloroplast damage. Antioxidant enzymes like SOD and APX were increased, together with several protein kinases and a Na^+/H^+ antiporter (*ZmNHX1*) responsible for decreasing Na^+ content in the roots. Also in maize, foliar application of Se at different concentrations resulted in increased total protein, reducing- and non-reducing sugar, phenol and flavonoid contents under control, and saline conditions (Gul et al., 2017). Moreover, in garlic, salt-stressed plants were treated with Se, which was able to improve chlorophyll index and carotenoid content, together with the increased relative water content. With the increasing amount of supplied Se, the leaves' K^+ concentration increased gradually while Na^+ content decreased (Astaneh et al., 2018).

Habibi and Sarvary (2015) found that the growth of *Melissa officinalis* treated with 40 mM NaCl suffered growth inhibition and their chlorophyll a and *b* content was reduced; 10 μM Se supplementation, however, was able to significantly improve growth rate, photosynthetic pigment, and amino acid content, and also reduce membrane damage by reduction of the activity of peroxidases and glutathione peroxidase, while phenylalanine ammonia-lyase activity was upregulated.

Foliar application of Se was able to compensate the effect of salt stress in lettuce (Shalaby et al., 2017). Biomass production increased, together with CAT and APX activity and decreased electrolyte leakage. Similarly, reduced growth and yield parameters of cowpea (*Vigna unguiculata*) treated with 50 mM NaCl was successfully reduced by the foliar application of 10 μM Se by the upregulation of antioxidant capacity (Manaf, 2016).

According to the recent results of Habibi (2017), the growth and photosynthetic parameters of *Petroselinum crispum* were also negatively affected by salt stress, but this was reduced by the application of 1 mg L^{-1} Se treatment. In the root-to-shoot translocation of Na^+, the Na^+ content of the shoot was decreased by Se due to the exclusion of Na^+ from the cell sap and binding Na^+ to the root cell walls.

The high amount of accumulated physiological data indicates the general alleviating effect of Se supply on salt-triggered oxidative stress mainly through the induction of the antioxidant system. Another seemingly general effect of Se is the alleviation of salt-induced chlorophyll loss and photosynthetic decline which positively affects the growth of salt-stresses plants. Moreover, several results point out that Se supplementation is able to increase K^+ and decreases Na^+ concentration. Recent molecular results indicate the possibility that Se regulates Na^+/H^+ antiporter at the transcriptional level and induces exclusion of Na^+ from the cell sap. In order to support these data and provide further explanation for the molecular mechanisms of Se effects, more experiments using molecular and genetic tools are needed.

16.4.2 Se and Drought

Drought has a severe effect on crop yield and production (Feng et al., 2013; Lesk et al., 2016; Saha et al., 2017). Drought stress may be the consequence of limited water supply caused by the drought of soil and in some cases the presence of excess salt ions. In plants, this usually means water deficit in the cells, coupled with diminished osmotic conditions (Sieprawska et al., 2015). Ultimately, limited water supply usually leads to significant weight loss, decrease in the efficiency of photosynthesis, and in the upregulation of antioxidant defense systems, which refers to the emergence of oxidative stress (Filek et al., 2015; Grzesiak et al., 2013).

The protective effect of Se on drought-stressed plants is reported in a number of previous articles. In some plants, Se is able to mitigate the negative effects of drought stress by the regulation of their water status (Kuznetsov et al., 2003). Another possibility for the protective role against drought-triggered damages is the contribution of Se ions to the regulation and maintenance of optimal water status (Proietti et al., 2013), through the accumulation of inorganic (P and Ca ions) or organic (carbohydrates and proteins) osmoprotectants to most likely improve water content as was reported in wheat tissues (Emam et al., 2014). Se can also maintain a lower level of ROS by inducing the biosynthesis of proline and peroxidases (Ahmad et al., 2016), while Ibrahim (2014) concluded that Se supplementation is able to improve the plants' oxidative stress-tolerance by inducing the antioxidant defense system (CAT, SOD, POD, AsA, GSH, alpha-tocopherol) in cases of drought stress. Interestingly, the application of Se—despite its stimulative effect on photosynthesis and the antioxidative system—did not improve the growth of the water stressed plants (Sieprawska et al., 2015).

Similarly, in the work of Yao et al. (2009), low doses (1, 2, and 3 mg Se kg^{-1}) of Se were able to significantly increase the chlorophyll and carotenoid content, antioxidant enzyme activity, and the ratio of reduced MDA in wheat submitted to drought stress. Also in wheat exposed to drought stress, foliar Se application improved turgor by lowered osmotic potential and transpiration rate, together with increased antioxidant capacity, accumulation of soluble sugars, and free amino acids (Nawaz et al., 2015). Recently, Tedeschini et al. (2015) reported that the germination rate of drought-stressed olive tree pollens was also improved by the Se supplementation; while foliar application of sodium selenite was able to upregulate the antioxidant defense system of drought-stressed barley (Habibi, 2013). As another possibility for rescuing drought-stressed plants, Se supplementation of water-stressed barley resulted in higher stomatal conductance and photosynthetic activity because of their better water management (Tadina et al., 2007).

Foliar application of Se improves green fodder yield of maize in cases of drought stress through restoring water potential and relative water content, enhancing the levels of photosynthetic pigments, free amino acids, and the activity of antioxidant enzymes such as SOD, CAT, POD, and APX, compared with maize plants treated with water stress without the addition of Se (Nawaz et al., 2016). According to recent findings of Nawaz et al. (2017), Se improves the drought tolerance of spring wheat through the significant increase of water retention by enhanced water uptake of the root system, but not through the reduction of the transpiration rate. On the other hand, Habibi and Alizade (2017) found that although 1 mg L^{-1} Se (foliar application) was able to enhance the vegetative and reproductive properties of *Melissa officinalis* (a drought tolerant but Se-sensitive plant), it could not countervail the drought-induced growth reduction and lipid peroxidation, despite the stimulated antioxidant system.

According to the current state of research, the role of Se in water-stressed plants is known mainly at the phenomenon level; however, more detailed molecular biological studies are needed in order to be able to explain the background mechanism of Se action.

16.4.3 Se and Heavy Metal Stress

Numerous studies revealed that the increased emission of heavy metals (HMs) in the environment due to different sources such as geological processes or anthropogenic activities (e.g., mining, industrial, application of fertilizers, and pesticides or usage of contaminated irrigation water) might result in several negative alterations in plants. The main consequences of HM stress and/or toxicity in plants are usually the disturbances in photosynthesis, reduced germination rate, and abnormal development of vegetative parts (root and/or shoot) (Emamverdian et al., 2015; Maksymiec, 2007). It is well known that not only the toxic metals like Cd, Pb, Hg, or As can cause oxidative stress in the plant cells impairing different tissues but at supra-optimal concentrations essential HMs (e.g., Fe, Ni, Zn, or Cu) also have stimulatory effect on the production of ROS and consequently on the activity of the antioxidant defense system (Anjum et al., 2011; Feng et al., 2016; Janas et al., 2010; Li et al., 2013).

Up to now plenty of experiments investigated the positive effects of Se at low concentrations (1–5 μM or 1–3 mg L^{-1} depending on plant species and circumstances of the experiment) on HM-stressed plants, and excellent reviews have summarized the results (Domokos-Szabolcsy et al., 2017; Feng et al., 2013). Here we focus on the differences between the general (including ROS-production and antioxidant system) and the HM stress-specific beneficial influences of low Se doses. Although aluminum (Al) is not a HM considering its density (2.70 g cm^{-3}), but because of its toxic effects, it is often regarded as HM (Cartes et al., 2010; Huang et al., 2014; Matsumoto and Motoda, 2012). Therefore, we discuss it as well.

The results of the studies that have been reviewed show that in plants excess HM generates oxidative burst which is realized in lipid peroxidation (LP), the accumulation of H_2O_2 and superoxide anion, and the induction of the antioxidant system, but generally it becomes well-balanced due to Se application at low doses both in root and shoot (Table 16.1). Thus, in most cases, low amounts of Se (regardless of its form selenite/selenate/selenium dioxide or the aspect of the HM) play a holistic protective role against HM-induced oxidative stress, namely it alleviates LP and ROS-overproduction, increases the activity of SOD and H_2O_2-eliminating enzymes like CAT, APX, GR, and the selenoenzyme GPOX (Alyemeni et al., 2018; Filek et al., 2008; Tang et al., 2015; Wu et al., 2016). Besides, some researcher found that Se supply may evoke the increase of non-enzymatic antioxidant reduced glutathione, ascorbic acid (Malik et al., 2012; Tang et al., 2015) or proline content (Alyemeni et al., 2018) in shoots while others observed the diminution of these components (Saidi et al., 2014; Singh et al., 2018). Thus, either the simultaneous application of

TABLE 16.1
The Effect of Se Supply in Heavy Metal-Stressed Plant Species

Name of HM	Plant Name	Concentration or degree of the HM Stress	Chemical form and Concentration of Se Applied	Time of Exposure	Plant organ Investigated	Effects of Se Addition at Low Doses*	References
Essential HMs							
Cu	*Pisum sativum* L. cv. Fenomen	5 μM (hydroponic)	Selenite or selenate: 5 μM	21 days	root	Cu uptake ↓	Landberg and Greger (1994)
					shoot	Cu uptake ↓	
	Sinapis alba L.	3 mg L^{-1} (hydroponic)	Selenium dioxide: 3 mg L^{-1}	8 days	root	Cu accumulation ↑ ns	Fargašová et al. (2006)
					shoot	Cu accumulation ↓; Chl and carotenoid content ↓	
	Triticum aestivum L. cv. Sunny	16 μM (hydroponic)	Selenite or selenate: 16 μM	21 days	root	Cu accumulation ↑ for selenate	Landberg and Greger (1994)
					shoot	Cu uptake ↓ ns	
Zn	*Sinapis alba* L.	15 mg L^{-1} (hydroponic)	Selenium dioxide: 3 mg L^{-1}	8 days	root	Zn accumulation ↓	Fargašová et al. (2006)
					shoot	Zn accumulation ↓; Chl and carotenoid content ↓	
Ni	*Lactuca sativa* L. *var. capitata cv.* Justyna	50 μM (hydroponic)	Selenite: 5 and 20 μM	14 days	leaves	Ni accumulation ↑ at 5 μM Se; Chl and carotenoid content ↑ at 5 μM Se	Hawrylak et al. (2007)
Non-essential HMs							
Cd							
	Boehmeria nivea (L.) Gaud.	5 mg L^{-1} (hydroponic)	Selenite: 1 μM	7 days	root	Cd uptake ↓ ns	Tang et al. (2015)
					stem	Cd uptake ↓	
					leaves	Cd uptake ↓; LP ↓; H_2O_2 content ↓; Chl content ↑; SOD activity ↑; POD activity ↑; APX activity ↑; GR activity ↑; GSH content ↑; AsA and Vit E content ↓	
	Brassica napus L. var. Górczański	400 and 600 μM (*in vitro*)	Selenate: 2 μM	14 days	root	SOD activity ↓; CAT activity ↓; APX activity ↑; no change in GSH-Px activity; H_2O_2 content ↓	Filek et al. (2008)

(*Continued*)

TABLE 16.1 (CONTINUED)
The Effect of Se Supply in Heavy Metal-Stressed Plant Species

Name of HM	Plant Name	Concentration or degree of the HM Stress	Chemical form and Concentration of Se Applied	Time of Exposure	Plant organ Investigated	Effects of Se Addition at Low Doses*	References
					shoot	SOD activity ↓ ns; CAT activity ↓; APX activity ↓ at 600 μM Cd; GSH-Px activity ↑; H_2O_2 content ↓	
	Brassica napus L. cv. Wanyou 18	1 and 5 mg kg^{-1} (soil)	Selenite: 1–20 mg kg^{-1} (soil)	40–100 days	root	Cd accumulation ↓	Wu et al. (2016)
					shoot	Cd accumulation ↓; GSH-Px activity ↑; SOD activity ↑; LP ↓; H_2O_2 content ↓; $•O_2^-$ content ↓	
	Cucumis sativus L. cv. 4200	5 and 7 μM (hydroponic)	Selenite: 2–6 mg L^{-1}	6 weeks	root	root growth ↑ but ↓ at 4–6 mg L^{-1} Se	Haghighi and da Silva (2016)
					shoot	Cd accumulation ↓	
	Helianthus annuus L.	20 μM (hydroponic)	Selenate: 5, 10 and 20 μM (pre-treatment for 24 h)	4 days	root	root growth ↑; Cd accumulation ↓ and ↓ ns;	Saidi et al. (2014)
					leaves	leaf biomass ↑; Cd accumulation ↓ and ↓ ns; Chl and carotenoid content ↑ but ↓ at 20 μM Se; LP ↓; H_2O_2 content ↓; GSH and AsA content ↓; SOD activity ↓; POD activity ↓; CAT and APX activity ↑; GR activity ↑	
	Lepidium sativum cv. Ogrodowa	0.5–5.0 mg L^{-1} (hydroponic)	Selenite: 0.2–2.0 mg L^{-1}	14 days	whole plant	Cd accumulation ↓	Elguera et al. (2013)
	Oryza sativa L. cv. Jiahua 1	1.03 ± 0.11 mg kg^{-1} (contaminated soil)	Selenite: 0.5–1 mg kg^{-1} (soil)	6 months	root	Cd accumulation ↓	Hu et al. (2014)
					shoot	Cd accumulation ↓	
					grain	Cd accumulation ↓	
	Pisum sativum L. cv. Fenomen	5 μM (hydroponic)	Selenite or selenate: 5 μM	21 days	root	Cd uptake ↑ ns and ↑	Landberg and Greger (1994)
					shoot	Cd uptake ↑ ns and ↑	
	Sinapis alba L.	6 mg L^{-1} (hydroponic)	Selenium dioxide: 3 mg L^{-1}	8 days	root	Cd accumulation ↑	Fargašová et al., (2006)

(Continued)

TABLE 16.1 (CONTINUED)
The Effect of Se Supply in Heavy Metal-Stressed Plant Species

Name of HM	Plant Name	Concentration or degree of the HM Stress	Chemical form and Concentration of Se Applied	Time of Exposure	Plant organ Investigated	Effects of Se Addition at Low Doses*	References
					shoot	Cd accumulation ↓; Chl and carotenoid content ↓	
	Solanum lycopersicon cv. K-21	150 mg L^{-1} (pot experiment)	Selenite: 10 μM	3 weeks	root	root growth ↑; Cd accumulation ↓	Alyemeni et al. (2018)
					shoot	shoot growth ↑; Cd accumulation ↓	
					top leaves	Cd accumulation ↓; LP ↓; H_2O_2 content ↓; SOD and CAT activity ↑; APX and GR activity ↑; Pro content ↑; Chl content ↑ and carotenoid content ↑ ns	
	Triticum aestivum L. cv. Sunny	16 μM (hydroponic)	Selenite or selenate: 16 μM	21 days	root	Cd accumulation ↑ ns for selenate but ↓ ns for selenite	Landberg and Greger (1994)
					shoot	Cd accumulation ↑ for selenate but ↓ for selenite	
Cr	*Brassica campestris* L. ssp. *Pekinensis*	1 mg L^{-1} (hydroponic)	Selenite: 1 mg L^{-1}	30 days	root	root growth ↑	Qing et al. (2015)
					shoot	shoot growth ↑; Cr accumulation ↑ ns	
					leaves	SOD and POD activity ↑; CAT activity ↑ ns; LP ↓; $\bullet O_2^-$ content ↓; Pro content ↓ ns	
	Iris pseudacorus L.	5 mg L^{-1} (hydroponic)	Selenite or selenate: 0.5 mg L^{-1} (pre-treatment)	2 weeks + 1 week	root	root growth ↑ ns; Cr uptake ↑ for selenite	Xu et al. (2017)
					shoot	shoot growth ↑ ns; Cr uptake ↓ for selenite	
	Spinacia oleracea L.	Cr (III) or Cr (VI): 2 and 5 mg L^{-1} (pot culture)	Selenite or selenate: 0.5–6.0 mg L^{-1}	60 days		Cr uptake ↓	Srivastava et al. (1998)
						Cr uptake ↓	

(*Continued*)

TABLE 16.1 (CONTINUED)
The Effect of Se Supply in Heavy Metal-Stressed Plant Species

Name of HM	Plant Name	Concentration or degree of the HM Stress	Chemical form and Concentration of Se Applied	Time of Exposure	Plant organ Investigated	Effects of Se Addition at Low Doses*	References
Hg	*Allium sativum* L.	0.01–100 mg L^{-1} (hydroponic)	Selenite or selenate: 0.01–100 mg L^{-1}	28 days	root	Hg uptake ↑ at lower Hg conc. and 10–100 mg L^{-1} Se, but ↓ at 100 mg L^{-1} Hg and 1–100 mg L^{-1} Se	Zhao et al. (2013)
					stem (bulb)	Hg uptake ↑ at 10–100 mg L^{-1} Se	
					leaves	Hg uptake ↑ at lower Hg conc. and 100 mg L^{-1} Se, but ↓ at 100 mg L^{-1} Hg and 1–100 mg L^{-1} Se	
	Oryza sativa L.cv. *japonica* ('Zixiang')	5 μM (hydroponic)	Selenite: 1 and 10 μM	14 days	root	Hg uptake ↓	Wang et al. (2014)
					shoot	Hg uptake ↓	
		46 mg kg^{-1} (contaminated soil)	Selenite: 1 and 5 mg kg^{-1} (soil)	90 days	brown rice	Hg accumulation ↓ ns and ↓	
	Oryza sativa L.cv. *indica* ('Nanfeng')	5 μM (hydroponic)	Selenite: 1 and 10 μM	14 days	root	Hg uptake ↓	
					shoot	Hg uptake ↓ ns and ↓	
		46 mg kg^{-1} (contaminated soil)	Selenite: 1 and 5 mg kg^{-1} (soil)	90 days	brown rice	Hg accumulation ↓	
	Oryza sativa L.	2.5 μM (hydroponic)	Selenite: 2.5 μM	21 days	root	Hg accumulation ↓	Li et al. (2018)
Pb	*Brassica napus* L. cv. Wanyou 18	300 and 500 mg kg^{-1} (soil)	Selenite: 1–20 mg kg^{-1} (soil)	40–100 days	root	Pb accumulation ↓	Wu et al. (2016)
					shoot	Pb accumulation ↓; GSH-Px activity ↑; SOD activity ↑; LP ↓; H_2O_2 content ↓; $O_2\bullet^-$ content ↓	
	Oryza sativa L. cv. Jiahua 1	98.1 ± 6.47 mg kg^{-1} (contaminated soil)	Selenite: 0.5–1 mg kg^{-1} (soil)	6 months	root	Pb accumulation ↓ ns	Hu et al. (2014)
					shoot	Pb accumulation ↓	

(*Continued*)

TABLE 16.1 (CONTINUED)
The Effect of Se Supply in Heavy Metal-Stressed Plant Species

Name of HM	Plant Name	Concentration or degree of the HM Stress	Chemical form and Concentration of Se Applied	Time of Exposure	Plant organ Investigated	Effects of Se Addition at Low Doses*	References
					grain	Pb accumulation ↓ (in husk) and no change (in brown rice)	
	Sinapis alba L.	100 mg L^{-1} (hydroponic)	Selenium dioxide: 3 mg L^{-1}	8 days	root	Pb accumulation ↓	Fargašová et al. (2006)
					shoot	Pb accumulation ↓; Chl and carotenoid content ↓	
	Triticum aestivum L. cv. Triso	50 and 100 mg kg^{-1} (soil)	Selenate: 0.4 and 0.8 mg kg^{-1} (soil)	14 days	root	root growth ↑ at lower but ↓ at higher Pb conc.	Balakhnina and Nadezhkina (2017)
					leaves	shoot growth and Chl content ↑ at lower but ↓ at higher Pb conc.; LP ↓ at Pb100; APX activity ↑ but ↓ at Pb100; GR activity ↓ at Pb100; POD activity ↑ at Pb50 but ↓ at Pb100	
	Vicia faba L. *minor* cv. Nadwiślański	50 μM (hydroponic)	Selenite: 1.5 and 6 μM	14 days	root	root growth ↑ at 6 μM Se; Pb accumulation ↓ ns; LP ↓ ns but ↑ at 6 μM Se; total thiol content ↑; CAT activity ↓ ns but ↑ ns at 6 μM Se; POD activity ↑; GSH-Px activity ↑ at 6 μM Se	Mroczek-Zdyrska and Wójcik (2012)
As	*Phaseolus aureus* Roxb.	2.5–10 μM (hydroponic)	Selenium: 2.5 and 5 μM	10 days	root	root growth ↑	Malik et al. (2012)
					shoot	shoot growth ↑; As uptake ↓; electrolyte leakage ↓; chlorophyll content ↑ ns and ↑; LP ↓; H_2O_2 content ↓; SOD activity ↓ and ↑; CAT activity ↑ns and ↑; APX activity ↑ns and ↑; GR activity ↓ ns and ↑; Asc and GSH content ↑; GST activity ↑	
	Oryza sativa L. cv. Minaksi	60 μM (hydroponic)	Selenate: 10 and 20 μM	7 days	root	As accumulation ↓	Singh et al. (2018)

(*Continued*)

TABLE 16.1 (CONTINUED)
The Effect of Se Supply in Heavy Metal-Stressed Plant Species

Name of HM	Plant Name	Concentration or degree of the HM Stress	Chemical form and Concentration of Se Applied	Time of Exposure	Plant organ Investigated	Effects of Se Addition at Low Doses*	References
					shoot/leaves	shoot growth ↑ ns; As uptake ↓ and ↓ ns; LP ↓; H_2O_2 content ↓; SOD and CAT activity ↓; APX activity ↓; Asc content ↓	
Al	*Lolium perenne* L. cv. Nui	0.2 mM (hydroponic)	Selenite: 1–10 µM	20 days	root	Al accumulation ↑; LP ↓ but ↑ at 5–10 µM Se; POD activity ↑ but ↓ at 5–10 µM Se; APX activity ↓ but ↑ at 5 µM Se; SOD activity ↓ but ↑ at 5–10 µM Se	Cartes et al. (2010)
					shoot	Al accumulation ↑ at 10 µM Se; LP ↓ but ↑ at 10 µM Se; POD activity ↑; APX activity ↓ but ↑ at 5–10 µM Se; SOD activity ↓ but ↑ at 10 µM Se	

* ↑ indicates significant and ↑ ns indicates non-significant increase, while ↓ refers to significant decrease and ↓ ns to non-significant reduction.

Se, either Se-pretreatment seems to be protective against HM-induced oxidative stress in plants. Another general effect of most HMs is the induction of chlorophyll degradation and the consequent decline of photosynthesis (Myśliwa-Kurdziel et al., 2004). Albeit, application of Se in Cu or Zn stressed plants resulted in a decrease of chlorophyll (Chl) and carotenoid content in the shoots (Fargašová et al., 2006; Landberg and Greger, 1994); Se seems to have a beneficial effect in this aspect in plants exposed to Cd or As (Malik et al., 2012; Tang et al., 2015).

Regarding HM uptake and accumulation, Se application at low doses can be advantageous for the plant, although there are some contradicting data. Namely, generally the presence of Se prevented both the underground and the aboveground plant parts from the uptake/accumulation of essential HM Cu, Zn in pea and white mustard (Landberg and Greger, 1994) or toxic HM Cd in sunflower and rapeseed (Saidi et al., 2014; Wu et al., 2016), Pb in white mustard and rice (Fargašová et al., 2006; Hu et al., 2014), Hg in rice (Li et al., 2018; Wang et al., 2014), and As (Malik et al., 2012). At the same time, despite the application of Se, for example, intense Ni accumulation was observed in lettuce (Hawrylak et al., 2007), elevated Cd uptake was found in pea and white mustard (Fargašová et al., 2006; Landberg and Greger, 1994), and Al accumulated in the roots of ryegrass (Cartes et al., 2010).

From the available literature data (presented in Table 16.1), it can be seen that low Se doses primarily protect plants from the oxidative damages induced by several different (essential or non-essential) HMs. Moreover, Se application was found to be beneficiary in reducing uptake and accumulation of some HMs, which is not a general rather it is a HM-specific process. However, in this case, molecular mechanisms of Se effect on HM transport have not been examined.

16.4.4 Se and Temperature Stresses, High Light Stress

Damages induced by low temperatures are considered to be a major threat that limits vegetative (Yadav, 2010) and reproductive (Thakur et al., 2010) growth, as well as the distribution of plants, especially in sub-tropical and temperate regions. Principally, cold affects the lipid composition and consequently the fluidity of membranes, but no related Se effects have been described yet. Suboptimal temperatures are known to decrease the biosynthesis of photosynthetic pigments; however, selenium was able to ameliorate chlorophyll loss in cold-treated wheat and *Sorghum* plants (Abbas, 2012; Chu et al., 2010) thus possibly improving photosynthetic capacity and carbohydrate metabolism. The higher level of free proline in cold-stressed plants has been suggested as a factor that improves chilling tolerance; however, it can also be considered as a symptom of injury. Anyway, low Se doses intensified proline accumulation in cold-stressed cucumber and sorghum (Abbas, 2012; Hawrylak-Nowak et al., 2010); although its connection with improved chilling tolerance has not been confirmed. The indirect effect of cold on plant physiology is realized through the induction of secondary oxidative stress involving ROS overproduction and lipid peroxidation. Selenium application reportedly enhanced the contents of antioxidant compounds such as anthocyanins, carotenoids, flavonoids, phenolic compounds, ascorbate, and antioxidant enzyme activities like APX, POX, and CAT (Abbas, 2012; Chu et al., 2010), suggesting that both enzymatic and non-enzymatic antioxidants are Se-responsive and contribute to the alleviating effect of Se on chilling-induced oxidative stress. It shades the picture that Se-increased chlorophyll synthesis was not associated with enhanced biomass production in wheat (Chu et al., 2010), or in Se-treated cold-stressed cucumber reduced lipid peroxidation, or proline accumulation was not accompanied by biomass gain (Hawrylak-Nowak et al., 2010). In order to clarify these relationships, as well as to explore molecular mechanisms of Se-triggered cold endurance, further experiments are needed.

Supraoptimal temperature is also a limiting stress factor for plants affecting membrane fluidity, ROS production, chlorophyll biosynthesis, and photosynthetic capacity. Photosynthesis is greatly sensitive to high temperature and is often inhibited before other cell functions are impaired (Mathur et al., 2014). Application of selenium as a foliar spray (75 mg L^{-1}) increased chlorophyll content, photosynthetic rate, stomatal conductance, and transpiration rate in heat-stressed (~40°C) wheat (Djanaguiraman et al., 2010). Furthermore, Se successfully diminished the heat-induced accumulation of H_2O_2 and superoxide while enhancing the activity of SOD, CAT, and POD enzymes. Similarly, antioxidant enzymes like CAT, APX, and non-enzymatic components like carotenoids, anthocyanins, and ascorbate were induced and H_2O_2 and MDA contents were decreased by Se treatment (2 and 4 mg L^{-1}) in high temperature-exposed wheat cultivars (38±2°C) suggesting that Se prevents heat-induced oxidative stress thus delaying premature leaf senescence (Iqbal et al., 2015). Low Se doses had very similar effects on heat-tolerant and sensitive wheat cultivars (Iqbal et al., 2015) indicating that the heat stress ameliorating the effect of Se is independent of genetically determined stress resistance.

The high temperature is often accompanied by enhanced light exposure, damaging photosynthetic pigments and photosystem II (Powles, 1984). Se supplementation (0.07 or 0.3 mg L^{-1}) to potato improved the recovery of chlorophyll content following light stress suggesting that Se can improve adaptation to photoinhibitory light intensities (Seppanen et al., 2003). The high light-triggered reduction of Fv/Fm in Se-treated plants was milder and their photosynthesis was slightly more tolerant to paraquat compared with non-treated plants. Interestingly, Se treatments decreased transcript accumulation of chloroplast CuZnSOD and GPOX in high light-stressed plants, which has not been examined in detail yet. In contrast, the MnSOD and *psbA* transcript levels were unresponsive to Se application. The lack of Se effect on Dl protein coding *psbA* in spite of the improved recovery from light stress may be explained by the fact that the expression of chloroplast-coded genes is mainly regulated at the post-transcriptional and translational levels (Seppanen et al., 2003).

16.4.5 Se and UV-B Radiation

Increased ultraviolet-B (UV-B) radiation due to the depletion of the stratospheric ozone layer in the atmosphere negatively affects plants since it causes ROS production, DNA damage and decline in photosynthetic activities, reduces growth and development, and alters plant morphology (Robson et al., 2015). During a field experiment using UV-B± Se treated Tatary buckwheat, a slight ameliorating effect of Se on underground biomass production could be observed, while other morphological parameters such as branching, node number, and petiole length were negatively affected by Se addition (Breznik et al., 2004). According to the early results of Pennanen et al. (2001), UV-B treatment induced the formation of a thicker epidermal layer in lettuce leaves and this event was more pronounced in plants treated with Se. The beneficiary effect of Se was more apparent in the aging leaves where it stabilized Fv/Fm values and prevented the increase of F0 parameter reflecting slighter UV-B-induced photoinhibitory damage. Furthermore, the protective effect of Se was found to be related to increased GPOX activities (Pennanen et al., 2001). In flagellate *Euglena gracilis*, photosynthetic rate and light-enhanced dark respiration increased upon UV-B radiation, while these parameters significantly decreased in the presence of Se indicating that Se may regulate repair processes during UV-B stress (Ekelund and Danilov, 2001). In wheat seedlings, Se application (1 mg kg^{-1} selenite) alleviated UV-B-induced reduction of biomass, chlorophyll, and antioxidant (proline, flavonoids, phenolic compounds) contents (Yao et al., 2010a, b, 2011, 2013). In the same experimental system, low Se dose significantly increased POD, SOD, and CAT activities, maintaining the superoxide level and MDA content; consequently the rate of oxidative stress was low. At the ultrastructural level, the enlargement of peroxisomes in barley cells was observed suggesting the activation of antioxidative enzymes, possibly catalase, in response to combined treatments of UV-B and Se (Valkama et al., 2003). In the case of selenium supplemented ryegrass and lettuce, 0.1 mg kg^{-1} Se exerted positive effects on plant growth and biomass production only in combination with UV-B stress indicating that UV light may act as a trigger for the growth-promoting effect of Se. Furthermore, the toxic effects of higher Se dosage were alleviated by the presence of UV-B radiation which suggests synergistic action of Se and UV light on plants (Hartikainen et al., 2000; Xue et al., 2001). In a recent study of Golob et al. (2017), it was found that Se applied in UV-B-stressed wheat leaves strengthens the protection of plants against stress by increasing cuticle thickness and stomatal density, enhancing the amount of UV absorbing compounds, light reflectance, and transmittance. These data provide new evidence for the inducing effect of Se on secondary plant metabolism; however, further experiments are needed to substantiate these effects.

16.5 CONCLUSIONS AND FUTURE PERSPECTIVES

As depicted in Figure 16.3, the major common effects of diverse abiotic stresses are secondary oxidative stress and photosynthetic damage. Selenium at low concentrations is undoubtedly an antioxidant that effectively counteracts the prooxidant effects of salinity, drought, heavy metals, extreme temperatures, high light, and UV-B radiation. Experimental data shows that the antioxidant effect of Se is achieved by the induction of antioxidant capacity; however, we have little knowledge about the effect of Se at the molecular level. Similarly, the effect of Se on stress-specific processes like ion uptake and accumulation are barely known.

Despite the huge amount of valuable experimental work, we feel that the number of experiments attempting to understand the nature of processes is small. Therefore, we propose an approach that not just describes but also explains the beneficial effects of Se. Possibly, the future identification of Se-responsive genes and proteins could contribute to the better understanding of the molecular effects of low Se doses.

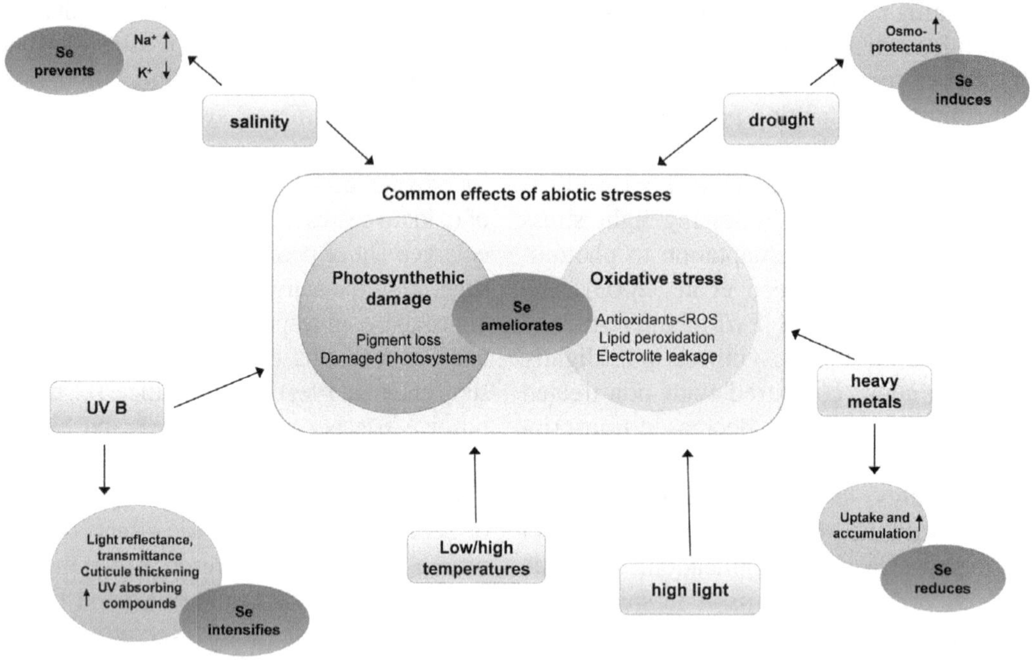

FIGURE 16.3 Schematic representation of the effects of Se on common and specific processes of abiotic stresses discussed in the chapter. Stress-specific processes are indicated by gray circles.

ACKNOWLEDGMENTS

This work was supported by the János Bolyai Research Scholarship of the Hungarian Academy of Sciences (Grant no. BO/00751/16/8) by the National Research, Development and Innovation Fund (Grant no. NKFI-6, K120383 and PD120962), and by the EU-funded Hungarian grant EFOP-3.6.116-2016-00008. Z.K. was supported by UNKP-17-4 New National Excellence Program of the Ministry of Human Capacities.

REFERENCES

Abbas, S.M. (2012) Effects of low temperature and selenium application on growth and the physiological changes in sorghum seedlings. *Journal of Stress Physiology and Biochemistry* 8: 268–286.

Ahmad, P., Prasad, M.N.V. (2012) *Abiotic Stress Responses in Plants: Metabolism, Productivity and Sustainability*. New York, NY: Springer.

Ahmad, P., Umar, S. (2011) *Antioxidants: Oxidative Stress Management in Plants*. New Delhi, India: Studium Press.

Ahmad, R., Waraich, E.A., Nawaz, F., Ashraf, M.Y., Khalid, M. (2016) Selenium (Se) improves drought tolerance in crop plants–a myth or fact? *Journal of the Science of Food and Agriculture* 96: 372–380.

Alyemeni, M.N., Ahanger, M.A., Wijaya, L., Alam, P., Bhardwaj, R., Ahmad, P. (2018) Selenium mitigates cadmium-induced oxidative stress in tomato (*Solanum lycopersicum* L.) plants by modulating chlorophyll fluorescence, osmolyte accumulation, and antioxidant system. *Protoplasma* 255: 459–469.

Anjum, N.A., Umar, S., Iqbal, M., Khan, N.A. (2011) Cadmium causes oxidative stress in mung bean by affecting the antioxidant enzyme system and ascorbate-glutathione cycle metabolism. *Russian Journal of Plant Physiology* 58: 92–99.

Astaneh, R.K., Bolandnazar, S., Nahandi, F.Z., Oustan, S. (2018) The effects of selenium on some physiological traits and K, Na concentration of garlic (*Allium sativum* L.) under NaCl stress. *Information Processing in Agriculture* 5: 156–161.

Balakhnina, T.I., Nadezhkina, E.S. (2017) Effect of selenium on growth and antioxidant capacity of *Triticum aestivum* L. during development of lead-induced oxidative stress. *Russian Journal of Plant Physiology* 64: 215–223.

Barros, R.S., Freitas, A.W.P. (2001) Selenomethionine as a dormancy-breaking agent in seeds of *Stylosanthes humilis*. *Acta Physiologiae Plantarum* 23: 279–284.

Bhatnagar-Mathur, P., Vadez, V., Sharma, K.K. (2008) Transgenic approaches for abiotic stress tolerance in plants: Retrospect and prospects. *Plant Cell Reports* 27: 411–424.

Boldrin, P.F., de Figueiredo, M.A., Yang, Y., Luo, H., Giri, S., Hart, J.J., Faquin, V., et al. (2016) Selenium promotes sulfur accumulation and plant growth in wheat (*Triticum aestivum*). *Physiologia Plantarum* 158: 80–91.

Breznik, B., Germ, M., Gaberscik, A., Kreft, I. (2004) The combined effects of elevated UV-B radiation and selenium on Tartary buckwheat (*Fagopyrum tataricum*) habitus. *Fagopyrum* 21: 59–64.

Cartes, P., Jara, A.A., Pinilla, L., Rosas, A., Mora, M.L. (2010) Selenium improves the antioxidant ability against aluminium-induced oxidative stress in ryegrass roots. *Annals of Applied Biology* 156: 297–307.

Chen, C.C., Sung, J.M. (2001) Priming bitter gourd seeds with selenium solution enhances germinability and antioxidative responses under sub-optimal temperature. *Physiologia Plantarum* 111: 9–16.

Chu, J., Yao, X., Zhang, Z. (2010) Responses of wheat seedlings to exogenous selenium supply under cold stress. *Biological Trace Element Research* 136: 355–363.

Das, K., Roychoudhury, A. (2014) Reactive oxygen species (ROS) and response of antioxidants as ROS-scavengers during environmental stress in plants. *Frontiers in Environmental Science* 2: 53.

Diao, M., Ma, L., Wang, J., Cui, J., Fu, A., Liu, H.Y. (2014) Selenium promotes the growth and photosynthesis of tomato seedlings under salt stress by enhancing chloroplast antioxidant defense system. *Journal of Plant Growth Regulation* 33: 671–682.

Djanaguiraman, M., Devi, D.D., Shanker, A.K., Sheeba, J.A., Bangarusamy, U. (2004) Impact of selenium spray on monocarpic senescence of soybean (*Glycine max* L.). *Journal of Food, Agriculture and Environment* 2: 44–47.

Djanaguiraman, M., Devi, D.D., Shanker, A.K., Sheeba, J.A., Bangarusamy, U. (2005) Selenium – An antioxidative protectant in soybean during senescence. *Plant and Soil* 272: 77–86.

Djanaguiraman, M., Prasad, P.V.V., Seppanen, M. (2010) Selenium protects sorghum leaves from oxidative damage under high temperature stress by enhancing antioxidant defense system. *Plant Physiology and Biochemistry: PPB* 48: 999–1007.

Domokos-Szabolcsy, É., Alshaal, T., Elhawat, N., Abdalla, N., Reis, A., El-Ramady, H. (2017) The interactions between selenium, nutrients and heavy metals in higher plants under abiotic stresses. *Environment Biodiversity and Soil Security* 1: 5–31.

Ekelund, N.G.A., Danilov, R.A. (2001) The influence of selenium on photosynthesis and "light-enhanced dark respiration" (LEDR) in the flagellate *Euglena gracilis* after exposure to ultraviolet radiation. *Aquatic Sciences* 63: 457–465.

Elguera, J.C.T., Barrientos, E.Y., Wrobel, K., Wrobel, K. (2013) Effect of cadmium (Cd (II)), selenium (Se (IV)) and their mixtures on phenolic compounds and antioxidant capacity in Lepidium sativum. *Acta Physiologiae Plantarum* 35: 431–441.

Emam, M.M., Khattab, H.E., Helal, N.M., Deraz, A.E. (2014) Effect of selenium and silicon on yield quality of rice plant grown under drought stress. *Australian Journal of Crop Science* 8: 596.

Emamverdian, A., Ding, Y., Mokhberdoran, F., Xie, Y. (2015) Heavy metal stress and some mechanisms of plant defense response. *The Scientific World Journal* 2015: 756120.

Fargašová, A., Pastierová, J., Svetková, K. (2006) Effect of Se-metal pair combinations (Cd, Zn, Cu, Pb) on photosynthetic pigments production and metal accumulation in *Sinapis alba* L. seedlings. *Plant, Soil and Environment* 52: 8–15.

Feng, R., Liao, G., Guo, J., Wang, R., Xu, Y., Ding, Y., Mo, L., Fan, Z., Li, N. (2016) Responses of root growth and antioxidative systems of paddy rice exposed to antimony and selenium. *Environmental and Experimental Botany* 122: 29–38.

Feng, R., Wei, C., Tu, S. (2013) The roles of selenium in protecting plants against abiotic stresses. *Environmental and Experimental Botany* 87: 58–68.

Filek, M., Keskinen, R., Hartikainen, H., Szarejko, I., Janiak, A., Miszalski, Z., Golda, A. (2008) The protective role of selenium in rape seedlings subjected to cadmium stress. *Journal of Plant Physiology* 165: 833–844.

Filek, M., Łabanowska, M., Kościelniak, J., Biesaga-Kościelniak, J., Kurdziel, M., Szarejko, I., Hartikainen, H. (2015) Characterization of barley leaf tolerance to drought stress by chlorophyll fluorescence and electron paramagnetic resonance studies. *Journal of Agronomy and Crop Science* 201: 228–240.

Golob, A., Kavčič, J., Stibilj, V., Gaberščik, A., Vogel-Mikuš, K., Germ, M. (2017) The effect of selenium and UV radiation on leaf traits and biomass production in *Triticum aestivum* L. *Ecotoxicology and Environmental Safety* 136: 142–149.

Grzesiak, M., Filek, M., Barbasz, A., Kreczmer, B., Hartikainen, H. (2013) Relationships between polyamines, ethylene, osmoprotectants and antioxidant enzymes activities in wheat seedlings after short-term PEG-and NaCl-induced stresses. *Plant Growth Regulation* 69: 177–189.

Gul, H., Kinza, S., Shinwari, Z.K., Hamayun, M. (2017) Effect of selenium on the biochemistry of *Zea mays* under salt stress. *Pakistan Journal of Botany* 49: 25–32.

Gupta, B., Huang, B. (2014) Mechanism of salinity tolerance in plants: physiological, biochemical, and molecular characterization. International Journal of Genomics 2014: 701596.

Habibi, G. (2013) Effect of drought stress and selenium spraying on photosynthesis and antioxidant activity of spring barley. *Acta Agriculturae Slovenica* 101: 31–39.

Habibi, G. (2014) Role of trace elements in alleviating environmental stress. In *Emerging Technologies and Management of Crop Stress Tolerance Biological Techniques*, eds. Ahmad, P Rasool, S., 313–331, Oxford: Elsevier.

Habibi, G. (2017) Selenium ameliorates salinity stress in Petroselinum crispum by modulation of photosynthesis and by reducing shoot Na accumulation. *Russian Journal of Plant Physiology* 64: 368–374.

Habibi, G., Alizade, Z. (2017) Selenium in lemon balm plants: Productivity, phytotoxicity and drought alleviation. *Journal of Plant Nutrition* 40: 1557–1568.

Habibi, G., Sarvary, S. (2015) The roles of selenium in protecting lemon balm against salt stress. Iranian. *Journal of Plant Physiology* 5: 1425–1433.

Haghighi, M., da Silva, J.A.T. (2016) Influence of selenium on cadmium toxicity in cucumber (*Cucumis sativus* cv. 4200) at an early growth stage in a hydroponic system. *Communications in Soil Science and Plant Analysis* 47: 142–155.

Hajiboland, R., Keivanfar, N. (2012) Selenium supplementation stimulates vegetative and reproductive growth in canola (*Brassica napus* L.) plants. *Acta Agriculturae Slovenica* 99: 13–19.

Han, D., Li, X., Xiong, S., Tu, S., Chen, Z., Li, J., Xie, Z. (2013) Selenium uptake, speciation, and stress response of *Nicotiana tabacum* L. *Environmental and Experimental Botany* 95: 6–14.

Hartikainen, H., Xue, T., Piironen, V. (2000) Selenium as an anti-oxidant and pro-oxidant in ryegrass. *Plant and Soil* 225: 193–200.

Hasanuzzaman, M., Hossain, M.A., da Silva, J.A.T., Fujita, M. (2012) Plant responses and tolerance to abiotic oxidative stress: Antioxidant defense is a key factor. In *Crop Stress and Its Management: Perspectives and Strategies*, eds. Bandi, V, Shanker, AK, Shanker, C, Mandapaka, M., 261–316. Berlin: Springer.

Hasanuzzaman, M., Hossain, M.A., Fujita, M. (2011) Selenium-induced up-regulation of the antioxidant defense and methylglyoxal detoxification system reduces salinity-induced damage in rapeseed seedlings. *Biological Trace Element Research* 143: 1704–1721.

Hasanuzzaman, M., Nahar, K., Fujita, M. (2013) Plant response to salt stress and role of exogenous protectants to mitigate salt-induced damages. In *Ecophysiology and Responses of Plants under Salt Stress*, eds. Ahmad P, Azooz M, Prasad M., 25–86. New York: Springer.

Hasegawa, P.M., Bressan, R.A., Zhu, J.K., Bohnert, H.J. (2000) Plant cellular and molecular responses to high salinity. *Annual Review of Plant Physiology and Plant Molecular Biology* 51: 463–499.

Hawrylak, B., Matraszek, R., Szymańska, M. (2007) Response of lettuce (*Lactuca sativa* L.) to selenium in nutrient solution contaminated with nickel. *Vegetable Crops Research Bulletin* 67: 63–70.

Hawrylak-Nowak, B. (2009) Beneficial effects of exogenous selenium in cucumber seedlings subjected to salt stress. *Biological Trace Element Research* 132: 259–269.

Hawrylak-Nowak, B. (2015) Selenite is more efficient than selenate in alleviation of salt stress in lettuce plants. *Acta Biologica Cracoviensia s. Botanica* 57: 49–54.

Hawrylak-Nowak, B., Matraszek, R., Szymańska, M. (2010) Selenium modifies the effect of short-term chilling stress on cucumber plants. *Biological Trace Element Research* 138: 307–315.

Hu, Y., Norton, G.J., Duan, G., Huang, Y., Liu, Y. (2014) Effect of selenium fertilization on the accumulation of cadmium and lead in rice plants. *Plant and Soil* 384: 131–140.

Huang, W., Yang, X., Yao, S., LwinOo, T., He, H., Wang, A., Li, C., He, L. (2014) Reactive oxygen species burst induced by aluminum stress triggers mitochondria-dependent programmed cell death in peanut root tip cells. *Plant Physiology and Biochemistry: PPB* 82: 76–84.

Ibrahim, H.M. (2014) Selenium pretreatment regulates the antioxidant defense system and reduces oxidative stress on drought—Stressed wheat (*Triticum aestivum* L.) plants. *Asian Journal of Plant Sciences* 13: 120–128.

Iqbal, M., Hussain, I., Liaqat, H., Ashraf, M.A., Rasheed, R., Rehman, A.U. (2015) Exogenously applied selenium reduces oxidative stress and induces heat tolerance in spring wheat. *Plant Physiology and Biochemistry: PPB* 94: 95–103.

Janas, K.M., Zielińska-Tomaszewska, J., Rybaczek, D., Maszewski, J., Posmyk,M.M., Amarowicz, R., Kosińska, A. (2010) The impact of copper ions on growth, lipid peroxidation, and phenolic compound accumulation and localization in lentil (*Lens culinaris* Medic.) seedlings. *Journal of Plant Physiology* 167: 270–276.

Jiang, C., Zu, C., Lu, D., Zheng, Q., Shen, J., Wang, H., Li, D. (2017) Effect of exogenous selenium supply on photosynthesis, Na+ accumulation and antioxidative capacity of maize (*Zea mays* L.) under salinity stress. *Scientific Reports* 7: 42039.

Johnson, C.C., Fordyce, F.M., Rayman, M.P. (2010) Symposium on 'Geographical and geological influences on nutrition' factors controlling the distribution of selenium in the environment and their impact on health and nutrition. *The Proceedings of the Nutrition Society* 69: 119–132.

Khaliq, A., Aslam, F., Matloob, A., Hussain, S., Geng, M., Wahid, A., Rehman, H. (2015) Seed priming with selenium: Consequences for emergence, seedling growth, and biochemical attributes of rice. *Biological Trace Element Research* 166: 236–244.

Kong, L.G., Wang, M., Bi, D.L. (2005) Selenium modulates the activities of antioxidant enzymes, osmotic homeostasis and promotes the growth of sorrel seedlings under salt stress. *Plant Growth Regulation* 45: 155–163.

Kronzucker, H.J., Britto, D.T. (2011) Sodium transport in plants: A critical review. *The New Phytologist* 189: 54–81.

Kuznetsov, V.V., Kholodova, V.P., Kuznetsov, V.V., Yagodin, B.A. (2003) Selenium regulates the water status of plants exposed to drought. *The Journal of Biological Sciences* 390: 266–268.

Landberg, T., Greger, M. (1994) Influence of selenium on uptake and toxicity of copper and cadmium in pea (*Pisum sativum*) and wheat (*Triticum aestivum*). *Physiologia Plantarum* 90: 637–644.

Lehotai, N., Feigl, G., Koós, Á., Molnár, Á., Ördög, A., Pető, A., Erdei, L., Kolbert, Zs. (2016a) Nitric oxide-cytokinin interplay influences selenite sensitivity in *Arabidopsis*. *Plant Cell Reports* 35: 2181–2195.

Lehotai, N., Lyubenova, L., Schröder, P., Feigl, G., Ördög, A., Szilágyi, K., Erdei, L., Kolbert, Z. (2016b) Nitro-oxidative stress contributes to selenite toxicity in pea (*Pisum sativum* L.). *Plant and Soil* 400: 107–122.

Leshem, Y.Y. (1988) Plant senescence processes and free radicals. *Free Radical Biology & Medicine* 5: 39–49.

Lesk, C., Rowhani, P., Ramankutty, N. (2016) Influence of extreme weather disasters on global crop production. *Nature* 529: 84–87.

Levine, V.E. (1925) The effect of selenium compounds upon growth and germination in plants. *American Journal of Botany* 12: 82–90.

Li, Y., Li, H., Li, Y.F., Zhao, J., Guo, J., Wang, R., Li, B., Zhang, Z., Gao, Y. (2018) Evidence for molecular antagonistic mechanism between mercury and selenium in rice (*Oryza sativa* L.): A combined study using 1,2-dimensional electrophoresis and SR-XRF techniques. *Journal of Trace Elements in Medicine and Biology* 50: 435–440.

Li, H.F., McGrath, S.P., Zhao, F.J. (2008) Selenium uptake, translocation and speciation in wheat supplied with selenate or selenite. *The New Phytologist* 178: 92–102.

Li, X., Yang, Y., Jia, L., Chen, H., Wei, X. (2013) Zinc-induced oxidative damage, antioxidant enzyme response and proline metabolism in roots and leaves of wheat plants. *Ecotoxicology and Environmental Safety* 89: 150–157.

Linchuan, Z., Binggao, Y. (1996) Regulation of selenium in the senescence of corn leaves. *Journal of Nanjing Agricultural University* 19: 22–25.

Lintschinger, J., Fuchs, N., Moser, J., Kuehnelt, D., Goessler, W. (2000) Selenium-enriched sprouts. A raw material for fortified cereal-based diets. *Journal of Agricultural and Food Chemistry* 48: 5362–5368.

Lyons, G.H., Genc, Y., Soole, K., Stangoulis, J.C.R., Liu, F., Graham, R.D. (2009) Selenium increases seed production in Brassica. *Plant and Soil* 318: 73–80.

Maathuis, F.J.M., Ahmad, I., Patishtan, J. (2014) Regulation of Na+ fluxes in plants. *Frontiers in Plant Science* 5: 467.

Mahajan, S., Tuteja, N. (2005) Cold, salinity and drought stresses: An overview. *Archives of Biochemistry and Biophysics* 444: 139–158.

Maksymiec, W. (2007) Signaling responses in plants to heavy metal stress. *Acta Physiologiae Plantarum* 29: 177–187.

Malik, J.A., Goel, S., Kaur, N., Sharma, S., Singh, I., Nayyar, H. (2012) Selenium antagonises the toxic effects of arsenic on mungbean (*Phaseolus aureus* Roxb.) plants by restricting its uptake and enhancing the antioxidative and detoxification mechanisms. *Environmental and Experimental Botany* 77: 242–248.

Malik, J.A., Kumar, S., Thakur, P., Sharma, S., Kaur, N., Kaur, R., Pathania, D.S., *et al.* (2011) Promotion of growth in mungbean (*Phaseolus aureus* Roxb.) by selenium is associated with stimulation of carbohydrate metabolism. *Biological Trace Element Research* 143: 530–539.

Manaf, H.H. (2016) Beneficial effects of exogenous selenium, glycine betaine and seaweed extract on salt stressed cowpea plant. *Annals of Agricultural Sciences* 61: 41–48.

Martens, D.A. (2003) Selenium. In *Encyclopedia of Water Science*, eds. Stewart, B.A., Howell, T., 840–842. Basel: Marcel Decker Inc.

Mathur, S., Agrawal, D., Jajoo, A. (2014) Photosynthesis: Response to high temperature stress. *Journal of Photochemistry and Photobiology. B, Biology* 137: 116–126.

Matsumoto, H., Motoda, H. (2012) Aluminum toxicity recovery processes in root apices. Possible association with oxidative stress. *Plant Science: An International Journal of Experimental Plant Biology* 185: 1–8.

Molnár, Á., Feigl, G., Trifán, V., Ördög, A., Szőllősi, R., Erdei, L., Kolbert, Z. (2018a) The intensity of tyrosine nitration is associated with selenite and selenate toxicity in *Brassica juncea* L. *Ecotoxicology and Environmental Safety* 147: 93–101.

Molnár, Á., Kolbert, Zs., Kéri, K., Feigl, G., Ördög, A., Szőllősi, R., Erdei, L. (2018b) Selenite-induced nitro-oxidative stress processes in *Arabidopsis thaliana* and *Brassica juncea*. *Ecotoxicology and Environmental Safety* 148: 664–674.

Mozafariyan, M., Kamelmanesh, M.M., Hawrylak-Nowak, B. (2016) Ameliorative effect of selenium on tomato plants grown under salinity stress. *Archives of Agronomy and Soil Science* 62: 1368–1380.

Mroczek-Zdyrska, M., Wójcik, M. (2012) The influence of selenium on root growth and oxidative stress induced by lead in *Vicia faba* L. minor plants. *Biological Trace Element Research* 147: 320–328.

Munns, R., Tester, M. (2008) Mechanisms of salinity tolerance. *Annual Review of Plant Biology* 59: 651–681.

Myśliwa-Kurdziel, B., Prasad, M.N.V., Strzałtka, K. (2004) Photosynthesis in heavy metal stressed plants. In *Heavy Metal Stress in Plants*, ed. Prasad, M.N.V., 146–181. Berlin: Springer.

Nawaz, F., Ahmad, R., Ashraf, M.Y., Waraich, E.A., Khan, S.Z. (2015) Effect of selenium foliar spray on physiological and biochemical processes and chemical constituents of wheat under drought stress. *Ecotoxicology and Environmental Safety* 113: 191–200.

Nawaz, F., Ashraf, M.Y., Ahmad, R., Waraich, E.A., Shabbir, R.N., Hussain, R.A. (2017) Selenium supply methods and time of application influence spring wheat (*Triticum aestivum* L.) yield under water deficit conditions. *The Journal of Agricultural Science* 155: 643–656.

Nawaz, F., Naeem, M., Ashraf, M.Y., Tahir, M.N., Zulfiqar, B., Salahuddin, M., Shabbir, R.N., Aslam, M. (2016) Selenium supplementation affects physiological and biochemical processes to improve fodder yield and quality of maize (*Zea mays* L.) under water deficit conditions. *Frontiers in Plant Science* 7: 1438.

Peng, A., Xu, Y., Wang, Z.J. (2001) The effect of fulvic acid on the dose effect of selenite on the growth of wheat. *Biological Trace Element Research* 83: 275–279.

Pennanen, A., Hartikainen, H., Lukkari, K., Ollilainen, V. (2001) Acclimation of *Lactuca sativa* to increased UV irradiation at various selenium levels. *Journal of Photosynthesis Research Abstract of 12th Congress on Photosynthesis* 69: 30.

Pezzarossa, B., Malorgio, F., Tonutti, P. (1999) Effects of selenium uptake by tomato plants on senescence, fruit ripening and ethylene evolution. In *Biology and Biotechnology of the Plant Hormone Ethylene II*, eds. Kanellis, A.K., Chang, C., Klee, H., Bleecker, A.B., Pech, J.C., Grierson, D., 275–276. Dordrecht: Springer.

Pilon-Smits, E.A.H., Quinn, C.F. (2010) Selenium metabolism in plants. In *Cell Biology of Metals and Nutrients*, Plant Cell Monographs 17, eds. Hell, R., Mendell, R.R., 225–241. Berlin: Springer.

Pinheiro, F.J.A., Barros, R.S., Ribeiro, D.M., De Lana Souza, B.M., Coelho, T.G. (2008) Efficiency of selenium compounds in breaking dormancy of Townsville stylo seeds. *Seed Science and Technology* 36: 271–282.

Powles, S.B. (1984) Photoinhibition of photosynthesis induced by visible-light. *Annual Review of Plant Physiology and Plant Molecular Biology* 35: 15–44.

Proietti, P., Nasini, L., Del Buono, D., D'Amato, R., Tedeschini, E., Businelli, D. (2013) Selenium protects olive (*Olea europaea* L.) from drought stress. *Scientia Horticulturae* 164: 165–171.

Qing, X., Zhao, X., Hu, C., Wang, P., Zhang, Y., Zhang, X., Wang, P., et al. (2015) Selenium alleviates chromium toxicity by preventing oxidative stress in cabbage (*Brassica campestris* L. ssp. *Pekinensis*) leaves. *Ecotoxicology and Environmental Safety* 114: 179–189.

Rahmat, S., Hajiboland, R., Sadeghzade, N. (2017) Selenium delays leaf senescence in oilseed rape plants. *Photosynthetica* 55: 338–350.

Robson, T.M., Klem, K., Urban, O., Jansen, M.A.K. (2015) Re-interpreting plant morphological responses to UV-B radiation. *Plant, Cell and Environment* 38: 856–866.

Roy, S.J., Negrão, S., Tester, M. (2014) Salt resistant crop plants. *Current Opinion in Biotechnology* 26: 115–124.

Saha, U., Fayiga, A., Sonon, L. (2017) Selenium in the soil-plant environment: A review. *International Journal of Applied Agricultural Sciences* 3: 1–18.

Saidi, I., Chtourou, Y., Djebali, W. (2014) Selenium alleviates cadmium toxicity by preventing oxidative stress in sunflower (*Helianthus annuus*) seedlings. *Journal of Plant Physiology* 171: 85–91.

Schiavon, M., Pilon-Smits, E.A.H. (2017) The fascinating facets of plant selenium accumulation – Biochemistry, physiology, evolution and ecology. *The New Phytologist* 213: 1582–1596.

Seppanen, M., Turakainen, M., Hartikainen, H. (2003) Selenium effects on oxidative stress in potato. *Plant Science* 165: 311–319.

Shalaby, T., Bayoumi, Y., Alshaal, T., Elhawat, N., Sztrik, A., El-Ramady, H. (2017) Selenium fortification induces growth, antioxidant activity, yield and nutritional quality of lettuce in salt-affected soil using foliar and soil applications. *Plant and Soil* 421: 245–258.

Shekari, F., Abbasi, A., Mustafavi, S.H. (2017) Effect of silicon and selenium on enzymatic changes and productivity of dill in saline condition. *Journal of the Saudi Society of Agricultural Sciences* 16(4): 367–374.

Shrift, A., Ulrich, J.M. (1976) Transport of selenate and selenite into astragalus roots. *Plant Physiology* 44: 893–896.

Sieprawska, A., Kornaś, A., Filek, M. (2015) Involvement of selenium in protective mechanisms of plants under environmental stress conditions–review. *Acta Biologica Cracoviensia s. Botanica* 57: 9–20.

Singh, R., Upadhyay, A.K., Singh, D.P. (2018) Regulation of oxidative stress and mineral nutrient status by selenium in arsenic treated crop plant *Oryza sativa*. *Ecotoxicology and Environmental Safety* 148: 105–113.

Sors, T.G., Els, D.R., Salt, D.E. (2005) Selenium uptake, translocation, assimilation and metabolic fate in plants. *Photosynthesis Research* 86: 373–389.

Srivastava, S., Shanker, K., Srivastava, S., Shrivastav, R., Das, S., Prakash, S., Srivastava, M.M. (1998) Effect of selenium supplementation on the uptake and translocation of chromium by spinach (*Spinacea oleracea*). *Bulletin of Environmental Contamination and Toxicology* 60: 750–758.

Tadina, N., Germ, M., Kreft, I., Breznik, B., Gaber, A. (2007) Effects of water deficit and selenium on common buckwheat (*Fagopyrum esculentum* Moench.) plants. *Photosynthetica* 45: 472–476.

Tang, H., Liu, Y., Gong, X., Zeng, G., Zheng, B., Wang, D., Sun, Z., Zhou, L., Zeng, X. (2015) Effects of selenium and silicon on enhancing antioxidative capacity in ramie (*Boehmeria nivea* (L.) Gaud.) under cadmium stress. *Environmental Science and Pollution Research International* 22: 9999–10008.

Tanou, G., Molassiotis, A., Diamantidis, G. (2009) Induction of reactive oxygen species and necrotic death-like destruction in strawberry leaves by salinity. *Environmental and Experimental Botany* 65: 270–281.

Tavakkoli, E., Rengasamy, P., McDonald, G.K. (2010) High concentrations of Na+ and Cl– ions in soil solution have simultaneous detrimental effects on growth of faba bean under salinity stress. *Journal of Experimental Botany* 61: 4449–4459.

Tedeschini, E., Proietti, P., Timorato, V., D'Amato, R., Nasini, L., Del Buono, D., Businelli, D., Frenguelli, G. (2015) Selenium as stressor and antioxidant affects pollen performance in Olea europaea. *Flora - Morphology, Distribution, Functional Ecology of Plants* 215: 16–22.

Terry, N., Zayed, A.M., de Souza, M.P., Tarun, A.S. (2000) Selenium in higher plants. *Annual Review of Plant Physiology and Plant Molecular Biology* 51: 401–432.

Thakur, P., Kumar, S., Malik, J.A., Berger, J.D., Nayyar, H. (2010) Cold stress effects on reproductive development in grain crops: An overview. *Environmental and Experimental Botany* 67: 429–443.

Valkama, E., Kivimäenpää, M., Hartikainen, H., Wulff, A. (2003) The combined effects of enhanced UV-B radiation and selenium on growth, chlorophyll fluorescence and ultrastructure in strawberry (Fragaria × ananassa) and barley (*Hordeum vulgare*) treated in the field. *Agricultural and Forest Meteorology* 120: 267–278.

Walaa, A.E., Shatlah, M.A., Atteia, M.H., Sror, H.A.M. (2010) Selenium induces antioxidant defensive enzymes and promotes tolerance against salinity stress in cucumber seedlings (*Cucumis sativus*). *Arab Universities Journal of Agricultural Sciences* 18: 65–76.

Wang, X., Tam, N.F.Y., Fu, S., Ametkhan, A., Ouyang, Y., Ye, Z. (2014) Selenium addition alters mercury uptake, bioavailability in the rhizosphere and root anatomy of rice (*Oryza sativa*). *Annals of Botany* 114: 271–278.

Wu, Z., Yin, X., Bañuelos, G.S., Lin, Z.Q., Liu, Y., Li, M., Yuan, L. (2016) Indications of selenium protection against cadmium and lead toxicity in oilseed rape (*Brassica napus* L.). *Frontiers in Plant Science* 7: 1875. doi:10.3389/fpls.2016.01875

Xu, B., Yu, J., Xie, T., Li, Y., Yang, L., Yu, Y., Chen, Y., Wang, G. (2017) Effect of brassinosteroid and selenium on uptake and accumulation of chromium in yellow flag (*Iris pseudacorus*). *International Journal of Agriculture and Biology* 19: 621–628.

Xue, T., Hartikainen, H., Piironen, V. (2001) Antioxidative and growth-promoting effect of selenium on senescing lettuce. *Plant and Soil* 237: 55–61.

Yadav, S.K. (2010) Cold stress tolerance mechanisms in plants. A review. *Agronomy for Sustainable Development* 30: 515–527.

Yao, X., Chu, J., Ba, C. (2010a) Antioxidant responses of wheat seedlings to exogenous selenium supply under enhanced ultraviolet-B. *Biological Trace Element Research* 136: 96–105.

Yao, X., Chu, J., Ba, C. (2010b) Responses of wheat roots to exogenous selenium supply under enhanced ultraviolet-B. *Biological Trace Element Research* 137: 244–252.

Yao, X., Chu, J., He, X., Ba, C. (2011) Protective role of selenium in wheat seedlings subjected to enhanced UV-B radiation. *Russian Journal of Plant Physiology* 58: 283–289.

Yao, X., Chu, J., Wang, G. (2009) Effects of selenium on wheat seedlings under drought stress. *Biological Trace Element Research* 130: 283–290.

Yao, X., Jianzhou, C., Xueli, H., Binbin, L., Jingmin, L., Zhaowei, Y. (2013) Effects of selenium on agronomical characters of winter wheat exposed to enhanced ultraviolet-B. *Ecotoxicology and Environmental Safety* 92: 320–326.

Zayed, A., Lytle, C.M., Terry, N. (1998) Accumulation and volatilization of different chemical species of selenium by plants. *Planta* 206: 284–292.

Zhang, L., Hu, B., Li, W., Che, R., Deng, K., Li,H., Yu, F., et al. (2014) OsPT2, a phosphate transporter, is involved in the active uptake of selenite in rice. *The New Phytologist* 201: 1183–1191.

Zhang, J.L., Shi, H.Z. (2013) Physiological and molecular mechanisms of plant salt tolerance. *Photosynthesis Research* 115: 1–22.

Zhao, J., Gao, Y., Li, Y.F., Hu, Y., Peng, X., Dong, Y., Li, B., Chen, C., Chai, Z. (2013) Selenium inhibits the phytotoxicity of mercury in garlic (*Allium sativum*). *Environmental Research* 125: 75–81.

Zhu, Z., Chen, Y., Shi, G., Zhang, X. (2017) Selenium delays tomato fruit ripening by inhibiting ethylene biosynthesis and enhancing the antioxidant defense system. *Food Chemistry* 219: 179–184.

17 Bio-Organic Fertilizer in Stress Mitigation in Plants

Fatima Bibi and Noshin Ilyas

CONTENTS

17.1 INTRODUCTION

Plants' accessibility to organic materials is improved by bio-organic fertilizers more than any ordinary organic fertilizer. Application of biofertilizers is an environmentally friendly biotechnological approach that offers an alternative to harmful chemical fertilizers. The word 'microbial inoculants' can be used instead of biofertilizer, or it can be called a preparation that contains live or latent cells of effectual strains of phosphate solubilizing cellulytic microbes or nitrogen-fixing used for application in soil, seed or cultivated areas with the aim of raising the numbers of useful microbes and increasing the definite microbial process to boost the degree of accessibility of nutrients in a range that can be incorporated by plants for life processes (Mohammadi and Sohrabi, 2012).

One of the main constituents of integrated nutrient management is biofertilizers. These bio-organic fertilizers not only play an essential function in sustainability and productivity of soil but also play a vital role in protecting the environment as environmentally friendly and commercial inputs for farmers. Biofertilizers are good for the environment, easily available and considered as a renewable source for plant uptake of nutrients to addendum chemical fertilizers in an agricultural system. When biofertilizers are applied to soil, plant surface or seed, microorganisms colonize the rhizosphere or the plant's interior and enhance growth by transferring nutritionally essential elements because they contain living cells of different kinds of microorganism. By using different biological processes, namely solubilization of rock phosphate or fixing of nitrogen, they convert phosphorus and nitrogen to an available form (Rokhzadi et al., 2008).

Propitious microbes in biofertilizers speed up and defend plants from pests and diseases and improve plant growth (El-Yazeid et al., 2007). The sustainable development of the agriculture role played by soil microorganisms in biofertilizers has been reviewed many times (Lee and Pankhurst, 1992; Wani et al., 1995). In a broader sense, we can use this word to include all organic resources for the better growth of plants, which are delivered in an accessible form for plant assimilation through plant associations or microbes (Mohammadi and Sohrabi, 2012).

17.2 TYPES OF BIO-ORGANIC FERTILIZER

Bio-organic fertilizers can be used in many forms, such as manure, and compost-based products are popular. On the other hand, algal extracts, microbial inoculation, mined mineral supplements and a mixture of the above can be used for good results. Many forms of slaughterhouse by-products are used such as powders, dry pellets and in liquid forms. These products contain carbon and other important micronutrients such as Ca, Fe (quantities may vary) and microorganisms. Nutrient release not only depends on soil conditions but also on temperature, moisture and crop demand.

Extracts of algae or plant-based can be used as dry biomass or in liquid form. It also contains carbon, nitrogen and many other important micronutrients. Similarly, different forms of microbial inoculants can be used in pellet form or powders, granules and pure or mixed liquid fermentation depending upon the formulation; biofertilizers may contain carbon, micronutrients and phytohormones. Biofertilizers may consist of nitrogen, phosphorus and potassium (NPK). It always contains different microbes but their activity, viability and the type may differ significantly. Conditions of soil and activity of microorganisms in the formulation may affect nutrient discharge. The mined mineral supplement can be used in the different form usually as granules or bagged powders. On the other hand, NPK composition sometimes contains microbes, but a considerable quantity of definite nutrients is present. Nutrient discharge greatly depends on soil condition as well as temperature, humidity and crop demands.

17.3 DIFFERENCES BETWEEN ORGANIC FERTILIZERS AND BIO-ORGANIC FERTILIZERS

Organic manure is one of the best sources of fertilizers because these materials are natural products applied by farmers to supply food to crop plants, such as green manures, farmyard wastes and from a crop residue prepared from compost and other agricultural wastes, slaughterhouse refuse and bones of animals. These organic manures not only play a vital role in enhancing growth and productivity but also yield components of different crops. Abd El-Rahman and Hosny (2001) reported that application of organic manure considerably improved the yield as well as components of yields in eggplants. The amount of organic matter increased in the soil by using organic manure. As a result, plant food is released by organic matter for the use of crops in an available form. On the other hand, organic manures should not be considered as a carrier of plant food. These organic manures not only facilitate a soil to hold more water but also help to develop the drainage in clay soils. An organic acid is also provided by the organic manure that aids in dissolving soil nutrients to make them available for plants. TahaRyan et al. (2011) reported that organic manures considerably affected tomato leaf area, plant height and fruit number.

17.4 CHEMICAL FERTILIZERS

Industrially produced fertilizers are chemical or inorganic fertilizers and these are prepared by mixing micro and macronutrients in proportions to get nutritionally balanced fertilizers based on the requirements of plants. There is a range of different types of chemical fertilizers, for example,

- **Phosphorous-rich**. These promote root growth, and are thus appropriate for the early growth of the plant. Mostly for flowering plants.
- **Nitrogen-rich**. These stimulate leaf and stem growth, and are suitable for evergreens and grasses.
- **Potassium-rich**. These promote not only flowers but also fruit growth, and the development of the lignifying protecting layers of stems (Coyne, 1999).

17.5 ORGANIC FERTILIZERS

By definition, organic fertilizers are derived from animal or plant sources. Generally, if they are properly utilized, they consist of inclusive substances that are nutritionally important and these are required for microbes in the soil to directly offer the minerals essential for the healthy growth of crops. On the other hand, it is important to consider that organic fertilizers may not always be the 'best' ones because even the uncontrolled applications of biofertilizers may cause environmental damage because of a greater amount of nitrogen released into the soil.

Actually, when fertilizer is applied to soil from different sources, the nitrogen distribution originated as follows: ureic (urea and a few mixed fertilizers); organic (compost, organic, manure and mineral fertilizers); nitric (calcium nitrate, ammonium nitrate and some mixed fertilizers) or ammonia-based (ammonium nitrate, ammonium sulfate and some mixed fertilizers) (Coyne, 1999).

17.6 RAW MATERIALS FOR BIO-ORGANIC FERTILIZERS

Selection of raw materials for making bio-organic fertilizers can be natural such as algae and seaweed, or it can be agro-industrial wastes such as feathers or husks of cereals.

17.7 PREPARATION METHOD OF BIO-ORGANIC FERTILIZERS

There are several things that need to be considered in making biofertilizer; the most important are the growth profile of the microbe, the types of organisms and their most suitable state and the inoculum formulation method. To successfully achieve biological product formulation of inocula, storage of the product and the method of application are both significant. There are generally six main steps in the preparation of biofertilizer. These are as follows:

- Selection of active microorganisms
- Separation and choosing of target microorganisms
- Choosing the method and the carrier material
- Selection method of propagation
- Testing of prototype
- Testing on a large scale

Initially, active microbes must be determined. For instance, we must decide whether to use the combination of some organisms, nitrogen fixer or organic acid bacteria. Then to isolate target microorganism from their habitants. Generally, organisms are isolated from roots of plants. The next step is growing isolated microorganisms on Petri plates, the isolated microbes will be grown in shake flasks and then the suitable applicant is chosen and moved to a glasshouse. It is also essential to determine the right carrier material by using biofertilizers sensibly. If biofertilizers are to be produced in powder form, then the correct carrier materials are peat or tapioca flour. To find the optimum conditions for microorganisms, the propagation method is the main consideration. This can be done by obtaining a growth profile under unusual conditions and parameters. After this step, a prototype (normally in different appearance) is prepared and tested. Finally, to estimate its efficacy and limit its ability in different surroundings, the biofertilizer is tested on a large-scale at different environmental conditions. To sustain a greater number of microorganisms for long-term storage on the carrier material, a critical step is the sterilization of this material. Biofertilizers are frequently prepared as carrier-based inoculants that contain efficient microbes. Actually merging microbes in carrier substances facilitates easy handling, storage for long time periods, and high and long-lasting efficiency of biofertilizers. Autoclaving or gamma irradiation can be used as methods for sterilization. A variety of forms of material can be used as a carrier for seed or soil inoculation. The characteristic of a good quality carrier material for seed inoculation is that it must be obtainable in abundant amounts and be low-cost. It must be non-toxic to the plant itself as well as non-toxic to the inoculant's bacterial strain. It should have an excellent moisture absorption capacity and bond best to seeds because it operates as a carrier for the inoculation of the seed. The final step is that the carrier must have excellent pH buffering capability, sterilized by using an autoclave or by using gamma radiation and easy to process (Mohammadi and Sohrabi, 2012).

17.8 COMPONENTS OF BIO-ORGANIC FERTILIZERS

Organic fertilizers, which chiefly come from animal or plant waste residues as spent mushroom compost and cow dung or from municipal solid waste compost (MSWC), are often well known as proper local bio-organic fertilizers. Biofertilizers have more amounts of nutrients, such as P and N, and a high level of organic matter (Olfati et al., 2009; Peyvast et al., 2007; Peyvast et al., 2008; Shabani et al., 2011).

17.9 EFFECTS OF BIO-ORGANIC FERTILIZERS ON PLANTS

17.9.1 Improve Productivity

Symbiotic phosphate solubilizing and nitrogen fixer microorganisms play a vital function by permitting a sustainable use of phosphate and nitrogen fertilizers by supplementing phosphorus and nitrogen to the plant for growth (Tambekar et al., 2009). Maize when inoculated with *Azospirillum* not only improved 59% dry weight of seed but also significantly improved the yield, which was comparable to 60 kg urea N ha^{-1}. A considerable encouraging effect on the yield of grain was achieved as a result of collective inoculation of *Pseudomonas striata* and *Azospirillum* in maize (Fulchieri and Frioni, 1994) and cotton (Radhakrishnan, 1996). Zaddy et al. (1993) showed that inoculation with single cell and aggregated suspensions of *A. brasilense* enhanced the growth of the plant. They also stated that inoculation of single cell suspensions of *Azospirillum* (prepared with fructose) extensively increased the foliage dry weight of the maize seedling and surface area of the root, as compared with controls or the plants inoculated with malate grown *Azospirillum*. Sorghum and pearl millet, which are grown as dryland crops, inoculated with *Azospirilla* and *Azotobacter* showed an 11–12% increased return. On the other hand, wheat, maize and rice received improved inputs and management than sorghum and pearl millet due to inoculation and showed 15–20% better yields (Wani et al., 1995).

17.9.2 Improve the Quality of Soil

There are so many benefits and values of applying organic materials in the production of crops, such as they not only enhance and sustain the quality of soil but also increase soil health as well as soil fertility. It is also noticed that they increase soil air and water qualities, and usually re-establish crucial nutrients required by plants and soil (FAO, 2005; Usman and Kundiri, 2016). Bio-inoculants, traditionally called biofertilizer, are biologically active strains of bacteria, fungi and algae. Basically, biofertilizers are preparations that are carrier based and chiefly contain efficient microorganism strains in adequate numbers which are helpful for nitrogen fixation in plants, uptake of phosphorus solubilization and synthesis of growth promoting substances like hormones, auxins and vitamins. Biofertilizers maintain the soil fertility and improve soil properties. With the use of organic agriculture, elevated total nitrogen (N) soil and organic matter are reported by comparing research soils of chemically and organically managed farming systems (Alvarez et al., 1988; Drinkwater et al., 1995; Reganold, 1988). Plant-available nutrient concentrations become higher, soil pH may be higher and under organic management the overall population of the microbe increases (Clark et al., 1998; Dinesh et al., 2000; Lee, 2010; Reganold, 1988).

17.9.3 Increase Soil Fertility

One of the main necessities for crop yield in agricultural systems is maintaining soil sustainability. Activities of soil enzymes, soil microbes and soil organic matter (SOM) are considered important signs of soil productivity and fertility because the biochemical properties of the soil are determined by them (Bandick and Dick, 1999; Olk and Gregorich, 2006; van der Heijden et al., 2008).

17.9.4 Enhance Enzyme Activity

Soil microorganisms and enzyme activity have been recommended as possible indicators of soil quality. Soil enzymes are not only involved in catalyzing diverse reactions and metabolic processes occurring in organic matter metabolism but also maintaining soil structure, producing energy for both microorganisms and plants and cycling nutrients (Khan et al., 2010; Kızılkaya et al., 2004). The outcome of fertilizer microbial composition and soil enzyme activity has been intensively revised (Kandeler et al., 1999; Saha et al., 2008a; Zhao et al., 2014). Various studies have indicated that the increases of overall enzyme activity are because of the application of organic fertilizers (García-Ruiz et al., 2008; Mäder et al., 2002; Moeskops et al., 2010), on the other hand, activities of particular enzymes possibly will change depending on the relative accessibility of nutrients and the type of the organic fertilizer. To meet microbial metabolic demands, enzymatic activity might be estimated to improve the availability of most restrictive nutrients (Allison et al., 2011; Sinsabaugh et al., 2008). For example, lifelong N fertilization increased the activity of soil enzymes involved in labile C breakdown in usually managed agricultural soils. The input of an ordinary organic fertilizer might increase enzymatic activity of the soil, but the reaction of enzymatic activity to the accumulation of bio-organic fertilizer is not completely understood (Bandick and Dick, 1999; Piotrowska and Wilczewski, 2012).

17.10 ADVANTAGES OVER CHEMICAL FERTILIZER

Based on natural inputs (decaying remains of organic matter).

17.10.1 Lower Environmental Pollution

If organic matter is completely decayed in the soil, it will help in retaining soil health as well as reducing soil hazards and soil contamination. As compared with inorganic chemical fertilizers, organic materials are quite safe for the environment. They also control soil pollution and contamination of soil (Usman and Kundiri, 2016).

17.10.2 Cheaper than Conventional Techniques

It means that the application of organic fertilizer in crop production would reduce the usage of inorganic fertilizer, which may result in increase in production. Therefore, the long-term effect of organic improvement will certainly improve conditions for countryside farmers. Bio organic fertilizers will prove to be more cost effective than chemical fertilizers and thus yield more profit (Diacono and Montemurro, 2010).

17.10.3 Reduce Use of Chemical Fertilizers

Verified chemical fertilizers can successfully increase the yield of many crops, but the significant application of chemical fertilizers can cause harmful effects on the surroundings. They also build up substances that contaminate the environment and soil and decrease the levels of some trace elements, which are required by the plant for its growth. Biofertilizers play a vital role in maintaining physical, chemical and biological characteristics of the soil. On the other hand, mostly favorable organic and biofertilizer dosage in addition to their usefulness for upland crops have not been studied in detail. In this regard, an effort has been made to study the influence of organic and biofertilizers compared with chemical fertilizers on rice production (Farah et al., 2014). Existing bio-organic fertilizer products in the market may lessen the dependence on inorganic fertilizers (Usman and Kundiri, 2016).

17.10.4 Improve Availability of Nutrients

Organic matter not only boosts nutrient accessibility in soils but also supplies them to crops (Usman and Kundiri, 2016).

17.11 STRESS MITIGATION

17.11.1 Saline Stress

Several interactions between plants and advantageous bacteria show a defensive response under limited ecological circumstances. When wheat and faba beans were inoculated with *Azospirillum* and subjected to salt stress, more growth was shown as compared with other non-inoculated plants (Bacilio et al., 2004; Hamaoui et al., 2001). This admiring effect can be attributable directly to the bacteria or indirectly to the influence on plant physiology. Different mechanisms are used by microorganisms to mitigate the salinity stress problems in crops. Some plant growth-promoting rhizobacteria (PGPR) rhizobacterial strains change the growth and growth of beans, tomatoes, lettuce and peppers grown in salty environments (Grover et al., 2010; Yildirium and Taylor, 2005). Wheat seedlings inoculated with bacteria that produce exopolysaccharates (EPS) affect the limitation of sodium uptake and prompt plant growth under conditions of stress caused by high salt stress (Grover et al., 2010). Corn, clover and beans inoculated with arbuscular mycorrhiza (AM) fungi enhanced their osmoregulation and increased proline accumulation which resulted in salinity resistance (Feng et al., 2002, cit. Grover et al., 2010).

17.11.2 Drought Stress

In drought stress, the production of microbial metabolites like polysaccharides change the soil structure and shows a helpful result on plant growth. Sunflower plant growth parameters under drought stress inoculated with an EPS generates *Rhizobium* sp. more than in uninoculated plants (Alami et al., 2000). In wheat plants inoculated with *Paenibacillus polymyxa*, the aggregation of rhizospheric soil depended on a bacterial polysaccharide that enlarged the amount of soil adhering to the roots (Bezzate et al., 2000; Gouzou et al., 1993). After inoculation with an EPS-producing *Pantoea agglomerans* separate, a better effect in wheat plants was observed (Amelal et al., 1998).

Certain microbial types may alleviate the impact of soil drought through the production of EPS, induction of resistance genes, increased circulation of water in the plant and the synthesis of ACC-deaminase, indole-acetic acid and proline, as well as drought stress bound crop growth and productivity, particularly in arid and semi-arid regions. Some microbial species and/or strains that reside in plant rhizosphere use different mechanisms to mitigate negative effects of drought on plants (Grover et al., 2010). Bacteria can also arouse the plant to turn on a particular metabolic activity like escalating its exudates, and consequently, improve rhizospheric soil qualities (Heulin et al., 1987).

PGPR lessens the impact of drought on plants through a process of so-called induced systemic tolerance which includes: (1) production of cytokinins; (2) production of antioxidants; and (3) degradation of the ethylene precursor ACC by bacterial ACC-deaminase. The production of cytokinins causes the accumulation of abscisic acid (ABA) in leaves, which in its turn results in the closing of stomata (Cowan, 1999; cit. Figueiredo et al., 2008; Yang et al., 2009).

17.12 CONCLUSION

Against pathogenic colonization inoculated plants show an enhanced reaction and drought stress as compared with control plants. Therefore, it appears that inoculation stimulates protection beside biotic agents and against abiotic ones. In general, PGPR can shield a plant against destructive environmental and particularly hostile soil conditions through the bacterial discharge of soil structure-improving substances and by inducing the plant to trigger reactive stress mechanisms. In aggressive soils, the use of bacteria that permit plants to flourish is probably the best option to acquire good production at minor ecological expenses.

REFERENCES

Abd El-Rahman, S. Z., Hosney, F. (2001) Effect of organic and inorganic fertilizer on growth and yield, fruitility and storability of eggplant. *The Journal of Agricultural Science* 26(10): 6307–6321.

Alami, Y., Achouak, W., Marol, C., Heulin, T. (2000) Rhizosphere soil aggregation and plant growth promotion of sunflowers by an exopolysaccharide-producing *Rhizobium* sp. strain isolated from sunflower roots. *Applied and Environmental Microbiology* 66(8): 3393–3398.

Allison, S. D., Weintraub, M. N., Gartner, T. B., Waldrop, M. P. (2011) Evolutionary economic principles as regulators of soil enzyme production and ecosystem function. In *Soil Enzymology*, ed. Shukla, G., Varma, A., 229–243. Berlin: Springer.

Alvarez, C. E., Garcia, C., Carracedo, A. E. (1988) Soil fertility and mineral nutrition of organic banana plantation in Tenerife. *Biological Agriculture and Horticulture* 5(4): 313–323.

Amelal, N., Burtan, G., Bartoli, F., Heulin, T. (1998) Colonization of wheat roots by an exopolysaccharide-producing *Pantoea agglomerans* strain and its effect on rhizosphere soil aggregation. *Applied and Environmental Microbiology* 64(10): 3740–3747.

Bacilio, M., Rodríguez, H., Moreno, M., Hernández, J. P., Bashan, Y. (2004) Mitigation of salt stress in wheat seedlings by a gfp-tagged *Azospirillum lipoferum*. *Biology and Fertility of Soils* 40(3): 188–193.

Bandick, A. K., Dick, R. P. (1999) Field management effects on soil enzyme activities. *Soil Biology and Biochemistry* 31(11): 1471–1479.

Bezzate, S., Aymerich, S., Chambert, R., Czarnes, S., Berge, O., Heulin, T. (2000) Disruption of the *Bacillus polymyxa* levansucrase gene impairs its ability to aggregate soil in the wheat rhizosphere. *Environmental Microbiology* 2(3): 333–342.

Clark, M. S., Horwath, W. R., Shennan, C., Scow, K. M. (1998) Changes in soil chemical properties resulting from organic and low-input farming practices. *Agronomy Journal* 90(5): 662–671.

Cowan, M. M. (1999) Plant product as antimicrobial agents. *Clinical Microbiology Reviews* 12(4): 564–582.

Coyne, M. S. (1999) *Soil Microbiology: Exploratory Approach*. Albany, NY: Delmar.

Diacono, M., Montemurro, F. (2010) Long-term effects of organic amendments on soil fertility. A review. *Agronomy for Sustainable Development* 30(2): 401–422.

Dinesh, R., Dubey, R. P., Ganeshamurthy, A. N., Prasad, G. S. (2000) Organic manuring in rice based cropping system: Effects on soil microbial biomass and selected enzyme activities. *Current Science* 79: 1716–1720.

Drinkwater, L. E., Letourneau, D. K., Workneh, F., Bruggen, A. H. C., Shennan, C. (1995) Fundamental difference between conventional and organic tomato agro ecosystems in California. *Ecological Applications* 5(4): 1098–1112.

El-Yazeid, A. A., Abou-Aly, H. A., Mady, M. A., Moussa, S. A. M. (2007) Enhancing growth, productivity and quality of squash plants using phosphate dissolving microorganisms (bio phosphor) combined with boron foliar spray. *Research Journal of Agriculture and Biological Sciences* 3(4): 274–286.

FAO. (2005) Current world fertilizer trends and outlook to 2009/2010. FAO, Rome.

Farah, G. A., Dagash, Y. M. I., Yagoob, S. O. (2014) Effect of different fertilizers (bio, organic and inorganic fertilizers) on some yield components of rice (*Oryza sativa* L.). *Universal Journal of Agricultural Research* 2(2): 67–70.

Feng, G., Zhang, F. S., Li, X., Tian, C. Y., Tang, C., Rengel, Z. (2002) Improved tolerance of maize plants to salt stress by arbuscular mycorrhiza is related to higher accumulation of soluble sugars in roots. *Mycorrhiza* 12: 185–190.

Figueiredo, M. V. B., Burity, H. A., Martinez, C. R., Chanway, C. P. (2008) Alleviation of drought stress in the common bean (*Phaseolus vulgaris* L.) by co-inoculation with *Paenibacillus polymyxa* and *Rhizobium tropici*. *Applied Soil Ecology* 40: 182–188.

Fulchieri, M., Frioni, L. (1994) Azospirillum inoculation on maize (*Zea mays*): Effect of yield in a field experiment in central Argentina. *Soil Biology and Biochemistry* 26: 921–924.

García-Ruiz, R., Ochoa, V., Hinojosa, M. B., Carreira, J. A. (2008) Suitability of enzyme activities for the monitoring of soil quality improvement in organic agricultural systems. *Soil Biology and Biochemistry* 40(9): 2137–2145.

Gouzou, L., Burtin, R., Philippy, R., Bartoli, F., Heulin, T. (1993) Effect of inoculation with *Bacillus polymyxa* on soil aggregation in wheat rhizosphere: Preliminary examination. *Geoderma* 56: 479–491.

Grover, M., Ali, S. Z., Sandhya, V., Rasul, A., Venkateswarlu, B. (2010) Role of microorganisms in adaptation of agriculture crops to abiotic stresses. *World Journal of Microbiology and Biotechnology* 27 (5): 1231–1240.

Hamaoui, B., Abbadi, J. M., Burdman, S., Rashid, A., Sarig, S., Okon, Y. (2001) Effects of inoculation with *Azospirillum brasilense* on chickpeas (*Cicer arietinum*) and faba beans (*Vicia faba*) under different growth conditions. *Agronomie* 21(6–7): 553–560.

Heulin, T., Gucker, A., Balandeau, J. (1987) Stimulation of root exudation of rice seedlings by *Azospirillum* strains: Carbon budget under gnotobiotic conditions. *Biology and Fertility of Soils* 4: 9–14.

Kandeler, E., Stemmer, M., Klimanek, E. M. (1999) Response of soil microbial biomass, urease and xylanase within particle size fractions to long-term soil management. *Soil Biology and Biochemistry* 31(2): 261–273.

Khan, S., Hesham, A. E. L., Qiao, M., Rehman, S., He, J. Z. (2010) Effects of Cd and Pb on soil microbial community structure and activities. *Environmental Science and Pollution Research* 17: 288–296.

Kızılkaya, R. A., Aşkın, T., Bayraklı, B., Sağlam, M. (2004) Microbiological characteristics of soils contaminated with heavy metals. *European Journal of Soil Biology* 40(2): 95–102.

Lee, J. (2010) Effect of application methods of organic fertilizer on growth, soil chemical properties and microbial densities in organic bulb onion production. *Scientia Horticulturae* 124(3): 299–305.

Lee, K. E., Pankhurst, C. E. (1992) Soil organisms and sustainable productivity. *Australian Journal of Soil Research* 30(6): 855–892.

Mäder, P. D. A., Fliebbach, A., Dubois, D., Gunst, L., Fried, P., Niggli, U. (2002) Soil fertility and biodiversity in organic farming. *Science* 296(5573): 1694–1697.

Moeskops, B., Sukristiyonubowo, Buchan, D., Sleutel, S., Herawaty, L., Husen, E., Saraswati, R., Setyorini, D., De Neve S. (2010) Soil microbial communities and activities under intensive organic and conventional vegetable farming in West Java, Indonesia. *Applied Soil Ecology* 45(2): 112–120.

Mohammadi, K., Sohrabi, Y. (2012) Bacterial biofertilizers for sustainable crop production: A review. *ARPN. Journal of Agricultural and Biological Science* 7(5): 307–316.

Olfati, J. A., Peyvast, G., Nosrati-Rad, Z., Saliqedar, F., Rezaie, F. (2009) Application of municipal solid waste compost on lettuce yield. *International Journal of Vegetable Science* 15(2): 168–172.

Olk, D. C., Gregorich, E. G. (2006) Overview of the symposium proceedings, 'Meaningful pools in determining soil carbon and nitrogen dynamics'. *Soil Science Society of America Journal* 70(3): 967–974.

Peyvast, G., Ramezani Kharazi, P., Tahernia, S., Nosratierad, Z., Olfati, J. A. (2008) Municipal solid waste compost increased yield and decreased nitrate amount of broccoli (*Brassica oleracea* var. Italica). *Journal of Applied Horticulture* 10(2): 129–132.

Peyvast, G., Sedghi Moghaddam, M., Olfati, J. A. (2007) Effect of municipal solid waste compost on weed control, yield and some quality indices of green pepper (*Capsicum annuum* L.). *Biosciences, Biotechnology Research Asia* 4(2): 449–456.

Piotrowska, A., Wilczewski, E. (2012) Effects of catch crops cultivated for green manure and mineral nitrogen fertilization on soil enzyme activities and chemical properties. *Geoderma* 189–190:72–80.

Radhakrishnan, K. C. (1996) Role of biofertilizers in cotton productivity. *National Seminar on Biofertilizer Production, Problems and Constraints*, TNAU, Coimbatore, 24–25 January, p. 17.

Reganold, J. P. (1988) Comparison of soil properties as influenced by organic and conventional farming systems. *American Journal of Alternative Agriculture* 3(4): 144–155.

Rokhzadi, A., Asgharzadeh, A., Darvish, F., Nourmohammadi, G., Majidi, E. (2008) Influence of plant growth-promoting rhizobacteria on dry matter accumulation and yield of chickpea (*Cicer arietinum* L.) under field condition. *American-Eurasian Journal of Agricultural & Environmental Sciences* 3(2): 253–257.

Saha, S., Gopinath, K. A., Mina, B. L., Gupta, H. S. (2008a) Influence of continuous application of inorganic nutrients to a maize-wheat rotation on soil enzyme activity and grain quality in a rain fed Indian soil. *European Journal of Soil Biology* 44(5–6): 521–531.

Shabani, H., Peyvast, Gh., Olfati, J. A., Ramezani Kharrazi, P. (2011) Effect of municipal solid waste compost on yield and quality of eggplant. *Comunicata Scientiae* 2(2): 85–90.

Sinsabaugh, R. L., Lauber, C. L., Weintraub, M. N., Ahmed, B., Allison, S. D., Crenshaw, C., Contosta, A. R., *et al.* (2008) Stoichiometry of soil enzyme activity at global scale. *Ecology Letters* 11(11): 1252–1264.

Tambekar, D. H., Gulhane, S. R., Somkuwar, D. O., Ingle, K. B., Kanchalwar, S. P. (2009) Potential Rhizobium and phosphate solubilizers as biofertilizers from saline belt of Akola and Buldhana district, India. *Research Journal of Biological Sciences. Introduction* 5: 578–582.

Usman, S., Kundiri A. M. (2016) Values of organic materials as fertilizers to northern Nigerian crop production systems. *Journal of Soil Science and Environmental Management* 7(12): 204–211.

van der Heijden, M. G. A., Bardgett, R. D., van Straalen, N. M. (2008) The unseen majority, soil microbes as drivers of plant diversity and productivity in terrestrial ecosystems. *Ecology Letters* 11(3): 296–310.

Wani, S. P., Rego, T. J., Rajeshwari, S., Lee, K. K. (1995) Effect of legume-based cropping systems on nitrogen mineralization potential of vertisol. *Plant and Soil* 175(2): 265–274.

Yang, J., Kloepper, J. W., Ryu, C. M. (2009) Rhizosphere bacteria help plants tolerate abiotic stress. *Trends in Plant Science* 14(1): 1–4.

Yildirium, E., Taylor, A. G. (2005) Effect of biological treatment on growth of bean plants under salt stress. *Annual Report on Bean Improvement Cooperative* 48: 176–177.

Zaddy E., Perevolosky, A., Okon, Y. (1993) Promotion of plant growth by inoculation with aggregated and single cell suspension by *Azospirillum brasilense. Soil Biology and Biochemistry* 25: 819–823.

Zhao, J., Ni, T., Li, Y., Xiong, W., Ran, W., Shen, B., Shen, Q., Zhang, R. (2014) Responses of bacterial communities in arable soils in a rice-wheat cropping system to different fertilizer regimes and sampling times. *PLOS ONE* 9(1): e85301.

18 Application of Biochar for the Mitigation of Abiotic Stress-Induced Damages in Plants

Muhammad Riaz, Muhammad Saleem Arif, Qaiser Hussain, Shahbaz Ali Khan, Hafiz Muhammad Tauqeer, Tahira Yasmeen, Muhammad Arslan Ashraf, Muhammad Arif Ali, Muhammad Iqbal, Sher Muhammad Shehzad, Samar Fatima, Afia Zia, Najam Abbas, Muhammad Siddique, and Muhammad Sajjad Haider

CONTENTS

18.1 INTRODUCTION

In the beginning of the 1960s, a significant jump in the population of the world remarkably increased agricultural growth and production. Since then, during the last 40 to 50 years, agricultural production has increased multiple times depending upon the country or region. In Asia, agricultural production has increased by about 300% (FAO et al., 2015). Similarly, in those four to five decades, the population of the world also increased from 3 billion to more than 6 billion presently. Since the beginning of agriculture, crop production has changed dramatically in terms of domestication, varietal selection, irrigation techniques, cropping patterns, planting methods, climate change mitigation and fertilization are continuously in evolution to improve crop production. In recent times, agricultural production has increased significantly, resulting in the generation of more food for the increasing global population. However, in a large number of countries, intensive cropping has ruined fertile lands, reduced groundwater availability, increased pests and their resistance, reduced biodiversity and contamination of land resources like air, water and soil. Thus, high-intensity crop production is no longer sustainable and the novel "Save and Grow" model of growing crops should be adopted (FAO et al., 2015).

At present, sustainability focuses on the need to develop technologies and practices in agriculture that do not have any noxious effects on the environment, approachable to every farmer and efficient and effective for agricultural productivity. In the agriculture system, the idea of sustainability includes the capacity of systems to buffer any negative effects of biotic and abiotic stresses as well as the capacity of systems to function over long periods. Achieving the goal of sustainable agriculture is under serious concern due to several anthropogenic and environmental factors, including climate change. The management of agroecosystems to produce fuel, fodder and food and the management of agroecosystems for the adoption of climate changes and climate change mitigation have the same basic principles that can operate simultaneously for obtaining the same goal. This goal can only be achieved through the sustainable management of ecosystems and ecosystem services (FAO et al., 2015). Crop production by conservation agriculture also provides environmental service by achieving competitive crop yields while maintaining the reduction of degradation in natural resources. For example, undisturbed soil with an ample supply of organic matter is a better habitat for soil biota, which generally improves soil health by increasing soil aggregation, soil porosity, aeration and soil structure. Increasing soil organic matter and carbon sequestration through conservation agriculture or through other human interventions are very helpful for better drainage, less erosion, improved infiltration of water and water storage capacity of the soil. These soil characteristics can have a great significance for the management of crops under drought or flooding conditions.

The changing climatic conditions, however, affect several biotic and abiotic factors that are or will significantly hamper crop productivity. Among these factors, abiotic factors probably have direct effects on crop growth and yield. For example, higher CO_2 has been believed to increase crop growth. However, other factors like changing temperature, availability of water and mineral nutrients, uneven precipitation, unpredictable drought and flooding events can counteract with potential yield response (Hatfield et al., 2014). Therefore, for the sustainability of the agroecosystems, it is important to reduce CO_2 in the atmosphere produced as a result of energy production (Beerling et al., 2018). Biochar is gaining attraction as a result of its applicability in sustainable crop production due to several advantages. It increases soil carbon sequestration, improves soil health and nutrient bioavailability and reduces gas emissions, which ultimately increase crop production. Several studies revealed that the applications of biochar significantly increased plant growth, biomass and the activities of key enzymes against abiotic stress. To date, less information is available in the literature about the interaction and applications of biochar in reducing heat stress in plants.

18.2 BIOCHAR FOR SUSTAINABLE CROP PRODUCTION AND SOIL MANAGEMENT

Biochar is known to have variable effects on crop production depending on the biochar types and properties, soil conditions and plant types. Several studies have reported the positive impact of biochars prepared from various feedstocks, with and without inorganic and organic fertilizers, on the crop yield both in acidic and alkaline soils (Jeffery et al., 2011). A meta-analysis of 782 publications conducted by the Jeffery et al. (2011) revealed that overall biochar application increased the crop yield by approximately 10% under various soil conditions. In another meta-analysis, Jeffery et al. (2017) found a 25% increase in crop yield in the tropics area. In several other studies, biochar derived from rice husk increased crop yields of rice (Haefele et al., 2011), lentil (Abrishamkesh et al., 2015), spinach (Milla et al., 2013), maize (Nguyen et al., 2012), lettuce and cabbage (Carter et al., 2013).

In a long term-field experiment, Griffin et al. (2017) reported that walnut shell-biochar increased the corn yield by 8% in only the second season out of four seasons, suggesting the delayed and short-lived effects of walnut-shell biochar on crop productivity. They explained that the beneficial effects of biochar on crop yield could be due to the increased soil concentration of P, K and Ca^{2+} contributed by the biochar. Further, across all years, they found that application of a mineral fertilizer gave higher crop yield than biochar and compost alone; this might be due to the timely provision of nutrients by mineral fertilizers as per the demand of plant. Overall, the positive impact of biochar could be due to enhanced colonization and activities of mycorrhizal fungi around crop roots, thus facilitating efficient nutrient uptake (Atkinson et al., 2010), alleviation of nutrients and moisture stress in acidic low-input soil (Pandit et al., 2018), facilitated seed germination by altering the thermal dynamics through developing the dark color of the soil (Genesio et al., 2012), improved soil water holding capacity (Laird et al., 2010), liming effect in acidic soils (Wang et al., 2014) and increased nutrient (N, P and K) availability (Dempster et al., 2012; Lehmann et al., 2003).

Most of the studies showed that biochar applications rates ranging from 5 to 50 t ha^{-1} gave positive results regarding crop yield; however, the high biochar applications rates are economically not feasible for farming systems due to high biochar production cost and transportation. Now research focus has been on the development of nutrient-enriched biochar fertilizers by blending the biochar with synthetic/chemical fertilizers or by modifying the biochar properties and coating with biopolymers. These biochar-based fertilizers could be low cost and nutrient efficient, and their application rate is relatively less, which is suitable for the agriculture system. Tian et al. (2018) found increase in cotton lint yield after biochar application under field conditions for three years which increased with rate of biochar application. The increased cotton yield could be attributed to nutrient availability from the NPK fertilizer and biochar, enhanced soil water retention and improved soil structure. In another study based on a two-year experiment, application of *Acacia* tree-biochar in combination with poultry manure or farmyard manure or chemical fertilizer increased biological and grain yield of maize and wheat significantly in nutrient deficient alkaline soil (Arif et al., 2017). The positive impact of the combined application of organic and inorganic fertilizer is due to the slow release and availability of nutrients from organic sources, which were less prone to losses as compared with nutrients from mineral fertilizers. The role of biochar in biochar-based fertilizers is to ameliorate the soil conditions and thus provide a conducive environment for plants to uptake fertilizer efficiently for better crop production, improve soil chemical properties such as CEC for better nutrient retention and reduced nutrients losses and act as a carrier matrix for nutrients and then slowly releasing nutrients to increase fertilizer use efficiency (Huang et al., 2014; Lee et al., 2013).

There is evidence that biochar application did not affect crop yields, even in several cases crop yield was reduced (Borchard et al., 2014; Crane-Droesch et al., 2013; Liang et al., 2014; Tammeorg et al., 2014). In a 3-year mesocosm experiment, Borchard et al. (2014) reported that biochars derived from hardwood, when applied at 15 g biochar kg^{-1} soil, did not show any significant effect on maize yield in sandy and silty soils, but biochar application at 100 g biochar kg^{-1} soil resulted in reduced maize yield. These negative effects of biochar on maize yield could be due to N-immobilization and nutrient imbalance. Under a long-term field experiment, Liang et al. (2014) applied biochar derived from shell cottonseed and rice husk mixture to the calcareous soil and found that annual yield of winter wheat and summer maize was not increased significantly. However, soil pH was increased by a maximum of 0.35 units after 2 years of biochar application. Biochar addition also increased the soil water holding capacity and exchangeable K, reduced the bulk density and alkaline hydrolyzable N and had no impact on Olsen P. Moreover, biochar application with excessive rate (100–120 t ha^{-1}) is reported to decrease the yield by 10–20% (Baronti et al., 2010; Laghari et al., 2016). The negative effects of different biochars on crop productivity could be due to the application of biochars that have been produced from toxic and contaminated feedstocks (municipal waste, sewage sludge and food waste) and these biochars contain Na and ultimately their high amount in the addition to soils may cause salinity, negative liming effect in alkaline and calcareous soils, high volatile matter contents (e.g. PAHs, PCBs) in biochars and nutrient deficient wood based-feedstock from which biochar is produced (Clough et al., 2013; Laghari et al., 2016; Rajkovich et al., 2012).

Biochar effects on soil physicochemical (water retention, bulk density, water infiltration rate, hydraulic conductivity, CEC, nutrients availability, pH) and biological properties (microbial biomass, microbial communities and structure, enzyme activities) are well documented (Bailey et al., 2011; Chintala et al., 2014; Gomez et al., 2014; Kuzyakov et al., 2009; Uzoma et al., 2011). Biochar has the potential to increase nutrient availability in various soils under cropping systems. Biochar incorporation to highly weathered soils, strongly acidic soils, nutrients deficient soils and low clay content soils/sandy soils improved the soil quality by increasing nutrients availability organic carbon content, microbial biomass and enzyme activities (Jien and Wang, 2013; Steinbeiss et al., 2009), which could be due to the increase of labile nutrient (C,P, N and K) release and availability in soil (Ouyang et al., 2014). Biochar application increased micronutrients (Mn, Zn and Cu) except available Fe and macronutrients (P and K) availability in calcareous soil and also their uptake by crops (bean and maize) (Gunes et al., 2015). Biochar has also been reported to contain significant amounts of P, which contributes to increased P availability in soil and uptake by plants (Arif et al., 2017). The possible mechanisms behind the availability of nutrients to plants are the elevation of pH of the acidic soil, nutrient retention due to an increase in surface area and CEC and direct release of nutrients from biochar (Alling et al., 2014; Subedi et al., 2016). The majority of biochars produced are alkaline and generally are suitable as a liming material for tropical acidic soils (Ameloot et al., 2013b; Domene et al., 2014; Subedi et al., 2016). Biochars, derived from manure and sludge, have strong liming effects because these biochars contain a high content of basic cations and alkali metals (Ca, Na, K, and Mg) (Domene et al., 2014). Biochar is generally a sterilized material as it is prepared at a high temperature and its application has no direct contribution to the stimulation of microbial populations (Lehmann et al., 2011); However, indirectly, biochar could stimulate microbial biomass and activity as it is a highly porous material and may provide a conducive habitat for microbial growth (Subedi et al., 2016). Biochar has the ability to directly influence the soil microbial community structure by providing inorganic and organic nutrients. The large surface area enables the biochar to hold more organic and inorganic nutrients that supply the necessary nutrition and energy to soil microbes for their prolific metabolic growth (Ameloot et al., 2013a; Lehmann et al., 2011).

18.3 TYPES AND CLASSIFICATION OF ABIOTIC STRESS FACTORS IN CROP PRODUCTION

Stress is the condition when environmental conditions exceed the ranges required for normal growth and development of living organisms (Roelofs et al., 2008). Plants respond to the various kinds of stresses by altering their response ranging from the morphological level to the molecular level. A single stress can have effects on various underlying processes while considering the whole plant. Abiotic stresses are the key factors limiting crop production worldwide and, because of the immobile nature of plants, they are prone to interact immediately with any abiotic environmental factor. These factors include drought, heat, cold, salinity, metals/metalloids, antioxidants, exposure to chemicals and nutrient imbalances. According to the estimates, these stresses can have a severe impact on growth and development of crops, reducing ca. 70% of overall yield (Ramegowda and Senthil-Kumar, 2015). Considering the expected population growth of the world and its food demand, finding the ways to improve crop management and/or tolerance concerning abiotic stress factors will be crucial to further improving agricultural production and enhancing food security.

18.3.1 Salinity Stress

The stress caused due to the accumulation of salts is a major abiotic stress. Salinity stress decreases growth rate, leaf expansion rate, vegetative development, leaf area index and net assimilation (Hasanuzzaman et al., 2013a). It also restricts photosynthesis and elevates production of reactive oxygen species (ROS). The damage produced due to ROS to proteins, lipids and nucleic acids halts metabolic processes (Ahmad et al., 2016; Hasanuzzaman et al., 2013b).

18.3.2 Drought Stress

Climate changes have resulted in elevated temperature and CO_2 levels with uneven rainfall patterns over recent decades. An extended period of inadequate rainfall causes drought and intense drought condition reduces the availability of water in the soil for plant uptake and brings the plant to a permanent wilting point or plant death. Whereas, in contrast, sporadic drought conditions do not affect the plant growth and development to lethal levels. Under such conditions, the ability of plants to produce biomass and yield is much less compared with the hydrated condition is called drought tolerance. This type of condition is called drought stress condition (Hasanuzzaman et al., 2013a; Moussa and Abdel-Aziz, 2008). The fundamental responses of the plant under drought include a reduction in metabolic demands and mobilize the metabolites in order to produce the protective compounds for osmotic adjustments. In addition, severe transpiration losses and induction of oxidative stress also occur under drought stress (Impa et al., 2012).

18.3.3 Cold Stress

Cold stress is one of the major abiotic stresses that decreases or limits crop productivity by affecting the quality of produce and the post-harvest shelf life (Thakur and Nayyar, 2013). Because the plants are sessile, they have to adjust their metabolism to survive under cold stress. Many agronomic crop plants cannot acclimatize to cold stress, however, most of the temperate plant can acquire cold tolerance under sub-lethal cold stress. Under cold stress, almost all cellular functions such as signal transduction pathways, which controls several enzyme activities, namely protein kinase and phosphatase, are negatively affected. In addition, other physiological changes under cold stress include plasma membrane solidification (Chen et al., 2013), reduced leaf expansion and stunted growth (Thakur and Nayyar, 2013) and decreased photosynthetic activity and efficiency (Farooq et al., 2009).

18.3.4 Heat Stress

Elevated temperature due to climatic changes has become a global concern for its severe negative effects on crop production, and the situation is worsening consistently and gradually. High vulnerability of crop plants to rising temperature or temperature variations has become a serious problem. In the last century, a 0.5°C rise in ambient temperature was recorded, which has been projected to rise by 0.2°C in the next two decades and the rise in temperature is expected to reach 1°C to 3.4°C in 2100 (IPCC, 2007). Heat stress affects seed germination, photosynthetic efficiency and plant development by anther indehiscence and pollen grain swelling causing yield reduction (Hasanuzzaman et al., 2013b; Wahid et al., 2007).

18.3.5 Metals and Mineral Nutrients

Agriculture in the current modern and industrialized period is highly dependent on the use of chemical fertilizers, sewage waste irrigation, pesticides and insecticides. A huge amount of toxic materials like metals are being introduced into agroecosystems from industrialization causing serious harmful effects on the soil-plant-environment system. For example, cadmium (Cd) contamination is considered a big concern to the agroecosystems because it resides in soil over many years. Accumulation of such toxic metals in plants has several consequences, including poor seed germination, impaired root growth, reduced photosynthesis rate, low transpiration and leaf chlorosis and senescence (Drzewiecka et al., 2012). Deficiency of mineral nutrients has widely been reported to affect plant growth and physiological responses coupled with other biotic and abiotic stresses (Hajiboland, 2012). Abiotic stresses affect crop productivity more directly and significantly throughout the world, reducing average yields of almost all the major crop plants. Importantly, all major abiotic stresses are related and interconnected to each other and produce specific and general effects on crops.

18.4 BIOCHAR FOR SUSTAINABLE CROP PRODUCTION AND SOIL MANAGEMENT

Biochar can be defined as a stable, recalcitrant organic material produced under low or zero oxygen levels at various temperatures in a controlled condition usually through a technique called pyrolysis. Biochar acts as a soil conditioner and increases the fertility and quality of soil by improving soil properties and its processes via enhancing physicochemical and biological characteristics of soil, including enhancing CEC, water and nutrient retention capacity and declining the rate of nutrient deficiency via adsorption, decreased groundwater contamination and supporting microbial activity like reducing the bioaccessibility of pesticides undergoing helpful soil functioning via decomposition (Major et al., 2010).

Biochar is a carbon-rich source produced from woody and agriculture biomass under a limited supply of oxygen at high temperatures (Verheijen et al., 2004). Biochar consists of hydrogen (H), carbon (C), oxygen (O), sulfur (S) and nitrogen (N) (Lehmann and Joseph, 2015). The chemical characteristics of biochar mainly depend upon the type of feedstock as well as pyrolysis temperature. The adsorption characteristic of biochar is a key property, which increases the adsorption capacity of biochar to hold nutrients as well as various environmental toxicants and thus makes it a suitable candidate for the management of degraded soils (Inyang et al., 2016).

Biochar is a heterogeneous material due to the wide range of biomass that can be utilized for the production of biochar such as forest waste, i.e., tree bark, wood pellets and wood chops; crop residues like wheat straw, corn stover, rice husk, nut shells; switch grass, bamboo and organic waste such as paper pulp, municipal green waste, sawmill waste, manure, sewage sludge, sugarcane bagasse and even human waste. Generation of energy from the pyrolysis process depends on biomass stock and its concentration (Clough et al., 2010; Maraseni, 2010). Suitability of each biomass depends upon ecological, climatic, fiscal, logistic, physical and chemical factors (Verheijen et al., 2004).

Physically, biochar is a porous material and has a high surface area due to which it inhibits or slows down the process of nutrients leaching by adsorption phenomenon and makes them available for vegetation. Biochar has a strong ability to adsorb toxic materials such as heavy metals like cadmium and arsenic in contaminated soils, as well as manganese from acidic soils owing to its surface ability that has several chemically reactive groups like ketones, OH and COOH (Berek et al., 2011). It has been proven that biochar lowers the rate of nitrous oxide emissions, NH_3 volatilization and nitrogen leaching (Clough et al., 2013). It has a high C/N ratio, low mass content and is stable due to long chain carbon (aromatic structure), shows persistence due to recalcitrant nature and acts as a carbon sink in the soil (Biederman and Harpole, 2013). Biochar has neutral to alkaline pH, favorable in acidic soil and usually increases the soil pH. Biochar production at low temperature can manage the soil pH. All biochar has similar characteristics to some extent. Physical and chemical composition of biochar depends on its temperature limits or conditions of pyrolysis and nature of feedstock used (Zheng et al., 2010).

18.5 BIOCHAR TO REDUCE DROUGHT STRESS ON PLANTS

Drought is becoming a rising concern across the world as water scarcity, changing climate patterns and the need to feed a growing population continues to dominate the landscape of pressing global issues. Drought stress is characterized by a reduction of water content, diminished leaf water potential and turgor loss, closure of stomata and decrease in cell enlargement and growth. Severe water stress may result in the arrest of photosynthesis, disturbance of metabolism and finally the death of the plant (Jaleel et al., 2008). Biochar is pyrolyzed organic material intended for use as a soil amendment to sustainably sequester C and concurrently improve soil function, while avoiding any adverse effects in both the short and long term (Lehmann and Joseph, 2015). Biochar as a miraculous soil conditioner has gained attention during the last decade, a highly porous material—a characteristic that when combined with its unique chemical makeup, allows it to capture and hold minute particles. This not only allows it to hang on to nutrients, but it also allows it to hold on to moisture and provide a habitat for microorganisms (Viger et al., 2014). However, responses of microbial communities to biochar addition in particular in relation to abiotic disturbances are seldom documented (Liang et al., 2014). An example of these disturbances, which is predicted to be exacerbated by global warming, is regional drought. It has been known that fungal-based food webs are more resistant to drought than their bacterial counterparts (de Vries et al., 2012). Biochar addition can increase the resistance of both bacterial and fungal networks to drought. Contrary to expectations, this result was not related to a change in the dominance of fungal or bacteria. In general, soil amended with biochar was characterized by a faster recovery of soil microbial properties to its basal values (Liang et al., 2014).

Drought is characterized by frequent drying and wetting cycles that strongly affect soil properties such as soil microbiological properties, organic matter decomposition and plant growth (Geng et al., 2015; Mariotte et al., 2015). Ohashi et al. (2015) reported non-significant effects of short-term drought stress on soil respiration in tropical seasonal rainforest soil. Drought stress reduces plant growth and biomass production by limiting leaf expansion and stomatal conductance, and resultantly, the photosynthetic rate is decreased (Osakabe et al., 2014; Tardieu et al., 2014). The other processes include plant hydraulics, phytohormone production, osmotic adjustment and signaling of reactive oxygen species (Farooq et al., 2009; Khan et al., 2015; Tardieu et al., 2014). Experimental evidence so far shows that incorporation of biochar to soil enhanced soil water-holding capacity (Asai et al., 2009; Karhu et al., 2011; Laird et al., 2010), improved soil water permeability (Asai et al., 2009), improved saturated hydraulic conductivity (Asai et al., 2009), reduced soil strength (Busscher et al., 2010; Chan et al., 2007, 2008), modified soil bulk density (Laird et al., 2011) and modified aggregate stability (Busscher et al., 2010; Peng et al., 2011). Due to its physical properties, biochar helps increase water holding capacity and reduces nutrient leaching. Biochar is very porous which increases adsorption properties allowing a greater retention of water and nutrients in the soil solution (Adams et al., 2013). One greenhouse study found that the water holding capacity nearly doubled when 15% poultry litter biochar by weight was added to 9 kg of sandy loam soil. The ability to retain a relatively large quantity of water aids plant growth when under water stress. In another greenhouse study, Artiola et al., (2012) found that soil amended with 2% and 4% biochar by weight (15 kg of loamy sand used) had higher yields than control plants after undergoing water stress.

Plant physiological response (other than crop yield) related to water uptake biochar amended soils remains poorly understood and has seldom been investigated (Elad et al., 2010; Graber et al., 2010). The addition of biochar to sandy soil changes soil characteristics such as its texture and porosity. Hypothetically, finer textured soils (after biochar addition) in arid climates should be associated with more negative plant and soil water potentials during drought, inducing a greater resistance of xylem to cavitation and shallower root systems than coarse soils (Sperry and Hacke, 2002). Biochar application can increase growth, drought tolerance and leaf-N– and water use efficiency of quinoa despite larger plant–leaf areas (Kammann et al., 2011). Biochar enhanced plant growth and biomass production under drought stress such as in okra (Batool et al., 2015) and maize (Haider et al., 2015). Similar results were observed for the effects of biochar on wheat grown under drought in the field (Olmo et al., 2014) and biomass and yield of tomato (Akhtar et al., 2014). Application of biochar increased resistance of tomato seedlings to drought in sand and silt clay soil (Mulcahy et al., 2013; Vaccari et al., 2015). Moreover, Githinji (2014) also observed an increased leaf quality after biochar application. Several researchers have found improved photosynthetic activity under drought (e.g. Lyu et al., 2016; Xiao et al., 2016). For example, Akhtar et al. (2014) recorded significantly higher chlorophyll contents, stomatal conductance, photosynthetic rate, water use efficiency and relative water content after biochar application. Similar changes in maize physiological attributed were observed after biochar application by Haider et al. (2015) and Uzoma et al. (2011) in sandy soil. Application biochar has extensively been linked to increased water use efficiencies and improved water relations under drought stress (Afshar et al., 2016; Baronti et al., 2014; Batool et al., 2015). However, these observations indicated that the extent of biochar effects on crop growth and development under drought strongly depend on soil type, biochar characteristics and crop species.

Artiola et al. (2012) investigated the ameliorative effects of pine waste-derived biochar on the growth of drought-stressed romaine lettuce (*Lactuca sativa* L.) and Bermuda grass (*Cynodon dactylon*) in the alkaline loamy sand under greenhouse conditions. The results revealed the positive effects of biochar on the growth, development and biomass production of both crops at 2% and 4% rates of biochar application, however, the growth of lettuce was initially decreased at the 4% biochar application rate, which was attributed to an increase in soil pH, indicating the requirement of biochar adjustment before getting positive effects of biochar application. Addition of biochar to a vineyard also increased grape productivity and quality during the low rainfall years. Biochar application to the vineyard field increased the vine yield especially in the years receiving lowest rainfall, however, no significant effects were observed on the grape quality parameters such as brisk, total acidity and anthocyanins (Genesio et al., 2015). There is also compelling evidence that biochar could reduce the drought stress when applied in combination with microbial inoculants (Liu et al., 2017; Mickan et al., 2016). For example, *Bradyrhizobium* sp. inoculation to biochar increased growth, biomass production, N and P uptake and nodulation in lupin (*Lupinus angustifolius* L.) under limited water availability (Egamberdieva et al., 2017). Similarly, in another study involving the birch wood-derived biochar inoculation with *Rhizophagus irregularis*, Liu et al. (2017) found that biochar application influenced the potato leaf area, root biomass, WUE and soil pH. However, Pressler et al. (2017) did not find significant effects of wood-derived biochar on soil fauna including protozoa, nematodes, bacteria, fungi and arthropods under drought conditions. This discussion may lead to the facts that biochar application with other bioresources such as organic amendments and microbial inoculants may reduce drought stress to plants, however, these effects vary with biochar type, soil characteristics, crop types and climatic conditions.

18.6 USE OF BIOCHAR TO ALLEVIATE SALINITY STRESS ON PLANTS

18.6.1 Effects of Salinity on Plant Growth and Development

Globally, soil salinization is recognized as the most threatening issue for crop production and food security. Approximately, 800 × 106 million hectares (ha) of soil is loaded with salt content and about 20% of the total productive land is influenced by salinity with an annual increase of 1–2% (Ali et al., 2017; Munns and Tester, 2008). The accumulation of free salts in soils is the key factor responsible for soil salinization. Generally, salt-affected soils are categorized as sodic, saline or sodic-saline soils, which are primarily based on the criterion of electrical conductivity (EC), exchangeable sodium percentage, sodium absorption and depend on the saturated paste extracts of the soil (Richards, 1954). Saline soils are deficient in C:N ratio as well as organic matter. The main sources of soil salinity are weathering of salt containing minerals as well as the release of salts from mineral deposits (Rath and Rousk, 2015). Moreover, soils from arid zones are also enriched with salts due to shallow saline ground water. This shallow groundwater when moved upward evaporates, leaving salts behind on or below the soil surface (Ali et al., 2017; Rath and Rousk, 2015). Another source of salinization is the application of brackish water in agricultural soils. Additional factors of soil salinity may include movement of the water table, climatic conditions, rainfall patterns, land use practices and seepage salting (Ali et al., 2017).

Saline soils adversely affect physiochemical properties of soils, reduce organic matter, microbial activity and microbial mass (Yan et al., 2015). Salt stress has been reported to significantly reduce soil respiration under short and long-term salt stress due to a reduction in soil enzymatic activities and alterations in soil microbial communities (Rath and Rousk, 2015). While reviewing the effects of salinity on soil C cycling and associated microbial processes, Rath and Rousk (2015) concluded that changes in soil respiration were linked with the effects of salinity on soil microbial communities. However, the quality of the organic matter also determines the susceptibility to decomposition by soil microorganisms (Hasbullah and Marschner, 2015). Similarly, plants grown on salt-affected soils are more vulnerable to salt toxicity in several ways, that is, reduced plant growth, biomass, yield, leaf water content, nutrient uptake, damage photosynthesis machinery, nutritional quality and further lead to death (Noman et al., 2015; Siddiqui et al., 2015). Moreover, salt stress also produced abiotic stress in plants by producing hydrogen peroxide (H_2O_2), which damages the activities of key enzymes including catalase (CAT), ascorbate peroxidase (APX) and superoxidase dismutase (SOD) (Abbasi et al., 2015; Mohamed et al., 2017).

18.6.2 Mechanisms of Biochar-Induced Salinity Alleviation

The use of biochar amendments has been found useful in mitigating salt stress to plants by modifying soil physical, chemical and biological properties, which are directly linked with Na dynamics in the soil solution phase (e.g., Chaganti and Crohn, 2015; Drake et al., 2016; Oo et al., 2015). Rizwan et al. (2016) also reviewed the extensive literature to conclude that biochar improved soil physicochemical and biological properties under salinity stress of differential origin. Luo et al. (2017) studied the effects of poultry-manure derived biochar compost on soil biological indicators and found that biochar enhanced microbial biomass carbon and the activities of urease, invertase and phosphatase enzymes in saline soil under maize cultivation. Similarly, Bhaduri et al. (2016) also found varying but positive effects of biochar on soil enzymatic activities in saline soil depending on the rate of biochar application and incubation conditions. Organic amendments other than biochar have also been effective to improve the soil physicochemical properties under saline soils (Wang et al., 2014), however, the data about the effects of biochar on soil properties under saline conditions is not consistent (Thomas et al., 2013; Wu et al., 2014). For example, biochar at the application rate of 30 g m^{-2} increased soil EC but resulted in non-significant effects on soil pH under salt stress (Thomas et al., 2013). In another study, application of biochar in saline soil reduced soil pH and increased SOC contents, CEC and available P (Wu et al., 2014). Similar results were observed when composted biochar was applied to the saline soil which improved soil organic matter and CEC but reduced the exchangeable Na and soil pH (Luo et al., 2017). These investigations indicate that the biochar amendments can improve soil properties and plant growth under saline conditions.

Biochar has multiple binding sites due to the presence/or modification of various functional groups produced during the pyrolysis process depending on the type of feedstock and temperature. After the addition of biochar to salt-affected soils, it was observed that sodium content and soil pH was significantly reduced compared with control treatments (Lashari et al., 2013). It was also observed that soil organic carbon and available phosphorus was also increased in biochar amended soils. It was suggested that biochar addition to salt-affected soils could be an alternative solution due to its high adsorption capacity of Na^+ (Lashari et al., 2013). Similarly, when pot grown wheat plants were exposed to salt stress, it was observed that yield and growth of plants were increased, and significant reduction of K^+, Na^+, Ca^{2+} and Mg^{2+} were observed in leachate. They suggested that the enhancement in exchangeable sites of Ca^{2+} and Mg^{2+} have promisingly reduced exchangeable Na^+ in soils, which resulted in the improved in soil health and physical properties in salt-rich soils (Akhtar et al., 2015).

Biochar amendments also stimulate biological activity in salt-affected soils. Due to its high content of C in biochar, C amount was increased in biochar amended soils. Reportedly, the increase in soil microbial activity might be due to excess amount of C, a sole source of energy required for the enhanced microbial activity in salt-affected soils. Under salt stress, these microbes could provide resistance to plants against salt uptake by enhancing soil carbon, further relieving salt stress in plants (Amini et al., 2016). Addition of biochar to salt-affected soils also increased plant growth, photosynthesis, biomass and nutritional quality of cereal crops. For example, when biochar derived from rice hull significantly increased maize growth in tidal land remediated soils enriched with high amounts of exchangeable sodium and high concentrations of dissolved soils (Kim et al., 2016). Results suggested that biochar addition to salt-affected soils could reduce salt stress by enhancing nutrient uptake in plants. Reportedly, biochar amendments increased P levels in maize tissues in a dose-dependent manner (Kim et al., 2016). Similarly, results from the study of Usman et al. (2016) also revealed that biochar increased K, P, Fe, Zn, Mn and Cu in tomato plants irrigated with saline water compared with non-saline irrigation.

18.7 BIOCHAR TO MITIGATE HEAVY METAL STRESS ON PLANTS

18.7.1 Heavy Metal Pollution

Heavy metals are composed of an ill-defined group of inorganic chemical hazards. These metals are chromium (Cr), nickel (Ni), cadmium (Cd), zinc (Zn), arsenic (As), lead (Pb) and mercury (Hg), which are commonly prevailed in contaminated soils (Evanko and Dzombak, 1997). Unfortunately, heavy metal pollution in the soils is a critical issue worldwide due to their abundant occurrence, non-biodegradable, toxic and accumulative behaviors (Azimi et al., 2017; Hu et al., 2017; Kayastha, 2015). Exposure of these toxic heavy metals to the soil environment is originated by various anthropogenic (humans related) sources like industrial effluents, burning of fossil fuels, metals mining or smelter and usage of agricultural products, i.e. pesticides and fertilizers (Akoto et al., 2017; Yan et al., 2018), as well as natural sources like weathering of rocks, volcanic eruption, forest fire and wind-blown dust (Cyraniak and Draszawka-Bołzan, 2014; Dixit et al., 2015; Nagajyoti et al., 2010).

18.7.2 Effects of Heavy Metals on Plants

Elevated levels of heavy metal pollution in soil also reduces growth and dry biomass like shoots and roots lengths (Ibrahim et al., 2017; Sheetal et al., 2016; Stambulska et al., 2018) and grain yield in plants and crops (Alia et al., 2015; Ashraf et al., 2017; Ramzani et al., 2016). Higher concentrations of heavy metals in plants adversely influence the contents of biochemical compounds like starch, protein, fat, amino acid and fiber in plants (Pant and Tripathi, 2014; Sharma and Dhiman, 2013), which, as a result, deteriorate overall food quality (Zhong et al., 2018).

The most important effects of heavy metal toxicity in plants is the excessive production of ROS such as malondialdehyde (MDA), superoxide anion (O_2^-) and H_2O_2 which directly affects metabolism (Aydoğan et al., 2017; Ramzani et al., 2017) and normal functionality of a cell through degrading proteins, enzymes and nucleic acids. While in parallel to this oxidative stress, the enzymatic [like SOD, CAT, glutathione reductase (GR), guaiacol peroxidase (GPX) and APX] and non-enzymatic [ascorbic acid (AsA) and dehydroascorbate reductase (DHAR)] components in plants are also produced as defense mechanisms in plants against HMs stress (Chandrakar et al., 2016; Kotapati et al., 2017) but the higher concentrations of these metals in soil suppressed the enzymatic activities in plants (Branco-Neves et al., 2017; Rabêlo and Borgo, 2016).

18.7.3 Role of Biochar in Mitigating Heavy Metal Stress

Biochar is a black carbon having exclusive attributes like a large surface area, porous structure, a high CEC, active functional groups and a mineral phase for adsorption of heavy metals (Luo et al., 2016; Qi et al., 2017; Xie et al., 2015). Biochar has its potential role for immobilization of heavy metals in the soil (Vilvanathan and Shanthakumar, 2018; Zhou et al., 2018). Several factors like soil pH, CEC and sorption mechanism are primary factors involved in this sorption process of heavy metals (Park et al., 2011). The large specific surface area of biochar contains a large number of negative exchange sites. Therefore, incorporation of biochar in the soil increases negative exchange sites for adsorption of heavy metals in the soil which resultantly increases the soil pH (Li et al., 2018). Biochar also increases CEC of the soil due to exclusive attributes like a porous structure and a large surface area (Jien and Wang, 2013; Lehmann, 2007) as well as aging mainly because of increased carboxylation of C by abiotic oxidation (Cheng et al., 2006; Lehmann, 2007). Therefore, this increase in both the pH and CEC of BC amended soil reduced the bioavailability of heavy metals to plants (Huang et al., 2017; Mohamed et al., 2017).

During the process of sorption of heavy metals, precipitation is the formation of solids either in solution or on the surface. Precipitation has been an important mechanism held responsible for immobilization of HMs in soil using biochar sorbent. The pH of soil solution is the most significant indicator during the adsorption process, influencing on charge surface area, speciation of adsorbent and degree of ionization. Moreover, biochar contains various surface functional groups [(hydroxyl (AOH) and carboxylate (ACOOH)] which alters with increasing pH of the solution (Kołodyńska et al., 2012; Patra et al., 2017; Zhang et al., 2013). The large surface area of biochar also contributes to ion exchange and surface complexation during heavy metals immobility in the soil (Patra et al., 2017). This efficient potential of biochar for immobilizing heavy metals in the soil also reduces metals uptake (Moreno-Barriga et al., 2017; Sizmur et al., 2016) and resultantly alleviates metals toxicity (Seneviratne et al., 2017) to plants. The presence of biochar in soil can be an efficient source of essential nutrients (Liu et al., 2014) and alleviation of heavy metals toxicity in plants through adsorption of metals on to large surfaces of BC (Patra et al., 2017), which consequently improves growth, biomass (Al-Wabel et al., 2015), yields (Hussain et al., 2017), biochemistry (Różyło et al., 2017), oxidative stress and antioxidant defense mechanisms (Ali et al., 2017; Ramzani et al., 2017) in plants. Biochar application in the soil improves overall soil health, enzymes and microbial biomass (Lwin et al., 2018).

Biochar consists of oxides, carbonates and hydroxides that can act as liming agents in soil and improve soil quality and productivity (Krishnakumar et al., 2014). Biochar has a porous carbonaceous structure, various surface functional groups and a charged surface. The porous structure of biochar contains significant quantities of extractable humic and fulvic type of substances (Lin et al., 2012). Furthermore, biochar is a low-density substance that can reduce soil bulk density of soil (Laird et al., 2010; Rogovska et al., 2011) that as a result, increases water infiltration, soil aeration, root penetration and improved soil aggregate stability (Glaser et al., 2002). Biochar mineralization occurs in the soil which improves soil aggregates through an increase in the quantity of oxidized functional groups (Cheng et al., 2006), which consequently facilitates flocculation of the soil particles and biochar. Consequently, the application of biochar in soil efficiently improved overall soil health and immobility of heavy metals causing less phytotoxicity (Al-Wabel et al., 2015).

Biochar improves the activities of soil enzymes like β-glucosaminidase, β-glucosidase, urease, phosphomonoesterase, acid phosphatase and catalase in heavy metal polluted soils (Cui et al., 2013) through increasing soil organic matter content and improving soil properties (Akça and Namli, 2015). In addition, biochar in the soil also alleviates heavy metals toxicity via adsorption of heavy metals on surfaces of biochar (Lwin et al., 2018). Heavy metals pollutants in soil negatively influence soil properties and functionality through disturbing soil biological and physiochemical attributes like poor soil health (structure and productivity) and low soil microbial activities (Lwin et al., 2018). Incorporation of a large amount of biochar in the soil improved the habitat for microbial population via increased porosity (Hairani et al., 2016; Wuddivira et al., 2009). Moreover, biochar has the ability to donate or accept electrons in their environments via biological pathways (Klüpfel et al., 2014; Saquing et al., 2016). Biochar in soil also promotes microbial electron shuttling processes analogous to soil organic matter with redox active functional groups (Graber et al., 2014; Yuan et al., 2017). Several other factors like pH, decreased the solubility of heavy metals, improved allochthonous microbial biomass and available nutrients, also contributing to the variation in soil microbial population (Alburquerque et al., 2011; Chen et al., 2015).

18.8 BIOCHAR TO REDUCE TEMPERATURE AND HEAT STRESS ON PLANTS

In the present era, agriculture and sustainable crop production are one of the key issues for population growth under varying environmental conditions. Abiotic stresses are recognized as one of the most important environmental threat for food production worldwide (Fahad et al., 2017). Among abiotic stresses, heat stress is one of the most important stresses which have adverse effects on growth, productivity, and yield of crops. Heat stress is not only affecting morphological, physiological and biochemical attributes but also reduced the nutritional quality of various cereal crops. It has been reported that the global wheat production was declined by 6% for each degree Celsius rise in temperature (Fahad et al., 2017). Though high temperatures are also good for sustainable crop production in some cooler areas around the globe, but the overall impacts on global food security are adverse (Challinor et al., 2014). Rising high temperatures severely disturbed photosynthesis production, proteins, inactivate key enzymes, damage proteins and produced ROS. All these factors are responsible for the reduction of plant growth and favor oxidative damage. Moreover, under high-temperature stress, during seed filling could also result in accelerated filling, which will further result in poor quality and yield reduction.

18.8.1 Plant Responses to Heat Stress

Elevated temperatures and excessive radiation are considered as major factors affecting plant growth and development in tropical regions. Plants exposed to high temperatures showed various symptoms of toxicity including leaves and twig scorching, sunburn, stunted growth and yellowing of leaves and fruits (Vollenweider and Günthardt-Goerg, 2005). In addition, high temperatures also resulted in poor seed germination which further leads to poor stand and establishment of plants. High temperatures also adversely affect cereal crops, but toxicity depends upon time and duration of exposure and severity of heat stress (Fahad et al., 2016a). High-temperature stress significantly reduced seed-set in sorghum and the number of spikes and florets per plant in rice (Fahad et al., 2016b). It has been observed that pollens and anthers were adversely affected by inside floret under heat stress (Fahad et al., 2015b). Similarly, a substantial growth inhabitation was observed in sugarcane and maize under heat stress (Wahid and Close, 2007).

High-temperature shocks are responsible for yield reduction in major cereals under varying climatic conditions (Camejo et al., 2006). Temperature shocks significantly reduced rice yield by damaging different growth and yield traits. Reportedly, high-temperature stress significantly reduced the yield of peanut (*Arachis hypogea* L.) and common beans (*Phaseolus vulgaris* L.) (Rainey and Griffiths, 2005). Similarly, fertilization, meiosis and growth of fertilized embryos were also adversely affected in tomato (*Lycopersicum esculentum* Mill.) by heat stress further leading to yield reduction (Camejo et al., 2005).

Some key factors including leaf and canopy temperature, transpiration rate, leaf water potential and stomatal conductance were also affected (Simoes-Araujo et al., 2003). On exposure to high temperatures, a decline in leaf tissue water content was noticed in sugarcane despite the fact that a sufficient supply of water was provided to the soil. In addition, it may be noted that it also had a negative impact on root conductance (Wahid et al., 2007; Wahid and Close, 2007). In general, water losses are extensively greater in day times due to increased transpiration rate which leads to further damaging key physiological processes in plants under heat stress. Similarly, temperature shocks also significantly reduced root growth, which ultimately reduced the translocation of water and nutrients (Huang et al., 2012). Similarly, heat stress also affects the activity of nitrate reductase involved in nutrient metabolism (Rennenberg et al., 2006). This reduction in nutrient uptake might be due to reduced root mass and nutrient uptake per unit root area (Basirirad, 2000).

In plants, exposure to rising temperatures was also responsible for the reduction and/or degradation of chlorophyll biosynthesis which resulted in a decrease in plant growth, biomass, yield and poor nutritional quality of major cereals (Bita and Gerats, 2013). This reduction in chlorophyll synthesis might be due to enhanced denaturing of key enzymes (Dutta et al., 2009). For example, 5-aminolevulinate dehydratase is an important enzyme actively participating in biosynthesis pathways of pyrrole. The activity of this key enzyme was reduced in wheat plants when exposed to temperature shocks (Mohanty et al., 2006). High temperatures also adversely affect sucrose and starch production by reducing the activity of sucrose and phosphate synthase, invertase and ADP-glucose pyrophosphorylase (Hasanuzzaman et al., 2013a; Rodríguez et al., 2005). Similarly, 60% degradation of chlorophyll synthesis was also observed in *Cucumis sativus* L when exposed to a temperature at 42°C due to inhabitation of production of 5-aminolevulinate at high-temperature shocks (Tewari and Tripathy, 1998).

The production of ROS is a common phenomenon that was greatly observed in plants when exposed to abiotic stress. Similarly, under heat stress, ROS are produced which affect growth, reduced yield, poor nutritional quality of major cereals and biomass (Fahad et al., 2015a). The production of the ROS has also been reported under high-temperature stress (Fahad et al., 2015a; Wahid et al., 2007). Plants have developed counteracting mechanisms to cope with high-temperature shocks. This mechanism is involved in the generation of key enzymatic and non-enzymatic activities. It has been reported that some key signaling molecules are responsible for enhancing the activities of key enzymes in plants. It was noticed that the activities of APX, SOD and CAT was significantly reduced at 50°C, whereas the activities of GR and peroxidase (POX) were decreased at all temperature levels ranging from 20–50°C (Almeselmani et al., 2006; Chakraborty and Pradhan, 2011).

18.8.2 Biochar and Heat Stress Alleviation

18.8.2.1 Improved Water Holding Capacity

Potentially, biochar amendments significantly modify soil hydrology which results in a change in water cycling processes driven by water. In addition, improving the water holding capacity of soil is an important parameter to increase crop production under abiotic stress. Water movement in soils mainly depends on spaces between or/within soil particles. Primarily, any change in soil hydrological characteristics also alters the size, volume and connectivity of these particles in soils. Moreover, biochar is a porous material and its porosity mainly depends upon the type of feedstock and pyrolysis temperature. Simply, biochar addition increased the water holding capacity of the soil due to its porous structure. When biochar derived from birch was applied in silt loam soil, water holding capacity was increased by 11% (Karhu et al., 2011). Similarly, an increase in water holding capacity of 7% was observed in Typic Kandiudult soil when amended with pecan shell derived biochar, (Busscher et al., 2011). It has been reported that biochar applications significantly increased tomato productivity by increasing water holding capacity (Akhtar et al., 2014). Regarding plant available water, various results have been published by numerous authors. The amount of plant available water was doubled in a soil amended with both biochar and composts (Liu et al., 2012). Similar results were also found when a sandy soil was amended with biochar to grow turf grass (Brockhoff et al., 2010). Although limited information is available on the application of biochar under heat stress, there might be a possible mechanism in the alleviation of heat stress in plants by providing available water for plant uptake.

18.8.2.2 Availability of Plant Nutrients and Improved Growth

Pyrolysis temperature and feedstock type are also vital in achieving desired characteristics of biochar. Generally, increase in pyrolysis temperature reduced biochar production but increased total K, C and Mg contents, pH and surface area and reduced cation exchange capacity. Biochar produced at slow pyrolysis temperature is enriched with S, N, Ca, available P, Mg, cation exchange capacity and surface area compared with biochar produced at fast pyrolysis temperature. Plant growth was also significantly altered by the addition of biochar, which was observed with the addition of biochar enhanced K and P concentrations in plant tissues (Taghizadeh-Toosi et al., 2012). The introduction of these nutrients might be due to the availability of labile organic compounds connected with biochar (Biederman and Harpole, 2013). Another possible mechanism of plant growth under various abiotic stresses might be due to the control of nutrient leaching (Biederman and Harpole, 2013). The porous structure of biochar, a large surface area with more negative binding sites, enhanced cation exchange capacity and retained nutrients in the long run (Major et al., 2012). In addition, the adsorption of P to biochar's surface slows down P leaching, which results in enhanced plant nutrient availability and uptake (Beck et al., 2011; Biederman and Harpole, 2013).

Biochar additions to soil also significantly increased plant growth characteristics in various cereal crops. It was observed that the addition of biochar significantly enhanced rice pollen development and anther dehiscence and protected rice plants from heat stress (Fahad et al., 2015b). Similarly, when biochar was applied alone and in combination with P, the growth of rice plants was significantly increased in plants treated with combined applications of biochar and P (Fahad et al., 2016a). Under heat stress, the availability of nutrients to plants may significantly enhance root elongation and nutrient uptake, which results in relief from the heat stress.

18.8.2.3 Improvement in Antioxidant Activities

Under abiotic stress, activities of key enzymes include APX, CAT, SOD and POX under heat stress (Bita and Gerats, 2013; Prasad and Djanaguiraman, 2011; Suriyasak et al., 2017). The addition of biochar to soils not only significantly increased the activities of key enzymes but also reduced the production of ROS under various abiotic stresses. Under heat stress, the activities of these enzymes could be enhanced due to the availability of nutrients in soils and improved soil physiochemical properties. This will, in turn, reduce the production of H_2O_2, a key component in producing abiotic stress in plants (Kanwal et al., 2018; Ramzani et al., 2017).

18.8.2.4 Positive Effects on Soil Biota

Soil microbes are very sensitive to many soil changes occurring due to anthropogenic as well as natural processes. The addition of organic matter and biochar significantly increases soil biota for several reasons such as biochar serving as a habitat for microorganisms due to its porous structure. This porous property protects microorganisms from predation (Lehmann et al., 2011) and biochar acts as a source of nutrients and energy by microbes (Ameloot et al., 2013a) and enhances soil-microbes-plant interaction. Improvement in soil microbes could also change solubilization of nutrients by microbes.

18.9 CONCLUSIONS AND THE WAY FORWARD

Biochar is an organic amendment consisting principally of recalcitrant organic C and offering multiple benefits to soil-plant systems. The use of biochar has recently been popularized for its role in mitigating abiotic stress-induced changes in plants due to its multifarious positive effect on soil physical, chemical, and biological properties. Extensive research has begun to emerge on the ameliorative role of biochar to reduce heat/temperature, salinity, drought and heavy metal stress on plants. However, the degree of these effects of biochar depends strongly on the nature and rate of biochar amendments as well as the physicochemical properties and climatic conditions of the soils. Moreover, the majority of studies reporting positive effects of biochar on plants under abiotic stresses were confined to laboratories and pot experiments under controlled conditions and field-scale investigation are scarce. More research, therefore, is required to validate the short-term findings from laboratory studies under field conditions.

REFERENCES

Abbasi, G.H., Akhtar, J., Ahmad, R., Jamil, M., Anwar-ul-Haq, M., Ali, S., Ijaz, M. (2015) Potassium application mitigates salt stress differentially at different growth stages in tolerant and sensitive maize hybrids. *Plant Growth Regulation* 76 No. 1: 111–125.

Abrishamkesh, S., Gorji, M., Asadi, H., Bagheri-Marandi, G.H., Pourbabaee, A.A. (2015) Effects of rice husk biochar application on the properties of alkaline soil and lentil growth. *Plant, Soil and Environment* 62: 475–482.

Adams, M.M., Benjamin, T.J., Emery, N.C., Brouder, S.J., Gibson, K.D. (2013) The effect of biochar on native and invasive prairie plant species. *Invasive Plant Science and Management* 6 No. 2: 197–207.

Afshar, R.K., Hashemi, M., DaCosta, M., Spargo, J., Sadeghbour, A. (2016) Biochar application and drought stress effects on physiological characteristics of *Silybum marianum. Communications in Soil Science and Plant Analysis* 47 No. 6: 743–752.

Ahmad, P., Abdul Latef, A.A., Hashem, A., Abd-Allah, E.F., Gucel, S., Tran, L.-S.P. (2016) Nitric oxide mitigates salt stress by regulating levels of osmolytes and antioxidant enzymes in chickpea. *Frontiers in Plant Sciences* 7: 347.

Akça, M.O., Namli, A. (2015) Effects of poultry litter biochar on soil enzyme activities and tomato, pepper and lettuce plants growth. *Eurasian Journal of Soil Science* 4: 161–168.

Akhtar, S.S., Andersen, M.N., Liu, F. (2015) Residual effects of biochar on improving growth, physiology and yield of wheat under salt stress. *Agricultural Water Management* 158: 61–68.

Akhtar, S.S., Li, G., Andersen, M.N., Liu, F. (2014) Biochar enhances yield and quality of tomato under reduced irrigation. *Agricultural Water Management* 138: 37–44.

Akoto, O., Bortey-Sam, N., Ikenaka, Y., Nakayama, S.M.M., Baido, E., Yohannes, Y.B., Ishizuka, M. (2017) Contamination levels and sources of heavy metals and a metalloid in surface soils in the Kumasi Metropolis, Ghana. *Journal of Health and Pollution* 7 No. 15: 28–39.

Alburquerque, J.A., De La Fuente, C., Bernal, M.P. (2011) Improvement of soil quality after "alperujo" compost application to two contaminated soils characterised by differing heavy metal solubility. *Journal of Environmental Management* 92 No. 3: 733–741.

Ali, S., Rizwan, M., Qayyum, M.F., Ok, Y.S., Ibrahim, M., Riaz, M., Arif, M.S., et al. (2017) Biochar soil amendment on alleviation of drought and salt stress in plants: A critical review. *Environmental Science and Pollution Research International* 24 No. 14: 12700–12712.

Alia, N., Sardar, K., Said, M., Salma, K., Sadia, A., Sadaf, S., Topeer, A., Miklas, S. (2015) Toxicity and bioaccumulation of heavy metals in spinach (*Spinacia oleracea*) grown in a controlled environment. *International Journal of Environmental Research and Public Health* 12 No. 7: 7400–7416.

Alling, V., Hale, S.E., Martinsen, V., Mulder, J., Smebye, A., Breedveld, G.D., Cornelissen, G. (2014) The role of biochar in retaining nutrients in amended tropical soils. *Journal of Plant Nutrition and Soil Science* 177 No. 5: 671–680.

Almeselmani, M., Deshmukh, P.S., Sairam, R.K., Kushwaha, S.R., Singh., T.P. (2006) Protective role of antioxidant enzymes under high temperature stress. *Plant Science: an International Journal of Experimental Plant Biology* 171 No. 3: 382–388.

Al-Wabel, M.I., Usman, A.R., El-Naggar, A.H., Aly, A.A., Ibrahim, H.M., Elmaghraby, S., Al-Omran, A. (2015) Conocarpus biochar as a soil amendment for reducing heavy metal availability and uptake by maize plants. *Saudi Journal of Biological Sciences* 22 No. 4: 503–511.

Ameloot, N., Graber, E.R., Verheije, F.G.A., De Neve, S. (2013a) Interactions between biochar stability and soil organisms: Review and research needs. *European Journal of Soil Science* 64 No. 4: 379–390.

Ameloot, N., Sleutel, S., Das, K.C., Kanagaratnam, J., de Neve, S. (2013b) Biochar amendment to soils with contrasting organic matter level: Effects on N mineralization and biological soil properties. *Global Change Biology Bioenergy* 7: 135–144.

Amini, S., Ghadiri, H., Chen, C., Marschner, P. (2016) Salt-affected soils, reclamation, carbon dynamics, and biochar: A review. *Journal of Soils and Sediments* 16 No. 3: 939–953.

Arif, M., Ilyas, M., Riaz, M., Ali, K., Shah, K., Haq, I.U., Fahad, S. (2017) Biochar improves phosphorus use efficiency of organic-inorganic fertilizers, maize-wheat productivity and soil quality in a low fertility alkaline soil. *Field Crops Research* 214: 25–37.

Artiola, J.F., Rasmussen, C., Freitas, R. (2012) Effects of a biochar-amended alkaline soil on the growth of romaine lettuce and Bermudagrass. *Soil Science* 177 No. 9: 561–570.

Asai, H., Samson, B.K., Stephan, H.M., Songyikhangsuthor, K., Homma, K., Kiyono, Y., Inoue, Y., et al. (2009) Biochar amendment techniques for upland rice production in Northern Laos. 1. Soil physical properties, leaf SPAD and grain yield. *Field Crops Research* 111 No. 1–2: 81–84.

Ashraf, U., Kanu, A.S., Deng, Q., Mo, Z., Pan, S., Tian, H., Tang, X. (2017) Lead (Pb) toxicity; physio-biochemical mechanisms, grain yield, quality, and Pb distribution proportions in scented rice. *Frontiers in Plant Science* 8: 259.

Atkinson, C.J., Fitzgerald, J.D., Hipps, N.A. (2010) Potential mechanisms for achieving agricultural benefits from biochar application to temperate soils: A review. *Plant and Soil* 337 No. 1–2: 1–18.

Aydoğan, S., Erdağ, B., Aktaş, L. (2017) Bioaccumulation and oxidative stress impact of Pb, Ni, Cu, and Cr heavy metals in two bryophyte species, *Pleurochaete squarrosa* and *Timmiella barbuloides*. *Turkish Journal of Botany* 41: 464–475.

Azimi, A., Azari, A., Rezakazemi, M., Ansarpour, M. (2017) Removal of heavy metals from industrial wastewaters: A review. *ChemBioEng Reviews* 4: 34–59.

Bailey, V.L., Fansler, S.J., Smith, J.L., Bolton, H. (2011) Reconciling apparent variability in effects of biochar amendment on soil enzyme activities by assay optimization. *Soil Biology and Biochemistry* 43 No. 2: 296–301.

Baronti, S., Alberti, G., Delle Vedove, G., Di Gennaro, F., Fellet, G., Genesio, L., Miglietta, F., et al. (2010) The biochar option to improve plant yields: First results from some field and pot experiments in Italy. *Italian Journal of Agronomy* 5 No. 1: 3–11.

Baronti, S., Vaccari, F.P., Miglietta, F., Calzolari, C., Lugato, E., Orlandini, S., Pini, R., et al. (2014) Impact of biochar application on plant water relations in *Vitis vinifera* (L.). *European Journal of Agronomy* 53: 38–44.

Basirirad, H. (2000) Kinetics of nutrient uptake by roots: Responses to global change. *New Phytologist* 147 No. 1: 155–169.

Batool, A., Taj, S., Rashid, A., Khalid, A., Qadeer, S., Saleem, A.R., Ghufran, M.A. (2015) Potential of soil amendments (Biochar and Gypsum) in increasing water use efficiency of *Abelmoschus esculentus* L. Moench. *Frontiers in Plant Science* 6: 733.

Beck, D.A., Johnson, G.R., Spelok, G.A. (2011) Amending greenroof soil with biochar to affect runoff water quantity and quality. *Environmental Pollution* 159: 2111–2118.

Beerling, D.J., Leake, J.R., Long, S.P., Scholes, J.D., Ton, J., Nelson, P.N., Bird, M., et al. (2018) Farming with crops and rocks to address global climate, food and soil security. *Nature Plants* 4 No. 3: 138–147.

Berek, A.K., Hue, N., Ahmad, A. (2011) Beneficial use of biochar to correct soil acidity. Available on line at www.ctahr.hawaii.edu/huen/nvh/biochar.pdf. Accessed on July 20, 2015.

Bhaduri, D., Saha, A., Desai, D., Meena, H.N. (2016) Restoration of carbon and microbial activity in salt-induced soil by application of peanut shell biochar during short-term incubation study. *Chemosphere* 148: 86–98.

Biederman, L.A., Harpole, W.S. (2013) Biochar and its effects on plant productivity and nutrient cycling: A meta-analysis. *GCB Bioenergy* 5 No. 2: 202–214.

Bita, C.E., Gerats, T. (2013) Plant tolerance to high temperature in a changing environment: Scientific fundamentals and production of heat stress-tolerant crops. *Frontiers in Plant Science* 4: 273.

Borchard, N., Siemens, J., Ladd, B., Möller, A., Amelung, W. (2014) Application of biochars to sandy and silty soil failed to increase maize yield under common agricultural practice. *Soil and Tillage Research* 144: 184–194.

Branco-Neves, S., Soares, C., Sousa, A., Martins, V., Azenha, M., Gerso, H., Fidalgo, F. (2017) An efficient antioxidant system and heavy metal exclusion from leaves make

Solanum cheesmaniae more tolerant to Cu than its cultivated counterpart. *Food and Energy Security* 6 No. 3: 123–133.

Brockhoff, S.R., Christians, N.E., Killorn, R.J., Horton, R., Davis, D.D. (2010) Physical and mineral-nutrition properties of sand-based turfgrass root zones amended with biochar. *Agronomy Journal* 102 No. 6: 1627–1631.

Busscher, W.J., Novak, J.M., Ahmedna, M. (2011) Physical effects of organic matter amendment of a southeastern US coastal loamy sand. *Soil Science* 176: 661–667.

Busscher, W.J., Novak, J.M., Evans, D.E., Watts, D.W., Niandou, M.A.S., Ahmedna, M. (2010) Influence of pecan biochar on physical properties of a Norfolk loamy sand. *Soil Science* 175 No. 1: 10–14.

Camejo, D., Jimenez, A., Alarcon, J.J., Torres, W., Gomez, J.M., Sevilla, F. (2006) Changes in photosynthetic parameters and antioxidant activities following heat-shock treatment in tomato plants. *Functional Plant Biology* 33 No. 2: 177–187.

Camejo, D., Rodriguez, P., Morales, M.A., Dell'amico, J.M., Torrecillas, A., Alarcon, J.J. (2005) High temperature effects on photosynthetic activity of two tomato cultivars with different heat susceptibility. *Journal of Plant Physiology* 162 No. 3: 281–289.

Carter, S., Shackley, S., Sohi, S., Suy, T.B., Haefele, S. (2013) The impact of biochar application on soil properties and plant growth of pot grown lettuce (*Lactuca sativa*) and cabbage (*Brassica chinensis*). *Agronomy* 3 No. 2: 404–418.

Chaganti, V.N., Crohn, D.M. (2015) Evaluating the relative contribution of physiochemical and biological factors in ameliorating a saline–sodic soil amended with composts and biochar and leached with reclaimed water. *Geoderma* 259–260: 45–55.

Chakraborty, U., Pradhan, D. (2011) High temperature-induced oxidative stress in *Lens culinaris*, role of antioxidants and amelioration of stress by chemical pretreatments. *Journal of Plant Interactions* 6 No. 1: 43–52.

Challinor, A.J., Watson, J., Lobell, D.B., Howden, S.M., Smith, D.R., Chhetri, N. (2014) A meta-analysis of crop yield under climate change and adaptation. *Nature Climate Change* 4 No. 4: 287–291.

Chan, K.Y., van Zwieten, L., Meszaros, I., Downie, A., Joseph, S. (2007) Agronomic values of greenwaste biochar as a soil amendment. *Soil Research* 45: 629–634.

Chan, K.Y., van Zwieten, L., Meszaros, I., Downie, A., Joseph, S. (2008) Using poultry litter biochars as soil amendments. *Soil Research* 46: 437–444.

Chandrakar, V., Dubey, A., Keshavkant, S. (2016) Modulation of antioxidant enzymes by salicylic acid in arsenic exposed *Glycine max* L. *Journal of Soil Science and Plant Nutrition* 16: 662–676.

Chen, S., Jin, W., Liu, A., Zhang, S., Liu, D., Wang, F., Lin, X., He, C. (2013) Arbuscular mycorrhizal fungi (AMF) increase growth and secondary metabolism in cucumber subjected to low temperature stress. *Scientia Horticulturae* 160: 222–229.

Chen, M., Xu, P., Zeng, G., Yang, C., Huang, D., Zhang, J. (2015) Bioremediation of soils contaminated with polycyclic aromatic hydrocarbons, petroleum, pesticides, chlorophenols and heavy metals by composting: Applications, microbes and future research needs. *Biotechnology Advances* 33 No. 6 Pt 1: 745–755.

Cheng, C.H., Lehmann, J., Thies, J.E., Burton, S.D., Engelhard, M.H. (2006) Oxidation of black carbon by biotic and abiotic processes. *Organic Geochemistry* 37 No. 11: 1477–1488.

Chintala, R., Mollinedo, J., Schumacher, T.E., Malo, D.D., Julson, J.L. (2014) Effect of biochar on chemical properties of acidic soil. *Archives of Agronomy and Soil Science* 60 No. 3: 393–404.

Clough, T.J., Bertram, J.E., Ray, J.L., Condron, L.M., O'Callaghan, M., Sherlock, R.R., Wells, N.S. (2010) Unweathered wood biochar impact on nitrous oxide emissions from a bovine-urine-amended pasture soil. *Soil Science Society of America Journal* 74 No. 3: 852–860.

Clough, T., Condron, L., Kammann, C., Müller, C. (2013) A review of biochar and soil nitrogen dynamics. *Agronomy* 3 No. 2: 275–293.

Crane-Droesch, A., Abiven, S., Jeffery, S., Torn, M.S. (2013) Heterogeneous global crop yield response to biochar: A meta-regression analysis. *Environmental Research Letters* 8 No. 4: 044049.

Cui, L., Yan, J., Yang, Y., Li, L., Quan, G., Ding, C., Chen, T., Fu, Q., Chang, A. (2013) Influence of biochar on microbial activities of heavy metals contaminated paddy fields. *BioResources* 8 No. 4: 5536–5548.

Cyraniak, E., Draszawka-Bołzan, B. (2014) Heavy metals in circulation biogeochemical. *World Scientific News* 4: 30–36.

de Vries, F.T., Liiri, M.E., Bjørnlund, L., Bowker, M.A., Christensen, S., Setälä, H.M., Bardgett, R.D. (2012) Land use alters the resistance and resilience of soil food webs to drought. *Nature Climate Change* 2 No. 4: 276–280.

Dempster, D.N., Jones, D.L., Murphy, D.V. (2012) Organic nitrogen mineralisation in two contrasting agro-ecosystems is unchanged by biochar addition. *Soil Biology and Biochemistry* 48: 47–50.

Dixit, R., Wasiullah, Malaviya, D., Pandiyan, K., Singh, U., Sahu, A., Shukla, R., et al. (2015) Bioremediation of heavy metals from soil and aquatic environment: An overview of principles and criteria of fundamental processes. *Sustainability* 7 No. 2: 2189–2212.

Domene, X., Mattana, S., Hanley, K., Enders, A., Lehmann, J. (2014) Medium-term effects of corn biochar addition on soil biota activities and functions in a temperate soil cropped to corn. *Soil Biology and Biochemistry* 72: 152–162.

Drake, J.A., Cavagnaro, T.R., Cunningham, S.C., Jackson, W.R., Patti, A.F. (2016) Does biochar improve establishment of tree seedlings in saline sodic soils? *Land Degradation and Development* 27 No. 1: 52–59.

Drzewiecka, K., Mleczek, M., Waśkiewicz, A., Goliński, P. (2012) Oxidative stress and phytoremediation. In *Abiotic Stress Responses in Plants: Metabolism, Productivity and Sustainability*, eds. Ahmad, P., Prasad, M.N.V., 425–449. Berlin: Springer.

Dutta, S., Mohanty, S., Tripathy, B.C. (2009) Role of temperature stress on chloroplast biogenesis and protein import in pea. *Plant Physiology* 150 No. 2: 1050–1061.

Egamberdieva, D., Reckling, M., Wirth, S. (2017) Biochar-based *Bradyrhizobium inoculum* improves growth of lupin (*Lupinus angustifolius* L.) under drought stress. *European Journal of Soil Biology* 78: 38–42.

Elad, Y., David, D.R., Harel, Y.M., Borenshtein, M., Kalifa, H.B., Silber, A., Graber, E.R. (2010) Induction of systemic resistance in plants by biochar, a soil-applied carbon sequestering agent. *Phytopathology* 100 No. 9: 913–921.

Evanko, C.R., Dzombak, D.A. (1997) Remediation of metals-contaminated soils and groundwater. Technology Evaluation Report, TE-97-01. Ground-Water Remediation Technologies Analysis Center, Pittsburgh, PA.

Fahad, S., Bajwa, A.A., Nazir, U., Anjum, S.A., Farooq, A., Zohaib, A., Sadia, S., et al. (2017) Crop production under drought and heat stress: Plant responses and management options. *Frontiers in Plant Science* 8: 1147.

Fahad, S., Hussain, S., Matloob, A., Khan, F.A., Khaliq, A., Saud, S., Hassan, S., et al. (2015a) Phytohormones and plant responses to salinity stress: A review. *Plant Growth Regulation* 75 No. 2: 391–404.

Fahad, S., Hussain, S., Saud, S., Hassan, S., Tanveer, M., Ihsan, M.Z., Shah, A.N., et al. (2016a) A combined application of biochar and phosphorus alleviates heat-induced adversities on physiological, agronomical and quality attributes of rice. *Plant Physiology and Biochemistry: PPB* 103: 191–198.

Fahad, S., Hussain, S., Saud, S., Khan, F., Hassan, S., Amanullah, Nasim, W., et al. (2016b) Exogenously applied plant growth regulators affect heat-stressed rice pollens. *Journal of Agronomy and Crop Science* 202 No. 2: 139–150.

Fahad, S., Hussain, S., Saud, S., Tanveer, M., Bajwa, A.A., Hassan, S., Shah, A.N., et al. (2015b) A biochar application protects rice pollen from high-temperature stress. *Plant Physiology and Biochemistry* 96: 281–287.

FAO, IFAD, WFP. (2015) The state of food insecurity in the world 2015. Meeting the 2015 international hunger targets: Taking stock of uneven progress. FAO, Rome, Italy.

Farooq, M., Aziz, T., Wahid, A., Lee, D.J., Siddique, K.H.M. (2009) Chilling tolerance in maize: Agronomic and physiological approaches. *Crop and Pasture Science* 60 No. 6: 501–516.

Genesio, L., Miglietta, F., Baronti, S., Vaccari, F.P. (2015) Biochar increases vineyard productivity without affecting grape quality: Results from a four years field experiment in Tuscany. *Agriculture, Ecosystems and Environment* 201: 20–25.

Genesio, L., Miglietta, F., Lugato, E., Baronti, S., Pieri, M., Vaccari, F.P. (2012) Surface albedo following biochar application in durum wheat. *Environmental Research Letters* 7 No. 1: 14025.

Geng, S.M., Yan, D.H., Zhang, T.X., Weng, B.S., Zhang, Z.B., Qin, T.L. (2015) Effects of drought stress on agriculture soil. *Natural Hazards* 75 No. 2: 1997–2011.

Githinji, L. (2014) Effect of biochar application rate on soil physical and hydraulic properties of a sandy loam. *Archives of Agronomy and Soil Science* 60 No. 4: 457–470.

Glaser, B., Lehmann, J., Zech, W. (2002) Ameliorating physical and chemical properties of highly weathered soils in the tropics with charcoal: A review. *Biology and Fertility of Soils* 35 No. 4: 219–230.

Gomez, J.D., Denef, K., Stewart, C.E., Zheng, J., Cotrufo, M.F. (2014) Biochar addition rate influences soil microbial abundance and activity in temperate soils. *European Journal of Soil Science* 65 No. 1: 28–39.

Graber, E.R., Meller Harel, Y., Kolton, M., Cytryn, E., Silber, A., Rav David, D., Tsechansky, L., et al. (2010) Biochar impact on development and productivity of pepper and tomato grown in fertigated soilless media. *Plant and Soil* 337 No. 1–2: 481–496.

Graber, E.R., Tsechansky, L., Lew, B., Cohen, E. (2014) Reducing capacity of water extracts of biochars and their solubilization of soil Mn and Fe. *European Journal of Soil Science* 65 No. 1: 162–172.

Griffin, D.E., Wang, D., Parikh, S.J., Scow, K.M. (2017) Short-lived effects of walnut shell biochar on soils and crop yields in a long-term field experiment. *Agriculture, Ecosystems and Environment* 236: 21–29.

Gunes, A., Inal, A., Sahin, O., Taskin, M.B., Atakol, O., Yilmaz, N. (2015) Variations in mineral element concentrations of poultry manure biochar obtained at different pyrolysis temperatures, and their effects on crop growth and mineral nutrition. *Soil Use and Management* 31 No. 4: 429–437.

Haefele, S.M., Konboon, Y., Wongboon, W., Amarante, S., Maarifat, A.A., Pfeiffer, E.M., Knoblauch, C. (2011) Effects and fate of biochar from rice residues in rice-based systems. *Fields Crop Research* 121: 430–440.

Haider, G., Koyro, H.W., Azam, F., Steffens, D., Müller, C., Kammann, C. (2015) Biochar but not humic acid product amendment affected maize yields via improving plant-soil moisture relations. *Plant and Soil* 395 No. 1–2: 141–157.

Hairani, A., Osaki, M., Watanabe, T. (2016) Effect of biochar application on mineral and microbial properties of soils growing different plant species. *Soil Science and Plant Nutrition* 62 No. 5–6: 519–525.

Hajiboland, R. (2012) Effect of micronutrient deficiencies on plants stress responses. In *Abiotic Stress Responses in Plants: Metabolism, Productivity and Sustainability*, eds. Ahmad, P., Prasad, M.N.V., 283–329. Berlin: Springer.

Hasanuzzaman, M., Gill, S.S., Fujita, M. (2013a) Physiological role of nitric oxide in plants grown under adverse environmental conditions. In *Plant Acclimation to Environmental Stress*, eds. Tuteja, N., Gill, S.S., 269–322. New York: Springer.

Hasanuzzaman, M., Nahar, K., Alam, M.M., Roychowdhury, R., Fujita, M. (2013b) Physiological, biochemical, and molecular mechanisms of heat stress tolerance in plants. *International Journal of Molecular Sciences* 14 No. 5: 9643–9684.

Hasbullah, H., Marschner, P. (2015) Residue properties influence the impact of salinity on soil respiration. *Biology and Fertility of Soils* 51 No. 1: 99–111.

Hatfield, J., Takle, G., Grotjahn, R., Holden, P., Izaurralde, R.C., Mader, T., Marshall, E., Liverman, D. (2014) Agriculture. Climate change impacts in the United States: The third national climate assessment. In: J.M. Melillo, T.C. Richmond, and G.W. Yohe, editors, *U.S. Global Change Research Program*, pp. 150–174.

Hu, W., Huang, B., Tian, K., Holm, P.E., Zhang, Y. (2017) Heavy metals in intensive greenhouse vegetable production systems along Yellow Sea of China: Levels, transfer and health risk. *Chemosphere* 167: 82–90.

Huang, Z., Lu, Q., Wang, J., Chen, X., Mao, X., He, Z. (2017) Inhibition of the bioavailability of heavy metals in sewage sludge biochar by adding two stabilizers. *PLOS ONE* 12 No. 8: e0183617.

Huang, B., Rachmilevitch, S., Xu, J. (2012) Root carbon and protein metabolism associated with heat tolerance. *Journal of Experimental Botany* 63 No. 9: 3455–3465.

Huang, M., Yang, L., Qin, H., Jiang, L., Zou, Y. (2014) Fertilizer nitrogen uptake by rice increased by biochar application. *Biology and Fertility of Soils* 50 No. 6: 997–1000.

Hussain, M., Farooq, M., Nawaz, A., Al-Sadi, A.M., Solaiman, Z.M., Alghamdi, S.S., Ammara, U., et al. (2017) Biochar for crop production: Potential benefits and risks. *Journal of Soils and Sediments* 17 No. 3: 685–716.

Ibrahim, M.H., Kong, Y.C., Mohd Zain, N.A. (2017) Effect of cadmium and copper exposure on growth, secondary metabolites and antioxidant activity in the medicinal plant Sambung Nyawa (*Gynura procumbens* (Lour.) Merr). *Molecules* 22 No. 10: 1623.

Impa, S.M., Nadaradjan, S., Jagadish, S.V.K. (2012) Drought stress induced reactive oxygen species and anti-oxidants in plants. In *Abiotic Stress Responses in Plants: Metabolism, Productivity and Sustainability*, eds. Ahmad, P., Prasad, M.N.V., 131–147. Berlin: Springer.

Inyang, M.I., Gao, B., Yao, Y., Xue, Y., Zimmerman, A., Mosa, A., Pullammanappallil, P., Ok, Y.S., Cao, X. (2016) A review of biochar as a low-cost adsorbent for aqueous heavy metal removal. *Critical Reviews in Environmental Science and Technology* 46 No. 4: 406–433.

IPCC (Intergovernmental Panel on Climatic Change). (2007) The physical science basis: Summary for policy makers. IPCC WG 14th Assessment Report.

Jaleel, C.A., Gopi, R., Sankar, B., Gomathinayagam, M., Panneerselvam, R. (2008) Differential responses in water use efficiency in two varieties of *Catharanthus roseus* under drought stress. *Comptes Rendus Biologies* 331 No. 1: 42–47.

Jeffery, S., Abalos, D., Prodana, M., Bastos, A.C., van Groenigen, J.W., Hungate, B.A., Verheijen, F. (2017) Biochar boosts tropical but not temperate crop yields. *Environmental Research Letters* 12 No. 5: 053001.

Jeffery, S., Verheijen, F.G.A., van der Velde, M., Bastos, A.C. (2011) A quantitative review of the effects of biochar application to soils on crop productivity using meta-analysis. *Agriculture, Ecosystems and Environment* 144 No. 1: 175–187.

Jien, S.H., Wang, C.S. (2013) Effects of biochar on soil properties and erosion potential in a highly weathered soil. *CATENA* 110: 225–233.

Kammann, C.I., Linsel, S., Gößling, J.W., Koyro, H.W. (2011) Influence of biochar on drought tolerance of *Chenopodium quinoa* Willd and on soil-plant relations. *Plant and Soil* 345 No. 1–2: 195–210.

Kanwal, S., Ilyas, N., Shabir, S., Saeed, M., Gul, R., Zahoor, M., Batool, N., Mazhar, R. (2018) Application of biochar in mitigation of negative effects of salinity stress in wheat (*Triticum aestivum* L.). *Journal of Plant Nutrition* 41 No. 4: 526–538.

Karhu, K., Mattila, T., Bergström, I., Regina, K. (2011) Biochar addition to agricultural soil increased CH_4 uptake and water holding capacity–results from a short-term pilot field study. *Agriculture, Ecosystems & Environment* 140 No. 1–2: 309–313.

Kayastha, S.P. (2015) Heavy metals Fractionation in Bagmati River Sediments, Nepal. *Journal of Hydrology and Meteorology* 9 No. 1: 119–128.

Khan, M.I.R., Asgher, M., Fatma, M., Per, T.S., Khan, N.A. (2015) Drought stress vis a vis plant functions in the era of climate change. *Climate Change and Environmental Sustainability* 3 No. 1: 13–25.

Kim, H.S., Kim, K.R., Yang, J.E., Ok, Y.S., Owens, G., Nehls, T., Wessolek, G., Kim, K.-H. (2016) Effect of biochar on reclaimed tidal land soil properties and maize (*Zea mays* L.) response. *Chemosphere* 142: 153–159.

Klüpfel, L., Keiluweit, M., Kleber, M., Sander, M. (2014) Redox properties of plant biomass-derived black carbon (biochar). *Environmental Science & Technology* 48 No. 10: 5601–5611.

Kołodyńska, D., Wnętrzak, R., Leahy, J.J., Hayes, M.H.B., Kwapiński, W., Hubicki, Z. (2012) Kinetic and adsorptive characterization of biochar in metal ions removal. *Chemical Engineering Journal* 197: 295–305.

Kotapati, K.V., Palaka, B.K., Ampasala, D.R. (2017) Alleviation of nickel toxicity in finger millet (*Eleusine coracana* L.) germinating seedlings by exogenous application of salicylic acid and nitric oxide. *The Crop Journal* 5 No. 3: 240–250.

Krishnakumar, S., Rajalakshmi, A.G., Balaganesh, B., Manikandan, P., Vinoth, C., Rajendran, V. (2014) Impact of biochar on soil health. *International Journal of Advanced Research* 2: 933–950.

Kumar Tewari, A., Charan Tripathy, B. (1998) Temperature-stress-induced impairment of chlorophyll biosynthetic reactions in cucumber and wheat. *Plant Physiology* 117 No. 3: 851–858.

Kuzyakov, Y., Subbotina, I., Chen, H., Bogomolova, I., Xu, X. (2009) Black carbon decomposition and incorporation into soil microbial biomass estimated by ^{14}C labeling. *Soil Biology and Biochemistry* 41 No. 2: 210–219.

Laghari, M., Hu, Z., Mirjat, M.S., Xiao, B., Tagar, A.A., Hu, M. (2016) Fast pyrolysis biochar from sawdust improves quality of desert soils and enhances plant growth. *Journal of the Science of Food and Agriculture* 96 No. 1: 199–206.

Laird, D.A., Fleming, P., Davis, D.D., Horton, R., Wang, B., Karlen, D.L. (2010) Impact of biochar amendments on the quality of a typical Midwestern agricultural soil. *Geoderma* 158 No. 3–4: 443–449.

Lashari, M.S., Liu, Y., Li, L., Pan, W., Fu, J., Pan, G., Zheng, J., et al. (2013) Effects of amendment of biochar-manure compost in conjunction with pyroligneous solution on soil quality and wheat yield of a salt-stressed cropland from Central China Great Plain. *Field Crops Research* 144: 113–118.

Lee, J.W., Hawkins, B., Li, X., Day, D.M. (2013) Biochar fertilizer for soil amendment and carbon sequestration. In *Advanced Biofuels and Bioproducts*, ed. James, W.L., 57–68. New York: Springer.

Lehmann, J. (2007) A handful of carbon. *Nature* 447 No. 7141: 143–144.

Lehmann, J., Da Silva, J.P., Steiner, C., Nehls, T., Zech, W., Glaser, B. (2003) Nutrient availability and leaching in an archaeological Anthrosol and a Ferralsol of the Central Amazon Basin: Fertilizer, manure and charcoal amendments. *Plant and Soil* 249 No. 2: 343–357.

Lehmann, J., Joseph, S. (2015) *Biochar for Environmental Management: Science, Technology and Implementation*. London: Routledge.

Lehmann, J., Rillig, M.C., Thies, J., Masiello, C.A., Hockaday, W.C., Crowley, D. (2011) Biochar effects on soil biota: A review. *Soil Biology and Biochemistry* 43 No. 9: 1812–1836.

Li, Y., Hu, S., Chen, J., Muller, K., Li, Y., Fu, W., Lin, Z., Wang, H. (2018) Effects of biochar application in forest ecosystems on soil properties and greenhouse gas emissions: A review. *Journal of Soils and Sediments* 18 No. 2: 546–563.

Liang, F., Li, G., Lin, Q., Zhao, X. (2014) Crop yield and soil properties in the first 3 years after biochar application to a calcareous soil. *Journal of Integrative Agriculture* 13 No. 3: 525–532.

Lin, Y., Munroe, P., Joseph, S., Kimber, S., Zwieten, L.V. (2012) Nanoscale organomineral reactions of biochars in ferrosol: An investigation using microscopy. *Plant and Soil* 357: 369–380.

Liu, C., Liu, F., Ravnskov, S., Rubæk, G.H., Sun, Z., Andersen, M.N. (2017) Impact of wood biochar and its interactions with mycorrhizal fungi, phosphorus fertilization and irrigation strategies on potato growth. *Journal of Agronomy and Crop Science* 203 No. 2: 131–145.

Liu, J., Schulz, H., Brandl, S., Miehtke, H., Huwe, B., Glaser, B. (2012) Short-term effect of biochar and compost on soil fertility and water status of a dystric cambisol in NE Germany under field conditions. *Journal of Plant Nutrition and Soil Science* 175 No. 5: 698–707.

Liu, T., Liu, B., Zhang, W. (2014) Nutrients and heavy metals in biochar produced by sewage sludge pyrolysis: Its application in soil amendment. *Polish Journal of Environmental Studies* 23: 271–275.

Luo, X., Liu, G., Xia, Y., Chen, L., Jiang, Z., Zheng, H., Wang, Z. (2017) Use of biochar-compost to improve properties and productivity of the degraded coastal soil in the Yellow River Delta, China. *Journal of Soils and Sediments* 17 No. 3: 780–789.

Luo, Y., Yu, Z., Zhang, K., Xu, J., Brookes, P.C. (2016) The properties and functions of biochars in forest ecosystems. *Journal of Soils and Sediments* 16 No. 8: 2005–2020.

Lwin, C.S., Seo, B.H., Kim, H.U., Owens, G., Kim, K.R. (2018) Application of soil amendments to contaminated soils for heavy metal immobilization and improved soil quality: A critical review. *Soil Science and Plant Nutrition* 64 No. 2: 156–167.

Lyu, S., Du, G., Liu, Z., Zhao, L., Lyu, D. (2016) Effects of biochar on photosystem function and activities of protective enzymes in *Pyrus ussuriensis* Maxim. under drought stress. *Acta Physiologiae Plantarum* 38 No. 9: 220.

Major, J., Rondon, M., Molina, D., Riha, S.J., Lehmann, J. (2010) Maize yield and nutrition during 4 years after biochar application to a *Colombian savanna oxisol*. *Plant and Soil* 333 No. 1–2: 117–128.

Major, J., Rondon, M., Molina, D., Riha, S.J., Lehmann, J. (2012) Nutrient leaching in a Colombian savanna oxisol amended with biochar. *Journal of Environmental Quality* 41 No. 4: 1076–1086.

Maraseni, T.N. (2010) Biochar: Maximising the benefits. *International Journal of Environmental Studies* 67 No. 3: 319–327.

Mariotte, P., Robroek, B.J.M., Jassey, V.E.J., Buttler, A. (2015) Subordinate plants mitigate drought effects on soil ecosystem processes by stimulating fungi. *Functional Ecology* 29 No. 12: 1578–1586.

Mickan, B.S., Abbott, L.K., Stefanova, K., Solaiman, Z.M. (2016) Interactions between biochar and mycorrhizal fungi in a water-stressed agricultural soil. *Mycorrhiza* 26 No. 6: 565–574.

Milla, O.V., Rivera, E.B., Huang, W.J., Chien, C.C., Wang, Y.M. (2013) Agronomic properties and characterization of rice husk and wood biochars and their effect on the growth of water spinach in a field test. *Journal of Soil Science and Plant Nutrition* 13: 251–266.

Mohamed, A.K.S.H., Qayyum, M.F., Abdel-Hadi, A.M., Rehman, R.A., Ali, S., Rizwan, M. (2017) Interactive effect of salinity and silver nanoparticles on photosynthetic and biochemical parameters of wheat. *Archives of Agronomy and Soil Science* 63 No. 12: 1736–1747.

Mohanty, S., Baishna, B.G., Tripathy, B.C. (2006) Light and dark modulation of chlorophyll biosynthetic genes in response to temperature. *Planta* 224 No. 3: 692–699.

Moreno-Barriga, F., Díaz, V., Acosta, J.A., Muñoz, M.Á., Faz, Á., Zornoza, R. (2017) Creation of Technosols to decrease metal availability in pyritic tailings with addition of biochar and marble waste. *Land Degradation & Development* 28 No. 7: 1943–1951.

Moussa, H.R., Abdel-Aziz, S.M. (2008) Comparative response of drought tolerant and drought sensitive maize genotypes to water stress. *Australian Journal of Crop Science* 1: 31–36.

Mulcahy, D.N., Mulcahy, D.L., Dietz, D. (2013) Biochar soil amendment increases tomato seedling resistance to drought in sandy soils. *Journal of Arid Environments* 88: 222–225.

Munns, R., Tester, M. (2008) Mechanisms of salinity tolerance. *Annual Review of Plant Biology* 59: 651–681.

Nagajyoti, P.C., Lee, K.D., Sreekanth, T.V.M. (2010) Heavy metals, occurrence and toxicity for plants: A review. *Environmental Chemistry Letters* 8 No. 3: 199–216.

Nguyen, H., Blair, G., Guppy, C. (2012) Effect of rice husk biochar and nitrification inhibitor treated urea on N and other macronutrient uptake by maize. 16th ASA Conference, Armidale, Australia.

Noman, A., Ali, S., Naheed, F., Ali, Q., Farid, M., Rizwan, M., Irshad, M.K. (2015) Foliar application of ascorbate enhances the physiological and biochemical attributes of maize (*Zea mays* L.) cultivars under drought stress. *Archives of Agronomy and Soil Science* 61 No. 12: 1659–1672.

Ohashi, M., Kume, T., Yoshifuji, N., Kho, L.K., Nakagawa, M., Nakashizuka, T. (2015) The effects of an induced short-term drought period on the spatial variations in soil respiration measured around emergent trees in a typical Bornean tropical forest, Malaysia. *Plant and Soil* 387 No. 1–2: 337–349.

Olmo, M., Alburquerque, J.A., Barrón, V., del Campillo, M.C., Gallardo, A., Fuentes, M., Villar, R. (2014) Wheat growth and yield responses to biochar addition under Mediterranean climate conditions. *Biology and Fertility of Soils* 50 No. 8: 1177–1187.

Oo, A.N., Iwai, C.B., Saenjan, P. (2015) Soil properties and maize growth in saline and nonsaline soils using cassava-industrial waste compost and vermicompost with or without earthworms. *Land Degradation and Development* 26 No. 3: 300–310.

Osakabe, Y., Osakabe, K., Shinozaki, K., Tran, L.P. (2014) Response of plants to water stress. *Frontiers in Plant Science* 5: 1–19.

Ouyang, L., Yu, L., Zhang, R. (2014) Effects of amendment of different biochars on soil carbon mineralisation and sequestration. *Soil Research* 52 No. 1: 46–54.

Pandit, N.R., Mulder, J., Hale, S.E., Martinsen, V., Schmidt, H.P., Cornelissen, G. (2018) Biochar improves maize growth by alleviation of nutrient stress in a moderately acidic low-input Nepalese soil. *The Science of the Total Environment* 625: 1380–1389.

Pant, P.P., Tripathi, A.K. (2014) Impact of heavy metals on morphological and biochemical parameters of *Shorea robusta* plant. *Ekologia* 33 No. 2: 116–126.

Park, J.H., Choppala, G.K., Bolan, N.S., Chung, J.W., Chuasavathi, T. (2011) Biochar reduces the bioavailability and phytotoxicity of heavy metals. *Plant and Soil* 348 No. 1–2: 439–451.

Patra, J.M., Panda, S.S., Dhal, N.K. (2017) Biochar as a low-cost adsorbent for heavy metal removal: A review. *International Journal of Research in Biosciences* 6: 1–7.

Peng, X., Ye, L.L., Wang, C.H., Zhou, H., Sun, B. (2011) Temperature- and duration-dependent rice straw-derived biochar: Characteristics and its effects on soil properties of an ultisol in southern China. *Soil and Tillage Research* 112 No. 2: 159?166.

Prasad, P.V.V., Djanaguiraman, M. (2011) High night temperature decreases leaf photosynthesis and pollen function in grain sorghum. *Functional Plant Biology* 38 No. 12: 993–1003.

Pressler, Y., Foster, E.J., Moore, J.C., Cotrufo, M.F. (2017) Coupled biochar amendment and limited irrigation strategies do not affect a degraded soil food web in a maize agroecosystem, compared to the native grassland. *GCB Bioenergy* 9 No. 8: 1344–1355.

Qi, F., Kuppusamy, S., Naidu, R., Bolan, N.S., Ok, Y.S., Lamb, D., Li, Y., et al. (2017) Pyrogenic carbon and its role in contaminant immobilization in soils. *Critical Reviews in Environmental Science and Technology* 47 No. 10: 795–876.

Rabêlo, F.H.S., Borgo, L. (2016) Changes caused by heavy metals in micronutrient content and antioxidant system of forage grasses used for phytoremediation: An overview. *Ciência Rural* 46 No. 8: 1368–1375.

Rainey, K.M., Griffiths, P.D. (2005) Evaluation of *Phaseolus acutifolius* A. Gray plant introductions under high temperatures in a controlled environment. *Genetic Resources and Crop Evolution* 52 No. 2: 117–120.

Rajkovich, S., Enders, A., Hanley, K., Hyland, C., Zimmerman, A.R., Lehmann, J. (2012) Corn growth and nitrogen nutrition after additions of biochars with varying properties to a temperate soil. *Biology and Fertility of Soils* 48 No. 3: 271–284.

Ramegowda, V., Senthil-Kumar, M. (2015) The interactive effects of simultaneous biotic and abiotic stresses on plants: Mechanistic understanding from drought and pathogen combination. *Journal of Plant Physiology* 176: 47–54.

Ramzani, P.M.A., Coyne, M.S., Anjum, S., Khan, W.U., Iqbal, M. (2016) In situ immobilization of Cd by organic amendments and their effect on antioxidant enzyme defense mechanism in mung bean (*Vigna radiata* L.) seedlings. *Plant Physiology and Biochemistry: PPB* 118: 561–570.

Ramzani, P.M.A., Shan, L., Anjum, S., Khan, W.U., Ronggui, H., Iqbal, M., Virk, Z.A., Kausar, S. (2017) Improved quinoa growth, physiological response, and seed nutritional quality in three soils having different stresses by the application of acidified biochar and compost. *Plant Physiology and Biochemistry: PPB* 116: 127–138.

Rath, K.M., Rousk, J. (2015) Salt effects on the soil microbial decomposer community and their role in organic carbon cycling: A review. *Soil Biology and Biochemistry* 81: 108–123.

Rennenberg, H., Loreto, F., Polle, A., Brilli, F., Fares, S., Beniwal, R.S., Gessler, A. (2006) Physiological responses of forest trees to heat and drought. *Plant Biology* 8 No. 5: 556–571.

Richards, L.A. (1954) Diagnosis and improvement of saline and alkali soils. In *Soil Science. Agricultural Handbook* No. 60. Riverside, CA: US Salinity Laboratory.

Rizwan, M., Meunier, J.-D., Davidian, J.-C., Pokrovsky, O.S., Bovet, N., Keller, C. (2016) Silicon alleviates Cd stress of wheat seedlings (*Triticum turgidum* L. cv. Claudio) grown in hydroponics. *Environmental Science and Pollution Research* 23 No. 2: 1414–1427.

Rodríguez, M., Canales, E., Borrás-Hidalgo, O. (2005) Molecular aspects of abiotic stress in plants. *Biotecnología Aplicada* 22: 1–10.

Roelofs, D., Aarts, M.G.M., Schat, H., van Straalen, N.M. (2008) Functional ecological genomics to demonstrate general and specific responses to abiotic stress. *Functional Ecology* 22: 8–18.

Rogovska, N., Laird, D., Cruse, R., Fleming, P., Parkin, T., Meek, D. (2011) Impact of biochar on manure carbon stabilization and greenhouse gas emissions. *Soil Science Society of America Journal* 75 No. 3: 871–879.

Różyło, K., Świeca, M., Gawlik-Dziki, U., Stefaniuk, M., Oleszczuk, P. (2017) The potential of biochar for reducing the negative effects of soil contamination on the phytochemical properties and heavy metal accumulation in wheat grain. *Agricultural and Food Science* 26 No. 1: 34–46.

Saquing, J.M., Yu, Y.H., Chiu, P.C. (2016) Wood-derived black carbon (biochar) as a microbial electron donor and acceptor. *Environmental Science & Technology Letters* 3 No. 2: 62–66.

Seneviratne, M., Weerasundara, L., Ok, Y.S., Rinklebe, J., Vithanage, M. (2017) Phytotoxicity attenuation in Vigna radiata under heavy metal stress at the presence of biochar and N fixing bacteria. *Journal of Environmental Management* 186 No. 2: 293–300.

Sharma, A., Dhiman, A. (2013) Nickel and cadmium toxicity in plants. *Journal of Pharmaceutical and Scientific Innovation* 2 No. 2: 20–24.

Sheetal, K.R., Singh, S.D., Anand, A., Prasad, S. (2016) Heavy metal accumulation and effects on growth, biomass and physiological processes in mustard. *Indian Journal of Plant Physiology* 21 No. 2: 219–223.

Siddiqui, M.H., Al-Khaishany, M.Y., Al-Qutami, M.A., Al-Whaibi, M.H., Grover, A., Ali, H.M., Al-Wahibi, M.S., Bukhari, N.A. (2015) Response of different genotypes of faba bean plant to drought stress. *International Journal of Molecular Sciences* 16 No. 5: 10214–10227.

Simoes-Araujo, J.L., Rumjanek, N.G., Margis-Pinheiro, M. (2003) Small heat shock proteins genes are differentially expressed in distinct varieties of common bean. *Brazilian Journal of Plant Physiology* 15 No. 1: 33–41.

Sizmur, T., Quilliam, R., Puga, A.P., Moreno-Jiménez, E., Beesley, L., Gomez-Eyles, J.L. (2016) Application of biochar for soil remediation. In *Agricultural and Environmental Applications of Biochar: Advances and Barriers*, eds. Guo, M., He, Z., Uchimiya, S.M., 295–324. Madison, WI: Soil Science Society of America, Inc.

Sperry, J.S., Hacke, U.G. (2002) Desert shrub water relations with respect to soil characteristics and plant functional type. *Functional Ecology* 16 No. 3: 367–378.

Stambulska, U.Y., Bayliak, M.M., Lushchak, V.I. (2018) Chromium (VI) toxicity in legume plants: Modulation effects of rhizobial symbiosis. *BioMed Research International* 2018: 8031213.

Steinbeiss, S., Gleixner, G., Antonietti, M. (2009) Effect of biochar amendment on soil carbon balance and soil microbial activity. *Soil Biology and Biochemistry* 41 No. 6: 1301–1310.

Subedi, R., Taupe, N., Ikoyi, I., Bertora, C., Zavattaro, L., Schmalenberger, A., Leahy, J.J., Grignani, C. (2016) Chemically and biologically-mediated fertilizing value of manure-derived biochar. *The Science of the Total Environment* 550: 924–933.

Suriyasak, C., Harano, K., Tanamachi, K., Matsuo, K., Tamada, A., Iwaya-Inoue, M., Ishihbashi, Y. (2017) Reactive oxygen species induced by heat stress during grain filling of rice (*Oryza sativa* L.) are involved in occurrence of grain chalkiness. *Journal of Plant Physiology* 216: 52–57.

Taghizadeh-Toosi, A., Clough, T.J., Sherlock, R.R., Condron, L.M. (2012) Biochar adsorbed ammonia is bioavailable. *Plant and Soil* 350 No. 1–2: 57–69.

Tammeorg, P., Simojoki, A., Mäkelä, P., Stoddard, F.L., Alakukku, L., Helenius, J. (2014) Short-term effects of biochar on soil properties and wheat yield formation with meat bone meal and inorganic fertiliser on a boreal loamy sand. *Agriculture, Ecosystems & Environment* 191: 108–116.

Tardieu, F., Parent, B., Caldeira, C.F., Welcker, C. (2014) Genetic and physiological controls of growth under water deficit. *Plant Physiology* 164 No. 4: 1628–1635.

Thakur, P., Nayyar, H. (2013) Facing the cold stress by plants in the changing environment: Sensing, signaling, and defending mechanisms. In *Plant Acclimation to Environmental Stress*, eds. Tuteja, N., Gill, S.S., 29–69. New York: Springer.

Thomas, S.C., Frye, S., Gale, N., Garmon, M., Launchbury, R., Machado, N., Melamed, S., et al. (2013) Biochar mitigates negative effects of salt additions on two herbaceous plant species. *Journal of Environmental Management* 129: 62–68.

Tian, X., Li, C., Zhang, M., Wan, Y., Xie, Z., Chen, B., Li, W. (2018) Biochar derived from corn straw affected availability and distribution of soil nutrients and cotton yield. *PLOS ONE* 13 No. 1: e0189924.

Usman, A.R.A., Al-Wabel, M.I., Ok, Y.S., Al-Harbi, A., Wahb-Allah, M., El-Naggar, A.H., Ahmad, M., et al. (2016) Conocarpus biochar induces changes in soil nutrient availability and tomato growth under saline irrigation. *Pedosphere* 26 No. 1: 27–38.

Uzoma, K.C., Inoue, M., Andry, H., Fujimaki, H., Zahoor, A., Nishihara, E. (2011) Effect of cow manure biochar on maize productivity under sandy soil condition. *Soil Use and Management* 27 No. 2: 205–212.

Vaccari, F.P., Maienza, A., Miglietta, F., Baronti, S., Di Lonardo, S., Giagnoni, L., Lagomarsino, A., et al. (2015) Biochar stimulates plant growth but not fruit yield of processing tomato in a fertile soil. *Agriculture, Ecosystems and Environment* 207: 163–170.

Verheijen, F., Jeffery, S., Bastos, A.C., van der Velde, M., Diafas, I. (2004) Biochar application to soils – A critical scientific review of effects on soil properties, processes and functions. EUR, 24099 EN, Office for the Official Publications of the European Communities, Luxembourg.

Viger, M., Hancock, R.D., Miglietta, F., Taylor, G. (2014) More plant growth but less plant defence? First global gene expression data for plants grown in soil amended with biochar. *Global Change Biology Bioenergy* 7: 658–672.

Vilvanathan, S., Shanthakumar, S. (2018) Ni^{2+} and Co^{2+} adsorption using Tectona grandis biochar: Kinetics, equilibrium and desorption studies. *Environmental Technology* 39 No. 4: 464–478.

Vollenweider, P., Günthardt-Goerg, M.S. (2005) Diagnosis of abiotic and biotic stress factors using the visible symptoms in foliage. *Environmental Pollution* 137 No. 3: 455–465.

Wahid, A., Close, T.J. (2007) Expression of dehydrins under heat stress and their relationship with water relations of sugarcane leaves. *Biologia Plantarum* 51 No. 1: 104–109.

Wahid, A., Gelani, S., Ashraf, M., Foolad, M.R. (2007) Heat tolerance in plants: An overview. *Environmental and Experimental Botany* 61 No. 3: 199–223.

Wang, L., Butterly, C.R., Wang, Y., Herath, H.M.S.K., Xi, Y.G., Xiao, X.J. (2014) Effect of crop residue biochar on soil acidity amelioration in strongly acidic tea garden soils. *Soil Use and Management* 30 No. 1: 119–128.

Wu, Y., Xu, G., Shao, H.B. (2014) Furfural and its biochar improve the general properties of a saline soil. *Solid Earth* 5 No. 2: 665–671.

Wuddivira, M.N., Stone, R.J., Ekwue, E.I. (2009) Structural stability of humid tropical soils as influenced by manure incorporation and incubation duration. *Soil Science Society of America Journal* 73 No. 4: 1353–1360.

Xiao, Q., Zhu, L.-X., Shen, Y.-F., Li, S.-Q. (2016) Sensitivity of soil water retention and availability to biochar addition in rainfed semi-arid farmland during a three-year field experiment. *Field Crops Research* 196: 284–293.

Xie, T., Reddy, K.R., Wang, C., Yargicoglu, E., Spokas, K. (2015) Characteristics and applications of biochar for environmental remediation: A review. *Critical Reviews in Environmental Science and Technology* 45 No. 9: 939–969.

Yan, N., Marschner, P., Cao, W., Zuo, C., Qin, W. (2015) Influence of salinity and water content on soil microorganisms. *International Soil and Water Conservation* 3 No. 4: 316–323.

Yan, X., Liu, M., Zhong, J., Guo, J., Wu, W. (2018) How human activities affect heavy metal contamination of soil and sediment in a long-term reclaimed area of the Liaohe river delta, North China. *Sustainability* 10 No. 2: 338.

Yuan, Y., Bolan, N., Prévoteau, A., Vithanage, M., Biswas, J.K., Ok, Y.S., Wang, H. (2017) Applications of biochar in redox-mediated reactions. *Bioresource Technology* 246: 271–281.

Zhang, P., Sun, H., Yu, L., Sun, T. (2013) Adsorption and catalytic hydrolysis of carbaryl and atrazine on pig manure-derived biochars: Impact of structural properties of biochars. *Journal of Hazardous Materials* 244–245: 217–224.

Zheng, W., Guo, M., Chow, T., Bennett, D.N., Rajagopalan, N. (2010) Sorption properties of greenwaste biochar for two triazine pesticides. *Journal of Hazardous Materials* 181 No. 1–3: 121–126.

Zhong, T., Xue, D., Zhao, L., Zhang, X. (2018) Concentration of heavy metals in vegetables and potential health risk assessment in China. *Environmental Geochemistry and Health* 40 No. 1: 313–322.

Zhou, Q., Liao, B., Lin, L., Qiu, W., Song, Z. (2018) Adsorption of Cu (II) and Cd (II) from aqueous solutions by ferromanganese binary oxide–biochar composites. *The Science of the Total Environment* 615: 115–122.

19 Exploring and Harnessing Plant Microbiomes for Abiotic Stress Tolerance and Yield Stability in Crop Plants

Syed Sarfraz Hussain

CONTENTS

19.1 INTRODUCTION: ABIOTIC STRESSES – CHALLENGES AND OPPORTUNITIES

The world population is 6.5 billion and is expected to increase to at least 9 billion by 2050 (Hussain et al., 2012, 2014). It is estimated that over 800 million people are experiencing food insecurity and malnutrition worldwide. Global food production is limited by a multitude of factors, primarily environmental stresses. Drought and salinity stresses are the major cause of historic and modern agricultural productivity losses throughout the world (Cushman and Bohnert, 2000). The green revolution is no more fruitful due to ever-increasing global population and different environmental stresses. Therefore, it is conceived that conventional agriculture alone cannot keep pace with future needs of the world population. To meet food requirements, all major crops must be improved either through selective breeding or genetic modifications to ensure productivity in rapidly changing climatic conditions (Bartels and Hussain, 2008; Hussain et al., 2012). Transgenic approaches have resulted in improvement of different biotic and abiotic stresses which significantly contribute to plant yield improvement (Hussain et al., 2016). Current engineering strategies rely on the transfer of one or several genes that are either involved in signaling and regulatory pathways, or that encode enzymes present in pathways leading to the synthesis of functional and structural protectants, such as osmolytes and antioxidants, or that encode stress tolerance conferring proteins (reviewed by Wang et al., 2003; Vinocur and Altman, 2005; Valliyodan and Nguyen, 2006; Sreenivasulu et al., 2007; Kathuria et al., 2007; Bartels and Hussain, 2008; Hussain et al., 2012, 2014; Marasco et al., 2016; Thao and Tran, 2016). Similarly, a combination of both transgenic strategies and conventional breeding can be used to create stress tolerant crop plants (Capell et al., 2004). However, identification and isolation of key genes and social acceptance of transgenic products pose the main bottleneck of this strategy.

A growing body of evidence shows that plant-associated microbiomes including symbiotic associations help significantly in plant yield sustainability (Gouda et al., 2018; Wagg et al., 2014). Integration and utilization of emerging microbe associated technologies offer a potential increase in plant growth and health, biotic/abiotic stress tolerance, and nutrient use efficiency leading to a significant increase in crop yields. However, this represents a vast but still largely an unexplored area, and intensive research efforts are required to fully utilize its potential in increasing crop yields in an environmentally friendly and sustainable way

(Hussain et al., 2018). Very recently, molecular and other omic-based tools have resulted in several diverse and unexpected research discoveries related to plant associated microbiome (Berg et al., 2016; Bulgarelli et al., 2012; Hussain et al., 2018; Lundberg et al., 2012; Mendes et al., 2011; Timmusk et al., 2017; White III et al., 2017). Several studies have reported many agriculturally important microbes which are being exploited for plant growth and disease management. Similarly, researchers also expect that plant-associated microbiomes can significantly contribute to tackling abiotic stresses in plants (Hayat et al., 2010; Lakshmanan et al., 2012; Mapelli et al., 2013; Vejan et al., 2016). Manipulation of plant microbiome holds great potential which can serve as a valuable tool and key determinants in promoting plant growth and plant fitness (Rolli et al., 2015; Wallenstein, 2017), managing plant health under biotic and abiotic stresses (Mapelli et al., 2013; Vejan et al., 2016), and increasing productivity for sustainable agriculture (Berg et al., 2016; Celebi et al., 2010; Lugtenberg and Kamilova, 2009; Marasco et al., 2016; Mengual et al., 2014; Rolli et al., 2015). The identification, characterization and utilization of beneficial microbiomes which enhance abiotic stress tolerance in plants by diverse mechanisms would help to sustain the next generation in agriculture worldwide (Jorquera et al., 2012; Nadeem et al., 2014). Diverse mechanisms which these microbes use to confer stress have been reviewed elsewhere (Lugtenberg and Kamilova, 2009; Nadeem et al., 2014; Yang et al., 2009; Zelicourt et al., 2013). Overall, sustainable agriculture challenged by abiotic stresses needs non-conventional solutions such as the use of plant-related microbiomes (Schlaeppi and Bulgarelli, 2015). Research efforts must be directed toward strengthening microbial traits beneficial to both plants and the environment; this offers a promising avenue for the development of sustainable agriculture (Lally et al., 2017). This chapter highlights advantages of the plant-related microbiome approach, in particular, increasing plant tolerance to different abiotic stresses which pose a serious threat to global crop productivity.

19.2 EXPLORING PLANT MICROBIOME DIVERSITY: TECHNICAL CHALLENGES

Ample evidence suggests that virtually every plant part is colonized by an astounding number of microorganisms (Quiza et al., 2015). These microorganisms have been categorized depending on the plant part into endophyte (inside plant part), epiphytic (aboveground plant part: leaves and twigs), and rhizospheric (belowground plant part: closely associated with roots). Several lines of research highlighted that these microbes heavily influence many plant characters like seed germination and vigor, growth and development, plant health and productivity (Mendes et al., 2013; Quiza et al., 2015). Historically, use of microbes in agriculture dates back to 1800, when rhizobium bacteria was recommended for inoculation of legume crops to promote growth, development, and uptake of nitrogen and phosphorous (Jones et al., 2014). However, these research efforts have focused on the functional roles of individual microbial groups (e.g., specific species or organisms from the same genera) associated with plants have met with limited success mostly because of technological limitations to assess non-culturable microbial groups (Amann et al., 1995; Andreote et al., 2009). Examples of these inferences are related to specific microbial groups able to promote plant growth, such as nitrogen-fixing bacteria (de Bruijin, 2015; Olivares et al., 2013) and mycorrhiza-forming fungi (Chagnon et al., 2013; Smith and Read, 2008; van der Heijden et al., 2015). However, a holistic view of the microbial system recognizes the importance of saprophytic or symbiotic interactions with plants spanning from beneficial to pathogenic (Mendes et al., 2013), depending upon several factors like plant species and nutrient availability (Quiza et al., 2015). The plant microbiome term, first used by Joshua Lederberg (Lederberg and McCray, 2001), has received substantial attention and addresses both plant health and plant productivity. However, the concept of microbiome has been broadly applied to microbe composition and the impact in specific hosts or environment (Boon et al., 2014; Lakshmanan et al., 2012; Ofek et al., 2014). However, the vast majority of microbes in any microbiome system have not been identified and characterized. Therefore, limited information is available on their involvement for increased plant productivity (Mendes et al., 2013). The current focus of plant-microbe interaction research includes three strategies. These include microbes involved in nutrient acquisition by symbiosis between plants and arbuscular mycorrhiza (Sessitsch and Mitter, 2014; Smith and Smith, 2011), atmospheric nitrogen-fixing rhizobia (Lundberg et al., 2012; Oldroyd et al., 2011), microbes promoting stress tolerance (Doornbos et al., 2011; Ferrara et al., 2012; Kavamura et al., 2013; Marasco et al., 2012; Zolla et al., 2013), and disease-causing microbes (Kachroo and Robin, 2013; Mendes et al., 2011, 2013; Quecine et al., 2014; Wirthmueller et al., 2013).

Mendes et al. (2013) revealed that the association of microbes with plants initially related to plant diseases. However, advance research in this field highlighted that the vast majority of microbial organisms are rendering beneficial services to plants and not always causal agents of damage (Mendes et al., 2013). Apart from well-known mutualistic interactions among plant and microbes, other beneficial microbes have been rarely included in field-based plant production strategies. Several functions have been attributed to microbial cells in close association with plants. A plant microbiome is a highly complex and dynamic component comprising several diverse microbial strains (Farrar et al., 2014). Therefore, recent studies have partitioned plant microbiomes and targeted different fractions separately. Three major compartments where microbial cells can establish and develop include: the so-called rhizosphere, endosphere, and phyllosphere (Hardoim et al., 2008; Hirsch and Mauchline, 2012). Currently, our knowledge of plant microbiome comprising of diverse microbial communities is limited, mainly due to methodological constraints. Therefore, development and implementation of protocols are essential for exploring the whole plant microbiome diversity. With the advent of next-generation sequencing and other molecular techniques like florescent tagging especially for studying unculturable species (endophytes) are now gradually routine in research (Bulgarelli et al., 2012). A huge body of data has been accumulated as a consequence of technological advancements in this field. On the other hand, integration of different computational models is also essential for dissection of this complex and dynamic hidden treasure (Farrar et al., 2014) with the aim to search for new beneficial microbes and effectively manipulate plant microbiomes for increasing plant productivity. Taken together, investment in research aimed at exploring microbial traits that are beneficial to plants or the environment or both presents a promising strategy toward the next generation of sustainable agriculture (Schlaeppi and Bulgarelli, 2015).

19.3 MICROBIOME REVOLUTION IN AGRICULTURE: A WAY FORWARD

It is noteworthy that advanced scientific technologies required for sustainable practices in agriculture are essential for continuous food supply in future. The idea of interactions between plants and associated microbes (plant microbiome) is not new. In fact, new developments and technical advances resulted in enhanced research in this unexplored field (Bakker et al., 2013; Berendsen et al., 2012; Berg et al., 2014; Bulgarelli et al., 2013; Guttman et al., 2014; Knief, 2014; Lebeis, 2014; Philippot et al., 2013; Porras-Alfaro and Bayman, 2011; Schlaeppi et al., 2013; Schlaeppi and Bulgarelli, 2015; Turner et al., 2013). Considering the importance of essential plant nutrients, it would be logical to discover microbes that affect macro and micronutrient uptake for plants under different deficient and toxic soil conditions (Lally et al., 2017; Leveau et al., 2010; Mapelli et al., 2012). Scientific literature provides many examples of well-characterized microbes like plant growth promoting rhizobacteria (PGPR) and plant growth promoting fungi (PGPF) which produce hormones (e.g., auxin), playing critical roles in host nutrition, growth, health, and protect plants from environmental stresses (Berendsen et al., 2012; Berg et al., 2014; Bulgarelli et al., 2013; Mendes et al., 2013; Prashar et al., 2014; Rastogi et al., 2013). Well explored systems for mutualistic microbes include *Rhizobia* spp. and arbuscular mycorrhizae (AM) that exchange plant carbohydrates and amino acids (Moe, 2013) for atmospheric nitrogen fixation and insoluble phosphate bioavailability (Leite et al., 2014; Luvizotto et al., 2010; Spaink, 2000) for plants. Microbes inhabiting the rhizosphere also facilitate the uptake of several trace elements such as iron (Marschner et al., 2011; Shirley et al., 2011; Zhang et al., 2009) and calcium (Lee et al., 2010). On the other hand, the plant microbiome also plays essential functions in degrading organic compounds which are required not only for their survival but also for plant growth and development in nutrient poor and contaminated soils (Bhattacharyya et al., 2015; Leveau et al., 2010; Mapelli et al., 2012; Turner et al., 2013).

Plant protection against pathogens represents an important and often ignored feature of plant-associated microbes. In fact, PGPR and PGPF are also involved in induction of immune "priming," by secreting signaling compounds, which does not result in direct immune activation, but activates and governs the subsequent defense responses to pathogens (Badri et al., 2009; Conrath, 2006; Dangl et al., 2013; De-la-Peña et al., 2012), even in distal tissues. Thus, protective rhizobacteria trigger induced systemic resistance (ISR; Ortiz-Castro et al., 2008) and AM can produce mycorrhizal induced resistance (MIR; Pozo and Azcon-Aguilar, 2007; Zamioudis and Pieterse, 2012) suggesting that microbial exploitation is common which allows the plants to endure pathogen attacks. Several studies have demonstrated that root inoculation with much different PGPR rendered the entire plant tolerant to lethal pathogens (Choudhary et al., 2007; Schuhegger et al., 2006; Tarkka et al., 2008). Zamioudis et al. (2013) demonstrated that *Pseudomonas fluorescens* WCS417 is able to promote increased leaf and root biomass in *Arabidopsis thaliana* which further revealed that stimulation of lateral root and root hair development occurs via an auxin-dependent and JA-independent (ISR) mechanism (Zamioudis et al., 2013). Thus, interactions between PGPR/PGPF and their plant host illustrate the power to unravel mechanisms which acts as the prime barrier of plant defense (Badri et al., 2009; Dangl et al., 2013; De-la-Peña et al., 2012). Despite this enormous progress in the description of plant microbiomes, numerous important crop species and their associated microbes have not yet been studied, which necessitates intensive research efforts.

19.4 ENGINEERING A FUNCTIONAL PLANT-ASSOCIATED MICROBIOME: KNOWLEDGE-DRIVEN SMART APPROACH

Several research efforts have highlighted the ability of microbes to promote plant growth and productivity under environmental stresses (Bhattacharyya and Jha, 2012; Coleman-Derr and Tringe, 2014; Goh et al., 2013; Lebeis et al., 2015; Schlaeppi et al., 2014; Tkacz et al., 2015; Yeoh et al., 2016). It is crucial to understand microbe–microbe and plant–microbe interactions for engineering a beneficial soil microbiome (rhizosphere). Many reports have revealed the genetic and molecular basis of these interactions (Bloemberg and Lugtenberg, 2001; Busby et al., 2017; Iannucci et al., 2017; Kim et al., 2015; Lally et al., 2017; Lim and Kim, 2013; Timmusk et al., 2014; Vargas et al., 2014; Wang et al., 2005), which can be used for the purpose of enhancing plant growth and productivity using genetic engineering strategies. Microbiome interactions are complex and often depend on several factors including soil type, plant species, and local environment which usually determine the composition and association of microbial communities with plant roots. These factors have been shown to play vital roles in triggering plant-species dependent physiological responses, resulting in different exudation patterns (Bais et al., 2006; Dumbrell et al., 2010; Hamel et al., 2005; Hartmann et al., 2009; Oburger et al., 2013). Consequently, the influence of soil type, plant species, and microbial communication has been reviewed extensively on the rhizomicrobiome (Berg and Smalla, 2009; Bulgarelli et al., 2013, 2015; De-la-Peña et al., 2012; Lareen et al., 2016; Philippot et al., 2013; Tarkka et al., 2008). Taken together, soil type and plant species are important players which coordinate the establishment and recruitment of diverse microbial communities for the establishment of a specific rhizobiome with the potential to increase crop productivity and reduce losses to environmental stresses (Bulgarelli et al., 2013, 2015; Lebeis et al., 2015; Peiffer et al., 2013; Philippot et al., 2013; Schlaeppi et al., 2014; Tkacz et al., 2015; Yeoh et al., 2016). These factors significantly contribute to the selective enrichment of beneficial microbes in the rhizobiome (Berendsen et al., 2012; Miller and Oldroyd, 2012; Morel and Castro-Sowinski, 2013; Oldroyd, 2013; Schenk et al., 2012; Sugiyama and Yazaki, 2012), which may help to identify heritable traits to improve plant health and productivity (Su et al., 2015; Yadav et al., 2015).

In fact, using the microbial communities in plant production is not a new strategy. Numerous studies have shown the potential of engineering/reconstructing rhizosphere microbiomes to achieve maximum benefits in crop production including plant growth, yield, disease resistance, and tolerance to environmental challenges offering a unique opportunity (Bainard et al., 2013; Bakker et al., 2012; Berendsen et al., 2012; Bulgarelli et al., 2015; Qiu et al., 2014; Yadav et al., 2015). While the rhizosphere microbiome plays a key role in plant health and productivity (Berendsen et al., 2012; Chaparro et al., 2014; Ziegler et al., 2013), engineering a sustainable synthetic microbial community has received considerable attention in recent years (Bakker et al., 2012; Bulgarelli et al., 2013; Lebeis et al., 2012; Su et al., 2015). Hundreds of bacterial/fungal strains with beneficial effects have been identified, isolated, and are currently being utilized as part of a plant microbiome (Dong and Zhang, 2014; Kim and Timmusk, 2013; Patel and Sinha, 2011). This sustainable approach has huge potential to increase tolerance to biotic and abiotic stresses (Barka et al., 2006; Berg et al., 2013; Jha et al., 2012; Jorquera et al., 2012), increase agricultural production (Bakker et al., 2012; Turner et al., 2013; Yang et al., 2009), reduce chemical inputs (Adesemoye et al., 2009; Adesemoye and Egamberdieva, 2013), and reduce greenhouse gas emissions (Singh et al., 2010), resulting in more sustainable agricultural productivity (Egamberdieva et al., 2017). This is vital for sustaining the ever-growing global population. Furthermore, identified naturally occurring beneficial microbes are now being used in agriculture for significant improvement of crop plant performance (Nadeem et al., 2014; Zolla et al., 2013). Consequently, this mechanistic approach can leverage knowledge from naturally existing microbes with publicly available genome sequences and has the potential to create a microbiome that can improve plant traits following species or genotype-driven selection in the composition of rhizobiome structure as documented in maize, wheat, barley, potato, rice, *Arabidopsis*, *Brassica rapa*, and sugarcane (Bulgarelli et al., 2013, 2015; Lebeis et al., 2015; Lundberg et al., 2012; Panke-Buisse et al., 2015; Peiffer et al, 2013; Raajimakers, 2015; Rasche et al., 2006; Yeoh et al., 2016).

In the light of the above discussion, some interesting strategies have been worked out to reshape the rhizobiome and redirect microbial activity by bringing about change in root exudates using conventional and modern breeding approaches (Bakker et al., 2012). On the other hand, development of PGPB and/or PGPF consortia using knowledge from plant ecosystem for mimicking or partially reconstructing the plant microbiome/rhizobiome are in progress. Adesemoye et al. (2009) have shown that tomato plants achieved full yield potential with 30% fewer inputs when inoculated with PGP consortia (*Bacillus amyloliquefaciens* IN937a, *Bacillus pumilus* T4, AMF *Glomus intraradices*) under greenhouse conditions (Adesemoye et al., 2009). Similarly, two soybean cultivars have significantly increased biomass after inoculation with *Bradyrhizobium japonicum* 532C, RCR3407 and *Bacillus subtilus* MIB600 (Atieno et al., 2012) compared with control plants. Masciarelli et al. (2014) investigated the interaction between microbes and nodulation formation in soybean. In this study, co-inoculation of soybean with two strains viz *B. japonicum* E109 and *Bacillus amyloliquefaciens* LL2012 have resulted in improved soybean nodulation efficiency due to phytohormones produced by co-inoculants. Mengual et al. (2014) employed a consortium of *B. megaterium*, *Enterobacter* sp., *Bacillus thuringiensis*, and *Bacillus* sp. along with composted sugar beet residues on *Lavandula dentate* L. to help restore soils by increasing phosphorus bioavailability, soil nitrogen fixation, and foliar NPK contents. Hence, the success of a rational design of a plant microbiome depends on several factors including identification of the genetic components of the microbiome control and smart integration of all players in the system.

19.5 PLANT MICROBIOME FOR ABIOTIC STRESS ALLEVIATION IN CROP PLANTS: SUCCESS STORY SO FAR

Plant microbiomes, particularly rhizosphere, has currently received increased attention for enhancing crop productivity and alleviation of plant stress by a variety of mechanisms (Bainard et al., 2013; Berg et al., 2014; Lau and Lennon, 2012; Marasco et al., 2012, 2013; Marulanda et al., 2009; Mendes et al., 2011; Panke-Buisse et al., 2015; Premachandra et al., 2016; Rolli et al., 2015; Sugiyama et al., 2013; Yang et al., 2009). The most well characterized of these microbial communities include the mycorrhizal fungi (Aroca and Ruíz-Lozano, 2012; Azcon et al., 2013; Bashan et al., 2012; Khan et al., 2008; Ruiz-Lozano et al., 2011; Sheng et al., 2011; Singh et al., 2011), symbiotic bacteria (Leite et al., 2014; Lugtenberg and Kamilova, 2009; Luvizotto et al., 2010; Spaink, 2000), and PGP rhizobacteria (Glick, 2012; Kloepper et al., 2004). PGPR contains a wide range of well-studied root-colonizing bacteria, which have the capacity to produce a wide range of enzymes and metabolites that play critical roles in host nutrition, growth, and health, and protect plants from biotic and abiotic stresses (Berendsen et al., 2012; Berg et al., 2014; Bulgarelli et al., 2013; Chauhan et al., 2015; Dimpka et al., 2009; Ding et al., 2013; Grover et al., 2011; Kim et al., 2009; Kim et al., 2013; Mendes et al., 2013; Pineda et al., 2013; Prashar et al., 2014; Rastogi et al., 2013; Timmusk et al., 2014; Timmusk and Nevo, 2011; Yang et al., 2009). Currently, efforts have been directed at exploring and utilizing naturally occurring, beneficial soil microbes for enhanced crop productivity under a changing climate (Bhattacharyya et al., 2016; Nadeem et al., 2014; Yang et al., 2009). The advantages of using root associated PGPR to help plants tolerate stress include their ability to confer abiotic tolerance to a wide range of crop species (Kasim et al., 2013; Mayak et al., 2004; Sandhya et al., 2010; Timmusk and Wagner, 1999; Tkacz and Poole, 2015) and also their ability to simultaneously tackle several biotic and/or abiotic stresses.

19.5.1 DROUGHT STRESS

Recent data have revealed that the plant-associated microbiome can influence several plant traits like root system architecture (Huang et al., 2014) to endure growth and biotic and abiotic stress tolerance (Bainard et al., 2013; Berg et al., 2014; Edwards et al., 2015; Lakshmanan et al., 2012; Marasco et al., 2012, 2013; Mendes et al., 2011; Ngumbi, 2011; Panke-Buisse et al., 2015; Rolli et al., 2015; Sugiyama et al., 2013). Drought stress represents a serious threat to agriculture worldwide (Hussain et al., 2012, 2014; Naveed et al., 2014a; Vinocur and Altman, 2005). The only approach to date is to develop powerful strategies to mitigate the drought associated negative effects on crop productivity. Use of plant-associated microbial communities offers a sustainable solution to abiotic stresses such as drought (Budak et al., 2013; Cooper et al., 2014). Two maize cultivars (Mazurka and Kaleo) showed 70% and 58% increases in root biomass respectively compared with controls under drought when inoculated with *Burkholderia phytofirmans* strain PsJN. Similarly, the same maize cultivars inoculated with *Enterobacter* sp. strain FD resulted in 47% and 40% increases in root biomass under drought condition (Naveed et al., 2014b). Several other researchers have shown the positive effects of plant-associated microbes on roots in different plants like maize and wheat (Timmusk et al., 2013, 2014; Yasmin et al., 2013). Inoculated plants performed better compared with control plants under drought, leading to the conclusion that an increase in root biomass resulted in enhanced water uptake by plants under drought stress. Two studies have also demonstrated the positive effects on shoot biomass in corn and wheat under drought when inoculated with different PGPR (Timmusk et al., 2014).

Crop plants treated with PGPR demonstrated enhanced growth, increased photosynthesis, better water use efficiency, nutrient management, high chlorophyll content biocontrol activity and grain yield due to alteration in root and shoot, phytohormonal activity, maintenance of high relative water content, EPS production, osmotic adjustment due to osmolyte accumulation, ACC deaminase activity, and antioxidant defense (Bano et al., 2013; Huang et al., 2014; Kasim et al., 2013; Marasco et al., 2013; Naveed et al., 2014b; Rolli et al., 2015; Sarma and Saikia, 2014; Timmusk et al., 2014; Yang et al., 2016a). PGPR treatment has improved growth in crop plants like wheat, maize, sunflower, pea, sorghum, tomato, pepper, rice, common bean, and lettuce under drought conditions (Alami et al., 2000; Arshad et al., 2008; Castillo et al., 2013; Cho et al., 2006; Creus et al., 2004; Dodd et al., 2005; Figueiredo et al., 2008; Kasim et al., 2013; Kim et al., 2013; Kohler et al., 2008; Lim and Kim, 2013; Marasco et al., 2016; Marquez et al., 2007; Mayak et al., 2004; Naseem and Bano, 2014; Perez-Montano et al., 2014; Sandhya et al., 2010; Sarma and Saikia, 2014; Timmusk et al., 2014, 2017).

19.5.2 Salinity Stress

Over the past few decades, salinity appears to have become a major stress and one of the most serious limiting factors to plant growth and productivity (Hussain et al., 2014; Wicke et al., 2011). Several strategies have been used in the past for tackling salinity problems including molecular (genetic engineering) and agronomic (soil reclamation and management practices). However, these methods are expensive, inefficient, and not practically sustainable due to an incomplete understanding of the fundamental mechanisms of stress tolerance in plants (Hussain et al. 2011). In contrast, the use of natural PGPR and PGPF has demonstrated promising success under salinity stress (Bharti et al., 2016; Shukla et al., 2012; Upadhyay et al., 2011; Yang et al., 2016b). These are being used as inoculants for crop plants growing on salt-affected land (Paul and Lade, 2014; Qin et al., 2014; Ruiz et al., 2016; Shabala et al., 2013; Tiwari et al., 2011). A growing body of research has shown that plant-associated microbial communities enhance productivity and improve plant health following different environmental stresses (Berendsen et al., 2012; Mahmood et al., 2016; Sharma et al., 2016; Sloan and Lebeis, 2015; Timmusk et al., 2017; Zuppinger-Dingley et al., 2014). Recently, several researchers have demonstrated the practical utility of PGPRs where plants like wheat, mung bean, and peanuts have shown significantly higher biomass under salinity stress compared to non-inoculated plants (Bharti et al., 2016; Mahmood et al., 2016; Sharma et al., 2016).

It is well documented that microbes inhabiting harsh environments develop adaptive tolerant traits and are potential candidates as plant growth promoters under stress conditions (Rodriguez et al., 2008; Timmusk et al., 2014). For example, halotolerant microbes thrive under high soil salinity and express traits to help plants to cope with salinity stress. A team of researchers isolated 130 rhizobacterial strains from roots of wheat plants growing under saline conditions. Furthermore, results revealed that 24 isolates grew well at relatively high levels (8%) of NaCl stress in culture (Upadhyay et al., 2009). It is reported that 17 of 20 bacteria isolated from different salt-tolerant plant species showed growth in 7.5% NaCl in culture and two of these grew well in 10% NaCl (Arora et al., 2014). Various PGPRs tackle stress using different mechanisms like halotolerant bacterial strains isolated from Korea enhanced plant growth under salinity stress whereby bacterial ACC deaminase activity negatively affected ethylene production under stress (Siddikee et al., 2010). Wheat inoculated with EPS-producing PGPRs (like *Bacillus* spp, *Enterobacter* spp, etc.) demonstrated high biomass production under high salinity stress (Upadhyay et al., 2011). Plant-associated microbiomes have improved growth in canola, pepper, tomato, bean, wheat, and lettuce (Ali et al., 2014; Bharti et al., 2016; Jha et al., 2012; Leite et al., 2014; Mahmood et al., 2016; Mayak et al., 2004; Sharma et al., 2016; Shukla et al., 2012; Timmusk et al., 2014; Upadhyay et al., 2009; Zhao et al., 2016).

Several reports have highlighted the involvement of PGPF resulting in enhanced host plant tolerance to high salinity stress (Giri and Mukerji, 2004; Velazquez-Hernandez et al., 2011) using several different mechanisms like osmotic adjustment, increased phosphate, and decreased Na^+ concentration in shoots, antioxidant systems, and reduced ROS compared with uninoculated controls. Therefore, maize, mung bean, clover, tomato, and cucumber have revealed improved salt tolerance after PGPF treatment possibly due to P acquisition, improved osmoregulation by proline accumulation, and reduced NaCl concentration (Al-Karaki et al., 2001; Ben Khaled et al., 2003; Feng et al., 2002; Jindal et al., 1993; Velazquez-Hernandez et al., 2011; Yang et al., 2009). These reports suggest that plants may readily recruit diverse microbes under different stresses with broad implications for plants grown particularly under salinity stress.

19.5.3 Extreme Temperature Stress

Plants engineer their own rhizobiome to improve nutrient availability and address macro and micro environments by recruiting specific microbes. There are reports that global temperatures are predicted to increase by 1.8°C–3.6°C by the end of this century due to climate changes (IPCC, 2007). High temperatures are a major obstacle in crop production as well as microbial colonization (Carson et al., 2010), resulting in major cellular damage such as protein degradation and aggregation. Both plants and microbes respond to heat stress by producing a specific group of polypeptides known as heat-shock proteins (HSPs). Stress adaptation in microorganisms represents a complex multilevel regulatory process that may involve several genes (Srivastava et al., 2008), such that microbes develop different adaptation strategies to combat the stress (Kumar and Verma, 2018; Yang et al., 2016a). High soil temperature negatively affects the performance of plant-associated microbes. However, certain microbes perform better at high temperatures, and these microbes may be important for crop plants under high temperature. Abd El-Daim et al. (2014) treated seeds of the cultivars Olivin and Sids1 with *Bacillus amyloliquefaciens* UCMB5113 or *Azospirillum brasilense* NO40 and young seedlings tested for management of short-term heat stress. Heat stress raised transcript levels of several stress-related genes in the leaves. However, expression was lower in inoculated plants but elevated compared with the control. Wheat seedling demonstrated improvement of heat tolerance by bacteria priming as revealed by reduced generation of reactive oxygen species, small changes in the metabolome. Furthermore, pre-activation of certain heat shock transcription factors seems important in conferring heat stress tolerance in wheat (Abd El-Daim et al., 2014). Srivastava et al. (2008) isolated *Pseudomonas putida* strain NBR10987, which exhibited thermotolerance in the drought-stressed rhizosphere of chickpea and was attributed to the stress sigma factor δs overexpression and thick biofilm formation. Certain bacterial strains combat stress by producing exopolysaccharides (EPS), which possess unique water holding and cementing characteristics and play vital roles in stress tolerance by water retention and biofilm formation. Redman et al. (2002) demonstrated high temperature tolerance in *Dichanthelium lanuginosum* by a symbiotic *Curvularia* spp. Similarly, *Pseudomonas* AKM-P6 and NBR10987 strains used to inoculate sorghum seedling had improved thermotolerance through enhanced physiological and metabolic performance indicating a unique interaction of inducible proteins in heat tolerance using microbes (Ali et al., 2009). In another study, *Paraphaeosphaeria quadriseptata*, a rhizosphere fungus also enhanced heat stress tolerance to *Arabidopsis* plants through induction of small heat shock HSP101 and HSP70 proteins (McLellan et al., 2007).

Cold (low-temperature) stress is an important limiting factor to crop productivity because it negatively affects plant growth and development. Several researchers addressed cold stress and documented bacterial strains for enhanced cold stress tolerance in plants (Selvakumar et al., 2008a, b, 2009, 2010a, b). *B. phytofirmans* PsJN confers increased tolerance to low non-freezing temperatures and resistance to gray mold in grapevine plants. While Barka et al. (2006) noted that grapevine roots inoculated with *Burkholderia phytofirmans* PsJN resulted in better root growth, higher plant biomass, and increased physiological activity at low temperature (4°C). Comprehensive physiological analysis revealed that bacterized plantlets significantly increased proline, starch, and phenolic levels compared with uninoculated control plantlets, which enhanced grapevine plantlets to tolerate low temperature. On the other hand, endophyte inoculation resulted in higher and faster accumulation of stress-related proteins and metabolites, which lead to more effective resistance to low temperature, indicating a positive priming effect on plants (Theocharis et al., 2012). Low temperature usually inhibits soybean symbiotic activities (nodule infection and nitrogen fixation), but inoculation of soybean with both *Bradyrhizobium japonicum* and *Serratia proteamaculans* resulted in faster growth at 15°C (Zhang et al., 1995, 1996). Switchgrass inoculated with *B. phytofirmans* PsJN had enhanced growth under glasshouse conditions (Kim et al., 2012). According to Mishra et al. (2009), wheat seedlings inoculated with *Pseudomonas* sp. strain PPERs23 highly improved root and shoot lengths resulting in dry root/shoot biomass, and total phenolics, chlorophyll, and amino acid contents. Furthermore, inoculated wheat seedlings had enhanced physiologically available iron, anthocyanins, proline, protein, and relative water contents, and reduced Na^+/K^+ ratio and electrolyte leakage, resulting in enhanced cold tolerance (Mishra et al., 2009). It is apparent from the above studies that *Burkholderia phytofirmans* PsJN has a wide host spectrum, which includes grapevines, maize, soybean, sorghum, wheat, and switchgrass with promising results at low-temperature stress.

19.5.4 Heavy Metal Stress

Industrialized economies often suffer from heavy metal contamination which has recently received attention globally due to the non-degradable nature of these contaminants (Kidd et al., 2009; Ma et al., 2011; Rajkumar et al., 2012). Common heavy metals include mercury (Hg), manganese (Mn), chromium (Cr), cadmium (Cd), lead (Pb), chromium (Cr), zinc (Zn), nickel (Ni), aluminum (Al), and copper (Cu). Some metalloids also show toxicity such as antimony (Sb) and arsenic (As) (Duruibe et al., 2007; Pandey, 2012; Park, 2010; Wuana and Okieimen, 2011). Heavy metals represent a significant threat to living beings when elevated above tolerance levels. Many physiochemical and biological techniques developed to reclaim contaminated soils have failed due to being non-sustainable, environmentally unsafe, and publicly unacceptable (Boopathy, 2000; Doble and Kumar, 2005; Vidali, 2001). Phytoremediation using plants to eliminate soil contaminants is cost-effective and environmentally friendly technology with high public acceptance (Afzal et al., 2011; Arslan et al., 2017; Beskoski et al., 2011; Fester et al., 2014; Hadi and Bano, 2010). However, high soil contamination negatively affects plant growth including root/shoot growth and expansion mainly due to oxidative stress, which limits the scope of soil reclamation by phytoremediation (Gerhardt et al., 2009; Hu et al., 2016). Other factors like shortage of plant nutrients and reduced microbial activity also limit phytoremediation (Gerhardt et al., 2009). Microbe-assisted phytoremediation represents a viable and promising alternative (Jamil et al., 2014) whereby microbe-associated activities in the rhizosphere increase plant metal uptake using several unique ways to use or transform by altering mobility and bioavailability of metals or render them inactive (Aafi et al., 2012; Ma et al., 2011; Rajkumar et al., 2010; Yang et al., 2012). Several reports have shown that microbes produce substances, such as plant growth hormones (IAA, cytokinins & gibberellins), siderophores, and ACC deaminase, which help to improve plant growth in contaminated soils (Babu and Reddy, 2011; Bisht et al., 2014; Ijaz et al., 2016; Kukla et al., 2014; Luo et al., 2011, 2012; Santoyo et al., 2016; Wang et al., 2011; Waqas et al., 2015).

Waqas et al. (2015) have described several PGPRs genera which deserve close attention among rhizosphere microbes involved in phytoremediation. These can directly enhance process efficiency by altering soil pH and oxidation/reduction reactions (Khan et al., 2009; Kidd et al., 2009; Ma et al., 2011; Rajkumar et al., 2010; Uroz et al., 2009; Wenzel, 2009). Rice inoculated with *Bacillus licheniformis* strain NCCP-59 showed improved seed germination under Ni stress compared with uninoculated control (Jamil et al., 2014), pointing toward the ability to confer protection against the toxic effects of Ni. Sheng et al. (2008) showed that two microbes namely *Microbacterium* sp. G16 and *Pseudomonas fluorescens* G10 significantly increased the solubility of lead (Pb) in *Brassica napus* and was mainly attributed to a variety of mechanisms including IAA, siderophores, ACC deaminase, and phosphate solubilization. Similarly, *Zea mays* co-inoculated with *Azotobacter chrococcum* or *Rhizobium leguminosarum* improved plant growth and biomass in lead-contaminated soil (Hadi and Bano, 2010; Hussain et al., 2018). A wide diversity of PGPRs included in *Bacillus* sp., *Serratia*, *Enterobacter*, *Burkholderia* sp., and *Agrobacterium* genera have increased the phytoremediation rate and biomass production in problem soils (Afzal et al., 2014; Feng et al., 2017; Glick, 2014, 2015; Hardoim et al., 2015; Ijaz et al., 2016; Jha et al., 2015; Kumar et al., 2009; Luo et al., 2012; Mastretta et al., 2009; Nonnoi et al., 2012; Singh et al., 2016; Wani et al., 2008; Zheng et al., 2016).

19.5.5 Nutrient Deficiency Stress

Beneficial microbes represent key determinants to enhance sustainable plant productivity in current agricultural systems (Berendsen et al., 2012), and their role in plant nutrition has received substantial attention only recently (Bulgarelli et al., 2013; Lebeis et al., 2012; Turner et al., 2013). Microbes of the rhizomicrobiome render significant services for maintaining adequate plant nutrient and high productivity (Adhya et al., 2015). Well-known microbes involved in solubilization and increased uptake of nitrogen include rhizobia and bradyrhizobia (nitrogen uptake) and mycorrhizal fungi for phosphorus uptake (Hawkins et al., 2000; Miransari, 2011; Richardson et al., 2009). Plants usually get nutrients from the rhizosphere and phyllosphere (Turner et al., 2013). Several factors involved in plant nutrient management are the optimal use of soil, water, atmospheric factors, and NPK fertilizers (Adesemoye and Egamberdieva, 2013; Miao et al., 2011), along with a beneficial rhizomicrobiome (Adesemoye et al., 2009). Similarly, recent work reported the usefulness of symbionts such as mycorrhizal fungi to many plant species for channeling nutrients and minerals such as phosphorous, water, and other essential elements from soil to host plants (Adeleke et al., 2012; Carvalhais et al., 2013; Gianinazzi et al., 2010; Hartmann et al., 2009; Johnson and Graham, 2013; Lareen et al., 2016; Salvioli et al., 2016), suggesting a role in modeling and improved soil structure and microbial communities (Bulgarelli et al.,

2012; Peiffer et al., 2013; Tkacz et al., 2015) in many crop plants belonging to several genera for maintaining their nutritional requirements (Johnson et al., 2012; Philippot et al., 2013; Salvioli and Bonfante, 2013; Schlaeppi et al., 2014). It has been demonstrated that apart from known nitrogen-fixing bacteria (*Rhizobium* and *Bradyrhizobium*), several other bacterial endophytes establish a symbiosis or symbiosis-like relationship with plants for bioavailable nitrogen fixation in unspecialized host tissues without using a nodule system (Gaby and Buckley, 2011; Guimaraes et al., 2012; Santi et al., 2013; Zehr et al., 2003). For example, *Cyanobacteria* establish a symbiotic association with a range of plants and form heterocysts suitable for biological nitrogen fixation (BNF) with nitrogenase (Berman-Frank et al., 2003; Santi et al., 2013). Another study revealed that sugarcane root-associated bacteria successfully fix nitrogen and solubilize phosphorus, respectively (Leite et al., 2014). Similarly, Guimaraes et al. (2012) based on analysis of the cowpea rhizosphere using 16 S rRNA sequencing reported that *Burkholderia* and *Achromobacter* species along with *Rhizobium* and *Bradyrhizobium* can nodulate cowpea and support BNF. Furthermore, some reports have argued that algal genera such as *Anabaena*, *Aphanocapsa*, and *Phormidium* can successfully fix available atmospheric nitrogen in paddy fields (Hasan, 2013; Shridhar, 2012). Recently, Symanczik et al. (2017) demonstrated that use of AMF in Naranjilla (*Solanum quitoense*) led to the significantly better acquisition of phosphorous (up to 104%) compared with the control resulting in better plant growth, improved nutrition, and soil water retention. Furthermore, this study highlighted that highly diverse belowground systems like AMF are essential for maintaining sustainable and increased the productivity of soil ecosystems (De Vries et al., 2013; Van der Heijden et al., 2008; Wagg et al., 2014). Similarly, many in-AMF are also reported to bring symbiotic like benefits to plants (Cai et al., 2015; Ghanem et al., 2014; Pandey et al., 2016).

Given the importance of nutrients in plant growth and health, it is important to identify bacterial strains that effectively increase macro and micronutrient uptake in crop plants under deficient and toxic soil conditions (Leveau et al., 2010; Mapelli et al., 2012). Similarly, several rhizospheric microbes can also facilitate the uptake of many trace elements such as iron (Marschner et al., 2011; Shirley et al., 2011; Zhang et al., 2009) and calcium (Lee et al., 2010) with improved root growth from the soil (Cummings and Orr, 2010). Taken together, it is safe to conclude that microbes of any plant microbiome play essential roles in degrading insoluble organic compounds required not only for their survival but also for promoting plant growth in nutrient-deficit soils (Bhattacharyya et al., 2015; Leveau et al., 2010; Mapelli et al., 2012; Turner et al., 2013).

19.6 CONCLUSION AND FUTURE PERSPECTIVE

Feeding a growing population requires high and stable yields from efficient crop production. Modern agriculture is mainly based on the cultivation of high-yield varieties combined with the use of agrochemicals. It is not surprising that abiotic stresses, especially drought and salinity, are considered by researchers the most significant threats to future agriculture (Busby et al., 2017). Given this, we need to either improve abiotic stress tolerance of crop plants or look for alternative and more sustainable agricultural practices (Bulgarelli et al., 2012; Mengual et al., 2014). Developing more sustainable solutions to agricultural problems becomes more important under climate change and uncontrolled growth of the human population (Hussain et al., 2014). Opportunities for exploiting the plant microbiome for raising crops are numerous and diverse, which can play a promising role for stress management in sustainable next-generation agriculture (Hussain et al., 2018; Vandenkoornhuyse et al., 2015).

The development of next-generation agricultural tools and practices will depend on the smart integration of all co-variates in the system. Progress has been made toward the establishment of model host-microbiome systems for populus, rice, sorghum, maize, miscanthus, tomato, and *Medicago truncatula* (Edwards et al., 2015; Hacquard and Schadt, 2015; Hussain et al., 2018; Johnston-Monje and Raizada, 2011; Knief et al., 2012; Lakshmanan, 2015; Li et al., 2016; Peiffer et al., 2013; Ramond et al., 2013; Sessitsch et al., 2011; Spence et al., 2014; Tian et al., 2015). However, there is a great variation and success depends on several factors including individual plant species and genotype, the native soil microbiota, and the microbiome and interplay between these players with their specific traits that interact with one or both under the given climatic conditions. Under such circumstances, it is recommended that native microbiomes are likely best suited to generate diverse and functionally variable host-microbiome associations to select on initially (Mueller and Sachs, 2015). Hence, novel methods to exploit the plant microbiome at work and next-generation agriculture could lead to crop production that is more resilient in the face of stresses (Bakker et al., 2012; Marasco et al., 2012; Prudent et al., 2015). Several recent reports have revealed significant improvements to stress tolerance using PGPM to crops under field conditions (Celebi et al., 2010; Mengual et al., 2014; Rolli et al., 2015), others have shown inconsistent or negative results (Nadeem et al., 2014). One promising strategy for a stable beneficial outcome is to use a microbial consortium in the field to tailor the rhizobiome to collectively respond to specific biotic and abiotic stresses without compromising plant growth and productivity (Trabelsi and Mhamdi, 2013). Development and application of multispecies consortia have the potential to address the inconsistency in performance. Therefore, the mechanisms by which microbes confer stress tolerance to their hosts warrant further research to develop novel microbial consortia for use under different biotic and abiotic stresses. However, concerted efforts are required at interdisciplinary levels from microbiologists, molecular biologists, plant physiologists, plant breeders, soil scientists, and agronomists. Recent developments like the use of omics approaches (White III et al., 2017) in this field provide powerful insights to understand how the microbe–microbe and plant–microbe interactions mediate the functional relationship between different players.

REFERENCES

Aafi, N.E., Brhada, F., Dary, M., Maltouf, A.F., Pajuelo, E. (2012) Rhizostabilization of metals in soils using *Lupinus luteus* inoculated with the metal resistant rhizobacterium *Serratia* sp. MSMC 541. *International Journal of Phytoremediation* 14: 26174.

Abd El-Daim, I.A., Bejai, S., Meijer, J. (2014) Improved heat stress tolerance of wheat seedling by bacterial seed treatment. *Plant and Soil* 379 No. 1–2: 337–350.

Adeleke, R.A., Cloete, T.E., Bertrand, A., Khasa, D.P. (2012) Iron ore weathering potentials of ectomycorrhizal plants. *Mycorrhiza* 22 No. 7: 535–544.

Adesemoye, A.O., Egamberdieva, D. (2013) Beneficial effects of plant growth-promoting rhizobacteria on improved crop production: Prospects for developing economies. In *Bacteria in Agrobiology: Crop Productivity*, ed. Maheshwari, D.K., Saraf, M., Aeron, A., 45–63. Berlin: Springer.

Adesemoye, A.O., Torbert, H.A., Kloepper, J.W. (2009) Plant growth promoting rhizobacteria allow reduced application rates of chemical fertilizers. *Microbial Ecology* 58 No. 4: 921–929.

Adhya, T.K., Kumar, N., Reddy, G., Podile, A.R., Bee, H., Bindiya, S. (2015) Microbial mobilization of soil phosphorus and sustainable P management in agricultural soils. *Current Science* 108: 1280–1287.

Afzal, M., Khan, Q.M., Sessitsch, A. (2014) Endophytic bacteria: Prospects and applications for the phytoremediation of organic pollutants. *Chemosphere* 117: 232–242.

Afzal, M., Yousaf, S., Reichenauer, T.G., Kuffner, M., Sessitsch, A. (2011) Soil type affects plant colonization, activity and catabolic gene expression of inoculated bacterial strains during phytoremediation of diesel. *Journal of Hazardous Materials* 186 No. 2–3: 1568–1575.

Alami, Y., Achouak, W., Marol, C., Heulin, T. (2000) Rhizosphere soil aggregation and plant growth promotion of sunflowers by exopolysaccharide producing Rhizobium sp. strain isolated from sunflower roots. *Applied and Environmental Microbiology* 66 No. 8: 3393–3398.

Ali, S., Duan, J., Charles, T.C., Glick, B.R. (2014) A bioinformatics approach to the determination of genes involved in endophytic behavior in Burkholderia spp. *Journal of Theoretical Biology* 343: 193–198.

Ali, S.Z., Sandhya, V., Grover, M., Kishore, N., Rao, L.V., Venkateswarlu, B. (2009) Pseudomonas sp. strain AKM-P6 enhances tolerance of sorghum seedlings to elevated temperatures. *Biology and Fertility of Soils* 46 No. 1: 45–55.

Al-Karaki, G.N., Ammad, R., Rusan, M. (2001) Response of two tomato cultivars differing in salt tolerance to inoculation with mycorrhizal fungi under salt stress. *Mycorrhiza* 11 No. 1: 43–47.

Amann, R.I., Ludwing, W., Schleifer, K.H. (1995) Phylogenetic identification and in situ detection of individual microbial cells without cultivation. *Microbiological Reviews* 59 No. 1: 143–169.

Andreote, F.D., Azevedo, J.L., Araújo, W.L. (2009) Assessing the diversity of bacterial communities associated with plants. *Brazilian Journal of Microbiology: [Publication of the Brazilian Society for Microbiology]* 40 No. 3: 417–432.

Aroca, R., Ruíz-Lozano, J.M. (2012) Regulation of root water uptake under drought stress conditions. In *Plant Responses to Drought Stress*, ed. Aroca, A., 113–128. Berlin: Springer.

Arshad, M., Sharoona, B., Mahmood, T. (2008) Inoculation with Pseudomonas spp. containing ACC deaminase partially eliminate the effects of drought stress on growth, yield and ripening of pea (*Pisum sativum* L.). *Pedosphere* 18 No. 5: 611–620.

Arslan, M., Imran, A., Khan, Q.M., Afzal, M. (2017) Plant-bacteria partnerships for the remediation of persistent organic pollutants. *Environmental Science and Pollution Research International* 24 No. 5: 4322–4336.

Atieno, M., Herrmann, L., Okalebo, R., Lesueur, D. (2012) Efficiency of different formulations of *Bradyrhizobium japonicum* and effect of co-inoculation of *Bacillus subtilis* with two different strains of *Bradyrhizobium japonicum*. *World Journal of Microbiology and Biotechnology* 28 No. 7: 2541–2550.

Azarias Guimarães, A., Duque Jaramillo, P.M., Simão Abrahão Nóbrega, R., Florentino, L.A., Barroso Silva, K., de Souza Moreira, F.M. (2012) Genetic and symbiotic diversity of nitrogen-fixing bacteria isolated from agricultural soils in the western Amazon by using cowpea as the trap plant. *Applied and Environmental Microbiology* 78 No. 18: 6726–6733.

Azcon, R., Medina, A., Aroca, R., Ruiz-Lozano, J.M. (2013) Abiotic stress remediation by the arbuscular mycorrhizal symbiosis and rhizosphere bacteria/yeast interactions. In *Molecular Microbial Ecology of the Rhizosphere*, ed. de Bruijin, F.J., 991–1002. Hoboken: Wiley-Blackwell.

Babu, A.G., Reddy, M.S. (2011) Dual inoculation of arbuscular mycorrhizal and phosphate solubilizing fungi contributes in sustainable maintenance of plant health in fly ash ponds. *Water, Air, and Soil Pollution* 219 No. 1–4: 3–10.

Badri, D.V., Weir, T.L., Van Der Lelie, D., Vivanco, J.M. (2009) Rhizosphere chemical dialogues: Plant-microbe interactions. *Current Opinion in Biotechnology* 20 No. 6: 642–650.

Bainard, L.D., Koch, A.M., Gordon, A.M., Klironomos, J.N. (2013) Growth response of crops to soil microbial communities from conventional monocropping and tree-based intercropping systems. *Plant and Soil* 363 No. 1–2: 345–356.

Bais, H.P., Weir, T.L., Perry, L.G., Gilroy, S., Vivanco, J.M. (2006) The role of root exudates in rhizosphere interactions with plants and other organisms. *Annual Review of Plant Biology* 57: 233–266.

Bakker, P.A., Berendsen, R.L., Doornbos, R.F., Wintermans, P.C., Pieterse, C.M. (2013) The rhizosphere is revisited: Root microbiomics. *Frontiers in Plant Science* 4: 165

Bakker, M.G., Manter, D.K., Sheflin, A.M., Weir, T.L., Vivanco, J.M. (2012) Harnessing the rhizosphere microbiome through plant breeding and agricultural management. *Plant and Soil* 360 No. 1–2: 1–13.

Bano, Q., Ilyas, N., Bano, A., Zafar, N., Akram, A., Hassan, F. (2013) Effect of Azospirillum inoculation on maize (*Zea mays* L.) under drought stress. *Pakistan Journal of Botany* 45: 13–20.

Barka, E.A., Nowak, J., Clement, C. (2006) Enhancement of chilling resistance of inoculated grapevine plantlets with a plant growth-promoting rhizobacterium, *Burkholderia phytofirmans* strain PsJN. *Applied and Environmental Microbiology* 72 No. 11: 7246–7252.

Bartels, D., Hussain, S.S. (2008) Current status and implications of engineering drought tolerance in plants using transgenic approaches. *CAB Reviews: Perspectives in Agriculture, Veterinary Science, Nutrition and Natural Resources* 3 No. 20: 020.

Bashan, Y., Salazar, B.G., Moreno, M., Lopez, B.R., Lindermann, R.G. (2012) Restoration of eroded soil in the sonorant desert with native leguminous trees using plant growth promoting microorganisms and limited amounts of compost and water. *Journal of Environmental Management* 102: 26–36.

Ben Khaled, L., Gomez, A.M., Ourraqi, E.M., Oihabi, A. (2003) Physiological and biochemical responses to salt stress of mycorrhized and/or nodulated clover seedlings (*Trifolium alexandrinum* L.). *Agronomie* 23 No. 7: 571–580.

Berendsen, R.L., Pieterse, C.M.J., Bakker, P.A. (2012) The rhizosphere microbiome and plant health. *Trends in Plant Science* 17 No. 8: 478–486.

Berg, G., Grube, M., Schloter, M., Small, K. (2014) Unraveling the plant microbe: Looking back and future perspectives. *Frontiers in Microbiology* 5: 148

Berg, G., Rybakova, D., Grube, M., Koberl, M. (2016) The plant microbiome explored: Implications for experimental botany. *Journal of Experimental Botany* 67 No. 4: 995–1002.

Berg, G., Smalla, K. (2009) Plant species and soil type cooperatively shape the structure and function of microbial communities in the rhizosphere. *FEMS Microbiology Ecology* 68 No. 1: 1–13.

Berg, G., Zachow, C., Müller, H., Philipps, J., Tilcher, R. (2013) Next-generation bio-products sowing the seeds of success for sustainable agriculture. *Agronomy* 3 No. 4: 648–656.

Berman-Frank, I., Lundgren, P., Falkowski, P. (2003) Nitrogen fixation and photosynthetic oxygen evolution in Cyanobacteria. *Research in Microbiology* 154 No. 3: 157–164.

Beskoski, V.P., Gojgic-Cvijovic, G., Milic, J., Ilic, M., Miletic, S., Solevic, T., Vrvic, M.M. (2011) Ex-situ bioremediation of a soil contaminated by mazut (heavy residual fuel oil), a field experiment. *Chemosphere* 83 No. 1: 34–40.

Bharti, N., Pandey, S.S., Barnawal, D., Patel, V.K., Kalra, A. (2016) Plant growth promoting rhizobacteria *Dietzia natronolimnaea* modulates the expression of stress responsive genes providing protection of wheat from salinity stress. *Scientific Reports* 6: 34768.

Bhattacharyya, P.N., Goswani, M.P., Bhattacharyya, L.H. (2016) Perspective of beneficial microbes in agriculture under changing climatic scenario: A review. *Journal of Phytology* 8: 26–41.

Bhattacharyya, P.N., Jha, D.K. (2012) Plant growth-promoting rhizobacteria (PGPR): Emergence in agriculture. *World Journal of Microbiology and Biotechnology* 28 No. 4: 1327–1350.

Bhattacharyya, P.N., Sarmah, S.R., Dutta, P., Tanti, A.J. (2015) Emergence in mapping microbial diversity in tea (*Camellia sinensis* L.) soil of Assam, North-East India: A novel approach. *European Journal of Biotechnology and Biosciences* 3: 20–25.

Bisht, S., Pandey, P., Kaur, G., Aggarwal, H., Sood, A., Sharma, S., Kumar, V., Bisht, N.S. (2014) Utilization of endophytic strain Bacillus sp. SBER3 for biodegradation of polyaromatic hydrocarbons (PAH) in soil model system. *European Journal of Soil Biology* 60: 67–76.

Bloemberg, G.V., Lugtenberg, B.J.J. (2001) Molecular basis of plant growth promotion and biocontrol by rhizobacteria. *Current Opinion in Plant Biology* 4 No. 4: 343–350.

Boon, E., Meehan, C.J., Whidden, C., Wong, D.H.J., Langille, M.G.I., Beiko, R.G. (2014) Interactions in the microbiome: Communities of organisms and communities of genes. *FEMS Microbiology Reviews* 38 No. 1: 90–118.

Boopathy, R. (2000) Factors limiting bioremediation technologies. *Bioresource Technology* 74 No. 1: 63–67.

Budak, H., Kantar, M., Yucebilgili Kurtoglu, K. (2013) Drought tolerance in modern and wild wheat. *The Scientific World Journal* 2013: 548246.

Bulgarelli, D., Garrido-Oter, R., Munch, P.C., Weiman, A., Droge, J., Pan, Y., McHardy, A.C., Schulze-Lefert, P. (2015) Structure and function of the bacterial root microbiota in wild and domesticated barley. *Cell Host & Microbe* 17 No. 3: 392–403.

Bulgarelli, D., Rott, M., Schlaeppi, K., Ver Loren van Themaat, E., Ahmadinejad, N., Assenza, F., Rauf, P., *et al.* (2012) Revealing structure and assembly cues for *Arabidopsis* root-inhabiting bacterial microbiota. *Nature* 488 No. 7409: 91–95.

Bulgarelli, D., Schlaeppi, K., Spaepen, S., Loren, van, V., Themaat, A.N., Schulze-Lefert, P. (2013) Structure and functions of the bacterial microbiota of plants. *Annual Reviews in Plant Biology* 64: 807–838.

Busby, P.E., Soman, C., Wagner, M.R., Friesen, M.L., Kremer, J., Bennett, A., Morsy, M., *et al.* (2017) Research priorities for harnessing plant microbiomes in sustainable agriculture. *PLOS Biology* 15 No. 3: e2001793.

Cai, F., Chen, W., Wei, Z., Pang, G., Li, R., Ran, W., Shen, Q. (2015) Colonization of *Trichoderma harzianum* strain SQR-T037 on tomato roots and its relationship to plant growth, nutrient availability and soil microflora. *Plant and Soil* 388 No. 1–2: 337–350.

Capell, T., Bassie, L., Christou, P. (2004) Modulation of the polyamine biosynthetic pathway in transgenic rice confers tolerance to drought stress. *Proceedings of the National Academy of Sciences of the United States of America* 101 No. 26: 9909–9914.

Carson, J.K., Gonzalez-Quinones, V., Murphy, D.V., Hinz, C., Shaw, J.A., Gleeson, D.B. (2010) Low pore connectivity increases bacterial diversity in soil. *Applied and Environmental Microbiology* 76 No. 12: 3936–3942.

Carvalhais, L.C., Dennis, P.G., Fan, B., Fedoseyenko, D., Kierul, K., Becker, A., von Wiren, N., Borriss, R. (2013) Linking plant nutritional status to plant-microbe interactions. *PLoS ONE* 8 No. 7: e68555.

Castillo, P., Escalante, M., Gallardo, M., Alemano, S., Abdala, G. (2013) Effects of bacterial single inoculation and co-inoculation on growth and phytohormone production of sunflower seedlings under water stress. *Acta Physiologiae Plantarum* 35 No. 7: 2299–2309.

Celebi, S.Z., Demir, S., Celebi, R., Durak, E.D., Yilmaz, I.H. (2010) The effect of arbuscular mycorrhizal fungi (AMF) applications on the silage maize (*Zea mays* L.) yield in different irrigation regimes. *European Journal of Soil Biology* 46 No. 5: 302–305.

Chagnon, P.L., Bradley, R.L., Maherali, H., Klironomos, J.N. (2013) A trait-based framework to understand life history of mycorrhizal fungi. *Trends in Plant Science* 18 No. 9: 484–491.

Chaparro, J.M., Badri, D.V., Vivanco, J.M. (2014) Rhizosphere microbiome assemblage is affected by plant development. *The ISME Journal* 8 No. 4: 790–803.

Chauhan, H., Bagyaraj, D.J., Selvakumar, G., Sundaram, S.P. (2015) Novel plant growth promoting rhizobacteria prospects. *Applied Soil Ecology* 95: 38–53.

Cho, K., Toler, H., Lee, J., Ownley, B., Stutz, J.C., Moore, J.L., Auge, R.M. (2006) Mycorrhizal symbiosis and response of sorghum plants to combined drought and salinity stresses. *Journal of Plant Physiology* 163 No. 5: 517–528.

Choudhary, D.K., Prakash, A., Johri, B.N. (2007) Induced systemic resistance (ISR) in plants: Mechanism of action. *Indian Journal of Microbiology* 47 No. 4: 289–297.

Coleman-Derr, D., Tringe, S.G. (2014) Building the crops of tomorrow: Advantages of symbiont-based approaches to improving abiotic stress tolerance. *Frontiers in Microbiology* 5: 283.

Conrath, U. (2006) Systemic acquired resistance. *Plant Signaling and Behavior* 1 No. 4: 179–184.

Cooper, M., Gho, C., Leafgren, R., Tang, T., Messina, C. (2014) Breeding drought-tolerant maize hybrids for the US corn-belt: Discovery to product. *Journal of Experimental Botany* 65 No. 21: 6191–6204.

Creus, C.M., Sueldo, R.J., Barassi, C.A. (2004) Water relations and yield in Azospirillum-inoculated wheat exposed to drought in the field. *Canadian Journal of Botany* 82 No. 2: 273–281.

Cummings, S.P., Orr, C. (2010) The role of plant growth promoting rhizobacteria in sustainable and low graminaceous crop production. In *Plant Growth and Health Promoting Bacteria*, ed. Maheshwari, D.K., 297–315. Berlin: Springer.

Cushman, J.C., Bohnert, H.J. (2000) Genomic approaches to plant stress tolerance. *Current Opinion in Plant Biology* 3 No. 2: 117–124.

Dangl, J.L., Horvath, D.M., Staskawicz, B.J. (2013) Pivoting the plant immune system from dissection to deployment. *Science* 341 No. 6147: 746–751.

de Bruijin, F.J. (2015) Biological nitrogen fixation. In *Principles of Plant-Microbe Interactions*, ed. Lugtenberg, B., 215–224. Heidelberg: Springer.

De Vries, F.T., Thebault, E., Liiri, M., Birkhofer, K., Tsiafouli, M.A., Bjornlund, L., Jorgensen, H.B., *et al.* (2013) Soil food web properties explain ecosystem sevices across European land use systems. *Proceedings of the National Academy of Sciences of the United States of America* 110 No. 35: 14296–14301.

de Zelicourt, A., Al-Yousif, M., Hirt, H. (2013) Rhizosphere microbes as essential partners for plant stress tolerance. *Molecular Plant* 6 No. 2: 242–245.

De-la-Peña, C., Badri, D., Loyola-Vargas, V. (2012) Plant root secretions and their interactions with neighbors. In *Secretions and Exudates in Biological Systems*, ed. Vivanco, J.M. and Baluška, F., 1–26. Berlin: Springer.

Dimpka, C., Weinard, T., Asch, F. (2009) Plant-rhizobacteria interactions alleviate abiotic stress conditions. *Plant, Cell and Environment* 32 No. 12: 1682–1694.

Ding, G.C., Piceno, Y.M., Heuer, H., Weinert, N., Dohrmann, A.B., Carrillo, A., Andersen, G.L., *et al.* (2013) Changes of soil bacterial diversity as a consequence of agricultural land use in a semi-arid ecosystem. *PLoS ONE* 8 No. 3: e59497.

Doble, M., Kumar, A. (2005) Biotreatment of industrial effluents: Introduction. In *Biotreatment of Industrial Effluents*, ed. Doble, M., Kumar, A., 1–9. Butterworth Heinemann, Elsevier.

Dodd, I.C., Belimov, A.A., Sobeih, W.Y., Safronova, V.I., Grierson, D., Davies, W.J. (2005) Will modifying plant ethylene status improve plant productivity in water-limited environments? 4th International Crop Science Congress. au/icsc2004/poster/1/3/4/510_doddicref.htm. www.cropscience.org. Accessed 21 August 2017.

Dong, H.N., Zhang, D.W. (2014) Current development in genetic engineering strategies of Bacillus species. *Microbial Cell Factories* 13: 63.

Doornbos, R.F., Loon, L.C., van Bakker, P.H.M. (2011) Impact of root exudates and plant defense signaling on bacterial communities in the rhizosphere: A review. *Agronomy for Sustainable Development* 32: 227–243.

Dumbrell, A.J., Nelson, M., Helgason, T., Dytham, C., Fitter, A.H. (2010) Relative roles of niche and neutral processes in structuring a soil microbial community. *The ISME Journal* 4 No. 3: 337–345.

Duruibe, J.O., Ogwuegbu, M.O.C., Egwurugwu, J.N. (2007) Heavy metal pollution and human biotoxic effects. *International Journal of Physical Sciences* 2: 112–118.

Edwards, J., Johnson, C., Santos-Medellín, C., Lurie, E., Podishetty, N.K., Bhatnagar, S., Eisen, J.A., Sundaresan, V. (2015) Structure, variation, and assembly of the root-associated microbiomes of rice. *Proceedings of the National Academy of Sciences of the United States of America* 112 No. 8: E911–E920.

Egamberdieva, D., Wirth, S.J., Alqarawi, A.A., Abd-Allah, E.F., Hashem, A. (2017) Phytohormones and beneficial microbes: Essential components for plant to balance stress and fitness. *Frontiers in Microbiology* 8: 2104.

Farrar, K., Bryant, D., Cope-Selby, N. (2014) Understanding and engineering beneficial plant-microbe interactions: Plant growth promotion in energy crops. *Plant Biotechnology Journal* 12 No. 9: 1193–1206.

Feng, N.X., Yu, J., Zhao, H.M., Cheng, Y.T., Mo, C.H., Cai, Q.Y., Li, Y.W., Li, H., Wang, M.H. (2017) Efficient phytoremediation of organic contaminants in soils using plant-endophyte partnerships. *The Science of the Total Environment* 583: 352–368.

Feng, G., Zhang, F.S., Li, X.L., Tian, C.Y., Tang, C., Renegal, Z. (2002) Improved tolerance of maize plants to salt stress by arbuscular mycorrhiza is related to higher accumulation of leaf P-concentration of soluble sugars in roots. *Mycorrhiza* 12: 185–190.

Ferrara, F.I.S., Oliveira, Z.M., Gonzales, H.H.S., Floh, E.I.S., Barbosa, H.R. (2012) Endophytic and rhizospheric enterobacteria isolated from sugar cane have different potentials for producing plant growth-promoting substances. *Plant and Soil* 353 No. 1–2: 409–417.

Fester, T., Giebler, J., Wick, L.Y., Schlosser, D., Kästner, M. (2014) Plant–microbe interactions as drivers of ecosystem functions relevant for the biodegradation of organic contaminants. *Current Opinion in Biotechnology* 27: 168–175.

Figueiredo, M.V.B., Burity, H.A., Martinez, C.R., Chanway, C.P. (2008) Alleviation of drought stress in common bean (*Phaseolus vulgaris* L.) by co-inoculation with *Paenibacillus polymyxa* and *Rhizobium tropici*. *Applied Soil Ecology* 40 No. 1: 182–188.

Gaby, J.C., Buckley, D.H. (2011) A global census of nitrogenase diversity. *Environmental Microbiology* 13 No. 7: 1790–1799.

Gerhardt, K.E., Huang, X.D., Glick, B.R., Greenberg, B.M. (2009) Phytoremediation and rhizoremediation of organic soil contaminants: Potential and challenges. *Plant Science* 176 No. 1: 20–30.

Ghanem, G., Ewald, A., Zerche, S., Hennig, F. (2014) Effect of root colonization with *Piriformospra indica* and phosphate availability on the growth and reproductive biology of a *Cyclamen persicum* cultivar. *Scientia Horticulturae* 172: 233–241.

Gianinazzi, S., Gollotte, A., Binet, M.N., van Tuinen, D., Redecker, D., Wipf, D. (2010) Agroecology: The key role of arbuscular mycorrhizae in ecosystem services. *Mycorrhiza* 20 No. 8: 519–530.

Giri, B., Mukerji, K.G. (2004) Mycorrhizal inoculant alleviate salt stress in *Sesbania aegyptiaca* and *Sesbania grandiflora* under field conditions: Evidence for reduced sodium and improved magnesium uptake. *Mycorrhiza* 14 No. 5: 307–312.

Glick, B.R. (2012) Plant growth-promoting bacteria: Mechanisms and applications. *Scientifica* 2012: 963401.

Glick, B.R. (2014) Bacteria with ACC deaminase can promote plant growth and help to feed the world. *Microbiological Research* 169 No. 1: 30–39.

Glick, B.R. (2015) Phytoremediation. In *Beneficial Plant-Bacterial Interactions*, ed. Glick B.R., 191–221. Heidelberg: Springer.

Goh, C.H., Valiz Vallejos, D.F., Nicotra, A.B., Mathesius, U. (2013) The impact of beneficial plant-associated microbes on plant phenotypic plasticity. *Journal of Chemical Ecology* 39 No. 7: 826–839.

Gouda, S., Kerry, R.G., Das, G., Paramithiotis, S., Shin, H.S., Patra, J.K. (2018) Revitalization of plant growth promoting rhizobacteria for sustainable development in agriculture. *Microbiological Research* 206: 131–140.

Grover, M., Ali, S.Z., Sandhya, V., Rasul, A., Venkateswarlu, B. (2011) Role of microorganisms in adaptation of agriculture crop to abiotic stresses. *World Journal of Microbiology and Biotechnology* 27 No. 5: 1231–1240.

Guttman, D.S., McHardy, A.C., Schulze-Lefert, P. (2014) Microbial genome-enabled insights into plant-microorganism interactions. *Nature Reviews Genetics* 15 No. 12: 797–813.

Hacquard, S., Schadt, C.W. (2015) Towards a holistic understanding of the beneficial interactions across the Populus microbiome. *The New Phytologist* 205 No. 4: 1424–1430.

Hadi, F., Bano, A. (2010) Effect of diazotrophs (Rhizobium and Azotobacter) on growth of maize (*Zea mays* L.) and accumulation of lead (Pb) in different plant parts. *Pakistan Journal of Botany* 42: 4363–4370.

Hamel, C., Vujanovic, V., Jeannotte, R., Nakano-Hylander, A., St-Arnaud, M. (2005) Negative feedback on perennial crop: Fusarium crown and root rot of asparagus is related to changes in soil microbial community structure. *Plant and Soil* 268 No. 1: 75–87.

Hardoim, P.R., van Overbeek, L.S., Berg, G., Pirtilla, A.M., Compant, S., Campisano, A., Doring, M., Sessitsch, A. (2015) The hidden world within plants: Ecological and evolutionary considerations for defining functioning of microbial endophytes. *Microbiology and Molecular Biology Reviews: MMBR* 79 No. 3: 293–320.

Hardoim, P.R., van Overbeek, L.S., van Elsas, J.D. (2008) Properties of bacterial endophytes and their proposed role in plant growth. *Trends in Microbiology* 16 No. 10: 463–471.

Hartmann, A., Schmid, M., van Tuinen, D., Berg, G. (2009) Plant-driven selection of microbes. *Plant Science* 268: 75–87.

Hasan, M.A. (2013) Investigation on the nitrogen fixing Cyanobacteria (BGA) in rice fields of North-West region of Bangladesh. I: Nonfilamentous. *Journal of Environmental Science and Natural Resources* 5 No. 2: 253–259.

Hawkins, H.J., Johansen, A., George, E. (2000) Uptake and transport of organic and inorganic nitrogen by arbuscular mycorrhizal fungi. *Plant and Soil* 226 No. 2: 275–285.

Hayat, R., Ali, S., Amara, U., Khalid, R., Ahmed, I. (2010) Soil beneficial bacteria and their role in plant growth promotion: A review. *Annals of Microbiology* 60 No. 4: 579–598.

Hirsch, P.R., Mauchline, T.H. (2012) Who's who in the plant root microbiome? *Nature Biotechnology* 30 No. 10: 961–962.

Hu, S., Gu, H., Cui, C., Ji, R. (2016) Toxicity of combined chromium (VI). *Environmental Science and Pollution Research International* 23 No. 15: 15227–15235.

Huang, B., DaCosta, M., Jiang, Y. (2014) Research advances in mechanisms of turfgrass tolerance to abiotic stresses: From physiology to molecular biology. *Critical Reviews in Plant Sciences* 33 No. 2–3: 141–189.

Hussain, S.S., Asif, M.A., Sornaraj, P., Ali, M., Shi, B.J. (2016) Towards integration of system based approach for understanding drought stress in plants. In *Water Stress and Crop Plants: A Sustainable Approach*, ed. Ahmad, P., Rasool, S., 227–247. Hoboken: John-Wiley & Sons.

Hussain, S.S., Iqbal, M.T., Arif, M.A., Amjad, M. (2011) Beyond osmolytes and transcription factors: Drought tolerance in plants via protective proteins and aquaporins. *Biologia Plantarum* 55 No. 3: 401–413.

Hussain, S.S., Mehnaz, S., Siddique, K.M. (2018) Harnessing the plant microbiome for improved abiotic stress tolerance. In *Plant Microbiome: Stress Response and Microorganisms for Sustainability*, ed. Ahmad, P, Egamberdieva, D., 21–43. Singapore: Springer.

Hussain, S.S., Raza, H., Afzal, I., Kayani, M.A. (2012) Transgenic plants for abiotic stress tolerance: Current status. *Archives of Agronomy and Soil Science* 58 No. 7: 693–721.

Hussain, S.S., Siddique, K.H.M., Lopato, S. (2014) Towards integration of bacterial genomics in plants for enhanced abiotic stress tolerance: Clues from transgenics. *Advances in Environmental Research* 33: 65–122.

Iannucci, A., Fragasso, M., Beleggic, R., Nigo, F., Papa, R. (2017) Evolution of crop rhizosphere: Impact of domestication on root exudates in tetraploid wheat (*Triticum turgidum* L.). *Frontiers in Plant Science* 8: 2124.

Ijaz, A., Imran, A., ul Haq, M.A., Khan, Q.M., Afzal, M. (2016) Phytoremediation: Recent advances in plant-endophytic synergistic interactions. *Plant and Soil* 405 No. 1–2: 179–195.

IPCC. (2007) *Climate Change 2007. The Physical Science Basis. Contribution of Working Group I to the Fourth Assessment Report of the Intergovernmental Panel on Climate Change.* Cambridge, NY: Cambridge University Press.

Jamil, M., Zeb, S., Anees, M., Roohi, A., Ahmed, I., Rehman, S.U., Rha, E.S. (2014) Role of *Bacillus licheniformis* in phytoremediation of nickel contaminated soil cultivated with rice. *International Journal of Phytoremediation* 16 No. 6: 554–571.

Jha, B., Gontia, I., Hartmann, A. (2012) The roots of the halophyte *Salicornia brachiata* are a source of new halotolerant diazotrophic bacteria with plant growth-promoting potential. *Plant and Soil* 356 No. 1–2: 265–277.

Jha, P., Panwar, J., Jha, P.N. (2015) Secondary plant metabolites and root exudates: Guiding tools for polychlorinated biphenyl biodegradation. *International Journal of Environmental Science and Technology* 12 No. 2: 789–802.

Jindal, V., Atwal, A., Sekhon, B.S., Rattan, S., Singh, R. (1993) Effect of vesicular-arbuscular mycorrhiza on metabolism of moong plants under salinity. *Plant Physiology and Biochemistry* 31: 475–481.

Johnson, N.C., Graham, J.H. (2013) The continuum concept remains a useful framework for studying mycorrhizal functioning. *Plant and Soil* 363 No. 1–2: 411–419.

Johnson, D., Martin, F., Cairney, J.W.G., Anderson, I.C. (2012) The importance of individuals: Intraspecific diversity of mycorrhizal plants and fungi in ecosystems. *The New Phytologist* 194 No. 3: 614–628.

Johnston-Monje, D., Raizada, M.N. (2011) Conservation and diversity of seed associated endophytes in zea across boundaries of evolution, ethnography and ecology. *PLOS ONE* 6 No. 6: e20396.

Jones, M.B., Finnan, J., Hodkinson, T.R. (2014) Morphological and physiological traits for higher biomass production in perennial rhizomatous grasses grown on marginal land. *Global Change Biology Bioenergy* 7: 375–385

Jorquera, M.A., Shaharoona, B., Nadeem, S.M., de la Luz Mora, M., Crowley, D.E. (2012) Plant growth-promoting rhizobacteria associated with ancient clones of creosote bush (*Larrea tridentata*). *Microbial Ecology* 64 No. 4: 1008–1017.

Kachroo, A., Robin, G.P. (2013) Systemic signaling during plant defense. *Current Opinion in Plant Biology* 16 No. 4: 527–533.

Kasim, W.A., Osman, M.E., Omar, M.N., Abd El-Daim, I.A., Bejai, S., Meijer, J. (2013) Control of drought stress in wheat using plant-growth promoting rhizobacteria. *Journal of Plant Growth Regulation* 32 No. 1: 122–130.

Kathuria, H., Giri, J., Tyagi, H., Tyagi, A.K. (2007) Advances in transgenic rice biotechnology. *Critical Reviews in Plant Sciences* 26 No. 2: 65–103.

Kavamura, V.N., Santos, S.N., Silva, J.L., Parma, M.M., Avila, L.A., Visconti, A., Zucchi, T.D., *et al.* (2013) Screening of Brazilian cacti rhizobacteria for plant growth promotion under drought. *Microbiological Research* 168 No. 4: 183–191.

Khan, I.A., Ayub, N., Mirza, S.N., Nizami, S.N., Azam, M. (2008) Yield and water use efficiency (WUE) of *Cenchrus ciliaris* as influenced by vesicular arbuscular mycorrhizae (VAM). *Pakistan Journal of Botany* 40: 931–937.

Khan, A., Jilani, G., Akhtar, M.S., Naqvi, S.M.S., Rasheed, M. (2009) Phosphorus solubilizing bacteria: Occurrence, mechanisms and their role in crop production. *Journal of Agriculture and Biological Sciences* 1: 48–58.

Kidd, P., Barcelo, J., Bernal, M.P., Navari-Izzo, F., Poschenrieder, C., Shilev, S., Clemente, R., Monterroso, C. (2009) Trace element behavior at the root-soil interface: Implications in phytoremediation. *Environmental and Experimental Botany* 67 No. 1: 243–259.

Kim, Y.C., Glick, B.R., Bashan, Y., Ryu, C.M. (2009) Enhancement of plant drought tolerance by microbes. In *Plant Responses to Drought Stress*, ed. Aroca, A., 382–412. Berlin: Springer.

Kim, Y.C., Glick, B., Bashan, Y., Ryu, C.M. (2013) Enhancement of plant drought tolerance by microbes. In *Plant Responses to Drought Stress*, ed. Aroca, R., 383–413. Berlin: Springer.

Kim, J.S., Lee, J., Seo, S.G., Lee, C., Woo, S.Y., Kim, S.H. (2015) Gene expression profile affected by volatiles of new plant growth promoting rhizobacteria, Bacillus subtilis strain JS, in tobacco. *Genes and Genomics* 37 No. 4: 387–397.

Kim, S., Lowman, S., Hou, G., Nowak, J., Flinn, B., Mei, C. (2012) Growth promotion and colonization of switchgrass (*Panicum virgatum*) cv. Alamo by bacterial endophyte *Burkholderia phytofirmans* strain PsJN. *Biotechnology for Biofuels* 5 No. 1: 37.

Kim, S.B., Timmusk, S. (2013) A simplified method for gene knockout and direct screening of recombinant clones for application in *Paenibacillus polymyxa*. *PLOS ONE* 8 No. 6: e68092.

Kloepper, J.W., Ryu, C.M., Zhang, S. (2004) Induced systemic resistance and promotion of plant growth by *Bacillus* spp. *Phytopathology* 94 No. 11: 1259–1266.

Knief, C. (2014) Analysis of plant microbe interactions in the era of next generation sequencing technologies. *Frontiers in Plant Science* 5: 216.

Knief, C., Delmotte, N., Chaffron, S., Stark, M., Innerebner, G., Wassmann, R., von Mering, C., Vorholt, J.A. (2012) Metaproteogenomic analysis of microbial communities in the phyllosphere and rhizosphere of rice. *The ISME Journal* 6 No. 7: 1378–1390.

Kohler, J., Hernandez, J.A., Caravaca, F., Rolden, A. (2008) Plant growth promoting rhizobacteria and arbuscular mycorrhizal fungi modify alleviation biochemical mechanisms in water stressed plants. *Functional Plant Biology* 35 No. 2: 141–151.

Kukla, M., Płociniczak, T., Piotrowska-Seget, Z. (2014) Diversity of endophytic bacteria in *Lolium perenne* and their potential to degrade petroleum hydrocarbons and promote plant growth. *Chemosphere* 117: 40–46.

Kumar, K.V., Srivastava, S., Singh, N., Behl, H.M. (2009) Role of metal resistant plant growth promoting bacteria in ameliorating fly ash to the growth of *Brassica juncea*. *Journal of Hazardous Materials* 170 No. 1: 51–57.

Kumar, A., Verma, J.P. (2018) Does plant-microbe interaction confer stress tolerance in plants: A review? *Microbiological Research* 207: 41–52.

Lakshmanan, V. (2015) Root microbiome assemblage is modulated by plant host factors. *Advances in Botanical Research* 75: 57–79.

Lakshmanan, V., Kitto, S.L., Caplan, J.L., Hsueh, Y.H., Kearns, D.B., Wu, Y.S., Bais, H.P. (2012) Microbe-associated molecular patterns-triggered root responses mediate beneficial rhizobacterial recruitment in *Arabidopsis*. *Plant Physiology* 160 No. 3: 1642–1661.

Lally, R.D., Galbally, P., Moreira, A.S., Spink, J., Ryan, D., Germaine, K.J., Dowling, D.N. (2017) Application of endophytic *Pseudomonas fluorescens* and a bacterial consortium to *Brassica napus* can increase plant height and biomass under greenhouse and field conditions. *Frontiers in Plant Science* 8: 2193.

Lareen, A., Burton, F., Schäfer, P. (2016) Plant root-microbe communication in shaping root microbiomes. *Plant Molecular Biology* 90 No. 6: 575–587.

Lau, J.A., Lennon, J.T. (2012) Rapid responses of soil microorganisms improve plant fitness in novel environments. *Proceedings of the National Academy of Sciences of the United States of America* 109 No. 35: 14058–14062.

Lebeis, S.L. (2014) The potential for give and take in plant-microbiome relationships. *Frontiers in Plant Science* 5: 287.

Lebeis, S.L., Paredes, S.H., Lundberg, D.S., Breakfield, N., Gehring, J., McDonald, M., Malfatti, S., *et al.* (2015) Plant microbiome. Salicylic acid modulates colonization of the root microbiome by specific bacterial taxa. *Science* 349 No. 6250: 860–864.

Lebeis, S.L., Rott, M., Dangl, J.L., Schulze-Lefert, P. (2012) Culturing a plant microbiome community at the cross-Rhodes. *The New Phytologist* 196 No. 2: 341–344.

Lederberg, I., McCray, A.T. (2001) Ome sweet omics: A genealogical treasury of words. *Scientist* 15: 8.

Lee, S.W., Ahn, P.I., Sim, S., Lee, S.Y., Seo, M.W., Kim, S., Park, S., Lee, Y.H., Kang, S. (2010) Pseudomonas sp. LSW25R antagonistic to plant pathogens promoted plant growth and reduced blossom red rot of tomato roots in a hydroponic system. *European Journal of Plant Pathology* 126: 1–11.

Leite, M.C.B.S., de Farias, A.R.B., Freire, F.J., Andreote, F.D., Sobral, J.K., Freire, M.B.G.S. (2014) Isolation, bioprospecting and diversity of salt-tolerant bacteria associated with sugarcane in soils of Pernambuco, Brazil. *Revista Brasileira de Engenharia Agrıcola e Ambiental* 18: S73–S79.

Leveau, J.H.J., Uroz, S., de Boer, W. (2010) The bacterial genus *Collimonas*: Mycophagy, weathering and other adaptive solutions to life in oligotrophic soil environments. *Environmental Microbiology* 12 No. 2: 281–292.

Li, D., Voigt, T.B., Kent, A.D. (2016) Plant and soil effects on bacterial communities associated with Miscanthus × giganteus rhizosphere and rhizomes. *GCB Bioenergy* 8 No. 1: 183–193.

Lim, J.H., Kim, S.D. (2013) Induction of drought stress resistance by multifunctional PGPR *Bacillus licheniformis* K11 in pepper. *The Plant Pathology Journal* 29 No. 2: 201–208.

Lugtenberg, B., Kamilova, F. (2009) Plant-growth-promoting rhizobacteria. *Annual Review of Microbiology* 63: 541–556.

Lundberg, D.S., Lebeis, S.L., Paredes, S.H., Yourstone, S., Gehring, J., Malfatti, S., Tremblay, J., *et al.* (2012) Defining the core *Arabidopsis thaliana* root microbiome. *Nature* 488 No. 7409: 86–90.

Luo, S., Xu, T., Chen, L., Chen, J., Rao, C., Xiao, X., Wan, Y., *et al.* (2012) Endophyte-assisted promotion of biomass production and metal-uptake of energy crop sweet sorghum by plant-growth-promoting endophyte *Bacillus* sp. SLS18. *Applied Microbiology and Biotechnology* 93 No. 4: 1745–1753.

Luo, S.L., Chen, L., Chen, J.L., Xiao, X., Xu, T.Y., Wan, Y., Rao, C., *et al.* (2011) Analysis and characterization of cultivable heavy metal-resistant bacterial endophytes isolated from Cd-hyperaccumulator *Solanum nigrum* L. and their potential use for phytoremediation. *Chemosphere* 85 No. 7: 1130–1138.

Luvizotto, D.M., Marcon, J., Andreote, F.D., Dini-Andreote, F., Neves, A.A.C., Araújo, W.L., Pizzirani-Kleiner, A.A. (2010) Genetic diversity and plant-growth related features of *Burkholderia* spp. from sugarcane roots. *World Journal of Microbiology and Biotechnology* 26 No. 10: 1829–1836.

Ma, Y., Prasad, M.N.V., Rajkumar, M., Freitas, H. (2011) Plant growth promoting rhizobacteria and endophytes accelerate phytoremediation of metalliferous soils. *Biotechnology Advances* 29 No. 2: 248–258.

Mahmood, S., Daur, I., Al-Solaimani, S.G., Ahmad, S., Madkour, M.H., Yasir, M., Hirt, H., Ali, S., Ali, Z. (2016) Plant growth promoting rhizobacteria and silicon synergistically enhance salinity tolerance of mung bean. *Frontiers in Plant Science* 7: 876.

Mapelli, F., Marasco, R., Balloi, A., Rolli, E., Cappitelli, F., Daffonchio, D., Borin, S. (2012) Mineral-microbe interactions: Biotechnological potential of bio-weathering. *Journal of Biotechnology* 157 No. 4: 473–481.

Mapelli, F., Marasco, R., Rolli, E., Barbato, M., Cherif, H., Guesmi, A., Ouzari, I., Daffonchio, D., Borin, S. (2013) Potential for plant growth promotion of rhizobacteria associated with Salicornia growing in Tunisian hypersaline soils. *BioMed Research International* 2013: 248078.

Marasco, R., Mapelli, F., Rolli, E., Mosqueira, M.J., Fusi, M., Bariselli, P., Reddy, M., *et al.* (2016) *Salicornia strobilacea* (synonym of *Halocnemum strobilaceum*) growth under different tidal regimes selects rhizosphere bacteria capable of promoting plant growth. *Frontiers in Microbiology* 7: 1286.

Marasco, R., Rolli, E., Attoumi, B., Vigani, G., Mapelli, F., Borin, S., Abou-Hadid, A.F., *et al.* (2012) A drought resistance-promoting microbiome is selected by root system under desert farming. *PLOS ONE* 7 No. 10: e48479.

Marasco, R., Rolli, E., Vigani, G., Borin, S., Sorlini, C., Ouzari, H., Zocchi, G., Daffonchio, D. (2013) Are drought-resistance promoting bacteria cross-compatible with different plant models? *Plant Signaling and Behavior* 8 No. 10: e26741.

Marquez, L.M., Redman, R.S., Rodriguez, R.J., Roosinck, M.J. (2007) A virus in a fungus in a plant: Three-way symbiosis required for thermal tolerance. *Science* 315 No. 5811: 513–515.

Marschner, P., Crowley, D., Rengel, Z. (2011) Rhizosphere interactions between microorganisms and plants govern iron and phosphorus acquisition along the root axis–model and research methods. *Soil Biology and Biochemistry* 43 No. 5: 883–894.

Marulanda, A., Barea, J.M., Azcón, R. (2009) Stimulation of plant growth and drought tolerance by native microorganisms (AM fungi and bacteria) from dry environments: Mechanisms related to bacterial effectiveness. *Journal of Plant Growth Regulation* 28 No. 2: 115–124.

Masciarelli, O., Llanes, A., Luna, V. (2014) A new PGPR co-inoculated with *Bradyrhizobium japonicum* enhances soybean nodulation. *Microbiological Research* 169 No. 7–8: 609–615.

Mastretta, C., Taghavi, S., Van Der Lelie, D., Mengoni, A., Galardi, F., Gonnelli, C., Barac, T., *et al.* (2009) Endophytic bacteria from seeds of *Nicotiana tabacum* can reduce cadmium phytotoxicity. *International Journal of Phytoremediation* 11 No. 3: 251–267.

Mayak, S., Tirosh, T., Glick, B.R. (2004) Plant growth promoting bacteria that confer resistance to water stress in tomato and pepper. *Plant Science* 166 No. 2: 525–530.

McLellan, C.A., Turbyville, T.J., Wijeratne, E.M., Kerschen, A., Vierling, E., Queitsch, C., Whitesell, L., Gunatilaka, A.A. (2007) A rhizosphere fungus enhances *Arabidopsis* thermotolerance through production of an HSP90 inhibitor. *Plant Physiology* 145 No. 1: 174–182.

Mendes, R., Garbeva, P., Raajimakers, J.M. (2013) The rhizosphere microbiome: Significance of plant beneficial, plant pathogenic and human pathogenic microorganisms. *FEMS Microbiology Reviews* 37 No. 5: 634–663.

Mendes, R., Kruijt, M., de Bruijn, I., Dekkers, E., van der Voort, M., Schneider, J.H., Piceno, Y.M., *et al.* (2011) Deciphering the rhizosphere microbiome for disease-suppressive bacteria. *Science* 332 No. 6033: 1097–1100.

Mengual, C., Schoebitz, M., Azcón, R., Roldán, A. (2014) Microbial inoculants and organic amendment improves plant establishment and soil rehabilitation under semi-arid conditions. *Journal of Environmental Management* 134: 1–7.

Miao, Y., Stewart, B.A., Zhang, F. (2011) Long-term experiments for sustainable nutrient management in China: A review. *Agronomy for Sustainable Development* 31 No. 2: 397–414.

Miller, J.B., Oldroyd, G.D. (2012) The role of diffusible signals in the establishment of rhizobial and mycorrhizal symbioses. In *Signaling and Communication in Plant Symbiosis*, ed. Perotto, S, Baluška, F., 1–30. Berlin: Springer.

Miransari, M. (2011) Hyperaccumulators, arbuscular mycorrhizal fungi and stress of heavy metals. *Biotechnology Advances* 29 No. 6: 645–653.

Mishra, P.K., Mishra, S., Selvakumar, G., Kundub, S., Gupta, H.S. (2009) Enhanced soybean (*Glycine max* L.) plant growth and nodulation by *Bradyrhizobium japonicum*-SB1 in presence of *Bacillus thuringiensis*-KR1. *Acta Agriculturae Scandinavica, Section-B: Soil and Plant Science* 59: 189–196.

Moe, L.A. (2013) Amino acids in the rhizosphere: From plants to microbes. *American Journal of Botany* 100 No. 9: 1692–1705.

Morel, M., Castro-Sowinski, S. (2013) The complex molecular signaling network in microbe-plant interaction. In *Plant Microbe Symbiosis: Fundamentals and Advances*, ed. Arora, N.K., 169–199. New Delhi: Springer.

Mueller, U.G., Sachs, J.L. (2015) Engineering microbiome to improve plant and animal health. *Trends in Microbiology* 23 No. 10: 606–617.

Nadeem, S.M., Ahmad, M., Zahir, Z.A., Javaid, A., Ashraf, M. (2014) The role of mycorrhizae and plant growth promoting rhizobacteria (PGPR) in improving crop productivity under stressful environments. *Biotechnology Advances* 32 No. 2: 429–448.

Naseem, H., Bano, A. (2014) Role of plant growth-promoting rhizobacteria and their exopolysaccharide in drought tolerance in maize. *Journal of Plant Interactions* 9 No. 1: 689–701.

Naveed, M., Hussain, M.B., Zahir, Z.A., Mitter, B., Sessitsch, A. (2014a) Drought stress amelioration in wheat through inoculation with *Burkholderia phytofirmans* strain PsJN. *Plant Growth Regulation* 73 No. 2: 121–131.

Naveed, M., Mitter, B., Reichenauer, T.G., Wieczorek, K., Sessitsch, A. (2014b) Increased drought stress resilience of maize through endophytic colonization by *Burkholderia phytofirmans* PsJN and Enterobacter sp. FD17. *Environmental and Experimental Botany* 97: 30–39.

Ngumbi, E.N. (2011) Mechanisms of olfaction in parasitic wasps: Analytical and behavioral studies of response of a specialist (*Microplitis croceipes*) and a generalist (*Cotesia marginiventris*) parasitoid to host-related odor. PhD Dissertation. Auburn University, Auburn.

Nonnoi, F., Chinnaswamy, A., García de la Torre, V.S., Coba de la Peña, T., Lucas, M.M., Pueyo, J.J. (2012) Metal tolerance of rhizobial strains isolated from nodules of herbaceous legumes *Medicago* sp. and *Trifolium* sp. growing in mercury-contaminated soils. *Applied Soil Ecology* 61: 49–59.

Oburger, E., Dell'Mour, M., Hann, S., Wieshammer, G., Puschenreiter, M., Wenzel, W.W. (2013) Evaluation of a novel tool for sampling root exudates from soil-grown plants compared to conventional techniques. *Environmental and Experimental Botany* 87: 235–247.

Ofek, M., Voronov-Goldman, M., Hadar, Y., Minz, D. (2014) Host signature effect on plant root-associated microbiomes revealed through analyses of resident vs active communities. *Environmental Microbiology* 16 No. 7: 2157–2167.

Oldroyd, G.E.D. (2013) Speak, friend, and enter: Signaling systems that promote beneficial symbiotic associations in plants. *Nature Reviews Microbiology* 11 No. 4: 252–263.

Oldroyd, G.E., Murray, J.D., Poole, P.S., Downie, J.A. (2011) The rules of engagement in the legume-rhizobial symbiosis. *Annual Review of Genetics* 45: 119–144.

Olivares, J., Bedmar, E.J., Sanjuan, J. (2013) Biological nitrogen fixation in the context of global change. *Molecular Plant-Microbe Interactions: MPMI* 26 No. 5: 486–494.

Ortiz-Castro, R., Martinez-Trujillo, M., Lopez-Bucio, J. (2008) N-acyl-L- homoserinelactones: a class of bacterial quorum-sensing signals alter post-embryonic root development in *Arabidopsis thaliana*. *Plant, Cell and Environment* 31 No. 10: 1497–1509.

Pandey, V.C. (2012) Phytoremediation of heavy metals from fly ash pond by *Azolla caroliniana*. *Ecotoxicology and Environmental Safety* 82: 8–12.

Pandey, V.C., Ansari, M.W., Tula, S., Yadav, S., Sahoo, R.K., Shukla, N., Bains, G., *et al.* (2016) Dose dependent response of *Trichoderma harzianum* in improving drought tolerance in rice genotypes. *Planta* 243 No. 5: 1251–1264.

Panke-Buisse, K., Poole, A.C., Goodrich, J.K., Ley, R.E., Kao-Kniffin, J. (2015) Selection on soil microbiomes reveals reproducible impacts on plant function. *The ISME Journal* 9 No. 4: 980–989.

Park, J.D. (2010) Heavy metal poisoning. *Hanyang Medical Reviews* 30 No. 4: 319–325.

Patel, U., Sinha, S. (2011) Rhizobia species: A boon for "Plant Genetic Engineering." *Indian Journal of Microbiology* 51 No. 4: 521–527.

Paul, D., Lade, H. (2014) Plant-growth-promoting rhizobacteria to improve crop growth in saline soils: A review. *Agronomy for Sustainable Development* 34 No. 4: 737–752.

Peiffer, J.A., Spor, A., Koren, O., Jin, Z., Tringe, S.G., Dangl, J.L., Buckler, E.S., Ley, R.E. (2013) Diversity and heritability of the maize rhizosphere microbiome under field conditions. *Proceedings of the National Academy of Sciences of the United States of America* 110 No. 16: 6548–6553.

Perez-Montano, F., Alías-Villegas, C., Bellogin, R.A., del Cerro, P., Espuny, M.R., Jimenez-Guerrero, I., Lopez-Baena, F.J., Ollero, F.J., Cubo, T. (2014) Plant growth promotion in cereal and leguminous agricultural important plants: from microorganism capacities to crop production. *Microbiological Research* 169 No. 5–6: 325–336.

Philippot, L., Raaijmakers, J.M., Lemanceau, P., van der Putten, W.H. (2013) Going back to the roots: The microbial ecology of the rhizosphere. *Nature Reviews. Microbiology* 11 No. 11: 789–799.

Pineda, A., Dicke, M., Pieterse, C.M.J., Pozo, M.J. (2013) Beneficial microbes in a changing environment: Are they always helping plants deal with insects? *Functional Ecology* 27 No. 3: 574–586.

Porras-Alfaro, A., Bayman, P. (2011) Hidden fungi, emergent properties: Endophytes and microbiomes. *Annual Review of Phytopathology* 49: 291–315.

Pozo, M.J., Azcon-Aguilar, C. (2007) Unraveling mycorrhiza-induced resistance. *Current Opinion in Plant Biology* 10 No. 4: 393–398.

Prashar, P., Kapoor, N., Sachdeva, S. (2014) Rhizosphere: Its structure, bacterial diversity and significance. *Reviews in Environmental Science and Bio/Technology* 13 No. 1: 63–77.

Premachandra, D., Hudek, L., Brau, L. (2016) Bacterial modes of action for enhancing plant growth. *Journal of Biotechnology and Biomaterials* 6: 3.

Prudent, M., Salon, C., Souleimanov, A., Emery, R.J.N., Smith, D.L. (2015) Soybean is less impacted by water stress using *Bradyrhizobium japonicum* and thuricin-17 from *Bacillus thuringiensis*. *Agronomy for Sustainable Development* 35 No. 2: 749–757.

Qin, S., Zhang, Y.J., Yuan, B., Xu, P.Y., Xing, K., Wang, J., Jiang, J.H. (2014) Isolation of ACC deaminase-producing habitat-adapted symbiotic bacteria associated with halophyte *Limonium sinense* (Girard) Kuntze and evaluating their plant growth promoting activity under salt stress. *Plant and Soil* 374 No. 1–2: 753–766.

Qiu, M., Li, S., Zhou, X., Cui, X., Vivanco, J.M., Zhang, N., Shen, Q., Zhang, R. (2014) De-coupling of root-microbiome associations followed by antagonist inoculation improves rhizosphere soil suppressiveness. *Biology and Fertility of Soils* 50 No. 2: 217–224.

Quecine, M.C., Araujo, W.L., Tsui, S., Parra, J.R.P., Azevedo, J.L., Pizzirani-Kleiner, A.A. (2014) Control of *Diatraea saccharalis* by the endophytic *Pantoea agglomerans* 33.1 expressing cry1Ac7. *Archives of Microbiology* 196 No. 4: 227–234.

Quiza, L., St-Arnaud, M., Yergeau, E. (2015) Harnessing phytomicrobiome signaling for rhizosphere microbiome engineering. *Frontiers in Plant Science* 6: 507.

Raajimakers, J.M. (2015) The minimal rhizosphere microbiome. In *Principles of Plant-Microbe Interactions*, ed. Lugtenberg, B., 411–417. Heidelberg: Springer.

Rajkumar, M., Ae, N., Prasad, M.N.V., Freitas, H. (2010) Potential of siderophore-producing bacteria for improving heavy metal phytoextraction. *Trends in Biotechnology* 28 No. 3: 142–149.

Rajkumar, M., Sandhya, S., Prasad, M.N.V., Freitas, H. (2012) Perspectives of plant-associated microbes in heavy metal phytoremediation. *Biotechnology Advances* 30 No. 6: 1562–1574.

Ramond, J.B., Tshabuse, F., Bopda, C.W., Cowan, D.A., Tuffin, M.I. (2013) Evidence of variability in the structure and recruitment of rhizospheric and endophytic bacterial communities associated with arable sweet sorghum (*Sorghum bicolor* (L.) Moench). *Plant and Soil* 372 No. 1–2: 265–278.

Rasche, F., Velvis, H., Zachow, C., Berg, G., van Elsas, J.D., Sessitsch, A. (2006) Impact of transgenic potatoes expressing antibacterial agents on bacterial endophytes is comparable with the effects of plant genotype, soil type and pathogen infection. *Journal of Applied Ecology* 43 No. 3: 555–566.

Rastogi, G., Coaker, G.L., Leaveu, J.H.J. (2013) New insights into the structure and function of phyllosphere microbiota through high-throughput molecular approaches. *FEMS Microbiology Letters* 348 No. 1: 1–10.

Redman, R.S., Sheehan, K.B., Stout, R.G., Rodriguez, R.J., Henson, J.M. (2002) Thermotolerance generated by plant/fungal symbiosis. *Science* 298 No. 5598: 1581.

Richardson, A.E., Barea, J.M., McNeill, A.M., Prigent-Combaret, C. (2009) Acquisition of phosphorus and nitrogen in the rhizosphere and plant growth promotion by microorganisms. *Plant and Soil* 321 No. 1–2: 305–339.

Rodriguez, H., Vessely, S., Shah, S., Glick, B.R. (2008) Effect of a nickel-tolerant ACC deaminase-producing pseudomonas strain on growth of nontransformed and transgenic canola plants. *Current Microbiology* 57 No. 2: 170–174.

Rolli, E., Marasco, R., Vigani, C., Ettoumi, B., Mapelli, F., Deangelis, M.I., Gandolfi, C., *et al.* (2015) Improved plant resistance to drought is promoted by the root-promoted microbiome as a water stress dependent trait. *Environmental Microbiology* 17: 316–331.

Ruiz, K.B., Biondi, S., Martínez, E.A., Orsini, F., Antognoni, F., Jacobsen, S.-E. (2016) Quinoa – A model crop for understanding salt-tolerance mechanisms in halophytes. *Plant Biosystems – an International Journal Dealing with All Aspects of Plant Biology* 150 No. 2: 357–371.

Ruiz-Lozano, J.M., Peralvarez, M.C., Aroca, R., Azcon, R. (2011) The application of a treated sugar beet waste residue to soil modifies the responses of mycorrhizal and non mycorrhizal lettuce plants to drought stress. *Plant and Soil* 346 No. 1–2: 153–166.

Salvioli, A., Bonfante, P. (2013) Systems biology and "omics" tools: A cooperation for next-generation mycorrhizal studies. *Plant Science: an International Journal of Experimental Plant Biology* 203–204: 107–114.

Salvioli, A., Ghignone, S., Novero, M., Navazio, L., Venice, F., Bagnaresi, P., Bonfante, P. (2016) Symbiosis with an endobacterium increases the fitness of a mycorrhizal fungus, raising its bioenergetic potential. *The ISME Journal* 10 No. 1: 130–144.

Sandhya, V., Ali, S.Z., Grover, M., Kishore, N., Venkateswarlu, B. (2010) Pseudomonas sp. strain P45 protects sunflowers seedlings from drought stress through improved soil structure. *Journal of Oilseed Research* 26: 600–601.

Sanjay, A., Purvi, N.P., Meghna, J.V., G., G.R. (2014) Isolation and characterization of endophytic bacteria colonizing halophyte and other salt tolerant plant species from coastal Gujarat. *African Journal of Microbiology Research* 8 No. 17: 1779–1788.

Santi, C., Bogusz, D., Franche, C. (2013) Biological nitrogen fixation in non-legume plants. *Annals of Botany* 111 No. 5: 743–767.

Santoyo, G., Moreno-Hagelsieb, G., Orozco-Mosqueda, Mdel C., Glick, B.R. (2016) Plant growth-promoting bacterial endophytes. *Microbiological Research* 183: 92–99.

Sarma, R.K., Saikia, R. (2014) Alleviation of drought stress in mung bean by strain *Pseudomonas aeruginosa* GGRJ21. *Plant and Soil* 377 No. 1–2: 111–126.

Schenk, S.T., Stein, E., Kogel, K.H., Schikora, A. (2012) *Arabidopsis* growth and defense are modulated by bacterial quorum sensing molecules. *Plant Signaling and Behavior* 7 No. 2: 178–181.

Schlaeppi, K., Bulgarelli, D. (2015) The plant microbiome at work. *Molecular Plant-Microbe Interactions* 28 No. 3: 212–217.

Schlaeppi, K., Dombrowski, N., Oter, R.G., van Themaat, E.V.L., Schulze-Lefert, P. (2014) Quantitative divergence of the bacterial root microbiota in *Arabidopsis thaliana* relatives. *Proceeding of the National Academy of Sciences of the United States of America* 111: 585–592.

Schlaeppi, K., van Themaat, E.V.L., Bulgarelli, D., Schulze-Lefert, P. (2013) *Arabidopsis thaliana* as model for studies on the bacterial root microbiota. In *Molecular Microbial Ecology of the Rhizosphere*, ed. de Bruijin, F.J., 243–256. Dordrecht: Kluwer Academic Publishers.

Schuhegger, R., Ihring, A., Gantner, S., Bahnweg, G., Knappe, C., Vogg, G., Hutzler, P., *et al.* (2006) Induction of systemic resistance in tomato by N-acyl-L-homoserine lactone-producing rhizosphere bacteria. *Plant, Cell and Environment* 29 No. 5: 909–918.

Selvakumar, G., Joshi, P., Nazim, S., Mishra, P.K., Bisht, J.K., Gupta, H.S. (2009) Phosphate solubilization and growth promotion by *Pseudomonas fragi* CS11RH1 (MTCC 8984) a psychrotolerant bacterium isolated from a high altitude Himalayan rhizosphere. *Biologia* 64 No. 2: 239–245.

Selvakumar, G., Joshi, P., Suyal, P., Mishra, P.K., Joshi, G.K., Bisht, J.K., Bhatt, J.C., Gupta, H.S. (2010a) *Pseudomonas lurida* M2RH3 (MTCC 9245), a psychrotolerant bacterium from the Uttarakhand Himalayas, solubilizes phosphate and promotes wheat seedling growth. *World Journal of Microbiology and Biotechnology* 5: 1129–1135.

Selvakumar, G., Kundu, S., Joshi, P., Nazim, S., Gupta, A.D., Gupta, H.S. (2010b) Growth promotion of wheat seedlings by *Exiguobacterium acetylicum* 1P (MTCC 8707) a cold tolerant bacterial strain from the Uttarakhand Himalayas. *Indian Journal of Microbiology* 50 No. 1: 50–56.

Selvakumar, G., Kundu, S., Joshi, P., Nazim, S., Gupta, A.D., Mishra, P.K., Gupta, H.S. (2008a) Characterization of a cold-tolerant plant growth-promoting bacterium *Pantoea dispersa* 1A isolated from a sub-alpine soil in the North Western Indian Himalayas. *World Journal of Microbiology and Biotechnology* 24 No. 7: 955–960.

Selvakumar, G., Mohan, M., Kundu, S., Gupta, A.D., Joshi, P., Nazim, S., Gupta, H.S. (2008b) Cold tolerance and plant growth promotion potential of Serratia marcescens strain SRM (MTCC 8708) isolated from flowers of summer squash (Cucurbita pepo). *Letters in Applied Microbiology* 46: 171–175.

Sessitsch, A., Hardoim, P., Döring, J., Weilharter, A., Krause, A., Woyke, T., Mitter, B., *et al.* (2011) Functional characteristics of an endophyte community colonizing rice roots as revealed by metagenomic analysis. *Molecular Plant–Microbe Interaction* 25: 28–36.

Sessitsch, A., Mitter, B. (2014) 21st century agriculture: Integration of plant microbiome for improved crop production and food safety. *Microbial Biotechnology* 8: 32–33.

Shabala, S., Hariadi, Y., Jacobsen, S.E. (2013) Genotypic difference in salinity tolerance in quinoa is determined by differential control of xylem Na^+ loading and stomatal density. *Journal of Plant Physiology* 170 No. 10: 906–914.

Sharma, S., Kulkarni, J., Jha, B. (2016) Halotolerant rhizobacteria promote growth and enhance salinity tolerance in peanut. *Frontiers in Microbiology* 7: 1600.

Sheng, X., He, L., Wang, Q., Ye, H., Jiang, C. (2008) Effects of inoculation of biosurfactant-producing Bacillus sp. J119 on plant growth and cadmium uptake in a cadmium-amended soil. *Journal of Hazardous Materials* 155 No. 1–2: 17–22.

Sheng, M., Tang, M., Zhang, F., Huang, Y. (2011) Influence of arbuscular mycorrhiza on organic solutes in maize leaves under salt stress. *Mycorrhiza* 21 No. 5: 423–430.

Shirley, M., Avoscan, L., Bernuad, E., Vansuyt, G., Lemanceau, P. (2011) Comparison of iron acquisition from Fe-pyoverdine by strategy I and strategy II plants. *Botany* 89 No. 10: 731–735.

Shridhar, B.S. (2012) Review: Nitrogen fixing microorganisms. *International Journal of Microbiological Research* 3: 46–52.

Shukla, P.S., Agarwal, P.K., Jha, B. (2012) Improved salinity tolerance of *Arachis hypogaea* (L.) by the interaction of halotolerant plant-growth-promoting rhizobacteria. *Journal of Plant Growth Regulation* 31 No. 2: 195–206.

Siddikee, M.A., Chauhan, P.S., Anandham, R., Han, G.H., SA, T. (2010) Isolation, characterization and use for plant growth promotion under salt stress, of ACC deaminase producing halotolerant bacteria derived from coastal soil. *Journal of Microbiology and Biotechnology* 20 No. 11: 1577–1584.

Singh, B.K., Bardgett, R.D., Smith, P., Reay, D.S. (2010) Microorganisms and climate change: Terrestrial feedbacks and mitigation options. *Nature Reviews Microbiology* 8 No. 11: 779–790.

Singh, L.P., Gill, S.S., Tuteja, N. (2011) Unraveling the role of fungal symbionts in plant abiotic stress tolerance. *Plant Signaling and Behavior* 6 No. 2: 175–191.

Singh, B., Kaur, T., Kaur, S., Manhas, R.K., Kaur, A. (2016) Insecticidal potential of an endophytic *Cladosporium velox* against *Spodoptera litura* mediated through inhibition of alpha glycosidases. *Pesticide Biochemistry and Physiology* 131: 46–52.

Sloan, S.S., Lebeis, S.L. (2015) Exercising influence: Distinct biotic interactions shape root microbiomes. *Current Opinion in Plant Biology* 26: 32–36.

Smith, R.E., Read, D.J. (2008) Mycorrhizal symbiosis-Introduction. In *Mycorrhizal Symbiosis*, ed. Smith, R.E., Read, D.J., 1–9. New York: Elsevier.

Smith, S.E., Smith, F.A. (2011) Roles of arbuscular mycorrhizas in plant nutrition and growth: New paradigms from cellular to ecosystem scales. *Annual Review of Plant Biology* 62: 227–250.

Spaink, H.P. (2000) Root nodulation and infection factors produced by rhizobial bacteria. *Annual Review of Microbiology* 54: 257–288.

Spence, C., Alff, E., Johnson, C., Ramos, C., Donofrio, N., Sundaresan, V., Bais, H. (2014) Natural rice rhizospheric microbes suppress rice blast infections. *BMC Plant Biology* 14: 130.

Sreenivasulu, N., Sopory, S.K., Kavi Kishor, P.B. (2007) Deciphering the regulatory mechanisms of abiotic stress tolerance in plants by genomic approaches. *Gene* 388 No. 1–2: 1–13.

Srivastava, S., Yadav, A., Seem, K., Mishra, S., Chaudhary, V., Srivastava, C.S. (2008) Effect of high temperature on *Pseudomonas putida* NBRI0987 biofilm formation and expression of stress sigma factor RpoS. *Current Microbiology* 56 No. 5: 453–457.

Su, J., Hu, C., Yan, X., Jin, Y., Chen, Z., Guan, Q., Wang, Y., *et al.* (2015) Expression of barley SUSIBA2 transcription factor yields high-starch low-methane rice. *Nature* 523 No. 7562: 602–606.

Sugiyama, A., Bakker, M.G., Badri, D.V., Manter, D.K., Vivanco, J.M. (2013) Relationships between *Arabidopsis* genotype-specific biomass accumulation and associated soil microbial communities. *Botany* 91 No. 2: 123–126.

Sugiyama, A., Yazaki, K. (2012) Root exudates of legume plants and their involvement in interactions with soil microbes. In *Secretions and Exudates in Biological Systems*, ed. Vivano, J.M., Baluška, F., 27–48. Berlin: Springer.

Symanczik, S., Gisler, M., Thonar, C., Schlaeppi, K., Van der Heijden, M.G., Kahmen, A., Boller, T., Maeder, P. (2017) Application of mycorrhiza and soil from a permaculture system improved phosphorus acquisition in Nanranjilla. *Frontiers in Plant Science* 8: 1263.

Tarkka, M., Schrey, S., Hamp, P. (2008) Plant associated soil microorganisms. In *Molecular Mechanisms of Plant and Microbe Coexistence*, ed. Nautiyal, C., Dion, P, 3–51. Berlin: Springer.

Thao, N.P., Tran, L.S. (2016) Enhancement of plant productivity in the post genomic era. *Current Genomics* 17 No. 4: 295–296.

Theocharis, A., Bordiec, S., Fernandez, O., Paquis, S., Dhondt-Cordelier, S., Baillieul, F., Clement, C., Barka, E.A. (2012) *Burkholderia phytofirmans* PsJN primes *Vitis vinifera* L. and confers a better tolerance to low nonfreezing temperatures. *Molecular Plant-Microbe Interactions: MPMI* 25 No. 2: 241–249.

Tian, B.Y., Cao, Y., Zhang, K.Q. (2015) Metagenomic insights into communities, functions of endophytes, and their associates with infection by root-knot nematode, *Meloidogyne incognita*, in tomato roots. *Scientific Reports* 5: 17087.

Timmusk, S., Behers, L., Muthoni, J., Aronsson, A.C. (2017) Perspectives and challenges of microbe application for crop improvement. *Frontiers in Plant Science* 8: 49.

Timmusk, S., El-Daim, I.A., Cpolovici, L., Tanilas, T., Kannaste, A., Behers, L., Nevo, E., *et al.* (2014) Drought-tolerance of wheat improved y rhizosphere bacteria from harsh environments: Enhanced biomass production and reduced emissions of stress volatiles. *PLOS ONE* 9 No. 5: e96086.

Timmusk, S., Nevo, E. (2011) Plant root associated biofilms. In *Bacteria in Agrobiology: Plant Nutrient Management*, ed. Meshwari, D.K., 285–300. Berlin: Springer.

Timmusk, S., Timmusk, K., Behers, L. (2013) Rhizobacterial plant drought stress tolerance enhancement towards sustainable water resource management and food security. *Journal of Food Security* 1: 6–9.

Timmusk, S., Wagner, E.G.H. (1999) The plant growth-promoting rhizobacterium *Paenibacillus polymyxa* induces changes in *Arabidopsis thaliana* gene expression: A possible connection between biotic and abiotic stress responses. *Molecular Plant-Microbe Interactions: MPMI* 12 No. 11: 951–959.

Tiwari, S., Singh, P., Tiwari, R., Meena, K.K., Yandigeri, M., Singh, D.P., Arora, D.K. (2011) Salt-tolerant rhizobacteria-mediated induced tolerance in wheat (*Triticum aestivum* L.) and chemical diversity in rhizosphere enhance plant growth. *Biology and Fertility of Soils* 47 No. 8: 907–916.

Tkacz, A., Cheema, J., Chandra, G., Grant, A., Poole, P.S. (2015) Stability and succession of the rhizosphere microbiota depends upon plant type and soil composition. *The ISME Journal* 9 No. 11: 2349–2359.

Tkacz, A., Poole, P. (2015) Role of root microbiota in plant productivity. *Journal of Experimental Botany* 66 No. 8: 2167–2175.

Trabelsi, D., Mhamdi, R. (2013) Microbial inoculants and their impact on soil microbial communities: A review. *BioMed Research International* 2013: 863240.

Turner, T.R., James, E.K., Poole, P.S. (2013) The plant microbiome. *Genome Biology* 14 No. 6: 209.

Upadhyay, S.K., Singh, D.P., Saikia, R. (2009) Genetic diversity of plant growth promoting rhizobacteria from rhizospheric soil of wheat under saline conditions. *Current Microbiology* 59 No. 5: 489–496.

Upadhyay, S.K., Singh, J.S., Singh, D.P. (2011) Exopolysaccharide-producing plant growth-promoting rhizobacteria under salinity condition. *Pedosphere* 21 No. 2: 214–222.

Uroz, S., Dessaux, Y., Oger, P. (2009) Quorum sensing and quorum quenching: The Yin and Yang of bacterial communication. *Chembiochem: A European Journal of Chemical Biology* 10 No. 2: 205–216.

Valliyodan, B., Nguyen, H.T. (2006) Understanding regulatory networks and engineering for enhanced drought tolerance in plants. *Current Opinion in Plant Biology* 9 No. 2: 189–195.

Van der Heijden, M.G., Bardgett, R.D., van Straalen, N.M. (2008) The unseen majority: Soil microbes as drivers of plant diversity and productivity in terrestrial ecosystems. *Ecology Letters* 11 No. 3: 296–310.

van der Heijden, M.G.A., Martin, F.M., Selosse, M.A., Sanders, I.R. (2015) Mycorrhizal ecology and evolution: The past, the present and the future. *The New Phytologist* 205 No. 4: 1406–1423.

Vandenkoornhuyse, P., Quaiser, A., Duhamel, M., LeVan, A., Dufresne, A. (2015) The importance of the microbiome of the plant holobiont. *The New Phytologist* 206 No. 4: 1196–1206.

Vargas, L., Santa Brigida, A.B., Mota-Filho, J.P., de Carvalho, T.G., Rojas, C.A., Vaneechoutte, D., Van Bel, M., *et al.* (2014) Drought tolerance conferred to sugarcane by association with *Gluconacetobactor diazotrophicus*: A transcriptomic view of hormone pathways. *PLOS ONE* 9 No. 12: e114744.

Vejan, P., Abdullah, R., Khadiran, T., Ismail, S., Nasrulhaq Boyce, A. (2016) Role of plant growth promoting rhizobacteria in agricultural sustainability—A review. *Molecules* 21 No. 5: 573.

Velazquez-Hernandez, M.L., Baizabal-Aguirre, V.M., Cruz-Vazquez, F., Trejo-Contreras, M.J., Fuentes-Ramırez, L.E., Bravo-Patino, A., Valdez-Alarcon, J.J. (2011) *Gluconacetobacter diazotrophicus* levansucrase is involved in tolerance to NaCl, sucrose and desiccation, and in biofilm formation. *Archives in Microbiology* 193: 137–149.

Vidali, M. (2001) Bioremediation: An overview. *Pure and Applied Chemistry* 73 No. 7: 1163–1172.

Vinocur, B., Altman, A. (2005) Recent advances in engineering plant tolerance to abiotic stress: Achievements and limitations. *Current Opinion in Biotechnology* 16 No. 2: 123–132.

Wagg, C., Bender, S.F., Widmer, F., Van der Heijden, M.G. (2014) Soil biodiversity and soil community composition determine ecosystem multifunctionality. *Proceedings of the National Academy of Sciences of the United States of America* 111 No. 14: 5266–5270.

Wallenstein, M.D. (2017) Managing and manipulating the rhizosphere microbiome for plant health: A system approach. *Rhizosphere* 3: 230–232.

Wang, Y., Ohara, Y., Nakayashiki, H., Tosa, Y., Mayama, S. (2005) Microarray analysis of the gene expression profile induced by the endophytic plant growth-promoting rhizobacteria, *Pseudomonas fluorescens* FPT9601-T5 in *Arabidopsis. Molecular Plant-Microbe Interactions: MPMI* 18 No. 5: 385–396.

Wang, W.X., Vinocur, B., Altman, A. (2003) Plant responses to drought, salinity and extreme temperatures: Towards genetic engineering for stress tolerance. *Planta* 218 No. 1: 1–14.

Wang, H.B., Zhang, Z.X., Li, H., He, H.B., Fang, C.X., Zhang, A.J., Li, Q.S., *et al.* (2011) Characterization of metaproteomics in crop rhizospheric soil. *Journal of Proteome Research* 10 No. 3: 932–940.

Wani, P.A., Khan, M.S., Zaidi, A. (2008) Effect of metal tolerant plant growth-promoting Rhizobium on the performance of pea grown in metal-amended soil. *Archives of Environmental Contamination and Toxicology* 55 No. 1: 33–42.

Waqas, M., Khan, A.L., Hamayun, M., Shahzad, R., Kim, Y.H., Choi, K.S., Lee, I.J. (2015) Endophytic infection alleviates biotic stress in sunflower through regulation of defense hormones, antioxidants and functional amino acids. *European Journal of Plant Pathology* 141 No. 4: 803–824.

Wenzel, W.W. (2009) Rhizosphere processes and management in plant-assisted bioremediation (phytoremediation) of soils. *Plant and Soil* 321 No. 1–2: 385–408.

White III, R.A., Borkum, M.I., Rivas-Ubach, A., Bilbao, A., Wendler, J.P., Colby, S.M., Koberl, M., Jansson, C. (2017) From data to knowledge: The future of multi-omics data analysis for the rhizosphere. *Rhizosphere* 3: 222–229.

Wicke, B., Smeets, E., Dornburg, V., Vashev, B., Gaiser, T., Turkenburg, W., Faij, A. (2011) The global technical and economic potential of bioenergy from salt-affected soils. *Energy & Environmental Science* 4 No. 8: 2669–2681.

Wirthmueller, L., Maqbool, A., Banfield, M.J. (2013) On the front line: Structural insights into plant–pathogen interactions. *Nature Reviews Microbiology* 11 No. 11: 761–776.

Wuana, R.A., Okieimen, F.E. (2011) Heavy metals in contaminated soil: A review of sources, chemistry, risks and best available strategies for bioremediation. *ISRN Ecology* 2011: 402647.

Yadav, U.P., Ayre, B.G., Bush, D.R. (2015) Transgenic approaches to altering carbon and nitrogen partitioning in whole plants: Assessing the potential to improve crop yields and nutritional quality. *Frontiers in Plant Science* 6: 275.

Yang, A.Z., Akhtar, S.S., Amjad, M., Iqbal, S., Jacobsen, S.-E. (2016a) Growth and physiological responses of quinoa to drought and temperature stress. *Journal of Agronomy and Crop Science* 202 No. 6: 445–453.

Yang, A.Z., Akhtar, S.S., Iqbal, S., Amjad, M., Naveed, M., Zahir, Z.A., Jacobsen, S.E. (2016b) Enhancing salt tolerance in quinoa by halotolerant bacterial inoculum. *Functional Plant Biology* 43 No. 7: 632–642.

Yang, J., Kloepper, J.W., Ryu, C.M. (2009) Rhizosphere bacteria help plants tolerate abiotic stress. *Trends in Plant Science* 14 No. 1: 1–4.

Yang, Q., Tu, S., Wang, G., Liao, X., Yan, X. (2012) Effectiveness of applying arsenate reducing bacteria to enhance arsenic removal from polluted soils by *Pteris vittata* L. *International Journal of Phytoremediation* 14 No. 1: 89–99.

Yasmin, H., Bano, A., Samiullah, A. (2013) Screening of PGPR isolates from semi-arid region and their implication to alleviate drought stress. *Pakistan Journal of Botany* 45: 51–58.

Yeoh, Y.K., Paungfoo-Lonhienne, C., Dennis, P.G., Robinson, N., Ragan, M.A., Schmidt, S., Hugenholtz, P. (2016) The core root microbiome of sugarcanes cultivated under varying nitrogen fertilizer application. *Environmental Microbiology* 18 No. 5: 1338–1351.

Zamioudis, C., Mastranesti, P., Donukshe, P., Blilou, I., Pieterse, C.M.J. (2013) Unraveling root development programs initiated by beneficial *Pseudomonas* spp. Bacteria. *Plant Physiology* 162 No. 1: 304–318.

Zamioudis, C., Pieterse, C.M.J. (2012) Modulation of host immunity by beneficial microbes. *Molecular Plant-Microbe Interactions: MPMI* 25 No. 2: 139–150.

Zehr, J.P., Jenkins, B.D., Short, S.M., Steward, G.F. (2003) Nitrogenase gene diversity and microbial community structure: A cross-system comparison. *Environmental Microbiology* 5 No. 7: 539–554.

Zhang, F., Dashti, N., Hynes, R., Smith, D.L. (1996) Plant growth promoting rhizobacteria and soybean [*Glycine max* (L.) Merr.] nodulation and nitrogen fixation at suboptimal root zone temperatures. *Annals of Botany* 77 No. 5: 453–460.

Zhang, F., Lynch, D.H., Smith, D.L. (1995) Impact of low root temperatures in soybean [*Glycine max* (L.) Merr.] on nodulation and nitrogen fixation. *Environmental and Experimental Botany* 35 No. 3: 279–285.

Zhang, H., Sun, Y., Xie, X., Kim, M.S., Dowd, S.E., Pare, P.W. (2009) A soil bacterium regulates plant acquisition of iron via deficiency-inducible mechanisms. *The Plant Journal: for Cell and Molecular Biology* 58 No. 4: 568–577.

Zhao, S., Zhou, N., Zhao, Z.Y., Zhang, K., Wu, G.H., Tian, C.Y. (2016) Isolation of endophytic plant growth-promoting bacteria associated with the halophyte *Salicornia europea* and evaluating their promoting activity under salt stress. *Current Microbiology* 73 No. 4: 574–581.

Zheng, Y.K., Qiao, X.G., Miao, C.P., Liu, K., Chen, Y.W., Xu, L.H., Zhao, L.X. (2016) Diversity, distribution and biotechnological potential of endophytic fungi. *Annals of Microbiology* 66 No. 2: 529–542.

Ziegler, M., Engel, M., Welzl, G., Schloter, M. (2013) Development of a simple root model to study the effects of single exudates on the development of bacterial community structure. *Journal of Microbiological Methods* 94 No. 1: 30–36.

Zolla, G., Badri, D.V., Bakker, M.G., Manter, D.K., Vivanco, J.M. (2013) Soil microbiomes vary in their ability to confer drought tolerance to *Arabidopsis*. *Applied Soil Ecology* 68: 1–9.

Zuppinger-Dingley, D., Schmid, B., Petermann, J.S., Yadav, V., De Deyn, G.B., Flynn, D.F. (2014) Selection for niche differentiation in plant communities increases biodiversity effects. *Nature* 515 No. 7525: 108–111.

20 Role of Beneficial Microorganisms in Abiotic Stress Tolerance in Plants

Kanika Khanna, Ravdeep Kaur, Shagun Bali, Anket Sharma, Palak Bakshi, Poonam Saini, A.K Thukral, Puja Ohri, Bilal Ahmad Mir, Sikander Pal Choudhary, and Renu Bhardwaj

CONTENTS

20.1 INTRODUCTION

Plants are exposed to different types of stresses in their life span and they develop different mechanisms to cope with these stresses in order to survive. Various environmental stresses such as high and low temperature, flooding, strong light, drought, UV, heavy metals, salinity, change in pH and so on are predominant and they adversely affect crop production, plant metabolism and their physiology (Chakraborty et al., 2015). The stresses are needed to be managed on a priority basis in plants by keeping in mind eco-friendly as well as cost-effective technologies. Although research is being carried out for this challenging task to develop the strategies to cope with the abiotic stresses through the production of stress-tolerant varieties, resource management practices and so on (Venkateswarlu and Shanker, 2009). Nevertheless, due to their high cost, these technologies are not very useful; therefore, cost-effective strategies are gaining importance. These techniques mainly involve the use of different microbes possessing multi-faceted traits including nutrient management, plant growth promotion and disease control that enable plants to withstand abiotic stresses. In addition to these, numerous reports are available on the use of these microbes in abiotic stress tolerance in plants (Yang et al., 2009). It has been confirmed that inoculation of microorganisms in plants exposed to different stresses may lead to enhanced growth, improved yield and overall plant health (Sturz et al., 2000). Consequently, many commercially available bacterial inoculates are being used in the form of biofertilizers and biopesticides against different biotic stresses (Dimkpa et al., 2009). Moreover, plant growth promoting rhizobacteria caused changes in plants in response to abiotic stresses which can be explained by the term induced systemic tolerance (IST) (Yang et al., 2009). Plants mainly possess adaptation mechanisms toward different types of stresses such as heavy metals, drought, salt, nutrient deficiency, temperature and so on in which they modulate the morphological changes with the help of phytohormones (Potters et al., 2009). Auxins, for example, play an important role in this process and it is also produced by a wide variety of root-associated bacteria, which when inoculated to various plant species results in improved root growth, root hair and lateral root formation (Long et al., 2008). It shows positive effects on nutrient uptake and water acquisition. Another characteristic possessed by the microbes is ACC deaminase activity and its regulation, which turns out to be the basic mechanism through which rhizobacteria exerts positive effects in abiotically stressed plants (Saleem et al., 2007).

Microorganisms producing this enzyme uses ACC (ethylene precursor) as a source of nitrogen, which in turn decreases ethylene levels of plants and improves growth (Belimov et al., 2007). Nonetheless, alteration of ethylene generally modifies the stressed condition of plants due to the fact that ethylene plays an important role in stress-related signal transduction pathways (Glick, 2005). Apart from the phytohormones, certain metabolites like proline, amino acids, osmolytes and so on tend to increase in response to stress that further mediates ROS scavenging, osmotic adjustment and membrane stabilization (Hare and Cress, 1997). It has also been reported that proline level was enhanced in abiotically stressed plants upon inoculation of beneficial bacteria such as *Burkholderia* (Barka et al., 2006), *Arthobacter* and *Bacillus* (Sziderics et al., 2007). Additionally, these beneficial microbes are also involved in maintaining ionic homeostasis in plants under abiotic stress conditions. Several ions such as Na^+ ions are enhanced during stress conditions, along with the reduction of K^+ ions. This could be overcome by inoculation of certain microbes which help in maintaining high Na^+:K^+ ratios in stressed plants (Giri et al., 2007). Moreover, many reports of plants associated with beneficial microorganisms have been evolved in modulating antioxidative defense mechanism within plants. For instance, Kohler et al. (2010) showed that inoculation with *Pseudomonas memdocina* in *Lactua sativa* L. plants altered CAT and peroxidase (POD) activities under salinity stress. Similar reports indicate that maize plants when exposed to drought stress and inoculated with *Pseudomonas* sp. lowered the activities of enzymatic antioxidants thereby alleviating drought stress (Sandhya et al., 2010).

However, recent studies have demonstrated the role of arbuscular mycorrhizal fungi (AMF) in increasing tolerance toward abiotic stresses as well as improved nutrient uptake ability (Ruíz-Lozano, 2003). The interaction of plant-AMF could be successfully exploited in abiotic stress tolerance. Several studies have been revealed in which AMF population tends to increase during abiotic stress, providing them with the enhanced mechanism of abiotic stress tolerance (Estrada et al., 2013a). However, they possess various mechanisms in improving their responses toward different types of stresses. For example, *Trichoderma* sp. enhances water holding capacity, mineral uptake and improves root growth in the presence of different types of stresses (Mastouri et al., 2010). They also enhance secretion of different plant hormones such as auxins that improve their growth and alleviate stress. Moreover, they also enhance accumulation of different metabolites such as amino acids, proteins, polyamines, amides, glycinebetaines and so on, which protect the plants from unfavorable conditions. The present chapter intends to look at the role of microorganisms in enhancing abiotic stress tolerance in plants and the mechanisms underlying the increased tolerance to abiotic stress via microbial inoculations.

20.2 ROLE OF MICROORGANISMS IN ABIOTIC STRESS TOLERANCE IN PLANTS

20.2.1 Plant Growth Promoting Rhizobacteria

The beneficial bacteria that colonize the rhizosphere or roots of plants, and directly or indirectly promote plant growth, are commonly known as plant growth promoting rhizobacteria (PGPR) (Lugtenberg and Kamilova, 2009). PGPR includes a wide range of bacterial genera such as *Agrobacterium, Arthrobacter, Azotobacter, Azospirillum, Allorhizobium, Azorhizobium, Bacillus, Bradyrhizobium, Burkholderia, Caulobacter, Chromobacterium, Enterobacter, Erwinia, Flavobacterium, Mesorhizobium, Micrococcus, Pseudomonas, Rhizobium, Serratia* and *Streptomyces* (Bhattacharyya and Jha, 2012; Dimkpa et al., 2009). They may be free-living, endophytic, i.e. colonize inside plant tissues, symbiotic (e.g. *Rhizobia* sp. and *Frankia* spp.) or cyanobacteria (Glick, 2012). They can promote growth of plants directly by regulating nutrient acquisition (by solubilizing phosphate, nitrogen fixation and sequestering iron, etc.) and hormone levels [by synthesizing indole-3-acetic acid (IAA), gibberellins, cytokinins, 1-aminocyclopropane-1-carboxylate (ACC) deaminase for lowering stress induced ethylene] and indirectly by providing immunity to plant against environmental stresses (Glick, 2012). PGPR induce physical and chemical changes in plants for suppression of diseases caused by pathogenic agents and tolerance to various abiotic stresses which are also referred as induced systemic resistance (IST) in plants (Yang et al., 2009). Bacteria secrete organic acids, compatible solutes, siderophores, enzymes, antibiotics and nitric oxide which indirectly promote the growth of plants by providing protection against pathogens, improving resource acquisition, stimulation of phytohormone production by plants and providing abiotic stress tolerance (Dimkpa et al., 2009).

Due to the beneficial effects of PGRRs on crop productivity and the ability to augment abiotic and biotic stress resistance, they are gaining huge attention as an eco-friendly strategy for management of abiotic and biotic stresses. Table 20.1 summarizes recent studies of the beneficial effects of inoculation with PGPRs to plants under various abiotic stresses.

TABLE 20.1
PGPRs under Abiotic Stress in Plants

Abiotic Stress	PGPR Strain	Plant Species	Plant Response to Inoculation with PGPR	References
Drought stress	*Achromobacter piechaudii*	*Lycopersicum esculentum* and *Capsicum annuum*	Facilitated growth during and after stress. Bacteria also reduced ethylene production in stressed seedlings.	Mayak et al. (2004)
	Azospirillum brasilense and *Herbaspirillum seropedicae*	*Zea mays*	Promoted high biomass, carbon, nitrogen and chlorophyll contents and low ABA and ethylene levels.	Cura et al. (2017)
	Bacillus subtilis	*Triticum aestivum*	Enhanced photosynthetic efficiency and IAA content while lowered ABA and ACC contents.	Barnawal et al. (2017)
	Bacillus thuringiensis	*Lavandula dentata*	Enhance growth and drought avoidance.	Armada et al. (2014)
	Enterobacter ludwigii and *Flavobacterium* sp.	*Triticum aestivum*	Plant growth, membrane integrity, water status and accumulation osmolytes was improved. PGPRs also altered expression of stress-responsive genes and showed better recovery from drought stress.	Gontia-Mishra et al. (2016)
	Phyllobacterium brassicacearum (STM196)	*Arabidopsis thaliana*	Promoted gain of biomass by delaying reproductive development and also improved water use efficiency and photosynthetic efficiency.	Bresson et al. (2014)
Moisture stress	*Bacillus* spp.	*Sorghum* spp.	Improved growth biomass, chlorophyll content and induced accumulation of osmolytes. Leaf water content and soil moisture content was also improved.	Grover et al. (2014)
Salinity	*Arthrobacter* and *Bacillus megaterium* strains	*Lycopersicon esculentum* Mill.	Seedling germination and growth attributes like shoot length, fresh and dry weights, and vigor index were improved under salt stress.	Fan et al. (2016)
	Arthrobacter protophormiae and *Dietziana tronolimnaea*	*Triticum aestivum*	Both the strains improved photosynthetic efficiency and IAA contents. *A. protophormiae* also reduced ABA and ACC contents.	Barnawal et al. (2017)
	Arthrobacter sp. and *Bacillus subtilis*	*Triticum aestivum*	Both strains promoted seedling dry weight, increase in total soluble sugars and proline contents. Co-inoculation also reduced Na^+ levels.	Upadhyay et al. (2012)
	Azotobacter vinellandii (SRIAz3)	*Oryza sativa*	Restored growth, nutrient status and improved compatible solutes.	Sahoo et al. (2014a)
	Bacillus amyloliquefaciens (SQR9)	*Zea mays*	Enhanced growth, chlorophyll, total sugar and strengthened antioxidative defense system. Reduced Na^+ levels.	Chen et al. (2016)
	Bacillus licheniformis	*Chrysanthemum* sp.	Enhance plant survival rates, biomass production, photosynthesis and iron accumulation. Reduced Na^+ contents.	Zhou et al. (2017b)
	Burkholderia phytofirmans	*Arabidopsis thaliana* Col-0 plants (salt sensitive)	Enhanced tolerance after long-term salt stress and exhibited fast recovery from salt stress.	Pinedo et al. (2015)
	Dietzia natronolimnaea	*Triticum aestivum*	Enhanced salt stress tolerance by inducing expression of plethora salt-stress-responsive genes.	Bharti et al. (2016)

(Continued)

TABLE 20.1 (CONTINUED)
Role of PGPRs in Plants under Abiotic Stress Conditions

Abiotic Stress	PGPR Strain	Plant Species	Plant Response to Inoculation with PGPR	References
	Enterobacter sp.	*Abelmoschus esculentus*	Improved seed germination rate, growth, chlorophyll content and activities of antioxidative enzymes.	Habib et al. (2016)
	Enterococcus sp. and *Pantoea* sp.	*Vigna radiata*	Lowered stress injuries like lipid peroxidation and Na^+ content, and enhanced antioxidative defense system thereby improved growth and yield of plants.	Panwar et al. (2016)
	Pseudomonas syringae, *Pseudomonas fluorescens* and *Rhizobium phaseoli*	*Vigna radiata*	Both PGPR and rhizobia improved growth and modulation, but co-application was more effective.	Ahmad et al. (2011)
	Pseudomonas putida and *Pseudomonas chlororaphis*	*Gossypium hirsutum*	Seedling germination and plant growth both improved by inducing IAA production.	Egamberdieva et al. (2015)
	Planococcus rifietoensis, *Micrococcus yunnanensis* and *Variovorax paradoxus*	*Beta vulgaris*	Improved seed germination, photosynthetic capacity and lowered ethylene content resulting in high plant biomass production.	Zhou et al. (2017a)
	Staphylococcus sciuri (SAT-17)	*Zea mays*	Alleviated salt-induced oxidative stress and improved growth.	Akram et al. (2016)
	Serratia sp. and *Rhizobium* sp.	*Lactuca sativa*	Improved growth, photosynthesis and strengthened antioxidative system.	Han and Lee (2005)
Extreme temperature	*Mycobacterium phlei* and *Mycoplana bullata*	*Triticum aestivum*	Facilitated root and shoot growth in both nutrient rich and deficient soils, and at 38°C and 16°C.	Egamberdiyeva and Höflich (2003)
High temperature	*Ochrobactrum intermedium*	*Arachis hypogaea*	Sustained growth, dry weight and alterations in lipid composition.	Paulucci et al. (2015)
Chilling stress	*Burkholderia phytofirmans* strain	*Vitis vinifera*	Improved root growth, plant biomass, CO_2 fixation and O_2 evolution, and increased contents of phenolics, proline and starch.	Barka et al. (2006)
Heavy metal stress				
Cu and Zn	*Acinetobacter* sp. (FQ-44)	*Brassica napus*	Promoted growth and increased bioavailability of Cu and Zn. Bacterial strain also accumulated appreciable amount of Cu and Zn.	Fang et al. (2016)
Cu, Cr, Co, Cd, Ni, Mn and Pb	*Bacillus cereus* and *Pseudomonas moraviensis*	*Triticum aestivum*	Decreased metal content in rhizospheric soil, bio-accumulation coefficient and translocation factor.	Hassan et al. (2017)
Ni	*Bacillus* sp. CIK-516 and *Stenotrophomonas* sp. CIK-517Y	*Raphanus sativus*	Improves plant growth, fresh and dry biomass, chlorophyll and nitrogen contents. It also enabled plants to accumulate more Ni by improving Ni tolerance.	Akhtar et al. (2018)
Hg	*Bradyrhizobium canariense* (L-7AH)	*Lupinus albus*	Inoculation promoted root Hg accumulation and reversal of Hg toxicity by increasing photosynthetic efficiency and maintaining photosynthetic pigments and nitrogenase activity.	Quinones et al. (2013)

(Continued)

TABLE 20.1 (CONTINUED)
Role of PGPRs in Plants under Abiotic Stress Conditions

Abiotic Stress	PGPR Strain	Plant Species	Plant Response to Inoculation with PGPR	References
Cr	*Bacillus*, *Staphylococcus* and *Aerococcus*	*Lolium multiflorum*	Promoted plant growth and Cr removal from soil irrigated with tannery effluents.	Khan et al. (2014)
Cu	*Enterobacter* sp.	*Vigna radiata*	Reduced Cu accumulation and improved plant growth.	Sharaff et al. (2017)
Cu and Zn	*Pantoea* sp. (Y4-4)	*Medicago sativa*	Improved plant dry biomass in metal contaminated soils and increased Cu and Zn accumulation by 15% and 30.3% respectively.	Li et al. (2017)
Zn	*Pseudomonas aeruginosa*	*Triticum aestivum*	Improved uptake of minerals, chlorophyll content, total soluble proteins content, antioxidative defense system and plant biomass. Lowered malondialdehyde, hydrogen peroxide and Zn accumulation in roots and shoots.	Islam et al. (2014)
Nutrient stress	*Paenibacillus polymyxa* (BFKC01)	*Arabidopsis*	Increased Fe uptake, improved root system and photosynthetic capacity.	Zhou et al. (2016)
Pesticide stress	*Bradyrhizobium* sp. MRM6 and *Pseudomonas aeruginosa* strain PS1	*Vigna radiata*	Increased growth, nodulation, nutrient profile and seed yield.	Ahemad and Khan (2011b, 2012)
	Rhizobium sp. strain MRL3	*Lens esculentus*	Increased growth, nodulation, nutrient profile and seed yield.	Ahemad and Khan (2011a)

20.2.2 Fungi (Including Mycorrhizal Fungi)

The root system of terrestrial plants shows symbiotic association with AMF which enhances plants ability to tolerate different stresses while plants provide photosynthate and niche to fungus (Smith and Read, 2008). Studies were revealed by Rodriguez et al. (2008) that showed interactions between many plant species and mycorrhizal or endophytic fungi that enhanced tolerance toward different types of abiotic stresses. They provide specificity in stress tolerance, for example, *Curvularia protuberata* provided resistance from heat in panic grass and enhanced the thriving ability of grass in warmer compared with normal conditions (Redman et al., 2002). On the other hand, Redman et al. (2002) reported that in the absence of associated microflora, plants are not able to survive in environmental stress conditions which shows habitat-specific features in stress tolerance. In various symbiotic endophytes, different mechanisms are involved in providing tolerance against stress conditions like in the *Colletotrichum* strain, which is non-pathogenic in nature, but has the ability to increase disease resistance in its host but it does not activate the host defense without pathogen presence (Redman et al., 1999; de Zelicourt et al., 2013), whereas disease resistance to host is not systemic, it localizes where fungus is colonized. Colonization of *Piriformospora indica* in the host triggers cell signaling by synthesizing phosphatidic acid (PA) with the stimulation of Phospholipase D, which further activates the MAPK pathway (mitogen-activated protein kinase) and the OXI1-MAPK pathway resulting to plant defense as shown in Figure 20.1.

Properties of AMF show one of the eco-friendly strategies to withstand stress conditions of the environment. Table 20.2 summarizes the studies showing the positive effect of fungal colonization in plants during stressed conditions.

20.3 MECHANISMS OF ABIOTIC STRESS AMELIORATION BY MICROBES

Plant interaction with rhizospheric soil microorganisms under abiotic stress is most important for tolerance and survival of stress conditions. Microorganisms have been reported to have impending basic genetic and

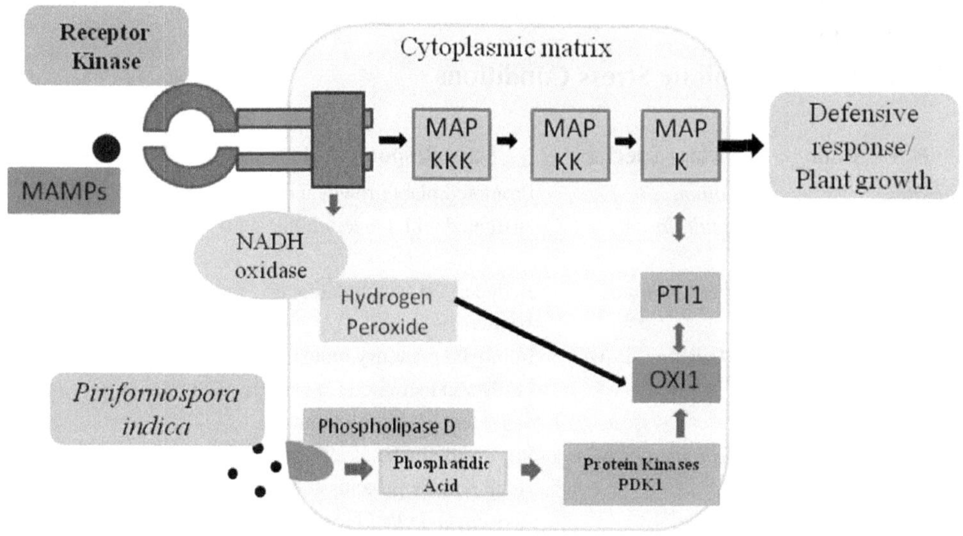

FIGURE 20.1 *Piriformospora indica*, a basidiomycetes, synthesizes phosphatidic acid by stimulating phospholipase D with further activation of PDK1, OXI1 and MAPK pathway. MAMPs (microbe-associated molecular patterns) activate OXI1 and MAPKs, and hydrogen peroxide generation, which all results in the OXI1-MAPK pathway, resulting in the activation of plant defence and growth mode. Activation and inactivation of the host defence pathways is maintained by production of fungal auxin which interferes with the activation of plant defense responses or the growth of the plant.

metabolic potential in alleviating abiotic stress in plants (Gopalakrishnan et al., 2015), resulting in the activation of molecular mechanisms and generation of reactive oxygen species (ROS) (Sharma et al., 2012). The synthesis of ROS is a common feature in plants under stress, and their role has been now explored in developing defense mechanisms. Generally, plants have their antioxidative defense mechanisms to protect themselves from oxidative damage (Jiang and Zhang, 2002; Yamane et al., 2004). The role of microorganisms in modulating the plant defense system by enhancing antioxidative enzymes, mineral uptake, maintaining ionic homeostasis and producing different hormones under abiotic stress has been studied in the recent past (de Zelicourt et al., 2013; Souza et al., 2015). Various rhizospheric bacteria like *Rhizobium* (Remans et al., 2008; Sorty et al., 2016), *Pseudomonas* (Ali et al., 2009; Sorty et al., 2016), *Azotobacter* (Sahoo et al., 2014a,b), *Bacillus* (Sorty et al., 2016; Tiwari et al., 2011; Vardharajula et al., 2011), *Burkholderia* (Barka et al., 2006; Oliveira et al., 2009), *Trichoderma* (Ahmad et al., 2015) and *cyanobacteria* (Singh et al., 2011) have been reported to show an important role in remediation of various pollution and stress tolerance in plants. Presence or inoculation of these microorganisms in the rhizosphere effects the production of various metabolites such as amino acids, proteins, amides, polyamines, sugars and glycine betaines in the root region as well as inside the plants. Their effect on some of the important metabolites during plant-microbe interaction is discussed below.

20.3.1 Microbes and Secondary Metabolites

Secondary metabolites such as polyamines, sugars, amino acids, proteins, amino acids and glycine betaines are accumulated in plants under different stress conditions (Shukla et al., 2012b). It has been reported that proline gets accumulated in plant cells under salt stress and it leads to cytoplasmic osmotic adjustment within plants (Leigh et al., 1981). Moreover, proline also enables stabilization of subcellular structures such as cell membranes, free radical scavenging and redox homeostasis under abiotic stress conditions (Ashraf and Foolad, 2007). The accumulation of proline in plants exposed to stressed conditions might be due to a decrease in their degradation or enhanced biosynthesis. Proline metabolism is therefore very complex and it enables plants to survive under stress conditions. It functions as an osmolyte and possesses an antioxidant property for ROS scavenging (Verbruggen and Hermans, 2008). It also acts as a molecular chaperone and functions in protein structure stabilization by modulating enzymatic activities (Kavi Kishor et al., 2005). Further studies were reported in which PGPR inoculated plants caused decreased proline content under NaCl stress as compared with control plants in which proline contents were higher. After which they suggested that PGPR-inoculated plants had much more tolerance toward salt stress than in its absence (Shukla et al., 2012a). In addition, *Sinorhizobium meliloti* inoculated *Medicago truncatula* plants showed enhanced proline content, which is positively correlated with

TABLE 20.2
Role of Fungi in Plants under Abiotic Stress Conditions

Abiotic Stress	Fungi	Plant Species	Plant Response to Inoculation with Fungi	References
Salt stress	*Piriformospora indica*	*Hordeum vulgare*	Effect of moderate salt stress completely abolished. Enhanced biomass.	Waller et al. (2005)
	Piriformospora indica	*Hordeum vulgare*	Prevents oxidative damage caused by the stress by modulating defense system of plants.	Baltruschat et al. (2008)
	Glomus etunicatum	*Carthamus tinctorius*	Increased leaf area, root and shoot weight. Phosphate uptake also enhanced.	Abbaspour (2010)
	Tricoderma harzianum	*Lycopersicum esculentum*	Increased seed germination ability as well as seedling vigor	Mastouri et al. (2010)
	Glomus intraradices	*Zea mays*	Upregulation of shoot dry mass. Elevated K^+ ion accumulation as compared to Na^+ ion.	Estrada et al. (2013b)
	Piriformospora indica	*Oryza sativa*	Induced photosynthetic pigments Chl *a*, Chl *b* and carotenoids in addition to the stimulated levels of osmolytes.	Jogawat et al. (2013)
	Glomus viscosum	*Medicago sativa* L.	Leaf area, root density resulted increased in biomass of plant.	Campanelli et al. (2013)
Heavy metal stress	*Glomus mosseae*	*Piper nigrum*	Stimulated levels of P, K, Ca and Mg and decrease in oxidative stress via Cu amelioration.	Abdel and Hamed (2011)
	Glomus macrocarpum	*Zea mays*	Enhanced plant biomass under Cd stress.	de Andrade and da Silveira (2008)
	Piriformospora indica	*Arabidopsis*	Promoted plant growth as well as modification of plasma proteins endoplasmic reticulum proteins under Cd stress.	Peskan-Berghöfer et al. (2004)
	Piriformospora indica, Glomus mosseae	*Triticum aestivum*	Increased plant growth, chlorophyll content, performance index.	Shahabivand et al. (2012)
	Piriformospora indica	*Triticum aestivum* L.	Increased shoot and root biomass as compared with non-inoculated controls were examined	Abadi and Sepehri (2016)
Water stress	*Tricoderma harzianum*	*Lycopersicum esculentum*	Increased seed germination and seedling vigor.	Mastouri et al. (2010)
Polyethylene glycol to mimic drought stress	*Piriformospora indica*	*Brassica rapa*	Overexpression of drought-related genes increases whereas upregulation of peroxidase, catalase, superoxide dismutase.	Sun et al. (2010)
	Tricoderma	*Oryza sativa*	Induced physiological changes and delayed wilting.	Shukla et al. (2012b)
	Piriformospora indica	*Hordeum vulgare*	Increased root and shoot biomass under well-watered and water-deficit condition.	Ghabooli et al. (2013)
	AMF	*Olea europaea*	Enhanced drought tolerance as compared with non-colonized and reduced ROS and MDA accumulation.	Fouad et al. (2014)
	AMF	*Phoenix dactylifera*	Improved antioxidative defence system and stress alleviation.	Benhiba et al. (2015)
	AMF	*Citrus reticulata*	Induced antioxidative defense system resulting in drought stress amelioration.	Sarkar et al. (2016)
Temperature stress	AMF	*Dichanthelium lanuginosum*	Enhanced total plant biomass.	Bunn et al. (2009)

(Continued)

TABLE 20.2 (CONTINUED)
Role of Fungi in Plants under Abiotic Stress Conditions

Abiotic Stress	Fungi	Plant Species	Plant Response to Inoculation with Fungi	References
Low temperature	*Glomus etunicatum*, *Glomus intraradices*	*Zea mays*	Stimulated tolerance in seedlings stress and increased Malondialdehyde and soluble sugar content.	Zhu et al. (2010)
Temperature Stress (Chilling stress)	*Funneliformis mosseae*	*Cucumis sativus*	Upregulation in secondary metabolites including phenols, flavonoids, lignin and phenolic activity as well as enhanced the expression of stress marker genes.	Chen et al. (2013)
	Glomus etunicatum	*Zea mays*	Reduced membrane lipid peroxidation and permeability. Accumulation of osmoprotectants and enhanced antioxidative enzymatic activities.	Zhu et al. (2010)

decreased salt stress symptoms such as wilting, necrosis and chlorosis (Bianco and Defez, 2009). Further, *Arabidopsis thaliana* plants showed osmotic stress tolerance by enhancing proline levels in *Bacillus subtilis* inoculated plants (Chen et al., 2007). Along with PGPR, AM fungi has also been reported to enhance proline levels under abiotic stress conditions. It was reported that NaCl treated soybean plants leads to higher proline accumulation upon AMF inoculation than that in uninoculated plants (Sharifi et al., 2007).

Besides proline, other secondary metabolites such as betaines, amino acids and quaternary ammonium compounds also act as osmolytes and accumulate under abiotic stresses. They actively participate in stabilizing cellular structures, enzymatic activities and membrane integrity under abiotic stresses. It was observed that AMF inoculated *Phragmites australis* plants enhanced glycine betaine contents under salt stress (Al-Garni, 2006). Further, the role of amino acids in drought stress have been elucidated. It was suggested that cacao seedlings, when subjected to drought stress, enhanced the levels of certain amino acids such as valine, arginine, leucine and histidine, which upon inoculation of drought-tolerant strain *T. hamatium* D1S219b showed no further increase in amino acid contents (Bae et al., 2009). Similar results were observed by Shukla et al. (2012b) in drought exposed rice plants in which drought-induced metabolites were reduced as compared with the plants treated by *T. harzianum*.

Another group of protective metabolites that are reported to be involved in plant tolerance against different abiotic stresses are polyamines (putrescine, spermidine and spermine). An increase in polyamines content under salt-stressed *Lotus glaber* mycorrhizal inoculated plants were reported. It was suggested that alteration of polyamines levels could be one of the mechanisms involved in host-AMF interactions against various abiotic stresses (Sannazzaro et al., 2007).

Other metabolites such as soluble sugars (trehalose, mannitol) play an important role in maintaining the osmotic potential of plants against various abiotic stresses (drought, salinity, heavy metals). Trehalose is the major storage material of AMF. Therefore, inoculation with AMF enhances its accumulation within plants so as to protect them from abiotic stresses (Becard et al., 1991). It has been found that inoculation of *Phaseolus vulgaris* with *Rhizobium etli*, overexpressing trehalose-6-phosphate synthase gene enhanced nitrogenase activity as well as induced drought tolerance (Shukla et al., 2012a). Microorganisms have the ability to accumulate certain metabolites while protecting them against different abiotic stresses. Table 20.3 summarizes the reports of alterations in secondary metabolites of plants under different abiotic stresses upon microbe inoculations.

20.3.2 Microbes and Mineral Uptake

All plants are known to perceive and respond to stress signals such as drought, heat, salinity, herbivory and pathogens (Hirt, 2009). Beneficial microbes such as plant growth promoting bacteria and fungi can stimulate plant growth and enhanced resistance to biotic and abiotic stresses (Lugtenberg and Kamilova, 2009). The knowledge of the interaction between plants and associated microbes could lead to the development of novel agricultural applications. Plants produce a broad range of organic compounds which can be used as nutrient or signals by microbial populations. Sugars, vitamins and organic acids are examples of organic compounds (Ortíz-Castro et al., 2009). In nutrient mobilization and

TABLE 20.3
Alterations in Secondary Metabolites of Plants Exposed to Abiotic Stresses upon Microbe Inoculations

Microorganisms	Plant Species	Type of Stress	Alterations of Secondary Metabolites	References
Pseudomonas oryzihabitans, *Variovorax paradoxus*, *Achromobacter xylosoxidans*	*Solanum tuberosum*	Drought	Decreased concentration of different amino acids such as valine, threonine, leucine, isoleucine, tryptophan, glutamic acid, tyrosine, histidine, phenylalanine and serine that is positively correlated with bacterial inoculation.	Belimov et al. (2015)
Achromobacter xylosoxidans	*Ocimum sanctum*	Waterlogging	Two-fold increase in proline content in water-logged plants which decreases to 18% upon bacterial inoculation.	Barnawal et al. (2012)
Pseudomonas pseudoalcaligenes, *Bacillus pumilus*	*Oryza sativa*	Salinity	Proline accumulation under saline condition but tends to decrease in plants inoculated with PGPRs.	Jha et al. (2011)
Bacillus subtilis, *Arthrobacter*	*Triticum aestivum*	Salinity	Increased dry mass, soluble sugars and proline content.	Upadhyay et al. (2012)
Bacillus	*Zea mays*	Drought	Increased levels of osmolytes such as proline, free amino acids along with decreased electrolyte leakage.	Vardharajula et al. (2011)
Pseudomonas	*Zea mays*	Drought	Higher levels of proline, sugars and amino acids upon bacterial inoculations.	Sandhya et al. (2010)
Glomus intraradices/Glomus mosseae	*Lactuca sativa*	Salinity	Elevated sugar and proline accumulation.	Kohler et al. (2009)
Rhizobium *Pseudomonas*	*Zea mays*	Salinity	Enhanced proline content with reduced electrolyte leakage, maintained relative water content as a result of inoculation of bacterial strains.	Bano and Fatima (2009)
Sinorhizobium meliloti	*Medicago truncatula*	Salinity	Accumulation of higher levels of trehalose as a result of osmoregulatory effect.	Bianco and Defez (2009)
Pseudomonas	*Sorghum bicolor*	High-temperature stress	Microbial inoculation induced the levels of cellular metabolites such as amino acids, proteins, proline and sugars.	Ali et al. (2009)
Pseudomonas mendocina, *Glomus intraradices*	*Lactuca sativa*	Drought	Proline accumulation in leaves under drought stress and microbial inoculations.	Kohler et al. (2008)
Burkholderia phytofirmans	*Vitis vinifera*	Chilling	Enhanced total phenols and proline contents upon bacterial inoculations.	Barka et al. (2006)
Glomus intraradices	*Acacia auriculiformis*	Salinity	Proline contents induced via mycorrhizal colonization in NaCl treated plants.	Diouf et al. (2005)
Bradyrhizobium	*Acacia mangium*	Salinity	Induced proline concentration in plants upon NaCl treatment which further increased upon microbial inoculation.	Diouf et al. (2005)
Glomus mosseae	*Zea mays*	Salinity	Mycorrhizal plants showed higher soluble sugar accumulation and electrolyte concentration, enhancing their osmoregulation capacity.	Feng et al. (2002)
Azospirillum brasilense	*Zea mays*	Drought	Bacterial inoculations prevented water potential drop by increasing proline contents in leaves and roots.	Casanovas et al. (2002)

mineral uptake, different classes of microorganisms like fungi, bacteria and cyanobacteria play an important role. They also promote plant growth and suppress diseases by different activities. Some well recognized microbial mediated processes are phosphate and sulfate solubilization, nitrogen fixation, siderophores production, denitrification and so on (Prakash et al., 2015). Different direct and indirect mechanisms are used by the microbes that help the plants in mineral uptake. For example, in phosphate solubilization, most of phosphorous occurs in an insoluble form like inositol phosphate, phosphomonoesters and phosphotriesters (Ma et al., 2003). Various soil microbes help in the solubilization of inorganic phosphorous by synthesizing low molecular weight organic acids such as a gluconic and citric acid (Rodríguez and Fraga, 1999. Rodriguez et al., 2004). In a direct process, microbes solubilize P by lowering the pH of the external medium and produce low molecular weight organic anions like citric, gluconic, α-ketogluconic, oxalic and succinic acids (Chen et al., 2006). Hydroxyl and carboxyl group of these acids chelate the cations bound to phosphate thereby converting it to a soluble form (Miller et al., 2010). It was reported that maize plants colonized by *Glomus mosseae* increased shoot phosphorous concentration under salt stress (Feng et al., 2002). Moreover, durum wheat (*Triticum durum*) inoculated by AM fungus *Glomus monosporum* had a higher content of P, Zn, Cu and Fe under water stress (Al-Karaki and Clark, 1998). In indirect processes, CO_2 released by microbes during respiration gets dissolved in water to form carbonic acid that decreased the pH of mycorrhizosphere and H^+ release during NH_4^+ assimilation lowered soil pH that enabled phosphorous solubilization (Illmer and Schinner, 1992).

Few microbial strains were isolated which could oxidize Fe^{2+} from primary phyllosilicates minerals to release iron and K from these minerals (Shelobolina et al., 2014). In addition to bacterial strains, fungi also solubilized the mineral form of potassium through the release of citric, malate and oxalate (Meena et al., 2014). A study showed that *Zea mays* when inoculated with *G. mosseae* and *G. intraradices* increased K uptake (Wu et al., 2005). Acidolysis is the mechanism involved in K mobilizing in which bacteria or fungi release succinic, gluconic, citric and oxalic acid that decreases local pH. Further, production of these organic acids and protons (H^+) increase chelation of cations which are bound to potassium (Rashid et al., 2016).

Siderophore producing microbes play an important role in alleviating the heavy metal stress on plants. Siderophores are low molecular mass proteins which have high affinity to form complexes with Fe^{3+} which are further taken by bacterial cell membrane where these complexes are reduced to Fe^{2+} (Machuca et al., 2007; Miethke and Marahiel, 2007). In iron-limited conditions, siderophores form complexes with heavy metals which undergo chelation of heavy metals (Schalk et al., 2011; Wang et al., 2002). Table 20.4 summarizes different reports of plants inoculated with different microbes enabling mineral uptake during abiotic stress is given in tabulated form.

20.3.3 Microbes and Hormones

Plant growth-stimulating bacteria are free-living bacteria in the rhizosphere that may directly or indirectly promote plant growth (Glick et al., 1995, 1998). Indirect activation of plant growth comprises an array of mechanisms by which the bacteria prohibit pathogens from hampering growth and development of plants. Direct activation might be comprised of biosynthesis of bacterial compounds which promote the uptake of micronutrients and essential nutrients from the soil as well as stimulate the synthesis of plant growth regulators, siderophore production, sequestration of iron and zinc, solubilization of potassium and phosphorus, atmospheric nitrogen fixation and stimulation of the enzyme ACC deaminase that may reduce the concentration of ethylene in plants (Glick and Bashan, 1997; Mayak et al., 2004; Sivan and Chet, 1992).

In the biosynthesis of ethylene, *S*-adenosylmethionine (S-AdoMet) is converted to 1-aminocyclopropane-1-carboxylate (ACC) by 1-aminocyclopropane-1-carboxylate synthase (ACS), and ACC is the immediate precursor of ethylene. Under stress, ethylene endogenously mediates plant homeostasis and causes reduction in shoot and root growth (Grover et al., 2011). The bacteria declined the adverse effects caused by ethylene and enhanced plant growth under stress (Glick, 2007). Under drought stress, *Pisum sativum* was inoculated with ACC deaminase containing bacteria, which stimulated longer roots that might be useful in the uptake of comparatively more water from the deep soil, thus consequently enhancing water use efficiency of plants under stress (Zahir et al., 2008).

It has been suggested that the function of ACC deaminase is to ensure that ethylene level should not be raised at the extent where plant growth (particularly roots) is inhibited (Glick et al., 1995, 1999; Glick and Bashan, 1997). ACC deaminase-containing bacteria (*Achromobacter piechaudii* ARV8) suppressed the synthesis of ethylene in tomato and pepper plants, which may be proficient in reducing harmful effects caused by

TABLE 20.4
Role of Microorganisms in Mineral Uptake in Plants under Abiotic Stress

Microorganisms	Plant Species	Type of Stress	Role of Microorganisms in Mineral Uptake	References
Plumeria pudica, Enterobacter cloacae, Serratia ficaria and *Pseudomonas fluorescens*	*Triticum aetivum*	Salinity	Enhanced uptake of nitrogen, phosphorous and potassium.	Nadeem et al. (2013)
Serratia ureilytica, Herbaspirillum seropedicae, Achromobacter xylosoxidans	*Ocimum sanctum*	Water logging	Increased phosphorous and nitrogen uptake.	Barnawal et al. (2012)
Brachybacterium saurashtrense, Brevibacterium casei and *Haerero halobacter*	*Arachis hypogaea*	Salt	Induced nitrogen phosphorous and calcium uptake.	Shukla et al. (2012a)
Azospirillum brasilense and *Pantoea dispersa*	*Capsicum annuum* L.	Saline	Upregulation of potassium and sodium ions.	del Amor and Cuadra-Crespo (2012)
Rhizobium and *Pseudomonas*	*Zea mays*	Salt	Microbial colonization enhanced potassium absorption under salt tress.	Bano and Fatima (2009)
Glomus intraradices or Glomus mosseae, Pseudomonas mendocina	*Lactuca sativa* L.	Salt	Increased phosphorous and potassium uptake.	Kohler et al. (2010)
Mycorrhiza	*Trifolium alexandrium*	Salt	Upregulation of phosphate solubilization via mycorrhizal colonization.	Shokri and Maadi (2009)
Bacillus sp.	*Lactuca sativa* L.	Drought	Induced uptake of nitrogen, phosphorous and potassium.	Han and Lee (2005)
Piriformospora indica (basidiomycete fungus)	*Hordeum vulgare* L.	Salt	Enhanced nitrogen uptake.	Waller et al. (2005)

water stress (Mayak et al., 2004). Ethylene synthesis is stimulated under stress conditions. It affects turnover of phospholipids in membranes, decreases the fluidity of the membrane, elevates seepage of solutes from plant cells and hampers elongation of roots (Mayak et al., 2004). Reduction in the concentration of ethylene modifies the stress status of the plant, as ethylene plays an important part in stress-related responses.

However, bacterial ACC deaminase leads to the degradation of ethylene precursor ACC in plants, which consequently reduces the stress and improves normal plant growth (Glick et al., 2007). Regulation of ACC is a primary mechanism by which bacteria employs advantageous effects on plants against abiotic stress (Saleem et al., 2007). Bacteria acquiring this enzyme may utilize the ethylene precursor (ACC) as a source of nitrogen. Hydrolysis of ACC by bacteria leads to a reduction in plant ethylene concentration that consequently enhances root growth (Belimov et al., 2007, 2009; Long et al., 2008). Another extensive trait among rhizospheric and endophytic bacteria is ACC deaminase activity. *Achromobacter piechaudii* has ACC deaminase activity that shows tolerance against water-deficit conditions by improving fresh and dry weights in tomato and pepper plants. Ethylene production was declined in inoculated plants in contrast to uninoculated ones and helped in alleviating stress (Mayak et al., 2004). Under drought stress, the ripening of pea pods was hampered in the plants inoculated with rhizobacterial strains (*Pseudomonas putida* and *P. fluorescens*) containing ACC-deaminase, which demonstrated that this PGPR might have reduced the endogenous levels of ethylene in peas through their ACC-deaminase activity (Arshad et al., 2008).

Triticum aestvum seedlings were inoculated with *Pseudomonas* sp. which facilitated plant growth by elevating the concentration of auxin and production of stress-specific proteins under Cr stress (Hasnain and Sabri, 1996). Apart from the growth promoting characteristics, the osmotolerant rhizobacteria have the capability to produce IAA. Enhancement in root proliferation in inoculated rice plants is stimulated by IAA under drought stress and consequently increased water uptake (Yuwono et al., 2005). The accessibility of particular substrates as precursors for plant hormones such as L-tryptophan for IAA, thus, is the main factor determining the degree of bacteria prompting the plant growth.

The most common elucidation for the effect of PGPR on plants depends on the production of plant hormones that modify the morphology and metabolism of plants and consequently leads to improvement in water and mineral absorption. Application of PGPR enhanced the levels of auxin, cytokinin and gibberellins but declined the level of abscisic acid in soybean plants. PGPR helped in combating drought stress by improving plant growth and metabolism (Zahedi and Abbasi, 2015). They stimulate plant growth through various mechanisms which include the production of phytohormones such as auxins, gibberellins and cytokinins and reducing the level of ethylene, siderophores sequestration of iron and nitrogen fixation (Glick et al., 1999). Synthesis of gibberellins, indole acetic acid and a few unfamiliar elements by PGPR led to increased root surface area, root length, and number of root tips and improved uptake of essential nutrients under stress (Egamberdieva and Kucharova, 2009). Few PGPR strains synthesized cytokinin and antioxidants which caused accumulation of abscisic acid (ABA) and reduced the formation of reactive oxygen species. The rapid increase in the activities of antioxidative enzymes is associated with tolerance against stress (Štajner et al., 1997). Rhizobacteria-induced drought endurance and resilience (RIDER) that comprises alterations in phytohormones level, defense-related enzymes and proteins, epoxypolysaccharide and antioxidants have been examined for microbe-regulated plant responses. Such approaches make plants resistant against abiotic stresses (Kaushal and Wani, 2016). The prospective of interactions of microbes with plants has a multifaceted role. Microbes stimulate local or systemic stress mitigation responses in plants to endure under abiotic stress conditions and also assist plants to sustain their growth and development via mobilization or production of essential nutrients, hormones and organic phytostimulant compounds. Such multipronged action of microbes or their communities makes them potential and practical options for abiotic stress alleviation approaches in crop plants.

20.3.4 Microbes and Antioxidative Mechanisms

In major physiological processes like photosynthesis and respiration, there is release of ROS as a by-product in very low amounts (Apel and Hirt, 2004), whereas stress conditions, including excess salt, water stress and temperature variation, affects plants by producing free radicals and ROS in high concentrations which is toxic to cell and leads to cell damage. In order to cope with such stress conditions and to minimize the damage, plants have evolved different mechanisms to quench the ROS. Free radicals and ROS species are neutralized by enzymatic systems, which include superoxide dismutase (SOD), glutathione peroxidase (GPOX), glutathione reductase (GR), catalase, ascorbate peroxidase (APOX), polyphenol oxidase (PPO), dehydroascorbate reductase (DHAR) and monodehydroascorbate reductase (MDHAR). Various toxic effects of different stresses are ameliorated by these antioxidative enzymes (Bano and Ashfaq, 2013). However, one of the most efficient methods to ameliorate plant stresses includes the use of microbes. Microbial population present in rhizospheric soils play an essential role in plants life cycle to withstand environmental stresses and also results in upregulation of antioxidative defense system (Ma et al., 2011; Wang et al., 2011). PGPR is a specific category of rhizobacteria that plays a key role in phytoremediation by ameliorating the oxidative damage caused by the release of siderophores, phytohormones and chelators binding to metal ions (Ma et al., 2011).

A study conducted by Chakraborty et al. (2013) on wheat showed a decrease in SOD and catalase (CAT) activities under drought stress but on the application of *Bacillus safensis* or *Ochyobactrum pseudogyegnonense* enhanced the ability to withstand the drought conditions. This shows that one of the mechanisms of drought tolerance of plants is via elevation of antioxidative state of plants. Table 20.5 summarizes the reports of the altered antioxidative potential of plants under stressed abiotic conditions and microbe presence.

20.3.5 Microbes and Ionic Homeostasis

Various ions such as Na^+, Cl^- and K^+ are accumulated within plant tissues which are mainly responsible for plant senescence and its limiting growth. Environmental stresses can cause an imbalance of ionic homeostasis in plants (Chakraborty et al., 2015). It has been reported that excess of NaCl in the soil can cause an increase in Na^+ uptake while decreasing K^+ uptake. K^+ ions are vital for many metabolic processes like stomatal conductance as well as protein synthesis in which they enable tRNA binding to ribosomes (Blaha et al., 2000). On the other hand, Na^+ starts competing with K^+, therefore disturbing the whole process by elevating Na^+:K^+ ratios (Giri et al., 2007). Moreover, it was studied that plants are able to maintain the low concentration of salt via extrusion through the plasma membrane in the cytosol and this is achieved by scavenging through NHX1 antiporters in the vacuole or by the salt overlay sensitive (SOS) pathway. Although an enhanced level of K^+ ion concentration

TABLE 20.5
Role of Microorganisms in Altering Antioxidative Defence System of Plants under Abiotic Stress

Stress	Microorganisms	Plant Species	Alteration of Antioxidative Defense System	References
Metal stress	*Pseudomonas aureofaciens* and *Klebsiella oxytoca*	*Glycine max*	Activated antioxidative defense system through enhanced glutathione-s-transferase (GST) levels.	Zaets et al. (2010)
	Agrobacterium radiobacter	*Populus deltoids* LHO5-17	Increased activities of SOD and CAT.	Wang et al. (2011)
	Azospirillum brasilense and *Azotobacter chroococcum*	*Triticum aestivum*	Enhanced enzymatic activities of APX, SOD, CAT and improved membrane integrity.	Janmohammadi et al. (2013)
	Sinorhizobium meliloti	*Medicago lupulina*	Enhanced SOD, CAT, APOX and GR activities.	Zribi et al. (2015)
	Rhizobacterial isolates	*Triticum aesvitum* L.	Upregulation of enzymatic activities such as GST, POD and CAT.	Hassan et al. (2016)
Salt stress	*Pseudomonas mendocina*	*Lactuca sativa* L.	PGPR inoculation enhanced ability to withstand stress and increased plant growth up to 30%	Kohler et al. (2010)
	Sinorhizobium meliloti 1021 and *Sinorhizobium meliloti Mt-* RD64	*Medicago truncatula*	Induced activities of antioxidant enzymes such as peroxidase, ascorbate peroxidase, glutathione reductase and superoxide dismutase.	Bianco and Defez (2009)
	Bacillus amyloliquefaciens NBRISN13 (SN13)	*Oryza sativa*	Modulation of differential transcriptional genes resulted in upregulation of plants defense system.	Nautiyal et al. (2013)
	Bacillus cereus	*Vigna radiata*, *Cicer arietinum* and *Oryza sativa*	Stimulated enzymatic activities of superoxide dismutase, peroxidase, ascorbate peroxidase and catalase as well as chitinase, β-1, 3-glucanase and phenyl alanine ammonia lyase	Chakraborty et al. (2011)
Water stress	*Bacillus safensis* and *Ochrobactrum pseudogregnonense*	*Triticum aestivum*	Upregulation of enzymatic activities such as catalase, peroxidase, ascorbate peroxidase, superoxide dismutase and glutathione reductase as well as carotenoids and ascorbate.	Chakraborty et al. (2013)
	Bacillus species	*Zea mays*	PGPR mitigates drought stress by rhizobacterial-induced drought endurance and resilience (RIDER) by reducing activity of the antioxidant enzymes APX and GPX	Vardharajula et al. (2011)
	Proteus penneri (Pp1), *Pseudomonas aeruginosa* (Pa2) and *Alcaligenes faecalis* (AF3)	*Zea mays*	Increased in relative water content, protein and sugar but proline content and the activities of antioxidant enzymes were decreased.	Naseem and Bano (2014)

can ameliorate the toxic effects of salinity on growth, metabolism and yield of the plants (Giri et al., 2007). Rhizobacteria are suggested to be involved in toxic ion homeostasis and further improve the plant growth during abiotic stress (Vaishnav et al., 2016). The mechanism by which these microorganisms reduce toxic ion uptake is either through the formation of rhizosheaths produced by exopolysaccharides (EPS) or through regulation of transporter expression system in plants (Vaishnav et al., 2016). Furthermore, studies have been conducted in which NaCl exposed *Phaseolus vulgaris* plants showed reduced Na^+ and rise in K^+ content when inoculated with PGPR, therefore, maintaining higher K^+:Na^+ ratios, which is significant due to restricted Na^+ uptake and induced K^+ uptake (Shukla et al., 2012a). Similar studies were reported by Ashraf et al. (2004) in which *Aeromonas hydrophila* and *Bacillus* inoculated wheat plants reduced Na^+ accumulation in roots and prevented its further transport toward shoots. Also, EPS-producing *B. polymyxa* and *B. circulans* elevated K^+/Na^+ and Ca^{2+}/Na^+ ratios in NaCl treated wheat plants (Khodair et al., 2008). Additionally, it was observed that EPS-producing bacterial strains elevated K^+/Na^+ ratios in soybean plants under salt stress (Kumari et al., 2015). Apart from this, microorganisms possess another mechanism of maintaining ionic homeostasis through releasing volatile organic compounds (VOCs) that regulate Na^+ homeostasis pathway within plants (Chakraborty et al., 2015). A study has been reported in which VOCs released by *Bacillus subtilis* GB03, downregulates the HKT1

TABLE 20.6
Role of Microorganisms in Maintaining Ionic Homeostasis in Plants under Abiotic Stresses

Microorganisms	Plant Species	Type of Stress	Role of Microorganisms in Ionic Homeostasis	References
Aspergillus aculeatus	*Cynodon dactylon*	Salinity	Increased concentration of K^+ and decrease in Na^+ concentration upon fungal inoculation and enhanced activities of antioxidants for Na^+ absorption through fungal hyphae in order to prevent ionic homeostasis disruption in plants.	Xie et al. (2017)
Pseudomonas simiae	*Glycine max*	Salinity	VOC mediated expression of VSP (vegetative storage protein) is positively correlated with reduced Na^+ uptake resulting in the regulation of sodium transporter activity.	Vaishnav et al. (2016)
Brachybacterium saurashtrense, Brevibacterium casei, Haererohalobacter	*Arachis hypogaea*	Salinity	Rise in K^+/Na^+ ratios and induced Ca^{2+}, phosphorous and N content.	Shukla et al. (2012a)
Achromobacter xylosoxidans, Herbaspirillum seropedicae, Ochrobacterium rhizosphaerae	*Ocimum sanctum*	Flooding	K^+ ion concentration in waterlogged plants inoculated with bacterial strains.	Barnawal et al. (2012)
Azospirillum brasilense, Pantola dispersa	*Capsicum annum* L.	Salinity	Reduced Cl^- concentration in leaves of inoculated plants along with higher K^+/Na^+ ratios in stressed plants.	del Amor and Cuadra-Crespo (2012)
Glomus mosseae, Pseudomonas mendocina	*Lactuca sativa* L.	Salinity	Inoculated plants showed enhanced Na content.	Kohler et al. (2010)
Pseudomonas putida	*Gossypium*	Salinity	Reduced Na^+ uptake from soil and increase in the absorption of Mg^{2+}, Ca^{2+} and K^+ ions.	Yao et al. (2010)
Rhizobium, Pseudomonas	*Zea mays*	Salinity	Uptake of K^+ ions synergistically via microbial inoculation.	Bano and Fatima (2009)
Glomus mosseae/Glomus intraradices	*Lavandula spica*	Drought	Rise in N and K content in plants inoculated by mycorrhizae which play crucial role in osmoregulation.	Marulanda et al. (2007)
Pseudomonas syringae, Enterobacter aerogenes, P. fluorescens	*Zea mays*	Salinity	Higher K^+/Na^+ ratios along with high relative water and low proline contents.	Nadeem et al. (2007)
Bacillus insolitus, Aeromonas hydrophila	*Triticum aestivum*	Salinity	Restricted Na^+ uptake in inoculated roots via apoplastic flow of Na^+ into stele probably due to covering of root zones by inoculants.	Ashraf et al. (2004)
Azospirillum	*Zea mays*	Salinity	Downregulation of Na^+ uptake along with upregulation of K^+ and Ca^{2+} uptake. Induced nitrogenase and nitrate reductase activities.	Hamdia et al. (2004)
Glomus macrocarpum	*Sesbania aegyptiaca*	Salinity	Reduced Na uptake in mycorrhizal colonized plants together with increase in Mg, N and P absorption.	Giri and Mukerji (2004)

expression involved in Na^+ entry into *Arabidopsis* roots. It further alleviates the stress within plants and leads to uniform distribution of Na^+ in cells (Zhang et al., 2008). Similar studies were conducted in *Arabidopsis* plants colonized by *B. subtilis* which secreted VOCs that not only decreased AtHKT1 in roots but promoted Na^+ recirculation within roots (Shkolnik-Inbar et al., 2013).

Apart from rhizobacteria, mycorrhizal inoculation also enables K^+ uptake and Na^+ inhibition toward roots, which is also considered to be the most crucial mechanism for iron homeostasis (Zuccarini and Okurowska, 2008). AMF inoculation of plants leads to increased K^+ absorption that causes higher K^+/Na^+ ratios in roots and shoots (Giri et al., 2007). This could be achieved by K^+ pumps which together regulate the transport of different cations (Parida and Das, 2005). Moreover, enhanced Na^+ and reduced K^+ levels resulting from different abiotic stresses can also disrupt various enzymatic processes and protein synthesis that could be overcome by higher K^+/Na^+ ratios (Colla et al., 2008). Table 20.6 summarizes the role of microorganisms in maintaining ionic homeostasis within plants under abiotic stresses.

20.4 CONCLUSIONS

Microorganisms associated with plants are able to mediate improved resistance toward different abiotic stresses such as drought, heavy metals, salinity, temperature, flooding, moisture and light. Identification of microbial strains that have the potential of providing cross-protection against multiple abiotic stresses would be highly valuable. Efficiencies of different bacteria and fungi in plant protection and improved growth characteristics needs to be studied more comprehensively. They alleviate adverse effects of various stresses through the production of different plant hormones like auxins and gibberellins that enhance plant growth under stress. Along with this, they also produce certain metabolites such as proteins, peptides, amino acids, polyamines and glycine betaines, which get accumulated within plants under stressful conditions, therefore enhancing their defense system. However, they also possess the potential of activating plant enzymatic and non-enzymatic antioxidant systems which provide the plant with an enhanced defense mechanism. They also possess the property of inducing phosphate solubilization and another mineral uptake within plants. Furthermore, microbes have been found to be involved in ionic homeostasis by maintaining the level of osmoprotectants.

From all these viewpoints, it is very important to explore further plant-associated microbial communities including bacteria and fungi that contribute greatly toward providing resistance among different environmental stresses. Apart from rhizobacteria, fungal symbionts establish a mutualistic relationship with plants and confer abiotic stress tolerance. These fungal symbionts are able to thrive in high-stress environment by adopting different mechanisms and defensive strategies.

As discussed above, only a few studies so far have been reported on the interactions between microbial consortium and plants in detail. Furthermore, the exact mechanism by which the microorganisms contribute to plant abiotic stress tolerance is yet to be studied. The factors contributing toward abiotic stress tolerance within plants could be identified by increasing the studies of plant-microbe interactions. Such knowledge may enable us to manipulate the role of microbes in plant performance under stressed conditions and utilizing their potential in abiotic stress tolerance. We anticipate that in the coming years, microorganisms could be completely explored in the field of plant protection toward abiotic stresses as low-input cost-effective technologies.

REFERENCES

Abadi, V.A.J.M., Sepehri, M. (2016) Effect of *Piriformospora indica* and *Azotobacter chroococcum* on mitigation of zinc deficiency stress in wheat (*Triticum aestivum* L.). *Symbiosis* 69 (1): 9–19.

Abbaspour, H. (2010) Investigation of the effects of vesicular arbuscular mycorrhiza on mineral nutrition and growth of *Carthamus tinctorius* under salt stress conditions. *Russian Journal of Plant Physiology* 57 (4): 526–531.

Abdel, L., Hamed, B. (2011) RETRACTED: Influence of arbuscular mycorrhizal fungi and copper on growth, accumulation of osmolyte, mineral nutrition and antioxidant enzyme activity of pepper (*Capsicum annuum* L.). *Mycorrhiza* 21: 495–503.

Ahemad, M., Khan, M.S. (2011a) Insecticide-tolerant and plant-growth-promoting rhizobium improves the growth of lentil (*Lens esculentus*) in insecticide-stressed soils. *Pest Management Science* 67 (4): 423–429.

Ahemad, M., Khan, M.S. (2011b) *Pseudomonas aeruginosa* strain PS1 enhances growth parameters of greengram [*Vigna radiata* (L.) Wilczek] in insecticide-stressed soils. *Journal of Pesticide Science* 84: 23–131.

Ahemad, M., Khan, M.S. (2012) Productivity of green gram in tebuconazole-stressed soil, by using a tolerant and plant growth-promoting *Bradyrhizobium* sp. MRM6 strain. *Acta Physiologiae Plantarum* 34 (1): 245–254.

Ahmad, P., Hashem, A., Abd-Allah, E.F., Alqarawi, A.A., John, R., Egamberdieva, D., Gucel, S. (2015) Role of *Trichoderma harzianum* in mitigating NaCl stress in Indian mustard (*Brassica juncea* L.) through antioxidative defense system. *Frontiers in Plant Science* 6 (6): 868.

Ahmad, M., Zahir, Z.A., Asghar, H.N., Asghar, M. (2011) Inducing salt tolerance in mung bean through coinoculation with rhizobia and plant-growth-promoting rhizobacteria containing 1-aminocyclopropane-1-carboxylate deaminase. *Canadian Journal of Microbiology* 57 (7): 578–589.

Akhtar, M.J., Ullah, S., Ahmad, I., Rauf, A., Nadeem, S.M., Khan, M.Y., Hussain, S., Bulgariu, L. (2018) Nickel phytoextraction through bacterial inoculation in *Raphanus sativus*. *Chemosphere* 190: 234–242.

Akram, M.S., Shahid, M., Tariq, M., Azeem, M., Javed, M.T., Saleem, S., Riaz, S. (2016) Deciphering *Staphylococcus sciuri* SAT-17 mediated anti-oxidative defense mechanisms and growth modulations in salt stressed maize (*Zea mays* L.). *Frontiers in Microbiology* 7: 867.

Al-Garni, S.M.S. (2006) Increasing NaCl-salt tolerance of a halophytic plant *Phragmites australis* by mycorrhizal symbiosis. *American-Eurasian Journal of Agricultural and Environmental Science* 1: 119–126.

Ali, S.Z., Sandhya, V., Grover, M., Kishore, N., Venkateswar Rao, L.V., Venkateswarlu, B. (2009) *Pseudomonas* sp. strain AKM-P6 enhances tolerance of sorghum seedlings to elevated temperatures. *Biology and Fertility of Soils* 46 (1): 45–55.

Al-Karaki, G.N., Clark, R.B. (1998) Growth, mineral acquisition, and water use by mycorrhizal wheat grown under water stress. *Journal of Plant Nutrition* 21 (2): 263–276.

Andrade, S.A.Ld, Silveira, A.P.Dd (2008) Mycorrhiza influence on maize development under Cd stress and P supply. *Brazilian Journal of Plant Physiology* 20 (1): 39–50.

Apel, K., Hirt, H. (2004) Reactive oxygen species: Metabolism, oxidative stress, and signal transduction. *Annual Review of Plant Biology* 55: 373–399.

Armada, E., Roldan, A., Azcon, R. (2014) Differential activity of autochthonous bacteria in controlling drought stress in native *Lavandula* and *Salvia* plants species under drought conditions in natural arid soil. *Microbial Ecology* 67 (2): 410–420.

Arshad, M., Shaharoona, B., Mahmood, T. (2008) Inoculation with *Pseudomonas* spp. containing ACC-deaminase partially eliminates the effects of drought stress on growth, yield, and ripening of pea (*Pisum sativum* L.). *Pedosphere* 18 (5): 611–620.

Ashraf, M., Foolad, M.R. (2007) Roles of glycine betaine and proline in improving plant abiotic stress resistance. *Environmental and Experimental Botany* 59 (2): 206–216.

Ashraf, M., Hasnain, S., Berge, O., Mahmood, T. (2004) Inoculating wheat seedlings with exopolysaccharide-producing bacteria restricts sodium uptake and stimulates plant growth under salt stress. *Biology and Fertility of Soils* 40 (3): 157–162.

Bae, H., Sicher, R.C., Kim, M.S., Kim, S.H., Strem, M.D., Melnick, R.L., Bailey, B.A. (2009) The beneficial endophyte *Trichoderma hamatum* isolate DIS 219b promotes growth and delays the onset of the drought response in *Theobroma cacao. Journal of Experimental Botany* 60 (11): 3279–3295.

Baltruschat, H., Fodor, J., Harrach, B.D., Niemczyk, E., Barna, B., Gullner, G., Janeczko, A., *et al.* (2008) Salt tolerance of barley induced by the root endophyte *Piriformospora indica* is associated with a strong increase in antioxidants. *The New Phytologist* 180 (2): 501–510.

Bano, S.A., Ashfaq, D. (2013) Role of mycorrhiza to reduce heavy metal stress. *Natural Science* 5 (12): 16–20.

Bano, A., Fatima, M. (2009) Salt tolerance in *Zea mays* (L.) following inoculation with Rhizobium and Pseudomonas. *Biology and Fertility of Soils* 45 (4): 405–413.

Barka, E.A., Nowak, J., Clement, C. (2006) Enhancement of chilling resistance of inoculated grapevine plantlets with a plant growth-promoting rhizobacterium, *Burkholderia phytofirmans* strain PsJN. *Applied and Environmental Microbiology* 72 (11): 7246–7252.

Barnawal, D., Bharti, N., Maji, D., Chanotiya, C.S., Kalra, A. (2012) 1-Aminocyclopropane-1-carboxylic acid (ACC) deaminase-containing rhizobacteria protect Ocimum sanctum plants during waterlogging stress via reduced ethylene generation. *Plant Physiology and Biochemistry: PPB* 58: 227–235.

Barnawal, D., Bharti, N., Pandey, S.S., Pandey, A., Chanotiya, C.S., Kalra, A. (2017) Plant growth-promoting rhizobacteria enhance wheat salt and drought stress tolerance by altering endogenous phytohormone levels and TaCTR1/TaDREB2 expression. *Physiologia Plantarum* 161 (4): 502–514.

Becard, G., Doner, L.W., Rolin, D.B., Douds, D.D., Pfeffer, P.E. (1991) Identification and quantification of trehalose in vesicular arbuscular mycorrhizal fungi in vivo 13C NMR and HPLC analyses. *New Phytologist* 118: 547–552.

Belimov, A.A., Dodd, I.C., Hontzeas, N., Theobald, J.C., Safronova, V.I., Davies, W.J. (2009) Rhizosphere bacteria containing 1-aminocyclopropane-1-carboxylate deaminase increase yield of plants grown in drying soil via both local and systemic hormone signalling. *The New Phytologist* 181 (2): 413–423.

Belimov, A.A., Dodd, I.C., Safronova, V.I., Hontzeas, N., Davies, W.J. (2007) *Pseudomonas brassicacearum* strain Am3 containing 1-aminocyclopropane-1-carboxylate deaminase can show both pathogenic and growth-promoting properties in its interaction with tomato. *Journal of Experimental Botany* 58 (6): 1485–1495.

Belimov, A.A., Dodd, I.C., Safronova, V.I., Shaposhnikov, A.I., Azarova, T.S., Makarova, N.M., Davies, W.J., Tikhonovich, I.A. (2015) Rhizobacteria that produce auxins and contain 1-amino-cyclopropane-1-carboxylic acid deaminase decrease amino acid concentrations in the rhizosphere and improve growth and yield of well-watered and water-limited potato (*Solanum tuberosum*). *Annals of Applied Biology* 167 (1): 11–25.

Benhiba, L., Fouad, M.O., Essahibi, A., Ghoulam, C., Qaddoury, A. (2015) Arbuscular mycorrhizal symbiosis enhanced growth and antioxidant metabolism in date palm subjected to long-term drought. *Trees* 29 (6): 1725–1733.

Bharti, N., Pandey, S.S., Barnawal, D., Patel, V.K., Kalra, A. (2016) Plant growth promoting rhizobacteria *Dietziana tronolimnaea* modulates the expression of stress responsive genes providing protection of wheat from salinity stress. *Scientific Reports* 6: 34768.

Bhattacharyya, P.N., Jha, D.K. (2012) Plant growth-promoting rhizobacteria (PGPR): Emergence in agriculture. *World Journal of Microbiology and Biotechnology* 28 (4): 1327–1350.

Bianco, C., Defez, R. (2009) *Medicago truncatula* improves salt tolerance when nodulated by an indole-3-acetic acid-overproducing *Sinorhizobium meliloti* strain. *Journal of Experimental Botany* 60 (11): 3097–3107.

Blaha, G., Stelzl, U., Spahn, C.M.T., Agrawal, R.K., Frank, J., Nierhaus, K.H. (2000) Preparation of functional ribosomal complexes and effect of buffer conditions on tRNA positions observed by cryoelectron microscopy. *Methods in Enzymology* 317: 292–309.

Bresson, J., Vasseur, F., Dauzat, M., Labadie, M., Varoquaux, F., Touraine, B., Vile, D. (2014) Interact to survive: *Phyllobacterium brassicacearum* improves *Arabidopsis* tolerance to severe water deficit and growth recovery. *PLOS ONE* 9 (9): e107607.

Bunn, R., Lekberg, Y., Zabinski, C. (2009) Arbuscular mycorrhizal fungi ameliorate temperature stress in thermophilic plants. *Ecology* 90 (5): 1378–1388.

Campanelli, A., Ruta, C., Mastro, G.D., Morone-Fortunato, I. (2013) The role of arbuscular mycorrhizal fungi in alleviating salt stress in *Medicago sativa* L. var. icon. *Symbiosis* 59 (2): 65–76.

Casanovas, E.M., Barassi, C.A., Sueldo, R.J. (2002) Azospirillum inoculation mitigates water stress effects in maize seedlings. *Cereal Research Communications* 30: 343–350.

Chakraborty, U., Chakraborty, B.N., Chakraborty, A.P., Dey, P.L. (2013) Water stress amelioration and plant growth promotion in wheat plants by osmotic stress tolerant bacteria. *World Journal of Microbiology and Biotechnology* 29 (5): 789–803.

Chakraborty, U., Chakraborty, B., Deyy, P., Chakraborty, A.P. (2015) Role of microorganisms in alleviation of abiotic stresses for sustainable agriculture. In *Microorganisms in Alleviation of Abiotic Stresses*, eds. Chakraborty, U., Chakraborty, B., 232–253. CABI Publishing, Wallingford.

Chakraborty, U., Roy, S., Chakraborty, A.P., Dey, P., Chakraborty, B. (2011) Plant growth promotion and amelioration of salinity stress in crop plants by a salt-tolerant bacterium. *Recent Research in Science and Technology* 3: 11.

Chen, S., Jin, W., Liu, A., Zhang, S., Liu, D., Wang, F., Lin, X., He, C. (2013) Arbuscular mycorrhizal fungi (AMF) increase growth and secondary metabolism in cucumber subjected to low temperature stress. *Scientia Horticulturae* 160: 222–229.

Chen, L., Liu, Y., Wu, G., Veronican Njeri, K., Shen, Q., Zhang, N., Zhang, R. (2016) Induced maize salt tolerance by rhizosphere inoculation of *Bacillus amyloliquefaciens* SQR9. *Physiologia Plantarum* 158 (1): 34–44.

Chen., Y.P., Rekha, P.D., Arun, A.B., Shen, F.T., Lai, W.-A., Young, C.C. (2006) Phosphate solubilizing bacteria from subtropical soil and their tricalcium phosphate solubilizing abilities. *Applied Soil Ecology* 34 (1): 33–41.

Chen, M., Wei, H., Cao, J., Liu, R., Wang, Y., Zheng, C. (2007) Expression of Bacillus subtilis proAB genes and reduction of feedback inhibition of proline synthesis increases proline production and confers osmotolerance in transgenic Arabidopsis. *Journal of Biochemistry and Molecular Biology* 40 (3): 396–403.

Colla, G., Rouphael, Y., Cardarelli, M., Tullio, M., Rivera, C.M., Rea, E. (2008) Alleviation of salt stress by arbuscular mycorrhizal in zucchini plants grown at low and high phosphorus concentration. *Biology and Fertility of Soils* 44 (3): 501–509.

Cura, J.A., Franz, D.R., Filosofia, J.E., Balestrasse, K.B., Burgueno, L.E. (2017) Inoculation with *Azospirillum* sp. and *Herbaspirillum* sp. bacteria increases the tolerance of maize to drought stress. *Microorganisms* 5(3): 41.

de Zelicourt, A., Al-Yousif, M., Hirt, H. (2013) Rhizosphere microbes as essential partners for plant stress tolerance. *Molecular Plant* 6 (2): 242–245.

del Amor, F.M., Cuadra-Crespo, P. (2012) Plant growth-promoting bacteria as a tool to improve salinity tolerance in sweet pepper. *Functional Plant Biology* 39(1): 82–90.

Dimkpa, C., Weinand, T., Asch, F. (2009) Plant–rhizobacteria interactions alleviate abiotic stress conditions. *Plant, Cell and Environment* 32 (12): 1682–1694.

Diouf, D., Duponnois, R., Ba, A.T., Neyra, M., Lesueur, D. (2005) Symbiosis of *Acacia auriculiformis* and *Acacia mangium* with mycorrhizal fungi and *Bradyrhizobium* spp. improves salt tolerance in greenhouse conditions. *Functional Plant Biology* 32 (12): 1143–1152.

Egamberdieva, D., Jabborova, D., Hashem, A. (2015) Pseudomonas induces salinity tolerance in cotton (Gossypium hirsutum) and resistance to Fusarium root rot through the modulation of indole-3-acetic acid. *Saudi Journal of Biological Sciences* 22 (6): 773–779.

Egamberdieva, D., Kucharova, Z. (2009) Selection for root colonising bacteria stimulating wheat growth in saline soils. *Biology and Fertility of Soils* 45 (6): 563–571.

Egamberdiyeva, D., Höflich, G. (2003) Influence of growth-promoting bacteria on the growth of wheat in different soils and temperatures. *Soil Biology and Biochemistry* 35 (7): 973–978.

Estrada, B., Aroca, R., Barea, J.M., Ruiz-Lozano, J.M. (2013a) Native arbuscular mycorrhizal fungi isolated from a saline habitat improved maize antioxidant systems and plant tolerance to salinity. *Plant Science: An International Journal of Experimental Plant Biology* 201–202:42–51.

Estrada, B., Barea, J.M., Aroca, R., Ruiz-Lozano, J.M. (2013b) A native Glomus intraradices strain from a Mediterranean saline area exhibits salt tolerance and enhanced symbiotic efficiency with maize plants under salt stress conditions. *Plant and Soil* 366 (1–2): 333–349.

Fan, P., Chen, D., He, Y., Zhou, Q., Tian, Y., Gao, L. (2016) Alleviating salt stress in tomato seedlings using *Arthrobacter* and *Bacillus megaterium* isolated from the rhizosphere of wild plants grown on saline-alkaline lands. *International Journal of Phytoremediation* 18 (11): 1113–1121.

Fang, Q., Fan, Z., Xie, Y., Wang, X., Li, X., Liu, Y. (2016) Screening and evaluation of the bioremediation potential of Cu/Zn-resistant, autochthonous *Acinetobacter* sp. FQ-44 from *Sonchusoleraceus* L. *Frontiers in Plant Science* 7: 1487.

Feng, G., Zhang, F.S., Li, X.L., Tian, C.Y., Tang, C., Rengel, Z. (2002) Improved tolerance of maize plants to salt stress by arbuscular mycorrhiza is related to higher accumulation of soluble sugars in roots. *Mycorrhiza* 12 (4): 185–190.

Fouad, M.O., Essahibi, A., Benhiba, L., Qaddoury, A. (2014) Effectiveness of arbuscular mycorrhizal fungi in the protection of olive plants against oxidative stress induced by drought. *Spanish Journal of Agricultural Research* 12 (3): 763–771.

Ghabooli, M., Khatabi, B., Ahmadi, F.S., Sepehri, M., Mirzaei, M., Amirkhani, A., Jorrín-Novo, J.V., Salekdeh, G.H. (2013) Proteomics study reveals the molecular mechanisms underlying water stress tolerance induced by *Piriformospora indica* in barley. *Journal of Proteomics* 94: 289–301.

Giri, B., Kapoor, R., Mukerji, K.G. (2007) Improved tolerance of *Acacia nilotica* to salt stress by arbuscular mycorrhiza, *Glomus fasciculatum*, may be partly related to elevated K^+/Na^+ ratios in root and shoot tissues. *Microbial Ecology* 54 (4): 753–760.

Giri, B., Mukerji, K.G. (2004) Mycorrhizal inoculant alleviates salt stress in *Sesbania aegyptiaca* and *Sesbania grandiflora* under field conditions: Evidence for reduced sodium and improved magnesium uptake. *Mycorrhiza* 14 (5): 307–312.

Glick, B.R. (2005) Modulation of plant ethylene levels by the bacterial enzyme ACC deaminase. *FEMS Microbiology Letters* 251 (1): 1–7.

Glick, B.R. (2012) Plant growth-promoting bacteria: Mechanisms and applications. *Scientifica (Cairo)* 2012: 963401.

Glick, B.R., Bashan, Y. (1997) Genetic manipulation of plant growth promoting bacteria to enhance biocontrol of fungal phytopathogens. *Biotechnology Advances* 15 (2): 353–378.

Glick, B.R., Karaturovic, D.M., Newell, P.C. (1995) A novel procedure for rapid isolation of plant growth-promoting pseudomonads. *Canadian Journal of Microbiology* 41 (6): 533–536.

Glick, B.R., Patten, C.L., Holguin, G., Penrose, D.M. (1999) *Biochemical and Genetic Mechanisms Used by Plant Growth Promoting Bacteria.* Imperial College Press, London.

Glick, B.R., Penrose, D.M., Li, J. (1998) A model for the lowering of plant ethylene concentrations by plant growth promoting bacteria. *Journal of Theoretical Biology* 190 (1): 63–68.

Glick, B.R., Todorovic, B., Czarny, J., Cheng, Z., Duan, J., McConkey, B. (2007) Promotion of plant growth by bacterial ACC deaminase. *Critical Reviews in Plant Sciences* 26 (5–6): 227–242.

Gontia-Mishra, I., Sapre, S., Sharma, A., Tiwari, S. (2016) Amelioration of drought tolerance in wheat by the interaction of plant growth-promoting rhizobacteria. *Plant Biology* 18 (6): 992–1000.

Gopalakrishnan, S., Srinivas, V., Alekhya, G., Prakash, B., Kudapa, H., Rathore, A., Varshney, R.K. (2015) The extent of grain yield and plant growth enhancement by plant growth-promoting broad-spectrum *Streptomyces* sp. in chickpea. *SpringerPlus* 4: 31.

Grover, M., Ali, S.Z., Sandhya, V., Rasul, A., Venkateswarlu, B. (2011) Role of microorganisms in adaptation of agriculture crops to abiotic stresses. *World Journal of Microbiology and Biotechnology* 27 (5): 1231–1240.

Grover, M., Madhubala, R., Ali, S.Z., Yadav, S.K., Venkateswarlu, B. (2014) Influence of *Bacillus* spp. strains on seedling growth and physiological parameters of sorghum under moisture stress conditions. *Journal of Basic Microbiology* 54 (9): 951–961.

Habib, S.H., Kausar, H., Saud, H.M. (2016) Plant growth-promoting rhizobacteria enhance salinity stress tolerance in okra through ROS-scavenging enzymes. *BioMed Research International* 2016: 6284547.

Hamdia, M.A.E., Shaddad, M.A.K., Doaa, M.M. (2004) Mechanism of salt tolerance and interactive effect of *Azospirillum brasilense* inoculation on maize cultivars grown under salt stress conditions. *Plant Growth Regulation* 44 (2): 165–174.

Han, H.S., Lee, K.D. (2005) Plant growth promoting rhizobacteria effect on antioxidant status, photosynthesis, mineral uptake and growth of lettuce under soil salinity. *Research Journal of Agriculture and Biological Sciences* 1: 210–215.

Hare, P.D., Cress, W.A. (1997) Metabolic implications of stress induced proline accumulation in plants. *Plant Growth Regulation* 21 (2): 79–102.

Hasnain, S., Sabri, A.N. (1996) Growth stimulation of *Triticum aestivum* seedlings under Cr-stress by non-rhizospheric pseudomonad strains. In Abstract book of 7th International Symposium on Nitrogen Fixation with Non-legumes, Faisalabad, pp 36.

Hassan, T.U., Bano, A., Naz, I. (2017) Alleviation of heavy metals toxicity by the application of plant growth promoting rhizobacteria and effects on wheat grown in saline sodic field. *International Journal of Phytoremediation* 19 (6): 522–529.

Hassan, W., Bashir, S., Ali, F., Ijaz, M., Hussain, M., David, J. (2016) Role of ACC-deaminase and/or nitrogen fixing rhizobacteria in growth promotion of wheat (*Triticum aestivum* L.) under cadmium pollution. *Environmental Earth Sciences* 75 (3): 267.

Hirt, H. (2009) *Plant Stress Biology: From Genomics to Systems Biology.* Wiley, Chichester.

Illmer, P., Schinner, F. (1992) Solubilization of inorganic phosphates by microorganisms isolated from forest soils. *Soil Biology and Biochemistry* 24 (4): 389–395.

Islam, F., Yasmeen, T., Ali, Q., Ali, S., Arif, M.S., Hussain, S., Rizvi, H. (2014) Influence of *Pseudomonas aeruginosa* as PGPR on oxidative stress tolerance in wheat under Zn stress. *Ecotoxicology and Environmental Safety* 104: 285–293.

Janmohammadi, M., Bihamta, M.R., Ghasemzadeh, F. (2013) Influence of rhizobacteria inoculation and lead stress on the physiological and biochemical attributes of wheat genotypes. *Cercetari Agronomice in Moldova* 46 (1): 49–67.

Jha, Y., Subramanian, R.B., Patel, S. (2011) Combination of endophytic and rhizospheric plant growth promoting rhizobacteria in *Oryza sativa* shows higher accumulation of osmoprotectant against saline stress. *Acta Physiologiae Plantarum* 33 (3): 797–802.

Jiang, M., Zhang, J. (2002) Water stress-induced abscisic acid accumulation triggers the increased generation of reactive oxygen species and up-regulates the activities of antioxidant enzymes in maize leaves. *Journal of Experimental Botany* 53 (379): 2401–2410.

Jogawat, A., Saha, S., Bakshi, M., Dayaman, V., Kumar, M., Dua, M., Varma, A., *et al.* (2013) *Piriformospora indica* rescues growth diminution of rice seedlings during high salt stress. *Plant Signaling & Behavior* 8 (10): e26891.

Kaushal, M., Wani, S.P. (2016) Plant-growth-promoting rhizobacteria: Drought stress alleviators to ameliorate crop production in drylands. *Annals of Microbiology* 66 (1): 35–42.

Kavi Kishor, P.B., Sangam, S., Amrutha, R.B. (2005) Regulation of proline biosynthesis, degradation, uptake and transport in higher plants: Its implications in plant growth and abiotic stress tolerance. *Current Science* 88: 424–438.

Khan, M.U., Sessitsch, A., Harris, M., Fatima, K., Imran, A., Arslan, M., Shabir, G., Khan, Q.M., Afzal, M. (2014) Cr-resistant rhizo- and endophytic bacteria associated with *Prosopis juliflora* and their potential as phytoremediation enhancing agents in metal-degraded soils. *Frontiers in Plant Science* 5: 755.

Khodair, T.A., Gehan, F., Galaland, T., El-Tayeb, S. (2008) Effect of inoculating wheat seedlings with exopolysaccharide-producing bacteria in saline soil. *Journal of Applied Sciences Research* 4: 2065–2017.

Kohler, J., Caravaca, F., Roldán, A. (2010) An AM fungus and a PGPR intensify the adverse effects of salinity on the stability of rhizosphere soil aggregates of *Lactuca sativa*. *Soil Biology and Biochemistry* 42 (3): 429–434.

Kohler, J., Hernández, J.A., Caravaca, F., Roldán, A. (2008) Plant-growth-promoting rhizobacteria and arbuscular mycorrhizal fungi modify alleviation biochemical mechanisms in water-stressed plants. *Functional Plant Biology* 35 (2): 141–151.

Kohler, J., Hernandez, J.A., Caravaca, F., Roldàn, A. (2009) Induction of antioxidant enzymes is involved in the greater effectiveness of a PGPR versus AM fungi with respect to increasing the tolerance of lettuce to severe salt stress. *Environmental and Experimental Botany* 65 (2–3): 245–252.

Kumari, S., Vaishnav, A., Jain, S., Varma, A., Choudhary, D.K. (2015) Bacterial-mediated induction of systemic tolerance to salinity with expression of stress alleviating enzymes in soybean (*Glycine max* L. Merrill). *Journal of Plant Growth Regulation* 34 (3): 558–573.

Leigh, R.A., Ahmad, N., Wyn-Jones, R.G. (1981) Assessment of glycine betaine and proline compartmentation by analysis isolated beet vacuoles. *Planta* 153: 34–41.

Li, S., Wang, J., Gao, N., Liu, L., Chen, Y. (2017) The effects of *Pantoea* sp. strain Y4-4 on alfalfa in the remediation of heavy-metal-contaminated soil, and auxiliary impacts of plant residues on the remediation of saline-alkali soils. *Canadian Journal of Microbiology* 63 (4): 278–286.

Long, H.H., Schmidt, D.D., Baldwin, I.T. (2008) Native bacterial endophytes promote host growth in a species-specific manner; phytohormone manipulations do not result in common growth responses. *PLOS ONE* 3 (7): e2702.

Lugtenberg, B., Kamilova, F. (2009) Plant-growth-promoting rhizobacteria. *Annual Review of Microbiology* 63: 541–556.

Ma, W., Guinel, F.C., Glick, B.R. (2003) *Rhizobium leguminosarum* biovar viciae 1-aminocyclopropane-1-carboxylate deaminase promotes nodulation of pea plants. *Applied and Environmental Microbiology* 69 (8): 4396–4402.

Ma, Y., Prasad, M.N.V., Rajkumar, M., Freitas, H. (2011) Plant growth promoting rhizobacteria and endophytes accelerate phytoremediation of metalliferous soils. *Biotechnology Advances* 29 (2): 248–258.

Ma, Y., Rajkumar, M., Vicente, J.A.F., Freitas, H. (2011) Inoculation of Ni-resistant plant growth promoting bacterium *Psychrobacter* sp. strain SRS8 for the improvement of nickel phytoextraction by energy crops. *International Journal of Phytoremediation* 13 (2): 126–139.

Machuca, A., Pereira, G., Aguiar, A., Milagres, A.M. (2007) Metal-chelating compounds produced by ectomycorrhizal fungi collected from pine plantations. *Letters in Applied Microbiology* 44 (1): 7–12.

Marulanda, A., Porcel, R., Barea, J.M., Azcon, R. (2007) Drought tolerance and antioxidant activities in lavender plants colonized by native drought-tolerant or drought-sensitive Glomus species. *Microbial Ecology* 54 (3): 543–552.

Mastouri, F., Björkman, T., Harman, G.E. (2010) Seed treatment with *Trichoderma harzianum* alleviates biotic, abiotic, and physiological stresses in germinating seeds and seedlings. *Phytopathology* 100 (11): 1213–1221.

Mayak, S., Tirosh, T., Glick, B.R. (2004) Plant growth-promoting bacteria that confer resistance to water stress in tomatoes and peppers. *Plant Science* 166 (2): 525–530.

Meena, V.S., Maurya, B.R., Verma, J.P. (2014) Does a rhizospheric microorganism enhance K^+ availability in agricultural soils? *Microbiological Research* 169 (5–6): 337–347.

Miethke, M., Marahiel, M.A. (2007) Siderophore-based iron acquisition and pathogen control. *Microbiology and Molecular Biology Reviews: MMBR* 71 (3): 413–451.

Miller, S.H., Browne, P., Prigent-Combaret, C., Combes-Meynet, E., Morrissey, J.P., Gara, F.O. (2010) Biochemical and genomic comparison of inorganic phosphate solubilization in Pseudomonas species. *Environmental Microbiology Reports* 2 (3): 403–411.

Nadeem, S.M., Zahir, Z.A., Naveed, M., Arshad, M. (2007) Preliminary investigations on inducing salt tolerance in maize through inoculation with rhizobacteria containing ACC deaminase activity. *Canadian Journal of Microbiology* 53 (10): 1141–1149.

Nadeem, S.M., Zaheer, Z.A., Naveed, M., Nawaz, S. (2013) Mitigation of salinity-induced negative impact on the growth and yield of wheat by plant growth-promoting rhizobacteria in naturally saline conditions. *Annals of Microbiology* 63 (1): 225–232.

Naseem, H., Bano, A. (2014) Role of plant growth-promoting rhizobacteria and their exopolysaccharide in drought tolerance of maize. *Journal of Plant Interactions* 9 (1): 689–701.

Nautiyal, C.S., Srivastava, S.S., Chauhan, P.S., Seem, K., Mishra, A., Sopory, S.K. (2013) Plant growth-promoting bacteria *Bacillus amyloliquefaciens* NBRISN13 modulates gene expression profile of leaf and rhizosphere community in rice during salt stress. *Plant Physiology and Biochemistry: PPB* 66: 1–9.

Oliveira, C.A., Alves, V.M.C., Marriel, I.E., Gomes, E.A., Scotti, M.R., Carneiro, N.P., Guimaraes, C.T., Schaffert, R.E., Sa, N.M.H. (2009) Phosphate solubilizing microorganisms isolated from rhizosphere of maize cultivated in an oxisol of the Brazilian Cerrado Biome. *Soil Biology and Biochemistry* 41 (9): 1782–1787.

Ortíz-Castro, R., Contreras-Cornejo, H.A., Macías-Rodríguez, L., López-Bucio, J. (2009) The role of microbial signals in plant growth and development. *Plant Signaling & Behavior* 4 (8): 701–712.

Panwar, M., Tewari, R., Nayya, H. (2016) Native halo-tolerant plant growth promoting rhizobacteria *Enterococcus* and *Pantoea* sp. improve seed yield of Mungbean (*Vigna radiata* L.) under soil salinity by reducing sodium uptake and stress injury. *Physiology and Molecular Biology of Plants: an International Journal of Functional Plant Biology* 22 (4): 445–459.

Parida, A.K., Das, A.B. (2005) Salt tolerance and salinity effects on plants: A review. *Ecotoxicology and Environmental Safety* 60 (3): 324–349.

Paulucci, N.S., Gallarato, L.A., Reguera, Y.B., Vicario, J.C., Cesari, A.B., García de Lema, M.B., Dardanelli, M.S. (2015) Arachis hypogaea PGPR isolated from Argentine soil modifies its lipids components in response to temperature and salinity. *Microbiological Research* 173: 1–9.

Peskan-Berghöfer, T., Shahollari, B., Giong, P.H., Hehl, S., Markert, C., Blanke, V., Kost, G., Varma, A., Oelmuller, R. (2004) Association of *Piriformospora indica* with *Arabidopsis thaliana* roots represents a novel system to study beneficial plant–microbe interactions and involves early plant protein modifications in the endoplasmic reticulum and at the plasma membrane. *Physiologia Plantarum* 122 (4): 465–477.

Pinedo, I., Ledger, T., Greve, M., Poupin, M.J. (2015) *Burkholderia phytofirmans* PsJN induces long-term metabolic and transcriptional changes involved in *Arabidopsis thaliana* salt tolerance. *Frontiers in Plant Science* 6: 466.

Potters, G., Pasternak, T.P., Guisez, Y., Jansen, M.A.K. (2009) Different stresses, similar morphogenic responses: Integrating a plethora of pathways. *Plant, Cell and Environment* 32 (2): 158–169.

Prakash, O., Sharma, R., Rahi, P., Karthikeyan, N. (2015) Role of microorganisms in plant nutrition and health. In *Nutrient Use Efficiency: from Basics to Advances*, ed.Rakshit A., Sigh H. B., Sen A., 125–161. Springer, Delhi.

Quinones, M.A., Ruiz-Diez, B., Fajardo, S., Lopez-Berdonces, M.A., Higueras, P.L., Fernandez-Pascual, M. (2013) Lupinusalbus plants acquire mercury tolerance when inoculated with an Hg-resistant Bradyrhizobium strain. *Plant Physiology and Biochemistry: PPB* 73: 168–175.

Rashid, M.I., Mujawara, L.H., Shahzade, T., Almeelbia, T., Ismaila, I.M.I., Ovesa, M. (2016) Bacteria and fungi can contribute to nutrients bioavailability and aggregate formation in degraded soils. *Microbiological Research* 183: 26–41.

Redman, R.S., Freeman, S., Clifton, D.R., Morrel, J., Brown, G., Rodriguez, R.J. (1999) Biochemical analysis of plant protection afforded by a nonpathogenic endophytic mutant of *Colletotrichum magna*. *Plant Physiology* 119 (2): 795–804.

Redman, R.S., Sheehan, K.B., Stout, R.J., Rodriguez, R.J., Henson, J.M. (2002) Thermotolerance conferred to plant host and fungal endophyte during mutualistic symbiosis. *Science* 298: 1581.

Remans, R., Ramaekers, L., Schelkens, S., Hernandez, G., Garcia, A., Reyes, J.L., Mendez, N., *et al.* (2008) Effect of Rhizobium–Azospirillum coinoculation on nitrogen fixation and yield of two contrasting *Phaseolus vulgaris* L. genotypes cultivated across different environments in Cuba. *Plant and Soil* 312 (1–2): 25–37.

Rodríguez, H., Fraga, R. (1999) Phosphate solubilizing bacteria and their role in plant growth promotion. *Biotechnology Advances* 17 (4–5): 319–339.

Rodriguez, H., Gonzalez, T., Goire, I., Bashan, Y. (2004) Gluconic acid production and phosphate solubilization by the plant growth-promoting bacterium *Azospirillum* spp. *Die Naturwissenschaften* 91 (11): 552–555.

Rodriguez, R.J., Henson, J., Van Volkenburgh, E., Hoy, M., Wright, L., Beckwith, F., Kim, Y.O., Redman, R.S. (2008) Stress tolerance in plants via habitat-adapted symbiosis. *The ISME Journal* 2 (4): 404–416.

Ruíz-Lozano, J.M. (2003) Arbuscular mycorrhizal symbiosis and alleviation of osmotic stress, new perspectives for molecular studies. *Mycorrhiza* 13 (6): 309–317.

Sahoo, R.K., Ansari, M.W., Dangar, T.K., Mohanty, S., Tuteja, N. (2014a) Phenotypic and molecular characterisation of efficient nitrogen-fixing Azotobacter strains from rice fields for crop improvement. *Protoplasma* 251 (3): 511–523.

Sahoo, R.K., Ansari, M.W., Pradhan, M., Dangar, T.K., Mohanty, S., Tuteja, N. (2014b) A novel *Azotobacter vinellandii* (SRI Az 3) functions in salinity stress tolerance in rice. *Plant Signaling & Behavior* 9: 511–523.

Saleem, M., Arshad, M., Hussain, S., Bhatti, A.S. (2007) Perspective of plant growth promoting rhizobacteria (PGPR) containing ACC deaminase in stress agriculture. *Journal of Industrial Microbiology & Biotechnology* 34 (10): 635–648.

Sandhya, V., Ali, S.Z., Grover, M., Reddy, G., Venkateswarlu, B. (2010) Effect of plant growth promoting *Pseudomonas* spp. on compatible solutes, antioxidant status and plant growth of maize under drought stress. *Plant Growth Regulation* 62 (1): 21–30.

Sannazzaro, A.I., Echeverria, M., Alberto, E.O., Ruiz, O.A., Menéndez, A.B. (2007) Modulation of polyamine balance in *Lotus glaber* by salinity and arbuscular mycorrhiza. *Plant Physiology and Biochemistry: PPB* 45 (1): 39–46.

Sarkar, J., Ray, A., Chakraborty, B., Chakraborty, U. (2016) Antioxidative changes in *Citrus reticulata* L. induced by drought stress and its effect on root colonization by arbuscular mycorrhizal fungi. *European Journal of Biological Research* 6: 1–13.

Schalk, I.J., Hannauer, M., Braud, A. (2011) New roles for bacterial siderophores in metal transport and tolerance. *Environmental Microbiology* 13 (11): 2844–2854.

Shahabivand, S., Maivan, H.Z., Goltapeh, E.M., Sharifi, M., Aliloo, A.A. (2012) The effects of root endophyte and arbuscular mycorrhizal fungi on growth and cadmium accumulation in wheat under cadmium toxicity. *Plant Physiology and Biochemistry: PPB* 60: 53–58.

Sharaff, M., Kamat, S., Archana, G. (2017) Analysis of copper tolerant rhizobacteria from the industrial belt of Gujarat, western India for plant growth promotion in metal polluted agriculture soils. *Ecotoxicology and Environmental Safety* 138: 113–121.

Sharifi, M., Ghorbanli, M., Ebrahimzadeh, H. (2007) Improved growth of salinity-stressed soybean after inoculation with salt pre-treated mycorrhizal fungi. *Journal of Plant Physiology* 164 (9): 1144–1151.

Sharma, P., Jha, A.B., Dubey, R.S., Pessarakli, M. (2012) Reactive oxygen species, oxidative damage, and antioxidative defense mechanism in plants under stressful conditions. *Journal of Botany* 2012: 1–26.

Shelobolina, E., Roden, E., Benzine, J., Xiong, M.Y. (2014) Using phyllosilicate-fe (ii)-oxidizing soil bacteria to improve fe and k plant nutrition. U.S. Patent Application 14/209,509, filed September 18, 2014.

Shkolnik-Inbar, D., Adler, G., Bar-Zvi, D. (2013) ABI4 downregulates expression of the sodium transporter HKT1;1 in Arabidopsis roots and affects salt tolerance. *The Plant Journal: for Cell and Molecular Biology* 73 (6): 993–1005.

Shokri, S., Maadi, B. (2009) Effects of arbuscular mycorrhizal fungus on the mineral nutrition and yield of *Trifolium alexandrium* plants under salinity stress. *Journal of Agronomy* 8 (2): 79–83.

Shukla, P.S., Agarwal, P.K., Jha, B. (2012a) Improved Salinity tolerance of *Arachis hypogaea* (L.) by the interaction of halotolerant plant growth-promoting rhizobacteria. *Journal of Plant Growth Regulation* 31 (2): 195–206.

Shukla, N., Awasthi, R.P., Rawat, L.J.K., Kumar, J. (2012b) Biochemical and physiological responses of rice (*Oryza sativa* L.) as influenced by *Trichoderma harzianum* under drought stress. *Plant Physiology and Biochemistry: PPB* 54: 78–88.

Singh, D.P., Prabha, R., Yandigeri, M.S., Arora, D.K. (2011) Cyanobacteria-mediated phenylpropanoids and phytohormones in rice (*Oryza sativa*) enhance plant growth and stress tolerance. *Antonie van Leeuwenhoek* 100 (4): 557–568.

Sivan, A., Chet, I. (1992) Microbial control of plant diseases. In *Environmental Microbiology*, ed. Mitchell, R., 335–354. Wiely-Liss Inc., New York.

Smith, S.E., Read, D.J. (2008) *Mycorrhizal Symbiosis*, 3rd edn. Academic Press, London.

Sorty, A.M., Meena, K.K., Choudhary, K., Bitla, U.M., Minhas, P.S., Krishnani, K.K. (2016) Effect of plant growth promoting bacteria associated with halophytic weed (*Psoralea corylifolia* L.) on germination and seedling growth of wheat under saline conditions. *Applied Biochemistry and Biotechnology* 180 (5): 872–882.

Souza, Rd, Ambrosini, A., Passaglia, L.M.P. (2015) Plant growth-promoting bacteria as inoculants in agricultural soils. *Genetics and Molecular Biology* 38 (4): 401–419.

Štajner, D., Kevrešan, S., Gašić, O., Mimica-Dukić, N., Zongli, H. (1997) Nitrogen and *Azotobacter chroococcum* enhance oxidative stress tolerance in sugar beet. *Biologia Plantarum* 39 (3): 441–445.

Sturz, A.V., Christie, B.R., Nowak, J. (2000) Bacterial endophytes: Potential role in developing sustainable systems of crop production. *Critical Reviews in Plant Sciences* 19 (1): 1–30.

Sun, C., Johnson, J.M., Cai, D., Sherameti, I., Oelmüller, R., Lou, B. (2010) *Piriformospora indica* confers drought tolerance in Chinese cabbage leaves by stimulating antioxidant enzymes, the expression of drought-related genes and the plastid-localized CAS protein. *Journal of Plant Physiology* 167 (12): 1009–1017.

Sziderics, A.H., Rasche, F., Trognitz, F., Sessitsch, A., Wilhelm, E. (2007) Bacterial endophytes contribute to abiotic stress adaptation in pepper plants (*Capsicum annuum* L.). *Canadian Journal of Microbiology* 53 (11): 1195–1202.

Tiwari, S., Singh, P., Tiwari, R., Meena, K.K., Yandigeri, M., Singh, D.P., Arora, D.K. (2011) Salt-tolerant rhizobacteria-mediated induced tolerance in wheat (*Triticum aestivum*) and chemical diversity in rhizosphere enhance plant growth. *Biology and Fertility of Soils* 47 (8): 907–916.

Upadhyay, S.K., Singh, J.S., Saxena, A.K., Singh, D.P. (2012) Impact of PGPR inoculation on growth and antioxidant status of wheat under saline conditions. *Plant Biology* 14 (4): 605–611.

Vaishnav, A., Varma, A., Tuteja, N., Choudhary, D.K. (2016) PGPR-mediated amelioration of crops under salt stress. In *Plant-Microbe Interaction: An Approach to Sustainable Agriculture*, eds. Choudhary D., Varma, A., Tuteja, N. Springer, Singapore, pp. 205–226.

Vardharajula, S., Ali, S.Z., Grover, M., Reddy, G., Bandi, V. (2011) Drought-tolerant plant growth promoting *Bacillus* spp.: Effect on growth, osmolytes, and antioxidant status of maize under drought stress. *Journal of Plant Interactions* 6 (1): 1–14.

Venkateswarlu, B., Shanker, A.K. (2009) Climate change and agriculture: Adaptation and mitigation strategies. *Indian Journal of Agronomy* 54: 226–230.

Verbruggen, N., Hermans, C. (2008) Proline accumulation in plants: A review. *Amino Acids* 35 (4): 753–759.

Waller, F., Achatz, B., Baltruschat, H., Fodor, J., Becker, K., Fischer, M., Heier, T., *et al.* (2005) The endophytic fungus *Piriformospora indica* reprograms barley to salt-stress tolerance, disease resistance, and higher yield. *Proceedings of the National Academy of Sciences of the United States of America* 102 (38): 13386–13391.

Wang, Q., Xiong, D., Zhao, P., Yu, X., Tu, B., Wang, G. (2011) Effect of applying an arsenic-resistant and plant growth–promoting rhizobacterium to enhance soil arsenic phytoremediation by Populus deltoides LH05-17. *Journal of Applied Microbiology* 111 (5): 1065–1074.

Wang, J., Zhao, F.J., Meharg, A.A., Raab, A., Feldmann, J., McGrath, S.P. (2002) Mechanisms of arsenic hyperaccumulation in *Pteris vittata*. Uptake kinetics, interactions with phosphate, and arsenic speciation. *Plant Physiology* 130 (3): 1552–1561.

Wu, S.C., Cao, Z.H., Li, Z.G., Cheung, K.C., Wong, M.H. (2005) Effects of biofertilizer containing N-fixer, P and K solubilizers and AM fungi on maize growth: A greenhouse trial. *Geoderma* 125 (1–2): 155–166.

Xie, Y., Han, S., Li, X., Amombo, E., Fu, J. (2017) Amelioration of salt stress on Bermudagrass by the fungus *Aspergillus aculeatus. Molecular Plant-Microbe Interactions: MPMI* 30 (3): 245–254.

Yamane, K., Rahman, S., Kawasaki, M., Taniguchi, M., Miyake, H. (2004) Pretreatment with antioxidants decreases the effects of salt stress on chloroplast ultrastructure in rice leaf segments (*Oryza sativa* L.). *Plant Production Science* 7 (3): 292–300.

Yang, J., Kloepper, J.W., Ryu, C.M. (2009) Rhizosphere bacteria help plants tolerate abiotic stress. *Trends in Plant Science* 14 (1): 1–4.

Yao, L., Wu, Z., Zheng, Y., Kaleem, I., Li, C. (2010) Growth promotion and protection against salt stress by *Pseudomonas putida* on cotton. *European Journal of Soil Biology* 46 (1): 49–54.

Yuwono, T., Handayani, D., Soedarsono, J. (2005) The role of osmotolerant rhizobacteria in rice growth under different drought conditions. *Australian Journal of Agricultural Research* 56 (7): 715–721.

Zaets, I., Kramarev, S., Kozyrovska, N. (2010) Inoculation with a bacterial consortium alleviates the effect of cadmium overdose in soybean plants. *Open Life Sciences* 5: 481–490.

Zahedi, H., Abbasi, S. (2015) Effect of plant growth promoting rhizobacteria (PGPR) and water stress on phytohormones and polyamines of soybean. *Indian Journal of Agricultural Research* 49 (5): 427–431.

Zahir, Z.A., Munir, A., Asghar, H.N., Shaharoona, B., Arshad, M. (2008) Effectiveness of rhizobacteria containing ACC deaminase for growth promotion of peas (*Pisum sativum*) under drought conditions. *Journal of Microbiology and Biotechnology* 18 (5): 958–963.

Zhang, Y., Zhao, L., Wang, Y., Yang, B., Chen, S. (2008) Enhancement of heavy metal accumulation by tissue specific co-expression of iaaM and ACC deaminase genes in plants. *Chemosphere* 72 (4): 564–571.

Zhou, C., Guo, J., Zhu, L., Xiao, X., Xie, Y., Zhu, J., Ma, Z., Wang, J. (2016) *Paenibacillus polymyxa* BFKC01 enhances plant iron absorption via improved root systems and activated iron acquisition mechanisms. *Plant Physiology and Biochemistry: PPB* 105: 162–173.

Zhou, N., Zhao, S., Tian, C. Y. (2017a) Effect of halotolerant rhizobacteria isolated from halophytes on the growth of sugar beet (Beta vulgaris L.) under salt stress. *FEMS Microbiology Letters* 364 (11): fnx091.

Zhou, C., Zhu, L., Xie, Y., Li, F., Xiao, X., Ma, Z., Wang, J. (2017b) *Bacillus licheniformis* SA03 confers increased saline-alkaline tolerance in chrysanthemum plants by induction of abscisic acid accumulation. *Frontiers in Plant Science* 8: 1143.

Zhu, X., Song, F., Xu, H. (2010) Influence of arbuscular mycorrhiza on lipid peroxidation and antioxidant enzyme activity of maize plants under temperature stress. *Mycorrhiza* 20 (5): 325–332.

Zribi, K., Nouairi, I., Slama, I., Talbi-Zribi, O., Mhadhbi, H. (2015) Medicago sativa- *Sinorhizobium meliloti* symbiosis promotes the bioaccumulation of zinc in nodulated roots. *International Journal of Phytoremediation* 17 (1–6): 49–55.

Zuccarini, P., Okurowska, P. (2008) Effects of mycorrhizal colonization and fertilization on growth and photosynthesis of sweet basil under salt stress. *Journal of Plant Nutrition* 31 (3): 497–513.

21 Enhancement of Temperature Stress Tolerance in Plants: Use of Multifaceted Bacteria and Phytoprotectants

Usha Chakraborty, Bishwanath Chakraborty, and Jayanwita Sarkar

CONTENTS

21.1 INTRODUCTION

Plants which are constantly subjected to variations in environmental conditions are under extreme stress leading to alterations in their metabolism, growth, and development and may even lead to death under long-term exposure (Boguszewska and Zagdanska, 2012). Among all the abiotic stresses such as drought, high and low temperatures, salinity, heavy metals, and so on, temperature stresses (cold or heat) can have devastating effects on plant growth and metabolism, which in turn leads to alterations in redox state of the plant cell which has been considered as one of the important consequences of the fluctuating environment conditions (Bita and Gerats, 2013; Suzuki and Mittler, 2006; Suzuki et al., 2012). Every plant species has optimum temperature limits for its growth and development, and abnormal temperatures have devastating effects on plant growth and metabolism (Hasanuzzaman et al., 2012, 2013; Kumar et al., 2013a,b; Suzuki et al., 2012; Yadav, 2010). Plants growing in colder climates when exposed to higher temperatures will suffer from elevated temperature stress, whereas the same temperature would be optimal for those growing in warmer climates. With global warming and other related phenomena, the earth is now witnessing severe changes in temperature conditions, varying from very high to very low, and this is now a major concern in agriculture.

Low temperatures, defined as low but not freezing temperatures (0–15°C), are common, especially in the temperate regions, and can damage many plant species. In order to cope with such conditions, many plants have the ability to increase their degree of freezing tolerance in response to low, non-freezing temperatures, a phenomenon known as cold acclimation. It is well established that some of the molecular and physiological changes that occur during cold acclimation are important for plant cold tolerance (Hsieh et al., 2002; Zhu et al., 2007). Accordingly, it has been concluded that cold tolerance that develops in initially insensitive plants is not entirely constitutive and at least some of it is developed during exposure to low temperatures (Theocharis et al., 2011).

21.2 PLANTS' PERCEPTION OF COLD TEMPERATURE

As a result of exposure to low temperatures, plants exhibit several visible symptoms such as wilting, chlorosis, or necrosis and many physiological and biochemical cell functions have been correlated with these (Ruelland and Zachowski, 2010). Often, these adverse effects are accompanied by changes in cellular changes such as cell membrane structure and lipid composition (Matteucci et al., 2011; Uemura and Steponkus, 1999), cellular leakage of electrolytes and amino acids, alterations in protoplasmic streaming and redistribution of intracellular calcium ions (Knight et al., 1998), as well as a diversion of electron flow to alternate pathways (Seo et al., 2010). They also involve changes in protein content and enzyme activities (Ruelland and Zachowski, 2010) as well as ultrastructural changes in a wide range of cell components, including plastids, thylakoid membranes and the phosphorylation of thylakoid proteins, and mitochondria (Zhang et al., 2011). It has been observed that brief exposures to low temperatures may only cause transitory changes, and plants generally survive, but prolonged exposure causes plant necrosis or death.

The exact identity of the plant sensors of low temperature has not yet been established (Chinnusamy et al., 2006). It is probable that multiple primary sensors may be involved, with each perceiving a specific aspect of the stress, and each involved in a distinct branch of the cold signaling pathway (Xiong et al., 2002). Potential sensors include Ca^{2+} influx channels, two-component histidine kinase, and receptors associated with G-proteins (Xiong et al., 2002). Certain cytoskeletal components (microtubules and actin filaments) have also been reported to participate in cold sensing by modulating the activity of Ca^{2+} channels following membrane rigidification (Abdrakhamanova et al., 2003; Nick, 2000). Because of its basic role in separating the internal from the external environment, the plasma membrane has been considered as a site for the perception of temperature change (Uemura et al., 2006; Vaultier et al., 2006; Wang et al., 2006). Protein phosphorylation, together with the suppression of protein phosphatase activity, may also provide a means for the plant to sense low temperatures (Rajashekar, 2000). Triggering of signaling pathways, including secondary messengers, ROS, Ca^{2+}-dependent protein kinases (CDPKs), mitogen-activated protein kinase (MAPK) cascades, and the activation of transcription factors (TFs) promote the production of cold-responsive proteins. These proteins can be divided into two distinct groups: regulatory ones controlling the transduction of the cold stress signal and those functionally involved in the tolerance response. The latter include LEA (late embryogenesis abundant) proteins, antifreeze proteins, mRNA-binding proteins, chaperones, detoxification enzymes, proteinase inhibitors, transporters, lipid-transfer proteins, and enzymes required for osmoprotectant biosynthesis (Grennan, 2006; Nakashima and Yamaguchi-Shinozaki, 2006; Shinozaki and Yamaguchi-Shinozaki, 2000; Wang et al., 2003; Xiong and Zhu, 2001; Yamaguchi-Shinozaki and Shinozaki, 2006).

21.3 MECHANISMS OF COLD TOLERANCE

Cold tolerance depends on several factors and has been identified as a multigene trait. Some interspecific variation has been identified among relevant gene products (Rajashekar, 2000), but the regulation of many of the genes induced during cold acclimation is conserved between species (Chen and Zhu, 2004; Chinnusamy et al., 2006; Nakashima and Yamaguchi-Shinozaki, 2006). It has been suggested that histone acetylation/deacetylation is an important player in gene activation and repression during cold acclimation, and in particular, HOS15 gene product, a nuclear-localized repressor protein that functions as a histone deacetylator, specifically interacts with histone H4 (Zhu et al., 2008). MicroRNAs (miRNAs) may also be involved as regulators of stress responses. Several stress related elements are present in the promoter regions of certain miRNAs, and some of these are known to be inducible by abiotic stress. miRNA expression profiling has also been used to demonstrate the existence of cross-talk between the salinity, cold, and drought stress signaling pathways (Liu et al., 2008).

The response of plants to low-temperature stress can be divided into three distinct phases. The first is cold acclimation (pre-hardening), which occurs at low, but above zero temperatures. The second stage (hardening), during which the full degree of tolerance is achieved, and which requires exposure to a period of sub-zero temperatures. The final phase is plant recovery after winter (Li et al., 2008). Some plants (especially trees) need a combination of short photoperiod and low temperature to fully develop their cold tolerance. In these cases, tolerance can be lost if the temperature is raised above zero and the photoperiod is lengthened (Juntilla and Robberecht, 1999; Kacperska, 1999). Plant organs differ in their level of tolerance—typically the roots are much more sensitive than the crown (McKersie and Leshem, 1994), which is understandable given that the crown is the site of the major meristem responsible for the production of new roots and shoots at the end of the cold period.

21.3.1 Cold Acclimation

Cold acclimation is the phenomenon whereby plants can tolerate very low freezing temperatures by exposure to low non-freezing temperatures, triggering a cascade of events that cause changes in gene expression and thus induce biochemical and physiological modifications that enhance their tolerance (Zhu et al., 2007). This can be compared with heat shock responses as a result of exposure to sublethal high temperatures prior to the lethal temperature, enhancing the ability of the plant to withstand high, lethal temperatures.

The primary mechanisms involved in cold acclimation are related to a number of processes which include molecular and physiological modifications occurring in plant membranes, the accumulation of cytosolic Ca^{2+}, increased levels of ROS and the activation of antioxidant systems, changes in the expression of cold-related genes and transcription factors, alterations in protein and sugar synthesis, proline accumulation, and biochemical changes that affect photosynthesis (Theocharis et al., 2011).

21.3.1.1 Cell Membrane Modifications

Being the first site of cold-induced injury, plasma membranes alterations during cold acclimation have been well characterized using biochemical and physiological approaches and recognized as a critical adaptation mechanism to low temperature (Steponkus, 1984; Uemura et al., 2006; Uemura and Yoshida, 1984). More specifically, during cold acclimation, increased P-type ATPase activity, disassembly of microtubules, and accumulation of several dehydrin family proteins occur on the plasma membrane (Abdrakhamanova et al., 2003; Ishikawa and Yoshida, 1985; Kosová et al., 2007). Further, several studies have also demonstrated that membrane rigidification, coupled with cytoskeletal rearrangements, calcium influxes, and the activation of MAPK cascades are involved in cold acclimation (Sangwan et al., 2002b; Uemura and Steponkus, 1999). The lipid composition of the plasma membrane and chloroplast envelopes in acclimated plants changes such that the threshold temperature for membrane damage is lowered relative to that for non-acclimated plants (Uemura and Steponkus, 1999). This lowering of threshold temperature is achieved by increasing the cold-adapted membranes' unsaturated fatty acid content, which makes them more fluid (Sangwan et al., 2002a; Vogg et al., 1998).

Proteins being important components of membranes research has also focused on elucidating the relationship between plasma membrane proteins and cold acclimation (Catala and Salinas, 2010; Li et al., 2012). Kawamura and Uemura (2003) first applied mass spectrometric technology and successfully identified 38 proteins, the levels of which changed during 3 days of cold acclimation treatment in *Arabidopsis*. Among these cold-responsive plasma membrane proteins are the early responsive to dehydration proteins (ERD10 and ERD14), which are members of the dehydrins and may protect proteins and membranes against freeze-induced dehydration (Kosová et al., 2007; Uemura et al., 2006). In addition, a novel cold acclimation-induced protein, plant synaptotagmin 1 (SYT1), was identified which may be involved in repairing the plasma membrane when it is disrupted during the freeze-thaw cycle. Yamazaki et al. (2008) later confirmed that synaptotagmin-associated membrane resealing is eventually involved in the plant freezing tolerance mechanism.

Alterations of the *Arabidopsis* microdomain during cold acclimation was also characterized by Minami et al. (2009) by proteome profiling. The microdomain is a lateral membrane subdomain composed of specific lipids and proteins in the plasma membrane (Takahashi et al., 2013). Studies have shown that many functional proteins that are associated with cold acclimation, such as P-type ATPases, aquaporins, tubulins (Abdrakhamanova et al., 2003; Peng et al., 2008), clathrins, and dynamin-related proteins (Minami et al., 2009) accumulate in microdomain fractions. Minami et al. (2009) predicted that functional changes of endocytosis activity during cold acclimation are regulated by microdomain-enriched clathrins and dynamin-related proteins.

21.3.1.2 Role of Calcium

Calcium is known to act as a mediator of stimulus-response coupling involved in the regulation of plant growth, development, and responses to environmental stimuli (Du and Poovaiah, 2005). Cold stress-induced rigidification of plasma membrane microdomains can cause actin cytoskeletal rearrangement. This may be followed by the activation of Ca^{2+} channels and increased cytosolic Ca^{2+} levels, which in turn may play a role in the cold acclimation process (Catalá et al., 2003).

Within seconds of a cold shock, a transient increase in cytosolic Ca^{2+} levels is observed (Knight et al., 1991). Membrane rigidification induces cytosolic Ca^{2+} signatures and the transient increase in Ca^{2+} regulates expression of COR (cold-responsive) gene. The Ca^{2+} signal can be transduced into the nucleus. Nuclear [Ca^{2+}], which is monitored by a chimera protein, formed by the fusion of aequorin to nucleoplasmin, is also transiently increased after cold shock, and the peak of nuclear [Ca^{2+}] is delayed at 5 to 10 s compared with the peak of cytosolic [Ca^{2+}] (Van der Luit et al., 1999). Which indicates that the increase in nuclear [Ca^{2+}] is probably propagated by cytosolic Ca^{2+} transients via the nuclear pore complexes (Mauger, 2012).

The cold stress-induced Ca^{2+} signature can be decoded by different pathways. Plants possess groups of Ca^{2+} sensors, including CaM (calmodulin) and CMLs (CaM-like), CDPKs (Ca^{2+}-dependent protein kinases), CCaMK (Ca^{2+}-and Ca^{2+}/CaM-dependent protein kinase), CAMTA

(CaM-binding transcription activator), CBLs (calcineurin B-like proteins), and CIPKs (CBL-interacting protein kinases) (Miura and Furumoto, 2013). While CDPKs work as positive regulators (Saijo et al., 2000) calmodulin3 is a negative regulator of gene expression and cold tolerance in plants (Townley and Knight, 2002). CBLs relay the Ca^{2+} signal by interacting with and regulating the family of CIPKs. CAMTA3 has been identified as a positive regulator of CBF2/DREB1C expression through binding to a regulatory element (CG-1 element, vCGCGb) in its promoter (Doherty et al., 2009). The camta2 camta3 double mutant plants are sensitive to freezing temperatures. The expression of CBF3/DREB1A is not regulated by CAMTA, because there is no CG-1 element in its promoter (Doherty et al., 2009).

21.3.1.3 ROS and Antioxidants in Cold Acclimation

The role of ROS in abiotic stress management has become a subject of leading to plant stress acclimation (Suzuki et al., 2012). Contrary to earlier perceptions, it is now known that ROS are not simply toxic by-products of metabolism, but act as signaling molecules that modulate the expression of various genes, including those encoding antioxidant enzymes and modulators of H_2O_2 production (Gechev et al., 2003; Neill et al., 2002; Suzuki et al., 2012). Low-temperature stress has been reported to cause significant increases in the levels of the soluble non-enzymatic antioxidants ascorbate and glutathione, as well as the activity of the main NADPH-generating dehydrogenases (Airaki et al., 2012). Also, the increased activity of the antioxidative enzymes such as superoxide dismutase, glutathione peroxidase, glutathione reductase, ascorbate peroxidase, and catalase, as well as the presence of a series of non-enzymatic antioxidants, such as tripeptidthiol, glutathione, ascorbic acid (vitamin C), and alpha-tocopherol (vitamin E) plays an important role in cold acclimation and maintenance of cellular redox homeostasis (Chen and Li, 2002). During cold acclimation, changes in H_2O_2 concentrations and GSH/GSSG ratio alter the redox state of cells and activate special defense mechanisms through the redox signaling chain (Kocsy et al., 2001). H_2O_2 generated by NADPH oxidase in the apoplast of plant cells plays a crucial role in cold acclimation induced chilling tolerance in tomato (*Lycopersicon esculentum*) (Zhou et al., 2012).

21.3.1.4 Regulation of Transcription in Cold Acclimation

Cold acclimation ultimately depends on changes in the expression of specific genes encoding products that confer increased cold tolerance (Doherty et al., 2009). This is achieved by modifying pre-existing proteins by up- or down-regulation of gene expression leading to modification of protein synthesis. It has been suggested that activity of cold/chilling-induced genes may facilitate the metabolic changes that confer cold tolerance (Doherty et al., 2009; Gilmour et al., 1998) and may also be involved in the signal transduction of the stress-response (Thomashow, 2010). Zhu et al. (2008) showed that HOS15, one of 237 predicted WD40-repeat proteins in *Arabidopsis*, functions as a repressor that modifies chromatin and controls the expression of genes involved in cold tolerance. There are several transcriptional regulatory pathways which play important roles in cold acclimation. Some of these pathways are discussed below.

21.3.1.4.1 The CRT/DREB1 Regulatory Pathway

A wide variety of COR genes have been isolated from cold-acclimated plants (Svensson et al., 2006). The cloned genes' products can be classified as (i) proteins that protect cells against environmental cold/chilling stress and (ii) proteins that regulate gene expression during the adaptation response (Fowler and Thomashow, 2002). One of the most important routes for the production of cold-responsive proteins is provided by the CBF/DREB responsive pathway. The major cis-acting element involved in this is DRE (dehydration-responsive element)/CRT. Two major groups of transcription factors bind to DRE/CRT sequences: CBF/DREB1 (CRT-binding factor/DRE-binding protein) in low-temperature signaling and DREB2 during osmotic stress (Nakashima and Yamaguchi-Shinozaki, 2006). CBF1, 2, and 3 are all responsive to low temperature, and their encoding genes are present in tandem on *Arabidopsis thaliana* chromosome 4. CBF2/DREB1C is a negative regulator of both CBF1/DREB1B and CBF3/DREB1A. CBF3 probably regulates the expression level of CBF2 (Chinnusamy et al., 2006; Novillo et al., 2004). Thus, the function(s) of CBF1 and CBF3 differ from those of CBF2, and act additively to induce the set of CBF-responsive genes required to complete the process of cold acclimation. ICE1 (inducer of CBF expression), a positive regulator of CBF3, and HOS1 (high expression of osmotically sensitive), a negative regulator of ICE1, lie upstream of CBF. The HOS1 product is a RING E3 ligase targeting ICE1 for degradation in the proteasome (Dong et al., 2006). Because of the rapid (within a few minutes) inductions of CBF transcripts following plant exposure to low temperature, ICE1 is unlikely to require de novo synthesis, but rather is already present in the absence of cold stress and is only activated when the temperature decreases (Chinnusamy et al., 2003). The LOS1 (low expression of osmotically responsive genes) product is a translation elongation factor 2-like protein, which negatively regulates CBF expression.

21.3.1.4.2 Abscisic Acid (ABA) and Other Phytohormones-Dependent Cold Signal Pathway

As is the case with several environmental stresses, ABA serves as a secondary signal in the transduction of cold signals too via second messengers, such as H_2O_2 and Ca^{2+}. Experimental evidence for this has been obtained by the use of los5 (low expression of osmotically responsive genes) mutant, which exhibits significantly decreased cold- and salt/drought-induced expression of COR. ABA enhances antioxidant defense and slows down the accumulation of ROS caused by low temperatures (Liu et al., 2011). ABA can also induce the expression of the CBF1, CBF2, and CBF3 genes, but to a significantly lower level than that caused by cold (Knight et al., 2004).

It has been reported that putrescine accumulation under cold stress is essential for proper cold acclimation and survival at freezing temperatures (Cuevas et al., 2008). *Arabidopsis* mutants defective in putrescine biosynthesis (adc1, adc2) exhibit reduced freezing tolerance compared with wild-type plants. This suggests that the detrimental consequences of putrescine depletion during cold stress are at least partially due to changes in the concentration of ABA. Polyamines have also been reported to have a role in alleviating oxidative stress: inhibiting polyamine synthesis causes increased oxidative damage in cold-treated plants (Groppa and Benavides, 2008). A number of genes involved in the biosynthesis or signaling of plant hormones, such as gibberellic acid and auxin, may also be regulated by cold stress which might be important in coordinating cold tolerance with growth and development. It has been reported that there is a role for cytokinin in mediating plant growth rates in the cold (Xia et al., 2009). Mutants with elevated cytokinin levels (ampl) displayed enhanced cell division at 40°C. Moreover, no changes in CBF expression were recorded in ampl or NahG plants at LT, suggesting that the effects of cytokinin temperature-regulated growth are independent of the CBF regulon (Clarke et al., 2004).

21.3.1.5 Involvement of Proteins in Cold Acclimation

21.3.1.5.1 CSPs and HSPs

Cold shock proteins (CSPs) play important roles in the development and stress adaptation in a variety of organisms, ranging from bacteria to mammals (Chaikam and Karlson, 2008). In higher plants these proteins are involved in the cold response (Nakaminami et al., 2006; Sasaki et al., 2007). It has been further demonstrated that CSP *Arabidopsis* 3 (AtCSP3), which shares a cold shock domain with the CSPs, is involved in the acquisition of freezing tolerance in plants (Kim et al., 2009).

Though generally involved in providing thermotolerance during high-temperature stress, some HSPs (in particular HSP90, HSP70, several small HSPs, and chaperonins 60 and 20) increase in abundance following exposure to low temperature, and unlike the HSPs produced in response to high-temperature stress, which function as molecular chaperones, these have a strong cryoprotective effect, participating in membrane protection in the refolding of denatured proteins and in preventing their aggregation (Renaut et al., 2006; Timperio et al., 2008). Small HSPs are not themselves able to refold non-native proteins but do facilitate refolding effected by HSP70 and HSP100 (Mogk et al., 2003; Sun et al., 2002).

21.3.1.5.2 LEA Proteins

The dehydrins belong to the group of heat-stable, glycine-rich LEA proteins considered to be important for membrane stabilization and the protection of proteins from denaturation when the cytoplasm becomes dehydrated. Nakayama et al. (2008) have suggested that some of them, especially COR15am, function as a protectant by preventing protein aggregation. On the other hand, the dehydrins ERD10 (early response to dehydration) and ERD14 function as chaperones and interact with phospholipid vesicles through electrostatic forces (Kovacs et al., 2008). Microarray experiments have shown that the expression profile of specific combinations of dehydrin genes can provide a reliable indication of low temperature and drought stress (Tommasini et al., 2008). Okawa et al. (2008) identified COR413im as an integral membrane protein targeted to the inner chloroplast envelope in response to low temperatures, where it contributes to plant freezing tolerance. However, the SFR2 protein, which is protective of the chloroplast during freezing, is localized in the chloroplast outer envelope membrane (Fourrier et al., 2008).

21.4 LOW-TEMPERATURE TOLERANCE ENHANCEMENT BY BENEFICIAL MICROBES

Despite the ability of plants to adapt partially to low-temperature stress in temperate climates (Saltveit, 2000), plant growth and overall productivity generally decline under chilling conditions (Haldiman, 1998). The extent of a plant's ability to withstand such stress is determined by metabolic alterations (Seki et al., 2003).

While cold acclimation provides a degree of tolerance to plants under freezing and below freezing temperatures, enhancement of tolerance by other means have also been studied extensively. It is now clear that cold resistance is not a quality conferred by the product of single gene, but has turned out as a syndrome (Fowler and Thomashow, 2002), comprising of many quite different traits, such as fluidity of the biomembranes, synthesis, accumulation of low and high molecular weight cryoprotectants, and increase in the potential to cope with oxidative stress, and most importantly, expression of cold tolerance related genes.

In this context, the use of beneficial microorganisms, specially growth-promoting rhizobacteria (PGPR), have a high potential for agriculture because they can improve plant growth, especially under limiting or stress conditions (Nabti et al., 2010). Attempts are now being made to use cold-tolerant plant growth-promoting rhizobacterial strains to improve cold stress in plants, and several workers have been successful in this attempt (Barka et al., 2006; Cheng et al., 2007). Barka et al. (2006) demonstrated that the plant growth-promoting bacteria colonizing grape plantlets could significantly influence the plantlets' resistance to chilling. Plantlet bacterization with *Burkholderia phytofirmans* (Bp) strain PsJN had a pronounced effect on grapevine growth, development, and responses to low temperatures, i.e., diminished rates of biomass reduction and electrolyte leakage during chilling and stimulated post-chilling recovery, as also enhanced photosynthesis, starch deposition, and accumulation of proline and phenolics. The presence of BpPsJN reduces the impact of chilling via a non-stomatal dependent pattern. Bacteria BpPsJN affects photosynthetic pigment accumulation and RbcLgene and protein accumulation after freezing treatment. Beneficial effect was shown on chlorophyll contents after three freezing nights whereas a significant reduction of RbcL expression and protein content was visible upon cold night exposure in bacterized plants. In contrast, Fernandez et al. (2012) have shown that BpPsJN presence did not modify pigment concentration (chlorophyll and carotenoid) in grapevine after five cold days. They concluded that these changes, combined with an enlarged root system and improved sucrose uptake from the medium, may have contributed to the stimulation of growth, development, and adaptation to stress (Li et al., 2013).

In a further study, Su et al. (2015) also evaluated the role of the endophytic PGPR, BpPsJN, on *Arabidopsis thaliana* cold tolerance. Beneficial effects triggered by PGPR colonization against cold are associated with photosynthesis, carbohydrates, and related metabolites. Poupin et al. (2013) reported that BpPsJN inoculation modified several gene regulations in *Arabidopsis*, including genes involved in defense or biotic or abiotic stimulus responses.

It is well established now that one of the mechanisms by which PGPR act is through the secretion of hormones such as IAA which can directly facilitate the proliferation of their root system by stimulating the density and length of root hairs, which in turn improves the plant uptake potential of water and mineral nutrients from a large volume of soil. The result of PGPR inoculation varies considerably depending on the bacteria, plant species, soil types, inoculum density, and environmental conditions. Hence, an effective PGPR strain isolated from one region may not perform the same way in other soil and climatic conditions (Principe et al., 2007)

Twelve psychrotolerant Pseudomonad strains were selected by Mishra et al. (2011) on the basis of various plant growth-promoting (PGP) activities at cold temperature (40°C). They studied the effect of inoculation with Pseudomonad strains on cold alleviation and growth of wheat seedlings at cold temperature (80°C) and reported that inoculation with Pseudomonad strains significantly enhanced root/shoot biomass and nutrients uptake as compared to non-bacterized control at 60 days of plant growth. Bacterization also significantly improved the level of cellular metabolites like chlorophyll, anthocyanin, free proline, total phenolics, starch content, physiologically available iron, proteins, and amino acids that are signs of alleviation of cold stress in wheat plants.

There are reports of a large number of psychrotolerant bacteria, which, due to their ability to grow under cold conditions, play a role in imparting cold tolerance ability to the various crops (Table 21.1). Some of these include several species of Pseudomonas, *Pantoea dispersa*, *Acinetobacter rhizosphaerae*, and *Exiguobacterium acetylicum*. Such bacteria also improve the plants' growth under cold conditions (Subramanian et al., 2011). In further studies, Subramanian et al. (2016) further consolidated the culturable diversity of psychrotolerant bacteria from soil and their cellular and physiological adaptations to low temperatures. They also reported that inoculation of seeds of tomato with plant growth promoting psychrotolerant bacteria significantly improved plant height, root length, and membrane damage in leaf tissues as evidenced through electrolyte leakage and malondialdehyde content. Further, the antioxidant enzyme activity was also improved in plants.

It is thus clear that several bacteria, which are tolerant to cold stress, have the ability to modify the plant's metabolism under chilling stress and make the plants

TABLE 21.1
Psychrotolerant Plant Growth Promoting Bacterial Association with Crops and their Induced Responses in Plants

Microorganisms	Source	Function	References
Pseudomonas lurida M2RH3	Rhizosphere of radish plants	P-solubilization, root and shoot length increased and N, P, K uptake	Selvakumar et al. (2011)
Pseudomonas putida (B0)	Soil from central Himalayas	P-solubilization, antagonistic to *Alternaria alternata, Fusarium oxysporum*	Pandey et al. (2006)
Pseudomonas lurida	Rhizosphere of Himalayan plants	Protects plants from chilling stress.	Bisht et al. (2013)
Pseudomonas sp. PGERs17	Garlic root	P-solublization, antagonistic to pathogens	Mishra et al. (2008)
Pseudomonas sp. NARs9	Rhizospheric soil of Amaranth, NW Indian Himalayas.	Increase germination rate, shoot, and root lengths in *Triticum* sp.	Mishra et al. (2009)
Pantoea dispersa 1A	NW Indian Himalaya	Involved in P-solubilization, IAA production, HCN production, increase in root and shoot lengths in *Triticum* sp.	Selvakumar et al. (2008a)
Acinetobacter rhizosphaerae BIHB 723	Rhizosphere of *Hippophae rhamnoides*	P-solubilization, IAA, ACC deaminase production *Hordeum vulgare*	Gulati et al. (2009)
Exiguobacterium acetylicum 1P	Rhizosphere of apple tree	P-solubilization, IAA production, HCN production. Increase root and shoot lengths and N, P, K uptake in *Triticum* sp	Selvakumar et al. (2010)
Burkholderia phytofirmans PsJN	*Glomus vesiculiferum*-infected onion roots	Tolerance to cold stress; increase in total phenolics, photosynthetic activity in *Vitis vinifera*	Barka et al. (2006)
Serratia marcescens SRM (MTCC 8708)	Flowers of summer squash (Cucurbita pepo).	Increase root and shoot lengths and N, P, K uptake in *Triticum* sp	Selvakumar et al. (2008b)
Mycobacterium sp. 44 *Mycobacterium phlei* MbP18 *Mycoplana bullata* MpB46	Soil from Müncheberg and Tashkent region	Increase root and shoot dry mass *Triticum aestivum* cv. Bussard and enhance N, P, K uptake	Egamberdiyeva and Höflich (2003)

tolerant to stress. It is further interesting that most of these bacteria also have the ability to improve plant growth and are thus extremely beneficial.

21.5 EXOGENOUS APPLICATION OF PHYTOPROTECTANTS FOR ENHANCING COLD TOLERANCE

Another approach for enhancing cold tolerance in plants has been to use exogenous chemicals that can modify certain metabolic responses to induce tolerance against stresses including cold stress (Table 21.2). Among such compounds, polyamines (PAs) are aliphatic polycation compounds ubiquitously found in living organisms which include putrescine (Put), spermidine (Spd), and spermine (Spm). PAs are considered as a class of plant growth regulators due to their fundamental roles in physiological processes, including protein synthesis, cell-cycle regulation, ion-channel regulation, free-radical scavenging, signal transduction, gene expression, and many other metabolic processes (Takahashi and Kakehi, 2010). These have a range of protective functions in low temperature affected plants such as enhancing cellular antioxidant capacity and decreasing the peroxidation of unsaturated fatty acids. PAs improve membrane stabilization and increase the fluidity (Alcázar et al., 2010). The polycationic PAs can combine with cellular membrane and help to avoid intracellular ice formation (Nahar et al., 2015). They reported that exogenous Spd improved the biochemical and physiological status of low temperature affected mung bean seedlings. Spd protected chl from oxidative damage or breakdown. Reduction of oxidative damage in mung bean seedlings was the prime advantageous effect among the studied parameters of their experiment, which was due to the enhanced AsA-GSH cycle components including AsA and GSH contents and activities of APX, MDHAR, DHAR, and GR. The activity of GPX also increased by Spd pretreatment of LT affected seedlings.

Melatonin is another ubiquitous chemical involved in regulation of plant growth and responses to abiotic

TABLE 21.2
Role of Phytoprotectants in Alleviation of Chilling Stress

Phytoprotectants	Crop	Effect	References
Melatonin -N-acetyl-5-methoxytryptamine	*Arabidopsis thaliana*	Up-regulation of transcription factors like CBFs, DREBs, CAMTA1, ZAT10, and ZAT12 following cold stress	Bajwa et al. (2014)
Spermidine	*Vitis vinifera* *Zea mays* *Cucumis sativus*	Reduced chilling injury and maintain of fruit quality and shelf life Improves cold tolerance Improve antioxidative response during cold stress	Harindra Champa et al. (2015), Saeidnejad et al. (2012), Zhang et al. (2009)
Putrescine	*Nicotiana tabacum* *Cucumis sativus*	Increase germination percentage, germination index, seedling length, and dry weight Improves cold tolerance	Xu et al. (2011), Zhang et al. (2009)
Cinnamic acid	*Cucumis sativus*	Increases superoxide dismutase catalase and protects plant from oxidative injury during chilling stress	Li et al. (2011)
Abscisic acid	*Chorispora bungeana* *Cicer arietinum*	Improves cold stress tolerance	Liu et al. (2011), Kumar et al. (2008)
ABA1 mimic (AM1)	*Cynodon dactylon*	Improves cold tolerance	Cheng et al. (2016)
Salicylic acid	Eggplant *Hordeum vulgare*	Protect plants from chilling stress	Chen et al. (2011), Mutlu et al. (2016)
Glycine betaine	*Lycopersicon esculentum*	Enhance catalase expression and catalase activity	Eung-Jun et al. (2006)
Gallic acid	Glycine max	Protect plants from chilling stress	Yildiztugay et al. (2017)
24-Epibrassinolide	Peach fruit	Protect plants from chilling stress and increase phenolic and proline metabolisms	Gao et al. (2016)

stresses. It has been reported recently that application of melatonin can induce cold tolerance not on at the site of application, but also at distal sites (Li et al., 2017). However, in their study, they observed that exogenous melatonin promoted cold-induced up-regulation of a set of regulatory genes in signal transduction and transcriptional regulation in leaves, but not in roots. They suggested that this may be due to divergence in gene expression between leaves and roots during cold acclimation, since 86% of cold-induced genes are not shared between rot and leaves (Liu et al., 2010). Melatonin may play important roles in the regulation of gene networks of leaves during cold stress. It has been shown that melatonin can activate defense systems by regulating the expression of cold-responsive genes such as ZAT10, ZAT 12, CBFs, COR15, and CAMTA1 (Doherty et al., 2009; Shi and Chan, 2014; Teige et al., 2004)

Exogenous application of other phenolic compounds such as salicylic and gallic acid, which have strong antioxidant activities, have also been shown to induce tolerance against abiotic stresses including low temperature. Yildiztugay et al. (2017) reported that application of gallic acid reduced cold injury in tomato leaves by scavenging of ROS through antioxidant mechanisms

21.6 CONCLUSION

It is thus clear that chilling stress, like other abiotic stresses causes metabolic disbalance in the plant system leading to several physiological and biochemical factors. Depending on several external and internal factors, plants have evolved cold-adaptive mechanisms which makes them tolerant to the cold to some extent (Figure 21.1). Cold acclimation processes are when the plant is initially exposed to a slightly higher temperature and then the freezing temperature upregulates several metabolic processes which confer tolerance. Besides constitutive cold acclimation processes, cold tolerance can also be induced in the plant by either treatment with microbes or phytoprotectants. Use of plant growth promoting bacteria, which are themselves cold tolerant for induction of cold tolerance, in bacteria is now being considered very useful because such bacteria, besides promoting the growth of plants, also provide them the ability to survive under freezing conditions. Several species of Pseudomonas have been tested and have shown a positive response. Besides the use of phytoprotectant chemicals such as spermidine, putrescine, melatonin, glycine betaine, salicylic acid, and so on have also shown the potential of cold acclimation in plants.

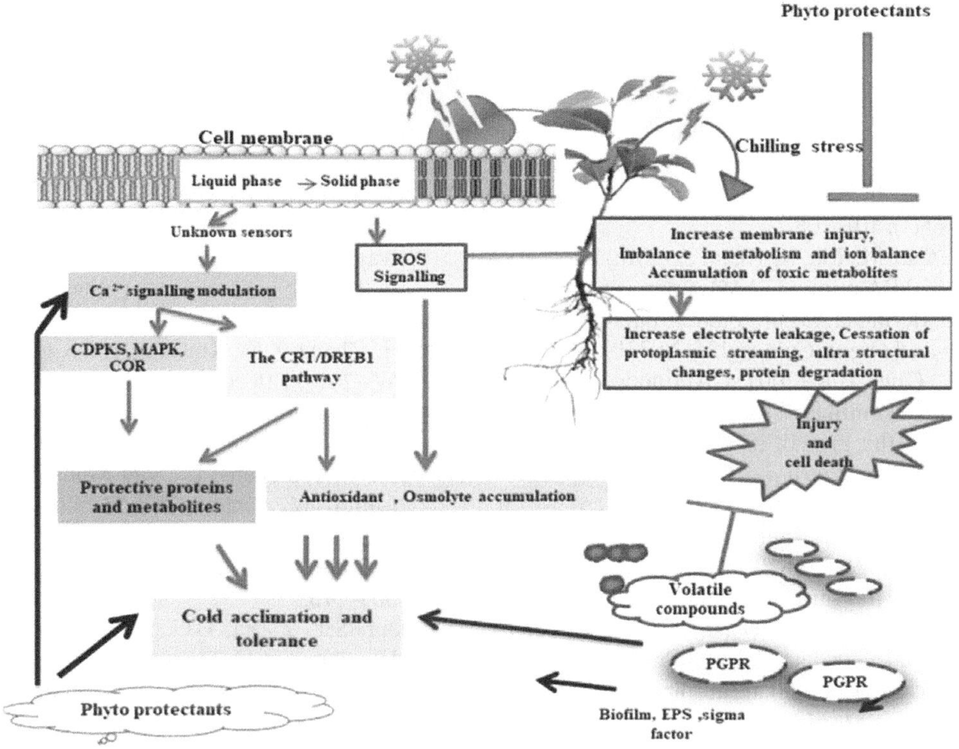

FIGURE 21.1 Schematic representation of events during chilling stress in plants and its amelioration by microbes and phytoprotectants.

REFERENCES

Abdrakhamanova, A., Wang, Q.Y., Khokhlova, L., Nick, P. (2003) Is microtubule disassembly a trigger for cold acclimation? *Plant & Cell Physiology* 44(7): 676–686.

Airaki, M., Leterrier, M., Mateos, R.M., Valderrama, R., Chaki, M., Barroso, J.B., Del Río, L.A., Palma, J.M., Corpas, F.J. (2012) Metabolism of reactive oxygen species and reactive nitrogen species in pepper (*Capsicum annuum* L.) plants under low temperature stress. *Plant, Cell and Environment* 35(2): 281–295.

Alcázar, R., Altabella, T., Marco, F., Bortolotti, C., Reymond, M., Koncz, C., Carrasco, P., Tiburcio, A.F. (2010) Polyamines: Molecules with regulatory functions in plant abiotic stress tolerance. *Planta* 231(6): 1237–1249.

Bajwa, V.S., Shukla, M.R., Sherif, S.M., Murch, S.J., Saxena, P.K. (2014) Role of melatonin in alleviating cold stress in *Arabidopsis thaliana*. *Journal of Pineal Research* 56(3): 238–245.

Barka, E., Nowk, J., Clement, C. (2006) Enhancement of chilling resistance of inoculated grapevine plantlets with plant growth promoting rhizobacteria *Burkholderia phytofermans* strain PsJN. *Applied and Environmental Microbiology* 72: 7246–7252.

Bisht, S.C., Joshi, G.K., Haque, S., Mishra, P.K. (2013) Cryotolerance strategies of pseudomonads isolated from the rhizosphere of Himalayan plants. *SpringerPlus* 2: 667.

Bita, C.E., Gerats, T. (2013) Plant tolerance to high temperature in a changing environment: Scientific fundamentals and production of heat stress-tolerant crops. *Frontiers in Plant Science* 4: 273.

Boguszewska, D., Zagdanska, B. (2012) ROS as signalling molecules and enzymes of plant response to unfavorable environmental conditions, oxidative stress-molecular mechanisms and biological effects. In *Oxidative Stress—Molecular Mechanisms and Biological Effects*, eds. Lushchak, V., Semchyshyn, H.M., 341–362. Janeza Trdine, Croatia: InTech.

Catala, R., Salinas, J. (2010) Temperature-perception, molecules and mechanism. *Journal of Applied Biomedicine* 8(4): 189–198.

Catalá, R., Santos, E., Alonso, J.M., Ecker, J.R., Martínez-Zapater, J.M., Salinas, J. (2003) Mutations in the Ca^{2+}/H^+ transporter CAX1 increase CBF/DREB1 expression and the cold-acclimation response in *Arabidopsis*. *The Plant Cell* 15(12): 2940–2951.

Chaikam, V., Karlson, D. (2008) Functional characterization of two cold shock domain proteins from *Oryza sativa*. *Plant, Cell and Environment* 31(7): 995–1006.

Chen, W.P., Li, P.H. (2002) Attenuation of reactive oxygen production during chilling in ABA-treated maize cultured cells. In *Plant Cold Hardiness*, eds. Li, C., Palva, E.T., 223–233. Dordrecht: Kluwer Academic Publishers.

Chen, W.J., Zhu, T. (2004) Networks of transcription factors with roles in environmental stress response. *Trends in Plant Science* 9(12): 591–596.

Chen, S., Zimei, L., Cui, J., Jiangang, D., Xia, X., Liu, D., Yu, J. (2011) Alleviation of chilling-induced oxidative damage by salicylic acid pretreatment and related gene expression in eggplant seedlings. *Plant Growth Regulation* 65(1): 101–108.

Cheng, Z., Jin, R., Cao, M., Liu, X., Chan, Z. (2016) Exogenous application of ABA mimic 1 (AM1) improves cold stress tolerance in Bermudagrass (*Cynodon dactylon*). *Plant Cell, Tissue and Organ Culture* 125(2): 231–240.

Cheng, Z., Park, E., Glick, B.R. (2007) 1-Aminocyclopropape-1-carboxylate deaminase from *Pseudomonas putida* UW4 facilitate the growth of canola in the presence of sat. *Canadian Journal of Microbiology* 53(7): 912–918.

Chinnusamy, V., Ohta, M., Kanrar, S., Lee, B.H., Hong, X., Agarwal, M., Zhu, J.K. (2003) ICE1: A regulator of cold-induced transcriptome and cold tolerance in *Arabidopsis. Genes & Development* 17(8): 1043–1054.

Chinnusamy, V., Zhu, J., Zhu, J.K. (2006) Gene regulation during cold acclimation in plants. *Physiologia Plantarum* 126(1): 52–61.

Clarke, S.M., Mur, L.A., Wood, J.E., Scott, I.M. (2004) Salicylic acid dependent signaling promotes basal thermotolerance but is not essential for acquired thermotolerance in *Arabidopsis thaliana. The Plant Journal: for Cell and Molecular Biology* 38(3): 432–447.

Cuevas, J.C., Lopez-Cobollo, R., Alcazar, R., Zarza, X., Koncz, C., Altabella, T., Salinas, J., Tiburcio, A.F., Ferrando, A. (2008) Putrescine is involved in *Arabidopsis* freezing tolerance and cold acclimation by regulating abscisic acid levels in response to low temperature. *Plant Physiology* 148(2): 1094–1105.

Doherty, C.J., Van Buskirk, H.A., Myers, S.J., Thomashow, M.F. (2009) Roles for *Arabidopsis* CAMTA transcription factors in cold regulated gene expression and freezing tolerance. *The Plant Cell* 21(3): 972–984.

Dong, C.H., Agarwal, M., Zhang, Y., Xie, Q., Zhu, J.K. (2006) The negative regulator of plant cold responses, HOS1, is a RING E3 ligase that mediates the ubiquitination and degradation of ICE1. *Proceedings of the National Academy of Sciences of the United States of America* 103(21): 8281–8286.

Du, L., Poovaiah, B.W. (2005) Ca^{2+}/calmodulin is critical for brassinosteroid biosynthesis and plant growth. *Nature* 437(7059): 741–745.

Egamberdiyeva, D., Höflich, G. (2003) Influence of growth-promoting bacteria on the growth of wheat in different soils and temperatures. *Soil Biology and Biochemistry* 35(7): 973–978.

Fernandez, O., Theocharis, A., Bordiec, S., Feil, R., Jacquens, L., Clément, C., Fontaine, F., Barka, E.A. (2012) *Burkholderia phytofirmans* PsJN acclimates grapevine to cold by modulating carbohydrate metabolism. *Molecular Plant-Microbe Interactions: MPMI* 25(4): 496–504.

Fourrier, N., Bédard, J., Lopez-Juez, E., Barbrook, A., Bowyer, J., Jarvis, P., Warren, G., Thorlby, G. (2008) A role for sensitive to freezing in protecting chloroplasts against freeze induced damage in *Arabidopsis. The Plant Journal: for Cell and Molecular Biology* 55(5): 734–745.

Fowler, S., Thomashow, M.F. (2002) *Arabidopsis* transcriptome profiling indicates that multiple regulatory pathways are activated during cold acclimation in addition to the CBF cold response pathway. *The Plant Cell* 14(8): 1675–1690.

Gao, H., Zhang, Z.K., XinGang, L.v., Cheng, N., Peng, B., Cao, W. (2016) Effect of 24-epibrassinolide on chilling injury of peach fruit in relation to phenolic and proline metabolisms. *Postharvest Biology and Technology* 111: 390–397.

Gechev, T., Willekens, H., Van Montagu, M., Inzé, D., Van Camp, W., Toneva, V., Minkov, I. (2003) Different responses of tobacco antioxidant enzymes to light and chilling stress. *Journal of Plant Physiology* 160(5): 509–515.

Gilmour, S.J., Zarka, D.G., Stockinger, E.J., Salazar, M.P., Houghton, J.M.,Thomashow, M.F. (1998) Low temperature regulation of the *Arabidopsis* CBF family of AP2 transcriptional activators as an early step in cold-induced COR gene expression. *The Plant Journal: for Cell and Molecular Biology* 16(4): 433– 442.

Grennan, A.K. (2006) Abiotic stress in rice. An "omic" approach. *Plant Physiology* 140(4): 1139–1141.

Groppa, M.D., Benavides, M.P. (2008) Polyamines and abiotic stress: Recent advances. *Amino Acids* 34(1): 35–45.

Gulati, A.P., Vyas, P.R., Kasana, R.C. (2009) Plant growth-promoting and rhizosphere-competent *Acinetobacter rhizosphaerae* strain BIHB 723 from the cold deserts of the Himalayas. *Current Microbiology* 58: 71–377.

Haldiman, P. (1998) Low growth temperature induced changes to pigment composition and photosynthesis in *Zea mays* genotypes differing in chilling sensitivity. *Plant, Cell and Environment* 21(2): 200–208.

Harindra Champa, W.A., Gilla, M.I.S., Mahajanb, B.V.C., Arora, N.K. (2015) Exogenous treatment of spermine to maintain quality and extend postharvest life of table grapes (*Vitis vinifera* L.) cv. flame seedless under low temperature storage. *Food Science and Technology* 60: 412–419.

Hasanuzzaman, M., Hossain, M.A., Teixeira da Silva, J.A., Fujita, M. (2012) Plant responses and tolerance to abiotic oxidative stress: Antioxidant defences is a key factor. In *Crop Stress and Its Management: Perspectives and Strategies*, eds. Bandi, V., Shanker, A.K., Shanker, C., Mandapaka, M., 261–316. Berlin: Springer.

Hasanuzzaman, M., Nahar, K., Alam, M.M., Roychowdhury, R., Fujita, M. (2013) Physiological, biochemical, and molecular mechanisms of heat stress tolerance in plants. *International Journal of Molecular Sciences* 14(5): 9643–9684.

Hsieh, T.H., Lee, J.T., Yang, P.T., Chiu, L.H., Charng, Y.Y., Wang, Y.C., Chan, M.T. (2002) Heterology expression of the *Arabidopsis* C-repeat/dehydration response element binding factor 1 gene confers elevated tolerance to chilling and oxidative stresses in transgenic tomato. *Plant Physiology* 129(3): 1086–1094.

Ishikawa, M., Yoshida, S. (1985) Seasonal changes in plasma membranes and mitochondria isolated from Jerusalem artichoke tubers: Possible relationship to cold hardiness. *Plant and Cell Physiology* 26: 1331–1344.

Juntilla, O., Robberecht, R. (1999) Ecological aspects of cold-adapted plants with special emphasis on environmental control of cold hardening and dehardening. In *Cold-Adapted Organisms – Ecology, Physiology, Enzymology and Molecular Biology*, eds. Margesin, R., Schinner, F., 57–77. Berlin, Germany: Springer-Verlag.

Kacperska, A. (1999) Plant response to low temperature: Signaling pathways involved in plant acclimation. In *Cold-Adapted Organisms – Ecology, Physiology, Enzymology and Molecular Biology*, eds. Margesin, R., Schinner, F., 79–103. Berlin, Germany: Springer-Verlag.

Kawamura,Y., Uemura, M. (2003) Mass spectrometric approach for identifying putative plasma membrane proteins of *Arabidopsis* leaves associated with cold acclimation. *The Plant Journal: for Cell and Molecular Biology* 36(2): 141–154.

Kim, M.H., Sasaki, K., Imai, R. (2009) Cold shock domain protein 3 regulates freezing tolerance in *Arabidopsis thaliana*. *The Journal of Biological Chemistry* 284(35): 23454–23460.

Kim, W.Y., Jung, H.J., Kwak, K.J., Kim, M.K., Oh, S.H., Han, Y.S., Kang, H. (2010) The *Arabidopsis* U12-type spliceosomal protein U11/U12-31K is involved in U12 intron splicing via RNA chaperone activity and affects plant development. *The Plant Cell* 22(12): 3951–3962.

Knight, H., Brandt, S., Knight, M.R. (1998) A history of stress alters drought calcium signalling pathways in *Arabidopsis*. *The Plant Journal: for Cell and Molecular Biology* 16(6): 681–687.

Knight, M.R., Campbell, A.K., Smith, S.M., Trewavas, A.J. (1991) Transgenic plant aequorin reports the effects of touch and cold-shock and elicitors on cytoplasmic calcium. *Nature* 352(6335): 524–526.

Knight, H., Zarka, D.G., Okamoto, H., Thomashow, M.F., Knight, M.R. (2004) Abscisic acid induces CBF gene transcription and subsequent induction of cold-regulated genes via the CRT promoter element. *Plant Physiology* 135(3): 1710–1717.

Kocsy, G.,Galiba, G., Brunold, C. (2001) Role of glutathione in adaptation and signalling during chilling and cold acclimationin plants. *Physiologia Plantarum* 113(2): 158–164.

Kosová, K., Vítámvás, P., Prášil, I.T. (2007) The role of dehydrins in plant response to cold. *Biologia Plantarum* 51(4): 601–617.

Kovacs, D., Kalmar, E., Torok, Z., Tompa, P. (2008) Chaperone activity of ERD10 and ERD14, two disordered stress-related plant proteins. *Plant Physiology* 147(1): 381–390.

Kumar, R.R., Goswami, S., Singh, K., Rai, G.K., Rai, R.D. (2013a) Modulation of redox signal transduction in plant system through induction of free radical/ROS scavenging redox-sensitive enzymes and metabolites. *Australian Journal of Crop Science* 7: 1744–1751.

Kumar, S., Kaur, G., Nayyar, H. (2008) Exogenous application of abscisic acid improves cold tolerance in chickpea (*Cicer arietinum* L.). *Journal of Agronomy and Crop Science* 194(6): 449–456.

Kumar, S., Kaushal, N., Nayyar, H., Gaur, P. (2013b) Abscisic acid induces heat tolerance in chickpea (*Cicer arietinum* L.) seedlings by facilitated accumulation of osmoprotectants. *Acta Physiology Plantarum* 34: 1651–1658.

Li, H., Chang, J., Zheng, J., Dong, Y., Liu, Q., Yang, X., Wei, C., *et al.* (2017) Local melatonin application induces cold tolerance in distant organs of *Citrullus lanatus* L. via long distance transport. *Scientific Reports* 7: 40858.

Li, X., Jiang, H., Liu, F., Cai, J., Dai, T., Cao, W., Jiang, D. (2013) Induction of chilling tolerance in wheat during germination by pre-soaking seed with nitric oxide and gibberellins. *Plant Growth Regulation* 71(1): 31–40.

Li, B., Takahashi, D., Kawamura, Y., Uemura, M. (2012) Comparison of plasma membrane proteomic changes of *Arabidopsis* suspension cultured cells (T87 line) after cold and ABA treatment in association with freezing tolerance development. *Plant and Cell Physiology* 53(3): 543–554.

Li, W., Wang, R., Li, M., Li, L., Wang, C., Welti, R., Wang, X. (2008) Differential degradation of extraplastidic and plastidic lipids during freezing and post-freezing recovery in *Arabidopsis thaliana*. *The Journal of Biological Chemistry* 283(1): 461–468.

Li, Q., Yu, B., Gao, Y., Dai, A.H., Bai, J.G. (2011) Cinnamic acid pretreatment mitigates chilling stress of cucumber leaves through altering antioxidant enzyme activity. *Journal of Plant Physiology* 168(9): 927–934.

Liu, L.Y., Duan, L.S., Zhang, J.C., Zhang, Z.X., Mi, G., Ren, H.Z. (2010) Cucumber (*Cucumis sativus* L.) overexpressing cold-induced transcriptome regulator ICE1 exhibits changed morphological characters and enhances chilling tolerance. *Scientia Horticulturae* 124(1): 29–33.

Liu, Y., Jiang, H., Zhao, Z., An, L. (2011) Abscisic acid is involved in brassinosteroids-induced chilling tolerance in the suspension cultured cells from *Chorispora bungeana*. *Journal of Plant Physiology* 168(9): 853–862.

Liu, H.H., Tian, X., Li, Y.J., Wu, C.A., Zheng, C.C. (2008) Microarray based analysis of stress-regulated microRNAs in *Arabidopsis thaliana*. *RNA* 14(5): 836–843.

Matteucci, M., D'Angeli, S., Errico, S., Lamanna, R., Perrotta, G., Altamura, M.M. (2011) Cold affects the transcription of fatty acid desaturases and oil quality in the fruit of *Olea europaea* L. genotypes with different cold hardiness. *Journal of Experimental Botany* 62(10): 3403–3420.

Mauger, J.P. (2012) Role of the nuclear envelope in calcium signalling. *Biology of the Cell* 104(2): 70–83.

McKersie, B.D., Leshem, Y.Y. (1994) *Stress and Stress Coping in Cultivated Plants*. Dordrecht, The Netherlands: Kluwer Academic Publishers.

Minami, A., Fujiwara, M., Furuto, A., Fukao, Y., Yamashita, T., Kamo, M., Kawamura, Y., Uemura, M. (2009) Alterations in detergent-resistant plasma membrane microdomains in *Arabidopsis thaliana* during cold acclimation. *Plant and Cell Physiology* 50(2): 341–359.

Mishra, P.K., Bisht, S.C., Ruwari, P., Selvakumar, G., Joshi, G.K., Bisht, J.K., Bhatt, J.C., Gupta, H.S. (2011) Alleviation of cold stress in inoculated wheat (*Triticum aestivum* L.) seedlings with psychrotolerant pseudomonads from NW Himalayas. *Archives of Microbiology* 193(7): 497–513. doi:10.1007/s00203-011-0693-x.

Mishra, P.K., Mishra, S., Selvakumar, G., Bisht, S.C., Bisht, J.K., Kundu, S., Gupta, H.S. (2008) Characterization of a psychrotolerant plant growth promoting *Pseudomonas* sp. strain PGERs17 (MTCC 9000) isolated from north western Indian Himalayas. *Annals of Microbiology* 58(4): 561–568.

Mishra, P.K., Mishra, S., Bisht, S.C., Selvakumar, G., Kundu, S., Bisht, J.K., Gupta, H.S. (2009) Isolation, molecular characterization and growth-promotion activities of a cold tolerant bacterium *Pseudomonas* sp. NARs9 (MTCC9002) from the Indian Himalayas. *Biological Research* 42(3): 305–313.

Miura, K., Furumoto, T. (2013) Cold signaling and cold response in plants. *International Journal of Molecular Sciences* 14(3): 5312–5337.

Mogk, A., Schlieker, C., Friedrich, K.L., Schönfeld, H.J., Vierling, E., Bukau, B. (2003) Refolding of substrates bound to small Hsps relies on a disaggregation reaction mediated most efficiently by ClpB/DnaK. *The Journal of Biological Chemistry* 278(33): 31033–31042.

Mutlu, S., Atici, Ö., Nalbantoğlu, B., Mete, E. (2016) Exogenous salicylic acid alleviates cold damage by regulating antioxidative system in two barley (*Hordeum vulgare* L.) cultivars. *Frontiers in Life Science*: 9(2): 99–109.

Nabti, E., Sahnoune, M., Ghoul, M., Fischer, D., Hofmann, A., Rothballer, M., Schmid, M., Hartmann, A. (2010) Restoration of growth of durum wheat (*Triticum durum* var. waha) under saline conditions due to inoculation with the rhizosphere bacterium *Azospirillum brasilense* NH and extracts of the marine alga *Ulva lactuca*. *Journal of Plant Growth Regulation* 29(1): 6–22.

Nahar, K., Hasanuzzaman, M., Alam, M.M., Fujita, M. (2015) Exogenous spermidine alleviates low temperature injury in mung bean (*Vigna radiata* L.) seedlings by modulating ascorbate-glutathione and glyoxalase pathway. *The International Journal of Molecular Sciences* 16(12): 30117–30132.

Nakaminami, K., Karlson, D.T., Imai, R. (2006) Functional conservation of cold shock domains in bacteria and higher plants. *Proceedings of the National Academy of Sciences of the United States of America* 103(26): 10122–10127.

Nakashima, K., Yamaguchi-Shinozaki, K. (2006) Regulons involved in osmotic stress-responsive and cold stress-responsive gene expression in plants. *Physiologia Plantarum* 126(1): 62–71.

Nakayama, K., Okawa, K., Kakizaki, T., Inaba, T. (2008) Evaluation of the protective activities of a late embryogenesis abundant (LEA) related protein, Cor15am, during various stresses in vitro. *Bioscience, Biotechnology, and Biochemistry* 72(6): 1642–1645.

Neill, S.J., Desikan, R., Clarke, A., Hurst, R.D., Hancock, J.T. (2002) Hydrogen peroxide and nitric oxide as signalling molecules in plants. *Journal of Experimental Botany* 53(372): 1237–1247.

Nick, P. (2000) Control of the response to low temperatures. In *Plant Microtubules: Potential for Biotechnology*, ed. Nick, P., 121–135. Berlin, Germany: Springer.

Novillo, F., Alonso, J.M., Ecker, J.R., Salinas, J. (2004) CBF2/DREB1C is a negative regulator of CBF1/DREB1B and CBF3/DREB1A expression and plays a central role in stress tolerance in *Arabidopsis. Proceedings of the National Academy of Sciences of the United States of America* 101(11): 3985–3990.

Okawa, K., Nakayama, K., Kakizaki, T., Yamashita, T., Inaba, T. (2008) Identification and characterization of Cor413im proteins as novel components of the chloroplast inner envelope. *Plant, Cell and Environment* 31(10): 1470–1483.

Pandey, A., Trivedi, P., Kumar, B., Palni, L.M. (2006) Characterization of a phosphate solublizing and antagonistic strain of *Pseudomonas putida* (B0) isolated from a sub-alpine location in the Indian central Himalaya. *Current Microbiology* 53(2): 102–107.

Park, E.J., Jeknic, Z., Chen, T.H. (2006) Exogenous application of glycine betaine increases chilling tolerance in tomato plants. *Plant and Cell Physiology* 47(6): 706–714.

Peng, Y., Arora, R., Li, G., Wang, X., Fessehaie, A. (2008) *Rhododendron catawbiense* plasma membrane intrinsic proteins are aquaporins, and their over-expression compromises constitutive freezing tolerance and cold acclimation ability of transgenic *Arabidopsis* plants. *Plant, Cell and Environment* 31(9): 1275–1289.

Poupin, M.J., Timmermann, T., Vega, A., Zuñiga, A., González, B. (2013) Effects of the plant growth-promoting bacterium *Burkholderia phytofirmans* PsJN throughout the life cycle of *Arabidopsis thaliana. PLOS ONE* 8(7): e69435.

Principe, A., Alvarez, F., Castro, M.G., Zacchi, L.F., Fischer, S.E., Mori, G.B., Jofré, E. (2007) Biocontrol and PGPR feature in native strains isolated from saline soils of Argentina. *Current Microbiology* 55(4): 314–322.

Rajashekar, C.B. (2000) Cold response and freezing tolerance in plants. In *Plant–Environment Interactions*, 2nd edn, ed. Wilkinson, R.E., 321–341. New York, NY: Marcel Dekker, Inc.

Renaut, J., Hausman, J.F., Wisniewski, M.E. (2006) Proteomics and low temperature studies: Bridging the gap between gene expression and metabolism. *Physiologia Plantarum* 126(1): 97–109.

Ruelland, E., Zachowski, A. (2010) How plants sense temperature. *Environmental and Experimental Botany* 69(3): 225–232.

Saeidnejad, A.H., Pouramir, F., Naghizadeh, M. (2012) Improving chilling tolerance of maize seedlings under cold conditions by spermine application. *Notulae Scientia Biologicae* 4(3): 110–117.

Saijo, Y., Hata, S., Kyozuka, J., Shimamoto, K., Izui, K. (2000) Over-expression of a single Ca^{2+}-dependent protein kinase confers both cold and salt/drought tolerance on rice plants. *The Plant Journal: for Cell and Molecular Biology* 23(3): 319–327.

Saltveit, M.E. (2000) Discovery of chilling injury. In *Discoveries in Plant Bology*, 3rd edn, eds. Kung, S.D., Yang, S.F., 423–448. Singapore: World Scientific Publishing.

Sangwan, V., Orvar, B.L., Beyerly, J., Hirt, H., Dhindsa, R.S. (2002a) Opposite changes in membrane fluidity mimic cold and heat stress activation of distinct plant MAP kinase pathways. *The Plant Journal* 31(5): 629–638.

Sangwan, V., Örvar, B.L., Dhindsa, R.S. (2002b) Early events during low temperature signaling. In *Plant Cold Hardiness*, eds. Li, C., Palva, E.T., 43–53. Dordrecht, The Netherlands: Kluwer Academic Publishers.

Sasaki, K., Kim, M.H., Imai, R. (2007) *Arabidopsis* COLD SHOCK DOMAIN PROTEIN2 is a RNA chaperone that is regulated by cold and developmental signals. *Biochemical and Biophysical Research Communications* 364(3): 633–638.

Seki, M., Kamei, A., Yamaguchi-Shinozaki, K., Shinozaki, K. (2003) Molecular responses to drought, salinity and frost: Common and different paths for plant protection. *Current Opinion in Biotechnology* 14(2): 194–199.

Selvakumar, G.P., Joshi, P., Suyal, P.K., Mishra, P.K., Joshi, G.K., Bisht, J.K., Bhatt, J.C., Gupta, H.S. (2011) *Pseudomonas lurida* M2RH3 (MTCC 9245), a psychrotolerant bacterium from the Uttarakhand Himalayas, solubilizes phosphate and promotes wheat seedling growth. *World Journal of Microbiology and Biotechnology* 27(5): 1129–1135.

Selvakumar, G.S., Kundu, P., Joshi, S., Nazim, S., Gupta, A.D., Gupta, H.S. (2010) Growth promotion of wheat seedlings by *Exiguobacterium acetylicum* 1P (MTCC 8707) a cold tolerant bacterial strain from Uttarakhand Himalayas. *Indian Journal of Microbiology* 50: 50–56.

Selvakumar, G., Kundu, S., Joshi, P., Nazim, S., Gupta, A.D., Mishra, P.K., Gupta, H.S. (2008a) Characterization of a cold-tolerant plant growth–promoting bacterium Pantoea dispersa 1A isolated from a sub-alpine soil in the north western Indian Himalayas. *World Journal of Microbiology and Biotechnology* 24(7): 955–960.

Selvakumar, G., Mohan, M., Kundu, S., Gupta, A.D., Joshi, P., Nazim, S., Gupta, H.S. (2008b) Cold tolerance and plant growth promotion potential of *Serratia marcescens* strain SRM (MTCC 8708) isolated from flowers of summer squash (*Cucurbita pepo*). *Letters in Applied Microbiology* 46(2): 171–175.

Seo, P.J., Kim, M.J., Park, J.Y., Kim, S.Y., Jeon, J., Lee, Y.H., Kim, J., Park, C.M. (2010) Cold activation of a plasma membrane-tethered NAC transcription factor induces a pathogen resistance response in *Arabidopsis*. *The Plant Journal: for Cell and Molecular Biology* 61(4): 661–671.

Shi, H., Chan, Z. (2014) The cysteine2/histidine2-type transcription factor zinc finger of *Arabidopsis thaliana* 6-activated C-REPEAT-BINDING FACTOR pathway is essential for melatonin-mediated freezing stress resistance in *Arabidopsis*. *Journal of Pineal Research* 57(2): 185–191.

Shinozaki, K., Yamaguchi-Shinozaki, K. (2000) Molecular responses to dehydration and low temperature: Differences and cross-talk between two stress signaling pathways. *Current Opinion in Plant Biology* 3(3): 217–223.

Steponkus, P.L. (1984) Role of the plasma membrane in freezing injury and cold acclimation. *Annual Review of Plant Physiology* 35(1): 543–584.

Su, F., Jacquard, C., Villaume, S., Michel, J., Rabenoelina, F., Clément, C., Barka, E.A., Dhondt-Cordelier, S., Vaillant-Gaveau, N. (2015) *Burkholderia phytofirmans* PsJN reduces impact of freezing temperatures on photosynthesis in *Arabidopsis thaliana*. *Frontiers in Plant Science* 6: 810.

Subramanian, P., Kim, K., Krishnamoorthy, R., Mageswari, A., Selvakumar, G., Sa, T. (2016) Cold stress tolerance in psychrotolerant soil bacteria and their conferred chilling resistance in tomato (*Solanum lycopersicum* mill.) under low temperatures. *PLOS ONE* 11(8): e0161592.

Subramanian, P., Melvin Joe, M.M., Yim, W., Hong, B., Tipayno, S.C., Saravanan, V.S., Yoo, J., et al. (2011) Psychrotolerance mechanisms in cold-adapted bacteria and their perspectives as plant growth-promoting bacteria in temperate agriculture. *Korean Journal of Soil Science and Fertilizer* 44(4): 625–636.

Sun, W., Van Montagu, M., Verbruggen, N. (2002) Small heat shock proteins and stress tolerance in plants. *Biochimica et Biophysica Acta* 1577(1): 1–9.

Suzuki, N., Koussevitzky, S., Mittler, R., Miller, G. (2012) ROS and redox signaling in the response of plants to abiotic stress. *Plant, Cell and Environment* 35(2): 259–270.

Suzuki, N., Mittler, R. (2006) Reactive oxygen species and temperature stresses: A delicate balance between signalling and destruction. *Physiologia Plantarum* 126(1): 45–51.

Svensson, J.T., Crosatti, C., Campoli, C., Bassi, R., Stanca, A.M., Close, T.J., Cattivelli, L. (2006) Transcriptome analysis of cold acclimation in barley albina and xantha mutants. *Plant Physiology* 141(1): 257–270.

Takahashi, T., Kakehi, J.I. (2010) Polyamines: Ubiquitous polycations with uniqueroles in growth and stress responses. *Annals of Botany* 105(1): 1–6.

Takahashi, D., Kawamura, Y., Yamashita, T., Uemura, M. (2013) Detergent-resistant plasma membrane proteome in oat and rye: Similarities and dissimilarities between two monocotyledonous plants. *Journal of Proteomic Research* 11: 1654–1665.

Teige, M., Scheikl, E., Eulgem, T., Dóczi, R., Ichimura, K., Shinozaki, K., Dangl, J.L., Hirt, H. (2004) The MKK2 pathway mediates cold and salt stress signaling in *Arabidopsis*. *Molecular Cell* 15(1): 141–152.

Theocharis, A., Bordiec, S., Fernandez, O., Paquis, S., Dhondt-Cordelier, S., Baillieul, F., Clément, C., Barka, E.A. (2011) *Burkholderia phytofirmans* strain PsJN

primes *Vitis vinifera* L. and confers a better tolerance to low non-freezing temperatures. *Molecular Plant–Microbe Interaction* 25: 241–249.
Thomashow, M.F. (2010) Molecular basis of plant cold acclimation: Insights gained from studying the CBF cold response pathway. *Plant Physiology* 154(2): 571–577.
Timperio, A.M., Egidi, M.G., Zolla, L. (2008) Proteomics applied on plant abiotic stresses: Role of heat shock proteins (HSP). *Journal of Proteomics* 71(4): 391–411.
Tommasini, L., Svensson, J.T., Rodriguez, E.M., Wahid, A., Malatrasi, M., Kato, K., Wanamaker, S., Resnik, J., Close, T.J. (2008) Dehydrin gene expression provides an indicator of low temperature and drought stress: Transcriptome-based analysis of barley (*Hordeum vulgare* L.). *Functional and Integrative Genomics* 8(4): 387–405.
Townley, H.E., Knight, M.R. (2002) Calmodulin as a potential negative regulator of *Arabidopsis* COR gene expression. *Plant Physiology* 128(4): 1169–1172.
Uemura, M., Steponkus, P.L. (1999) Cold acclimation in plants: Relationship between the lipid composition and the cryostability of the plasma membrane. *Journal of Plant Research* 112(2): 245–254.
Uemura, M., Tominaga, Y., Nakagawara, C., Shigematsu, S., Minami, A., Kawamura, Y. (2006) Responses of the plasma membrane to low temperatures. *Physiologia Plantarum* 126(1): 81–89.
Uemura, M., Yoshida, S. (1984) Involvement of plasma membrane alterations in cold acclimation of winter rye seedlings (*Secale cereale* L. cv Puma). *Plant Physiology* 75(3): 818–826.
Van der Luit, A.H., Olivari, C., Haley, A., Knight,M.R., Trewavas, A.J. (1999) Distinct calcium signaling pathways regulate calmodulin gene expression in tobacco. *Plant Physiology* 121(3): 705–714.
Vaultier, M.N., Cantrel, C., Vergnolle, C., Justin, A.M., Demandre, C., Benhassaine-Kesri, G., Çiçek, D., Zachowski, A., Ruelland, E. (2006) Desaturase mutants reveal that membrane rigidification acts as a cold perception mechanism upstream of the diacylglycerol kinase pathway in *Arabidopsis* cells. *FEBS Letters* 580(17): 4218–4223.
Vogg, G., Heim, R., Gotschy, B., Beck, E., Hansen, J. (1998) Frost hardening and photosynthetic performance of Scots pine (*Pinus sylvestris* L.). II. Seasonal changes in the fluidity of thylakoid membranes. *Planta* 204(2): 201–206.
Wang, X., Li, W., Li, M., Welti, R. (2006) Profiling lipid changes in plant response to low temperatures. *Physiologia Plantarum* 126(1): 90–96.
Wang, W., Vinocur, B., Altman, A. (2003) Plant responses to drought, salinity and extreme temperatures: Towards genetic engineering for stress tolerance. *Planta* 218(1): 1–14.
Xia, X.J., Wang, Y.J., Zhou, Y.H., Tao, Y., Mao, W.H., Shi, K., Asami, T., Chen, Z., Yu, J.Q. (2009) Reactive oxygen species are involved in brassinosteroid-induced stress tolerance in cucumber. *Plant Physiology* 150(2): 801–814.
Xiong, L., Schumaker, K.S., Zhu, J.K. (2002) Cell signaling during cold, drought and salt stress. *Plant Cell* 14(Suppl): s165–s183.
Xiong, L., Zhu, J.K. (2001) Abiotic stress signal transduction in plants. Molecular and genetic perspectives. *Physiologia Plantarum* 112(2): 152–166.
Xu, S., Hu, J., Li, Y., Ma, W., Zheng, Y., Zhu, S. (2011) Chilling tolerance in *Nicotiana tabacum* induced by seed priming with putrescine. *Plant Growth Regulation* 63(3): 279–290.
Yadav, S.K. (2010) Cold stress tolerance mechanisms in plants: A review. *Agronomy for Sustainable Development* 30(3): 515–527.
Yamaguchi-Shinozaki, K., Shinozaki, K. (2006) Transcriptional regulatory networks in cellular responses and tolerance to dehydration and cold stresses. *Annual Review of Plant Biology* 57: 781–803.
Yamazaki, T., Kawamura, Y., Minami, A., Uemura, M. (2008) Calcium dependent freezing tolerance in *Arabidopsis* involves membrane resealing via synaptotagmin SYT1. *The Plant Cell* 20(12): 3389–3404.
Yildiztugay, E., Ozfindan-Konakiciz, C., Kucukoduk, M. (2017) Improvement of cold stress resistance via free radical scavenging ability and promoted water stress status and photosynthetic activity of gallic acid in soybean leaves. *Journal of Soil Science and Plant Nutrition* 17(2). doi:dx.doi.org/10.4067/S0718-95162017005000027.
Zhang, W.P., Jiang, B., Li, W.G., Song, H., Yu, Y., Chen, J. (2009) Polyamines enhance chilling tolerance of cucumber (*Cucumis sativus* L.) through modulating antioxidative system. *Scientia Horticulturae* 122(2): 200–208.
Zhang, S., Jiang, H., Peng, S., Korpelainen, H., Li, C. (2011) Sex-related differences in morphological, physiological, and ultrastructural responses of Populus cathayana to chilling. *Journal of Experimental Botany* 62(2): 675– 686.
Zhou, J., Wang, J., Shi, K., Xia, X.J., Zhou, Y.H., Yu, J.Q. (2012) Hydrogen peroxide is involved in the cold acclimation-induced chilling tolerance of tomato plants. *Plant Physiology and Biochemistry: PPB* 60: 141–149.
Zhu, J.H., Dong, C.H., Zhu, J.K. (2007) Interplay between cold-responsive gene regulation, metabolism and RNA processing during plant cold acclimation. *Current Opinion in Plant Biology* 10(3): 290–295.
Zhu, J., Jeong, J.C., Zhu, Y., Sokolchik, I., Miyazaki, S., Zhu, J.K., Hasegawa, P.M., *et al.* (2008) Involvement of *Arabidopsis* HOS15 in histone deacetylation and cold tolerance. *Proceedings of the National Academy of Sciences of the United States of America* 105(12): 4945–4950.

22 Effect of Biostimulants on Plant Responses to Salt Stress

José Ramón Acosta-Motos, Pedro Diaz-Vivancos, Manuel Acosta, and José Antonio Hernandez

CONTENTS

22.1 INTRODUCTION

Salt stress causes osmotic stress as well as ion toxicity in plants, leading to an alteration of ionic homeostasis. In addition to toxic and osmotic effects, salt stress also induces oxidative stress at the subcellular level. These three factors can contribute to the detrimental effects produced by salinity in plants (Acosta-Motos et al., 2015a, b; Barba-Espín et al., 2011). Some authors have studied the exogenous application of different molecules to induce salt tolerance in different plant species. Different works, for instance, have shown that the use of humic acids, certain amino acids like proline (Pro) or glycine betaine (GB), hydrogen peroxide (H_2O_2), polyamines (PAs), salicylic acid (SA), and melatonin improve the response of plants to salt stress. In addition, the use of arbuscular mycorrhizal fungi is another strategy for improving salt tolerance. In this chapter, we have reviewed the use of these different strategies for improving the growth and adaptation of plants of agronomic interest in saline conditions.

22.2 HUMIC ACIDS AND FULVIC ACIDS

In agriculture, the term humic acid (HA) is often used to refer to both humic acids (HAs) and fulvic acids (FAs). HAs are present in soils and they are the most active part of organic matter. They are a mixture of complex organic molecules formed by the decomposition and oxidation of organic matter. This progressive process is called humification. Among other uses, these products are employed as raw materials for the preparation of agricultural biostimulants. HAs originated in diverse sources, such as peat or plant residues, although most HAs are obtained from leonardite. Due to the characteristics of leonardite, these HAs are considered to be of the best quality and are thought to have better agronomic properties. Leonardite is a humified plant substance that is very rich in organic matter and is at an intermediate state of transformation between peat and lignite. It comes from plant materials that have been buried for millions of years and it is usually found in the upper layers of open-cast lignite (coal) mines.

As mentioned above, when we talk about HAs, we also include FAs in this concept. These biomaterials differ in how they behave in basic and acidic media. Both HAs and FAs are soluble in basic media and this chemical property is used to extract them in liquid form using an alkaline extractant, usually potassium hydroxide. However, in an acidic media, the HAs precipitate, while the FAs remain soluble. This different behavior is the base of the official method of analysis for separating and quantifying the HAs and FAs of a product. In addition, HAs have a higher molecular weight, greater cation exchange capacity, and greater water retention capacity than FAs. Furthermore, HAs have a slower and more durable effect on the structure of the soil, while the FAs have a more rapid effect on the plant but are less persistent in the soil. HAs positively influence the fertility of soil by favoring microbial activity and performing various actions depending on the soil type. In general, they contribute to the unblocking of nutrients and act as natural complexing agents, facilitating the assimilation of the nutrients by the plant.

Applications of HAs and calcium nitrate have been found to significantly influence the growth of pepper seedlings in a concentration-dependent manner (Gulser et al., 2010). Humic acid (1000 and 2000 mg kg^{-1}) and calcium nitrate (50 mg kg^{-1}) applications increased fresh and dry leaf and root weight, stem diameter, and root and shoot length. However, higher rates of HAs (4000 mg kg^{-1}) and calcium nitrate (100 and 150 mg kg^{-1}) had no beneficial effects on pepper growth. Çimrin et al. (2010) described that HA treatment could ameliorate the deleterious effects of salt stress on pepper plants, offering an economical and simple way to reduce salinity problems in pepper production in moderately saline soil. The application of HAs significantly increased N, P, K, Ca, Mg, S, Mn, Fe, and Cu concentrations in the shoots and roots of pepper seedlings and decreased the Na concentrations in both plant organs. It can be concluded that high doses of HAs have positive effects on salt tolerance based on the plant growth parameters and nutrient contents.

In bean plants, Aydin et al. (2012) described that total chlorophyll and nitrate concentrations decreased with increasing salt doses. The addition of HAs to the saline soil significantly improved these bean plant variables affected by high salinity. Furthermore, the HAs also increased nitrates, nitrogen, and phosphorus in the plants and reduced soil electric conductivity, Pro and electrolyte leakage, therefore enhancing the plant root and shoot dry weight by allowing nutrients and water to be released to the plant as needed. The results suggest that HAs have great potential in alleviating the effects of salinity stress on plant growth in saline soils in arid and semiarid areas. In this study, HAs were highly effective soil conditioners regarding vegetable growth, improving crop tolerance and growth under saline conditions (Aydin et al., 2012).

Jarošová et al. (2016) described relatively high sensitivity to NaCl (100 mM) in barley plants, manifested by a considerable decline in plant growth, tissue water depletion, and high sodium accumulation. Humic acids typically increased the concentrations of organic metabolites [syringic acid, alanine, Pro, ascorbic acid, glutathione (GSH), and phytochelatin], whereas NaCl provoked the opposite effect (except in the case of Pro). The combined treatment ($NaCl^{+}$ HAs), however, mostly demonstrated the positive impact of HAs. Humic acids suppressed the NaCl-induced increase in Na, while the impact on other nutrients was not extensive. Moreover, foliar and hydroponic application HAs revealed similar mitigating effects on NaCl stress. Overall, these data indicate the potential of HAs to protect barley against NaCl stress by limiting Na uptake and positively impacting the number of certain metabolites.

In tomato, the fruit yield decreased significantly due to NaCl, but the yield was not affected after HA application (Korkmaz et al., 2016). In addition, HAs significantly increased the canopy (stem plus leaves) biomass under NaCl stress. At a low dose, HAs also increased root dry matter, but a negative effect was produced at higher doses. Under saline conditions, HAs increased root dry matter content and provided tomato plants with tolerance to NaCl stress. Humic acid treatments also improved fruit quality regarding dry matter content and the pH of the fruit juice and reduced the blossom end rot symptoms induced by salinity.

In cotton plants, the application of HAs has also been shown to enhance tolerance to NaCl stress (Rady et al., 2016). HA-treated plants showed improved photosynthetic efficiency, water use efficiency (WUE), nutritional status, and seed and fiber quality compared with untreated plants. Furthermore, the application of HAs to the soil led to significant decreases in leaf Na concentrations, total soluble sugars, and free Pro. HAs were determined to have a pronounced positive effect on the growth, yields, fiber quality, and WUE of salt-stressed cotton plants.

Matuszak-Slamani et al. (2017) studied the influence of molecular fractions of HA (HA $<$ 30 kDa and HA $>$ 30 kDa) on enhancing the tolerance of soybean seedlings to salt stress (50 mM NaCl). These authors demonstrated that HA $>$ 30 kDa was more effective than HA $<$ 30 kDa in alleviating the negative effects of salt stress. In conclusion, HAs have the potential to be used as a soil amendment for different plant species to overcome the adverse effects of soil salinity.

22.3 ARBUSCULAR MYCORRHIZAL FUNGI

The word mycorrhiza was derived from a Greek word MYKOS (mushroom) and a Latin word RHIZA (root). The German botanist Albert Bernard Frank first coined the term Mycorrhiza in 1985 to designate the association that took place between the hyphae of some soil fungi with the roots of the vast majority of higher plants. The fungus, by extending the root area, helps the plant by increasing its capacity to sustain itself physically in the soil, improving its resistance and adaptability. An arbuscular mycorrhiza is a type of endomycorrhiza in which the fungus penetrates the cortical cells of the roots of a vascular plant. Arbuscular mycorrhizae are characterized by the formation of unique structures, arbuscules, and vesicles of the Glomeromycota phylum fungi (Bolan, 2005). In this symbiotic association, where a dual organism is formed (Paul and Clark, 1989), the fungus helps the plant capture nutrients, such as phosphorus, sulfur, nitrogen, and soil micronutrients, and improves the plant's water relations, affording protection against pathogenic agents and abiotic stress conditions. The fungus-plant association is of considerable ecological importance since this interaction seems to play a key role in the succession of species in natural plant communities. In return, the fungus receives carbohydrates (mainly hexoses) needed for sustenance, which come from the photosynthesis of the plant.

It is believed that the development of symbiosis with arbuscular mycorrhizae played a crucial role in the initial colonization of the soil by plants and in the evolution of vascular plants (Brundrett, 2002). Arbuscular mycorrhizae are found in 80% of the families of the known vascular plants (Schüßler et al., 2001). The beneficial effects of mycorrhizae on the soil are closely related to their effects on plants. Mycorrhizae perform various functions in the soil that greatly increase its agro-productive potential and its possibilities of sustaining and maintaining different plant species. The mycorrhizae in the natural ecosystems have the following functions:

1. They prolong the root system of plants, which facilitates a greater physical retention of soil particles and limits the harmful effects of soil erosion.
2. Mycorrhizae regenerate degraded soils by facilitating the improvement of the soil structure, increasing the possibilities of moisture retention, aeration, and the decomposition of organic matter.
3. The presence of mycorrhizae in soils mobilizes a large amount of nutrients that were previously not available to plants, and thus increases soil fertility. In less fertile soils, more fungal structures are needed to achieve greater mycorrhizal efficiency.
4. Mycorrhizae improve the productive capacity of unproductive soils, such as those affected by desertification, wind erosion, salinization, and water stress.
5. Another interesting effect of mycorrhizae in soil is the role they play in relation to the ecosystem in which they develop. They interact with various microorganisms in the soil, for instance, establishing beneficial cooperation with some and competition with others (generally of pathogenic type). Mycorrhizae even interact with the microfauna of the rhizosphere (nematodes, aphids and mites, among others).
6. Mycorrhizae prolong the life of productive agricultural soils, contributing to a more diverse, economic, and ecological use of the soil.
7. In arid and semiarid zones, mycorrhizae can help symbiotic plants collect water to tolerate water stress or to cope with salt stress.

The application of arbuscular mycorrhizal fungi (AMF) could, therefore, offer a cost-effective alternative for counteracting the problem of salinity (Saxena et al., 2017). AMF establish a direct physical link between plant roots and soils and help host plants acquire mineral nutrients from soils, particularly under nutrient stress conditions; these fungi also modify the rhizosphere environment, thereby alleviating the adverse effects of salinity stress (Evelin et al., 2009; Jahromi et al., 2008; Smith and Read, 1997). AMF have been found to improve salt tolerance in different plant species, including lettuce, tomato, maize, acacia, sesbania, fenugreek, and citrus (Al-Karaki, 2000; Evelin et al., 2012, 2013; Feng and Zhang, 2002; Giri et al., 2003, 2007; Giri and Mukerji, 2004; Navarro et al., 2014; Ruiz-Lozano et al., 1996). Furthermore, AMF strongly influences the uptake of P under saline conditions (Giri et al., 2003; Ojala et al., 1983). AMF also allows extra-radical hyphae to explore more soil volume for improved absorption of other nutrients, particularly under stress conditions (Ruiz-Lozano and Azcón, 2000). Regarding this last point, other AMF (*G. iranicum* var. *tenuihypharum*) have been found to have a protective effect on soil and *Viburnum tinus* irrigated with reclaimed water. Good AMF-soil-plant interactions could make it possible to reuse reclaimed water, particularly when the roots are in a saline soil (Gómez-Bellot et al., 2015).

A high concentration of salts in soil may induce three different types of stresses: osmotic, ionic, and oxidative, all of which drastically affect plant growth and productivity (Acosta-Motos et al., 2017). Experiments carried out to understand the AMF-salinity interaction have revealed that mycorrhizal fungi reduce the negative effects of these stresses and promote plant growth (Evelin et al., 2009; Wu et al., 2010a, b; Zuccarini and Okurowska, 2008). It has widely been accepted, for instance, that AMF improves plant water use efficiency and nutrient uptake under saline conditions, thus helping reduce the negative impact of salt stress. Moreover,

AMF diminishes the detrimental effects of toxic ions on membrane permeability and cell organelles, maintains the level of compatible organic solutes, increases antioxidant production (both enzymatic and non-enzymatic), and positively controls the expression of salt-related genes. Researchers have presented several physiological, biochemical, and molecular approaches by which AM plants can alleviate salt stress (Augé et al., 2014; Evelin et al., 2009; Kumar et al., 2015; Ruiz-Lozano et al., 2012).

The mechanisms of salt stress alleviation in plants produced by AMF can be summarized as follows: i) improved ion uptake by roots, ion compartmentation, and ion transport into plant tissues, to maintain nutrient homeostasis; ii) greater accumulation of osmolytes such as Pro and glycine-betaine; iii) selective uptake or extrusion of salts by the roots; iv) improved uptake of water by the roots and water distribution to plant tissues, in which aquaporins play a key role; v) increased production of antioxidants, which help protect against oxidative stress by maintaining membrane structure and its integrity; vi) improved photosynthetic rates in leaves for optimal plant growth; vii) controlled root-to-shoot signaling of phytohormones; and viii) avoidance of ultrastructure damage. These strategies seem to develop integrated responses in a concerted manner to improve plant salinity tolerance. In sum, the use of AMF is an inexpensive and effective approach for coping with the deleterious effects of salinity and improving plant growth and production.

22.4 REACTIVE OXYGEN SPECIES METABOLISM UNDER SALT STRESS

ROS are produced under normal conditions as a by-product of aerobic metabolism in plants (Niu and Liao, 2016). However, it is well known that ROS can be over-generated under environmental stress conditions, including salinity (Hernández et al., 1993, 1995, 2001). Salt stress interferes with carbon metabolism and affects the different electron transport chains in different cell compartments as well as different redox reactions leading to ROS production at the subcellular level (Corpas et al., 1993; Hernández et al., 1993, 1995, 2001; Hossain and Dietz, 2016).

Although ROS are considered toxic molecules, they also play a role as signaling molecules. This role has been the subject of many studies within the last 25 to 30 years, particularly in the case of H_2O_2. For instance, several researchers have investigated the role of H_2O_2 in the regulation of different physiological processes like plant growth, seed dormancy, senescence, flowering, and stomatal closure (Barba-Espín et al., 2011; Gapper and Dolan, 2006; Ishibashi et al., 2011; Niu and Liao, 2016). Furthermore, the positive effects of H_2O_2-priming on seed germination and early seedling growth have been reported in different plant species in the absence (Barba-Espín et al., 2010; Ogawa and Iwabuchi, 2001) or presence of NaCl (Gondim et al., 2010). In addition, the pre-treatment of maize seeds with 100 mM H_2O_2 has been found to increase seedling tolerance to 80 mM NaCl (Gondim et al., 2010).

Regarding the role of H_2O_2 as a signaling molecule in the stress response of plants, some authors have investigated the beneficial effect of exogenous H_2O_2 on the induction of tolerance to different environmental stresses, including salinity (Hossain et al., 2015). An H_2O_2 pre-treatment can induce abiotic stress tolerance by regulating and/or modulating different physiological and metabolic processes such as antioxidant metabolism, photosynthesis, gas exchange parameters, and Pro accumulation, leading to improved plant growth and development (Hossain et al., 2015). Different studies have shown that seedlings treated with low H_2O_2 levels protected the photosynthetic pigments and showed better photosynthetic performance than untreated plants under saline stress (Uchida et al., 2002; Wahid et al., 2007).

In barley, for instance, treatment with low H_2O_2 levels (1–5 μM) induced salt tolerance in seedlings (Fedina et al., 2009). This response was correlated with decreased Cl- uptake in roots and leaves as well as with the induction of three new superoxide dismutase (SOD) isoenzymes in H_2O_2-pretreated seedlings (Table 22.1).

Low H_2O_2 levels (0.05 μM) were also effective in enhancing tolerance to salt stress in wheat seedlings (Li et al., 2011). In this case, H_2O_2 treatment reduced salt-induced damage to membranes, which occurred in parallel to an increase in certain antioxidant enzymes [SOD, peroxidase (POX), ascorbate peroxidase (APX), and catalase (CAT)] and the non-enzymatic antioxidants GSH and carotenoids. In addition, these authors reported that H_2O_2 promoted seedling growth (Table 22.1).

The soaking of maize seeds with 100 mM H_2O_2 increased the germination percentage when seeds were incubated in the presence of 80 mM NaCl, and the treatment stimulated CAT, glutathione peroxidase (GPX), and APX activities with the soaking time (Gondim et al., 2010). This H_2O_2 pre-treatment induced salt tolerance up to 80 mM NaCl in terms of plant growth parameters (leaf area, shoot and root dry mass, and total dry mass). In addition, H_2O_2 priming stimulated SOD, APX, CAT, and POX activity in leaves. The stimulation of the antioxidant defenses can partially explain the increased tolerance to salt stress of the maize seedlings after the H_2O_2 seed priming (Gondim et al., 2010) (Table 22.1).

TABLE 22.1
Effects of Exogenous H_2O_2 Treatments in Plants Subjected to Salt Stress

Stress and Plant Species	H_2O_2 Concentration	Effect on Phenotype	Effect on Metabolism	References
150 mMNaCl Barley	5 μM	Mitigate the photosynthetic pigments decrease. Improved CO_2 fixation	Lower H_2O_2 accumulation Increased SOD activity	Fedina et al. (2009)
80 mM NaCl Maize	100 mM (by soaking seeds)	Accelerated germination (in the absence of NaCl) Increases biomass (seedlings)	Increased CAT, POX, and APX in seeds Increased SOD, CAT, POX, and APX in leaves	Gondim et al. (2010)
150 mM NaCl Wheat	0.05 μM	Improved plant growth (plant height, biomass, and root length)	Lower $O_2^{\bullet-}$ generation Lower Lipid peroxidation Increased antioxidant defenses (SOD, POX, CAT, APX, GSH, carotenoids)	Li et al. (2011)
80 mM NaCl Maize	10 mM (foliar spray)	Improved plant growth (shoot and root DW)	Increased CAT and APX in leaves	Gondim et al. (2012)
100 mM NaCl Wheat	50–100 nM (in water irrigation)	Improved plant growth	Reduced Na^+ and Cl^- uptake Increased K^+/Na^+ ratio Increased Pro content and N assimilation Increase in photosynthetic pigments and PN	Ashfaque et al. (2014)
120 mM NaCl *Vigna radiata*	0.25–1 mM (foliar spray)	Not described	Increased antioxidant defenses (APX, DHAR, GR, γ-ECS, ASC, GSH) Reduced oxidative damage	Shan and Liu (2017)

In maize plants, a foliar spray treatment with 10 mM H_2O_2 reduced the effects of salt stress (80 mM NaCl), and this response correlated with increases in the antioxidant enzymes CAT and APX in leaves (Gondim et al., 2012). These authors also found that spraying maize plants with H_2O_2 improved their gas exchange parameters and increased chlorophyll concentrations and the relative water content (RWC) under salt stress, compared with control plants subjected to the same stress (Gondim et al., 2013). In addition, H_2O_2 treatment has been found to improve salt tolerance by regulating ion homeostasis and increasing Pro levels and N assimilation (Ashfaque et al., 2014) (Table 22.1).

Rejeb et al. (2015) studied the role of H_2O_2 produced by the plasma membrane-localized NADPH oxidase in the response to NaCl stress in *Arabidopsis* plants. These authors found a link between the transient NaCl-induced increase in H_2O_2 and the induction of Pro accumulation. This effect was prevented by dimethylthiourea, an H_2O_2 scavenger, and diphenyleneiodonium (DPI), an inhibitor of H_2O_2 production by NADPH oxidase. These results demonstrate that the H_2O_2, generated by NADPH oxidase can act as a signal to increase the accumulation of Pro in salt-stressed plants.

More recently, Shan and Liu (2017) indicated that exogenous H_2O_2 could enhance salt tolerance in *Vigna radiata* plants by regulating the ascorbate and glutathione metabolism. These authors observed that salt-stress-induced oxidative stress, as observed by the increase in electrolyte leakage (EL) and the levels of malondialdehyde (MDA), a biochemical marker of the lipid peroxidation of membranes. However, pre-treatment with exogenous H_2O_2 plus NaCl significantly decreased the oxidative stress parameters (EL and MDA). At the same time, these authors also reported an increase in the antioxidant enzymes APX, glutathione reductase (GR), and dehydreascorbate reductase (DHAR) as well as in the key enzyme for GSH biosynthesis, ϒ-glutamyl-cysteine synthetase (ϒ-ECS). The increase in DHAR, GR, and ϒ-ECS correlated with significant increases in ascorbate and glutathione, suggesting that H_2O_2 can also regulate the metabolism of both non-enzymatic antioxidants (Shan and Liu, 2017) (Table 22.1). In sum, the exogenous H_2O_2 application can improve the growth of crop plants under NaCl stress by protecting the photosynthesis process and enhancing plant antioxidant defenses.

22.5 OSMOLYTES AND SALT STRESS

Plants undergoing a process of salt stress immediately activate their physiological mechanisms to regulate the osmotic potential of the cells, especially to prevent the loss of water and the entry of toxic ions. In this process, some plants synthesize certain osmoprotectant organic compounds like Pro and GB, among others (Nounjana et al., 2012; Singh et al., 2014; Wutipraditkul et al., 2015). Not all plants have developed these adaptations, however, and the vast majority of crop plants do not have the capacity to synthesize and accumulate such protective compounds.

Most studies on the role of Pro in salt stress have focused on its ability to mediate osmotic adjustment, stabilize subcellular structures, and scavenge ROS. However, high levels of Pro synthesized during stress conditions can also maintain the $NAD(P)^+/NAD(P)H$ ratio Most studies on the role of Pro in salt stress have focused on its ability to mediate osmotic adjustment, stabilize subcellular structures, and scavenge ROS. The biosynthesis and degradation of Pro and its accumulation in plants is regulated by different abiotic stresses, including salinity, which is one of the most significant forms of abiotic stress that plants encounter (Yang et al., 2009). Proline synthesis in plants consists in two different cycles. The first is the glutamate cycle, in which glutamate is phosphorylated to ϒ-glutamyl phosphate and reduced to glutamate-ϒ-semialdehyde (GSA), which is spontaneously cyclized to Δ1-pyrroline-5-carboxylate (P5C). The second is the ornithine cycle, in which ornithine is transaminated to GSA by ornithine ϒ-aminotransferase (OAT) (Yang et al., 2009). Proline biosynthesis from glutamate consists of two enzyme reactions involving Δ1-pyrroline-5-carboxylate synthetase (P5CS) and glutamate dehydrogenase (GDH). Proline accumulation depends on its degradation rate, which is catalyzed by the mitochondrial enzyme proline dehydrogenase (PDH) (Yang et al., 2009). In plants, both PDH and Δ1-pyrroline-5-carboxylate dehydrogenase (P5CDH) are attached to the matrix side of the inner mitochondrial membrane (Hare and Cress, 1997). Proline synthesis initiates the generation of $NADP^+$, which acts as the backbone for ribose 5-phosphate required for purine synthesis, and Pro oxidation yields the reduced electron carriers, which provide energy for different biochemical reactions such as nitrogen fixation (Kim and Nam, 2013). Researchers have suggested that the exogenous application of Pro may be a good approach for decreasing the undesirable effects of salt stress on plants (Yang et al., 2009).

Researchers have also studied the role that the organic compound GB plays in salt stress. GB is found in bacteria, cyanobacteria, algae, animals, and several plant families. It is a quaternary ammonium compound that is synthesized and accumulated in some higher plants in response to adverse conditions such as lack of water or high concentrations of salt in the soil. Betaine aldehyde dehydrogenase (BADH, EC 1.2.1.8) belongs to the super family of aldehyde dehydrogenases (ALDHs) (EC 1.2.1), enzymes that catalyze the irreversible oxidation of aldehydes to their corresponding carboxylic acids, with the concomitant reduction of NAD^+ or $NADP^+$. Specifically, BADHs catalyze the irreversible oxidation of betaine aldehyde to GB. GB falls into the category of compatible solutes that form a group of small organic metabolites that are easily soluble in water and are not toxic at high concentration. The synthesis of GB in plants has the purpose of adjusting the internal osmotic potential to compensate for the external osmotic potential and in this way avoid the loss of turgor. In other words, the plant diminishes its internal osmotic potential by accumulating solutes at the cytosol level and in the organelles to compensate for the external osmotic potential. This helps stabilize macromolecules and certain valuable proteins and maintain the integrity of the cell membrane in plants.

Different tests in both laboratory field conditions have shown that the exogenous application of GB increases tolerance to salinity by protecting cells from dehydration in rice, sunflower, maize, wheat, rape, eggplant, pepper, ryegrass, *Arabidopsis*, lettuce, safflower, pigeon pea, soybean, and tomato (Abbas et al.,2010; Alasvandyari et al., 2017; Ashraf et al., 2008; Athar et al., 2009; Demiral and Türkan, 2004; Hu et al., 2012; Ibrahim et al., 2006; Korkmaz and Sirikci, 2011; Kumar et al., 2017; Lai et al., 2014; Nawaz and Ashraf, 2007; Raza et al., 2007; Sabagh et al., 2017; Shams et al., 2016; Tian et al., 2017; Wei et al., 2017; Yildirim et al., 2015). Furthermore, different studies have shown that GB acts in the protection of the extrinsic protein structure from the photosynthetic complex, particularly in photosystem II. GB also maintains photosynthetic activity by increasing stomatal conductance and maintaining Rubisco activity and chloroplast stability. Moreover, researchers have suggested a link between GB and nutrients or phytohormones, which interact together to confer tolerance to abiotic stress in plants. In summary, exogenous application of Pro and GB can be a good means to increase salt tolerance in different crop species by regulating osmotic potential, reducing the accumulation of phytotoxic ions, and protecting essential macromolecules such as proteins from the photosynthetic complex.

22.6 POLYAMINES AND THE SALT STRESS RESPONSE

Plant polyamines (PAs) play important roles in different physiological processes such as morphogenesis, growth, embryogenesis, organ development, leaf senescence, and the abiotic and biotic stress response (Kusano et al., 2008). The PAs [agmatine (Agm), cadaverine (Cad), diaminopropane (Dap), putrescine (Put), spermidine (Spd), and spermine (Spm)] are osmolytes of low molecular weight commonly present in higher plants. They are known as ROS scavengers and are considered as mediators in protective reactions against different stresses (Kovacs et al., 2010). PAs are present in all cellular compartments, including the nucleus, highlighting their participation in diverse fundamental processes in the cell such as plant growth, development, and response to stress (Bouchereau et al., 1999; Gill and Tuteja, 2010). PAs have been reported to play a role in the protection of plants to abiotic stresses, including salinity (Ikbal et al., 2014; Roy et al., 2005; Santa-Cruz et al., 1997a, b, 1998, 1999; Zhao and Qin, 2004).

PAs have different mechanisms to boost salt tolerance in plants. They can protect cell membrane integrity, ion transport processes, plant growth, photosynthesis, antioxidative metabolism, and carbohydrate metabolism through the modulation of glycolysis and the Krebs cycle pathways (Dobrovinskaya et al., 1999; Ikbal et al., 2014; Nahar et al., 2016; Shu et al., 2013; Zhong et al., 2016).

Foliar spraying with Spm was found to improve the growth of *Cucumis sativus* L. plants in the presence of 75 mM NaCl regarding plant height, stem diameter, and fresh and dry plant weight (Shu et al., 2013). In addition, the photosynthesis process was protected in Spm-treated plants under salt stress by a decrease in chlorophyll loss and an increase in some chlorophyll fluorescence parameters. This response also correlated with a decrease in superoxide ion generation and lipid peroxidation and increases in SOD and the ASC-GSH cycle components in chloroplasts. All of these responses led to a protection of the chloroplasts as observed by electron microscopy (Shu et al., 2013).

In a recent study, Ikbal et al. (2014) studied the role of PAs in the response to salinity in grapevine plantlets. These authors observed that the deleterious effect of salinity was enhanced in the presence of MGBG [Methylglyoxal-bis (guanylhydrazone)], an inhibitor of the enzyme S-adenosylmethionine decarboxylase (SAMDC), thus affecting the endogenous PA levels (Bouchereau et al., 1999). The combined treatment of grapevine plantlets with 100 mM NaCl plus 1 mM MGBG resulted in more severe chlorotic symptoms in the leaf margin, and leaves showed a higher decrease in Fv/Fm compared with the salt-treated plants. Both NaCl and MGBG induced oxidative stress, as shown by the increase in lipid peroxidation in leaves. Furthermore, a synergistic effect in this oxidative stress parameter was produced in the combined treatment (NaCl + MGBG). Likewise, higher ROS accumulation (H_2O_2, $O_2^{\cdot -}$) was observed by histochemical staining with 3,3'-diaminobenzidine (DAB) or nitroblue tetrazolium (NBT), respectively. Salt stress affected the levels of the free and conjugate forms of the polyamines agmatine and Put, and this effect was noticeable in the presence of MGBG. The authors suggested that MGBG treatment contributes to the deleterious effect of oxidative stress in grapevine plantlets grown in the presence of NaCl, disturbing different physiological and biochemical processes, including plant growth, Put levels, photosynthesis, and the redox state of the cells. They highlighted the possible protective role of PAs, including Agm and Put, in plants subjected to salt stress (Ikbal et al., 2014; Table 22.2).

In addition to photosynthesis, salinity can affect the glycolysis and the Krebs cycle pathways (Zhong et al., 2016). In the presence of 75 mM NaCl, cucumber plants showed altered carbohydrate metabolism in leaves, displaying an accumulation of starch, sucrose, glucose, and total soluble sugar contents. This response correlated with a decrease in the activity of key glycolytic and Krebs cycle enzymes, such as phosphofructokinase (PFK), pyruvate kinase (PK), and phosphoenolpyruvate pyruvate kinase (PEPC) or isocitrate dehydrogenase (IDH), malate dehydrogenase (MDH,) and succinate dehydrogenase (SDH), respectively. However, foliar treatment with 8 mM of Put improved plant growth, reduced the accumulation of carbohydrates, and produced an increase in the aforementioned enzymes. It is important to remark that the accumulation of sugars in leaves can result in the feedback inhibition of photosynthesis (Roitsch, 1999). These authors indicate that Put can regulate carbohydrate metabolism through the modulation of glycolysis metabolism and the TCA cycle, thus leading to a decrease in starch accumulation in leaves. This, in turn, promotes respiratory metabolism and mitochondrial electron transport, thus reducing the effects of salt damage (Zhong et al., 2016) (Table 22.2).

In mung bean seedlings, the exogenous application of PAs (0.2 mM of Put, Spm or Spd) alleviated the negative effects of salt stress (200 mM) on plant growth and relative water content (RWC) (Nahar et al., 2016). This effect was linked to different changes, including a decrease in Na content and an increase in K, Ca, Mg, and Zn levels in roots and shoots and a decrease in Pro in leaves. PA treatments also reduced ROS accumulation in leaves, which correlated with an increase in the reduced forms of ascorbate (ASC) and GSH, but a decrease in the corresponding oxidized forms (DHA and GSSG, respectively), thus increasing the redox state of both antioxidants. In addition, PAs produced an increase in catalase,

TABLE 22.2
Effects of Exogenous Polyamines (PAs) Treatments in Plants Subjected to Salt Stress

Stress and Plant Species	PAs Concentration	Effect on Phenotype	Effect on Metabolism	References
75 mM NaCl *Cucumis sativus*	Spd, Spm (0.3 mM). Foliar spray	Improved plant growth Improved photosynthesis Chloroplast protection	Reduced oxidative stress Induction of antioxidant defenses Chlorophyll loos alleviation	Shu et al. (2013)
100 mMNaCl Grape vine plantlets	None. Use of MGBG (SAMDC inhibitor)	More severe chlorotic symptoms	Increased ROS accumulation Increased lipid peroxidation	Ikbal et al. (2014)
150 mM NaCl Ginseng seedlings	Spm (0.01, 0.1 1 mM) MS glass bottles	Improved plant growth	Increase in expression and activity of CAT, APX, and GPX Reduced ROS accumulation and lipid peroxidation. Increased photosynthetic pigments	Parvin et al. (2014)
100 mM NaCl *Citrus aurantium* L.	Put, Spd, Spm (1 mM). Hydroponic culture	Not described	Reduced oxidative stress Induction of antioxidant defenses Improved photosynthesis	Tanou et al. (2014)
75 mM NaCl cucumber	8 mM Put	Improved plant growth	Stimulated glycolysis and Krebs cycle pathways	Zhong et al. (2016)
200 mM Mung bean	0.2 mM (Put, Smp, or Spd)	Improved plant growth and nutrient homeostasis	Decreased ROS accumulation Increase in antioxidant defenses (ASC, GSH, DHAR, GR, GPX)	Nahar et al. (2016)
200 mM NaCl *Petunia hybrida*	Low concentrations (μM level of Put, Spm, or Spd)	Improved plant growth	Increase of antioxidant enzymes (CAT, POX) Increased photosynthetic pigments	Arun et al. (2016)

DHAR, GR, and GPX activities and restored the activity levels of some antioxidant enzymes that were affected by the salt treatment, such as glutathione-S-transferase (GST). PAs also produced an increase in the chlorophyll content (Nahar et al., 2016). The authors concluded that the application of PAs prevented Na toxicity and improved the nutrient homeostasis in salt-treated plants, probably by regulating the plasma membrane ion channels. Moreover, the increase in Pro content can favor improved osmoregulation, facilitating the increase in RWC. Finally, by preventing the degradation of photosynthetic pigments and increasing the levels of some antioxidant defenses, PAs can protect the photosynthetic process and induce oxidative stress tolerance that can lead to better plant growth under saline stress (Nahar et al., 2016) (Table 22.2).

Arun et al. (2016) studied the effect of PA treatments in the response of salt-sensitive petunia (*Petunia hybrida*) plantlets to NaCl stress. The salt treatment significantly affected seed germination and seedling growth, increasing adventitious rooting. In the absence of NaCl, the addition of Put, Spm, or Spd to the culture medium at low concentrations improved plant growth. In the presence of 200 mM NaCl, a decrease in chlorophyll content and an increase in lipid peroxidation occurred in the leaves of the petunia seedlings. These effects were mitigated in the presence of the different PAs. Similarly, salinity significantly reduced the activity of catalase, peroxidase, and polyphenol oxidase (PPO). However, the plantlets rooted using PAs showed higher activity in the previously mentioned antioxidant enzymes. The authors suggested that the application of exogenous PAs in the salt-sensitive *P. hybrida* induced the growth of roots and shoots and also increased the activity of antioxidant defenses and the photosynthetic pigments. All of these responses favored the response of these plants to salinity. Petunia growers can, therefore, use this information to develop healthy cultivars with improved tolerance to abiotic stresses, including salt stress (Arun et al., 2016) (Table 22.2).

In conclusion, results seem to indicate that PAs exhibit various positive effects on antioxidant machinery, photosynthesis, glycolysis, and the Krebs cycle pathways of plants in response to abiotic stresses, including salt stress. PA treatments, therefore, lead to improvements in plant growth and adaptation in saline conditions.

22.7 SALICYLIC ACID AND THE RESPONSE TO SALT STRESS

Salicylic acid is a phenolic plant hormone widely distributed in plants, although its basal levels differ among species. It plays an important role in the regulation of a multitude of physiological processes, including seed germination, vegetative growth, photosynthesis, respiration, thermogenesis, flower formation, seed production, and senescence (Hernández et al., 2017). SA is mainly known for its central role in plant-pathogen interactions. During the last two decades, hundreds of papers regarding the implication of SA in the plant response to biotic stress have been published. Recent works have also indicated that SA plays an important role in the response of higher plants to abiotic stresses, including saline stress (Takatsuji and Jiang, 2014).

SA is synthesized via two distinct pathways, the isochorismate (IC) pathway and the phenylalanine ammonia-lyase (PAL) pathway (Miura and Tada, 2014). Furthermore, our team recently reported that the cyanogenic glycoside (CNgls) pathway might be involved in a new SA biosynthetic pathway in peach with mandelonitrile (MD) at the hub, controlling both amygdalin (one of the main CNglcs in peach) and SA biosynthesis (Diaz-Vivancos et al., 2017). Although the biosynthesis of SA from MD is still functional under stress conditions, the physiological functions of this new SA biosynthetic pathway remain to be elucidated in further works (Bernal-Vicente et al., 2018).

The role of SA in the plant response under salinity conditions remains to be fully unraveled. Nevertheless, different authors have studied the effect of exogenous SA treatments on the response to NaCl stress in different plant systems. This effect seems to be dependent on the SA concentrations used, the plant species, the application mode, the physiological state of the plant at the time of application, the level of salinity, and the exposure time to NaCl (Barba-Espín et al., 2011). Some authors have found that exogenous SA treatments improved plant growth and seed germination under saline stress (Bastam et al., 2013; He and Zhu, 2008; Lee et al., 2010; Liu et al., 2014; Rajjou et al., 2006) (Table 22.3).

The effects of SA on seed germination under salt stress are dependent on the concentration used (Lee et al., 2010). The germination of the transgenic NahG *Arabidopsis* line was delayed under salinity conditions (100 mM NaCl), but the addition of exogenous SA (0.5 mM) improved the germination vigor under the stress conditions (Rajjou et al., 2006). Other authors, however, did not find any negative effects of NaCl stress on the germination of NahG plants (Borsani et al., 2001; Lee et al., 2010). It has been suggested that SA regulates seed germination through the modulation of the ROS balance (Lee et al., 2010). This effect can be explained by the fact that SA induces H_2O_2 accumulation by inhibiting H_2O_2-scavenging enzymes, while at the same time H_2O_2 also induces SA accumulation (Rao and Davis, 1999). Accordingly, different works have shown that exogenous H_2O_2 can stimulate seed germination and early seedling growth (Barba-Espín et al., 2010, 2012).

SA could be also related to plant acclimation in saline conditions. NaCl-adapted tomato cells, for instance, were found to contain a lower concentration of SA than un-adapted cells (Molina et al., 2002). The NaCl-adaptation process has also been linked to a higher antioxidative capacity, because salt-adapted cells were found to contain a higher basal level of APX and GR activities (Molina et al., 2002).

Bastam et al. (2013) reported that the exogenous application of SA improved the tolerance of pistachio seedlings to NaCl-stress (up to 90 mM NaCl). The SA-treated plants showed lower NaCl-induced injury symptoms, a better growth rate, higher chlorophyll content, and improved photosynthetic capacity than the non-treated plants (Table 22.3). In this case, the authors used SA concentrations ranging from 0 to 1 mM, and the treatments were applied by foliar spray. At low concentrations (less than 10 μM), SA alleviated the salt-induced decrease in photosynthesis by increasing net photosynthesis (PN), CO_2 fixation, transpiration, stomatal conductance, and antioxidant activity, but an opposite effect was observed at high concentrations (1–5 mM SA) (Jayakannan et al., 2015). The foliar application of 0.1 mM SA also improved the growth of cotton seedlings in the presence of 100 mM NaCl. The SA-treated plants displayed better growth and photosynthetic rates and showed low ROS accumulation ($O_2^{\cdot-}$ and H_2O_2) and lipid peroxidation, which correlated with a significant increase in CAT activity (Liu et al., 2014) (Table 22.3).

SA application (1 mM) by foliar treatment also mitigated the deleterious effects of NaCl in tomato plants (He and Zhu, 2008). These authors observed an SA-linked alleviation in NaCl-induced oxidative stress, reflected in reduced levels of lipid peroxidation and H_2O_2 accumulation and an increase in the antioxidant capacity of the tomato plants due to increases in some antioxidant defenses (He and Zhu, 2008) (Table 22.3). However, Barba-Espín et al. (2011) observed that SA negatively affects the response of pea plants to NaCl stress. In this work, pea seeds and seedlings were treated with different SA levels (25, 50, and 100 μM). In the absence of NaCl, 100 μM SA significantly reduced plant growth and the effect was more evident in roots than in shoots.

TABLE 22.3
Effects of Exogenous Application of Salicylic Acid (SA) on Plants Under Salt or Drought Stress Conditions

Stress and Name of Plant	SA Concentration	Effect on Phenotype	Effect on Metabolism	References
100 mM NaCl Tomato	10^{-4} M Hydroponic culture	Improved acclimation	Increased APX and GPX	Szepesi (2006)
100 mM NaCl Tomato	1 mM SA, foliar application	Improved plant growth	Increased ascorbate, GSH, CAT, APX, DHAR	He and Zhu (2008)
150 mM *Arabidopsis*	1–10 μM MS agar plates	Germination promotion	Decreased H_2O_2 generation	Lee et al. (2010)
100 mM NaCl Tomato	10^{-3}–10^{2} M 10^{7}–10^{-4} M	Cell death, reduced cell viability Acclimation to NaCl stress, higher cell viability	H_2O_2 and NO accumulation No ROS and/or NO accumulation	Gémes et al. (2011)
70 mM NaCl pea plants	25–100 μM, foliar application	Reduction of plant growth	Increased ascorbate, GSH, CAT, SOD Decreased APX and GR	Barba-Espín et al. (2011)
30–90 mM NaCl Pistachio	0.5–1 mM, foliar application	Improved plant growth Increased PN	Reduced electrolyte leakage, Increased chlorophyll contents	Bastam et al. (2013)
100 mM NaCl Cotton	0.1 mM, foliar application	Improved plant growth Increased PN	Reduced ROS accumulation and Lipid peroxidation. Increased CAT	Liu et al. (2014)
300 mM *Caralluma tuberculata*	200 μM	Increased calli fresh weight and protein contents	Increased SOD and GR	Rehman et al. (2014)
3 and 6 g/L *Dianthus superbus* L.	0.5 mM, foliar application	Improved plant growth Increased PN, chloroplast protection	Reduced ROS accumulation and Lipid peroxidation. Increased SOD, POX, CAT	Ma et al. (2017)

SA treatment also had an effect on the antioxidative machinery of pea plants. For example, in the absence of NaCl, 100 μM SA increased APX and catalase activities, whereas in the presence of NaCl a decrease in APX, as well as increases in SOD and GST activities, took place, and this response correlated with an accumulation of H_2O_2 in leaves (Barba-Espín et al., 2011) (Table 22.3). Low SA levels led to the induction of the PR-1b gene in leaves from NaCl-stressed pea plants. These authors suggested that the induction of the PR-1b gene could be an adaptive response in order to prevent a possible opportunistic fungal or bacterial infection in a situation of weakness (Barba-Espín et al., 2011).

Cross-talk between SA and NaCl-generated ROS and NO in tomato has been described during acclimation to high salinity (Gémes et al., 2011). However, this response was dependent on the SA concentration used in the experiments. At high concentrations (10^{-3}–10^{-2} M), SA induced massive H_2O_2 accumulation, which eventually produced cell death. In root tips, the same SA concentration provoked the overgeneration of ROS and NO, with a concomitant decrease in cell viability (Gémes et al., 2011). At lower concentrations (10^{-7}–10^{-4} M), however, SA alleviated the effect of salt stress regarding ROS and NO production in the root apex. The same response was observed in leaf protoplasts prepared from control plants when exposed to low SA concentrations in the presence of 100 mM NaCl. In this case, the authors detected less production of ROS and NO and higher cell viability than in the salt-treated samples in the absence of SA (Gémes et al., 2011). The authors suggested that treatments with sublethal SA concentrations can be considered as an eustress that leads to limited ROS and NO accumulation in the root tips of tomato plants subjected to salinity. However, higher SA concentrations can cause oxidative burst, damaging the root meristems (Gémes et al., 2011) (Table 22.3).

In a more recent work, Ma et al. (2017) described that exogenous SA could effectively counteract the adverse effects of moderate salt stress on the growth and development of *Dianthus superbus* L. plants. The presence of NaCl (3 or 6 g/L) reduced plant growth, chlorophyll content, and net photosynthesis (PN), but produced an increase in the oxidative stress parameters, including ROS accumulation, lipid peroxidation, and relative electric conductivity. The treatment with SA (0.5 mM) reduced the decrease in growth in salt-stressed *D. superbus* plants and increased the chlorophyll content and PN. These effects were parallel to a decrease in all the oxidative stress parameters analyzed and to an increase in SOD, POX, and CAT activities. Salt stress also caused an alteration in the chloroplast ultrastructure, which showed irregular and disordered stroma thylakoids. However, the plants grown under SA treatment mostly had grana containing more and better-organized thylakoids than the plants grown under salt stress (Ma et al., 2017) (Table 22.3). These results suggested that SA can protect the chloroplast, alleviating the damage caused by salt stress in this cell organelle and contributing to the resistance of *D. superbus* to salinity.

SA has also been described as playing a role in ionic homeostasis. Accordingly, SA pretreatments reduce Na^+ concentrations in shoots during prolonged salt stress, but can also prevent salt-induced K^+ leakage from roots and increase K^+ concentrations in shoots under salinity (Jayakannan et al., 2015). Although the effect of SA in response to NaCl seems to be dependent on different factors, its exogenous application, in some cases, ameliorates the acclimation of plants to NaCl stress by improving plant growth and increasing PN and antioxidant defenses, thus reducing ROS accumulation and damage to biological membranes.

22.8 MELATONIN AND SALT STRESS TOLERANCE

Melatonin (N-acetyl-5-methoxytryptamine) is a potent, naturally occurring antioxidant that effectively scavenges both ROS and reactive nitrogen species (RNS) in animals and plants (Tan et al., 2012). The effects of melatonin have been documented in studies with various abiotic stressors (Arnao and Hernández-Ruiz, 2014), including salt stress (Kostopoulou et al., 2015; Liang et al., 2015; Wei et al., 2015). Melatonin has been found to enhance tolerance to salt stress in different plant species.

NaCl stress (100 mM) has been shown to significantly inhibit the growth of *Malus hupehensis* seedlings and decrease chlorophyll concentrations and the photosynthesis rate. However, exogenous melatonin (0.1 μM) significantly alleviated these responses (Li et al., 2012). In addition, the melatonin treatment relieved salt-induced oxidative stress, as observed by the decrease in H_2O_2 accumulation under salinity. This reduction in H_2O_2 accumulation was probably due to an enhancement of the H_2O_2-scavenging enzymes APX, catalase and POX (Li et al., 2012). Moreover, it seems that melatonin can control the ion-homeostasis response of *M. hupehensis* seedlings under salt stress through the expression of ion-channel genes. MdNHX1 and MdAKT, encoding Na^+ and K^+ transporters, were greatly up-regulated in the leaves from melatonin-treated plants. This response possibly contributed to the maintenance of ion homeostasis, therefore improving salinity resistance in plants exposed to exogenous melatonin (Li et al., 2012) (Table 22.4).

The treatment of rice plants with 10–20 μM melatonin was found to delay leaf senescence and cell death and enhance salt stress tolerance by directly or indirectly counteracting the cellular accumulation of H_2O_2. Melatonin treatment also reduced the expression of genes involved in chlorophyll degradation, as well as others senescence-associated genes, and up-regulated genes linked to oxidation/reduction that might play an important role in ROS scavenging. Moreover, melatonin also induced genes linked to chlorophyll and pigment biosynthesis and stress response. Melatonin directly or indirectly induces the expression of some transcription factors (TFs), such as bZIP, MAC, and MYB. These TFs activate certain gene-encoding enzymes related to H_2O_2 scavenging. In addition, melatonin possibly regulates the expression of DREB and HSF TFs, thereby establish in GA transcriptional cascade to reduce intracellular H_2O_2 levels. All of these responses demonstrate that melatonin delays leaf senescence, thereby enhancing salt stress tolerance (Liang et al., 2015) (Table 22.4).

Melatonin can also regulate plant growth and has been found to increase germination and plant yield in soybean. Melatonin can also increase the salt and drought tolerance of soybean plants (Wei et al., 2015). When soybean seedlings were treated with NaCl (10 g/L), the presence of melatonin (100 μM) improved different plant growth parameters (plant height, leaf size, and biomass production) compared with salinized plants, and melatonin-treated plants also presented unchanged chlorophyll levels (Wei et al., 2015). Salt-stress-induced oxidative stress in soybean plants, as observed by the accumulation of H_2O_2 and the increase in electrolyte leakage. However, melatonin-treated plants presented a decrease in these oxidative stress parameters (Wei et al., 2015). Gene ontology analysis showed that melatonin promoted the expression of many genes that can help inhibit the deleterious effects of salt stress in plants. The authors observed that melatonin up-regulates the expression of genes related with photosynthesis, the Calvin cycle, starch and sucrose metabolism, ascorbate synthesis and metabolism, glycolysis, and the Krebs cycle (Wei et al., 2015).

Melatonin has also been found to regulate GR activity and reduce ROS accumulation in sunflower cotyledons in response to salt stress. The presence of melatonin in salt-treated (120 mM NaCl) sunflower seeds increased

TABLE 22.4
Described Effects of Exogenous Melatonin Treatments in Plants Subjected to Salt or Drought Stress

Stress and Plant Species	Melatonin Concentration	Effect on Phenotype	Effects on Metabolism	References
100 mM NaCl *Malus halepensis*	0.1 μM	Improved plant growth parameters	Improved photosynthesis Alleviated loos of chlorophyll Better ion-homeostasis control Reduced H_2O_2 accumulation Increased APX, CAT, POX	Li et al. (2012)
5 g/L NaCl Rice	10–20 μM	Delayed salt-induced leaf senescence and cell death	Reduced H_2O_2 accumulation Alleviation of chlorophyll degradation	Liang et al. (2015)
10 g/L NaCl Soybean	100 μM	Germination and plant growth promotion	Induction of genes related with photosynthesis, Calvin cycle, carbohydrate metabolism, ascorbate synthesis and metabolism, glycolysis, and Krebs cycle Reduced H_2O_2 accumulation	
100 mM NaCl *Citrus aurantium* L.	1 μM	Reduced toxicity symptoms in leaves (chlorosis, dehydration, necrosis, twisted leaves)	Lipid peroxidation mitigation Reduced electrolyte leakage Increased photosynthetic pigments Higher Cl accumulation in leaves and roots. Increased GSH levels in roots. PPO increase in roots and leaves SOD increase in roots	Kostopoulou et al. (2015)
Diluted sea water (3.85 to 7.69 dS/m Wheat	100–500 μM	Improved plant growth parameters and yield	Increased photosynthetic pigments, IAA levels and antioxidant capacity	Sadak (2016)
300 mM NaCl Watermelon	50–100 μM	Improved plant growth	Improved redox homeostasis Photosynthesis protection Increased antioxidant defenses	Li et al. (2017)

PPO: polyphenol oxidase.

GR activity and glutathione contents, but also reduced the ROS accumulation in cotyledons (Kaur and Bhatla, 2016).

Sadak (2016) described that salinity stress caused marked decreases in wheat plant growth parameters (shoot height, number of leaves/plant, fresh and dry shoot weights), as well as in photosynthetic pigments and indole acetic acid (IAA) concentrations. In response to different salinity levels, yield as well as and carbohydrate, protein, nitrogen, phosphorous, and potassium levels decreased. Flavonoid and phenolic contents, on the other hand, increased under salinity stress. Melatonin treatments proved to be effective in enhancing growth parameters, photosynthetic pigments, and IAA levels in salt-stressed plants. At different levels, melatonin treatments also produced significant increases in yield and carbohydrate, protein, nitrogen, phosphorous, potassium, flavonoid, and phenolic contents. Furthermore, melatonin increased the antioxidant activity of the yielded seeds in both non-stressed and salinity-stressed plants relative to the corresponding controls. Generally, 500 μM of melatonin was the most effective treatment in alleviating the deleterious effect of salinity stress on wheat plants.

Li et al. (2017) described the effects of melatonin on leaf photosynthesis and redox homeostasis in watermelon under salt stress (300 mM NaCl). Salt stress inhibited photosynthesis and increased ROS accumulation and membrane damage in the leaves of watermelon seedlings. However, pretreatment with melatonin on roots alleviated the NaCl-induced decrease in the photosynthetic rate and oxidative stress in a dose-dependent manner (50–150–500 μM). The protection of photosynthesis by melatonin was closely associated with the inhibition of stomatal closure and improved light energy absorption and electron transport in photosystem II. The decrease in oxidative stress due to melatonin was attributed to improved redox homeostasis coupled with enhanced antioxidant enzyme activity.

Zheng et al. (2017) described that chloroplasts produce melatonin, a recently discovered plant antioxidant molecule. Importantly, the MzASMT9 gene involved in melatonin synthesis was found to be upregulated by the high light intensity and salt stress. Increased melatonin due to the highly expressed MzASMT9 resulted in *Arabidopsis* lines with enhanced salt tolerance compared to wild-type plants, as indicated by reduced ROS, lower lipid peroxidation, and enhanced photosynthesis.

All of this research shows that melatonin has a beneficial effect on plant growth and improves plant performance under salt stress. Given this and its low price, low toxicity, and the absence of side effects, melatonin can be used as a potential environmentally friendly treatment for improving plant growth and yield in agriculture (Liang et al., 2015).

22.9 CONCLUSION

Biostimulants are natural substances capable of improving plant metabolism and the plant response to different environmental stresses, including salt stress. HAs can positively influence the soil microbiome and contribute to the unblocking of nutrients, facilitating their assimilation by the root system. This property can influence plant growth and the plant response in saline soils. The application of AMF constitutes an inexpensive and effective approach for coping with the deleterious effects of salinity and improving plant growth and production. This fungus helps plants mobilize a large number of nutrients and improves the soil structure and the plant-water relations, affording protection against pathogenic agents and abiotic stress conditions.

Although ROS are considered as toxic molecules, their role as signaling molecules has also been described, especially in the case of H_2O_2. Accordingly, the exogenous H_2O_2 application can improve seed germination and early seedling growth under NaCl stress by protecting the photosynthesis process and enhancing the antioxidant defenses. On the other hand, exogenous application of Pro and GB can be good strategies for increasing salt tolerance in different crop species by regulating osmotic potential, reducing the accumulation of phytotoxic ions, and protecting essential macromolecules like proteins from the photosynthetic complex.

PAs also exhibit various positive effects on the antioxidant machinery, photosynthesis, glycolysis, and the Krebs cycle pathways of plants in response to abiotic stresses, including salt stress. This leads to improvements in plant growth and adaptation in saline conditions.

Although the effect of SA in response to NaCl seems to be dependent on different factors, its exogenous application, in some cases, ameliorates the acclimation of plants to NaCl stress by improving plant growth and increasing PN and antioxidant defenses. This in turn reduces ROS accumulation and damage to biological membranes.

Different works have shown that melatonin has a beneficial effect on plant growth and improves plant performance under salt stress. Given this and its low price, low toxicity and the absence of side effects, melatonin can be used as a potential environmentally friendly treatment for improving plant growth and yield in agriculture.

In conclusion, the implementation of these different strategies can improve the growth and adaptation of plants of agronomic interest in saline conditions.

REFERENCES

Abbas, W., Ashraf, M., Akram, N.A. (2010) Alleviation of salt-induced adverse effects in eggplant (*Solanum melongena* L.) by glycinebetaine and sugarbeet extracts. *Scientia Horticulturae* 125:188–195.

Acosta-Motos, J.R., Díaz-Vivancos, P., Álvarez, S., Fernández-García, N., Sánchez-Blanco, M.J., Hernández, J.A. (2015a) NaCl-induced physiological and biochemical adaptative mechanism in the ornamental *Myrtus cummunis* L. plants. *Journal of Plant Physiology* 183:41–51.

Acosta-Motos, J.R., Díaz-Vivancos, P., Álvarez, S., Fernández-García, N., Sánchez-Blanco, M.J., Hernández, J.A. (2015b) Physiological and biochemical mechanisms of the ornamental *Eugenia myrtifolia* L. plants for coping with NaCl stress and recovery. *Planta* 242:829–846.

Acosta-Motos, J.R., Ortuño, M.F., Bernal-Vicente, A., Diaz-Vivancos, P., Sanchez-Blanco, M.J., Hernandez, J.A. (2017) Plant responses to salt stress: Adaptive mechanisms. *Agronomy* 7:18.

Alasvandyari, F., Mahdavi, B., Madah Hosseini, S. (2017) Glycine betaine affects the antioxidant system and ion accumulation and reduces salinity-induced damage in safflower seedlings. *Archives of Biological Sciences* 69:139–147.

Al-Karaki, G.N. (2000) Growth of mycorrhizal tomato and mineral acquisition under salt stress. *Mycorrhiza* 10:51–54

Arnao, M.B., Hernández-Ruiz, J. (2014) Melatonin: Plant growth regulator and/or biostimulator during stress? *Trends in Plant Science* 19:789–797.

Arun, M., Radhakrishnan, R., Ai, T.N., Naing, A.H., Lee, I.J., Kim, C.K. (2016) Nitrogenous compounds enhance the growth of petunia and reprogram biochemical changes against the adverse effect of salinity. *The Journal of Horticultural Science and Biotechnology* 91:562–572.

Ashfaque, F., Khan, M.I.R., Khan, N.A. (2014) Exogenously applied H_2O_2 promotes proline accumulation, water relations, photosynthetic efficiency and growth of wheat (*Triticum aestivum* L.) under salt stress. *Annual Research & Review in Biology* 4:105–120.

Ashraf, M., Athar, H.R., Harris, P.J.C., Kwon, T.R. (2008) Some prospective strategies for improving crop salt tolerance. *Advances in Agronomy* 97:45–110

Athar, H.R., Ashraf, M., Wahid, A., Jamil, A. (2009) Inducing salt tolerance in Canola (*Brassica napus* L.) by exogenous application of glycine beanie and proline: Response at the initial growth stages. *Pakistan Journal of Botany* 41:1311–1319.

Augé, R.M., Toler, H.D., Saxton, A.M. (2014) Arbuscular mycorrhizal symbiosis and osmotic adjustment in response to NaCl stress: A meta-analysis. *Frontiers in Plant Science* 5:562.

Aydin, A., Kant, C., Turan, M. (2012) Humic acid application alleviate salinity stress of bean (*Phaseolus vulgaris* L.) plants decreasing membrane leakage. *African Journal of Agricultural Research* 7:1073–1086.

Barba-Espín, G., Clemente-Moreno, M.J., Álvarez, S., García-Legaz, M.F., Hernández, J.A., Díaz-Vivancos, P. (2011) Salicylic acid negatively affects the response to salt stress in pea plants. *Plant Biology* 13:909–917.

Barba-Espín, G., Diaz-Vivancos, P., Clemente-Moreno, M.J., Albacete, A., Faize, L., Faize, M., Pérez-Alfocea, F., Hernández, J.A. (2010) Interaction between hydrogen peroxide and plant hormones during germination and the early growth of pea seedlings. *Plant, Cell and Environment* 33:981–994.

Barba-Espín, G., Hernández, J.A., Diaz-Vivancos, P. (2012) Role of H_2O_2 in pea seed germination. *Plant Signaling & Behavior* 7:193–195.

Bastam, N., Baninasab, B., Ghobadi, C. (2013) Improving salt tolerance by exogenous application of salicylic acid in seedlings of pistachio. *Plant Growth Regulation* 69:275–284.

Bernal-Vicente, A., Petri, C., Hernández, J.A., Diaz-Vivancos, P. (2018) The effect of abiotic and biotic stress on the salicylic acid biosynthetic pathway from mandelonitrile in peach. *Plant Biology* 20:986–994.

Bolan, F. (2005) *A Critical Review on the Role of Mycorrhizal Fungi in Microbiology*. Academic Press Ltd., London.

Borsani, O., Valpuesta, V., Botella, M.A. (2001) Evidence for a role of salicylic acid in the oxidative damage generated by NaCl and osmotic stress in *Arabidopsis* seedlings. *Plant Physiology* 126:1024–1030.

Bouchereau, A., Aziz, A., Larher, F., Martin-Tanguy, J. (1999) Polyamines and environment challenges: Recent advances. *Plant Science* 140:103–125.

Brundrett, M.C. (2002) Coevolution of roots and mycorrhizas of land plants. *New Phytologist* 154:275–304.

Çimrin, K.M., Türkmen, Ö., Turan, M., Tuncer, B. (2010) Phosphorus and humic acid application alleviate salinity stress of pepper seedling. *African Journal of Biotechnology* 9:5845–5851.

Corpas, F.J., Gómez, M., Hernández, J.A., del Río, L.A. (1993) Metabolism of activated oxygen in peroxisomes from two *Pisumsativum* L. cultivars with different sensitivity to sodium chloride. *Journal of Plant Physiology* 141:160–165.

Demiral, T., Türkan, I. (2004) Does exogenous glycine betaine affect antioxidative system of rice seedlings under NaCl treatment? *Journal of Plant Physiology* 161:1089–1100.

Diaz-Vivancos, P., Bernal-Vicente, A., Cantabella, D., Petri, C., Hernández, J.A. (2017) Salicylic acid biosynthesis is linked to the cyanogenic glycosides pathway in peach plants. *Plant and Cell Physiology* 58:2057–2066.

Dobrovinskaya, O.R., Muñiz, J., Pottosin, I.I. (1999) Inhibition of vacuolar ion channels by polyamines. *The Journal of Membrane Biology* 167:127–140.

Evelin, H., Giri, B., Kapoor, R. (2012) Contribution of *Glomus intraradices* inoculation to nutrient acquisition and mitigation of ionic imbalance in NaCl-stressed *Trigonellafoenum graecum*. *Mycorrhiza* 22:203–217.

Evelin, H., Giri, B., Kapoor, R. (2013) Ultrastructural evidence for AMF mediated salt stress mitigation in *Trigonellafoenum graecum*. *Mycorrhiza* 23:71–86.

Evelin, H., Kapoor, R., Giri, B. (2009) Arbuscular mycorrhizal fungi in alleviation of salt stress: A review. *Annals of Botany* 104:1263–1280.

Fedina, I.S., Nedeva, D., Çiçek, N. (2009) Pre-treatment with H_2O_2 induces salt tolerance in barley seedlings. *Biologia Plantarum* 53:321–324.

Feng, G., Zhang, F.S., Li, X.L., Tian, C.Y., Tang, C., Rengel, Z. (2002) Improved tolerance of maize plants to salt stress by arbuscular mycorrhiza is related to higher accumulation of soluble sugars in roots. *Mycorrhiza* 12:185–190.

Gapper, C., Dolan, L. (2006) Control of plant development by reactive oxygen species. *Plant Physiology* 141:341–345.

Gémes, K., Poór, P., Horváth, E., Kolbert, Z., Szopkó, D., Szepesi, A., Tari, I. (2011) Cross-talk between salicylic acid and NaCl-generated reactive oxygen species and nitric oxide in tomato during acclimation to high salinity. *Physiologia Plantarum* 142:179–192.

Gill, S.S., Tuteja, N. (2010) Polyamines and abiotic stress tolerance in plants. *Plant Signaling & Behavior* 5:26–33.

Giri, B., Kapoor, R., Mukerji, K.G. (2003) Influence of arbuscular mycorrhizal fungi and salinity on growth, biomass and mineral nutrition of *Acacia auriculiformis*. *Biology and Fertility of Soils* 38:170–175.

Giri, B., Kapoor, R., Mukerji, K.G. (2007) Improved tolerance of *Acacia nilotica* to salt stress by arbuscular mycorrhiza, *Glomus fasciculatum*, may be partly related to elevated K/Na ratios in root and shoot tissues. *Microbial Ecology* 54:753–760.

Giri, B., Mukerji, K.G. (2004) Mycorrhizal inoculant alleviates salt stress in *Sesbania aegyptiaca* and *Sesbania grandiflora* under field conditions: Evidence for reduced sodium and improved magnesium uptake. *Mycorrhiza* 14:307–312.

Gómez-Bellot, M.J., Ortuño, M.F., Nortes, P.A., Vicente-Sánchez, J., Martín, F.F., Bañón, S., Sánchez-Blanco, M.J. (2015) Protective effects of *Glomus iranicum* var. *tenuihypharum* on soil and *Viburnum tinus* plants irrigated with treated wastewater under field conditions. *Mycorrhiza* 25:399–409.

Gondim, F.A., Gomes-Filho, E., Costa, J.H., Alencar, N.L.M., Priso, J.T. (2012) Catalase plays a key role in salt stress acclimation induced by hydrogen peroxide pretreatment in maize. *Plant Physiology and Biochemistry: PPB* 56:62–71.

Gondim, F.A., Gomes-Filho, E., Lacerda, C.F., Prisco, J.T., Neto, A.D.A., Marques, E.C. (2010) Pretreatment with H_2O_2 in maize seeds: Effects on germination and seedling acclimation to salt stress. *Brazilian Journal of Plant Physiology* 22:103–112.

Gondim, F.A., Miranda, R. de S., Gomes-Filho, E., Prisco, J.T. (2013) Enhanced salt tolerance in maize plants induced by H_2O_2 leaf spraying is associated with improved gas exchange rather than with non-enzymatic antioxidant system. *Theoretical and Experimental Plant Physiology* 25:251–260.

Gulser, F., Sonmez, F., Boysan, S. (2010) Effects of calcium nitrate and humic acid on pepper seedling growth under saline condition. *Journal of Environmental Biology* 31:873–876.

Hare, P.D., Cress, W.A. (1997) Metabolic implications of stress-induced proline accumulation in plants. *Plant Growth Regulation* 21:79–102.

He, Y., Zhu, Z.J. (2008) Exogenous salicylic acid alleviates NaCl toxicity and increases antioxidative enzyme activity in *Lycopersicun esculentum*. *Biologia Plantarum* 52:792–795.

Hernández, J.A., Corpas, F.J., Gómez, M., del Río, L.A., Sevilla, F. (1993) Salt induced oxidative stress mediated by activated oxygen species in pea leaf mitochondria. *Physiologia Plantarum* 89:103–110.

Hernández, J.A., Díaz-Vivancos, P., Barba-Espín, G., Clemente-Moreno, M.J. (2017) On the role of salicylic acid in plant responses to environmental stresses. In: Nazar, R., Iqbal, N., Khan, N. (eds.) *Salicylic Acid: A Multifaceted Hormone*. Springer, Singapore, 17–34.

Hernández, J.A., Ferrer, M.A., Jiménez, A., Ros-Barceló, A., Sevilla, F. (2001) Antioxidant systems and $O_2^{\cdot-}/H_2O_2$ production in the apoplast of *Pisumsativum* L. leaves: Its relation with NaCl-induced necrotic lesions in minor veins. *Plant Physiology* 127:817–831.

Hernández, J.A., Olmos, E., Corpas, F.J., Sevilla, F., del Río, L.A. (1995) Salt-induced oxidative stress in chloroplast of pea plants. *Plant Science* 105:151–167.

Hossain, M.A., Bhattacharjee, S., Armin, S.M., Qian, P., Xin, W., Li, H.Y., Burritt, D.J., Fujita, M., Tran, L.S.P. (2015) Hydrogen peroxide priming modulates abiotic oxidative stress tolerance: Insights from ROS detoxification and scavenging. *Frontiers in Plant Science* 6:420. doi:10.3389/fpls.2015.00420

Hossain, M.S., Dietz, K.J. (2016) Tuning of redox regulatory mechanisms, reactive oxygen species and redox homeostasis under salinity stress. *Frontiers in Plant Science* 7:548. doi:10.3389/fpls.2016.00548

Hu, L., Hu, T., Zhang, X., Pang, H., Fu, J. (2012) Exogenous glycine betaine ameliorates adverse effects of salt stress on perennial ryegrass. *Journal of the American Society for Horticultural Science* 137:38–46.

Ibrahim, M., Anjum, A., Khaliq, N., Iqbal, M., Athar, H.U.R. (2006) Four foliar applications of glycine betaine did not alleviate adverse effects of salt stress on growth of sunflower. *Pakistan Journal of Botany* 38:1561–1570.

Ikbal, F.E., Hernández, J.A., Barba-Espín, G., Koussa, T., Aziz, A., Faize, M., Diaz-Vivancos, P. (2014) Enhanced salt-induced antioxidative responses involve a contribution of polyamine biosynthesis in grapevine plants. *Journal of Plant Physiology* 171:779–788.

Ishibashi, Y., Yamaguchi, H., Yuasa, T., Iwaya-Inoue, M., Arima, S., Zheng, S.H. (2011) Hydrogen peroxide spraying alleviates drought stress in soybean plants. *Journal of Plant Physiology* 168:1562–1567.

Jahromi, F., Aroca, R., Porcel, R., Ruiz-Lozano, J.M. (2008) Influence of salinity on the *in vitro* development of *Glomus intraradices* and on the *in vivo* physiological and molecular responses of mycorrhizal lettuce plants. *Microbial Ecology* 55:45–53.

Jarošová, M., Klejdus, B., Kováčik, J., Babula, P., Hedbavny, J. (2016) Humic acid protects barley against salinity. *Acta Physiologiae Plantarum* 38:161. doi:10.1007/s11738-016-2181-z

Jayakannan, M., Bose, J., Babourina, O., Rengel, Z., Shabala, S. (2015) Salicylic acid in plant salinity stress signalling and tolerance. *Plant Growth Regulation* 76:25–40.

Kaur, H., Bhatla, S.C. (2016) Melatonin and nitric oxide modulate glutathione content and glutathione reductase activity in sunflower seedling cotyledons accompanying salt stress. *Nitric Oxide: Biology and Chemistry* 59:42–53.

Kim, G.B., Nam, Y.W. (2013) A novel Δ1-pyrroline-5-carboxylate synthetase gene of *Medicago truncatula* plays a predominant role in stress-induced proline accumulation during symbiotic nitrogen fixation. *Journal of Plant Physiology* 170:291–302.

Korkmaz, A., Karagöl, A., Horuz, A. (2016) The effects of humic acid added into the nutrient solution on yield and some fruit quality properties of tomato plant under the increasing NaCl stress conditions. *Anadolu Tarım Bilimleri Dergisi* 31:275–282.

Korkmaz, A., Sirikci, R. (2011) Improving salinity tolerance of germinating seeds by exogenous application of glycine betaine in pepper. *Seed Science and Technology* 39:377–388.

Kostopoulou, Z., Therios, I., Roumeliotis, E., Kanellis, A.K., Molassiotis, A. (2015) Melatonin combined with ascorbic acid provides salt adaptation in *Citrus aurantium* L. seedlings. *Plant Physiology and Biochemistry: PPB* 86:155–165.

Kovacs, Z., Simon-Sarkadi, L., Szücs, A., Kocsy, G. (2010) Different effects of cold, osmotic stress and abscisic acid on polyamine accumulation in wheat. *Amino Acids* 38:623–631.

Kumar, A., Dames, J.F., Gupta, A., Sharma, S., Gilbert, J.A., Ahmad, P. (2015) Current developments in arbuscular mycorrhizal fungi research and its role in salinity stress alleviation: A biotechnological perspective. *Critical Reviews in Biotechnology* 35:461–474.

Kumar, P., Sharma, V., Atmaram, C.K., Singh, B. (2017) Regulated partitioning of fixed carbon (^{14}C), sodium (Na^+), potassium (K^+) and glycine betaine determined salinity stress tolerance of gamma irradiated pigeon pea [*Cajanuscajan* (L.) Millsp]. *Environmental Science and Pollution Research International* 24:7285–7297. doi:10.1007/s11356-017-8406-x

Kusano, T., Berberich, T., Tateda, C., Takahashi, Y. (2008) Polyamines: Essential factors for growth and survival. *Planta* 228:367–381.

Lai, S.J., Lai, M.C., Lee, R.J., Chen, Y.H., Yen, H.E. (2014) Transgenic *Arabidopsis* expressing osmolyte glycine betaine synthesizing enzymes from halophilic methanogen promote tolerance to drought and salt stress. *Plant Molecular Biology* 85:429–441

Lee, S., Kim, S.G., Park, C.M. (2010) Salicylic acid promotes seed germination under high salinity by modulating antioxidant activity in *Arabidopsis. The New Phytologist* 188:626–637.

Li, H., Chang, J.J., Chen, H.J., Wang, Z.Y., Gu, X.R., Wei, C.H., Zhang, Y., *et al.* (2017) Exogenous melatonin confers salt stress tolerance to watermelon by improving photosynthesis and redox homeostasis. *Frontiers in Plant Science* 8:295. doi:10.3389/fpls.2017.00295

Li, J.T., Qiu, Z.B., Zhang, X.W., Wang, L.S. (2011) Exogenous hydrogen peroxide can enhance tolerance of wheat seedlings to salt stress. *Acta Physiologiae Plantarum* 33:835–842.

Li, C., Wang, P., Wei, Z., Liang, D., Liu, C., Yin, L., Jia, D., Fu, M., Ma, F. (2012) The mitigation effects of exogenous melatonin on salinity-induced stress in *Malus hupehensis. Journal of Pineal Research* 53:298–306.

Liang, C., Zheng, G., Li, W., Wang, Y., Hu, B., Wang, H., Wu, H., *et al.* (2015) Melatonin delays leaf senescence and enhances salt stress tolerance in rice. *Journal of Pineal Research* 59:91–101.

Liu, S., Dong, Y., Xu, L., Kong, J. (2014) Effects of foliar application of nitric oxide and salicylic acid on salt-induced changes in photosynthesis and antioxidative metabolism of cotton seedlings. *Plant Growth Regulation* 73:67–78.

Ma, X., Zheng, J., Zhang, X., Hu, Q., Qian, R. (2017) Salicylic acid alleviates the adverse effects of salt stress on *Dianthus superbus* (*Caryophyllaceae*) by activating photosynthesis, protecting morphological structure, and enhancing the antioxidant system. *Frontiers in Plant Science* 8:600. doi:10.3389/fpls.2017.00600

Matuszak-Slamani, R., Bejger, R., Cieśla, J., Bieganowski, A., Koczańska, M., Gawlik, A., Kulpa, D., *et al.* (2017) Influence of humic acid molecular fractions on growth and development of soybean seedlings under salt stress. *Plant Growth Regulation* 83:465–477.

Miura, K., Tada, Y. (2014) Regulation of water, salinity, and cold stress responses by salicylic acid. *Frontiers in Plant Science* 5:4. doi:10.3389/fpls.2014.00004

Molina, A., Bueno, P., Marín, M.C., Rodriguez-Rosales, M.P., Belver, A., Venema, K., Donaire, J.P. (2002) Involvement of endogenous salicylic acid content, lipoxygenase and antioxidant enzyme activities in the response of tomato cell suspension cultures to NaCl. *New Phytologist* 156:409–415.

Nahar, K., Hasanuzzaman, M., Rahman, A., Alam, M.M., Al Mahmud, J.A., Suzuki, T., Fujita, M. (2016) Polyamines confer salt tolerance in mung bean (*Vigna radiata* L.) by reducing sodium uptake, improving nutrient homeostasis, antioxidant defense, and methylglyoxal detoxification systems. *Frontiers in Plant Science* 7:1104. doi:10.3389/fpls.2016.01104

Navarro, J.M., Pérez-Tornero, O., Morte, A. (2014) Alleviation of salt stress in citrus seedlings inoculated with arbuscular mycorrhizal fungi depends on the rootstock salt tolerance. *Journal of Plant Physiology* 171:76–85.

Nawaz, K., Ashraf, M. (2007) Improvement in salt tolerance of maize by exogenous application of glycine betaine: Growth and water relations. *Pakistan Journal of Botany* 39:1647–1653.

Niu, L., Liao, W. (2016) Hydrogen peroxide signaling in plants development and abiotic responses: Crosstalk with nitric oxide and calcium. *Frontiers in Plant Science* 7:230. doi:10.3389/fpls.2016.00230

Nounjana, N., Nghiab, P.T., Theerakulpisut, P. (2012) Exogenous proline and trehalose promote recovery of rice seedlings from salt-stress and differentially modulate antioxidant enzymes and expression of related genes. *Journal of Plant Physiology* 169:596–604.

Ogawa, K., Iwabuchi, M. (2001) A mechanism for promoting the germination of *Zinnia elegans* seeds by hydrogen peroxide. *Plant and Cell Physiology* 42:286–291.

Ojala, J.C., Jarrell, W.M., Menge, J.A., Johnson, E.L.V. (1983) Influence of mycorrhizal fungi on the mineral nutrition and yield of onion in saline soil. *Agronomy Journal* 75:255–259.

Parvin, S., Lee, O.R., Sathiyaraj, G., Khorolragchaa, A., Kim, Y.J., Yang, D.C. (2014) Spermidine alleviates the growth of saline-stressed ginseng seedlings through antioxidative defense system. *Gene* 537:70–78.

Paul, E.A., Clark, F.E. (1989) *Soil Microbiology and Biochemistry.* Academic Press, San Diego.

Rady, M.M., Abd El-Mageed, T.A., Abdurrahman, H.A., Mahdi, A.H. (2016) Humic acid application improves field performance of cotton (*Gossypium barbadense* L.) under saline conditions. *Journal of Animal and Plant Sciences* 26:487–493.

Rajjou, L., Belghazu, M., Huget, R., Robin, C., Moreau, A., Job, C., Job, D. (2006) Proteomic investigation of the effect of salicylic acid on *Arabidopsis* seed germination and establishment of early defense mechanisms. *Plant Physiology* 141:910–923.

Rao, M.V., Davis, K.R. (1999) Ozone-induced cell death occurs via two distinct mechanisms in *Arabidopsis*: The role of salicylic acid. *The Plant Journal: for Cell and Molecular Biology* 17:603–614

Raza, S.H., Athar, H.R., Ashraf, M., Hameed, A. (2007) Glycine betaine induced modulation of antioxidant enzymes activities and ion accumulation in two wheat cultivars differing in salt tolerance. *Environmental and Experimental Botany* 60:368–376.

Rehman, R.U., Zia, M., Abbasi, B.H., Lu, G., Chaudhary, M.F. (2014) Ascorbic acid and salicylic acid mitigate NaCl Stress in *Carallum atuberculata* Calli. *Applied Biochemistry and Biotechnology* 173:968–979.

Rejeb, K.B., Lefevre-De Vos, D., Le Disquet, I., Leprince, A.S., Bordenave, M., Maldiney, R., Jdey, A., Abdelly, C., Savouré, A. (2015) Hydrogen peroxide produced by NADPH oxidases increases proline accumulation during salt or mannitol stress in *Arabidopsis thaliana*. *The New Phytologist* 208:1138–1148.

Roitsch, T. (1999) Source-sink regulation by sugar and stress. *Current Opinion in Plant Biology* 2:198–206.

Roy, P., Niyogi, K., SenGupta, D.N., Ghosh, B. (2005) Spermidine treatment to rice seedlings recovers salinity stress-induced damage of plasma membrane and PM-bound Hþ-ATPase in salt-tolerant and salt-sensitive rice cultivars. *Plant Science* 168:583–591.

Ruiz-Lozano, J.M., Azcón, R. (2000) Symbiotic efficiency and infectivity of an autochthonous arbuscular mycorrhizal *Glomus* sp. from saline soils and *Glomus deserticola* under salinity. *Mycorrhiza* 10:137–143.

Ruiz-Lozano, J.M., Azcón, R., Gómez, M. (1996) Alleviation of salt stress by arbuscular-mycorrhizal *Glomus* species in *Lactuca sativa* plants. *Physiologia Plantarum* 98:767–772.

Ruiz-Lozano, J.M., Porcel, R., Azcón, C., Aroca, R. (2012) Regulation by arbuscular mycorrhizae of the integrated physiological response to salinity in plants: New challenges in physiological and molecular studies. *Journal of Experimental Botany* 63:4033–4044.

Sabagh, A.E., Sorour, S., Ragab, A., Saneoka, H., Islam, M.S. (2017) The effect of exogenous application of proline and glycine betaine on the nodule activity of soybean under saline condition. *Journal of Agriculture Biotechnology* 2:01–05.

Sadak, M.S. (2016) Mitigation of salinity adverse effects of on wheat by grain priming with melatonin. *International Journal of ChemTech Research* 9:85–97.

Santa-Cruz, A., Acosta, M., Pérez-Alfocea, F., Bolarin, M.C. (1997a) Changes in free polyamine levels induced by salt stress in leaves of cultivated and wild tomato species. *Physiologia Plantarum* 101:341–346.

Santa-Cruz, A., Acosta, M., Rus, A., Bolarin, M.C. (1999) Short-term salt tolerance mechanisms in differentially salt tolerant tomato species. *Plant Physiology and Biochemistry* 37:65–71.

Santa-Cruz, A., Estañ, M.T., Rus, A., Bolarin, M.C., Acosta, M. (1997b) Effects of NaCl and mannitoliso-osmotic stresses on the free polyamine levels in leaf discs of tomato species differing in salt tolerance. *Journal of Plant Physiology* 151:754–758.

Santa-Cruz, A., Perez-Alfocea, F., Caro, M., Acosta, M. (1998) Polyamines as short-term salt tolerance traits in tomato. *Plant Science* 138:9–16.

Saxena, B., Shukla, K., Giri, B. (2017) Arbuscular mycorrhizal fungi and tolerance of salt stress in plants. In: Wu, Q.S. (ed.) *Arbuscular Mycorrhizas and Stress Tolerance of Plants*. Springer, Singapore, 67–98.

Schüßler, A., Schwarzott, D., Walker, C. (2001) A new fungal phylum, the Glomeromycota: Phylogeny and evolution. *Mycological Research* 105:1413–1421.

Shams, M., Yildirim, E., Ekinci, M., Turan, M., Dursun, A., Parlakova, F., Kul, R. (2016) Exogenously applied glycine betaine regulates some chemical characteristics and antioxidative defence systems in lettuce under salt stress. *Horticulture, Environment, and Biotechnology* 57:225–231. doi:10.1007/s13580-016-0021-0

Shan, C., Liu, R. (2017) Exogenous hydrogen peroxide up-regulates the contents of ascorbate and glutathione in the leaves of *Vigna radiata* (Linn.) Wilczek. exposed to salt stress. *Brazilian Journal of Botany* 40:583–589.

Shu, S., Yuan, L.Y., Guo, S.R., Sun, J., Yuan, Y.H. (2013) Effects of exogenous spermine on chlorophyll fluorescence, antioxident system and ultrastructure of chloroplasts in *Cucumis sativus* L. under salt stress. *Plant Physiology and Biochemistry: PPB* 63:209–216.

Singh, M., Kumar, J., Singh, V.P., Prasad, S.M. (2014) Proline and salinity tolerance in plants. *Biochemical Pharmacology* 3:6. doi:10.4172/2167-0501.1000e170

Smith, S.E., Read, D.J. (1997) *Mycorrhizal Symbiosis*. Academic Press, San Diego.

Szepesi, A. (2006) Salicylic acid improves the acclimation of *Lycopersicon esculentum* Mill. L. to high salinity by approximating its salt stress response to that of the wild species *L. pennellii*. *Acta Biologica Szegediensis* 50:177.

Takatsuji, H., Jiang, C.J. (2014) Plant hormone crosstalks under biotic stresses. In: Tran, L.S., Pal, S. (eds.) *Phytohormones: A Window to Metabolism, Signaling and Biotechnological Applications*. Springer, New York, 323–350.

Tan, D.X., Hardeland, R., Manchester, L.C., Korkmaz, A., Ma, S., Rosales-Corral, S., Reiter, R.J. (2012) Functional roles of melatonin in plants, and perspectives in nutritional and agricultural science. *Journal of Experimental Botany* 63:577–597.

Tanou, G., Ziogas, V., Belghazi, M., Christou, A., Filippou, P., Job, D., Fotopoulos, V., Molassiotis, A. (2014) Polyamines reprogram oxidative and nitrosative status and the proteome of citrus plants exposed to salinity stress. *Plant, Cell and Environment* 37:864–885.

Tian, F., Wang, W., Liang, C., Wang, X., Wang, G., Wang, W. (2017) Over accumulation of glycine betaine makes the function of the thylakoid membrane better in wheat under salt stress. *The Crop Journal* 7:73 –82.

Uchida, A., Jagendorf, A.T., Hibino, T., Takabe, T., Takabe, T. (2002) Effects of hydrogen peroxide and nitric oxide on both salt and heat stress tolerance in rice. *Plant Science* 163:515–523.

Wahid, A., Perveen, M., Gelani, S., Basra, S.M.A. (2007) Pretreatment of seed with H_2O_2 improves salt tolerance of wheat seedlings by alleviation of oxidative damage and expression of stress proteins. *Journal of Plant Physiology* 164:283–294

Wei, W., Li, Q.T., Chu, Y.N., Reiter, R.J., Yu, X.M., Zhu, D.H., Zhang, W.K., *et al.* (2015) Melatonin enhances plant growth and abiotic stress tolerance in soybean plants. *Journal of Experimental Botany* 66:695–707.

Wei, D., Zhang, W., Wang, C., Meng, Q., Li, G., Chen, T.H.H., Yang, X. (2017) Genetic engineering of the biosynthesis of glycine betaine leads to alleviate salt-induced potassium efflux and enhances salt tolerance in tomato plants. *Plant Science: an International Journal of Experimental Plant Biology* 257:74–83.

Wu, Q.S., Zou, Y.N., He, X.H. (2010a) Contributions of arbuscular mycorrhizal fungi to growth, photosynthesis, root morphology and ionic balance of citrus seedlings under salt stress. *Acta Physiologiae Plantarum* 32:297–304.

Wu, Q.S., Zon, Y.N., Liu, W., Ye, X.F., Zai, H.F., Zhao, L.J. (2010b) Alleviation of salt stress in citrus seedlings inoculated with mycorrhiza: Changes in leaf antioxidant defense systems. *Plant, Soil and Environment* 56:470–475.

Wutipraditkul, N., Wongwean, P., Buaboocha, T. (2015) Alleviation of salt-induced oxidative stress in rice seedlings by proline and/or glycine betaine. *Biologia Plantarum* 59:547–553.

Yang, S.L., Lan, S.S., Gong, M. (2009) Hydrogen peroxide-induced proline and metabolic pathway of its accumulation in maize seedlings. *Journal of Plant Physiology* 166:1694–1699.

Yildirim, E., Ekinci, M., Turan, M., Dursun, A., Kul, R., Parlakova, F. (2015) Roles of glycine betaine in mitigating deleterious effect of salt stress on lettuce (*Lactuca sativa* L.). *Archives of Agronomy and Soil Science* 61:1673–1689.

Zhao, F.G., Qin, P. (2004) Protective effect of exogenous polyamines on root tonoplast function against salt stress in barley seedlings. *Plant Growth Regulation* 42:97–103.

Zheng, X.D., Tan, D.X., Allan, A.C., Zuo, B.X., Zhao, Y., Reiter, R.J., Wang, L., *et al.* (2017) Chloroplastic biosynthesis of melatonin and its involvement in protection of plants from salt stress. *Scientific Reports* 7:41236. doi:10.1038/srep41236

Zhong, M., Yuan, Y., Shu, S., Sun, J., Guo, S., Yuan, R., Tang, Y. (2016) Effects of exogenous putrescine on glycolysis and Krebs cycle metabolism in cucumber leaves subjected to salt stress. *Plant Growth Regulation* 79:319–330.

Zuccarini, P., Okurowska, P. (2008) Effects of mycorrhizal colonization and fertilization on growth and photosynthesis of sweet basil under salt stress. *Journal of Plant Nutrition* 31:497–513.

23 Enhancement of Abiotic Stress Tolerance in Plants by Probiotic Bacteria

Md. Mohibul Alam Khan, Patrick Michael Finnegan, Sajid Mahmood, Yasir Anwar, Saleh M. S. Al-Garni, Ahmed Bahieldin, and Md. Tofazzal Islam

CONTENTS

23.1 INTRODUCTION

Plants are sessile organisms, so are constantly subjected to a wide range of abiotic stress conditions such as salinity, drought, flooding, extreme temperatures, heavy metals, and pollutants. These abiotic stresses may negatively affect both natural environments and in cropping systems of plants. These stresses negatively impact yields of many crops worldwide (Forni et al., 2016). To ensure food security for a fast-growing world population, it is essential to significantly increase agricultural productivity under normal as well as suboptimal conditions within the next few decades (Glick, 2014; Forni et al., 2016).

23.1.1 Abiotic Stressors and their Effects on Plant

Soil salinity is a major environmental stress that adversely affects the soil quality of agricultural lands (Radhakrishnan and Baek, 2017). When the electrical conductivity of a saturated paste soil extract (ECe) exceeds 4 dS m^{-1} (approximately 40 mM NaCl) at 25°C and has exchangeable sodium ions (Na^+) of about 15%, the soil is usually classified as saline (Shrivastava and Kumar, 2015). At present, about 20% of total cultivated lands and 50% of irrigated agricultural lands across the globe suffer from a high salt concentration (Cheng et al., 2012). Moreover, the salt-affected arable land is increasing by about 10% each year. It has been estimated that about half of the agricultural land will be salinized by 2050 (Jamil et al., 2011). Salinity stress detrimentally affects almost every aspect of plant growth and development including seed germination, water and nutrient uptake, photosynthetic efficiency, total biomass, and yields. These effects are primarily triggering by ion imbalances, ion-toxicity-induced metabolic imbalances and hyperosmotic stress-induced water deficits (Bharti et al., 2016; Cardinale et al., 2015; Li et al., 2017). Salinity stress induces the generation of reactive oxygen species (ROS) such as superoxide radical (O_2^-), hydrogen peroxide (H_2O_2), and hydroxyl radical (OH·) that cause oxidative damage to proteins, lipids, and deoxyribonucleic acid (DNA), and impair normal functions in plant cells (Chen et al., 2016; Habib et al., 2016; Islam et al., 2016; Radhakrishnan and Baek, 2017). ROS are also responsible for the synthesis of stress-induced ethylene, which is often involved in initiating responses such as premature senescence, chlorosis, and leaf abscission (Glick, 2014).

Drought is another major constraint on agricultural productivity worldwide and is likely to affect more than 50% of arable lands by 2050 (Vurukonda et al., 2016). It affects plant-water potential, turgor pressure and photosynthesis as well as availability and transport of soil nutrients, resulting in stunted plant growth and reduced yield (Kang et al., 2014a; Rolli et al., 2015; Vurukonda et al., 2016). Like salinity, drought stress also stimulates the production of ROS and stress-induced ethylene in plants (Kumar et al., 2016; Naveed et al., 2014; Xu et al., 2012). In fact, drought is multidimensional stress, which is more pervasive and devastating than salinity; however, the response of plants to both stresses are closely related (Hussain et al., 2008), largely because salinity imposes physiological drought of cells.

Flooding decreases the availability of oxygen and light to plants, resulting in an energy and carbohydrate crisis (Voesenek and Bailey-Serres, 2015). Due to global climatic change, extreme flooding events are predicted to increase in many areas around the world. Under submergence conditions, there is an increased accumulation of 1-aminocyclopropane-1-carboxylate (ACC) in root tissues, which is transported to the shoots and converted into ethylene (Barnawal et al., 2012; Glick et al., 2007). This higher concentration of ethylene causes epinasty, leaf chlorosis, necrosis, and reduced crop yield (Glick et al., 2007).

Low temperatures encourage the growth of saprophytic fungi and reduce soil fertility by disturbing the natural soil nutrient cycling that often limits geographical distribution and agricultural productivity of crops worldwide (Subramanian et al., 2016; Templer, 2012). At low, non-freezing temperatures, cell membranes of chilling-sensitive plants harden, which perturbs various cellular processes such as the opening of ion channels and membrane-associated electron transfer reactions (Uemura and Steponkus, 1999). Exposure to chilling conditions also disrupts cellular homeostasis and photosynthesis, and increases the generation of ROS in plants, and may lead to cell death (Barka et al., 2006; Gill and Tuteja, 2010; Ruelland et al., 2009). On the other hand, elevated temperatures decrease water status, photosynthesis rate, and increase ROS production in plants, which may severely damage plant cells and inhibit developmental processes within a short period (El-Daim et al., 2014). High temperatures may also detrimentally affect the integrity and functions of biological membranes by altering the tertiary and quaternary structures of membrane proteins (Redondo-Gómez, 2013).

Heavy metal pollution due to industrial and agricultural activities, for example, mining and smelting of metalliferous ores, wastewater irrigation, and abuse of synthetic fertilizers and pesticides, is increasing in the environment. In the presence of high levels of toxic metals such as aluminum (Al), arsenic (As), cadmium (Cd), copper (Cu), lead (Pb),nickel (Ni), zinc (Zn), and so on, and organic contaminants such as oil, polycyclic aromatic hydrocarbons (PAHs), and polycyclic biphenyls (PCBs), most plants produce ethylene and ROS, and also become severely iron (Fe) depleted. Together, these conditions lead to chlorosis, necrosis, root system damage, photosynthesis inhibition, plasma membrane permeability damage, and reduced biomass production (Adrees et al., 2015; Ahmad et al., 2014; Al-Khateeb and Al-Qwasemeh, 2014; Glick et al., 2007). Heavy metal pollution, together with the other abiotic stressors mentioned here, adversely affects plant growth, physiology, and yield. Therefore, as a group, these stressors are considered to be a major threat to the agricultural industry and global food security.

23.1.2 Mechanisms of Plant Adaptation under Adverse Environmental Conditions

Several mechanisms have evolved in plants that provide tolerance to abiotic stress. These comprise physiological, biochemical, molecular, and genetic changes that allow the plant to cope with the negative effects of various abiotic stresses (Figure 23.1; Redondo-Gómez, 2013). However, the degree of tolerance is variable from plant species to species. Some plants may be highly sensitive, while others may be relatively immune to particular abiotic stress. To adapt to the saline environment, plants primarily have three strategies: (i) active Na^+ efflux; (ii) Na^+ influx prevention; and (iii) intracellular ion

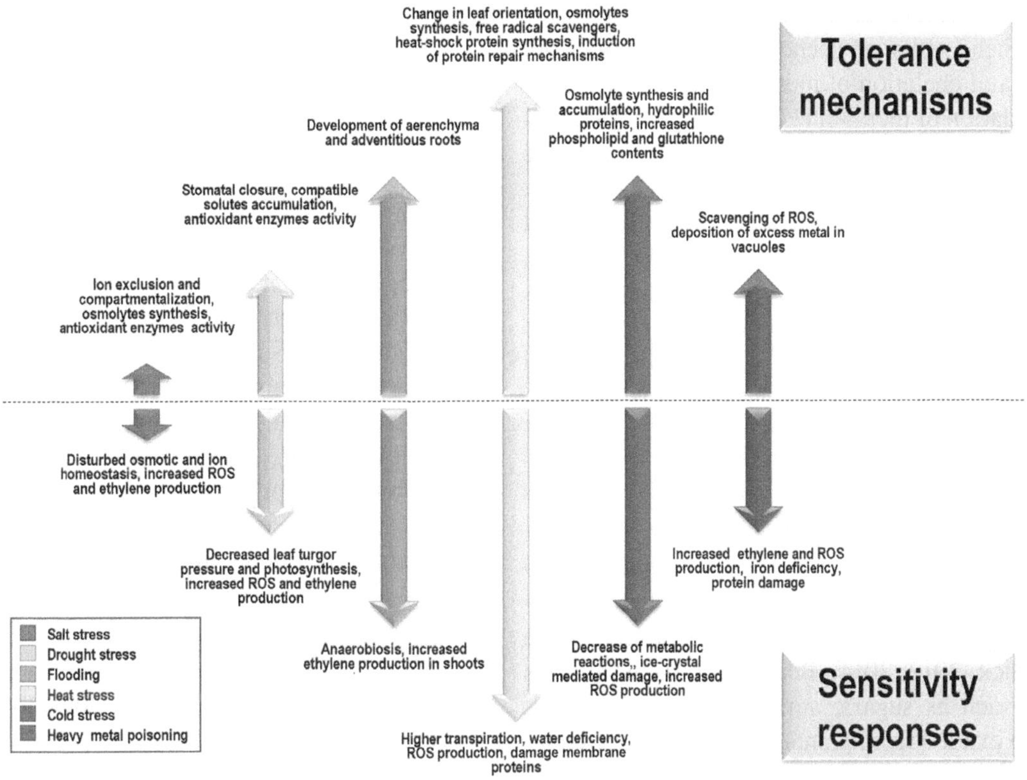

FIGURE 23.1 Effects of various abiotic stresses on plants and their mechanisms of tolerance.

compartmentalization (Rajendran et al., 2009). Sodium exclusion is accomplished by H^+-ATPase pumps and Na^+/H^+ antiporters, while salt glands and bladders are the principal salt-accumulating structures in plants (Redondo-Gómez, 2013). In addition, plants develop antioxidant defense systems that play key roles in balancing and preventing oxidative damage caused by ROS produced as a result of various stresses. These defense systems comprise both enzymatic components such as superoxide dismutases (SOD), catalase (CAT), and ascorbate peroxidases (APX), and non-enzymatic components such as cysteine, glutathione (GSH), ascorbic acid, and flavonoids (Foyer et al., 1994). Under drought conditions, plants reduce transpirational water loss by closing their stomata, where abscisic acid (ABA) plays the pivotal role (Assmann et al., 2000; Voesenek and Van der Veen, 1994). They also synthesize and accumulate compatible solutes such as proline (Pro), sorbitol, glycine betaine (GB), and soluble carbohydrates that help maintain turgor pressure and cell water content (Orcutt and Nilsen, 2000; Yokota et al., 2006). Osmolytes also help in maintaining the structural integrity of biological membranes by replacing the hydroxyl (OH)group of water with the OH group of sugar alcohols and by maintaining the hydrophilic interaction with membrane lipids and proteins (Hoekstra and Buitink, 2001; Tamura et al., 2003). Plants tolerant to flood-induced water-logging stress have structural adaptations such as the formation of aerenchyma and adventitious roots (Ashraf, 2012). Change in leaf orientation, closure of stomata, transpirational cooling, alteration of membrane lipid composition, and production of ion transporters, osmoprotectants, and free-radical scavengers are some of the major plant tolerance mechanisms that counteract the effects of heat stress (Wahid et al., 2007; Wang et al., 2004). However, the production of heat-shock proteins (HSPs) that improve photosynthesis, assimilate partitioning, water and nutrient use efficiency, and membrane stability is also considered to be an important adaptive strategy for heat stress (Feder and Hoffman, 1999; Wahid et al., 2007). In contrast, an increase in phospholipids, especially phosphatidyl ethanolamine, was reported to be associated with freezing tolerance in plants (Yoshida and Uemura, 1984). In addition, glutathione may protect membrane proteins from denaturation during cold stress (Redondo-Gómez, 2013). Soluble sugars and other osmolytes can protect cell membranes and organelles by decreasing the degree of freeze-induced dehydration (Rajashekar, 2000; Trischuk et al., 2006). The sequestering of metals in the cell walls or cellular compartments like vacuoles has been described as a tolerance mechanism to heavy metal toxicity in plants (Weis and Weis, 2004).

Additionally, excretion of metals, for instance, Cd^{2+} through the head cells of trichomes in tobacco (Choi et al., 2001), and chelation of metals in the extracellular space by exudation of metal-chelating substances including organic (e.g., carbohydrates, organic acids [e.g., citric acid, succinic acid]) as well as inorganic (e.g., Cl^-, SO_4^{2-}, NH_4^+, CO_3^{2-}) ligands provide tolerance to plants against various heavy metals (Bertrand et al., 2001; Dong et al., 2007). Furthermore, cysteine-rich polypeptides such as metallothioneins (MTs) and phytochelatins (PCs) have been reported to play an important role in cellular metal homeostasis and protection (Cobbett and Goldsbrough, 2002; Steffens, 1990).

23.1.3 Plant Probiotic Bacteria

Most soils contain an enormous diversity of microorganisms including bacteria, fungi, algae, and protozoa; however, bacteria are by far the most common among these microbes (Glick, 2012, 2014). Because of the high levels of nutrients such as sugars, amino acids, and organic acids that are exuded from plant roots, the bacterial concentration in the rhizosphere zone is typically higher than in the bulk soil (Glick, 2012). The interactions between a plant and the bacteria in its rhizosphere may be beneficial, harmful, or neutral for the plant (Forni et al., 2016). The terms "plant probiotic bacteria" and "plant growth promoting bacteria (PGPB)" have been used to describe rhizospheric, phyllospheric, and endophytic associations between a plant and bacteria that directly or indirectly facilitate plant growth and development (Glick, 2012; Islam and Hossain, 2012). Interestingly, plant probiotic bacteria belonging to a wide range of genera such as *Achromobacter, Acinetobacter, Azospirillum, Bacillus, Burkholderia, Enterobacter, Klebsiella, Ochrobactrum, Paenibacillus, Pseudomonas, Rhizobium, Serratia*, and *Sphingomonas* have been found to enhance tolerance of host plants to various abiotic stresses (Tables 23.1 through 23.3). The term "induced systemic tolerance" (IST) is used to define this phenomenon (Yang et al., 2009). Bacteria that naturally inhabit sites that frequently experience conditions that are stressful to plants are likely to be more tolerant to that stress and may exhibit better performance in terms of augmenting plant growth and yield under that stress condition (Lifshitz et al., 1986; Timmusk et al., 2011). Therefore, the use of probiotic bacteria as bioinoculants is considered to be potentially cost-effective and environmentally friendly strategy to increase crop productivity in marginalized agricultural lands (Glick, 2014; Grover et al., 2011; Radhakrishnan and Baek, 2017; Yang et al., 2009).

23.2 ALLEVIATION OF ABIOTIC STRESSES IN PLANTS BY PROBIOTIC BACTERIA

There is ample evidence that shows plant probiotic bacteria can adapt themselves to adverse environmental conditions and effectively alleviate the negative impacts on plants of various environmental stresses like salinity (Li et al., 2017; Mahmood et al., 2016), drought (Belimov et al., 2015; Raheem et al., 2017), flooding (Barnawal et al., 2012; Ravanbakhsh et al., 2017), extreme temperatures (Kang et al., 2015; Subramanian et al., 2016), heavy metals and organic contaminants (Burd et al., 2000; Gurska et al., 2009; Ma et al., 2017), and nutrient deficiency (İpek et al., 2017; Meldau et al., 2013). These enhancements improve plant growth and productivity. Tables 23.1 through 23.3 summarize reports on the inoculation effects of probiotic bacteria on plants under diverse abiotic stresses.

23.2.1 Salinity

There are many reports where inoculation with plant probiotic bacteria enhanced tolerance to salinity stress, and augmented growth, physiology, and yield in a variety of plants (Table 23.1). For example, Mahmood et al. (2016) recorded a significant increase in stomatal conductance, transpiration rates, relative water content (RWC), photosynthetic pigments (chlorophyll a, chlorophyll b and carotenoids), plant height, leaf area, dry biomass, seed yield, and salt tolerance index in mung bean plants inoculated with bacterial strains *Bacillus pretenses* P16 and *Enterobacter cloacae* P6 compared with non-inoculated (control) plants under salt-stressed field conditions. Similarly, hydroponically grown wheat cultivar (cv.) HD 2285 plants inoculated with *Dietzia natronolimnaea* STR1 exhibited better performance regarding photosynthetic pigments, plant height, shoot and root length, and biomass production in comparison with control treatments under salinity stress (Bharti et al., 2016). Recently, Radhakrishnan, and Baek (2017) reviewed the effects of probiotic bacteria on non-salt-tolerant glycophytic plants grown in salt-stressed soils and concluded that the use of probiotic bacteria could reprogram the expression of salt-stress-responsive genes and proteins in salt-affected plants, and thereby can improve plant defense mechanism.

23.2.2 Drought

Use of plant-associated bacteria has been documented to be a useful tool for ameliorating drought stress in host plants (Barnawal et al., 2017; Yang et al., 2009). For

TABLE 23.1
Effects of Probiotic Bacteria and Their Mechanisms to Improve Tolerance in Host Plants to Salinity Stress

Plant Species	Bacterial Inoculate	Effects on Host Plant	Suggested Mechanism(s)	References
Arabidopsis thaliana L.	*Paenibacillus yonginensis* DCY84	Improved plant growth.	IAA, P solubilization, siderophores. Up-regulating plant salinity responsive genes *AtRSA1* and *AtWRKY8*, and down-regulating *AtVQ9*.	Sukweenadhi et al. (2015)
Arabidopsis thaliana L.	*Pseudomonas lini* KBEcto4	Enhanced plant growth and biomass production.	ACC deaminase, IAA, P solubilization, siderophores.	Palacio-Rodríguez et al. (2017)
Barley (*Hordeum vulgare* L.)	*Curtobacterium flaccumfaciens* E108	Increased seed germination rate, RWC, and plant biomass production.	IAA, P solubilization.	Cardinale et al. (2015)
Canola (*Brassica napus* L.)	*Enterobacter cloacae* HSNJ4	Increased seed germination rate, plant growth and biomass production, chlorophyll, and Pro contents; decreased malondialdehyde (MDA) and ethylene contents.	ACC deaminase, IAA. Increasing antioxidant enzymes activity and IAA content in plants.	Li et al. (2017)
Chinese cabbage (*Brassica rapa* L.)	*Herbaspirillum* sp. GW103	Increased K^+/Na^+ ratio in roots and plant biomass production.	ACC deaminase, IAA, siderophores.	Lee et al. (2016)
Cotton (*Gossypium hirsutum* L.)	*Klebsiella oxytoca* Rs-5	Increased seed germination rate, plant growth, and biomass production; reduced MDA and Pro contents.	ACC deaminase, IAA. Increasing IAA content in plants.	Liu et al. (2013a)
Cucumber (*Cucumis sativus* L.)	*Acinetobacter calcoaceticus* SE370, *Burkholderia cepacia* SE4, *Promicromonospora* sp. SE188	Increased plant growth and biomass production, water potential, chlorophyll, P and K contents; decreased electrolytic leakage and Na^+ content.	Gibberellins, P solubilization. Regulating plant endogenous ABA, gibberellins, and SA production.	Kang et al. (2014a)
Maize (*Zea mays* L.)	*Bacillus amyloliquefaciens* SQR9	Increased plant growth and biomass production, chlorophyll content; decreased Na^+ content.	IAA. Increasing total soluble sugar and glutathione contents, peroxidase/catalase activity, decreasing ABA content, up-regulating *RBCS*, *RBCL*, *H^+-PPase*, *HKT1*, *NHX1*, *NHX2* and *NHX3*, and down-regulating *NCED* genes expression in plants.	Chen et al. (2016)
Mung bean (*Vigna radiata* L.)	*Bacillus cereus* Pb25	Increased plant growth, biomass, and yield; chlorophyll content; N, P and K contents; decreased Na^+, MDA and H_2O_2 contents.	ACC deaminase, IAA, P solubilization, siderophores. Increasing plant antioxidant enzymes activity, and improving soil fertility.	Islam et al. (2016)
Oats (*Avena sativa*)	*Acinetobacter* sp. CMH2 and *Pseudomonas corrugate* CMH3	Enhanced plant biomass production.	ACC deaminase, IAA.	Chang et al. (2014)

(Continued)

TABLE 23.1 (CONTINUED)
Effects of Probiotic Bacteria and Their Mechanisms to Improve Tolerance in Host Plants to Salinity Stress

Plant Species	Bacterial Inoculate	Effects on Host Plant	Suggested Mechanism(s)	References
Okra (*Abelmoschus esculentus* L.)	*Bacillus megaterium* UPMR2, *Enterobacter* sp. UPMR18	Increased seed germination rate, plant growth and biomass production, chlorophyll content.	ACC deaminase, IAA, N-fixation, P solubilization. Up-regulating the expression of ROS pathway genes (*CAT*, *APX*, *GR*, and *DHAR*) in plants.	Habib et al. (2016)
Pea (*Pisum sativum* L.)	*Variovorax paradoxus 5C-2*	Increased plant growth and biomass production, photosynthetic efficiency, maximal electron transport rate, and K content; decreased stomatal resistance, xylem balancing pressure, and Na^+ content.	ACC deaminase.	Wang et al. (2016)
Peanut (*Arachishypogaea* L.)	*Agrobacterium tumefaciens* MBE01, *Klebsiella* sp. MBE02, *Ochrobactrum anthropi* MBE03, *Pseudomonas stutzeri* MBE04, *Pseudomonas* sp. MBE05	Increased plant growth and biomass production.	ACC deaminase, IAA, N-fixation, P solubilization. Inducing *CAT*, *APX*, and *SOD* genes expression in plants.	Sharma et al. (2016)
Quinoa (*Chenopodium quinoa*)	*Bacillus* sp. MN54, *Enterobacter* sp. MN17	Increased plant growth, biomass and grain yield, photosynthetic efficiency, stomatal conductance, plant water relations, and K content; decreased Na^+ content.	ACC deaminase, EPS, IAA. Reducing ABA content in leaves.	Yang et al. (2016)
Rice (*Oryza sativa* L.)	*Bacillus pumilus*, *Pseudomonas pseudoalcaligenes*	Increased plant growth and biomass production; decreased lipid peroxidation and SOD activity.	Reducing plant cell membrane index, cell caspase-like protease activity, and programmed cell death.	Jha and Subramanian (2014)
Tomato (*Solanum lycopersicum* L.)	*Sphingomonas* sp. LK11	Increased plant growth and biomass production; reduced lipid peroxidation.	Gibberellins. Regulating antioxidant enzymes activity in plants.	Halo et al. (2015)

instance, Naseem and Bano (2014) showed improved drought tolerance in maize (*Zea mays*) cv. Agaiti-2002 when the seed was inoculated with the exopolysaccharide (EPS)-producing bacteria *Proteus penneri* (Pp1), *Pseudomonas aeruginosa* (Pa2), and *Alcaligenes faecalis* (AF3) alone and in combination with their respective EPS. The treatments improved soil moisture content, RWC, root and shoot length, leaf area, and plant biomass production compared with control plants. Recently, Raheem et al. (2017) reported that wheat cv. FD2006 plants treated with *Bacillus amyloliquefaciens* S-134 had remarkably increased shoot length over control plants under drought of 10% field water-holding capacity (FC). In addition, they observed a 34% increase in spike length when *B. muralis* D-5 was the treatment and a 2-fold increase in seed weight when *Enterobacter aerogenes* S-10 was the treatment. Moreover, two mixed culture combinations displayed significant improvement in the number of tillers and spikelets (Raheem et al., 2017). Enhanced drought stress tolerance after bacterial treatment has also been documented in many other crops (Table 23.2). The ability of plant-associated bacteria to confer drought stress tolerance to host plants has been recently discussed by Vurukonda et al. (2016).

23.2.3 Flooding

Several studies have mentioned enhanced tolerance to flooding stress, and improved growth and biomass production in various plants pre-treated with probiotic bacteria (Table 23.2). A recent investigation by Ravanbakhsh et al. (2017) demonstrated that inoculation of marsh dock (*RumexpalustrisSm*) with *Pseudomonas putida* UW4 significantly reduced stress-induced ethylene production and improved submergence-induced young leaf and petiole elongation, shoot and root fresh and dry weight, and plant height compared with non-treated control plants exposed to short-term (3 days) and long-term (17 days) submergence. These results suggest that pre-treatment with bacteria could be an effective strategy to ameliorate flooding stress. However, more studies are required to assess the impact of bacterial pre-treatment on water-logging stress, as very little work has been done in this area.

23.2.4 Extreme Temperatures

Inoculation with plant-associated bacteria was found to augment tolerance to extremes of both high and low temperatures in several plants (Table 23.3). As an example, at normal temperatures, Kang et al. (2015) recorded about 55% and 15% increases in the shoot and root lengths, respectively, and a 1.8-fold increase in fresh biomass in pepper plants inoculated with *Serratia nematodiphila* PEJ1011 compared with uninoculated control plants. They also assessed the effect of low-temperature stress (5°C for 4 h) on randomly selected plants from both inoculated and control treatments. They found that plants treated with bacteria-maintained vigor and performed better regarding the shoot and root lengths and biomass production than uninoculated plants. In addition, El-Daim et al. (2014) observed greater than 40% higher survival rates to short-term heat stress (45°C for 24 h) in wheat cvs. Olivine and Sids1 plants pre-treated with *B. amyloliquefaciens* UCMB5113 or *Azospirillum brasilense* NO40 compared with untreated control plants. They also noted increased water content and biomass production in the bacteria-treated plants compared with untreated controls.

23.2.5 Other Abiotic Stresses

Treatment of plants with probiotic bacteria augmented plant growth and health, and increased plant tolerance to stresses imposed by toxic metals such as As, Cd, Cu, Pb, Ni, and Zn, among others (Burd et al., 2000; Han et al., 2015; Ma et al., 2016a, 2017); organic contaminants such as oil and PAHs (Glick, 2014; Gurska et al., 2009); and nutrient deficiency (Dimkpa et al., 2009a; Ípek et al., 2017; Khan et al., 2017) (Table 23.3). Hence, the use of plant-associated bacteria to improve phytoremediation efficiency (Ma et al., 2016b) and enhancement of crop production in nutrient deficient soils (Dimkpa et al., 2009a; Ípek et al., 2017) appears to be an effective, environmentally friendly and potentially cost-effective solution to abiotic stresses. For instance, Pishchik et al. (2002) reported that barley plants inoculated with commercially available *Klebsiella mobilis* CIAM 880 growing on Cd-contaminated field (5 mg Cd Kg^{-1} soil) soil had 120% higher grain yield and 2-fold reduced content (0.25 g Cd Kg^{-1} grain) in grains compared with non-inoculated plants. Likewise, in a three-year field trial at a site contaminated with petroleum hydrocarbons, Gurska et al. (2009) observed improved photosynthetic activity, growth and biomass production in annual ryegrass, tall fescue, barley, and fall rye treated with *Pseudomonas putida* UW4 and *Pseudomonas* sp. UW3 compared with untreated plants. Recently, in a study with rice (*Oryza sativa* L.) cv. BRRIdhan29 (Khan et al., 2017) demonstrated that bacteria treatment could enhance plant nutrient use efficiency. In this study, plants inoculated with *Burkholderia* sp. BRRh-4 or *Pseudomonas aeruginosa* BRRh-5 and given 50% of the recommended amount of N-P-K fertilizer produced an equivalent or higher yield than uninoculated plants that were given the full recommended amount of fertilizer. Similarly, inoculation with six probiotic bacteria increased plant growth and Fe nutrition in pear grown on calcareous soil that locked up Fe (Ípek et al., 2017).

23.3 MECHANISMS FOR ABIOTIC STRESS TOLERANCE IN PLANTS INDUCED BY PROBIOTIC BACTERIA

Plant probiotic bacteria have been reported to enhance abiotic stress tolerance in plants through a variety of mechanisms (Figure 23.2; Tables 23.1 through 23.3; Glick, 2014; Liu and Zhang, 2015; Yang et al., 2009; Forni et al., 2016; Radhakrishnan and Baek, 2017). Bacterial traits such as the production of phytohormones (Egamberdieva, 2009; Raheem et al., 2017), ACC deaminase (Etesami et al., 2014; Ravanbakhsh et al., 2017),

TABLE 23.2
Effects of Probiotic Bacteria and their Mechanisms to Enhance Tolerance in host Plants Under Drought and Flooding Conditions

Stress Type	Plant Species	Bacterial Inoculate	Effects on Host Plant	Suggested Mechanism(s)	References
Drought	Alfalfa (*Medicago sativa*)	*Sinorhizobium meliloti* LMG202	Increased plant survival rate, fresh shoot and root weight, and RWC; delayed premature senescence, reduced H_2O_2 content in leaves.	Cytokinin up-regulating the expression of *SOD*, *CAT*, *sAPX*, *thylAPX*, *DHAR*, *MDHAR*, *GR*, and *GPX* genes in plants.	Xu et al. (2012)
Drought	Chickpea (*Cicer arietinum* L.)	*Bacillus amyloliquefaciens* NBRISN13, *Pseudomonas putida* NBRIRA	Increased number of nodules, plant growth, and biomass production.	ACC deaminase, IAA, P solubilization, siderophores, biofilm formation. Modulating plant antioxidant enzymes activity, soil enzymes, and microbial diversity.	Kumar et al. (2016)
Drought	Common beans (*Phaseolus vulgaris*)	*Rhizobium etli* Ox	Increased plant survival rates and biomass production, number of nodules, nitrogenase activity, and RWC.	Trehalose	Suarez et al. (2008)
Drought	Cucumber (*Cucumis sativus* L.)	*Acinetobacter calcoaceticus* SE370, *Burkholderia cepacia* SE4, *Promicromonospora* sp. SE188	Increased plant growth and biomass production, leaf water potential, chlorophyll content, P and K contents; decreased electrolytic leakage.	Gibberellins, P solubilization. Regulating plant endogenous ABA, SA, and gibberellins production.	Kang et al. (2014a)
Drought	Grapevine (*Vitis vinifera* L.)	*Acinetobacter* sp. S2, *Pseudomonas* sp. S1 and S3	Increased plant growth and biomass production, photosynthetic efficiency, transpiration rate, stomatal conductance, water use efficiency; decreased internal CO_2.	ACC deaminase, EPS, IAA, P solubilization, siderophores.	Rolli et al. (2015)
Drought	Lettuce (*Lactuca sativa* L.)	*Bacillus subtilis* IB-22	Enhanced shoot biomass.	Cytokinin. Modulating plant endogenous cytokinin and ABA contents.	Arkhipova et al. (2007)
Drought	Maize (*Zea mays* L.)	*Burkholderia phytofirmans*PsJN, *Enterobacter* sp. FD17	Increased plant growth and biomass production, chlorophyll content, photosynthetic efficiency, RWC, stomatal conductance, transpiration rate, vapor pressure deficit, number of leaves per plant, leaf area; decreased relative membrane permeability, ethylene, and H_2O_2 contents.	ACC deaminase, IAA.	Naveed et al. (2014)
Drought	Mung bean (*Vigna radiata* L.)	*Pseudomonas aeruginosa* GGRJ21	Increased plant growth, biomass and yield, chlorophyll and RWC contents.	ACC deaminase, IAA, P solubilization, siderophores. Increasing antioxidant enzymes activity, osmolytes accumulation, and up-regulating *DREB2A*, *CAT1*, *DHN* genes expression in plants.	Sarma and Saikia (2014)

(*Continued*)

TABLE 23.2 (CONTINUED)
Effects of Probiotic Bacteria and their Mechanisms to Enhance Tolerance in host Plants Under Drought and Flooding Conditions

Stress Type	Plant Species	Bacterial Inoculate	Effects on Host Plant	Suggested Mechanism(s)	References
Drought	Pepper (*Capsicum annuum* L.)	*Bacillus licheniformis* K11	Increased plant survival rate, growth, and biomass production.	ACC deaminase, IAA. Up-regulating the expression of *Cadhn*, *VA*, *sHSP*, *CaPR-10* in plants.	Lim and Kim (2013)
Drought	Potato (*Solanum tuberosum* L.)	*Achromobacter xylosoxidans* Cm4, *Pseudomonas oxyzihabitans* Ep4, *Variovorax paradoxus* 5C-2	Increased water use efficiency, root biomass, and tuber yield.	ACC deaminase, IAA.	Belimov et al. (2015)
Drought	Rice (*Oryza sativa* L.)	*Bacillus amyloliquefaciens* Bk7, *Brevibacillus laterosporus* B4	Increased plant survival rate and growth, chlorophyll and Pro content; decreased MDA content and electrolyte leakage.	IAA, P solubilization, siderophores. Modulating antioxidant enzymes activity and the expression of *OsDREB1A*, *OsAP37*, *OsGADPH*, *OsWRKY11*, *OsDIL*, *OsCNGC10* genes in plants.	Kakar et al. (2016)
Drought	Wheat (*Triticum aestivum L.*)	*Bacillus subtilis* (LDR2)	Increased plant growth and biomass production, photosynthetic efficiency; reduced ethylene content.	ACC deaminase, IAA, P solubilization. Increasing endogenous IAA content, decreasing ABA content, and up-regulating *TaCTR1* and *TaDREB2* genes expression in plants.	Barnawal et al. (2017)
Flooding	Cucumber (*Cucumis sativus* L.),	*Pseudomonas putida* UW4	Increased plant growth and biomass production.	ACC deaminase, IAA, siderophores. Modulating the protein expression in plants.	Li et al. (2013)
Flooding	Holy basil (*Ocimum sanctum*)	*Achromobacter xylosoxidans* FD2, *Herbaspirillum seropedicae* Oci9, *Ochrobactrum rhizosphaerae* Oci13, *Serratia ureilytica* Bac5	Increased plant growth, biomass and yield, number of leaves and nodes, chlorophyll content, N, P and K contents; reduced ethylene and MDA contents.	ACC deaminase, IAA, P solubilization, siderophores.	Barnawal et al. (2012)
Flooding	Rice (*Oryza sativa* L.)	*Pseudomonas fluorescens* REN1	Increased root length.	ACC deaminase, IAA, siderophores.	Etesami et al. (2014)
Flooding	Tomato (*Lycopersiconesculentum*)	*Enterobacter cloacae* UW4 and CAL2	Increased plant growth and biomass production, chlorophyll content, decreased ethylene content.	ACC deaminase.	Grichko and Glick (2001)
Flooding	Marsh dock (*Rumexpalustris*)	*Pseudomonas putida* UW4	Improved sub-mergence induced young leaf and petiole elongation, plant growth and biomass production; decreased ethylene content.	ACC deaminase.	Ravanbakhsh et al. (2017)

TABLE 23.3
Effects of Probiotic Bacteria and their Mechanisms to Enhance Tolerance in Host Plants to Extreme Temperatures, and Heavy Metals Stresses

Stress Type	Plant Species	Bacterial Inoculate	Effects on Host Plant	Suggested Mechanism(s)	References
Cold	*Arabidopsis thaliana*	*Burkholderia phytofirmans* PsJN	Enhanced photosynthesis, chlorophyll content, and cell wall strengthening, decreased plasmalemma disruption in leaf mesophyll.	Differential accumulation of pigments, and reduced expression of *RbcL* and *COR78* genes in plants.	Su et al. (2015)
Cold	Barley (*Hordeum vulgare*)	*Azospirillum brasilense* Sp245, *Bacillus megaterium* M3, *B. subtilis* OSU142, *Raoultella terrigena*	Increased shoot and root dry matters, decreased H_2O_2 content.	IAA, N-fixation, P solubilization. Increasing antioxidant activities in plants.	Turan et al. (2013)
Cold	Grapevine (*Vitis vinifera* L.)	*Burkholderia phytofirmans*PsJN	Increased Pro content, decreased MDA and H_2O_2 contents.	Up-regulating *StSy*, *Chit4c*, *Gluc*, *PAL*, *Chit1b*, *LOX*, and *CBF4* genes expression in plants.	Theocharis et al. (2012)
Cold	Rice (*Oryza sativa* L.)	*Bacillus amyloliquefaciens* Bk7, *Brevibacillus laterosporus* B4	Increased plant survival rate and growth, chlorophyll and proline content; decreased MDA content and electrolyte leakage.	IAA, P solubilization, siderophores. Modulating antioxidant enzymes activity and the expression of *OsDREB1A*, *OsNAC6*, *OsMYB3R-2*, *OsGolS1*, *OsCNGC6*, *OsCNGC10*, and *OsCDPK13* genes in plants.	Kakar et al. (2016)
Cold	Tomato (*Solanum lycopersicum*)	*Pseudomonas frederiksbergensis* OS261, *P. vancouverensis* OB155	Increased seed germination rate and plant growth; reduced electrolyte leakage and MDA content.	ACC deaminase, IAA, P solubilization, salicylic acid, siderophores. Increasing antioxidant enzymes activity and Pro synthesis in plants.	Subramanian et al. (2016)
Heat	Sorghum	*Pseudomonas* sp. AKM-P6	Increased plant survival rates, growth and biomass production, chlorophyll, Pro, total sugars, and starch contents; decreased electrolyte leakage.	EPS, gibberellins, IAA, P solubilization, siderophores.	Ali et al. (2009)
Heat	Wheat (*Triticum aestivum* L.)	*Azospirillum brasilense* NO40, *Bacillus amyloliquefaciens* UCMB5113	Increased plant survival rates, growth and biomass production, and water content.	IAA, N-fixation, siderophores. Down-regulating *APX1*, *SAMS1* genes expression in plants.	El-Daim et al. (2014)
Heavy metals	*Brassica oxyrrhina*	*Pseudomonas libanensis* TR1, *P. reactans* Ph3R3	Increased plant growth and biomass production, RWC, chlorophyll content, copper and zinc content; decreased proline and MDA contents, lipid peroxidation.	ACC deaminase, IAA, siderophores.	Ma et al. (2016a)
Heavy metals	Maize (*Zea mays*), wheat (*Triticum aestivum*)	*Klebsiella* sp. CIK-502	Increased plant growth and biomass production; decreased electrolyte leakage, and Cd content.	ACC deaminase, EPS, IAA, siderophores.	Ahmad et al. (2014)
Heavy metals	Rice (*Oryza sativa* L.)	*Pseudomonas stutzeri* A1501	Increased plant growth and biomass production.	ACC deaminase.	Han et al. (2015)
Heavy metals	*Trifolium arvense*	*Pseudomonas azotoformans* ASS1	Increased plant growth and biomass production, RWC, chlorophyll content, Cu, Ni, and Zn contents, decreased Pro and MDA contents.	ACC deaminase, N-fixation, P solubilization, siderophores. Increasing antioxidant activities in plants.	Ma et al. (2017)

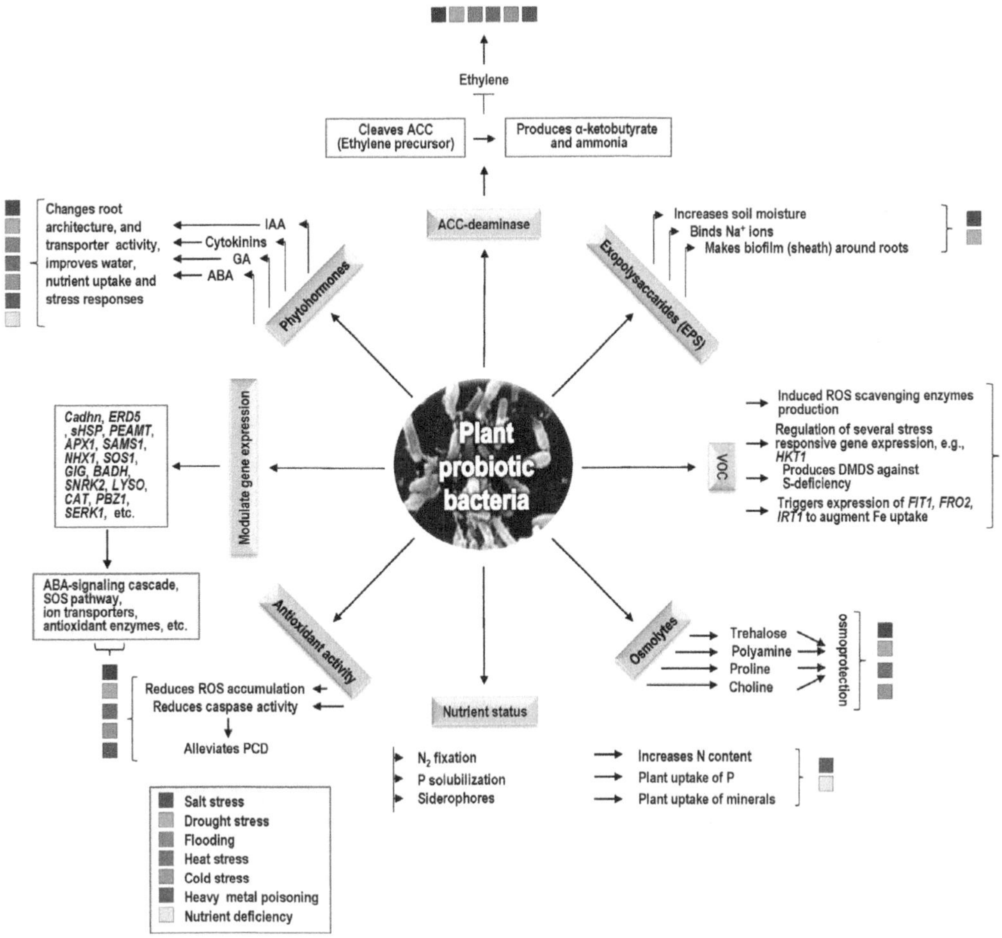

FIGURE 23.2 Diverse mechanisms of abiotic stress tolerance in plants conferred by probiotic bacteria. Some probiotic bacteria synthesize phytohormones such as IAA, cytokinins, GA, and ABA that alter plant root architecture and improve overall nutrient uptake and responses to stress (Radhakrishnan and Baek, 2017; Vurukonda et al., 2016; Yang et al., 2009). Degradation of the ethylene precursor ACC by bacterial ACC deaminase rescues normal plant growth under multiple stress conditions (Glick et al., 2007; Glick, 2014). Bacterial EPS protect plant roots from soil hardness by forming biofilms as well as retaining soil moisture content under drought conditions (Vurukonda et al., 2016). EPS may also improve plant salinity tolerance by binding Na^+ and reducing its availability for plant uptake (Ashraf et al., 2004). Increased osmolyte accumulation in plants due to bacterial inoculation augments plant tolerance to salinity, drought, and extreme temperature (Dimkpa et al., 2009a; Radhakrishnan and Baek, 2017; Vurukonda et al., 2016). Probiotic bacteria also improve plant health under nutrient deficient conditions by fixing atmospheric N_2, solubilizing mineral P, and synthesizing siderophores. They can induce enzymes that scavenge ROS, reducing ROS accumulation and caspase activity (Habib et al., 2016; Jha and Subramanian, 2014; Yang et al., 2009). Plant-associated bacteria can regulate the expression of stress-responsive plant genes by producing VOCs and/or by other unknown mechanisms that enhance plant tolerance against abiotic stresses (Liu and Zhang, 2015; Yang et al., 2009).

EPS (Naseem and Bano, 2014; Yang et al., 2016), volatile organic compounds (VOCs) (Ledger et al., 2016; Vaishnav et al., 2015), trehalose (Rodriguez-Salazar et al., 2009; Suarez et al., 2008), and polyamines (Cassan et al., 2009), as well as atmospheric nitrogen (N_2) fixation (Sharma et al., 2016), phosphorus (P) solubilization (Ma et al., 2017), and siderophore synthesis (Dimkpa et al., 2008; Palacio-Rodríguez et al., 2017) have been suggested to be involved in conferring stress tolerance to treated plants. Plant-associated bacteria may also improve plant responses under stress conditions by stimulating osmolyte accumulation (Sandhya et al., 2010; Upadhyay and Singh, 2015) and antioxidant defense systems (Habib et al., 2016; Li et al., 2017). Many of these plant-based responses would involve changes in gene expression (Barnawal et al., 2017; Bharti et al., 2016). These various mechanisms are largely interconnected and affect one another (Forni et al., 2016).

23.3.1 Production of Phytohormones

Plant hormones play pivotal roles in plant growth and development, and plants often adjust their endogenous phytohormone levels under adverse conditions, which may negatively affect the physiology, growth, and yield of crop plants (Glick, 2012; Vurukonda et al., 2016). However, many plant probiotic bacteria have the ability to produce a variety of phytohormones including auxins, gibberellins (GAs), cytokinins, and ABA (Kang et al., 2014a; Li et al., 2017; Liu et al., 2013b; Palacio-Rodríguez et al., 2017; Sukweenadhi et al., 2015; Xu et al., 2012). Indole acetic acid (IAA) is the physiologically most-active auxin in plant growth and development. Its production by many plant-associated bacteria can increase root length, root surface area, and number of root tips, leading to improved water and nutrient uptake by plants, and thereby enhanced plant tolerance to environmental stresses (Grover et al., 2011; Radhakrishnan and Baek, 2017; Vurukonda et al., 2016; Yang et al., 2009). Bacterial IAA can also induce the activity of ACC deaminase through a signaling cascade (Glick, 2012, 2014). This important trait of probiotic bacteria is discussed later. Bacterial strains synthesizing cytokinins (Arkhipova et al., 2007; Liu et al., 2013b; Xu et al., 2012), GAs (Cohen et al., 2009; Halo et al., 2015; Kang et al., 2014a, b) and ABA (Cohen et al., 2008, 2009) have been found to ameliorate plant resistance to abiotic stresses. Apart from producing phytohormones, plant-beneficial bacteria also modulate endogenous plant production of phytohormones under adverse environmental conditions. Examples of such modulation have been reported for IAA (Li et al., 2017; Liu et al., 2013a), GA (Kang et al., 2014a, b), ABA (Arkhipova et al., 2007; Chen et al., 2016; Cohen et al., 2008; Kang et al., 2014a, b; Yang et al., 2016), salicylic acid (SA), and jasmonic acid (JA) (Kang et al., 2014a). It is not yet clear exactly how bacterial-mediated alterations of these phytohormones confer stress tolerance to plants. The biosynthetic pathways of IAA, GAs, and cytokinins in bacteria have recently been discussed by Khan et al. (2016).

23.3.2 Synthesis of 1-Aminocyclopropane-1-Carboxylate (ACC) Deaminase

The synthesis of the phytohormone ethylene, an important modulator of normal plant growth and development, is affected by a wide range of abiotic and biotic factors (Abeles et al., 1992). Under stress conditions, plants often produce a high level of ethylene, termed stress ethylene, that can remarkably exacerbate the deleterious effects of the stress inhibiting plant growth (Glick, 2014; Singh et al., 2015). A wide range of plant-associated bacteria, including members of the genera *Achromobacter, Alcaligenes, Bacillus, Burkholderia, Enterobacter, Pseudomonas, Rhizobium*, and *Mesorhizobium* have been reported to have ACC deaminase activity. This pyridoxal phosphate-dependent enzyme provides tolerance to host plants to a variety of abiotic stresses including salinity (Li et al., 2017; Mayak et al., 2004a), drought (Barnawal et al., 2017; Mayak et al., 2004b), flooding (Barnawal et al., 2012; Ravanbakhsh et al., 2017), chilling (Subramanian et al., 2015, 2016), and heavy metal toxicity (Han et al., 2015). The ACC deaminase enzyme cleaves ACC, an immediate precursor of plant ethylene, to α-ketobutyrate and ammonia, decreasing the ability of the plant to produce ethylene. The effect is to improve plant tolerance to environmental stressors by lowering the level of stress ethylene in the plant (Glick, 2014; Honma and Shimomura, 1978). Recently, Glick (2014) and Singh et al. (2015) have reviewed the detailed mechanism of bacterial ACC deaminase-mediated stress tolerance in plants.

23.3.3 Production of Exopolysaccharides

EPS are high molecular weight carbohydrate compounds attached to the outer surface of bacteria as a capsule or slime. They are responsible for allowing bacteria to form biofilms and to attach to surfaces, including plant roots and soil particles (Forni et al., 2016). Bacteria that normally live in harsh environments produce different polysaccharides that protect the cells from their extreme environment (Demirjian et al., 2001; Konnova et al., 2001; Vurukonda et al., 2016). These polysaccharides also improve the physicochemical and biological properties of environmentally affected soils (Amellal et al., 1998; Ashraf et al., 2013; Naseem and Bano, 2014). Bacteria that produce EPS enhance water and nutrient uptake by plants during water stress conditions by increasing soil aggregation, macroporosity of root-adhering soils (RAS), and maintaining higher water potential around the roots (Alami et al., 2000; Amellal et al., 1998; Naseem and Bano, 2014; Sandhya et al., 2009). Bacterial EPS were also found to form a hydrophilic biofilm around the roots that acts as an additional sheath to protect the root system from soil hardness during drought and excessive Na^+ in salinity soils (Ashraf et al., 2004; Dimkpa et al., 2009a; Rolli et al., 2015). Moreover, EPS synthesized by probiotic bacteria might bind cations including Na^+, decreasing Na^+ availability for plant uptake and consequently improving the salinity tolerance of the plant (Ashraf et al., 2004; Geddie and Sutherland, 1993; Nunkaew et al., 2015; Siddikee et al., 2011).

23.3.4 Emission of Volatile Organic Compounds

Plant probiotic bacteria emit VOCs that can augment plant tolerance to various abiotic stresses, primarily by stimulating plant genes that encode ROS scavenging enzymes, such as glutathione reductase (GR), monodehydroascorbate reductase (MDHAR), SOD, and CAT (Timmusk et al., 2014; Liu and Zhang, 2015; Forni et al., 2016). For example, *Bacillus amyloliquefaciens* GB03 (Zhang et al., 2008) and *Paraburkholderia phytofirmans* PsJN (Ledger et al., 2016) was reported to induce tolerance to salinity in *Arabidopsis thaliana* through the emission of volatile metabolites. The VOCs emitted by *amyloliquefaciens* GB03 improved plant salt tolerance by modulating Na^+ homeostasis via tissue-specific regulation of *HKT1* (Zhang et al., 2008). In addition, volatile emissions from *Pseudomonas simiae* AU protected soybean plants from salt-induced osmotic stress by inducing vegetative storage proteins (VSP) and several other proteins synthesis (Vaishnav et al., 2015). The VOCs from *Pseudomonas chlororaphis* strain O6 contain 2,3-butanediol, which can induce stress tolerance in *Arabidopsis* during drought (Cho et al., 2008). This VOC-induced drought tolerance might be linked with the SA pathway, and enhanced accumulation of H_2O_2 and nitric oxide (NO) (Cho et al., 2008,2013; Liu and Zhang, 2015).

Bacterial VOCs can also improve plant tolerance to nutrient stress (Liu and Zhang, 2015). *Bacillus* sp. B55 protected *Nicotiana attenuata* plants from growth retardation caused by S starvation under sulfur (S)-deficient conditions by producing dimethyl disulfide (DMDS), an S-containing volatile metabolite that the plant may have used directly (Meldau et al., 2013). Moreover, *B. amyloliquefaciens* GB03 was reported to augment Fe uptake in *Arabidopsis* during iron deprivation by triggering the expression of *FIT1, FRO2*, and *IRT1* (Zhang et al., 2009), genes necessary for the modulation of iron uptake. Further research is warranted for a better understanding of the mechanisms through which bacterial VOCs enhance abiotic stress tolerance in plants.

23.3.5 Synthesis of Compatible Solutes/Osmolytes

Some plant-beneficial bacteria synthesize osmolytes in response to different stresses, which act synergistically with plant-produced osmolytes, resulting in improved plant stress tolerance (Vurukonda et al., 2016). Probiotic bacteria have been reported to secrete trehalose, a highly stable non-reducing glucoside consisting of two molecules of α-glucose, which can act as an osmoprotectant (Forni et al., 2016; Vurukonda et al., 2016). Suarez et al. (2008) reported that *Phaseolus vulgaris* plants inoculated with a trehalose-overproducing mutant strain of *Rhizobium etli* displayed better growth performance under drought conditions compared with plants inoculated with the wild strain, indicating that higher levels of trehalose can confer drought tolerance in plants. Similar effects were observed by Rodriguez-Salazar et al. (2009) in maize plants treated with a mutant strain of *Azospirillum brasilense* that overproduced trehalose. *Azospirillum brasilense* strain Az39, which produces the polyamine cadaverine, was reported to alleviate osmotic stress in rice seedlings (Cassan et al., 2009). However, more studies are required to understand the role and mechanisms of bacterial trehalose and polyamines in improving abiotic stress tolerance in plants.

Probiotic bacteria were also found to induce accumulation of compatible solutes such as Pro (Bharti et al., 2016; Kakar et al., 2016; Li et al., 2017; Sandhya et al., 2010) and choline, a precursor of GB (Bashan et al., 2014; Gou et al., 2015; Upadhyay and Singh, 2015; Zhang et al., 2010) in plants, and thus increased plant resistance against osmotic stress, as well as salinity-induced oxidative stress.

23.3.6 Modification of Plant Nutrient and Metal Acquisition by Probiotic Bacteria

Probiotic bacteria can improve plant response to abiotic stresses, especially under nutrient deficient conditions, by improving both the macronutrient (N, P, and K) and micronutrient (Zn, Fe, Cu, and Mn) status of the host plants (Choudhary et al., 2016; Dimkpa et al., 2009a; Glick, 2012). It is well known that plant-beneficial bacteria such as *Rhizobium* spp., *Azospirillum* spp., and *Pantoea agglomerans* can provide plants with atmospheric N that is fixed either symbiotically (Esitken et al., 2006) or non-symbiotically (Bashan and de-Bashan, 2010). Many bacterial strains have been reported to have the ability to solubilize inorganic phosphate through the secretion of low molecular weight organic acids, such as gluconic acid and citric acid. They also mineralize organic P through the synthesis of a variety of phosphatases. Both processes make P available to plants, resulting in better plant growth, especially under P-deficient conditions (Glick, 2012; Islam and Hossain, 2012). They may also increase plant uptake of mineral ions by stimulating proton-pumping ATPases (Mantelin and Touraine, 2004). Furthermore, some bacteria can release metal-chelating substances, such as iron-chelating siderophores, and have been reported to stimulate plant uptake of diverse metals including Fe, Zn, and Cu (Carrillo-Castañeda et al., 2002, 2005; Dimkpa et al., 2009b, c; Egamberdiyeva, 2007). Microbial siderophores were also documented to alleviate the stresses associated with heavy metal pollution, primarily by chelating and reducing toxic metal concentrations in the rhizosphere (Belimov et al., 2005; Burd et al., 2000; Dimkpa et al., 2008, 2009b, c; Glick, 2012; Ma et al., 2016b). In addition, the cell-wall composition of some probiotic bacteria confers them with metal-binding properties (Beveridge et al., 1982), which might be associated with the reduction of metal uptake by plants (Ganesan, 2008; Pishchik et al., 2002), enhancing their tolerance to heavy metals.

23.3.7 Altering Antioxidant Defense Systems

As discussed earlier, the production of ROS increases in plants subjected to abiotic stresses. This increase causes oxidative damage to proteins, lipids, and DNA, and impairs normal cell function (Choudhury et al., 2013; Miller et al., 2010; Radhakrishnan and Baek, 2017; Vurukonda et al., 2016). However, inoculation with bacteria can stimulate ROS-scavenging enzyme activities in plants and enhance tolerance to environmental stresses by reducing ROS accumulation (Jha and Subramanian, 2014; Habib et al., 2016; Kaushal and Wani, 2016; Li et al., 2017; Tables 23.1 through 23.3). Some bacteria such as *Pseudomonas pseudoalcaligenes* and *Bacillus pumulis* were found to reduce the salinity-induced elevation of cell caspase-like protease activity, which is involved in programmed cell death (PCD) (Keyster et al., 2012); thus, helping plants to survive in stress conditions (Jha and Subramanian, 2014).

23.3.8 Modulation of Gene Expression in Plants by Probiotic Bacteria

Probiotic bacteria may also improve plant tolerance against environmental stresses by modulating stress-related gene expression patterns in plants (Barnawal et al., 2017; Chen et al., 2016; Kakar et al., 2016; Sarma and Saikia, 2014; Sukweenadhi et al., 2015; Yang et al., 2009). For instance, Timmusk and Wagner (1999) reported that inoculation of *Arabidopsis thaliana* with the *Paenibacillus polymyxa* induced transcription of a drought response gene, *EARLY RESPONSIVE TO DEHYDRATION 15* (*ERD15*), which enhanced drought tolerance. Increased expression of the *phosphoethanolamine N-methyl transferase* (*PEAMT*) gene, which is associated with a key enzyme involved in the synthesis of the plant osmolyte choline, was observed by Zhang et al. (2010) under drought in *A. thaliana* treated with *bacillus amyloliquefaciens* GB03. Pepper plants primed with *B. licheniformis* K11 and subjected to osmotic stress showed a greater than 1.5-fold increase in expression of specific genes related to stress, including genes encoding *Capsicum annum* dehydrin (Cadhn) and small heat-shock protein (sHSP), when compared with non-inoculated plants (Lim and Kim, 2013). Similarly, Kasim et al. (2013) reported improved plant tolerance to drought stress and induction of stress-related genes *ascorbate peroxidase 1* (*APX1*), *S-adenosylmethionine synthase 1* (*SAMS1*), and *heat-shock* protein 17.8 (*HSP17.8*) in leaves of wheat plants inoculated with *B. amyloliquefaciens* 5113 and *Azospirillum brasilense* NO40. Furthermore, *Gluconacetobacter diazotrophicus* inoculation of sugarcane cv. SP70-1143 activated ABA-dependent signaling genes and conferred drought resistance to the inoculated plants (Vargas et al., 2014).

Furthermore, the treatment of rice with *B. amyloliquefaciens* NBRISN13 ameliorated the salinity-induced down-regulation of a suite of genes allowing the plants to better acclimate to salinity (Nautiyal et al., 2013). Likewise, *Burkholderia phytofirmans* PsJN inoculated *Arabidopsis* plants displayed increased transcript abundance for a suite of genes related to abscisic acid signaling (*RD29A, RD29B*), ROS scavenging (*APX2*), and detoxification (*GLYI7*) and decreased expression of *LOX2* (related to jasmonic acid biosynthesis), thereby enhanced plant tolerance to salt-stress (Pinedo et al., 2015). Recently, Bharti et al. (2016) demonstrated that increased salinity tolerance in wheat plants treated with *D. natronolimnaea* STR1 involved modulation of transcriptional machinery associated with the ABA-signaling cascade, SOS pathway, ion transporters, and antioxidant enzymes.

23.4 CONCLUSION AND FUTURE PERSPECTIVES

This review indicates that probiotic bacteria have great potential to enhance the protection of plants from a wide variety of environmental stresses and enhance plant growth and yield; however, more research effort is required before this strategy becomes a mainstay of agricultural practice. Although considerable information has been accumulated regarding the impact of bacterial inoculation on plant responses to abiotic stress, the underlying mechanisms and cross-talk between bacteria and plants remain largely speculative. Hence, further research is needed to understand the molecular mechanisms of plant-bacteria interactions underlying the induction of stress tolerance in host plants by probiotic bacteria. In addition, most studies evaluated the plant-bacteria interactions based on a single plant species interacting with one or a mixture of a few bacterial strains under controlled conditions. However, in real-life situations, a plant population is interacting with many different types of soil microorganisms along with many environmental signals. Therefore, scientists need to make the leap from greenhouse or growth chamber experiments to field evaluation. To date, most of the plant-associated bacteria that have been studied are rhizospheric; however, researchers should also give more focus in future to endophytic bacteria, as they are protected by plant tissues from many biotic and abiotic environmental challenges. There is a growing interest among researchers in using genetic engineering tools to develop more efficacious bacterial strains. In this regard, scientists will need to prove that genetically modified probiotic bacteria do not present new threats to the environment. Overall, it is expected that continued research on probiotic bacteria-mediated abiotic stress tolerance in plants will bring this technology to use as an environmentally friendly tool for sustainable management of stressed agriculture in near future.

ACKNOWLEDGMENTS

The authors are thankful to the Deanship of Scientific Research (DSR), King Abdulaziz University (KAU), Saudi Arabia, for financial support for this work through project no. 517/130/438G.

REFERENCES

Abeles, F.B., Morgan, P.W., Saltveit, M.E. Jr. (1992) *Ethylene in Plant Biology*. New York: Academic Press.

Adrees, M., Ali, S., Rizwan, M., Ibrahim, M., Abbas, F., Farid, M., Zia-Ur-Rehman, M., Irshad, M.K., Bharwana, S.A. (2015) The effect of excess copper on growth and physiology of important food crops: A review. *Environmental Science and Pollution Research International* 22:8148–8162.

Ahmad, I., Akhtar, M.J., Zahir, Z.A., Naveed, M., Mitter, B., Sessitsch, A. (2014) Cadmium-tolerant bacteria induce metal stress tolerance in cereals. *Environmental Science and Pollution Research International* 21:11054–11065.

Alami, Y., Achouak, W., Marol, C., Heulin, T. (2000) Rhizosphere soil aggregation and plant growth promotion of sunflowers by an exopolysaccharide-producing Rhizobium sp. strain isolated from sunflower roots. *Applied and Environmental Microbiology* 66:3393–3398.

Ali, S.Z., Sandhya, V., Grover, M., Kishore, N., Rao, L.V., Venkateswarlu, B. (2009) *Pseudomonas* sp. strain AKM-P6 enhances tolerance of Sorghum seedlings to elevated temperatures. *Biology and Fertility of Soils* 46:45–55.

Al-Khateeb, W., Al-Qwasemeh, H. (2014) Cadmium, copper and zinc toxicity effects on growth, proline content and genetic stability of Solanum nigrum L., a crop wild relative for tomato; comparative study. *Physiology and Molecular Biology of Plants: An International Journal of Functional Plant Biology* 20:31–39.

Amellal, N., Burtin, G., Bartoli, F., Heulin, T. (1998) Colonization of wheat roots by an exopolysaccharide-producing *Pantoea agglomerans* strain and its effect on rhizosphere soil aggregation. *Applied and Environmental Microbiology* 64:3740–3747.

Arkhipova, T.N., Prinsen, E., Veselov, S.U., Martinenko, E.V., Melentiev, A.I., Kudoyarova, G.R. (2007) Cytokinin producing bacteria enhance plant growth in drying soil. *Plant and Soil* 292:305–315.

Ashraf, M.A. (2012) Waterlogging stress in plants: A review. *African Journal of Agricultural Research* 7:1976–1981.

Ashraf, M., Hasnain, S., Berge, O. (2013) Bacterial exopolysaccharides: A biological tool for the reclamation of salt-affected soils. In *Developments in Soil Salinity Assessment and Reclamation*, eds. Shahid, S., Abdelfattah, M., Taha, F., 641–658. Dordrecht: Springer.

Ashraf, M., Hasnain, S., Berge, O., Mahmood, T. (2004) Inoculating wheat seedlings with exopolysaccharide producing bacteria restricts sodium uptake and stimulates plant growth under salt stress. *Biology and Fertility of Soils* 40:157–162.

Assmann, S.M., Snyder, J.A., Lee, Y.J. (2000) ABA-deficient (aba1) and ABA-insensitive (abi1-1, abi2-1) mutants of *Arabidopsis* have a wild-type stomatal response to humidity. *Plant, Cell and Environment* 23:387–395.

Barka, E.A., Nowak, J., Clement, C. (2006) Enhancement of chilling resistance of inoculated grapevine plantlets with a plant growth-promoting rhizobacterium, *Burkholderia phytofirmans* strain PsJN. *Applied and Environmental Microbiology* 72:7246–7252.

Barnawal, D., Bharti, N., Maji, D., Chanotiya, C.S., Kalra, A. (2012) 1-aminocyclopropane-1-carboxylic acid (ACC) deaminase-containing rhizobacteria protect *Ocimum sanctum* plants during water-logging stress via reduced ethylene generation. *Plant Physiology and Biochemistry: PPB* 58:227–235.

Barnawal, D., Bharti, N., Pandey, S.S., Pandey, A., Chanotiya, C.S., Kalra, A. (2017) Plant growth promoting rhizobacteria enhances wheat salt and drought stress tolerance by altering endogenous phytohormone levels and TaCTR1/TaDREB2 expression. *Physiologia Plantarum* 161:502–514. doi:10.1111/ppl.12614

Bashan, Y., de-Bashan, L.E. (2010) Chapter two - How the plant growth-promoting bacterium *Azospirillum* promotes plant growth–a critical assessment. *Advances in Agronomy* 108:77–136.

Bashan, Y., de-Bashan, L.E., Prabhu, S.R., Hernandez, J.P. (2014) Advances in plant growth-promoting bacterial inoculant technology: Formulations and practical perspectives (1998–2013). *Plant and Soil* 378:1–33.

Belimov, A.A., Dodd, I.C., Safronova, V.I., Shaposhnikov, A.I., Azarova, T.S., Makarova, N.M., Davies, W.J., Tikhonovich, I.A. (2015) Rhizobacteria that produce auxins and contain 1-amino-cyclopropane-1-carboxylic acid deaminase decrease amino acid concentrations in the rhizosphere and improve growth and yield of well-watered and water-limited potato (*Solanum tuberosum*). *Annals of Applied Biology* 167:11–25.

Belimov, A.A., Hontzeas, N., Safronova, V.I., Demchinskaya, S.V., Piluzza, G., Bullitta, S., Glick, B.R. (2005) Cadmium-tolerant plant growth-promoting bacteria associated with the roots of Indian mustard (*Brassica juncea* L. Czern.). *Soil Biology and Biochemistry* 37:241–250.

Bertrand, M., Guary, J.C., Schoefs, B. (2001) How plants adapt their physiology to an excess of metals. In *Handbook of Plant and Crop Physiology*, ed. Pessarakli, M., 751–762. New York: Marcel Dekker.

Beveridge, T.J., Forsberg, C.W., Doyle, R.J. (1982) Major sites for metal binding in *Bacillus licheniformis* walls. *Journal of Bacteriology* 150:1438–1448.

Bharti, N., Pandey, S.S., Barnawal, D., Patel, V.K., Kalra, A. (2016) Plant growth promoting rhizobacteria *Dietzia natronolimnaea* modulates the expression of stress responsive genes providing protection of wheat from salinity stress. *Scientific Reports* 6:34768.

Burd, G.I., Dixon, D.G., Glick, B.R. (2000) Plant growth-promoting bacteria that decrease heavy metal toxicity in plants. *Canadian Journal of Microbiology* 46:237–245.

Cardinale, M., Ratering, S., Suarez, C., Montoya, A.M.Z., Geissler-Plaum, R., Schnell, S. (2015) Paradox of plant growth promotion potential of rhizobacteria and their actual promotion effect on growth of barley (*Hordeum vulgare* L.) under salt stress. *Microbiological Research* 181:22–32.

Carrillo-Castañeda, G., Juárez Muñoz, J., Ramón Peralta-Videa, J., Gomez, E., Gardea-Torresdey, J.L. (2002) Plant growth-promoting bacteria promote copper and iron translocation from root to shoot in alfalfa seedlings. *Journal of Plant Nutrition* 26:1801–1814.

Carrillo-Castañeda, G., Juárez Munoz, J.J., Ramón Peralta-Videa, J., Gomez, E., Gardea-Torresdey, J.L. (2005) Modulation of uptake and translocation of iron and copper from root to shoot in common bean by siderophore-producing microorganisms. *Journal of Plant Nutrition* 28:1853–1865.

Cassan, F., Maiale, S., Masciarelli, O., Vidal, A., Luna, V., Ruiz, O. (2009) Cadaverine production by *Azospirillum brasilense* and its possible role in plant growth promotion and osmotic stress mitigation. *European Journal of Soil Biology* 45:12–19.

Chang, P., Gerhardt, K.E., Huang, X.D., Yu, X.M., Glick, B.R., Gerwing, P.D., Greenberg, B.M. (2014) Plant growth-promoting bacteria facilitate the growth of barley and oats in salt-impacted soil: Implications for phytoremediation of saline soils. *International Journal of Phytoremediation* 16:1133–1147.

Chen, L., Liu, Y., Wu, G., Veronican Njeri, K., Shen, Q., Zhang, N., Zhang, R. (2016) Induced maize salt tolerance by rhizosphere inoculation of *Bacillus amyloliquefaciens* SQR9. *Physiologia Plantarum* 158:34–44.

Cheng, Z., Woody, O.Z., McConkey, B.J., Glick, B.R. (2012) Combined effects of the plant growth promoting bacterium *Pseudomonas putida* UW4 and salinity stress on the *Brassica napus* proteome. *Applied Soil Ecology* 61:255–263.

Cho, S.M., Kang, B.R., Han, S.H., Anderson, A.J., Park, J.Y., Lee, Y.H., Cho, B.H., *et al.* (2008) 2R, 3R-butanediol, a bacterial volatile produced by *Pseudomonas chlororaphis* O6, is involved in induction of systemic tolerance to drought in *Arabidopsis thaliana*. *Molecular Plant-Microbe Interactions: MPMI* 21:1067–1075.

Cho, S.M., Kim, Y.H., Anderson, A.J., Kim, Y.C. (2013) Nitric oxide and hydrogen peroxide production are involved in systemic drought tolerance induced by 2R,3R-butanediol in *Arabidopsis thaliana*. *The Plant Pathology Journal* 29:427–434.

Choi, Y.E., Harada, E., Wada, M., Tsuboi, H., Morita, Y., Kusano, T., Sano, H. (2001) Detoxification of cadmium in tobacco plants: Formation and active excretion of crystals containing cadmium and calcium through trichomes. *Planta* 213:45–50.

Choudhary, D.K., Kasotia, A., Jain, S., Vaishnav, A., Kumari, S., Sharma, K.P., Varma, A. (2016) Bacterial-mediated tolerance and resistance to plants under abiotic and biotic stresses. *Journal of Plant Growth Regulation* 35:276–300.

Choudhury, S., Panda, P., Sahoo, L., Panda, S.K. (2013) Reactive oxygen species signaling in plants under abiotic stress. *Plant Signaling and Behavior* 8:e23681.

Cobbett, C.S., Goldsbrough, P. (2002) Phytochelatins and metallothioneins: Roles in heavy metal detoxification and homeostasis. *Annual Review of Plant Biology* 53:159–182.

Cohen, A.C., Bottini, R., Piccoli, P.N. (2008) *Azospirillum brasilense* sp. 245 produces ABA in chemically defined culture medium and increases ABA content in *Arabidopsis* plants. *Plant Growth Regulation* 54:97–103.

Cohen, A.C., Travaglia, C.N., Bottini, R., Piccoli, P.N. (2009) Participation of abscisic acid and gibberellins produced by endophytic *Azospirillum* in the alleviation of drought effects in maize. *Botany* 87:455–462.

Demirjian, D.C., Moris-Varas, F., Cassidy, C.S. (2001) Enzymes from extremophiles. *Current Opinion in Chemical Biology* 5:144–151.

Dimkpa, C.O., Merten, D., Svatoš, A., Büchel, G., Kothe, E. (2009b) Metal-induced oxidative stress impacting plant growth in contaminated soil is alleviated by microbial siderophores. *Soil Biology and Biochemistry* 41:154–162.

Dimkpa, C.O., Merten, D., Svatoš, A., Büchel, G., Kothe, E. (2009c) Siderophores mediate reduced and increased uptake of cadmium by Streptomyces tendaeF4 and sunflower (*Helianthus annuus*), respectively. *Journal of Applied Microbiology* 107:1687–1696.

Dimkpa, C., Svatoš, A., Merten, D., Büchel, G., Kothe, E. (2008) Hydroxamate siderophores produced by *Streptomyces acidiscabies* E13 bind nickel and promote growth in cowpea (*Vigna unguiculata* L.) under nickel stress. *Canadian Journal of Microbiology* 54:163–172.

Dimkpa, C., Weinand, T., Asch, F. (2009a) Plant-rhizobacteria interactions alleviate abiotic stress conditions. *Plant, Cell and Environment* 32 (12):1682–1694.

Dong, J., Mao, W.H., Zhang, G.P., Wu, F.B., Cai, Y. (2007) Root excretion and plant tolerance to cadmium toxicity – A review. *Plant, Soil and Environment* 53:193–200.

Egamberdieva, D. (2009) Alleviation of salt stress by plant growth regulators and IAA producing bacteria in wheat. *Acta Physiologiae Plantarum* 31:861–864.

Egamberdiyeva, D. (2007) The effect of plant growth promoting bacteria on growth and nutrient uptake of maize in two different soils. *Applied Soil Ecology* 36:184–189.

El-Daim, I.A., Bejai, S., Meijer, J. (2014) Improved heat stress tolerance of wheat seedlings by bacterial seed treatment. *Plant and Soil* 379:337–350.

Esitken, A., Pirlak, L., Turan, M., Sahin, F. (2006) Effects of floral and foliar application of plant growth promoting rhizobacteria (PGPR) on yield, growth and nutrition of sweet cherry. *Scientia Horticulturae* 110:324–327.

Etesami, H., Hosseini, H.M., Alikhani, H.A. (2014) Bacterial biosynthesis of 1-aminocyclopropane-1-carboxylate (ACC) deaminase, a useful trait to elongation and endophytic colonization of the roots of rice under constant flooded conditions. *Physiology and Molecular Biology of Plants: An International Journal of Functional Plant Biology* 20:425–434.

Feder, M.E., Hoffman, G.E. (1999) Heat-shock proteins, molecular chaperones, and the stress response: Evolutionary and ecological physiology. *Annual Review of Physiology* 61:243–282.

Forni, C., Duca, D., Glick, B.R. (2016) Mechanisms of plant response to salt and drought stress and their alteration by rhizobacteria. *Plant and Soil* 410:335–356.

Foyer, C.H., Lelandais, M., Kunert, K.J. (1994) Photooxidative stress in plants. *Physiologia Plantarum* 92:696–717.

Ganesan, V. (2008) Rhizoremediation of cadmium soil using a cadmium-resistant plant growth-promoting rhizopseudomonad. *Current Microbiology* 56:403–407.

Geddie, J.L., Sutherland, I.W. (1993) Uptake of metals by bacterial polysaccharides. *Journal of Applied Bacteriology* 74:467–472.

Gill, S.S., Tuteja, N. (2010) Reactive oxygen species and antioxidant machinery in abiotic stress tolerance in crop plants. *Plant Physiology and Biochemistry: PPB* 48:909–930.

Glick, B.R. (2012) Plant growth-promoting bacteria: Mechanisms and applications. *Scientifica* 2012:963401. doi:10.6064/2012/963401

Glick, B.R. (2014) Bacteria with ACC deaminase can promote plant growth and help to feed the world. *Microbiological Research* 169:30–39.

Glick, B.R., Cheng, Z., Czarny, J., Duan, J. (2007) Promotion of plant growth by ACC deaminase containing soil bacteria. *European Journal of Plant Pathology* 119:329–339.

Gou, W., Tian, L., Ruan, Z., Zheng, P., Chen, F., Zhang, L., Cui, Z., *et al.* (2015) Accumulation of choline and glycinebetaine and drought stress tolerance induced in maize (*Zea mays*) by three plant growth promoting rhizobacteria (pgpr) strains. *Pakistan Journal of Botany* 47:581–586.

Grichko, V.P., Glick, B.R. (2001) Amelioration of flooding stress by ACC deaminase-containing plant growth-promoting bacteria. *Plant Physiology and Biochemistry* 39:11–17.

Grover, M., Ali, Sk.Z., Sandhya, V. Rasul, A., Venkateswarlu, B. (2011) Role of microorganisms in adaptation of agriculture crops to abiotic stresses. *World Journal of Microbiology and Biotechnology* 27:1231–1240.

Gurska, J., Wang, W., Gerhardt, K.E., Khalid, A.M., Isherwood, D.M., Huang, X.D., Glick, B.R., Greenberg, B.M. (2009) Three year field test of a plant growth promoting rhizobacteria enhanced phytoremediation system at a land farm for treatment of hydrocarbon waste. *Environmental Science and Technology* 43:4472–4479.

Habib, S.H., Kausar, H., Saud, H.M. (2016) Plant growth-promoting rhizobacteria enhance salinity stress tolerance in okra through ROS-scavenging enzymes. *BioMed Research International* 2016:6284547. doi:10.1155/2016/6284547.

Halo, B.A., Khan, A.L., Waqas, M., Al-Harrasi, A., Hussain, J., Ali, L., Adnan, M., Lee, I. (2015) Endophytic bacteria (*Sphingomonas* sp. LK11) and gibberellin can improve *Solanum lycopersicum* growth and oxidative stress under salinity. *Journal of Plant Interactions* 10:117–125.

Han, Y., Wang, R., Yang, Z., Zhan, Y., Ma, Y., Ping, S., Zhang, L., Lin, M., Yan, Y. (2015) 1-Aminocyclopropane-1-carboxylate deaminase from *Pseudomonas stutzeri* A1501 facilitates the growth of rice in the presence of salt or heavy metals. *Journal of Microbiology and Biotechnology* 25:1119–1128.

Hoekstra, F.A., Buitink, J. (2001) Mechanisms of plant desiccation tolerance. *Trends in Plant Science* 8:431–438.

Honma, M., Shimomura, T. (1978) Metabolism of 1-aminocyclopropane-1-carboxylic acid. *Agricultural and Biological Chemistry* 43:1825–1831.

Hussain, T.M., Chandrasekhar, T., Hazara, M., Sultan, Z., Saleh, B.K., Gopal, G.R. (2008) Recent advances in salt stress biology-a review. *Biotechnology and Molecular Biology Reviews* 3:8–13.

İpek, M., Aras, S., Arikan, Ş., Esitken, A., Pırlak, L., Dönmez, M.F., Turan, M. (2017) Root plant growth promoting rhizobacteria inoculations increase ferric chelate reductase (FC-R) activity and Fe nutrition in pear under calcareous soil conditions. *Scientia Horticulturae* 219:144–151.

Islam, M.T., Hossain, M.M. (2012) Plant probiotics in phosphorus nutrition in crops, with special reference to rice. In *Bacteria in Agrobiology: Plant Probiotics*, ed. Maheshwari, D.K., 325–363. Berlin, Heidelberg: Springer.

Islam, F., Yasmeen, T., Arif, M.S., Ali, S., Ali, B., Hameed, S., Zhou, W. (2016) Plant growth promoting bacteria confer salt tolerance in *Vigna radiata* by up-regulating antioxidant defense and biological soil fertility. *Plant Growth Regulation* 80 (1):23–36.

Jamil, A., Riaz, S., Ashraf, M., Foolad, M.R. (2011) Gene expression profiling of plants under salt stress. *Critical Reviews in Plant Sciences* 30:435–458.

Jha, Y., Subramanian, R.B. (2014) PGPR regulate caspase-like activity, programmed cell death, and antioxidant enzyme activity in paddy under salinity. *Physiology and Molecular Biology of Plants: An International Journal of Functional Plant Biology* 20:201–207.

Kakar, K.U., Ren, X.L., Nawaz, Z., Cui, Z.Q., Li, B., Xie, G.L., Hassan, M.A., Ali, E., Sun, G.C. (2016) A consortium of rhizobacterial strains and biochemical growth elicitors improve cold and drought stress tolerance in rice (*Oryza sativa* L.). *Plant Biology* 18:471–483.

Kang, S.-M., Khan, A.L., Waqas, M., You, Y., Hamayun, M., Joo, G., Shahzad, R., Choi, K., Lee, I. (2015) Gibberellin-producing *Serratia nematodiphila* PEJ1011 ameliorates low temperature stress in *Capsicum annuum* L. *European Journal of Soil Biology* 68:85–93.

Kang, S.M., Khan, A.L., Waqas, M., You, Y., Kim, J., Kim, J., Hamayun, M., Lee, I. (2014a) Plant growth promoting rhizobacteria reduce adverse effects of salinity and osmotic stress by regulating phytohormones and antioxidants in *Cucumis sativus. Journal of Plant Interactions* 9:673–682.

Kang, S.M., Radhakrishnan, R., Khan, A.L., Kim, M.J., Park, J.M., Kim, B.R., Shin, D.H., Lee, I.J. (2014b) Gibberellin secreting rhizobacterium, *Pseudomonas putida* H-2-3 modulates the hormonal and stress physiology of soybean to improve the plant growth under saline and drought conditions. *Plant Physiology and Biochemistry: PPB* 84:115–124.

Kasim, W.A., Osman, M.E., Omar, M.N., Abd El-Daim, I.A., Bejai, S., Meijer, J. (2013) Control of drought stress in wheat using plant growth promoting bacteria. *Journal of Plant Growth Regulation* 32:122–130.

Kaushal, M., Wani, S.P. (2016) Plant-growth-promoting rhizobacteria: Drought stress alleviators to ameliorate crop production in drylands. *Annals of Microbiology* 66:35–42.

Keyster, M., Klein, A., Ludidi, N. (2012) Caspase-like enzymatic activity and the ascorbate-glutathione cycle participate in salt stress tolerance of maize conferred by exogenously applied nitric oxide. *Plant Signaling and Behavior* 7:349–360.

Khan, M.M.A., Haque, E., Paul, N.C., Khaleque, M.A., Al-Garni, S.M.S., Rahman, M., Islam, M.T. (2017) Enhancement of growth and grain yield of rice in nutrient deficient soils by rice probiotic bacteria. *Rice Science* 24:264–273.

Khan, M.M.A., Khatun, A., Islam, M.T. (2016) Promotion of plant growth by phytohormone producing bacteria. In *Microbes in Action*, eds. Garg, N., Aeron, A., 45–76. New York: Nova Sci. Pub.

Konnova, S.A., Brykova, O.S., Sachkova, O.A., Egorenkova, I.V., Ignatov, V.V. (2001) Protective role of the polysaccharide containing capsular components of *Azospirillum brasilense. Microbiology* 70:436–440.

Kumar, M., Mishra, S., Dixit, V., Kumar, M., Agarwal, L., Chauhan, P.S., Nautiyal, C.S. (2016) Synergistic effect of *Pseudomonas putida* and *Bacillus amyloliquefaciens* ameliorates drought stress in chickpea (*Cicer arietinum* L.). *Plant Signaling and Behavior* 11:e1071004.

Ledger, T., Rojas, S., Timmermann, T., Pinedo, I., Poupin, M.J., Garrido, T., Richter, P., Tamayo, J., Donoso, R. (2016) Volatile-mediated effects predominate in *Paraburkholderia phytofirmans* growth promotion and salt stress tolerance of *Arabidopsis thaliana. Frontiers in Microbiology* 7:1838.

Lee, G.W., Lee, K.J., Chae, J.C. (2016) Herbaspirillum sp. strain GW103 alleviates salt stress in *Brassica rapa* L. ssp. Pekinensis. *Protoplasma* 253:655–661.

Li, H., Lei, P., Pang, X., Li, S., Xu, H., Xu, Z., Feng, X. (2017) Enhanced tolerance to salt stress in canola (*Brassica napus* L.) seedlings inoculated with the halotolerant Enterobacter cloacae HSNJ4. *Applied Soil Ecology* 119:26–34.

Li, J., McConkey, B.J., Cheng, Z., Guo, S., Glick, B.R. (2013) Identification of plant growth-promoting rhizobacteria-responsive proteins in cucumber roots under hypoxic stress using a proteomic approach. *Journal of Proteomics* 84:119–131.

Lifshitz, R., Kloepper, J.W., Scher, F.M., Tipping, E.M., Laliberte, M. (1986) Nitrogen-fixing pseudomonads isolated from roots of plants grown in the Canadian high arctic. *Applied and Environmental Microbiology* 51:251–255.

Lim, J.H., Kim, S.D. (2013) Induction of drought stress resistance by multi-functional PGPR *Bacillus licheniformis* K11 in Pepper. *The Plant Pathology Journal* 29:201–208.

Liu, Y., Shi, Z., Yao, L., Yue, H., Li, H., Li, C. (2013a) Effect of IAA produced by *Klebsiella oxytoca* Rs-5 on cotton growth under salt stress. *The Journal of General and Applied Microbiology* 59:59–65.

Liu, F., Xing, S., Ma, H., Du, Z., Ma, B. (2013b) Cytokinin producing, plant growth promoting rhizobacteria that confer resistance to drought stress in *Platycladus orientalis* container seedlings. *Applied Microbiology and Biotechnology* 97:9155–9164.

Liu, X.M., Zhang, H. (2015) The effects of bacterial volatile emissions on plant abiotic stress tolerance. *Frontiers in Plant Science* 6:774.

Ma, Y., Rajkumar, M., Moreno, A., Zhang, C., Freitas, H. (2017) Serpentine endophytic bacterium *Pseudomonas azotoformans* ASS1 accelerates phytoremediation of soil metals under drought stress. *Chemosphere* 185:75–85.

Ma, Y., Rajkumar, M., Zhang, C., Freitas, H. (2016a) Inoculation of *Brassica oxyrrhina* with plant growth promoting bacteria for the improvement of heavy metal phytoremediation under drought conditions. *Journal of Hazardous Materials* 320:36–44.

Ma, Y., Rajkumar, M., Zhang, C., Freitas, H. (2016b) Beneficial role of bacterial endophytes in heavy metal phytoremediation. *Journal of Environmental Management* 174:14–25.

Mahmood, S., Daur, I., Al-Solaimani, S.G., Ahmad, S., Madkour, M.H., Yasir, M., Hirt, H., Ali, S., Ali, Z. (2016) Plant growth promoting rhizobacteria and silicon synergistically enhance salinity tolerance of mung bean. *Frontiers in Plant Science* 7:876.

Mantelin, S., Touraine, B. (2004) Plant growth-promoting bacteria and nitrate availability impacts on root development and nitrate uptake. *Journal of Experimental Botany* 55:27–34.

Mayak, S., Tirosh, T., Glick, B.R. (2004a) Plant growth-promoting bacteria that confer resistance in tomato to salt stress. *Plant Physiology and Biochemistry* 42:565–572.

Mayak, S., Tirosh, T., Glick, B.R. (2004b) Plant growth-promoting bacteria that confer resistance to water stress in tomato and pepper. *Plant Science* 166:525–530.

Meldau, D.G., Meldau, S., Hoang, L.H., Underberg, S., Wunsche, H., Baldwin, I.T. (2013) Dimethyl disulfide produced by the naturally associated bacterium *Bacillus* sp B55 promotes *Nicotiana attenuata* growth by enhancing sulfur nutrition. *The Plant Cell* 25:2731–2747.

Miller, G., Susuki, N., Ciftci-Yilmaz, S., Mittler, R. (2010) Reactive oxygen species homeostasis and signalling during drought and salinity stresses. *Plant, Cell & Environment* 33:453–467.

Naseem, H., Bano, A. (2014) Role of plant growth-promoting rhizobacteria and their exopolysaccharide in drought tolerance of maize. *Journal of Plant Interactions* 9:689–701.

Nautiyal, C.S., Srivastava, S., Chauhan, P.S., Seem, K., Mishra, A., Sopory, S.K. (2013) Plant growth-promoting bacteria *Bacillus amyloliquefaciens* NBRISN13 modulates gene expression profile of leaf and rhizosphere community in rice during salt stress. *Plant Physiology and Biochemistry: PPB* 66:1–9.

Naveed, M., Mitter, B., Reichenauer, T.G., Wieczorek, K., Sessitsch, A. (2014) Increased drought stress resilience of maize through endophytic colonization by *Burkholderia phytofirmans* PsJN and *Enterobacter* sp. FD17. *Environmental and Experimental Botany* 97:30–39.

Nunkaew, T., Kantachote, D., Nitoda, T., Kanzaki, H., Ritchie, R.J. (2015) Characterization of exopolymeric substances from selected *Rhodopseudomonas palustris* strains and their ability to adsorb sodium ions. *Carbohydrate Polymers* 115:334–341.

Orcutt, D.M., Nilsen, E.T. (2000) *The Physiology of Plants under Stress: Soil and Biotic Factors*. New York: Wiley.

Palacio-Rodríguez, R., Coria-Arellano, J.L., López-Bucio, J., Sánchez-Salas, J., Muro-Pérez, G., Castañeda-Gaytán, G., Sáenz-Mata, J. (2017) Halophilic rhizobacteria from *Distichlis spicata* promote growth and improve salt tolerance in heterologous plant hosts. *Symbiosis* 73:179–189. doi:10.1007/s13199-017-0481-8.

Pinedo, I., Ledger, T., Greve, M., Poupin, M.J. (2015) *Burkholderia phytofirmans* PsJN induces long-term metabolic and transcriptional changes involved in *Arabidopsis thaliana* salt tolerance. *Frontiers in Plant Science* 6:466.

Pishchik, V.N., Vorobyev, N.I., Chernyaeva, I.I., Timofeeva, S.V., Kozhemyakov, A.P., Alexeev, Y.V., Lukin, S.M. (2002) Experimental and mathematical simulation of plant growth promoting rhizobacteria and plant interaction under cadmium stress. *Plant and Soil* 243:173–186.

Radhakrishnan, R., Baek, K.H. (2017) Physiological and biochemical perspectives of non-salt tolerant plants during bacterial interaction against soil salinity. *Plant Physiology and Biochemistry: PPB* 116:116–126.

Raheem, A., Shaposhnikov, A., Belimov, A.A., Dodd, I.C., Ali, B. (2017) Auxin production by rhizobacteria was associated with improved yield of wheat (*Triticum aestivum* L.) under drought stress. *Archives of Agronomy and Soil Science* 64:574–587.

Rajashekar, C.B. (2000) Cold response and freezing tolerance in plants. In *Plant Environment Interaction*, ed. Wilkinson, R.E., 321–342. New York: Marcel Dekker.

Rajendran, K., Tester, M., Roy, S.J. (2009) Quantifying the three main components of salinity tolerance in cereals. *Plant, Cell & Environment* 32:237–249.

Ravanbakhsh, M., Sasidharan, R., Voesenek, L.A.C.J., Kowalchuk, G.A., Jousset, A. (2017) ACC deaminase-producing rhizosphere bacteria modulate plant responses to flooding. *Journal of Ecology* 105:979–986.

Redondo-Gómez, S. (2013) Abiotic and biotic stress tolerance in plants. In *Molecular Stress Physiology of Plants*, eds. Rout, G.R., Das, A.B., 1–20. India: Springer.

Rodriguez-Salazar, J., Suarez, R., Caballero-Mellado, J., Iturriaga, G. (2009) Trehalose accumulation in *Azospirillum brasilense* improves drought tolerance and biomass in maize plants. *FEMS Microbiology Letters* 296:52–59.

Rolli, E., Marasco, R., Vigani, G., Ettoumi, B., Mapelli, F., Deangelis, M.L., Gandolfi, C., *et al.* (2015) Improved plant resistance to drought is promoted by the root-associated microbiome as a water stress-dependent trait. *Environmental Microbiology* 17:316–331.

Ruelland, E., Vaultier, M.N., Zachowski, A., Hurry, V. (2009) Cold signalling and cold acclimation in plants. In *Advances in Botanical Research*, eds. Kader, J.C., Delseny, M., 35–150. New York: Academic Press.

Sandhya, V., Ali, Sk. Z., Grover, M., Reddy, G., Venkateswarlu, B. (2009) Alleviation of drought stress effects in sunflower seedlings by the exopolysaccharides producing *Pseudomonas putida* strain GAP-P45. *Biology and Fertility of Soils* 46:17–26.

Sandhya, V., Ali, Sk. Z., Grover, M., Reddy, G., Venkateswarlu, B. (2010) Effect of plant growth promoting *Pseudomonas* spp. on compatible solutes antioxidant status and plant growth of maize under drought stress. *Plant Growth Regulation* 62:21–30.

Sarma, R.K., Saikia, R. (2014) Alleviation of drought stress in mung bean by strain *Pseudomonas aeruginosa* GGRJ21. *Plant and Soil* 377:111–126.

Sharma, S., Kulkarni, J., Jha, B. (2016) Halotolerant rhizobacteria promote growth and enhance salinity tolerance in peanut. *Frontiers in Microbiology* 7:1600.

Shrivastava, P., Kumar, R. (2015) Soil salinity: A serious environmental issue and plant growth promoting bacteria as one of the tools for its alleviation. *Saudi Journal of Biological Sciences* 22:123–131.

Siddikee, M.A., Glick, B.R., Chauhan, P.S., Yim, W.J., Sa, T. (2011) Enhancement of growth and salt tolerance of red pepper seedlings (*Capsicum annum* L.) by regulating stress ethylene synthesis with halotolerant bacteria containing 1-aminocyclopropane-1-carboxylic acid deaminase activity. *Plant Physiology and Biochemistry: PPB* 49:427–434.

Singh, R.P., Shelke, G.M., Kumar, A., Jha, P.N. (2015) Biochemistry and genetics of ACC deaminase: A weapon to "stress ethylene" produced in plants. *Frontiers in Microbiology* 6:937.

Steffens, J.C. (1990) The heavy metal-binding peptides of plants. *Annual Review of Plant Physiology and Plant Molecular Biology* 41:553–575.

Su, F., Jacquard, C., Villaume, S., Michel, J., Rabenoelina, F., Clément, C., Barka, E.A., Dhondt-Cordelier, S., Vaillant-Gaveau, N. (2015) *Burkholderia phytofirmans PsJN* reduces impact of freezing temperatures on photosynthesis in *Arabidopsis thaliana. Frontiers in Plant Science* 6:810.

Suarez, R., Wong, A., Ramirez, M., Barraza, A., Orozco, Mdel C., Cevallos, M.A., Lara, M., Hernández, G., Iturriaga, G. (2008) Improvement of drought tolerance and grain yield in common bean by overexpressing trehalose-6-phosphate synthase in rhizobia. *Molecular Plant-Microbe Interactions: MPMI* 21:958–966.

Subramanian, P., Kim, K., Krishnamoorthy, R., Mageswari, A., Selvakumar, G., Sa, T. (2016) Cold stress tolerance in psychrotolerant soil bacteria and their conferred chilling resistance in tomato (*Solanum lycopersicum* Mill.) under low temperatures. *PLoS One* 11(8):e0161592.

Subramanian, P., Mageswari, A., Kim, K., Lee, Y., Sa, T. (2015) Psychrotolerant endophytic *Pseudomonas* sp. strains OB155 and OS261 induced chilling resistance in tomato plants (*Solanum lycopersicum* Mill.) by activation of their antioxidant capacity. *Molecular Plant-Microbe Interactions: MPMI* 28:1073–1081.

Sukweenadhi, J., Kim, Y.J., Choi, E.S., Koh, S.C., Lee, S.W., Kim, Y.J., Yang, D.C. (2015) *Paenibacillus yonginensis* DCY84T induces changes in *Arabidopsis thaliana* gene expression against aluminum, drought, and salt stress. *Microbiological Research* 172:7–15.

Tamura, T., Hara, K., Yamaguchi, Y., Koizumi, N., Sano, H. (2003) Osmotic stress tolerance of transgenic tobacco expressing a gene encoding a membrane-located receptor-like protein from tobacco plants. *Plant Physiology* 131:454–462.

Templer, P.H. (2012) Changes in winter climate: Soil frost, root injury, and fungal communities. *Plant and Soil* 353:15–17.

Theocharis, A., Bordiec, S., Fernandez, O., Paquis, S., Dhondt-Cordelier, S., Baillieul, F., Clément, C., Barka, E.A. (2012) *Burkholderia phytofirmans* PsJN primes *Vitis vinifera* L. and confers a better tolerance to low nonfreezing temperatures. *Molecular Plant-Microbe Interactions: MPMI* 25:241–249.

Timmusk, S., Abd El-Daim, I.A., Copolovici, L., Tanilas, T., Kännaste, A., Behers, L., Nevo, E., *et al.* (2014) Drought-tolerance of wheat improved by rhizosphere bacteria from harsh environments: Enhanced biomass production and reduced emissions of stress volatiles. *PLoS ONE* 9:e96086.

Timmusk, S., Paalme, V., Pavlicek, T., Bergquist, J., Vangala, A., Danilas, T., Nevo, E. (2011) Bacterial distribution in the rhizosphere of wild barley under contrasting microclimates. *PLoS ONE* 6:e17968.

Timmusk, S., Wagner, G.H. (1999) The plant-growth-promoting rhizobacterium *Paenibacillus polymyxa* induces changes in *Arabidopsis thaliana* gene expression: A possible connection between biotic and abiotic stress responses. *Molecular Plant–Microbe Interactions* 12:951–959.

Trischuk, R.G., Schilling, B.S., Wisniewski, M., Gusta, L.V. (2006) Freezing stress: Systems biology to study cold tolerance. In *Physiology and Molecular Biology of Stress Tolerance in Plants*, eds. Rao, K.V.M., Raghavendra, A.S., Reddy, K.J., 131–156. Dordrecht: Springer.

Turan, M., Güllüce, M., Çakmak, R., Şahin, F. (2013) Effect of plant growth-promoting rhizobacteria strain on freezing injury and antioxidant enzyme activity of wheat and barley. *Journal of Plant Nutrition* 36:731–748.

Uemura, M., Steponkus, P.L. (1999) Cold acclimation in plants: Relationship between the lipid composition and the cryostability of the plasma membrane. *Journal of Plant Research* 112:245–254.

Upadhyay, S.K., Singh, D.P. (2015) Effect of salt-tolerant plant growth-promoting rhizobacteria on wheat plants and soil health in a saline environment. *Plant Biology* 17:288–293.

Vaishnav, A., Kumari, S., Jain, S., Varma, A., Choudhary, D.K. (2015) Putative bacterial volatile-mediated growth in soybean (*Glycine max* L. Merrill) and expression of induced proteins under salt stress. *Journal of Applied Microbiology* 119:539–551.

Vargas, L., Santa Brígida, A.B., Mota Filho, J.P., de Carvalho, T.G., Rojas, C.A., Vaneechoutte, D., Van Bel, M., *et al.* (2014) Drought tolerance conferred to sugarcane by association with *Gluconacetobacter diazotrophicus*: A transcriptomic view of hormone pathways. *PLoS One* 9:e114744.

Voesenek, L.A.C.J., Bailey-Serres, J. (2015) Flood adaptive traits and processes: An overview. *The New Phytologist* 206:57–73.

Voesenek, L.A.C.J., Van der Veen, R. (1994) The role of phytohormones in plant stress: Too much or too little water. *Acta Botanica Neerlandica* 43:91–127.

Vurukonda, S.S.K.P., Vardharajula, S., Shrivastava, M., SkZ, A. (2016) Enhancement of drought stress tolerance in crops by plant growth promoting rhizobacteria. *Microbiological Research* 184:13–24.

Wahid, A., Gelani, S., Ashraf, M., Foolad, M.R. (2007) Heat tolerance in plants: An overview. *Environmental and Experimental Botany* 61:199–223.

Wang, Q., Dodd, I.C., Belimov, A.A., Jiang, F. (2016) Rhizosphere bacteria containing 1-aminocyclopropane-1-carboxylate deaminase increase growth and photosynthesis of pea plants under salt stress by limiting Na^+ accumulation. *Functional Plant Biology* 43:161–172.

Wang, W., Vinocur, B., Shoseyov, O., Altman, A. (2004) Role of plant heat-shock proteins and molecular chaperones in the abiotic stress response. *Trends in Plant Science* 9:244–252.

Weis, J.S., Weis, P. (2004) Metal uptake, transport and release by wetland plants: Implications for phytoremediation and restoration. *Environment International* 30:685–700.

Xu, J., Li, X.L., Luo, L. (2012) Effects of engineered *Sinorhizobium meliloti* on cytokinin synthesis and tolerance of alfalfa to extreme drought stress. *Applied and Environmental Microbiology* 78:8056–8061.

Yang, A., Akhtar, S.S., Iqbal, S., Amjad, M., Naveed, M., Zahir, Z.A., Jacobsen, S. (2016) Enhancing salt tolerance in quinoa by halotolerant bacterial inoculation. *Functional Plant Biology* 43:632–642.

Yang, J., Kloepper, J.W., Ryu, C.M. (2009) Rhizosphere bacteria help plants tolerate abiotic stress. *Trends in Plant Science* 14:1–4.

Yokota, A., Takahara, K., Akashi, K. (2006) Water stress. In *Physiology and Molecular Biology of Stress Tolerance in Plants*, eds. Rao, K.V.M., Raghavendra, A.S., Reddy, K.J., 15–40. Dordrecht: Springer.

Yoshida, S., Uemura, M. (1984) Protein and lipid composition of isolated plasma membranes from orchard grass (*Dactylisgl omerata* L.) and changes during cold acclimation. *Plant Physiology* 75:31–37.

Zhang, H., Kim, M.S., Sun, Y., Dowd, S.E., Shi, H., Paré, P.W. (2008) Soil bacteria confer plant salt tolerance by tissue-specific regulation of the sodium transporter HKT1. *Molecular Plant-Microbe Interactions: MPMI* 21:737–744.

Zhang, H., Murzello, C., Sun, Y., Kim, M.S., Xie, X., Jeter, R.M., Zak, J.C., Dowd, S.E., Paré, P.W. (2010) Choline and osmotic-stress tolerance induced in *Arabidopsis* by the soil microbe *Bacillus subtilis* (GB03). *Molecular Plant-Microbe Interactions: MPMI* 23:1097–1104.

Zhang, H., Sun, Y., Xie, X., Kim, M.S., Dowd, S.E., Paré, P.W. (2009) A soil bacterium regulates plant acquisition of iron via deficiency-inducible mechanisms. *The Plant Journal* 58:568–577.

24 The Long Road to Develop Novel Priming Products to Increase Crop Yield under Stressful Environments

Andrés A. Borges, Estefanía Carrillo-Perdomo, David Jiménez-Arias, Francisco J. García-Machado, Francisco Valdés-González, and Juan C. Luis

CONTENTS

24.1 INTRODUCTION

For many years, chemical pesticides have been the most effective pathogen and pest management tools. However, their use has become very controversial as these chemicals are toxic substances that contaminate the environment and negatively affect living species (Guedes et al., 2016). As a consequence of the growing demand for food production worldwide, new strategies are required to permit sustainable production with minimal impact on natural resources. The agrochemical industry is continually challenged to discover, develop and produce new, more effective and safer plant protection products (PPPs) that also need to be cost- and time-effective. These novel products could help farmers to more effectively prevent crop losses that globally cost billions of dollars. For instance, in 2013, the estimated losses (Figure 24.1) due to plant diseases affecting the four major crops wheat, potato, maize and soybean were 10–20% of potential production, while those due to abiotic stresses such as salinity, heat, chilling and drought prevented at least 70% of the potential production (FAOSTAT, 2013; Bray et al., 2002).

Novel crop protection agents have to be effective as well as safe for users, consumers and the environment. Thus, it is urgently necessary to find new strategies to increase yields, preferably with minimal impact on natural ecosystems, including a reduction in the use of chemical pesticides. Among the most promising alternatives that do not involve the application of pesticides is the implementation of plant resistance inducers capable of limiting pathogen and pest damage, along with tolerance inducers that mitigate abiotic stresses such as salinity or drought (Table 24.1). Induced resistance is a non-specific form of disease resistance in plants, acting against a wide range of pathogens and pests. It can be activated by a range of inducers, which may be biotic (elicitors) or abiotic (natural or synthetic chemicals). Many plant species are able to pre-activate or prime their defence mechanisms once exposed to these treatments that induce genetic and/or biochemical modifications, resulting in an enhanced systemic acquired resistance in response to a subsequent stress exposure (Conrath, 2006). So, higher plants can show some stress 'memory', or stress imprinting (Bruce et al., 2007; Conrath et al., 2015). After receiving the exogenous priming treatment, the plant enters a primed state in which its defenses are not yet fully mobilized. Then as soon as its safety is challenged, faster and stronger basal resistance mechanisms are induced, leading to a more efficient and effective resistance response (Conrath et al., 2015; Pastor et al., 2013). Preliminary stress exposure is indeed known to enhance the stress tolerance of the plant through induction of adaptation responses (Conrath et al., 2015; Pastor et al., 2013). Such tolerance can be linked to an array of morphological, physiological and biochemical

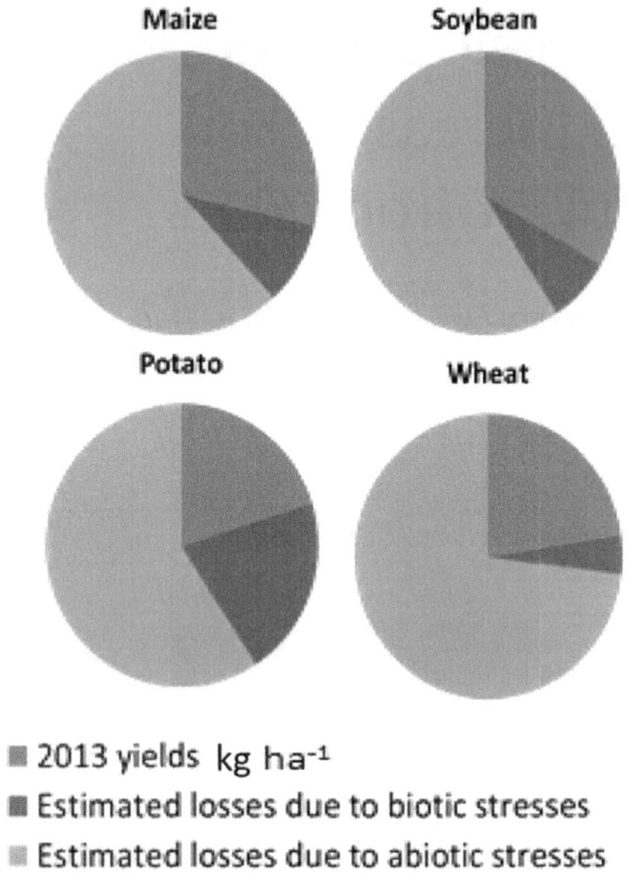

FIGURE 24.1 Percentages of estimated yield losses due to biotic and abiotic stresses. Source: FAOSTATS 2013.

responses that decrease stress damage or facilitate repair in damaged systems (Potters et al., 2007). Primed plants subsequently attacked by pathogens and/or subjected to environmental threats tend to grow more vigorously, bloom earlier and have higher yields than unprimed (Farooq et al., 2008). Consequently, priming could provide a low-cost protection against relatively high-stress environmental conditions (van Hulten et al., 2006).

The primed state can also be triggered naturally by plants in response to stress, working as an alert warning that helps them face stress in the future (Filippou et al., 2013). It is now widely accepted that priming is an intrinsic part of induced resistance because when plants are challenged by a potential hazard, they activate their defense pathways to protect themselves. They also prepare their defense system to improve their immune response against later environmental stress. Therefore, exploiting their natural defense system appears to be the key to developing novel bioactive compounds that are environmentally friendly and contribute to sustainable agriculture (Figure 24.2).

24.2 THE MAIN APPROACHES TO DISCOVERY AND DEVELOPMENT OF NOVEL PRIMING COMPOUNDS FOR COMMERCIALIZATION AS PPPs

Simplistically, the way the agrochemical industry has traditionally searched for novel plant protection agents is through mass screenings designed to cover most of the spectrum of diseases that affect the crop. Nowadays, the chance of identifying a candidate molecule has been enhanced by the development of high-throughput screening methods that have achieved the screening of almost 160,000 molecules from 2010 to 2014 (McDougall, 2016). Once the most effective active compounds are identified, the next steps are to characterize them and determine their mode of action regarding spatio-temporal effects (local or systemic, protective and/or curative) designed to demonstrate the biological activity of the new molecule. Normally, this information is assembled once the biochemical mode of action of the PPP is identified. Indeed, the rapid advance in understanding systemic acquired resistance (SAR) and induced systemic resistance (ISR) mechanisms in the last few decades has brought deeper knowledge of the mode of action of bioactive substances in general (Leadbeater and Staub, 2017).

After progress through the previous steps, the decision whether the new compound is worth full development will also depend on the patentability of the molecule, good toxicological and environmental properties and good business prospects. So, for example, preliminary safety testing in biological systems (plant, animal, human, soil and water systems) and market studies are performed. Thus, the process requires the establishment of a pilot plant to produce sufficient quantities of the molecule for testing. While almost 160,000 molecules are screened during the research stage, on average, only 1.5 molecules pass on to the development stage and finally 1 molecule will reach the registration stage (McDougall, 2016). The candidate compounds will first be subject to the formulation. Although in rare cases the screenings single out the final molecule, modifications of the primary molecule –semi-synthesis – or even an overall reconfiguration of the core structure – synthetic mimic – are required.

Then dosage, mode of employment and production optimization are focused on, which often results in high costs (Leadbeater and Staub, 2017). These are incurred in testing the compound in the control of the target pests, weeds or diseases on a number of crops, under a variety of environmental situations including small and

TABLE 24.1
List of Some Biotic (Elicitors) and Abiotic (Chemical) Resistance-Induced Compounds

Biotic Inducers (Elicitors from Microorganisms)	Plant–Pathogen Interaction	References
Fungal, bacterial and plant growth promoting rhizobacteria products	Carnation vs. *Fusarium* Barley vs. *Blumeria graminis* f.sp. *hordei*	Van Peer and Schippers (1992), Nelson (2005)
Chitin	Tobacco vs. tobacco mosaic virus (TMV)	Zhang et al. (2004)
Glucans	Tobacco vs. viruses Barley vs. *Blumeria graminis* f.sp. *hordei*	Kopp et al. (1989), Reglinski et al. (1994)
Lipopolysaccharides	Tobacco vs. *Phytophtora nicotianae*	Coventry and Dubery (2001)
Proteins and peptides	Tobacco vs. *Pseudomonas syringae* pv. *tabaci*	Lotan and Fluhr (1990)
Plant extracts		
Brassinosteroid	Rice and tobacco vs. TMV and *Pseudomonas syringae* pv. *tabaci*	Nakashita et al. (2003)
Spermine	Tobacco vs. TMV	Yamakawa et al. (1998)
Hedera helix	*Cotoneaster salicifolius* vs. *Erwinia amylovora*	Baysal et al. (2002)
Abiotic inducers (natural or synthetic chemicals)		
B-aminobutyric acid (BABA)	*Arabidopsis* vs. *Pseudomonas syringae* pv. Tomato DC3000, *Alternaria brassicicola*	Flors et al. (2008)
Benzothiadiazole (BTH)	Wheat vs. *Erysiphe graaminis* f sp *tritic* Tomato vs. *Bemisia tabaci*	Görlach et al. (1996), Nombela et al. (2005)
Fungicides		
Probenazole	Tobacco vs. TMV and *Pseudomonas syringae* pv. *tabaci*	Nakashita et al. (2002)
Strobilurin	Tobacco vs. TMV and *P. syringae* pv. *tabaci*	Herms et al. (2002)
Recently identified natural activators		
Azelaic acid (AZA)	*Arabidopsis* vs. *P. syringae* pv. *maculicola*	Jung et al. (2009)
Pipecolic acid	*Arabidopsis* vs. *P. syringae* pv. *maculicola*	Návarová et al. (2012)
Vitamins		
Vitamin B_1	Rice and vegetable crops vs. fungal, bacterial and viral infections	Ahn et al. (2005)
Vitamin K_3 or menadione	Banana vs. *Fusarium oxysporum* f. sp. cubense Oilseed rape vs. *Leptosphaeria maculans* *Arabidopsis* vs. *Pseudomonas syringae* pv. Tomato DC3000	Borges et al. (2003, 2004, 2009)
Vitamin B_2 or riboflavin	*Arabidopsis* vs. *Phytophtora pasasitica* and *Pseudomonas syringae* pv. Tomato DC3000	Dong and Beer (2000)

large-scale field trials. Between the years 2010 and 2014, this cost has been estimated at \$253 million/€190 million (88.5% of the budget), involving the discovery and research and developmental (R&D) phases (McDougall, 2016). Although the complete cost and time required to launch a new PPP onto the market has increased (from around \$152 million and 8.3 years in 1995 to \$286 million and 11.3 years for 2010–2014), the number of new commercial products has remained at only 1 from 1995 to 2014. The number of potential candidates to be screened has increased over time (from 52,500 in 1995 to almost 160,000 from 2010–2014), but the agrochemical industry has reduced the number of molecules it gives the opportunity to pass along the developmental stage. These restrictions are probably due to the increasing difficulty of finding novel compounds that meet all the plant-protection, production-cost and market demands. Among the explanations for the longer period needed to place the PPP on the shelves are the increase in the complexity and volume of data required by registration agencies and the time necessary to carry out the related experiments, along with the requirements of regulatory bodies to ensure that the registration dossiers are absolutely complete prior to authorization, rather than issuing a provisional approval. As an example, Loso et al. (2017) have detailed the strategy addressed by Dow

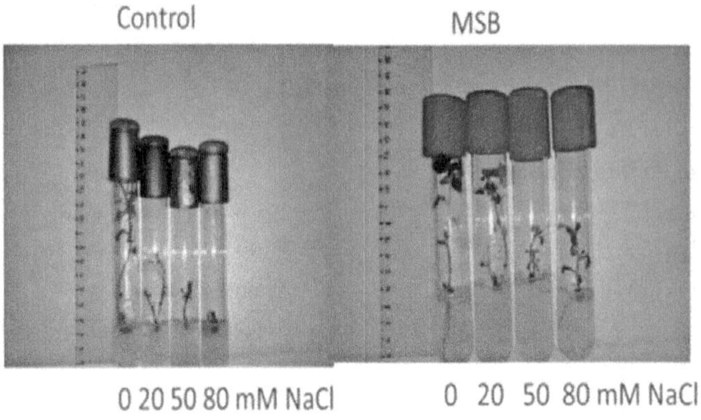

FIGURE 24.2 Effect of seed-soaking in menadione sodium bisulfite (MSB), a priming agent, on salt stress tolerance in potato.

AgroSciences (DAS) for the generation of novel PPPs and provide a couple of case studies to provide further insight into some of the approaches employed by that company.

The agrochemical industry (led by Europe, spending 40.1% of total world funding for PPPs) plans to increase its budget for developing novel PPPs in 2019. It is expected 30.5% of the budget will be spent on discovering new compounds and 28.3% on their development. This represents \$979 million and \$907 million invested by the agrochemical industry, respectively. Together with product marketing (32.5%, \$1041 million), this amount accounts for most of the agrochemical industry budget (McDougall, 2016). So far during the 21st century, the growing awareness of climate change and sustainable development has led to a rapid increase in public demand for PPPs that are safe for the environment and human health. Interest in biocontrol products research is expanding, so in 2019 it is expected that almost \$270 million (9.2%) will be invested in their development (however, chemical control still leads the investment). This is possible thanks to rapid advancements in genomics, genetics and chemistry that have also been exploited in the field of priming compounds research. Among the advantages of these compounds is that they are usually eco-friendly and low in cost. Moreover, they are capable of imprinting long-term resistance to stresses, and sometimes only a single application is required (Jiménez-Arias et al., 2017). Here we present two of the main approaches adopted to find novel priming agents for plant protection.

24.2.1 Genomics

Genomics, the study of genomes and the effect of chemicals on biological targets using large-scale and high-throughput methods, has recently led to the discovery of uncovered aspects of plant biology. Such methods are providing potent information about the intricate nature of living systems at an unprecedented rate. In the recent past, the application of genomic technologies has similarly provided revolutionary new insights into how plants defend themselves from pathogen attack. The plant model *Arabidopsis thaliana* was the first plant genome ever sequenced, and research on this model has paved the way for plant genomics research. Several large-scale gene expression profiling studies followed by functional analyses of genes altered in expression have discovered unexpectedly complex interactions between plant signaling pathways. Thus, scientists are now slowly moving beyond the limitations of understanding single processes such as the behavior of an individual transcription factor in plant defense response (Kazan and Schenk, 2007). Reverse and genetic approaches have permitted the discovery of genes encoding master switches that could be promising candidates for use in engineering disease resistance. A master switch is a regulatory protein, for instance, a transcription factor (TF), acting upstream in a signaling cascade pathway. These TFs regulate gene expression by activating TFs, which, in turn, directly bind to the promoter region of target genes. In some cases, these regulatory proteins with a role as the master switch can be constitutively present in the cell or be activated by protein-protein interactions and/or phosphorylation/dephosphorylation. For instance, protein kinase and mitogen-activated (MAP) kinase kinase kinase (MAPKKK) act upstream from MAPKK and MAPK pathways and function in activating downstream MAPKs in response to a stimulus (Pedley and Martin, 2005). Most of the master regulators have so far been identified by reverse/forward genetic approaches. However, during the last decade, advances in transcriptome analyses and bioinformatics tools allowed the identification of a number of TFs, based on their response to pathogens. The *Arabidopsis* genome contains more than 1500 TFs belonging to different TF gene families, including AP2/ERFs, MYBs and WRKYs involved in plant defense. These were identified using large-scale identification (microarray) from pathogen-inoculated versus uninoculated (control) plants. Once identified, these genes were confirmed by RT-Q-PCR, and selected genes were either inactivated by T-DNA, transposon insertions or RNAi or overexpressed, in order to study the effect of phenotypes generated from such manipulation in disease resistance (Chen and Chen, 2002; MacGrath et al., 2005).

The use of the biological system approach to induce plant defenses in model plant species such as *Arabidopsis* has provided a wealth of information about the genes involved in such defense mechanisms. It is expected that such information will help speed up research in crop species. However, the current challenge in transferring this knowledge to crop species is the identification of crop genes that are functional orthologues of those identified in the plant model. Bioinformatics, molecular mapping, cloning and functional complementation approaches make this task relatively easy (Kazan and Schenk, 2007). Plant resistance is multigenic and genomic analysis of induced defense has established that an induced resistance response requires coordinated action of many genes and/or defense-signaling pathways. These latter are mainly dependent on the phytohormones salicylic acid (SA), jasmonic acid (JA) and ethylene (ET). The inherent complexity associated with individual defense-signaling pathways can be further complicated by the extensive crosstalk that occurs between multiple stress and/or defense-signaling pathways. Therefore, a holistic approach needs to be developed to integrate the effects of such multiple parameters on the whole plant system and to assess the responses of plants sometimes exposed to both types of stresses simultaneously: biotic and abiotic.

24.2.2 Chemical Genetics

Another approach to studying the effect of chemicals on biological targets is known as chemical genetics. This discipline studies biological systems at cell level or in the living organism by applying small bioactive molecules to them. These molecules bind reversibly or irreversibly to the protein target of a particular cellular process to alter its functioning over a certain period and produce a specific phenotype (Schreiber, 1998). Thus, chemical genetics is considered multidisciplinary, since it unites the field of chemical synthesis with genetics and diverse analytical and bioinformatic techniques. Indeed, the combination of these different approaches facilitates its success (Smukste and Stockwell, 2005). The genetic foundations of chemical genetics lie in the properties of the molecules used. Among them should be highlighted the specificity of the ligands for their protein targets in numerous metabolic processes, the permeability of the cell toward them and the rapidity at which they activate or inhibit these targets (or receptors). Furthermore, the biological activity of the molecules depends on the timing and dose applied, so, due to which the resulting phenotype changes qualitatively and quantitatively over time. AS a result of these factors, chemical genetics is considered a suitable tool for studying multifunctional proteins and carrying out high-performance large-scale analyses in a wide variety of organisms. Moreover, this discipline aids greatly in the study of proteins belonging to multigenic families, among which there may be functional redundancy, as well as in identifying the functions of essential genes whose knock-out genotype induces lethality (O'Connor et al., 2011; Stockwell, 2000; Walsh and Chang, 2006).

The two types of strategies used in chemical genetics are called direct and reverse (Figure 24.3). The direct strategy applies bioactive compounds in a biological system, in order to select candidates that modulate the phenotype of interest. Later it is necessary to identify the target protein and its function in a particular signaling pathway. This identification is one of the key steps and at the same time one of the factors limiting the success of this strategy. In contrast, reverse chemical genetics focuses directly on searching for candidate molecules with a specific protein interaction. These are fixed into a matrix to study the phenotype that induces the molecule in the organism and then deciphers the molecular function it performs in this protein (O'Connor et al., 2011). Both strategies require optimization in order to find the most efficient and effective methodology for screening bioactive molecules. Another key point is to use the chemical library most suitable to the aim of the project, selected from chemically synthesized compounds or natural product derivatives. The effectiveness of the technique depends on the diversity of the molecules used (Hicks and Raikhel, 2012; Kumar et al., 2011; Toth and van der Hoorn, 2010).

Chemical genetics has mainly pharmaceutical and agrochemical applications. Aside from industrial screenings based on the search for pesticides, the use of chemical genetics in basic research into plants has also recently built up a decade of experience. This has resulted in success stories in clarifying novel mechanisms in the study areas of endomembrane trafficking, hormonal signaling, cell-walls, immunity and small RNAs (Hicks and Raikhel, 2014). One of these chemicals is the pyrabactin, a novel analogue of ABA, with which the receptor PYR1 was identified within the receptor family PYR/PYL involved in the signaling and the response to ABA. This was possible because this compound induces a quantified phenotype through its selective action on one of the receptors of the multigenic family, avoiding functional redundancy (Park et al., 2009). In most cases, the target was identified through a strategy based on searching for mutants insensitive to the compound. The use of chemical genetics in modulating the immune response is another field of study, resulting in the identification of new signaling pathways besides novel synthetic defense inducers of great commercial potential (Betkas and Eulgem, 2014). Other examples of synthetic defense inducers are: i) functional analogues of SA such

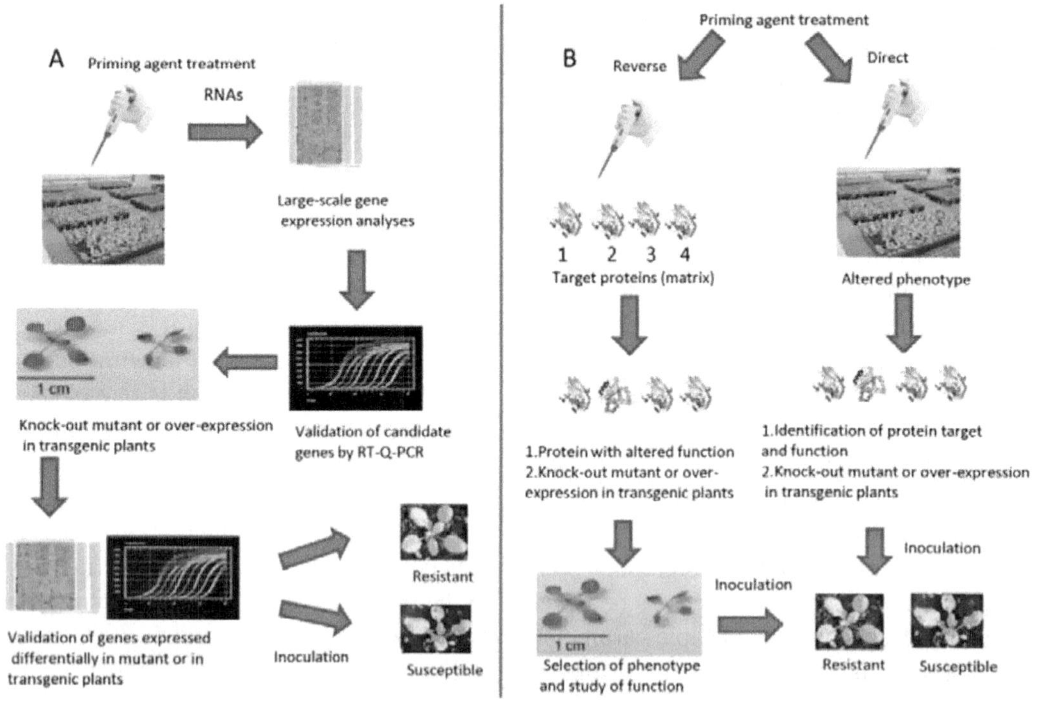

FIGURE 24.3 Different approaches to the search for novel priming compounds. Genomics (A) and chemical genetics (B).

as isonicotinic acid, which induces a long-term higher expression of pathogenesis-related (PR) genes without affecting SA and also resistance in *Arabidopsis* against virulent strains of *Hyaloperonospora arabidopsidis* or *Pseudomonas syringae* pv. tomato DC3000 (Uknes, 1996); ii) a set of compounds of different chemical origins defined as imprimatins, due to inducing a priming mechanism in *Arabidopsis* cells that promotes an immune response (Noutoshi et al., 2012a); iii) sulphonamides able to strengthen the resistance of *Arabidopsis* against *P. syringae* DC3000 (Noutoshi et al., 2012b); iv) a water-soluble vitamin K_3 derivative (MSB, menadione sodium bisulfite) that primes *Arabidopsis* against *P. syringae* DC3000, as well as other species against various pathogens (Borges et al., 2009). However, the mode of action of these synthetic defense activators is still unclear and highlights the importance of implementing successful strategies for the identification of targets receptive to low molecular weight molecules.

In the last few decades, several chemical plant defense activators have been marketed. For instance, the inducer of systemic acquired resistance (SAR), acibenzolar-S-methyl (benzo-1,2,3-thiadiazole-7-carbothioic acid S-methyl ester), formally named benzothiadiazol (BTH) was commercialized by Syngenta as a plant strengthener, a low-risk compound not marketed as a PPP and only in Germany. Plant strengtheners are compounds or mixtures exclusively intended to serve the well-being of the plant, given they are not PPPs as defined in Art. 2(1) of Regulation (EC) 1107/2009.

24.3 PROTECTING INTELLECTUAL PROPERTY AND MARKETING THE NOVEL PPP PRODUCT

After discovering the effectiveness of the novel priming compound and studying its action mechanism in the plant, the next step is to patent and register it as a PPP to safeguard the intellectual property and obtain authorization for its commercial use (plant heath registration). Registration *in sensu lato* (including patent) takes up approximately 11.5% of the budget ($33 million/€25 million), but previously mandatory toxicity and environmental studies (especially residue analysis) that require around 22.4% of the development budget ($64 million/€48 million) are performed as requirements of the registration dossier. In recent years, fast-track registration procedures have been implemented, but as stated above, the time to launch a PPP far from shortening has lengthened, probably because registration policies are more demanding than before (particularly the hazard assessment). This also reflects rising registration costs from the beginning of the 21st century to date (McDougall, 2016). Patent legislation is different for each country; it is possible to reserve the right of exploitation at a national level (submitted at the national office of patents and brands of each country), Europe-wide or internationally. Thus, the holders will be given temporary privileges for industrial or commercial exploitation in those territories where they register the invention for approximately 20 years from the submission date. If greater protection is intended, an international 'guard' can be requested. A time reservation must be requested from the World Intellectual Property Organization (WIPO), so that the owner can apply for the patent in the desired territories without losing the priority date. Here we briefly detail the procedure for a European Union (EU) patent registration, but for a detailed review of the issue consult Carroll (2016).

The European Patent Office (EPO, www.epo.org/) is the agency in charge of administering the European Patent Convention, EPC or EU patent directive that controls the patents protected in the 28 countries of the European Union (EU). It also administers patents for the countries of the Munich Convention (Albania, Iceland, Liechtenstein, the former Yugoslav Republic of Macedonia, Monaco, Norway, San Marino, Serbia, Switzerland and Turkey). To date, Bosnia and Herzegovina, and Montenegro also hold extension agreements with the EPO. The applicant must submit a validation request in each of the designated countries within three or six months, to avoid the risk of losing the patent right there. After the expiration of the patent's time limit, the applicant/holder can also ask for a supplementary protection certificate (SPC), which extends the patented intellectual property right when entering the market is postponed due to bureaucratic delays. Nevertheless, in 2018, a unitary patent will be also available, that will make it possible to obtain patent protection in up to 26 EU Member States (Austria, Belgium, Bulgaria, Cyprus, Czech Republic, Estonia, France, Germany, Greece, Ireland, Latvia, Lithuania, Luxembourg, Malta, Poland, Portugal, Romania, Slovakia, United Kingdom, Slovenia) by submitting a single request to the EPO, making the procedure much simpler and more cost-effective for applicants. Before marketing, the PPP needs to be validated by the European Union (EU) (regulation (EC) No 1107/2009 in the form in which it is to be supplied to the user. Before any PPP can be placed on the EU market or used, it must be authorized (https://ec.europa.eu/food/plant/pesticides/authorisation_of_ppp_en) in the Member State(s) concerned. In this context, the EU defines a PPP as follows:

> PPPs (also referred to as 'pesticides') are products in the form in which they are supplied to the user, consisting of or containing one or more approved active substances, safeners or synergists, and intended for one of the following uses: (a) protecting plants or plant products against harmful organisms or preventing the action of such organisms, unless the main purpose of these products is considered to be for reasons of hygiene rather than for the protection of plants or plant products (e.g. fungicides, insecticides); (b) influencing the life processes of plants, such as substances influencing their growth, other than as a nutrient (e.g. plant growth regulators, rooting hormones); (c) preserving plant products, in so far as such substances or products are not subject to special Community provisions on preservatives (e.g. extending the life of cut flowers); (d) destroying undesired plants or parts of plants, except algae unless the products are applied on soil or water to protect plants (e.g. herbicides/weed-killers to kill actively growing weeds); (e) checking or preventing undesired growth of plants, except algae unless the products are applied on soil or water to protect plants (e.g. herbicides/weed-killers preventing the growth of weeds; https://ec.europa.eu).

In addition, the active compound will be checked for its presence in the 'List of Candidates for Substitution'. If so, it must be replaced (substituted) by other appropriate solutions (chemical and non-chemical). For the first authorization of a new PPP, the EPO processes other applications: mutual recognition, amendment, renewal, emergency authorization, minor uses, parallel trade permits or assessment of technical equivalence. Whatever the type of application; it has to be submitted through the Plant Protection Products Application Management System (PPPAMS). PPPAMS manages the workflow of applications, enabling applicants and Member States to communicate with each other during the application process, primarily by changing the status of the application. The registration dossier must contain a part A dealing with risk management, part B with data evaluation and risk assessment and a part C detailing the confidential information. This report is submitted to the zonal *Rapporteur Member State* (*RMS*) (not to the PPPAMS), which evaluates the proposal and prepares a report, after which the European Food Safety Authority (EFSA) issues a decision. Its final status will depend on approval by the Standing Committee for Food Chain and Animal Health, which, if it votes in favor, will authorize the PPP in up to 1.5 years (communicated via the PPPAMS) after submitting the application. After 2.5 to 3.5 years, the regulation is published approving the new active substance (depending on how complex and complete is the registration dossier).

In countries such as the USA, the process is quite different, and there is no possibility of applying for SPC.

The United States Environmental Protection Agency (EPA) is in charge of the evaluation process and contrast to Europe, the EPA sets the specific requirements for validating the novel PPP (Carroll, 2016). Once on the market, work refining it will continue. The industry needs to follow up its PPPs, undertaking post-introduction studies and controls required by regulatory authorities. Research and development should carry on progressing to improve product knowledge and find new applications in controlling pests and diseases in the same or other crops. Also, the industry could try to work its way into new countries and conquer new markets in order to recover the costs incurred in developing the novel PPPs.

ACKNOWLEDGMENTS

This work was supported by Programa Estatal de Investigación, Desarrollo e Innovación Orientada a los Retos de la Sociedad SAF2013-48399-R from Ministerio de Economía y Competitividad (Spain). The authors also thank Mr. Guido Jones, who endeavored to edit the English translation of the manuscript.

REFERENCES

Ahn, I.P., Kim, S., Lee, Y.H. (2005) Vitamin B1 functions as an activator of plant disease resistance. *Plant Physiology* 138(3):1505–1515.

Baysal, O., Laux, P., Zeller, W. (2002) Further studies on the induced resistance (IR) effect of plant extract from *Hedera helix* against fire blight (*Erwinia amylovora*). *Acta Horticulturae* 590(590):273–277.

Betkas, Y., Eulgem, T. (2014) Synthetic plant defence elicitors. *Frontiers in Plant Science* 5:804.

Borges, A.A., Borges-Perez, A., Fernandez-Falcon, M.J. (2004) Induced resistance to Fusarium wilt of banana by menadione sodium bisulphite treatments. *Crop Protection* 23(12):1245–1247.

Borges, A.A., Cools, H.J., Lucas, J.A. (2003) Menadione sodium bisulphite: A novel plant defence activator which enhances local and systemic resistance to infection by *Leptosphaeria maculans* in oilseed rape. *Plant Pathology* 52(4):429–436.

Borges, A.A., Dobon, A., Expósito-Rodríguez, M., Jiménez-Arias, D., Borges-Pérez, A., Casañas-Sánchez, V., Pérez, J.A., et al. (2009) Molecular analysis of menadione-induced resistance against biotic stress in *Arabidopsis*. *Plant Biotechnology Journal* 7(8):744–762.

Bruce, T.J.A., Matthes, M.C., Napier, J.A., Pickett, J.A. (2007) Stressful 'memories' of plants: Evidence and possible mechanisms. *Plant Science* 173(6):603–608.

Carroll, M.J. (2016) The importance of regulatory data protection or exclusive use and other forms of intellectual property rights in the crop protection industry. *Pest Management Science* 72(9):1631–1637.

Chen, C., Chen, Z. (2002) Potentiation of developmentally regulated plant defence response by AtWRKY18, a pathogen-induced *Arabidopsis* transcription factor. *Plant Physiology* 129(2):706–716.

Conrath, U. (2006) Systemic acquired resistance. *Plant Signaling & Behavior* 1(4):179–184.

Conrath, U., Beckers, G.J., Langenbach, C.J., Jaskiewicz, M.R. (2015) Priming for enhanced defense. *Annual Review of Phytopathology* 53:97–119.

Coventry, H.S., Dubery, I.A. (2001) Lipopolysaccharides from *Burkholderia ceparia* contribute to an enhanced defensive capacity and the induction of pathogenesis-related proteins in Nicotiana tabacum. *Physiol Molecular Plant Biology* 58: 149–158.

Dong, H., Beer, S.V. (2000) Riboflavin induces resistance in plants by activating a novel signal transduction pathway. *Phytopathology* 90(8):801–811.

FAOSTAT. (2013) http://www.fao.org/statistics/en.

Farooq, M., Basra, S.M.A., Hafeez-u-Rehman, Saleem, B.A. (2008) Seed priming enhances the performance of late sown wheat (*Triticum aestivum* L.) by improving chilling tolerance. *Journal of Agronomy Crop Science* 194:55–60.

Filippou, P., Tanou, G., Molassiotis, A., Fotopoulos, V. (2013) Plant acclimation to environmental stress using priming agents. In. *Plant Acclimation to Environmental Stress*, eds. Tuteja, N., Gill, S.S., 1–27. New York: Springer.

Flors, V., Ton, J., van Doorn, R., García-Agustín, P., Mauch-Mani, B. (2008) Interplay between JA, SA and ABA signalling during basal and induced resistance against *Pseudomonas syringae* and *Alternaria brassicicola*. *Plant Journal* 44:81–92.

Görlach, J., Voirath, S., Knauf-Beiter, G., Hengy, G., Beckhove., U., Kogel, K.H., Oostendorp, M., et al. (1996) Benzothiadiazole, a novel class of inducers of systemic acquired resistance, activates gene expression and disease resistance in wheat. *The Plant Cell* 8(4):629–643.

Guedes, R.N.C., Smagghe, G., Stark, J.D., Desneux, N. (2016) Pesticides-induced stress in arthropod pests for optimized integrated pest management programs. *Annual Review of Entomology* 61:43–62.

Herms, S., Seehaus, K., Koehle, H., Conrath, U. (2002) A strobilurin fungicide enhances the resistance of tobacco against tobacco mosaic virus and *Pseudomonas syringae* pv. tabaci. *Plant Physiology* 130(1):120–127.

Hicks, G.R., Raikhel, N.V. (2012) Small molecules present large opportunities in plant biology. *Annual Review of Plant Biology* 63:261–282.

Hicks, G.R., Raikhel, N.V. (2014) Plant chemical biology: Are we meeting the promise? *Frontiers in Plant Science* 5:455.

Jiménez-Arias, D., Carrillo-Perdomo, E., Garcia-Machado, F.J., Luis-Jorge, J.C., Borges, A.A. (2017) Priming crops to cope with stress: Advances in seed-priming approach. In *Agricultural Research Updates*, Vol. 15, 1–29. New York: NOVA Science Publisher.

Jung, H.W., Tschaplinski, T.J., Wang, L., Glazebrook, J., Greenberg, J.T. (2009) Priming in systemic plant immunity. *Science* 324(5923):89–91.

Kazan, K., Schenk, P.M. (2007) Genomics in induced resistance. In *Induced Resistance for Plant Defence. A Sustainable Approach to Crop Protection*, eds. Walters, D., Newton, A., Lyon, G., Chapter 2, 9–29. Oxford: Blackwell Publishing.

Kopp, M., Rouster, J., Fritig, B., Darvill, A., Albersheim, P. (1989) Host-pathogen interactions XXXII. A fungal glucan preparation protects Nicotinae against infection by viruses. *Plant Physiology* 90(1):208–216.

Kumar, R., Das, I., Samanta, A., Ghosh, K., Duanting, Z., Wang, X., Dongdong, S., Cheryl, L., Young, T.C. (2011) Target identification: A challenging step in forward chemical genetics. *Interdisciplinary Bio Central* 3(3):1–18.

Leadbeater, A., Staub, T. (2017) Exploitation of induced resistance: A commercial perspective. In *Induced Resistance for Plant Defence. A Sustainable Approach to Crop Protection*, eds. Walters, D., Newton, A., Lyon, G., Chapter 2, 9–29. Oxford: Blackwell Publishing.

Loso, M.R., Garizi, N., Hegde, V.B., Hunter, J.E., Sparks, T.C. (2017) Lead generation in crop protection research: A portfolio approach to agrochemical discovery. *Pest Management Science* 73(4):678–685.

Lotan, T., Fluhr, R. (1990) Xylanase, a novel elicitor of pathogenesis-related proteins in tobacco, uses a non-ethylene pathway for induction. *Plant Physiology* 93(2):811–817.

MacGrath, K.C., Dombrecht, B., Manners, J.M., Schenk, P.M., Edgar, C.I., Maclean, D.J., et al. (2005) Repressor- and activator-type ethylene response factors functioning in jasmonate signalling and disease resistance identified via a genome-wide screen of *Arabidopsis* transcription factor gene expression. *Plant Physiology* 139(2):949–959.

McDougall, P. (2016) The cost of new agrochemical product discovery, development and registration, 1995, 2000, 2005–8 and 2010 to 2014. Report, CropLife America, Washington, DC, March

Nakashita, H., Yasuda, M., Nishioka, M., Asami, T., Fujioka, S., Arai, Y., Sekimata, K., et al. (2003) Brassinosteroid functions in a broad range of disease resistance in tobacco and rice. *Plant and Cell Physiology* 43:887–898.

Nakashita, H., Yoshioka, K., Yasuda, M., Nitta, T., Arai, Y., Yoshida, S., Yamaguchi, I. (2002) Probenazole induces systemic acquired resistance in tobacco through salicylic acid accumulation. *Physiological and Molecular Plant Pathology* 61(4):197–203.

Návarová, H., Bernsdorff, F., Döring, A.C., Zeier, J. (2012) Pipecolic acid, an endogenous mediator of defense amplification and priming, is a critical regulator of inducible plant immunity. *The Plant Cell* 24(12):5123–5141.

Nelson, H.E. (2005) *Fusarium oxysporum f. sp.radicis-lycopersici* can induce systemic resistance in barley against powdery mildew. *Journal of Phytopathology* 153(6):366–370.

Nombela, G., Pascual, S., Aviles, M., Guillard, E., Muñiz, M. (2005) Benzothiadiazole induces local resistance to *Bemisia tabaci* in tomato plants. *Journal of Economy and Entomology* 98:2266–2271.

Noutoshi, Y., Ikeda, M., Saito, T., Osada, H., Shirasu, K. (2012b) Sulfonamides identified as plant immune-priming compounds in high-throughput chemical screening increase disease resistance in *Arabidopsis thaliana. Frontiers in Plant Science* 3:245.

Noutoshi, Y., Okazaki, M., Kida, T., Nishina, Y., Morishita, Y., Ogawa, T., Suzuki, H., et al. (2012a) Novel plant immune-priming compounds identified via high-throughput chemical screening target salicylic acid glucosyltransferases in *Arabidopsis. The Plant Cell* 24(9):3795–3804.

O'Connor, C.J., Lairaia, L., Spring, D.R. (2011) Chemical genetics. *Chemical Society Reviews* 40(8):4332–4345.

Park, S.Y., Fung, P., Nishimura, N., Jensen, D.R., Fujii, H., Zhao, Y., Lumba, S., *et al.* (2009) Abscisic acid inhibits type 2c protein phosphatases via the PYR/PYL family of START proteins. *Science* 324(5930):1068–1071.

Pastor, V., Luna, E., Mauch-Mani, B., Ton, J., Flors, V. (2013) Primed plants do not forget. *Environmental and Experimental Botany* 94:46–56.

Pedley, K.F., Martin, G.B. (2005) Role of mitogen-activated protein kinases in plant immunity. *Current Opinion in Plant Biology* 8(5):541–547.

Potters, G., Pasternak, T.P., Guisez, Y., Palme, K.J., Jansen, A.K. (2007) Stress-inducedmorphogenic responses: growing out of trouble? *Trends in Plant Science.* 12:98–105.

Reglinski, T., Newton, A.C., Lyon, G.D. (1994) Assessment of the ability of yeast-derived elicitors to control powdery mildew in the field. *Journal of Plant Disease and Protection* 101:1–10.

Schreiber, S.L. (1998) Chemical genetics resulting from a passion for synthetic organic chemistry. *Bioorganic & Medicinal Chemistry* 6(8):1127–1152.

Smukste, I., Stockwell, B.R. (2005) Advances in chemical genetics. *Annual Review of Genomics and Human Genetics* 6:261–286.

Stockwell, B.R. (2000) Chemical genetics: Ligand-based discovery of gene function. *Nature Reviews Genetics* 1(2):116–125.

Toth, R., van der Hoorn, R.A. (2010) Emerging principles in plant chemical genetics. *Trends in Plant Science* 15(2):81–88.

Uknes, S. (1996) The role of benzoic acid derivatives in systemic acquired resistance. In *Phytochemical Diversity and Redundancy in Ecological Interactions*, eds. Romeo, J.T., Saunders, J.A., Barbosa, P., 253–263. Boston, MA: Springer.

van Hulten, M., Pelser, M., Van Loon, L.C., Pieterse, C.M., Ton, J. (2006) Costs and benefits of priming for defense in *Arabidopsis. Proceedings of the National Academy of Sciences of the United States of America* 103(14):5602–5607.

van Peer, R., Schippers, B. (1992) Lipopolysaccharides of plant-growth promoting Pseudomonas sp. strain WCS417r induce resistance in carnation to Fusarium wilt. *Netherlands Journal of Plant Pathology* 98:129–139.

Walsh, D.P., Chang, Y.T. (2006) Chemical genetics. *Chemical Reviews* 106(6):2476–2530.

Yamakawa, H., Kamada, H., Satoh, M., Yuko, O. (1998) Spermine is a salicylate-dependent endogenous inducer for both tobacco acidic pathogenesis-related proteins and resistance against tobacco mosaic virus infection. *Plant Physiology* 118:1213–1222.

Zhang, Z.-G., Wang, Y.-C., Li, J., Jia, R., Shen, G., Wang, S.-C., Xie Zhou, X., Zheng, X.-B. (2004) The role of SA in the hypersensitive response and systemic acquired resistance induced by elicitor PB90 from *Phytophthora boehmeriae. Physiological and Molecular Plant Pathology* 65(1):31–38.

25 Role of Plant-Derived Smoke in Amelioration of Abiotic Stress in Plants

Sumera Shabir and Noshin Ilyas

CONTENTS

25.1 INTRODUCTION

Some plant species depend on fire for completion of their life cycle (Bradshaw et al., 2011). Fire has multiple effects on plant species including removal of leaf litter and canopy that ultimately changes quantity as well as quality of light, nutrient availability and soil moisture. It also removes allelochemicals that cause inhibition (Dixon et al., 2009; Nelson et al., 2012). It has been investigated that traditional African farmers place maize cobs in a hut where smoke can reach. The smoking of plant material contains components that stimulate seed germination in fire-exposed seeds as compared with untreated seeds (Modi, 2002). De Lange and Boucher (1990) stated that smoking of plant materials contains germination stimulants.

It has been established that smoke has positive and negative impacts (Daws et al., 2007; Drewes et al., 1995; Light et al., 2002). According to the *International Union for Conservation of Nature and Natural Resources*, there are 1200 plant species that have a pronounced effect on their physiological and ecological roles due to smoke compounds. However, out of a total of 315,000 (www.iucnredlist.org/documents/summarystatistics/2010 _1RL_Stats_Table_1.pdf), that is just 0.38% of all known plant species. It is exclusively seen in plant species that have tough and impermeable to water seed coats such as members of the family Fabaceae (Keeley and Fotheringham, 2000).

Since 1990, the study of smoke compounds has focused on its roles in breaking seed dormancy and increasing germination rate in some plant species. It was first reported by De Lange and Boucher (1990) that seed germination was stimulated by compounds found in plant smoke. After this early work, many researches have shown that definite water-soluble compounds produced by burning plant material is effective in breaking seed dormancy (Brown and Van Staden, 1997; Dixon et al., 1995; Jefferson et al., 2008; Keeley and Fotheringham, 1998; Landis, 2000; Lindon and Menges, 2008; Pennacchio et al., 2007; Tieu et al., 1999). Egerton-Warburton (1998) showed that smoke acts as a scarifying agent to the seed surface, but according to Briggs and Morris (2008), this mechanism is not universal in all plant species. Different responses of various plant species are shown (Table 25.1).

25.2 MODES OF APPLICATION OF SMOKE

Jefferson et al. (2014) established that plant-derived smoke solutions prove effective in increasing seed germination and breaking seed dormancy in 1335 plant

TABLE 25.1
Different Plant Species Response Toward Effects of Plant-Derived Smoke

Plant Species	Response	References
Lactuca sativa	Stimulates germination, substitute light requirement of lettuce	Drewes et al. (1995)
Oryza sativa	Stimulates germination, release of dormancy	Doherty and Cohn (2000)
Avena fatua	Stimulates germination	Adkins and Peters (2001)
Nicotiana attenuata	Stimulates germination	Baldwin et al. (1994)
Themeda triandra	Stimulates germination	Van Staden et al. (1995)
Solanum esculentum	Stimulates germination, stimulate root growth	Van Staden et al. (2004); Jain and Van Staden (2006); Kulkarni et al. (2006); Van Staden et al. (2006); Taylor and Van Staden (1998).
Abelmoschus esculentus cv. Clemson spineless	Stimulates germination	Van Staden et al. (2004); Jain and Van Staden (2006); Kulkarni et al. (2006); Van Staden et al. (2006).
Phaseolus vulgaris cv. Dwarfimbali	Stimulates germination	Van Staden et al. (2004); Jain and Van Staden (2006); Kulkarni et al. (2006)
Zea mays	Both germination and seedling Growth increase, exhibited the highest vigour index	Van Staden et al. (2004); Jain and Van Staden (2006); Kulkarni et al. (2006); Van Staden et al. (2006).
Oryza sativa	Induces germination	Van Staden et al. (2004); Jain and Van Staden (2006); Kulkarni et al. (2006); Van Staden et al. (2006)
Apium graveolens	Induces germination	Thomas and Van Staden (1995)
Silybum marianum L.	Germination and seedling growth increased significantly with parameters (germination percentage, germination rate, seedling length and seedling vigor index)	Abdollahi et al. (2011)
Solanum elongena cv. Kemer	Rapid emergence and uniform seedlings	Demir et al. (2009)
Cucumis melo cv. Kirkagac	Increases seed vigor	Mavi et al. (2010)
Cyrtanthus ventricosus	Induces flowering	Keeley (1993)
Watsonia borbonica	Induces flowering	Keeley (1993)
Avena sterilis ssp.	Induces germination	Adkins and Peters (2001)
Artemisia ludoviciana	Induces germination	Adkins and Peters (2001)
Alopecurus myosuroides	Induces germination	Adkins and Peters (20010
Sorghum Halepense	Induces germination	Adkins and Peters (2001)
Phalaris paradoxa	Induce germination	Adkins and Peters (2001)
Bromus sterilis	Effective in enhancing seedlings	Stevens et al. (2007)
Capsella bursa-pastoris	Effective in enhancing seedlings	Stevens et al. (2007)
Galium aparine	Effective in enhancing seedlings	Stevens et al. (2007)
Arctotheca calendula	Effective in enhancing seedlings	Stevens et al. (2007)
Brassica tournefortii	Effective in enhancing seedlings	Stevens et al. (2007)
Raphanus raphanistrum	Effective in enhancing seedlings	Stevens et al. (2007)
Avena fatua	Effective in enhancing seedlings	Stevens et al. (2007)
Solanum viarum	Seed germination markedly stimulated	Kandari et al. (2011)
Carica papaya cv. Tainung No. 2	Increased germination rate	Chumpookam et al. (2012)
Emmenanthe penduliflora	Promotes germination	Preston et al. (2004)
Orobanche minor	Promotes germination	Flematti et al. (2004)
Striga hermonthica	Promotes germination	Flematti et al. (2004)
Brassica napus	Improving the plantlet regeneration percentage, root length and shoot length	Govindaraj and Masilamani (2016)
Syncarpha vestita	Improves germination	Drewes et al. (1995)
Tersonia cyathiflora	Improves germination	Downes et al. (2010)

TABLE 25.2
Different Plant Material used for Producing Smoke and Smoke Solution

Plant Species	Scientific Names	Effects	References
Cotton palm/fan palm and rice straw	*Washingtonia filifera*, *Oryza sativa*	Positive (high germination percentage, vigor index and breaks seed dormancy, at the same time it contains karrikins as KAR1 and KAR2)	Govindaraj and Masilamani (2016)
Combustion of wood plus mixture of odoriferous and medicinal herbs	–	Helpful in eliminating bacteria hazardous for agricultural crops	Kulkarni et al. (2011)
Climax grass	*Themeda triandra*	Stimulates germination	Jager et al. (1996)
Extracts prepared by heating agar and cellulose	–	Increase germination of light-sensitive seeds	Jager et al. (1996)
Eucalyptus and tissue paper smoke extracts	*Eucalyptus globulus*	Germination inhibitory activity	Jager et al. (1996)
Dry rice straw smoke water	*Oryza sativa*	Improves germination	Chumpookam et al. (2012)
Cymbopogon	*Cymbopogon jwarancusa*	Improves germination	Khan et al. (2017)
Bauhinia and *Cymbopogon*	*Bauhinia variegata*, *Cymbopogon jwarancusa*	Increase seed germination and seedling vigor	Jamil et al. (2014)

species from 120 families including crop plants and arable weeds in fire- and non-fire-prone ecosystems. Many plant materials are being utilized for the production of smoke (Table 25.2) and the experimental setup for production of smoke from plant material is shown in Figure 25.1. There are different ways by which smoke is applied to plants.

25.2.1 Fumigation or Aerosol

Considered as the simplest and easiest method, smoke is applied to soil or seeds are given prior treatment before sowing. It is done by partially burning plant material, and as a result, smoke is produced by which seeds, soil or potting media (e.g. sand, vermiculite, perlite, coco peat) can be sown (Abdollahi et al., 2011; Keeley and Fotheringham, 1998; Sparg et al., 2006). It is established by various studies that direct smoke exposure of seeds enhance seed germination (Baxter et al., 1995; Blank and Young, 1998; Brown et al., 1993; De Lange and Boucher, 1990; Dixon et al., 1995; Jefferson et al., 2008; Lindon and Menges, 2008; Pennacchio et al., 2007). It is effective for the promotion of seed germination of some species when directly fumigated to smoke. Light et al. (2010) stated that seeds should be rinsed with distilled water after exposure to smoke because smoke contains certain inhibitory chemicals that could negatively affect seed germination and the emergence of the seedling. It is advisable that intense exposure to smoke should be avoided.

25.2.2 Plant-Derived Smoke Solution or Smoke Water

Smoke solution application is also an easy and economical method of application. Brown and Van Staden (1997) reported that a large number of plant materials have been used for producing a smoke solution. It is done by producing smoke by incomplete burning of plant material in a drum and then bubbling smoke into the water. Smoke water has biologically active compounds that get readily dissolved in water. This smoke water can be used for treating seeds prior to sowing. De Lange and Boucher (1990) first reported this phenomenon of producing smoke extract. Jäger et al. (1996) stated that all plant materials could be used for the preparation of smoke extracts. Several studies have shown that concentrated smoke solution is further diluted with water (1:250; 1:500; 1:1000; 1:1500 and 1:2000 (v/v)). Van Staden et al. (2004) found that different dilutions prove to be more effective in encouraging seed germination, however, it effectivity differs in species to species. The smoke solution is more advantageous as far as its application is concerned because it is easier to apply on agricultural lands and to be sprayed on crops. Smoke has certain antimicrobial properties due to which it is beneficial in

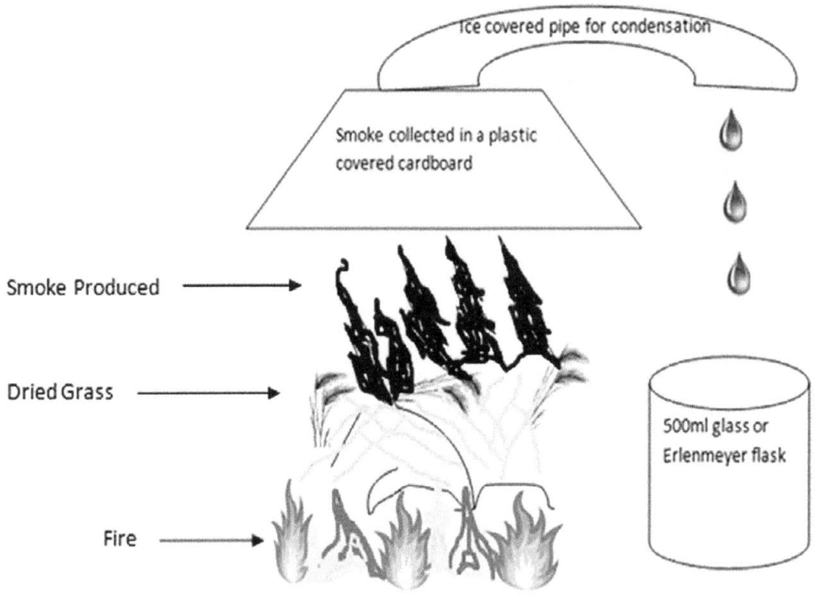

FIGURE 25.1 Diagrammatically showing the experimental setup for production of smoke from plant material (grasses). smoke reaches the U-turn in the pipe that is cooled by ice on the outer side of the pipe where its condensation takes place. Condensed smoke-saturated water (SSW) is collected in a glass vessel.

protecting seeds against a number of diseases, as well as protecting seeds and seedlings from detrimental microbial contaminations (Nautiyal et al., 2007).

25.2.3 Commercial Smoke Preparations

Commercial smoke preparations include '*Seed Starter–Australian Smoky Water*' from the Friends of Kings Park and Botanic Garden in Perth, Western Australia; '*Kirstenbosch Instant Smoke Plus*' seed primer from the National Botanical Institute in Cape Town, South Africa; and '*Super Smoke Plus*' Seed Germinator Primer from Seedman.com. Standardized smoke solutions could be generated from burning cellulose filter paper (Flematti et al., 2004).

25.3 CHEMICAL CHARACTERIZATION OF SMOKE AND SMOKE SOLUTION

There are about 5000 different components found in plant-derived smoke (Smith et al., 2003). Baldwin et al. (1994) identified 71 compounds, GC-MS and atomic absorption (AA) spectrometry in active fractions of smoke and tested a total 233 compounds in *Nicotiana attenuata* seeds. It was interpreted that the difficulty in the isolation of active compounds was a large number of compounds found in the smoke extract (up to several thousand) (Maga, 1988) or it may be due to a fewer quantity of active constituents as compared with other components. In the beginning, it was established that germination promoting chemicals are water-soluble chemicals that are thermostable, long-lasting in solution and highly active at very low concentrations (Baldwin et al., 1994; Van Staden et al., 2000).

Among different components found in smoke nitric oxides and butenolide, derivatives play an active role in inducing a stimulating effect on plant germination. Butenolide derivatives play a prominent role in physiological and ecological attributes of plants (Flematti et al., 2004, 2005). Chiwocha et al. (2009) stated that they are termed as karrikinolides or karrikins (KAR). A compound named 3-methyl-2H-furo[2,3-c]pyran-2-one, obtained from plant-derived smoke has been referred to KAR1 and was discovered in 2004 (Flematti et al., 2004). Flematti et al. (2005) reported that 3-methyl-2H-furo [2,3-c]pyran-2-one synthesized from pyromeconic acid. Its synthesis and several analogous compounds from d-xylose is described by Goddard-Borger et al. (2007). The synthetic material displayed germination activity at sub-nanomolar (10^{-9}-M) concentrations (Flematti et al., 2004).

Light et al. (2005) found that butenolide formed as a product during a heating reaction between d-xylose and glycine. Greatest germination response is shown by the extracts prepared by the reaction between d-xylose or d-ribose (aldopentose sugars) with the amino acids arginine, asparagine, aspartic acid, glycine, serine, tyrosine or valine. Observed germination activity was probably not due to ethylene (Van Staden et al., 1995). Flematti et al. (2011) identified a new bioactive compound containing nitrogen and was termed cyanohydrin glyceronitrile. Most researchers agree that one of the smoke active components, butenolide, interacts with hormones including gibberellin, ABA and auxin pathways in seeds (Chiwocha et al., 2009; Light et al., 2005). Jain et al. (2008) inferred that it might be due to up-regulation of protein expansins that disrupt the hydrogen bonds within the cell wall. Smoke components, due to their oxidative role, may interfere with the cell redox status (Light et al., 2009). Chemical structures of the naturally occurring germination stimulants are shown in Table 25.3.

25.3.1 Karrikins: A Butenolide Role

Light et al. (2009) suggested that butenolides, a key component of smoke, act as a plant growth regulator. Active butenolide derivatives can be synthesized in laboratory conditions physiologically (Flematti et al., 2005; Light et al., 2005; Nagase et al., 2008; Sun et al., 2008). Pepperman and Cutler (1991) identified the bioactivity of 'butenolides' that are related with butenolides of smoke by conduction bioassay on wheat coleoptiles. After initial discovery, closely related butenolides have been found in smoke (Flematti et al., 2009). Such a closely related butenolide family has been termed 'karrikins' (Chiwocha et al., 2009; Flematti et al., 2009). It is a term that is based on the native word 'karrik' for smoke, used by the Noongar people of south-western Australia. As 'karrikins' are analogous to gibberellin or auxin so '-in' suffix denotes a family of these biologically active molecules. The original parent molecule is denoted as karrikin-1 (KAR1) or 'karrikinolide' (Dixon et al., 2009). Karrikins are thermostable, water-soluble, long lasting in solution and greatly active at very low concentrations of 1–9 M in darkness (Flematti et al., 2004; Light et al., 2005). Physiologically active butenolide derivatives can be even synthesized in laboratory conditions (Flematti et al., 2005; Light et al., 2005; Nagase et al., 2008; Sun et al., 2008).

25.4 BENEFICIAL AND HARMFUL ASPECTS OF PDS SOLUTION

Active components of smoke respond similarly to plant growth regulators (Senaratna et al., 1999). These components of smoke act via endogenous modification of receptor molecules (Thomas and Van Staden, 1995; Van Staden et al., 1995). Kępczyński and Van Staden (2012) demonstrated KAR1 requires ethylene action and gibberellin biosynthesis to stimulate germination of *Avena fatua* dormant caryopses. Cembrowska-Lech et al. (2015) and Cembrowska-Lech and Kępczyński (2016) demonstrated that the release of *A. fatua* caryopsis dormancy is mainly associated with control of the ABA content, cell cycle, metabolic activity and homeostasis between ROS and antioxidants in the embryo. Butenolides, for example, peagol, a dehydrocostus lactone that is formed by pea as well as sunflower roots, increases germination of some parasitic weed seeds (Evidente et al., 2009; Joel et al., 2011). Around other butenolides, leaf-closing cause (from *Phyllanthus urinaria*) called phyllanthurinolactone (Ueda and Nakamura, 2006) and the mycotoxin patulin (Brase et al., 2009) bring a significant similarity with karrikins. Flematti et al. (2009, 2011) accounted for seed germination enhances toward KAR1 also repressed by smoke. The stimulatory impact of smoke will be principally attributed to KAR1. Certain other germination stimulators include KAR2 and KAR3 and/or glyceronitrile found in smoke (Flematti et al., 2009, 2011). Some of the beneficial roles are given in Figure 25.2.

Smoke may be a highly complex mixture holding a few thousand compounds (Maga, 1988). Because of this certainty, it will be troublesome to calibrate optimum conditions; higher germination should be considered as a good clue of better crop stand (Adkins and Peters, 2001; Boucher and Meets, 2004), utilizing pure butenolide compound. Smoke additionally contains components that might negatively influence germination (Light et al., 2002). Drewes et al. (1995), Adkins and Peters (2001), Light et al. (2002) and Daws et al. (2007) accounted antagonistic impacts of smoke in seed germination. There are few inhibitors which would have inhibitory impacts, for example, 3,4,5-trimethylfuran-2(5H)-one in smoke (Pošta et al., 2013). Mennan et al. (2012) reported that the germination and root development inhibited in rice, for example, saturated butenolides (or butanolides).

In the investigation by Downes et al. (2010) that purified active compounds found in smoke, it became apparent that some smoke-responsive plant species do not react. The fire ephemeral plant *Tersonia cyathiflora* (Gyrostemonaceae) has been discovered to germinate because of smoke but not, in any case, will karrikinolide. An investigation by Sparg et al. (2006) showed that treating maize seeds with smoke water brought about restraint of germination.

25.5 MODE OF ACTION OF PDS SOLUTION

Pepperman and Cutler (1991) reported that 'butenolides' are structurally similar to the butenolide from smoke while conducting bioassay on coleoptile of wheat. Activities of these compounds attributed to the structural

TABLE 25.3
Chemical Structures of the Naturally Occurring Germination Stimulants, KAR1 (Karrikinolide) and Lactones

Name of Compound	Structure	References
Strigol	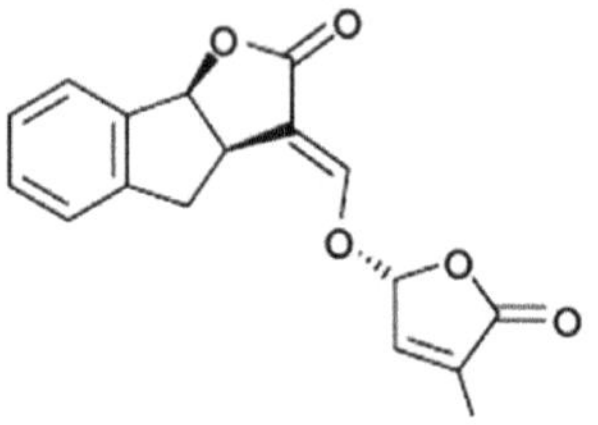	Cook et al. (1972); Mangnus and Zwanenburg (1992)
GR 24*	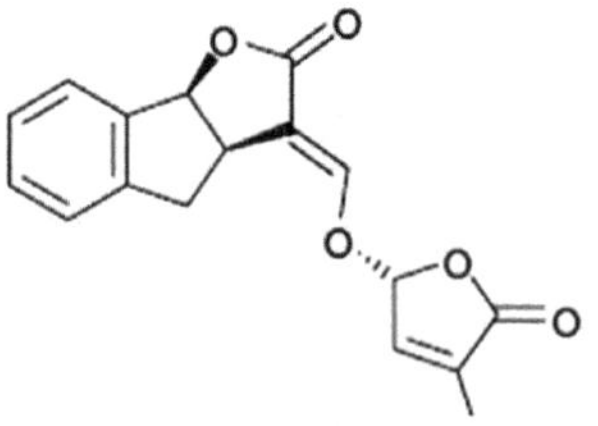	Johnson et al. (1981); Nefkens et al. (1997)
Nijmegan-1	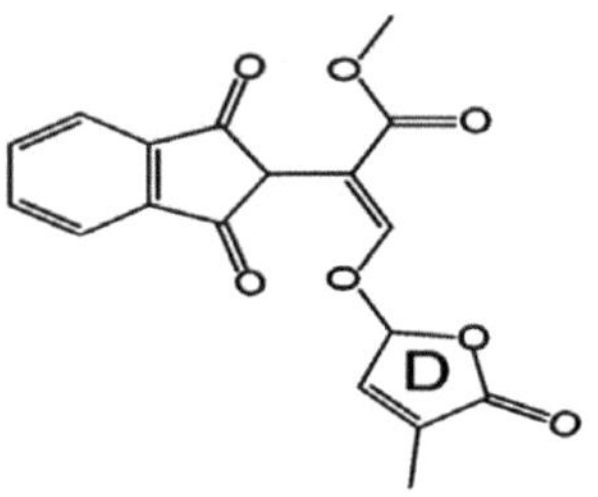	Johnson et al. (1981); Nefkens et al. (1997)
Butenolide	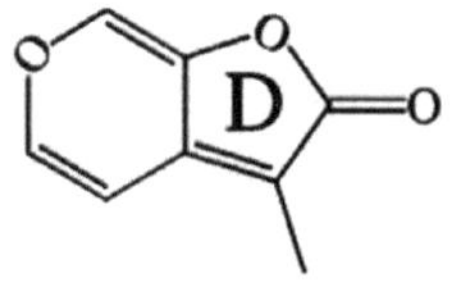	Van Staden et al. (2004)
Karrikinolide	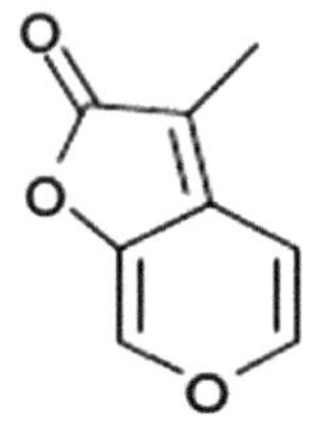	Zwanenburg and Pospisil (2013)
Peagol	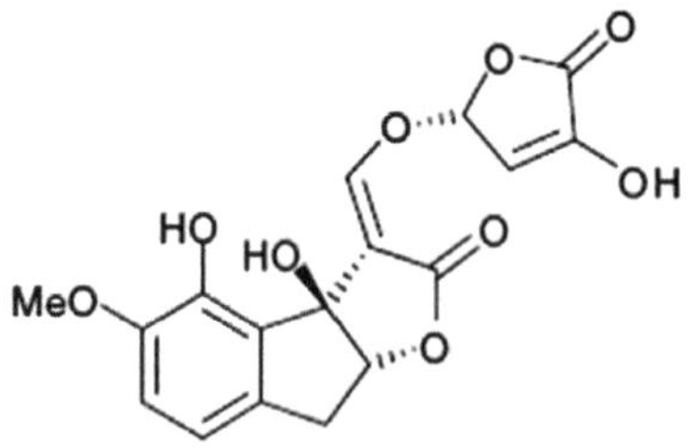	Zwanenburg and Pospisil (2013)
Dehydrocostus lactone	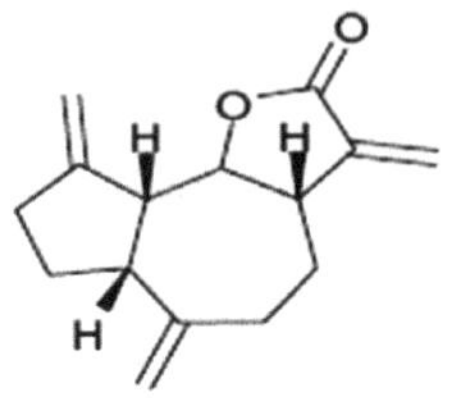	Zwanenburg and Pospisil (2013)

(Continued)

TABLE 25.3 (CONTINUED)
Chemical Structures of the Naturally Occurring Germination Stimulants, KAR1 (Karrikinolide) and Lactones

Name of Compound	Structure	References
Phyllanthurinolactone	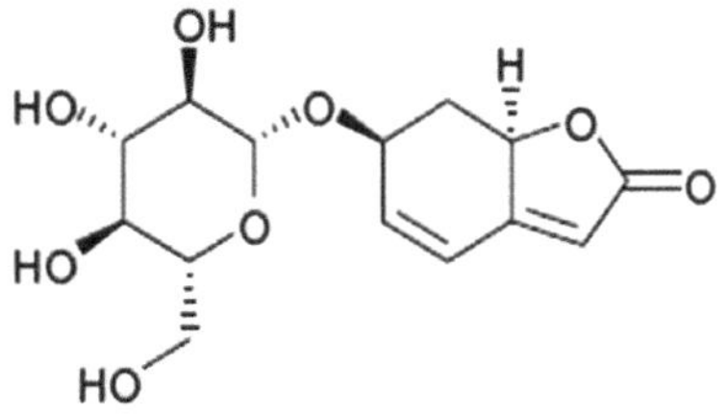	Zwanenburg and Pospisil (2013)
Patullin	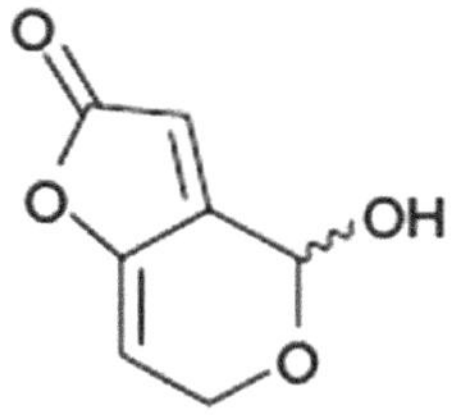	Zwanenburg and Pospisil (2013)
Momilactone B	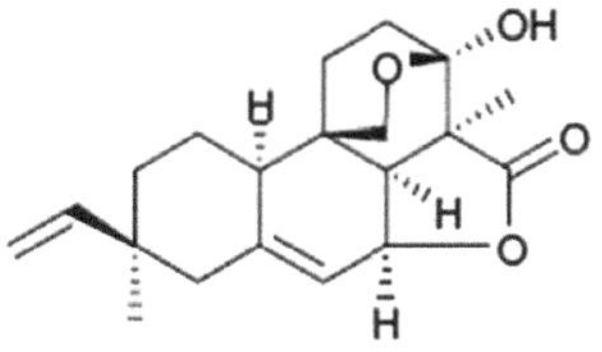	Zwanenburg and Pospisil (2013)
Mandelonitrile	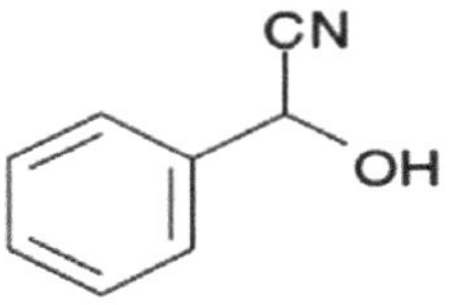	Flematti et al. (2011)
Glycolonitrile	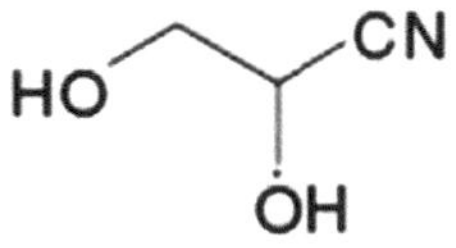	Flematti et al. (2011)
Acetone cyanohydrin		Flematti et al. (2011)
Glycolonitrile		Flematti et al. (2011)
Racemic 3,4,5-trimethylfuran-2(5H)-one (2,3,4-trimethylbut-2-enolide)	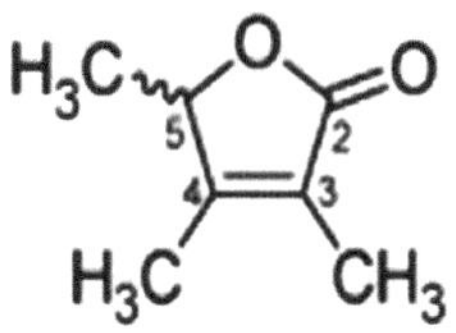	Light et al. (2010)

similarities to strigolactones (e.g. strigol) that act as a germination stimulant for parasitic weeds (Butler, 1995). However, their mode of action on seed germination of parasitic species is not yet fully understood (Humphrey and Beale, 2006). The bioactiphore is located in the lactone-enol ether D-ring portion of the molecule that is common with butenolide (Mangnus and Zwanenburg, 1992; Wigchert et al., 1999).

Pepperman and Cutler (1991) accounted that 'butenolides' are structurally comparative of the butenolide from smoke by directing bioassay on the coleoptile of wheat. Activities about these compounds have shown structural resemblances with strigolactones (e.g. strigol) that act as a germination stimulant to parasitic weeds (Butler, 1995). However, their mode of action about seed germination of parasitic species has not yet been completely

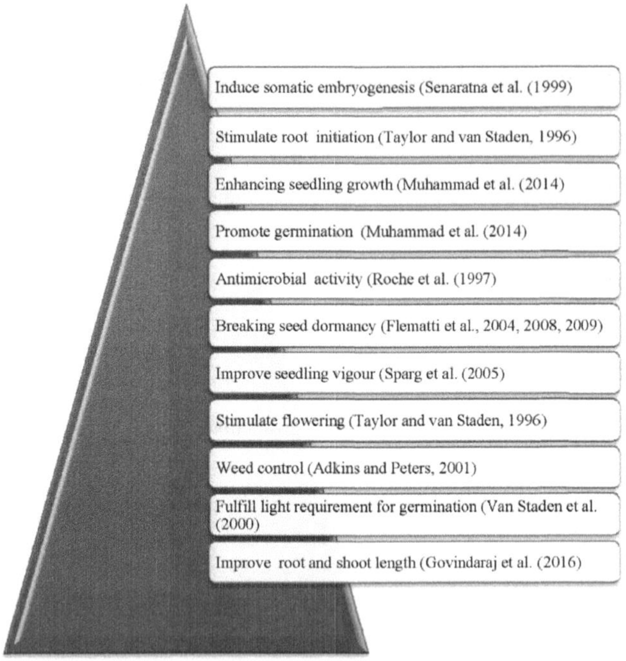

FIGURE 25.2 Different aspects of plant-derived smoke activities.

studied (Humphrey and Beale, 2006). The bioactiphore located in the lactone-enol ether D-ring portion is common with butenolide.

Keeley and Fotheringham (1998) concluded that prior to the identification of active smoke constituents, nitrogen oxides and nitric oxide (NO) in smoke were more probably seed germination stimulants. Light and Van Staden (2003) identified that NO-specific scavenger 2-(4-carboxyphenyl)-4,4,5,5-tetramethylimidazoline-1-oxyl-3-oxide potassium (c-PTIO) is not able to reduce germination as it is observed in smoke solution. Light and Van Staden (2003) identified the effects of two NO-releasing smoke germination components (N-tert-butyl-α-phenylnitrone (PBN) and sodium nitroprusside (SNP)) in increasing germination in Grand Rapids lettuce seeds. These two components also stimulate these seeds to germinate in darkness. Light and Van Staden (2003) observed that NO-specific scavenger 2-(4-carboxyphenyl)-4,4,5,5-tetramethylimidazoline-1-oxyl-3-oxide potassium (c-PTIO) remains unable to reduce seed germination in Grand Rapids lettuce seeds by applying smoke solutions.

This compound has stimulatory effects on a wide range of plant species regarding increasing seed germination at concentrations less than 10^{-9} M (in Grand Rapids lettuce seeds) and 10^{-7} M (in *Conostylis aculeata* and *Stylidium affine*) (Flematti et al., 2004). In a study, Grand Rapids lettuce seeds shown effectivity at 10^{-4} M down to 10^{-9} M (Van Staden et al., 2004). Drewes et al. (1995) and Van Staden et al. (1995) have shown in Grand Rapids lettuce germination that smoke has effects similar to GA3 in substituting for red light (640nm). Butenolide has positive effects in increasing germination and substituting light requirements in Australian Asteraceae (Merritt et al., 2006) and in arable weeds as well (Daws et al., 2007).

Gardner et al. (2001) and Schwachtje and Baldwin (2004) reported that smoke components affect endogenous GA and ABA synthesis. Auxins have shown a stimulatory role in seedling development and embryogenesis. These phytohormones are considered as important components in inducing normal growth and development in *in vitro* cultures by coordinating cell patterns in different stages (Fischer-Iglesias and Neuhaus, 2001; Teale et al., 2006). Butenolide functions in a similar way to auxins as in somatic embryogenesis of *Baloskion tetraphyllum*, it substitutes the role of 2,4-D (a synthetic auxin).

On the other hand, the effects of butenolide on the levels of endogenous plant hormones have not been fully investigated yet. However, there are few similarities in the chemical structure of butenolide from smoke and GA3. Table 25.3 shows that there are structural similarities in butenolide and the strigolactones. Flematti et al. (2004) showed that butenolide and the strigolactones stimulate germination in parasitic plant species (*Orobanche* and *Striga*). It might be inferred that there are similarities between butenolide and gibberellins, auxin and strigolactones. One molecule might have comparable properties to a range of plant growth regulating compounds (e.g. Jain et al., 2008) as certain plants have proteins that work in several pathways.

25.6 ABIOTIC STRESSES AND PLANT-DERIVED SMOKE

Smoke has been used as an important factor in the understanding of vegetation dynamics in some fire-prone environments (Table 25.4). It has diversified uses in horticulture, agriculture, ecological management and the rehabilitation of disturbed areas. Plant growth is highly affected by various environmental stresses including temperature, drought or high salinity. From an ecological point of view, such stress conditions have a damaging effect on crops and results in excessive loss of yield. Overcoming abiotic stresses is a major challenge faced by agronomists and crop physiologists, which reduces crop productivity and yield and ultimately economic losses as an outcome.

25.6.1 Salinity Stress, Consequences and Reversal

High salinity (mainly due to NaCl) results in a decrease in water potential, and it leads to an osmotic imbalance in plants by generating oxidative stresses in plants. High

TABLE 25.4
Fire Prone Environments

Fire Prone Environment	References
Australian kwongan	Roche et al. (1997)
Californian chaparral	Keeley and Fotheringham (1998)
Western Cape fynbos	Brown et al. (2003)
Mediterranean basin	Crosti et al. (2006)

salinity also results in ion toxicity and nutrient imbalance (Serrano et al., 1999). High salinity is considered as a worldwide environmental problem of agricultural lands.

Khan et al. (2017) examined linked impacts of plant growth-promoting microscopic organisms (*Bacillus safensis*) and plant-derived smoke (*Cymbopogon jwarancusa*) looking into rice under separate saltiness levels such as 50, 100 and 150 mM (cv. Basmati-385). Seeds were soaked (10 h) in smoke water being diluted (C-500 and C-1000) in bacterial cultures. Salt has negative impacts on germination percentage, plant vegetative growth, ion contents (K^+ and Ca^{2+}) and photosynthetic pigments (Chl 'a', Chl 'b' and carotene), and an increase in Na^+, aggregate solvent protein (TSP) and total soluble sugar (TSS), proline, catalase (CAT) and peroxidase (POD) contents. *B. safensis*, as well as smoke primed seeds, have indicated an increase in seed germination percentage, seedling growth, ion content (K^+ and Ca^{2+}) and photosynthetic pigments (Chl 'a', Chl 'b' and carotene), while lessened Na^+ ion content, total soluble protein, proline content, total soluble sugar, catalase and peroxidase activity by bringing down those intense impact of salinity.

Khan et al. (2017) reported that smoke and PGPR have more stimulating effects on germination then their individual effects. Plant-derived smoke solutions have been shown to have growth-promoting effects after being studied for two decades. Jamil et al. (2014) conducted a study to investigate the effect of the plant-derived smoke solution with different dilutions (1:100, 1:500 and 1:1000) over various concentrations of salinity (0, 50, 100 and 150 mM NaCl) on rice variety (NIAB-IR-9). It was observed that germination percentage (23.3%), seedling vigor and fresh and dry weight increased in smoke primed seeds as compared with non-primed seeds at high salinity. Seeds primed with different smoke dilutions have shown an increase in K^+ and Ca^{2+} with decreased Na+ content as compared with plant seeds that are hydro primed seeds under saline conditions. Primed seeds had higher chlorophyll a and b, carotenoids contents and total nitrogen and protein contents as compared with plants of non-smoke treated seeds as the level of salt increased. The most significant effect was shown by *Bauhinia* (1:500) dilution that significantly alleviated salt stress both at the physiological and biochemical level. It could be inferred that smoke has a mitigatory effect in reducing adverse effects of salinity.

25.6.2 Heavy Metals, Consequences and Reversal

Heavy metal accumulation in the environment results from a large number of anthropogenic activities. Plants show a diverse response as a result of their exposure to heavy metals. Cobalt is among one of the heavy metals having toxic effects. It remains in the environment for a long time and causes serious environmental stresses and public health problems (Jayakumar et al., 2008). Jayakumar and Vijayarengan (2006) stated that its high concentration (up to 750 ig L^{-1}) is found in some plants. Seed germination and radical length of ragi and paddy have been adversely affected by higher concentrations of cobalt (Jayakumar et al., 2008). Aslam et al. (2014) stated that heavy metals have serious effects on the environment as well as human health including cancer of the bladder, lung and skin due to arsenic (As) being consumed by As-contaminated rice. Chromium can lead to cancer in humans through exposure to smoking and use of Cr-laden vegetables. Lead and mercury are well-known neurotoxins being consumed via seafood, vegetables and rice while cadmium has deleterious effects on the liver, bones and the female reproduction system. Plant-derived smoke acts as a growth regulator and it has a mitigatory effect over a number of compounds. Aslam et al. (2014) studied the effect of $COCl_2$ (10, 20 and 30 ppm) on the roots of *Ipomoea marguerite* along with the effect of *Cymbopogon*-derived smoke solution (1:500 dilution). $CoCl_2$ has an inhibitory effect on a number of adventitious roots, lateral roots and adventitious root length with increasing concentrations. It was found that 1:500 dilution proved to be effective in an increasing number of adventitious rot as well as lateral roots. An investigation by Okem et al. (2015) investigated the stimulatory impacts for smoke-water (SW), KAR1 and different known plant growth regulators (PGRs) with impacts of Cd treatment to Hoagland solution. KAR1 exhibited certain positive impacts in enhancing shoot as well as root dry weight (at the most increased conc. Cd 10 mg L^{-1}) as contrasted with different treatments.

25.7 CONCLUSION AND RECOMMENDATIONS

Plant-derived smoke has stimulatory effects on various plant morphological and physiological attributes. It was found that smoke been has applied in various ways, for example, fumigation, aerosol sprays and various dilutions of the smoke solution. Smoke has a positive effect on the plant if applied at low concentrations (1:500 smoke water dilutions). Various abiotic stresses have negative effects on crop productivity and yield so ultimately results in economic losses. It was observed that smoke has mitigatory effects in reducing adverse effects of salinity as well as metal stress. There is a need for applications of smoke at field level as well as undertaking a molecular study of smoke so that mechanisms of its mode of action can be found.

REFERENCES

Abdollahi, M.R., Mehrshad, B., Moosavi, S.S. (2011) Effect of method of seed treatment with plant derived smoke solutions on germination and seedling growth of milk thistle (*Silybum marianum* L.). *Seed Science and Technology* 39(1): 225–229.

Adkins, S.W., Peters, N.C.B. (2001) Smoke derived from burnt vegetation stimulates germination of arable weeds. *Seed Science Research* 11: 213–222.

Aslam, M.M., Akhter, A., Jamil, M., Khatoon, A., Malook, I., Rehman, S. (2014) Effect of plant-derived smoke solution on root of *Ipomoea marguerite* cuttings under cobalt stress. *Journal of Bio-Molecular Sciences* 2(1): 6–11.

Baldwin, I.T., Staszak-Kozinski, L., Davidson, R. (1994) Up in smoke. I. Smoke-derived germination cues for post-fire annual, *Nicotiana attenuate* Torr ex Watson. *Journal of Chemical Ecology* 20(9): 2345–2371.

Baxter, B.J.M., Granger, J.E., Van Staden, J. (1995) Plant-derived smoke and seed germination: Is all smoke good smoke? That is the burning question. *South African Journal of Botany* 61(5): 275–277.

Blank, R.R., Young, J.A. (1998) Heated substrate and smoke: Influence on seed emergence and plant growth. *Journal of Range Management* 51(5): 577–583.

Boucher, C., Meets, M., Brown, N.A.C. (2004) Determination of the relative activity of aqueous plant-derived smoke solutions used in seed germination. *South African Journal of Botany* 70(2): 313–318.

Bradshaw, S.D., Dixon, K.W., Hopper, S.D., Lambers, H., Turner, S.R. (2011) Little evidence for fire-adapted plant traits in Mediterranean climate regions. *Trends in Plant Science* 16(2): 69–76.

Brase, S., Encinas, A., Keck, J., Nising, C.F. (2009) Chemistry and biology of mycotoxins and related fungal metabolites. *Chemical Reviews* 109(9): 3903–3990.

Briggs, C.L., Morris, E.C. (2008) Seed-coat dormancy in *Grevillea linearifolia*: Little change in permeability to an apoplastic tracer after treatment with smoke and heat. *Annals of Botany* 101(5): 623–632.

Brown, N.A.C., Kotze, G., Botha, P.A. (1993) The promotion of seed germination of Cape Erica species by plant derived smoke. *Seed Science and Technology* 21: 573–580.

Brown N.A.C., Van Staden J. (1997) Smoke as a germination cue: A review. *Plant Growth Regulation* 22: 115–124.

Brown, S., Kozinets, R. V., Sherry, J. (2003) Teaching old brands new tricks: Retro branding and the revival of brand meaning. *Journal of Marketing* 67(3): 19–33.

Butler, L.G. (1995) Chemical communication between the parasitic weed *Striga* and its crop host: A new dimension of allelochemistry. In: Inderjit, K., Dakshini, M.M., Einhellig, F.A. (Eds.), *Allelopathy: Organism Processes and Application. ACS Symposium Series*, 158–168.

Cembrowska-Lech, D., Kępczyński, J. (2016) Gibberellin-like effects of KAR1 on dormancy release of *Avena fatua* caryopses include participation of non-enzymatic antioxidants and cell cycle activation in embryos. *Planta* 243(2): 531–548.

Cembrowska-Lech, D., Koprowski, M., Kępczyński, J. (2015) Germination induction of dormant Avena fatua caryopses by KAR1 and GA3 involving the control of reactive oxygen species (H_2O_2 and $O_2^{\bullet-}$) and enzymatic antioxidants (superoxide dismutase and catalase) both in the embryo and the aleurone layers. *Journal of Plant Physiology* 176: 169–179.

Chiwocha, S.D.S., Dixon, K.W., Flematti, G.R., Ghisalberti, E.L., Merritt, D.J., Nelson, D.C., Riseborough, J.M., Smith, S.M., Stevens, J.C. (2009) Karrikins: A new family of plant growth regulators in smoke. *Plant Science* 177(4): 252–256.

Chumpookam, J., Lin, H.L., Shiesh, C.C. (2012) Effect of smoke-water on seed germination and seedling growth of papaya (*Carica papaya* cv. Tainung No. 2). *Horticultural Science* 47: 741–744.

Cook, C.E., Whichard, L.P., Wall, M.E., Egley, G.H., Coggon, P., Luhan, P.A., McPhail, A.T. (1972) Germination stimulants. II. The structure of strigol—A potent seed germination stimulant for witchweed (*Striga lutea* Lour.). *Journal of the American Chemical Society* 94(17): 6198–6199.

Crosti, R., Ladd, P.G., Dixon, K.W., Piotto, B. (2006) Post-fire germination: The effect of smoke on seeds of selected species from the central Mediterranean basin. *Forest Ecology and Management* 221(1–3): 306–312.

Daws, M.I., Davies, J., Pritchard, H.W., Brown, N.A.C., Van Staden, J. (2007) Butenolide from plant-derived smoke enhances germination and seedling growth of arable weed species. *Plant Growth Regulation* 51(1): 73–82.

De Lange, J.H., Boucher, C. (1990) Autecological studies on *Audouinia capitata*. Bruniaceae. I. Plant-derived smoke as a seed germination cue. *South African Journal of Botany* 56(6): 700–703.

Demir, I., Light, M.E., Van Staden, J., Kenanoglu, B.B., Celikkol, T. (2009) Improving seedling growth of unaged and aged aubergine seeds with smoke-derived butenolide. *Seed Science and Technology* 37(1): 255–260.

Dixon, K.W., Merritt, D.J., Flematti, G.R., Ghisalberti, E.L. (2009) Karrikinolide – A phytoreactive compound derived from smoke with applications in horticulture, ecological restoration and agriculture. *Acta Horticulturae* 813: 155–170.

Dixon, K.W., Roche, S., Pate, J.S. (1995) The promotive effect of smoke derived from burnt native vegetation on seed germination of Western Australian plants. *Oecologia* 101(2): 185–192.

Doherty, L.C., Cohn, M.A. (2000) Seed dormancy in red rice (*Oryza sativa*). XI. Commercial liquid smoke elicits germination. *Seed Science Research* 10: 415–421.

Downes, K.S., Lamont, B.B., Light, M.E., Van Staden, J. (2010) The fire ephemeral *Tersonia cyathiflora* (Gyrostemonaceae) germinates in response to smoke but not the butenolide 3-methyl-2H-furo[2,3-c]pyran-2-one. *Annals of Botany* 106(2): 381–384.

Drewes, F.E., Smith, M.T., Van Staden, J. (1995) The effect of a plant-derived smoke extract on the germination of light-sensitive lettuce seed. *Plant Growth Regulation* 16(2): 205–209.

Egerton-Warburton, L.M. (1998) A smoke-induced alteration of the sub-testa cuticle in seeds of the post-fire recruiter, *Emmenanthe penduliflora* Benth (Hydrophyllaceae). *Journal of Experimental Botany* 49(325): 1317–1327.

Evidente, A., Fernandez-Aparicio, M., Cimmino, A., Rubiales, D., Andolfi, A., Motta, A. (2009) Peagol and peagoldione, two new strigolactone-like metabolites isolated from pea root exudates. *Tetrahedron Letters* 50(50): 6955–6958.

Fischer-Iglesias, C., Neuhaus, G. (2001) Zygotic embryogenesis-hormonal control of embryo development. In: Bhojwani, S.S., Soh, W.Y. (Eds.), *Current Trends in the Embryology of Angiosperms*, 223–247. Kluwer Academic, Dordrecht, The Netherlands.

Flematti, G.R., Ghisalberti, E.L., Dixon, K.W., Trengove, R.D. (2009) Identification of alkyl substituted 2 H-furo [2,3-c] pyran-2-ones as germination stimulants present in smoke. *Journal of Agricultural and Food Chemistry* 57(20): 9475–9480.

Flematti, G.R., Ghisalberti, E.L., Dixon, K.W., Trengove, R.D. (2004) A compound from smoke that promotes seed germination. *Science* 305(5686): 977.

Flematti, G.R., Ghisalberti, E.L., Dixon, K.W., Trengove, R.D. (2005) Synthesis of the seed germination stimulant 3-methyl-2H-furo[2,3-c]pyran-2-one. *Tetrahedron Letters* 46(34): 5719–5721.

Flematti, G.R., Merritt, D.J., Piggott, M.J., Trengove, R.D., Smith, S.M., Dixon, K.W., Ghisalberti, E.L. (2011) Burning vegetation produces cyanohydrins that liberate cyanide and promote seed germination. *Nature Communications* 2: 360.

Gardner, M.J., Dalling, K.J., Light, M.E., Jager, A.K., Van Staden, J. (2001) Does smoke substitute for red light in the germination of light-sensitive lettuce seeds by affecting gibberellin metabolism? *South African Journal of Botany* 67(4): 636–640.

Goddard-Borger, E.D., Ghisalberti, E.L., Stick, R.V. (2007) Synthesis of the germination stimulant 3-methyl-2H-furo[2,3-c]pyran-2-one and analogous compounds from carbohydrates. *European Journal of Organic Chemistry* 2007(23): 3925–3934.

Govindaraj, M., Masilamani, P., Alex Albert, V., Bhaskaran, M. (2016) Plant derived smoke stimulation for seed germination and enhancement of crop growth: A review. *Agricultural Reviews* 37(2): 87–100.

Humphrey, A.J., Beale, M.H. (2006) Strigol: Biogenesis and physiological activity. *Phytochemistry* 67(7): 636–640.

Jäger, A.K., Light, M.E., Van Staden, J. (1996) Effects of source of plant material and temperature on the production of smoke extracts that promote germination of light-sensitive lettuce seeds. *Environmental and Experimental Botany* 36(4): 421–429.

Jain, N., Stirk, W.A., Van Staden, J. (2008) Cytokinin-and auxin-like activity of a butenolide isolated from plant-derived smoke. *South African Journal of Botany* 74(2): 327–331.

Jain, N., Van Staden, J. (2006) A smoke-derived butenolide improves early growth of tomato seedlings. *Plant Growth Regulation* 50(2–3): 139–148.

Jamil, M., Rehman, S., Rha, E.S. (2014) Response of Growth, PSII Photochemistry and Chlorophyll Content to Salt Stress in Four Brassica Species. *Life Science Journal* 11(3): 139–145.

Jayakumar K., Jaleel C.A., Azooz M.M. (2008) Impact of cobalt on germination and seedling growth of *Eleusine coracana* L. and *Oryza sativa* L. under hydroponic culture. *Global Journal of Molecular Science* 3(1): 18–20.

Jayakumar, K., Vijayarengan, P. (2006) Influence of zinc on seed germination and seedling growth of Vigna mungo (L.) Hepper. *South African Journal of Botany* 61: 27–29.

Jefferson, L.V., Pennacchio, M., Havens, K., Forsberg, B., Sollenberger, D., Ault, J. (2008) Ex situ germination responses of Midwestern USA prairie species to plant-derived smoke. *The American Midland Naturalist* 159(1): 251–256.

Jefferson, L.V., Pennacchio, M., Havens-Young, K. (2014) *Ecology of Plant-Derived Smoke: Its Use in Seed Germination*. Oxford University Press, Oxford.

Joel, D.M., Chaudhuri, S.K., Plakhine, D., Ziadna, H., Steffens, J.C. (2011) Dehydrocostus lactone is exuded from sunflower roots and stimulates germination of the root parasite *Orobanche cumana*. *Phytochemistry* 72(7): 624–634.

Johnson, A.W., Gowda, G., Hassanali, A., Knox, J., Monaco, S., Razawi, Z., Roseberry, G. (1981) The preparation of synthetic analogues of strigol. *Journal of the Chemical Society, Perkin Transactions 1* 1: 1734–1743.

Kandari, L.S., Kulkarni, M.G., Van Staden, J. (2011) Effect of nutrients and smoke solutions on seed germination and seedling growth of tropical soda apple (*Solanum viarum*). *Weed Science* 59(4): 470–475.

Keeley, J.E. (1993) Smoke-induced flowering in the fire-lily *Cyrtanthus ventricosus*. South African. *Journal of Biology* 59: 638.

Keeley, J.E., Fotheringham, C.J. (1998) Smoke-induced seed germination in California chaparral. *Ecology* 79(7): 2320–2336.

Keeley, J.E., Fotheringham, C.J. (2000) Role of fire in regeneration from seed. In: Fenner, M. (Ed.), *Seeds: The Ecology of Regeneration in Plant Communities*, 2nd ed, 311–330. CABI Publishing, Wallingford.

Kępczyński J., Białecka B., Light M. E., Van Staden, J. (2006) Regulation of *Avena fatua* seed germination by smoke solutions, Gibberellin A3 and ethylene. *Plant Growth Regulation Plant Growth Regulation* 49(1): 9–16.

Kępczyński J., Van Staden J. (2012) Interaction of karrikinolide and ethylene in controlling germination of dormant *Avena fatua* L. caryopses. *Plant Growth Regulation* 67: 185–190.

Khan, I., Ibrahim, A.A.M., Sohail, M., Qurashi, A. (2017) Sonochemical assisted synthesis of RGO/ZnO nanowire arrays for photoelectrochemical water splitting. *Ultrasonic Sonochemistry* 37: 669–675.

Kulkarni, M.G., Light, M.E., Van Staden, J. (2011) Plant-derived smoke: Old technology with possibilities for economic applications in agriculture and horticulture. *South African Journal of Botany* 77(4): 972–979.

Kulkarni, M.G., Sparg, S.G., Light, M.E., Van Staden, J. (2006) Stimulation of rice (*Oryza sativa* L.) seedling vigour by smoke water and butenolide. *Journal of Agronomy and Crop Science* 192(5): 395–398.

Landis, T.D. (2000) Where there's smoke. There's germination? *Native Plants Journal* 1(1): 25–29.

Light, M.E., Burger, B.V., Staerk, D., Kohout, L., Staden, J.V. (2010) Butenolides from plant-derived smoke: Natural plant-growth regulators with antagonistic actions on seed germination. *Journal of Natural Products* 73(2): 267–269.

Light, M.E., Burger, B.V., Van Staden, J. (2005) Formation of a seed germination promoter from carbohydrates and amino acids. *Journal of Agricultural and Food Chemistry* 53(15): 5936–5942.

Light, M.E., Daws, M.I., Van Staden, J. (2009) Smoke-derived butenolide: Towards understanding its biological effects. *South African Journal of Botany* 75(1): 1–7.

Light, M.E., Gardner, M.J., Jäger, A.K., Van Staden, J. (2002) Dual regulation of seed germination by smoke solutions. *Plant Growth Regulation* 37(2): 135–141.

Light, M.E., Van Staden, J. (2003) The nitric oxide specific scavenger carboxy-PTIO does not inhibit smoke stimulated germination of Grand Rapids lettuce seeds. *South African Journal of Botany* 69(2): 217–219.

Lindon, H.L., Menges, E. (2008) Effects of smoke on seed germination of twenty species of fire-prone habitats in Florida. *Castanea* 73(2): 106–110.

Maga, J.A. (1988) *Smoke in Food Processing*, 1–160. CRC Press, Boca Raton FL, USA.

Mangnus, E.M., Zwanenburg, B. (1992) Tentative molecular mechanism for germination stimulation of *Striga* and *Orobanche* seeds by strigol and its synthetic analogues. *Journal of Agricultural and Food Chemistry* 40(6): 1066–1070.

Mavi, K., Light, M.E., Demir, I., Van Staden, J., Yasar, F. (2010) Positive effect of smoke-derived butenolide priming on melon seedling emergence and growth. *New Zealand Journal of Crop and Horticultural Science* 38(2): 147–155.

Mennan, H., Ngouajio, M., Sahin, M., Isik, D., Altop, E.K. (2012) Quantification of momilactone B in rice hulls and the phytotoxic potential of rice extracts on the seed germination of *Alisma plantago-aquatica*. *Weed Biology and Management* 12(1): 29–39.

Merritt, D.J., Kristiansen, M., Flematti, G.R., Turner, S.R., Ghisalberti, E.L., Trengove, R.D., Dixon, K.W. (2006) Effects of a butenolide present in smoke on light-mediated germination of Australian Asteraceae. *Seed Science Research* 16(1): 29–35.

Modi, A.T. (2002) Indigenous storage method enhances seed vigour of traditional maize. *South African Journal of Science* 98: 138–139.

Nagase, R., Katayama, M., Mura, H.H., Matsuo, N., Tanabe, Y. (2008) Synthesis of the seed germination stimulant 3-methyl-2H-furo[2,3-c]pyran-2-one utilizing direct and regioselective Ti-crossed aldol addition. *Tetrahedron Letters* 49(29–30): 4509–4512.

Nautiyal, C.S., Chauhan, P.S., Nene, Y.L. (2007) Medicinal smoke reduces airborne bacteria. *Journal of Ethnopharmacology* 114(3): 446–451.

Nefkens, G.H.L., Thuring, J.W.J.F., Beenakkers, M.F.M., Zwanenburg, B. (1997) Synthesis of a phtaloylglycine-derived strigol analogue and its germination stimulatory activity towards seeds of the parasitic weeds *Striga hermonthica* and *Orobanche crenata*. *Journal of Agricultural and Food Chemistry* 45(6): 2273–2277.

Nelson, D.C., Flematti, G.R., Ghisalberti, E.L., Dixon, K.W., Smith, S.M. (2012) Regulation of seed germination and seedling growth by chemical signals from burning vegetation. *Annual Review of Plant Biology* 63: 107–130.

Okem A., Stirk W. A., Street R. A., Van Staden J. (2015) Effects of Cd and Al stress on secondary metabolites, antioxidant and antibacterial activity of Hypoxis hemerocallidea Fisch. and C.A. Mey. *Plant Physiology and Biochemistry* 97: 147–155.

Pennacchio, M., Jefferson, L.V., Havens, K. (2007) Where there's smoke, there's germination. *The Illinois Steward* 16: 24–28.

Pepperman, A.B., Cutler, H.G. (1991) Plant-growth-inhibiting properties of some 5-alkoxy-3-methyl-2(5H)-furanones related to strigol. *ACS Symposium Series* 443: 278–287.

Pošta, M., Light, M.E., Papenfus, H.B., Van Staden, J., Kohout, L. (2013) Structure-activity relationships of analogs of 3,4,5-trimethylfuran-2(5H)-one with germination inhibitory activities. *Journal of Plant Physiology* 170: 1235–1242.

Preston, C.A., Becker, R., Baldwin, I.T. (2004) Is 'NO' news good news? Nitrogen oxides are not components of smoke that elicits germination in two smoke-stimulated species, *Nicotiana attenuata* and *Emmenanthe penduliflora*. *Seed Science Research* 14(1): 73–79.

Roche, S., Dixon, K.W., Pate, J.S. (1997) Seed ageing and smoke: Partner Cuesin the amelioration of seed Dormancyin selected Australian native species. *Australian Journal of Botany* 45(5): 783–815.

Schwachtje, J., Baldwin, I.T. (2004) Smoke exposure alters endogenous gibberellin and abscisic acid pools and gibberellin sensitivity while eliciting germination in the post-fire annual, *Nicotiana attenuata*. *Seed Science Research* 14(1): 51–60.

Senaratna, T., Dixon, K., Bunn, E., Touchell, D. (1999) Smoke-saturated water promotes somatic embryogenesis in geranium. *Plant Growth Regulation* 28(2): 95–99.

Serrano R., Macia F.C., Moreno V. (1999) Genetic engineering of salt and drought tolerance with yeast regulatory genes. *Horticultural Science* 78: 261–269.

Smith, C.J., Perfetti, T.A., Garg, R., Hansch, C. (2003) IARC carcinogens reported in cigarette mainstream smoke and their calculated log P values. *Food and Chemical Toxicology: an International Journal Published for the British Industrial Biological Research Association* 41(6): 807–817.

Sparg, S.G., Kulkarni, M.G., Light, M.E., Van Staden, J. (2005) Improving seedling vigour of indigenous medicinal plants with smoke. *Bioresource Technology* 96(12): 1323–1330.

Sparg, S.G., Kulkarni, M.G., Van Staden, J. (2006) Aerosol smoke and smoke-water stimulation of seedling vigor of a commercial maize cultivar. *Crop Science* 46(3): 1336–1340.

Stevens, J.C., Merritt, D.J., Flematti, G.R., Ghisalberti, E.L., Dixon, K.W. (2007) Seed germination of agricultural weeds is promoted by the butenolide 3-methyl-2H-furo [2, 3-c] pyran-2-one under laboratory and field conditions. *Plant and Soil* 298(1–2): 113–124.

Sun J., Smith L., Armento A., Deng W.M. (2008) Regulation of the endocycle/gene amplification switch by Notch and ecdysone signaling. *The Journal of Cell Biology* 182(5): 885–896.

Taylor, J.L.S., Van Staden, J. (1996) Root initiation in *Vigna radiata* (L.) Wilczek hypocotyl cuttings is stimulated by smoke-derived extracts. *Plant Growth Regulation* 18(3): 165–168.

Taylor, J.L.S., Van Staden, J. (1998) Plant-derived smoke solutions stimulate the growth of *Lycopersicon esculentum* roots in vitro. *Plant Growth Regulation* 26(2): 77–83.

Teale, W.D., Paponov, I.A., Palme, K. (2006) Auxin in action: Signalling, transport and the control of plant growth and development. *Nature Reviews. Molecular Cell Biology* 7(11): 847–859.

Thomas T.H., Van Staden, J. (1995) Dormancy break of celery (*Apium graveolens* L.) seeds by plant-derived smoke extracts. *Plant Growth Regulation* 17: 195–198.

Tieu, A., Dixon, K.A., Sivasithamparam, K., Plummer, J.A., Sieler, I.M. (1999) Germination of four species of native Western Australian plants using plant-derived smoke. *Australian Journal of Botany* 47(2): 207–219.

Ueda, M., Nakamura, Y. (2006) Metabolites involved in plant movement and 'memory': Nyctinasty of legumes and trap movement in the Venus flytrap. *Natural Product Reports* 23(4): 548–557.

Van Staden, J.V., Brown, N.A.C., Jäger, A.K., Johnson, T.A. (2000) Smoke as a germination cue. *Plant Species Biology* 15(2): 167–178.

Van Staden, J., Drewes, F.E., Jager, A.K. (1995) The search for germination stimulants in plant-derived smoke extracts. *South African Journal of Botany* 61(5): 260–263.

Van Staden J., Jäger A. K., Light M. E., Burger B.V. (2004) Isolation of the major germination cue from plant-derived smoke. *South African Journal of Botany* 70: 654–659.

Van Staden, J., Sparg, S.G., Kulkarni, M.G., Light, M.E. (2006) Post-germination effects of the smoke-derived compound 3-methyl-2H-furo [2, 3-c] pyran-2-one and its potential as a preconditioning agent. *Field Crops Research* 98(2–3): 98–105.

Wigchert, S.C.M., Kuiper, E., Boelhouwer, G.J., Nefkens, G.H.L., Verkleij, J.A.C., Zwanenburg, B. (1999) Dose-response of seeds of the parasitic weeds *Striga* and *Orobanche* towards the synthetic germination stimulants GR24 and Nijmegen-1. *Journal of Agricultural and Food Chemistry* 47(4): 1705–1710.

Zwanenburg, B., Pospisil, T. (2013) Structure and activity of strigolactones: new plant hormones with a rich future. *Molecular Plant* 6: 38–62.

26 Magnetopriming Alleviates Adverse Effects of Abiotic Stresses in Plants

Sunita Kataria and Meeta Jain

CONTENTS

26.1 INTRODUCTION

Various abiotic factors affect plants throughout the course of their growth and development (Zhao et al., 2007). Extreme temperature, drought, high salinity, and UV-B are the major abiotic stresses to which plants are exposed. In agriculture, these stresses are the most significant factors leading to a substantial and unpredictable loss in crop production (Jakab et al., 2005). All the growth stages in plants are controlled by environmental conditions, but the stages of crop plants which are most influenced by environmental factors are seedling emergence, establishment, and early vegetative growth. Rapid and uniform field emergence is critical to optimize stand establishment, especially under suboptimum conditions. Under all environmental conditions, the rapid and uniform field emergence is required to compete with weeds and optimize stand establishment. Physiological constraints like poor germination and vigor, low seedling emergence, and stand establishment results in uneven and sparse plant population (Singh and Afria, 1990). Modern agriculture strategies aim at enhancing harvest yields per acreage and reducing pre-harvest and post-harvest losses caused by detrimental abiotic cues (Gust et al., 2010). Due to drought, salinity, and UV-B stress, a variety of biochemical, physiological, and metabolic changes occur in plants (Kataria et al., 2014, 2017a; Xiong and Zhu, 2002), which may result in oxidative stress and affect plant metabolism, performance, and thereby the yield (Baghel et al., 2016, 2018; Kataria et al., 2017b).

Soil salinity may affect the germination of seeds either by creating a lower osmotic potential external to the seed, preventing water uptake or through the toxic effects of Na^+ and Cl^- ions on the germinating seed (Khajeh-Hosseini et al., 2003). The physiological mechanisms through which plants respond to salinity and drought show high similarity, suggesting that both the stresses must be perceived by the plant cell as a deprivation of water (Tavili et al., 2011). Other abiotic stress factors such as heat, cold, UV-B, or light stress are also known to adversely affect the growth and yield of the crops (Kataria et al., 2017a; Reyes and Cisneros-Zevallos, 2007).

During critical periods of germination and seedling establishment, the seed priming treatments improve the rate and synchronization of field emergence and shorten the exposure time of seeds to adverse environmental and biotic factors in soil and thus construct a successful crop stand (Chang and Sung, 1998; Harris et al., 2007; Shehzad et al., 2012). In a number of crop species, various seed priming treatments have been devised to improve the rate and uniformity of emergence as well as seed viability under normal and stress conditions (Afzal et al., 2006; Ghobadi et al., 2012; Punjabi et al., 1982). Basra et al. (2003) reported that primed crops grew more vigorously, flowered earlier, and yielded higher. It has also been reported that seed priming improves emergence, stand establishment, tillering, grain and straw yields, and harvest index (Farooq et al., 2008). Harris et al. (2007) reported that seed priming led to better establishment and growth, earlier flowering, increased seed tolerance to the adverse environment, and greater yield in maize.

Most often, chemical methods consisting of seed dressing, priming with various chemical substances, are used in the pre-sowing seed treatment. The various approaches include hydropriming, osmopriming, chemical priming, hormonal priming, biological priming, redox priming, solid matrix priming, and so on. Although priming improves the rate and uniformity of seedling emergence and growth particularly under stress conditions (Parera and Cantliffe, 1992), the effectiveness of different priming agents varies under different stresses and with different crop species (Iqbal and Ashraf, 2005; Vaishnav and Jain, 2015). Seed priming is the induction of a particular physiological state in plants by the treatment of natural and synthetic compounds to the seeds before germination. Such methods are considered as very effective in vigor improvement but may or may not be eco-friendly and have handling problems. Interest has therefore shifted to physical treatments like gamma rays, laser, electron beam, microwave, magnetic field, and radiofrequency energies to bring about biostimulation of seeds, which leads to increased vigor and contributes to the improved development of the plants. Physical methods moreover provide significant yield improvement without the hazard of toxic fertilizer and management cost. Therefore, the practical applicability of physical seed treatment for enhancing the seed performance should be standardized for commercial use.

Magnetic seed treatment/magnetopriming is one of the physical pre-sowing seed treatments especially worth attention since its impact on the seeds can change the processes taking place in the seed and stimulate plant growth and development. Magnetic field (MF) treatment of seeds is potentially a safe and affordable physical method that has been reported to greatly speed up the release of seeds from the dormant state (Reina et al., 2001), to improve seed germination and vigor (Kataria et al., 2015a, 2017b; Shine et al., 2011), plant growth, and biomass (Baghel et al., 2016, 2018; Kataria et al., 2015b, 2017a). Also, static magnetic field (SMF) induces an increase in seed water uptake (Kataria et al., 2015a, 2017b; Reina et al., 2001), enzymatic activity of seeds (Bhatnagar and Deb, 1977; Kataria et al., 2015a, 2017b), essential nutrient uptake into leaves (Esitken and Turan, 2004), and chlorophyll pigment content (Kataria et al., 2015b, 2017a; Novitskaya et al., 2001).

SMF provokes protection against heat stress (Ruzic and Jerman, 2002), salt stress (Baghel et al., 2016; Kataria et al., 2017a,b; Thomas et al., 2013), drought (Anand et al., 2012; Baghel et al., 2018), deficit soil moisture stress (Mridha et al., 2016), UV-B stress (Kataria et al., 2017a), and pathogens (De Souza et al., 2006; Pál, 2005) without adversely affecting the environment. Previous studies also reported that MF treatment helps in overcoming the unfavorable consequence of weeding and adequate use of solar radiation in plants (Aladjadjiyan and Ylieva, 2003). SMF-pretreatment with 0.2–0.3 T for 10 min enhanced saline-alkali tolerance of *Leymus chinensis* seedlings (Xia and Guo, 2000). The low static magnetic field (4 or 7 mT) caused significant enhancement in early growth and dry biomass accumulation of bean and wheat seeds in different osmotic conditions (Cakmak et al., 2010). This chapter discusses the alleviation of the salt stress by magnetopriming in plants.

26.2 EFFECT OF MAGNETOPRIMING ON THE EARLY SEEDLING CHARACTERISTICS UNDER ABIOTIC STRESSES

26.2.1 Germination and Early Seedling Vigor

The available literature revealed the effects of salinity on the seed germination of various crops like *Oryza sativa* (Xu et al., 2011), *Triticum aestivum* (Akbarimoghaddam et al., 2011), *Zea mays* (Carpici et al., 2009; Khodarahmpour et al., 2012), *Brassica* spp. (Ibrar et al., 2003; Ulfat et al., 2007), *Glycine max* (Essa, 2002), *Vigna* spp., (Jabeen et al., 2003), and *Helianthus annuus* (Mutlu and Buzcuk, 2007). It is well established that salt stress has a negative correlation with seed germination and vigor (Rehman et al., 2000). Salinity has many-fold effects on the germination process; it alters the imbibition of water by seeds due to lower osmotic potential of germination media (Khan and Weber, 2008), causes toxicity which changes the activity of enzymes of nucleic acid metabolism (Gomes-Filho et al., 2008), alters protein metabolism (Dantas et al., 2007 ; Yupsanis et al., 1994), disturbs hormonal balance (Khan and Rizvi, 1994), and reduces the utilization of seed reserves (Othman et al., 2006; Promila and Kumar, 2000). To improve the rate and synchronization of seed germination, seed priming is an alternative method of seed treatment that is applied to seeds, which are sown under abiotic stress conditions. MF treatment is a non-invasive physical stimulant used for improving vigor and field emergence in seeds. Several reports are available on the stimulating effects of magnetopriming on seed germination, seed vigor, seedling length, plant growth, and yield of a number of plant species under non-stressed conditions (Galland and Pazur, 2005; Harichand et al., 2002; Kataria et al., 2015a; Phirke et al., 1996; Radhakrishnan and Kumari, 2012; Shine et al., 2017; Vashisth and Nagarajan, 2010).

Salinity causes reduction in germination, length of root and shoots, and reduced the vigor index-I and II while on the other hand the stimulatory effects of SMF was found on the performance of the seedlings under saline as well as non-saline conditions in chickpea, maize, and soybean (Kataria et al., 2017b; Thomas et al., 2013). Figure 26.1 illustrates the dramatic increase in early seedling characteristics of maize and soybean by SMF pre-treatment under salt stress. There are a few reports that showed that magnetopriming enhanced germination, the speed of germination, vigor indices of the seeds as compared with their untreated control under abiotic stress conditions (Baghel et al., 2016; Kataria et al., 2017b; Thomas et al., 2013). In maize and chickpea, seedlings could withstand moisture stress due to improved soil–water relations through seed SMF treatment (Anand et al., 2012; Mridha et al., 2016), indicating that the advantage of seed pre-treatment gets augmented under stress compared with normal conditions (Khaliq et al., 2013).

In leguminous crops, Hadas (1970) observed that germination occurs at soil moisture content depending on the seed capabilities, seed moisture potential, and conductivity. This indicated that the rate of water uptake by seeds depended mainly on the internal water potential of the

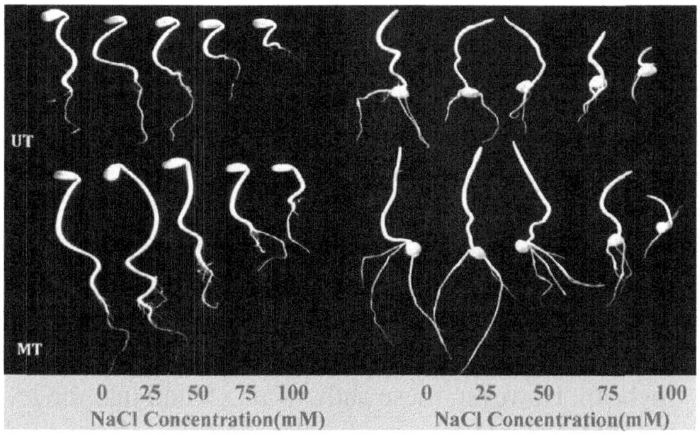

FIGURE 26.1 Effect of pre-sowing exposure of soybean and maize seeds to SMF (200 mT for 1 h) on early growth characteristics (length of root and shoot) under salt stress (unpublished data of the authors).

seed. Pittman and Ormrod (1970) reported that in wheat, seedlings grown from magnetically treated seeds absorbed more moisture, respired more slowly, released less heat energy, and grew faster than the untreated controls. Seeds exposed to the magnetic field had increased capacity to absorb moisture in soybean (Kavi, 1977). The influence of very low-frequency alternating magnetic field on the ionic permeability of cell membrane has been reported earlier (Khizhenkov et al., 2001) and it is envisaged that magnetic field influences the structure of the cell. The previous findings suggested that under non-stress and stress conditions, SMF pre-treatment to the seeds increased the water uptake as compared with their untreated ones (Kataria et al., 2017b; Shine et al., 2011; Shine and Guruprasad, 2012; Thomas et al., 2013). An increase in seed water content as a function of time was observed in magnetoprimed seeds of Pusa 1053 under non-saline and saline conditions (Thomas et al., 2013). Kataria et al. (2017b) also found out a remarkable decrease in water uptake with the increasing concentration of NaCl from 0 to 50 mM in untreated seeds of maize and soybean. On the other hand, they had found a significant increase in water uptake when the SMF-treated seeds were kept for imbibition under saline as well as non-saline conditions as compared with their untreated controls. Thus, positive effects of SMF toward stress factors have been reported in few studies, including under cadmium stress in mung bean (Chen et al., 2011), salt and drought stresses in maize seedling (Javed et al., 2011), and salt stress in soybean and maize (Kataria et al., 2017b).

The percentage germination, early growth characteristics, and vigor indices were reduced in chickpea, maize, and soybean under salt stress; this may be due to the ions in high concentration in the external solution (Na^+ and Cl^-) that are taken up by radicals at high rates which may lead to excessive accumulation in the tissue (Baghel et al., 2016). These ions may inhibit the uptake of other ions into the root (such as K^+) and their transport to the shoot, eventually leading to a deficiency in the tissue. Thus, there is the potential for many nutrient interactions in salt-stressed plants which may lead to important consequences for reduced growth (Gadallah, 1996). Rathod and Anand (2016) found that under salinity, the Na^+/K^+ ratio was less in plants emerged from SMF-treated seeds compared with untreated seeds and the sodium exclusion in SMF-treated seeds may be beneficial in imparting tolerance to wheat genotypes under salt stress. The higher nutrient content of cotton plants treated with pulse magnetic field (PMF) resulted in higher root growth which in turn increases the intake of water (Bilalis et al., 2013) transport in the ion-channels which affects the metabolic pathway activity (Chung et al., 2008).

Magnetic field exposure cause changes in the intracellular level of Ca^+ and ionic current density across the cellular membrane, altering the osmotic pressure and the capacity of the cellular tissue to absorb water (García-Reina and Arza-Pascual, 2001). The increased physiological activity due to greater absorption of moisture by SMF-treated seeds may be responsible for the increase in germination and seedling vigor under salt stress.

26.2.2 Biochemical Changes in Germinating Magnetoprimed Seeds Under Abiotic Stresses

26.2.2.1 Enzymatic Activities

A number of repair mechanisms are activated during seed imbibition, including repairing of cells and organelles and protein and enzyme activation to break down food reserves (McDonald, 2000). The most well-studied process in seedling development is the mobilization of complex polymers such as starch that serves as an energy source and building block for seedling development (Srivastava, 2002). During germination, α and β-amylases results in the breakdown of stored carbohydrate reserves of the seeds and the produced monosaccharides subsequently are utilized by the growing seedlings. Seed reserve utilization, seedling growth, and weight of mobilized seed reserve decreased with increasing drought and salt intensity (Ansari et al., 2012; Soltani et al., 2006). Seeds that have been treated by SMF for a short time generate more extensive root systems as well as more vigorous shoots compared with untreated seeds (Florez et al., 2007; Vashisth and Nagarajan, 2010). SMF treatments have shown to enhance seed germination, seed vigor, and productivity (Radhakrishnan and Kumari, 2012; Vashisth and Nagarajan, 2010). Such effects may be due to the promotion of gene expression, protein biosynthesis, enzymes activity, cell reproduction, and overall metabolism of plants (Atak et al., 2007; Stange et al., 2002; Vashisth and Nagarajan, 2010). Rapid germination of seedlings is linked with the enhancement of activities of α-amylase, dehydrogenase, and protease in seeds following SMF exposure under non-stress as well as abiotic stress conditions (Kataria et al., 2017b; Radhakrishnan and Kumari, 2012; Thomas et al., 2013; Vashisth and Nagarajan, 2010).

Vashisth and Nagarajan (2010) observed that wheat and sunflower seed, treated with SMF had significantly higher α-amylase activity than their untreated controls. He et al. (2002) found that rice seeds primed with mixed salt solution resulted in significantly increased activity of α-amylase, β-amylase, and root dehydrogenase and a moderate increase in the activity of shoot catalase under salt stress. The increased rate of germination in SMF-treated seeds of chickpea, maize, and soybean was associated with the higher activity of amylase and protease under salinity (Kataria et al., 2017b; Thomas et al., 2013).

Proteases are involved in the hydrolysis of proteins in the germinating seeds; the reduction being initiated by endoproteases, which convert the water-insoluble storage proteins into soluble peptides that can be further hydrolyzed in amino acids by exopeptidases (Callis, 1995). An increase in protease activity of soybean and maize represented the increase in hydrolyzed protein per gram weight of seeds. Kataria et al. (2017b) found that hydrolyzed protein per gram was decreased with the increased concentration of salt (NaCl from 0 to 50 mM) in both SMF-treated seeds and untreated seeds after 48 h of imbibitions. Whereas they have also found that SMF-treated seeds of maize and soybean maintained higher protease activity than the untreated seeds, with the difference being greater at later stages of imbibitions; 48 h and 72 h under non-saline (0 mM NaCl) as well as saline conditions (25 and 50 mM NaCl). Seed germination stimulation might be attributed to a combined effect of biochemical, physiological, metabolic, and enhanced enzymatic activities. It is considered that that SMF treatment alters cell membrane permeability and enables to transfer water and energy signals into the cell (Reina et al., 2001) and resultantly, metabolic pathways may be enhanced (Iqbal et al., 2012a,b) under abiotic stress conditions.

26.2.2.2 Reactive Oxygen Species (ROS) and Antioxidative Enzymes

In the achievement of major events of seed life, such as germination or dormancy release, ROS plays a key signaling role. In the germinating seeds, ROS such as superoxide radicals ($O_2^{\cdot-}$), hydrogen peroxide (H_2O_2) and hydroxyl radicals ($OH^{\cdot}$) are generated as a result of aerobic metabolism in mitochondria, peroxisomes, and the apoplastic space. For ROS to act as cellular messengers, seeds have evolved specific ROS removing mechanisms to protect against the overproduction of ROS.

Abiotic stresses increased production of ROS like $O_2^{\cdot-}$ and H_2O_2 during the germination of the seeds. Earlier studies (Kataria et al., 2017b; Thomas et al., 2013) showed that salt stress increased the production of $O_2^{\cdot-}$ and H_2O_2 during germination in chickpea, maize, and soybean. However, these authors also found that the levels of $O_2^{\cdot-}$ and H_2O_2 were further increased in the seedlings emerged from SMF pre-treatments as compared with their untreated seeds grown under saline conditions. Enhanced germination and seedling growth in SMF-treated seeds of soybean, chickpea, and maize were found to relate with the increased production of ROS under non-saline as well as saline conditions (Kataria et al., 2017b; Shine et al., 2012; Thomas et al., 2013). Elevated ROS may serve as signaling molecules to bring about the mobilization of reserve food material and support rapid axis growth (Verma and Sharma, 2010). ROS generation during seed germination helps in endosperm weakening, protection against the pathogen, Ca^{+2} signaling, gene expression, redox regulation, hormone signaling, cell wall elongation, and so on (El-Maarouf-Bouteau and Bailly, 2008).

SMF also activates the antioxidative defense, improving plant tolerance to biotic and abiotic stresses (Rochalska and Grabowska, 2007). This might be the result of stimulation of enzymes activities implicated in the detoxification of $O_2^{\cdot-}$ (SOD) and H_2O_2 (catalases, peroxidases) (Atak et al., 2007; Celik et al., 2009). The significant enhancement in the production of free radicals after treating the seeds of soybean and maize with SMF, the EPR spectra of adducts produced with $O_2^{\cdot-}$ and OH^- radicals at an early stage (just germinated and 8-day-old seedlings) provide evidence for the important role of ROS in enhancing the performance of seeds at this stage (Shine et al., 2012, 2017).

Although the ROS content enhanced during germination of SMF-treated seeds, the ROS scavenging enzymes such as catalase (CAT) and superoxide dismutase (SOD) were reduced (Shine et al., 2012; Shine et al., 2017; Shine and Guruprasad, 2012), whereas the activity of peroxidase enzyme was enhanced by SMF treatment which is predicted to be the result of enhanced NADPH activity (Radhakrishnan and Kumari, 2013a). SOD is an important enzyme to deactivate $O_2^{\cdot-}$ by converting the radical to H_2O_2 (Scandalios, 1993). SMF treatment although enhanced $O_2^{\cdot-}$, the activity of SOD was reduced in soybean and maize seedlings (Shine et al., 2012, 2017).

26.3 EFFECT OF MAGNETOPRIMING ON GROWTH, PHOTOSYNTHESIS, AND YIELD OF PLANTS UNDER ABIOTIC STRESSES

26.3.1 Growth and Photosynthesis

Abiotic stress factors, including drought, salinity, UV-B, heavy metal, and so on, reduce the plant growth, photosynthetic performance, and yield performances (Baghel et al., 2016, 2018; Kataria et al., 2015, 2017; Kataria and Verma, 2018). Salinity stress causes adverse effects on growth and yield of many crop plants (Baghel et al., 2016; Golezani et al., 2009). Muraji et al. (1998) found that an alternating magnetic field of 10 and 20 Hz resulted in 20% greater root growth than control corn seedlings. Moreover, root length and root surface area are used as important physiological parameters for evaluation of chemical elements uptake (Wang and Ritz, 2006). Plants that emerged from SMF treated seeds of soybean showed better root characteristics (root length, fresh weight, and dry weight) when compared with the untreated ones. Number and weight of root nodules per plant were also enhanced after SMF treatment (200 mT for 60 min.) under non-stress as well as stress conditions (Baghel et al., 2016, 2017). Vashisth and Nagarajan (2010) found that root length and root surface area showed significant increases in sunflower seedlings exposed to SMF strength from 0 to 250 mT. Martínez et al. (2000, 2002) observed similar effects on SMF-treated wheat and barley seeds. The stimulatory effect of the application of SMF on the seedling growth is reported by many researchers like Flórez et al. (2004), who observed an increase for the initial growth stages and early sprouting of rice seeds exposed to SMF (125 and 250 mT) for different time durations. The seedlings of maize and soybean from SMF pre-treated seeds grew taller and heavier than untreated controls in non-stress conditions (Shine et al., 2011, 2012). Baghel et al. (2016) indicated that salt stress considerably suppressed the shoot and root fresh and dry weights, and leaf area as well as photosynthetic efficiency and yield of soybean. However, they have also observed that SMF treatment of 200 MT for 1 h was effective to enhance the growth, biomass accumulation, and carbon and nitrogen metabolism in soybean under saline as well as non-saline conditions. Positive effects of SMF toward stress factors have been reported in a number of studies, including under cadmium stress in mung bean (Chen et al., 2011) and the combination of SMF pre-treatment with 100 mM NaCl or 60 g/L PEG-6000 also showed that SMF applications moderated negative effects of salt and drought stresses in wheat mature embryo explants (Sen and Alikamanoglu, 2014). Vashisth and Nagarajan (2010) have shown electromagnetic treatment-induced growth improvement in chickpea (*Cicer arietinum*) under water-deficit conditions. SMF could alleviate the toxic effects of cadmium salts through increasing the photosynthetic rate and reducing the lipid peroxidation in mung bean seedlings (Chen et al., 2011). Under salt-stressed conditions, PMF pre-treatment helped in the regeneration of soybean (Radhakrishnan and Kumari, 2013b; Radhakrishnan and Kumari, 2012). SMF treatment of soybean and maize seeds (100 MT for 2 h and 200 MT for 1 h), enhanced the growth and photosynthesis under abiotic stress conditions like water (Anand et al., 2012), ambient UV-B (Kataria et al., 2015b, 2017a), and salt (Baghel et al., 2016). The positive impact of SMF of 200 mT for 1 h was observed on carbon and nitrogen metabolism under saline as well as non-saline conditions in soybean (Baghel et al., 2016) and maize (unpublished data of authors). The alleviation of the adverse effect of salt stress on soybean and maize plants by the seed pre-treatment with SMF is shown in Figure 26.2. All the above-ground parameters like plant height, leaf area, and plant biomass of soybean (Baghel et al., 2016) and maize have been found to be promoted by SMF in saline as well as non-saline conditions suggesting that this addition of biomass may be due to the enhanced fixation of carbon and nitrogen. Similar results of enhanced growth and biomass have also been recorded after SMF pre-treatment under ambient conditions in many plant species like soybean (Shine et al., 2011), maize (Shine et al., 2017; Shine and Guruprasad, 2012), cucumber (Bhardwaj et al., 2012), cotton (Bilalis et al., 2012), sunflower (Vashisth and Nagarajan, 2010), and chickpea (Vashisth and Nagarajan, 2008) along with increased germination percentage and speed of germination.

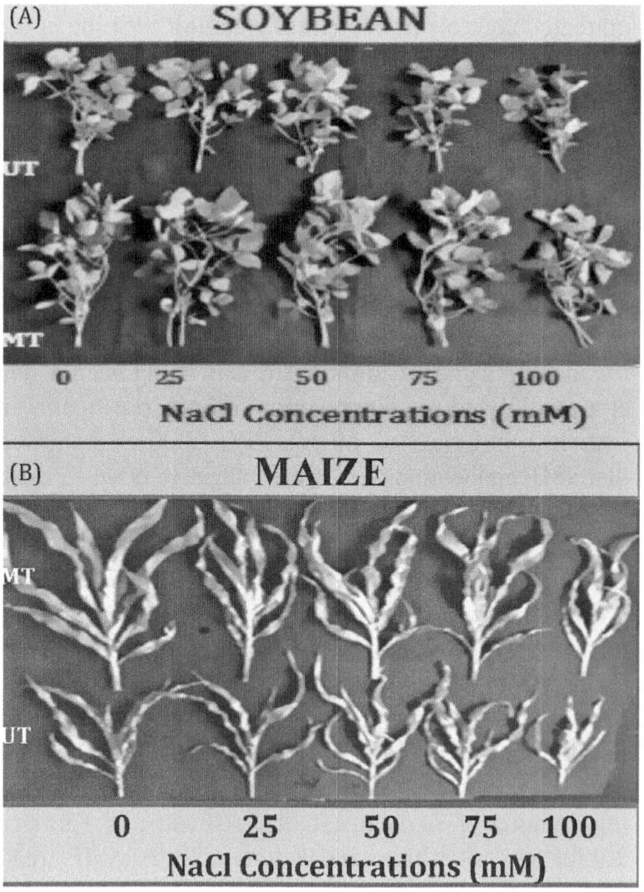

FIGURE 26.2 Effect of pre-sowing exposure of soybean (A) and maize (B) seeds to SMF (200 mT for 1 h) on growth attributes (plant height, leaf area, leaf number, and biomass) under salt stress (unpublished data of authors).

Abiotic stresses like salt, UV-B, and water stress caused a prominent decrease in the photosynthetic pigments, which could affect the photosynthesis and reduced biomass accumulation and economic yield of the crop plants (Baghel et al., 2016, 2018; Kataria et al., 2017a). On the other hand, the significant increase in Chl *a*, Chl *b*, total Chl, and carotenoid concentrations were observed in plants emerged from SMF pre-treatment even in the presence of ambient UV stress and salt and water stress (Baghel et al., 2016, 2018; Kataria et al., 2017a). Similarly it was confirmed by several authors for different plants but under non-stress conditions; whereas SMF treatment increased the chlorophyll content in sugar beet leaves (Rochalska and Orzeszko-Rywka, 2005) and content of chlorophyll a, *b* and carotenoids in potato (Rakosy-Tican et al., 2005), soybean, and maize (Kataria et al., 2015b; Shine et al., 2011). Plants from MF-treated seeds had a marginal advantage over those from untreated seeds. Magnetically treated seeds had a stimulatory effect, which enabled soybean plants for more efficient sunlight harvesting, leading to increased crop biomass (Shine et al., 2012). Other general effects on SMF and PMF application on chlorophyll content have been documented for several plant species (Radhakrishnan and Kumari, 2013a; Rochalska and Orzeszko-Rywka, 2005; Turker et al., 2007). Similarly, pre-sowing different electromagnetic treatments enhanced the chlorophyll contents (*a* and *b*) under non-stress and drought stress conditions (Javed et al., 2011). The electromagnetic treatment-induced improvement in chlorophyll pigments could be due to the presence of paramagnetic properties of chloroplast which can increase the rate of seed metabolism (Aladjadjiyan and Ylieva, 2003; Rochalska and Orzeszko-Rywka, 2005). In view of a number of studies, it is evident that electromagnetic treatments alter relative growth rate, seed germination, chlorophyll contents, root growth, cell membrane characteristics, and cell division by inducing changes in plant cell metabolism (Aladjadjiyan, 2002; Baghel et al., 2016; Florez et al., 2007; Shine et al., 2011). Hozayn and Qados (2010) demonstrated a marked increase in chlorophyll *a* and *b* contents in wheat while using electromagnetically treated water, and they ascribed this increase to electromagnetic treatment induced better ion mobility as well as uptake during growth.

In maize cultivars, pre-treated seeds with different MFs were seen to overcome the effect of drought by improving chlorophyll *a*, photosynthetic rate, photochemical, and non-photochemical quenching (Javed et al., 2011). Anand et al. (2012) found that pre-treatment of seeds with SMF reduced soil water stress in maize seedlings and maintained enhanced growth, leaf water status, and chlorophyll content. Chen et al. (2012) hypothesized that priming-induced activation of pre-germinative metabolism might imprint a sort of "stress memory" in seeds. Consequently, the cellular protective system, activated during priming, makes seedlings more tolerant to stresses. Chickpea seeds treated with mannitol alleviated effects of water deficit and salt stress on the seedling growth (Kaur et al., 2003). Mridha et al. (2016) found that chickpea from SMF-treated (strength: 100 mT for 1 h) seeds had better water utilization and radiation use efficiencies, higher biomass, root volume, and surface area as compared with the untreated control under deficit soil moisture conditions. The improvement in radiation use efficiency and water use efficiency in SMF-treated plants was attributed to better shoot and root development in chickpea to alleviate soil moisture stress (Mridha et al., 2016). Among the tools used to study the effects of environmental changes on the photosynthetic apparatus, chlorophyll *a* fluorescence is often proposed as a simple, rapid, and sensitive method (Kalaji et al., 2016), which has been successfully used to monitor the changing physiological states of photosynthetic system under abiotic stress like UV-B radiation (Wang et al., 2010), salt stress (Baghel et al., 2016), and water stress (Baghel et al., 2018). The polyphasic chlorophyll *a* fluorescence (O-J-I-P) transients from SMF-treated plants (200 mT for 1 h) gave a higher fluorescence yield at the J-I-P phase under non-stress as well as under abiotic stress conditions (salt, water, and UV-B) as compared with untreated plants (Baghel et al., 2016, 2018; Kataria et al., 2017a). Figure 26.3 depicts the rise in the fluorescence curve after SMF treatment in the leaves of soybean plants that emerged from SMF-treated (A) and untreated seeds (B) under saline as well as non-saline conditions. The rise in fluorescence curve after SMF treatment is the result of the faster reduction of electron acceptors in the photosynthetic pathway downstream of PS II, notably plastoquinone and in particular Q_A (Maxwell and Johnson, 2000). In the O-J-I-P curve, the J step represents the momentary maximum of Q_A^- and I step suggested to be related to heterogeneity in the filling up of the plastoquinone pool (Govindjee, 1995; Strasser and Strasser, 1995).

The results of Baghel et al. (2016, 2018) showed that the maximal quantum efficiency of PSII (calculated from Fv/Fm) and the efficiency of the water-splitting complex on the donor side of PSII (as inferred from Fv/Fo) slightly reduced while both Fv/Fm and Fv/Fo were significantly increased with SMF treatment under saline and water stress conditions. Electron transport per leaf CS (ETo/CSm) showed a 10% decrease at 50 mM NaCl which was increased by 50% after SMF treatment at 50 mM NaCl as compared to untreated ones.

A phenomenological leaf model (generated by Biolyzer HP3 software) of the leaves of plants that emerged from SMF-treated seeds of soybean depicts more active reaction centers per unit area (Figure 26.4). In this model, open circles represent the active reaction center and in leaves of plants emerged from SMF-treated plants had more active reaction centers combined with the higher efficiency of electron transport indicated by the broader width of the arrow (Figure 26.4B). Under salinity, the active reaction center and electron transport per leaf CS (ETo/CSm) decreased with the increasing concentration of NaCl and it was drastically reduced at 100 mM NaCl in leaves of untreated soybean plants (Figure 26.4A), while SMF treatment caused significant promotion in both active reaction center and (ETo/CSm) as compared with the plants that emerged from untreated seeds under salinity (Figure 26.4B). Performance index (PI_{ABS}) is a complex parameter that is related to the ratio

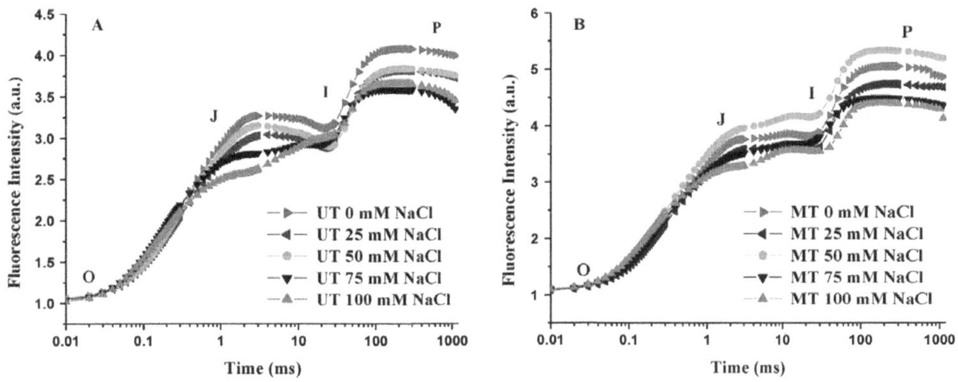

FIGURE 26.3 Fluorescence emission transient (normalized at F_0) of leaves of plants emerged from SMF pretreatment of 200 mT for 1 h and from untreated seeds under salt stress. (O-J-I-P are fluorescence yield at 20 μs, 2 ms, 30 ms, and maximum fluorescence, respectively). UT (A) and MT (B).

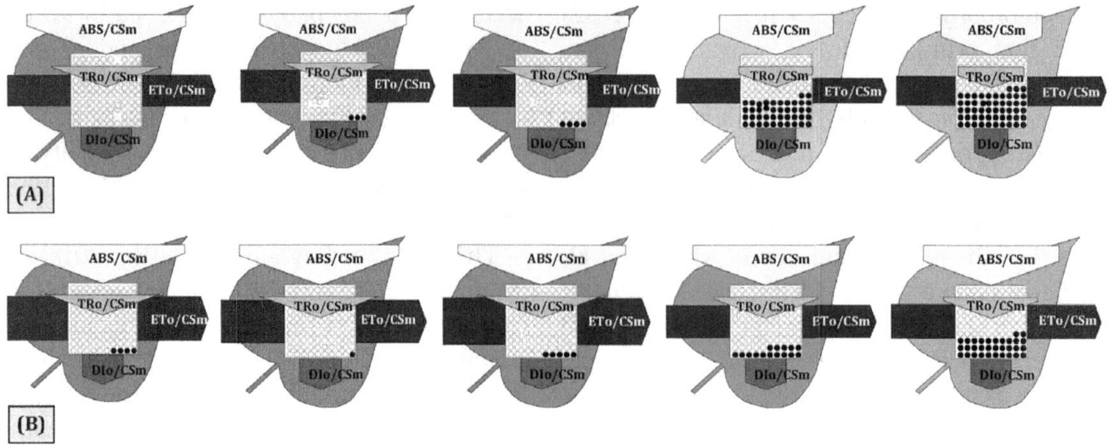

FIGURE 26.4 Impact of pre-sowing exposure of soybean seeds to SMF (200 mT for 1h) on Leaf model showing phenomenological energy fluxes per excited cross-section (CS) of soybean leaf. ABS/CSm, absorption flux CS approximately by Fm; TRo/CSM, trapped energy per CS, ETo/CSm, electron transport flux per CS; DI0/CSM, dissipated energy per CS. Each relative value is represented by the size of the proper parameters (arrow), empty circles represent reducing QA reaction centers (active), full black circles represent non-reducing QA reaction centers (inactive or silent). UT (A), MT (B). (Unpublished data of authors).

of reaction center per (light) absorption flux, the maximal quantum yield for primary photochemistry, and the quantum yield for electron transport. Baghel et al. (2016) found the largest decrease (18%) upon salt treatment at 50 mM NaCl salinity and the priming of soybean seeds with SMF resulted with enhanced PI_{ABS} by 85% over the untreated controls (0 mM NaCl) under non-saline conditions while under saline conditions SMF treatment enormously enhanced the PI_{ABS} by 70% and 134% respectively at 25 and 50 mM NaCl as compared with their untreated controls. This parameter encompasses fluorescent changes associated with changes in antenna conformation and energy fluctuations. Therefore, the PI_{ABS} helps to estimate the vitality of the plants with high resolution. Similarly, the changes in OJIP parameters and PI_{ABS} were found after SMF treatment of soybean seeds under water and UV-B stress (Baghel et al., 2018; Kataria et al., 2017a).

Fisarakis et al. (2001) reported a positive growth inhibition caused by salinity associated with a marked inhibition of photosynthesis. Salinity affects plant growth due to changes in many physiological processes including photosynthesis (Kalaji and Guo, 2008; Kalaji and Pietkiewicz, 1993). A reduction in chlorophyll content, stomatal conductance, ribulose-1,5-bisphosphate carboxylase/oxygenase (Rubisco) activity and an increase in the chlorophyll *a*/*b* ratio had been observed earlier under salt stress (Delfine et al., 1999; Kalaji and Nalborczyk, 1991). A magnetic flux density of around 4 mT had beneficial effects, regardless of the direction of MF, on the growth promotion and enhancement of CO_2 uptake of potato plantlets *in vitro* (Iimoto et al., 1996). Javed et al. (2011) found that pre-treated corn seeds with different electromagnetic treatments particularly 100 and 150 mT for 10 min significantly alleviated the drought-induced adverse effects on growth by improving photosynthesis, transpiration rate, and stomatal conductance. In a similar experiment under greenhouse conditions, it was found that photosynthesis, stomatal conductance, and chlorophyll content increased in maize plants exposed to SMFs of 100 mT for 2 h and 200 for 1 h, compared with control under irrigated and mild stress condition (Anand et al., 2012). Under salt stress conditions, PMF treatment of soybean seeds has the potential to counteract the adverse effects of salt stress on calli growth by improving primary and secondary metabolites (Radhakrishnan and Kumari, 2012).

Baghel et al. (2016) reported that the rate of net photosynthesis (*Pn*) along with stomatal conductance, intercellular CO_2 concentration, and rate of transpiration was decreased with the increased level of salinity while SMF treatment (200 mT for 1 h) caused a significant increase in all of these gas exchange parameters under saline as well as non-saline conditions except inter-cellular CO_2 concentration which was decreased slightly. Similarly, SMF pre-treatment to the soybean seeds increased the rate of photosynthesis, stomatal conductance, and rate of transpiration under water and UV-B stress conditions (Baghel et al., 2018; Kataria et al., 2017a) due to the increase in the activity of Rubisco and Carbonic anhydrase (Joshi, 2015; Shine et al., 2011).

26.3.2 Nitrogen Metabolism

Biochemical analysis of nodules has revealed specific enhancement in the level of Lb; a protein, which plays an important part in the fixation of nitrogen in the root nodules. The higher the content of Lb, the more efficient the plant regarding its capacity to fix atmospheric nitrogen (Gurumoorthi et al., 2003). The previous literature indicates a positive effect of SMF treatment on the activity of nitrogen fixation by an enhancement in the content of Lb and hemechrome content in the root nodule of soybean in non-saline and saline conditions (Baghel et al., 2016). Nitrogen requirement of legumes can be met by inorganic N assimilation and symbiotic N_2 fixation; in practice, they obtain N through both processes. A reduction of nitrate to nitrite (NO_2^-) is catalyzed by nitrate reductase. Changes in nitrate reductase activities in germinating wheat seeds treated by electromagnets of different field strengths were also observed by Bhatnagar and Deb (1977). Rhadhakrisnan and Kumari (2013b) found that the presence of increasing concentrations of NaCl decreased the regeneration potential of cotyledonary nodal soybean explants; they have demonstrated that PMF exposure was suitable to enhance *in vitro* salt tolerance in soybean. Radhakrishnan and Kumari (2013a) also reported enhancement in nitrate reductase in 10-day-old seedlings of soybean that emerged from seeds treated with PMF for 5 h per day for 20 days at different frequencies under laboratory conditions. SMF pre-treatment to the seeds significantly enhances the activity of nitrate reductase in the leaves of soybean in salt, water, and UV-B stress and non-stress conditions (Baghel et al., 2016, 2018; Kataria et al., 2017a). The increase in nitrate reductase activity by SMF treatment may be due to the increase of ATP supply through the enhancement in the activity of PS II under water, salt, and UV-B stress as well as well under non-stress conditions.

26.3.3 Antioxidant Defense System

Salt stress can lead to stomatal closure, which reduces CO_2 availability in the leaves and inhibits carbon fixation, exposing chloroplasts to excessive excitation energy which in turn increases the generation of ROS such as superoxide ($O_2^{\cdot -}$), hydrogen peroxide (H_2O_2), hydroxyl radical ($OH^{\cdot}$), and singlet oxygen ($^1O_2$) (Ahmad et al., 2010; Ahmad and Sharma, 2008; Parida and Das, 2005). However, plants contain numerous antioxidant compounds, both enzymatic and non-enzymatic, which acts to prevent oxidative damage by scavenging free radicals before they attack membranes (Merritt et al., 2003; Sharma et al., 2012). Some protective mechanisms involving free radical and peroxide scavenging enzymes, such as CAT, guaiacol peroxidase (POX), ascorbate peroxidase (APX), and SOD, have been evaluated within the mechanism of seed aging (Pukacka and Ratajczak, 2007; Sharma et al., 2012).

Decreased free radical production (H_2O_2), antioxidant content (ascorbic acid and α-tocopherol), and antioxidant enzyme activities (SOD, POD, APX, GR) in the leaves from 45-day-old maize plants raised from SMF-treated seeds were reported under salt, water, and UV-B stress conditions (Anand et al., 2012; Baghel et al., 2015; Kataria et al., 2017a). These studies suggested that MF treatment ameliorated the adverse effect of salt, water, and UV-B stress, and therefore, plants did not have to divert their metabolic energy in detoxification of free radicals that are generally produced under abiotic stress conditions. This has been collaborated by actual measurement of superoxide radicals in leaves of field-grown soybean plants from SMF-treated seeds (Shine et al., 2011).

According to Ashraf and Rauf (2001), accumulations of high levels of Na^+ and Cl^- in sweetcorn probably triggers the injuries to the plasma membrane by affecting ion homeostasis in a cell under salt stress. Physiological resonance causes ROS overproduction and induces lipid peroxidation in a cell under this condition, which eventually leads to plasma membrane injuries and malfunction (Foyer and Shigeoka, 2011). SMF treatment also increases the cell membrane permeability and ion transport in the ion channels, which then affects some metabolic pathways activity (Chung et al., 2008; Iqbal et al., 2012a,b; Labes, 1993); this leads to improving the growth and development of crop plants. Free movement of ions activates the metabolic pathways by enhancing biochemical and physiological feedback (ul Haq et al., 2012). The results of the Kataria et al. (2017a) revealed that in addition to the increase in H_2O_2 content, the activities of antioxidant enzymes like SOD, APx, GR, and GPx were also higher under ambient UV-B stress. ROS (H_2O_2) content and antioxidant defense system enzymes activities were lower in SMF-treated plants under saline as well as under non-saline conditions (unpublished data of authors). These results indicate that the presence of the H_2O_2 content activates the antioxidant enzymes in untreated ones and SMF pre-treatment to the seeds eliminates the need for the defense against harmful abiotic stresses like salinity stress and UV-B stress and leads to enhancement in the primary metabolism.

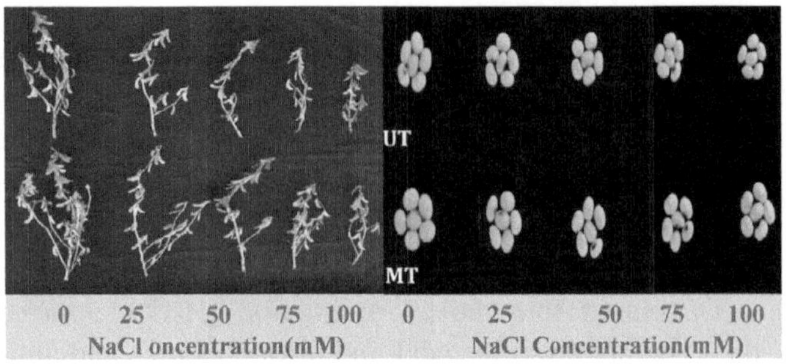

FIGURE 26.5 Effect of pre-sowing exposure of soybean seeds to SMF (200 mT for 1 h) on yield attributes (length of pods, number of pods/plants, size of seeds) under salt stress (unpublished data of authors).

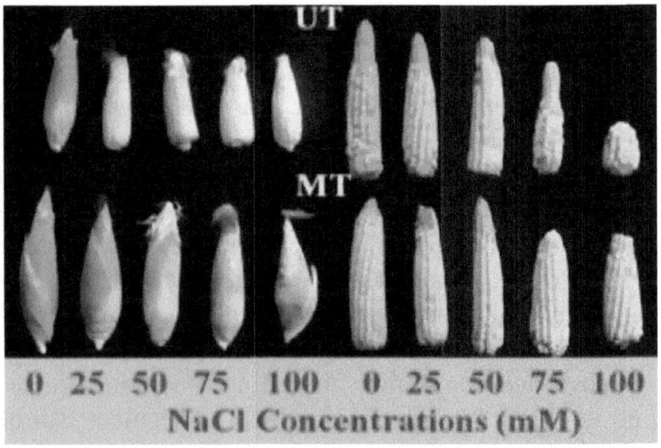

FIGURE 26.6 Effect of pre-sowing exposure of maize seeds to SMF (200 mT for 1 h) on yield attributes (length of cob/arrangement of the seeds per cob) under salt stress.

26.3.4 Crop Yield

The above-mentioned effects of salt stress on plants ultimately leads to a reduction of yield of the crop which is the most countable effect of salt stress in agriculture. The effect of SMF on the yield parameters of soybean plants like number of pods and size of seeds are shown in Figure 26.5. Figure 26.5 illustrates that the soybean plants that emerged from SMF(200 mT for 1 h) treated seeds have a large number of pods and higher seed size even in the presence of salt stress as well as under non-saline conditions (unpublished data of authors). Figure 26.6 represents the maize yield parameters like the length of cobs and number and size of seeds per cob; it shows that all of these parameters are increased by the SMF pre-treatment under salt stress as compared to their untreated ones (unpublished data of authors). Naz et al. (2012) found a significant increase in number of flowers per plant, leaf area, plant height, number of fruits per plant, pod mass per plant, and number of seeds per plant after 99 mT for 11 min exposure to okra (*Abelmoschus esculentus*) seeds as compared with the controls. PMF exposure significantly altered the number of leaves, pods per plant, length of pods, and number and weight of seeds in soybean (Radhakrishnan and Kumari, 2012) under non-stress conditions. The plants emerged after SMF treatment acquired an ability to reduce the ROS level in the leaves and enhance the efficiency of PSII in harvesting light and increase the rate of photosynthesis and nitrogen fixation which resulted in the improved yield of crop plants under non-stress as well as under abiotic stress conditions like salt, water, and UV-B (Baghel et al., 2016, 2018; Kataria et al., 2017a).

The salt stress reduced growth, biomass accumulation, nitrogen metabolism, photosynthetic performance, and yield of crop plants like soybean and maize may be due to the excess osmotic potential of the soil due to the presence of higher concentrations of salts which reduces the ability of plants to take up water and minerals like K^+ and Ca^{2+} (Munns et al., 2006). Na^+ and Cl^- ions can enter into the cells and have their direct toxic effects on cell membranes, as well as on metabolic activities in the cytosol (Hasegawa et al., 2000). On the other hand, SMF treatment was able to ameliorate the effect of salt stress to some extent which may be attributed to the maintenance of better plant water status by osmotic adjustment and greater root growth than the untreated controls. SMF treatment may also increase the cell membrane permeability and ion transport in the ion channels which then affects some metabolic pathways activity leading to the improved growth of plants (Galland and Pazur, 2005). In most cases, a magnetic field can affect the growth processes at the cellular and subcellular levels and alter the Ca_2^+ balance, enzyme activities, and various metabolic processes (Celik et al., 2009).

26.4 CONCLUSION AND FUTURE PERSPECTIVES

In conclusion, perusal of the literature indicated that abiotic stresses like salt, water, and UV-B reduced the speed of germination, vigor of the seedlings, growth, nodulation, biomass, and carbon and nitrogen metabolism, which ultimately reduced the yield of crop plants, whereas SMF-treated seeds showed a significant increase in all of these parameters under abiotic stresses as well as under non-stress conditions. Figure 26.7 illustrates the hypothetical model proposing the central role of ROS and hydrolytic enzymes in the increased seedling vigor of magnetoprimed seeds and the increased

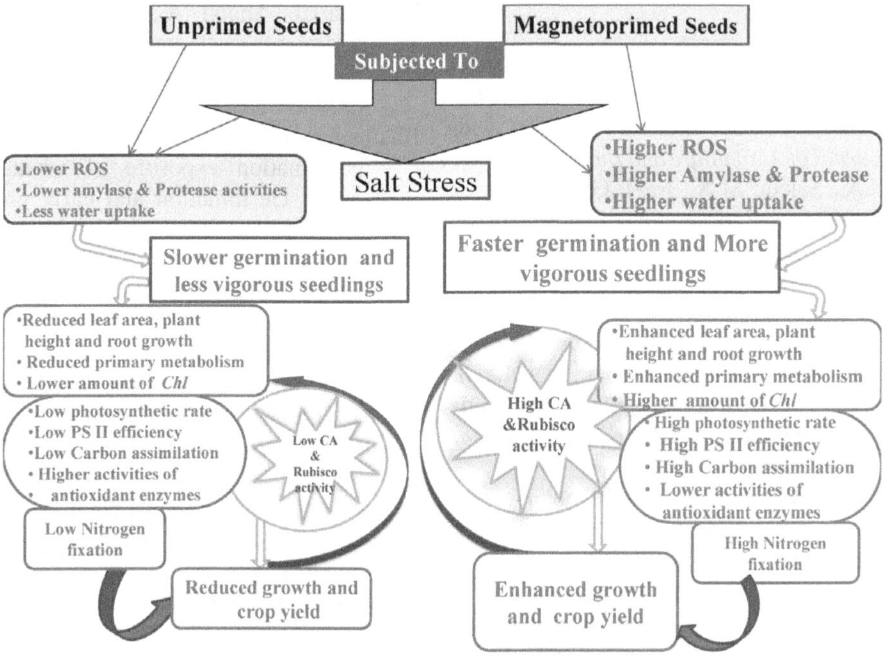

FIGURE 26.7 Hypothetical model proposing a central role of ROS and hydrolytic enzymes in increased seedling vigor of magnetoprimed seeds and the stimulatory effect of pre-sowing MF treatment of seeds also persisted in plants until its maturity and stimulated plant growth, leaf area, biomass, photosynthetic performance, improved the crop yield, and reduced antioxidative response under salt stress conditions.

growth and yield by SMF pre-treatment under salt stress appears to be due to the lower level of ROS and scavenging enzymes in plants that emerged after SMF-treated seeds. This suggested that SMF ameliorated the adverse effect of stress by restricting the production of ROS. The metabolic energy that would have been utilized for scavenging these ROS was efficiently utilized toward maintaining growth and yield of the plant under stress as well as under non-stress conditions (Figure 26.7). The magnetoprimed seeds had a long-lasting stimulatory effect on plants as reduced superoxide, and hydrogen peroxide content and higher performance index of photosystem II contributed to the higher efficiency of light harvesting that consequently increased biomass and photosynthetic performance in plants in stress as well as non-stress conditions. This suggests that a signal transduction pathway may be operative, but the underlying molecular mechanism needs further study to explain plant response.

Thus, pre-sowing SMF treatment of seeds has revealed beneficial effects in several crops. Exposure of seeds to SMF is one of the potential safe and affordable physical pre-sowing treatments to enhance post-germination plant development and crop stand. This has a potential to increase crop production per unit area of land without having any damaging effect toward any environmental component. Pre-seed SMF treatments could be used to enhance the growth and yield production by minimizing the salt-induced adverse effects on different crop plants. SMF may be a good tool for improving agricultural production in future because of the moderating effect toward drought and salinity stress. Thus, magnetopriming of the seeds alleviates the oxidative stress caused by the salinity. Pre-seed SMF treatments could be used to enhance the growth and yield production by minimizing the salt-induced adverse effects on different crop plants. The mechanism by which plants perceive MFs and regulate the signal transduction pathway is not fully understood yet and needs further study. SMF may be a good tool for improving agricultural production in the future because of the moderating effects toward abiotic stresses. This chapter has presented many results from recent research, however, further studies are necessary to explain how the physiological and biochemical changes caused by magnetopriming improve the plant adaptations to stressful environmental conditions.

ACKNOWLEDGMENTS

Financial support by the Department of Science Technology Women Scientists-A Scheme (SR/WOS-A/LS-674/2012-G and SR/WOS-A/LS-17/2017-G) to Dr. Sunita Kataria is thankfully acknowledged.

REFERENCES

Afzal, I., Basra, S.M.A., Farooq, M., Nawaz, A. (2006) Alleviation of salinity stress in spring wheat by hormonal priming with ABA, salicylic acid and ascorbic acid. *International Journal of Agriculture and Biology* 8:23–28.

Ahmad, P., Jaleel, C.A., Salem, M.A., Nabi, G., Sharma, S. (2010) Roles of enzymatic and non-enzymatic antioxidants in plants during abiotic stress. *Critical Reviews in Biotechnology* 30(3):161–175.

Ahmad, P., Sharma, S. (2008) Salt stress and phytobiochemical responses of plants. *Plant, Soil and Environment* 54:89–99.

Akbarimoghaddam, H., Galavi, M., Ghanbari, A., Panjehkeh, N. (2011) Salinity effects on seed germination and seedling growth of bread wheat cultivars. *Trakia Journal of Sciences* 9:43–50.

Aladjadjiyan, A. (2002) Study of the influence of magnetic field on some biological characteristics of *Zea mays. Journal of Central European Agriculture* 4:90–94.

Aladjadjiyan, A., Ylieva, T. (2003) Influence of stationary magnetic field on the early stages of the development of tobacco seeds (*Nicotiana tabacum* L.). *Journal of Central European Agriculture* 4:131–135.

Anand, A., Nagarajan, S., Verma, A.P.S., Joshi, D.K., Pathak, P.C., Bhardwaj, J. (2012) Pre-treatment of seeds with static magnetic field ameliorates soil water stress in seedlings of maize (*Zea mays* L.). *Indian Journal of Biochemistry and Biophysics* 49(1):63–70.

Ansari, O., Choghazardi, H.R., SharifZadeh, F., Nazarli, H. (2012) Seed reserve utilization and seedling growth of treated seeds of mountain ray (*Secale montanum*) as affected by drought stress. *Cercetări Agronomice în Moldova* 2:43–48.

Ashraf, M., Rauf, H. (2001) Inducing salt tolerance in maize (*Zea mays* L.) through seed priming with chloride salts: Growth and ion transport at early growth stages. *Acta Physiologiae Plantarum* 23(4):407–414.

Atak, Ç., Çelik, Ö., Olgum, A., Alikamanoğlu, S., Rzakoulieva, A. (2007) Effect of magnetic field on peroxidase activities of soybean tissue culture. *Biotechnology & Biotechnological Equipment* 21(2):166–171.

Baghel, L., Kataria, S., Guruprasad, K.N. (2015) Impact of seed pretreatment by static magnetic field on antioxidant defense of the maize seedlings against ambient ultraviolet (280–400 nm) radiation. *International Journal of Tropical Agriculture* 33:3631–3636.

Baghel, L., Kataria, S., Guruprasad, K.N. (2016) Static magnetic field treatment of seeds improves carbon and nitrogen metabolism under salinity stress in soybean. *Bioelectromagnetics* 37(7):455–470.

Baghel, L., Kataria, S., Guruprasad, K.N. (2018) Effect of SMF pretreatment on growth, photosynthetic performance and yield of soybean under water stress. *Photosynthetica* 55:718–730.

Basra, S.M.A., Ehsanullah, E.A., Warraich, M.A., Afzal, I. (2003) Effect of Storage on growth and yield of primed canola (*Brassica napus*) seeds. *International Journal of Agriculture and Biology* 5:117–120.

Bhardwaj, J., Anand, A., Nagarajan, S. (2012) Biochemical and biophysical changes associated with magnetopriming in germinating cucumber seeds. *Plant Physiology and Biochemistry: PPB* 57:67–73.

Bhatnagar, D., Deb, A.R. (1977) Some aspects of pregermination exposure of wheat seeds to magnetic fields: Germination and early growth. *Seed Research* 5:129–137.

Bilalis, D.J., Katsenios, N., Efthimiadou, A., Karkanis, A., Efthimiadis, P. (2012) Investigation of pulsed electromagnetic field as a novel organic pre-sowing method on germination and initial growth stages of cotton. *Electromagnetic Biology and Medicine* 31(2):143–150.

Bilalis, D.J., Katsenios, N., Efthimiadou, A., Karkanis, A., Khah, E.M., Mitsis, T. (2013) Magnetic field pre-sowing treatment as an organic friendly technique to promote plant growth and chemical elements accumulation in early stages of cotton. *Australian Journal of Crop Science* 7:46–50.

Cakmak, T., Dumlupinar, R., Erdal, S. (2010) Acceleration of germination and early growth of wheat and bean seedlings grown under various magnetic field and osmotic conditions. *Bioelectromagnetics* 31(2):120–129.

Callis, J. (1995) Regulation of protein degradation. *The Plant Cell* 7(7):845–857.

Carpici, E.B., Celık, N., Bayram, G. (2009) Effects of salt stress on germination of some maize (*Zea mays* L.) cultivars. *African Journal of Biotechnology* 8:4918–4922.

Celik, O., Bűyukuslu, N., Atak, C., Rzakoulieva, A. (2009) Effects of magnetic field on activity of superoxide dismutase and catalase in *Glycine max* (L.) Merr. Roots. *Polish Journal of Environmental Studies* 18:175–182.

Chang, S.M., Sung, J.M. (1998) Deteriorative changes in primed sweet corn seeds during storage. *Seed Science and Technology* 26:613–626.

Chen, K., Fessehaie, A., Arora, R. (2012) Dehydrin metabolism is altered during seed osmopriming and subsequent germination under chilling and desiccation in *Spinacia oleracea* L. cv. Bloomsdale: Possible role in stress tolerance. *Plant Science: an International Journal of Experimental Plant Biology* 183:27–36.

Chen, Y.P., Li, R., He, J.M. (2011) Magnetic field can alleviate toxicological effect induced by cadmium in mungbean seedlings. *Ecotoxicology* 20(4):760–769.

Chung, J.S., Zhu, J.K., Bressan, R.A., Hasegawa, P.M., Shi, H. (2008) Reactive oxygen species mediate Na^+ induced SOS1 mRNA stability in *Arabidopsis. The Plant Journal: for Cell and Molecular Biology* 53(3):554–565.

Dantas, B.F., De-Sa Ribeiro, L., Aragao, C.A. (2007) Germination, initial growth and cotyledon protein content of bean cultivars under salinity stress. *Revista Brasileira de Sementes* 29(2):106–110.

De Souza, A., García, D., Sueiro, L., Gilart, F., Porras, E., Licea, L. (2006) Presowing magnetic treatments of tomato seeds increase the growth and yield of plants. *Bioelectromagnetics* 27(4):247–257.

Delfine, S., Alvino, A., Villani, M.C., Loreto, F. (1999) Restrictions to carbon dioxide conductance and photosynthesis in spinach leaves recovering from salt stress. *Plant Physiology* 119(3):1101–1106.

El-Maarouf-Bouteau, H., Bailly, C. (2008) Oxidative signaling in seed germination and dormancy. *Plant Signaling and Behavior* 3(3):175–182.

Esitken, A., Turan, M. (2004) Alternating magnetic field effects on yield and plant nutrient element composition of strawberry (*Fragaria ananassa* cv. camarosa). *Acta Agriculturae Scandinavica, Section B - Soil & Plant Science* 54(3):135–139.

Essa, T.A. (2002) Effect of salinity stress on growth and nutrient composition of three soybean (*Glycine max* L. Merrill) cultivars. *Journal of Agronomy and Crop Science* 188(2):86–93.

Farooq, M., Basra, S.M.A., Rehman, H., Saleem, B.A. (2008) Seed priming enhances the performance of late sown wheat (*Triticum aestivum* L.) by improving chilling tolerance. *Journal of Agronomy and Crop Science* 194(1):55–60.

Fisarakis, I., Chartzoulakis, K., Stavrakas, D. (2001) Response of Sultana vines (*V. vinifera* L.) on six rootstocks to NaCl salinity exposure and recovery. *Agricultural Water Management* 51(1):13–27.

Flórez, M., Carbonell, M.V., Martínez, E. (2004) Early sprouting and first stages of growth of rice seeds exposed to a magnetic field. *Electromagnetic Biology and Medicine* 23(2):157–166.

Florez, M., Carbonell, M.V., Martinez, E. (2007) Exposure of maize seeds to stationary magnetic fields: Effects on germination and early growth. *Environmental and Experimental Botany* 59(1):68–75.

Foyer, C.H., Shigeoka, S. (2011) Understanding oxidative stress and antioxidant functions to enhance photosynthesis. *Plant Physiology* 155(1):93–100.

Gadallah, M.A. (1996) Effect of proline and glycinebetaine on Vicia faba responses to salt stress. *Biologia Plantarum* 42:249–257.

Galland, P., Pazur, A. (2005) Magnetoreception in plants. *Journal of Plant Research* 118(6):371–389.

García-Reina, F., Arza-Pascual, L. (2001) Influence of a stationary magnetic field on water relations in lettuce seeds. Part I. Theoretical considerations. *Bioelectromagnetics* 22:589–595.

Ghobadi, M., Shafiei Abnavi, M., Honarmand, S.J., Ghobadi, M.E., Reza Mohammadi, G. (2012) Effect of hormonal priming (GA3) and osmopriming on behavior of seed germination in wheat (*Triticum aestivum* L.). *Journal of Agricultural Science* 4(9):244–250.

Golezani, K.G., Taifeh-Noori, M., Oustan, S., Moghaddam, M. (2009) Response of soybean cultivars to salinity stress. *Journal of Food, Agriculture and Environment* 7:401–404.

Gomes-Filho, E., Lima, C.R., Costa, J.H., da Silva, A.C., da Guia Silva Lima, M., de Lacerda, C.F., Prisco, J.T. (2008) Cowpea ribonuclease: Properties and effect of NaCl-salinity on its activation during seed germination and seedling establishment. *Plant Cell Reports* 27(1):147–157.

Govindjee. (1995) Sixty-three years since Kautsky: Chlorophyll a fluorescence. *Australian Journal of Plant Physiology* 22(2):131–160.

Gurumoorthi, P., Senthil, K.S., Vadivel, V., Janardhnan, K. (2003) Studies on agro botanical characters of different accessions of velvet bean collected from Western Ghats, South India. *Tropical and Subtropical Agroecology* 2:105–115.

Gust, A.A., Brunner, F., Nürnberger, T. (2010) Biotechnological concepts for improving plant innate immunity. *Current Opinion in Biotechnology* 21(2):204–210.

Hadas, A. (1970) Factors affecting seed germination under soil moisture stress. *Israel Journal of Agricultural Research* 20:3–14.

Harichand, K.S., Narula, V., Raj, D., Singh, G. (2002) Effect of magnetic field on germination, vigour, and seed yield of wheat. *Seed Research* 30:289–293.

Harris, D., Rashid, A., Miraj, G., Arif, M., Shah, H. (2007) On-farm seed priming with zinc sulphate solution, a cost effective way to increase the maize yields of resource poor farmers. *Field Crop Research* 102:119–127.

Hasegawa, P.M., Bressan, R.A., Zhu, J.K., Bohnert, H.J. (2000) Plant cellular and molecular responses to high salinity. *Annual Review of Plant Physiology and Plant Molecular Biology* 51:463–499.

He, C.Z., Hu, J., Zhu, Z.Y., Ruan, S.L., Song, W.J. (2002) Effect of seed priming with mixed- salt solution on germination and physiological characteristics of seedling in rice (*Oryza sativa* L.) under stress conditions. *Journal of Zhejiang University: (Agriculture and Life Sciences)* 28:175–178.

Hozayn, M., Qados, A.M.S. (2010) Magnetic water application for improving wheat (*Triticum aestivum* L.) crop production. *Agriculture and Biology Journal of North America* 1:677–682.

Ibrar, M., Jabeen, M., Tabassum, J., Hussain, F., Ilahi, I. (2003) Salt tolerance potential of *Brassica juncea* Linn. *Journal of Science and Technology of University of Peshawar* 27:79–84.

Iimoto, M.,Watanabe, K., Fujiwara, K. (1996) Effects of magnetic flux density and direction of the magnetic field on growth and CO_2 exchange rate of potato plantlets in vitro. *Acta Horticulturae* 440:606–610.

Iqbal, M., Ashraf, M. (2005) Changes in growth, photosynthetic capacity and ionic relations in spring wheat (*Triticum aestivum* L.) due to pre-sowing seed treatment with polyamines. *Plant Growth Regulation* 46(1):19–30.

Iqbal, M., Haq, Z., Jamil, Y., Ahmad, M. (2012a) Effect of presowing magnetic treatment on properties of pea. *International Agrophysics* 26(1):25–31.

Iqbal, M., Muhammad, D., ul-Haq, Z., Jamil, Y., Ahmad, M.R. (2012b) Effect of presowing magnetic field treatment to garden pea (*Pisum sativum* L.) seed on germination and seedling growth. *Pakistan Journal of Botany* 44:1851–1856.

Jabeen, M., Ibrar, M., Azim, F., Hussain, F., Ilahi, I. (2003) The effect of sodium chloride salinity on germination and productivity of Mung bean (*Vigna mungo* Linn.). *Journal of Science and Technology* 27:1–5.

Jakab, G., Ton, J., Flors, V., Zimmerli, L., Metraux, J.P., Mauch-Mani, B. (2005) Enhancing Arabidopsis salt and drought stress tolerance by chemical priming for its abscisic acid responses. *Plant Physiology* 139(1):267–274.

Javed, N., Ashraf, M., Akram, N.A., Al-Qurainy, F. (2011) Alleviation of adverse effects of drought stress on growth and some potential physiological attributes in maize (*Zea mays*) by seed electromagnetic treatment. *Photochemistry and Photobiology* 87(6):1354–1362.

Joshi, J. (2014) Physiological and biochemical changes in soybean after treatment with magnetic field and strobilurin F-500. Ph.D thesis, School of Life Sciences, DAVV, Indore (M.P.), India.

Kalaji, M.H., Guo, P. (2008) Chlorophyll fluorescence: A useful tool in barley plant breeding programs. In *Photochemistry Research Progress*, eds. Sanchez A., Gutierrez S.J., 439–463. Nova Press Publishers, Hauppauge.

Kalaji, H.M., Jajoo, A., Oukarroum, A., Brestic, M., Zivcak, M., Samborska, I.A., Cetner, M.D., *et al.* (2016) Chlorophyll a fluorescence as a tool to monitor physiological status of plants under abiotic stress conditions. *Acta Physiologiae Plantarum* 38(4):102.

Kalaji, M.H., Nalborczyk, E. (1991) Gas exchange of barley seedlings growing under salinity stress. *Photosynthetica* 25:197–202.

Kalaji, H.M., Pietkiewicz, S. (1993) Salinity effects on plant growth and other physiological processes. *Acta Physiologia Plantatarum* 15:89–124.

Kataria, S., Baghel, L., Guruprasad, K.N. (2015a) Acceleration of germination and early growth characteristics of soybean and maize after pre-treatment of seeds with static magnetic field. *International Journal of Tropical Agriculture* 33:985–992.

Kataria, S., Baghel, L., Guruprasad, K.N. (2015b) Effect of seed pre-treatment by magnetic field on the sensitivity of maize seedlings to ambient ultraviolet radiation (280–400 nm). *International Journal of Tropical Agriculture* 33:3645–3652.

Kataria, S., Baghel, L., Guruprasad, K.N. (2017a) Alleviation of adverse effects of ambient UV stress on growth and some potential physiological attributes in soybean (*Glycine max*) by seed pre-treatment with static magnetic field. *Journal of Plant Growth Regulation* 36(3):550–565.

Kataria, S., Baghel, L., Guruprasad, K.N. (2017b) Pre-treatment of seeds with static magnetic field improves germination and early growth characteristics under salt stress in maize and soybean. *Biocatalysis and Agricultural Biotechnology* 10:83–90.

Kataria, S., Jajoo, A., Guruprasad, K.N. (2014) Impact of increasing ultraviolet-B radiation on photosynthetic processes. *Journal of Photochemistry and Photobiology B: Biology* 137:55–66.

Kataria, S., Verma S.K. (2018) Salinity stress responses and adaptive mechanisms in major glycophytic crops: The story so far. In *Salinity Responses and Tolerance in Plants, Volume 1, Targeting Sensory, Transport and Signaling Mechanisms,* eds. Kumar V., Hussain W.S., Suprasanna P., Phan Tran L.-S., 1–40. Springer, Cham.

Kaur, S., Gupta, A.K., Kaur, N. (2003) Priming of chickpea seeds with water and mannitol can over-come the effect of salt stress on seedling growth. *International Chickpea Pigeon Pea Newsletter* 10:18–268.

Kavi, P.S. (1977) The effect of magnetic treatment of soybean seed on its moisture absorbing capacity. *Scientific Culture* 43:405–406.

Khajeh-Hosseini, M., Powell, A.A., Bimgham, I.J. (2003) The interaction between salinity stress and seed vigor during germination of soybean seeds. *Seed Science and Technology* 31(3):715–725.

Khaliq, A., Matloob, A., Mahmood, S., Wahid, A. (2013) Seed pretreatments help improve maize performance under sorghum allelopathic stress. *Journal of Crop Improvement* 27(5):586–605.

Khan, M.A., Rizvi, Y. (1994) Effect of salinity, temperature and growth regulators on the germination and early seedling growth of *Atriplex griffithii* var. Stocksii. *Canadian Journal of Botany* 72(4):475–479.

Khan, M.A., Weber, D.J. (2008) *Ecophysiology of High Salinity Tolerant Plants (Tasks for Vegetation Science)*, 1st edn. Springer, Amsterdam.

Khizhenkov, P.K., Dobritsa, N.V., Netsvetov, M.V., Driban, V.M. (2001) Influence of low- and super low-frequency alternating magnetic fields on ionic permeability of cell membranes. *Dopovidi Natsionalnoi Akademii Nauk Ukraini* 4:161–164.

Khodarahmpour, Z., Ifar, M., Motamedi, M. (2012) Effects of NaCl salinity on maize (*Zea mays* L.) at germination and early seedling stage. *African Journal of Biotechnology* 11(2):298–230.

Labes, M.M. (1993) A possible explanation for the effect of magnetic fields on biological systems. *Nature* 211:969.

Martínez, E., Carbonell, M.V., Amaya, J. (2000) Static magnetic field of 125 mT stimulates de initial growth stages of barley (*Hordeum vulgare* L.). *Electro and Magnetobiology* 19:271–277.

Martínez, E., Carbonell, M.V., Flórez, M. (2002) Magnetic biostimulation of initial growth stages of wheat (*Triticum aestivum* L.). *Electromagnetic Biology and Medicine* 21(1):43–53.

Maxwell, K., Johnson, G.N. (2000) Chlorophyll fluorescence—A practical guide. *Journal of Experimental Botany* 51(345):659–668.

McDonald, M.B. (2000) Seed priming. In *Seed Technology and Its Biological Basis*, eds. Black M., Bewley J.D., 287–325. Sheffield Academic Press Ltd, Sheffield.

Merritt, D.J., Senaratna, T., Touchell, D.H., Dixon, K.W., Sivasithamparam, K. (2003) Seed ageing of four Western Australian species in relation to storage environment and seed antioxidant activity. *Seed Science Research* 13(2):155–165.

Mridha, N., Chattaraj, S., Chakraborty, D., Anand, A., Aggarwal, P., Nagarajan, S. (2016) Pre-sowing static magnetic field treatment for improving water and radiation use efficiency in chickpea (*Cicer arietinum* L.) under soil moisture stress. *Bioelectromagnetics* 37(6):400–408.

Munns, R., James, R.A., Lauchli, A. (2006) Approaches to increasing the salt tolerance of wheat and other cereals. *Journal of Experimental Botany* 57(5):1025–1043.

Muraji, M., Asai, T., Tatebe, W. (1998) Primary root growth rate of Zea mays seedlings grown in an alternating magnetic field of different frequencies. *Bioelectrochemistry and Bioenergetics* 44(2):271–273.

Mutlu, F., Buzcuk, S. (2007) Salinity induced changes of free and bound polyamine levels in Sun flower (*Helianthus annuus* L.) roots differing in salt tolerance. *Pakistan Journal of Botany* 39:1097–1102.

Naz, A., Jamil, Y., ul Haq, Z., Iqbal, M., Ahmad, M.R., Ashraf, M.I., Ahmad, R. (2012) Enhancement in the germination, growth and yield of Okra (Abelmoschus esculentus) using pre-sowing magnetic treatment of seeds. *Indian Journal of Biochemistry and Biophysics* 49:211–214.

Novitskaya, G.V., Tulinova, E.A., Kocheshkova, T.K., Novitsky, I. (2001) The effect of weak permanent magnetic field on cotyledon emergence and neutral lipid content in 5-day-old radish seedlings. In *Plant under Environmental Stress*, 212–213. Publishing House of Peoples Friendship, University of Russia, Moscow.

Othman, Y., Al-Karaki, G., Al-Tawaha, A.R., Al-Horani, A. (2006) Variation in germination and ion uptake in barley genotypes under salinity conditions. *World Journal of Agricultural Sciences* 2:11–15.

Pál, N. (2005) The effect of low inductivity static magnetic field on some plant pathogen fungi. *Journal of Central European Agriculture* 6:167–171.

Parera, C.A., Cantliffe, D.J. (1992) Enhanced emergence and seedling vigor in shrunken-2 sweet corn via seed disinfection and solid matrix priming. *Journal of American Society for Horticultural Science* 117:400–403.

Parida, A.K., Das, A.B. (2005) Salt tolerance and salinity effect on plants: A review. *Ecotoxicology and Environmental Safety* 60(3):324–349.

Phirke, P.S., Kubde, A.B., Umbarkar, S.P. (1996) The influence of magnetic field on plant growth. *Seed Science and Technology* 24:375–392.

Pittman, U.J., Ormrod, D.P. (1970) Physiological and chemical features of magnetically treated winter wheat seeds and resultant seedlings. *Canadian Journal of Plant Science* 50(3):211–217.

Promila, K., Kumar, S. (2000) *Vigna radiata* seed germination under salinity. *Biologia Plantarum* 43(3):423–426.

Pukacka, S., Ratajczak, E. (2007) Age-related biochemical changes during storage of beech (*Fagus sylvatica* L.) seeds. *Seed Science Research* 17(1):45–53.

Punjabi, B., Mandal, A.K., Basu, R.N. (1982) Maintenance of vigour, viability and productivity of stored barley seed. *Seed Research* 19:69–71.

Radhakrishnan, R., Kumari, B.R. (2012) Pulsed magnetic field: A contemporary approach offers to enhance plant growth and yield of soybean. *Plant Physiology and Biochemistry: PPB* 51:139–144.

Radhakrishnan, R., Kumari, B.R. (2013a) Influence of pulsed magnetic field on soybean (*Glycine max* L.) seed germination, seedling growth and soil microbial population. *Indian Journal of Biochemistry and Biophysics* 50(4):312–317.

Radhakrishnan, R., Kumari, B.R. (2013b) Protective role of pulsed magnetic field against salt stress effects in soybean organ culture. *Plant Biosystems - an International Journal Dealing with All Aspects of Plant Biology* 147(1):135–140.

Rakosy-Tican, L., Aurori, C.M., Morariu, V.V. (2005) Influence of near null magnetic field on in vitro growth of potato and wild Solanum species. *Bioelectromagnetics* 26(7):548–557.

Rathod, G.R., Anand, A. (2016) Effect of seed magneto-priming on growth, yield and Na/K ratio in wheat (*Triticum aestivum* L.) under salt stress. *Indian Journal of Plant Physiology* 21(1):15–22.

Rehman, S., Harris, P.J.C., Bourne, W.F., Wilkin, J. (2000) The relationship between ions, vigour and salinity tolerance of acacia seeds. *Plant and Soil* 220(1/2):229–233.

Reina, F.G., Pascual, L.A., Fundora, I.A. (2001) Influence of a stationary magnetic field on water relations in lettuce seeds. Part II: Experimental results. *Bioelectromagnetics* 22(8):596–602.

Reyes, L.F., Cisneros-Zevallos, L. (2007) Electron-beam ionizing radiation stress effects on mango fruit (*Mangifera indica* L.) antioxidant constituents before and during post-harvest storage. *Journal of Agricultural and Food Chemistry* 55(15):6132–6139.

Rochalska, M., Grabowska, K. (2007) Influence of magnetic fields on the activity of enzymes α- and β-amylase and glutathione S-transferase (GST) in wheat plants. *International Agrophysics* 21:185–188.

Rochalska, M., Orzeszko-Rywka, A. (2005) Magnetic field treatment improves seed performance. *Seed Science and Technology* 33(3):669–674.

Ruzic, R., Jerman, I. (2002) Weak magnetic field decreases heat stress in cress seedlings. *Electromagnetic Biology and Medicine* 21(1):69–80.

Scandalios, J.G. (1993) Oxygen stress and superoxide dismutases. *Plant Physiology* 101(1):7–12.

Sen, A., Alikamanoglu, S. (2014) Effects of static magnetic field pretreatment with and without peg 6000 or NaCl exposure on wheat biochemical parameters. *Russian Journal of Plant Physiology* 61(5):646–655.

Sharma, P., Jha, A.B., Dubey, R.S., Pessarakli, M. (2012) Reactive oxygen species, oxidative damage and antioxidative defence mechanisms in plants under stressful conditions. *Journal of Botany* 2012:1–26.

Shehzad, M., Ayub, M., Ahmad, A.H.U., Yaseen, M. (2012) Influence of priming techniques on emergence and seedling growth of forage sorghum (*Sorghum bicolour* L.). *The Journal of Animal and Plant Sciences* 22:154–158.

Shine, M.B., Guruprasad, K.N. (2012) Impact of pre-sowing magnetic field exposure of seeds to stationary magnetic field on growth, reactive oxygen species and photosynthesis of maize under field conditions. *Acta Physiologiae Plantarum* 34(1):255–265.

Shine, M.B., Guruprasad, K.N., Anand, A. (2011) Enhancement of germination, growth, and photosynthesis in soybean by pre-treatment of seeds with magnetic field. *Bioelectromagnetics* 32(6):474–484.

Shine, M.B., Guruprasad, K.N., Anand, A. (2012) Effect of stationary magnetic field strengths of 150 and 200mT on reactive oxygen species production in soybean. *Bioelectromagnetics* 33(5):428–437.

Shine, M.B., Kataria, S., Guruprasad, K.N., Anand, A. (2017) Enhancement of maize seeds germination by magnetopriming in perspective with reactive oxygen species. *Journal of Agriculture and Crop Research* 5(4):66–76.

Singh, K., Afria, B.S. (1990) Seed germination seedling growth, emergence and establishment responses of cotton cultivators as regulated by growth substances. *Seed Research* 18(1):25–30.

Soltani, A., Gholipoor, M., Zeinali, E. (2006) Seed reserve utilization and seedling growth of wheat as affected by drought and salinity. *Environmental and Experimental Botany* 55(1–2):195–200.

Srivastava, L.M. (2002) *Seed Germination, Mobilization of Food Reserves, and Seed Dormancy*. Academic Press, London, 447–471.

Stange, B.C., Rowland, R.E., Rapley, B.I., Podd, J.V. (2002) ELF magnetic fields increase amino acid uptake into *Vicia faba* L. roots and alter ion movement across the plasmamembrane. *Bioelectromagnetics* 23(5):347–354.

Strasser, B.J., Strasser, R.J. (1995) Measuring fast fluorescence transients to address environmental questions: The JIP Test. In *Photosynthesis: From Light to Biosphere*, ed. Mathis, P., 977–980. Kluwer Academic Publishers, Dordrecht, The Netherlands.

Tavili, A., Zare, S., Moosavi, S.A., Enayati, A. (2011) Effects of seed priming on germination characteristics of Bromus species under salt and drought conditions. *American-Eurasian Journal of Agriculture and Environmental Science* 10:163–168.

Thomas, S., Anand, A., Chinnusamy, V., Dahuja, A., Basu, S. (2013) Magnetopriming circumvents the effect of salinity stress on germination in chickpea seeds. *Acta Physiologiae Plantarum* 35(12):3401–3411.

Turker, M.,Temirci, C., Battal, P., Erez, M.E. (2007) The effects of an artificial and static magnetic field on plant growth, chlorophyll and phytohormone levels in maize and sunflower plants. *Phyton; Annales Rei Botanicae* 46:271–284.

ul Haq, Z., Jamil, Y., Irum, S., Randhawa, M.A., Iqbal, M., Amin, N. (2012) Enhancement in the germination, seedling growth and yield of radish (*Raphanus sativus*) using seed pre-sowing magnetic field treatment. *Polish Journal of Environmental Studies* 21:369–374.

Ulfat, M., Athar, H., Ashraf, M., Akram, N.A., Jamil, A. (2007) Appraisal of physiological and biochemical selection criteria for evaluation of salt tolerance in canola (*Brassica napus* L.). *Pakistan Journal of Botany* 39:1593–1608.

Vaishnav J., Jain M. (2015) Influence of halopriming and hydropriming on seed germination and growth characteristics of *Zea mays* L. cv. GSF-2 under salt stress. *Research Journal of Chemistry and Environment* 19: 1–6.

Vashisth, A., Nagarajan, S. (2008) Exposure of seeds to static magnetic field enhances germination and early growth characteristics in chickpea (*Cicer arietinum* L.). *Bioelectromagnetics* 29(7):571–578.

Vashisth, A., Nagarajan, S. (2010) Effect on germination and early growth characteristics in sunflower (*Helianthus annus*) seeds exposed to static magnetic field. *Journal of Plant Physiology* 167(2):149–156.

Verma, G., Sharma, S. (2010) Role of H_2O_2 and cell wall monoamine in germination of Vigna radiata seeds. *Indian Journal of Biochemistry and Biophysics* 47(4):249–253.

Wang, G., Hao, Z., Anken, R.H., Lu, J., Liu, Y. (2010) Effects of UV-B radiation on photosynthesis activity of *Wolffia arrhiza* as probed by chlorophyll fluorescence transients. *Advances in Space Research* 45(7):839–845.

Wang, K.F., Ritz, T. (2006) Zeeman resonances for radical-pair reactions in weak static magnetic fields. *Molecular Physics* 104(10–11):1649–1658.

Xia, L., Guo, J. (2000) Effect of magnetic field on peroxidase activation and isozyme in Leymus chinensis. *Ying Yong Sheng Tai Xue Bao = the Journal of Applied Ecology* 11(5):699–702.

Xiong, L., Zhu, J.K. (2002) Molecular and genetic aspects of plant responses to osmotic stress. *Plant, Cell and Environment* 25(2):131–139.

Xu, S., Hu, J., Li, Y., Ma, W., Zheng, Y., Zhu, S. (2011) Chilling tolerance in *Nicotiana tabacum* induced by seed priming with putrescine. *Plant Growth Regulation* 63(3):279–290.

Yupsanis, T., Moustakas, M., Domiandou, K. (1994) Protein phosphorylation-dephosphorylation in alfalfa seeds germinating under salt stress. *Journal of Plant Physiology* 38:57–61.

Zhao, T.J., Liu, Y., Yan, Y.B., Feng, F., Liu, W.Q., Zhou, H.M. (2007) Identification of the aminoacids crucial for the activities of drought responsive element binding factors (DREBs) of Brassica napus. *FEBS Letters* 581(16):3044–3050.

Index

R

S

T

V

W

X